Entomology

Third Edition

Entomology

Third Edition

Cedric Gillott
University of Saskatchewan
Saskatoon, Saskatchewan, Canada

 Springer

A C.I.P. Catalogue record for this book is available from the Library of Congress.

ISBN-10 1-4020-3182-3 (PB)
ISBN-13 978-1-4020-3182-3 (PB)
ISBN-10 1-4020-3184-X (HB)
ISBN-13 978-1-4020-3184-7 (HB)
ISBN-10 1-4020-3183-1 (e-book)
ISBN-13 978-1-4020-3183-0 (e-book)

Published by Springer,
P.O. Box 17, 3300 AA Dordrecht, The Netherlands.

www.springeronline.com

Printed on acid-free paper

Cover image:
Bee flies and a blister beetle feeding on pollen of Echinacea
(courtesy of Jason Wolfe and Tyler Wist)

Contents

I. Evolution and Diversity

1

**Arthropod
Evolution**

2

**Insect
Diversity**

II. Anatomy and Physiology

12

Sensory Systems

13

Nervous and Chemical Integration

14
Muscles and Locomotion

15
Gas Exchange

16

**Food Uptake
and
Utilization**

17

**The
Circulatory
System**

18

**Nitrogenous
Excretion and
Salt and
Water
Balance**

III. Reproduction and Development

19

Reproduction

20

Embryonic Development

21

Postembryonic Development

IV. Ecology

22

The Abiotic Environment

23

The Biotic Environment

24

**Insects and
Humans**

Preface

The strongly favorable reception accorded previous versions of this book, together with the not infrequent urgings of colleagues and students, encouraged me to take on the task of preparing a third edition of *Entomology*. My early retirement, in 1999, freed up the time necessary for a project of this size, and for the past 2 years my effort has been almost entirely focused in this direction. Obviously, all chapters have been updated; this includes not only the addition of new information and concepts (some of which are highlighted below), but also the reduction or exclusion of material no longer considered 'mainstream' so as to keep the book at a reasonable size.

My strong belief that an introductory entomology course should present a balanced treatment of the subject still holds and is reflected in the retention of the format of earlier editions, namely, arrangement of the book into four sections: Evolution and Diversity, Anatomy and Physiology, Reproduction and Development, and Ecology.

Section I (Evolution and Diversity) has again undergone a great reworking, mainly because the last decade has seen the uncovering of significant new fossil evidence, and the application of molecular and cladistic analyses to extant groups. As a result, ideas both on the relationships of insects to other arthropods and on the higher classification of many orders have changed drastically. However, as in previous editions, I have stressed that most phylogenies are not 'embedded in stone' but represent the consensus based on existing information; thus, they are liable to refinement as additional data are forthcoming. Chapter 1 discusses the evolution of Insecta in relation to other arthropods, emphasizing the ageless debate on whether arthropods form a monophyletic or polyphyletic group, and the relationship of insects to other hexapodous arthropods. Evolutionary relationships within the Insecta are considered in Chapter 2, together with discussion of the factors that contributed to the overwhelming success of the group. Chapter 3 serves two purposes: It provides a description of external structure, which remains the principal basis on which insects can be classified and identified, while stressing diversity with reference to mouthpart and appendage modifications. In Chapter 4 the principles of classification and identification are outlined, and a key to the orders of insects is provided. Diversity of form and habits is again emphasized in Chapters 5 to 10, which deal with the orders of insects, including the Mantophasmatodea, established only in 2002. For many orders, new proposed phylogenies are presented, and the text has undergone significant rearrangement to reflect modern ideas on the classification of these taxa.

The chapters in Section II (Anatomy and Physiology) deal with the homeostatic systems of insects; that is, those systems that keep insects 'in tune' with their environment, enabling them to develop and reproduce optimally. The section begins with a discussion of the integument (Chapter 11), as this has had such a profound influence on the success of insects. Chapter 12 examines sensory systems, whose form and function are greatly influenced by the cuticular nature of the integument. In Chapter 13, where neural and chemical integration are discussed, new sections on kairomones and allomones have been included. Chapter 14 considers muscle structure and function, including locomotion. In this chapter the section on flight has been significantly revised, especially with respect to recent proposals for the generation of lift using non-steady-state aerodynamics. Chapter 15 reveals the remarkable efficiency of the tracheal system in gaseous exchange, and Chapter 16 deals with the acquisition and utilization of food. Chapter 17 describes the structure and functions of the circulatory system, including the immune response of insects about which much has been learned in the past decade. New to this chapter is a section on how parasites and parasitoids are able to defend themselves against the host insect's immune system. Chapter 18 concludes this section with a discussion of nitrogenous waste removal and salt/water balance.

In Section III reproduction (Chapter 19), embryonic development (Chapter 20), and postembryonic development (Chapter 21) are discussed. Chapter 19 includes additional information on behavioral aspects of reproduction (courtship, mate guarding and sexual selection), as well as sperm precedence. Chapter 21 has been revised to provide an updated account of the endocrine regulation of development and molting.

Section IV (Ecology) examines those factors that affect the distribution and abundance of insects. In Chapter 22 abiotic (physical) factors in an insect's environment are considered. Chapter 23 deals with the biotic factors that influence insect populations and serves as a basis for the final chapter, in which the specific interactions of insects and humans are discussed. Of all of the chapters, Chapter 24 has received the most drastic overhaul; such has been the 'progress' (and the costs of such progress) in the battle against insect pests.

As may be inferred from the opening paragraph of this Preface, the book is intended as a text for senior undergraduates taking their first course in entomology. Such students probably will have an elementary knowledge of insects acquired from an earlier course in general zoology, as well as a basic understanding of animal physiology and ecological principles. With such a background, students should have no difficulty understanding the text.

Preparation of the third edition has benefited, not only from both published and unsolicited reviews of previous editions, but also from my solicitation of comments on the content of specific chapters from experts in those areas. Of course, any errors that remain, and I hope these are extremely few, are my responsibility. I have enjoyed preparing this third edition, for it has given me, once again, the opportunity to delve into aspects of entomology that are well outside the range of an 'insect sexologist'. For example, I never cease to be impressed by the remarkable discoveries and insights of those entomologists who deal with fossil insects, by those who develop integrated strategies for the management of insect pest populations, and by the patience and dedication (*and* imagination—see Chapter 4, Section 2) of insect taxonomists. Hopefully, readers of the new edition will receive the same enjoyment.

Cedric Gillott
Professor Emeritus

University of Saskatchewan,
Saskatoon, Saskatchewan, Canada

Acknowledgments

Though the book has single authorship, its preparation would not have been possible but for my colleagues, too numerous to mention individually, who provided information and answered specific questions that improved the book's content and currency. To these people I am most grateful.

Mr. Dennis Dyck and Mrs. Shirley Brodsky are thanked for their considerable assistance with preparation of the original figures. For this edition, all figures were converted into electronic format and, when necessary, reworked by Mr. Dyck to achieve greater uniformity of style.

Thanks are also extended to a large number of publishers, editors, and private individuals who allowed me to use materials for which they hold copyright. The source of each figure is acknowledged individually in the text.

I am grateful to the University of Saskatchewan, which granted me the facilities necessary to bring this project to fruition. I specifically acknowledge the assistance given by staff in the Library's inter-library loans department; so numerous were my requests for material that I felt, at times, as though they were my personal assistants! The confidence, patience, and assistance of Kluwer Academic Publishers, especially Zuzana Bernhart (Publishing Editor, Life Sciences), Ineke Ravesloot (Assistant to the Publishing Editor), and Tonny van Eekelen (Production Supervisor, Books) are also appreciated.

Finally, the enormous help given me by my wife, Anne, is acknowledged. To her fell the major task of proofreading to ensure that the revised text was coherent, figures were correctly numbered, labeled and cited, reference lists were accurate, and tables were complete. She also checked copyright approvals and assisted in preparation of the index. It is to her that this book is dedicated.

I

Evolution and Diversity

1

Arthropod Evolution

1. Introduction

Despite their remarkable diversity of form and habits, insects possess several common features by which the group as a whole can be distinguished. They are generally small arthropods whose bodies are divisible into cephalic, thoracic, and abdominal regions. The head carries one pair of antennae, one pair of mandibles, and two pairs of maxillae (the hind pair fused to form the labium). Each of three thoracic segments bears a pair of legs and, in the adult, the meso- and/or metathoracic segments usually have a pair of wings. Abdominal appendages, when present, generally do not have a locomotory function. The genital aperture is located posteriorly on the abdomen. With few exceptions eggs are laid, and the young form may be quite different from the adult; most insects undergo some degree of metamorphosis.

Although these may seem initially to be an inauspicious set of characters, when they are examined in relation to the environment it can be seen quite readily why the Insecta have become the most successful group of living organisms. This aspect will be discussed in Chapter 2.

In the present chapter we shall examine the possible origins of the Insecta, that is, the evolutionary relationships of this group with other arthropods. In order to do this meaningfully it is useful first to review the features of the other groups of arthropods. As will become apparent below, the question of arthropod phylogeny is controversial, and various theories have been proposed.

2. Arthropod Diversity

Arthropods share certain features with which they can be defined. These features are: segmented body covered with a chitinous exoskeleton that may be locally hardened and is periodically shed, tagmosis (the grouping of segments into functional units, for example, head, thorax, and abdomen in insects), presence of preoral segments, paired jointed appendages on a varied number of segments, hemocoelic body cavity containing ostiate heart enclosed within a pericardium, nervous system comprising dorsal brain and ventral ganglionated nerve cord, muscles almost always striated, and epithelial tissue almost always non-ciliated.

Though the "true" arthropods fit readily within this definition, three small groups, the Onychophora, Tardigrada, and Pentastoma, whose members are soft-bodied, wormlike animals with unjointed appendages, are less obviously arthropodan and each is usually given separate phylum status.

2.1. Onychophora, Tardigrada, and Pentastoma

The approximately 200 extant species of Onychophora (Figure 1.1A) are terrestrial animals living on land masses derived from the Gondwanan supercontinent: Africa, Central and South America, and Australasia (Tait, 2001). They are generally confined to moist habitats and are found beneath stones in rotting logs and leaf mold, etc. They possess a combination of annelidan and arthropodan characters and, as a result, are always prominent in discussions of arthropod evolution. Although covered by a thin arthropodlike cuticle (comprising procuticle and epicuticle, but no outer wax layer—see Chapter 11), the body wall is annelidan, as are the method of locomotion, unjointed legs, the excretory system, and the nervous system. Their arthropodan features include a hemocoelic body cavity, the development and structure of the jaws, the possession of salivary glands, an open circulatory system, a tracheal respiratory system, and claws at the tips of the legs. Among living arthropods, myriapods resemble the Onychophora most closely: their body form is similar, tagmosis is restricted to the three-segmented head, exsertile vesicles are present in Diplopoda and Symphyla as well as in some onychophorans, a digestive gland is absent, the midgut is similar, the genital tracts of Onychophora resemble those of myriapods, the gonopore is subterminal, and certain features of embryonic development are common to both groups (Tiegs and Manton, 1958). However, this resemblance is superficial. Recent onychophorans are but the remnants of a more widespread fauna (fossils from the Carboniferous are very similar to modern forms) that may have evolved from marine lobopods in the Cambrian period.

Tardigrades are mostly very small (<0.5 mm long) animals, commonly known as water bears (Figure 1.1B). The majority of the 800 extant species are found in the temporary water films that coat mosses and lichens. A few live in permanent aquatic habitats, either marine or freshwater, or in water films in soil and forest litter (Kinchin, 1994; Nelson, 2001). Their body is covered with a chitinized cuticle and bears four pairs of unjointed legs, each

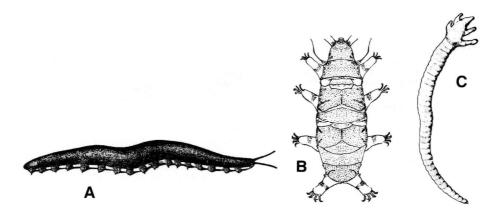

FIGURE 1.1. (A) *Peripatopsis* sp. (Onychophora); (B) *Pseudechiniscus suillus* (Tardigrada); and (C) *Cephalobaena tetrapoda* (Pentastomida). [A, from A. Sedgewick, 1909, *A Student's Textbook of Zoology*, Vol. III, Swan, Sonnenhein and Co., Ltd. B, C, from P.-P. Grassé, 1968, *Traité de Zoologie*, Vol. 6. By permission of Masson et Cie.]

pair being innervated from a segmental ganglion in the ventral nerve cord. The fluid-filled, hemocoelic body cavity serves as a hydrostatic skeleton. The affinities of the tardigrades remain unclear. They were traditionally aligned with pseudocoelomates. However, they have a number of onychophoran and arthropodan structural features, and the modern view is that they are closely related to these groups. Recent molecular evidence supports this proposal.

Pentastomids (tongue worms) (Figure 1.1C), of which about 100 species are known, are parasitic in the nasal and pulmonary cavities of vertebrates, principally reptiles but including birds and mammals. The body of these worms, which range from 2 to 13 cm long, is covered with a cuticle and has two pairs of anterior unjointed legs. Internally, there is a fluid-filled cavity (debatably either a hemocoel or a pseudocoelom) containing a paired ventral nerve cord with segmental ganglia innervating each leg. Larval development occurs in the tissues of an intermediate host, which may be an omnivorous or herbivorous insect, fish, or mammal. Though pentastomids are highly modified as a result of their parasitic life, they are undoubtedly arthropods. Their exact position remains controversial, relationships with Acarina, myriapods, and branchiuran crustaceans having been suggested.

2.2. Trilobita

The trilobites (Figure 1.2), of which almost 4000 species have been described, are marine fossils that reached their peak diversity in the Cambrian and Ordovician periods (500–600 million years ago) (Whittington, 1992). Despite their antiquity they were, however, not primitive but highly specialized arthropods. In contrast to modern arthropods the trilobites as a whole show a remarkable uniformity of body structure. The body, usually

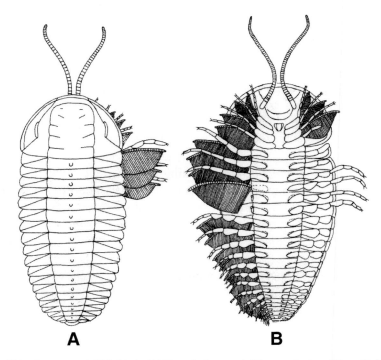

A   **B**

FIGURE 1.2. *Triarthrus eatoni* (Trilobita). (A) Dorsal view; and (B) ventral view. [From R. D. Barnes, 1968, *Invertebrate Zoology*, 2nd ed. By permission of the W. B. Saunders Co., Philadelphia.]

oval and dorsoventrally flattened, is divided transversely into three tagmata (head, thorax, and pygidium) and longitudinally into three lobes (two lateral pleura and a median axis). The head, which bears a pair of antennae, compound eyes, and four pairs of biramous appendages, is covered by a carapace. A pair of identical biramous appendages is found on each thoracic segment. The basal segment of each limb bears a small, inwardly projecting endite that is used to direct food toward the mouth.

Much about the habits of trilobites can be surmised from examination of their remains and the deposits in which these are found. Most trilobites lived near or on the sea floor. While some species preyed upon small, soft-bodied animals, the majority were scavengers. However, like earthworms, a few smaller trilobites took in mud and digested the organic matter from it. On the basis of X-ray studies of pyritized trilobite specimens, which show that trilobites possess a combination of chelicerate and crustacean characteristics, Cisne (1974) concluded that the Trilobita, Chelicerata, and Crustacea form a natural group with a common ancestry. Their ancestor would have a body form similar to that of trilobites. Most authors dispute the proposed trilobite-crustacean link, and some even reject the association between trilobites and chelicerates. Indeed, there are those who suggest that the trilobites themselves are polyphyletic (Willmer, 1990).

Although the decline of trilobites (and their replacement by the crustaceans as the dominant aquatic arthropods) is a matter solely for speculation, Tiegs and Manton (1958) suggested that their basic, rather cumbersome body plan may have prohibited the evolution of fast movement at a time when highly motile predators such as fish and cephalopods were becoming common. In addition, the many identical limbs presumably moved in a metachronal manner, which is a rather inefficient method in large organisms.

2.3. The Chelicerate Arthropods

The next four groups are often placed together under the general heading of Chelicerata because their members possess a body that is divisible into cephalothorax and abdomen, the former usually bearing a pair of chelicerae (but lacking antennae), a pair of pedipalps, and four pairs of walking legs. Although there is little doubt of the close relationship between the Xiphosura, Eurypterida, and Arachnida, the position of the Pycnogonida is uncertain. Though they are usually included as a class of chelicerates, their affinities with other members of this group remain unclear, and there are some authors who consider they deserve more separated status (see King, 1973; Manton, 1978; Edgecombe, 1998; Fortey and Thomas, 1998).

Xiphosura. *Limulus polyphemus*, the king or horseshoe crab (Figure 1.3), is one of four surviving species of a class of arthropods that flourished in the Ordovician-Upper Devonian periods. King crabs occur in shallow water along the eastern coasts of North and Central America. Three species of *Tachypleus* and *Carcinoscorpius* occur along the coasts of China, Japan, and the East Indies. Like trilobites they are bottom feeders, stirring up the substrate and extracting the organic material from it. In *Limulus* the cephalothorax is covered with a horseshoe-shaped carapace. The abdomen articulates freely with the cephalothorax and at its posterior end carries a long telson. On the ventral side of the cephalothorax are six pairs of limbs. The most anterior pair are the chelicerae, and these are followed by five pairs of legs. Each leg has a large gnathobase, which serves to break up food and pass it forward to the mouth. Six pairs of appendages are found on the abdomen. The first pair fuse medially to form the operculum. This protects the remaining pairs, which bear gills on their posterior surface.

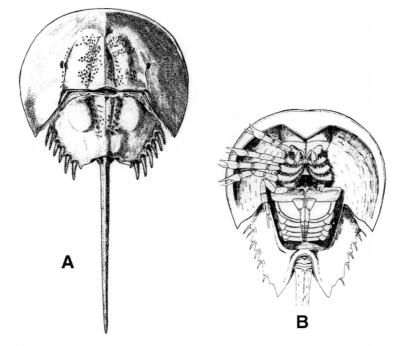

FIGURE 1.3. The horseshoe crab, *Limulus polyphemus*. (A) Dorsal view and (B) ventral view. [From R. D. Barnes, 1968, *Invertebrate Zoology*, 2nd ed. By permission of the W. B. Saunders Co., Philadelphia.]

Eurypterida. The Eurypterida (giant water scorpions) (Figure 1.4A) were formerly included with the Xiphosura in the class Merostomata. However, most recent studies have concluded that the two are not sister groups and that the Merostomata is a paraphyletic assemblage (authors in Edgecombe, 1998). More than 300 species of this entirely fossil group of predatory arthropods, which existed from the Ordovician to the Permian periods, are known. Because of their sometimes large size (up to 2.5 m) they are also known as Gigantostraca. They are believed to have been important predators of early fish, providing selection pressure for the evolution of dermal bone in the Agnatha. In body plan they were rather similar to the xiphosurids. Six pairs of limbs occur on the cephalothorax, but, in contrast to those of king crabs, the second pair is often greatly enlarged and chelate forming pedipalps, which presumably served in defense and to capture and tear up prey. The trunk of eurypterids can be divided into an anterior preabdomen on which appendages (concealed gills) are retained and a narrow taillike postabdomen from which appendages have been lost. Though the early eurypterids were marine, adaptive radiation into freshwater and perhaps even terrestrial habitats occurred. Indeed, it was from freshwater forms that arachnids are believed to have evolved.

Arachnida. Scorpions, spiders, ticks, and mites belong to the class Arachnida whose approximately 62,000 species are more easily recognized than defined. Living members of the group are terrestrial (although a few mites are secondarily aquatic) and have respiratory organs in the form of lung books or tracheae. In contrast to the two aquatic chelicerate groups described earlier, most arachnids take only liquid food, extracted from their prey by means of a pharyngeal sucking pump, often after extraoral digestion. Scorpions, of which there are about 1500 living species, are the oldest arachnids with fossils known from the Silurian. Some of these fossils were aquatic (Polis, 1990). With about 35,000 species, spiders form

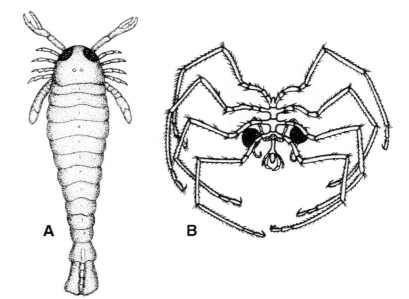

FIGURE 1.4. (A) Eurypterid and (B) *Nymphon rubrum* (Pycnogonida). [A, from D. T. Anderson (ed.), 2001, *Invertebrate Zoology*, 2nd ed. By permission of Oxford University Press. B, from R. D. Barnes. 1968, *Invertebrate Zoology*, 2nd ed. By permission of the W. B. Saunders Co., Philadelphia.]

an extremely diverse group. The earliest spider fossils are from the Devonian, and by the Tertiary the spider fauna was very similar to that seen today (Foelix, 1997).

Pycnogonida. The approximately 1000 species of living Pycnogonida (Pantopoda) are the remnants of a group that originated in the Devonian. They are commonly known as sea spiders because of their superficial similarity to these arachnids (Figure 1.4B). They are found at varying depths in all oceans of the world, but are particularly common in the shallower waters of the Arctic and Antarctic Oceans. They live on the sea floor and feed on coelenterates, bryozoans, and sponges. On the cephalothorax is a large proboscis, a raised tubercle bearing four simple eyes, a pair of chelicerae and an associated pair of palps, and five pairs of legs. The legs of the first pair differ from the rest in that they are small and positioned ventrally. These ovigerous legs are used in the male for carrying the eggs. The abdomen is very small and lacks appendages.

As noted above, the precise relationships of the pycnogonids to other arthropods remain controversial. Although the presence of chelicerae, the structure of the brain, and the nature of the sense organs are chelicerate characters, the structure and innervation of the proboscis, the similarity between the intestinal diverticula and those of annelids, the multiple paired gonopores, and the suggestion that the pycnogonids have a true coelom show that they must have left the main line of arthropod evolution at a very early date (Sharov, 1966). Other non-chelicerate features that they possess are (1) the partial segmentation of the leg-bearing part of the body, (2) the reduction of the opisthosoma to a small abdominal component, and (3) the presence, in the male, of ovigerous legs.

2.4. The Mandibulate Arthropods

The remaining groups of arthropods (crustaceans, myriapods and hexapods) were originally grouped together as the Mandibulata by Snodgrass (1938) because their members possess a pair of mandibles as the primary masticatory organs. Though this view became widely accepted, some later authors, notably Manton, argued forcefully that the mandible of the crustaceans is not homologous with that of the myriapods and insects. That is, the term Mandibulata should not be used to imply a phylogenetic relationship but only a common level of advancement reached by several groups independently (Tiegs and Manton, 1958; Manton, 1977). The debate over whether the Mandibulata constitute a monophyletic group continues to be vigorous (see chapters in Edgecombe, 1998; Fortey and Thomas, 1998; also Section 3.3.1), and the conclusion reached typically hinges on the type of evidence presented. Evidence from comparative morphology, biochemistry, and molecular biology of living species tends to support monophyly, whereas data from fossils generally align the Crustacea with the Chelicerata. With their two pairs of antennae, the Crustacea would appear very distinct from the other two groups. Myriapods and hexapods have a single pair of antennae, a feature that led Sharov (1966) to unite these groups in the Atelocerata. Tiegs and Manton (1958) and Manton (1977) went a step further, placing the two groups with the Onychophora in the Uniramia. However, looks can be deceiving, and many modern phylogeneticists cannot accept the Atelocerata (see Section 3.3.1) and the Uniramia (see Sections 3.2.2 and 3.3) as monophyletic taxa.

Crustacea. To the Crustacea belong the crabs, lobsters, shrimps, prawns, barnacles, and woodlice. The Crustacea are a successful group of arthropods: some 40,000 living species have been described and there is an abundant fossil record. They are primarily aquatic, and few have managed to successfully conquer terrestrial habitats. They exhibit a remarkable diversity of form; indeed, many of the parasitic forms are unrecognizable in the adult stage. Typical Crustacea, however, usually possess the following features: body divided into cephalothorax and abdomen; cephalothorax with two pairs of antennae, three pairs of mouthparts (mandibles and first and second maxillae), and at least five pairs of legs; biramous appendages.

The reason for the success of Crustacea (and perhaps the reason why they replaced trilobites as the dominant aquatic arthropods) is their adaptability. Like their terrestrial counterparts, the insects, crustaceans have exploited to the full the advantages conferred by possession of a segmented body and jointed limbs. Primitive crustaceans, for example, the fairy shrimp (Figure 1.5), have a body that shows little sign of tagmosis and limb specialization. In contrast, in a highly organized crustacean such as the crayfish (Figure 1.6) the appendages have become specialized so that each performs only one or two functions, and the body is clearly divided into tagmata. In the larger (bottom-dwelling) Crustacea specialized defensive weapons have evolved (e.g., chelae, the ability to change color in relation to the environment, and the ability to move at high speed over short distances by snapping the flexible abdomen under the thorax). By contrast, smaller, planktonic Crustacea are often transparent and have evolved high reproductive capacities and short life cycles to facilitate survival.

Myriapoda. The members of four groups of mandibulate arthropods (Chilopoda, Diplopoda, Pauropoda, and Symphyla) share the following features: five- or six-segmented head, unique mandibular biting mechanism, single pair of antennae, absence of compound

FIGURE 1.5. *Branchinecta* sp., a fairy shrimp. [From R. D. Barnes, 1968, *Invertebrate Zoology*, 2nd ed. By permission of the W. B. Saunders Co., Philadelphia.]

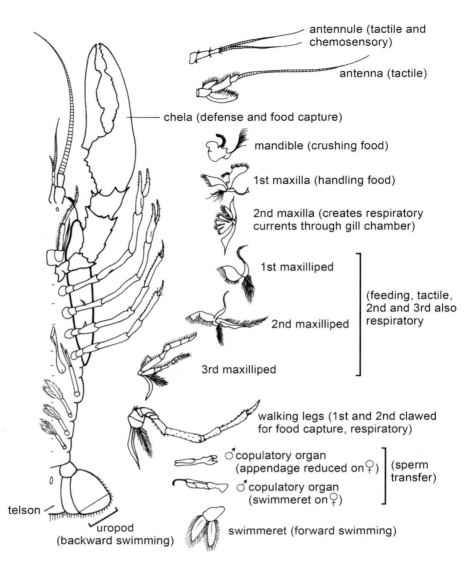

FIGURE 1.6. Crayfish. Ventral view of one side to show differentiation of appendages.

eyes, elongate trunk that bears many pairs of legs, articulation of the coxa with the sternum (rather than the pleuron as in hexapods), tracheal respiratory system, Malpighian tubules for excretion, absence of mesenteric ceca, and distinctive mechanism by which the animal exits the old cuticle during ecdysis. Further, they are found in similar habitats (e.g., leaf mold, loose soil, rotting logs).

For these reasons, they were traditionally placed in a single large taxon, the Myriapoda. The monophyletic nature of the myriapods has been supported by some, but not all, cladistic analyses of large data sets with a combination of morphological and molecular characters of living species (Wheeler *et al.*, 1993; authors in Fortey and Thomas, 1998). Yet other morphological and molecular studies indicate that the myriapods constitute a paraphyletic or even polyphyletic group. Determination of the relationships within the Myriapoda has proved difficult because potentially homologous characters are shared by different pairs of groups. For example, Diplopoda and Pauropoda have the same number of head segments and one pair of maxillae; Diplopoda and Chilopoda have segmental tracheae; and Symphyla

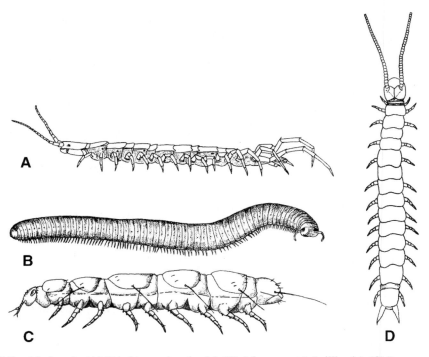

FIGURE 1.7. Myriapoda. (A) *Lithobius* sp. (centipede), (B) *Julus terrestris* (millipede), (C) *Pauropus silvaticus* (pauropod), and (D) *Scutigerella immaculata* (symphylan). [From R. D. Barnes, 1968, *Invertebrate Zoology*, 2nd ed. By permission of the W. B. Saunders Co., Philadelphia.]

and Pauropoda develop embryonic ventral organs that become eversible vesicles, as well as contributing to the ventral ganglia. Boudreaux's (1979) overall conclusion was that the Pauropoda-Diplopoda and Chilopoda-Symphyla are the two sister groups within the taxon (Figure 1.8). An alternative view, based on cladistic analysis of morphological characters of living forms (Kraus, in Fortey and Thomas, 1998), is that the myriapods are paraphyletic: the Chilopoda is the sister group to the other three. Unfortunately, though there is a rich fossil record of myriapods extending back to the Upper Silurian, insufficient study has been done to clarify the monophyletic nature or otherwise of this group (see Shear, in Fortey and Thomas, 1998).

Some 3000 species of chilopods (centipedes) (Figure 1.7A) have been described (Lewis, 1981). They are typically active, nocturnal predators whose bodies are flattened dorsoventrally. The first pair of trunk appendages (maxillipeds) are modified into poison claws that are used to catch prey. In most centipedes the legs increase in length from the anterior to the posterior of the animal to facilitate rapid movement. The earliest known fossil centipedes, from the Upper Silurian, are remarkably similar to some extant species, suggesting that the group may be considerably more ancient.

In contrast to the centipedes, the diplopods (millipedes) (Figure 1.7B) are slow-moving herbivorous animals. The distinguishing feature of the almost 10,000 species in the class is the presence of diplosegments, each bearing two pairs of legs, formed by fusion of two originally separate somites. It is believed that the diplosegmental condition enables the animal to exert a strong pushing force with its legs while retaining rigidity of the trunk region. As they cannot escape from would-be predators by speed, many millipedes have evolved such protective mechanisms as the ability to roll into a ball and the secretion of

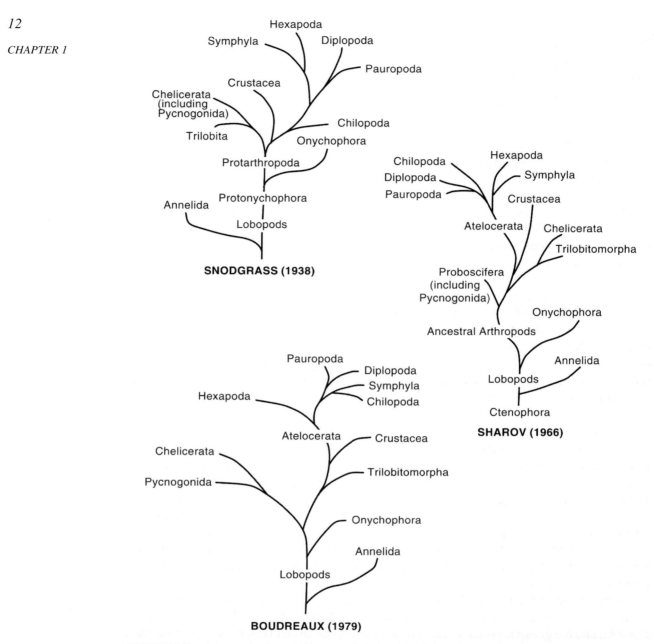

FIGURE 1.8. Schemes for the possible monophyletic origin of the arthropods as proposed by Snodgrass (1938), Sharov (1966), and Boudreaux (1979). Note also the differing relationships of the Annelida, Onychophora, and Arthropoda.

defensive chemicals (Hopkins and Read, 1992). Fossil millipedes are known from the Lower Devonian.

Pauropoda (500 species) are minute arthropods (0.5–2 mm long) that live in soil and leaf mold. Superficially they resemble centipedes, but detailed examination reveals that they are likely the sister group to the millipedes. This affinity is confirmed by such common features as the position of the gonopore, the number of head segments, and the absence of appendages on the first trunk segment (Sharov, 1966). A characteristic feature are the large tergal plates on the trunk, which overlap adjacent segments (Figure 1.7C). It is believed that these large structures prevent lateral undulations during locomotion.

Symphylans (Figure 1.7D) are small arthropods that differ from other myriapods in the possession of a labium (the fused second maxillae) and the position of the gonopore (on the 11th body segment). Although forming only a very small class of arthropods (160 species), the Symphyla have stimulated special interest among entomologists because of the several features they share with insects, leading to the suggestion that the two groups may have had a common ancestry. The symphylan and insectan heads have an identical number of segments and, according to some zoologists, the mouthparts of symphylans are insectan in character. At the base of the legs of symphylans are eversible vesicles and coxal styli. Similar structures are found in some apterygote insects.

Hexapoda.   Five groups of six-legged arthropods (hexapods) are recognized: the wingless Collembola, Protura, Diplura, and thysanurans, and the winged insects (Pterygota). All the wingless forms were traditionally included in the subclass Apterygota (Ametabola) within the class Insecta (= Hexapoda). Although most recent studies indicate that the hexapods are monophyletic, the nature of the relationships of the constituent groups has proved controversial, with perhaps only the thysanurans (now arranged in two orders Microcoryphia and Zygentoma) having a close affinity with the Pterygota.

The Collembola, Protura, and Diplura are often placed in the taxon Entognatha(ta) principally because of the unique arrangement of their mouthparts enclosed within the ventrolateral extensions of the head. Other possible synapomorphies of the entognathans include protrusible mandibles, reduced Malpighian tubules, and reduced or absent compound eyes. However, Bitsch and Bitsch (2000) argue strongly that most of these similarities are due to convergence; that is, the Entognatha is not a monophyletic group. Some classifications unite the Collembola and Protura as sister groups within the Ellipura(ta) based on the following synapomorphies: small body size (8 mm or less), absence of cerci, antennae with four or fewer segments, maxillary palps with three or fewer segments, one-segmented labial palps, and possibly the coiled immotile sperm and absence of abdominal spiracles (Boudreaux, 1979; Kristensen, 1991). Despite these similarities, the Collembola and Protura are quite distinct both from each other and from other hexapods. Collembola have a six-segmented abdomen bearing specialized appendages (see Chapter 5, Section 2), total cleavage in the egg, a long (composite tibiotarsal?) penultimate segment in the legs, and spiracles that either open in the neck region or are absent. Protura lack a tentorium, eyes, and antennae, have 11 abdominal segments (3 of which are added by anamorphosis, and have vestigial appendages on the first three abdominal segments. Indeed, the extensive cladistic analysis of Bitsch and Bitsch (2000) rejects the monophyly of the Ellipura. The position of the Diplura is questionable, and the group is probably not monophyletic (Bitsch and Bitsch, 2000). Some early authors included them in the same order as the thysanurans; Kukalová-Peck (1991) argued that the Diplura are Insecta, forming the sister group to the thysanurans + pterygotes; and many other authors, have placed them in their own class.

It will be readily apparent that a variety of schemes have been devised to show the possible relationships of the hexapod groups (Figure 1.9). The "lumpers" (e.g., Boudreaux) use the terms Hexapoda and Insecta synonymously, so that the Collembola, Protura, and Diplura are considered orders of insects. The "splitters" (e.g., Kristensen), on the other hand, assign each of these groups the rank of class, on a par therefore with the Insecta. As their taxonomic status is controversial, the Protura, Collembola, and Diplura have been included with the thysanurans in Chapter 5 where details of their biology are presented.

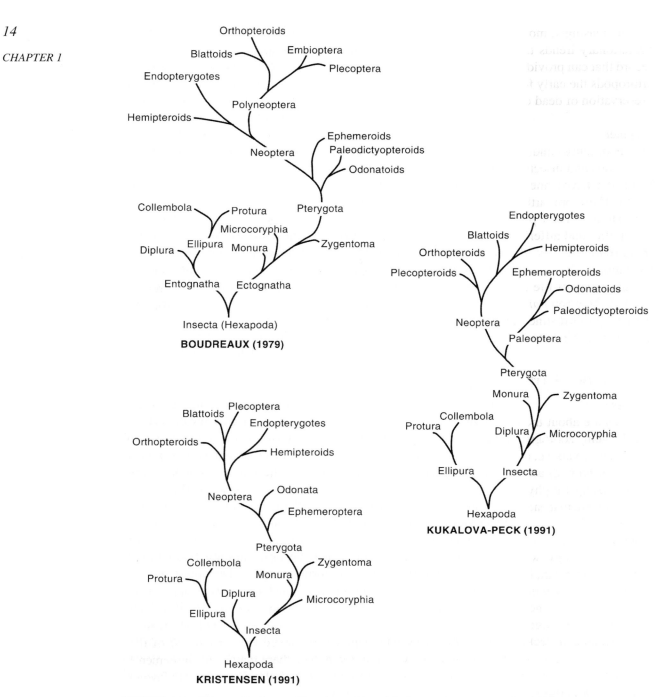

FIGURE 1.9. Schemes for the possible relationships of the hexapod groups as envisaged by Boudreaux (1979), Kristensen (1991), and Kukalová-Peck (1991).

3. Evolutionary Relationships of Arthropods

3.1. The Problem

In determining the evolutionary relationships of animals zoologists use evidence from a variety of sources. The comparative morphology, embryology, physiology, biochemistry and, increasingly, molecular biology of living members of a group provide clues about the evolutionary trends that have occurred within that group. It is, however, only the fossil record that can provide the *direct* evidence for such processes. Unfortunately, in the case of arthropods the early fossil record is poor. By the time the earth's crust became suitable for preservation of dead organisms, in the Cambrian period (about 600 million years ago), the arthropods had already undergone a wide adaptive radiation. Trilobites, crustaceans, and eurypterids were abundant at this time. Even after this time the fossil record is incomplete mainly because conditions were unsuitable for preserving rather delicate organisms such as myriapods and insects. The remains of such organisms are only preserved satisfactorily in media that have a fine texture, for example, mud, volcanic ash, fine humus, and resins (Ross, 1965). Therefore, arthropod phylogeneticists have had to rely almost entirely on comparative studies. Their problem then becomes one of determining the relative importance of similarities and differences that exist between organisms and whether apparently identical, shared characters are homologous (synapomorphic) or analogous (see Chapter 4, Section 3). Evolution is a process of divergence, and yet, paradoxically, organisms may evolve toward a similar way of life (and hence develop similar structures). A distinction must therefore be made between *parallel* and *convergent* evolution. As we shall see below, the difficulty in making this distinction led to the development of very different theories for the origin of and relationship between various arthropod groups.

3.2. Theories of Arthropod Evolution

As Manton (1973, p. 111) noted, "it has been a zoological pastime for a century or more to speculate about the origin, evolution, and relationships of Arthropoda, both living and fossil." Many zoologists have expounded their views on this subject. Not surprisingly, for the reasons noted above, these views have been widely divergent. Some authors have suggested that the arthropods are monophyletic, that is, have a common ancestor; others have proposed that the group is diphyletic (two major subgroups evolved from a common ancestor), and yet others believe that each major subgroup evolved independently of the others (a polyphyletic origin). Within the last 50 years, much evidence has been accumulated in the areas of functional morphology and comparative embryology but especially in paleontology and molecular biology, which has been brought to bear on the matter of arthropod phylogeny. This does not mean, however, that the problem has been solved! On the contrary, vigorous debate continues, with the proponents of each viewpoint pressing their claims, typically by using a particular methodology or a specific kind of evidence (for examples, see authors in Gupta, 1979; Edgecombe, 1998; Fortey and Thomas, 1998; also Emerson and Schram, 1990; Kukalová-Peck, 1992). Only rarely have authors attempted to marshall *all* of the evidence in order to arrive at an *overall* conclusion. Even then, there may be no agreement! For example, the analyses of Boudreaux (1979) and Wheeler *et al.* (1993) led them to favor a monophyletic origin whereas Willmer (1990) concluded that, for the present, a polyphyletic origin for the arthropods is more likely. In outlining the pros and cons of these theories it is useful to separate the mono- and diphyletic theories from the polyphyletic theory and to present them in a historical context showing the gradual development of evidence in support of one view or the other.

3.2.1. Mono- and Diphyletic Theories

In a nutshell, proponents of the monophyletic theory simply point to the abundance of features common to arthropods (Section 2) and argue that so many similarities could

FIGURE 1.10. *Aysheaia pedunculata.*

not have been achieved other than through a common origin. However, their argument goes beyond simply noting the presence of these features; rather, as a result of improved technology and knowledge, the monophyleticists can now point to the highly conserved nature of key arthropod structures and the processes by which they are formed, for example, cuticle chemistry and molting, the development and fine structure of compound eyes, and embryonic head development (see Gupta, 1979). To this can be added ever-increasing evidence from molecular biology, most (but not all) of which supports monophyly. This should not be interpreted to mean that there is agreement among the monophyleticists as to a general scheme for arthropod evolution. On the contrary, there are quite divergent views with respect to the relationships of the various arthropod groups (Figure 1.8).

Space does not permit a detailed account of the early history of monophyletic proposals and readers interested in this should consult Tiegs and Manton (1958). Nevertheless, a few very early schemes should be noted to show how ideas changed as new information became available. The first monophyletic scheme for arthropod evolution was devised by Haeckel (1866).[*] Though believing that arthropods had evolved from a common ancestor, he divided them into the Carides (Crustacea, which included Xiphosura, Eurypterida, and Trilobita) and the Tracheata (Myriapoda, Insecta, and Arachnida). After recognizing that *Peripatus* (Onychophora) had a number of arthropodan features (including a tracheal system), Moseley (1894) envisaged it as being the ancestor of the Tracheata, with the Crustacea having evolved independently. Here, then, was the first diphyletic theory for the origin of arthropods.

At about the same time, after the realization that *Limulus* is an aquatic arachnid, not a crustacean, it was proposed that the aquatic Eurypterida were the ancestors of all terrestrial arachnids. As a result the eurypterid-xiphosuran-arachnid group emerged as an evolutionary line entirely separate from the myriapod-insect line and having perhaps only very slight affinities with the crustaceans. Thus emerged the first example of convergence in the Arthropoda, namely, a twofold origin of the tracheal system.

Handlirsch (1908, 1925, 1937)[*] saw the Trilobita as the group from which all other arthropod classes arose separately. *Peripatus* was placed in the Annelida, its several arthropod features presumed to be the result of convergence. The greatest difficulty with Handlirsch's scheme is the idea that the pleura of trilobites became the wings of insects. This means that the apterygote insects must have evolved from winged forms, which is contrary to all available evidence.

It was at about this time that the Cambrian lobopod fossil *Aysheaia pedunculata* (Figure 1.10) was discovered. This *Peripatus*-like creature had a number of primitive features (six claws at the tip of each leg, a terminal mouth, first appendages postoral, second and third appendages are legs). The associated fauna suggested that this creature was from a marine or amphibious habitat. This and other discoveries led Snodgrass (1938) to suggest another monophyletic scheme of arthropod evolution (Figure 1.8). In this scheme the hypothetical ancestral group were the lobopods (so-called because of the lobelike

[*] Cited from Tiegs and Manton (1958).

outgrowths of the body wall that served as legs). After chitinization of the cuticle and loss of all except one pair of tentacles (which formed the antennae), the lobopods gave rise to the Protonychophora. From the protonychophorans developed, on the one hand, the Onychophora and, on the other, the Protarthropoda in which the cuticle became sclerotized and thickened. Such organisms lived in shallow water near the shore or in the littoral zone. The Protarthropoda gave rise to the Protrilobita (from which the trilobite—chelicerate line developed) and the Protomandibulata (Crustacea and Protomyriapoda). From the protomyriapods arose the myriapods and hexapods. In other words, two essential features of Snodgrass' scheme are that the Onychophora play no part in arthropod evolution and that the mandibulate arthropods (Crustacea, Myriapoda, and Hexapoda) form a natural group, the Mandibulata.

Originally, the major drawback to the scheme was a lack of supporting evidence, especially from the fossil record. Specifically, there were no protomandibulate fossils in the Cambrian period. A second difficulty is that all mandibulate arthropods are united on the basis of a single character. Manton (see Section 3.2.2), especially, argued strongly that the mandible has evolved convergently in Crustacea and the Myriapoda-Hexapoda line. The monophyleticists, on the other hand, believe that the mandibles of crustaceans, myriapods, and hexapods are homologous. Indeed, Kukalová-Peck (1992) insisted that the jaws of *all* arthropod groups are homologous, being formed from the same five original segments. A third difficulty of the Snodgrass scheme is the implied homology of the seven- to nine-segmented biramous appendage of Crustacea with the five-segmented, uniramous appendage of Insecta. Supporters of the Mandibulata concept, for example, Matsuda (1970), derived the insect leg from the ancestral crustacean type by proposing that the extra segments were incorporated into the thorax as subcoxal components. This proposal may be somewhat close to reality as there is now fossil evidence that early insects had appendages with side branches, comparable to those crustaceans, and further, the ancestral insect leg included 11 segments (Kukalová-Peck, 1992, and in Edgecombe, 1998).

Over the 75 years since it was proposed, the merits or otherwise of Snodgrass' scheme have been debated vigorously, and there is still no consensus. Broadly speaking, evidence from morphological, biochemical, and molecular biological studies tend to support the scheme (e.g., Boudreaux, 1979; Wägele, 1993; Wheeler *et al.*, 1993; Wheeler, in Edgecombe, 1998; Bitsch, 2001a,b). For example, Wheeler and coworkers compared more than 100 morphological characters and the 18S rDNA, 28S rDNA and polyubiquitin sequences for almost 30 taxa of arthropods, onychophorans, annelids, and a tardigrade. They reached the unequivocal conclusion that the arthropods are monophyletic, and that the concept of the Uniramia (see below) is no longer tenable. Indeed, their results support Snodgrass' (1938) scheme in every way except that their data indicate the monophyletic nature of the Myriapoda (Snodgrass believed the myriapods to be paraphyletic—see Figure 1.8). By contrast, the Mandibulata concept is rejected by those who examine the fossil evidence (see authors in Edgecombe, 1998, and Fortey and Thomas, 1998). Rather, these workers favor a close relationship between Crustacea and Chelicerata.

3.2.2. The Polyphyletic Theory

Proponents of a polyphyletic origin of arthropods, who share the view that the members of different groups are simply too different to have had a common ancestor, must above all be prepared to make the case that the many similarities in body plan noted in the opening paragraph of Section 2 are the result of convergence. Polyphyleticists point out that convergence is a fairly common phenomenon in evolution and, on theoretical grounds alone, it could be expected that two unrelated groups of animals would evolve toward the same highly desirable situation and, as a result, develop almost identical structures serving the same purpose. (Even the monophyleticists have to accept some degree of convergence, for example, among the tracheae of insects and those of some arachnids and crustaceans.) Polyphyleticists argue that the similar features of arthropods are interrelated and interdependent; that is, they all result from the evolution of a rigid exoskeleton. Thus, in order to grow, arthropods must periodically molt; to move around, they must have articulated limbs and body; tagmosis is a logical consequence of segmentation and results in changes to the nervous and muscular systems; the presence of the cuticle demands changes in the gas exchange, sensory, and excretory systems; and the open circulatory system (hemocoel) is the result of an organism no longer requiring a body cavity with hydrostatic functions. In a sense, then, all the polyphyleticists need to demonstrate is the polyphyletic nature of the cuticular exoskeleton. As noted in Section 2.1, the onychophorans, tardigrades, and pentastomids have such an outer covering (and *some* other arthropod features) yet are generally considered distinct from true arthropods (their very existence tends to make life "uncomfortable" for the members of the monophyletic camp).

The second approach taken by the polyphyletic supporters is to criticize the evidence or the methodology used by those who favor monophyly. For example, they argue that the processes of cuticular hardening used by the three major arthropod groups are quite distinct; quinone-tanning in insects, disulfide bridges in arachnids, and impregnation with organic salts in crustaceans. Likewise, they claim that there are great differences in the structure of the compound eyes among the major arthropod groups. However, they also point out that the ommatidium (the eye's functional unit—see Chapter 12, Section 7.1) of some polychaete worms and bivalve mollusks is highly similar to its counterpart in arthropods, emphasizing the ease with which convergence occurs. Ironically, even the monophyleticists disagree over the number and homologies of the segments that make up the arthropod head.

The early polyphyleticists, including Tiegs and Manton (1958), Anderson (1973), and Manton (1973, 1977), also presented direct evidence to support their theory, largely from comparative embryology and functional morphology. More recently, proponents of polyphyly have added information from paleontology (see authors in Gupta, 1979, and Fortey and Thomas, 1998). According to these authors, the evidence weighs heavily in support of the division of the arthropods into at least three natural groups, each with the rank of phylum (Figure 1.11). The phyla are the Chelicerata, the Crustacea, and the Uniramia

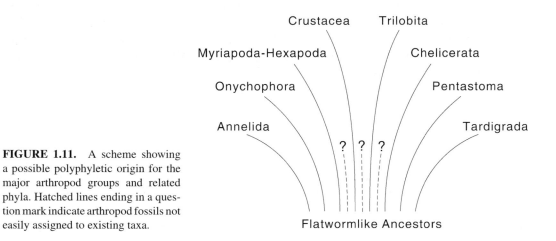

FIGURE 1.11. A scheme showing a possible polyphyletic origin for the major arthropod groups and related phyla. Hatched lines ending in a question mark indicate arthropod fossils not easily assigned to existing taxa.

(Onychophora-Myriapoda-Hexapoda). In some schemes the Trilobita are included as a sister group of the Chelicerata in the phylum Arachnomorpha; in others they are ranked as an independent phylum.

Manton, especially, argued that there are fundamental differences in the structure of the limbs in members of each phylum, related to the manner in which the animals move. In Crustacea the limb is biramous, bearing a branch (exopodite) on its second segment (basipodite); in Uniramia there is an unbranched limb; and in almost all Chelicerata there is a uniramous (unbranched) limb. However, in *Limulus* (and, incidentally, trilobites) the limb is biramous, though the branch originates on the first leg segment (the coxopodite), suggesting that chelicerates may have been initially biramous, losing the branch when the group became terrestrial. This view has been strongly disputed by Kukalová-Peck (1992, and in Fortey and Thomas, 1998) who sees the ancestral leg of *all* arthropods as being biramous and points out that many fossil insects have legs with several branches (i.e., they are "polyramous Uniramia"!). She has urged that the term Uniramia be abandoned. Manton (1973, 1977) made a strong case that the mandibles of the three major groups are not homologous. Based on their structure and mechanism of action, she suggested that the jaws of crustaceans and chelicerates are formed from the basal segment of the ancestral appendage (gnathobasic jaw), though in each group the mechanism of action is different. In Uniramia, however, Manton claimed that the mandible was formed from the entire ancestral appendage, and that members of this group bite with the tip of the limb. Manton pointed out that a segmented mandible is still evident in some myriapods, though the mandible of insects and onychophorans appears unsegmented. Again, this proposal has been severely criticized by Kukalová-Peck (1992, and in Fortey and Thomas, 1998) whose paleoentomological studies suggest the ancestral limb of all arthropods included 11 segments, 5 of which make up the jaw seen in extant species.

Anderson (1973, and in Gupta, 1979) drew evidence from embryology in support of the polyphyletic theory. He compared fate maps (figures indicating which embryonic cells give rise to which organs and structures) among the various groups and concluded that the pattern of development seen in Uniramia bears similarities to that in annelids, yet is very different from that of crustaceans (chelicerates show no generalized pattern, leading to speculation that they may themselves be polyphyletic). It should be noted that not all embryologists agree with Anderson's methods of analysis and, therefore, his conclusions (e.g., Weygoldt, in Gupta, 1979).

When the Mantonian viewpoint was initially presented, there was little supporting evidence from the arthropod fossil record. Within the last three decades, however, there has been considerable activity both in analyzing new species and in reinterpreting some specimens described earlier. Many of these fossils cannot be placed in extant arthropod groups or even along evolutionary lines leading to these groups (Whittington, 1985), indicating that arthropodization was experimented with many times and implying that arthropods had multiple origins. Most of the groups to which these Cambrian fossils belong rapidly became extinct. The arthropod groups seen today represent "successful attempts in applying a continuous, partially stiffened cuticle to a soft-bodied worm" (Willmer, 1990, p. 290).

Willmer (1990) drew attention to the very different methodology used by polyphyletic and monophyletic schools, by which they reach opposite conclusions regarding arthropod evolution. The approach taken by Manton and her supporters has been to search for differences among groups in the belief that they provided evidence for polyphyly. On the other hand, the modern monophyleticists, notably Boudreaux and Wheeler *et al.*, have attempted to determine similarities and use these as proof of a common origin for all arthropods.

The debate as to arthropod relationships and evolution continues to be vigorous (and polarized!) (see Edgecombe, 1998; Fortey and Thomas, 1998). Overall, the balance currently rests in favor of monophyly, with the major groups having had a common origin from a primitive segmented wormlike animal.

3.3. The Uniramians

In agreement with the 19th century zoologists Haeckel and Moseley, Tiegs and Manton (1958) and Manton (1973) made a forceful case for uniting the Onychophora, Myriapoda, and Hexapoda in the arthropod group Uniramia. In their view the many structural similarities between onychophorans and myriapods (see Section 2.1) indicated true affinity and were not the result of convergence. This view received support from the fate map analyses made by Anderson (1973, and in Gupta, 1979) showing the similarity of embryonic development in the three groups. These authors envisaged the evolution of myriapods and insects from onychophoranlike ancestors as a process of progressive cephalization. To the original three-segmented head (seen in modern Onychophora) were added progressively mandibular, first maxillary, and second maxillary (labial) segments, giving rise to the so-called monognathous, dignathous, and trignathous conditions, respectively. Of the monognathous condition there has been found no trace. The dignathous condition occurs in the Pauropoda and Diplopoda, and the trignathous condition is seen in the Chilopoda (in which the second maxillae remain leglike) and the Symphyla and Hexapoda (in which the second maxillae fuse to form the labium).

Few modern authors would support the idea of the onychophorans having common ancestry with the myriapods and insects, preferring to believe that the similarities are due to convergence. Indeed, some authors do not accept that the myriapods and hexapods are sister groups. For example, Friedrich and Tautz (1995) concluded from their comparison of ribosomal nuclear genes that the myriapods were the sister group to the chelicerates, while the crustaceans were the sister group to the hexapods. Unfortunately, the term Uniramia is still used in some texts (e.g., Barnes *et al.*, 1993; Barnes, 1994) to include only the Myriapoda and Hexapoda (i.e., as a synonym of the Atelocerata). As noted earlier, Kukalová-Peck (1992, and in Fortey and Thomas, 1998) has recommended that use of this word be discontinued as the group includes organisms with polyramous legs.

3.3.1. Myriapoda-Hexapoda Relationships

The sharing of features such as one pair of antennae, Malpighian tubules (though these may be secondarily reduced or lost in both groups), anterior tentorial arms, and a tracheal system gave rise to the traditional view that the Myriapoda and Hexapoda are sister groups, collectively forming the Atelocerata (Tracheata), with a common multilegged ancestor (Sharov, 1966; Boudreaux, 1979). Indeed, the existence of several shared features in Symphyla and Hexapoda (Section 2.4) led in the 1930s to the development of the Symphylan Theory for the origin of the hexapods. Within the last decade, however, a major change in opinion has occurred with respect to the relationship between myriapods and hexapods. It is now believed that their common features are the result of convergence or, at best, parallel evolution from a distant common ancestor. Much recent research, in molecular biology, neurobiology, and comparative morphology, often combined in extensive cladistic analyses, supports the hypothesis that hexapods are more closely related to crustaceans than to myriapods. Equally, the data suggest that myriapods are allied with the chelicerates.

Comparisons of mitochondrial and nuclear gene sequences and large hemolymph proteins, examination of eye and brain structure, and studies of nerve development have come out strongly in favor of insects and modern crustaceans as sister groups (Dohle, in Fortey and Thomas, 1998; Shultz and Regier, 2000; Giribet *et al.*, 2001; Hwang *et al.*, 2001; Cook *et al.*, 2001; Burmester, 2002). Some data even suggest that insects arose from the same crustacean lineage as the Malacostraca (crabs, lobsters, etc.), an idea for which Sharov (1966) had been criticized almost 40 years ago.

4. Summary

The arthropods are a very diverse group of organisms whose evolution and interrelationships have been vigorously debated for more than a century. Supporters of a monophyletic origin for the group rely heavily on the existence of numerous common features in the arthropod body plan. Their opponents, who must account for the extraordinary degree of convergent evolution inherent in any polyphyletic theory, argue that all of these features are essentially the result of a single phenomenon, the evolution of a hard exoskeleton, and that arthropodization could easily have been repeated several times among the various ancestral groups. In the polyphyletic theory, therefore, the four dominant groups of arthropods (Trilobita, Crustacea, Chelicerata, and Insecta), as well as several smaller groups both fossil and extant, originated from distinct, unrelated ancestors. The proponents of polyphyly use evidence from comparative morphology (notably studies of limb and mandible structure), comparative embryology (fate maps), and more recently the fossil record (which shows an abundance of arthropod types not easily assignable to already known groups). The monophyleticists claim, in turn, that these comparative embryological and morphological studies are of doubtful value because of the methodology employed and assumptions made. Overall, the current balance seems in favor of a monophyletic origin for the arthropods.

The uniting of Onychophora, Myriapoda, and Hexapoda as the clade Uniramia is highly questionable. Most modern authors agree that apparent similarities between onychophorans and members of the other two groups are due to convergent evolution. The Myriapoda, although including four rather distinct groups (Diplopoda, Chilopoda, Pauropoda, and Symphyla), are widely thought to be monophyletic. For many years, myriapods were considered the sister group to the Hexapoda. However, recent research indicates that myriapods may be allied more closely to the chelicerates, and hexapods to crustaceans. Five distinct groups of hexapods occur: collembolans, proturans, diplurans, thysanurans, and winged insects. On the basis of their entognathous mouthparts and other synapomorphies the first three groups are placed in the Entognatha and are distinct from the thysanurans and pterygotes which form the true Insecta.

5. Literature

Numerous general textbooks on invertebrates, as well as specialized treatises provide information on the biology of arthropods. Tiegs and Manton (1958) give a detailed historical account of schemes for the evolutionary relationships of arthropods. Other major contributors to this fascinating debate include Manton (1973, 1977), Sharov (1966), Anderson

(1973), Boudreaux (1979), Willmer (1990), Wheeler *et al.* (1993), Bitsch (2001a,b), and authors in Gupta (1979), Edgecombe (1998), and Fortey and Thomas (1998).

Anderson, D. T., 1973, *Embryology and Phylogeny of Annelids and Arthropods*, Pergamon Press, Elmsford, NY.

Barnes, R. D., 1994, *Invertebrate Zoology*, 6th ed., Saunders, Philadelphia.

Barnes, R. S. K., Calow, P., and Olive, P. J. W., 1993, *The Invertebrates: A New Synthesis*, 2nd ed., Blackwell, Oxford.

Bitsch, J., 2001a, The hexapod appendage: Basic structure, development and origin, *Ann. Soc. Entomol. Fr. (N.S.)* **37**:175–193.

Bitsch, J., 2001b, The arthropod mandible: Morphology and evolution. Phylogenetic implications, *Ann. Soc. Entomol. Fr. (N.S.)* **37**:305–321.

Bitsch, C., and Bitsch, J., 2000, The phylogenetic interrelationships of the higher taxa of apterygote arthropods, *Zool. Scrip.* **29**:131–156.

Boudreaux, H. B., 1979, *Arthropod Phylogeny, with Special Reference to Insects*, Wiley, New York.

Burmester, T., 2002, Origin and evolution of arthropod hemocyanins and related proteins, *J. Comp. Physiol. B* **172**:95–107.

Cisne, J. L., 1974, Trilobites and the origin of arthropods, *Science* **186**:13–18.

Cook, C. E., Smith, M. L., Telford, M. J., Bastianello, A., and Akam, L., 2001, Hox genes and the phylogeny of arthropods, *Curr. Biol.* **11**:759–763.

Edgecombe, G. D. (ed.), 1998, *Arthropod Fossils and Phylogeny*, Columbia University Press, New York.

Emerson, M. J., and Schram, F. R., 1990, The origin of crustacean biramous appendages and the evolution of Arthropoda, *Science* **250**:667–669.

Foelix, R. F., 1997, *Biology of Spiders*, 2nd ed., Harvard University Press, Cambridge, Massachusetts.

Fortey, R. A., and Thomas, R. H., 1998, *Arthropod Relationships*, Chapman and Hall, London.

Friedrich, M., and Tautz, D., 1995, Ribosomal DNA phylogeny of the major extant arthropod classes and the evolution of myriapods, *Nature* **376**:165–167.

Giribet, G., Edgecombe, G. D., and Wheeler, W. C., 2001, Arthropod phylogeny based on eight molecular loci and morphology, *Nature* **413**:157–161.

Gupta, A. P. (ed.), 1979, *Arthropod Phylogeny*, Van Nostrand-Reinhold, New York.

Hopkins, S. P., and Read, H. J., 1992, *The Biology of Millipedes*, Oxford University Press, London.

Hwang, U. W., Friedrich, M., Tautz, D., Park, C. J., and Kim, W., 2001, Mitochondrial protein phylogeny joins myriapods with chelicerates, *Nature* **413**:154–157.

Kinchin, I. M., 1994, *The Biology of the Tardigrada*, Portland Press, London.

King, P. E., 1973, *Pycnogonids*, St. Martin's Press, New York.

Kristensen, N. P., 1991, Phylogeny of extant hexapods, in: *The Insects of Australia*, 2nd ed., Vol. I (CSIRO, ed.), Melbourne University Press, Carlton, Victoria.

Kukalová-Peck, J., 1991, Fossil history and the evolution of hexapod structures, in: *The Insects of Australia,* 2nd ed., Vol. 1 (CSIRO, ed.), Melbourne University Press, Carlton, Victoria.

Kukalová-Peck, J., 1992, The "Uniramia" do not exist: The ground plan of the Pterygota as revealed by Permian Diaphanopterodea from Russia (Insecta: Paleodictyopteroidea), *Can. J. Zool.* **70**:236–255.

Lewis, J. G. E., 1981, *The Biology of Centipedes*, Cambridge University Press, Cambridge.

Manton, S. M., 1973, Arthropod phylogeny—A modern synthesis, *J. Zool. (London)* **171**: 111–130.

Manton, S. M., 1977, *The Arthropoda: Habits, Functional Morphology and Evolution.* Oxford University Press, London.

Manton, S. M., 1978, Habits, functional morphology and the evolution of pycnogonids, *Zool. J. Linn. Soc.* **63**:1–21.

Matsuda, R., 1970. Morphology and evolution of the insect thorax. *Mem. Entomol. Soc. Can.* **76**:431 pp.

Nelson, D. R., 2001, Tardigrada, in: *Ecology and Classification of North American Freshwater Invertebrates*, 2nd ed., Academic Press, San Diego.

Polis, G. A. (ed.), 1990, *The Biology of Scorpions*, Stanford University Press, Stanford.

Ross, H. H., 1965, *A Textbook of Entomology*, 3rd ed., Wiley, New York.

Sharov, A. G., 1966, *Basic Arthropodan Stock*, Pergamon Press, Elmsford, NY.

Shultz, J. W., and Regier, J. C., 2000, Phylogenetic analysis of arthropods using two nuclear protein-encoding genes supports a crustacean + hexapod clade, *Proc. R. Soc. Lond. Ser. B* **267**:1011–1019.

Snodgrass, R. E., 1938, Evolution of the annelida, onychophora, and arthropoda, *Smithson. Misc. Collect.* **97**:159 pp.

Tait, N. N., 2001, The Onychophora and Tardigrada, in: *Invertebrate Zoology*, 2nd ed. (D. T. Anderson, ed.), Oxford University Press, South Melbourne.

Tiegs, O. W., and Manton, S. M., 1958, The evolution of the Arthropoda, *Biol. Rev.* **33**:255–337.

Wägele, J. W., 1993, Rejection of the "Uniramia" hypothesis and implications of the Mandibulata concept, *Zool. Jahrb. Abt. Syst. kol. Geog. Tiere* **120**:253–288.

Wheeler, W. C., Cartwright. P., and Hayashi, C. Y., 1993, Arthropod phylogeny: A combined approach, *Cladistics* **9**:1–39.

Whittington, H. B., 1985, *The Burgess Shale*, Yale University Press, New Haven.

Whittington, H. B., 1992, *Trilobites*, The Boydell Press, Rochester, NY.

Willmer, P., 1990, *Invertebrate Relationships: Patterns in Animal Evolution*, Cambridge University Press, London.

Insect Diversity

1. Introduction

In this chapter, we shall examine the evolutionary development of the tremendous variety of insects that we see today. From the limited fossil record it would appear that the earliest insects were wingless, thysanuranlike forms that abounded in the Silurian and Devonian periods. The major advance made by their descendants was the evolution of wings, facilitating dispersal and, therefore, colonization of new habitats. During the Carboniferous and Permian periods there was a massive adaptive radiation of winged forms, and it was at this time that most of the modern orders had their beginnings. Although members of many of these orders retained a life history similar to that of their wingless ancestors, in which the change from juvenile to adult form was gradual (the hemimetabolous or exopterygote orders), in other orders a life history evolved in which the juvenile and adult phases are separated by a pupal stage (the holometabolous or endopterygote orders). The great advantage of having a pupal stage (although this is neither its original nor its only significance) is that the juvenile and adult stages can become very different from each other in their habits, thereby avoiding competition for the same resources. The evolution of wings and development of a pupal stage have had such a profound effect on the success of insects that they will be discussed as separate topics in some detail below.

2. Primitive Wingless Insects

The earliest wingless insects to appear in the fossil record are Microcoryphia (Archeognatha) (bristletails) from the Lower Devonian of Quebec (Labandeira *et al.*, 1988) and Middle Devonian of New York (Shear *et al.*, 1984). These, together with fossil Monura (Figure 2.1A) and Zygentoma (silverfish) (Figure 2.1B) from the Upper Carboniferous and Permian periods, constitute a few remnants of an originally extensive apterygote fauna that existed in the Silurian and Devonian periods. Primitive features of the microcoryphians include the monocondylous mandibles which exhibit segmental sutures, fully segmented (i.e., leglike) maxillary palps with two terminal claws, a distinct ringlike subcoxal segment on the meso- and metathorax (in all remaining Insecta this becomes flattened and forms part of the pleural wall), undivided cercal bases, and an ovipositor that has no gonangulum.

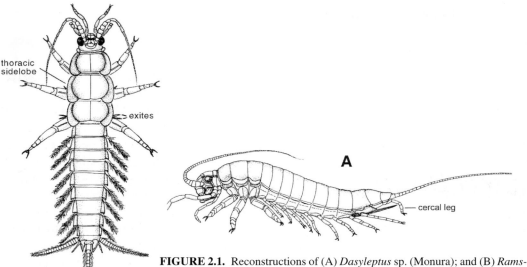

thoracic
sidelobe

exites

A

cercal leg

B

FIGURE 2.1. Reconstructions of (A) *Dasyleptus* sp. (Monura); and (B) *Ramsdelepidion schusteri* (Zygentoma). [From J. Kukalová-Peck, 1987, New Carboniferous Diplura, Monura, and Thysanura, the hexapod ground plan, and the role of thoracic side lobes in the origin of wings (Insecta), *Can. J. Zool.* **65**:2327–2345. By permission of the National Research Council of Canada and the author.]

The early bristletails, like their modern relatives, perhaps fed on algae, lichens, and debris. They escaped from predators by running and jumping, the latter achieved by abrupt flexing of the abdomen.

Monura are unique among Insecta in that they retain cercal legs (Kukalová-Peck, 1985). Other primitive features of this group are the segmented head, fully segmented maxillary and labial palps, lack of differentiation of the thoracic segments, segmented abdominal leglets, the long caudal filament, and the coating of sensory bristles over the body (Kukalová-Peck, 1991). Features they share with the Zygentoma and Pterygota are dicondylous mandibles, well-sclerotized thoracic pleura, and the gonangulum, leading Kukalová-Peck (1987) to suggest that the Monura are the sister group of the Zygentoma + Pterygota. Carpenter (1992), however, included the Monura as a suborder of the Microcoryphia. Shear and Kukalová-Peck (1990) suggested, on the basis of their morphology, that monurans probably lived in swamps, climbing on emergent vegetation, and feeding on soft matter. Escape from predators may have occurred, as in the Microcoryphia, by running and jumping.

In contrast to their rapidly running, modern relatives, the early silverfish, for example, the 6-cm-long. *Ramsdelepidion schusteri* (Figure 2.1B), with their weak legs, probably avoided predators by generally remaining concealed. When exposed, however, the numerous long bristles that covered the abdominal leglets, cerci, and median filament may have provided a highly sensitive, early warning system. Of particular interest in any discussion of apterygote relationships is the extant silverfish *Tricholepidion gertschi*, discovered in California in 1961. The species is sufficiently different from other recent Zygentoma that it is placed in a separate family Lepidotrichidae, to which some Oligocene fossils also belong. Indeed, *Tricholepidion* possesses a number of features common to both Microcoryphia and Monura (see Chapter 5, Section 6), leading Sharov (1966) to suggest that the family to which it belongs is closer than any other to the thysanuranlike ancestor of the Pterygota.

3.1. Origin and Evolution of Wings

The origin of insect wings has been one of the most debated subjects in entomology for close to two centuries, and even today the question remains far from being answered. Most authors agree, in view of the basic similarity of structure of the wings of insects, both fossil and extant, that wings are of monophyletic origin; that is, wings arose in a single group of ancestral apterygotes. Where disagreement occurs is with respect to (1) whether the wing precursors (pro-wings) were fused to the body or were articulated; (2) the position(s) on the body at which pro-wings developed (and, related to this, how many pairs of pro-wings originally existed); (3) the original functions of pro-wings; (4) the selection pressures that led to the formation of wings from pro-wings; and (5) the nature of the ancestral insects; that is, were they terrestrial or aquatic, were they larval or adult, and what was their size (Wootton, 1986, 2001; Brodsky, 1994; Kingsolver and Koehl, 1994).

At the core of all theories on the origin of wings is the matter of whether the pro-wings initially were outgrowths of the body wall (i.e., non-articulated structures) or were hinged flaps. Although there have been several proposals for wing origin based on non-articulated pro-wings (see Kukalová-Peck, 1978), undoubtedly the most popular of these is the Paranotal Theory, suggested by Woodward (1876, cited in Hamilton, 1971), and supported by Sharov (1966), Hamilton (1971), Wootton (1976), Rasnitsyn (1981), and others. The theory is based on three pieces of evidence: (1) the occurrence of rigid tergal outgrowths (wing pads) on modern larval exopterygotes (ontogeny recapitulating phylogeny); (2) the occurrence in fossil insects, both winged (Figure 2.5) and wingless (Figure 2.1B), of large paranotal lobes with a venation similar to that of modern wings; and (3) the assumed homology of wing pads and lateral abdominal expansions, both of which have rigid connections with the terga and, internally, are in direct communication with the hemolymph.

Essentially the theory states that wings arose from rigid, lateral outgrowths (paranota) of the thoracic terga that became enlarged and, eventually, articulated with the thorax. It presumes that, whereas three pairs of paranotal lobes were ideal for attitudinal control (see below), only two pairs of flapping wings were necessary to provide a mechanically efficient system for flight. (Indeed, as insects have evolved there has been a trend toward the reduction of the number of functional wings to one pair [see Chapter 3, Section 4.3.2]). This freed the prothorax for other functions such as protection of the membranous neck and serving as a base for attachment of the muscles that control head movement.

Various suggestions have been made to account for development of the paranota. For example, Alexander and Brown (1963) proposed that the lobes functioned originally as organs of epigamic display or as covers for pheromone-producing glands. Whalley (1979) and Douglas (1981) suggested a role in thermoregulation for the paranota, an idea that has received support from the experiments of Kingsolver and Koehl (1985) using models. Most authors, however, have traditionally believed that the paranota arose to protect the insect, especially, perhaps, its legs or spiracles. Enlargement and articulation of the paranotal lobes were associated with movement of the insect through the air. Packard (1898, cited in Wigglesworth, 1973) suggested that wings arose in surface-dwelling, jumping insects and served as gliding planes that would increase the length of the jump. However, the almost synchronous evolution of insect wings and tall plants supports the idea that wings evolved in insects living on plant foliage. Wigglesworth (1963a,b) proposed that wings arose in small aerial insects where light cuticular expansions would facilitate takeoff and dispersal.

The appearance later of muscles for moving these structures would help the insect to land the right way up. Hinton (1963a), on the other hand, argued that they evolved in somewhat larger insects and the original function of the paranota was to provide attitudinal control in falling insects. There is an obvious selective advantage for insects that can land "on their feet," over those that cannot, in the escape from predators. As the paranota increased in size, they would become secondarily important in enabling the insect to glide for a greater distance. Flower's (1964) theoretical study examined the hypotheses of both Wigglesworth and Hinton. Flower's calculations showed that small projections (rudimentary paranotal lobes) would have no significant advantage for very small insects in terms of aerial dispersal. However, such structures would confer great advantages in attitudinal control and, later, glide performance for insects 1–2 cm in length. Flower's proposals have been examined experimentally through the use of models (Kingsolver and Koehl, 1985; Wootton and Ellington, 1991; Ellington, 1991; Hasenfuss, 2002). These studies have served to emphasize the importance of the ancestral insect's body size, as well as confirming that even quite small projections could contribute to stability (a possible role for appendages such as antennae, legs, and cerci should not be ignored, however). Another consideration is the insect's speed on landing (and whether the insect might be damaged). Ellington's (1991) analysis suggested that the winglets might have been important in reducing this terminal velocity, and there would be strong selection pressure to increase their size as a means of further reducing landing speed.

In the Paranotal Theory a critical step in the transition from gliding to flapping flight would be the development of a hinge so that the winglets became articulated with the body. Most supporters would suggest that this would occur simply to improve the insect's control of attitude or landing speed, though various non-aerodynamic functions may also have been improved through the development of articulated winglets. For example, Kingsolver and Koehl (1985) noted the potential for more efficient thermoregulation that would arise from having movable winglets. Other authors have suggested that the hinge evolved initially in order that the projections could be folded along the side of the body, thereby enabling the insect to crawl into narrow spaces and thus avoid capture. Only later would the movements become sufficiently strong as to make the insect more or less independent of air currents for its distribution. In this hypothesis the earliest flying insects would rest with their wings spread at right angles to the body, as do modern dragonflies and mayflies. The final major step in wing evolution was the development of wing folding, that is, the ability to draw the wings when at rest over the back. This ability would be strongly selected for, as it would confer considerable advantage on insects that possessed it, enabling them to hide in vegetation, in crevices, under stones, etc., thereby avoiding predators and desiccation. An implicit part of the Paranotal Theory is that this ability evolved in the adult stage.

It was Oken (1811, cited in Wigglesworth, 1973) who made the first suggestion that wings evolved from an already articulated structure, namely gills. Woodworth (1906, cited in Wigglesworth, 1973), having noted that gills are soft, flexible structures perhaps not easily converted (in an evolutionary sense) into rigid wings, modified the Gill Theory by suggesting that wings were more likely formed from accessory gill structures, the movable gill plates which protect the gills and cause water to circulate around them. The gill plates, by their very functions, would already possess the necessary rigidity and strength. This proposal receives support from embryology, which has shown abdominal segmental gills of larval Ephemeroptera to be homologous with legs, not wings. Wigglesworth (1973, 1976) resurrected, and attempted to extend, the Gill Theory by proposing that in terrestrial apterygotes the homologues of the gill plates are the coxal styli, and it was from the thoracic coxal styli that wings evolved. Kukalová-Peck (1978) stated that the homology of the wings and styli as proposed by Wigglesworth was not acceptable and pointed out that wings are always located above the thoracic spiracles, whereas legs always articulate with the thorax below the spiracles. In support of Wigglesworth's proposal, it should be noted that primitively wings are moved by muscles attached to the coxae (see Chapter 14, Section 3.3.3) and are tracheated by branches of the leg tracheae.

Gradually, the "articulated pro-wings" proposal has gained support, drawing on evidence from paleontology, developmental biology, neurobiology, genetics, comparative anatomy, and transplant experiments. Among its leading proponents is Kukalová-Peck (1978, 1983, 1987) who not only presented a strong case for a wing origin from articulated pro-wings, but simultaneously cast major doubt on the paranotal theory and the evidence for it. She argued that the fossil record supports none of this evidence. Rather, it indicates just the opposite sequence of events, namely, that the primitive arrangement was one of freely movable pro-wings on all thoracic and abdominal segments of juvenile insects, and it was from this arrangement that the fixed wing-pad condition of modern juvenile exopterygotes evolved. According to Kukalová-Peck, numerous fossilized juvenile insects have been found with articulated thoracic pro-wings. However, with few exceptions even in the earliest fossil insects, both juvenile and adult, the abdominal pro-wings are already fused with the terga and frequently reduced in size. Some juvenile Protorthoptera with articulated abdominal pro-wings have been described, and in extant Ephemeroptera the abdominal pro-wings are retained as movable gill plates.

In proposing her ideas for the origin and evolution of wings, Kukalová-Peck emphasized that these events probably occurred in "semiaquatic" insects living in swampy areas and feeding on primitive terrestrial plants, algae, rotting vegetation, or, in some instances, other small animals. It was in such insects that pro-wings developed. The pro-wings developed on all thoracic and abdominal segments (specifically from the epicoxal exite at the base of each leg), were present in all instars, and at the outset were hinged to the pleura (not the terga). With regard to the selection pressures that led to the origin of pro-wings, Kukalová-Peck used ideas expressed by earlier authors. She suggested that pro-wings may have functioned initially as spiracular flaps to prevent entry of water into the tracheal system when the insects became submerged or to prevent loss of water via the tracheal system as the insects climbed vegetation in search of food. Alternatively, they may have been plates that protected the gills and/or created respiratory currents over them, or tactile organs comparable to (but not homologous with) the coxal styli of thysanurans. Initially, the pro-wings were saclike and internally confluent with the hemocoel. Improved mechanical strength and efficiency would be gained, however, by flattening and by restricting hemolymph flow to specific channels (vein formation). Kukalová-Peck speculated that eventually the pro-wings of the thorax and abdomen became structurally and functionally distinct, with the former growing large enough to assist in forward motion, probably in water. This new function of underwater rowing would create selection pressure leading to increased size and strength of pro-wings, improved muscular coordination, and better articulation of the pro-wings, making rotation possible. These improvements would also improve attitudinal control, gliding ability, and therefore survival and dispersal for the insects if they jumped or fell off vegetation when on land. The final phase would be the development of pro-wings of sufficient size and mobility that flight became possible.

A major difficulty in the theory that wings arose from articulated pro-wings in aquatic or amphibious ancestors is to explain satisfactorily the nature of the intermediate stages. That is, how could fliers evolve from swimmers? Marden and Kramer (1994) made the fascinating suggestion that surface skimming, as seen in some living stoneflies (Plecoptera) and the subadult stage of some mayflies (Ephemeroptera), may represent this intermediate phase. Essentially, surface skimming is running on the water surface, using the weak flapping movements of the wings to generate propulsion. Because the water supports the weight of the insect's body, the muscular demands of skimming are far less than those required in a fully airborne insect. Thus, stoneflies with quite small wings and weak flight muscles can surface skim. Thomas *et al.* (2000) combined a molecular phylogenetic analysis of the Plecoptera with an examination of locomotor behavior and wing structure in representatives of families across the order. Their study showed that surface skimming, along with weak flight, is a retained ancestral trait in stoneflies, supporting the hypothesis that the first winged insects were surface skimmers. Marden and Thomas (2003) have provided further support for Kukalová-Peck's proposals by studying the Chilean stonefly *Diamphipnopsis samali*. The weakly flying adults of *D. samali* use their forewings as oars to row across the water surface. Further, they retain abdominal gills. The larval stage is amphibious, living by day in fast-moving streams, but foraging at the water's edge by night. Thus, *D. samali* may represent a very early stage on the road to true flight: an amphibious lifestyle, the co-occurrence of wings and gills, and the ability to row on the water surface.

In addition to her views on wing origin, Kukalová-Peck has also speculated on the evolution of fused wing pads in juveniles and wing folding. Noting that the earliest flying insects had wings that stuck out at right angles to the body, Kukalová-Peck pointed out that, as they developed (in an ontogenetic sense), the insects would be subjected to two selection pressures. One, exerted in the adult stage, would be toward improvement of flying ability; the other, which acted on juvenile instars, would promote changes that enabled them to escape or hide more easily under vegetation, etc. In other words, it would lead to a streamlining of body shape in juveniles. In most Paleoptera streamlining was achieved through the evolution of wings that in early instars were curved so that the tips were directed backward. At each molt, the curvature of the wings became less until the "straight-out" position of the fully developed wings was achieved. Two other groups of paleopteran insects became more streamlined as juveniles through the evolution of a wing-folding mechanism, a feature that was also advantageous to, and was therefore retained in, the adult stage. The first of these groups, the fossil order Diaphanopterodea, remained primitive in other respects and is included therefore in the infraclass Paleoptera (Table 2.1 and Figure 2.6). The second group, whose wing-folding mechanism was different from that of Diaphanopterodea, contained the ancestors of the Neoptera. The greatest selection pressure would be exerted on the older juvenile instars, which could neither fly nor hide easily. In Kukalová-Peck's scheme, the older juvenile instars were eventually replaced by a single metamorphic instar in which the increasing change of form between juvenile and adult could be accomplished. To further aid streamlining and, in the final juvenile instar, to protect the increasingly more delicate wings developing within, the wings of juveniles became firmly fused with the terga and more sclerotized, that is, wing pads. This state is comparable to that in modern exopterygote (hemimetabolous) insects. Further reduction of adult structures to the point at which they exist until metamorphosis as undifferentiated embryonic tissues (imaginal discs) beneath the juvenile integument led to the endopterygote (holometabolous) condition, that is, the evolution of the pupal stage (Section 3.3).

Regardless of their origin, the wings of the earliest flying insects were presumably well-sclerotized, heavy structures with numerous ill-defined veins. Slight traces of fluting (the formation of alternating concave and convex longitudinal veins for added strength) may have been apparent (Hamilton, 1971). The wings (and flight efficiency) were improved

TABLE 2.1. The Major Groups of Pterygota

Infraclass	Orders	Divisions within Neoptera	
		Martynov's scheme	Hamilton's scheme
Paleoptera	Paleodictyoptera[a] Megasecoptera[a] Diaphanopterodea[a] Permothemistida[a] Protodonata[a] Odonata (dragonflies, damselflies) Ephemeroptera (mayflies)		
Neoptera	Protorthoptera[a] Dictyoptera (cockroaches, mantids) Isoptera (termites) Orthoptera (grasshoppers, locusts, crickets) Miomoptera[a] Protelytroptera[a] Dermaptera (earwigs) Grylloblattodea (grylloblattids) Mantophasmatodea Phasmida (stick and leaf insects) Embioptera (web spinners)	Polyneoptera	Pliconeoptera
	Paraplecoptera[a] Caloneurodea[a] Protoperlaria[a] Plecoptera (stoneflies) Zoraptera (zorapterans)		
	Glosselytrodea[a] Psocoptera (booklice) Phthiraptera (biting and sucking lice) Hemiptera (bugs) Thysanoptera (thrips)	Paraneoptera	Planoneoptera
	Megaloptera (dobsonflies, alderflies) Raphidioptera (snakeflies) Neuroptera (lacewings, mantispids) Mecoptera (scorpionflies) Lepidoptera (butterflies, moths) Trichoptera (caddis flies) Diptera (true flies) Siphonaptera (fleas) Hymenoptera (bees, wasps, ants, ichneumons) Coleoptera (beetles) Strepsiptera (stylopoids)	Oligoneoptera	

[a]Entirely fossil orders.

by a reduction in sclerotization, as seen in Paleoptera. Only the articulating sclerites at the base of the wing and the integument adjacent to the tracheae remained sclerotized, the latter giving rise to the veins. Fluting was accentuated in the Paleoptera, and the distal area of the wing was additionally strengthened by the formation of non-tracheated intercalary veins and numerous crossveins (Hamilton, 1971, 1972).

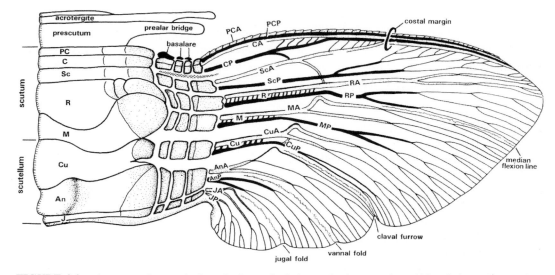

FIGURE 2.2. A proposed ground plan of wing articulation and wing venation. Abbreviations: A, anterior; An, anal; C, costa; Cu, cubitus; J, jugal; M, media; P, posterior; PC, precosta; R, radius; Sc, subcosta. [After J. Kukalová-Peck, 1983, Origin of the insect wing and wing articulation from the arthropodan leg, *Can. J. Zool.* **61**: 1618–1669. By permission of the National Research Council of Canada and the author.]

Kukalová-Peck (1983) argued that the ground plan of wing articulation included eight rows of four articulating sclerites (Figure 2.2). These sclerites were derived, in her view, from the epicoxa of the primitive leg and, as a result, were moved by ancestral leg muscles. Originating on the outer edge of each row was a wing vein. This articular arrangement, seen only in Diaphanopterodea, allowed the sclerites to be crowded and slanted by contraction of these muscles, so that a primitive form of wing folding could occur. In all other Paleoptera, fossil and extant, fusion of sclerites occurred to form axillary plates that, in turn, became united with some veins. Though the details of this process varied among the paleopteran groups, the end result was that, while it undoubtedly strengthened the wing attachment, it prevented wing folding. Essentially, in modern Paleoptera the base of each wing articulates at three points with the tergum, the three axillary sclerites running in a straight line along the body. In the evolution of Neoptera the axillary sclerites altered their alignment so that each wing articulated with the tergum at only two points. This alteration of alignment made wing folding possible.

A second important consequence of the altered articulation of the wing was a further improvement in flight efficiency. In Ephemeroptera and, presumably, most or all fossil Paleoptera the wing beat is essentially a simple up-and-down motion; in Neoptera each wing twists as it flaps and its tip traces a figure-eight path. In other words, the wing "rows" through the air, pushing against the air with its undersurface during the downstroke yet cutting through the air with its leading edge on the upstroke. To carry out this rowing motion effectively necessitated the loss of most of the wing fluting. Only the costal area (Figure 3.27) needs to be rigid as this leads the wing in its stroke, and fluting is retained here (Hamilton, 1971).

Another evolutionary trend, again leading to improved flight, was a reduction in wing weight, permitting both easier wing twisting and an increased rate of wing beating (see also Chapter 14, Section 3.3.4). Concomitant with this reduction in weight was a fusion or loss of some major veins and the loss of crossveins. The extent and nature of fusion or loss of veins followed certain patterns that, together with other structural features, for example, lines of flexion and lines of folding, are potentially important characters on which conclusions about the evolutionary relationships of neopteran insects can be based. Unfortunately, complicating this important tool has been a tendency for authors to use different terminologies when describing the veins and wing areas of different groups of insects, an aspect that is dealt with more fully in Chapter 3 (Section 4.3.2).

3.2. Phylogenetic Relationships of the Pterygota

There are some 25–30 orders of living pterygote insects and about 10 containing only fossil forms, the number varying according to the authority consulted. Clarification of the relationships of these groups may utilize fossil evidence, comparisons of extant forms, or a combination of both. Increasingly, morphological data and molecular information are being combined in massive cladistical analyses in an effort to resolve some long-standing arguments. For example, Wheeler *et al.* (2001) employed 275 morphological variables and 18S and 28S rDNA sequences from more than 120 species of hexapods, plus 6 outgroup representatives, to obtain a "best-fit" analysis of the relationships of the insect orders. Even so, none of these approaches is entirely satisfactory. For example, in extant species secondary modifications may mask the ancestral apomorphic characters. Equally, molecular studies may give spurious results if the sample size is too small. Fossils, on the other hand, are relatively scarce and often poorly or incompletely preserved,[*] especially from the Devonian and Lower Carboniferous periods during which a great adaptive radiation of insects occurred. By the Permian period, from which many more fossils are available, almost all of the modern orders had been established. Misidentification of fossils and misinterpretation of structures by early paleontologists led to incorrect conclusions about the phylogeny of certain groups and the development of confusing nomenclature. For example, *Eugereon*, a Lower Permian fossil with sucking mouthparts, was placed in the order Protohemiptera. It is now realised that this insect is a member of the order Paleodictyoptera and is not related to the modern order Hemiptera as was originally concluded. Likewise the Protohymenoptera, whose wing venation superficially resembles that of Hymenoptera, were thought originally to be ancestral to the Hymenoptera. It is now appreciated that these fossils are paleopteran insects, most of which belong to the order Megasecoptera (Hamilton, 1972). Carpenter (1992) published an authoritative account of the fossil Insecta in which he recognized nine orders of fossil pterygotes. With further work, some of these will undoubtedly require splitting (i.e., they are polyphyletic groups), for example, the Protorthoptera [described by Kukalová-Peck and Brauckmann (1992) as the "wastebasket taxon"!], and species now classified *incertae sedis* (of unknown affinity) will be placed in their correct taxon (Wootton, 1981).

To aid subsequent discussion of the evolutionary relationships within the Pterygota, the various orders referred to in the text are listed in Table 2.1.

It has generally been assumed that the Paleoptera and Neoptera had a common ancestor [in the hypothetical order Protoptera (Sharov, 1966)] in the Middle Devonian, although there is no fossil record of such an ancestor. Remarkably, a recent re-examination of a pair of mandibles first described in 1928 as *Rhyniognatha hirsti*, from the same Lower Devonian deposits as the collembolan *Rhyniella praecursor* (Chapter 5, Section 2), suggests that winged insects may have had a much earlier origin than previously thought (Engel and

---

[*] Many fossil orders were established on the basis of limited fossil evidence (e.g., a single wing). Carpenter (1977) recommended that at least the fore and hind wings, head, and mouthparts should be known before a specimen is assigned to an order.

Grimaldi, 2004). The mandibles are not only dicondylic (Chapter 3, Section 3.2.2) but have other features that are characteristic of mandibles of Pterygota. In other words, flying insects were already well established by the Lower Devonian, some 80 million years earlier than previously assumed. This conclusion agrees with a molecular clock study indicating that insects arose in the Early Silurian (about 430 million years ago), with neopteran forms present by about 390 million years ago (Gaunt and Miles, 2002).

By the Upper Carboniferous period, when conditions became suitable for fossilization, almost a dozen paleopteran and neopteran orders had evolved. Most authors, especially paleontologists, consider the Paleoptera to be monophyletic and the sister group to the Neoptera, and list a number of apomorphies in support of this view (Kukalová-Peck, 1991, 1998). Further, a recent study of 18S and 28S rDNA sequences from almost 30 species of Odonata, Ephemeroptera, and neopterans has provided strong support for the monophyly of the Paleoptera (Hovmöller *et al.*, 2002). However, there are those, notably Boudreaux (1979), Kristensen (1981, 1989, 1995) and Willmann (1998), who, having undertaken cladistic analyses of the extant Ephemeroptera (mayflies) and Odonata (damselflies and dragonflies), believe the Paleoptera to be paraphyletic. In Boudreaux's view the Ephemeroptera + Neoptera form the sister group to the Odonata, while according to Kristensen the best scenario has the Ephemeroptera as the sister group of the Odonata + Neoptera. This view is supported by Wheeler *et al.*'s (2001) analysis, though these authors examined only three species each of Odonata and Ephemeroptera. According to Kukalová-Peck (1991), within the Paleoptera, two major evolutionary lines appeared, one leading to the paleodictyopteroids (Paleodictyoptera, Diaphanopterodea, Megasecoptera, and Permothemistida), the other to the odonatoids + Ephemeroptera. All paleodictyopteroids (Upper Carboniferous-Permian) had a hypognathous head with piercing-sucking mouthparts (Figure 2.3). Adults and large juveniles used these to suck the contents of cones while younger instars probably ingested only fluids (Shear and Kukalová-Peck, 1990). Prothoracic extensions were

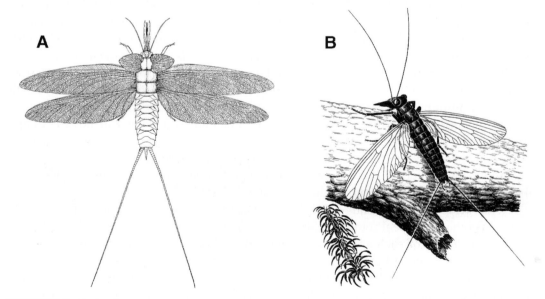

FIGURE 2.3.   Paleodictyopteroids. (A) *Stenodictya* sp. (Paleodictyoptera); and (B) *Permothemis* sp. (Permothemistida). [A, from J. Kukalová, 1970, Revisional study of the order Paleodictyoptera in the Upper Carboniferous shales of Commentry, France. Part III, *Psyche* **77**:1–44. B, from A. P. Rasnitsyn and D. L. J. Quicke (eds.), 2002, *History of Insects.* ⓒ Kluwer Academic Publishers, Dordrecht. With kind permission of Kluwer Academic Publishers and the authors.]

prominent in some paleodictyopteroids (Figure 2.3A) and Kukalová-Peck (1983, 1985) suggested that these were articulated. There was no metamorphic final instar as in modern exopterygotes; that is, wing development was gradual and older juveniles could probably fly. Their extinction at the end of the Permian may be correlated with the demise of the Paleozoic flora (see Section 4.2). Paleodictyoptera formed the largest order of paleodictyopteroids and included some very large species with wingspans up to 56 cm. As noted earlier, the Diaphanopterodea, which may be the sister group of Paleodictyoptera, were unique among Paleoptera in that they were able to fold their wings. Though most diaphanopterodeans were plant-juice feeders, Kukalová-Peck and Brauckmann (1990) observed that some Permian species were remarkably mosquitolike and speculated that these may have fed on blood. Megasecoptera had several features in common with Diaphanopterodea, though these were likely the result of convergence. Contrary to earlier opinions, the Megasecoptera were not carnivores but sucked plant material; a few may have caught other insects and sucked their body fluids. The Permothemistida [formerly the Archodonata and included in the Paleodictyoptera by Carpenter (1992)] were a small group, characterized by having greatly reduced or no metathoracic wings, short mouthparts, and unique wing venation (Figure 2.3B).

Early members of the Ephemeroptera + odonatoid group had biting mouthparts and aquatic juveniles with nine pairs of abdominal gill plates and leglets. Adults of early Ephemeroptera (Upper Carboniferous-Recent) (including the Protoephemeroptera, formerly separated because of their two pairs of identical wings) differed from extant forms in having functional mouthparts. Some very large forms evolved, for example, *Bojophlebia prokopi* with a wingspan of 45 cm. The nature of their mouthparts suggests that nymphs were probably predators, some perhaps feeding on amphibian tadpoles (Kukalová-Peck, 1985) (Figure 2.4A). The early odonatoids differed from Ephemeroptera in features of their venation and in having nymphs that lacked abdominal gill plates, using instead the rectal branchial chamber for gas exchange (Chapter 15, Section 4.1). The group includes two orders Protodonata (Meganisoptera) (Upper Carboniferous-Triassic) and Odonata (Triassic-Recent) that are evidently closely related, some authorities even including the former in the latter order. However, Kukalová-Peck (1991) presented five wing features, and features of the genitalia and cerci that justify their separation. The Protodonata were superb aerial predators, catching prey in flight or from its perch using their long, strong legs (Figure 2.4B). In this diverse and abundant group were the largest known insects (Meganeuridae), including

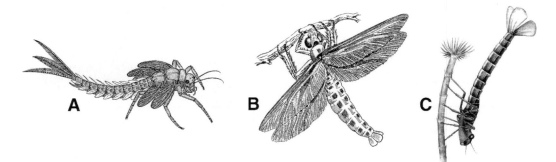

FIGURE 2.4. Early Paleoptera. (A) Juvenile of the Early Permian mayfly, *Kukalová americana*; (B) *Arctotypus* sp., a late Permian protodonatan; and (C) Early Jurassic dragonfly nymph, *Samamura gigantea*. Though the nymph had large anal flaps, reminiscent of the caudal lamellae of damselflies, it used a branchial chamber for gas exchange. [From A. P. Rasnitsyn and D. L. J. Quicke (eds.), 2002, *History of Insects.* © Kluwer Academic Publishers, Dordrecht. With kind permission of Kluwer Academic Publishers and the authors.]

Meganeuropsis permiana with a 71-cm wingspan. Only recently have protodonate juveniles been discovered (Kukalová-Peck, 1991); these had a mask similar to that of odonate larvae (see Figures 2.4C and 6.8). Some also had prominent wings, leading to the possibility that they could fly. A number of Permian fossils originally described as Odonata, specifically in the suborders Archizygoptera and Protanisoptera, have now been reassigned to the Protodonata (Kukalová-Peck, 1991) so that true Odonata are not known before the Triassic. These generally small predators already bore a strong resemblance to the extant Zygoptera and Anisoptera both in form and habits (Figure 2.4C).

In contrast to the Paleoptera, which were inhabitants of open spaces, the Neoptera evolved toward a life among overgrown vegetation where the ability to fold the wings over the back when not in use would be greatly advantageous. The early fossil record for Neoptera is poor, but from the great diversity of fossil forms discovered in Permian strata it appears that the major evolutionary lines had become established by the Upper Carboniferous period.

Two major schools of thought exist with regard to the origin and relationships of these evolutionary lines. The traditional view, proposed by Martynov (1938), is that, shortly after the separation of ancestral Neoptera from Paleoptera, three lines of Neoptera became distinct from each other (Table 2.1 and Figure 2.5A). Based on his studies of fossil wing venation Martynov arranged the Neoptera in three groups, Polyneoptera (plecopteroid, orthopteroid, and blattoid orders), Paraneoptera (hemipteroid orders), and Oligoneoptera (endopterygotes). In a modification of this view Sharov (1966) proposed that the Neoptera and Paleoptera had a common ancestor (i.e., the former did not arise from the latter) and, more importantly, that the Neoptera may be a polyphyletic group. In his scheme (Figure 2.5B) each of the three groups arose independently, a consequence of which must be the assumption that wing folding arose on three separate occasions.

Ross (1955), from studies of body structure, and Hamilton (1972), who examined the wing venation of a wide range of extant species as well as that of fossil forms, concluded that there are two primary evolutionary lines within the Neoptera, the Pliconeoptera and

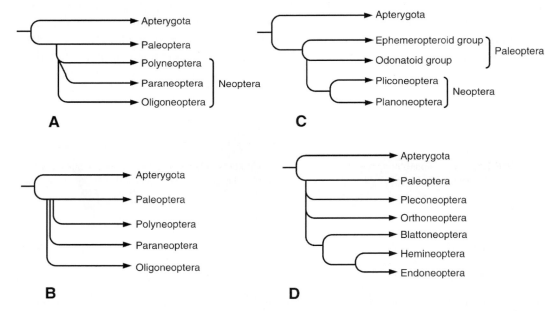

FIGURE 2.5. Schemes for the origin and relationships of the major groups of Neoptera. (A) Martynov's scheme; (B) Sharov's scheme; (C) Hamilton's scheme; and (D) Kukalova-Peck's scheme.

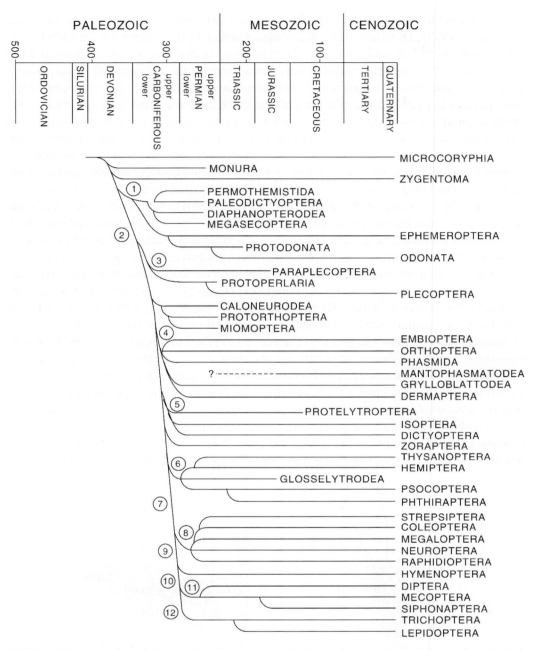

FIGURE 2.6. A possible phylogeny of the insect orders. Numbers indicate major evolutionary lines: (1) Paleoptera; (2) Neoptera; (3) Plecopteroids; (4) Orthopteroids; (5) Blattoids; (6) Hemipteroids; (7) Endopterygotes; (8) Neuropteroids-Coleoptera; (9) Panorpoids-Hymenoptera; (10) Panorpoids; (11) Antliophora; (12) Amphiesmenoptera.

the Planoneoptera (Figures 2.5C and 2.6). The Pliconeoptera corresponds approximately to the Polyneoptera of Martynov but excludes the plecopteroids and the Zoraptera, and a few fossil orders considered planoneopteran by Hamilton. The Planoneoptera includes the Paraneoptera and Oligoneoptera of Martynov's scheme, plus the plecopteroids and Zoraptera. In other words, both schools agree that there are three major groups within the Neoptera but differ with regard to the relationships among these groups.

The monophyletic nature of the Planoneoptera is now widely supported [e.g., see Boudreaux (1979), Hennig (1981), Kristensen (1981,1989), Kukalová-Peck (1991), and Kukalová-Peck and Brauckmann (1992)]. However, there is still some argument as to whether the Pliconeoptera constitutes its sister group (i.e., is monophyletic) or is a polyphyletic assemblage. In Martynov's view the features uniting the pliconeopteran orders included chewing mouthparts, a large anal lobe in the hind wing that folds like a fan along numerous anal veins, complex wing venation (typically including many crossveins) that differs between fore and hind wings, presence of cerci, numerous Malpighian tubules, and separate ganglia in the nerve cord. However, except for the first two, these features are no longer considered to be synapomorphic. The proposed sister-group relationship of the Paraneoptera and Oligoneoptera (i.e., the unity of the Pliconeoptera) has been given strong support by the extensive analysis of Wheeler *et al.* (2001). Kukalová-Peck (1991) and Kukalová-Peck and Brauckmann (1992) presented a new scheme for relationships among the Neoptera (Figure 2.5D), claiming several potential synapomorphic features of wing venation between plecopteroids and orthopteroids (implying a sister-group relationship). Yet, they found no apomorphies shared by orthopteroids and blattoids. Rather, the latter have possible synapomorphies with the Paraneoptera; that is, the two may be sister groups.

Generally included in the plecopteroids are the fossil orders Protoperlaria (Upper Carboniferous-Permian) and Paraplecoptera (Upper Carboniferous-Jurassic) [both of which are considered to be Protorthoptera by Carpenter (1992)], and the extant order Plecoptera (Permian-Recent). However, members of the two fossil orders are included in the Grylloblattodea by Storozhenko (1997) (and see below). The Protoperlaria may have been the ancestors of the Plecoptera. Early plecopteroids had well formed prothoracic winglets, chewing mouthparts, and long cerci. In some species there was no metamorphic final juvenile instar. In some species the young nymphs were semiaquatic, with articulated thoracic winglets and nine pairs of abdominal gills (Figure 2.7A). Older juveniles may have been terrestrial and able to fly. The Plecoptera (stoneflies) appear to have separated from the

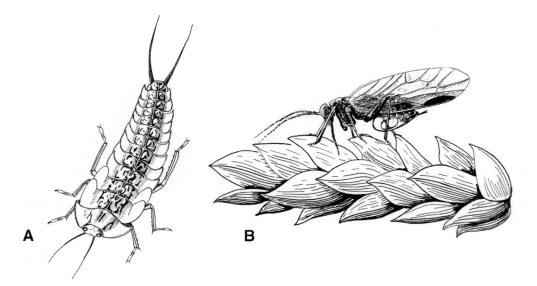

FIGURE 2.7. (A) Early Permian plecopteroid nymph, *Gurianovaella silphidoides*; and (B) The most primitive hemipteran, a member of the Archescytinidae, feeding on a cone of an Early Permian gymnosperm. [From A. P. Rasnitsyn and D. L. J. Quicke (eds.), 2002, *History of Insects.* © Kluwer Academic Publishers, Dordrecht. With kind permission of Kluwer Academic Publishers and the authors.]

remaining plecopteroids early and even by the time at which fossil stoneflies appear, some of these are assignable to extant families (Wootton, 1981).

As noted earlier, the Protorthoptera (Upper Carboniferous-Permian) is a "mixed bag" of fossils, almost certainly a polyphyletic group. Not surprisingly, it has often been suggested as the group from which the remaining orthopteroid orders evolved. The major difficulty in clarifying relationships within the group is that some 80% of Carboniferous protorthopterans are known only from fore wings or wing fragments. Permian forms are generally more completely preserved and superficially may resemble other groups (e.g., Plecoptera and Dictyoptera), though are obviously "too late" to be their ancestors (Wootton, 1981). A recent re-examination of the Protorthoptera by Kukalová-Peck and Brauckmann (1992) indicated that the majority of protorthopterans are primitive hemipteroids, though the group also includes plecopteroids, orthopteroids, blattoids, and even endopterygotes! The order Miomoptera (Upper Carboniferous-Permian) was erected to include a group of small, chewing insects with homonomous wings, simple venation, and short, distinct cerci, that were originally included in the Protorthoptera. The position of this order remains debatable; some authors (e.g., Carpenter, 1992) suggested that miomopterans may be hemipteroid, perhaps close to the Psocoptera, while others (e.g., Kukalová-Peck, 1991) believe that they may be endopterygotes, possibly close to the panorpoid-Hymenoptera stem group. Unfortunately, the immature stages are unknown. Another Upper Carboniferous-Permian group, the Caloneurodea, is also problematical. The chewing mouthparts seen in some fossils, short cerci, and wing venation led Carpenter (1977, 1992) to place them close to the Protorthoptera. Shear and Kukalová-Peck (1990) and Kukalová-Peck (1991), on the basis of the inflated clypeus (housing the sucking apparatus) and the chisellike laciniae, consider them hemipteroids, while some Russian paleontologists have suggested they are plecopteroids or even endopterygotes, perhaps close to the base of the neuropteroids and Coleoptera (Storozhenko, 1997).

The orthopteroid orders include the Orthoptera, Phasmida, Dermaptera, Grylloblattodea, probably the Mantophasmatodea, and possibly the Embioptera and Zoraptera. Orthoptera were widespread by the Upper Carboniferous, being easily recognizable by their modified hindlegs and particular wing venation. Early in the evolution of this order a split occurred, one line leading to the Ensifera (long-horned grasshoppers and crickets), the other to the Caelifera (short-horned grasshoppers and locusts). Indeed, Kevan (1986) and others have strongly urged that the two groups each be given ordinal status, an arrangement supported by those who claim, on the basis of dubious paleontological evidence, that the Caelifera and Phasmida (stick insects) may be sister groups. However, in addition to the two features already noted, the laterally extended pronotum covering the pleuron, the horizontally divided prothoracic spiracle, and the hind tibia with two rows of teeth appear to be synapomorphies confirming the unity of the Orthoptera. The fossil record of the Phasmida is poor, though specimens are known from the Upper Permian onward. Kamp's (1973) phenetic analysis of extant forms indicated that the phasmids are closest to the Dermaptera and Grylloblattodea, the three orders forming a natural group. Boudreaux (1979), on the other hand, listed a number of possible synapomorphies that would render the Phasmida and Orthoptera sister groups, (a view supported by Wheeler *et al.*'s (2001) study). Authorities still disagree on the affinities of the Dermaptera (earwigs), which do not appear in the fossil record until the Lower Jurassic. Some have placed them close to the Plecoptera, Orthoptera, Embioptera, and even the endopterygote Coleoptera, while others considered them to be only distantly related to any of the extant orthopteroid groups. Giles' (1963) comparative morphological study and the combined morphological-molecular analysis by Wheeler *et al.*

(2001) suggest that they are the sister group to the Grylloblattodea. In contrast, Boudreaux (1979) and Kukalová-Peck (1991) included the order in the blattoid group, though according to Kristensen (1981) the presumed synapomorphies are weak. Grylloblattodea (rock crawlers) show an interesting mixture of orthopteran, phasmid, dermapteran, and dictyopteran features, which led to an early suggestion that they are remnants of a primitive stock from which both orthopteroids and blattoids evolved. According to Storozhenko (1997), grylloblattid fossils are known from the Middle Carboniferous onward, and these insects were among the most abundant insects in the Permian. He believes that the group included the ancestors of Plecoptera, Embioptera and Dermaptera. As noted above, Kamp's analysis showed that considerable similarity exists between the grylloblattids and dermapterans, which supports the conclusion reached by Giles (1963) and Wheeler *et al.* (2001) that the two may be sister groups.

The fossil record of the Embioptera (web spinners) extends back to the Lower Permian, though even by this stage the wing venation was reduced and the asymmetric genitalia of males was evident. Web spinners share features with the Plecoptera, Dermaptera, and Zoraptera], though it is unclear whether these are primitive or derived. Wheeler *et al.* (2001) place them as the sister group to Plecoptera based on examination of two species. The phylogenetic position of the Zoraptera (zorapterans) is also uncertain. The order is not encountered in the fossil record until the Upper Eocene/Lower Miocene. As noted, zorapterans share features with the web spinners, earwigs, and stoneflies; however, the few Malpighian tubules, composite abdominal ganglia, and two-segmented tarsi are features that could align them with the hemipteroids. Wheeler *et al.*'s (2001) analysis suggests a sister-group relationship with the Dictyoptera + Isoptera.

Included in the blattoid group of orders are the Protelytroptera (Permian-Lower Cretaceous), Dictyoptera (Upper Carboniferous-Recent), and Isoptera (Lower Cretaceous-Recent). Protelytropterans were apparently an abundant group judging by the amount of fossil material discovered, though this may be somewhat artifactual because their highly sclerotized, elytralike fore wings were readily preserved. The latter are remarkably similar to the elytra of some early Coleoptera, and often it is only when other evidence is available (e.g., the hind wing) that the correct identification can be made (Wootton, 1981). The Protelytroptera appear to be an early branch off the line leading to the Dictyoptera, and in Kukalová-Peck's (1991) view were probably ancestral to the Dermaptera. The Dictyoptera (cockroaches and mantids) and Isoptera (termites) are clearly monophyletic, and some authors (e.g., Kristensen, 1981, 1991) see little point in giving each of these ordinal status. Cockroaches underwent a massive radiation in the Upper Carboniferous (often referred to as the Age of Cockroaches in view of the commonness of their remains) and the order remains extensive today. Female Paleozoic cockroaches had a long, well-developed ovipositor, and the evolution of the short, internal structure seen in modern forms apparently did not occur until the end of the Mesozoic. Reports of fossilized oothecae from the Upper Carboniferous are, according to Carpenter (1992), "not very convincing." Within the Dictyoptera two trends can be seen. The cockroaches became omnivorous, saprophagous, nocturnal, often secondarily wingless insects, whereas the mantids (not known as fossils until the Eocene) remained predaceous and diurnal. Although termites are known as fossils only from the Cretaceous onward, comparison of their structure and certain features of their biology with those of cockroaches (some of which are subsocial) indicates that they are derived from blattoidlike ancestors (Weesner, 1960). Indeed, certain venational features and the method of wing folding in the primitive termite *Mastotermes* resemble those of fossil rather than extant cockroaches.

The relationships of the recently erected order Mantophasmatodea remain unclear. Though unquestionably orthopteroid, members of this order possess a blend of features that suggests their closest relatives may be Grylloblattodea or Phasmida (Klass *et al.*, 2002).

The Paraneoptera (hemipteroid orders) share a number of features. They possess suctorial mouthparts, and the clypeal region of the head is enlarged to accommodate the cibarial sucking pump (see Figure 3.17). Their tarsi have three or fewer segments, the ganglia in the nerve cord are fused, there are six or fewer Malpighian tubules, and cerci are absent (at least in extant species). The anal lobe of the hind wing is reduced, never having more than five veins, and when the hind wing is drawn over the abdomen, it folds once along the anal or jugal fold, not between the anal veins as in the pliconeopterans. The wing venation of hemipteroids is much reduced as a result of fusion of primary veins and almost complete loss of crossveins and, when both fore and hind wings are present, is basically similar in each. Included in the hemipteroid assemblage are four extant orders and a few entirely fossil groups, though the number of these will surely increase as the Protorthoptera are reworked. As noted earlier, the Miomoptera and Caloneurodea are considered hemipteroid by some authors, but are included in the endopterygote or pliconeopteran groups by others. The Glosselytrodea (Permian-Jurassic) are considered hemipteroid by Hamilton (1972) and Kukalová-Peck (1991) on the basis of limited wing-venational features, though most authorities consider them endopterygotes close to the neuropteroids (Carpenter, 1977, 1992), based on different interpretation of the homologies of the wing venation. Unfortunately, the mouthparts and immature stages are not known. Hemiptera (true bugs) (Figure 2.7B), Psocoptera (psocids), and Thysanoptera (thrips) are known from as early as the Lower Permian period. The rich fossil record of the hemipterans indicates that the three major groups (Sternorrhyncha, Auchenorrhyncha, and Heteroptera) were already separated in the Permian (Wootton, 1981; Kukalová-Peck, 1991). Psocopterans may be the closest to the hemipteroid stem group, though their simplified wing venation and stubby, triangular mouthparts are derived features. The thrips are poorly represented in the Paleozoic fossil record, though interestingly the earliest specimens still have symmetrical mouthparts, unlike the extant forms in which the right mandible has been lost. Phthiraptera (chewing and sucking lice) are barely known as fossils; *Saurodectes*, a chewing louse from the Lower Cretaceous, may have parasitized pterosaurs. However, the many similarities between them and Psocoptera, notably the specialized preoral water-uptake mechanism, ovipositor structure, polytrophic ovarioles, hypopharynx (in primitive chewing lice), and nuclear rDNA sequences, suggest that the two are sister groups (Wheeler *et al.*, 2001). Indeed, some authorities include them in a single order Psocodea.

The main feature that unites members of the Oligoneoptera (endopterygotes) is the presence of the pupal stage between the larval and adult stages in the life history. Other probable synapomorphies include the absence of compound eyes in the immature stages (instead, stemmata occur), development of the wing rudiments in pouches beneath the larval integument, and the absence of external genitalia in immature stages. Despite these features, early investigators experienced some difficulty in deciding whether the group had a monophyletic or polyphyletic origin. The difficulty arose because, whereas the orders Mecoptera, Lepidoptera, Trichoptera, Diptera, and Siphonaptera show obvious affinities with each other and form the so-called panorpoid complex (Hinton, 1958), the remaining groups (neuropteroids, Coleoptera, Hymenoptera, and Strepsiptera) appear quite distinct, each apparently bearing little similarity to any other endopterygote group. The modern consensus, supported by both morphological and molecular data, is that the Oligoneoptera is a monophyletic taxon with the major subgroups forming at a very early date. However, opinions differ with respect to the constituent sister groups [see Boudreaux (1979), Kristensen (1981, 1989, 1995), Kukalová-Peck (1991, 1998), Wheeler *et al.* (2001), Kukalová-Peck and Lawrence (2004)]. Currently, the most favored view is that the two primary sister groups are the neuropteroids + Coleoptera and the panorpoids + Hymenoptera (temporarily setting aside the status of the Strepsiptera), but it must be emphasized that the supporting evidence is not strong. Putative synapomorphies of the former group include the absence of cruciate ventral neck muscles, prognathous head with a gula, female genitalia, and campodeiform larva; those of the panorpoids + Hymenoptera are orthognathous head without a gula, eruciform larva with single-clawed legs, and ability to produce silk from labial glands. The majority of molecular phylogenetic analyses support this arrangement (see Wheeler, 1989; Wheeler *et al.*, 2001).

The neuropteroid group includes three quite homogeneous orders—Neuroptera (lacewings), Megaloptera (alderflies and dobsonflies), and Raphidioptera (snakeflies)—which are sometimes included in a single order primarily on the basis of their very similar ovipositor (and the difficulty in determining good apomorphic characters for each). Neuroptera and Megaloptera were already well established in the Permian and probably reached their peak diversity in the Triassic/Jurassic. Fossil Raphidioptera are not known until the Jurassic [reports suggesting their earlier existence are dubious according to Kukalová-Peck (1991)] and never reached the abundance of the other neuropteroids.

Remains of genuine Coleoptera (beetles) are known from the Upper Permian period. Somewhat earlier elytralike remains, originally thought to be from beetles, are now known to belong to the Protelytroptera (see above). Though some early paleontologists suggested that the Coleoptera had protorthopteran ancestors, implying at least a diphyletic origin for the endopterygotes, Crowson (1960, 1981) and Kukalová-Peck (1991), among others, made a case for common ancestry with the neuropteroids. According to Crowson, this proposal is substantiated by the Lower Permian fossil *Tshekardocoleus*, which is intermediate in form between Coleoptera and Megaloptera. Crowson (1975) included *Tshekardocoleus* in the suborder Protocoleoptera, within the Coleoptera. Kukalová-Peck (1991), however, preferred to place it (and other beetlelike insects known from elytra in the same period) in a separate, probably paraphyletic order. The Coleoptera-neuropteroid sister-group relationship is strongly supported by the extensive analysis of Wheeler *et al.* (2001).

The position of the Strepsiptera (stylopoids), highly modified endoparasitic insects, remains controversial. The earliest fossils, from the Lower Cretaceous, are assignable to the extant family Elenchidae, so that speculation on their origin is based on comparative morphology. Kristensen (1981, 1989, 1991) has repeatedly noted that, based on the occurrence of instars with external wing buds and the carryover of larval eyes to the adult instar, the stylopoids could be considered exopterygotes. However, on the basis of many other features, they are unquestionably endopterygotes and on different occasions they have been allied with the panorpoids, Hymenoptera, and Coleoptera. Many authorities agree that the latter is the most likely arrangement, though opinions differ as to whether, for example, they are highly modified beetles [Crowson (1981) includes them as a family, Stylopidae, of Coleoptera] or are the sister order of the Coleoptera (Boudreaux, 1979; Kristensen, 1981; Kukalová-Peck, 1991, 1998; Kukalová-Peck and Lawrence, 1993, 2004). In support of the latter view the use of the hind wings only in flight and features of the hind wing venation are cited as synapomorphies. Other features taken to indicate a close association between the two groups are the extensive sclerotization of the sternum (rather than the tergum), the resemblance between the first instar larva of Strepsiptera and the triungulin larva of the beetle families Meloidae and Rhipiphoridae, and the similarity between the habits of endoparasitic forms of Rhipiphoridae and those of Strepsiptera. By contrast, a number of molecular or combined morphological/molecular analyses have come out strongly in support of a Strepsiptera-Diptera sister-group relationship (see references in Whiting, 1998 and Wheeler *et al.*, 2001).

Synapomorphic features of the orders that make up the panorpoid complex are "admittedly inconspicuous" according to Kristensen (1991). They include the vestigial or lost ovipositor, insertion of the pleural muscle on the first axillary plate, transversely divided larval stipes, and loss or addition of various muscles in the larval labium and maxilla. Within the panorpoid complex two well-substantiated sister groups (sometimes designated superorders) Antliophora (Mecoptera, Diptera, and Siphonaptera) and Amphiesmenoptera (Trichoptera and Lepidoptera) are recognized. An abundance of mecopteralike fossil wings have been recovered from Paleozoic strata, but it frequently has been difficult to determine whether these belong to Mecoptera, Diptera, or the stem group ancestral to both of these orders. Nevertheless, genuine Mecoptera (scorpionflies) are known from Upper Permian deposits and may have been the first endopterygotes to diversify widely. The close link between Diptera (true flies) and Mecoptera implied above, that is, by the inability to distinguish between fossil wings of the two groups, is supported by the existence of four-winged fossil "flies" (*Permotanyderus* and *Choristotanyderus*) in the Upper Permian. These may not be true Diptera but just off the main evolutionary line in a separate group Protodiptera. Interestingly, the only direct evidence for the existence of Paleozoic Diptera (a single wing of *Permotipula patricia,* collected from Upper Permian deposits in Australia during the 1920s) was lost for more than 50 years, only to be rediscovered in the British Museum (London) and redescribed in the late 1980s (Willmann, 1989). Though they may have originated from the mecopteroid stem group during the Carboniferous (Kukalová-Peck, 1991), the Siphonaptera (fleas) do not appear in the fossil record until the Lower Cretaceous. As their names indicate, some of these (*Saurophthirus* and *Saurophthiroides*) are thought to have possibly been parasites of flying reptiles. Because they are so highly modified for their ectoparasitic mode of life, comparative studies of living fleas must be interpreted cautiously. With their apodous larvae and adecticous pupae, fleas resemble Diptera, suggesting the two may be sister groups. However, most authors, noting similarities in sperm ultrastructure, thoracic skeleton, nervous system, foregut, and molecular genetic sequences believe that these indicate a sister-group relationship between the Mecoptera and Siphonaptera.

The monophyletic nature of the Amphiesmenoptera is unquestioned with more than 20 synapomorphies common to the Trichoptera (caddisflies) and Lepidoptera (butterflies and moths) (Kristensen, 1984). It is presumed that these orders had their origin in the Paleozoic from mecopteralike ancestors, though there is little in the fossil record to substantiate this claim. *Microptysmella* and related fossils from the Lower Permian may be members of the stem group from which the two orders are derived. From his comparison of primitive members of both orders, Ross (1967) suggested that the common ancestor was in the adult stage trichopteran and in the larval stage lepidopteran in character. In the evolution of Trichoptera the larva became specialized for an aquatic existence, but the adult remained primitive. Along the line leading to Lepidoptera the larva retained its primitive features, but the adult became specialized, especially in the development of the suctorial proboscis. The earliest genuine caddis fly fossils are from the Triassic and some of these are assignable to extant families. Most extant families probably originated in the Jurassic (Hennig, 1981), and caddis fly cases have been found in Lower Cretaceous deposits. Fossil Lepidoptera are known with certainty only from the Lower Jurassic onward, earlier specimens from the Triassic being more likely Trichoptera or Mecoptera. Like those of extant Micropterigidae, adults of the earliest Lepidoptera probably had chewing mouthparts and were pollen feeders;

the larvae probably fed on liverworts and mosses. The great adaptive radiation of the order probably came at the end of the Cretaceous period and beyond and was correlated with the evolution of the flowering plants.

Though generally aligned with the panorpoids, the Hymenoptera (bees, wasps, ants, sawflies) are quite distinct from all other endopterygotes. Indeed, the recent study by Kukalová-Peck and Lawrence (2004) has the Hymenoptera as the sister group to all other endopterygotes. Fossils are known from the Triassic period, but these were already quite specialized, clearly recognizable as belonging to the extant symphytan family Xyelidae. Members of the suborder Apocrita, which contains the parasitic and stinging forms, are not known as fossils until the Jurassic and Cretaceous periods. The great adaptive radiation of this suborder was, like that of the Lepidoptera, clearly associated with the evolution of the angiosperms (see Section 4.1).

The foregoing discussion of the evolutionary relationships within the Insecta is summarized in Figure 2.6.

3.3. Origin and Functions of the Pupa

As noted in the previous section the Oligoneoptera (endopterygote orders) are characterized by the presence of a pupal stage between the juvenile and adult phases in the life history. The development of this stage, which serves various functions, is a major reason for the success (i.e., diversity) of endopterygotes. Given the importance of the pupal stage, it is not surprising that several theories have been proposed for its origin (Figure 2.8).

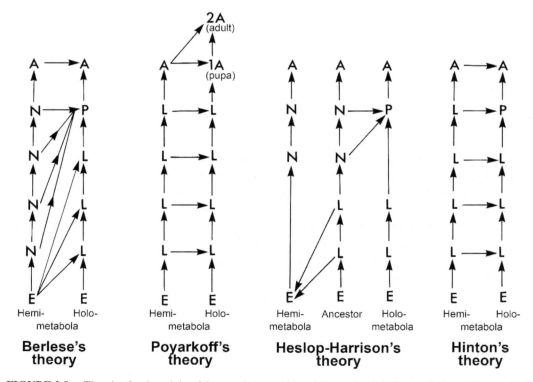

FIGURE 2.8. Theories for the origin of the pupal stage. Abbreviations: A, adult; E, egg; L, larva; N, nymph; P, pupa. [Partly after H. E. Hinton, 1963b, The origin and function of the pupal stage, *Proc. R. Entomol. Soc. Lond. Ser. A* **38**:77–85. By permission of the Royal Entomological Society.]

Among the earliest proposals was that of Berlese (1913, cited in Hinton, 1963b) who used the principle of "ontogeny recapitulates phylogeny" to develop his ideas. During its development an insect embryo passes through three distinct stages. In the first (protopod) stage no appendages are visible; this is followed by the polypod stage (in which appendages are present on most segments); and finally the oligopod stage (when the appendages on the abdomen have been resorbed) (see Chapter 20, Section 7.1). Berlese suggested that the eggs of exopterygotes, by virtue of their greater yolk reserves, hatch in a postoligopod stage of development, whereas the eggs of endopterygotes, which have less yolk, hatch in the polypod or oligopod stages. According to Berlese, the larvae of endopterygotes correspond to free-living embryonic stages, while the pupa represents the compression of the exopterygote nymphal stages into a single instar. The major fault of Berlese's idea is the absence of evidence that the eggs of exopterygotes have a better supply of yolk than those of endopterygotes (Hinton 1963b). Further, the theory implies that abdominal prolegs (see Chapter 3, Section 5.2) are homologous with thoracic legs. Hinton (1963b) argued that prolegs are secondary larval structures, though this claim is not justified given the multilegged nature of the ancestors of insects (Heslop-Harrison, 1958; Kukalová-Peck, 1991). Truman and Riddiford (1999, 2002) have resurrected interest in Berlese's proposal following their detailed studies of the endocrine control of embryonic development. Truman and Riddiford compared the subtle shifts in the timing of juvenile hormone activity in embryos of hemimetabolous and holometabolous insects. As well, they examined the effects of treating embryos with extra juvenile hormone or precocene (which destroys the corpora allata) at various stages of development. Truman and Riddiford argued that in ancestral hemimetabolous insects there were three postembryonic stages: pronymph, nymph and adult. These correspond to the larval, pupal, and adult stages, respectively, of holometabous forms. In modern hemimetabolous forms the pronymph has been retained as a short-lived, non-feeding stage, which typically is spent within the egg. By contrast, in holometabolous insects with earlier secretion of juvenile hormone the pronymph took on increasing importance, becoming a long-lived, multi-instar feeding stage while the nymphal instars, as proposed by Berlese, would be reduced to the single (pupal) stage.

Poyarkoff's (1914) theory (cited in Hinton, 1963b) offers a major advantage over Berlese's, namely, that it provides a causal explanation for the origin and function of the pupal stage. According to this theory, the eggs of both endopterygotes and exopterygotes hatch at a similar stage of development. The adult stage in the exopterygote ancestors of the endopterygotes became divided into two instars, the pupa and the imago. Poyarkoff suggested that the subimago of Ephemeroptera (see Chapter 6, Section 2) and the "pupal" stage of some exopterygotes (see Chapter 8, Sections 4 and 5) are equivalent to the endopterygote pupal stage. Further, the pupa (especially that of primitive endopterygotes such as Neuroptera) resembles the adult rather than the larva. In his view the pupal stage evolved in response to the need for a mold in which the adult systems, especially flight musculature, could be constructed. The second (pupal-imaginal) molt was then necessary in order that the new muscles could become attached to the exoskeleton. In Poyarkoff's theory there is no difference between the endopterygote larva and the exoptergote nymph.*

---

* Because they were considered originally to be quite distinct, the juvenile stages of exopterygotes and endopterygotes were referred to as "nymph" and "larva," respectively. The modern view (see Hinton's theory) is that nymphal and larval stages are homologous, and that pterygote juvenile stages should be called larvae. For clarity of discussion, however, in this chapter only, the traditional distinction has been retained.

Though it received support from some quarters, Poyarkoff's theory was strongly criticized by Heslop-Harrison (1958) and Hinton (1963b). Heslop-Harrison claimed that the explanation given for the origin of the pupal stage is teleological; in other words, Poyarkoff's explanation is that the pupal stage arose in fulfillment of a "need." Hinton stated that there is direct evidence against Poyarkoff's idea concerning the function of the pupal stage. First, it has been shown that tonofibrillae (microtubules within epidermal cells which attach muscles to the exoskeleton) (see Figure 14.1) can be formed long after the pupal-adult molt has occurred. Second, even in highly advanced endopterygotes the fiber rudiments of the wing muscles are present at the time of hatching. These develop in the larva in precisely the same way as the flight muscles of many primitive exopterygotes. In other words, no molt is required.

Implicit in the theories of Berlese, Poyarkoff, and Hinton (1963b) is the evolution of endopterygotes from exopterygote ancestors. Heslop-Harrison (1958) suggested, however, that the earliest forms of both groups were present at the same time and evolved from a common ancestor. This ancestor had a life history similar to that of modern Isoptera and Thysanoptera, namely, EGG → LARVAL INSTARS (showing no sign of wings) → NYMPHAL INSTARS (having external wing buds) → ADULT (see Chapter 7, Section 5; Chapter 8, Section 5; and Figure 21.15). Heslop-Harrison proposed that in the evolution of exopterygotes the larval instars were suppressed, and the modern free-living juvenile stages correspond to the nymphal instars of the ancestors. In endopterygote evolution the nymphal stages were compressed into the prepupal and pupal stages of modern forms. (The prepupal stage is a period of quiescence in the last larval instar prior to the molt to the pupa. It is not a distinct instar.) Thus, Berlese's original concept that the pupa comprised the ontogenetic counterparts of nymphal instars was supported by Heslop-Harrison. The basis of Heslop-Harrison's theory was his comparative study of the life history of various homopterans in which the last nymphal instar is divided into two phases. In the most primitive condition the first of these phases is an active one where the insect feeds and/or prepares its "pupal" chamber. In the most advanced condition both phases are inactive, and there are, for all intents and purposes, distinct prepupal and pupal stages, as in true endopterygotes.

The main general criticism of Heslop-Harrison's theory is that it lacks supporting evidence. More specific criticisms are that (1) the majority of early pterygote fossils (i.e., from the Carboniferous period) belong to exopterygote groups, endopterygotes mostly appearing for the first time in the Permian, leading most authorities to believe that endopterygotes came from exopterygote ancestors; (2) the Isoptera and Thysanoptera on which Heslop-Harrison's "primitive life history" was based are two highly specialized exopterygote orders; and (3) the implied homology of the endopterygote pupa and the last juvenile instars of the exopterygote homopterans studied by Heslop-Harrison is not justified [see discussion in Hinton (1963b)].

Perhaps the attraction of Hinton's (1963b) theory is its simplicity. It avoids the "suppression of larval," "compression of nymphal," and "expansion of imaginal" stages, found in the earlier theories and provides a simple functional explanation for the evolution of a pupal stage.

In Hinton's theory the pupa is homologous with the *final* nymphal instar of exopterygotes, and the terms "larva" and "nymph" are synonymous. Hinton proposed that, during the evolution of endopterygotes, the last juvenile stage (with external wings) was retained to complete the link between the earlier juvenile stages (larvae with internal wings) and the adult, hence the general resemblance between the pupa and adult in modern endopterygotes. Initially, the pupa would also resemble the earlier instars (just as the final instar nymph of modern exopterygotes resembles both the adult and the earlier nymphal stages). Once this intermediate stage had been established it is easy to visualize how the earlier juvenile stages could have become more and more specialized (for feeding and accumulating reserves) and quite different morphologically from both the pupa and the adult (the reproductive and dispersal stage). At the same time the pupa itself became more specialized. It ceased feeding actively, became less mobile, and was concerned solely with metamorphosis from the juvenile to the adult form.

Concerning the functional significance of the pupal stage, Hinton suggested that, as the endopterygote condition evolved, there was insufficient space in the thorax to accommodate both the "normal" contents (muscles and other organ systems) and the wing rudiments. Thus, the function of the larval-pupal molt was to evaginate the wings. This would permit not only considerable wing growth (as greatly folded structures within the pupal external wing cases) but also the enormous growth of the imaginal wing muscles within the thorax. The latter is facilitated, of course, by histolysis of the larval muscles (a process that is often not completed for many hours after the pupal-adult molt). The function of the pupal-adult molt is simply to effect release of the wings from the pupal case. The original function of the pupal stage was, then, to create space for wing and wing muscle development. But, once a stage had been developed in the life history in which structural rearrangement could take place, the way was open for increasing divergence of juvenile and adult habits and, subsequently, a decrease in the competition for food, space, etc. between the two stages. For many species the pupa has taken on a third function, namely as a stage in which the insect can pass through adverse climatic conditions, especially freezing temperatures.

4. The Success of Insects

The degree of success achieved by a group of organisms can be measured either as the total number of organisms within the group or, more commonly, as the number of different species of organisms that comprise the group. On either account the insects must be considered highly successful. Success is dependent on two interacting factors: (1) the potential of the group for adapting to new environmental conditions and (2) the degree to which the environmental conditions change. As success measured as the number of different species is a direct result of evolution, the environmental changes that must be considered are the long-term climatic changes that have occurred in different parts of the world over a period of several hundred million years.

4.1. The Adaptability of Insects

The basic feature of insects to which their success can be attributed must surely be that they are arthropods. As such they are endowed with a body plan that is superior to that of any other invertebrate group. Of the various arthropodan features the integument is the most important, as it serves a variety of functions. Its lightness and strength make it an excellent skeleton for attachment of muscles as well as a "shell" within which the tissues are protected. Its physical structure (usually including an outermost wax layer) makes it especially important in the water relations of arthropods. Because they are generally small organisms, arthropods in almost any environment face the problem of maintaining a suitable salt and water balance within their bodies. The magnitude of the problem (and, therefore, the energy expended in solving it) is greatly reduced by the impermeable cuticle (see Chapters

11 and 18). Arthropods are segmented animals and therefore have been able to exploit the advantages of tagmosis to the full extent. Directly related to this is the adaptability of the basic jointed limb, a feature used fully by different groups of arthropods (see Figure 1.6 and descriptions of segmental appendages in Chapter 3).

As all arthropods possess these advantageous features, the obvious question to ask is "Why have insects been especially successful?" or, put differently, "What features do insects have that other arthropods do not?" Answering this question will provide only a partial response for two reasons. First, as was stressed above, and is discussed more fully in Section 4.2, the environmental changes that take place are also very important in determining success. Consider, for example, the Crustacea. Compared to other invertebrate groups they must be regarded as successful (at least 40,000 extant species have been described), yet in comparison to the Insecta they come a very distant second. Although this is related partly to their different features, it must also reflect the different habitats in which they evolved. As a predominantly marine group, crustaceans evolved under relatively stable environmental conditions. Further, it is likely that when they were evolving the number of niches available to them would be quite limited because most were occupied by already established groups. Insects, on the other hand, evolved in a terrestrial environment subject to great changes in physical conditions. They were one of the earliest animal groups to "venture on land" and, therefore, had a vast number of niches available to them in this new adaptive zone. Second, the success of the Insecta as a whole is primarily related to the extraordinarily large number of species in a handful of orders, namely, the Coleoptera, Lepidoptera, Diptera, and Hymenoptera. Thus, the question ultimately becomes "What is it about these groups that allowed them to become so species-rich?" The answer to this is considered below.

Most insects, modern and fossil, are small animals. A few early forms achieved a large size but became extinct presumably because of climatic changes and their inability to compete successfully with other groups. Small size confers several advantages on an animal. It facilitates dispersal, it enables the animal to hide from potential predators, and it allows the animal to make use of food materials that are available in only very small amounts. The great disadvantage of small size in terrestrial organisms is the potentially high rate of water loss from the body. In insects this has been successfully overcome through the development of an impermeable exoskeleton.

The ability to fly was perhaps the single most important evolutionary development in insects. With this asset the possibilities for escape from predators and for dispersal were greatly enhanced. It led to colonization of new habitats, geographic isolation of populations, and, ultimately, formation of new species. Wide dispersal is particularly important for those species whose food and breeding sites are scattered and in limited supply.

Reproductive capacity and life history are two related factors that have contributed to the success of insects. Production of large numbers of eggs, combined with a short life history, means a greater amount of genetic variation can occur and be tested out rapidly within a population. This has two consequences. First, rapid adaptation to changes in environmental conditions will occur. This is best exemplified by the development of resistance to pesticides (see Chapter 24, Section 4.2). Second, there will be rapid attainment of genetic incompatibility between isolated populations and formation of new species. For example, the approximately 10,000 species of native Hawaiian insects are thought to have evolved from about 100 immigrant species. The evolution of a pupal stage between the larval and adult stages has led to a more specialized (and, in a sense, a more "efficient") life history. In some species this has led to the exploitation of different food sources by the larvae and adults (compare the foliage-feeding caterpillar with the nectar-drinking adult moth).

Further, it enables insects to use food sources that are available for only short periods of time. Eventually, the main function of the larva becomes the accumulation of metabolic reserves, whereas the adult is primarily concerned with reproduction and dispersal (and in some species does not feed). Although the pupa is primarily to allow transformation of the larva to the adult, in many species it has become a stage in which insects can resist unfavorable conditions. This development and the restriction of feeding activity to one phase of the life history have facilitated the expansion of insects into some of the world's most inhospitable habitats.

Four orders of insects have become extremely diverse: Coleoptera (300,000 species) Lepidoptera (200,000), Hymenoptera (130,000), and Diptera (110,000). Clearly, these must have particular features that allowed them to preferentially exploit new niches as these became available through evolutionary time.

For Coleoptera, the features were the development of elytra that protect the hind wings and cover the spiracles to reduce water loss, the "compact" body as a result of housing the coxal segments in cavities, and the increased proportion of the integument that was sclerotized. These allowed this group to occupy enclosed spaces and cryptic habitats such as soil and litter, and to invade arid environments. Within the Coleoptera two groups are especially diverse, the Curculionoidea and the Chrysomeloidea, which collectively total more than 130,000 species (see Chapter 10, Section 5). The ancestor of these groups likely fed on primitive plants such as pteridophytes, cycads and conifers. The species "explosion" that led to the modern curculionoids and chrysomeloids began in the post-Jurassic period and closely paralled the evolution of the angiosperms (Farrell, 1998).

The evolution of the proboscis enabled adult Lepidoptera to easily ingest water, hence avoid desiccation, and to obtain nectar, often stored cryptically by the plants with which they coevolved. Diptera, too, with their specialized mouthparts have been able to exploit particular liquid food sources, notably nectar, juices from decaying materials, and animal tissue fluids. It is generally considered that the radiation of the Lepidoptera and Diptera closely paralleled that of the flowering plants and while this may be correct, Labandeira and Sepkoski (1993) noted that the accelerated radiation of insects began 100 million years before that of the angiosperms, and that the great majority of mouthpart types were in existence by the Middle Jurassic. In Labandeira and Sepkoski's view it may have been the evolution of seed plants in general, and not specifically the angiosperms, that was the driving force behind the explosive evolution of the Insecta.

The Hymenoptera, the great majority of which are small to minute, have "piggybacked" on the success of other insect groups, by becoming parasitoids, especially on larvae or eggs, through the development of the ovipositor as a paralyzing organ. In a further step, many species evolved simple forms of parental care (e.g., by placing the prey in special cells along with their egg), leading eventually to cooperative nest care and true sociality.

Several features of insects have contributed therefore to their success (diversification). It is important to realize that these features have acted *in combination* to effect success, and, furthermore, little of this success would have been possible except for the changing environmental conditions in which the insects evolved.

4.2. The Importance of Environmental Changes

The importance of environmental changes in the process of evolution, acting through natural selection, is well known. These changes can be seen acting at the population or species level on a short-term basis, and many examples are known in insects, perhaps the best two being the development of resistance to pesticides and the formation of melanic forms of certain moths in areas of industrial air pollution. Of greater interest in the present context are the long-term climatic changes that have taken place over millions of years, for it is these that have controlled the evolution of insects both directly and indirectly through their influence on the evolution of other organisms, especially plants.

Although life began at least 2.5 billion years ago, it was not until the Upper Ordovician/ Lower Silurian periods (about 425–450 million years ago) that the first terrestrial organisms appeared, an event probably correlated with the formation of an ozone layer in the atmosphere, which reduced the amount of ultraviolet radiation reaching the earth's surface (Berkner and Marshall, 1965). The earliest terrestrial organisms were simple low-growing land plants that reproduced by means of spores. They were soon followed by mandibulate arthropods (scorpions, myriapods, and apterygotes) that presumably fed on the plants, their decaying remains, their spores, or on other small animals (Peck and Munroe, 1999). In these early plants, spores were produced on short side branches of upright stems. An important evolutionary development was the concentration of the sporangia (spore-producing structures) into a terminal spike. Whether these spikes were particularly attractive as food for insects and whether, therefore, they may have been important in the evolution of flight is a matter for speculation (Smart and Hughes, 1973).

During the Devonian and Lower Carboniferous periods, a wide radiation of plants occurred. Especially significant was the development of swamp forests that contained, for example, tree lycopods, calamites, and primitive gymnosperms. The evolution of treelike form, though in part related to the struggle for light, may also have enabled the plants to protect (temporarily) their reproductive structures against spore-feeding insects and other animals. In contrast to the humid or even wet conditions on the forest floor, the air several meters above the ground was probably relatively dry. Thus, the evolution of trees with terminal sporangia may have been an important stimulant to the evolution of a waterproof cuticle, spiracular closing mechanisms, and, eventually, flight (Kevan *et al.*, 1975).

The trees, together with the ground flora, would provide a wide range of food material. As noted earlier, winged insects appeared in the Lower Carboniferous. Though some of these were mandibulate and fed on soft parts of plants or litter on the forest floor, the great majority (>80%), notably the paleodictyopteroids and ancestral hemipteroids, had mouthparts in the form of a proboscis, which has led to the suggestion that these insects were adapted to feeding on either free-standing liquids or plant sap. Smart and Hughes (1973) believed, however, that the proboscis might have been used as a probe for extracting pollen and spores from the plants' reproductive structures. They argued that not until the Upper Carboniferous did plants evolve with phloem close enough to the stem surface that it was accessible to Hemiptera. Yet other insects such as the Protodonata and, later, Odonata, were predators, feeding on the sucking forms as well as early Ephemeroptera.

Possibly as a result of an "arms race" between predator and prey, some members of all these groups became very large (Peck and Munroe, 1999). Another view suggests that the evolution of large size (gigantism) was a result of competition between these insects and the earliest terrestrial vertebrates, the Amphibia. Certainly large size would be favored by the year-round, uniform growing conditions (Ross, 1965). A third factor related to the evolution of Late Paleozoic gigantism was the increase in oxygen concentration in the atmosphere that occurred at this time, with values possibly as high as 35%. This would improve the ability of the insects' tracheal system to supply tissues with oxygen (Chapter 15, Section 3.1), which is a major constraint to increased body size (Dudley, 1998).

In addition to the forest ecosystem, there were presumably other ecosystems, for example, the edges of swamps and higher ground, which had their complement of insects. However, such ecosystems did not apparently favor fossilization, and their insect fauna is practically unknown.

Toward the end of the Carboniferous and extending through the Permian periods significant climatic changes occurred that had a major effect on insect evolution. The Northern Hemisphere became progressively drier, extensive glaciation took place in the Southern Hemisphere, and many mountain ranges were formed, leading to distinct climatic zones and major alterations in plant and insect biota. Among the significant floral changes were the more restricted distribution or even extinction of many of the Carboniferous groups and their replacement by gymnosperms. These biogeographic changes created not only new habitats but also barriers that prevented gene flow among populations so that a veritable insect population explosion occurred, especially of endopterygotes, the class reaching its peak diversity in the Permian (Kukalová-Peck, 1991). However, near the end of this period, most of the Paleozoic orders (including many "experimental" side groups) became extinct, to be replaced by representatives of the modern orders. The reasons for the Great Permian Extinction remain obscure. Some authors suggest that a major catastrophe occurred, for example, the earth being struck by a very large meteorite or comet, while others believe that climatic changes were the cause (Erwin, 1990; Peck and Munroe, 1999).

In the Triassic and early Jurassic the gradual radiation of the extant orders continued but was largely overshadowed by evolution of the reptiles. The latter occurred in such large numbers and in such a variety of habitats that the Triassic period is generally known as the "Age of Reptiles." Many of them were insectivorous, and this may have acted as a selection pressure favoring small size (miniaturization), which is a general feature of fossil insects from this period. An alternative explanation for miniaturization has come from examination of the flight musculature across a range of insect groups, which suggests that asynchronous flight muscle evolved at this time. Asynchronous flight muscle is a prerequisite for the high wing-beat frequencies seen in the majority of good fliers (Chapter 14, Section 3.3.4). However, because wing-beat frequency is inversely proportional to body size, miniaturization was also necessary to achieve these high wing-beat frequencies (Dudley, 2001).

Early in the Triassic period the first bisexual flowers appeared. The occurrence together of male and female structures immediately led to the possibility of a role as pollinators for insects that fed on the reproductive parts of plants. Because of the risk of having their reproductive structures eaten, these early plants probably produced a large number of small ovules and much pollen (Smart and Hughes, 1973). The insect fauna of the Triassic period still included "orthodox" plant feeders such as Orthoptera and some Coleoptera, plant-sucking Hemiptera, and predaceous species (Odonata and Neuroptera). However, there were also large numbers of primitive endopterygotes, mostly belonging to the panorpoid complex, Hymenoptera, and Coleoptera, which as adults were mandibulate and therefore were potential "mess-and-soil" pollinators (Smart and Hughes, 1973).

By the middle of the Jurassic period the decline of the reptiles had begun and the earth's climate had become generally warmer. As a result the insect fauna increased both in mean body size and in variety. A good deal of mountain formation occurred at the end of the Jurassic, creating new climatic conditions in various parts of the world including, for the first time, winters. It was also about this time that the ancient world continent Pangea began to break apart and form the modern continents with their distinct insect faunas.

The Cretaceous period was an important phase in the evolution of insects, for it was during this time that adaptive radiations of several endopterygote orders took place. In some instances it must be assumed that these radiations directly paralleled the evolution of angiosperms, although, it must be emphasized, the fossil record of angiosperm flowers is sparse. Further, as noted above, there is a suggestion that much insect radiation preceded the angiosperm diversification by as much as 100 million years (Labandeira and Sepkoski, 1993). Nevertheless, some extremely close interrelationships have evolved between plants and their insect pollinators (see Chapter 23, Section 2.3). For other orders, the radiation was only indirectly related to plants; for example, a large variety of parasitic Hymenoptera appeared, correlated with the large increase in numbers of insect hosts. By the middle Cretaceous, 84% of the insect fauna belonged to families that exist today.

The Cretaceous period was also rather active, geologically speaking, for a good deal of mountain making, lowland flooding (by the sea), and formation and breakage of land bridges took place. All of these processes would assist in the isolation and diversification of the insect fauna.

The decline of reptiles, which became accelerated during the late Cretaceous period, was followed by the increase in numbers of mammals and birds. These groups became very widespread and diverse during the Tertiary period. Paralleling this diversification was the evolution of their insect parasites. Throughout the Tertiary period the climate seems to have alternated between warm and cold. In the Paleocene epoch the climate was cooler than that of the Cretaceous. Thus, cold-adapted groups became widely distributed. A warming trend followed in the Eocene so that cold-adapted organisms became restricted to high altitudes, while the warm-adapted types spread. It appears that by the end of the Eocene period (approximately 36 million years ago) most modern tribes or genera of insects had evolved (Ross, 1965). In the Oligocene the climate became cooler and remained so during the Miocene and Pliocene epochs. However, in these two epochs new mountain ranges were formed, and some already existing ones were pushed even higher. The Tertiary period ended about 1 million years ago and was followed by the Pleistocene epoch of the Quaternary period. Temperatures in the Pleistocene (Ice Age) were generally much lower than in the Tertiary, and four distinct periods of glaciation occurred, at which time most of the North American and European continents had a thick covering of ice. Between these periods warming trends caused the ice to recede northward. Accompanying these ice movements were parallel movements of the fauna and flora. However, the overall significance of the Ice Age in terms of insect evolution remains uncertain.

To conclude this section on the importance of environmental changes, the effects of humans on insect populations and biodiversity must be mentioned. Initially, the hunter-gatherer habits of humans would have had little effect. However, starting about 60,000 years ago, hunting by the growing human population began to drive to extinction many species of large mammals (and presumably their insect parasites). The trend began in Africa, was then evident in Europe, and has since spread (at an increasing rate) to North and South America, Australasia, and many oceanic islands. But by far the greatest human impact on ecosystems, hence on insect diversity and numbers, was the evolution of agriculture. Conversion of native grassland, deforestation, drainage of wetlands, and use of herbicides and insecticides have occurred at an ever-increasing rate, leading to massive declines in ecosystem diversity. While many species are in terminal decline or are already extinct as a result of these practices, others (both plant and animal) have thrived as a result of monoculture, to become major agricultural pests. Many organisms were introduced accidentally into new areas as a result of commerce. In the absence of natural enemies these, too, have become pests (see also Chapter 24). Some species were deliberately released under the misguided idea that they would enhance their new environment, with no thought given to their possible long-term impacts, for example, displacement of native species.

5. Summary

From their beginning as primitively wingless thysanuranlike creatures, the insects have undergone a vast adaptive radiation to become the world's most successful group of living organisms. Undoubtedly, a large measure of this success can be attributed to the evolution of wings. The Paranotal Theory suggests that wings evolved from rigid outgrowths of the terga in adult insects, their initial function being the protection of the legs. Increase in the size of the paranotal lobes was associated with improvement in attitudinal control as the insects dropped from vegetation in order to escape predation. In the Paranotal Theory articulation of these winglets was a later development associated with improving attitudinal control. An idea that is becoming more popular is that wings developed from already articulated structures. These were present on all thoracic and abdominal segments in aquatic juvenile insects and from the outset were hinged in association with their function as gills, spiracular flaps, or swimming organs. Before they became large and powerful enough for true flight, the wings may have been used to propel insects over the water surface (surface skimming), as seen in some extant stoneflies and mayflies.

The Pterygota had a monophyletic origin in the Devonian and soon split into two major evolutionary lines, the Paleoptera and the Neoptera. By the late Carboniferous several distinct neopteran groups were established. These were the pliconeopterans (plecopteroids, orthopteroids, and blattoids), which are perhaps polyphyletic, the paraneopterans (hemipteroids), and the oligoneopterans (endopterygotes), both of which are probably monophyletic. Of these, the endopterygotes have been by far the most successful. This is related, in large part, to the evolution of a pupal stage within the group. Various theories have been advanced for the origin of the pupal stage. The most likely theory proposes that it is equivalent to the final nymphal instar of the original exopterygote ancestor. Its initial function was to provide space for wing and wing muscle development. However, its evolution also facilitated divergence of adult and larval habits, so that the two stages no longer competed for the same resources. It also became a stage in which species could survive adverse climatic conditions, especially freezing temperatures.

Insect success (i.e., diversity) is related not only to the group's adaptability but also to the environment in which they have evolved. Being arthropods, insects possess a body plan that is superior to that of other invertebrates. They are generally small and able to fly. They usually have a high reproductive capacity, often coupled with a life history that is short and contains a pupal stage. Because of features peculiar to them, four orders (Coleoptera, Lepidoptera, Hymenoptera, and Diptera) are especially successful, comprising almost 75% of the described species. Insect evolution was coincident with the evolution of land plants. The insects were among the first invertebrates to establish themselves on land. By virtue of their adaptability they were able to colonize rapidly the new habitats formed as a result of climatic changes over the earth's surface.

6. Literature

For further information readers should consult Carpenter (1977, 1992), Boudreaux (1979), Hennig (1981), Kristensen (1981, 1989, 1991, 1995, 1998), Wootton (1981), Kukalová-Peck (1985, 1991, 1998), Wheeler *et al.* (2001), Rasnitsyn and Quicke (2002) [geological history and phylogenetic relationships]; Kukalová-Peck (1978, 1983, 1987), Wootton (1986), Ellington (1991), Brodsky (1994), Kingsolver and Koehl (1994), Marden and Kramer (1994) [origin and evolution of wings]; Hinton (1963b) and Truman and Riddiford (2002) [origin and function of the pupal stage]; and Becker (1965), Smart and Hughes (1973), Kevan *et al.* (1975), Kevan and Baker (1999), and Peck and Munroe (1999) [coevolution of insects and plants].

Alexander, R. D., and Brown, W. L., Jr., 1963, Mating behavior and the origin of insect wings, *Occas. Pap. Mus. Zool. Univ. Mich.* **628**:1–19.

Becker, H. F., 1965, Flowers, insects, and evolution, *Nat. Hist.* **74**:38–45.

Berkner, L. V., and Marshall, L. C., 1965, On the origin and rise of oxygen concentration in the earth's atmosphere, *J. Atmos. Sci.* **22**:225–261.

Boudreaux, H. B., 1979, *Arthropod Phylogeny with Special Reference to Insects*, Wiley, New York.

Brodsky, A. K., 1994, *The Evolution of Insect Flight*, Oxford University Press, Oxford.

Carpenter, F. M., 1977, Geological history and evolution of the insects, *Proc. XV Int. Congr. Entomol.*, pp. 63–70.

Carpenter, F. M., 1992, *Treatise on Invertebrate Paleontology. Part R. Arthropoda 4, Vols. 3 and 4 (Superclass Hexapoda)*, University of Kansas, Lawrence.

Crowson, R. A., 1960, The phylogeny of Coleoptera, *Annu. Rev. Entomol.* **5**:111–134.

Crowson, R. A., 1975, The evolutionary history of Coleoptera, as documented by fossil and comparative evidence, *Atti Congr. Naz. Ital. Ent.* **10**:47–90.

Crowson, R. A., 1981, *The Biology of the Coleoptera*, Academic Press, New York.

Douglas, M. M., 1981, Thermoregulatory significance of thoracic lobes in the evolution of insect wings, *Science* **211**:84–86.

Dudley, R., 1998, Atmospheric oxygen, giant Paleozoic insects and the evolution of aerial locomotor performance, *J. Exp. Biol.* **201**:1043–1050.

Dudley, R., 2001, The biomechanics and functional diversity of flight, in: *Insect Movement: Mechanisms and Consequences* (I. P. Woiwod, D. R. Reynolds, and C. D. Thomas, eds.), CAB International, Wallingford, U.K.

Ellington, C. P., 1991, Aerodynamics and the origin of insect flight, *Adv. Insect Physiol.* **23**:171–210.

Engel, M. S., and Grimaldi, D. A., 2004, New light shed on the oldest insect, *Nature* **427**:627–630.

Erwin, D.H., 1990, The end-Permian mass extinction, *Annu. Rev. Ecol. Syst.* **21**:69–91.

Farrell, B. D., 1998, "Inordinate fondness" explained: Why are there so many beetles?, *Science* **281**:555–558.

Flower, J. W., 1964, On the origin of flight in insects, *J. Insect Physiol.* **10**:81–88.

Gaunt, M. W., and Miles, M. A., 2002, An insect molecular clock dates the origin of insects and accords with palaeontological and biogeographic landmarks, *Mol. Biol. Evol.* **19**:748–761.

Giles, E. T., 1963, The comparative external morphology and affinities of the Dermaptera, *Trans. R. Entomol. Soc. Lond.* **115**:95–164.

Hamilton, K. G. A., 1971, 1972, The insect wing, Parts 1 and IV, *J. Kans. Entomol. Soc.* **44**:421–433; **45**:295–308.

Hasenfuss, I., 2002, A possible evolutionary pathway to insect flight starting from lepismatid organization, *J. Zool. Syst. Evol. Res.* **40**:65–81.

Hennig, W., 1981, *Insect Phylogeny*, Wiley, New York.

Heslop-Harrison, G., 1958, On the origin and function of the pupal stadia in holometabolous Insecta, *Proc. Univ. Durham Philos. Soc. Ser. A* **13**:59–79.

Hinton, H. E., 1958, The phylogeny of the panorpoid orders, *Annu. Rev. Entomol.* **3**:181–206.

Hinton, H. E., 1963a, Discussion: The origin of flight in insects, *Proc. R. Entomol. Soc. Lond. Ser. C* **28**:23–32.

Hinton, H. E., 1963b, The origin and function of the pupal stage, *Proc. R. Entomol. Soc. Lond. Ser. A* **38**:77–85.

Hovmöller, R., Pape, T., and Källersö, M., 2002, The Palaeoptera problem: Basal pterygote phylogeny inferred from 18S and 28S rDNA sequences, *Cladistics* **18**:313–323.

Kamp, J .W., 1973, Numerical classification of the orthopteroids, with special reference to the Grylloblattodea, *Can. Entomol.* **105**:1235–1249.

Kevan, D. K. McE., 1986, A rationale for the classification of orthopteroid insects—The saltatorial orthopteroids or grigs—one order or two? *Proc. Fourth Triennial Meet. Pan-Am. Acridol. Soc.*, pp. 49–67.

Kevan, P. G., and Baker, H. G., 1999, Insects on flowers, in: *Ecological Entomology*, 2nd ed. (C. B. Huffaker and A.P. Gutierrez, eds.), Wiley, New York.

Kevan, P. G., Chaloner, W. G., and Savile, D. B. O., 1975, Interrelationships of early terrestrial arthropods and plants, *Palaeontology* **18**:391–417.

Kingsolver, J. G., and Koehl, M. A. R., 1985, Aerodynamics, thermoregulation, and the evolution of insect wings: Differential scaling and evolutionary change, *Evolution* **39**:488–504.

Kingsolver, J. G., and Koehl, M. A. R., 1994, Selective factors in the evolution of insect wings, *Annu. Rev. Entomol.* **39**:425–451.

Klass, K.-D., Zompro, O., Kristensen, N. P., and Adis, J., 2002, Mantophasmatodea: A new insect order with extant members in the Afrotropics, *Science* **296**:1456–1459.

Kristensen, N. P., 1981, Phylogeny of insect orders, *Annu. Rev. Entomol.* **26**:135–157.

Kristensen, N. P., 1984, Studies on the morphology and systematics of primitive Lepidoptera (Insecta), *Steensrrupia* **10**:141–191.

Kristensen, N. P., 1989, Insect phylogeny based on morphological evidence, in: *The Hierarchy of Life* (B. Fernholm, K. Bremer, and H. Jomvall, eds.), Elsevier, Amsterdam.

Kristensen, N. P., 1991, Phylogeny of extant hexapods, in: *The Insects of Australia,* 2nd ed., Vol. I (CSIRO, ed.), Melbourne University Press, Carlton, Victoria.

Kristensen, N. P., 1995, Forty years insect phylogenetic systematics. Hennig's 'Kritische Bemerkungen . . . ' and subsequent developments, *Zoologische Beiträge NF* **36**:83–124.

Kristensen, N. P., 1998, The groundplan and basal diversification of the hexapods, in: *Arthropod Relationships* (R. A. Fortey and R. H. Thomas, eds.), Chapman and Hall, London.

Kukalová-Peck, J., 1978, Origin and evolution of insect wings and their relation to metamorphosis, as documented by the fossil record, *J. Morphol.* **156**:53–126.

Kukalová-Peck, J., 1983, Origin of the insect wing and wing articulation from the arthropodan leg, *Can. J. Zool.* **61**:1618–1669.

Kukalová-Peck, J., 1985, Ephemeroid wing venation based upon new gigantic Carboniferous mayflies and basic morphology, phylogeny, and metamorphosis of pterygote insects (Insecta, Ephemerida), *Can. J. Zool.* **63**:933–955.

Kukalová-Peck, J., 1987, New Carboniferous Diplura, Monura, and Thysanura, the hexapod ground plan, and the role of thoracic side lobes in the origin of wings (Insecta), *Can. J. Zool.* **65**:2327–2345.

Kukalová-Peck, J., 1991, Fossil history and the evolution of hexapod structures, in: *The Insects of Australia*, 2nd ed., Vol. I (CSIRO, ed.), Melbourne University Press, Carlton, Victoria.

Kukalová-Peck, J., 1998, Arthropod phylogeny and 'basal' morphological structures, in: *Arthropod Relationships* (R. A. Fortey and R. H. Thomas, eds.), Chapman and Hall, London.

Kukalová-Peck, J., and Brauckmann, C., 1990, Wing folding in pterygote insects, and the oldest Diaphanopterodea from the early Late Carboniferous of West Germany, *Can. J. Zool.* **68**:1104–1111.

Kukalová-Peck, J., and Brauckmann, C., 1992, Most Paleozoic Protorthoptera are ancestral hemipteroids: Major wing braces as clues to a new phylogeny of Neoptera (Insecta), *Can. J. Zool.* **70**:2452–2473.

Kukalová-Peck, J., and Lawrence, J. F., 1993, Evolution of the hind wing in Coleoptera, *Can. Entomol.* **125**:181–258.

Kukalová-Peck, J., and Lawrence, J. F., 2004, Relationships among coleopteran suborders and major endoneopteran lineages: Evidence from hind wing characters, *Eur. J. Entomol.* **101**:95–144.

Labandeira, C. C., and Sepkoski, J. J., Jr., 1993, Insect diversity in the fossil record, *Science* **261**:310–315.

Labandeira, C. C., Beall, B. S., and Heuber, F. M., 1988, Early insect diversification: Evidence from a Lower Devonian fossil from Quebec, *Science* **242**:913–916.

Marden, J. H., and Kramer, M. G., 1994, Surface-skimming stoneflies: A possible intermediate stage in insect flight evolution, *Science* **266**:427–430.

Marden, J. H., and Thomas, M. A., 2003, Rowing locomotion by a stonefly that possesses the ancestral pterygote condition of co-occurring wings and abdominal gills, *Biol. J. Linn. Soc.* **79**:341–349.

Martynov, A. V., 1938, Etudes sur l'histoire géologique et de phylogénie des ordres des insectes (Pterygota), 1-3e. partie, Palaeoptera et Neoptera-Polyneoptera, *Trav. Inst. Paléont. Acad. Sci. URSS*, pp. 1–150.

Peck, S. B., and Munroe, E., 1999, Biogeography and evolutionary history: Wide-scale and long-term patterns of insects, in: *Ecological Entomology*, 2nd ed. (C. B. Huffaker and A.P. Gutierrez, eds.), Wiley, New York.

Rasnitsyn, A. P., 1981, A modified paranotal theory of insect wing origin, *J. Morphol.* **168**:331–338.

Rasnitsyn, A. P., and Quicke, D. L. J. (eds.), 2002, *History of Insects*, Kluwer Academic Publishers, Dordrecht.

Ross, H. H., 1955, Evolution of the insect orders. *Entomol. News* **66**:197–208.

Ross, H. H., 1965, *A Textbook of Entomology*, 3rd ed., Wiley, New York.

Ross, H. H., 1967, The evolution and past dispersal of the Trichoptera, *Annu. Rev. Entomol.* **12**:169–206.

Sharov, A. G., 1966, *Basic Arthropodan Stock*, Pergamon Press, Elmsford, NY.

Shear, W. A., and Kukalová-Peck, J., 1990, The ecology of Paleozoic terrestrial arthropods: The fossil evidence, *Can. J. Zool.* **68**:1807–1834.

Shear, W. A., Bonamo, P. M., Grierson, J. D., Rolfe, W. D. I., Smith, E. L., and Norton, R. A., 1984, Early land animals in North America: Evidence from Devonian Age arthropods from Gilboa, New York, *Science* **224**:492–494.

Smart, J., and Hughes, N. F., 1973, The insect and the plant: Progressive palaeoecological integration, *Symp. R. Entomol. Soc.* **6**:143–155.

Storozhenko, S. Y., 1997, Fossil history and phylogeny of orthopteroid insects, in: *The Bionomics of Grasshoppers, Katydids and Their Kin* (S. K. Gangwere, M. C. Muralirangan, and M. Muralirangan, eds.), CAB International, Wallingford, U.K.

Thomas, M. A., Walsh, K. A., Wolf, M. R., McPheron, B. A., and Marden, J. H., 2000, Molecular phylogenetic analysis of evolutionary trends in stonefly wing structure and locomotor behavior, *Proc. Natl. Acad. Sci. U S A* **97**:13178–13183.

Truman, J. W., and Riddiford, L. M., 1999, The origins of insect metamorphosis, *Nature* **401**:447–452.

Truman, J. W., and Riddiford, L. M., 2002, Endocrine insights into the evolution of metamorphosis in insects, *Annu. Rev. Entomol.* **47**:467–500.

Weesner, F. M., 1960, Evolution and biology of termites, *Annu. Rev. Entomol.* **5**:153–170.

Whalley, P. E. S., 1979, New species of Protorthoptera and Protodonata (Insecta) from the Upper Carboniferous of Britain, with a comment on the origin of wings, *Bull. Br. Mus. Nat. Hist. (Geol.)* **32**:85–90.

Wheeler, W. C., 1989, The systematics of insect ribosomal DNA, in: *The Hierarchy of Life* (B. Fernholm, K. Bremer, and H. Jörnvall, eds.), Elsevier, Amsterdam.

Wheeler, W. C., Whiting, M., Wheeler, Q. D., and Carpenter, J. M., 2001, The phylogeny of the extant hexapod orders, *Cladistics* **17**:113–169.

Whiting, M., 1998, Phylogenetic position of the Strepsiptera: Review of molecular and morphological evidence, *Int. J. Insect Morphol. Embryol.* **27**:53–60.

Wigglesworth, V. B., 1963a, Origin of wings in insects, *Nature (London)* **197**:97–98.

Wigglesworth, V. B., 1963b, Discussion: The origin of flight in insects, *Proc. R. Entomol. Soc. Lond. Ser. C* **28**:23–32.

Wigglesworth, V. B., 1973, Evolution of insect wings and flight, *Nature (London)* **246**:127–129.

Wigglesworth, V. B., 1976, The evolution of insect flight, *Symp. R. Entomol. Soc. Lond.* **7**:255–269.

Willmann, R., 1989, Rediscovered: *Permotipula patricia*, the oldest known fly, *Naturwissenschaften* **76**:375–377.

Willmann, R., 1998, Advances and problems in insect phylogeny, in: *Arthropod Relationships* (R. A. Fortey and R. H. Thomas, eds.), Chapman and Hall, London.

Wootton, R. J., 1976, The fossil record and insect flight, *Symp. R. Entomol. Soc. Lond.* **7**:235–254.

Wootton, R. J., 1981, Palaeozoic insects, *Annu. Rev. Entomol.* **26**:319–344.

Wootton, R. J., 1986, The origin of insect flight: Where are we now? *Antenna* **10**:82–86.

Wootton, R. J., 2001, How insect wings evolved, in: *Insect Movement: Mechanisms and Consequences* (I. P. Woiwod, D. R. Reynolds, and C. D. Thomas, eds.), CAB International, Wallingford, U.K.

Wootton, R. J., and Ellington, C. P., 1991, Biomechanics and the origin of insect flight, in: *Biomechanics in Evolution* (J. M. Y. Rayner and R. J. Wootton, eds.), Cambridge University Press, London.

3

External Structure

1. Introduction

The extreme variety of external form seen in the Insecta is the most obvious manifestation of this group's adaptability. To the taxonomist who thrives on morphological differences, this variety is manna from Heaven; to the morphologist who likes to refer everything back to a basic type or ground plan, it can be a nightmare! Paralleling this variety is, unfortunately, a massive terminology, even the basics of which an elementary student may find difficult to absorb. Some consolation may be derived from the fact that "form reflects function." In other words, seemingly minor differences in structure may reflect important differences in functional capabilities. It is impossible to deal in a text of this kind with all of the variation in form that exists, and only the basic structure of an insect and its most important modifications will be described.

2. General Body Plan

Like other arthropods insects are segmented animals whose bodies are covered with cuticle. Over most regions of the body the outer layer of the cuticle becomes hardened (tanned) and forms the exocuticle (see Chapter 11, Section 3.3). These regions are separated by areas (joints) in which the exocuticular layer is missing, and the cuticle therefore remains membranous, flexible, and often folded. The presence of these cuticular membranes facilitates movement between adjacent hard parts (*sclerites*). The degree of movement at a joint depends on the extent of the cuticular membrane. In the case of intersegmental membranes there is complete separation of adjacent sclerites, and therefore movement is unrestricted. Usually, however, especially at appendage joints, movement is restricted by the development of one or two contiguous points between adjacent sclerites; that is, specific articulations are produced. A *monocondylic* joint has only one articulatory surface, and at this joint movement may be partially rotary (e.g., the articulation of the antennae with the head). In *dicondylic* joints (e.g., most leg joints) there are two articulations, and the joint operates like a hinge. The articulations may be either *intrinsic*, where the contiguous points lie within the membrane (Figure 3.1A), or *extrinsic*, in which case the articulating surfaces lie outside the skeletal parts (Figure 3.lB).

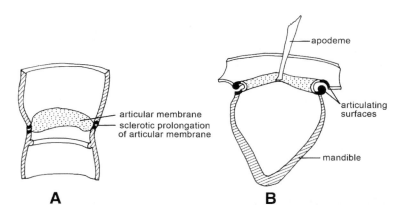

FIGURE 3.1. Articulations. (A) Intrinsic (leg joint); and (B) extrinsic (articulation of mandible with cranium). [From R. E. Snodgrass, *Principles of Insect Morphology.* Copyright 1935 by McGraw-Hill, Inc. Used with permission of McGraw-Hill Book Company.]

In many larval insects (as in annelids) the entire cuticle is thin and flexible, and segments are separated by invaginations of the integument (*intersegmental folds*) to which longitudinal muscles are attached (Figure 3.2A). Animals possessing this arrangement (known as *primary segmentation*) have almost unlimited freedom of body movement. In the majority of insects, however, there is heavy sclerotization of the cuticle to form a series of dorsal and ventral plates, the *terga* and *sterna*, respectively. As shown in Figure 3.2B,

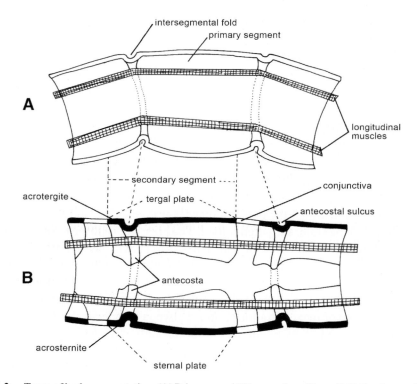

FIGURE 3.2. Types of body segmentation. (A) Primary; and (B) secondary. [From R. E. Snodgrass, *Principles of Insect Morphology.* Copyright 1935 by McGraw-Hill, Inc. Used with permission of McGraw-Hill Book Company.]

these regions of sclerotization do not correspond precisely with the primary segmental pattern. The tergal and sternal plates do not cover entirely the posterior part of the primary segment, yet they extend anteriorly slightly beyond the original intersegmental groove. Thus, the body is differentiated into a series of secondary segments (*scleromata*) separated by membranous areas (*conjunctivae*) that allow the body to remain flexible. This is termed *secondary segmentation*. Each secondary segment contains four exoskeletal components, a tergum and a sternum separated by lateral, primarily membranous, *pleural areas*. Each of the primary components may differentiate into several sclerites to which the general terms *tergites* and *sternites* are applied; small sclerites, generally termed *pleurites*, may also occur in the pleural areas. The primitive intersegmental fold becomes an internal ridge of cuticle, the *antecosta*, seen externally as a groove, the *antecostal sulcus*. The narrow strip of cuticle anterior to the sulcus is the *acrotergite* (when dorsal) or *acrosternite* (when ventraJ). The posterior part of both the tergum and sternum is primitively a simple cuticular plate, but this undergoes considerable modification in the thoracic region of the body. The pleurites are usually secondary sclerotizations but in fact may represent the basal segment of the appendages. The pleurites may become greatly enlarged and fused with the tergum and sternum in the thoracic segments. In the abdomen the pleurites may fuse with the sternal plates.

The basic segmental structure is frequently obscured as a result of tagmosis. In insects three tagmata are found: the head, the thorax, and the abdomen. In the head almost all signs of the original boundaries of the segments have disappeared, though, for most segments, the appendages remain. In the thorax the three segments can generally be distinguished, although they undergo profound modification associated with locomotion. The anterior abdominal segments are usually little different from the typical secondary segment described above. At the posterior end of the abdomen a varied number of segments may be modified, reduced, or lost, associated with the development of the external genitalia.

Examination of the exoskeleton reveals the presence of a number of lines or grooves whose origin is varied. If the line marks the union of two originally separate sclerites, it is known as a *suture*. If it indicates an invagination of the exoskeleton to form an internal ridge of cuticle (*apodeme*), the line is properly termed a *sulcus* (Snodgrass, 1960). Pits may also be seen on the exoskeleton. These pits mark the sites of internal, tubercular invaginations of the integument (*apophyses*). Secondary discontinuations of the exocuticular component of the cuticle may occur, for example, the ecdysial line along which the old cuticle splits during molting, and these are generally known as sutures.

Primitively each segment bore a pair of appendages. Traces of these can still be seen on almost all segments for a short time during embryonic development, but on many segments they soon disappear, and typical insects lack abdominal appendages on all except the posterior segments. According to Kukalová-Peck (1987), the ground plan of the insect segmental appendage included 11 *podites*, some of which carried inner or outer branches (*endites* and *exites*, respectively) (Figure 3.21A). All of these podites can be identified in some fossil insects (Kukalová-Peck, 1992), but in the great majority of extant forms only five or six podites at most are obvious, notably in the legs [*coxa, trochanter, femur, tibia, tarsus* (and *pretarsus*, which some authors do not consider to be a podite)]. The appendages of the head and abdomen have become so highly modified that homologizing their podites may be extremely difficult. Traces of exites can be seen as gills in some aquatic juvenile insects, and the endites remain as the exsertile vesicles of some apterygotes, but in the majority of insects these branches have completely disappeared.

3. The Head

The head, being the anterior tagma, bears the major sense organs and the mouthparts. Considerable controversy still surrounds the problem of segmentation of the insect head, especially concerning the number and nature of segments anterior to the mouth.* At various times it has been argued that there are from three to seven segments in the insect head, though it is now widely agreed that there are six. It is not feasible to discuss here the many theories concerning the segmental composition of the insect head, but the main points of contention should be noted. These are: (1) whether arthropods possess an *acron* (which is non-segmental and homologous with the annelid prostomium); (2) whether a *preantennal* segment occurs between the acron and the *antennal* segment and what appendages are associated with such a segment; and (3) whether the *antennae* are segmental appendages or merely outgrowths of the acron [see Rempel (1975), Bitsch (1994), Kukalová-Peck (1992), and Scholtz (1998) for reviews of the subject]. The embryological studies of Rempel and Church (1971) have demonstrated convincingly that an acron is present. However, it is never seen in fossil insects or other arthropods (Kukalová-Peck, 1992) because it moved dorsally to merge imperceptibly into the region between the compound eyes (Kukalová-Peck, 1998).

Both embryology and paleontology have confirmed that there are three preoral and three postoral segments The first preoral segment is preantennal; it is called the *protocerebral* or *clypeolabral* segment. The segment itself has disappeared but its appendages remain as the *clypeolabrum*. The second preoral (*antennal/deutocerebral*) segment bears the *antennae*, which are therefore true segmental appendages. The third preoral (*intercalary/tritocerebral*) segment appears briefly during embryogenesis, then is lost. Its appendages, however, remain as part of the *hypopharynx* (Kukalová-Peck, 1992). Head segments 4–6 are postoral and named the *mandibular*, *maxillary*, and *labial*, respectively. Their appendages form the mouthparts from which their names are derived. In addition, the sternum of the mandibular segment becomes part of the hypopharynx.

3.1. General Structure

Primitively the head is oriented so that the mouthparts lie ventrally (the *hypognathous* condition) (Figure 3.3B). In some insects, especially those that pursue their prey or use their mouthparts in burrowing, the head is *prognathous* in which the mouthparts are directed anteriorly (Figure 3.4A). In many Hemiptera the suctorial mouthparts are posteroventral in position (Figure 3.4B), a condition described as *opisthognathous* (*opisthorhynchous*).

The head takes the form of a heavily sclerotized capsule, and only the presence of the antennae and mouthparts provides any external indication of its segmental construction. In most adult insects and juvenile exopterygotes a pair of *compound eyes* is situated dorsolaterally on the cranium, with three *ocelli* between them on the anterior face (Figure 3.3A). The two posterior ocelli are somewhat lateral in position; the third ocellus is anterior and median. The *antennae* vary in location from a point close to the mandibles to a more median position between the compound eyes. On the posterior surface of the head capsule is an aperture, the *occipital foramen*, which leads into the neck. Of the mouthparts, the *labrum* hangs down from the ventral edge of the *clypeus*, the *labium* lies below the occipital foramen, and the

---

* Perhaps the most interesting conclusion was drawn by Snodgrass (1960, p. 51) who stated "it would be too bad if the question of head segmentation ever should be finally settled; it has been for so long such fertile ground for theorizing that arthropodists would miss it as a field for mental exercise"!

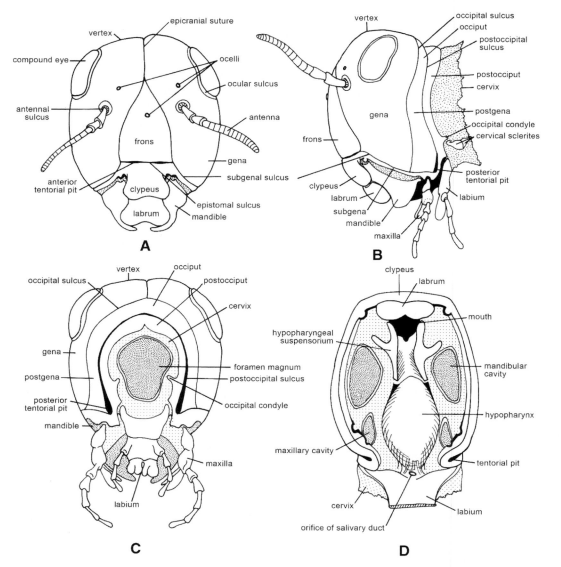

FIGURE 3.3. Structure of the typical pterygotan head. (A) Anterior; (B) lateral; (C) posterior; and (D) ventral (appendages removed). [From R. E. Snodgrass. *Principles of Insect Morphology.* Copyright 1935 by McGraw-Hill, Inc. Used with permission of McGraw-Hill Book Company.]

paired *mandibles* and *maxillae* occupy ventrolateral positions (Figure 3.3B). The mouth is situated behind the base of the labrum. The true ventral surface of the head capsule is the hypopharynx (Figure 3.3D), a membranous lobe that lies in the preoral cavity formed by the ventrally projecting mouthparts.

There are several grooves and pits on the head (Figure 3.3A–C), some of which, by virtue of their constancy of position within a particular insect group, constitute important taxonomic features. The grooves are almost all sulci. The *postoccipital sulcus* separates the maxillary and labial segments and internally forms a strong ridge to which are attached the muscles used in moving the head and from which the posterior arms of the *tentorium* arise (see following paragraph). The points of formation of these arms are seen externally as deep pits in the postoccipital groove, the *posterior tentorial pits*. The *epicranial suture*

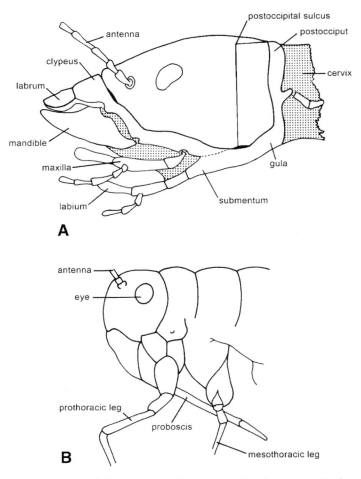

FIGURE 3.4. (A) Prognathous; and (B) opisthognathous types of head structure. [A, from R. E. Snodgrass, *Principles of Insect Morphology.* Copyright 1935 by McGraw-Hill, Inc. Used with permission of McGraw-Hill Book Company. B, after R. F. Chapman, 1971, *The Insects: Structure and Function.* By permission of Elsevier North-Holland, Inc., and the author.]

is a line of weakness occupying a median dorsal position on the head. It is also known as the *ecdysial line*, for it is along this groove that the cuticle splits during ecdysis. In many insects the epicranial suture is in the shape of an inverted Y whose arms diverge above the median ocellus and pass ventrally over the anterior part of the head. The *occipital sulcus*, which is commonly found in orthopteroid insects, runs transversely across the posterior part of the cranium. Internally it forms a ridge that strengthens this region of the head. The *subgenal sulcus* is a lateral groove in the cranial wall running slightly above the mouthpart articulations. That part of the subgenal sulcus lying directly above the mandible is known as the *pleurostomal sulcus*; that part lying behind is the *hypostomal sulcus*, which is usually continuous with the postoccipital suture. In many insects the pleurostomal sulcus is continued across the front of the cranium (above the labrum), where it is known as the *epistomal* (*frontoclypeal*) *sulcus*. Within this sulcus lie the *anterior tentorial pits*, which indicate the internal origin of the *anterior tentorial arms*. The *antennal* and *ocular sulci* indicate internal cuticular ridges bracing the antennae and compound eyes, respectively. A *subocular sulcus* running dorsolaterally beneath the compound eye is often present.

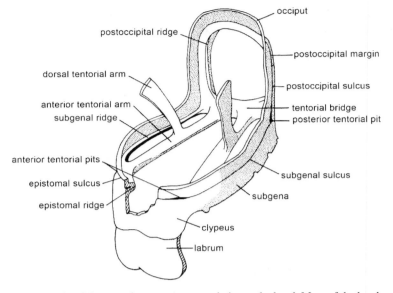

FIGURE 3.5. Relationship of the tentorium to grooves and pits on the head. Most of the head capsule has been cut away. [From R. E. Snodgrass. *Principles of Insect Morphology.* Copyright 1935 by McGraw-Hill, Inc. Used with permission of McGraw-Hill Book Company.]

The tentorium (Figure 3.5) is an internal, cranial-supporting structure whose morphology varies considerably among different insect groups. Like the *furca* of the thoracic segments (Section 4.2), with which it is homologous, it is produced by invagination of the exoskeleton. Generally, it is composed of the anterior and posterior tentorial arms that may meet and fuse within the head. Frequently, additional supports in the form of dorsal arms are found. The latter are secondary outgrowths of the anterior arms and not apodemes. The junction of the anterior and posterior arms is often enlarged and known as the *tentorial bridge* or *corporotentorium.* In addition to bracing the cranium, the tentorium is also a site for the insertion of muscles controlling movement of the mandibles, maxillae, labium, and hypopharynx.

The grooves described above delimit particular areas of the cranium that are useful in descriptive or taxonomic work. The major areas are as follows. The *frontoclypeal area* is the facial area of the head, between the antennae and the labrum. When the epistomal sulcus is present, the area becomes divided into the dorsal *frons* and the ventral clypeus. The latter is often divided into a *postclypeus* and an *anteclypeus.* The *vertex* is the dorsal surface of the head. It is usually delimited anteriorly by the arms of the epicranial suture and posteriorly by the occipital sulcus. The vertex extends laterally to merge with the *gena,* whose anterior, posterior, and ventral limits are the subocular, occipital, and subgenal sulci, respectively. The horseshoe-shaped area lying between the occipital sulcus and postoccipital sulcus is generally divided into the dorsal *occiput,* which merges laterally with the *postgenae.* The *postocciput* is the narrow posterior rim of the cranium surrounding the occipital foramen. It bears a pair of *occipital condyles* to which the *anterior cervical sclerites* are articulated. Below the gena is a narrow area, the *subgena,* on which the mandible and maxilla are articulated. The labium is usually articulated directly with the neck membrane (Figure 3.3C), but in some insects a sclerotized region separates the two. This sclerotized area develops in one of three ways: as extensions of the subgenae which fuse in the midline to form a *subgenal bridge,* as extensions of the hypostomal areas to form a *hypostomal bridge,* or (in most prognathous heads) through the extension ventrally and anteriorly of a ventral cervical sclerite to form the *gula*. At the same time the basal segment of the labium may also become elongated (Figure 3.4A).

3.2. Head Appendages

3.2.1. Antennae

A pair of antennae are found on the head of the pterygote insects and the apterygote groups with the exception of the Protura. However, in the larvae of many higher Hymenoptera and Diptera they are reduced to a slight swelling or disc.

In a typical antenna (Figure 3.6) there are three principal components: the basal *scape* by which the antenna is attached to the head, the *pedicel* containing Johnston's organ (Chapter 12, Section 3.1), and the *flagellum*, which is usually long and annulated. According to Kukalová-Peck (1992), the scape, pedicel, and flagellum are homologous with the subcoxa, coxa, and remaining segments, respectively, of the ancestral leg (Figure 3.21A). The annuli on the flagellum do not correspond with the ancestral leg joints; that is, the annuli are constrictions, not sutures. The scape is set in a membranous socket and surrounded by the antennal sclerite on which a single articulation may occur. In the majority of insects movement of the whole antenna is effected by muscles inserted on the scape and attached to the cranium or tentorium. However, in Collembola there is no Johnston's organ and each antennal segment is moved by a muscle inserted in the previous segment.

Although retaining the basic structure outlined above, the antennae take on a wide variety of forms (Figure 3.7) related to their varied functions. Generally, it is the flagellum that is modified. For example, in some male moths and beetles the flagellum is plumose and flabellate, respectively, providing a large surface area for the numerous chemosensilla that give these insects their remarkable sense of smell (see Chapter 12, Section 4). By contrast, the plumose nature of the antennae of male mosquitoes makes them highly sensitive to the sounds generated by the beating of the female's wings (Chapter 12, Section 3.1). Other functions of antennae include touching, temperature and humidity perception, grasping prey, and holding on to the female during mating (Schneider, 1964; Zacharuk, 1985). For taxonomists, this variety of form may be an important diagnostic feature.

3.2.2. Mouthparts

The mouthparts consist of the labrum, a pair of mandibles, a pair of maxillae, the labium, and the hypopharynx. In Collembola, Protura, and Diplura the mouthparts are

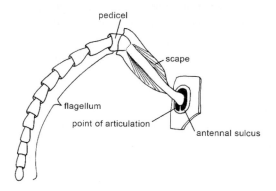

FIGURE 3.6. Structure of an antenna. [From R. E. Snodgrass, *Principles of Insect Morphology*. Copyright 1935 by McGraw-Hill, Inc. Used with permission of McGraw-Hill Book Company.]

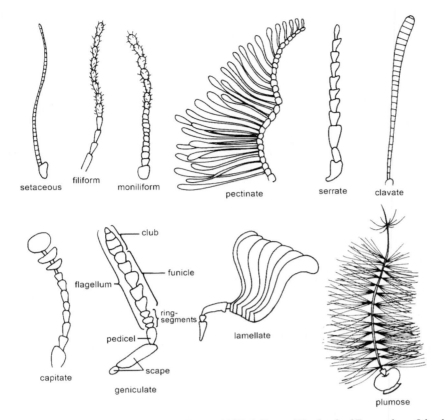

FIGURE 3.7. Types of antennae. [After A. D. Imms, 1957, *A General Textbook of Entomology*, 9th ed. (revised by O. W. Richards and R. G. Davies), Methuen and Co.]

enclosed within a cavity formed by the ventrolateral extension of the genae, which fuse in the midline (the *entognathous* condition). In Microcoryphia, Zygentoma, and Pterygota the mouthparts project freely from the head capsule, a condition described as *ectognathous*. The form of the mouthparts is extremely varied (see below), and it is appropriate to describe first their structure in the more primitive chewing condition.

Typical Chewing Mouthparts. In a typical chewing insect the labrum (Figure 3.3A) is a broadly flattened plate hinged to the clypeus. Its ventral (inner) surface is usually membranous and forms the lobe-like epipharynx, which bears mechano- and chemosensilla.

The mandible (Figure 3.8A) is a heavily sclerotized, rather compact structure having almost always a dicondylic articulation with the subgena. Its functional area varies according to the diet of the insect. In herbivorous forms there are both cutting edges and grinding surfaces on the mandible. The cutting edges are typically strengthened by the addition of zinc, manganese or, rarely, iron, in amounts up to about 4% of the dry weight. In carnivorous species the mandible possesses sharply pointed "teeth" for cutting and tearing. In Microcoryphia the mandible has a single articulation with the cranium and, as a result, much greater freedom of movement.

Of all of the mouthparts the maxilla (Figure 3.8B) retains most closely the structure of the primitive insectan limb. The basal segment is divided by a transverse line of flexure into two subsegments, a proximal *cardo* and a distal *stipes*. The cardo carries the single condyle with which the maxilla articulates with the head. Both the cardo and stipes are, however, attached on their entire inner surface to the membranous head pleuron. The stipes

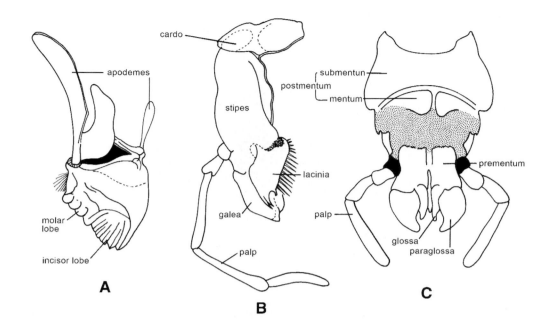

FIGURE 3.8. Structure of (A) mandible, (B) maxilla, and (C) labium of a typical chewing insect. [From R. E. Snodgrass, *Principles of Insect Morphology.* Copyright 1935 by McGraw-Hill, Inc. Used with permission of McGraw-Hill Book Company.]

bears an inner *lacinia* and outer *galea*, and a *maxillary palp*. This basic structure is found in both apterygotes and the majority of chewing pterygotes, although in some forms reduction or loss of the lacinia, galea, or palp occurs. In Kukalová-Peck's (1991) view the cardo and stipes correspond to the subcoxa and coxa + trochanter, respectively, of the ancestral appendage; the lacinia and the galea to the coxal and trochanteral endites, respectively; and the palp to the remaining segments. The laciniae assist in holding and masticating the food, while the galeae and palps are equipped with a variety of mechano- and chemosensilla.

The labium (Figure 3.8C) is formed by the medial fusion of the primitive appendages of the postmaxillary segment, together with, in its basal region, a small part of the sternum of that segment. The labium is divided into two primary regions, a proximal *postmentum* corresponding to the maxillary cardines plus the sternal component, and a distal *premen-tum* homologous with the maxillary stipites. The postmentum is usually subdivided into *submentum* and *mentum* regions. The prementum bears a pair of inner *glossae* and a pair of outer *paraglossae*, homologous with the maxillary laciniae and galeae, respectively, and a pair of *labial palps*. When the glossae and paraglossae are fused they form a single structure termed the *ligula*.

Arising as a median, mainly membranous, lobe from the floor of the head capsule and projecting ventrally into the preoral cavity is the hypopharynx (Figures 3.3D and 3.9). It is frequently fused to the labium. In a few insects (bristletails and mayfly larvae) a pair of lobes, the *superlinguae*, which arise embryonically in the mandibular segment, become associated with the hypopharynx. The hypopharynx divides the preoral cavity into anterior and posterior spaces, the upper parts of which are the *cibarium* (leading to the mouth) and *salivarium* (into which the salivary duct opens), respectively.

Mouthpart Modifications. The typical chewing mouthparts described above can be found with minor modifications in Odonata, Plecoptera, the orthopteroids and blattoids,

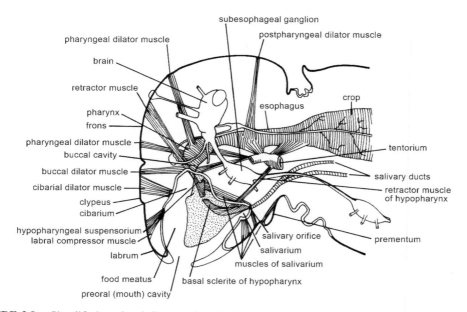

FIGURE 3.9. Simplified sectional diagram through the insect head showing the general arrangement of the parts. [From R. E. Snodgrass, *Principles of Insect Morphology*. Copyright 1935 by McGraw-Hill, Inc. Used with permission of McGraw-Hill Book Company.]

Neuroptera and Coleoptera (with the exceptions mentioned below), Mecoptera, primitive Hymenoptera, and larval Ephemeroptera, Trichoptera, and Lepidoptera. However, the basic arrangement may undergo great modification associated with specialized feeding habits (especially the uptake of liquid food) or other, nontrophic functions. Suctorial mouthparts are found in members of the hemipteroid orders, and adult Siphonaptera, Diptera, higher Hymenoptera, and Lepidoptera. The mouthparts are reduced or absent in non-feeding or endoparasitic forms.

Examination of the structure of the mouthparts provides information on an insect's diet and feeding habits, and is also of assistance in taxonomic studies. Some of the more important modifications for the uptake of liquid food are described below. It will be noted that all sucking insects have two features in common. Some components of their mouthparts are modified into tubular structures, and a sucking pump is developed for drawing the food into the mouth.

Coleoptera and Neuroptera. In certain species of Coleoptera and Neuroptera the mouthparts of the larvae are modified for grasping, injecting, and sucking. In the beetle *Dystiscus*, for example, the laterally placed mandibles are long, curved structures with a groove having confluent edges on their inner surface (Figure 3.10). The labrum and labium are closely apposed so that the cibarium is cut off from the exterior. When prey is grasped, digestive fluids from the midgut are forced along the mandibular grooves and into the body. After external digestion, liquefied material is sucked back into the cibarium. In *Dytiscus* the suctorial pump is constructed from the cibarium, the pharynx, and their dilator muscles (see Figure 3.9).

Hymenoptera. In adult Hymenoptera a range of specialization of mouthparts can be seen. In primitive forms, such as sawflies, the mandible is a typical biting structure, and the maxillae and labium, though united, still exhibit their component parts. In the advanced forms, such as bees, the mandibles become flattened and are used for grasping and molding

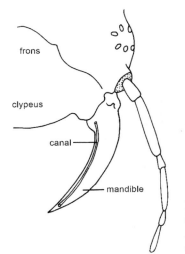

frons

clypeus

canal

mandible

FIGURE 3.10. Left mandible of *Dytiscus* larva, seen dorsally, showing the canal on its inner side. [From R. E. Snodgrass, *Principles of Insect Morphology.* Copyright 1935 by McGraw-Hill. Inc. Used with permission of McGraw-Hill Book Company.]

materials rather than biting and cutting. The maxillolabial complex is elongate and the glossae form a long flexible "tongue," a sucking tube capable of retraction and protraction (Figure 3.11). The laciniae are lost and the maxillary palps reduced, but the galeae are much enlarged, flattened structures, which in short-tongued bees are used to cut holes in the flower corolla to gain access to the nectary. When the food is easily accessible, the glossae, labial palps, and the galeae form a composite tube up which the liquid is drawn. When the food is confined in a narrow cavity such as a nectary, only the glossae are used to obtain it. The sucking mechanism of the Hymenoptera includes the pharynx, buccal cavity, and cibarium, and their dilator muscles.

   Lepidoptera.   Functional biting mouthparts are retained in the adults of only one family of Lepidoptera, the Micropterigidae. In all other groups the mouthparts (Figure 3.12) are considerably modified in conjunction with the diet of nectar. The mandibles are usually lost, the labrum is reduced to a narrow transverse sclerite, and the labium is a small flap

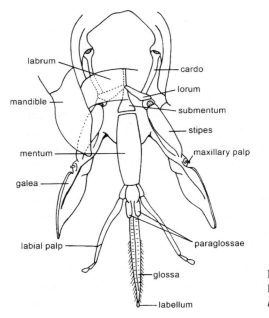

labrum

mandible

mentum

galea

labial palp

cardo

lorum

submentum

stipes

maxillary palp

paraglossae

glossa

labellum

FIGURE 3.11. Mouthparts of the honey bee. [After R. E. Snodgrass. 1925, *Anatomy and Physiology of the Honey bee*, McGraw-Hill Book Company.]

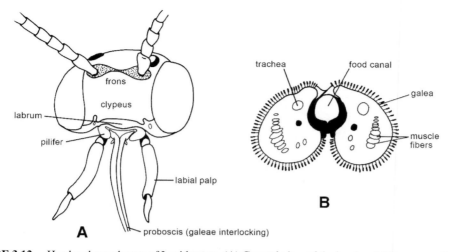

FIGURE 3.12. Head and mouthparts of Lepidoptera. (A) General view of the head and (B) cross-section of the proboscis. [From R. E. Snodgrass, *Principles of Insect Morphology.* Copyright 1935 by McGraw-Hill, Inc. Used with permission of McGraw-Hill Book Company.]

(though its palps remain quite large). The long, suctorial proboscis is formed from the interlocking galeae, whose outer walls comprise a succession of narrow sclerotized arcs alternating with thin membranous areas: presumably this arrangement facilitates coiling. Extension of the proboscis is brought about by a local increase in blood pressure. The sucking pump of Lepidoptera comprises the same elements as that of Hymenoptera. In Lepidoptera that do not feed as adults all mouthparts are greatly reduced and the pump is absent.

Diptera. In both larval and adult Diptera the form and function of the mouthparts have diverged considerably from the typical chewing condition. Indeed, in extreme cases [seen in some of the larvae (maggots) of Muscomorpha] it appears that not only a new feeding mechanism but an entirely new functional head and mouth have evolved, the true mouthparts of the adult fly being suppressed during the larval period. This remarkable modification of the head and its appendages is, of course, the result of the insect living entirely within its food.

Larva. In larvae of many orthorrhaphous flies the head is retracted into the thorax and enclosed within a sheath formed from the neck membrane. The mandibles and maxillae possess the typical biting structure (though the palps are small or absent). The labrum is large and overhanging. The labium is rudimentary and often confused with the *hypostoma*, a toothed, triangular sclerite on the neck membrane (Figure 3.13A–C).

In maggots the true head is completely invaginated into the thorax, and the conical "head" is, in fact, a sclerotized fold of the neck. The functional "mouth" is the inner end of the preoral cavity, the *atrium*, from which a pair of sclerotized hooks protrude. The cibarium is transformed into a massive sucking pump, and the true mouth is the posterior exit from the pump lumen (Figure 3.13D).

Adult. No adult Diptera have typical biting mouthparts, although, of course, many blood feeders are said to "bite" when they pierce the skin. The mouthparts can be divided functionally into those that only suck and those that first pierce and then suck. In the latter the piercing structure may be the mandibles, the labium, or the hypopharynx.

In Diptera that merely suck or "sponge" up their food (e.g., the house fly and blow fly) the mandibles have disappeared and the elongate feeding tube, the *proboscis*, is a composite structure that includes the labrum, hypopharynx, and labium (Figure 3.14). The proboscis

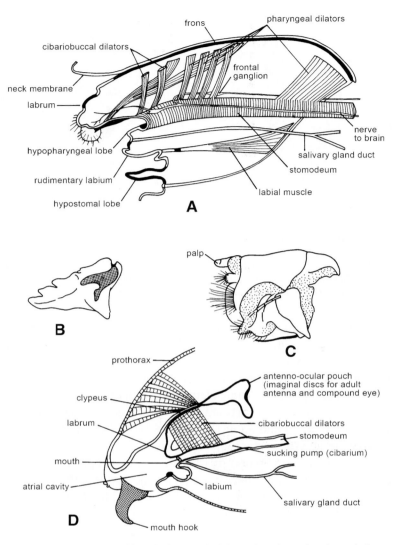

FIGURE 3.13. Head and mouthparts of larval Diptera. (A) Diagrammatic section through the retracted head of *Tipula*; (B) right mandible of *Tipula*; (C) left maxilla of *Tipula*; and (D) diagrammatic section through the anterior end of a maggot. [From R. E. Snodgrass, *Principles of Insect Morphology*. Copyright 1935 by McGraw-Hill, Inc. Used with permission of McGraw-Hill Book Company.]

is divisible into a basal *rostrum* bearing the maxillary palps, a median flexible *haustellum*, and two apical *labella*. The latter are broad sponging pads, equipped with *pseudotracheae* along which food passes to the oral aperture. The latter is not the true mouth, which lies at the upper end of the food canal. As in other Diptera, the sucking apparatus is formed from the cibarium and its dilator muscles that are inserted on the clypeus.

Many Diptera feed on blood. Some of these (e.g., the tsetse fly, stable fly, and horn fly), like their non-piercing relatives, have a composite proboscis. However, the haustellum is elongate and rigid, and the distal labellar lobes are small but bear rows of *prestomal teeth* on their inner walls. The labrum and labium interlock to form the food canal within which lies the hypopharynx enclosing the salivary duct (Figure 3.15).

Other blood-feeding flies (e.g., horse flies, deer flies, black flies, and mosquitoes) use the mandibles for piercing the host's skin. The mouthparts of the horse fly *Tabanus* may be

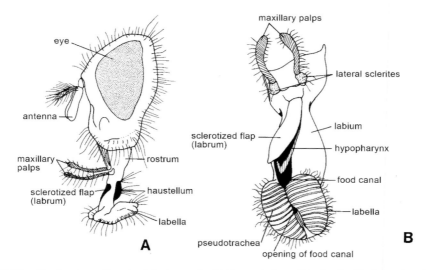

FIGURE 3.14. Head and mouthparts of the house fly. (A) Lateral view of the head with the proboscis extended; and (B) anterodistal view of the proboscis. [From R. E. Snodgrass, *Principles of Insect Morphology.* Copyright 1935 by McGraw-Hill, Inc. Used with permission of McGraw-Hill Book Company.]

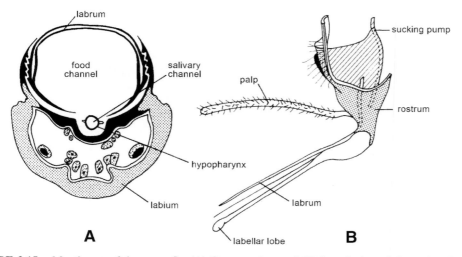

FIGURE 3.15. Mouthparts of the tsetse fly. (A) Cross-section; and (B) lateral view of the proboscis. [From R. E. Snodgrass, *Principles of Insect Morphology.* Copyright 1935 by McGraw-Hill, Inc. Used with permission of McGraw-Hill Book Company.]

taken as an example (Figure 3.16). The labrum is dagger-shaped but flexible and blunt at the tip. On its inner side is a groove closed posteriorly by the mandibles to form the food canal. The mandibles are long and sharply pointed. The maxillae retain most of the components of the typical biting form (except the laciniae) but the galeae are long bladelike structures. The hypopharynx is a styletlike structure and contains the salivary duct. The labium is a large, thick appendage with a deep anterior groove into which the other mouthparts normally fit. Distally it bears two large labellar lobes. Blood flows along the pseudotracheae to the tip of the food canal.

Hemiptera. The major contributor to the hemipteran proboscis (Figure 3.17) is the labium, a flexible segmented structure with a deep groove on its anterior surface. Within this groove are found the piercing organs, the *mandibular* and *maxillary bristles*. The two

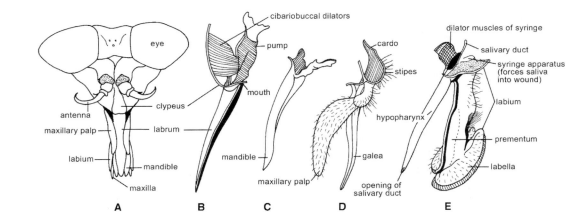

FIGURE 3.16. Head and mouthparts of the horse fly. (A) Anterior view of the head; and (B–E) lateral views of the separated mouthparts. [From R. E. Snodgrass, *Principles of Insect Morphology.* Copyright 1935 by McGraw-Hill, Inc. Used with permission of McGraw-Hill Book Company.]

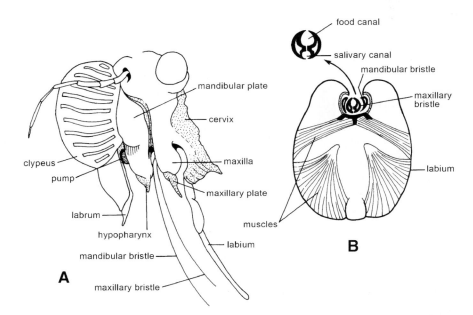

FIGURE 3.17. Head and mouthparts of Hemiptera. (A) Head with the mouthparts separated; and (B) cross-section of the proboscis. [From R. E. Snodgrass, *Principles of Insect Morphology.* Copyright 1935 by McGraw-Hill, Inc. Used with permission of McGraw-Hill Book Company.]

maxillary bristles are interlocked within the labial groove and form the food and salivary canals. Because of the great enlargement of the clypeal region of the head associated with the opisthognathous condition, the cibarial sucking pump is entirely within the head.

4. The Neck and Thorax

The thorax is the locomotory center of the insect. Typically each of its three segments (pro-, meso-, and metathorax) bears a pair of legs, and in the adult stage of the Pterygota the meso- and metathoracic segments each have a pair of wings. Between the head and thorax is the membranous neck (*cervix*).

4.1. The Neck

Study of its embryonic development shows that the neck contains both labial and prothoracic components and therefore the primary intersegmental line must be within the neck membrane. The muscles that control head movement arise on the postoccipital ridge and are attached to the antecosta of the prothoracic segment. Thus, they must include the fibers of two segments. This modification of the basic segmental structure was apparently necessary to provide sufficient freedom of head movement.

Usually supporting the head and articulating it with the prothorax are the cervical sclerites, a pair of which occur on each side of the neck (Figure 3.3B). Either one or both sclerites may be absent. When only one occurs, it is often fused with the prothorax. Occasionally, additional dorsal and ventral sclerites are found.

4.2. Structure of the Thorax

In the evolution of the typical insectan body plan there have been two phases associated with the development of the thorax as the locomotory center; in the first the walking legs became restricted to the three thoracic segments, and in the second articulated wings were formed on the meso- and metathoracic terga. Accompanying each of these developments were major changes in the basic structure of the secondary segments of the thoracic region. These changes were primarily to strengthen the region for increased muscular power.

In the apterygotes and many juvenile pterygotes the thoracic terga are little different from those of the typical secondary segment described in Section 2. In the adult the terga of the wing-bearing segments are enlarged and much modified (Figure 3.18). Although it may remain a single plate, the tergum (or *notum*, as it is called in the thoracic segments) is usually divided into the anterior wing-bearing *alinotum*, and the posterior *postnotum*. These are firmly supported on the pleural sclerotization by means of the *prealar* and *postalar arms*, respectively. The antecostae of the primitive segments become greatly enlarged forming *phragmata*, to which the large dorsal longitudinal muscles are attached. As wing movement is in part brought about by flexure of the terga (see Chapter 14, Section 3.3.3), which is itself caused by contraction and relaxation of the dorsal longitudinal muscles, it is clear that the connection between the *mesonotum* and *metanotum* and between the metanotum and first abdominal tergum must be rigid. The intersegmental membranes are therefore reduced or absent. Additional supporting ridges are developed on the meso- and metanota, the most common of which are the V-shaped *scutoscutellar ridge* and the *transverse (prescutal) ridge* (Figure 3.18A). The lateral margins of the alinotum are constructed for articulation of the wing. They possess both *anterior* and *posterior notal processes*, to which the *first* and *third axillary sclerites*, respectively, are attached. Further details of the wing articulation are given in Section 4.3.2.

The originally membranous pleura have been strengthened to varying degrees by sclerotization and the formation of internal cuticular ridges. In some apterygotes, for example, two small, crescent-shaped pleural sclerites may be seen above the coxa, though the rest of the pleuron is membranous. In the prothorax of Plecoptera there are likewise two sclerites, but these are much larger than those of apterygotes and occupy more than half the pleural area. In the thoracic segments of all other pterygotes the pleura are fully sclerotized and are

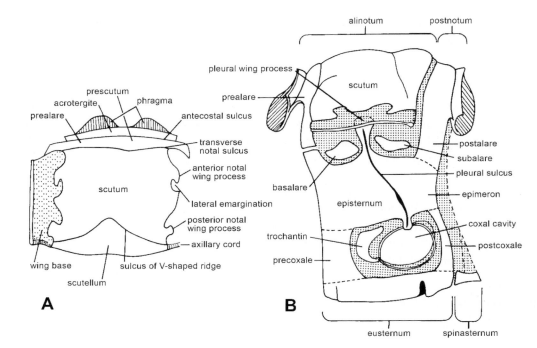

FIGURE 3.18. (A) Dorsal view of a generalized alinotum; and (B) lateral view of a typical wing-bearing segment. [From R. E. Snodgrass, *Principles of Insect Morphology.* Copyright 1935 by McGraw-Hill, Inc. Used with permission of McGraw-Hill Book Company.]

additionally strengthened by the formation of an internal *pleural ridge* that extends dorsally into the *pleural wing process* (Figure 3.18B). Articulating with this process is the *second axillary sclerite.* Each pleural ridge is extended inwardly as a *pleural arm* (Figure 3.19) that meets and may fuse with similar apophyses from the sternum. The pleural ridge is seen externally as the *pleural sulcus* (Figure 3.18B) above the coxa. This groove divides the pleuron into an anterior *episternum* and posterior *epimeron*. Often these sclerites are divided secondarily into dorsal and ventral areas, the *supraepisternum* and *infraepisternum*, and *supraepimeron* and *infraepimeron*. Derived from the episternum and epimeron and appearing above them usually as distinct, articulated sclerites are the *basalar* and *subalar*, to which important wing muscles are attached.

In the thorax the acrosternite of the typical secondary segment forms an independent intersegmental plate or *intersternite*. The intersternites, which are found between the pro- and mesothorax and between the meso- and metathorax, are known as *spinasterna* because each bears an internal spine to which a few ventral muscles are attached. Frequently, the spinasterna fuse with the segmental plate, *eusternum*, of the preceding segment. The eusternum is a composite structure, comprising the *primary sternal plate* and the *sternopleurite*. The eusternum may be divided secondarily into an anterior *basisternum* and posterior *sternellum* by the *sternacostal sulcus* (Figure 3.20). The latter is the result of an invagination to form the *sternacosta*, a ridge of cuticle that unites the *sternal apophyses* (Figure 3.19). In the higher pterygotes these apophyses are borne on a median internal ridge and form a Y-shaped furca (Figure 3.19). As noted earlier, these apophyses combine with the pleural arms to form a rigid internal support. The latter provides attachment for the major longitudinal ventral muscles and certain muscles of the leg.

It must be emphasized that many variations occur from the rather general description of a thoracic segment provided above. In all insects the prothorax is modified by the development

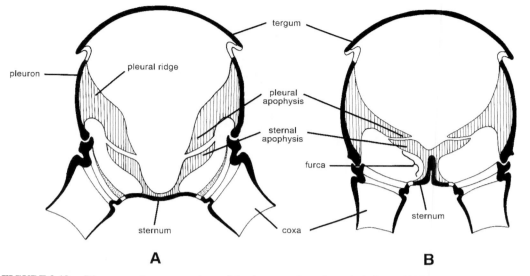

FIGURE 3.19. Diagrammatic cross-sections of the thorax to show the endoskeleton. (A) Normal condition; and (B) condition when furca present. [From R. E. Snodgrass, *Principles of Insect Morphology.* Copyright 1935 by McGraw-Hill, Inc. Used with permission of McGraw-Hill Book Company.]

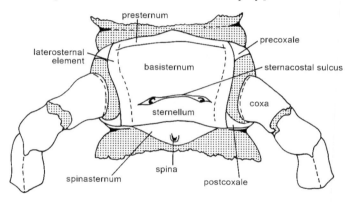

FIGURE 3.20. Ventral view of a generalized thoracic sternum. [From R. E. Snodgrass, *Principles of Insect Morphology.* Copyright 1935 by McGraw-Hill, Inc. Used with permission of McGraw-Hill Book Company.]

of the neck region. The *pronotum* especially is different, lacking the antecostal region and phragma through neck membranization. In some groups (e.g., Orthoptera, Hemiptera, and Coleoptera) the pronotum is greatly enlarged; in others it is reduced to a narrow band between the head and mesothorax. In those orders whose members have a single pair of functional wings, the tergal plates of the segment from which the wings are absent are usually reduced in size.

4.3. Thoracic Appendages

4.3.1. Legs

In the vast majority of insects each thoracic segment bears a pair of legs. In the cases where legs are absent, for example, in all dipteran, and many coleopteran and hymenopteran larvae, the condition is secondary. Typically, the legs are concerned with walking and running, but they may be specialized for a range of other physical functions, some of which are described below. In addition, for many insects they are important organs of taste

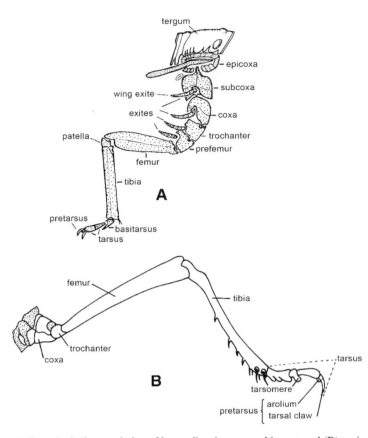

FIGURE 3.21. (A) Hypothetical ground plan of leg podites in ancestral insect; and (B) typical leg of a modern insect. [A, after J. Kukalová-Peck, 1987, New Carboniferous Diplura, Monura, and Thysanura, the hexapod ground plan, and the role of thoracic side lobes in the origin of wings (Insecta), *Can. J. Zool.* **65**:2327–2345. By permission of the National Research Council of Canada and the author. B, from R. E. Snodgrass, *Principles of Insect Morphology.* Copyright 1935 by McGraw-Hill, Inc. Used with permission of McGraw-Hill Book Company.]

(see Chapter 12, Section 4.1). As noted earlier, Kukalová-Peck (1987) suggested that the ancestral limb included 11 podites, as well as exites and endites (Figure 3.21A). Because of fusion of podites with the pleuron or with adjacent podites the full complement of podites in the leg is never seen, though in many fossils and a few extant Ephemeroptera and Odonata as many as eight podites can be identified.

Typical Walking Leg. The leg consists of six podites, the coxa, trochanter, femur, tibia, tarsus, and pretarsus (Figure 3.21B). Between adjacent parts are a narrow, annulated membrane, the *corium,* and usually a mono- or dicondylic articulation.

The coxa is a short, thick segment strengthened at its proximal end by an internal ridge, the *basicosta* (Figure 3.22). The coxa usually has a dicondylic articulation with the pleuron. In some orders the *basicostal sulcus* is U- or V-shaped over the posterior half of the coxa (Figure 3.22). The sclerite thus demarcated becomes thickened and is known as the *meron.* The trochanter is a small segment. It always has a dicondylic articulation with the coxa but is usually firmly fixed to the femur, which is generally the largest leg segment. Following the slender tibia is the tarsus, a segment that is usually subdivided into between two and five *tarsomeres* and a pretarsus. The pretarsus, in most insects, takes the form of a pair of *tarsal claws* and a median lobe, the *arolium* (Figure 3.23).

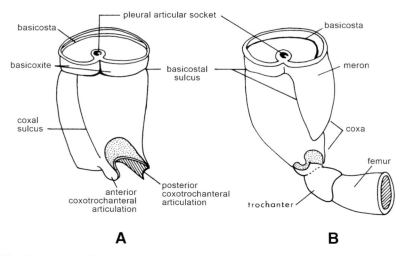

FIGURE 3.22. Structure of the coxa. (A) Lateral view; and (B) coxa with a well-developed meron. [From R. E. Snodgrass, *Principles of Insect Morphology.* Copyright 1935 by McGraw-Hill, Inc. Used with permission of McGraw-Hill Book Company.]

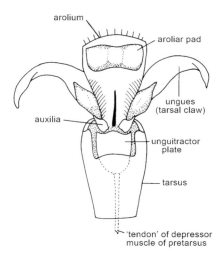

FIGURE 3.23. Distal part of a leg showing the arolium and claws. [From R. E. Snodgrass, *Principles of Insect Morphology.* Copyright 1935 by McGraw-Hill, Inc. Used with permission of McGraw-Hill Book Company.]

Leg Modifications. The functions for which the legs have become modified include jumping, swimming, grasping, digging, sound production, and cleaning.

In Orthoptera and a few Coleoptera (e.g., flea beetles) the femur on the hindleg is greatly enlarged to accommodate the extensor muscles of the tibia used in jumping. In swimming insects, the tibia and tarsus of the hindlegs (occasionally also the middle legs) are flattened and bear rigid hairs around the periphery (Figure 3.24A). Legs modified for grasping are found in predaceous insects such as the mantis and giant water bug, in ectoparasitic lice, and in males of various species where they are used for hanging onto the female during mating. In the mantis, the tibia and femur of the foreleg are equipped with spines and operate together as pincers (Figure 3.24B). The foreleg of a louse is short and thick and has at its tip a single, large tarsal claw that folds back against the tibial process (Figure 3.24C). Suctorial pads have been developed on the fore limbs in males of many beetle species. In *Dytiscus*, for example, the first three tarsomeres are flattened and possess large numbers of cuticular

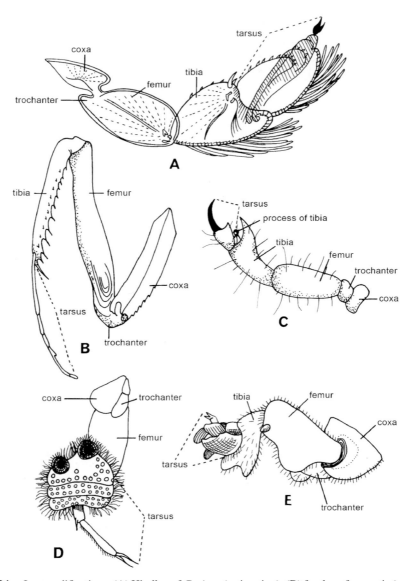

FIGURE 3.24. Leg modifications. (A) Hindleg of *Gyrinus* (swimming); (B) foreleg of a mantis (grasping prey); (C) foreleg of a louse (attachment to host); (D) foreleg of *Dytiscus* (holding onto female); and (E) foreleg of a mole cricket (digging). [A, after L. C. Miall, 1922, *The Natural History of Aquatic Insects*, published by Macmillan Ltd. B, D, E, after J. W. Folsom, 1906, *Entomology: With Special Reference to Its Biological and Economic Aspects*.]

cups, two of which are extremely enlarged (Figure 3.24D). The forelegs of soil-dwelling insects such as the mole cricket (Figure 3.24E), cicadas, and various beetles are modified for digging. The legs are large, heavily sclerotized, and possess stout claws. The tarsomeres are reduced in number or may disappear entirely in some forms. In many Orthoptera sounds are produced when the hind femora, which have a row of cuticular pegs on their inner surface, are rubbed against ridged veins on the fore wing. Modifications to the forelegs for cleaning purposes are found in many insects. In certain Coleoptera and Hymenoptera, for example, the honey bee (Figure 3.25A), a notch lined with hairs occurs on the metatarsus

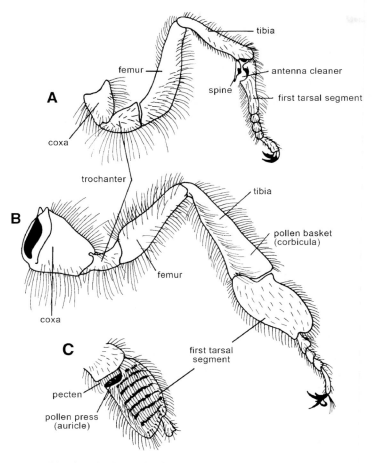

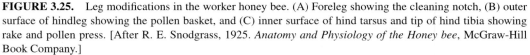

FIGURE 3.25. Leg modifications in the worker honey bee. (A) Foreleg showing the cleaning notch, (B) outer surface of hindleg showing the pollen basket, and (C) inner surface of hind tarsus and tip of hind tibia showing rake and pollen press. [After R. E. Snodgrass, 1925. *Anatomy and Physiology of the Honey bee*, McGraw-Hill Book Company.]

of the foreleg through which the antenna can be drawn and cleaned. The hindlegs of the bee are modified for pollen collection (Figure 3.25B). Rows of hairs, the *comb*, on the inner side of the first tarsomere scrape pollen off the abdomen. The *rake*, a fringe of hairs at the distal end of the tibia, then collects the pollen from the comb on the opposite leg and transfers it to the pollen *press*. When the press is closed, the pollen is pushed up into the pollen *basket*, where it is stored until the bee returns to its nest.

4.3.2. Wings

The majority of adult Pterygota have one or two pairs of functional wings. The complete absence of wings is a secondary condition, associated with the habits of the group concerned, for example, soil-dwelling or endoparasitism. The wings may be modified for a variety of purposes other than flight.

Development and General Structure.   Regardless of its evolutionary origin (Chapter 2, Section 3.1) a wing contains the usual integumental elements (cuticle, epidermis, and basal lamina) and its lumen, being an extension of the hemocoel, contains tracheae, nerves, and hemolymph. As the wing develops, the dorsal and ventral integumental layers become closely apposed over most of their area forming the *wing membrane*. The remaining areas form channels, the future *veins*, in which the nerves and tracheae may occur. The cuticle surrounding the veins becomes thickened to provide strength and rigidity to the wing. Hairs of two types may occur on the wings: *microtrichia*, which are small and irregularly scattered, and *macrotrichia*, which are larger, socketed, and may be restricted to veins. The scales of Lepidoptera and Trichoptera are highly modified macrotrichia. For a detailed review of wing morphology, see Wootton (1992).

It must be assumed that the extremely varied arrangement of veins found among the insect orders is derived from a primitive common pattern, the ground plan or archetype venation. Details of the latter will never be known with certainty because even the earliest fossilized wings had a highly complex venation. In order to develop a basic plan of wing venation, it is necessary to compare (and homologize) not only the veins, but also flexion lines (lines along which a wing creases during flight), fold lines (where wings crease during folding), and wing areas (areas delimited by specific veins and/or lines) (Wootton, 1979). Given the diversity of wings of insects, both fossil and extant, it is not surprising that authors have sometimes reached quite different conclusions with respect to the homologies of veins (compare Figures 2.2 and 3.26). This is unfortunate given the enormous importance of wing venation in phylogenetic and taxonomic studies.

The usual method of determining homology is direct comparison of the position and form of veins. In addition, there are other, more subtle features with which to assess potential homologies (Ragge, 1955; Hamilton, 1972a; Lawrence *et al.*, 1991). Thus, certain veins are always associated with particular axillary sclerites. Tillyard (1918) made use of the fact

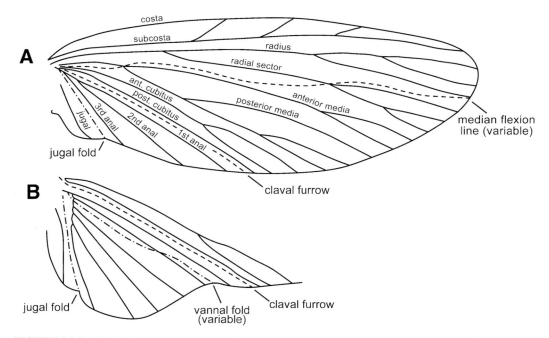

FIGURE 3.26. Basic scheme of wing venation, flexion lines and folding lines. (A) Fore wing or hind wing without vannus; and (B) vannal area of hind wing. [After R. J. Wootton, 1979, Function, homology and terminology in insect wings, *Syst. Entomol.* **4**:81–93. By permission of the Royal Entomological Society.]

that some veins have associated with them a row of macrotrichia. In many wings this row persists even when the original vein has disappeared. The widely used Comstock-Needham system of wing venation was based on the assumption that tracheae are always present in particular veins and that the pattern of tracheae develops in a characteristic manner and thereby determines the venational pattern (Comstock, 1918). In fact, the lacunae from which veins arise develop as hemolymph channels and it is only after veins have formed that tracheae and nerves extend into them. In addition, tracheal patterns may differ, for example, between pupal and adult instars, according to different functional needs. Other authors have based their studies of vein homology on wing fluting, the alternation of concave and convex veins. This approach is of limited use, however, and is not applicable to branched vein systems where fluting is practically non-existent, or to wings that are secondarily fluted (Hamilton, 1972a).

Wootton (1979) made a major contribution toward resolution of the confusion over the nomenclature of veins and other wing components. His scheme, presented in Figure 3.26, includes the following longitudinal veins (from anterior to posterior): *costa, subcosta, radius, radial sector, anterior media, posterior media, anterior cubitus, posterior cubitus,* and *anals* (of which there may be several). Any of these veins may be branched, though the costa and posterior cubitus rarely are because the former is at the leading edge of the wing while the latter is associated with the claval furrow. In wings with a *jugum* (see below), *jugal vein(s)* may occur. The major flexion lines are the *median flexion line*, which originates just behind the media and runs outward just behind the radial sector, and the *claval furrow*, which lies along the posterior cubitus (Figure 3.27). Though fold lines may be transverse, as in the hind wings of beetles and earwigs, they are normally radial to the base of the wing, allowing adjacent sections of a wing to be folded over or under each other. The commonest fold line is the *jugal fold*, situated just behind the third anal vein. In hind wings with an

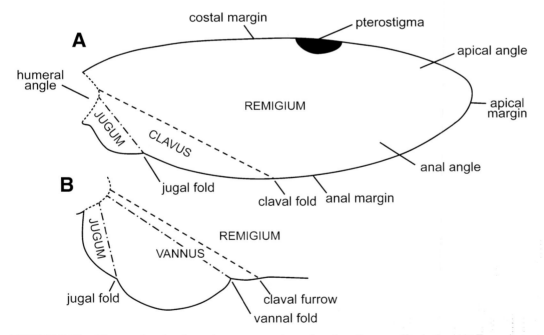

FIGURE 3.27. Diagram showing the major areas, margins and angles of a generalized wing. (A) Fore wing or hind wing without vannus; and (B) vannal area of hind wing [Partly after R. J. Wootton, 1979, Function, homology and terminology in insect wings, *Syst. Entomol.* **4**:81–93. By permission of the Royal Entomological Society.]

expanded posterior area (the *ano-jugal area* or *vannus*), several fold lines may occur. The principal wing areas are the jugum, behind the jugal fold, and the *clavus*, the area between the claval furrow and jugal fold in wings without a vannus. The vannus extends between the clavus and the jugum and therefore includes any vannal fold lines. The *remigium* is the area anterior to the claval furrow.

Various other terms may be encountered in wing descriptions (Figure 3.27). Because the wing is approximately triangular, it has three margins: the anterior *costal margin*, lateral *apical margin*, and posterior *anal margin*. These margins form three angles: the *humeral angle* at the wing base, the *apical angle* between the costal and apical margins, and the *anal angle* (*tornus*) between the apical and anal margins. In Psocoptera, Hymenoptera, and Odonata an opaque or pigmented area, the *pterostigma*, is found near the costal margin of the wing. The areas between adjacent veins are referred to as *cells*, which may be open (when extending to the wing margin) or closed (when entirely surrounded by veins). The cells are named after the longitudinal vein that forms their anterior edge.

Wing Modifications.   There are many modifications of the typical condition in which the insect has two pairs of triangle-shaped flapping wings. The modifications fall, however, into two broad categories: (1) Those that lead, directly or indirectly, to improved flight (see Wootton, 1992); and (2) those in which a wing takes on a function entirely unrelated to movement of the insect through the air. Both types of modification are possible in the same insect.

It seems that the two-winged condition is aerodynamically more efficient than the four-winged condition, and in a number of insect orders wing-coupling mechanisms have evolved that link together the fore and hind wings on the same side of the body (Tillyard, 1918). The precise nature of the coupling mechanism varies, but usually it consists of groups of hairs, the *frenulum*, on the anterior basal margin of the hind wing that interlocks with hairs or curved spines, the *retinaculum*, attached to various veins of the fore wing (Figure 3.28). The other way in which the two-winged condition has been achieved is through the loss, functionally speaking, of either the fore wings (in Coleoptera and male

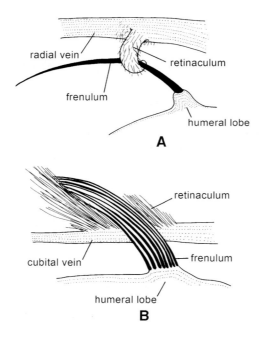

FIGURE 3.28.   Wing-coupling mechanism in *Hippotion scrota* (Lepidoptera). (A) Male; and (B) female. [After R. J. Tillyard, 1918, The panorpoid complex. Part I. The wing-coupling apparatus, with special reference to the Lepidoptera. *Proc. Linn. Soc. N.S.W.* **43**:286–319. By permission of the Linnean Society, N.S.W.]

Strepsiptera) or the hind wings (Diptera, and some Ephemeroptera and Hemiptera). In these insects the wings, no longer used directly in flight (i.e., to create lift through flapping), are still present but modified for other functions. The modified hind wings, *halteres*, of Diptera and male Coccoidea (Hemiptera) are highly developed sensory structures used in attitudinal control (see Figure 14.15). The heavily sclerotized *elytra* (fore wings) of Coleoptera are mainly protective in function, though they may be secondarily important in control of attitude.

Even in insects that retain two functional pairs of wings there are frequently modifications of these structures for other functions. In many Lepidoptera, for example, the wing margins are irregular and the wings appropriately colored so that when the insect is at rest it is camouflaged. The wings of male Orthoptera are commonly modified for sound production. In crickets and long-horned grasshoppers the hardened fore wings possess a toothed *file* (the modified cubital vein) and a *scraper* (a sclerotized ridge at the wing margin). Rapid opening and closing of the wings causes the file on one wing to be dragged over the scraper of the other wing and sound to be produced. In short-horned grasshoppers that sing the file is on the hind femur; the scraper takes the form of ridged veins on the fore wings.

5. The Abdomen

The abdomen differs from the head and thorax in that its segments generally have a rather simple structure, they are usually quite distinct from each other, and most of them lack appendages. The number of abdominal segments varies. The primitive number appears to be 12, though this number is found today in only the Protura. Most insects have 10 or 11 abdominal segments, but several of these are reduced. The reduction occurs primarily at the posterior end, but in some endopterygotes the first segment is reduced and intimately fused with the metathorax. For the purpose of discussion the abdominal segments may be considered to form three groups: pregenital segments, genital segments, and postgenital segments.

5.1. General Structure

The more anterior (pregenital) segments are little different from the typical secondary segment described in Section 2. In the first segment, however, the antecosta of the tergum bears internally a pair of phragmata to which the metathoracic dorsal longitudinal muscles are attached. Furthermore, the acrotergite of this segment forms the postnotal plate of the metathorax. Frequently the antecostal region and acrotergite are clearly separated from the rest of the tergum and form part of the metanotum. In the higher Hymenoptera the entire first segment, the *propodeum*, is fused with the metathorax and the conspicuous "waist" of these insects occurs between the first and second abdominal segments.

The genital opening (*gonopore*) is located, in the majority of Pterygota, on or behind the eighth or ninth sternum in the female, and behind the ninth sternum in the male. In Ephemeroptera and most male Dermaptera paired gonopores occur. The genital segments are modified in various ways for oviposition or sperm transfer. In female Diptera, Lepidoptera, and Coleoptera the posterior segments lack appendages and form smooth cuticular cylinders often telescoped into the anterior part of the abdomen. When extended (Figure 3.29) they form a long narrow tube that facilitates egg laying in inaccessible places. Sometimes the tip of the abdomen is sclerotized for piercing tissues. In other orders the tergum and sternum

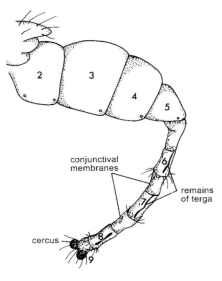

FIGURE 3.29. Abdomen of the house fly extended. The segments are numbered. [From L. S. West, 1951, *The Housefly*, Comstock Publishing Co., Inc.]

of the genital segments remain as distinct cuticular plates. In the female the sternum of the eighth segment is sometimes enlarged to form the *subgenital plate*, in which case the sternum of the ninth segment is reduced to a membranous sheet. In the male the tergum and sternum of the ninth segment are distinct but may be greatly modified. The genital segments retain their appendages, which are modified to serve in the reproductive process. In the female they form the *ovipositor*, and in the male, *clasping* and *intromittent organs*.

The postgenital segments include the 10th and, when present, the 11th abdominal segments. In the lower orders where both postgenital segments are present, the 10th segment is usually united with the 9th or 11th segments and it never bears appendages. The 11th segment comprises a somewhat triangular tergal plate, the *epiproct*, and a pair of ventrolateral plates, the *paraprocts*. It bears appendages, the *cerci*, inserted in the membranous area between the 10th and 11th segments on each side of the body (Figure 3.30).

In the majority of endopterygotes there is only one postgenital segment, the 10th abdominal segment. It is usually much modified and has no true appendages. When appendiculate structures are present, they are secondary developments.

5.2. Abdominal Appendages

Generally the only appendages seen on the abdomen of an adult pterygote insect are those on the genital segments (the *external genitalia*) and the cerci. In the apterygote groups

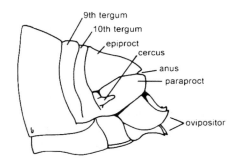

FIGURE 3.30. Postgenital segments of a female grasshopper. [From R. E. Snodgrass, *Principles of Insect Morphology.* Copyright 1935 by McGraw-Hill, Inc. Used with permission of McGraw-Hill Book Company.]

there are, in addition, appendages on a varied number of pregenital segments. All of these may be considered as "primary" appendages, that is, derived from the primitive insect limb.

Many larvae possess segmentally arranged *prolegs* or *gills* on the abdomen. According to some authorities (e.g., Hinton, 1955), these are secondary structures and are not derived from the primitive limb. However, others, notably Kukalová-Peck (1991), have argued that these are homologues of the thoracic legs, that is, are direct descendants of the ancestral paired appendages present on all segments. Finally, a number of insects possess non-segmental appendages on the abdomen, which clearly are of secondary origin.

5.2.1. External Genitalia

The morphology of the external genitalia is highly varied, especially in the male, and it is sometimes extremely difficult to homologize the different components. In the female the appendages combine to form an ovipositor, used for placing eggs in specific sites. The appendages of the male are used as clasping organs during copulation. Because of their specific form, the external genitalia are widely used by taxonomists for identification purposes.

Female. According to Scudder (1961) the primitive structure of the pterygote insect ovipositor can be seen in the silverfish *Lepisma* (Figure 3.31A). The ovipositor typically has a basal part and a shaft. The basal part consists of two pairs of *gonocoxae* (*valvifers*) on the eighth and ninth segments, homologous with the coxa of the leg. Projecting ventroposteriorly from each gonocoxa is an elongate process, the *gonapophysis* (*valvula*). Together the four gonapophyses make up the shaft. The first gonapophyses are ventral, the second dorsal. In most pterygote orders the second gonocoxa has a second posterior process, the *third valvula* or *gonoplac*, attached to it (Figure 3.31B). The gonoplac may form part of the ovipositor or serve as a protective sheath. The *gonangulum* is a small sclerite, articulated with the second gonocoxa (from which it is derived) and the ninth tergum.

Among Pterygota an ovipositor is found in Grylloblattodea, Dictyoptera, Orthoptera, and Hymenoptera, some Odonata and most Hemiptera, Thysanoptera, and Psocoptera. In each of these groups it is more or less modified from the general condition described above. The ovipositor of Odonata and Grylloblattodea is almost identical with that of *Lepisma*, except that a gonoplac is present. In Dictyoptera and Orthoptera the gonangulum fuses with the first gonocoxa. In viviparous cockroaches the ovipositor is considerably modified. The gonoplac, in Orthoptera, is a large sclerotized plate that forms part of the ovipositor shaft. The second gonapophyses, when present, are usually median and concealed. These structures are much reduced in Acrididae.

In Hemiptera, Thysanoptera, and Psocoptera, an ovipositor may or may not be present. When present, the gonangulum is fused with the ninth tergum, and the second gonapophyses are fused, making the shaft a three-part structure. The gonoplacs normally ensheath the shaft.

The ovipositor of Hymenoptera may be considerably modified for boring, piercing, sawing, and stinging (Figure 3.32). Only when modified for stinging does it no longer participate in egg laying, though it retains the basic components, with the exception of the first gonocoxae, which have disappeared. The first gonapophyses are attached to the gonangulum, which is articulated with the ninth tergum and second gonocoxa. As in Hemiptera, the second gonapophyses are fused. The shaft is typically very elongate, and the eggs are considerably compressed as they are squeezed along it. In the stinging Hymenoptera the eggs are released at the base of the ovipositor. The fused second valvulae form an inverted

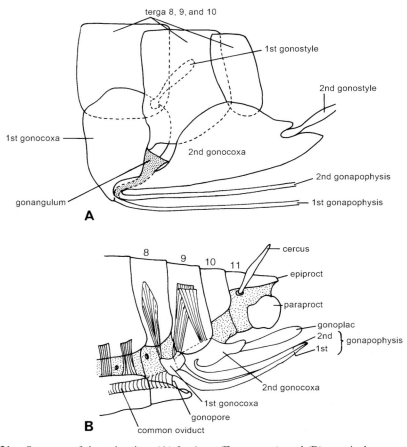

FIGURE 3.31. Structure of the ovipositor. (A) *Lepisma* (Zygentoma); and (B) a typical pterygote insect. [A, after G. G. E. Scudder, 1961, The comparative morphology of the insect ovipositor, *Trans. R. Entomol. Soc. Lond.* **113**:25–40. By permission of the Royal Entomological Society. B, from R. E. Snodgrass, *Principles of Insect Morphology.* Copyright 1935 by McGraw-Hill, Inc. Used with permission of McGraw-Hill Book Company.]

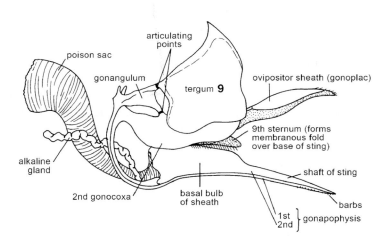

FIGURE 3.32. Sting of the honey bee. [After R. E. Snodgrass, 1925. *Anatomy and Physiology of the Honey bee.* McGraw-Hill Book Company.]

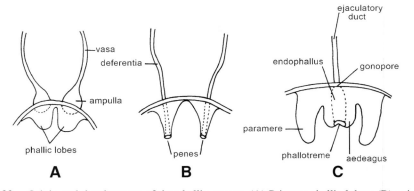

FIGURE 3.33. Origin and development of the phallic organs. (A) Primary phallic lobes; (B) paired penes of Ephemeroptera; and (C) formation of the aedeagus. [Reproduced by permission of the Smithsonian Institution Press from *Smithsonian Miscellaneous Collections,* Volume 135, "A revised interpretation of the external reproductive organs of male insects," Number 6, December 3, 1957, 60 pages, by R. E. Snodgrass: Figures 1A–C, page 3. Washington, D.C., 1958, Smithsonian Institution.]

groove that is enlarged proximally into the *basal bulb*. When an insect stings the *poison gland* releases its fluid into the basal bulb. The poison is caused to flow along the groove by back and forth movements of the *lancets* (modified first gonapophyses).

Male. "The great structural diversity in the male genitalia of insects is the delight of taxonomists, the despair of morphologists," wrote Snodgrass (1957, p. 11). Paralleling this diversity of structure is a mass of taxonomic terms (see Tuxen, 1970). The genitalia are composed of two sets of structures. First, there are basic structures that, though varied in form, are common to all insects, and second, there are structures that are peculiar to a group or even a species. Space does not permit description of the latter, or even the variation that occurs within the former, set of structures. The following paragraphs will, therefore, be limited to a discussion of the basic form of the genitalia.

Comparative study indicates that in all insects the basic genitalia are derived from a pair of *primary phallic lobes*, ectodermal outgrowths belonging to the tenth segment (Figure 3.33A). However, only in Ephemeroptera do they remain as separate lobes, through the posterior tips of which open the gonopores (Figure 3.33B). In Zygentoma the lobes meet in the midline to form a short, tubular structure, the *"penis."* The latter is, in fact, a misnomer, since this structure is not an intromittent organ. In Odonata secondary copulatory structures are developed on the ventral surface of the anterior abdominal segments (see Chapter 6, Section 3), and the genitalia on the tenth abdominal segment are much reduced. In all of the remaining orders the primary lobes are each divided into two secondary lobes, the *phallomeres*. Between the median pair, *mesomeres*, the ectoderm invaginates to form the *ejaculatory duct*. The outer pair, *parameres*, elongate and develop into clasping organs. The mesomeres unite medially to form a tubular intromittent organ, the *aedeagus*, whose inner passage is termed the *endophallus* (Figure 3.33C). The distal opening of the endophallus is the *phallotreme*. Occasionally the parameres and aedeagus are united basally, this region being known as the *phallobase*. Usually, however, the parameres are placed laterally on the ninth segment.

Two assumptions that are generally made concerning the male genitalia are (1) that they are modified limb appendages and (2) that they belong to the ninth abdominal segment. Snodgrass (1957) questioned the validity of both assumptions. He presented a convincing argument against assumption (1) and suggested, instead, that the genitalia are derived from primitive paired penes. This view contrasts, of course, with that of Kukalová-Peck (1991) who identified the various segments of the ancestral limb that make up the genitalia. According to Snodgrass (1957), the evidence from embryonic and postembryonic development does not support assumption (2) but indicates that the genitalia arose primitively on the tenth abdominal segment.

5.2.2. Other Appendages

Cerci. Paired cerci occur in Diplura, Microcoryphia, Zygentoma, Ephemeroptera, Zygoptera (a suborder of the Odonata), Plecoptera, the orthopteroids and blattoids, and Mecoptera. It is generally agreed that the cerci (Figure 3.31) are true appendages of the 11th segment, although frequently all traces of the latter have disappeared. Typically, they are elongate multisegmented structures that function as sense organs. They may, however, be considerably modified. In nearly all Dermaptera the cerci form unjointed *forceps*. The cerci of nymphs of Zygoptera are modified to form the *lateral caudal lamellae* (Figure 6.11C). The latter, along with the *median caudal lamella* (developed from the epiproct), are accessory respiratory structures. In adult male Zygoptera the cerci form *claspers* for grasping the female during copulation. In most male Embioptera the basal segment of the left cercus forms a hook with which the insect can clasp his mate.

Styli and Eversible Vesicles. *Styli* occur on most abdominal segments of Microcoryphia, Zygentoma, and Diplura, and on the ninth sternum of some male orthopteroids. In some bristletails the styli are articulated with a distinct coxal plate (Figure 3.34), but generally the original coxal segment is fused with the sternum. In bristletails, at least, the styli serve to raise the abdomen off the ground during locomotion.

Eversible vesicles (Figure 3.34) are short cylindrical structures found on some pregenital segments of apterygotes. They are closely associated with the styli when present, but their homology is unclear. They are believed to have the ability to take up water from the environment.

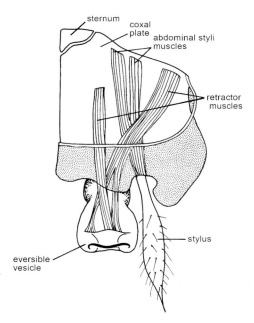

FIGURE 3.34. Stylus and eversible vesicle of a thysanuran. Part of the wall of the plate has been removed to show the musculature. [From R. E. Snodgrass, *Principles of Insect Morphology.* Copyright 1935 by McGraw-Hill, Inc. Used with permission of McGraw-Hill Book Company.]

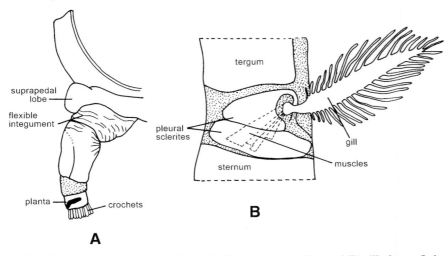

FIGURE 3.35. Secondary segmental appendages. (A) Proleg of a caterpillar; and (B) gill of a mayfly larva. [From R. E. Snodgrass, *Principles of Insect Morphology.* Copyright 1935 by McGraw-Hill, Inc. Used with permission of McGraw-Hill Book Company.]

Prolegs. Segmentally arranged, leglike structures are present on the abdomen of many endopterygote larvae (Figure 3.35A). They are known as prolegs (*pseudopods* or *larvapods*). Their structure is varied, though typically (e.g., in caterpillars) three regions can be distinguished: a basal membranous articulation, followed by a longer section having a sclerotized plate on the outer wall, and an apical protractile lobe, the *planta*, which bears claws, *crochets*, peripherally. The planta is protracted by means of blood pressure. Immediately above the leg there is a swollen area in the body wall, the *suprapedal lobe*.

Gills. A large number of aquatic larvae possess segmentally arranged gills on a varied number of abdominal segments. These are flattened, filamentous structures, frequently articulated at the base (Figure 3.35B).

Non-Segmental Appendages. In many insects non-segmental structures are present. These are typically a mediodorsal projection on the last abdominal segment. Examples of such structures are the median lamella of zygopteran larvae and the caudal filament of Microcoryphia, Zygentoma, and Ephemeroptera. Occasionally these structures are paired (e.g., the urogomphi of some larval Coleoptera) and are easily mistaken for cerci. The anal papillae of certain dipteran larvae also fit in this category.

6. Literature

Many books and review articles deal with more or less specific aspects of external structure, for example, Snodgrass (1935) [general morphology]; DuPorte (1957), Snodgrass (1960), Matsuda (1965) [head]; Matsuda (1963) [thorax]; Schneider (1964), Zacharuk (1985) [antennae]; Matsuda (1976) [abdomen]; Snodgrass (1957), Scudder (1971) [male external genitalia]; Scudder (1961, 1971) [ovipositor]. Care must be taken when reading papers that deal with morphology because authors may use differing terminology and have differing views on the homology of structures.

Bitsch, J., 1994, The morphological groundplan of Hexapoda: Critical review of recent concepts, *Ann. Soc. Entomol. Fr. (N.S.)* **30**:103–129.

Comstock, J. H., 1918, *The Wings of Insects*, Comstock, Ithaca, NY.

DuPorte, E. M., 1957, The comparative morphology of the insect head, *Annu. Rev. Entomol.* **2**:55–70.

Hamilton, K. G. A., 1972a,b, The insect wing, Parts II and III, *J. Kans. Entomol. Soc.* **45**:54–58, 145–162.

Hinton, H. E., 1955, On the structure, function, and distribution of the prolegs of Panorpoidea, with a criticism of the Berlese-Imms theory, *Trans. R. Entomol. Soc. Lond.* **106**:455–545.

Kukalová-Peck, J., 1987, New Carboniferous Diplura, Monura, and Thysanura, the hexapod ground plan, and the role of thoracic side lobes in the origin of wings (Insecta), *Can. J. Zool.* **65**:2327–2345.

Kukalová-Peck, J., 1991, Fossil history and the evolution of hexapod structures, in: *The Insects of Australia*, 2nd ed., Vol. I (CSIRO, ed.), Melbourne University Press, Carlton, Victoria.

Kukalová-Peck, J., 1992, The "Uniramia" do not exist: The ground plan of the Pterygota as revealed by Permian Diaphanopterodea from Russia (Insecta: Paleodictyopteroidea), *Can. J. Zool.* **70**:236–255.

Kukalová-Peck, J., 1998, Arthropod phylogeny and 'basal' morphological structures, in: *Arthropod Relationships* (R. A. Fortey and R. H. Thomas, eds.), Chapman and Hall, London.

Lawrence, J. F., Nielsen, E. S., and Mackerras, I. M., 1991, Skeletal anatomy and key to orders, in: *The Insects of Australia*, 2nd ed., Vol. 1 (CSIRO, ed.), Melbourne University Press, Carlton, Victoria.

Matsuda, R., 1963, Some evolutionary aspects of the insect thorax, *Annu. Rev. Entomol.* **8**:59–76.

Matsuda, R., 1965, Morphology and evolution of the insect head, *Mem. Am. Entomol. Inst. (Ann Arbor)* **4**:334 pp.

Matsuda, R., 1976, *Morphology and Evolution of the Insect Abdomen*, Pergamon Press, Elmsford, NY.

Ragge, D. R., 1955. *The Wing-Venation of the Orthoptera Saltatoria*, British Museum, London.

Rempel, J. G., 1975, The evolution of the insect head: The endless dispute, *Quaest. Entomol.* **11**:7–25.

Rempel, J. G., and Church, N. S., 1971, The embryology of *Lytta viridana* Le Conte (Coleoptera: Meloidae) VII. Eighty-eight to 132 h: The appendages, the cephalic apodemes, and head segmentation, *Can. J. Zool.* **49**:1571–1581.

Schneider, D., 1964, Insect antennae, *Annu. Rev. Entomol.* **9**:103–122.

Scholtz, G., 1998, Cleavage, germ band formation and head segmentation: The ground pattern of the Euarthropoda, in: *Arthropod Relationships* (R. A. Fortey and R. H. Thomas, eds.), Chapman and Hall, London.

Scudder, G. G. E., 1961, The comparative morphology of the insect ovipositor, *Trans. R. Entomol. Soc. Lond.* **113**:25–40.

Scudder, G. G. E., 1971, Comparative morphology of insect genitalia, *Annu. Rev. Entomol.* **16**:379–406.

Snodgrass, R. E., 1935, *Principles of Insect Morphology*, McGraw-Hill, New York. (Reprinted 1993, Cornell University Press.)

Snodgrass, R. E., 1957, A revised interpretation of the external reproductive organs of male insects, *Smithson. Misc. Collect.* **135**:60 pp.

Snodgrass, R. E., 1960, Facts and theories concerning the insect head, *Smithson. Misc. Collect.* **142**:61 pp.

Tillyard, R. J., 1918, The panorpoid complex. Part I—The wing-coupling apparatus, with special reference to the Lepidoptera, *Proc. Linn. Soc. N.S.W.* **43**:286–319.

Tuxen, S. L., 1970, *Taxonomist's Glossary of Genitalia in Insects*, 2nd ed., Munksgaard, Copenhagen.

Wootton, R. J., 1979, Function, homology and terminology in insect wings, *Syst. Entomol.* **4**:81–93.

Wootton, R. J., 1992, Functional morphology of wings, *Annu. Rev. Entomol.* **37**:113–140.

Zacharuk, R. Y., 1985, Antennae and sensilla, in: *Comprehensive Insect Physiology, Biochemistry, and Pharmacology*, Vol. 6 (G. A. Kerkut and L. I. Gilbert, eds.), Pergamon Press. Elmsford, NY.

4

Systematics and Taxonomy

1. Introduction

Systematics may be defined as the study of the kinds and diversity of organisms and the relationships among them. Taxonomy, the theory and practice of identifying, describing, naming, and classifying organisms, is an integral part of systematics. Classification is the arrangement of organisms into groups (*taxa*, singular *taxon*) on the basis of their relationships. It follows that identification can take place only after a classification has been established. It should be emphasized that not all authors adopt these definitions. Taxonomy is often used as a synonym of systematics (as defined above), while classification is sometimes used rather loosely (and incorrectly) as a synonym of identification.

Systematics is an activity that impinges on most other areas of biological endeavor. Yet, its importance (and fiscal support for it) seem to have diminished in recent years. To some extent, this may be the fault of systematists who tend to work in isolation, often focusing on some small and obscure group of organisms. This may be especially true of entomological systematists who, faced with the enormous diversity of the Insecta, tend to be seen as "counters of bristles," "measurers of head width" and performers of other activities of little relevance to the outside world. In fact, as Danks (1988) elegantly pointed out, nothing could be further from the truth. Systematics has played, and continues to play, a major role in fundamental evolutionary and ecological studies, for example faunistic surveys, zoogeographic work, life-history investigations and studies of associations between insects and other organisms. In applied entomology good systematic work is the basis for decisions on the management of pests. Indeed, Danks (1988) provided examples of pest-management projects in which inadequate or faulty systematics resulted in failure, sometimes with great economic and social cost (and see Section 2).

The taxonomy of insects, like that of most other groups of living organisms, continues to be based primarily on external structure, though limited use has also been made (sometimes of necessity, especially between species) of physiological, developmental, behavioral, and cytogenetic data. Molecular biological analyses of problems in insect systematics have increased exponentially over the past two decades (Caterino *et al.*, 2000). These analyses, principally using mtDNA sequences, have principally focused on the resolution of relationships at lower taxonomic levels, for example, among subspecies, species and species groups. Molecular phylogenetic studies of higher insect taxa (e.g., relationships among orders), though far fewer, have nevertheless generated important, sometimes even controversial, conclusions (see Chapter 2 for examples).

The purpose of this chapter is to provide a short introduction to the systematics of insects, including some of the technical terms applied by workers in these fields, as a basis for Chapters 5–10 inclusive, which deal with individual insect orders.

2. Naming and Describing Insects

For a variety of reasons but most obviously the enormous diversity within the class Insecta and economic considerations, insect taxonomists usually work within fairly narrow boundaries. Only by doing this can they acquire the necessary familiarity with a particular group (including knowledge of the relevant literature) to determine whether the specimen they are examining has been described and named or may be new to science. Even after a particular group has been chosen for study, there are typically superimposed biogeographic constraints, that is, taxonomists restrict their studies to particular geographic regions.

Many frequently encountered insects, especially pests, have a "common name" by which they are known. The name may refer to a particular species (e.g., house fly) or to a larger group (e.g., scorpionflies) and reflects a characteristic feature of the insect's appearance or habits. Unfortunately, insects of widely different groups may have similar habits (e.g., so-called "leaf miners" may be larvae of Diptera, Lepidoptera, or Hymenoptera) or the same common name may refer to different species of insects in different parts of the world. Thus, to avoid possible confusion, each insect species, like all other organisms both fossil and extant, is given a unique latinized binomial (two-part) name, a system introduced by Linnaeus in the early 1700s. In the Latin name, which is always italicized, the first word denotes the genus, the second the species (e.g., *Musca domestica* for the house fly). Rarely, the name has three parts, the third indicating the subspecies. (It should be noted, however, that some national entomological societies such as those of the United States and Canada publish lists of the *approved* common names for species in order to allow their use, yet avoid possible misunderstanding.)

Species are normally distinguished on the basis of a small number of key features (*characters*) that exist in a specific *character state* in each species (e.g., "number of tarsal segments" is a character, and "five tarsal segments" is a character state). Thus, a taxonomist will base the description of a new species on the characters already established for other species in the same group to facilitate comparison with them. Careful collection and curation (preparation, preservation, and maintenance) of specimens are critical to taxonomy to ensure that potentially important characters (which may be minute and delicate) are not damaged. The specimens must be properly labeled with the date and place of collection (preferably using map coordinates) and the collector's name. To facilitate proper maintenance, as well as accessibility for further studies, specimens are usually submitted to a central repository, the name of which is included in the published description of the species, to become part of the reference collection.

The specimens whose description leads to the establishment of a new species form the *type series*, one only of which becomes the standard reference specimen, the *holotype*, the others in the series being *paratypes*. The name given to a new species must follow the rules and universal nomenclatural system laid down by the International Commission on Zoological Nomenclature (published in the International Code of Zoological

Nomenclature). The species-specific part of the name may be a genuine Latin word, as in the dragonfly *Hemicordulia flava* (from the Latin "flavus" meaning yellow, referring to the extensive yellow coloration on the body), or may be a latinized form of a word, for example, a name of a person or place, as in the damselfly *Neosticta fraseri*, named for the Australian amateur odonatologist, F. C. Fraser. Sometimes, authors show remarkable imagination in naming a species, making study of the derivation of insect names ("entomological etymology"?) a fascinating subject in its own right. Take, for example, the Australian katydid *Kawanaphila lexceni* Rentz 1993 (in Rentz, 1993), the generic name of which is derived from the aboriginal word "kawana" meaning flower, a reference to the fact that all known species frequent flowers, while the species is named in honor of Ben Lexcen, designer of the Americas Cup challenger *Australia II*, in which the keel is similar to a structure (the subgenital plate) on the female katydid! Similarly, the damselfly *Pseudagrion jedda* Watson and Theischinger 1991 (in Watson *et al.*, 1991) receives its name from the 1955 film *Jedda*, parts of which were shot in Katherine Gorge, Northern Territory, Australia, the *type locality* (place of collection of the holotype) for the species! In publications, a species' name when first mentioned is given in full, and may be followed by the name of the original describer (authority), which may be abbreviated, and sometimes the year the description was published as in the two preceding examples. In some cases, the name of the authority (and date) appears in parentheses as, for example, in the termite *Porotermes adamsoni* (Froggatt, 1897), showing that the species was described first under a different genus, subsequently shown to be incorrect. In this example, Froggatt originally placed the species in the genus *Calotermes*.

As noted above, most species are described on the basis of their structure, especially external characters. However, on occasion such "morphospecies" are not equivalent to biological species (reproductively isolated populations); that is, groups that cannot be differentiated structurally may nevertheless be true biological species and are said to be "sibling species." Such species have been detected by a variety of means, including their different host preferences (e.g., some mosquitoes), mating behavior (courtship songs in some katydids), and cytogenetics (karyotypes of some black flies). The recognition of sibling species and their host specificity are critically important in biological control programs. For example, in the control of prickly pear (*Opuntia* spp.) by caterpillars of *Cactoblastis* (see Chapter 24, Section 2.3), it is now believed that the "slow" start made by the insects may have been due to introduction of the "wrong" sibling species which failed to establish themselves, not an unsuitable climate as suggested earlier (McFadyen, 1985).

If a new species is sufficiently different that it cannot be assigned to an existing genus, a new genus is proposed, following the same considerations as for species with respect to name, authority, and date as, for example, *Anax* Leach 1815, and this species is then denoted as the *type species* for this genus. Since 1930, it has been a requirement for a type species to be selected for any new genus. For genera described before this time and lacking a type species, the Code specifies how the type species should be determined. Within a genus, especially one with many species, there may be clearly defined groups of species, and each group may be given its own subgeneric name placed parenthetically after the genus; e.g., *Aedes (Chaetocruiomyia)* spp. for a species group of mosquitoes endemic to Australia. Each taxon above the genus level will also have its authority and date, and for each family (but not for taxa higher than this) there is a *type genus*, which by definition must have a name that is incorporated into the family name (e.g., *Apis* in the bee family Apidae).

3. Classification

Biological systems of classification are hierarchical, that is, the largest taxa are subdivided into successively smaller taxa. Thus, each taxon has a particular level (rank) within the system. Groups of the same rank are said to belong to the same taxonomic category, to which a particular name is given. Some of these categories are obligatory (capitalized in the example below), while others are optional. To show the hierarchical arrangement and to introduce the names of the various categories, let us take as an example the classification of the honey bee, *Apis mellifera*:

KINGDOM	Animalia
PHYLUM	Uniramia
Subphylum	Hexapoda
CLASS	Insecta
Subclass	Pterygota
Infraclass	Neoptera
Division	Oligoneoptera
ORDER	Hymenoptera
Suborder	Apocrita
Superfamily	Apoidea
FAMILY	Apidae
Subfamily	Apinae
Tribe	Apini
Subtribe	—
GENUS	*Apis*
Subgenus	—
SPECIES	*Apis mellifera*
Subspecies	

[In zoology, the subspecies is the lowest category considered valid; in botany, variety, form, and subform are recognized (and given latinized names).]

Classification, then, is a means of more efficiently storing (and retrieving) information about organisms. In other words, it is not necessary to describe all of the characteristics of a species each time that species is referred to. For example, as standard practice, a large proportion of entomological research articles include in their titles, after the name of the species being studied, the family, (superfamily), and order to which the species belongs. In this way, a reader can immediately gain some insight into the nature of the insect being studied, even though he or she may not be familiar with the species. Related to this last point, classification is also important in that it enables predictions to be made about incompletely studied organisms. For example, organisms are almost always classified first on the basis of their external structure. However, once an organism has been assigned to a particular taxon using structural criteria, it may then be possible to predict, in general terms, its habits (including life history), internal features, and physiology, on the basis of what is known concerning other, better studied, members of the taxon.

A classification may be either artificial or natural. It is possible, for example, to arrange organisms in groups according to their habitat or their economic importance. Such classifications may even be hierarchical in their arrangement. Artificial classifications are usually designed so that organisms belonging to different taxa within the system can be separated on the basis of single characters. As a result, such schemes have extremely restricted value and, usually, can be used only for the purpose for which they were initially designed. More importantly, artificial classifications provide no indication of the "true" or "natural" relationships of the constituent species.

Almost all modern classifications are natural, that is, they indicate the affinity (degree of similarity) between the organisms within the classification. Organisms placed in the same taxon (showing the greatest affinity) are said to form a natural group. There is, however, considerable controversy among systematists over the meaning of "degree of similarity," "natural group," and "natural classification." Essentially systematists fall into three major groups, according to their interpretation of the above terms. These are the phyleticists, cladists, and pheneticists. To the cladistic group, led by Hennig (see Hennig, 1965, 1966, 1981), belong those systematists who base classification entirely on *genealogy*, the recency of common ancestry. Critical to the *modus operandi* of cladists are the distinction between primitive and advanced homologous characters (so-called "*character polarity*") and the recognition of *sister groups* (see below for further discussion of these terms). Among the various ways used by cladists to assign character polarity are paleontology, ontogeny, and outgroup comparison. In theory, the study of fossils should clearly show when a character first appears, making the separation of primitive and advanced characters an easy task. However, the fossil record is typically discontinuous and preservation imperfect so that vital characters are missing. The idea that "ontogeny recapitulates phylogeny," suggested by Haeckel in 1866, proposes that an organism's development will reflect its evolution, giving clues therefore as to which of its features are primitive and which are advanced. Ontogeny has been relatively little used by cladists, however, perhaps because in development evolutionary steps are compressed, omitted, or masked. Outgroup comparison, which is the method most used, is a comparison of character states in the group under study with those in increasingly distant sister groups. The character state common to the largest sister groups is generally taken to be the primitive condition. This method requires, of course, some previous knowledge of a group's phylogeny and has been criticized because of its circularity. As a result of their studies, cladists usually express their results in the form of a *cladogram*.

Beginning in the 1950s, some taxonomists, dissatisfied with the perceived subjective approach to classification, began to devise schemes based on the number of common characters among organisms, regardless of whether these were primitive or advanced. The pheneticists (originally known as numerical taxonomists), led by Sokal and Sneath (see Sokal and Sneath, 1963; Sneath and Sokal, 1973), have as their major principles: (1) the more characters studied the better; (2) all characters are of equal weight; and (3) the greater the proportion of similar characters, the closer are two groups related. Pheneticists usually present the results of their analyses as *phenograms* or *scatter diagrams*.

Phyleticists such as Simpson (1961) and Mayr (1969, 1981) may be considered as forming a "middle-of-the-road" group, employing both cladistic and phenetic information on which to base their classifications. The proportions of cladistic and phenetic information used may vary significantly depending, for example, on the extent of the fossil record; in other words, in contrast to the cladistic and phenetic methods, the phyletic system does not follow a set of carefully established rules.

An implicit point of natural classifications, regardless of how they are derived, is that they are based on genealogy (i.e., relationship by descent). In other words, they show evolutionary relationships among taxa. Thus, the key step in any natural classification is the determination of *homology* (whether features common to groups were derived from the same feature in the most recent common ancestor of the groups). Similar, but non-homologous, features are said to show *homoplasy* (analogy) and are the result of either *parallelism* (the features had a distant, common ancestor) or *convergence* (the features are derived from entirely unrelated ancestral conditions). Once homology is established, it is then a matter of determining whether the character states under consideration are advanced (derived) or primitive (ancestral) (*apomorphies* or *plesiomorphies*, respectively). Both comparative morphology of extant forms and the fossil record have been used extensively in such determinations. Apomorphies shared by taxa are said to be *synapomorphies*, while those unique to a taxon are described as *autapomorphies*. Neither autapomorphies nor plesiomorphies can show relationships between groups. Broadly speaking, the greater the number of synapomorphies, the closer will be the relationship between taxa. Each taxon, regardless of rank, will have a sister group—its closest relative—so that the development of classifications and phylogenies is the establishment of successively larger sister groups, often depicted as a branching diagram known as a *phylogenetic tree* (see the section on Phylogeny and Classification under each order for examples). An ancestor and all of its descendants form a *monophyletic* group; when some of the descendants are lacking, the remaining descendants are said to be *paraphyletic*. Groups derived from more than one ancestor are said to be *polyphyletic*. It must be emphasized that the actual ancestor of two taxa is rarely known, though its general features (the so-called "*ground plan*") will be defined by the plesiomorphic characters of its descendants. The term *stem group* refers to collections of fossils that have some plesiomorphic characters of a more recent group; they may be close to, but are not directly on, the group's line of descent.

As the following section (and comparison of the current with previous editions of this book) shows, ideas on relationships among insect groups change with time, sometimes quite significantly. Though partly related to the acquisition of new knowledge, it is also because taxonomists differ in their analysis and interpretation of data, or use different data sets on which to base their conclusions.

3.1. The History of Insect Classification

Wilson and Doner (1937) have fully documented the many schemes that have been devised for the classification of insects, and it is from their account that the following short history is mainly compiled. (Papers marked with an asterisk are cited from Wilson and Doner's review.) Only the major developments (i.e., those that have had a direct bearing on modern schemes) have been included, though it should be realized that a good many more systems have been proposed.

Insect systematics may be considered to have begun with the work of Aristotle, who, according to Kirby and Spence (1815–1826),* included the Entoma as a subdivision of the Anaima (invertebrates). Within the Entoma Aristotle placed the Arthropoda (excluding Crustacea), Echinodermata, and Annelida. Authors who have examined Aristotle's writings differ in their conclusions regarding this author's classification of the insects, but it does appear clear that Aristotle realized that there were both winged and wingless insects and that they had two basic types of mouthparts, namely, chewing and sucking.

Amazingly, it was not for almost another 2000 years that further serious attempts to classify insects were made. Aldrovanus (1602)* divided the so-called "insects" into terrestrial and aquatic forms and subdivided these according to the number of legs they possessed and on the presence or absence and the nature of the wings. In Aldrovanus' classification the term "insect" encompassed other arthropods, annelids, and some mollusks. The work of Swammerdam (1669)* is of particular interest because it represents the first attempt to classify insects according to the degree of change that they undergo during development. Although Swammerdam's concept of development was inaccurate, he distinguished clearly between ametabolous, hemimetabolous, and holometabolous insects. A more elaborate scheme of classification, still based primarily on the degree of metamorphosis but also incorporating such features as number of legs, presence or absence of wings, and habitat, was that of Ray and Willughby (1705).* Ray was the first naturalist to form a concept of a "species," a term that was to take on more significance following the introduction, by Linnaeus, of the binomial system some 30 years later. Between 1735 and 1758, Linnaeus* gradually improved on his system for the classification of insects, based entirely on features of the wings. Linnaeus recognized seven orders of "insects," namely, the Aptera, Neuroptera, Coleoptera, Hemiptera, Lepidoptera, Diptera, and Hymenoptera. Of the seven, the first four orders each contained a heterogeneous group of insects (and other arthropods) that today are separated into many different orders. The Diptera, Lepidoptera, and Hymenoptera have remained, however, more or less as Linnaeus envisaged them more than 200 years ago. Like earlier authors, Linnaeus included in the Aptera (wingless forms) spiders, woodlice, myriapods, and some non-arthropodan animals. He failed also to distinguish between primitively and secondarily wingless insect groups.

Surprisingly, perhaps, up to this time no one had made a serious attempt to classify insects on the basis of their mouthparts. However, the Danish entomologist Fabricius, who was a student of Linnaeus, produced several "cibarian" or "maxillary" systems for classification during the period 1775–1798.* The primary subdivision was into forms with biting mouthparts and forms with sucking mouthparts. Like Linnaeus, however, Fabricius included a variety of non-insectan arthropods in his system and, furthermore, based his systems on a single anatomical feature.

De Geer (1778),* who also studied under Linnaeus, appears to have been one of the earliest systematists to realize the importance of using a combination of features as a basis for classification. Such an approach was used by the French entomologist Latreille, who, during the period 1796–1831,* gradually produced what he considered to be a natural arrangement of the Insecta. In 1810 Latreille separated the Crustacea and Arachnida from the "Insecta," in which he included still the Myriapoda. The latter group was not given class status until 1825. In the final version of his system Latreille distinguished 12 insect orders. The Linnaean order Aptera was split into the orders Thysanura, Parasita (= Anoplura), and Siphonaptera, although Latreille did not appreciate that the first group was primitively wingless, while the other two were secondarily so. The order Coleoptera of Linnaeus was subdivided into Coleoptera (*sensu stricto*), Dermaptera, and Orthoptera. The Phiphiptera (= Strepsiptera), believed to be related to the Diptera in which order they had been included, were separated as a distinct group by Latreille. The Frenchman was also among the earliest systematists to appreciate the heterogeneity of the Linnaean order Neuroptera, splitting the group into three tribes, the Subulicarnes (= modern Odonata and Ephemeroptera), Planipennes (= modern Plecoptera, Isoptera, Mecoptera, and neuropteroid insects[1]) and Plicipennes (= modern Trichoptera).

During the first half of the 19th century a large number of systematists produced their version of how insects should be classified. A majority argued, like Latreille, that the wings (presence or absence, number, and nature) were the primary feature on which a classification should be established. Yet others, such as Leach (1815)* and von Siebold

---

[1] Insects that are included in the modern orders Neuroptera, Megaloptera, and Raphidioptera.

(1848),* considered that the nature of metamorphosis was the first-order character, with wings, mouthparts, etc. of secondary importance. If nothing else, the use of metamorphosis as a separating character drew further attention to the heterogeneity of the neuropteroid group, which contained both hemi- and holometabolous forms. Indeed, in his classification von Siebold adopted Erichson's (1839)* arrangement in which the termites, psocids, embiids, mayflies, dragonflies, and damselflies were removed from the Neuroptera and placed together as the suborder Pseudoneuroptera in the order Orthoptera.

The foundations of modern systems of classification were laid by Brauer (1885),* who appears to have been greatly influenced by the principles of comparative anatomy and paleontology established by the French zoologist Cuvier, and by the work of Darwin. Brauer divided the Insecta into two subclasses, the Apterygogenea, containing the primitively wingless Thysanura and Collembola, the latter having been given ordinal status by Lubbock (1873),* and the Pterygogenea, containing 16 orders, in which he placed the winged and secondarily wingless forms. Three major divisions were established in the Pterygogenea: (1) Menognatha ametabola and hemimetabola (insects with biting mouthparts in both juvenile and adult stages, or mouthparts atrophied in the adult and with no or partial metamorphosis) containing the orders Dermaptera, Ephemerida, Odonata, Plecoptera, Orthoptera (including Embioptera), Corrodentia (which included the termites, psocids, and lice), and Thysanoptera; (2) Menorhyncha (insects with sucking mouthparts in both the juvenile and adult stages), containing the order Rhynchota (= Hemiptera); and (3) Menognatha metabola and Metagnatha metabola (insects having a complete metamorphosis, and with biting mouthparts in the juvenile stage and biting, sucking, or atrophied mouthparts in the adult), containing the neuropteroid insects, and the orders Panorpatae (= Mecoptera), Trichoptera, Lepidoptera, Diptera, Siphonaptera, Coleoptera, and Hymenoptera. Thus, Brauer appreciated the heterogeneity of the "Neuroptera" and correctly separated the Plecoptera, Odonata, and Ephemerida from the neuropteroids, Mecoptera, and Trichoptera. He failed, however, to recognize the heterogeneity of the orders Orthoptera and Corrodentia.

Between 1885 and 1900, a number of modifications to Brauer's system were suggested. Most of these were concerned solely with the subdivision or aggregation of orders according to the author's views on the affinity of the groups. There were, however, two proposals that have a more direct bearing on modern systems. In 1888 Lang* proposed that the terms Apterygota and Pterygota be substituted for Apterygogenea and Pterygogenea, respectively. Sharp (1899) refocused attention on the importance of metamorphosis, but, claiming that the terms Ametabola, Hemimetabola, and Holometabola were not sufficiently definite for taxonomic purposes, proposed new terms describing whether the wings developed internally or externally. His arrangement was as follows: Apterygota (primitively wingless forms); Anapterygota (secondarily wingless forms); Exopterygota (forms in which the wings develop externally); Endopterygota (forms in which the wings develop internally). Sharp was criticized for grouping together the secondarily wingless orders (Mallophaga, Anoplura, Siphonaptera), as these contained both hemi- and holometabolous forms, and the term Anapterygota was discarded. The terms Exopterygota and Endopterygota were widely accepted, however, and became synonymous with Hemimetabola and Holometabola, respectively. It was not until the work of Crampton and Martynov in the 1920s (see below) that it was realized that these terms had no phylogenetic significance but were merely descriptive, indicating "grades of organization." Sharp recognized 21 orders of insects. His system improved on Brauer's mainly in the splitting of the Corrodentia and Orthoptera, thereby giving ordinal status to the Isoptera, Embioptera, Psocoptera, Mallophaga, and Siphunculata.

Toward the end of the 19th century the full force of Darwin's ideas on evolution and the importance and usefulness of fossils began to make themselves felt in insect classification. Gone was the old idea that evolution was a single progressive series of events, and in its place came the appreciation that evolution was a process of branching. Thus, insect classification entered, at the beginning of the 20th century, the phylogenetic phase of its development, although Haeckel (1866)* had been the first to use a phylogenetic tree to indicate the relationships of the Insecta. Unfortunately his ideas on genealogy were incorrect. Most recent systems have been influenced to some degree by the work of an Austrian paleoentomologist, Handlirsch, who criticized earlier workers for their one-sided systems, in which a single character was used for separation of the major subdivisions. Another failure of the 19th century authors was, he claimed, their inability to distinguish between parallel and convergent evolution of similar features. Finally, he pointed out that almost no one had taken into account fossil evidence. Handlirsch's first scheme, produced in 1903, was, at the time, regarded as revolutionary. He raised the Collembola, Campodeoidea (= Diplura), and Thysanura each to the level of class. (Prior to this the Diplura had been considered usually as a suborder of the Thysanura.) He also raised the Pterygogenea of Brauer to the level of class and arranged the 28 orders of winged insects in 11 subclasses. His second scheme, published in 1908, was identical with the first except for some slight changes in the names of orders. In 1925 Handlirsch published his modified views on insect classification. In this scheme he reintroduced Brauer's two subclasses, Apterygogenea and Pterygogenea. In the former group he placed the orders Thysanura, Collembola, Diplura, and the recently discovered Protura. In the Pterygogenea he listed 29 orders (including the Zoraptera, first described in 1913) arranged in 11 superorders (his former subclasses). The most significant point in Handlirsch's work was his recognition of the heterogeneous nature of the Orthoptera, the contents of which he split into orders and regrouped with other orders in two superorders, Orthoptera (containing the orders Saltatoria, Phasmida, Dermaptera, Diploglossata, and Thysanoptera) and Blattaeformia (containing the Blattariae, Mantodea, Isoptera, Zoraptera, Corrodentia, Mallophaga, and Siphunculata). He did not appreciate, however, the orthopteroid nature of the Plecoptera and placed the group in a superorder of its own. Handlirsch was also in error in regarding the Corrodentia, Mallophaga, and Siphunculata as orthopteroid groups. They are undoubtedly more closely related to the Hemiptera. Handlirsch's arrangement was strongly criticized by Börner (1904), who said that it did not express the true phylogenetic relationships of the Insecta. Börner considered that fossil wings did not have much value in insect systematics, and, in any case, there were far too few fossils for paleontology to have much bearing on classification. Comparative anatomical studies of recent forms, Börner argued, would give a more accurate picture. Börner, whose system was widely accepted, arranged the 19 orders of winged insects that he recognized in five sections. Three of these correspond with the "paleopteran orders," "orthopteroid orders," and "hemipteroid orders" recognized today. In other words, Börner correctly assigned the Corrodentia, Mallophaga, and Siphunculata with the Hemiptera. The two remaining sections contained the endopterygote orders, though Börner's ideas on their affinities were to be shown by Tillyard (see below) to be incorrect.

Comstock (1918, and earlier), an American entomologist, supported Brauer's arrangement as a result of his comparative studies of the wing venation of living insects. Comstock was the first person to make extensive use of wing venation in determining affinities. He emphasized, however, that classifications should be based on many characters and not wings alone.

During a period of more than 20 years, beginning in 1917, Tillyard expounded his views on insect phylogeny, stemming from his extensive research into the fossil insects of Australia and North America. Although he made important contributions concerning the origin and relationships of many insect orders, Tillyard's (1918–1920) work on the endopterygotes is particularly well known. In this work he showed that the Hymenoptera and Coleoptera (with the Strepsiptera) form two rather distinct orders, only distantly related to the other endopterygote groups which collectively formed the panorpoid complex. Within the complex, the Mecoptera, Trichoptera, Lepidoptera, Diptera, and Siphonaptera form a well defined group, with the neuropteroid orders clearly distinct from them. In fact, as noted in Chapter 2, Hinton (1958) made a strong case for excluding these orders entirely from the panorpoid complex and placing them closer to the Coleoptera.

While Tillyard was concentrating on the phylogeny of the endopterygotes, his American contemporary, Crampton, was directing his efforts toward solution of the problems of exopterygote relationships, especially the position of the Zoraptera, Embioptera, Grylloblattidae, and Dermaptera. Following his anatomical study on the newly discovered winged zorapteran *Zorotypus hubbardi*, Crampton (1920) concluded that the Zoraptera were related to the orthopteroid orders, and he placed them in a group (superorder Panisoptera) that also contained the Isoptera, Blattida, and Mantida. However, the following year Crampton revised his views and transferred the Zoraptera to the psocoid (hemipteroid) superorder, after consideration of their wing venation. In 1922 Crampton placed the Zoraptera in the order Psocoptera and suggested that it was from psocoidlike ancestors that the modern hemipteroid orders evolved. Originally, Crampton (1915) had placed the Grylloblattidae in a separate order, Notoptera, in the orthopteroid group. Five years later he concluded that the grylloblattids were closer to the Orthoptera (*sensu stricto*) than the blattoid groups and made the Grylloblattodea a suborder of the Orthoptera. The modern view is that the grylloblattids are probably survivors of the protothopteran stock from which both the orthopteran and blattoid lines developed. Crampton considered that the closest relatives of the Embioptera were the Plecoptera, placing the two groups in the superorder Panplecoptera. In his early schemes Crampton also placed the Dermaptera in the Panplecoptera. He later changed this view and included them in the orthopteroid superorder, at the same time pointing out that the Diploglossata (Hemimerida) are parasitic Dermaptera.

Almost simultaneously in 1924 Crampton and the Russian paleoentomologist Martynov proposed an apparently natural division of the winged insects on the basis of the ability to flex the wings horizontally over the body when at rest. In the Paleoptera (= Paleopterygota = Archipterygota) are the orders Ephemeroptera and Odonata whose members do not possess a wing-folding mechanism. It must be emphasized, however, that the two orders are only very distantly related through their paleodictyopteran ancestry. The remaining orders, whose members are able to fold their wings over the body, are placed in the Neoptera (= Neopterygota). The latter contains three natural subdivisions, the Polyneoptera (orthopteroid orders), Paraneoptera (hemipteroid orders), and Oligoneoptera (endopterygote orders).

Even recently, vigorous debate has continued over the taxonomic rank of, and nature of the evolutionary relationships among, hexapod groups (see Chapter 1, Section 3.3.1 [apterygotes], and Chapter 2, Section 3.2 [pterygotes] for a fuller discussion). For example, most authors consider the Collembola and Protura to be sister groups and sometimes unite them in the class Ellipura (= Parainsecta). However, the position of the Diplura is less clear; Kristensen (1991) placed them close to the Ellipura principally on the basis of the entognathous condition, whereas Kukalová-Peck (1991), putting more emphasis on features of the thorax, suggested that they are true Insecta. Again, the monophyletic nature, or otherwise, of the Paleoptera is controversial. Sharov (1966) and Kukalová-Peck (1985, 1991) argued strongly that Ephemeroptera and Odonata had a common ancestor, whereas Kristensen (1991) lumped the Odonata with the Neoptera, this assemblage thereby becoming the sister group of the Ephemeroptera. The status of the Polyneoptera likewise remains questionable. Some workers believe that this is a monophyletic group, while others insist that the group is polyphyletic, the term "polyneopterous" simply describing a grade of organization. Certainly the position of the Zoraptera is enigmatic, this small order having a mixture of orthopteroid and hemipteroid characters. One recent suggestion is that zorapterans may be the sister group of the Embioptera, itself an order of uncertain affinity showing similarities with Plecoptera, Dermaptera, and Phasmida! Of all the major groups, the Paraneoptera is the one that is widely accepted to be monophyletic, though there is argument over whether the Psocoptera and Phthiraptera should be linked as a single order (Psocodea) or remain separate. Most modern authors also consider the endopterygote orders (except for the Strepsiptera) to be monophyletic, the two major sister groups being the Coleoptera-neuropteroids and the Hymenoptera-panorpoids. However, members of the small Southern Hemisphere family Nannochoristidae are clearly set apart from the other scorpionflies, with which they have been traditionally grouped in the order Mecoptera, and further study may result in the family being placed in its own order (Nannomecoptera) as suggested by Hinton (1981). Likewise, the primitive thysanuran *Tricholepidion gertschi* is considered by Boudreaux (1979) to be distinct enough to warrant its own order. The system adopted in the present volume is given below:

Superclass Hexapoda.
    1. CLASS. Collembola
          ORDERS. Arthropleona, Neelipleona, and Symphypleona
    2. CLASS AND ORDER. Protura
    3. CLASS AND ORDER. Diplura
    4. CLASS. Insecta
      I. SUBCLASS. Apterygota
          ORDERS. Microcoryphia and Zygentoma
     II. SUBCLASS. Pterygota
       A. INFRACLASS. Paleoptera
       ORDERS. Ephemeroptera and Odonata
       B. INFRACLASS. Neoptera
         a. DIVISION. Polyneoptera (orthopteroid orders)
         ORDERS. Orthoptera, Grylloblattodea, Dermaptera, Plecoptera,
              Embioptera. Dictyoptera, Isoptera, Phasmida,
              Mantophasmatodea, and Zoraptera
         b. DIVISION. Paraneoptera (hemipteroid orders)
         ORDERS. Psocoptera, Phthiraptera, Hemiptera, and Thysanoptera
         c. DIVISION. Oligoneoptera (endopterygote orders)
         ORDERS. Mecoptera, Lepidoptera, Trichoptera, Diptera. Siphonaptera,
              Neuroptera, Megaloptera, Raphidioptera, Coleoptera,
              Strepsiptera, and Hymenoptera

4. Identification

In principle the identification of insects is the same as that of any other animal. In practice it is more difficult, for two major reasons. First, the enormous number of species that occur means that often very minor differences in structure must be used to distinguish between forms, and second, the small size of most insects frequently means that the identifying characters are not easily seen. There are various methods for identifying organisms: (1) the specimen may be sent to an expert, (2) it may be compared with the specimens in a labeled collection, (3) it may be compared with pictures or descriptions, or (4) it may be identified by use of a key. Pictorial keys, which can be valuable to both specialists and non-specialists, include not only printed material but also user-friendly computer-based interactive systems such as those developed by Bishop *et al.* (1989), Lawrence *et al.* (1993), and Weeks *et al.* (1999). Most often, however, only conventional written keys are available. A tentative identification from a key should be confirmed by comparing the specimen's characters with the diagnosis or description for the species.

There are different ways of arranging a key, though all involve the same general principle, namely, the stepwise elimination of characters until a name is reached. Keys may be devised so as to reflect the evolutionary relationships between the taxa identified. However, because character state differences between closely related taxa may be slight, the use of a phylogenetic key with "weak" or "difficult" couplets may make identification difficult. Thus, most keys are quite arbitrary, as they have as their only objective, ease of identification. In this arbitrary system the same taxon may key out at several points in the key, whereas in a phylogenetic key, the taxon would appear only once.

Typically, a key is in the form of a series of couplets (occasionally triplets may be included) of contrasting character states. For maximum usefulness, the couplets should present clear-cut alternatives for the characters under consideration. The simplest form of sequence within a key is one in which each couplet includes only a single character. The drawback of such monothetic keys is that they do not work for organisms in which a character does not follow the norm. The alternative is a polythetic key in which at least some couplets include several statements, each about a different character. Sneath and Sokal (1973) suggest three reasons for using polythetic keys: (1) one or more characters may not be observable (e.g., if the specimen is incomplete, damaged, or at the "wrong" life stage), (2) some species may be exceptional for a particular character, and (3) the user of a key may err in deciding about a character. By having several characters in each couplet with which to work, a user can operate on a "majority vote" basis, that is, select the branch of the couplet that overall most closely describes the characters of the specimen. A disadvantage of such an arrangement is that a decision on which branch to select may not be clear-cut (especially if the specimen is exceptional in one of the characters listed). Further, the "rules" to be followed in a polythetic key must be carefully stated [i.e., do all characters in a couplet have equal value, or does one (the first) or more carry greater weight—and, if so, how much?].

A serious drawback to many keys is that in order not to become unwieldy they are constructed either specifically for identification of specimens in a particular geographic area or for identification of specimens to a higher taxonomic level only, typically to family. This is especially true of insect keys because of the great diversity of the insect fauna. In short, their use may be rather limited. The arrangement in this text is the provision of a polythetic key for identification of insects to the level of order, rarely the suborder. A list of keys for identification beyond the ordinal level is then provided under the description of each order (see Chapters 5–10). This list is by no means exhaustive, and it is anticipated that instructors will direct students to useful keys for the geographic area or insect group of interest.

4.1. Key to the Orders of Insects

The following key, modified from Brues *et al.* (1954), is in accordance with the classification used in this book. A few comments are necessary regarding its use. The key is suitable for use with the adult and most larval forms of insects. However, the larval forms of the endopterygote orders are often difficult to identify and, if at all possible, they should be allowed to metamorphose to the adult stage. In some cases it is important to know the original habitat of the specimen, and care should be taken to note this when the collection is made. Though not insects, the Collembola, Protura, and Diplura are included for the sake of completeness. Orders marked with an asterisk are unlikely to be encountered in a general collection. Because of its novelty and size [three species, each known only from single specimens (Zompro *et al.*, 2002)], the order Mantophasmatodea has not been included.

Key to the Orders of Insecta

1. Wings developed..2
   Wingless, or with vestigial wings, or with rudimentary wings not
      suitable for flight (wingless adults and immature stages)........................31
2. Fore wings horny, leathery, or parchmentlike, at least at base; hind wings
      membranous (occasionally absent). Prothorax large and not fused with
      mesothorax (except in Strepsiptera)..3
   Fore wings membranous..11
3. Fore wings containing veins, or at least hind wings not folded crossways
      when hidden under fore wings..4
   Fore wings veinless, of uniform horny consistency; hind wings, when
      present, folded crossways as well as lengthwise when at rest and
      hidden beneath fore wings; mouthparts mandibulate 10
4. Mouthparts forming a jointed beak, fitted for piercing and sucking.
      Bugs...HEMIPTERA   (Page 210)
   Mouthparts with mandibles fitted for chewing and moving laterally...............5
5. Hind wings not folded, similar to fore wings; thickened basal part of
      wings very short, separated from rest of wing by a preformed
      transverse suture; social species, living in colonies. Termites.....................
      .... ISOPTERA   (Page 163)
   Hind wings folding, fanlike, broader than fore wings...............................6
6. Usually rather large or moderately large species; antennae usually
      lengthened and threadlike; prothorax large and free from mesothorax;
      cerci present; fore wings rarely minute, usually long..............................7
   Very small active species; antennae short with few joints, at least one
      joint bearing a long lateral process; no cerci; fore wings minute;
      prothorax small. Rare, short-lived insects, parasites of other insects,
      usually wasps and bees................Males of STREPSIPTERA*   (Page 326)

7. Hind femora not larger than fore femora; body more or less flattened
with wings superposed when at rest; tergites and sternites subequal.............8
Hind femora almost always much larger than fore femora, jumping
species, if not (Gryllotalpidae) front legs broadened for burrowing;
species usually capable of chirping or making a creaking noise;
body more or less cylindrical, wings held sloping against sides
of the body when at rest, tergites usually larger than sternites.
Grasshoppers, katydids, crickets .... ORTHOPTERA   (Page 184)

8. Body elongate; head free, not concealed from above by the prothorax;
deliberate movers...9
Body oval, much flattened; head nearly concealed beneath the oval
pronotum; legs identical, coxae large and tibiae noticeably spiny or bristly......
Cockroaches..
.... DICTYOPTERA, Suborder BLATTODEA   (Page 160)

9. Prothorax much longer than mesothorax; front legs almost always
heavily spined, formed for seizing prey; cerci usually with several
joints. Mantids .... DICTYOPTERA, Suborder MANTODEA   (Page 161)
Prothorax short; legs similar, formed for walking; cerci unjointed.
Stick and leaf insects....................................PHASMIDA   (Page 179)

10. Abdomen terminated by movable, almost always heavily chitinized
forceps; antennae long and slender; fore wings short, leaving most
of abdomen uncovered, hind wings nearly circular, delicate,
radially folded from near the center; elongate insects. Earwigs..................
.... DERMAPTERA   (Page 175)
Abdomen not terminated by forceps; antennae of various forms but
usually with 11 subdivisions; fore wings usually completely
sheathing the abdomen; generally hard-bodied species. Beetles.................
.... COLEOPTERA   (Page 305)

11. With four wings...12
With only mesothoracic wings, usually outspread when at rest.................... 29

12. Wings long, very narrow, the margins fringed with long hairs, almost
veinless; tarsi 1- or 2-jointed, with swollen tips; mouthparts
asymmetrical without biting mandibles, fitted for lacerating and
sucking plant tissues; no cerci; minute species.
Thrips .... THYSANOPTERA   (Page 233)
Wings broader and most often with veins; if wings rarely somewhat
linear, tarsi have more than two joints and last tarsal joint is not
swollen...13

13. Wings, legs, and body covered, at least in part, with elongate flattened
scales (often intermixed with hairs) that nearly always form a color
pattern on the wings; mouthparts (rarely vestigial) forming a coiled
tongue composed of the maxillae; biting mandibles present only in
Micropterigidae. Moths and butterflies........... LEPIDOPTERA   (Page 276)
Wings, legs, and body not covered with scales, although sometimes
hairy and having a few scales intermixed; sometimes covered with
bristles, especially on legs, or rarely with wax flakes or dust; color
pattern when present extending to wing membrane 14

14. Hind wings with anal area separated, folded fanlike when at rest, nearly
    always wider and noticeably larger than fore wings; antennae prominent;
    wing veins usually numerous 15
    Hind wings without a separated anal area, not folded and not larger
    than fore wings 17
15. Tarsi five-jointed; cerci not pronounced 16
    Tarsi three-jointed; cerci well developed, usually long and many
    jointed; prothorax large, free; species of moderate to large size.
    Stoneflies .... PLECOPTERA   (Page 147)
16. Wings with a number of subcostal crossveins; prothorax rather large;
    species of moderate to large size. Alderflies ....
    .... MEGALOPTERA   (Page 297)
    Wings without subcostal crossveins, with surface hairy; pro thorax
    small; species of small to moderate size. Caddisflies ....
    .... TRICHOPTERA   (Page 268)
17. Antennae short and inconspicuous; wings netveined with numerous
    crossveins; mouthparts mandibulate 18
    Antennae large; if antennae small, wings have few crossveins or
    mouthparts form a jointed sucking beak 19
18. Hind wings much smaller than fore wings; abdomen ending in long
    threadlike processes; tarsi normally four- or five-jointed; sluggish
    fliers. Mayflies .... EPHEMEROPTERA   (Page 127)
    Hind wings nearly like fore wings; no caudal setae; tarsi three-jointed;
    vigorous, active fliers, often of large size. Dragonflies, damselflies ....
    .... ODONATA   (Page 136)
19. Head elongated ventrally forming a rostrum, at tip of which are
    mandibulate mouthparts; hind wings not folded; wings usually with
    color pattern, crossveins numerous; male genitalia usually greatly
    swollen, forming a reflexed bulb. Scorpionflies ....
    .... MECOPTERA   (Page 239)
    Head not drawn out as a mandibulate rostrum; male abdomen
    not forcipate 20
20. Mouthparts modified for sucking (occasionally reduced or absent);
    mandibles absent or in form of long bristles; no cerci; crossveins few 21
    Mouthparts for biting [occasionally for sucking (higher
    Hymenoptera)]; mandibles always present and having typical
    biting form 22
21. Wings not covered with scales, not outspread when at rest; prothorax
    large; antennae with few subdivisions; mouthparts forming a jointed
    piercing beak. Bugs .... HEMIPTERA   (Page 210)
    Wings and body covered with colored scales that form a definite
    pattern on wings; antennae greatly subdivided; mouthparts when
    present forming a coiled tongue. Moths and butterflies ....
    .... LEPIDOPTERA   (Page 276)
22. Tarsi five-jointed; if rarely three- or four-jointed, hind wings are
    smaller than front ones and wings lie flat over body; no cerci 23
    Tarsi two-, three-, or four-jointed; veins and crossveins not numerous 26

23. Prothorax small or only moderately long. (In Mantispidae prothorax
    is very long, but front legs are strongly raptorial.)..............................24
    Prothorax very long and cylindrical, much longer than head; front legs
    normal; antennae with more than 11 subdivisions; crossveins numerous.
    Snakeflies...RAPHIDIOPTERA   (Page 299)

24. Wings similar, with many veins and crossveins; prothorax more
    or less free...25
    Wings with relatively few angular cells, costal cell without crossveins;
    hind wings smaller than fore pair; prothorax fused with mesothorax;
    abdomen frequently constricted at base and ending in a sting or
    specialized ovipositor. Ants, wasps, bees, etc.........................
    ....HYMENOPTERA   (Page 330)

25. Costal cell, at least in fore wing, almost always with
    many crossveins. Lacewings, antlions.............NEUROPTERA   (Page 301)
    Costal cell without crossveins. Scorpionflies.........MECOPTERA   (Page 239)

26. Wings equal in size, or rarely hind wings larger, held superposed on
    top of abdomen when at rest; media fused with radial sector for a
    short distance near middle of wing; tarsi three-, four-, or five-jointed.........27
    Hind wings smaller than fore wings; wings held at rest folded back
    against abdomen; radius and media not fusing; tarsi two- or
    three-jointed...28

27. Tarsi apparently four-jointed; cerci usually minute; wings with a
    transverse preformed suture near the base; social species, living
    in colonies. Termites......................................ISOPTERA   (Page 163)
    Tarsi three-jointed, front metatarsi swollen; cerci conspicuous; usually
    solitary species. Webspinners.....................EMBIOPTERA*   (Page 153)

28. Cerci absent; tarsi two- or three-jointed; wings remaining attached
    throughout life; radial sector and media branched, except when
    fore wings are much thickened. Book lice........PSOCOPTERA   (Page 199)
    Cerci present; tarsi two-jointed; wings shed at maturity, venation
    greatly reduced; radial sector and media simple, unbranched.....................
    ....ZORAPTERA*   (Page 195)

29. Mouthparts not functional; abdomen with a pair of caudal filaments............30
    Mouthparts forming a proboscis, only exceptionally vestigial; abdomen
    without caudal filaments; hind wings replaced by knobbed halteres.
    True flies....................................DIPTERA   (Page 243)

30. No halteres; antennae inconspicuous; cross veins abundant. A few
    rare mayflies...................................EPHEMEROPTERA   (Page 127)
    Hind wings represented by minute hooklike halteres; antennae
    evident; venation reduced to a forked vein; crossveins lacking;
    minute delicate insects. Males of scale insects.....HEMIPTERA   (Page 210)

31. Body with more or less distinct head, thorax and abdomen, and
    jointed legs; capable of locomotion...32
    Without distinct body parts or without jointed legs, or incapable
    of locomotion...75

32. Terrestrial, breathing through spiracles; rarely without special
    respiratory organs...33
    Aquatic, usually gill-breathing, larval forms.......................................62

Parasites on warm-blooded animals 70

33. Mouthparts retracted into head and scarcely or not at all visible;
    underside of abdomen with styles or other appendages; less than
    three joints on maxillary palps if antennae present; delicate,
    small or minute insects...34
    Mouthparts conspicuously visible externally; if mouthparts mandibulate,
    maxillary palps more than two-jointed; antennae always present;
    underside of abdomen rarely with styles...36

34. Antennae absent; no long cerci, pincers, springing apparatus, or anterior
    ventral sucker on abdomen; head pear-shaped.........PROTURA   (Page 118)
    Antennae conspicuous; pincers, long cerci, or basal ventral sucker
    present on abdomen..35

35. Abdomen consisting of six segments or less, with a forked sucker
    at base of abdomen below; no terminal pincers or long cerci; usually
    with conspicuous springing apparatus near end of abdomen.
    Springtails .... COLLEMBOLA   (Page 114)
    Abdomen consisting of more than eight visible segments, with long
    multi-jointed cerci or strong pincers at the end; eyes and ocelli
    absent.. DIPLURA   (Page 120)

36. Mouthparts mandibulate, formed for chewing.......................................37
    Mouthparts haustellate, formed for sucking..59

37. Body usually covered with scales; abdomen with three prominent
    caudal filaments and bearing at least two pairs of ventral styles 38
    Body never covered with scales; never with three caudal filaments;
    ventral styles absent on abdomen 39

38. Head with large compound eyes and ocelli; legs with three tarsal
    segments; paired styli present on each abdominal segment.
    Bristletails......................................MICROCORYPHIA   (Page 122)
    Compound eyes small or absent; legs with two to four tarsal segments;
    paired styli on abdominal segments 7–9 (rarely 2–9). Silverfish,
    firebrats .... ZYGENTOMA   (Page 123)

39. Underside of abdomen entirely without legs..40
    Abdomen bearing false legs beneath, which differ from those of
    thorax; body caterpillarlike, cylindrical; thorax and abdomen not
    distinctly separated; larval forms..57

40. Antennae long and distinct 41
    Antennae short, not pronounced; larval forms......................................54

41. Abdomen terminated by strong movable forceps; prothorax free.
    Earwigs..DERMAPTERA   (Page 175)
    Abdomen not ending in forceps 42

42 Abdomen strongly constricted at base; prothorax fused with
    mesothorax. Ants, etc..............................HYMENOPTERA   (Page 330)
    Abdomen not strongly constricted at base; broadly joined to thorax.............. 43

43. Head not elongated ventrally..44
    Head elongated ventrally forming a rostrum, at tip of which are
    mandibulate mouthparts. Scorpionflies.............MECOPTERA   (Page 239)

44. Very small species; body soft and weakly sclerotized; tarsi two- or
    three-jointed...45

Usually much larger species; tarsi usually with more than three joints, or, if not, body is hard and heavily sclerotized and cerci are absent...46

45. Cerci absent. Book lice ....PSOCOPTERA   (Page 199)

Cerci unjointed, prominent...............................ZORAPTERA*   (Page 195)

46. Hind femora enlarged; wing pads of larva when present in inverse position, that is, metathoracic overlapping mesothoracic..ORTHOPTERA   (Page 184)

Hind legs not enlarged for jumping; wing pads, if present, in normal position...47

47. Prothorax much longer than mesothorax; front legs fitted for grasping prey. Mantids DICTYOPTERA, Suborder MANTODEA   (Page 161)

Prothorax not greatly lengthened...48

48. Cerci present; antennae usually with more than 15 subdivisions, often multiply subdivided...49

No cerci; body often hard-shelled; antennae usually with 11 subdivisions. Beetles..COLEOPTERA   (Page 305)

49. Cerci with more than three joints..50

Cerci short, with one to three joints...52

50. Body flattened and oval; head inflexed; prothorax oval. Cockroaches .... DICTYOPTERA, Suborder BLATTODEA   (Page 160)

Body elongate; head nearly horizontal...51

51. Cerci long; ovipositor rigid and exserted; tarsi five-jointed..................................GRYLLOBLATTODEA*   (Page 173)

Cerci short; no ovipositor; tarsi four-jointed; social forms, living in colonies. Termites..ISOPTERA   (Page 163)

52. Tarsi five-jointed; body usually very slender and long. Stick insects..PHASMIDA   (Page 179)

Tarsi two- or three-jointed; body not elongate..53

53. Front tarsi with first joint swollen, containing a silk-spinning gland, producing a web in which the insects live; body long and slender. Webspinners...EMBIOPTERA*   (Page 153)

Front tarsi not swollen, without silk-spinning gland; body much stouter; social species. Termites.........................ISOPTERA   (Page 163)

54. Body cylindrical, caterpillarlike...55

Body more or less depressed, not caterpillarlike 56

55. Head with six ocelli on each side; labium with spinnerets; antennae inserted in membranous area at base of mandibles .... Larvae of some LEPIDOPTERA   (Page 276)

Head with more than six ocelli on each side; metathoracic legs distinctly larger than prothoracic legs .... Larvae of Boreidae (MECOPTERA)   (Page 239)

56. Mandibles united with corresponding maxillae to form sucking organs......................................Larvae of NEUROPTERA   (Page 301)

Mandibles almost always separate from maxillae.................................... .... Larvae of COLEOPTERA   (Page 305);

RAPHIDIOPTERA   (Page 299); STREPSIPTERA *   (Page 326);

DIPTERA..   (Page 243)

57. False legs (prolegs) numbering five pairs or fewer, located on various abdominal segments, but not on first, second, or seventh; false legs tipped with many minute hooks (hookless prolegs rarely on second and seventh segments)............Larvae of most LEPIDOPTERA (Page 276)

    False legs numbering from six to ten pairs, one pair of which occurs on second abdominal segment; prolegs not tipped with minute hooks 58

58. Head with a single ocellus on each side..
    ....Larvae of some HYMENOPTERA (Page 330)

    Head with several ocelli on each side ..... Larvae of MECOPTERA (Page 239)

59. Body bare or with few scattered hairs, or with waxy coating.......................60

    Body densely covered with hair or scales; proboscis if present coiled under head. Moths............................LEPIDOPTERA (Page 276)

60. Last tarsal joint swollen; mouth consisting of a triangular unjointed beak; minute species. Thrips....................THYSANOPTERA (Page 233)

    Tarsi not bladderlike at tip, and with distinct claws...................................61

61. Prothorax distinct. Bugs...................................HEMIPTERA (Page 210)

    Prothorax small, hidden when viewed from above. True flies.........................
    .... DIPTERA (Page 243)

62. Mouthparts mandibulate..63

    Mouthparts suctorial, forming a strong pointed inflexed beak..Larvae of HEMIPTERA (Page 210)

63. Mandibles exserted straight forward and united with the corresponding maxillae to form piercing jaws.........Larvae of NEUROPTERA (Page 301)

    Mandibles normal, moving laterally to function as biting jaws.....................64

64. Body not encased in a shell made of sand, pebbles, leaves, etc....................65

    Case-bearing forms; tracheal gills usually present......................................
    .... Larvae of TRICHOPTERA (Page 268)

65. Abdomen furnished with external lateral gills of respiratory processes (a few Coleoptera and Trichoptera larvae here also)..................66

    Abdomen without external gills...67

66. Abdomen terminated by two or three long caudal filaments............................ Larvae of EPHEMEROPTERA (Page 127)

    Abdomen with short end processes....Larvae of MEGALOPTERA (Page 297)

67. Labium strong, extensible, and furnished with a pair of opposable hooks.. Larvae of ODONATA (Page 136)

    Labium not capable of being thrust forward and not hooked 68

68. Abdomen without false legs..69

    Abdomen bearing paired false legs on several segments...........................A few larvae of LEPIDOPTERA (Page 276)

69. The three divisions of thorax loosely united; antennae and caudal filaments long and slender...............Larvae of PLECOPTERA (Page 147)

    Thoracic divisions not constricted; antennae and caudal filaments short (also some aquatic larvae of Diptera and a few Trichoptera here)......................................Larvae of COLEOPTERA (Page 305)

70. Body flattened (or larval maggots)..71

    Body strongly compressed; mouth formed as a sharp inflexed beak; jumping species. Fleas............................SIPHONAPTERA (Page 264)

71. Mandibulate mouthparts 72
    Mouthparts formed for piercing and sucking 73
72. Mouth inferior; cerci long; ectoparasites of bats or rodents ....
    .... Rare DERMAPTERA*   (Page 175)
    Mouth anterior; no cerci; generally elongate-oval insects with somewhat
    triangular head; ectoparasites of birds (occasionally mammals).
    Chewing lice .... PHTHIRAPTERA, in part   (Page 203)
73. Antennae exserted, visible, though rather short 74
    Antennae inserted in pits, not visible from above (also larval maggots,
    without antennae) .... Pupiparous DIPTERA   (Page 243)
74. Beak (mouthparts) unjointed; tarsi formed as a hook for grasping
    hairs of the host (Figure 3.24C); permanent parasites. Sucking lice ....
    .... PHTHIRAPTERA, in part   (Page 203)
    Beak jointed; tarsi not hooked; temporary
    parasites .... Some HEMIPTERA   (Page 210)
75. Legless grubs, maggots or borers; locomotion effected by a squirming
    motion .... Larvae of STREPSIPTERA*   (Page 326);
    SIPHONAPTERA (Page 264); and of some COLEOPTERA   (Page 305)
    (see also couplet 56); DIPTERA (Page 243); LEPIDOPTERA   (Page 276);
    and HYMENOPTERA (Page 330). (If living in body
    of wasps and bees, with flattened head exposed, compare females
    of STREPSIPTERA* (Page 326); if aquatic wrigglers, see
    larvae and pupae of mosquitoes, etc.) Sedentary forms, incapable
    of locomotion 76
76. Small degraded forms bearing little superficial resemblance to insects,
    with long slender beak, and usually covered with a waxy scale, powder,
    or cottony tufts; living on various plants. Scale insects ....
    .... HEMIPTERA   (Page 210)
    Body quiescent, but able to bend from side to side; not capable of feeding,
    enclosed in a skin which is tightly drawn over all appendages, or which
    leaves limbs free but folded against body; sometimes free; sometimes
    enclosed in cocoon or in shell formed from dried larval skins 77
77. Skin encasing legs, wings, etc., holding appendages tightly against body;
    prothorax small; proboscis showing 78
    Legs, wings, etc., more or less free from body; biting mouthparts
    showing 79
78. Proboscis usually long, rarely absent; four wing cases; sometimes in
    cocoon .... Pupae of LEPIDOPTERA   (Page 276)
    Proboscis short; two wing cases, pupa often enclosed in oval shell
    (puparium) formed of hardened larval skin ....
    .... Pupae of DIPTERA   (Page 243)
79. Prothorax small, fused into one piece with mesothorax; sometimes
    enclosed in loose cocoon .... Pupae of HYMENOPTERA   (Page 330)
    Prothorax larger and not closely fused with mesothorax 80
80. Wing cases with few or no veins .... Pupae of COLEOPTERA   (Page 305)
    Wing cases with several branched veins ....
    .... Pupae of NEUROPTERA   (Page 301)

5. Literature

For an excellent introduction to the basics of taxonomy and systematics, readers should consult Cranston *et al.* (1991). The arguments for and against the cladistic, phenetic, and phyletic approaches to classification are discussed by Simpson (1961), Skal and Sneath (1963), Hennig (1965, 1966, 1981), Mayr (1969, 1981), Wagner (1969), Michener (1970), Sneath and Sokal (1973), Ross (1974), and Goto (1982). Boudreaux (1979), Hennig (1981), Kristensen (1981, 1989, 1991), and Kukalová-Peck (1985, 1987, 1991) provide extensive reviews of the phylogeny of insects.

Bishop, A. L., Holtkamp, R. H., Waterhouse, D. B., and Lowes, D., 1989, PESTKEY: An expert system based key for identification of the pests and beneficial arthropods in lucerne, *Gen. Appl. Entomol.* **21**:25–32.

Börner, C., 1904, Zur Systematik der Hexapoda, *Zool. Anz.* **27**:511–533.

Boudreaux, H. B., 1979, *Arthropod Phylogeny, with Special Reference to Insects*, Wiley, New York.

Brues, C. T., Melander, A. L., and Carpenter, F. M., 1954, *Classification of Insects*, 2nd ed., Museum of Comparative Zoology, Cambridge, MA.

Caterino, M. S., Cho, S., and Sperling, F. A. H., 2000, The current state of insect molecular systematics: A thriving Tower of Babel, *Annu. Rev. Entomol.* **45**:1–54.

Comstock, J. H., 1918, *The Wings of Insects*, Comstock, New York.

Crampton, G. C., 1915, The thoracic sclerites and the systematic position of *Grylloblatta campodeiformis* Walker, a remarkable annectant orthopteroid insect, *Entomol. News* **26**:337–350.

Crampton, G. C., 1920, Some anatomical details of the remarkable winged zorapteran, *Zorotypus hubbardi* Caudell, with notes on its relationships, *Proc. Entomol. Soc. Wash.* **22**:98–106.

Crampton, G. C., 1922, Evidences of relationship indicated by the venation of the fore wings of certain insects with especial reference to the Hemiptera-Homoptera, *Psyche* **29**:23–41.

Crampton, G. C., 1924, The phylogeny and classification of insects, *Pomona J. Entomol. Zool.* **16**:33–47.

Cranston, P. S.,Gullan, P. J., and Taylor, R. W., 1991, Principles and practice of systematics, in: *The Insects of Australia*, 2nd ed., Vol. I (CSIRO, ed.), Melbourne University Press, Carlton, Victoria.

Danks, H. V., 1988, Systematics in support of entomology, *Annu. Rev. Entomol.* **33**:271–296.

Goto, H. E., 1982, *Animal Taxonomy*, Arnold, London.

Handlirsch, A., 1903, Zur Phylogenie der Hexapoden, *Sitz. Nat. Klasse Acad. Wiss* **112**:716–738.

Handlirsch, A., 1906–1908, *Die fossilen Insecten und die Phylogenie der rezenten Formen*, Engelman, Leipzig.

Handlirsch, A., 1925, Geschichte, Litteratur, Technik, Palaontologie, Phylogenie und Systematik der Insecten, in: *Handbuch der Entomologie* (C. Schröder, ed.), Vol. 3, pp. 1–1201, Fischer, Jena.

Hennig, W., 1965, Phylogenetic systematics, *Annu. Rev. Entomol.* **10**:97–116.

Hennig, W., 1966, *Phylogenetic Systematics*, University of Illinois Press, Urbana.

Hennig, W., 1981, *Insect Phylogeny*, Wiley, New York.

Hinton, H. E., 1958, The phylogeny of the panorpoid orders, *Annu. Rev. Entomol.* **3**:181–206.

Hinton, H. E., 1981, *Biology of Insect Eggs*, 3 volumes, Pergamon Press, Elmsford, NY.

Kristensen, N. P., 1981, Phylogeny of insect orders, *Annu. Rev. Entomol.* **26**:135–157.

Kristensen, N. P., 1989, Insect phylogeny based on morphological evidence, in: *The Hierarchy of Life* (B. Fernholm, K. Bremer, and H. Jornvall, eds.), Elsevier, Amsterdam.

Kristensen, N. P., 1991, Phylogeny of extant hexapods, in: *The Insects of Australia*, 2nd ed., Vol. I (CSIRO, ed.), Melbourne University Press, Carlton, Victoria.

Kukalová-Peck, J., 1985, Ephemeroid wing venation based upon new gigantic Carboniferous mayflies and basic morphology, phylogeny, and metamorphosis of pterygote insects (Insecta, Ephemerida), *Can. J. Zool.* **63**:933–955.

Kukalová-Peck, J., 1987, New Carboniferous Diplura, Monura, and Thysanura, the hexapod ground plan, and the role of thoracic side lobes in the origin of wings (Insecta), *Can J. Zool.* **65**:2327–2345.

Kukalová-Peck, J., 1991, Fossil history and the evolution of hexapod structures, in: *The Insects of Australia*, 2nd ed., Vol. 1 (CSIRO, ed.), Melbourne University Press, Carlton, Victoria.

Lawrence, J. F., Hastings, A. M., Dallwitz, M. J., and Paine, T. A., 1993, *Beetle Larvae of the World: Interactive Identification and Information Retrieval for Families and Subfamilies*. CD-ROM; Version 1.0 for MS-DOS, CSIRO, Melbourne.

Martynov, A. V., 1924, The interpretation of the wing venation and tracheation of the Odonata and Agnatha, *Russ. Ent. Obozl:* **18**:145–174. [Original in Russian. Translated, with an introductory note, by F. M. Carpenter in *Psyche* **37**(1930):245–280.]

Mayr, E., 1969, *Principles of Systematic Zoology*, McGraw-Hill, New York.

Mayr, E., 1981, Biological classification: Toward a synthesis of opposing methodologies, *Science* **214**:510–516.

McFadyen, R. E., 1985, Larval characteristics of *Cactoblastis* spp. (Lepidoptera: Pyralidae) and the selection of species for biological control of prickly pears (*Opuntia* spp.), *Bull. Entomol. Res.* **75**:159–168.

Michener, C. D., 1970, Diverse approaches to systematics, *Evol. Biol.* **4**:1–38.

Rentz, D. C. F., 1993, A *Monograph of the Tettigoniidae of Australia*, Vol. 2 (The Austrosaginae, Zaprochilinae and Phasmodinae), CSIRO, Melbourne.

Ross, H. H., 1974, *Biological Systematics*, Addison-Wesley, Reading, MA.

Sharov, A. G., 1966, *Basic Arthropodan Stock*, Pergamon Press, Elmsford, NY.

Sharp, D., 1899, Some points in the classification of the Insecta Hexapoda, *Proc. 4th Int. Congr. Zool.*, pp. 246–249.

Simpson, G. G., 1961, *Principles of Animal Taxonomy*, Columbia University Press, New York.

Sneath, P. H. A., and Sokal, R. R., 1973, *Numerical Taxonomy—The Principles and Practice of Numerical Classification*, Freeman, San Francisco.

Sokal, R. R., and Sneath, P. H. A., 1963, *Principles of Numerical Taxonomy*, Freeman, San Francisco.

Tillyard, R. J., 1918–1920, The panorpoid complex. A study of the holometabolous insects with special reference to the sub-classes Panorpoidea and Neuropteroidea, *Proc. Linn. Soc. N.S.W.* **43**:265–284, 395–408, 626–657; **44**:533–718; **45**:214–217.

Wagner, W. H., Jr., 1969, The construction of a classification, in: *Systematic Biology*, National Academy of Science, Washington, D.C., Publ. No. 1692.

Watson, J. A. L., Theischinger, G., and Abbey, H. M., 1991, *The Australian Dragonflies*, CSIRO, Melbourne.

Weeks, P. J. D., O'Neill, M. A., Gaston, K. J., and Gauld, I. D., 1999, Automating insect identification: Exploring the limitations of a prototype system, *J. Appl. Entomol.* **123**:1–8.

Wilson, H. F., and Doner, M. H., 1937, *The Historical Development of Insect Classification*, Madison, WI.

Zompro, O., Adis, J., and Weitschat, W., 2002, A review of the order Mantophasmatodea, *Zool. Anz.* **241**: 269–279.

5

Apterygote Hexapods

1. Introduction

Traditionally, the groups included in the term "apterygote hexapods," namely, the Collembola, Protura, Diplura, and Thysanura (including Microcoryphia and Zygentoma), were considered orders of primitively wingless insects and placed in the subclass Apterygota (Ametabola). They show the following common features: lack of wings, lack of a pleural sulcus on the thoracic segments, presence of pregenital abdominal appendages, slight or absent metamorphosis, and indirect sperm transfer. As more information on their structure and habits has become available, it has become apparent that (1) their status as insects (except for Thysanura) is doubtful and (2) the relationship of the groups with each other is more distant than originally believed. Several authors have therefore recommended that the insectan subclass Apterygota be reserved solely for the Thysanura and that the Collembola, Protura, and Diplura each be given the rank of class, with the Collembola and Protura considered as sister groups within the Ellipura (see Figure 1.11). These three groups differ fundamentally from insects in several features; for example, they are entognathous, have intrinsic musculature in the antennae, and lack compound eyes which are characteristic of most insects, at least in the adult stage. Thus, the Ellipura and Diplura are sometimes considered sister groups within the Entognatha. However, other analyses indicate that the Entognatha is a paraphyletic assemblage (see Chapter 1, Section 3.3.1).

The Collembola are probably furthest removed from the winged insects. They possess only six abdominal segments, a postantennal sensory organ similar to the organ of Tömösvary found in myriapods, gonads with lateral (rather than apical) germaria, and eggs in which there is total cleavage. Non-insectan features of the Protura are the absence of antennae (perhaps a secondary condition associated with their soil-dwelling habit), the occurrence of anamorphosis, and a genital aperture that opens behind the 11th segment. Diplura are superficially similar to Thysanura, with which some authors group them. However, in addition to the features mentioned above, they differ from typical insects in having unusual respiratory and reproductive systems. Even though all Thysanura are considered insectan, it is now apparent that the group contains two distinct subgroups, the Microcoryphia and the Zygentoma (= Thysanura *sensu stricto*), to which some authors accord ordinal status. The primary basis for this distinction concerns the mouthparts. In the Microcoryphia (such as *Machilis* and its allies, the bristletails) the mandibles have a single articulation with the head and bite with a rolling motion. In the Zygentoma (which includes silverfish and firebrats), on the other hand, there is a dicondylic articulation of the mandible, which thus bites transversely as in pterygote insects. As noted in Chapter 1, differences in the structure and operation of the mouthparts are of fundamental phylogenetic importance.

2. Collembola

SYNONYMS: Oligentoma, Oligoentomata   COMMON NAME: springtails

Small to minute wingless hexapods; head pro- or hypognathous, antennae segmented, compound eyes absent, mouthparts entognathous; abdomen 6-segmented, typically with three medially situated pregenital appendages (collophore on segment 1, retinaculum on segment 2, furcula on segment 4), gonopore on 5th segment.

Collembola are abundant on every continent, including Antarctica. About 6500 species have been described, including more than 1600 from Australia, 300 from the United Kingdom, and about 840 from North America. Individual species may be quite cosmopolitan, sometimes as a result of human activity when they may become pests.

Structure

Collembola vary in length from about 0.2 to 10 mm. They are generally dark, but many species are whitish, green, or yellowish, and some are striped or mottled. The body is either elongate (Arthropleona) (Figure 5.1B) or more or less globular (Symphypleona and Neelipleona) (Figure 5.1A). The head is primitively prognathous, but is hypognathous

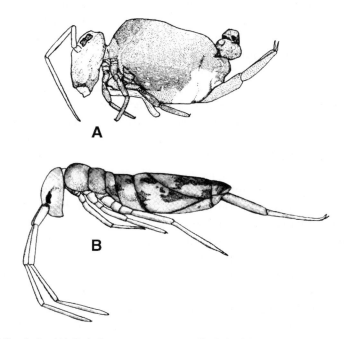

FIGURE 5.1. Collembola. (A) *Sminthurus purpurescens* (Sminthuridae); and (B) *Entomobrya nivalis* (Entomobryidae). [Reprinted from Elliott A. Maynard, 1951, *A Monograph of the Collembola or Springtail Insects of New York State*, Comstock Publishing Co., Inc.]

in the Symphypleona and Neelipleona, and the mouthparts are enclosed within a pouch formed by the ventrolateral extension of the head capsule. The mouthparts are typically of the chewing type, though in many species they are rasping or suctorial. The 4-segmented antennae vary greatly in length and each segment may be subdivided into two (segments 1 and 2) or numerous (segments 3 and 4) subsegments. Immediately behind the antennae is a structure of varied form, the postantennal organ, which appears to have an olfactory function. Compound eyes never occur, but a varied number of ocelli (up to eight) are found on each side of the head. The thoracic segments are distinct in Arthropleona, but not in Symphypleona and Neelipleona; in all species the prothorax is small or vestigial. In the Symphypleona and Neelipleona the thorax is fused with the abdomen and individual segments are not easily distinguished except at the posterior end. The legs have no true tarsus but terminate in one or two claws that arise from the tibia. No more than six abdominal segments can be distinguished at any time (even during embryonic development). The first abdominal segment bears the collophore (ventral tube), which arises by fusion and differentiation of the embryonic appendages. The tube contains a pair of vesicles that can be extruded by hemolymph pressure. The tube appears to have several functions, though it was originally named because it was thought to be adhesive (Greek *colle*, glue). Other likely functions include gaseous exchange and, especially, salt-water balance. Most Collembola have a springing organ (furcula) on the fourth abdominal segment, held under tension beneath the body by a hooklike structure, the retinaculum, formed from the appendages of the third abdominal segment. When released from the retinaculum, the furcula is forced downward and backward by both muscular action and hemolymph pressure. As it strikes the substrate, the animal is thrown through the air, sometimes a significant distance (e.g., up to 30 cm in some sminthurids). Abdominal appendages may be greatly reduced in small subterranean forms. Cerci are absent in Collembola. Some species are ecomorphic, their form changing from instar to instar as a result of unusual environmental conditions, others are cyclomorphic (having seasonally different forms and habits, usually in summer and winter), and some show epitoky in which reproductive instars are morphologically different from non-reproductive (feeding instars) with which they alternate.

Noteworthy features of the internal structure of Collembola are the absence of Malpighian tubules and, in most species, tracheal system. However, a pair of spiracles between the head and thorax, leading to tracheae in the head, sometimes also the body, have been reported for a few Symphypleona. The nervous system is specialized and includes brain, subesophageal ganglion, and three ventral ganglia, the ganglia of the abdominal segments having fused with the metathoracic ganglion.

Life History and Habits

Springtails are almost ubiquitous, being found in high latitudes (e.g., 84° south in Antarctica) and altitudes (above 7700 m in the Himalayas), deserts, and glaciers, in addition to more conventional biomes. Within these regions they occupy a wide range of habitats, though they are most abundant and diverse in moist soil, leaf litter, and rotting wood. Others live in dung, in the fleshy parts of fungi, in the nests of termites, ants, and vertebrates, on grasses, and in flowers. Several species occur in caves, on the surface of standing water (including, rarely, tidal pools), and in both marine and freshwater intertidal zones, though very few are truly aquatic. Occasionally, species form large aggregations on the surface of snow, though the significance of this is not clear. Most Collembola cannot tolerate desiccation, and those that colonize dry habitats show various morphological, physiological, and behavioral adaptations, such as scales or hairs, specific desiccation-resistant stages (including more or less completely dried out "anhydrobiotic" forms), and activity restricted to periods when the relative humidity is high. Collembola are saprophagous, fungivorous (including some spore feeders), or phytophagous (including some pollen feeders), and some species appear to have an intrinsic gut cellulase. Only rarely are they predaceous. Though a few species are occasionally of economic importance because of the damage they cause to plants, overall Collembola may be considered beneficial. By feeding on fungal hyphae and decaying material, they may influence mycorrhizal growth and control fungal diseases (Hopkin, 1997).

In adult Collembola reproductive and feeding instars alternate. Most species of Collembola reproduce sexually though a few soil-dwelling species are parthenogenetic. Sperm transfer is usually indirect, males placing their stalked spermatophores on the substrate, to be found by females as a result of aggregation (perhaps involving pheromones) or after an elaborate courtship dance in which the male grasps the female using spines on his antennae, head, or legs, and draws her over the spermatophore. The pale, spherical eggs are laid singly or in small clusters (sometimes several females oviposit collectively) and may be covered with freshly eaten soil and fecal material, or cleaned by the adults, to prevent fungal infection. There is little change in external form as the young animal develops, and sexual maturity is reached usually after 5–10 molts. As many as 50 molts may occur during the adult phase.

Phylogeny and Classification

According to Kukalová-Peck (1991), the early Collembola were semiaquatic. The earliest fossil collembolan, *Rhyniella praecursor*, from the Lower Devonian of Scotland resembled modern isotomids. Other fossils are known from the Lower Permian of South Africa, the Upper Cretaceous of Canada, and Tertiary amber deposits, the latter being assignable to extant genera. Currently, the two most primitive families are thought to be the Hypogastruridae and Isotomidae, with the Entomobryidae and Sminthuridae among the most advanced. Early classifications [e.g., that of Gisin (1960)] arranged extant species of the order Collembola in two suborders, Arthropleona and Symphypleona, principally on the basis of the striking difference in body form, and five families. Nowadays, up to 14 families are recognized, and these are allocated among three orders within the class Collembola, the round-bodied Neelidae being separated from the Arthropleona in their own order Neelipleona. However, lack of information prevents the construction of a phylogenetic tree. Of the extant families, the largest and most common are the Hypogastruridae, Neanuridae, Onychiuridae, Entomobryidae, Isotomidae, and Sminthuridae.

Order Arthropleona

Superfamily Poduroidea

The HYPOGASTRURIDAE (580 species) are generally 1–3 mm in length, whitish, pinkish, or darkly colored Collembola with a predominantly holarctic distribution. They have an obvious prothorax, a granular cuticle, and short antennae, but a postantennal organ may or may not be present. They are found in a wide range of habitats though most species live among decaying vegetation, in soil, in cracks in the bark of trees, or in fungi. Some hypogastrurids are known as "snow fleas" through their being found, sometimes in immense numbers, jumping about on snow, usually shortly after a period of mild weather. One widely distributed form, *Hypogastrura (Ceratophysella) denticulata* (likely a complex of several species), sometimes becomes a pest in mushroom cellars.

The majority of NEANURIDAE (1160 species) are found under stones and bark, or in soil and leaf litter where they feed on fungal hyphae that they pierce with their sharp mouthparts. Other species, however, are predaceous, eating rotifers, other Collembola, and their eggs. ONYCHIURIDAE (600 species) are soil- and litter-dwelling collembolans that lack ocelli and a furcula.

Superfamily Entomobryoidea

The characteristics of the ISOTOMIDAE (over 1000 species) are highly varied, and future work may well result in its being subdivided into several additional families. Isotomids are found worldwide, in a range of biomes, including the polar regions, arid areas, and seashores, though many are conventional inhabitants of soil or leaf litter. Many species, especially the soil dwellers, are cosmopolitan. The family ENTOMOBRYIDAE (Figure 5.1B), with 1365 species, includes many of the larger Collembola that reach 5 mm or more in length. Species have a greatly reduced prothorax and a smooth cuticle; a postantennal organ may or may not be present. They may be found in soil or leaf litter, under bark, in moss, and on vegetation. Some species are naturally cosmopolitan, and others have been transferred around the world by human activity.

Order Neelipleona

The approximately 25 species in this order are included in a single family NEELIDAE. These tiny collembolans (0.5 mm or less long) live in soil and leaf litter. They differ from Symphypleona in that their rounded body is formed from expansion of thoracic rather than abdominal segments.

Order Symphypleona

Superfamily Sminthuroidea

Most of the 890 species of SMINTHURIDAE (Figure 5.1A) are 1–3 mm in length and have a roundish body, hypognathous head, and conspicuous ocelli. A postantennal organ is absent. Often there is sexual dimorphism, with the antennae of males having hooks and spines. Most sminthurids are epigaeic, living near the surface of leaf litter, or on grasses or other low-growing vegetation. A number of species are economically important. For example, *Sminthurus viridis*, the lucerne flea, a European species introduced into Australia, has become an important pest on alfalfa (lucerne) and other leguminous crops. Other species may do considerable damage in greenhouses and to many garden vegetables at the seedling stage.

Literature

Accounts of the biology of Collembola are provided by Fjellberg (1985), Greenslade (1991, 1994), Hopkin (1997), and Christiansen and Bellinger (1998). Keys for their identification are to be found in Christiansen and Bellinger (1998) [North American species], Gisin (1960) [European species] (in German), and Greenslade (1991) [Australian families].

Christiansen, K., and Bellinger, P., 1998, *The Collembola of North America North of the Rio Grande*, 2nd ed., Grinnell College, Grinnell, IA.

Fjellberg, A., 1985, Recent advances and future needs in the study of Collembola biology and systematics, *Quaest. Entomol.* **21**:559–570.

Gisin, H., 1960, *Collembolenfauna Europas*, Museum d'histoire naturelle, Geneva.

Greenslade, P., 1994, Collembola, in: *Zoological Catalogue of Australia*, Vol. 22 (W. W. K. Houston, ed.), CSIRO, Melbourne.

Greenslade, P. J., 1991, Collembola, in: *The Insects of Australia*, 2nd ed., Vol. I (CSIRO, ed.), Melbourne University Press, Carlton, Victoria.

Hopkin, S. P., 1997, *Biology of the Springtails (Insecta: Collembola)*, Oxford University Press, Oxford.

Kukalová-Peck, J., 1991, Fossil history and the evolution of hexapod structures, in: *The Insects of Australia*, 2nd ed., Vol. I (CSIRO, ed.), Melbourne University Press, Carlton, Victoria.

3. Protura

SYNONYM: Myrientomata   COMMON NAME: proturans

Minute wingless hexapods; head cone-shaped, compound eyes, ocelli, and antennae absent, mouthparts entognathous and suctorial; foreleg modified as a sense organ; abdomen 12-segmented in adult with appendages on first three segments; gonopore (two in males) behind 11th segment; cerci absent.

About 660 species of proturans have been described worldwide. Of these, about 80 are European (including 20 in Britain), 30 Australian, and about 20 North American, though the latter figure may be misleading, reflecting the lack of taxonomic work done on this group.

Structure

Proturans (Figure 5.2) are elongate, generally pale arthropods 2 mm or less in length. The head is cone-shaped and bears anteriorly the styliform entognathous mouthparts. Photoreceptor organs are absent from the head, as are typical antennae. However, a pair of "pseudoculi" occur dorsolaterally that may be humidity receptors or chemosensory. The thoracic segments are distinct, though the first is greatly reduced. The six identical legs have an unsegmented tarsus. The forelegs are generally not used in locomotion but are held aloft

FIGURE 5.2.   *Acerella barberi*, a proturan. [From H. E. Ewing, 1940, The Protura of North America, *Ann. Entomol. Soc. Am.* **33**:495–551. By permission of the Entomological Society of America.]

and probably act as sense organs. In adult proturans the abdomen is 12-segmented; in the newly hatched animal there are only 9 abdominal segments, 3 being added anamorphically during postembryonic development. Short, unsegmented or 2-segmented appendages with eversible vesicles are found on the first three abdominal segments. Cerci are absent.

Internally there are no distinct Malpighian tubules, but six papillae occur at the junction of the midgut and hindgut, and these may serve an excretory function. A tracheal system is present in Eosentomoidea, originating from paired meso- and metatergal spiracles, but not in other groups. The tracheae do not anastomose. The nervous system is generalized, with discrete ganglia in the first seven abdominal segments.

Life History and Habits

Like springtails, proturans are found in a variety of moist habitats. Although frequently overlooked because of their small size, they are quite numerous in certain situations particularly in soil and litter. Few details of their biology are available. They are thought to be fungivorous. It is reported that four juvenile stages are passed through before sexual maturity is reached (five in male Acerentomidae which have a preimago instar), but it is not known whether proturans molt when adult. There are probably several generations per year with, in cooler climates, the adults spending the winter in a dormant condition.

Phylogeny and Classification

Tuxen (1964) recognized two suborders and three families within the class and order Protura, namely, the Eosentomoidea (family EOSENTOMIDAE) and Acerentomoidea (families ACERENTOMIDAE and PROTENTOMIDAE). However, following the description of *Sinentomon erythranum* by Yin (1965), which shows morphological features of all three families and may be the most primitive living proturan, some authors (e.g., Nosek, 1973) have placed this species in its own suborder Sinentomoidea (family SINENTOMIDAE). Though most workers consider the Eosentomidae and Sinentomidae as the most primitive proturans, their possession of a tracheal system and certain features of their sperm have led Yin (1984) to propose that they are more specialized than the Acerentomoidea.

Literature

Most literature on Protura is taxonomic. Accounts of the biology of this group are given by Ewing (1940), Nosek (1973), and Imidaté (1991). Nosek (1973) provides a key to the European species and Nosek (1978) a key to the world genera; Ewing (1940) deals with the North American forms; and Imidaté (1991) with the Australian families.

Ewing, H. E., 1940, The Protura of North America, *Ann. Entomol. Soc. Am.* **33**:495–551.
Imidaté, G., 1991, Protura, in: *The Insects of Australia*, 2nd ed., Vol. I (CSIRO, ed.), Melbourne University Press, Carlton, Victoria.
Nosek, J., 1973, *The European Protura*, Museum d'histoire naturelle, Geneva.
Nosek, J., 1978, Key and diagnosis of Protura genera of the world, *Annot. Zool. Bot. Bratislava* **122**:1–59.
Tuxen, S. L., 1964, *The Protura. A Revision of the Species of the World with Keys for Determination*, Hermann, Paris.
Yin, W.-Y., 1965, Studies on Chinese Protura. II. A new family of the suborder Eosentomoidea, *Acta Entomol. Sin.* **14**:186–195. (In Chinese with English summary.)
Yin, W.-Y., 1984, A new idea on phylogeny of Protura with approach to its origin and position, *Sci. Sin. (B)* **27**:149–160.

4. Diplura

SYNONYMS: Dicellura, Entotrophi, Entognatha   COMMON NAME: diplurans

Elongate apterygotes; head with long many-segmented antennae, compound eyes, and ocelli absent, mouthparts entognathous; thoracic segments distinct, legs with unsegmented tarsus; abdomen 10-segmented, most pregenital segments with styli, eversible vesicles on some abdominal segments, cerci present as either long multiannulate or forcepslike structures, gonopore between segments 8 and 9.

More than 800 species of this widely distributed, though mainly tropical or subtropical, order have been described. Most holarctic forms belong to the family Campodeidae, including the 11 British species. Just over 60 species have been described from North America, and about 30 from Australia.

Structure

In general form Diplura (Figure 5.3) resemble Thysanura but differ in being entognathous and lacking a median process on the last abdominal segment. Most species are a few millimeters long, but a few may reach almost 60 mm. The roundish or oval head carries the multisegmented antennae, whose flagellar segments (except the most distal) are provided with muscles. The reduced biting mouthparts occupy a pouch on the ventral surface. The six identical legs have an unsegmented tarsus. Two to four lateral spiracles occur on the thorax. Ten abdominal segments are distinguishable. The sterna of segments 2–7 bear styli, and eversible vesicles occur on a varied number of segments. The conspicuous cerci vary in structure and are an important taxonomic feature. In most Diplura (but not Campodeidae, which have none) there are seven pairs of abdominal spiracles.

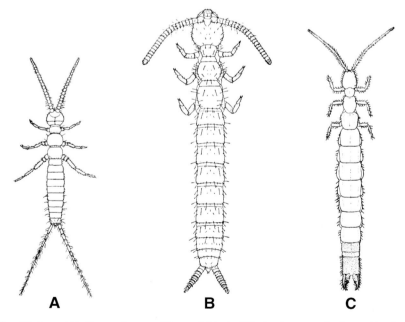

FIGURE 5.3. Diplura. (A) *Campodea* sp. (Campodeidae); (B) *Anajapyx vesiculosus* (Anajapygidae); and (C) *Heterojapyx* sp. (Heterojapygidae). [From A. D. Imms, 1957, *A General Textbook of Entomology*, 9th ed. (revised by O. W. Richards and R. G. Davies), Methuen and Co.]

As in Protura, Malpighian tubules are represented usually by a varied number of papillae at the junction of the midgut and hindgut. The tracheal system is developed to a varied extent. Tracheae leading from one spiracle never anastomose with those from other spiracles, and they lack cuticular supporting rings characteristic of insectan tracheae. The nervous system is not specialized, the ventral nerve cord containing eight (Japygidae) or seven (other Diplura) abdominal ganglia. The reproductive system is greatly varied within the Diplura, although in all species the germaria are apical. In *Japyx* (Japygidae) there are seven pairs of segmentally arranged ovarioles, in *Anajapyx* (Anajapygidae) two, and in *Campodea* (Campodeidae) one. One or two pairs of testes occur in the male.

Life History and Habits

Because of their rather secretive habits little is known of the life history of diplurans. They are found in damp habitats, for example, leaf litter, under stones and logs, and in soil. The campodeids are herbivorous, but other Diplura are carnivorous, catching prey with their forceps (modified cerci) or with their maxillae. Like male springtails, male diplurans deposit stalked spermatophores but make no attempt to attract females to them. The eggs are laid in groups in a chamber dug by the female. In some species the female guards the eggs and young. Development is slow and molting (up to 30 times in *Campodea*) continues through life.

Phylogeny and Classification

Fossil Diplura are known from the Upper Carboniferous of Illinois (*Testajapyx thomasi*) though at this stage the appendages were still fully segmented, eyes were present, and the entognathous condition was not fully developed; that is, the lateral margins of the head were not fused with the labium to form a pouch around the mandibles and maxillae. Extant Diplura were arranged by Paclt (1957) in three families [CAMPODEIDAE, PROJAPY-GIDAE (= ANAJAPYGIDAE), and JAPYGIDAE]. However, more recent classifications [e.g., that of Condé and Pagés (1991)] tend to raise the subfamilies within these groups to the rank of family. In campodeids (Figure 5.3A), which may reach 4 mm at maturity, the cerci are multiannulate and usually as long as the abdomen. In contrast, the projapygids (Figure 5.3B) are minute arthropods with relatively short cerci (less than half the length of the abdomen and having fewer than 10 subdivisions). The cerci of japygids (in the sense of Paclt, 1957) take the form of strongly sclerotized, undivided forceps (Figure 5.3C).

Literature

General information on Diplura is given by Wallwork (1970) and Condé and Pagés (1991). Most North American species can be identified from Smith (1960), the British species from Delany (1954), and the Australian families from Condé and Pagés (1991).

Condé, B., and Pagés, J., 1991, Diplura, in: *The Insects of Australia*, 2nd ed., Vol. I (CSIRO, ed.), Melbourne University Press, Carlton, Victoria.

Delany, M. J., 1954, Thysanura and Diplura, *R. Entomol. Soc. Handb. Ident. Br. Insects* **1**(2):1–7.

Paclt, J.,1957, Diplura, *Genera Insect.* **212**:1–123.

Smith, L. M., 1960, The family Projapygidae and Anajapygidae (Diplura) in North America, *Ann. Entomol. Soc. Am.* **53**:575–583.

Wallwork, J. A., 1970, *Ecology of Soil Animals*, McGraw-Hill, New York.

5. Microcoryphia

SYNONYMS: Archeognatha, Ectotrophi (in part),   COMMON NAME: bristletails
        Ectognatha (in part)

Small or moderately sized apterygote insects; head with long multiannulate antennae, large contiguous compound eyes, ocelli, ectognathous chewing mouthparts, mandibles with single articulation, maxillary palps 7-segmented; thorax strongly arched with terga extending over pleura, legs with 3 (rarely 2) tarsal segments; abdomen 11-segmented, though 10th segment reduced and tergum of 11th forming median caudal filament, paired styli present on each abdominal segment, long cerci with multiple subdivisions present.

As noted in the Introduction to this chapter, Microcoryphia and Zygentoma (see Section 6) were originally united in the order Thysanura. However, fundamental differences in their structure (compare the definitions of the orders) have led to their separation as distinct orders. The Microcoryphia form a small (about 350 species) but cosmopolitan group, with about 35 species in North America (mostly Machilidae), 7 in Australia (all Meinertellidae), and 7 in Britain (all Machilidae).

Structure

Microcoryphia (Figure 5.4A) are elongate insects up to 20 mm in length. Their body is strongly convex dorsally (with the large terga extending around the sides to cover the pleura), generally tapered posteriorly, and covered with scales. The head is hypognathous, in some species prognathous, and carries prominent chewing mouthparts, the long, 7-segmented maxillary palps being particularly conspicuous. Each mandible has a single articulation with the head. The antennae are filiform and comprise 30 or more subdivisions that lack intrinsic musculature. Compound eyes are well developed and contiguous (meet in a middorsal position). Median and paired lateral ocelli are also present. The legs have three tarsal segments and, in some species, those of the mesothorax and metathorax bear coxal styli. Abdominal styli occur on segments 2–9, and eversible vesicles are almost always found on abdominal segments 1–7. In females an ovipositor is formed from the appendages of

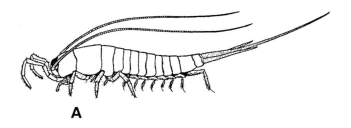

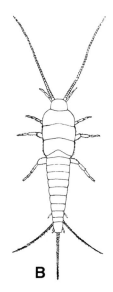

FIGURE 5.4. Microcoryphia and Zygentoma. (A) *Machilis* sp. (Machilidae); and (B) *Lepismodes inquilinus* (Lepismatidae). [A, reprinted from R. E. Snodgrass, 1952, *A Textbook of Arthropod Anatomy*, Cornell University Press. B, from A. D. Imms, 1957, *A General Textbook of Entomology*, 9th ed. (revised by O. W. Richards and R. G. Davies), Methuen and Co.]

abdominal segments 8 and 9. In males the appendages of the ninth abdominal segment fuse to form a median penis.

Internally Microcoryphia exhibit features that might be expected of primitive insects. They have 12–20 Malpighian tubules, a nervous system that includes paired longitudinal connectives, 3 thoracic and 8 abdominal ganglia, 9 pairs of spiracles and tracheae that do not anastomose with those of adjacent segments, and a primitive reproductive system. In females there are typically seven or eight panoistic ovarioles on each side of the body, and in some species these are arranged in a more or less segmental manner. Females lack a spermatheca. In males each testis comprises three or four follicles, and the vas deferens on each side is double, the two channels being connected by several transverse tubes.

Life History and Habits

Microcoryphia are usually nocturnal, hiding by day in rotting wood, among leaves, in crevices, under stones, etc., and are restricted to such habitats by their inability to resist desiccation. They are phytophagous or omnivorous, feeding on algae, lichens, leaf litter, and sometimes other arthropods. Bristletails can run fairly quickly and sometimes jump (up to 10 cm).

Details of the life history of Microcoryphia are not well known. Though parthenogenesis occurs in some species, reproduction is usually sexual, males leaving stalked spermatophores on the substrate or placing a sperm droplet on the ovipositor. Females lay eggs singly or in batches of up to 30 in crevices or holes dug with the ovipositor. The absence of a spermatheca suggests that mating must occur frequently. Postembryonic development is slow (generally taking several months to a year from hatching to adulthood) and includes at least five juvenile instars in *Machilis*. Molting continues in adults.

Phylogeny and Classification

Fossil machilids are known from as early as the Lower Devonian of Quebec. Extant species are placed in the single superfamily Machiloidea, containing two families, the MACHILIDAE and the more primitive MEINERTELLIDAE. The former is principally a Northern Hemisphere group while the latter is mainly from the Southern Hemisphere.

Literature

See under Section 6 (Zygentoma).

6. Zygentoma

SYNONYMS: Thysanura (*sensu stricto*),   COMMON NAME: silverfish, firebrats
          Ectotrophi (in part),
          Ectognatha (in part)

Small or moderately sized, dorsoventrally flattened apterygote insects; head with long multiannulate antennae, compound eyes small or absent, ocelli absent (except Lepidotrichidae), ectognathous chewing mouthparts, mandible with dicondylic articulation, maxillary palp 5-segmented; legs with 2–4 tarsal segments; abdomen 11-segmented but with 10th segment reduced and tergum of 11th segment forming median caudal filament, paired styli on abdominal segments 7–9 (rarely 2–9), cerci generally long with multiple subdivisions, but sometimes quite short, strongly diverging from body.

The Zygentoma is a small, worldwide order of about 370 species that are both structurally and ecologically more diverse than microcoryphians. About 30 species have been described from North America, with about the same number from Australia, including some that are cosmopolitan and occasionally pests, and 2 from Britain (both Lepismatidae).

Structure

Zygentoma (Figure 5.4B) are broadly similar to Microcoryphia, and only the more important differences in structure will be noted here. The body of Zygentoma is dorsoventrally flattened and may or may not have scales. Compound eyes are reduced or absent and ocelli do not occur (except in Lepidotrichidae). The maxillary palps are 5-segmented, and the mandibles have a dicondylic articulation with the head, as do those of pterygote insects. The legs include 2–4 tarsal segments. Abdominal styli normally occur only on segments 7–9, though in Nicoletiidae they are found on segments 2–9. Eversible vesicles are absent in many Lepismatidae but occur on abdominal segments 2–7 in Nicoletiidae and Lepidotrichidae. In most species the cerci are long, multiannulate structures that diverge sharply from the body (Figure 5.4B). However, in some inquiline Nicoletiidae the cerci are very short.

Four to eight Malpighian tubules are present. Ten pairs of spiracles occur and the tracheal system is well developed compared to that of Microcoryphia. On each side of the body a longitudinal trunk links the tracheae originating from the spiracles, and transverse segmental tracheae unite one side with the other. The female reproductive system includes 2–7 panoistic ovarioles, a spermatheca, and accessory glands. The number of lobes comprising each testis is greater than in Microcoryphia (eight lobes in Lepismatidae, many in Nicoletiidae and Lepidotrichidae).

Life History and Habits

Though many Zygentoma resemble Microcoryphia in their general habits, living in leaf litter, under bark, and in other places with high humidity, others are extremely resistant to desiccation and have the ability to take up water from the atmosphere. All are very agile insects. Zygentoma are omnivorous or phytophagous.

Eggs are laid singly or in groups, in crevices, etc. Individuals may live for several years, and as many as 60 molts have been recorded. About a dozen molts occur before sexual maturity is reached. Because the lining of the spermatheca is shed at each molt, females must mate between molts in order to produce fertilized eggs. The majority of species reproduce sexually, males producing stalked spermatophores that are deposited on the substrate for females to pick up as in Microcoryphia. However, parthenogenesis probably occurs in a few species where males are very scarce.

Phylogeny and Classification

The earliest known fossil zygentoman is *Ramsdelepidion schusteri* from the Upper Carboniferous of Illinois. This was a large (6 cm long) insect with such primitive features as a full set of long, abdominal, fully segmented leglets, and two pairs of vesicles on the ventral side of abdominal segments 1–7. Extant species fall into four families (LEPISMATIDAE, NICOLETIIDAE, LEPIDOTRICHIDAE, and MAINDRONIIDAE), of which only the first two are of any size. Though most Lepismatidae live in litter, under bark, etc., the family includes a number of domiciliary species (found in buildings) that have been transported worldwide by human activity, including the common silverfish, *Lepismodes inquilinus,* and the firebrat, *Thermobia domestica.* The former prefers warm and humid environments and is often found in places such as bookcases, cupboards, and bathrooms. Firebrats, in contrast, live in hot, dry environments, for example, in the vicinity of fireplaces, furnaces and boilers, and in bakeries. They are highly resistant to desiccation. Both species may cause considerable damage to books, clothing, and foods that contain starch or cellulose, and they are among the few animals that produce an intrinsic gut cellulase. The Nicoletiidae, which are distinguished from Lepismatidae by not having compound eyes, live principally in caves or underground though some are inquilines in the nests of ants and termites, stealing the hosts' food or causing the host to regurgitate fluids on which they feed.

The Lepidotrichidae should be mentioned in view of this family's important phylogenetic position. Prior to 1961 the family was known only as fossils. However, in 1961 Wygodzinsky discovered a living representative, *Tricholepidion gertschi,* in California, that possesses a large number of primitive characters. This species is clearly the most archaic living apterygote insect discovered to date and is likely to be similar to the common ancestor of other modern groups.

Literature

Accounts of the biology of Microcoryphia and Zygentoma are provided by Delaney (1957) and Sharov (1966). Keys for identification are given in the papers by Delany (1954) [British species], Wygodzinsky (1972) [North and Central American Lepismatidae], and Wygodzinsky and Schmidt (1980) [Microcoryphia of eastern North America].

Delany, M. J., 1954, Thysanura and Diplura, *R. Entomol. Soc. Handb. Ident. Br. Insects* **1**(2):1–7.

Delany, M. J., 1957, Life histories in the Thysanura, *Acta Zool. Cracov.* **2**:61–90.

Sharov, A. G., 1966, *Basic Arthropodan Stock*, Pergamon Press, Elmsford, NY.

Wygodzinsky, P., 1961, On a surviving representative of the Lepidotrichidae (Thysanura), *Ann. Entomol. Soc. Am.* **54**:621–627.

Wygodzinsky, P., 1972, A review of the silverfish (Lepismatidae, Thysanura) of the United States and the Caribbean Area, *Am. Mus. Novit.* **2481**:26.

Wygodzinsky, P., and Schmidt, K., 1980, Survey of the Microcoryphia (Insecta) of the northeastern United States and adjacent provinces of Canada, *Am. Mus. Novit.* **2701**:17.

6

Paleoptera

1. Introduction

In the infraclass Paleoptera are the orders Ephemeroptera (mayflies) and Odonata (drag-onflies and damselflies), the living species of which represent the few remains of two formerly very extensive groups. Although both are placed in the Paleoptera, authorities disagree on whether the two orders are monophyletic or have separate origins (see Chap-ter 2, Section 3.2). Even if monophyletic, the Ephemeroptera and Odonata are two very different groups that must have diverged at a very early stage in the evolution of winged insects. They possess the following common features that unite them as Paleoptera: wings that cannot be folded back against the body when not in use, retention of the anterior median wing vein, netlike arrangement of wing veins (many crossveins), aquatic juve-nile stage, and considerable change from juvenile to adult form. In members of both orders, wing development is external, though this feature is not, of course, restricted to Paleoptera.

2. Ephemeroptera

SYNONYMS: Plectoptera, Ephemerida   COMMON NAMES: mayflies, shadflies

Adults small- to medium-sized elongate fragile insects; antennae short and setaceous, mouthparts vestigial, compound eyes large, three ocelli present; generally two pairs of membranous wings (though hind pair greatly reduced) held vertically over body when at rest, with many crossveins; abdomen terminated with two very long cerci and frequently a median caudal filament; with subimaginal and imaginal winged stages.

Larvae aquatic; body campodeiform; antennae short, compound eyes well-developed, biting mouthparts; abdomen usually with long cerci and a median caudal filament, and four to seven pairs of segmental tracheal gills.

Approximately 2100 species of this widely distributed order have been described, though this may represent only about one-third of the extant species. Of the de-scribed species, about 675 occur in North America, 84 in Australia, and about 50 in Britain.

Adult. The head is triangular in shape when viewed from above. The compound eyes are large, especially in males where they often meet middorsally and typically are divided horizontally into an upper region with large facets and a lower region with smaller facets (Figure 6.1). This arrangement provides a male with both high acuity and good sensitivity, allowing him to detect and capture an individual female in a swarm at low light intensity. Three ocelli are present, the two laterals often large. The antennae are small, multiannulate, setaceous structures. The mouthparts are vestigial. The thoracic region is dominated by the large mesothoracic segment. Pleural sulci are poorly developed or absent even on the pterothorax. Two pairs of fragile wings are generally present, though the hind pair is always reduced or absent. The wing venation is primitive, the median vein being divided into anterior and posterior branches. The legs are sometimes reduced, associated with the habit of passing the entire adult life on the wing. However, the forelegs of males are usually enlarged and used to grip a female during mating. Primitively there are five tarsal segments, but the basal one or two segments may fuse with the tibia in higher families. The apex of the abdomen has three, usually very long, multiannulate caudal filaments, consisting of the two lateral cerci and a median filament (this is sometimes reduced or absent). In females paired gonopores open behind the seventh abdominal sternum. A typical ovipositor is absent. In males a pair of claspers occurs on the ninth sternum. Between these claspers lies a pair of penes.

The most noteworthy internal feature is the modification of the gut as an aerostatic organ to reduce the specific gravity of the insect. The esophagus is a narrow tube equipped with muscles that regulate the amount of air in the gut. Swallowed air is held in the midgut, which no longer has a digestive function and is lined with pavement rather than columnar epithelium. The hindgut also has a valve to prevent loss of air. The reproductive organs are very primitive; accessory glands are absent, and the gonoducts are paired in both sexes.

Larva. Mayfly larvae exhibit a wide range of body form associated with the diverse habitats in which they are found. The body is of varied shape but is often flattened dorsoventrally. The antennae, compound eyes, and ocelli differ little from those of adults. Larvae possess well-developed biting mouthparts. The structure of the legs varies according to whether a larva is a swimming, burrowing, or clinging form. The abdomen is terminated with a pair of long cerci and usually a median caudal filament. Between four and seven pairs of tracheal gills occur on the abdomen. In open-water forms the gills are usually lamellate; in burrowing species they tend to be plumose. In some species gills may not be directly important in gaseous exchange. They are capable of coordinated flapping movements and may serve simply to create a current of water flowing over the body. In some species accessory gill-like respiratory structures develop on the thorax and head.

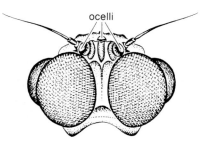

FIGURE 6.1. Dorsal view of head of male *Atalophlebia* (Leptophlebiidae) showing large compound eye divided into upper part with large facets and lower part with small facets. [From W. L. Peters and I. C. Campbell, 1991, Ephemeroptera, in: *The Insects of Australia*, 2nd ed., Vol. I (CSIRO, ed.), Melbourne University Press. By permission of the Division of Entomology, CSIRO.]

Adult mayflies are commonly found in the vicinity of water, often in huge mating swarms. They are short-lived creatures, existing for only a few hours (mostly nocturnal species) or a few days. A swarm consists generally only of males, often in the thousands, flying in an up-and-down pattern over water or a specific marker such as a rock, bush, or shoreline. Swarming commonly occurs at dusk in temperate species, light intensity and temperature being the major determinants of when it occurs. Females enter a swarm, and mating usually occurs immediately and lasts for less than a minute. Parthenogenesis has been reported for about 50 species, though it is rarely obligate. The egg-laying habits are quite varied, as is the number of eggs laid (generally from 500 to 3000). In some short-lived species eggs are laid *en masse* on the water surface. The clutch breaks up and the eggs sink, becoming scattered over the substrate. In species that survive for several days the eggs may be laid in small batches; *Baetis* spp. females descend below the water surface to secure the eggs on the substrate. Eggs often have special structures that serve to anchor them in position. They usually hatch within 10–20 days, but in a few species the eggs enter a diapause to overcome low winter temperatures. Consequently, they do not hatch until the following spring. In some species that have a relatively long adult life (up to 3 weeks) ovoviviparity occurs, females retaining fertilized eggs in the genital tract for several days prior to oviposition. Embryos then hatch from the eggs within a few minutes of deposition.

In most mayfly species the larval life span is 2–4 months; however, some mayfly larvae are long-lived, with a development time of at least a year and, in some instances, of 2 or 3 years. During this period they molt many times (15–30 is most common, but as many as 50 have been recorded). Larvae occupy a wide range of habitats, though each one is characteristic for a particular species. They may burrow into the substrate, hide beneath stones and logs, clamber about among water plants, or cling to the upper surface of rocks and stones in fast-flowing streams. With the exception of a few carnivorous forms, larvae feed on algae, or plant detritus, and thus play a key role in energy flow and nutrient recycling in freshwater ecosystems. Populations of larval mayflies show characteristic movements at specific times during their life. These may be diurnal, seasonal, and/or directional. For example, species in running water may have daily migrations into and out of the substrate, or they may move into the substrate during periods of heavy water flow. Typically, in both still and moving waters larvae move toward the shore during the later stages of their existence. And some species, especially of *Baetis*, have characteristic nocturnal rhythms of downstream drift. For other species, drift is influenced by both larval characters (e.g., age and population density) and environmental factors such as temperature, oxygen, current velocity, sediment, and food. How species compensate for the potential decrease in population upstream is not well understood, though for some upstream movement of larvae has been demonstrated, while for others the imagos undertake upstream flights before oviposition.

Mayflies are unique among living Pterygota in that they molt in the adult stage. A mature larva, on leaving its aquatic environment, molts into a subimago, a winged adult form (but see Chapter 2, Section 3.2), often capable of flight. A subimago can be distinguished from the imago into which it molts by its duller coloration and by the translucent wings, which are often fringed with hairs. A subimago exists usually for about 24 hours before molting to the imago. Under adverse conditions, however, a subimago may survive for many days. In a few exceptional species the subimago never molts but is the reproductive stage. It has been speculated that the adult molt may be a primitive trait retained because of a lack of selection pressure on the short-lived stages to have a single adult instar. An alternative suggestion is that an adult molt became necessary to complete elongation of the caudal filaments and adult forelegs (Maiorana, 1979). In populations of subtropical and tropical mayflies emergence tends to occur over a considerable length of time, whereas in species from cooler climates it is often highly synchronized, leading to the production of enormous swarms for short periods of the year. On a day-to-day basis, however, emergence may show distinct rhythmicity or require environmental cues such as a minimum water temperature or full moon for its initiation.

Phylogeny and Classification

Although the basic groups within the Ephemeroptera have been recognized since the work of Eaton (1883–1888, cited in Edmunds, 1962), differences of opinion continue to exist with regard to the taxonomic rank that should be assigned these groups, and the relationships among the groups. The primary obstacle to determining these relationships is the high degree of parallel evolution that has occurred among members of different groups. In many insect groups this problem can be overcome usually by comparing a number of different characters from all stages of the life history (Edmunds, 1972). Unfortunately, many mayflies are known only from the juvenile or the adult form.

The scheme used here is based on McCafferty (1991), Wang and McCafferty (1995), Bae and McCafferty (1995), and McCafferty (personal communication). It proposes that the order be arranged in 23 families shared among four suborders. A proposed phylogeny of these groups is depicted in Figure 6.2. According to McCafferty and Edmunds (1979), the ancestor of modern mayflies may have resembled members of the extant family Siphlonuridae

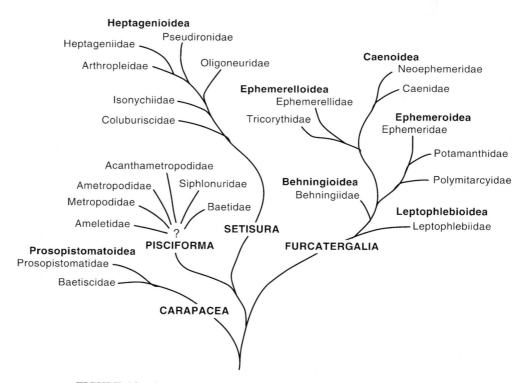

FIGURE 6.2. Proposed phylogenetic relationships within the Ephemeroptera.

that show a large number of primitive features; that is, they have evolved relatively little compared to other mayfly groups. From this siphlonuridlike ancestor, two major lines evolved. One led to the suborder Carapacea, the other to the suborders Furcatergalia, Setisura (which are sister groups) and Pisciforma. Though relationships of the families in the first three suborders are reasonably clear, those for the Pisciforma remain to be established.

Suborder Carapacea

Members of the Carapacea (an allusion to the carapacelike enlargement of the larval mesonotum) are included in a single superfamily Prosopistomatoidea.

Superfamily Prosopistomatoidea

The two small families in this group, the BAETISCIDAE [12 North American species of *Baetisca* (Figure 6.3)] and PROSOPISTOMATIDAE (11 species of *Prosopistoma* with a wide distribution including Africa, Australia, Europe, and southern Asia), show considerable parallel evolution in the larval stage. Indeed, their larvae are remarkable in having an enormous, posteriorly projecting mesonotal shield that protects the gills so that they superficially resemble notostracan crustacea, into which group *Prosopistoma* was originally placed by the French zoologist Latreille in 1833 (Berner and Pescador, 1980). Larvae of most species live in moving water, from streams to large rivers, where the bottom has sand, fine gravel, or small stones. Adult baetiscids, which are medium-sized insects, have an unusually large mesothorax; the eyes of males are large and almost contiguous but not divided horizontally. Prosopistomatid adults of both sexes have small, widely separated eyes; males have relatively short forelegs; the legs of females are vestigial; and females do not have an adult molt.

Suborder Pisciforma

McCafferty (1991) introduced the suborder Pisciforma (the name refers to the minnowlike body and actions of the larvae) for a group of families whose relationships remain unclear. For this reason, no arrangement into superfamilies is undertaken, though in earlier

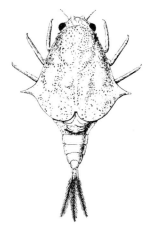

FIGURE 6.3. Larva of *Baetisca bajkovi* (Baetiscidae). [From B. D. Burks, 1953, The mayflies, or Ephemeroptera, of Illinois, *Bull. Ill. Nat. Hist. Surv.* **26**(1). By permission of the Illinois Natural History Survey.]

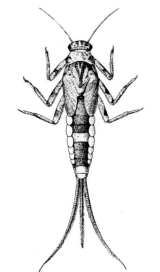

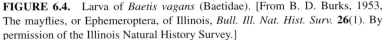

FIGURE 6.4. Larva of *Baetis vagans* (Baetidae). [From B. D. Burks, 1953, The mayflies, or Ephemeroptera, of Illinois, *Bull. Ill. Nat. Hist. Surv.* **26**(1). By permission of the Illinois Natural History Survey.]

schemes the families were lumped in a single superfamily Baetoidea. Only the major families are outlined here.

The SIPHLONURIDAE is a fairly large, probably paraphyletic, family containing about 160 described species with a worldwide distribution but especially diverse in the holarctic region. The streamlined, active larvae are found on the bottom of fast-flowing streams or among vegetation in still-water habitats. Some are predaceous. Adults are medium- to large-sized mayflies, and the sexes are similar in coloration. In both sexes the compound eyes are large and have a transverse band dividing the upper and lower regions. In males the eyes are usually contiguous.

The BAETIDAE (Figure 6.4) is easily the largest family of Ephemeroptera (>500 species) and has a worldwide distribution. The torpedo-shaped larvae are found in a variety of habitats but commonly on the bottom of fast-flowing streams where they may be well camouflaged. Adults are generally small and sexually dimorphic. The hind wings are greatly reduced or absent. The compound eyes of males are large and divided horizontally into distinct parts; in females the eyes are small and simple.

Suborder Setisura

Included in this suborder are the families listed under the superfamily Heptagenioidea in older classifications. The major family is the HEPTAGENIIDAE (Figure 6.5) which ranks next to the Baetidae in terms of number of described species (380). Heptageniids are an almost entirely holarctic and oriental group and are not represented in the Australasian region. The generally darkly colored larvae are typically found clinging to the underside (occasionally the exposed face) of stones in fast-flowing streams and on wave-washed shores of large lakes. They are remarkably well adapted for this life. Their body is extremely flattened dorsoventrally; the femora are broad and flat; the tarsal claws have denticles on the lower side; the gills are strengthened on their anterior margin; in some species the entire body takes on the shape (and function) of a sucking disc. Some larvae have fore tarsi with numerous setae that filter algae, etc. from the water and give the

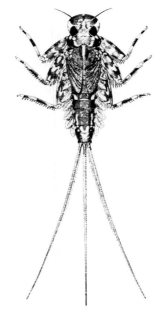

FIGURE 6.5. Larva of *Heptagenia flavescens* (Heptageniidae). [From B. D. Burks, 1953, The mayflies, or Ephemeroptera, of Illinois, *Bull. Ill. Nat. Hist. Surv.* **26**(1). By permission of the Illinois Natural History Survey.]

suborder its name. Adults are of varied size and color. The eyes of males are large but not contiguous.

Suborder Furcatergalia

The Furcatergalia is the largest mayfly suborder. It name derives from the forked nature of the larval gills. The group includes five superfamilies: Leptophlebioidea, Behningioidea, Ephemeroidea (burrowing mayflies), Ephemerelloidea, and Caenoidea. The last two superfamilies collectively form the pannote mayflies, so-called because of the fused fore wing pads of the larvae.

Superfamily Leptophlebioidea

The LEPTOPHLEBIIDAE (about 380 described species, representing perhaps only about 10% of the total) is another large and probably paraphyletic group of worldwide distribution but especially common in the Southern Hemisphere. A good deal of parallel evolution of habits and morphology appears to have taken place between the Leptophlebiidae in the Australasian region and the Baetidae and Heptageniidae in the holarctic region. Thus, many leptophlebiid species are found as larvae in still or slow-moving water, and, in some instances, the adults closely resemble baetids. Larvae of other species are found clinging to rocks in fast-flowing waters and resemble heptageniid larvae.

Superfamily Behningioidea

All members of this very small group (seven extant species) are included in the family BEHNINGIIDAE (tuskless burrowing mayflies). The family is holarctic, with representatives in eastern Europe, Siberia, and Thailand, plus one species in the eastern U.S.A. The larvae are predaceous and burrow in sand in rivers.

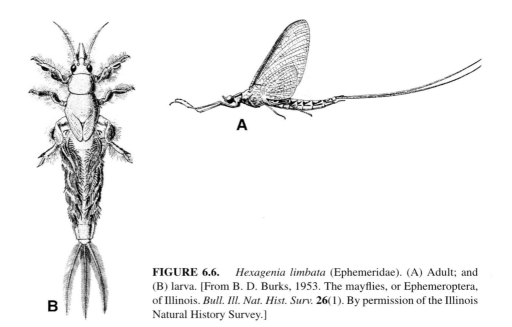

FIGURE 6.6. *Hexagenia limbata* (Ephemeridae). (A) Adult; and (B) larva. [From B. D. Burks, 1953. The mayflies, or Ephemeroptera, of Illinois. *Bull. Ill. Nat. Hist. Surv.* **26**(1). By permission of the Illinois Natural History Survey.]

Superfamily Ephemeroidea

Most species in this largely Northern Hemisphere group, the tusked mayflies, belong to the EPHEMERIDAE (about 100 species) (Figure 6.6) or the POLYMITARCYIDAE (about 70 species). Ephemerid larvae have the tibiae of the forelegs modified for burrowing in the mud or sand of large lakes, rivers, and streams. The mandibles (tusks) are long and used for lifting the roof of the burrow. Most of the body and the appendages are covered with fine hairs. These become coated with silt, and the insect is thereby well camouflaged. Adults are generally moderately sized to large insects. Their wings are hyaline, though they may be spotted in some species. In Polymitarcyidae the middle legs and hind legs of males, and all legs of females, are vestigial. Like ephemerids, polymitarcyid larvae have digging forelegs and tusks and are burrowers, usually in mud or fine sand, though some tunnel into clay on the banks of large rivers.

Superfamily Ephemerelloidea

With about 100 described species, the EPHEMERELLIDAE is widespread in the holarctic region, with genera also in South America, Asia, and southern Africa. Australia, by contrast, has but one species. Ephemerellid larvae are found in a wide variety of still- and moving-water habitats, especially cold, fast-flowing streams. Adults are small- to medium-sized mayflies. Members of the related family TRICORYTHIDAE (about 120 species), a predominantly Asian, African, and North American group, are generally similar in their habits to ephemerellids.

Superfamily Caenoidea

The CAENIDAE (Figure 6.7), with some 80 described species, is a widely distributed family of generally small mayflies. The hairy larvae sprawl on the surface of fine sediments in still or slow-moving water. The second pair of gills is enlarged and strengthened, forming a

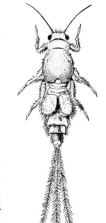

FIGURE 6.7. Larva of *Caenis simulans* (Caenidae). [From B. D. Burks, 1953. The mayflies, or Ephemeroptera, of Illinois, *Bull. Ill. Nat. Hist. Surv.* **26**(1). By permission of the Illinois Natural History Survey.]

plate that overlaps and protects the remaining four pairs of gills. The plate is alternately raised and lowered to effect water circulation. Male and female adults appear almost identical. The compound eyes are not especially large, but the lateral ocelli are about half the size of the compound eyes. The hind wings are absent.

Literature

General accounts of the structure and biology of mayflies are provided by Needham *et al.* (1935), Edmunds *et al.* (1976), Brittain (1982), Edmunds (1984), Harker (1989), and Peters and Campbell (1991). More specialized treatments, especially of life histories, are given by Clifford (1982) and in the volumes edited by Flannagan and Marshall (1980) and Campbell (1990). For an appreciation of the continuing controversy regarding the phylogeny and classification of Ephemeroptera, see McCafferty and Edmunds (1979), Landa and Soldan (1985), McCafferty (1991), Bae and McCafferty (1995), and Kluge (1998). Edmunds *et al.* (1976) and Edmunds (1984) provide keys for the North American genera, Macan (1970), Kimmins (1972), and Harker (1989) for the British species, and Peters and Campbell (1991) for the Australian families.

Bae, Y. J., and McCafferty, W. P., 1995, Ephemeroptera tusks and their evolution, in: *Current Directions in research on Ephemeroptera* (L. D. Corkum and J. J. H. Ciborowski, eds.), Canadian Scholars' Press, Toronto.

Berner, L., and Pescador, M. L., 1980, The mayfly family Baetiscidae (Ephemeroptera). Part I, in: *Advances in Ephemeroptera Biology* (J. F. Flannagan and K. E. Marshall, eds.), Plenum Press, New York.

Brittain, J. E., 1982, Biology of mayflies, *Annu. Rev. Entomol.* **27**:119–147.

Campbell, I. C., (ed.), 1990, *Mayflies and Stoneflies: Life Histories and Biology*, Kluwer, Dordrecht.

Clifford, H. F., 1982, Life cycles of mayflies (Ephemeroptera), with special reference to voltinism, *Quaest. Entomol.* **18**:15–90.

Edmunds, G. F., Jr., 1962, The principles applied in determining the hierarchic level of the higher categories of Ephemeroptera, *Syst. Zool.* **11**:22–31.

Edmunds, G. F., Jr., 1972, Biogeography and evolution of Ephemeroptera, *Annu. Rev. Entomol.* **17**:21–42.

Edmunds, G. F., Jr., 1984, Ephemeroptera, in: *An Introduction to the Aquatic Insects of North America*, 2nd ed. (R. W. Merritt and K. W., Cummins, eds.), Kendall/Hunt, Dubuque, IA.

Edmunds, G. F., Jr., Jensen, S. L., and Berner, L., 1976, *The Mayflies of North and Central America*, University of Minnesota Press, Minneapolis.

Flannagan, J. F., and Marshall, K. E., (eds.), 1980, *Advances in Ephemeroptera Biology*, Plenum Press, New York.

Harker, J., 1989, *Mayflies*, Richmond Publishing Co., Slough, U.K.

Kimmins, D. E., 1972, A revised key to the adults of the British species of Ephemeroptera with notes on their ecology (second revised edition), *Sci. Publ. F.W. Biol. Assoc.* **15**:75 pp.

Kluge, N. J., 1998, Phylogeny and higher classification of Ephemeroptera, *Zoosystematica* **7**:255–269.

Landa, V., and Soldan, T., 1985, Phylogeny and higher classification of the order Ephemeroptera: A discussion from the comparative anatomical point of view, *Studie Csl. Acad. Ved.* **4**:1–121.

Macan, T. T., 1970, A key to the nymphs of British species of Ephemeroptera (2nd revised ed.), *Sci. Publ. F.W. Biol. Assoc.* **20**:63 pp.

Maiorana, V. C., 1979, Why do adult insects not moult? *Biol. J. Linn. Soc.* **11**:253–258.

McCafferty, W. P., 1991, Toward a phylogenetic classification of the Ephemeroptera (Insecta): A commentary on systematics, *Ann. Entomol. Soc. Am.* **84**:343–360.

McCafferty, W. P., and Edmunds, G. F., Jr., 1979, The higher classification of the Ephemeroptera and its evolutionary basis, *Ann. Entomol. Soc. Am.* **72**:5–12.

Needham. J. G., Traver, J. R., and Hsu, Y.-C., 1935, *The Biology of Mayflies*, Comstock, New York.

Peters, W. L., and Campbell, I. C., 1991, Ephemeroptera, in: *The Insects of Australia*, 2nd ed., Vol. I (CSIRO, ed.), Melbourne University Press, Carlton, Victoria.

Wang, T. W., and McCafferty, W. P., 1995, Relationships of the Arthropleidae, Heptageniidae, and Pseudironidae (Ephemeroptera: Heptagenioidea), *Entomol News* **106**:251–256.

3. Odonata

SYNONYM: Paraneuroptera   COMMON NAMES: dragonflies and damselflies

Adults medium-sized to large elongate insects, frequently strikingly marked; head with antennae short and setaceous, compound eyes prominent, biting mouthparts; thorax with two pairs of membranous wings of approximately equal size and with netlike venation, pterostigma usually present; abdomen of male with copulatory organs on second and third sterna.

Larvae aquatic; body campodeiform; head equipped with extensible "mask" (modified labium) for catching prey, antennae small, compound eyes large; abdomen terminated with three processes, either short and stocky or extended into large lamellate structures.

Almost 5000 species of Odonata have been identified from different areas of the world. About 10% of these are from North America. Some 45 species are found in the British fauna, and 300 species have been described from Australia.

Structure

Adult.   The body of adult Odonata is remarkable for its colors, both pigmentary and structural, that frequently form a characteristic pattern over the dorsal region (Figure 6.12A). Most adults range from 30 to 90 mm in length and are sturdy, actively flying insects. The head is freely articulated with the thorax, and a large part of its surface, especially in Anisoptera (dragonflies), is occupied by the well-developed compound eyes. Three ocelli form a triangle on the vertex. The antennae are short, hairlike structures that apparently carry few sense organs. The mouthparts are powerful structures of the biting and chewing type. The thorax is somewhat parallelogram-shaped, with the legs placed anteroventrally and the wings situated posterodorsally. The prothorax is distinct but small, and in female Zygoptera (damselflies) is sculptured so as to articulate with the claspers of the male during mating. The mesothorax and metathorax are large and fused together. The pleura of these segments are very large and possess prominent sulci. The legs are weak and unsuitable for walking. They serve to grasp the prey and hold it to the mouth during feeding. In Zygoptera the fore and hind wings are almost identical; in Anisoptera the hind wing is somewhat broader near the base. A prominent pterostigma (see Figure 3.27) is present on each wing in all except a few species. The wing venation is a primitive netlike arrangement. Ten abdominal segments are visible with segments 1–8 bearing spiracles. In male Odonata the sternum of segment 2 is grooved and that of segment 3 bears copulatory structures used in the transfer of sperm to (sometimes also from) the female genital tract. The true genital opening is located behind the ninth sternum. In female Zygoptera and many Anisoptera that are endophytic (lay their eggs into plant tissues) a well-developed ovipositor is present. In exophytic Anisoptera it is reduced or absent.

Most internal organs are greatly elongated because of the narrow body. The testes extend from abdominal segments 4–8 and the ovaries occupy the whole length of the abdomen. Between 50 and 70 Malpighian tubules, united in groups of 5 or 6, enter the alimentary canal at the junction of the midgut and hindgut. The respiratory system is well developed and in many species includes a large number of air sacs in the thoracic region. The nervous system is generally primitive, although the brain is enlarged transversely due to the presence of large optic lobes.

Larva. Odonate larvae are usually shorter and stockier than adults. In general the larval head resembles that of the adult, though it differs in possession of the "mask," the elongated labium (Figure 6.8), used to capture prey. At rest the mask (so-called because it often covers the other mouthparts) is folded at the junction of the postmentum and prementum and held between the bases of the legs. It is extended extremely rapidly (in about 16–25 msec) by means of localized blood pressure changes, assisted by the release of tension in the labial locking muscles, and the prey is grasped by the labial palps. In contrast to those of adults, the legs are normally positioned on the thorax, well developed and quite long. At the tip of the abdomen there are three appendages, one mediodorsal and two lateral (the cerci). These are small in Anisoptera but enlarged to form caudal lamellae in Zygoptera.

Internally larvae differ from adults in several features. In the foregut there is a well-developed gizzard for breaking up food. There are initially only a few Malpighian tubules, though the number increases in each instar. In all odonate larvae, some gaseous exchange takes place directly across the body wall, including the wing pads, and via the wall of the rectum. In addition, larvae have special respiratory structures. In Anisoptera the wall of the rectum is greatly folded and well supplied with tracheae, forming "rectal gills." Water

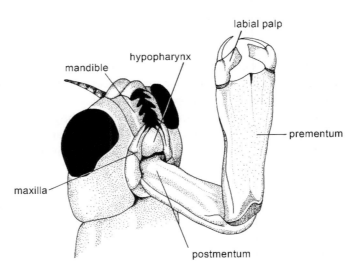

FIGURE 6.8. Lateroventral view of head of dragonfly larva showing mask. [After A. D. Imms, 1957, *A General Textbook of Entomology*, 9th ed. (revised by O. W. Richards and R. G. Davies), Methuen and Co.]

is continually pumped in and out of the rectum. Interestingly, the musculature used to ventilate the rectal gills serves also to jet propel the larva in swimming and to extend the labium for prey capture! In Zygoptera the caudal lamellae appear to supplement the surface area available for gaseous exchange (although in highly oxygenated water larvae appear to survive perfectly well when the lamellae are removed). In a few Zygoptera paired gills occur on most abdominal segments, while in Amphipterygidae filamentous perianal gills develop. (See also Chapter 15, Section 4.1.)

Life History and Habits

After emergence, immature adult Odonata spend some time away from water, usually among trees or tall grass where they hunt for prey and become sexually mature. It is during this maturation phase that some Odonata migrate over long distances. The maturation period, which lasts anywhere from a few days in smaller species to a month in large dragonflies, is followed by the reproductive phase in which mating and oviposition occur. Usually these two processes occur at the same site, although this is not necessarily so.

Mature adult Odonata generally can be classified as "perchers" or "fliers," the former spending most of their time perched and making only short flights, while the latter, when active, fly continuously. Adults feed throughout their life, Zygoptera often catching stationary prey whereas Anisoptera mostly capture their prey in flight, occasionally aggregating in large numbers where prey is concentrated. Males of many species are territorial, that is, they occupy and defend an area against other males. Perchers have a base from which they undertake patrol flights or sallies against intruders while fliers patrol the area continuously for extended periods. Other species show little or no spatial territoriality, though may behave aggressively against conspecifics they encounter, a feature that ensures the spacing out of males within a habitat. Some fliers are territorial for several short (10–40 minutes) periods throughout the day, between which they leave the area, allowing other conspecific males to occupy the space. Should a receptive female conspecific enter the territory and be recognized by the male (probably using visual cues), he will attempt to mate with her. Using his legs the male grasps the female on the pterothorax, then curls his abdomen around so as to be able to grip the female's prothorax (Zygoptera) or head (Anisoptera) with his claspers (the tandem position). As noted above, the compatibility of the male and female structures are key determinants of the conspecificity of the partners. The male's legs then release their grip, and the female bends her abdomen forward until its tip contacts the accessory genitalia on the male's second and third abdominal segments (the wheel position) (Figure 6.9). Earlier, it was assumed that the sole purpose to this was the transfer of sperm from male to female. However, in probably all species that settle after taking up the wheel position, the major portion of the time spent in copulation (which may last up to 30 minutes) is taken up by the male's penis removing much (40–100% depending on the species) of the sperm of a previous mating from the female's bursa copulatrix and spermatheca, before insemination occurs. Sperm displacement also occurs in the libellulid *Erythemis (Lepthemis) simplicollis* where copulation lasts less than 20 seconds. Whether the phenomenon is widespread among species that copulate for such brief periods (usually in flight) or whether sperm packing occurs (sperm from earlier matings is pushed more deeply into the spermatheca while that of the most recent mating remains adjacent to the spermathecal opening so that during oviposition it is used preferentially) requires further study. Certainly, however, the structures of both the penis and the female storage organs in some anisopteran families are very different from those of Zygoptera which have been

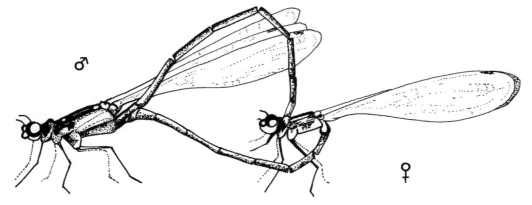

FIGURE 6.9. Copulating damselflies. [After E. M. Walker, 1953. *The Odonata of Canada and Alaska*, Vol. I. By permission of the University of Toronto Press.]

mostly studied, suggesting that sperm packing may be more common in Anisoptera (Waage, 1984).

Oviposition usually occurs soon after copulation, and in many Odonata the male remains close to (hovering or perching nearby) or in tandem with the female, the intensity of the association being correlated with the probability of other males disturbing the female. Territorial males usually adopt the first of these strategies, whereas non-territorial forms typically remain in tandem. Species that oviposit in leaves and stems of plants or woody material are typically quite selective in their choice of sites (a feature that may be correlated with their life history strategy) (see also Chapter 23, Section 3.2.1) and have elongate eggs. Often, the female climbs a considerable distance below the water surface before laying the eggs. Exophytic species that oviposit in ponds or swamps may simply release their eggs into the water, or stick them on or under a leaf or on mud. However, the eggs of stream- or river-dwelling species may be deposited above the water level or have hooks that catch on submerged objects.

Embryonic development is usually direct, the eggs hatching within 5–40 days; however, in some temperate species the eggs serve as the overwintering stage and undergo diapause. The first larval instar, known as the prolarva, does not feed and its sole purpose is to reach a suitable body of water. Immediately this is achieved (which may take from less than a minute up to several hours), the prolarva molts. Second-instar and older larvae are facultative predators, feeding on whatever animals of appropriate size are available. Detection of prey is achieved primitively through the use of both visual and contact sense organs. In advanced species the eyes become of primary importance. Odonate larvae are themselves preyed on by aquatic vertebrates and other aquatic insects, including larger members of the same order, though intraspecific cannibalism is very rare as a result of territorial behavior among similarly sized larvae and because age cohorts within a species tend to be spatially separated (e.g., by preferring perches of different diameters). Larvae of most Odonata inhabit permanent waters either still or flowing. Those of many species live in burrows in the substrate, whereas most others, especially Zygoptera, are generally found perched on detritus or aquatic plants where their color provides camouflage. When detected, these larvae can escape rapidly by either expelling water from the rectal cavity—a form of jet propulsion (Anisoptera)—or using rapid undulating movements of the abdomen and caudal lamellae (Zygoptera). A relatively few species have colonized temporary bodies of water through the use of such strategies as very rapid growth or burrowing into the substrate to avoid desiccation. Very rarely species have semiaquatic larvae or larvae that live in moist litter in rain forests. The majority of Odonata are warm-temperate or tropical species in which larval development is rapid so that one to several generations may be completed each year, with temperature and availability of food being the major determinants of larval growth rate. In higher latitudes species may be univoltine (one generation per year), semivoltine (taking 2 years to complete development), or take up to 6 years to develop (the duration of development can vary within a species over its range). Typically, these Odonata overwinter in a temperature and/or photoperiodically controlled diapause in the larval stage, though some species pass the winter as eggs. Diapause, together with instar-specific temperature thresholds for development, ensures that adult emergence the following spring is highly synchronized, thus improving the chances of successful reproduction. Larvae pass through 10–15 instars. Late in the final stadium, the larval gills cease to function and the pharate adult crawls to the water surface to breathe air through the mesothoracic spiracles. Immediately prior to ecdysis, the pharate adult climbs out of the water on a suitable support; it then swallows air to split the exuvium and expand the abdomen and wings. In warmer regions many larger Anisoptera emerge at night, perhaps to avoid predation; elsewhere, emergence occurs through the day and is dependent on a threshold temperature being achieved.

Phylogeny and Classification

In contrast to that of the other paleopteran order, the Ephemeroptera, the fossil record of the Odonata is remarkably extensive. Carpenter (1992) suggested that the aquatic juvenile stage and the tendency of adults to remain near water would favor the fossilization of these generally robust insects. According to Carpenter (1992), the earliest odonates (from the Permian) belonged to the entirely fossil suborders Protanisoptera and Archizygoptera. Other suborders include the Triadophlebiomorpha (Triassic), Anisozygoptera (Triassic—Cretaceous), Anisoptera (Jurassic—Recent) and Zygoptera (Jurassic—Recent). The Odonata underwent a rapid evolution during the Triassic period, and in the Jurassic Period, Anisozygoptera were especially abundant. Originally, two extant species of *Epiophlebia*, described from Japan and the Himalayas, were placed in this suborder. However, with more information it has become clear that *Epiophlebia* is an early anisopteran offshoot, in the superfamily Epiophlebioidea. By the Late Jurassic, representatives of recent families of Zygoptera and Anisoptera were already in existence.

Some early authorities suggested that ancient zygopterans were the group from which the remaining Odonata evolved. In this proposal, an early dichotomy led, on the one hand, to the modern groups of Zygoptera, and, on the other, through the Anisozygoptera to the Anisoptera. However, the abundant fossil evidence for the Odonata shows that these three suborders arose contemporaneously; that is, no one suborder gave rise to the others. The higher-level relationships of Odonata remain controversial, with markedly different hypotheses being generated, depending on the type and quantity of information used in the analysis (compare Trueman [1996], Bechly *et al.* [1998], Misof [2002], and Rehn [2003]). Thus, at one extreme, the question of whether the Zygoptera are monophyletic continues to be debated, while at the other, the make-up of some families (i.e., whether they are monophyletic, paraphyletic, or polyphyletic) remains unresolved. A suggested proposal for the relationships of the extant groups of Odonata is shown in Figure 6.10. In this scheme the monophyletic Zygoptera fall into three superfamilies, Calopterygoidea being the sister group to the Lestinoidea (paraphyletic) + Coenagrionoidea. In the Anisoptera, the

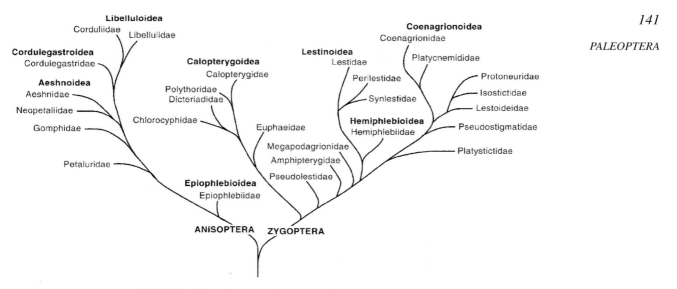

FIGURE 6.10. Proposed phylogeny of extant Odonata.

Epiophlebioidea form the sister group to all others. Of these, the very primitive Petaluridae form the sister group to the remaining true dragonflies.

Suborder Zygoptera (Damselflies)

Damselflies are characterized by the following structural features: fore and hind wings almost identical in shape and venation, quadrangular discoidal cell never longitudinally divided, eyes far apart, and larvae with three (rarely two) caudal lamellae.

Superfamily Coenagrionoidea

Most of the 1500 species of Coenagrionoidea fall into four families. The paraphyletic family COENAGRIONIDAE is the most successful zygopteran group, containing more than 1000 species. The family as a whole is cosmopolitan, and certain genera, for example, *Coenagrion* and *Ischnura* (Figure 6.11), are found throughout the world. Larvae are found among vegetation in still or slowly moving water. The generally small adults are weak fliers and rest with their narrow wings closely apposed over the body. The sexes are differently colored, with males usually much brighter. Commonly, the dorsal surface of males is a complex pattern of pale blue and black markings. Females are usually drab in color and in some species there may be two or more color forms. Pruinescence, the development of a waxy, whitish to pale blue secretion, is seen in older specimens of both sexes in some species. PLATYCNEMIDIDAE (150 species) are common in the palearctic, oriental, and tropical African regions where they breed in swamps, forests, streams, and fast-flowing water. The PROTONEURIDAE (220 species) are a widespread group, though absent from the palearctic region. They are most common in shaded localities, including forests, and breed in slowly moving water. Most of the 130 species of PLATYSTICTIDAE are oriental though some species occur in the New World tropics. Typically, they are found in forests, breeding in fast-flowing streams.

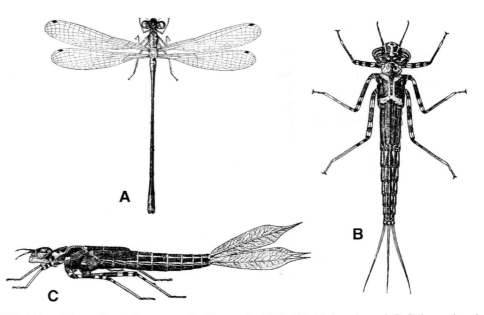

FIGURE 6.11. A damselfly, *Ischnura cervula* (Coenagrionidae). (A) Adult male; and (B,C) larva, dorsal and lateral views. [Reproduced by permission of the Smithsonian Institution Press from *Smithsonian Institution United States National Museum Proceedings*, Volume 49, 'Notes on the life history and ecology of the dragonflies (Odonata) of Washington and Oregon,' July 28, 1915, by C. H. Kennedy: Figures 77, 120, and 121. Washington, D.C., U.S. Government Printing Office, 1916.]

Superfamily Lestinoidea

About three quarters of the Lestinoidea are arranged in two families, the cosmopolitan LESTIDAE (140 species) and the primarily tropical MEGAPODAGRIONIDAE (200 species). Lestids are medium-sized, metallically colored insects that typically rest with their wings partially or completely outspread. They are found near still water or quiet streams. Eggs are laid in emergent vegetation and those of temperate species often show delayed development, an adaptation to overcome adverse climatic conditions such as drought or cold. Larvae are elongate, streamlined creatures, often well camouflaged. Megapodagrionids occur mainly in forests, breeding in streams, marshy places, and occasionally tree holes; however, some species breed in temporary swamps. Larvae are short and thick, with the caudal lamellae held horizontally, not vertically as in other Zygoptera.

Superfamily Calopterygoidea

Members of the cosmopolitan family CALOPTERYGIDAE (160 species) are medium to large, broad-winged damselflies characterized by the brilliant metallic coloring of their bodies, and, in males, the wings also. Larval Calopterygidae are found at the margins of fast-flowing water; they have relatively long and stout antennae, long spidery legs, and elongate caudal lamellae. CHLOROCYPHIDAE (120 species) are primarily restricted to tropical Africa and Asia, though there are old reports of their occurrence in northern Australia. In the larva the dorsal caudal lamella, sometimes all three lamellae, are spikelike.

Distinguishing features of dragonflies include fore and hind wings dissimilar in venation and, usually, shape; discoidal cell divided into two triangular areas; eyes contiguous or nearly so; and larvae stout and without caudal lamellae.

Superfamily Aeshnoidea

The south Australian and South American family NEOPETALIIDAE (nine species) contains the most primitive of recent dragonflies. The PETALURIDAE (11 species) is also an archaic family within which are found the world's largest extant species with wingspans of more than 16 cm. Their larvae are semiaquatic burrowers, living in swamps or beside steams. The AESHNIDAE form a large and cosmopolitan family of about 375 species of large, strongly flying insects characterized by their enormous eyes that meet broadly in the midline of the head. Larvae are mostly stout, elongate insects found among vegetation in a variety of still- or moving-water habitats; a few, however, are semi-terrestrial or terrestrial. The GOMPHIDAE (800 species) is another primitive family whose adults have widely separated eyes and are generally black and yellow, with one or the other color predominating according to the habitat in which they are found. They have a rudimentary ovipositor, and eggs are laid by simply dipping the tip of the abdomen into the water. Though some gomphids breed in still or sluggish waters, most breed in flowing water, and often their eggs have an adhesive exochorion or ropelike filaments that may prevent their being washed away. Gomphid larvae are burrowers or sprawlers in the substrate, and some burrowing species have a greatly elongated 10th abdominal segment in order to retain respiratory contact with the water and fossorial forelegs.

Superfamily Cordulegastroidea

This superfamily contains only one small family CORDULEGASTRIDAE (60 species), with a palearctic and oriental distribution. Its members carry a combination of aeshnoid and libelluloid characters. Some species are open-country forms that breed in small ponds or streams; others are associated with mountain streams.

Superfamily Libelluloidea

Both the CORDULIIDAE (Figure 6.12) and the LIBELLULIDAE are large, cosmopolitan families, though the former is a paraphyletic group. Corduliids (360 species) breed in a range of still- and moving-water habitats, including temporary pools and swamps, and the larvae of some species are able to withstand limited desiccation. A few species have terrestrial larvae. Libellulidae (900 species) principally breed in still-water habitats, though larvae of some species are stream dwellers. Larvae of most species are secretive, hiding among rotten vegetation at the bottom of the pond or lake; a few others have become secondarily adapted for a more active existence among growing vegetation. Adults vary greatly in size and coloration, the family including some of the most strikingly marked Anisoptera with pale wings bearing spots or bands of pigment, commonly dark but sometimes bright shades of orange or reddish brown.

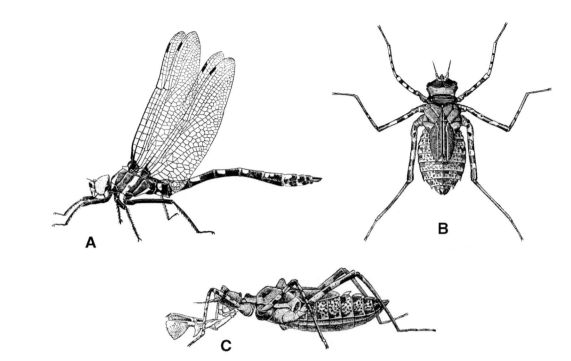

FIGURE 6.12. A dragonfly, *Macromia magnifica* (Corduliidae). (A) Adult male; (B) larva, dorsal view; and (C) larva, lateral view with labium extended. [Reproduced by permission of the Smithsonian Institution Press from *Smithsonian Institution United States National Museum Proceedings*, Volume 49, 'Notes on the life history and ecology of the dragonflies (Odonata) of Washington and Oregon,' July 28, 1915, by C. H. Kennedy: Figures 134, 146, and 147. Washington D.C., U.S. Government Printing Office, 1916.]

Literature

The Odonata have been one of the most popular insect groups for study, and the literature on them is abundant. General accounts of their biology are given by Tillyard (1917), Walker (1953), Corbet *et al.* (1960), Corbet (1962, 1980, 1999), Miller (1987), Watson and O'Farrell (1991), Needham *et al.* (2000), and Silsby (2001). The latter includes superb color photographs certain to stimulate interest in the order. The phylogeny and classification of the group, which remain controversial, are discussed by Trueman (1996), Bechly *et al.* (1998), Misof *et al.* (2001), Misof (2002), and Rehn (2003). Odonata may be identified through the keys of Walker (1953, 1958), Walker and Corbet (1975), and Westfall (1984) [North American forms]; Askew (1988) [European fauna]; Corbet *et al.* (1960) and Miller (1987) [British species]; and Watson *et al.* (1991) [Australian fauna].

Askew, R. R., 1988, *The Dragonflies of Europe*, Harley Books, Colchester, U.K.

Bechly, G., Nel, A., Martínez-Delclòs, X., and Fleck, G., 1998, Four new dragonflies from the Upper Jurassic of Germany and the Lower Cretaceous of Mongolia (Anisoptera: Hemeroscopidae, Sonidae, and Proterogomphidae fam. nov.), *Odonatologica* **27**:149–187.

Carpenter, F. M., 1992, Superclass Hexapoda, in: *Treatise on Invertebrate Paleontology*, Part R, Vol. 3 (R.L. Kaesler, ed.), Geological Society of America and University of Kansas, Boulder, CO and Lawrence, KA.

Corbet, P. S., 1962, A *Biology of Dragonflies*, Witherby, London. (Reprinted in 1983 by Classey, Faringdon, U.K.)

Corbet, P. S., 1980, Biology of Odonata, *Annu. Rev. Entomol.* **25**:189–217.

Corbet, P. S., 1999, *Dragonflies: Behavior and Ecology of Odonata*, Cornell University Press, Ithaca, NY.

Corbet, P. S., Longfield, C., and Moore, N. W., 1960, *Dragonflies*, Collins, London.

Miller, P. L., 1987, *Dragonflies*, Cambridge University Press, Cambridge.

Misof, B., 2002, Diversity of Anisoptera (Odonata): Infering speciation processes from patterns of morphological diversity, *Zoology* **105**:355–365.

Misof, B., Rickert, A. M., Buckley, T. R., Fleck, G., and Sauer, K. P., 2001, Phylogenetic signal and its decay in mitochondrial SSU and LSU rRNA gene fragments of Anisoptera, *Mol. Biol. Evol.* **18**:27–37.

Needham, J. G., Westfall, M. J., Jr., and May, M. L., 2000, *Dragonflies of North America*, rev. ed., Scientific Publishers, Gainesville, FL.

Rehn, A. C., 2003, Phylogenetic analysis of higher-level relationships of Odonata, *Syst. Entomol.* **28**:181–239.

Silsby, J., 2001, *Dragonflies of the World*, Smithsonian Institution Press, Washington, DC.

Tillyard, R. J., 1917, *The Biology of Dragonflies*, Cambridge University Press, London.

Trueman, J. W. H., 1996, A preliminary cladistic analysis of odonate wing venation, *Odonatologica* **25**:59–72.

Waage, J. K., 1984, Sperm competition and the evolution of odonate mating systems, in: *Sperm Competition and the Evolution of Animal Mating Systems* (R. L. Smith, ed.), Academic Press, New York.

Walker, E. M., 1953, *The Odonata of Canada and Alaska*, Vol. I: General and Zygoptera, University of Toronto Press, Toronto.

Walker, E. M., 1958, *The Odonata of Canada and Alaska*, Vol. 2: Anisoptera—4 Families, University of Toronto Press, Toronto.

Walker, E. M., and Corbet, P. S., 1975, *The Odonata of Canada and Alaska*, Vol. 3: Anisoptera—3 Families, University of Toronto Press, Toronto.

Watson, J. A. L., and O'Farrell, A. F., 1991, Odonata, in: *The Insects of Australia*, 2nd ed., Vol. I (CSIRO, ed.), Melbourne University Press, Carlton, Victoria.

Watson, J. A. L., Theischinger, G., and Abbey, H. M., 1991, *The Australian Dragonflies*, CSIRO, Melbourne.

Westfall, M. J., Jr., 1984, Odonata, in: *An Introduction to the Aquatic Insects of North America*, 2nd ed. (R. W. Merritt and K. W. Cummins, eds.), Kendall/Hunt, Dubuque, IA.

7

The Plecopteroid, Blattoid, and Orthopteroid Orders

1. Introduction

This chapter deals with the following 10 orders: Plecoptera, Embioptera, Dictyoptera, Isoptera, Grylloblattodea, Dermaptera, Phasmida, Orthoptera, Zoraptera, and the recently established Mantophasmatodea. Members of these orders can be distinguished from other exopterygotes [the hemipteroid orders (Chapter 8)] by the following features: generalized biting mouthparts, wing venation usually well developed with numerous crossveins (though less netlike than that of Paleoptera), cerci present, terminalia of male may be asymmetrical and reduced, many Malpighian tubules, and generalized nervous system with several discrete abdominal ganglia. However, as discussed in Chapter 2, the existence of these common features should not be taken as confirmation that these orders constitute a monophyletic group.

2. Plecoptera

SYNONYMS: Perlaria, Perlida   COMMON NAME: stoneflies

Moderate-sized to fairly large soft-bodied insects; head with long setaceous antennae, weak mandibulate mouthparts, well-developed compound eyes and two or three ocelli; thorax almost always with two pairs of membranous wings (sometimes reduced), hind pair in most species with a large anal lobe, venation frequently specialized, legs identical and with a three-segmented tarsus; abdomen of most species terminated by long multiannulate cerci, females lacking a true ovipositor, males without gonostyles and phallic organs on abdominal segment 9.

Larvae aquatic, generally resembling adults except for presence of a varied number of tracheal gills.

More than 2000 species of this very ancient order have been described, including just over 600 from North America, about 30 from Britain, and 200 from Australia. Though the order has representatives on all continents except Antarctica, most families have a rather restricted distribution.

147

Adult.   The plecopteran head is prognathous and bears a pair of elongate, multiannulate antennae, well-developed compound eyes, three (rarely two) ocelli, and weak, often non-functional biting-type mouthparts. Usually all the mouthparts are present, but in members of a few families the mandibles are vestigial. The thorax is primitive. Its segments are free and the prothorax is large. Two pairs of membranous wings are nearly always present, though brachypterous and apterous species occur at high altitudes and latitudes. The hind wing typically has a large anal fan, but this is reduced in the more advanced families. The wing venation is generally primitive, but considerable variation is seen within the order. In members of primitive families a typical archedictyon is developed to a greater or lesser degree; in those of advanced groups the number of branches of the longitudinal veins and the number of crossveins are greatly reduced. The abdomen contains 10 complete segments, with the 11th represented by the epiproct, paraprocts, and long cerci. In Nemouridae, however, the latter are reduced to an unsegmented structure used in copulation.

The esophagus is very long, the gizzard rudimentary, and midgut and hindgut short. There are between 20 and 100 Malpighian tubules. In primitive families the central nervous system includes three thoracic and eight abdominal ganglia, but in advanced groups the sixth to eighth abdominal ganglia fuse. The tracheal system opens to the exterior via two thoracic and eight abdominal spiracles. In males the testes meet in the midline, but their products are carried by separate vasa deferentia to a pair of seminal vesicles. Usually there is a median ejaculatory duct, but in some species the vasa deferentia remain separate until they reach the median gonopore located behind the ninth abdominal segment. In females the panoistic ovarioles arise from a common duct that joins the oviducts of each side. A spermatheca is usually present.

Larva.   In general form larvae resemble adults, except for the absence of wings and the presence, in most species, of several pairs of gills. Primitively there are five or six pairs of abdominal gills, but in members of more advanced groups these are reduced in number and secondary gill structures may appear on more anterior regions of the body (mentum, submentum, neck, thorax, and coxae) or may encircle the anus. In addition to gas exchange, the gills are important osmoregulators, equipped with chloride-uptake cells, as is also seen in larval Ephemeroptera. In many species the legs are fringed with hairs that assist swimming.

Life History and Habits

Adult stoneflies are weak flyers and seldom found far from the banks of streams or edges of lakes where they rest, often well camouflaged, on vegetation, rocks, logs, etc. Nocturnal species usually hide in crevices or among vegetation during the day. Many stoneflies do not feed as adults. Others feed on lichens, acellular algae, pollen, bark, and rotten wood.

Prior to mating, many Arctoperlaria tap the substrate with the tip of the abdomen (drumming). Males initiate the drumming and virgin females respond. The drumming is species-specific and serves to bring the partners together (Stewart and Maketon, 1990). Mating usually occurs in daylight, on the ground, though a few species are nocturnal. Large numbers of eggs are laid, singly or, more often, in batches of 100 or more. In flying species females hover over the water and dip the abdomen beneath the surface. Brachypterous and apterous forms crawl to the water's edge, or below the water surface, in order to oviposit. Eggs of many species develop adhesive properties on contact with water. Embryonic development is usually direct, though eggs of some species may survive drought conditions in

149

*THE
PLECOPTEROID,
BLATTOID, AND
ORTHOPTEROID
ORDERS*

diapause. A few species are ovoviviparous. Larvae are typically found in streams or lakes whose bottom is covered with stones under which they can hide. Development is slow, frequently taking more than a year in the larger species. Many molts occur, 33 having been recorded over a period of 3 years for one species. Most stonefly larvae are phytophagous, feeding on lichens, algae, moss, and diatoms. Typically these are the species that also feed in the adult stage. Juveniles of other species are carnivorous, living on other insects. These species do not feed as adults. Like that of Odonata and mayflies, emergence of stoneflies is frequently highly synchronized.

Phylogeny and Classification

Plecoptera, very primitive insects sometimes described as "flying Thysanura," probably had their origins in the Lower Permian period from a stem group, the plecopteroid assemblage, that included the extinct Paraplecoptera and Protoperlaria (Illies, 1965). Some paleoentomologists assigned some of the Permian fossil Plecoptera to recent families, though Zwick (1981) considered this incorrect, representatives of the latter not appearing in the fossil record until the Eocene (or possibly the Cretaceous).

Stoneflies traditionally were placed in two suborders, Filipalpia (Holognatha) and Setipalpia (Systellognatha). Illies (1965), however, considered the extremely primitive Southern Hemisphere families Eustheniidae and Diamphipnoidae sufficiently distinct from the remaining Filipalpia that they should be grouped in a separate suborder, the Archiperlaria. Both Hennig (1981) and Zwick (1981) argued that Illies' arrangement was not soundly based, and Zwick (1980, 1981, 2000), whose classification is followed here, proposed that the stoneflies could be divided into an exclusively Southern Hemisphere group (suborder Antarctoperlaria) and a predominantly Northern Hemisphere group (suborder Arctoperlaria), the separation and subsequent evolution of the two groups resulting from the breakup of the Pangean landmass (into Laurasia and Gondwanaland) during the Jurassic period. A few Arctoperlaria occur in the Southern Hemisphere, presumably as a result of secondary invasions. Figure 7.1 provides a suggested phylogeny for the order.

Suborder Antarctoperlaria

In Zwick's classification this suborder includes the superfamilies Eusthenioidea (families EUSTHENIIDAE and DIAMPHIPNOIDAE) and Gripopterygoidea (AUSTROPERLIDAE and GRIPOPTERYGIDAE). Illies (1965) considered members of the small family Eustheniidae, which is restricted to eastern Australia, New Zealand, and Chile, to represent the prototype of plecopteran organization. They are large, colorful insects having wings with numerous crossveins and an anal fan in the hind wing with eight or nine anal veins. Larvae are carnivorous and have four to six pairs of abdominal gills. The Gripopterygidae is a large family (about 150 species) mostly found in Australia, with a few species in New Zealand and South America. The adults are mostly dull in color; the larvae, which are sluggish and typically found under rocks and debris in fast-moving water, have a tuft of gills around the anus. Larvae of a few species are terrestrial and lack gills (Zwick, 2000).

Suborder Arctoperlaria

Zwick (2000) divided this suborder into the infraorders Systellognatha and Euholognatha. The former contains the superfamilies Perloidea (families PERLODIDAE,

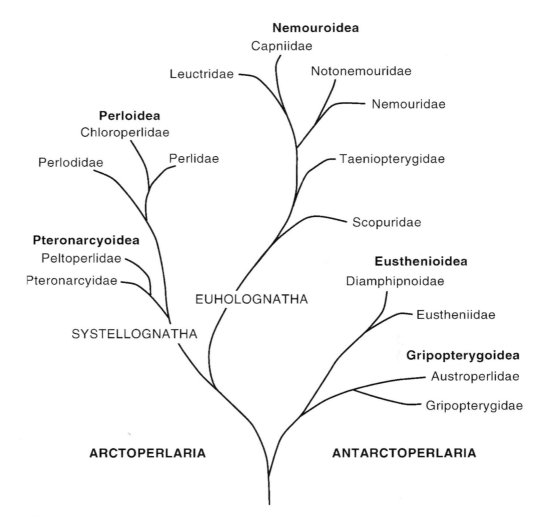

FIGURE 7.1. Proposed phylogeny of the Plecoptera. [Modified from P. Zwick, 1980, Plecoptera (Sternfliegen), in: *Handbuch der Zoologie*, Vol. IV, Insecta Lfg. **26**:1–115. By permission of Walter de Gruyter and Co.]

PERLIDAE, and CHLOROPERLIDAE) and Pteronarcyoidea (families PTERONARCYI-DAE and PELTOPERLIDAE). Included in the Euholognatha is a single superfamily Nemouroidea (families TAENIOPTERYGIDAE, NOTONEMOURIDAE, NEMOURI-DAE, CAPNIIDAE, and LEUCTRIDAE) and the very small family SCOPURIDAE. The Scopuridae forms the sister group to the nemuroids.

The Pteronarcyidae (Figure 7.2A) is a small, primitive family whose members include the largest stoneflies and have wings with numerous crossveins. It is primarily a North American group that has invaded eastern Asia in relatively recent times. The herbivorous or detritivorous larvae are found in medium- to large-sized rivers. Another small family, the Peltoperlidae, has a similar distribution to the Pteronarcyidae, and the larvae, which are somewhat cockroachlike in appearance, also feed on plant material or detritus. The Perlodidae (Figure 7.2B,C) is a large holarctic group (>200 species) of medium-sized stoneflies whose larvae are carnivorous, lack gills, and are typically found in slowly flowing rivers. The Perlidae is the largest family in the order with some 350 species. Though primarily a holarctic-oriental group, the family has representatives in South America and

151

*THE
PLECOPTEROID,
BLATTOID, AND
ORTHOPTEROID
ORDERS*

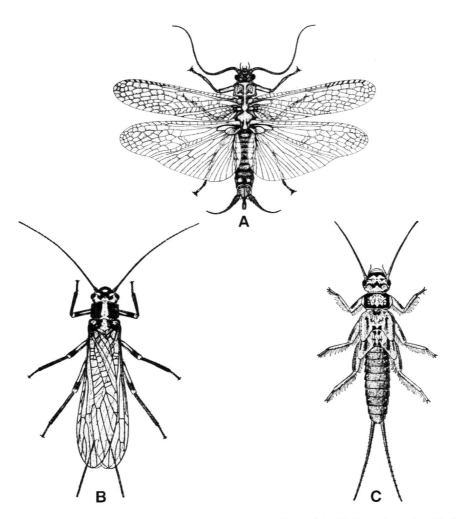

FIGURE 7.2. Plecoptera. (A) *Pteronarcys californica* (Pteronarcyidae) adult; (B) *Isoperla confusa* (Perlodidae) adult; and (C) *I. confusa* larva. [A, from A. R. Gaufin, W. E. Ricker, M. Miner, P. Milam, and R. A. Hayes, 1972, The stoneflies (Plecoptera) of Montana, *Trans. Am. Entomol. Soc.* **98**:1–161. By permission of the American Entomological Society. B, C, from T. H. Frison, 1935, The stoneflies, or Plecoptera, of Illinois, *Bull. Ill. Nat. Hist. Surv.* **20**(4). By permission of the Illinois Natural History Survey.]

Africa. That this is a rather advanced group is suggested by the reduced glossae, reduced first abdominal sternite, the fusion of the first two abdominal ganglia with that of the metathorax, and the absence of abdominal gills in larvae that are generally carnivorous. Containing more than 110 species, the holarctic family Chloroperlidae is considered to be the most specialized of the suborder by virtue of the reduced body size and wing venation (especially the absence of the anal fan in the hind wing) and the complex male reproductive system. Adults are often green (hence the family name); larvae of most species are predators though a few are detritivores or herbivores, they lack gills, and may show adaptations for burrowing in the substrate of the streams and small rivers where they are found.

Taeniopterygidae constitute the most primitive family of Nemouroidea as is indicated by the comparatively rich wing venation, large anal lobe in the hind wing, and five- or six-part cerci. Adults of this holarctic group, comprising about 70 species, are commonly known as winter stoneflies because of their habit of emerging between January and April. Some

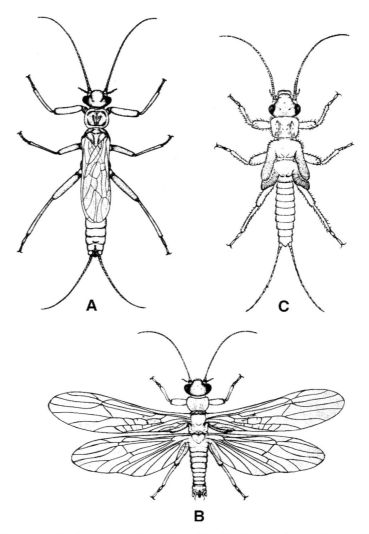

FIGURE 7.3. Plecoptera. (A) *Capnia nana* (Capniidae) adult; (B) *Nemoura flexura* (Nemouridae) adult; and (C) *N. flexura* larva. [From A. R. Gaufin, W. E. Ricker, M. Miner, P. Milam, and R. A. Hayes, 1972, The stoneflies (Plecoptera) of Montana, *Trans. Am. Entomol. Soc.* **98**:1–161. By permission of the American Entomological Society.]

adults feed on pollen. Larvae, commonly found in large streams and rivers, are herbivores or detritivores. In Capniidae (Figure 7.3A), a holarctic family of about 200 species, adults are generally small, their wings have few cross veins, and the size of the anal fan in the hind wing is reduced. The cerci, however, are long. Like Taeniopterygidae, capniids may emerge during the winter. The generally detritivorous larvae are mostly found in small rivers and streams, though a few species occur in alpine lakes. Leuctridae, which comprise a holarctic family of about 170 species, are recognized by their ability to roll their wings around the abdomen. The small anal area of the hind wings, the undivided cerci, and the specialized male genitalia suggest that this is an advanced family. Typically, the larvae are found in small mountain streams where they feed on detritus. With about 340 species, the holarctic family Nemouridae (Figure 7.3B,C) ranks next to the Perlidae in terms of size. Though the wing venation is primitive, the generally small size of the adults, the highly modified cerci and genitalia of the male, and the nerve cord with only five abdominal ganglia (due to fusion of

153

THE
PLECOPTEROID,
BLATTOID, AND
ORTHOPTEROID
ORDERS

posterior ones) make this perhaps the most advanced family in this group. Larvae are found in fast-moving streams, often with rocky substrates, where they feed on detritus or, rarely, growing plants and algae. The family Notonemouridae (about 60 species in Madagascar, South Africa, South America, Australia, and New Zealand) is likely a paraphyletic group. Zwick (1981) suggested, on the basis of differences in genitalia and internal structure, that the group arose as a result of several independent invasions from originally Northern Hemisphere stock. Larvae are found in a variety of habitats and are detritivores.

Literature

Hitchcock (1974), Hynes (1976), and Harper and Stewart (1984) provide much information on the general biology of stoneflies. The phylogeny of the order is discussed by Illies (1965) and Zwick (1980, 1981, 2000). Keys for identification are provided by Hynes (1967) [British species], Harper and Stewart (1984) and Stewart and Stark (1988) [North American genera], and Theischinger (1991) [Australian families].

Harper, P. P., and Stewart, K. W., 1984, Plecoptera, in: *An Introduction to the Aquatic Insects of North America*, 2nd ed. (R. W. Merritt and K. W. Cummins, eds.), Kendall/Hunt, Dubuque, IA.

Hennig, W., 1981, *Insect Phylogeny*, Wiley, New York.

Hitchcock, S. W., 1974, Guide to the insects of Connecticut. Part VII. The Plecoptera or stoneflies of Connecticut, *Conn. State Geol. Nat. Hist. Surv. Bull.* **107**:262 pp.

Hynes, H. B. N., 1967, A key to the adults and nymphs of British stoneflies (Plecoptera) (2nd ed.), *F.W. Biol. Assoc. Sci. Publ.* **17**:86 pp.

Hynes, H. B. N., 1976, Biology of Plecoptera, *Annu. Rev. Entomol.* **21**:135–154.

Illies, J., 1965, Phylogeny and zoogeography of the Plecoptera, *Annu. Rev. Entomol.* **10**:117–140.

Stewart, K. W., and Maketon, M., 1990, Intraspecific variation and information content of drumming in three Plecoptera species, in: *Mayflies and Stoneflies: Life Histories and Biology* (I. C. Campbell, ed.), Kluwer, Dordrecht.

Stewart, K. W., and Stark, B. P., 1988, *Nymphs of North American Stonefly Genera (Plecoptera)*, Thomas Say Foundation, Lanham, MD.

Theischinger, G., 1991, Plecoptera, in: *The Insects of Australia*, 2nd ed., Vol. I (CSIRO, ed.), Melbourne University Press, Carlton, Victoria.

Zwick, P., 1980, Plecoptera (Sternfliegen), in: *Handbuch der Zoologie*, Vol. IV, Insecta Lfg. **26**:1–115, de Gruyter, Berlin.

Zwick, P., 1981, Plecoptera, revisionary notes, in: *Insect Phylogeny* (W. Hennig), Wiley, New York.

Zwick, P., 2000, Phylogenetic system and zoogeography of the Plecoptera, *Annu. Rev. Entomol.* **45**:709–746.

3. Embioptera

SYNONYMS: Embiodea, Embiidina, Embiida   COMMON NAMES: webspinners, embiids, footspinners

Elongate, small, or moderately sized insects that live gregariously in silk tunnels; head with filiform antennae, compound eyes, and mandibulate mouthparts but lacking ocelli; males of almost all species with two pairs of nearly identical wings in which radial vein is thickened, females apterous, tarsi three-segmented and basal segment of fore tarsus greatly enlarged; cerci two-segmented and usually asymmetrical in males.

The Embioptera (Figure 7.4) are mostly confined to the larger land masses in tropical or subtropical areas of the world, though a few have found their way even to oceanic islands. Although fewer than 200 species have been described, including 13 species from North

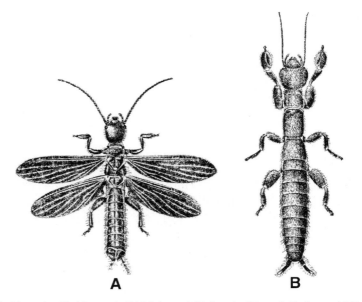

A   **B**

FIGURE 7.4.   *Embia major* (Embioptera). (A) Male; and (B) female. [From A. D. Imms, 1913, On *Embia major* n. sp. From the Himalayas, *Trans. Linn. Soc. Zool.* **11**:167–195. By permission of Blackwell Publishing Ltd.]

America and 65 from Australia, Ross (1970, 1991) suggested that this figure may represent only 10% of the world total.

Structure

As a group, the Embioptera are of remarkably uniform structure, a feature related to the widespread similarity of the tunnels in which they live. Webspinners are soft-bodied insects that fly only weakly or not at all. The prognathous head bears filiform antennae, compound eyes (often large and kidney-shaped in males, small in females), and mandibulate mouthparts. In males the mandibles are usually flattened and elongate. Ocelli are absent. No trace of wings can be seen in females; males may be apterous, brachypterous, or fully winged. In the latter the fore and hind wings are very similar. The radius is thickened; the other veins are reduced. The wings are flexible and able to fold at any point. This facilitates backward movement along the tunnels. For flight the wings are made more rigid by pumping blood into the radius. The forelegs are stout, and the basal tarsal segment is swollen to accommodate the silk glands, which number about 200. Ducts from the glands carry the product to the exterior via hairlike ejectors. The hind femur is also enlarged to contain a large tibial depressor muscle. This is correlated with the ability to run backward with great speed. There are 10 obvious abdominal segments. The cerci are two-segmented and tactile, serving as caudal "eyes" when the insect is running backward. In males the cerci are usually asymmetrical. .

The internal structure is generalized. The gut is straight, and 20–30 Malpighian tubules open into it. The ventral nerve cord is paired and includes three thoracic and seven abdominal ganglia. Each ovary consists of five panoistic ovarioles that are connected at intervals with the oviduct. A spermatheca is present. The five testis follicles on each side are also arranged serially along the vas deferens, which swells proximally into a seminal vesicle. Two pairs of accessory glands occur in males.

Life History and Habits

Both adult and juvenile Embioptera can produce silken tunnels that are just wide enough to permit the animals to move forward or backward along them. Generally, many embiids are found associated together in a "nest" of interconnected tubes. It must be emphasized, however, that this gregarious behavior is in no way social; that is, there is no caste system or division of labor. In humid regions an entire nest may be exposed, but in drier parts of the world it is usually partially subterranean as a protection against desiccation and fire. Nests are constructed in the immediate vicinity of a food source, and tunnels often extend directly into this. Embiids are phytophagous, with dead grass and leaves, lichens, moss, and bark constituting the main food. Early workers believed that males might be carnivorous because of the rather distinct mandibles. It is now known, however, that the structure of the latter is correlated with their use in grasping the female's head during copulation, and, in many species, mature males do not feed.

A typical nest contains a few mature females and their developing young. Mature males are generally short-lived and, in some species, are eaten by the female after mating. Parthenogenesis probably occurs in some species. Eggs are laid in a tunnel and guarded by the female. Parental care is extended to the young larvae, but these soon produce their own tunnels in which to develop. New colonies are formed in the vicinity of the old ones, and it is during this short migration to new sites that embiids are especially vulnerable. The absence of wings in females has more or less restricted the distribution of the Embioptera to the major land masses, though some species, perhaps transported by commerce, are found on remote Pacific islands.

Phylogeny and Classification

The fossil record of Embioptera is poor. Some authors (e.g., Hennig, 1981; Kukalová-Peck, 1991) believe that it is a very ancient insect order with a fossil record that extends to the Lower Permian period. It is claimed that these fossil remnants have a combination of primitive (e.g., wings in females, multisegmented cerci, and short ovipositor) and advanced characters (e.g., asymmetric cerci in males and reduced wing venation). However, the "embiid" nature of these Permian fragments is disputed by other workers (e.g., Carpenter, 1992; Rasnitsyn and Quicke, 2002), so that genuine embiopteran fossils do not appear before the Late Cretaceous-Early Eocene. The order is clearly orthopteroid but its strongly apomorphic character has hindered clarification of its position within the larger group. Relationships with Plecoptera, Dermaptera, Phasmida, and Zoraptera have been suggested by various authors (see Kristensen, 1991).

Because of the neotenous nature of females, identification and classification using morphological characters can be carried out with certainty only by examining mature males. Ross (1970) recognized eight families of living Embioptera, but it is not yet possible to draw many conclusions regarding their phylogenetic relationships because of the general structural uniformity of the order and the amount of parallel evolution that has taken place among families. The northern South American and West Indian family CLOTHODIDAE is the most primitive group. In this family, to which certain Miocene fossils are assigned, the cerci of the male are symmetrical and comprise two smooth segments. The largest family, EMBIIDAE, is a rather heterogeneous group of Old and New World forms. Szumik's (1996) cladistic analysis showed that, as presently constituted, the Embiidae is a paraphyletic group.

Another large and likely paraphyletic family is the OLIGOTOMIDAE, a rather primitive group with representatives in Asia, Australia, southern Europe, and possibly East Africa. Three species of *Oligotoma* have been introduced accidentally into the United States. Other families are the AUSTRALEMBIIDAE (restricted to eastern Australia and Tasmania), NOTOLIGOTOMIDAE (Southeast Asia and Australia), EMBONYCHIDAE (East Asia), TERATEMBIIDAE (South America and southern United States), and ANISEMBIIDAE (Central America and southern United States).

Literature

The biology of the Embioptera is dealt with by Ross (1970, 1991) who also (1984, 1991) has described and provided keys for the identification of North American genera and Australian families, respectively. Ross (1970) and Szumik (1996) have discussed the evolution and classification of the order.

Carpenter, F. M., 1992, *Treatise on Invertebrate Paleontology. Part R. Arthropoda 4, Vols. 3 and 4 (Superclass Hexapoda)*, University of Kansas, Lawrence.

Hennig, W., 1981, *Insect Phylogeny*, Wiley, New York.

Kristensen, N. P., 1991, Phylogeny of extant hexapods, in: *The Insects of Australia*, 2nd ed., Vol. I (CSIRO, ed.), Melbourne University Press, Carlton, Victoria.

Kukalová-Peck, J., 1991, Fossil history and the evolution of hexapod structures, in: *The Insects of Australia*, 2nd ed., Vol. I (CSIRO, ed.), Melbourne University Press, Carlton, Victoria.

Rasnitsyn, A. P., and Quicke, D. L. J. (eds.), 2002, *History of Insects*, Kluwer, Dordrecht.

Ross, E. S., 1970, Biosystematics of the Embioptera, *Annu. Rev. Entomol.* **15**:157–172.

Ross, E. S., 1984, A synopsis of the Embiidina of the United States, *Proc. Entomol. Soc. Wash.* **86**:82–93.

Ross, E. S., 1991, Embioptera, in: *The Insects of Australia*, 2nd ed., Vol. I (CSIRO, ed.), Melbourne University Press, Carlton, Victoria.

Szumik, C. A., 1996, The higher classification of the order Embioptera: A cladistic analysis, *Cladistics* **12**:41–64.

4. Dictyoptera

SYNONYMS: Dictuoptera, Oothecaria, Blattiformia, Blattopteriformia

COMMON NAMES: cockroaches and mantids

Small to very large terrestrial insects of varied form; head hypognathous with filiform, multisegmented antennae, mandibulate mouthparts and well developed compound eyes, ocelli present (Mantodea) or usually absent (Blattodea); pronotum large and disclike (Blattodea) or elongate (most Mantodea), legs with five-segmented tarsi, fore wings modified as tegmina, brachyptery, and aptery common; ovipositor reduced and hidden, male genitalia complex and concealed, cerci fairly short but multisegmented.

This mainly tropical to subtropical order contains some 5500 described species that fall into two clearly defined suborders, Blattodea (cockroaches), with at least 3500 species (including about 70 in North America, more than 400 in Australia, and 9 in Britain), and Mantodea (mantids), a predominantly Old World group of about 2000 species (including about 900 in Africa and 530 in Asia). About 20 mantid species occur in North America and 160 in Australia. Several species of cockroaches are important cosmopolitan pests.

Structure

157

*THE
PLECOPTEROID,
BLATTOID, AND
ORTHOPTEROID
ORDERS*

Cockroaches are typically flattened, oval-shaped insects whose head is covered by the large disclike pronotum. In contrast, mantids are elongate and easily recognized by their raptorial forelegs, prominent, movable head, and usually elongate pronotum. Almost all mantids are procryptically colored, though it is not known whether such camouflage is more important in concealing them from prey or from would-be predators. Some species show color polymorphism, the change from one color to another occurring either in individual insects over a few days or on a population-wide basis from season to season.

The head is hypognathous. Compound eyes are well developed in most forms but may be reduced or absent in cockroaches that live in caves, ants' nests, etc. Three ocelli are present in mantids, but in most cockroaches the ocelli have degenerated, being represented by a pair of transparent areas on the cuticle, the fenestrae. The antennae, which in some species are very long, are filiform and multisegmented. Well-developed mandibulate mouthparts are present. The legs are essentially similar in cockroaches, but in most mantids the forelegs are greatly enlarged and bear spines for catching prey. In both cockroaches and mantids wings may be fully developed, shortened, or absent. In some cockroach species both fully winged and short-winged forms occur. In mantids males are typically fully winged whereas females are frequently brachypterous or apterous. When present, the fore wings are moderately sclerotized and form tegmina. The hind wings have large anal areas. The venation is primitive, with the longitudinal veins much branched and large numbers of crossveins present. Ten obvious segments are present in the abdomen, with the 11th represented in both sexes by the paraprocts and short, multisegmented cerci. In males the ninth sternum forms the subgenital plate, which usually bears a pair of styli. The genitalia, which are partially hidden by the subgenital and supra-anal (10th tergal) plates, are membranous and asymmetrical. In females the subgenital plate is formed from the seventh sternum, which envelops the small ovipositor. Sterna 8–10 are reduced and internal.

The gut, which is long and coiled in cockroaches, short and straight in mantids, contains a large crop, well-developed gizzard, and a short midgut attached to which are eight ceca. Up to 100 or more Malpighian tubules originate at the anterior end of the hindgut. The nervous system is generalized, and three thoracic and six or seven abdominal ganglia are usually present. In some cockroaches only four or five abdominal ganglia can be seen, as a result of coalescence of the anterior ones with the metathoracic ganglion. The testes comprise four or more follicles enclosed in a peritoneal sheath. The vasa deferentia enter the ejaculatory duct, at the anterior end of which are the seminal vesicles and various accessory glands. A large conglobate gland of uncertain function opens separately to the exterior in male cockroaches. There are several panoistic ovarioles in each ovary. The lateral oviducts lead to the common oviduct, which opens into a large genital chamber. The spermatheca also enters this chamber on its dorsal side. Accessory glands, whose secretions form the ootheca, also open into the genital chamber. Subcutaneous glands, whose secretions may be either repugnatory or important in courtship (males only), occur in cockroaches.

Life History and Habits

As the differences are so great, the life history and habits of Blattodea and Mantodea are described separately.

Blattodea. Cockroaches are mostly secretive, primarily nocturnal, typically ground-dwelling insects that hide by day in cracks and crevices, under stones, in rotting logs, among decaying vegetation, etc. Some, however, live on foliage, etc. well above the ground and may be diurnal, even basking in the sun. Most species prefer a rather humid environment, though some are found in semidesert or even desert conditions and others in semiaquatic situations. A few live in caves, ants' nests, and similar places. Some species may be gregarious, insects at the same stage of development occupying the same hiding places and feeding together. Subsocial behavior occurs in a few species. Generally cockroaches are omnivorous but are rarely active predators. A few species feed on rotting wood, which is digested by symbiotic bacteria or protozoans in the cockroaches' gut. These microorganisms are very similar to those found in termites. However, it remains debatable whether these were inherited from a common ancestor or were originally in one of these groups, then transferred secondarily when members of one group preyed on members of the other (see Grandcolas and Deleporte, 1996).

Usually courtship precedes mating, which may take more than an hour to complete. The secretion of the male's tergal glands attracts the female into the appropriate position and serves as an aphrodisiac, allowing the male to mount and transfer a spermatophore. Surrounding the spermatophore produced by males of most Blattellidae is a layer of uric acid, which is subsequently eaten by the female and provides a source of nitrogen for use in ootheca construction. Cockroaches exhibit four types of reproductive strategy: (1) *oviparity* (all families except Blaberidae), the eggs being enclosed in a leathery or horny ootheca that may be seen protruding from a female's genital chamber prior to deposition; (2) *false ovoviviparity* (almost all Blaberidae and a few Blattellidae), the membranous ootheca being held internally within a brood sac during embryonic development; (3) *true ovoviviparity* (seen in only four genera of Blaberidae), in which an ootheca is not formed, the eggs passing directly from the oviduct into the brood sac where the embryonic development occurs; and (4) *viviparity* (known only in *Diploptera punctata* but probably occurs in other members of this genus), where the eggs are small and lack yolk, the embryos obtaining nourishment directly from secretions of the brood sac. Facultative parthenogenesis has been observed in some species, and obligate parthenogenesis occurs in *Pycnoscelis surinamensis*. Hatching from the ootheca requires collaborative effort on the part of the embryos, which swallow air, swell, and cause the ootheca to split open, the embryos escaping more or less synchronously. Larval development is often slow, taking up to a year and involving as many as 12 molts. Adults are frequently long-lived.

Mantodea. The life-style of mantids is in marked contrast with that of cockroaches. Mantids live a solitary, sometimes territorial existence, mostly in shrubs, trees, and other vegetation, where they wait motionless for the arrival of suitable prey, usually other insects, though anything of appropriate size is fair game. Occasionally, mantids will stalk their prey until they are within grasping distance. This is normally the situation with ground-living species (which are mostly found in arid regions).

Mating in mantids is sometimes risky for a male, as his partner, almost always larger, may regard him as being more desirable as a meal than as a lover! However, cannibalism of the male by the female, often seen in captivity, is probably rare under natural conditions. Eggs are laid in a mass of frothy material that hardens to form an ootheca. Usually this is attached to an object some distance from the ground, though a few species deposit the ootheca in the soil. Parental care of the eggs and even first-instar larvae is shown by females in a few species. Obligate parthenogenesis occurs, rarely, for example, in *Brunneria borealis* from the southern United States. As in cockroaches, the development time is rather long and there may be many molts.

Phylogeny and Classification

Numerous cockroaches broadly similar to those living today existed in the Carboniferous period, some 300 million years ago. The fossil record, including probable oothecae, is especially strong from the Upper Carboniferous onward from many regions of the world, and by the Paleocene species assignable to modern families occurred. Opinions differ on when the mantid and cockroach lines diverged. At one extreme, Carpenter (1992) suggested that the mantids may have evolved independently, from protorthopterous ancestors. In contrast, Rasnitsyn and Quicke (2002) indicate an evolution from a polyphagid cockroachlike ancestor in the Late Triassic. The earliest mantid fossils come from the Early Cretaceous, though these are generally only wing or foreleg fragments. By the Eocene, the group was well established; indeed, some fossils from this period can be placed in modern families.

Attempts to interpret the phylogeny of the suborder Blattodea have been hampered by the high degree of parallel evolution that has occurred within the group. McKittrick (1964) examined the external genitalia, oviposition behavior, and crop structure in a wide variety of extant species. She suggested that cockroach evolution proceeded along two lines, one leading to the superfamily Blattoidea (families Cryptocercidae and Blattidae), the other to the superfamily Blaberoidea (families Polyphagidae, Blattellidae (= Ectobiidae), and Blaberidae). Durden's (1969) study of Carboniferous cockroaches generally supported McKittrick's conclusions, though he recognized several additional superfamilies. More recent proposals have been based on extensive cladistic analysis of morphological and anatomical features of extant species (Grandcolas, 1996), mDNA sequences (Maekawa and Matsumoto, 2000), and fossils (Rasnitsyn and Quicke, 2002). Taken together, these studies show that the Blattidae and Cryptocercidae are not sister groups, the Polyphagidae and Cryptocercidae are closely related, and the Blattidae are the sister group to the Blattellidae + Blaberidae. A proposed phylogeny is shown in Figure 7.5.

Classification of the suborder Mantodea is also difficult because of parallel evolutionary trends among the constituent groups. Beier (1964) divided the suborder into eight families, contained within the single superfamily Mantoidea, of which the Amorphoscelidae and Mantidae are the largest. The six remaining families are small, tropical groups of restricted distribution.

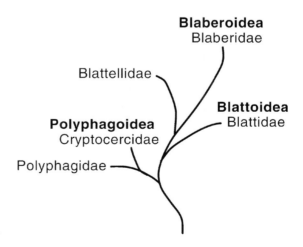

FIGURE 7.5. Proposed phylogeny of Blattodea.

Suborder Blattodea

In members of the suborder Blattodea the head is covered with a large, shield shaped pronotum; the legs are identical; and the gizzard is strongly dentate.

Superfamily Polyphagoidea

Included in this group are two families, POLYPHAGIDAE (about 190 species) and CRYPTOCERCIDAE (nine species in the genus *Cryptocercus*). The Polyphagidae is a widely distributed family that includes the most primitive living cockroaches. They are generally small (2 cm or less in length) and often have a hairy pronotum. Some inhabit arid regions, living in small burrows that they leave to forage at night, and a few species are inquilines in ants' nests. Until 1997, the Cryptocercidae was considered to include only three species, one being *C. punctulatus,* from mountainous regions in eastern and western United States. However, this number has now been increased to nine following recent discoveries in Eurasia (four species) and molecular biological analyses of the United States' populations which indicate that *C. punctulatus* is a complex of five species (Hossain and Kambhampati, 2001). Studies on *Cryptocercus* have been particularly important in discussions of evolutionary links between the Blattodea and termites. These cockroaches live in colonies containing individuals of all ages beneath rotting logs and show subsocial behavior. They feed on wood that, as in the "lower" termites, is digested by flagellate protozoans present in the hindgut. As the lining and contents of the hindgut are lost at each molt, insects must obtain a fresh supply of protozoans. This they do by eating fecal pellets.

Superfamily Blattoidea

This superfamily includes the BLATTIDAE, BLATTELLIDAE, and BLABERIDAE. The approximately 525 species in the cosmopolitan Blattidae are generally fairly large cockroaches (2–5 cm in length) and may be recognized by the numerous spines on the ventroposterior margin of the femora. The family contains several species that are closely associated with humans and do considerable damage to their property, as well as cause health hazards through contamination of food. *Blatta orientalis* (the Oriental cockroach) (Figure 7.6A) appears to be a native of the Mediterranean region but has been distributed through commerce to many parts of the world. It is the major cockroach pest in Britain and is widely distributed throughout North America. It prefers generally cool situations and is typically found in cellars, basements, toilets, bathrooms, and kitchens. It can tolerate warmer conditions provided that water is available. Four species of *Periplaneta, P. americana* (the American cockroach) (Figure 7.6B), *P. australasiae* (the Australian cockroach), *P. fuliginosa* (the smokey-brown cockroach), and *P. brunnea* (the brown cockroach), which are of African origin, are also found in and around human habitations. All four species prefer warmer, moister habitats than those enjoyed by *B. orientalis,* and are frequently found in outdoor habitats in subtropical regions.

Blattellidae are generally small cockroaches (not usually more than about 1 cm in length), with relatively long, slender legs. This, the largest cockroach family (about 1740 species), is widely distributed and contains two major pest species, *Blattella germanica* (the German cockroach) (Figure 7.6C) and *Supella longipalpa* (formerly *supellictilium)* (the brown-banded cockroach). The German cockroach ranks second to the Oriental cockroach in economic importance. It prefers warm, humid surroundings, such as are found in bakeries,

161

*THE
PLECOPTEROID,
BLATTOID, AND
ORTHOPTEROID
ORDERS*

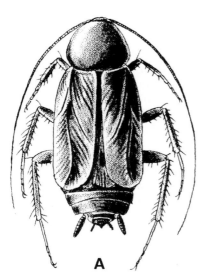

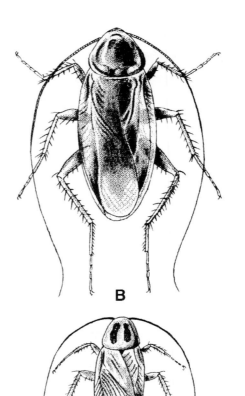

FIGURE 7.6. Cockroaches. (A) The Oriental cockroach, *Blatta orientalis* (Blattidae); (B) the American cockroach, *Periplaneta americana* (Blattidae); and (C) the German cockroach, *Blattella germanica* (Blattellidae). [From L. A. Swan and C. S. Papp, 1972, Copyright 1972 by L. A. Swan and C. S. Papp. Reprinted by permission of Harper & Row, Publishers, Inc.]

restaurants, and domestic kitchens. Like *B. germanica*, *S. longipalpa* is probably of African origin. In North America it became established in Florida at the beginning of the 20th century and has now been reported from all states. It is also common in some areas of Canada.

The family Blaberidae (1020 species) is the most recently evolved cockroach family and the one that has undergone the most extensive adaptive radiation. The group is primarily tropical and contains the largest cockroach species. Its members are generally found under logs, in humus, etc., though some species are arboreal. A few species may occasionally become associated with humans, for example, *Pycnoscelis surinamensis* (the Surinam cockroach), *Leucophaea maderae* (the Madeira cockroach), and *Nauphoeta cinerea* (the lobster cockroach). *P. surinamensis* may be found in greenhouses or, in warmer climates, outdoors where it can significantly damage roots of crops; it is also found in chicken houses and is known to be an intermediate host for the chicken eyeworm nematode (*Oxyspirura mansoni*).

Suborder Mantodea

In members of the suborder Mantodea the head is not covered with a pronotum; three ocelli are present; the forelegs are raptorial; and the gizzard is not well developed.

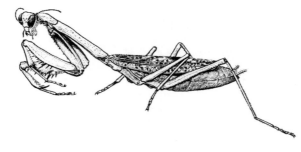

FIGURE 7.7. The Carolina mantid, *Stagmomantis carolina* (Mantodea). [From M. Hebard. 1934. The Dermaptera and Orthoptera of Illinois, *Bull. Ill. Nat. Hist. Surv.* **20**(3). By permission of the Illinois Natural History Survey.]

Superfamily Mantoidea

The family AMORPHOSCELIDAE is best represented in the Australasian region, though species are also found in Asia, Africa, and southern Europe. Two morphological features distinguish members of this family from other mantids. The pronotum is short, and the tibiae and femora of the raptorial forelegs lack spines. Many species are procryptically colored and have various spines and prominences on the head and pronotum. Amorphoscelids are generally small and live on the ground or on tree trunks. Apterous females of some Australian species mimic ants though it is unclear whether this assists the mantids in capturing the ants as prey or protects the mantids from predators.

The MANTIDAE (Figure 7.7) is easily the largest family of Mantodea, with almost 1500 species, and has a wide distribution throughout the tropical and warmer temperate regions of the world. Adults vary in size from just under 10 mm to over 15 cm. They are frequently well camouflaged, living among foliage, on tree trunks, or on the ground; in the latter case they actively pursue their prey. All mantids found in North America belong to this family, including four that have been introduced, for example, *Mantis religiosa,* the "soothsayer" or "praying mantis" of southern Europe.

Literature

Good accounts of the biology of cockroaches are given by Guthrie and Tindall (1968), Cornwell (1968, 1976), Bell and Adiyodi (1982), and Schal *et al.* (1984). Guthrie and Tindall, and Cornwell, also deal with their economic importance. The biotic associations of cockroaches are discussed by Roth and Willis (1960), and their medical and veterinary importance by the same authors (1957). The phylogeny and classification of Blattodea is dealt with by McKittrick (1964), Durden (1969), Roth (1970), Grandcolas (1996), and Maekawa and Matsumoto (2000). Accounts of mantid biology are provided by Gurney (1950) and Preston-Mafham (1990). North American Dictyoptera may be identified from Rehn (1950), Gurney (1950), Helfer (1987), and Arnett (2000). Harz and Kaltenbach (1976) provide keys to the European genera of Dictyoptera.

Arnett, R. H., Jr., 2000, *American Insects: A Handbook of the Insects of America North of Mexico*, 2nd ed., CRC Press, Boca Raton, FL.

Beier, M., 1964, Blattopteroidea. Ordnung Mantodea Burmeister 1838 (Raptoriae Latreille 1802; Mantoidea Handlirsch 1903; Mantidea auct.), *Bronn's Kl. Ordn. Tierreichs* **6**:849–870.

Bell, W. J., and Adiyodi, K. G. (eds.), 1982, *The American Cockroach*, Chapman & Hall, London.

Carpenter. F. M.. 1992, *Treatise on Invertebrate Paleontology. Part R. Arthropoda* 4, *Vols. 3 and 4 (Superclass Hexapoda),* University of Kansas, Lawrence.

Cornwell, P. B., 1968, 1976, *The Cockroach*, Vols. I and II, Hutchinson, London.

Durden, C. J., 1969, Pennsylvanian correlation using blattoid insects, *Can. J. Earth Sci.* **6**:1159–1177.

Grandcolas, P., 1996, The phylogeny of cockroach families: A cladistic appraisal of morpho-anatomical data, *Can. J. Zool.* **74**:508–527.

Grandcolas, P., and Deleporte, P., 1996, The origin of protistan symbionts in termites and cockroaches: A phylogenetic perspective, *Cladistics* **12**:93–98.

Gurney, A. B., 1950, Praying mantids of the United States, native and introduced, *Annu. Rep. Smithson. Inst.* **1950**:339–362.

Guthrie, D. M., and Tindall, A. R., 1968, *The Biology of the Cockroach*, Arnold, London.

Harz, K., and Kaltenbach, A., 1976, *The Orthoptera of Europe*, Vol. III, Junk, The Hague.

Helfer, J. R., 1987, *How to Know the Grasshoppers, Crickets, Cockroaches and Their Allies*, Dover, New York.

Hossain, S., and Kambhampati, S., 2001, Phylogeny of *Cryptocercus* species (Blattodea: Cryptocercidae) inferred from nuclear ribosomal DNA, *Mol. Phylog. Evol.* **21**:162–165.

Maekawa, K., and Matsumoto, T., 2000, Molecular phylogeny of cockroaches (Blattaria) based on mitochondrial COII gene sequences, *Syst. Entomol.* **25**:511–519.

McKittrick, F. A., 1964, Evolutionary studies of cockroaches, *Mem. Cornell Univ. Agric. Exp. Stn.* **389**:177 pp.

Preston-Mafham, K., 1990, *Grasshoppers and Mantids of the World*, Blandford, London.

Rasnitsyn, A. P., and Quicke, D. L. J. (eds.), 2002, *History of Insects*, Kluwer, Dordrecht.

Rehn, J. W. H., 1950, A key to the genera of North American Blattaria, including established adventives, *Entomol. News* **61**:64–67.

Roth, L. M., 1970, Evolution and taxonomic significance of reproduction in Blattaria, *Annu. Rev. Entomol.* **15**: 75–96.

Roth, L. M., and Willis, E. R., 1957, The medical and veterinary importance of cockroaches, *Smithson. Misc.Collect.* **134**(10):147 pp.

Roth, L. M., and Willis, E. R., 1960, The biotic associations of cockroaches, *Smithson. Misc. Collect.* **141**(4422): 470 pp.

Schal, C., Gautier, J.- Y., and Bell, W. J., 1984, Behavioral ecology of cockroaches, *Biol. Rev.* **59**:209–254.

5. Isoptera

SYNONYMS: Termitina, Termitida, Socialia   COMMON NAMES: termites, white ants

Polymorphic social insects living in colonies that comprise reproductives, soldiers, and workers; head with moniliform multisegmented antennae and mandibulate mouthparts, compound eyes present but frequently degenerate, ocelli often absent; wings when present almost identical (except *Mastotermes)* and membranous, lying horizontally over abdomen at rest, capable of being shed by a predetermined basal fracture, legs identical and with a large coxa, tarsi almost always four-segmented (five-segmented in *Mastotermes); cerci short and with few segments, external genitalia lacking in both sexes of most species.

More than 2300 species of termites are known, mainly from tropical to warm temperate areas, though a few species are found in cool temperate climates such as those of southern Europe and southern and western North America as far north as southern Canada. Several species have been transported to new areas by commerce, and some of these have become established in heated buildings (e.g., in Hamburg and Toronto) well outside their normal range of climatic tolerance.

Structure

In almost all species the mature termite colony contains individuals of remarkably different form and function. Each group of individuals that perform the same function is known as a caste. In most species three castes occur: reproductive (primary and secondary; both male and female), soldier (sterile adults of both sexes), and worker (also sterile adults of both sexes). Immature stages of all castes may also be present in the colony along with (occasionally) intercastes. As the castes are of different form, it is appropriate to describe them separately.

Reproductive. The body of primary reproductives (king and queen) is normally well sclerotized; however, in physogastric queens, that is, females whose abdomen becomes enormously swollen through hypertrophy of the ovaries and consequent stretching of the intersegmental membranes (Figure 7.11C), the abdomen is pale, and the original tergal and sternal plates are the only areas of sclerotization. The head is round or oval and carries well-developed compound eyes, moniliform antennae with a varied number of segments (generally fewer in more advanced termites), and mandibulate mouthparts. In Termitidae and Rhinotermitidae a small pore, the fontanelle, occurs in the midline between or behind the compound eyes. This marks the opening of the frontal gland. In the thorax the pronotum is distinctive; the thoracic sterna are membranous. Except in *Mastotermes*, the two pairs of wings are very similar in appearance, with strongly sclerotized veins in the anterior portion and a basal (humeral) suture along which fracture of the wing occurs. In *Mastotermes* the wings have a primitive venation; also, the hindwings have a large anal lobe as in cockroaches but lack a basal suture, though a line of weakness occurs to facilitate wing shedding. The legs are all very similar, having large coxae and four- (very rarely three-)segmented tarsi; in *Mastotermes* the tarsi are five-segmented. Ten obvious abdominal segments occur with the 11th tergum having fused with the 10th, and the 11th sternum being represented by the paraprocts. Except in Hodotermitidae, in females the seventh sternum forms a large subgenital plate that obscures the remaining sterna. Short cerci are present that are three- to eight-segmented in lower termites but are reduced to an unsegmented or two-segmented tubercle in higher forms. External genitalia are absent except in *Mastotermes* where females have a blattoid-type ovipositor and males a copulatory organ.

In neotenics (also called secondary, supplementary, and replacement reproductives) the body is less sclerotized than that of the primaries. The compound eyes are usually reduced. Neotenics may have wing buds or be wingless, their wings having been chewed off by workers. In some species female neotenics may become physogastric.

Soldier. Members of this caste are readily recognized by their large, well-sclerotized head that in some species may exceed the rest of the body in size. Though soldiers may be of either sex, the proportion of male to female individuals in this caste may vary. Primitively the mandibles are very large, sometimes enormous, and suited for biting. In other species in which the mandibles are large they may serve as pincers, or they may be asymmetrical and hinged so as to snap closed at great speed, thus delivering a powerful blow to an adversary. In Nasutitermitinae (Figure 7.12) the frons is enlarged to form a more or less pointed rostrum, at the tip of which opens the frontal gland, and the mandibles are reduced or vestigial. A large frontal gland occurs in both Rhinotermitidae and Termitidae, sometimes (in rhinotermitids) occupying most of the abdomen. The secretion of the gland may be toxic, repellent, or sticky; it is usually smeared on intruders but in some termites it can be ejected some distance. In some Kalotermitidae the head is phragmotic, that is, has a thick, sometimes sculptured frons designed to plug access holes to the nest and prevent entry of invaders. Generally soldiers are apterous though in Kalotermitidae and Termitidae they may develop from juveniles with wing buds.

Worker. In most species the body of workers is generally pale and weakly sclerotized. The head resembles that of a primary reproductive, except that the compound eyes are reduced or absent and the mandibles more powerful. Workers may be polymorphic according to their age and sex.

Except for the gut, which is modified with their mode of life (and of great use taxonomically), the internal structure of termites is generalized. The esophagus is a long, narrow tube and is followed by a scarcely differentiated crop. The gizzard wall is greatly folded longitudinally, each fold having cuticular thickenings and, often, teeth. The midgut is typically a short tube of uniform diameter though in physogastric queens it may be enormously enlarged, a development presumably associated with absorption of the large quantities of saliva fed to them by workers. Mesenteric ceca may or may not be present. The hindgut is well developed and differentiated into a number of regions, the most prominent of which is the large paunch containing bacterial or protozoan symbionts. The posterior wall of the paunch contains columnar epithelium and is probably a region of absorption. In lower termites (except *Mastotermes* with up to 15) 8 Malpighian tubules enter the gut at the junction of the midgut and hindgut; in Termitidae only 4 tubules occur. The central nervous system is orthopteroid, with three thoracic and six abdominal ganglia. In reproductive males each testis comprises up to 10 fingerlike follicles that enter the paired vasa deferentia. At the junction of the vasa deferentia and ejaculatory duct there is a pair of seminal vesicles. In reproductive females each ovary initially contains only a few panoistic ovarioles and this number remains in lower termites. However, in physogastric reproductives the number increases with maturity, reaching several thousand in some species. The paired oviducts enter the short common oviduct, which leads into the genital chamber. A spermatheca and accessory glands also enter this chamber. The reproductive organs are atrophied in workers and soldiers.

Life History and Habits

New colonies may be formed in various ways. By far the commonest method is swarming, in which large numbers of winged individuals (alates) leave the parent colony. The onset of swarming is closely correlated with climatic conditions. In tropical species it occurs typically at the onset of the rainy season, an adaptation that facilitates nest formation in the damp, soft earth for subterranean species. In species from temperate climates swarming occurs during the summer. Flights may occur at any time of the day, but for a given species there are frequently specific hours during which swarming takes place. Swarming may be temporarily postponed, however, if environmental conditions are unsuitable. The distance traveled by the alates is usually only a few hundred meters unless they are assisted by wind. It is at this time that termites are most susceptible to predators.

On landing individuals shed their wings, and a male is attracted (probably chemically) to a female, which he follows until she locates a suitable nesting site. After closing the entrance to the nest, the royal pair, as the founding pair is called, mate within a few hours or days. (Mating is, however, periodically repeated throughout the life of the pair.) Egg laying begins soon after the royal pair have become established, but the first batch of eggs is usually less than 20, and egg laying is not resumed until the young are capable of looking after themselves and feeding the queen. Initially only workers are produced but, as the number of individuals increases, soldiers differentiate. Alates are not produced until the colony is several years old while neotenics normally differentiate only if the primary reproductives are lost. The original royal pair may live for a considerable time (e.g., at least 17 years in *Mastotermes)* and, at maturity, a physogastric queen may produce up to 3000 eggs daily. It is likely, however, that such a high rate is not sustained on a year-round basis but is seasonal. The proportions of the different castes vary; for example, in *Nasutitermes* up to 15% of individuals may be soldiers whereas in many Kalotermitidae the fraction is less than

1%, and in *Invasitermes* spp. which live in the nests of other termites there is no soldier caste. The differentiation of the various castes and their maintenance in a fixed ratio to each other are complex phenomena, controlled by the interaction of pheromonal, nutritional, hormonal, and perhaps other factors (see Chapter 21, Section 7). A colony matures (i.e., begins producing winged reproductives) after several years, but it continues to increase in size after this time. It is obviously difficult to estimate the number of individuals in mature colonies, but in the lower termites the figure is usually several hundred or thousands, while in the higher termites it may be several million.

Two other methods of colony foundation are known. In some species, in which the nest is a rather diffuse structure, groups of individuals may become more or less isolated from the rest of the colony. In these groups neotenics differentiate, and the group becomes independent of the parent colony. This is described as budding. The foundation of new colonies by deliberate social fragmentation (sociotomy) has been reported for a few species. In this situation many individuals of all castes (often including the original royal pair) emerge from the parent colony and march to a new location. The original colony then becomes headed by neotenics.

Termite nests exhibit a wide range of form, the complexity of which parallels approximately the phylogeny of the order. In the primitive Kalotermitidae and Termopsidae the nest is simply a series of cavities and tunnels excavated in wood. Few partitions are constructed by these termites, and there is no differentiation of the nest into specific regions. In other lower termites the nest may be in wood or subterranean, but even in the former situation contact with the ground is maintained by a series of tunnels. This ensures that the humidity of the nest remains high. Most Hodotermitidae build completely subterranean nests, in which the beginnings of specialization are seen. Food is stored in chambers immediately below the surface of the ground. The main chamber, which is considerably subdivided by both horizontal and vertical walls, is several feet below the surface. However, in nests of this family there is no chamber specifically for the royal pair. Nests of Rhinotermitidae may be entirely in soil or in wood or in both of these media. In a few Hodotermitidae, some Rhinotermitidae in the genus *Coptotermes,* and many Termitidae epigeous (above-ground) nests are constructed (Figure 7.8), though it should be emphasized that even in these species a considerable portion of the nest may be subterranean. In the simplest epigeous nests little differentiation occurs between the peripheral and internal parts, which comprise a mass of interconnecting, uniform chambers; the royal chamber is either absent or located in the subterranean part of the nest. In more complex nests the above-ground component comprises a thick peripheral wall enclosed within which is the habitacle (nursery) and surrounding food chambers. The royal chamber is usually located near the base of the structure.

A major problem for all social insects is maintenance of a suitable nest climate. Regulation of relative humidity, temperature, and carbon dioxide concentration occurs (Korb, 2003). For termites that live in wetter regions humidity regulation is not a serious problem, and the relative humidity within the nest is generally 96% to 99%. In termites from regions with long dry spells various behavioral adaptations ensure the well-being of a colony. The commonest of these is for the termites to move more deeply into the ground where the moisture content is greater. Other species behave like honey bees and regurgitate saliva or crop contents onto the walls of the nest, especially in the nursery region. Some species burrow deeply into the ground to the level of the water table and bring moisture-laden particles up into the nest area.

Temperature is also regulated in some termite nests to a remarkable degree. To some extent this is facilitated by the location of the nests in wood and soil, which serve as excellent

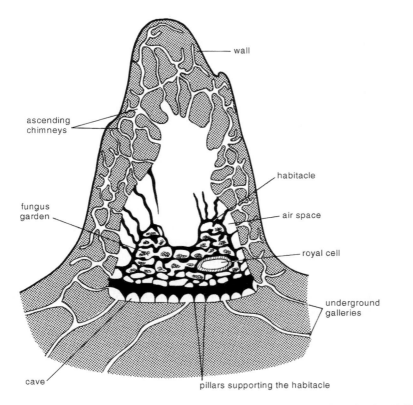

FIGURE 7.8. Mature nest of *Bellicositermes natalensis* (Termitidae). [After P.-P. Grassé (ed.), 1949, *Traité de Zoologie*, Vol. IX. By permission of Masson, Paris.]

buffers against sudden changes in external temperature. In cold weather, termites behave much like bees, clustering together in the center of the nest and effectively reducing the "operating space," whose temperature must be maintained by metabolic heat. In mound-building termites, whose nest may be fully exposed to the sun, the temperature in the center of the nest is held steady as a result of the excellent insulation provided by either thick walls or thin, cavity-bearing walls in which food is stored. However, as the degree of insulation from external temperature fluctuations increases, so does the problem of gas exchange. Although it has been shown experimentally that termites can withstand very high carbon dioxide concentrations, field studies have indicated that under natural conditions they do not face this problem because of the nest's air-conditioning system. Convection currents, created by the different temperatures at the center and periphery of the nest, are the basis of the system. In *Bellicositermes natalensis* the heat created in the central (nursery) area causes the air in this region to rise to the upper chamber (Figure 7.8). The air then moves along the radial ducts to the peripheral region of the nest, which comprises a system of thin-walled tubes. Carbon dioxide and oxygen can diffuse easily across these walls. As the "fresh" air in the peripheral tubes cools, it sinks into the "cellar" of the nest, eventually to be drawn by convection back into the central area.

Termites are primitively wood-eating insects, and this habit is retained in most lower termites and many of the higher forms. Others feed on dry grass, fungi, leaves, humus, rich soil, and herbivore dung. In most species food is consumed at its source, individuals remaining in the nest being fed by trophallaxis (see below). However, some termites travel considerable distances (e.g., 100 m or more in *Mastotermes)* from the nest to a food source.

Foraging may be done via underground tunnels or thin-walled surface tubes, or in the open at night or on humid, overcast days. Many species release trail-marking pheromones (see Chapter 13, Section 4.5).

Cellulose is the primary component used by the termites whose midgut produces cellulase. In addition, to facilitate breakdown and use of the food, complex relationships have evolved between termites and microorganisms (protozoa, bacteria, and fungi). In all families except Termitidae, protozoa in the paunch produce a range of enzymes (including cellulase) that degrade the food into organic acids such as acetate and butyrate. In Termitidae anaerobic bacteria replace protozoa in the paunch, though the bacteria do not themselves break down cellulose. Within the Termitidae, members of the subfamily Macrotermitinae also culture a basidiomycete fungus of the genus *Termitomyces* in special "fungus gardens." Although the occurrence of these structures has been known since 1779, it is only quite recently that the precise relationship between the termite and fungus has been established. In a typical fungus garden the fungus grows on sheets of reddish-brown "comb" (decaying vegetable material) and is visible as a whitish mycelium containing conidia and conidiophores. This latter observation led early authors to suggest that the young termites were fed on the fungus, though it soon became apparent that the small amount of fungus would not satisfy even their requirements. It was some time before it was realized that the comb was a dynamic structure, being removed from below and built up on its upper surface or in the space beneath. In other words, the comb forms the food of the termites. Using staining techniques, it has been shown that the primary role of the fungus is digestion of the lignin component of the comb, releasing material that is then broken down by bacteria in the termites' gut. Secondarily, however, the fungus also provides vitamins and a source of nitrogen. Another point of contention was the method of comb construction. It was believed originally that the termites regurgitated chewed-up food to produce comb, but more recent work has shown that the comb is derived from feces. Thus, the vegetable material passes twice through the gut of Termitidae, a situation that is comparable with that in other termite families in which proctodeal feeding is an important method of extracting the maximum nutrition from the food (see below).

Only workers are able to feed themselves. Members of other castes and very young stages must be fed. Furthermore, their diet, as in other social insects, is different to a greater or lesser degree from that of workers. Exchange of food material (trophallaxis) occurs either by anus-to-mouth transfer (proctodeal feeding) or by mouth-to-mouth transfer (stomodeal feeding). The former method takes place in all families except the Termitidae, and normally it occurs only between workers or larger juveniles, although occasionally soldiers may act as donors. Proctodeal food is a liquid containing protozoans, products of digestion, and undigested food. Stomodeal food is either a semisolid material comprising the regurgitated contents of the crop, which are fed to soldiers in the lower termite families, or saliva, which appears to be the only food received by reproductives of all families, very young stages of lower termites, and all juvenile stages and soldiers of Termitidae.

Phylogeny and Classification

There is little doubt that termites are derived from Paleozoic cockroachlike ancestors perhaps similar in some ways to *Cryptocercus punctulatus,* a subsocial, wood-eating cockroach. Indeed, some authorities consider the similarities between termites and cockroaches to be sufficiently great as to include the former as a suborder of the Dictyoptera; that is, the termites are eusocial cockroaches (Eggleton, 2001). The earliest fossil termites are from

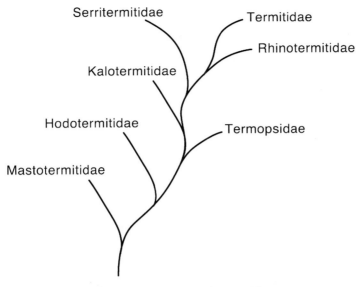

169

*THE
PLECOPTEROID,
BLATTOID, AND
ORTHOPTEROID
ORDERS*

FIGURE 7.9.   Proposed phylogeny of Isoptera.

the Lower Cretaceous and are assignable to the family Hodotermitidae. Strangely, fossil Mastotermitidae, widely accepted as the most primitive termite group, are known only from the Late Oligocene. It is anticipated that fossil remains of the order will eventually be found in Jurassic or Triassic deposits (Carpenter, 1992).

The possible relationships of the extant termite families are shown in Figure 7.9, though it should be emphasized that authorities still disagree over the status of some groups and relationships both within and between them, principally because of the high degree of convergence that has occurred as a result of their specialized mode of life (see Donovan *et al.*, 2000). The Mastotermitidae appear to be the sister group to other termites. The Kalotermitidae share a number of features (mandibular dentition, presence of ocelli, and an arolium on the tarsus) with the Mastotermitidae, and in some early schemes (e.g., that of Krishna, 1970) were shown as its sister group. These common features are now thought to have resulted from parallel evolution, with some studies indicating that the kalotermitids are the sister group to the (Rhinotermitidae + Serritermitidae + Termitidae). The Hodotermitidae constitute another primitive family, from an early form of which arose the Termopsidae. The termopsids, too, are considered primitive; for example, some genera have three teeth on the left mandible, a feature also found in cockroaches. The Rhinotermitidae, which appear to have evolved from an early kalotermitid ancestor, is a heterogeneous and likely polyphyletic group (Grassé, 1982–1986). The position of the single-species family Serritermitidae is questionable. Though included previously in the Rhinotermitidae or Termitidae because of its mixture of characters, it is probably best to consider the group as a distinct family that, like the Termitidae, evolved from early rhinotermitid stock. The families Mastotermitidae, Kalotermitidae, Hodotermitidae, Rhinotermitidae, and Serritermitidae are collectively known as the lower termites. Common to them all is a mutualistic relationship between the termite host and certain flagellate protozoans found in the hindgut. In the remaining termite family, Termitidae, often called higher termites, there are generally few or no protozoans in the hindgut, and the relationship between them and the host is never mutualistic. Where such a relationship exists it is between the termite and the bacteria of the hindgut.

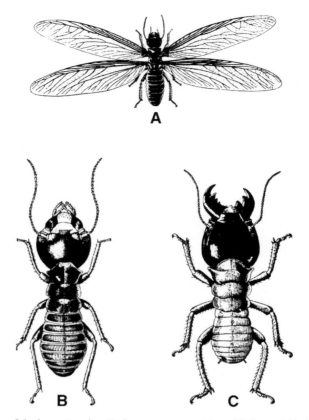

FIGURE 7.10. Castes of the lower termite, *Hodotermes mossambicus* (Hodotermitidae). (A) Alate; (B) pseud-ergate; and (C) soldier. [From W. H. G. Coaton, 1958, The hodotermitid harvester termites of South Africa, *Union of South Africa, Department of Agricultural Science Bulletin*, Vol. 375. By permission of the South African Department of Agricultural Technical Services.]

The family MASTOTERMITIDAE contains a single living species, *Mastotermes dar-winiensis,* endemic to tropical areas of northern Australia and introduced by commerce into New Guinea. *Mastotermes* has a large number of primitive characters that would support the idea of a close relationship between the cockroaches and termites. These include the five-segmented tarsi, long multisegmented antennae, well-developed compound eyes and ocelli, netlike wing venation, distinct anal lobe in the hind wing, absence of a basal suture in the hind wing, certain structural similarities in the gizzard and genitalia of both groups, and the laying of eggs in an ootheca, a feature found in no other termite family. *Mastotermes* normally exists in small colonies but after disturbances that lead to increased food supplies, colony size can increase rapidly to over a million individuals. The species is economi-cally very important through its destruction of structural timber, living plant material, and synthetic materials.

In the family HODOTERMITIDAE (Figure 7.10) are about 15 species of so-called harvester termites that forage above ground for grass, leaves, etc., which are then stored in special chambers in their predominantly underground nests. Hodotermitids typically occur in desert and steppe regions of the Old World, including northern and southern Africa, across the Near and Middle East to north India, Pakistan, and Afghanistan.

The TERMOPSIDAE is a very small (15 species), primitive termite family, com-monly known as the damp-wood termites, found especially in fungus-affected wood, either standingpt or fallen, occasionally in damp structural timbers. The group is primarily a northern warm temperate family, though some species are found in cool temperate regions in both Northern and Southern Hemispheres.

Members of the family KALOTERMITIDAE (300 species) are called dry-wood termites from their habit of living in sound, dry wood that is not in contact with the ground. The family is extremely widespread, with representatives in all tropical and some cool temperate regions. Several extant genera are also known from fossils. It is only in this family that soldiers with phragmotic heads occur. Some species are of major economic importance.

Most of the about 160 species in the widespread family RHINOTERMITIDAE are subterranean forms that live in buried, rotting wood. Some species, however, construct nests directly in the soil, or in rotting logs above ground, and yet others build a mound nest. All species are wood eaters, and many are extremely important economically including *Reticulitermes flavipes* and *R. hesperus* in the eastern and western United States, respectively. An interesting feature of some species is the occurrence of dimorphism in the soldiers. The larger form retains the large biting mandibles; the smaller form has reduced mandibles, but the labrum is elongate and grooved, enabling the insect to smear the noxious secretion from the frontal gland onto invaders.

Serritermes serrifer, from Brazil, is the only member of the family SERRITERMITIDAE. Among the smallest of termites (alates are about 4 mm long), *Serritermes* has a mixture of characters that led early authors to place it in the Rhinotermitidae or Termitidae. However, the existence of protozoa in the hindgut would seem to rule out the latter possibility. Colonies of this species have been found only in the outer wall of nests of *Cornitermes* (Nasutitermitinae, see below).

It is within the family TERMITIDAE (Figure 7.11), which contains about three quarters of the living termite species, that the greatest range of social development and specialization exists. Four subfamilies are recognized by Grassé (1982–1986) in this possibly polyphyletic group: (1) the cosmopolitan and largest subfamily (with about 800 species), TERMITINAE, in which the two major groups are the *Amitermes* group (soldiers with biting mandibles) and the *Termes* complex (soldiers with snapping mandibles); (2) APICOTERMITINAE, an almost entirely African subfamily (a few species of uncertain affinity occur in South America and eastern Asia) made up of two main groups, the *Apicotermes* group and the *Anoplotermes* group (there is no soldier caste in species in the latter group); (3) MACROTERMITINAE, which contains the Old World fungus-growing termites; and (4) NASUTITERMITINAE, the second largest subfamily with more than 500 species, characterized by the evolutionary development in soldiers of a rostrum (nasus) at the tip of which opens the frontal gland (Figure 7.12). They are consequently known as "nasute soldiers." The sticky, irritant fluid from the gland is either dribbled and smeared or forcefully ejected onto nest invaders. In many species the soldiers are di- or trimorphic, according to the age and sex of the stage from which they differentiate.

Literature

Krishna and Weesner (1969, 1970) have edited two volumes entitled "Biology of Termites" in which all aspects of termite biology are discussed. Grassé's (1982–1986) three-volume treatise also provides comprehensive coverage of the group. Other introductions to the biology of the order are given by Wilson (1971), Howse (1970), Harris (1971), and Pearce (1997). Termite phylogeny is discussed by McKittrick (1964), Krishna (1970),

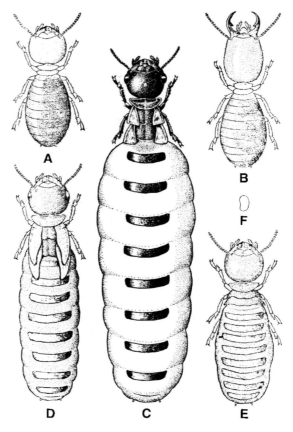

FIGURE 7.11. Castes of the higher termite, *Amitermes hastatus* (Termitidae). (A) Worker; (B) soldier; (C) physogastric queen; (D) secondary queen; (E) tertiary queen; and (F) egg. All figures are to same scale. The worker is about 5 mm long. [From S. H. Skaife, 1954, The black-mound termite of the Cape, *Amitermes atlanticus* Fuller, *Trans. R. Soc. S. Afr.* **34**:251–271. By permission of the Royal Society of South Africa.]

Kambhampati *et al.* (1996), Kambhampati and Eggleton (2000), Donovan *et al.* (2000), and Eggleton (2001). Their economic importance is reviewed by Harris (1969, 1971), Hickin (1971), and Pearce (1997). North American termites are discussed by Weesner (1965) (also 1970, in "Biology of Termites," Vol. II) and may be identified in Helfer (1987). Watson and Gay (1991) discuss the order from the Australian perspective and provide access to the literature on termites from this region. Caste differentiation in termites is considered briefly in Chapter 21, Section 7, and extensively in the volume edited by Watson *et al.* (1985).

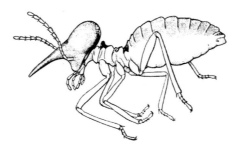

FIGURE 7.12. Nasute soldier of *Trinervitermes* sp. (Termitidae). [From P.-P. Grassé (ed.), 1949, *Traité de Zoologie*, Vol. IX. By permission of Masson, Paris.]

Carpenter, F. M., 1992, *Treatise on Invertebrate Paleontology. Part R. Arthropoda* 4, *Vols. 3 and 4 (Superclass Hexapoda),* University of Kansas, Lawrence.

Donovan, S. E., Jones, D. T., Sands, W. A., and Eggleton, P., 2000, Morphological phylogenetics of termites (Isoptera), *Biol. J. Linn. Soc.* **70**:467–513.

Eggleton, P., 2001, Termites and trees: A review of recent advances in termite phylogenetics, *Insectes Soc.* **48**:187–193.

Grassé, P. P., 1982–1986, *Termitologia,* Vols. I–III, Masson, Paris.

Harris, W. V., 1969, *Termites as Pests of Crops and Trees,* Commonwealth Institute of Entomology, London.

Harris, W. V., 1971, *Termites—Their Recognition and Control* (2nd ed.), Longmans-Green, London.

Helfer, J. R., 1987, *How to Know the Grasshoppers, Crickets, Cockroaches and Their Allies,* Dover, New York.

Hickin, N. E., 1971, *Termites: A World Problem,* Hutchinson, London.

Howse, P. E., 1970, *Termites,* Hutchinson, London.

Kambhampati, S., and Eggleton, P., 2000, Phylogenetics and taxonomy, in: *Termites: Evolution, Sociality, Symbioses, Ecology* (T. Abe, D. E. Bignell, and M. Higashi, eds.), Kluwer Academic Publishing, Dordrecht.

Kambhampati, S., Kjer, K. M., and Thorne, B. L., 1996, Phylogenetic relationships among termite families based on DNA sequence of mitochondrial 16S ribosomal RNA gene, *Insect Mol. Biol.* **5**:229–238.

Korb, J., 2003, Thermoregulation and ventilation of termite mounds, *Naturwissenschaften* **90**:212–219.

Krishna, K., 1970, Taxonomy, phylogeny and distribution, in: *Biology of Termites,* Vol. II (K. Krishna and F. M. Weesner, eds.), Academic Press, New York.

Krishna, K., and Weesner, F. M. (eds.), 1969, 1970, *Biology of Termites,* Vols. I and II, Academic Press, New York.

McKittrick, F. A., 1964, Evolutionary studies of cockroaches, *Mem. Cornell Univ. Agri. Exp. Stn.* **389**:177 pp.

Pearce, M. J., 1997, *Termites: Biology and Pest Management,* CAB International, Wallingford, U.K..

Watson, J. A. L., and Gay, F. J., 1991, Isoptera, in: *The Insects of Australia,* 2nd ed., Vol. I (CSIRO, ed.), Melbourne University Press, Carlton, Victoria.

Watson, J. A. L., Okot-Kotber, B. M., and Noirot, C. (eds.), 1985, *Caste Differentiation in Social Insects,* Pergamon Press, Elmsford, NY.

Weesner, F. M., 1965, *The Termites of the United States,* National Pest Control Association, Elisabeth, NJ.

Wilson, E. O., 1971, *The Insect Societies,* Harvard University Press, Cambridge, MA.

6. Grylloblattodea

SYNONYMS: Notoptera, Grylloblattaria, Grylloblattida   COMMON NAMES: rock crawlers, ice crawlers

Elongate insects; head prognathous, mouthparts mandibulate, compound eyes reduced or absent, ocelli absent, antennae long and filiform; thoracic segments similar and well distinguished, legs virtually identical with large coxae and five-segmented tarsi, wings and auditory organs absent; abdomen with long, segmented cerci, females with well-developed ovipositor, males with articulated coxites on ninth sternum and asymmetrical genitalia.

Walker (1914) described the first living representative of the order. Now totalling 26 species from western North America, eastern Siberia, Japan, Korea, and northeast China, living grylloblattids are the remnants of an abundant and diverse group of insects from the Permian.

Structure

Grylloblattodea (Figure 7.13) are elongate, wingless insects that as adults are 2–3.5 cm long. Their head is flattened and prognathous. It carries well-developed mandibulate mouthparts and long, filiform antennae. The compound eyes are reduced to a few ommatidia or are entirely absent. There are no ocelli. The thoracic segments are more or less identical, though the prothorax is slightly larger than the other two. The six legs are similar in structure,

FIGURE 7.13. *Grylloblatta campodeiformis* (Grylloblattodea). [From E. M. Walker, 1914, A new species of Orthoptera, forming a new genus and family, *Can. Entomol.* **46**:93–99. By permission of the Entomological Society of Canada.]

each with a large coxa and a five-segmented tarsus. The abdomen has 10 obvious segments and the 11th is represented by the epiproct and paraprocts. The ovipositor of females comprises three pairs of valves. The ninth sternite in males carries a pair of asymmetric coxites, each bearing a small style. The cerci are long, eight-segmented structures.

The internal structure is orthopteroid.

Life History and Habits

Grylloblattodea are cryptozoic insects, generally found in cold wet locations, often at relatively high altitudes, under stones, in rotting logs or leaf litter of cool temperate forests, etc. In Korea grylloblattids have also been taken in caves. Contrary to what is usually reported, the insects are active by both day and night (Rentz, 1991). They favor low temperatures (optimally around 4°C) and go underground during warmer months. In winter they probably remain active, occupying the air space between the snow and ground surface. Interestingly, they have very limited ability to withstand below-freezing temperatures. They are typically predaceous, eating a variety of other insects, dead or alive, but also ingest some plant material. The black eggs are laid singly, and embryonic development takes from a few months to 3 years. There are eight juvenile stages, and development may take up to 7 years.

Phylogeny and Classification

The relationship of the Grylloblattodea to other orthopteroid groups has been the subject of considerable discussion. Recent grylloblattids possess a combination of primitive "blattoid" and "orthopteroid" features, together with specialized features of their own. The multisegmented cerci, five-segmented tarsi, large coxae, and asymmetric male genitalia suggest a relationship with cockroaches, whereas the well-developed ovipositor and the structure of the tentorium are more orthopteroid in nature. Thus, some earlier authors suggested that living grylloblattids were specialized survivors of a primitive protorthopteran stock from which the Dictyoptera and Orthoptera evolved. However, new discoveries, plus the recognition that many so-called protorthopterans were, in fact, grylloblattids, have resulted in a significant change in thinking with respect to the phylogenetic position and importance of this order. Thus, although extant species are few and quite uniform, in the

Permian the order included the most abundant and diverse insects. The comparative studies of Giles (1963) and Kamp (1973) suggested that the closest relatives of grylloblattids are Dermaptera and that the two groups may have had a common ancestry. Storozhenko (1997, 2002) and Vrsansky *et al.* (2001) believe that it was from grylloblattids that Dermaptera, Plecoptera, and Embioptera evolved.

The living Grylloblattodea are placed in a single family, GRYLLOBLATTIDAE, containing four genera: *Grylloblatta* (11 western North American species); *Grylloblattina* (1 species from Siberia); *Galloisiana* (11 species from Japan, Korea, and China); and *Grylloblattella* (3 species from Siberia, China, and Korea).

Literature

The biology of Grylloblattodea is described by Walker (1937), Mills and Pepper (1937), and various authors in Ando (1982). Kamp (1979) and other authors in Ando (1982) discuss the taxonomy and distribution, and provide a bibliography, of the group. Giles (1963), Kamp (1973), and Storozhenko (1997, 2002) discuss their affinities with other orthopteroids. Storozhenko (1988) provides a key to living species.

Ando, H. (ed.), 1982, *Biology of the Notoptera*, Kashiyo-Insatsu, Nagano, Japan.

Giles, E. T., 1963, The comparative external morphology and affinities of the Dermaptera, *Trans. R. Entomol. Soc. Lond.* **115**:95–164.

Kamp, J. W., 1973, Numerical classification of the orthopteroids, with special reference to the Grylloblattodea, *Can. Entomol.* **105**:1235–1249.

Kamp, J. W., 1979, Taxonomy, distribution, and zoogeographic evolution of *Grylloblatta* in Canada (Insecta: Notoptera), *Can. Entomol.* **111**:27–38.

Mills, H. B., and Pepper, J. H., 1937, Observations on *Grylloblatta campodeiformis* Walker, *Ann. Entomol. Soc. Am.* **30**:269–274.

Rentz, D. C. F., 1991, Grylloblattodea, in: *The Insects of Australia*, 2nd ed., Vol. I (CSIRO, ed.), Melbourne University Press, Carlton, Victoria.

Storozhenko, S., 1988, A review of the family Grylloblattidae (Insecta), *Articulata* **3**:167–181.

Storozhenko, S. Y., 1997, Fossil history and phylogeny of orthopteroid insects, in: *The Bionomics of Grasshoppers, Katydids and Their Kin* (S. K. Gangwere, M. C. Muralirangan, and M. Muralirangan, eds.), CAB International, Wallingford, U.K.

Storozhenko, S. Y., 2002, Order Grylloblattida, in: *History of Insects* (A. P. Rasnitsyn and D. L. J. Quicke, eds.), Kluwer, Dordrecht.

Vrsansky, P., Storozhenko, S. Y., Labandeira, C. C., and Ihringova, P., 2001, *Galloisiana olgae* sp. nov. (Grylloblattodea: Grylloblattidae) and the paleobiology of a relict order of insects, *Ann. Entomol. Soc. Am.* **94**:179–184.

Walker, E. M., 1914, A new species of Orthoptera forming a new genus and family, *Can. Entomol.* **46**:93–99.

Walker, E. M., 1937, *Grylloblatta*, a living fossil, *Trans. R. Soc. Can.* **31**:1–10.

7. Dermaptera

SYNONYMS: Euplexoptera, Euplecoptera, Dermoptera,   COMMON NAME: earwigs
           Labiduroida, Forficulida

Generally elongate, dorsoventrally flattened insects; head prognathous with biting mouthparts and multisegmented antennae, compound eyes present (reduced or absent in epizoic forms), ocelli absent; wings generally present, fore wings modified into short smooth veinless tegmina, hind wings semicircular and membranous with veins arranged radially, legs subequal and with three-segmented tarsi; abdomen with unsegmented forcepslike cerci, ovipositor of females reduced or absent.

Dermaptera are found in all but the polar regions of the world, though they are most common in the warmer parts. Of the approximately 1800 described species, only about 25 occur in North America, 45 in Europe (including 7 in Britain), and 60 in Australia. Almost all are free-living, with about 15 species epizoic[*] on bats or rodents.

Structure

Dermaptera are dorsoventrally flattened, pale brown to black insects that range in length from about 7–50 mm, and are readily recognized by the forceps at the tip of the abdomen. The prognathous head carries a pair of multisegmented antennae, a pair of large compound eyes (except in epizoic species where they are reduced or absent), and mandibulate mouthparts. Ocelli are absent. The pronotum is enlarged somewhat and has a quadrangular shape. The metanotum is also large and has two lines of spines that lock the tegmina in the resting position. Both the tegmina and hind wings are absent in epizoic forms and some free-living species. In other species the degree of wing development is varied. The tegmina are smooth, veinless structures that do not extend beyond the metathorax. The hind wings are membranous, semicircular structures composed mainly of a very large anal area supported by 10 radiating branches of the first anal vein. The preanal area is reduced to a small anterior area and contains only the radius and cubitus veins. The wings fold both longitudinally and transversely and are stored beneath the tegmina. The legs are more or less similar in form, increasing in size posteriorly. They bear a 3-segmented tarsus. The 10-segmented abdomen is flattened dorsoventrally and telescopic (because of the overlapping of the tergal plates). At its posterior tip is a smooth pygidium (called the eleventh segment by some authors). The form of the cerci is varied. In Forficulina they are unsegmented, forcepslike structures, more curved in the male than in the female. They are used offensively and defensively, and for assisting in the opening and closing of the wings, and during copulation. In epizoic species they are hairy, stylelike structures. In females of some primitive Forficulina a reduced ovipositor is found, but other Dermaptera lack this structure. Paired penes are found in males of some species, but usually one of these organs is reduced or absent.

The gut resembles that of other orthopteroids, except that it lacks mesenteric ceca. Between 8 and 20 Malpighian tubules occur, usually in groups of 4 or 5. The nervous system is generalized, the ventral nerve cord containing three thoracic and six abdominal ganglia. The tracheal system, which lacks air sacs, opens to the exterior via two thoracic and eight abdominal spiracles. In both sexes the reproductive system shows considerable variation. In males there is a pair of testes, paired vasa deferentia, single or paired seminal vesicles, and one or two ejaculatory ducts. In females, the ovaries are paired and contain polytrophic ovarioles. The ovarioles are arranged in two ways, either in three rows along the length of the lateral oviduct, or as a group at the anterior end of the oviduct.

Life History and Habits

The free-living earwigs are secretive, nocturnal creatures that hide in crevices, under stones, in logs, etc. They are fast runners and, although many have wings, they use them only rarely. Most species are omnivorous, and a few may damage young shoots and buds of plants. When animal food is available, however, it seems to be preferred.

---

[*] Species of Arixeniina and Hemimerina are usually decribed as ectoparasitic. However, their parasitic nature has not been established.

A short courtship precedes mating in which the partners face in opposite directions with their genitalia in contact. In warmer regions reproduction occurs year-round whereas in temperate climates it is probably restricted to the summer. Except for epizoic species, Dermaptera are oviparous, eggs being laid in batches in a short tunnel constructed by the female. Females care for the eggs and first two instars, but then become noticeably cannibalistic. There are four or five juvenile instars. In warmer regions young earwigs develop rapidly, and there are several generations per year; in temperate species larval development is arrested with the onset of cold weather, to be completed the following spring.

The epizoic Dermaptera are viviparous, eggs developing within the follicle of the ovariole (see Figure 20.15). Approximately six embryos develop at a time, apparently being nourished via a placentalike structure attached to the head of each embryo.

Phylogeny and Classification

According to Giles (1963), the nearest relatives of the Dermaptera are the Grylloblattodea, the two orders having evolved from some common protorthopteran stock. This conclusion is supported by Kamp's (1973) numerical analysis of selected orthopteroids. However, other authors, notably Storozhenko (1997, 2002) claim that the Dermaptera have evolved from early Grylloblattodea. The earliest fossil Dermaptera, known from the Upper Jurassic, are assignable to the extinct suborder Archidermaptera and the Forficulina which includes most extant species. Usually, modern Dermaptera have been arranged in three suborders: Forficulina (free-living forms), Arixeniina, and Hemimerina (rare epizoic forms). The Arixeniina comprises five species in two genera, *Arixenia* (Figure 7.14B) and *Xeniaria*, that live in close association with Southeast Asian cave bats. The Hemimerina

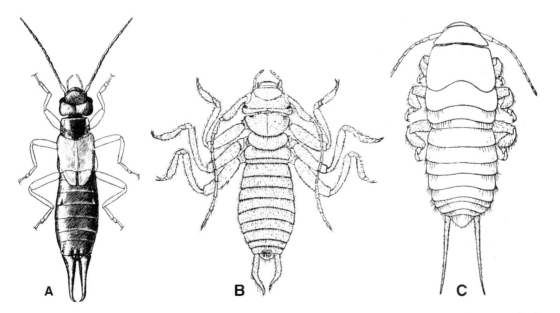

FIGURE 7.14. Dermaptera. (A) Female European earwig *Forficula auricularia*; (B) *Arixenia* sp.; and (C) *Hemimerus* sp. [B, C, from P.-P. Grassé (ed.), 1949, *Traité de Zoologie*, Vol. IX. By permission of Masson, Paris.]

includes about 10 species of *Hemimerus* (Figure 7.14C), all of which are epizoic on African giant rats of the genus *Cricetomys*. Some authors consider that the features by which *Arixenia* (Popham, 1965) and *Hemimerus* (Klass, 2001) differ from free-living forms are simply adaptations to their epizoic life and therefore include these in the Forficulina. The classification used here is that of Vickery and Kevan (1983, 1986).

Suborder Forficulina

Superfamily Pygidicranoidea

Most of the 200 or so species in this group are placed in the family PYGIDICRANIDAE, though this may be a paraphyletic arrangement. It includes the most primitive living earwigs, with representatives in Asia, Australia, South Africa, Madagascar, and South America.

Superfamily Karschielloidea

All members of this small group belong to the family KARSCHIELLIDAE. These large carnivorous Dermaptera are restricted to South Africa where they feed on ants.

Superfamily Spongiphoroidea (= Labioidea)

Included in this superfamily are the SPONGIPHORIDAE (= LABIIDAE) (240 species) and ANISOLABIDIDAE (= CARCINOPHORIDAE) (115 species). Members of both familiespt are principally found in warmer regions of the world, though in each group there are a few representatives in temperate regions and some species with a worldwide distribution (often as a result of human activity).

Superfamily Forficuloidea

Three families are included in the Forficuloidea, the CHELISOCHIDAE, LABIDURIDAE, and FORFICULIDAE. About 55 species of chelisochids are recognized, mostly from Southeast Asia, with a few from Africa and Australia. The family Labiduridae (60 species) has a worldwide distribution and includes some large earwigs such as the cosmopolitan *Labidura riparia* found under debris along riverbanks and beaches. Some 250 species of the worldwide family Forficulidae have been described, including the European earwig, *Forficula auricularia* (Figure 7.14A), which as a result of commerce is now cosmopolitan in cooler parts of the world. The species became established in North America in the early 1900s. Contrary to normal opinion, *F. auricularia* is generally a beneficial insect, feeding preferentially on other, potentially pestiferous, arthropods. However, when its numbers increase so that this kind of food becomes scarce, it will attack flowers, fruit, and vegetables causing severe damage.

Literature

Brown (1982), and Vickery and Kevan (1983, 1986) deal with the biology of earwigs. The evolution and classification of the order are discussed by Giles (1963), Popham (1965, 1985), and Vickery and Kevan (1983, 1986). Keys for identification are provided by Helfer (1987), Vickery and Kevan (1983, 1986) [North American forms], Harz and Kaltenbach

179

THE
PLECOPTEROID,
BLATTOID, AND
ORTHOPTEROID
ORDERS

(1976) [European genera], Marshall and Haes (1988) [British species], and Rentz and Kevan (1991) [Australian families].

Brown, W. L., 1982, Dermaptera, in: *Synopsis and Classification of Living Organisms*, Vol. 2 (S. P. Parker, ed.), McGraw-Hill, New York.

Giles, E. T., 1963, The comparative external morphology and affinities of the Dermaptera, *Trans. R. Entomol. Soc. London* **115**: 95–164.

Harz, K., and Kaltenbach, A., 1976, *The Orthoptera of Europe*, Vol. III, Junk, The Hague.

Helfer, J. R., 1987, *How to Know the Grasshoppers, Crickets, Cockroaches and Their Allies*, Dover, New York.

Kamp, J. W., 1973, Numerical classification of the orthopteroids, with special reference to the Grylloblattodea, *Can. Entomol.* **105**: 1235–1249.

Klass, K. D., 2001, The female abdomen of the viviparous earwig *Hemimerus vosseleri* (Insecta: Dermaptera: Hemimeridae), with a discussion of the postgenital abdomen of Insecta, *Zool. J. Linn. Soc.* **131**: 251–307.

Marshall, J. A., and Haes, E. C. M., 1988, *Grasshoppers and Allied Insects of Great Britain and Ireland*, Harley Books, Colchester, U.K.

Popham, E. J., 1965, The functional morphology of the reproductive organs of the common earwig *(Forficula auricularia)* and other Dermaptera with reference to the natural classification of the order, *J. Zool.* **146**:1–43.

Popham, E. J., 1985, The mutual affinities of the major earwig taxa (Insecta, Dermaptera), *Z. Zool. Syst. Evolutionsforsch.* **23**: 199–214.

Rentz, D. C. F., and Kevan, D. K. McE., 1991, Dermaptera, in: *The Insects of Australia*, 2nd ed., Vol. I (CSIRO, ed.), Melbourne University Press, Carlton, Victoria.

Storozhenko, S. Y., 1997, Fossil history and phylogeny of orthopteroid insects, in: *The Bionomics of Grasshoppers, Katydids and Their Kin* (S. K. Gangwere, M. C. Muralirangan, and M. Muralirangan, eds.), CAB International, Wallingford, U.K.

Storozhenko, S. Y., 2002, Order Grylloblattida, in: *History of Insects* (A. P. Rasnitsyn and D. L. J. Quicke, eds.), Kluwer, Dordrecht.

Vickery, V. R., and Kevan, D. K. McE., 1983, A monograph of the orthopteroid insects of Canada and adjacent regions, Vol. I, *Mem. Lyman Entomol. Mus. Res. Lab.* **13**: 1–679.

Vickery, V. R., and Kevan, D. K. McE., 1986, The grasshoppers, crickets, and related insects of Canada and adjacent regions. Ulonata: Dermaptera, Cheleutoptera, Notoptera, Dictuoptera, Grylloptera, and Orthoptera, *The Insects and Arachnids of Canada* **14**: 1–918.

8. Phasmida

SYNONYMS: Phasmodea, Cheleutoptera, Phasmatoptera, Phasmatodea, Gressoria

COMMON NAMES: stick insects, leaf insects, walking sticks, phasmids

Moderate-sized to very large insects, usually of elongate cylindrical form, occasionally leaflike; head with well developed compound eyes and mandibulate mouthparts, ocelli often absent; prothorax short, mesothorax and metathorax long, with or without wings, all legs very similar with small widely separated coxae and 3- to 5-segmented tarsi; ovipositor small and concealed, male genitalia asymmetrical and hidden, cerci unsegmented; specialized auditory and stridulatory structures absent in most species.[*]

Most of the approximately 2500 described species in this predominantly tropical group occur in the Indo-Malaysian region. About 150 species are found in Australia, 30 in North America, and 3 in Britain. The order includes the longest extant insect species, *Pharnacia serratipes*, from Borneo, females of which reach 33 cm, excluding the antennae and legs.

---

[*] *Phyllium* and related genera have a stridulatory apparatus on the antennae.

Most members of this order are remarkable for their close resemblance to the plants on which they are normally found (Figure 7.15). The body, which in some species may exceed 30 cm in length, is commonly elongate, wingless, or brachypterous and resembles a twig. In those species that retain wings the body may be dorsoventrally flattened and sculptured so as to resemble a leaf or a group of leaves.

The prognathous head bears a pair of antennae that may be short to very long. Compound eyes are always present, but ocelli are found only in some of the winged species, when they may be confined to males. In accord with the phytophagous habit, the mouthparts are of a generalized mandibulate form. The prothorax is small, while the mesothorax and metathorax are elongate with the latter usually firmly connected to the first abdominal segment. Wings may be present or absent, and all intermediate conditions of brachyptery are known. When wings are present they are typically fully developed in males but reduced in females. The fore wings take the form of tegmina, which in many species are much shorter than the hind wings. In leaflike species the venation is much modified to mimic the veins of a leaf. All legs are similar, with small, widely separated coxae and, in the leaflike forms, broadly flattened tibiae and femora. Eleven abdominal segments are present, though the 11th is represented only by the epiproct, paired paraprocts, and unsegmented cerci. In males the terminal abdominal segments and the aedeagus are of varied form; in some species the cerci are modified as claspers; and a sclerotized, prong-shaped process, the vomer, used in copulation, occurs in some groups. The ovipositor, which comprises three pairs of small valves, is covered by the operculum, a keel-shaped structure formed from the eighth sternum.

The gut is straight and comprises a large crop whose posterior part functions as a gizzard, a long midgut that bears numerous external papillae over its posterior part, and a short hindgut. Numerous Malpighian tubules, arranged in two groups, enter the gut via a common duct. The central nervous system is primitive and includes three thoracic and seven abdominal ganglia. The male reproductive system consists of a pair of tubular testes and short vasa deferentia that open together with various accessory glands into the ejaculatory

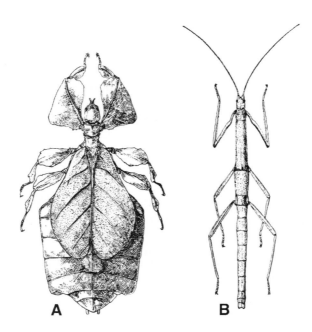

A   **B**

FIGURE 7.15. Phasmida. (A) A leaf insect, *Phyllium* sp.; and (B) a stick insect, *Carausius morosus*. [A, from P.-P. Grassé (ed.), 1949, *Traité de Zoologie*, Vol. IX. By permission of Masson, Paris. B, from H. Ling Roth, 1916, Observations on the growth and habits of the stick insect, *Carausius morosus* Br., *Trans. Entomol. Soc. Lond.* **1916**:345–386. By permission of the Royal Entomological Society.]

duct. In females several panoistic ovarioles lie alongside the lateral oviducts. One or two spermathecae and a pair of accessory glands open into the dorsally placed bursa copulatrix near its junction with the common oviduct. Paired prothoracic repugnatory glands, opening to the exterior in front of the fore coxae, occur in some species, for example, the North American two-striped walking stick (*Anisomorpha buprestoides*).

Life History and Habits

Phasmids are sedentary insects whose ability to escape from would-be predators is largely dependent on their close resemblance to twigs or leaves. However, they do possess a number of other devices that can be brought into operation should they be detected. These include falling to the ground and entering a cataleptic state, in which they may avoid further detection, secretion of an obnoxious material from the repugnatory glands, and reflex autotomy between the femur and trochanter should a leg be seized. Many phasmids are nocturnal, and during the day may remain completely immobile. Typically they are green or brown insects but the color of different populations of the same species may vary considerably. A few species exhibit density-dependent phase polymorphism analogous to that in locusts. It should be emphasized, however, that the behavioral differences found between the different phases of locusts are not seen in phasmids. In addition, a few species can change their color physiologically in a matter of hours. This color change results from the aggregation or dispersion of pigment in the epidermal cells and is controlled by the endocrine system.

Though phasmids are generally uncommon, on occasion population densities of some species are sufficiently high that they become economically important defoliators in hardwood forests (e.g., eucalyptus in Australia [Key, 1991]). Associated with their rarity is the development, in many species, of facultative parthenogenesis, the unfertilized eggs giving rise only to females (very rarely to individuals of both sexes). In a few species, for example, *Carausius* (= *Dixippus*) *morosus*, parthenogenesis is virtually obligate, males being extremely rare. Eggs, which frequently resemble seeds, are laid singly and usually allowed to simply fall to the ground. Development time is variable, even among eggs laid by the same female. Eggs may remain viable for 2 or more years. Postembryonic development takes several months, and there are on average six molts in females, one or two fewer in males.

Phylogeny and Classification

Fossil remains of phasmids, though not plentiful, extend back to the Upper Permian. Surprisingly, perhaps, these are remarkably similar to some of the less specialized modern forms. Phasmids assignable to the two extant families first appear in the Early Oligocene (Rasnitsyn and Quicke, 2002). Sharov's (1968) extensive study of these fossils suggested that phasmid affinities are closest to the Orthoptera (suborder Caelifera). In contrast, Kamp's (1973) comparative morphological analysis of living species led him to conclude that the Phasmida are most closely related to the Dermaptera and Grylloblattodea, and that perhaps the three form a natural group. Günther (1953), principally on the basis of a single feature on the tibial segment, divided the order into two families, PHYLLIIDAE (leaf insects and related sticklike forms) (650 species) and PHASMATIDAE (the remaining stick insects) (1850 species), though the two groups show much parallelism. More recent, though not necessarily more satisfactory classifications (Key, 1991) have tended to divide the group up into many more families (e.g., Bradley and Galil, 1977; Kevan, 1982). Key (1991) supported an earlier suggestion that the aberrant North American genus *Timema*, species of which have a very short (1–3 cm long) body, three-segmented tarsi, and the metanoturn not fused with the first abdominal segment, be placed in its own family TIMEMATIDAE.

Literature

Phasmid biology is reviewed by Clark (1974), Bedford (1978), Kevan (1982), and Key (1991). The taxonomy and distribution of the order are discussed by Günther (1953) and Kevan (1982), and phylogeny by Sharov (1968), Hennig (1981), and Mazzini and Scali (1987). Helfer (1987) provides a key to the North American forms, while Bradley and Galil (1977) have keys to subfamilies and tribes.

Bedford, G. E., 1978, Biology and ecology of the Phasmatodea, *Annu. Rev. Entomol.* **23**:125–149.

Bradley, J. C., and Galil, B. S., 1977, The taxonomic arrangement of the Phasmatodea with keys to the subfamilies and tribes, *Proc. Entomol. Soc. Wash.* **79**:176–208.

Clark, J. T., 1974, *Stick and Leaf Insects*, Sherlock, Winchester, Hants.

Günther, K., 1953, Über die taxonomische Gleiderung und die geographische Verbreitung der Insektenordnung der Phasmatodea, *Beitr. Entomol.* **3**:541–563.

Helfer, J. R., 1987, *How to Know the Grasshoppers, Crickets, Cockroaches and Their Allies*, Dover, New York.

Hennig, W., 1981, *Insect Phylogeny*, Wiley, New York.

Kamp, J. W., 1973, Numerical classification of the orthopteroids, with special reference to the Grylloblattodea, *Can. Entomol.* **105**:1235–1249.

Kevan, D. K. McE. 1982, Phasmatoptera, in: *Synopsis and Classification of Living Organisms*, Vol. 2 (S. P. Parker, ed.), McGraw-Hill, New York.

Key, K. H. L., 1991, Phasmatodea, in: *The Insects of Australia*, 2nd ed., Vol. 1 (CSIRO, ed.), Melbourne University Press, Carlton, Victoria.

Mazzini, M., and Scali, V. (eds.), 1987, *Stick Insects. Phylogeny and Reproduction*, Universities of Siena and Bologna, Italy.

Rasnitsyn, A. P., and Quicke, D. L. J. (eds.), 2002, *History of Insects*, Kluwer, Dordrecht.

Sharov, A. G., 1968, The phylogeny of the Orthopteroidea, *Tr. Paleontol. Inst. Akad. Nauk SSSR* **118**:1–213 (Engl. transl., 1971).

9. Mantophasmatodea

SYNONYMS: none   COMMON NAMES: heelwalkers, gladiators

Medium-sized, wingless insects; hypognathous head with mandibulate mouthparts; well-developed compound eyes, filiform multisegmented antennae, ocelli absent; prothoracic pleuron large and fully exposed, legs uniform except fore femora thickened and both fore and mid femora with ventral rows of short spines, coxae elongate, tarsi five-segmented; female with well-developed ovipositor, cerci unsegmented (shorter in male than in female) and modified for clasping.

This is the most recently established insect order. Its formation became necessary when previously unidentified specimens, including preserved extant forms and fossils in Baltic amber, could not be assigned to any recognized order (Klass *et al.*, 2002; Zompro *et al.*, 2002; Adis *et al.*, 2002). Extant species are known only from Africa south of the equator, though the fossils indicate that the order's earlier distribution included Europe.

183

*THE
PLECOPTEROID,
BLATTOID, AND
ORTHOPTEROID
ORDERS*

FIGURE 7.16. Mantophasmatodea. *Praedatophasma maraisi* female. [From O. Zompro, J. Adis, and W. Weitschat, 2002, A review of the order Mantophasmatodea, *Zool. Anz.* **241**:269–279. By permission of Urban and Fischer Verlag and Dr. Oliver Zompro.]

Structure

At first glance, these insects appear to be a "hybrid" of a mantis and a stick insect (hence the name of the order) (Figure 7.16). They range in length from about 11–25 mm. The hypognathous head has typical mandibulate mouthparts and compound eyes, but ocelli do not occur. The long antennae have a detailed structure different to that of other insects. The subgenal sulcus follows a very different course to that seen in other insects as a result of a dorsal shift in the position of the anterior tentorial arms. The prothoracic pleuron is large and fully exposed. Wings are absent. The legs are generally similar, except that the fore and mid tibiae (and to some extent the femora) have two ventral rows of short spines, the fore femora are thickened, and the hind femora are slightly elongated, enabling the insect to jump weakly. The coxae are elongate and the tarsi are five-segmented. Very typical of the Mantophasmatodea is a small dorsal projection in the membrane distal to the third tarsomere and a huge arolium that is elevated during walking, hence the order's common name "heelwalkers." The abdomen is 11-segmented. In the male the subgenital plate has a medioventral projection used for drumming, a feature seen in only one other insect order, the Plecoptera. Further, in almost all species the complicated phallic structures are asymmetric, the right phallomere being small and showing great similarity with that of Dictyoptera; only *Tanzaniophasma* has simple symmetrical genitalia. Above the genitalia there is a sclerotized projection, the vomeroid, which resembles the vomer of Phasmatodea. The female has a well developed but short ovipositor that protrudes beyond the subgenital lobe formed by parts of coxosternum 8. Distinct features of the mantophasmatodean ovipositor are the blunt gonapophyses on segment 8, the short, upcurved gonoplacs, and the gonapophyses of segment 9 fused to the latter. The unsegmented cerci are prominent (less so in females), clasping (in males) and not articulated with tergum 10.

The internal structure is not well known. The alimentary canal includes a large proventriculus, very similar to that of Grylloblattodea, and two mesenteric ceca. The heart lacks arteries in the midabdominal region, and a ventral diaphragm is present. Individual segmental ganglia occur in the thorax and abdomen. Females lack the ectodermal accessory glands seen, for example, in Dictyoptera.

Life History and Habits

Little is known about the biology of Mantophasmatodea. They are clearly hemimetabolous, with specimens at various stages of juvenile development collected in Baltic amber. In spring females deposit an egg pod containing 10–12 eggs in sandy soil. The eggs lie dormant through the hot, dry summer, hatching after the first autumn rains. Head capsule measurements and observations of changes in the structure of antennae and ovipositor suggest that there are five juvenile instars in each sex. Postembryonic development takes 1–1.5 months.

Living species have been taken in dry to moderately humid, stony areas in Namibia, Tanzania, and South Africa. They appear to be nocturnal, crawling slowly on vegetation in search of insect prey (including small moths, silverfish, cockroaches, psocopterans, and spiders). Small insects are captured using only the forelegs; larger prey is caught with the forelegs and middle legs, in the manner of some predatory katydids.

Phylogeny and Classification

Presently, this order comprises 11 extant species arranged in three families (AUSTROPHASMATIDAE [eight species], MANTOPHASMATIDAE [two species], and TANZANIOPHASMATIDAE [one species], plus *Praedatophasma maraisi* and *Tyrannophasma gladiator* (extant) and *Raptophasma kerneggeri* from Baltic amber (Eocene), which are of uncertain placement due to lack of data (Klass *et al.*, 2003a,b). As the above description of their structure indicates, they share unusual features with certain orders: Grylloblattodea, Plecoptera, Phasmatodea, and Dictyoptera. Klass *et al.* (2002) speculated that the closest relatives of the mantophasmatids might be the Grylloblattodea and/or Phasmatodea. A subsequent limited molecular analysis suggested that mantids (Dictyoptera; suborder Mantodea) form the sister group (Klass *et al.*, 2003a). However, it should be noted that this study did not include representation from the Grylloblattodea. Molecular studies in progress indicate that the grylloblattids are the closest relatives of heelwalkers (Terry and Whiting, unpublished).

Literature

Adis, J., Zompro, O., Moombolah-Goagoses, E., and Marais, E., 2002, Gladiators: A new order of insect, *Sci. Amer.* **287**(November):60–65.

Klass, K.-D., Zompro, O., Kristensen, N. P., and Adis, J., 2002, Mantophasmatodea: A new insect order with extant members in the Afrotropics, *Science* **296**:1456–1459.

Klass, K.-D., Picker, M. D., Damgaard, J., van Noort, S., and Tojo, K., 2003a, The taxonomy, genitalic morphology, and phylogenetic relationships of Southern African Mantophasmatodea, *Entomol. Abhandl.* **61**:3–67.

Klass, K.-D., Damgaard, J., and Picker, M. D., 2003b, Species diversity and intraordinal phylogenetic relationships of Mantophasmatodea, *Entomol. Abhandl.* **61**:144–146.

Zompro, O., Adis, J., and Weitschat, W., 2002, A review of the order Mantophasmatodea, *Zool. Anz.* **241**:269–279.

10. Orthoptera

SYNONYMS: Saltatoria, Saltatoptera, Orthopteroida COMMON NAMES: grasshoppers, locusts, katydids, crickets

Medium-sized to large, winged, brachypterous or apterous insects; head with mandibulate mouthparts; well-developed compound eyes, either long or relatively short antennae; prothorax large, hindlegs in almost all species enlarged for jumping, coxae small and widely spaced, tarsi usually three- or four-segmented, when present fore wings usually forming thickened tegmina; females usually with well developed exposed ovipositor, males with concealed copulatory structures, cerci short to moderately long and unsegmented; auditory and stridulatory organs very often present.

The Orthoptera is a large order of insects with more than 20,000 species having a worldwide, though mainly tropical, distribution. More than 1800 species are found in North America, about 2800 in Australia, and 33 in Britain.

Structure

Taken as a group, the Orthoptera, in keeping with their varied biology, exhibit a wide range of anatomical and morphological features. Though they are mostly medium-sized to large insects, the order includes some of the largest and the smallest members of the class. The head is usually hypognathous, but may be prognathous in some burrowing species, katydids, and tree crickets. The compound eyes are typically large, and the antennae vary from comparatively short (suborder Caelifera) to very long (suborder Ensifera). Eyes and antennae are reduced in size in burrowing or cave-dwelling forms. The mouthparts are mandibulate but show some modifications according to the diet of the insect and occasionally to other considerations. The pronotum is large and extends lateroventrally to cover the pleural region. The mesothoracic and metathoracic nota are closely associated, though the basic components can still be readily distinguished (see Chapter 3, Section 4.2). The legs are unequally developed. The forelegs of primarily digging forms are short but much enlarged. In some predaceous species the fore femora and/or tibiae are equipped with rows of long spines. In almost all Ensifera the fore tibiae bear auditory organs (see Chapter 12, Section 3.2). The hind femora of most species are greatly enlarged to accommodate the muscles used in jumping, but they may also be modified for production of sound by having a row of small pegs on their inner face. Typically, the two pairs of wings are well developed, but varying degrees of reduction of the fore and hind wings, or the latter alone, may occur, especially in females. In several groups, the wings may be modified to resemble leaves, grass blades, or stems so as to camouflage the insect. Whether fully developed or not, the fore wings are thickened and known as tegmina. The hind wings, unless reduced, are broad, because of the development of a large anal area. The wing venation, particularly of the tegmina, is varied and frequently modified in conjunction with stridulation. Paired auditory (tympanal) organs (see Chapter 12, Section 3.2), if present, occur in both sexes and are found either on the first abdominal segment (in Acridoidea) or on the fore tibiae (in Tettigonioidea and Grylloidea). Eleven abdominal segments are distinguishable, though the sternum of the first is closely associated with the metathorax, and the most posterior segments are modified in conjunction with reproduction. In females a well-developed ovipositor is usually found, except in primarily burrowing forms. It is made up of three pairs of valves, though the inner pair may be reduced (see Chapter 3, Section 5.2.1). In Ensifera the ovipositor is long and used to place eggs in crevices, soft ground, or plant tissues. In Caelifera the ovipositor valves are short and stout in accordance with the digging function that they perform. In most Orthoptera the "external" copulatory structures of males are enclosed within a pouch formed by the greatly enlarged ninth abdominal sternum. The cerci are unsegmented and, in most species, short and rigid. However, in many Ensifera (especially Grylloidea) they are rather long and flexible, and in males of many species (especially Tettigonioidea) they are modified to form clasping structures used during copulation.

The internal structure of Orthoptera is usually rather generalized. In the gut the crop is large and proventriculus fairly to very well developed. The anterior midgut possesses mesenteric ceca, two in Ensifera, usually six in Caelifera. There are many Malpighian tubules that enter the gut directly or via a common duct. In the central nervous system there are three thoracic and three to seven abdominal ganglia. The tracheal system, which communicates with the exterior by means of two thoracic and eight abdominal pairs of spiracles, is often modified, particularly in migratory forms, through the development of large, segmentally arranged air sacs. These serve to increase the volume of air exchanged during breathing movements (see Chapter 15, Section 3.3). The paired testes of males are typically united in a middorsal position. Accessory glands, often complex, open into the ejaculatory duct. In females the laterally placed ovaries comprise a varied number of panoistic ovarioles that enter the lateral oviducts serially or as a group at the anterior end. A single spermatheca is usual, and accessory glands occur either as anterior extensions of the lateral oviducts (Caelifera) or as tubules opening at the base of the ovipositor (Ensifera).

Life History and Habits

Most species are active, diurnal (most Caelifera), or nocturnal (many Ensifera) insects, capable of jumping in order to escape from would-be predators, or to launch themselves into flight. Flying ability is mostly rather limited, but a number of species (mostly members of the Acrididae) are capable of sustained flight for many hours. Members of a few groups are typically cryptozoic, living in humus or beneath logs and stones. A few species are true soil dwellers and some live in caves. Although normally "solitary," members of a number of species (especially of Caelifera) may become gregarious under certain conditions to form swarms that may reach enormous proportions (Superfamily Acridoidea). Most Orthoptera are phytophagous and consequently are extremely important because of the damage they may do to crops. Many species of Ensifera are omnivorous and a few are carnivorous, feeding on other insects, which in some instances they catch with their specialized forelegs.

Stridulation occurs in a large number of Orthoptera and serves to bring sexes together and to elicit certain behavioral responses leading to copulation. Stridulation is achieved in a great variety of ways, but, in most species, the principal method of sound production is either tegminal (crickets and katydids) or femoroalary (in many grasshoppers). In the former, the sound is typically produced when one tegmen is drawn across the other. The especially strengthened, toothed cubital vein (file) on one tegmen makes contact with the raised edge (scraper) of the other tegmen, and an adjacent area of wing, the mirror, is caused to resonate. In femoroalary stridulation it is normally the drawing of the hind femora across the tegmina that creates the sound. Another common method of stridulation, particularly among more primitive forms, is femoroabdominal, whereby the inside of the femur is rubbed over teeth or ridges on the side of the abdomen. Some grasshoppers make a clattering sound in flight either by striking the hind wing on the tegmen or by the rapid opening and closing of the fanlike hind wings. And there are numerous other less common means of sound production (Otte, 1977). It is generally the male that stridulates, though in some species the female also may perform this act. Stridulation is almost always species-specific, that is, each species has its own song, and this probably serves as an isolating mechanism between closely related species. Exceptions to this rule may sometimes occur, however, if there are other isolating mechanisms. For example, the songs of *Gryllus pennsylvanicus* and *G. veletis* are almost identical, but the former occurs as adults only in late summer and autumn and the latter in late spring and early summer.

Reproduction is almost always sexual, though a few facultatively parthenogenetic species are known. Sperm are transferred in spermatophores (except in some Eumastacidae) that vary widely in size and complexity. In many Ensifera the spermatophore is large and, after sperm evacuation, is eaten by the female, thereby making an important nutritive contribution to egg development (Gwynne, 2001). In Caelifera one to many small spermatophores may be transferred during each copulation. Eggs are laid singly or in small groups (in many Ensifera) or in batches (in Caelifera) in a variety of locations. The Ensifera typically lay them in or on stems or leaves, or in loose soil or humus. The Caelifera mostly dig into the soil, making alternate opening and closing movements of the ovipositor, and typically deposit the eggs in a mass of frothy material. This soon hardens to form a cylindrical mass known as the egg pod. Usually several batches of eggs are laid. The duration of embryonic development is varied, and many temperate species pass the winter in a diapausing egg stage, though some overwinter as juveniles. The pronymph (first-instar larva) is covered with a loose cuticle that serves to protect the insect as it emerges from the substrate. This cuticle is shed usually within a few minutes of hatching. In most species there are four to six additional larval instars in which the young form increasingly resembles the adult.

Phylogeny and Classification

Paleontology, comparative morphology, and molecular studies suggest that the sister group of the Orthoptera is the Phasmida (see Flook *et al.*, 1999). Quite possibly, the earliest orthopterans, belonging to the suborder Ensifera, evolved from phasmidlike ancestors early in the Carboniferous period. The early evolution of the Orthoptera led toward the specialization of the fore wings as tegmina, the development of stridulatory and tympanal organs, and the enlargement of the hind femora to accommodate the muscles for jumping (Carpenter, 1992). Even by the Upper Carboniferous period, there was a clear separation of the two major evolutionary lines, one of which led to the Ensifera (crickets, long-horned grasshoppers, etc.) and the other to the Caelifera (short-horned grasshoppers and their allies). This led to the suggestion that the two groups evolved independently from different protorthopteran ancestors (Ragge, 1955; Sharov, 1968; Hennig, 1981). Thus, some authorities, for example, Vickery *et al.* (1974) and Kevan (1986), have argued strongly that each group be given ordinal status, the ensiferans being placed in the order Grylloptera, with the caeliferans in the order Orthoptera *sensu stricto*. Most studies, including recent cladistic and molecular analyses, favor the view that the differences between each group do not warrant their separation into two orders and that the traditional subdivision of the Orthoptera into the suborders Ensifera and Caelifera be retained. The possible evolutionary relationships of the major families of Orthoptera are indicated in Figure 7.17.

Within the Ensifera, the Grylloidea may be the sister group to the other three superfamilies. Of these, the Hagloidea is an almost entirely fossil group, containing only four living species.

The Caelifera may be arranged in six superfamilies, with the Tridactyloidea being the sister group to the remaining five. Both the tridactyloids and the Tetrigoidea show a number of primitive characters that place them near the base of the caeliferan stem, and they feed on bacteria, algae, mosses, or roots of higher plants. By contrast, the generally grasshopperlike members of the remaining superfamilies (in older classifications collectively called the Acridomorpha [Dirsch, 1975]) feed on the foliage of higher plants.

Suborder Ensifera

Members of the suborder Ensifera are characterized by their multisegmented antennae that, with few exceptions, are as long as or longer than the body. Tympanal organs, when present, are located on the fore tibiae. The principal living superfamilies are the Gryllacridoidea (Stenopelmatoidea), Tettigonioidea, and Grylloidea.

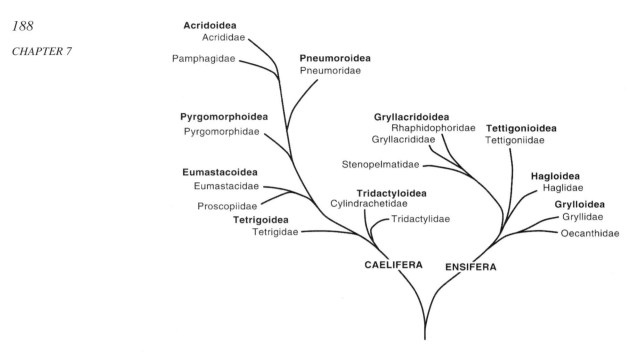

FIGURE 7.17. A proposed phylogeny of Orthoptera.

Superfamily Gryllacridoidea

The superfamily Gryllacridoidea is a primitive superfamily that contains nearly 1500 species. Its members are usually somewhat cricketlike in appearance, with longish cerci. However, they may be distinguished from true crickets because males lack a stridulatory mirror on the tegmen, and females have a laterally compressed, not needlelike, ovipositor. They may be winged, brachypterous, or apterous and may or may not have tympana on the fore tibiae. They are secretive, nocturnal creatures. The superfamily, which is mainly restricted to tropical or subtropical regions or to temperate parts of the Southern Hemisphere, includes three families: RHAPHIDOPHORIDAE, STENOPELMATIDAE, and GRYLLACRIDIDAE. The Rhaphidophoridae (cave or camel crickets) includes about 300 species of wingless, often darkly colored, humpbacked insects that live in caves, hollow logs, under stones, and in other humid situations; the genus *Ceuthophilus* (Figure 7.18A) is well known in the basements of North American houses, and *Tachycines asynamorus* is an oriental species commonly found in greenhouses. The 190 species of Stenopelmatidae (Jerusalem crickets, stone crickets, true wetas, and king crickets) are winged or apterous, large, omnivorous or carnivorous insects, found under stones, in rotting logs, etc. Some are subterranean and may have vestigial eyes and antennae, and enlarged forelegs. *Deinacrida heteracantha*, from New Zealand, may be the world's heaviest insect, gravid females weighing up to 70 g. The leaf-rolling crickets (Gryllacrididae) (Figure 7.18B) are so-called because many build shelters by rolling up a leaf and tying it with a silklike secretion of the salivary glands. Other members of this large family (600 species), from arid regions, live by day in burrows lined with silk and having a trap door so as to avoid desiccation. Species may be omnivorous, predaceous on other arthropods, or specialized herbivores feeding on grass seeds.

189

*THE
PLECOPTEROID,
BLATTOID, AND
ORTHOPTEROID
ORDERS*

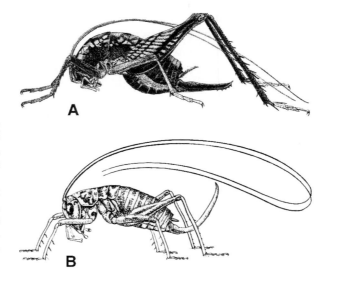

FIGURE 7.18. Gryllacridoidea. (A) A camel cricket, *Ceuthophilus maculatus*; and (B) a leaf roller, *Camptonotus carolinensis*, [A, from M. Hebard, 1934, The Dermaptera and Orthoptera of Illinois, *Bull. Ill. Nat. Hist. Surv.* **20**(3). By permission of the Illinoispt Naturalpt History Survey. B, from W. S. Blatchley, 1920, *Orthoptera of Northeastern America with Especial Reference to the Faunas of Indiana and Florida*, The Nature Publishing Co., Indianapolis, Indiana.]

Superfamily Tettigonioidea

The superfamily Tettigonioidea, which includes the common long-horned grasshoppers and katydids (bush crickets) (Figure 7.19), is the largest ensiferan superfamily, with more than 5000 described species, virtually all of which fall in the family TETTIGONIIDAE. The family is primarily tropical, though many species occur in temperate regions of the world. Tettigoniids range in size from a few millimeters to more than 13 cm in length (excluding antennae), and the largest may have wingspans exceeding 20 cm. They generally live among herbage or in trees and are commonly green. Males almost always possess a stridulatory apparatus, and in many species females also can stridulate. The species that live above

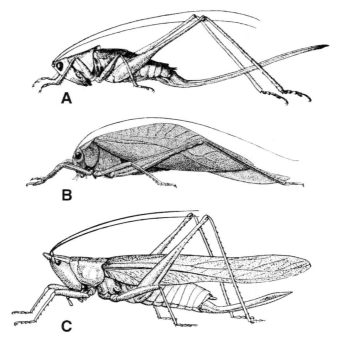

FIGURE 7.19. Tettigonioidea. (A) The meadow katydid, *Conocephalus strictus*; (B) a bush katydid, *Microcentrum rhombifolium*; and (C) a long-horned grasshopper, *Neoconocephalus palustris*. [A, B, from M. Hebard, 1934, The Dermaptera and Orthoptera of Illinois, *Bull. Ill. Nat. Hist. Surv.* **20**(3). By permission of the Illinois Natural History Survey. C, from W. S. Blatchley, 1920, *Orthoptera of Northeastern America, with Especial Reference to the Faunas of Indiana and Florida*, The Nature Publishing Co, Indianapolis, Indiana.]

the ground are usually fully winged, and their wings often resemble leaves. In ground-dwelling forms, which are less numerous, the wings are often reduced so that only the sound-producing area remains. The ovipositor is a laterally flattened, bladelike structure, either curved or straight, that is sometimes as long as the body itself or even longer. Tettigonioids in general are phytophagous or omnivorous, though a few are apparently entirely carnivorous. Occasionally its population density is sufficiently high that a species becomes economically important; for example, the Mormon cricket, *Anabrus simplex*, is frequently a serious pest of field crops and may reach "plague" proportions in western and midwestern North America. At least one African species of the genus *Ruspolia* (= *Homorocoryphus*) may form migratory flying swarms. It is widely collected as human food.

Superfamily Grylloidea

The Grylloidea (true crickets) form another very large superfamily, containing almost 3000 described species. They bear some resemblance to the Tettigonioidea but differ from the latter in having three rather than four tarsal segments and, in females, an ovipositor that is generally needlelike or cylindrical, usually straight or only slightly curved and comprising only two pairs of valves. They have rather long, flexible cerci, and a few species are green. Most species are typically nocturnal insects that hide in humid microhabitats during the day. Some species are commonly found adjacent to ponds and streams, and a number are cave dwellers. Like the tettigonioids, the grylloids exhibit the full range of wing development. They are generally omnivorous. The vast majority of species belong to the family GRYLLIDAE and are mainly drab brown to black, nocturnal insects. Occasionally, populations of some species (e.g., *Teleogryllus commodus*, the Australian field cricket) build up to economically important levels. A few species are common in buildings, for example, the house cricket (*Acheta domesticus*) (Figure 7.20) and tropical house cricket [*Gryllodes supplicans* (= *sigillatus*)]. OECANTHIDAE (tree crickets), a family of about 65 species, are generally delicate, pale-bodied insects less than 15 mm long. They oviposit in twigs, sometimes causing damage in orchards though this is perhaps offset by their habit of feeding on other insects, notably aphids. The approximately 45 species of MYRMECOPHILIDAE (ant crickets) are small (<2 mm), apterous insects with reduced eyes and short antennae, found in ant and termite nests where they apparently feed on secretions of their hosts. Some species are facultatively parthenogenetic, and females lay eggs about one third their body size! The family GRYLLOTALPIDAE (mole crickets) (Figure 7.21) includes about 50 very distinctive species. As their common name indicates, they live underground, and have greatly enlarged forelegs, reduced eyes, short antennae, short

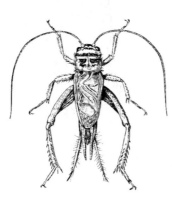

FIGURE 7.20. The house cricket, *Acheta domesticus* (Gryllidae). [From D. Sharp, 1901, *Peripatus, Myriapods and Insects* (The Cambridge Natural History Series, Vol. 5). Reprinted in 1970 as *Insects*, Vol. 1 by Dover Publications, New York.]

FIGURE 7.21. A mole cricket, *Gryllotalpa hexadactyla* (Gryllotalpidae). [From M. Hebard, 1934, The Dermaptera and Orthoptera of Illinois, *Bull. Ill. Nat. Hist. Surv.* **20**(3). By permission of the Illinois Natural History Survey.]

(non-jumping) hindlegs, and vestigial ovipositor. However, most species are fully winged and can fly strongly. Most species are omnivorous, and sometimes cause significant damage to field crops, other cultivated plants, and turf. Others feed on soil insects.

Suborder Caelifera

In members of the suborder Caelifera, the antennae are comparatively short with, in most species, less than 30 segments. Tympanal organs, when present, are on the first abdominal segment. The six superfamilies that comprise this group are the Tridactyloidea, Tetrigoidea, Eumastacoidea, Pyrgomorphoidea, Pneumoroidea, and Acridoidea.

Superfamily Tridactyloidea

Members of this superfamily were originally thought to be Ensifera, closely related to Gryllotalpidae, on account of the remarkable similarity in body and leg structure between the two groups. Included in the superfamily are two small families TRIDACTYLIDAE (pygmy mole crickets) (130 species) and CYLINDRACHETIDAE (sand gropers) (16 species). Tridactylids are semiaquatic, living in damp or wet sand adjacent to streams and lakes. They are reported to feed on algae. Sand gropers also live in burrows in soil where they feed on roots and perhaps other arthropods.

Superfamily Tetrigoidea

More than 1000 species are included in the superfamily Tetrigoidea. All are placed in a single family, TETRIGIDAE (Figure 7.22). They are commonly known as grouse locusts or pygmy grasshoppers. These grayish, often cryptically patterned insects are distinguished by an enormously enlarged pronotum that projects posteriorly to cover most of the abdomen, two-segmented tarsi, and, in most species, short antennae with 12 or fewer segments. In many species the wings (especially the tegmina) are reduced. Auditory and stridulatory structures are absent. The grouse locusts are most common in warmer areas of the world, though many species are found in temperate regions, where they overwinter in the adult stage. They prefer rather moist habitats and some species are semiaquatic, swimming freely

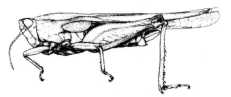

FIGURE 7.22. A grouse locust, *Tetrix subulata* (Tetrigidae). [From J. A. G. Rehn and H. J. Grant, Jr., 1961, *A Monograph of the Orthoptera of North America (North of Mexico)*, Vol. I. Monographs of the Academy of Natural Sciences of Philadelphia. By permission of the Academy of Natural Sciences of Philadelphia.]

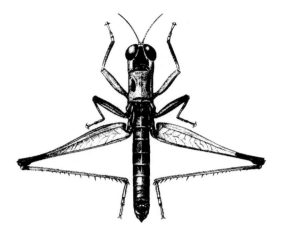

FIGURE 7.23. A monkey grasshopper, *Biroella* sp. (Eumastacidae). [From D. C. F. Rentz, 1991, Orthoptera, in: *The Insects of Australia*, 2nd ed., Vol. I (CSIRO, ed.), Melbourne University Press. By permission of the Division of Entomology, CSIRO.]

and able to remain submerged for some time. They feed on algae, moss, or mud from which they extract the organic matter.

Superfamily Eumastacoidea

Most of the more than 1100 species of Eumastacoidea are placed in the family EUMASTACIDAE (Figure 7.23), commonly known as monkey grasshoppers because of the supposed monkeylike appearance of their faces. The majority of species occur in tropical rainforests. Many are metallic or iridescent in color though some mimic leaves or twigs. The antennae are very short, wings may be present or absent, and the hindlegs are splayed out to the side when the insect is resting. Eumastacids are phytophagous and may be diurnally or nocturnally active. The remaining 130 species in this superfamily belong to the South American family PROSCOPIIDAE (stick grasshoppers), most of which are elongate, twiglike forms. They are wingless, lack jumping hindlegs, and have tiny antennae. They are found in a range of habitats, including tropical rainforests, lowland deserts, and mountainous regions.

Superfamily Pyrgomorphoidea

The single family (PYRGOMORPHIDAE) that makes up this group includes about 530 species with a worldwide distribution. The family includes some of the smallest and some of the largest grasshoppers. Generally, they can be easily distinguished from acridoids by their conical heads. Many are aposematically colored, reflecting their poisonous nature. Some exude irritant fluids or froth as a means of discouraging predators. In contrast to acridoids, pyrgomorphs tend to be sluggish and prefer to crawl rather than jump.

Superfamily Pneumoroidea

The approximately 20 species in this group, found in South and East Africa, are included in the PNEUMORIDAE. They appear to be among the most primitive living Caelifera. They are commonly known as bladder grasshoppers as males of some species can inflate their abdomen, which then serves as a sound amplifier during stridulation. Such males are fully winged and strong fliers. Non-inflating males and females are brachypterous or micropterous. Typically, individuals are cryptically colored, closely resembling their food plant.

Superfamily Acridoidea

The Acridoidea (short-horned grasshoppers and locusts) is the largest orthopteran superfamily with about 8000 described species, most of them in the family ACRIDIDAE (Figure 7.24). Although most species are found in warmer areas of the world, a large number occur in temperate climates, where usually they overwinter in the egg stage. Most acridoids inhabit grassland and other low vegetation, but a number live in trees. They are predominantly diurnally active insects. They are virtually exclusively phytophagous, eating mainly living plants. Most species are winged and some are strong fliers. There are several important subfamilies of Acrididae. The CYRTACANTHACRIDINAE are a subfamily of rather large forms confined mainly to tropical and subtropical regions of the world. The subfamily contains a number of economically very important species, including several locusts, for

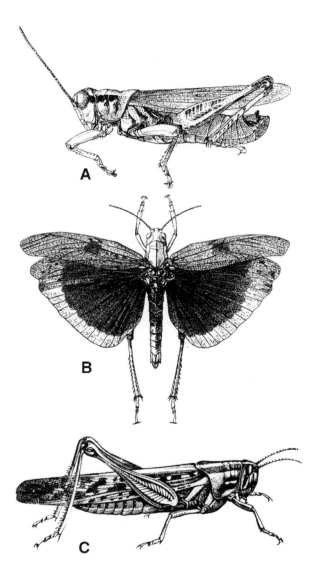

FIGURE 7.24. Acrididae. (A) The differential grasshopper, *Melanoplus differentialis*; (B) the Carolina grasshopper, *Dissosteira carolina*; and (C) the South American locust, *Schistocerca americana*. [A, B, from L. Chopard, 1938, *Encyclopedie Entomologique. XX. La Biologie des Orthoptères*. By permission of Lechevalierpt, Paris. C, from M. Hebard, 1934, The Dermaptera and Orthoptera of Illinois, *Bull. Ill. Nat. Hist. Surv.* **20**(3). By permission of the Illinois Natural History Survey.]

example, *Schistocerca gregaria gregaria* (the desert locust of Africa and southwest Asia), *S. americana* (the South American locust, comprising at least two major subspecies), and *Nomadacris septemfasciata* (the red locust). The OEDIPODINAE are another subfamily of worldwide distribution, but whose members are found mainly in warmer, drier areas. It contains many important pest species, including *Locusta migratoria* (the migratory locust of Africa, northern Australia, Asia, and southern Europe, with several distinct subspecies) and *Chortoicetes terminifera* (the Australian plague locust). The CATANTOPINAE constitute another large, widely distributed subfamily. As presently constituted, it is a rather heterogeneous assemblage and is in gradual process of revision. Usually included in this subfamily are the genera *Melanoplus* and *Dichropus*, several species of which are major pests, particularly in the grassland areas of North and South America, respectively.

The term "locust" is applied to about 20 species of Acrididae that are capable, under certain conditions, of aggregating in immense swarms that may migrate for considerable distances and cause massive damage to vegetation. Locusts (and also some Tettigonioidea, Phasmida, and Lepidoptera) can exist in more than one form or "phase." These phases differ not only in color and morphology, but also in their physiology, ecology, and behavior. Phase polymorphism is largely a density-dependent phenomenon. Under extended conditions of low population density, the locusts exist in the "solitary phase." If conditions change so that the population density increases, the locusts enter a "transition phase," and under high density conditions change to the "gregarious phase." It is the latter phase that is of importance with reference to swarming. Gregarious larvae (hoppers) are highly active and "march" in vast groups from place to place. As adults they may be carried for very considerable distances by wind currents (see Chapter 22, Section 5.2). New swarms in most species originate in more or less permanent breeding areas (outbreak areas), though desert locusts seem to be more strictly nomadic, even in the solitary phase. Though adults may breed in the regions to which they are distributed for a period of several years, the population gradually decreases because of unsuitable conditions. The effects of population density on phase development are mediated via the endocrine system (see Chapter 21, Section 7).

The PAMPHAGIDAE includes about 320 species of acridoids found in the palearctic region, south-west Asia, and Africa. Commonly known as toad grasshoppers on account of their rough, sometimes spiny appearance, they are highly camouflaged, mimicking foliage, stones, dead branches, etc. They typically occur in dry, semi-desert environments.

Literature

General accounts of the Orthoptera are given by Preston-Mafham (1990) and Rentz (1996), both of whom include many excellent photographs. Uvarov (1966, 1977) and authors in Chapman and Joern (1990) deal specifically with Acridoidea, while Bailey and Rentz (1990) and Gwynne (2001) focus on Tettigoniidae. Authors in the volume edited by Gangwere *et al.* (1997) cover a wide range of topics. The evolutionary relationships of the Orthoptera are discussed by Ragge (1955), Sharov (1968), Hennig (1981), Flook and Rowell (1997), and Flook *et al.* (1999). Dirsh (1975) and Kevan (1986) present their views on the classification of the group. Dirsh (1961) has discussed the relationships of the families of Acridoidea and provided keys for their separation. Other keys are those of Marshall and Haes (1988) [British species]; Harz (1969, 1975) [European genera]; Helfer (1987) and Arnett (2000) [most North American species]; and Rentz (1991) [Australian subfamilies].

Arnett, R. H., Jr., 2000, *American Insects: A Handbook of the Insects of America North of Mexico*, 2nd ed., CRC Press, Boca Raton, FL.

Bailey, W. J., and Rentz, D. C. F., 1990, *The Tettigoniidae: Biology, Systematics and Evolution*, Crawford House Press, Bathurst, Australia.

Carpenter, F. M., 1992, *Treatise on Invertebrate Paleontology. Part R. Arthropoda 4, Vols. 3 and 4 (Superclass Hexapoda)*, University of Kansas, Lawrence.

Chapman, R. F., and Joern, A. (eds.), 1990, *Biology of Grasshoppers*, Wiley, New York.

Dirsh, V. M., 1961, A preliminary revision of the families and subfamilies of Acridoidea (Orthoptera, Insecta), *Bull. Br. Mus. Nat. Hist. Entomol.* **10**:351–419.

Dirsh, V. M., 1975, *Classification of the Acridomorphoid Insects*, Classey, Faringdon, U.K.

Flook, P. K., and Rowell, C. H. F., 1997, The phylogeny of the Caelifera (Insecta, Orthoptera) as deduced from mtrRNA gene sequences, *Mol. Syst. Evol.* **8**:89–103.

Flook, P. K., Klee, S., and Rowell, C. H. F., 1999, Combined molecular phylogenetic analysis of the Orthoptera (Arthropoda, Insecta) and implications for their higher systematics, *Syst. Biol.* **48**:233–253.

Gangwere, S. K., Muralirangan, M. C., and Muralirangan, M (eds.), 1997, *The Bionomics of Grasshoppers, Katydids and Their Kin*, CAB International, Wallingford, U.K.

Gwynne, D. T., 2001, *Katydids and Bush Crickets: Reproductive Behavior and Evolution of the Tettigoniidae*, Cornell University Press, Ithaca, NY.

Harz, K., 1969, 1975, *The Orthoptera of Europe*, Vols. I and II, Junk, The Hague.

Helfer, J. R., 1987. *How to Know the Grasshoppers, Crickets, Cockroaches and Their Allies*, Dover, New York.

Hennig, W., 1981, *Insect Phylogeny*, Wiley, New York.

Kevan, D. K. McE., 1986, A rationale for the classification of orthopteroid insects—The saltatorial orthopteroids or grigs—One order or two? *Proc. 4th Triennial Meet. Pan Am. Acridol. Soc.*, pp. 49–67.

Preston-Mafham, K., 1990, *Grasshoppers and Mantids of the World*, Blandford, London.

Ragge, D. R., 1955, *The Wing-Venation of the Orthoptera Saltatoria*, British Museum of Natural History, London.

Rentz, D. C. F., 1991, Orthoptera, in: *The Insects of Australia*, 2nd ed., Vol. I (CSIRO. ed), Melbourne University Press, Carlton, Victoria.

Rentz, D. C. F., 1996, *Grasshopper Country: The Abundant Orthopteroid Insects of Australia*, University of New South Wales Press, Sydney.

Sharov, A. G., 1968, The phylogeny of the Orthopteroidea. *Tr. Paleontol. Inst. Akad. Nauk SSSR* **118**:1–213 (Engl. Transl., 1971).

Uvarov, B. P., 1966, 1977, *Grasshoppers and Lacusts. A Handbook of General Acridology*, Vols. I and 2, Cambridge University Press, London.

Vickery, V. R., Johnstone, D. E., and Kevan, D. K. McE., 1974, The orthopteroid insects of Quebec and the Atlantic provinces of Canada, *Mem. Lyman Entomol. Mus. Res. Lab.* **1**:204 pp.

11. Zoraptera

SYNONYMS: Zorotypida   COMMON NAME: angel insects

Minute gregarious insects; head with nine-segmented moniliform antennae and biting mouthparts; commonly apterous but occasionally winged, with greatly reduced venation, tarsi two-segmented; cerci short and unsegmented, ovipositor absent, male genitalia specialized and frequently asymmetrical.

The Zoraptera constitute a very small order of insects, containing about 32 living species in a single genus *Zorotypus* plus 6 fossil species (Engel and Grimaldi, 2002). The order has a worldwide but patchy distribution in warmer climates, including many oceanic islands. A few species occur in regions with a temperate climate, for example, the southern U.S.A. and Tibet.

Structure

First described in 1913, these minute insects (3 mm or less in length) are unusual in that in each species adults occur in two forms: a usually pale, eyeless and wingless morph, and a darker, eyed and winged morph (Figure 7.25). The hypognathous head bears a pair

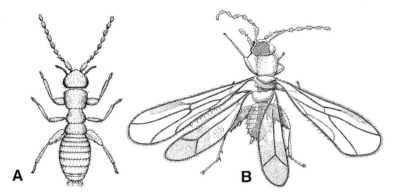

FIGURE 7.25. Zoraptera. (A) Wingless morph of male *Zorotypus hubbardi*, a species found in the United States; and (B) winged female of *Z. nascimbenei*, a Cretaceous species from Burmese amber. [A, from C. N. Smithers, 1991, Zoraptera, in: *The Insects of Australia*, 2nd ed., Vol. I (CSIRO, ed.), Melbourne University Press. By permission of the Division of Entomology, CSIRO. B, from M. S. Engel and D. A. Grimaldi, 2002, The first Mesozoic Zoraptera (Insecta), *Amer. Mus. Novit.* **3362**:20 pp.]

of moniliform antennae and biting mouthparts. Compound eyes and three ocelli occur only in the winged form. A prominent Y-shaped epicranial suture can be seen on the head. The prothorax is prominent. The mesonotum and metanotum are simple in apterous individuals. When present, the wings are membranous and have a much-reduced venation. They are frequently shed at the base, though there is no basal suture as occurs in termites. The legs have a large coxa and a two-segmented tarsus. The 11-segmented abdomen carries a pair of unsegmented cerci. There is no ovipositor, and, in males, the external genitalia are frequently asymmetrical. The homologies of the genitalia are uncertain.

The gut contains a very large crop, a short midregion, and a convoluted hind part. Six Malpighian tubules occur. The nervous system is specialized with only three thoracic and two abdominal ganglia. The reproductive organs are typically orthopteroid, with two follicles per testis and six panoistic ovarioles per ovary.

Life History and Habits

The life history and habits of Zoraptera are poorly understood. Species are found in rotting logs, sawdust piles, humus, termites' nests, etc., where they appear to feed principally on fungal hyphae and minute arthropods (e.g., mites and Collembola). Although they are gregarious and dimorphic, there appears to be little social organization. However, in colonies of *Zorotypus gurneyi* males show dominance hierarchy. Older males tend to dominate and secure most matings (Choe, 1994). Both apterous and alate forms are sexually functional. Winged forms apparently swarm, shedding their wings thereafter. Development is hemimetabolous, and juveniles are of two forms, one with and the other without wing buds. The number of juvenile instars is not known.

Phylogeny and Classification

The relationship of the Zoraptera with other insect orders is unclear, in part due to the poor fossil record. Until recently, only two fossil species were described, from Miocene amber deposits in the Dominican Republic (Poinar, 1988; Engel and Grimaldi, 2000). However, Engel and Grimaldi (2002) have described four additional species from Burmese amber

197

*THE
PLECOPTEROID,
BLATTOID, AND
ORTHOPTEROID
ORDERS*

(Cretaceous). Consequently, comparative anatomy of living species has provided the main basis for claims of phylogenetic relationships. The general form of the head, mandibulate mouthparts, structure of the thorax, presence of cerci, and nature of the male genitalia are orthopteroid or blattoid. The reduced wing venation, small number of Malpighian tubules, and concentrated abdominal ganglia, however, have been considered indicative of a hemipteroid relationship by some authors. Thus, at one time or another, Zoraptera have been designated as the sister group of Embioptera, Isoptera, Dictyoptera, Dermaptera, Paraneoptera (hemipteroid orders), even Holometabola (endopterygotes) (references in Engel and Grimaldi, 2002). Engel and Grimaldi (2000) suggested that the most strongly supported hypothesis is a sister-group relationship between Zoraptera and Embioptera.

Kukalová-Peck and Peck (1993) and Chao and Chen (2000) have argued that the extant New World species are quite distinct from those in the Old World and should be placed in seven new genera (and probably a new family). Engel and Grimaldi (2000, 2002) did not subscribe to this view and included all but one of the described species (fossil and extant) in the genus *Zorotypus*, family ZOROTYPIDAE.

Literature

The biology of this little-known order is summarized by Gurney (1938, 1974) and Riegel (1963). The possible phylogenetic relationships of the order are discussed by Engel and Grimaldi (2002) (and authors cited therein). Engel and Grimaldi (2002) provide a list of the described species and their distribution.

Chao, R. F., and Chen, C. S., 2000, *Formosozoros newi*, a new genus and species of Zoraptera (Insecta) from Taiwan, *Pan-Pac. Entomol.* **76**:24–27.

Choe, J. C., 1994, Sexual selection and mating system in *Zorotypus gurneyi* Choe (Insecta: Zoraptera): II. Determinants and dynamics of dominance, *Behav. Ecol. Sociobiol.* **34**:233–237.

Engel, M. S., and Grimaldi, D. A., 2000, A winged *Zorotypus* in Miocene amber from the Dominican Republic (Zoraptera: Zorotypidae), with discussion on relationships of and within the order, *Acta Geol. Hisp.* **35**:149–164.

Engel, M. S., and Grimaldi, D. A., 2002, The first Mesozoic Zoraptera (Insecta), *Amer. Mus. Novit.* **3362**:20 pp.

Gurney, A. B., 1938, A synopsis of the order Zoraptera, with notes on the biology of *Zorotypus hubbardi* Caudell, *Proc. Entomol. Soc. Wash.* **40**:57–87.

Gurney, A. B., 1974, Class Insecta. Order Zoraptera, *Dept. Agri. Tech. Serv. Rep. S. Afr.* **38**:32–34.

Kukalová-Peck, J., and Peck, S. B., 1993, Zoraptera wing structures: Evidence for new genera and relationship with the blattoid orders (Insecta: Blattoneoptera), *Syst. Entomol.* **18**:333–350.

Poinar, G. T., Jr., 1988, *Zorotypus palaeus*, new species, a fossil Zoraptera (Insecta) in Dominican amber, *J.N.Y. Entomol. Soc.* **96**:253–259.

Riegel, G. T., 1963, Distribution of *Zorotypus hubbardi* (Zoraptera), *Ann. Entomol. Soc. Am.* **56**:744–747.

8

The Hemipteroid Orders

1. Introduction

The four orders (Psocoptera, Phthiraptera, Hemiptera, and Thysanoptera) that constitute the hemipteroid group are united by the following features: specialized, usually suctorial, mouthparts; small anal lobe in hind wing; wing venation reduced; cerci absent; few Malpighian tubules; and ventral nerve cord with few discrete ganglia. On the whole, the hemipteroid group is more homogeneous than the orthopteroid group, although two evolutionary lines have developed, leading to the Psocoptera-Phthiraptera, on the one hand, and the Hemiptera-Thysanoptera, on the other.

2. Psocoptera

SYNONYMS: Corrodentia, Copeognatha   COMMON NAMES: barklice, booklice, psocids

Small or minute soft-bodied insects; mobile head with long filiform antennae and specialized chewing mouthparts, compound eyes usually prominent but reduced in some species; prothorax small, wings present or absent, legs with two- or three-segmented tarsi; external genitalia of both sexes concealed, cerci absent.

This order, containing about 3200 described species, has a worldwide, though predominantly tropical, distribution. Some 290 species occur in North America, about 80 in Britain, and 300 in Australia.

Structure

Psocoptera are stocky, soft-bodied insects whose length is usually less than 10 mm. The large, mobile head bears a swollen postclypeus, long filiform antennae, and, usually, prominent compound eyes, though the latter are reduced in some wingless species. Three ocelli are usually present in winged forms but absent in apterous species. The Y-shaped epicranial suture is prominent. The mouthparts, though retaining a chewing function, are specialized. The mandibles are dissimilar, though each has both grinding and cutting edges. In the maxillae the cardo and stipes are not always distinct. The galea is a large, fleshy lobe, whereas the lacinia is a narrow, sclerotized rod (the pick), which may be used to scrape food from the substrate. The hypopharynx, which is able to take up water from the atmosphere (Rudolph, 1982), has a characteristic structure. The lingua bears a pair of ventral sclerites that are connected to the median sitophore sclerite by five ligaments. The sitophore sclerite is situated on the ventral surface of the base of the cibarium. Opposite to it, on the dorsal surface of the cibarium wall, is a knoblike process that is believed to move against the sclerite in the manner of a mortar and pestle and facilitate the grinding up of food.

In winged forms the small prothorax is largely concealed by the pterothorax. The wings are membranous and have a prominent but reduced venation. The anterior pair is larger than the hind pair. At rest they are held rooflike over the body. The fore and hind wings are coupled both during flight and at rest. Varying degrees of brachyptery occur, even within the same species, and aptery is common, especially in females. The legs are usually slender and similar; in some species the hind coxae carry what is believed to be a stridulatory organ. The abdomen is 10-segmented and terminates in a dorsal epiproct and a pair of lateral paraprocts that may represent the 11th segment. The external genitalia of males are weakly developed, and their homologies are uncertain. A small ovipositor is usually present in females, though it is much reduced or absent in some forms. Cerci are never present.

Four Malpighian tubules originate at the posterior end of the midgut, which is long and convoluted. The nervous system is highly modified and comprises only five ganglionic centers: brain, subesophageal ganglion, prothoracic ganglion, a composite pterothoracic ganglion, and a composite abdominal ganglion. The tracheal system usually opens to the exterior by means of two pairs of thoracic and eight pairs of abdominal spiracles. Each ovary contains three to five polytrophic ovarioles. The lateral oviducts are short and open into a larger median duct. A spermatheca is present. The testes are roundish or three-lobed. The vasa deferentia lead into large seminal vesicles that appear to produce the material of the spermatophore.

Life History and Habits

Most Psocoptera are found on vegetation or under bark, though some live among leaf litter, under stones, or in caves. A few species are associated with humans and may be encountered in houses or buildings in which food materials are stored. Though they may occur in vast numbers, they are seldom of economic importance. They are primarily phytophagous, feeding on algae, lichens, fungi, pollen, and decaying fragments of higher plants; occasionally they eat dead animal matter. Species associated with humans live on cereal products or, in the case of the common booklice, molds that develop on old books. Many species are gregarious, with individuals of all ages living together, often beneath a silken web produced from secretions of the modified labial glands. Most outdoor species are fully winged, though brachyptery and aptery are common under certain environmental conditions.

Males are unknown in some species, while in others facultative parthenogenesis may occur. In dioecious species a male typically courts a female prior to mating. Eggs, between 20 and 100 at a time, are laid singly or in groups, usually on vegetation or under bark, and they may be covered with silk, particles of debris, or fecal material. They have a thin chorion and lack micropyles and aeropyles. A few species are viviparous. Larvae usually pass through six instars prior to metamorphosis, but this figure is often reduced in polymorphic species. Typically the apterous morphs have one fewer instar than the fully winged forms. Where climatic conditions permit, Psocoptera may have several generations per year; elsewhere, species are typically univoltine and may enter diapause to overcome adverse conditions.

Phylogeny and Classification

Modern Psocoptera are but the remnants of an order that had already undergone an extensive evolution by the end of the Permian period. Of the four hemipteroid orders, the Psocoptera are generally considered to be closest to the ancestral stock. The earliest fossil Psocoptera (from the Lower Permian of Kansas [U.S.A.] and Moravia [Czech Republic]) differ from modern species with regard to wing venation and mouthpart characters and are placed in a distinct suborder Permopsocina. However, many of the numerous Oligocene fossils, and even some Cretaceous species, can be assigned to extant families (some even to extant genera) indicating that the order has undergone relatively little change since the Mesozoic. Living Psocoptera are divisible into three well defined suborders: Trogiomorpha, Troctomorpha, and Psocomorpha (Eupsocida). The Trogiomorpha contains the most primitive and the Psocomorpha the most advanced Psocoptera. The Psocomorpha has been the dominant suborder since the Late Cretaceous. Unfortunately, insufficient systematic work has been done to permit firm conclusions to be reached with respect to the relationships of the many families. This is especially true for the Psocomorpha, to which at least 75% of the recent species belong. The classification of Badonnel (1951), which is used here, continues to be accepted as the one that most accurately reflects phylogenetic relationships within the order (Smithers, 1991; Mockford, 1993).

Suborder Trogiomorpha

Distinguishing characters of Trogiomorpha include antennae with more than 20 segments, never secondarily annulated; tarsi three-segmented; labial palps two-segmented; pterostigma not thickened, or absent; and paraprocts with strong posterior spine.

In this suborder the major families are the LEPIDOPSOCIDAE, TROGIIDAE, PSOQUILLIDAE, and PSYLLIPSOCIDAE. Lepidopsocidae (165 species) form a primarily tropical group of bark- and leaf litter-inhabiting forms recognized by their somewhat moth-like appearance produced by the scales on their body and wings. Though a small family (about 20 species), the Trogiidae is a cosmopolitan group that includes several species found in buildings. Examples are *Trogium pulsatorium* (Figure 8.1A), a common book-louse that feeds on paper, vegetable matter, and cereal products, and *Lepinotus inquilinus* that is found especially in granaries and warehouses. Psoquillidae and Psyllipsocidae are both small (about 20 species in each), widely distributed families. Psoquillids are found on bark, in bird nests, and litter. Psyllipsocids occur in caves and termite nests. Both families include some species found indoors.

Suborder Troctomorpha

Features of Troctomorpha are 12- to 17-segmented antennae, with some flagellar segments secondarily annulated; 2- or 3-segmented tarsi; 2-segmented labial palps; pterostigma not thickened; and paraprocts without a strong posterior spine.

The largest families in this suborder are the LIPOSCELIDAE (140 species), AMPHIENTOMIDAE (40 species), and PACHYTROCTIDAE (60 species). The

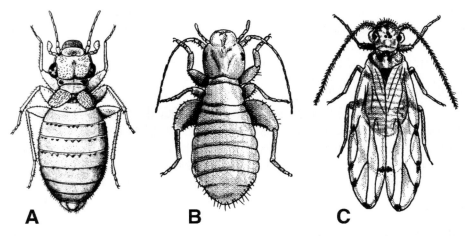

A   **B**   **C**

FIGURE 8.1. Psocoptera. (A) *Trogium pulsatorium* (Trogiidae) (distal antennal segments omitted); (B) *Liposcelis* sp. (Liposcelidae); and (C) *Ectopsocus californicus* (Ectopsocidae). [A, from P.-P. Grassé (ed.), 1951, *Traité de Zoologie*, Vol. X. By permission of Masson, Paris. B, C, from L. A. Swan and C. S. Papp, 1972, *The Common Insects of North America.* Copyright 1972 by L. A. Swan and C. S. Papp. Reprinted by permission of Harper & Row, Publishers, Inc.]

Liposcelidae is a cosmopolitan group whose members are recognized by their greatly enlarged hind femora. The family includes a number of common booklice (*Liposcelis* spp.) (Figure 8.1B) found in houses, warehouses, and ship holds. Outdoor species typically occur in litter and under bark. However, occasionally they have been taken in the nests of vertebrates, in the fur of mammals, and on birds' feathers, these associations possibly aiding dispersal of the psocopterans. Amphientomids are predominantly found in tropical regions of the Old World. They occur under bark and in litter. Pachytroctidae are also mainly tropical, occurring in both the Old and the New World. Typically, they are found under bark or in litter, occasionally in buildings.

Suborder Psocomorpha

Features of Psocomorpha include antennae almost always 13-segmented, with flagellar segments not secondarily annulated; tarsi 2- or 3-segmented; labial palps unsegmented; pterostigma not thickened; and paraprocts with strong posterior spine.

Judging by their common features, this very large suborder, containing more than 20 families, is probably derived from or had a common ancestry with the Troctomorpha. Most members of the suborder are found outdoors, on growing vegetation, in leaf litter, or on bark. The AMPHIPSOCIDAE (140 species) is a cosmopolitan family that includes some of the largest Psocoptera. They are generally found on broad-leaved foliage. ARCHIPSOCIDAE (60 species) form a tropical family, largely from South America and Africa, some species of which live in massive aggregations under sheets of webbing that may cover an entire tree. The CAECILIIDAE is a very large family (370 species mostly in the genus *Caecilius*) of worldwide distribution. Most species are foliage dwellers. ECTOPSOCIDAE (120 species), which form a cosmopolitan group, are typically found in dry foliage and leaf litter though a few species are found elsewhere; for example, *Ectopsocus californicus* (Figure 8.1C) is a common booklouse and *E. pumilis* is a cosmopolitan species found in granaries and warehouses. PERIPSOCIDAE (120 species) are bark dwellers with a cosmopolitan distribution. Members of the ELIPSOCIDAE and EPIPSOCIDAE (each with about 100 species) are bark and litter inhabitants, with worldwide (especially Southern Hemisphere) and primarily tropical distributions, respectively. The cosmopolitan LACHESILLIDAE (250 species) are primarily found on dry foliage or in leaf litter with a few widespread species (e.g., *Lachesilla pedicularia*) occurring in granaries, barns, and warehouses. MYOPSOCIDAE (125 species) are generally large, tropical Psocoptera found on bark. PHILOTARSIDAE (150 species) live on bark or low vegetation. The family is widely distributed but especially common in the southwest Pacific region. The PSEUDOCAECILIIDAE (120 species) is a mainly tropical group of foliage- and bark-dwelling species. The PSOCIDAE, a cosmopolitan group of about 520 species, is the largest family in the order. Its members are darkly colored and live on bark or, occasionally, on the ground.

Literature

A detailed account of the biology of the Psocoptera is given by New (1987) and Mockford (1993). Thornton (1985) discusses the zoogeography and ecology of the arboreal forms. The phylogeny and classification of the order are considered by Smithers (1972). New (1987) and Smithers (1990) provide keys to the world families and genera, respectively. New (1974) has a key to the British species, and Smithers (1991) and Mockford (1993) keys to the Australian families and North American species, respectively.

Badonne1, A., 1951, Ordre des Psocoptères, in: *Traité de Zoologie* (P.-P. Grassé, ed.), Vol. X, Masson, Paris.

Mockford, E. L., 1993, *North American Psocoptera (Insecta)*, Sandhill Crane Press, Gainesville, FL.

New, T. R., 1974, Psocoptera, *Handb. Ident. Br. Insects* **1**(7):1–102.

New, T. R., 1987, Biology of the Psocoptera, *Or. Insects* **21**:1–109.

Rudolph, D., 1982, Occurrence, properties and biological implications of the active uptake of water vapour from the atmosphere in Psocoptera, *J. Insect Physiol.* **28**:111–121.

Smithers, C. N., 1972, The classification and phylogeny of the Psocoptera, *Mem. Aust. Mus.* **14**:1–349.

Smithers, C. N., 1990, Keys to the families and genera of Psocoptera, *Tech. Rep. Aust. Mus.* **2**:1–82.

Smithers, C. N., 1991, Psocoptera, in: *The Insects of Australia*, 2nd ed., Vol. 1 (CSIRO, ed.), Melbourne University Press, Carlton, Victoria.

Thornton, I. W. B., 1985, The geographical and ecological distribution of arboreal Psocoptera, *Annu. Rev. Entomol.* **30**:175–196.

3. Phthiraptera

SYNONYMS: Pseudorhynchota, Mallophaga
(in part), Lipoptera (in part),
Siphunculata (in part),
Anoplura (in part)

COMMON NAMES: lice, sucking lice (in part),
chewing lice or bird lice
(in part)

Minute to small, apterous, dorsoventrally flattened ectoparasites of birds or mammals; head prognathous or hypognathous, compound eyes reduced or absent, ocelli absent, antennae 3- to 5-segmented, mouthparts of chewing or piercing-sucking type with palps of maxillae and labium reduced or absent; prothorax free or fused with pterothorax, legs with unsegmented or 2-segmented tarsi, more or less modified for clinging to hair or feathers, one, two, or no tarsal claws; abdomen 7- to 10-segmented, cerci absent.

The order, which has a worldwide distribution, includes about 3100 described species. The sucking lice (suborder Anoplura) are exclusively parasites of placental mammals, while the remainder are chewing lice ("Mallophaga") comprising the suborders Amblycera,

Ischnocera, and Rhynchophthirina. About 85% of the species of chewing lice are parasites of birds. Many Phthiraptera are important pests, in their role as disease vectors, of domesticated animals and humans.

Structure

The minute to small (0.35–10 mm long) body is dorsoventrally flattened and shows a varied degree of sclerotization. In chewing lice the head is relatively large and bears chewing mouthparts that show certain resemblances to those of Psocoptera. In sucking lice the relatively small head has partially retracted, suctorial mouthparts. The labrum forms a short eversible proboscis. Three stylets are contained within a pouch that runs ventrally off the cibarium. The ventral stylet represents the modified labium, the middle stylet probably is an extension of the opening from the salivary duct, and the dorsal stylet is either the modified maxillae or the hypopharynx. The mandibles disappear during embryogenesis. The head bears three- to five-segmented antennae that may be filiform (in Anoplura), capitate and in grooves (Amblycera), or modified for grasping (Ischnocera). Compound eyes are reduced or absent and ocelli are never present. In Anoplura the thoracic segments are fused, but in other lice the prothorax is distinct from the pterothroax. The well-developed legs include a two-segmented or unsegmented tarsus and usually one or two tarsal claws. Eight to ten visible abdominal segments occur. Male genitalia include a permanently everted endophallus. There is no true ovipositor though in all sucking and some chewing lice the gonapophyses on segment 8 are used to attach eggs to the host's hair. Cerci are absent.

In sucking lice the cibarium and pharynx form a strong sucking pump, but the crop and gizzard are poorly differentiated. In chewing lice the crop is large. In all Phthiraptera the midgut is large and has two mesenteric ceca. Four Malpighian tubules enter the midgut posteriorly. The nervous system is highly modified and includes a composite metathoracoabdominal or thoracoabdominal ganglion. The internal reproductive organs generally resemble those of Psocoptera except that female Phthiraptera may have accessory glands and lack a spermatheca.

Life History and Habits

Generally, lice are considered to be highly host-specific being restricted to a single or a few closely related species of host, often occupying specific regions on the host's body. However, Marshall (1981) questioned whether taxonomic problems with this group may be obscuring the real situation, citing the example of *Menacanthus eurysternus*, which, on the basis of a recent revision, now counts 118 passerine species (in 20 families) and 5 woodpecker species among its hosts! By contrast, a given bird or mammal may play host to several species of lice, up to 15 being recorded on a South American tinamou. In contrast to fleas, lice spend their entire life on the host. They are highly sensitive to changes in temperature and humidity and survive for only a few days should the host die.

Most species of chewing lice live among the feathers of birds where they feed on fragments of feathers and skin. A few species, for example, the chicken body louse, *Menacanthus stramineus*, feed on blood in addition to epidermal products and are able to pierce the skin or developing quills. The members of two small families of chewing lice are parasitic on mammals, though their general habits are like those of species found on birds. Some chewing lice, like Psocoptera, have the ability to take up moisture from the atmosphere via their hypopharynx, presumably an adaptation to prevent desiccation in the dry air among feathers or fur of their host (Rudolph, 1983).

The sucking lice feed exclusively on the blood of the host, always a placental mammal. It has been suggested that a possible reason for the high host specificity of sucking lice is the lethal effect an unsuitable host's blood might have on the symbiotic bacteria present in certain gut cells or in the mycetome, a structure closely associated with the gut (see Chapter 16, Section 5.1.2).

Heavy infestations of lice may render a host more susceptible to disease and cause economic loss related to reduction in quality. In addition, some sucking lice are important disease vectors (see Suborder Anoplura).

In some Anoplura, at least, mating occurs frequently, presumably because females lack a spermatheca. In many species of chewing lice, males are less common than females, and in some species, are rare or unknown so that parthenogenesis occurs. Eggs are usually cemented to hairs or feathers by means of a secretion from the female's accessory gland. Postembryonic development is rapid, juveniles passing through three molts, and adults become sexually mature within a few days of the final molt. Transfer to new hosts is by physical contact and occurs during mating, communal roosting, and brooding and feeding the young. Some bloodsucking Diptera may carry lice from host to host.

Phylogeny and Classification

It is generally accepted that the Phthiraptera are derived from a free-living Psocopter-alike ancestor, based on a number of synapomorphies (Chapter 2, Section 3.2). There is virtually no fossil record: the Early Cretaceous chewing louse *Saurodectes vrsanskyi* may have parasitized pterosaur reptiles, and a sucking louse, *Neohaematopinus relictus* has been found on a rodent, *Citellus*, from the Pleistocene of Siberia. However, this has not prevented some authors from speculating that the order may be quite ancient, possibly originating as early as the Upper Carboniferous/Lower Permian. Kim and Ludwig (1982) proposed this very early origin for the order to be consistent with their hypothesis that the group arose from an ancestor within the psocopteran suborder Permopsocida, an idea rejected by Lyal (1985). Rather, Lyal (1985) proposed, the lice may have evolved from liposcelidlike Pso-coptera that, as noted in the previous section, have occasionally been found in the nests and on the body of birds and mammals. Initially, this association perhaps aided dispersal of the insects, but in time they may have begun to feed on flakes of dead skin, bits of feathers, etc. At this stage, the association between the ancestral louse and its host presumably would have been facultatively parasitic and much less host-specific than is the case with modern Phthiraptera. The virtual absence of fossils has meant that proposals with respect to the origin and phylogeny of the order have come largely from examination of the host associations of lice, and the phylogeny and zoogeography of their mammalian and avian hosts. Such studies suggest that ancestral lice did not arise until the Cretaceous, with either birds or mammals as hosts. The earliest lice were chewing forms, and from these arose the modern Amblycera on the one hand, and a line leading to the remaining groups on the other. The latter itself split, giving rise to the Ischnocera, whose ancestor retained chewing mouthparts and had either a bird or a mammal as host, and the common ancestor of the Rhyncoph-thirina and Anoplura which, like all modern members of these two groups, presumably fed on mammals. While the Rhyncophthirina retained chewing mouthparts, the Anoplura evolved suctorial mouthparts in conjunction with their blood-feeding habit. Though this scheme may be a reasonable explanation of the evolution of Phthiraptera, Barker (1994)

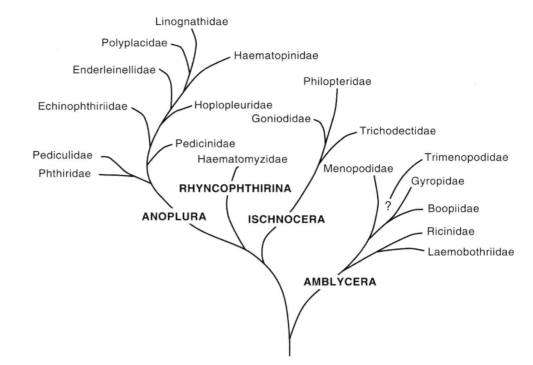

FIGURE 8.2. A suggested phylogeny of the Phthiraptera. The relationship of the Trimenopodidae to other Amblyceran families is uncertain.

has stressed that the coevolution of lice and their vertebrate hosts has not always occurred, and that host-switching by lice is fairly common. Thus, a phylogeny based on lice characters rather than characters of the host is preferable.

Some authors place the chewing lice and sucking lice in separate orders, the Mallophaga and Anoplura, respectively, the great differences in mouthpart structure and feeding habits being taken as sufficient justification for this separation. Generally, however, all lice are included in the order, Phthiraptera, divisible into four suborders, Amblycera, Ischnocera, Rhyncophthirina, and Anoplura. Thus, the Mallophaga, which comprises the first three suborders, is a paraphyletic group in this scheme, leading some authorities (e.g., Barker *et al.*, 2003) to urge that the term be abandoned. A suggested phylogeny of the Phthiraptera is presented in Figure 8.2.

Suborder Amblycera

Members of the suborder Amblycera have the following characteristics: capitate, four-segmented antennae, lying in grooves; mandibles horizontal, maxillary palps present; and mesothorax and metathorax usually separate.

Generally six families are recognized in this suborder, a group of about 850 species that is considered to contain the more primitive lice. Of the six families, three are restricted to avian hosts, two to marsupials, and one to placental mammals. The MENOPONIDAE form the largest family (650 species) and have a cosmopolitan distribution. Its members infest birds, and several species are important pests of poultry, for example, *Menacanthus stramineus* (the chicken body louse) and *Menopon gallinae* (the shaft louse) (Figure 8.3A). The LAEMOBOTHRIIDAE are a small family containing the single genus *Laemobothrion*,

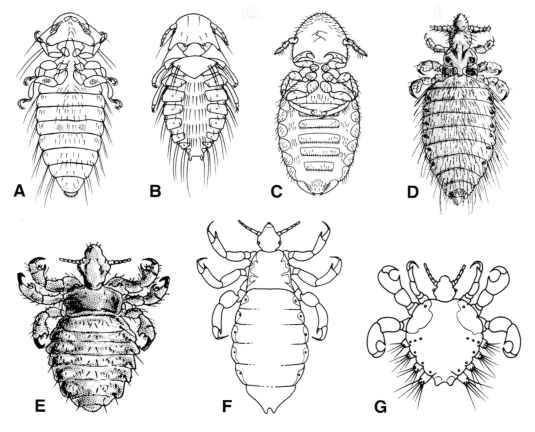

FIGURE 8.3. Phthiraptera. (A) The shaft louse, *Menopon gallinae* (Menoponidae); (B) the large turkey louse, *Chelopistes meleagridis* (Goniodidae); (C) the red cattle louse, *Damalinia bovis* (Trichodectidae); (D) the long-nosed cattle louse, *Linognathus vituli* (Linognathidae); (E) the short-nosed cattle louse, *Haematopinus eurysternus* (Haematopinidae); (F) the human body louse, *Pediculus humanus* (Pediculidae); and (G) the crab louse, *Phthirus pubis* (Phthiridae). [D, E, from L. A. Swan and C. S. Papp 1972, *The Common Insects of North America.* Copyright 1972 by L. A. Swan and C. S. Papp. Reprinted by permission of Harper & Row, Publishers, Inc.]

the 14 species of which are parasites of water birds and hawks. The RICINIDAE, containing 65 species in two genera, are found on hummingbirds and several families of passerines. Because of the sister-group relationship of their hosts, the BOOPIIDAE (35 species), found on Australian marsupials, and the TRIMENOPONIDAE (10 species), found on South American marsupials and histricomorph rodents (porcupines, guinea pigs, etc.), were formerly thought to have had a common ancestor. This view no longer appears tenable; rather, the Boopiidae appear to be the sister group of the GYROPIDAE (60 species). This family is endemic to South and Central America, though two species, *Gyropus ovalis* and *Gliricola porcelli*, which are found on guinea pigs, have been spread by commerce to other parts of the world. Most species of Gyropidae live on histricomorph rodents.

Suborder Ischnocera

Characteristics of Ischnocera are exposed, three- to five-segmented filiform antennae; mandibles vertical, maxillary palps absent; and mesothorax and metathorax usually fused.

Included in this probably paraphyletic or even polyphyletic suborder of about 1750 species are two large families. The paraphyletic PHILOPTERIDAE, a cosmopolitan group with about 1460 species, is the largest family of the order. Its members are parasitic on birds and include a number of pest species found on poultry, for example, *Cuclutogaster heterographus* (the chicken head louse). The cosmopolitan family TRICHODECTIDAE (290 species), which is restricted to placental mammals, contains a number of species found on domesticated animals, for example, *Damalinia (Bovicola) bovis* (cattle) (Figure 8.3C), *D. ovis* (sheep), *D. equi* (horses), *Felicola subrostratus* (cats), and *Trichodectes canis* (dogs), which can serve as an intermediate host for the dog tapeworm, *Dipylidium caninum*. The GONIODIDAE, formerly included in the Philopteridae, are found on galliform and columbiform birds. The family includes some major poultry pests, for example, *Chelopistes meleagridis* (large turkey louse) (Figure 8.3B), *Goniodes gigas* (large chicken louse) and *G. dissimilis* (brown chicken louse).

Suborder Rhyncophthirina

Members of the suborder Rhyncophthirina have the following characteristics: head prolonged into a rostrum, mandibles at apex of rostrum, and labium and maxillae vestigial.

This suborder contains only two species, *Haematomyzus elephantis,* a parasite of both Indian and African elephants, and *H. hopkinsi*, which infests warthogs. It has been suggested that this group may resemble the ancestors of the Anoplura. Certainly, the two groups share a number of primitive characters (Lyal, 1985).

Suborder Anoplura

Members of the suborder Anoplura are recognized by their relatively small head, styliform mouthparts, and lack of mandibles.

The approximately 530 species of Anoplura were arranged by Kim and Ludwig (1978) in 15 families, of which 8 contain 4 or fewer species, The features of the seven largest families, plus the Pediculidae and Phthiridae, which have special significance for humans, are summarized below, The ECHINOPHTHIRIIDAE (12 species) are parasites of Pinnipedia (seals, sea lions, and walruses) and the river otter *(Lutra canadensis)*. Their body is covered with strong spines or scales that retain a film of air over the body when submerged. It is evident that many of these lice must be very long-lived, as their hosts spend most of their life at sea, coming ashore to breed (when presumably transfer of parasites occurs) for only a short time each year. The cosmopolitan LINOGNATHIDAE (70 species) parasitize dogs *(Linognathus setosus)*, hyraxes, and ruminants, including sheep *(L. ovillus* and *L. pedalis)*, cattle *(L. vituli)* (Figure 8.3D), and goats *(L. stenopsis)*. The HAEMATOPINIDAE (22 species) form a cosmopolitan group whose members parasitize ungulates, including several domesticated forms on which they may become serious pests. *Haematopinus eurysternus* (Figure 8.3E) occurs on cattle, *H. suis* on pigs, and *H. asini* on horses. The HOPLOPLEURI-DAE (about 130 species) are also a large group whose hosts are mainly rodents and rabbits, but also include moles and shrews. The largest anopluran family is the POLYPLACIDAE, a cosmopolitan group whose 175 species very largely parasitize rodents, though some are found on primates (lemurs, lorises, and galagos), rabbits, moles, and shrews. The ENDER-LEINELLIDAE (50 species) are another rodent-infesting group, specifically parasitizing Sciuridae (squirrel family). The family is widespread though not represented in Australia, Madagascar, and southern South America. PEDICINIDAE (16 species) infest Old World monkeys. The PEDICULIDAE (2 species) live on primates. The genus *Pediculus* is found on humans (as *P. humanus* and *P. capitis*) and other hominoids as well as New World monkeys. *P. capitas* (the head louse) and *P. humanus* (the body louse) (Figure 8.3F) differ in size and habits. The smaller head louse attaches its eggs to hair, whereas the body louse or "cootie" lays its eggs on clothing to which it usually remains attached. Until recently, *P. humanus* and *P. capitis* were considered sibling species. However, a recent molecular study (Leo *et al.*, 2002) indicated that head and body lice are conspecific, despite their differences. Certainly, they interbreed readily in the laboratory. This discovery is significant given that the two forms are generally believed to be very different in their roles as disease vectors. Thus, the body louse is an important vector of diseases such as endemic typhus and trench fever, caused by blood-borne rickettsias. Relapsing fever, caused by a spirochete, is also spread by the louse. By contrast, the head louse is not usually considered a disease vector (but see Robinson *et al.*, 2003). Included in the PHTHIRIDAE are *Phthirus gorillae*, found on gorillas, and *P. pubis*, the pubic or crab louse (Figure 8.3G) found on humans. Unlike the body louse, the crab louse does not appear to be a disease vector.

Literature

Information on the biology of Phthiraptera is given by Askew (1971), Marshall (1981), and Price and Graham (1997). Ferris (1951) and Kim (1985) deal specifically with the Anoplura (though the former's systematic treatment of the group is outdated), while Emerson and Price (1985) consider the Mallophaga on mammals. Clay (1970), Lyal (1985), Barker (1994), Cruickshank *et al.* (2001), and Barker *et al.* (2003) discuss the phylogeny of the order. Arnett (2000) includes keys to North American families of Phthiraptera, Calaby and Murray (1991) to the Australian families of Phthiraptera, and Kim and Ludwig (1978) to world families of Anoplura. Kim *et al.* (1986) provide a key for North American Anoplura.

Arnett, R. H., Jr., 2000, *American Insects: A Handbook of the Insects of America North of Mexico*, 2nd ed., CRC Press, Boca Raton, FL.

Askew, R. R., 1971, *Parasitic Insects*, American Elsevier, New York.

Barker, S. C., 1994, Phylogeny and classification, origins, and evolution of host associations of lice, *Int. J. Parasitol.* **24**:1285–1291.

Barker, S. C., Whiting, M., Johnson, K. P., and Murrell, A., 2003, Phylogeny of the lice (Insecta: Phthiraptera) inferred from small subunit rRNA, *Zool. Scrip.* **32**:407–414.

Calaby, J. H., and Murray, M. D., 1991, Phthiraptera, in: *The Insects of Australia*, 2nd ed., Vol. I (CSIRO, ed.), Melbourne University Press, Carlton, Victoria.

Clay, T., 1970, The Amblycera (Phthiraptera: Insecta), *Bull. Br. Mus. Nat. Hist.* **25**:75–98.

Cruickshank, R. H., Johnson, K. P., Smith, V. S., Adams, R. J., Clayton, D. H., and Page, R. D. M., 2001, Phylogenetic analysis of partial sequences of elongation factor 1α identifies major groups of lice (Insecta: Phthiraptera), *Mol. Phylog. Evol.* **19**:202–215.

Emerson, K. C., and Price, R. D., 1985, Evolution of Mallophaga on mammals, in: *Coevolution of Parasitic Arthropods and Mammals* (K. C. Kim, ed.), Wiley, New York.

Ferris, G. F., 1951, The sucking lice, *Mem. Pac. Coast Entomol. Soc.* **1**:320 pp.

Kim, K. C., 1985, Evolution and host associations of Anoplura, in: *Coevolution of Parasitic Arthropods and Mammals* (K. C. Kim, ed.), Wiley, New York.

Kim, K. C., and Ludwig, H. W., 1978, The family classification of the Anoplura, *Syst. Entomol.* **3**:249–284.

Kim, K. C., and Ludwig, H. W., 1982, Parallel evolution, cladistics, and classification of parasitic Psocodea, *Ann. Entomol. Soc. Am.* **75**:537–548.

Kim, K. C., Pratt, H. D., and Stojanovich, C. J., 1986, *The Sucking Lice of North America: An Illustrated Manual for Identification*, Pennsylvania State University Press, University Park.

Leo, N. P., Campbell, N. J. H., Yang, X., Mumcuoglu, K., and Barker, S. C., 2002, Evidence from mitochondrial DNA that head lice and body lice of humans (Phthiraptera: Pediculidae) are conspecific, *J. Med. Entomol.* **39**:662–666.

Lyal, C. H. C., 1985, Phylogeny and classification of the Psocodea, with particular reference to the lice (Psocodea: Phthiraptera), *Syst. Entomol.* **10**:145–165.

Marshall, A. G., 1981, *The Ecology of Ectoparasitic Insects*, Academic Press, New York.

Price, M. A., and Graham, O. H., 1997, *Chewing and Sucking Lice as Parasites of Mammals and Birds*, U.S.D.A., Tech. Bull. **1849**:1–309.

Robinson, D., Leo, N., Prociv, P., and Barker, S. C., 2003, Potential role of head lice, *Pediculus humanus capitis*, as vectors of *Rickettsia prowazekii*, *Parasitol. Res.* **90**:209–211.

Rudolph, D., 1983, The water-vapour uptake system of the Phthiraptera, *J. Insect Physiol.* **29**:15–25.

4. Hemiptera

SYNONYM: Rhynchota   COMMON NAME: true bugs

Minute to large insects, head opisthognathous (homopterans) or prognathous (heteropterans), compound eyes usually well developed but rarely absent, two or three ocelli usually present, antennae with few segments, mouthparts suctorial with mandibles and maxillae in form of stylets enclosed within a labial sheath; two pairs of wings usually present with fore wings of harder consistency than hind pair; abdomen with 9–11 segments, external genitalia varied in both sexes, cerci absent.

This group is the largest and most heterogeneous order of exopterygotes, containing some 80,000 described species from all regions of the world, About 11,000 species have been described from North America, 5600 from Australia, and 1600 from Britain. The order traditionally has been divided into the Homoptera (Sternorrhyncha + Auchenorrhyncha) and Heteroptera (and some authors have raised these to the rank of order). However, there is a growing body of evidence, both morphological and molecular, to indicate that the Homoptera is a paraphyletic group, and that the Auchenorrhyncha is the sister group of the Heteroptera (see Phylogeny and Classification). Thus, in the description below the term *homopteran* is used to describe a level of organization, not a taxonomic rank.

Structure

Hemiptera range in size from about 1 mm to 11 cm. They are frequently procryptically colored in shades of green or brown, or aposematically colored in striking patterns, often in red and white. Yet others mimic ants and other insects with which they live, or resemble plant parts. Various forms of polymorphism (see Chapter 21, Section 7) are common throughout the order. The head is either opisthognathous and lacking a gula in homopterans or prognathous and with a sclerotized gula in most Heteroptera. The antennae comprise only four or five segments in most Hemiptera. The compound eyes are typically well developed and of varied shape and size. Two or three ocelli are usually present. The unifying feature of the order are the suctorial mouthparts, which are remarkably similar in all except a very few members of the group. The mandibles and maxillae form two pairs of piercing stylets that are contained within the flexible, segmented labium (see Chapter 3, Section 3.2.2 and Figure 3.17). The stylets are sometimes much longer than the labium and, when not in use, may be coiled within an integumental fold, the crumena. Maxillary and labial palps are absent, though the labium, which never enters the tissue that is pierced, has a number of sensory hairs at its tip. Mouthparts are absent in male and many female Coccoidea and the sexual forms of some aphids.

In Heteroptera the pronotum is large, the mesonotum and metanotum small. In most homopterans the pronotum is small and collarlike, the mesonotum is well developed, the metanotum somewhat less so. Usually two pairs of wings are present, though brachyptery and aptery are common, sometimes occurring in the same species. In homopterans the wings are generally held rooflike over the body; the fore wings are of uniform texture and are often of a harder consistency than the hind pair. In most Heteroptera the wings are held flat, and there is typically a marked difference in the consistency of the fore and hind wings. The fore wings (hemelytra) usually are well sclerotized basally, with only a small distal portion remaining membranous. The hind pair remain membranous and, at rest, are folded beneath the hemelytra. The legs are generally identical and suited for a cursorial habit. However, in certain groups they are modified for a variety of functions, such as catching prey, swimming, jumping, moving over the water surface, and production of sound, or they may be vestigial or absent in females that are sedentary.

In its least specialized condition the abdomen has 11 segments of which the first two may be modified in connection with sound production, the eighth and ninth possess the external genitalia, and the last two are extremely reduced. Frequently reduction of up to three anterior segments occurs. The ovipositor is complete in many homopterans and those Heteroptera that lay eggs in plant tissues, but reduced or absent in other groups. The homologies of the male genitalia are complex. Primitively, the enlarged ninth sternum bears a pair of lateral claspers, while the cavity formed by invagination of the membrane between the ninth and tenth sterna contains the aedeagus and a pair of parameres. However, either the claspers or both the claspers and parameres are absent in some groups.

The alimentary canal presents a wide variety of structural modifications associated with the liquid diet of the order. Posterior to the cibarial sucking pump, which is not part of the alimentary canal, is a short foregut. The midgut is large, frequently occupying a major part of the abdomen and usually differentiated into several regions, of which the most anterior is a croplike structure. The posterior region is tubular and from it, in many Heteroptera, arise many ceca that contain symbiotic bacteria. In Aphidoidea a special structure (mycetome) within the body cavity houses symbiotic bacteria (*Buchnera* spp.) that provide the host with the essential amino acids unavailable in the diet (see Chapter 16, Section 5.1.2). In many plantsucking Hemiptera the hindgut is long and convoluted. The anterior part of the hindgut and the Malpighian tubules may come to lie alongside the swollen midgut, thus providing a "shortcut" to the hindgut for the large volumes of liquid feces (see Figure 16.6). In a few Heteroptera the midgut ends blindly or is entirely separated from the hindgut. Four Malpighian tubules are the rule, but rarely there are only three or two. In Aphidoidea there are no Malpighian tubules, and the function of excretion is taken over by the midgut. The central nervous system is highly specialized, and discrete abdominal ganglia never occur. In most Hemiptera there is a single composite ventral ganglion; in others up to four ventral ganglia (subesophageal, prothoracic, mesothoracic, and composite metathoracic-abdominal) may be found. Usually each ovary comprises between four and eight acrotrophic ovarioles arranged in a group at the distal end of the lateral oviduct. In a few homopterans, for example, aphids, there may, however, be up to 100 ovarioles. Paired accessory glands and zero to three spermathecae are also present. Each testis contains from one to seven follicles, which may or may not be enclosed in a follicular sheath. Well developed accessory glands also occur in males. Various types of subcutaneous glands are common in Hemiptera. Most Heteroptera possess scent glands in both juvenile and adult stages. In juveniles these are typically on the dorsal surface of abdominal segments 4–7. Though in some heteropterans the glands of juveniles continue to function in the adult stage, in most species new, ventrally positioned metathoracic glands differentiate at eclosion. Mostly the glands secrete repugnatorial chemicals; in some species, however, aggregation, warning, or sexual pheromones are produced. Wax-secreting glands occur in many Sternorrhyncha and Fulgoroidea, usually on the dorsal surface of the abdomen. Their function is presumably protective.

Life History and Habits

Homopterans and the majority of Heteroptera are terrestrial insects. Other Heteroptera show varying degrees of adaptation to an aquatic existence. Some occur in the littoral or intertidal zone of the seashore, on marshy ground, or in damp moss. Others live on the surface of water. Among the truly aquatic forms there is a range of adaptation from those species that must periodically visit the surface to respire (and which periodically fly from one location to another) to those that remain submerged permanently and respire by means of a plastron (see Chapter 15, Section 4.2).

All homopterans and many Heteroptera feed on fluids from plants. All parts of the plant are attacked: roots, stem, leaves, flowers, and seeds (often when these have fallen to the ground). Sternorrhyncha and Fulgoroidea principally feed in phloem, Cicadoidea and Cercopoidea in xylem, and Heteroptera in parenchyma, while Cicadelloidea vary in their choice of tissue (Carver *et al.*, 1991; Novotny and Wilson, 1997). The remaining Heteroptera are predaceous, living on the body fluids of other arthropods and vertebrates. It is primarily because of these feeding habits, assisted in many cases by extraordinarily high rates of reproduction (see Superfamily Aphidoidea), that the order is considered by many people to be the most economically important insect group. The damaging effect on plants may be direct or indirect. When insect populations are large, the loss of sap results in stunted growth and poor yield and quality. Many species, when feeding, inject saliva that causes necrosis of plant tissue. Indirectly, the effects are to weaken the plant, making it more susceptible to attack by other pathogens, especially fungi and viruses. Most important of all, however, is the role of Hemiptera (especially aphids) as vectors of viruses that cause major plant diseases, for example, mosaic, leaf roll, yellows. Among the predaceous Hemiptera, a few species may act as vectors for the transmission of disease; for example, certain South American Reduviidae are carriers of *Trypanosoma cruzi*, a flagellate protozoan that causes trypanosomiasis (Chagas' disease), a form of sleeping sickness among humans. On the beneficial side, many homopterans play an important part in weed control; for example, *Dactylopius* species have been successfully used in the control of the prickly pear cactus (*Opuntia*), and many predaceous Heteroptera exert a major controlling effect on some arthropod pests.

The majority of Hemiptera are bisexual and oviparous. There are, however, species that are facultatively parthenogenetic, and some that are ovoviviparous or viviparous. In many aphids all of these reproductive conditions may be met within the same species in the course of a year. In aphids parthenogenesis and viviparity commonly occur together in the spring and early summer, thus enabling the insects to exploit fully the increased food available at this time. Insemination is typically of the usual intragenital type. However, in Cimicoidea various forms of hemocoelic insemination occur (see Chapter 19, Section

4.3.1). The eggs of homopterans are generally simple, ovoid structures; those of Heteroptera are very diverse in form and coloring. They are glued to plant surfaces, inserted in crevices or between adjacent parts of a plant, laid in litter or soil, or in the case of predaceous species, in the vicinity of a host if possible. In a few Heteroptera the eggs are stuck on the dorsal surface of males. Parental care is shown by a few species. Usually, there are between three and seven instars (almost always six in Heteroptera). The juvenile stages tend to feed on the same part of the plant as the adult, though in some species juveniles are found on the roots, while adults occur on the upper parts. Although most Hemiptera are typically exopterygote in their postembryonic development, there occurs in some Aleurodidae and winged male Coccoidea one or two resting instars, the pupal and prepupal stages. These instars generally do not feed but undergo some degree of metamorphosis. A similar phenomenon occurs in Thysanoptera.

Phylogeny and Classification

Like the Psocoptera, the Hemiptera are a very ancient group whose fossil record extends into the Lower Permian period. The earliest known Hemiptera are the Archescytinidae (Figure 2.9B) from the Lower Permian of Moravia (Czech Republic). These represent "stem-group" homopterans, close to but not within the Auchenorrhyncha, which appear slightly later. Sternorrhyncha and Heteroptera first appear in the fossil record of the Upper Permian. By the Upper Permian, the auchenorrhynchans were clearly divisible into Fulgoromorpha and Cicadomorpha, and fossils assignable to extant families have been obtained from Upper Triassic and Lower Jurassic deposits. Among the Sternorrhyncha, the Psylloidea and their sister group, Aleurodoidea, date from the Upper Permian, while the earliest Aphidoidea and Coccoidea occur in the Triassic. The Coleorrhyncha, whose fossil record extends back to the Lower Jurassic, share characters with both the Auchenorrhyncha and the Heteroptera, though the consensus is that they branched off the heteropteran line at an early date, perhaps even in the Permian. Some authors consider this group sufficiently distinct from other Hemiptera as to warrant the rank of suborder. In the Jurassic the Heteroptera underwent a wide radiation, and the comparatively rich fossil record from this period includes representatives of many extant families, especially Nepomorpha and Leptopodomorpha. Although some fossil Cimicomorpha have been obtained from the Upper Jurassic, this group and its sister group, Pentatomorpha, underwent a major expansion in the Cretaceous, paralleling the radiation of the angiosperms. A proposed phylogeny of the order is shown in Figure 8.4.

As noted in the introduction to this order, the traditional subdivision of the Hemiptera into Homoptera and Heteroptera is no longer tenable, the Sternorrhyncha now being considered the sister group to the rest of the Hemiptera (i.e., Auchenorrhyncha + Heteroptera, including Coleorrhyncha) (see Carver *et al.*, 1991; Campbell *et al.*, 1994; Sorensen *et al.*, 1995; von Dohlen and Moran, 1995). Until recently, based on morphological criteria, it was widely accepted that the three suborders were monophyletic. Molecular studies have supported the monophyly of the Sternorrhyncha and Heteroptera, but strongly suggest that the Auchenorrhyncha is a paraphyletic group. Thus, Sorensen *et al.* (1995) proposed the division of Auchenorrhyncha into two suborders, corresponding to the Fulgoromorpha and Cicadomorpha of morphology-based classifications. Within the Heteroptera, eight infraorders are recognized in the classification used here (largely from Carver *et al.*, 1991). Of these, the Gerromorpha correspond to the Amphibicorisae (amphibious bugs), the Nepomorpha to

FIGURE 8.4. A proposed phylogeny of the Hemiptera. [After M. Carver, G. F. Gross. and T. E. Woodward, 1991, Hemiptera, in: *The Insects of Australia*, 2nd ed., Vol. 1 (CSIRO, ed.), Melbourne University Press. By permission of the Division of Entomology, CSIRO.]

the Hydrocorisae (aquatic bugs), also known as Cryptocerata on the basis of their short, concealed antennae, and the Cimicomorpha + Pentatomorpha to the Geocorisae (terrestrial forms) of earlier systems.

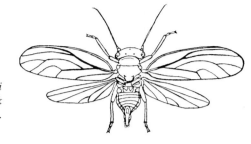

FIGURE 8.5. Psylloidea. The apple sucker, *Psylla mali* (Psyllidae). [From A. D. Imms, 1957, *A General Textbook of Entomology*, 9th ed. (revised by O. W. Richards and R. G. Davies), Methuen and Co.]

Suborder Sternorrhyncha

Like Auchenorrhyncha, Sternorrhyncha have an opisthognathous head without a gula, a small pronotum, and fore wings (when present) with a uniform texture and held rooflike over the body at rest. Derived features of recent sternorrhynchans include the position of the base of the proboscis (between or posterior to the fore coxae), two-segmented tarsi, no mesothoracic trochantin, a sawlike egg-burster, and various wing features, notably the absence of a vannus and vannal fold in the hind wing. Another feature that distinguishes them from auchenorrhynchans is their multisegmented, filiform antennae.

Superfamily Psylloidea

About 1400 of the 2200 species of Psylloidea are included in the family PSYLLIDAE. The jumping plant lice, as members of the family are commonly called, are small (2–5 mm long) and resemble miniature cicadas. They have strong hind legs, and wings are present in both sexes. The family contains many pest species, which may cause galls or a general stunting of host plants. The commonly held view that this stunting is caused by viruses for which psyllids serve as vectors is now believed to be wrong; it is the toxic saliva injected during feeding that causes the damage. Examples of pest species are *Psylla pyricola* and *P. mali* (Figure 8.5), two species introduced into North America from Europe, which feed on pear and apple, respectively. TRIOZIDAE (650 species worldwide) have similar habits to Psyllidae, in which family they are often included.

Superfamily Aleurodoidea

Included in the Aleurodoidea is a single family, ALEURODIDAE (ALEYRODIDAE), with approximately 1100 species, commonly known as whiteflies (Figure 8.6). They are small insects, usually 3 mm or less in length, generally covered with a whitish waxy secretion. They are commonly found on the underside of leaves. The group is mainly tropical or subtropical, though a few species are pests of greenhouse crops in temperate regions. The life cycle is complex, and parthenogenesis is commonly involved. Larvae are sedentary, and the final stage ("pupa") does not feed but undergoes a marked metamorphosis to the adult.

Superfamily Coccoidea

Some 20 families are included in the Coccoidea, a large, heterogeneous group of more than 7000 species. Despite this heterogeneity a unifying characteristic of the group is the more or less degenerate females. These are apterous, and may be scalelike, gall-like, or covered with a waxy or powdery coating. For this reason they are commonly known as scale

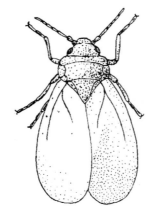

FIGURE 8.6. Aleurodoidea. The greenhouse whitefly, *Trialeurodes vaporariorum* (Aleurodidae). [From L. Lloyd, 1922, The control of the greenhouse whitefly *(Asterochiton vaporariorum)* with notes on its biology, *Ann. Appl. Biol.* 9:1–32. By permission of the Association of Applied Biologists.]

insects or mealybugs. Adult males are either apterous or have fore wings only, and have nonfunctional mouthparts. Females are oviparous (in which case the eggs are usually retained within the scaly covering of a female), ovoviviparous, or viviparous. Parthenogenesis is common and hermaphroditism is known to occur in one genus. Many species have become cosmopolitan as a result of distribution by trade. Notes on some of the commoner families are given below. The DIASPIDIDAE (armored scales) form the largest family of coccoids with 2500 species worldwide. Females are covered with a hard, waxy layer that is separate from the body. Included in the family are many pests of trees and shrubs, for example, *Lepidosaphes ulmi*, the oystershell scale (Figure 8.7A,B), and *Quadraspidiotus perniciosus*, the San Jose scale. Another large family is the COCCIDAE (soft scales, wax scales) (1200 species), the female members of which show a wide range of form. The family contains several pests that are now widespread through commerce; for example, *Pulvinaria innumerabilis*, the cottony maple scale, introduced from Europe, is now found throughout North America feeding on forest, shade, and fruit trees. The KERRIIDAE (LACCIFERIDAE), a small (70 species), mainly tropical and subtropical group, have females that are extremely degenerate and live in a resinous cell. *Laccifer lacca*, the Indian lac insect, produces a secretion from which shellac is prepared. The PSEUDOCOCCIDAE (2000 species worldwide) are the common mealybugs, so-called because females are covered with a mealy or filamentous waxy secretion. Several species of *Pseudococcus* (Figure 8.7C) are major pests as they are vectors of disease-causing viruses. The ERIOCOCCIDAE (felt scales), a cosmopolitan group of some 500 species, were formerly included in the Pseudococcidae because of their close resemblance to mealybugs. The family contains a number of pest species, including some gall formers, as well as some potentially beneficial ones. Various *Dactylopius* species, for example, have been introduced into Australia in an attempt to control the prickly pear cactus (*Opuntia*) (see Chapter 24, Section 2.3). The MARGARODIDAE (ground pearls) (250 species) form a widely distributed family, included in which is *Icerya purchasi*, the cottony-cushion scale (Figure 8.7D), an Australian species that was transplanted through commerce to many regions of the world where it became an important pest of citrus fruit. In California this pest has been controlled successfully following the introduction of the predaceous beetle *Rodolia cardinalis* (see Chapter 24, Section 2.3).

Superfamily Aphidoidea

More than 4700 species of Aphidoidea have been described, including some of the world's most important insect pests. Some authors include all Aphidoidea in one family

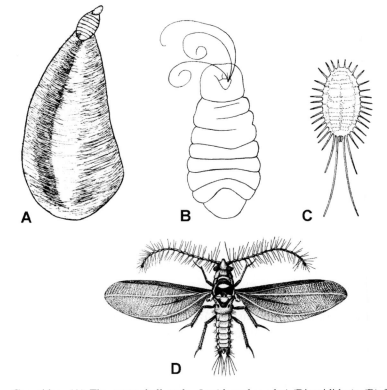

FIGURE 8.7. Coccoidea. (A) The oystershell scale, *Lepidosaphes ulmi* (Diaspididae); (B) female *L. ulmi*; (C) female long-tailed mealybug, *Pseudococcus longispinus* (Pseudococcidae); and (D) male cottony-cushion scale, *Icerya purchasi* (Margarodidae). [A, B, from D. J. Borror, D. M. Delong, and C. A. Triplehorn, 1976, *An Introduction to the Study of Insects*, 4th ed. By permission of Holt, Rinehart and Winston, Inc. C, D, from P.-P. Grassé (ed.), 1951, *Traité de Zoologie*, Vol. X. By permission of Masson, Paris.]

while others recognize up to 11 families. Members of the superfamily are characterized by their complex life cycles, in which the species takes on a variety of forms and frequently alternates between two taxonomically distinct host plants. By far the largest of the four families of Aphidoidea, with some 4300 species, is the APHIDIDAE (aphids, plant lice, greenfly, and black fly). Most aphids are found on leaves, shoots, or buds, though a few species live in rather specialized situations, for example, in unfolded leaves or in earth shelters especially constructed for them by ants with which they are associated. Most species are polymorphic in different generations and reproduce in a variety of ways. They may also show host alternation. A typical life cycle, that of *Dysaphis plantaginea*, the rosy apple aphid, is shown in Figure 8.8. Usually aphids overwinter in the egg stage. In spring the eggs hatch and give rise to wingless, viviparous parthenogenetic females. A variable number of such generations occur, and these are followed by the production of winged individuals that migrate to the alternate host on which reproduction continues. Later in the season sexual males and females are produced, and the aphids return to the original host. They mate and females lay the overwintering eggs. Parthenogenesis provides the means by which aphids can increase their population extremely rapidly. Fortunately, the occurrence of a large number of predators and adverse weather conditions usually keep their numbers in check.

Aphids in sufficient numbers may have a direct effect on their hosts, causing wilting and stunted growth. They are, however, economically more important through their role

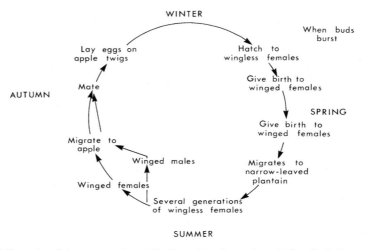

WINTER

Lay eggs on
apple twigs

When buds
burst

Hatch to
wingless females

Give birth to
winged females

AUTUMN

Mate

SPRING

Give birth to
winged females

Migrate to
apple

Winged males

Migrates to
narrow-leaved
plantain

Winged females

Several generations
of wingless females

SUMMER

FIGURE 8.8. Life cycle of the rosy apple aphid, *Dysaphis plantaginea.* [After D. J. Borror, D. M. Delong, and C. A. Triplehorn, 1976, *An Introduction to the Study of Insects*, 4th ed. By permission of Holt, Rinehart and Winston, Inc.]

as vectors of disease-producing viruses. Migratory species that are not particularly host-specific are especially important, since these can transmit diseases among a wide variety of plants. *Myzus persicae*, the green peach aphid (Figure 8.9A), is the classic example, being known as a vector for more than 100 virus diseases.

PEMPHIGIDAE (ERIOSOMATIDAE) are closely related to the Aphididae. The family is small (275 species), widely distributed (often by commerce), and includes both above- and below-ground feeders and many gall-making species. Many of the woolly aphids, so-called

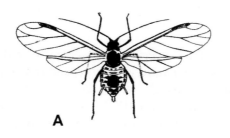

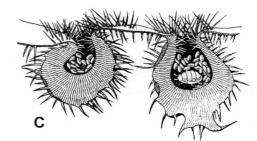

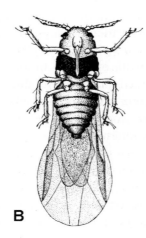

FIGURE 8.9. Aphidoidea. (A) The green peach aphid, *Myzus persicae* (Aphididae); (B) winged female of the grape phylloxera, *Phylloxera vitifoliae* (Phylloxeridae); and (C) section through galls on grape leaf showing wingless female and eggs. [A, reproduced by permission of the Smithsonian Institution Press from *Smithsonian Scientific Series*, Volume V *(Insects: Their Ways and Means of Living)* by Robert Evans Snodgrass: Fig. 101B, page 171. Copyright © 1930, Smithsonian Institution, New York. B, C, from P.-P. Grassé (ed.), 1951, *Traité de Zoologie*, Vol. X. By permission of Masson, Paris.]

because of the woollike waxy filaments that they produce, are included in this family. Most species undergo host alternation. The ADELGIDAE (CHERMIDAE) are a small family (50 species) whose members are confined to conifers. They feed on needles, twigs, or within galls. Most species alternate hosts, the primary one always being spruce. The PHYLLOX-ERIDAE (70 species) are a small but widely distributed family that includes a number of pests, for example, *Viteus* (= *Phylloxera*) *vitifolii*, the vine or grape phylloxera that devastated European vineyards in the 1870s and 1880s following its accidental introduction on North American vines. In North America winged females of *V. vitifolii* produce two types of offspring, gallicolae that form leaf galls (Figure 8.9B,C) and radicicolae that live underground feeding on the vine roots. In Europe only radicicolae are produced. Both types of offspring reproduce parthenogenetically throughout the summer. Roots of the natural North American hosts are resistant to attack by *V. vitifolii* whereas those of the European vine are highly susceptible. Successful control of the pest in Europe was thus achieved by growing the European vines on resistant North American rootstock.

Suborder Auchenorrhyncha

The two features that characterize this suborder are the complex tympanal organs and antennae whose flagellum is aristoid (hairlike). The labium originates from the posterior head region, and there is no gula. Except for those of Cicadoidea, the hind legs are spined and modified for jumping.

Infraorder Fulgoromorpha

Superfamily Fulgoroidea

In the large, heterogeneous superfamily Fulgoroidea, whose members are commonly known as plant hoppers, there are more than 10,000 species arranged in some 20 families. Plant hoppers are mainly phloem feeders on higher plants, but a few feed on fungi. The majority of species are 1 cm or less in length, though some tropical forms may reach a length of 5 cm or more. Some of the larger and commonly encountered families are as follows. The CIXIIDAE (Figure 8.10A), with more than 1300 species distributed throughout the world, are regarded as the most primitive fulgoroid family. Little is known of their biology, but the young stages are subterranean and typically feed on grass roots. Larvae of a few species have been encountered in ant nests while others are cave dwellers with reduced eyes. DELPHACIDAE (Figure 8.10B) are small (less than 1 cm in length) fulgoroids that frequently have reduced wings. This is the largest family of fulgoroids, and its more than 1500 species are recognized by the large spur on the tibia of the hind legs. A few members of this family are serious pests by acting as vectors of virus diseases. The DERBIDAE constitute a mainly tropical family, whose more than 800 species typically have very long wings and feed on fungi or higher plants. DICTYOPHARIDAE (Figure 8.10C) are medium-sized Fulgoroidea whose head bears a distinct anterior process. The family, which contains more than 600 species, is widely distributed through arid or semiarid areas of the world. The FULGORIDAE, with about 700 species, are a widely distributed group, many of whose members are known as lantern flies because the inflated anterior part of the head was believed originally to be luminous. The largest plant hoppers are members of this family. The ACHILIDAE form a widely distributed, though primarily tropical family containing about 380 species. Juveniles live beneath bark or in rotting wood. More than 1000 species

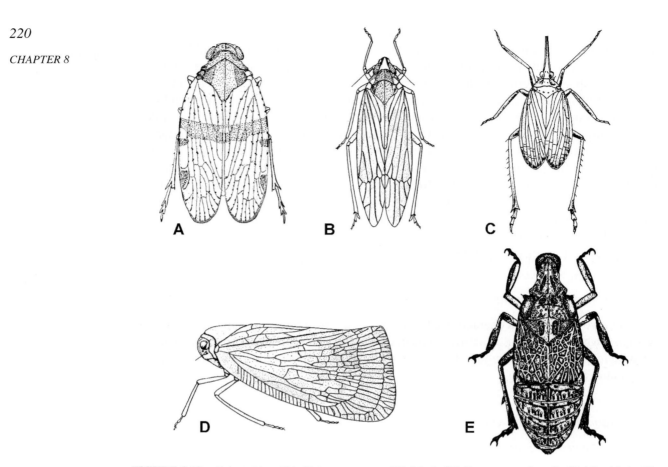

FIGURE 8.10. Fulgoroidea. (A) *Cixius angustatus* (Cixiidae); (B) *Stenocranus dorsalis* (Delphacidae); (C) *Scolops perdix* (Dictyopharidae); (D) *Anormenis septentrionalis* (Flatidae); and (E) *Fitchiella robertsoni* (Issidae). [From H. Osborn, 1938, The Fulgoridae of Ohio, *Ohio Biol. Surv. Bull.* **6**(6):283–349 (Bulletin #35). By permission of the Ohio Biological Survey.]

are included in the family FLATIDAE (Figure 8.10D), a highly specialized, mainly tropical, group. Members of this family resemble moths by virtue of their triangular, opaque tegmina, which are folded to form a steep roof over the body. The ISSIDAE (Figure 8.10E) are another large family, with more than 1100 species. Its members are mostly dull colored and frequently have a squat, beetlelike facies.

Infraorder Cicadomorpha

Superfamily Cercopoidea

The members of this small, rather homogeneous superfamily are arranged in three families, CERCOPIDAE, APHROPHORIDAE, and MACHAEROTIDAE. Species are seldom more than 15 mm in length and frequently strikingly colored as adults. The larvae of a few Cercopidae are subterranean, but mostly they live either in a mass of froth (Cercopidae and Aphrophoridae), when they are known as cuckoo-spit insects or spittlebugs, or in a calcareous tube (Machaerotidae). The function of these structures, which are affixed to plant stems, is to provide protection from predators and to prevent desiccation.

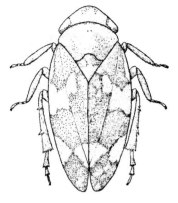

FIGURE 8.11. Cercopoidea. A froghopper, *Philaenus spumarius* (Cercopidae). [From D. J. Borror, D. M. Delong, and C. A. Triplehorn, 1976. *An Introduction to the Study of Insects*, 4th ed. By permission of Brooks/Cole, a division of Thomson Learning.]

Adults are active, hopping insects that in many cases bear a crude resemblance to a frog, hence the common name of froghopper. Aphrophoridae, with 1300 species, form the largest family with a wide distribution. Its members are usually found on herbaceous plants, but a few live on trees to which they sometimes do considerable damage. Cercopidae (1000 species) are also widespread but particularly common in the tropics. A few species are pests of grasses and clovers, for example, *Philaenus spumarius*, the meadow spittlebug (Figure 8.11). Machaerotidae (100 species) are restricted to Asia, tropical Africa, and Australia.

Superfamily Cicadoidea

Cicadas are common insects in all of the warmer regions of the world. They are generally between 2 and 5 cm in length and are particularly well known because of their sound-producing abilities and the length of time required for juvenile development. Virtually all of the 1500 or so species are placed in the family CICADIDAE. The sound-producing organs (tymbals) are located on the dorsal side of the first abdominal segment of males only. The auditory tympana are better developed in males than females and are found on the ventral side of the anterior abdominal segments. Most cicadas require several years for juvenile development. A well-known periodic cicada is *Magicicada septendecim* (Figure 8.12), which, in the eastern United States, requires 17 years for its development; in contrast, the southern form takes only 13 years to mature. A larva spends this entire period underground, feeding on roots, especially those of perennial plants. Prior to the final molt the larva leaves the soil to complete its metamorphosis on a tree or other object. Eggs are laid in twigs, a process that may cause considerable dieback of the tree.

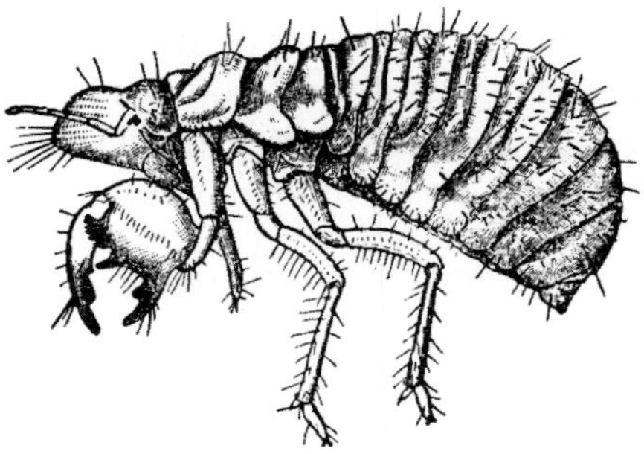

FIGURE 8.12. Cicadoidea. Juvenile periodic cicada. *Magicicada septendecim* (Cicadidae). [From C. L. Marlatt, 1907, The periodical cicada, *Bull. U.S. Dept Agr. Bur. Ent. (N. S.)* **71**:181 pp.]

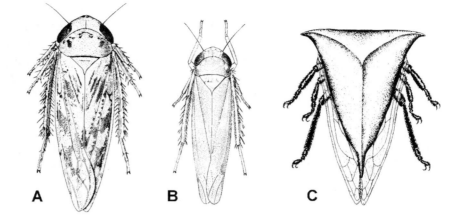

FIGURE 8.13. Cicadelloidea. (A) The beet leafhopper, *Circulifer tenellus* (Cicadellidae); (B) the potato leafhopper, *Empoasca fabae* (Cicadellidae); and (C) the buffalo treehopper, *Stictocephala bubalus* (Membracidae). [A, B, from D. M. Delong, 1948, The leafhoppers, or Cicadellidae, of Illinois, *Bull. Ill. Nat. Hist. Surv.* **24**(2):97–376. By permission of the Illinois Natural History Survey. C, from D. J. Borror, D. M. Delong, and C. A. Triplehorn, 1976, *An Introduction to the Study of Insects*, 4th ed. By permission of Brooks/Cole, a division of Thomson Learning.]

Superfamily Cicadelloidea

Most of the Cicadelloidea are placed in two very large families, CICADELLIDAE (JASSIDAE) (leafhoppers) and MEMBRACIDAE (treehoppers). The cosmopolitan Cicadellidae (20,000 species), which form the largest homopteran family, are found on almost all types of plants. They rank second only to the Aphididae in the enormity of their numbers and, in consequence, are major pests. They cause a wide variety of injuries to plants. They may remove large quantities of sap, block the phloem tubes, or destroy the chlorophyll so that growth is stunted. Many are vectors of viruses that cause disease. A few damage the plants by their oviposition habits. Two well-known pests are *Circulifer tenellus*, the beet leafhopper (Figure 8.13A), and *Empoasca fabae*, the potato leafhopper (Figure 8.13B), which feeds on solanaceous plants, beans, celery, alfalfa, and various flowers. Several other species of *Empoasca* are also important pests (see Swan and Papp, 1972). Membracidae (2400 species) are easily recognized by their enormous pronotum that projects backward over the abdomen and often assumes bizarre shapes. The family is primarily neotropical, They are generally gregarious and attended by ants for the honeydew they produce. They are seldom of economic importance; however, *Stictocephala bubalus*, the buffalo treehopper (Figure 8.13C), may damage young fruit trees and nursery stock as a result of its egg-laying activity.

Suborder Heteroptera

In Heteroptera the head is usually prognathous, almost always with a gula; the pronotum is well developed; and fore wings when present are in the form of hemelytra, with wings held flat over the body when at rest.

Infraorder Coleorrhyncha

Superfamily Peloridioidea

Contained in this superfamily is a single, Southern Hemisphere family, PELORIDIIDAE, whose 25 species are small, flattened, cryptically colored Hemiptera found among moss and liverworts or in caves, The family is all that remains of a formerly numerous and widespread Northern Hemisphere group, with fossils from as early as the Lower Jurassic period, Their current distribution is disjunct, with species occurring in South America and the Australian region. Peloridiids have a fascinating mixture of homopteran and heteropteran features, as well as such ancestral characters as the complete tentorium, discrete pro- and mesothoracic ganglia, and eight pairs of abdominal spiracles.

Infraorder Enicocephalomorpha

Superfamily Enicocephaloidea

The 260 or so species are usually arranged in a single family, ENICOCEPHALIDAE, which has a predominantly tropical to subtropical distribution. They are small- to medium-sized bugs, typically found among leaf litter, rotting logs, etc., where they prey on other arthropods. Many are brachypterous or apterous, though winged species often appear in large swarms at dusk, hence their common name of gnat bugs.

Infraorder Dipsocoromorpha

Superfamily Dipsocoroidea

About three quarters of the approximately 225 species of Dipsocoroidea are included in the cosmopolitan family SCHIZOPTERIDAE. These are minute, predaceous insects found in damp habitats such as moss, leaf litter, and ant nests.

Infraorder Gerromorpha

Some authors include all members of the infraorder in a single superfamily, Gerroidea. Others, including Carver *et al.* (1991), split the group into the following four superfamilies.

Superfamily Mesovelioidea

The 30 or so species are arranged in a single, widely distributed family, MESOVELI-IDAE. These long-legged, predaceous bugs are usually found at the margins of ponds, crawling among vegetation and debris; sometimes they move over the water surface. A few species occur away from water in damp places, including caves.

Superfamily Hebroidea

Hebroidea are very small, semiaquatic bugs found among moss or on ponds with an abundance of floating or emergent vegetation where they prey on small arthropods. The group has a cosmopolitan distribution and its 150 species are included in a single family, HEBRIDAE.

Superfamily Gerroidea

All but about a dozen of the approximately 1000 species in this superfamily are included in two equally sized families, GERRIDAE and VELIIDAE. Gerridae, commonly known as pond skaters and water striders (Figure 8.14A), are elongate bugs whose body is covered with a dense layer of waterproof hairs. Members of this widespread group are mainly found

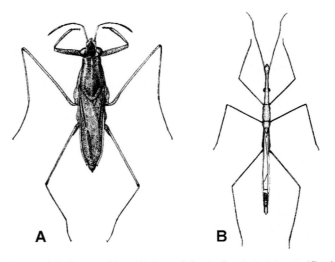

FIGURE 8.14. Gerroidea and Hydrometroidea. (A) A pond skater, *Gerris marginatus* (Gerridae); and (B) a water measurer, *Hydrometra martini* (Hydrometridae). [A, from R. C. Froeschner, 1962, Contributions to a synopsis of the Hemiptera of Missouri, *Am. Midl. Nat.* **67**(1):208–240. By permission of the American Midland Naturalist. B, from A. R. Brooks and L. A. Kelton, 1967, Aquatic and semiaquatic Heteroptera of Alberta, Saskatchewan and Manitoba (Hemiptera), *Mem. Entomol. Soc. Can.* **51**:92 pp. By permission of the Entomological Society of Canada.]

on freshwater surfaces, moving or still, but a few species are marine. They feed on insects that fall onto the water surface. Veliids (water crickets) are primarily neotropical and oriental in distribution. Like gerrids, they are predaceous and mainly occur on the surface of moving or still fresh water. A few species are marine and others live in moist forest soil or among wet rocks.

Superfamily Hydrometroidea

All but 5 of the approximately 120 species in this group are placed in the mainly tropical family HYDROMETRIDAE (water measurers), though the genus *Hydrometra* (Figure 8.14B) is cosmopolitan. These slow-moving insects, with a superficial resemblance to stick insects, mostly crawl over floating vegetation or still water, preying on insects and crustaceans; however, a few species are found on the floor of rain forests.

Infraorder Leptopodomorpha

Superfamily Leptopodoidea

Most members of this very small group are included in the family LEPTOPODIDAE (30 species). These are primarily found in the Eastern Hemisphere tropics, on vertical rock faces and boulders adjacent to streams where they prey on other arthropods.

Superfamily Saldoidea

Almost all of the 260 species of Saldoidea are included in the cosmopolitan family SALDIDAE (Figure 8.15), commonly called shore bugs because they are found in the intertidal zone, on mud flats, in salt marshes, and other open areas, where they prey or scavenge on other invertebrates. Their long legs facilitate rapid pursuit of potential prey.

FIGURE 8.15. Saldoidea. *Teloleuca pellucens* (Saldidae). [From A. R. Brooks and L. A. Kelton, 1967, Aquatic and semiaquatic Heteroptera of Alberta, Saskatchewan and Manitoba (Hemiptera), *Mem. Entomol. Soc. Can.* **51**:92 pp. By permission of the Entomological Society of Canada.]

Infraorder Nepomorpha

Superfamily Nepoidea

Two small but well-known families of aquatic bugs are included in this group, NEPIDAE (200 species) and BELOSTOMATIDAE (150 species). Nepidae are called water scorpions because of their posterior respiratory siphon through which gas exchange occurs during periodic visits to the water surface. The family is mainly tropical and subtropical in distribution, though some common genera, for example, *Ranatra* (Figure 8.16C) and *Nepa* (Figure 8.16D), are cosmopolitan. Among the Belostomatidae (giant water bugs) are some of the largest Heteroptera, some South American species reaching a length of 11 cm. *Lethocerus* (Figure 8.16B) has a worldwide distribution; its species are brown, oval-shaped bugs with raptorial forelegs used to catch prey, both invertebrate and vertebrate. Occasionally, these insects become a nuisance in fish hatcheries.

Superfamily Ochteroidea

The Ochteroidea (Gelastocoroidea) includes two families of small, flattened, "knobbly surfaced" bugs, the OCHTERIDAE (30 species) and GELASTOCORIDAE (150 species). Both families are widespread, and species are mostly found on the shores of lakes and ponds, mud flats, etc., where they feed on other arthropods. Their warty appearance, together with their ability to jump, has given them their common name of toadbugs.

Superfamily Naucoroidea

All 170 species of this widely distributed group of aquatic bugs are placed in the family NAUCORIDAE. These slow-moving insects, which bear some resemblance to belostomatids, are found among submerged vegetation in moving or still water where they use their raptorial forelegs to capture prey.

Superfamily Notonectoidea

The great majority of the 360 species of Notonectoidea are included in the widespread family NOTONECTIDAE (back swimmers) (Figure 8.16A). As their common name

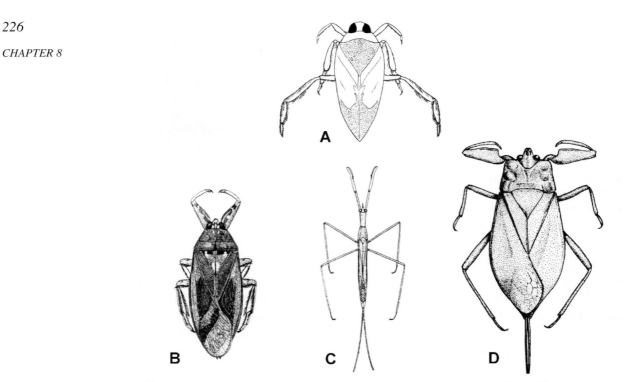

FIGURE 8.16. Notonectoidea and Nepoidea. (A) A back swimmer, *Notonecta undulata* (Notonectidae); (B) the giant water bug, *Lethocerus americanus* (Belostomatidae); (C) a water scorpion, *Ranatra fusca* (Nepidae); and (D) a water scorpion, *Nepa apiculata* (Nepidae). [B, C, from A. R. Brooks and L. A. Kelton, 1967, Aquatic and semiaquatic Heteroptera of Alberta, Saskatchewan and Manitoba (Hemiptera) *Mem. Entomol. Soc. Can.* **51**:92 pp. By permission of the Entomological Society of Canada. D, from R. C. Froeschner, 1962, Contributions to a synopsis of the Hemiptera of Missouri, *Am. Midl. Nat.* **67**(1):208–240. By permission of the American Midland Naturalist.]

indicates, the insects swim with the ventral surface uppermost. The hind legs are long and oarlike, with a fringe of bristles on the posterior margin. Back swimmers, which are predaceous, usually rest on the water surface and, when disturbed, swim actively downward and grasp onto a submerged object. Cosmopolitan genera are *Notonecta* and *Anisops*.

Superfamily Corixoidea

The 500 or so species of Corixoidea are placed in a single cosmopolitan family, CORIXIDAE (water boatmen) (Figure 8.17). They are typically microphagous bugs that feed on detritus, algae, etc., which they scoop up with their flattened, hairy fore tarsi. However, some species feed on other aquatic arthropods, using their fore tarsi to locate prey (e.g., chironomid larvae) buried in the substrate. Corixids generally cling onto the substrate or submerged vegetation with their middle legs, surfacing only to renew their air supply. The hind legs are enlarged, flattened, and fringed with hairs for swimming purposes but, in contrast to notonectids, corixids swim with the dorsal surface uppermost.

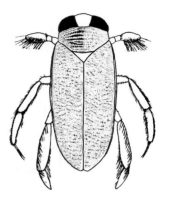

FIGURE 8.17. Corixoidea. A water boatman. *Sigara atropodonta* (Corixidae).

Infraorder Cimicomorpha

Superfamily Thaumastocoroidea

The 15 species of Thaumastocoroidea are arranged in a single family, THAUMASTO-CORIDAE. Most species of these phytophagous bugs are Australian, though others occur in India, South America, Cuba, and Florida.

Superfamily Tingoidea

About 1800 species of Tingoidea are known, the vast majority of which are included in the family TINGIDAE (lacebugs). Members of this cosmopolitan family are easily recognized by the lacelike pattern on the dorsal surface of the head and body, including the wings (Figure 8.18). They are phytophagous on a wide range of plants and occasionally become pests. A few species are gall formers and some show maternal care of the nymphs.

Superfamily Miroidea

All but about two dozen of the approximately 10,000 species of miroids are included in the worldwide family MIRIDAE (CAPSIDAE). The majority of species are phytophagous, and the group includes some important pest species, for example, *Lygus lineolaris*, the

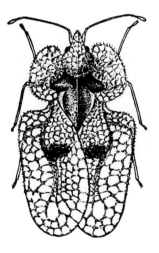

FIGURE 8.18. Tingoidea. The sycamore lacebug, *Corythuca ciliata* (Tingidae). [From R. C. Froeschner, 1944, Contributions to a synopsis of the Hemiptera of Missouri, *Am. Midl. Nat.* **31**(3):638–683. By permission of the American Midland Naturalist.]

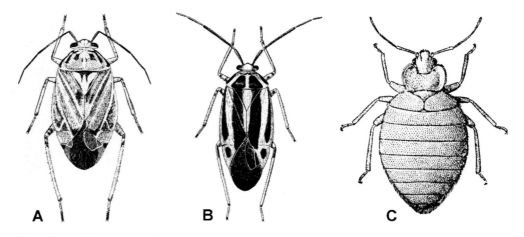

FIGURE 8.19. Miroidea and Cimicoidea. (A) The tarnished plant bug, *Lygus lineolaris* (Miridae); (B) the four-lined plant bug, *Poecilocapsus lineatus* (Miridae); and (C) the bedbug, *Cimex lectularius* (Cimicidae). [A, B, from H. H. Knight, 1941, The plant bugs, or Miridae, of Illinois, *Bull. Ill. Nat. Hist. Surv.* **22**(1):234 pp. By permission of the Illinois Natural History Survey.]

tarnished plant bug (Figure 8.19A), which feeds on cotton, alfalfa, hay, and various vegetables and fruits. Its relative, *Poecilocapsus lineatus*, the four-lined plant bug (Figure 8.l9B), feeds on gooseberry, currant, rose, and various annual flowers. Some species feed on fungi while others are omnivorous, sucking both plant and insect fluids (including those of pest species).

Superfamily Cimicoidea

Members of this superfamily of about 1000 species are united by the occurrence of hemocoelic insemination (see Chapter 19, Section 4.3.1), egg fertilization in the vitellarium of the ovary, and pre-ovipositional embryonic development. Most belong to three families, CIMICIDAE, NABIDAE, and ANTHOCORIDAE. Nabidae (300 species worldwide) were formerly included in the Reduvioidea, but the demonstration of hemocoelic insemination in members of one subfamily led to their transfer to this group. Nabids prey actively on other insects. Anthocorids (flower bugs) (500 species) form a widespread group of bugs that feed mostly on the blood or eggs of arthropods, sometimes on pollen, and rarely on the blood of mammals, including humans. They are found on flowers, under bark, or in leaf litter. The Cimicidae is a small (75 species) but widely distributed family of wingless bugs that are bloodsucking ectoparasites of birds and mammals. Included in the family are the bedbugs, *Cimex lectularius* (Figure 8.19C) (cosmopolitan) and *C. hemipterus* (mainly southern Asia and Africa). These are particularly common in unhygienic and/or overcrowded conditions. They hide during the day, and also lay their eggs, in crevices, coming out to feed at night. Though bedbugs have been implicated in the transmission of more than 20 diseases, conclusive evidence for this role is lacking.

Superfamily Reduvioidea

Almost all of the more than 5000 described species of Reduvioidea are placed in one family, the REDUVIIDAE, commonly known as assassin bugs (Figure 8.20). All species are predaceous, particularly on other arthropods, though a number feed on the blood of

FIGURE 8.20. Reduvioidea. The bloodsucking conenose, *Triatoma sanguisuga* (Reduviidae). [From R. C. Froeschner, 1944, Contributions to a synopsis of the Hemiptera of Missouri, *Am. Midl. Nat.* 31(3):638–683. By permission of the American Midland Naturalist.]

vertebrates, including humans. Most species inject saliva that paralyzes the tissue as well as assists in its digestion, thereby causing severe and painful bites. Species of *Triatoma* and *Rhodnius prolixus* carry *Trypanosoma cruzi*, which causes a fatal form of sleeping sickness (Chagas' disease) in humans. In many species the forelegs are raptorial. The family is cosmopolitan, but especially rich in the tropics and subtropics. Some species mimic phytophagous Hemiptera or are well camouflaged.

Infraorder Pentatomorpha

Superfamily Aradoidea

More than 1000 species of Aradoidea are known, all but about 10 of which are placed in the family ARADIDAE (Figure 8.21). Aradids are widely distributed and occur typically beneath bark, in leaf litter, or in rotting wood where they suck fluid from fungal mycelia or fruiting bodies. Their mouthparts are extremely long and at rest are coiled within the head.

Superfamily Idiostoloidea

Only a handful of species occur in this group, all being included in the family IDIOSTOLIDAE, found only in *Nothofagus* forests of southern South America and Australia. Their habits and food are unknown.

FIGURE 8.21. Aradoidea. *Aradus acutus* (Aradidae). [From R. C. Froeschner, 1942, Contributions to a synopsis of the Hemiptera of Missouri, *Am. Midl. Nat.* 27(3):591–609. By permission of the American Midland Naturalist.]

Superfamily Piesmatoidea

This widespread group contains only about 20 species, all in the family PIESMATI-DAE. Though these phytophagous bugs superficially resemble Tingoidea, morphological studies confirm that they are closely related to the Lygaeoidea (in which they are sometimes included).

Superfamily Lygaeoidea

About 80% of the more than 3600 species of Lygaeoidea belong to the family LY-GAEIDAE, a widely distributed group whose members are usually found on the ground, among vegetation, or under stones or low plants, where they feed usually on mature seeds that have fallen to the ground. Some species live off the ground, feeding on stems or immature seeds of a variety of plants, especially grasses. A few others are predaceous. Several species are pests, the best-known example being *Blissus leucopterus*, the chinch bug (Figure 8.22A), which attacks maturing cereal crops in the United States. A small but economically important family is the PYRRHOCORIDAE (300 species), whose members are widely distributed, commonly black and red bugs. Several species of *Dysdercus* (Figure 8.22B) are major pests of cotton and other Malvaceae on whose seeds (bolls) they feed. During this activity the bolls become contaminated with a fungus, which later stains the cotton fibers, hence the common name of cotton stainers for these insects. The BERYTIDAE (stilt bugs) (Figure 8.22C) are secretive, slow-moving bugs found chiefly on low-growing plants. The family is small (160 species) but widely distributed. Its members have narrow, elongate bodies and long slender legs and antennae. Though principally phytophagous, some species are facultative predators.

Superfamily Coreoidea

The Coreoidea are closely related to the Lygaeoidea, and the demarcating line between the two groups is often difficult to define. There is also disagreement over the higher classification of the superfamily, some authorities (e.g., Carver *et al.*, 1991) recognizing up to five families, others placing all species in the single family COREIDAE. The more than 2000 species of coreids are generally dark-colored bugs, most common in Asia, Africa, and South America. All are phytophagous, and some are pests, for example, the squash bug, *Anasa tristis* (Figure 8.23), on Cucurbitaceae in North America. Many species rival pentatomids in their abilities to produce foul smells. ALYDIDAE (250 species worldwide) feed on the vegetative parts of plants and on both ripe and unripe seeds. Some species are pests of rice and, occasionally, legumes.

Superfamily Pentatomoidea

Included among the 7000 species of Pentatomoidea are the familiar shield and stink bugs. The group is subdivided into as many as 15 families of which by far the largest, with more than 5000 species, is the PENTATOMIDAE. Most pentatomids are phytophagous, but a few are predaceous on other insects, especially caterpillars. They are typically brightly colored or conspicuously marked insects, capable of emitting a foul fluid from thoracic repugnatorial glands. A few species are economically important, for example, *Murgantia histrionica*, the harlequin cabbage bug (Figure 8.24) found on cabbage and other Cruciferae.

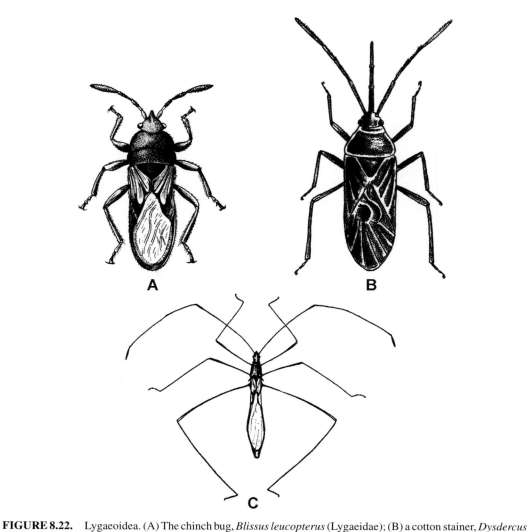

FIGURE 8.22. Lygaeoidea. (A) The chinch bug, *Blissus leucopterus* (Lygaeidae); (B) a cotton stainer, *Dysdercus suturellus* (Pyrrhocoridae); and (C) a stilt bug, *Jalysus wickhami* (Berytidae). [A, from P.-P. Grassé (ed.), 1951, *Traité de Zoologie*, Vol. X. By permission of Masson, Paris. B, from L. A. Swan and C. S. Papp, 1972, *The Common Insects of North America.* Copyright 1972 by L. A. Swan and C. S. Papp. Reprinted by permission of Harper & Row, Publishers, Inc. C, from R. C. Froeschner, 1942, Contributions to a synopsis of the Hemiptera of Missouri, *Am. Midl. Nat.* **27**(3):591–609. By permission of the American Midland Naturalist.]

FIGURE 8.23. Coreoidea. The squash bug, *Anasa tristis* (Coreidae). [From R. C. Froeschner, 1942, Contributions to a synopsis of the Hemiptera of Missouri, *Am. Midl. Nat.* **27**(3):591–609. By permission of the American Midland Naturalist.]

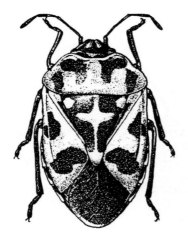

FIGURE 8.24. Pentatomoidea. The harlequin cabbage bug, *Murgantia histrionica* (Pentatomidae). [From R. C. Froeschner, 1941, Contributions to a synopsis of the Hemiptera of Missouri, *Am. Midl. Nat.* **26**(1):122–146. By permission of the American Midland Naturalist.]

The CYDNIDAE (negro bugs) (400 species worldwide) resemble the Pentatomidae in shape but are darkly colored and live under stones or dead leaves, or occasionally in ant nests. The SCUTERELLIDAE are easily recognized by their enormously enlarged mesoscutellum, which extends posteriorly to cover the entire abdomen and wings. Most of the 400 species are tropical or subtropical. They are phytophagous, sometimes becoming pests of grain crops in the Near East. PLATASPIDAE (500 species) mainly occur in tropical regions of the Eastern Hemisphere. They are mostly phytophagous, including species that are pests of legumes, though some feed on fungi and a few live in ant nests. The chiefly Old World tropical TESSARATOMIDAE (250 species) are sometimes called giant shield bugs as they are generally large to very large (>15 mm). These phytophagous bugs sometimes become pests of cultivated plants, for example, citrus in Australia.

Literature

Good general accounts of the Hemiptera are given by Pesson (1951), Poisson (1951), and Dolling (1991). Miller (1971), Schuh and Slater (1995), and Schaefer (2000) discuss the biology of the Heteroptera, while authors in Schaefer and Panizzi (2000) detail economically important Heteroptera. Phylogeny and classification are discussed by Hennig (1981), Hamilton (1981), Popov (1981), Schuh (1986), Wootton and Betts (1986), Wheeler *et al.* (1993), Campbell *et al.* (1994), Sorensen *et al.* (1995), and von Dohlen and Moran (1995). Keys to the North American families of Hemiptera are given by Arnett (2000) and Slater and Baranowski (1978) [Heteroptera]. Genera of the aquatic and semiaquatic Hemiptera of North America can be identified from the key provided by Polhemus (1984). Keys and descriptions of British Hemiptera are given by Southwood and Leston (1959) [Heteroptera], Macan (1965) [aquatic Heteroptera], Le Quesne (1960, 1965, 1969) [Fulgoromorpha and Cicadomorpha], and Dolling (1991) [all families]. Carver *et al.* (1991) provide a key to the Australian families (and some subfamilies) of Hemiptera. Stonedahl and Dolling (1991) have prepared a lengthy bibliography on Heteroptera of the world, while Schaefer and Kosztarab *et al.* (in Kosztarab and Schaefer, 1990) provide bibliographies to the Heteroptera and homopterans, respectively, of the North American region.

Arnett, R. H., Jr., 2000, *American Insects: A Handbook of the Insects of America North of Mexico*, 2nd ed., CRC Press, Boca Raton, FL.

Campbell, B. C., Steffen-Campbell, J. D., and Gill, R. J., 1994, Evolutionary origin of whiteflies (Hemiptera: Sternorrhyncha: Aleyrodidae) inferred from 18S rDNA sequences, *Insect Mol. Biol.* **3**:73–88.

Carver, M., Gross, G. F., and Woodward, T. E., 1991, Hemiptera, in: *The Insects of Australia*, 2nd ed., Vol. I (CSIRO, ed.), Melbourne University Press, Carlton, Victoria.

Dolling, W. R., 1991, *The Hemiptera*, Oxford University Press, Oxford.

Hamilton, K. G. A., 1981, Morphology and evolution of the rhynchotan head (Insecta: Hemiptera, Homoptera), *Can. Entomol.* **113**:953–974.

Hennig, W., 1981, *Insect Phylogeny*, Wiley, New York.

Kosztarab, M., and Schaefer, C. W. (eds.), 1990, *Systematics of the North American Insects and Arachnids: Status and Needs*, Virginia Polytechnic Institute and State University, Blacksburg.

Le Quesne, W. J., 1960, 1965, 1969, Hemiptera-Fulgoromorpha and Hemiptera-Cicadomorpha, *R. Entomol. Soc. Handb. Ident. Br. Insects* **2**(3):1–68; **2**(2):1–64, 65–148.

Macan, T. T., 1965, A revised key to the British water bugs (Hemiptera-Heteroptera), 2nd revised ed., *Sci. Publ. F.W. Biol. Assoc.* **16**:78 pp.

Miller, N. C. E., 1971, *The Biology of the Heteroptera*, 2nd ed., Hill, London.

Novotny, V., and Wilson, M. R., 1997, Why are there no small species among xylem-sucking insects? *Evol. Ecol.* **11**:419–437.

Pesson, P., 1951, Ordre des Homoptères, in: *Traité de Zoologie* (P.-P. Grassé, ed.), Vol. X, Masson, Paris.

Poisson, R., 1951, Ordre des Heteroptères, in: *Traité de Zoologie* (P.-P. Grassé, ed.), Vol. X, Masson, Paris.

Polhemus, J. T., 1984, Aquatic and semiaquatic Hemiptera, in: *An Introduction to the Aquatic Insects of North America*, 2nd ed. (R. W. Merritt and K. W. Cummins, eds.), Kendall/Hunt, Dubuque, IA.

Popov, Y. A., 1981, Historical development and some questions on the general classification of Hemiptera, *Rostria* **33**(Suppl.):85–99.

Schaefer, C. W., 2000, The Heteroptera, or true bugs, in: *Encyclopedia of Biodiversity* (S. Levin, ed.), Academic Press, San Diego.

Schaefer, C. W., and Panizzi, A. R. (eds.), 2000, *Heteroptera of Economic Importance*, CRC Press, Boca Raton, FL.

Schuh, R. T., 1986, The influence of cladistics on heteropteran classification, *Annu. Rev. Entomol.* **31**:67–93.

Schuh, R. T., and Slater, J. A., 1995, *True Bugs of the World (Hemiptera: Heteroptera)*, Cornell University Press, Ithaca, NY.

Slater, J. A., and Baranowski, R. M., 1978, *How to Know the True Bugs (Hemiptera-Heteroptera)*, William Brown, Dubuque, Iowa.

Sorensen, J. T., Campbell, B. C., Gill, R. J., and Steffen-Campbell, J. D., 1995, Non-monophyly of Auchenorrhyncha ("Homoptera"), based upon 18S rDNA phylogeny: Eco-evolutionary and cladistic implications within pre-Heteropterodea Hemiptera (s.l.) and a proposal for monophyletic suborders, *Pan-Pac. Entomol.* **71**:31–60.

Southwood, T. R. E., and Leston, D., 1959, *Land and Water Bugs of the British Isles*, Warne, London.

Stonedahl, G. M., and Dolling, W. R., 1991, Heteroptera identification: A reference guide, with special emphasis on economic groups, *J. Nat. Hist.* **25**:1027–1066.

Swan, L. A., and Papp, C. S., 1972, *The Common Insects of North America*, Harper & Row, New York.

Von Dohlen, C. D., and Moran, N. A., 1995, Molecular phylogeny of the Homoptera: A paraphyletic taxon, *J. Mol. Evol.* **41**:211–223.

Wheeler, W. C., Schuh, R. T., and Bang, R., 1993, Cladistic relationships among higher groups of Heteroptera: Congruence between morphological and molecular data sets, *Entomol. Scand.* **24**:121–137.

Wootton, R. J., and Betts, C. R., 1986, Homology and function in the wings of Heteroptera, *Syst. Entomol.* **11**:389–400.

5. Thysanoptera

SYNONYMS: Physapoda, Thripida   COMMON NAME: thrips

Small to minute slender-bodied insects; head with short four- to nine-segmented antennae, asymmetrical suctorial mouthparts, small but prominent compound eyes, ocelli present or absent; prothorax large and free, legs with unsegmented or two-segmented tarsi and terminal eversible vesicles, wings when present are narrow and fringed with long setae; external genitalia varied, cerci always absent.

Almost 5200 described species belong to this widely distributed order, including about 700 in North America, 420 in Australia, and 160 in Britain. Many species are very widespread either as a result of natural drift on wind currents or by trade. Several hundred species are economically important to varying degrees either as pests, especially as disease vectors, or as beneficial predators of other arthropod pests.

Structure

Thrips are generally yellowish, brown, or black elongate insects that range in length from about 0.5 to 15 mm. The hypognathous or opisthognathous head is devoid of sutures and sulci and bears a pair of prominent compound eyes, four- to nine-segmented antennae inserted close together on the head, and asymmetrical suctorial mouthparts. The labrum, labium, and maxillary stipites form a short cone-shaped rostrum that encloses the styletlike left mandible and paired laciniae. The right mandible disappears during embryogenesis. Labial and maxillary palps are present. Ocelli (three) are found only on winged adults. The prothorax is large and free, the mesothorax and metathorax fused. The forelegs are usually slender but may have enlarged tarsal teeth and swollen femora for gripping. The tarsi are unsegmented or two-segmented and at their tip carry a bladderlike structure (arolium) that can be everted by means of hemolymph pressure to enable the insect to walk on a variety of surfaces. Wings may be fully developed, shortened, or absent. When present they are membranous, narrow structures with few or no veins. Each possesses a fringe of long setae. The fore and hind wings are coupled during flight. The 11-segmented abdomen tapers posteriorly. Females of the suborder Tubulifera have a delicate, reversible, chutelike ovipositor, but in Terebrantia this is usually a strong, external structure with four sawlike valves. Both tubuliferan and terebrantian males have well-developed external genitalia.

The internal structure of Thysanoptera is generally similar to that of Hemiptera. There is a large cibarial pump anterior to the alimentary canal. The foregut is short and leads into a large midgut, which is differentiated into an anterior croplike region and tubular hind portion. There are four Malpighian tubules. The nervous system is highly specialized. The subesophageal and prothoracic ganglia are fused, those of the mesothorax and metathorax remain separate, and those of the abdomen have coalesced to form a single center. Each ovary contains four "neopanoistic" ovarioles (secondarily derived from the ancestral polytrophic condition). A spermatheca occurs, but accessory glands may or may not be present. In males the testes are fusiform and connect via short vasa deferentia to the ejaculatory duct. The latter also receives ducts from one or two pairs of large accessory glands.

Life History and Habits

Though thrips are commonly encountered on growing vegetation, particularly among flowers, many species live on the ground among rotting vegetation where they feed on fungal mycelia or spores, a habit established by their psocopteroid ancestors. Others feed on pollen, leaf and flower tissue, and small arthropods, including their eggs (Heming, 1993). Feeding is achieved by a "punch and suck" mechanism (Heming, 1978) whereby the left mandible first pierces the tissue, the maxillary stylets (which form a tube) are then inserted, and the liquid food is drawn into the oral cavity by the action of the cibarial pump. Not surprisingly, many thrips are important pests because of the damage they cause, either directly by weakening plants and perhaps preventing fruit formation, or indirectly by acting as transmitters of disease-causing viruses. Partially offsetting their importance as pests is the benefit they produce by assisting in pollination and, in predaceous species, killing harmful insects. Most thrips seem to be rather inactive insects, though a number can run rapidly on occasion. Many can fly but only rarely resort to this activity. Many species are readily dispersed on wind currents though, curiously, these are more often wingless rather than winged forms.

Though bisexual, most species probably reproduce by haplodiploid parthenogenesis (see Chapter 20, Section 8.1). Some species apparently are obligate parthenogens, males being rare or unknown, and a few are ovoviviparous or viviparous. The eggs of Terebrantia are somewhat kidney-shaped and laid in plant tissues; those of Tubulifera are oval and simply deposited on the surface of a plant. The postembryonic development of thrips is of interest because it parallels in some respects that found in endopterygote orders. The first two juvenile stages are typically exopterygote in that the insect generally resembles the adult except for the lack of wings and in the possession of fewer antennal segments. However, the remaining instars (two in Terebrantia, three in Tubulifera) are resting stages that do not feed and in which some degree of metamorphosis occurs. The first of these resting stages is called the prepupa (propupa), the remaining one or two, the pupal stage(s). Whether the latter is homologous with the pupa of endopterygotes is open to discussion (see Chapter 2, Section 3.3).

Phylogeny and Classification

The Thysanoptera appear to be most closely related to the Hemiptera, with which they may have had a common ancestor in the Upper Carboniferous or Lower Permian period. An earlier suggestion (Grinfel'd, 1959, cited in Stannard, 1968) that the order arose as a pollen-feeding group is no longer tenable in light of recent comparative morphological and ontogenetical studies of the mouthparts (Heming, 1993). This author suggested that thrips arose from a litter-dwelling, fungivorous ancestor, with the gradual loss of the left mandible perhaps resulting from selection pressure for small body size, a single structure being quite adequate for making a hole in the tissue. Subsequently, numerous small but significant changes in mouthpart structure enabled thrips (especially Tubulifera) to radiate into the numerous, cryptic microhabitats that they now occupy. Unfortunately, this radiation cannot be traced in the relatively poor, early fossil record for this order; indeed, some supposed early fossils (e.g., *Permothrips*) are now thought to belong to other hemipteroid orders (Mound *et al.*, 1980). From the Cretaceous period on, fossils become much more abundant, but most of these can be assigned to modem families. Recent Thysanoptera can be arranged in two suborders and nine families. Of the two suborders, the Terebrantia, whose members retain wing veins and, in females, a well-developed ovipositor, are undoubtedly more primitive than the Tubulifera, whose species have veinless wings and a greatly reduced ovipositor. A proposed phylogeny of the extant families is shown in Figure 8.25.

Suborder Terebrantia

Uzelothrips scabrosus, collected in Brazil and Singapore on dead twigs and litter, is the sole species in the family UZELOTHRIPIDAE. Its relationship with other Terebrantia remains conjectural but, on the basis of its fully developed tentorium, Mound *et al.* (1980) suggested that it might be a relict form. The MEROTHRIPIDAE form a small Neotropical group whose 15 species occur in leaf litter or on dead twigs. The family has several characters that suggest it may be the most primitive of the extant Thysanoptera. Most of the 260 species of AEOLOTHRIPIDAE live in temperate regions in both the Northern and the Southern

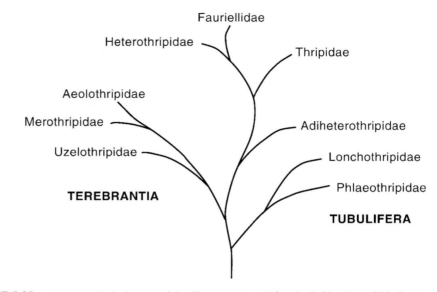

FIGURE 8.25. A proposed phylogeny of the Thysanoptera. [After B, S. Heming, 1993, Structure, function, ontogeny, and evolution of feeding in thrips, in: *Functional Morphology of Insect Feeding* (C. W. Schaefer and R. A. B. Leschen, eds.), Thomas Say Publications in Entomology: Proceedings. By permission of the Entomological Society of America.]

Hemisphere. They mostly feed on other arthropods though some species eat pollen or the tissues of vascular plants. The HETEROTHRIPIDAE (70 species) (as defined by Mound *et al.,* 1980) is a strictly New World group of pollen-feeding thrips found on trees and flowers. The FAURIELLIDAE and ADIHETEROTHRIPIDAE, with four and five species of pollen feeders, respectively, were originally included in the previous family. The fauriellids include two genera from southern Africa and one from southern Europe, the adiheterothripids one genus from western North America and a second that ranges from the eastern Mediterranean to India. By far the largest and most specialized group in the suborder is the cosmopolitan family THRIPIDAE (1710 species). Though some thripids are predaceous, pollenophagous, or fungivorous, the great majority feed on sap obtained from the epidermis or mesophyll of vascular plants. Most pest thrips belong to this family and cause damage either directly by generally weakening their hosts or indirectly by serving as vectors of disease-causing viruses. Several pest species have been transported through trade to many parts of the world. *Taeniothrips inconsequens*, the pear thrips (Figure 8.26A), is a European species now widespread through North America where it is a major pest of sugar maple. *Taeniothrips simplex*, the gladiolus thrips, is another widespread species. *Thrips tabaci*, the onion thrips, is a cosmopolitan species, which, although preferring onions, is known to feed also on many other plants, including tomatoes, tobacco, cotton, and beans. This, and other species of *Thrips* and *Frankliniella*, are known to transmit the virus that causes spotted wilt of tomatoes. Species of *Limothrips* (grain thrips) are important pests of cereal crops in various parts of the world.

Suborder Tubulifera

All but one species (*Lonchothrips linearis*; LONCHOTHRIPIDAE) in this suborder of about 3100 species are placed in the cosmopolitan family PHLAEOTHRIPIDAE

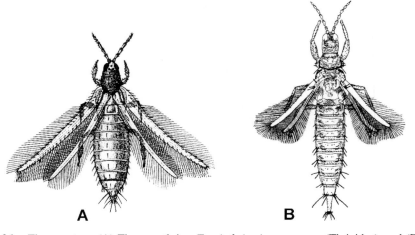

FIGURE 8.26. Thysanoptera. (A) The pear thrips, *Taeniothrips inconsequens* (Thripidae); and (B) *Liothrips citricornis* (Phlaeothripidae). [A, from P.-P. Grassé (ed.), 1951, *Traité de Zoologie*, Vol. X. By permission of Masson, Paris. B, from L. J. Stannard, Jr., 1968, The thrips, or Thysanoptera, of Illinois, *Bull. Ill. Nat. Hist. Surv.* **29**(4):215–552. By permission of the Illinois Natural History Survey.]

(Figure 8.25B). Phlaeothripids are generally large thrips, occasionally reaching more than 10 mm in length, that exhibit a wide range of life styles. Mound *et al.* (1980), on the basis of the generally dorsoventrally flattened body, suggested that the early members of the family lived as fungus feeders under bark. Most extant species continue to feed on fungi, including their spores, while others prefer lichens, mosses, flowers, leaves and stems of angiosperms (some causing galls), or small arthropods. Crespi (1992) has shown that some of the gall-forming species have evolved eusociality, the founding female producing both fully winged and short-winged offspring. The latter ("soldiers") have large, armed forelegs with which they defend the colony against would-be invaders, notably other thrips species, but including ants and caterpillars.

Literature

Stannard (1968), Lewis (1973), Mound *et al.* (1980), Ananthakrishnan (1979, 1984, 1993), and authors in Lewis (1997) give good accounts of the biology of thrips. Mound *et al.* (1980) and Heming (1993) discuss the phylogeny of the order, the former also including a key to the world families. Mound *et al.* (1976) provide a key to the British species, while the North American and Australian families may be identified from Arnett (2000) and Mound and Heming (1991), respectively.

Ananthakrishnan, T. N., 1979, Biosystematics of Thysanoptera, *Annu. Rev. Entomol.* **24**:159–183.

Ananthakrishnan, T. N., 1984, *Bioecology of Thrips*, Indira Publishing House, Oak Park, MI.

Ananthakrishnan, T. N., 1993, Bionomics of thrips, *Annu. Rev. Entomol.* **38**:71–92.

Arnett, R. H., Jr., 2000, *American Insects: A Handbook of the Insects of America North of Mexico*, 2nd ed., CRC Press, Boca Raton, FL.

Crespi, B. J., 1992, Eusociality in Australian gall thrips, *Nature* **359**:724–726.

Heming, B. S.,1978, Structure and function of the mouthparts in larvae of *Haplothrips verbasci* (Osborn) (Thysanoptera, Tubulifera, Phlaeothripidae), *J. Morphol.* **156**:1–37.

Heming, B. S.,1993, Structure, function, ontogeny, and evolution of feeding in thrips, in: *Functional Morphology of Insect Feeding* (C. W. Schaefer and R. A. B. Leschen, eds.), Thomas Say Publications in Entomology: Proceedings, Entomological Society of America, Lanham, MD.

Lewis, T., 1973, *Thrips, Their Biology, Ecology and Economic Importance*, Academic Press, New York.

Lewis, T. (ed.), 1997, *Thrips as Crop Pests*, CAB International, Wallingford, U.K.

Mound, L. A., and Heming, B. S.,1991, Thysanoptera, in: *The Insects of Australia*, 2nd ed., Vol. I (CSIRO, ed.), Melbourne University Press, Carlton, Victoria.

Mound, L. A., Morison, G. D., Pitkin, B. R., and Palmer, J. M., 1976, Thysanoptera, *Handb. Ident. Br. Insects* **1**(11):1–79.

Mound, L. A., Heming, B. S.,and Palmer, J. M., 1980, Phylogenetic relationships between the families of recent Thysanoptera (Insecta), *Zool. J. Linn, Soc.* **69**:111–141.

Stannard, L. J., Jr., 1968, The thrips, or Thysanoptera, of Illinois, *Bull. Ill. Nat. Hist. Surv.* **29**(4):215–552.

9

The Panorpoid Orders

1. Introduction

In this and the following chapter we shall deal with the endopterygote insects—those that have a distinct pupal instar in which the insect undergoes a drastic metamorphosis from the larval to the adult form. As noted in Chapter 2, Section 3.2, considerable difficulty has arisen in deciding whether the endopterygote orders have a common origin or are polyphyletic. The five orders considered in this chapter, Mecoptera, Diptera, Siphonaptera, Trichoptera, and Lepidoptera, show clear affinities that enable them to be grouped together as the panorpoid complex. Within the complex there are two sister lines of evolution: the Antliophora (first three orders) and the Amphiesmenoptera (Trichoptera and Lepidoptera). The remaining orders, Megaloptera, Raphidioptera, Neuroptera, Coleoptera, Hymenoptera, and Strepsiptera, dealt with in Chapter 10, show few affinities with the panorpoid group.

2. Mecoptera

SYNONYMS: Panorpatae, Panorpina, Panorpida   COMMON NAME: scorpionflies

Slender medium-sized insects; head usually prolonged ventrally into a broad rostrum with long filiform antennae, well-developed compound eyes, and biting mouthparts; usually with two pairs of identical membranous wings with primitive venation and carried horizontally at rest; abdomen with short cerci and, in males, prominent genitalia.

Larvae usually eruciform with simple eyes, biting mouthparts, and thoracic legs; abdominal legs present or absent. Pupae decticous and exarate.

This is a small order containing about 500 known species, about 90% of which belong to two families, Panorpidae and Bittacidae. The order is particularly common in the Northern Hemisphere and includes about 75 North American and 4 British species. About 30 species occur in Australia.

Structure

Adult.   A characteristic feature of most Mecoptera is the ventral prolongation of the head into a broad rostrum. Incorporated into this structure are the clypeus, labrum, and maxillae. Compound eyes are well developed, and in most species there are three ocelli. The antennae are multisegmented and filiform. The mouthparts are mandibulate, except in *Nannochorista*, where they are specialized and may be interpreted as foreshadowing the suctorial type seen in lower Diptera. The prothorax is small, the pterothorax well developed. The legs are long and thin and adapted for walking. They have a five-segmented tarsus. In Bittacidae the fifth tarsal segment folds back on the fourth and is used for catching prey. Two pairs of fully developed, identical, membranous wings are present in most species; the venation is primitive. In Boreidae the wings of females are small sclerotized pads while those of males are hooklike and used to grasp the female during mating. Wings are reduced in some female Panorpodidae and Bittacidae, and absent in female Apteropanorpidae. The abdomen of females is 11-segmented and usually carries 2-segmented cerci (unsegmented in Bittacidae and Boreidae). In female Boreidae the 10th tergum is prolonged and together with the pointed, sclerotized cerci forms a functional ovipositor. In males segment 9 is bifurcate and bears a pair of bulbous claspers. Segment 10 is inconspicuous and bears unsegmented cerci. The aedeagus lies at the base of the claspers. In Panorpidae the terminal segments are turned upward and resemble somewhat a scorpion's sting, hence the common name for the order.

The foregut has two interesting features. The esophagus contains two dilations that appear to form a sucking apparatus, and the crop is provided with long setae (acanthae) that may act as a filter. Six Malpighian tubules occur. The nervous system is generalized, with three thoracic and between five and eight abdominal ganglia (males usually with one more than females). Each testis comprises three or four follicles. The paired vasa deferentia open separately into a median seminal vesicle, which also receives paired accessory glands. In females each ovary contains 7–19 polytrophic ovarioles (panoistic ovarioles in Nannochoristidae and Boreidae). The paired oviducts unite before entering a genital pouch. The ducts from the spermatheca and accessory glands also lead into the pouch.

Larva and Pupa.   Larvae are typically caterpillarlike, with a distinct head capsule that bears simple eyes. Prolegs occur on the first eight abdominal segments, and the apex of the abdomen bears either a suction disc or a pair of hooks. In Boreidae and Panorpodidae larvae are grublike, lacking prolegs and a terminal suction disc. Larvae of Nannochoristidae are very elongate, lack prolegs, but have a pair of apical hooks. Pupae are decticous and exarate.

Life History and Habits

Adult scorpionflies are most frequently encountered in cool, shaded locations, especially among low vegetation, though a few species occur in semidesert habitats. They can fly actively when disturbed, though they normally rest on grass, under leaves, etc. Adult Panorpidae feed mostly on dead soft-bodied arthropods (including insects caught in spiders' webs); they also eat nectar, pollen, and fruit juices. Bittacidae, by contrast, are insect predators, catching their prey either in flight or by hanging under vegetation till it comes within range. In members of both of these families much of the food of females is provided in the form of a nuptial gift by a male during courtship (see Chapter 19, Section 4.2). The food item may be an arthropod recently obtained by the male or a mass of saliva secreted from the male's greatly enlarged salivary glands. Adult Boreidae feed on mosses,

and members of other families may be herbivorous, saprophagous, or carrion feeders. To attract females, male panorpids and bittacids secrete pheromones from glands on the posterior abdominal segments. Visual signals (wing movements and abdominal vibrations) are also important in close-range courtship interactions in these and other (non-pheromone-producing) mecopteran families. Eggs of bittacids are dropped randomly; those of panorpids and choristids are laid in batches in moist depressions in the ground. Boreidae deposit eggs singly or in small batches in soil adjacent to moss rhizoids. The egg stage may last from as little as a week up to several months in species where there is an egg diapause. Larvae are saprophagous, carnivorous, or moss-feeders and pass through four instars before entering a quiescent prepupal phase, usually in an earthen cell. The length of the prepupal phase is varied and may include a diapause. The pupal stadium usually lasts 14–50 days. Most species are univoltine; some are bivoltine; and boreids take 2 years to complete a generation.

Phylogeny and Classification

A rich array of mecopteralike fossils is known from the Lower Permian, and the modern consensus is that this includes a few genuine Mecoptera. By the Upper Permian, and extending through the Jurassic, the order was abundant and diverse. Some Upper Permian scorpionflies from deposits in Australia and Siberia are assignable to the extant family Nannochoristidae, while representatives of some other modern families appear in the Lower Jurassic. Recent Mecoptera are arranged by Willmann (1987) in two suborders, containing nine families whose possible evolutionary relationships are shown in Figure 9.1.

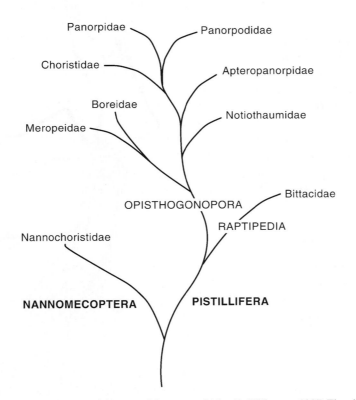

FIGURE 9.1. Proposed phylogeny of the extant Mecoptera. [After R. Willmann, 1987, The phylogenetic system of the Mecoptera, *Syst. Entomol.* **12**:519–524. By permission of Blackwell Scientific Publications Ltd.]

Suborder Nannomecoptera

This suborder contains the single, small, primitive Southern Hemisphere family NANNOCHORISTIDAE. The eight species in this group differ from other mecopterans in lacking a pump for transferring sperm from male to female, and in details of their external genitalia, wing venation, and mandibular structure. Adults are small and live in the vicinity of streams or lakes. Their larvae are aquatic and carnivorous, feeding on larval Diptera, especially chironomids.

Suborder Pistillifera

The suborder Pistillifera (literally "piston-bearers" in reference to the sperm pump of males) is divided by Willmann (1987) into two infraorders, Raptipedia and Opisthogonopora. The Raptipedia includes the large (145 species) and cosmopolitan family BITTACIDAE (Figure 9.2C) whose members have raptorial tarsi with which they grasp their prey, often hanging under vegetation by their forelegs. Among the Opisthogonopora, the BOREIDAE (Figure 9.2A) constitute the most distinct family; indeed, Hinton (1958) placed this holarctic group of about 25 species in a separate order, Neomecoptera.

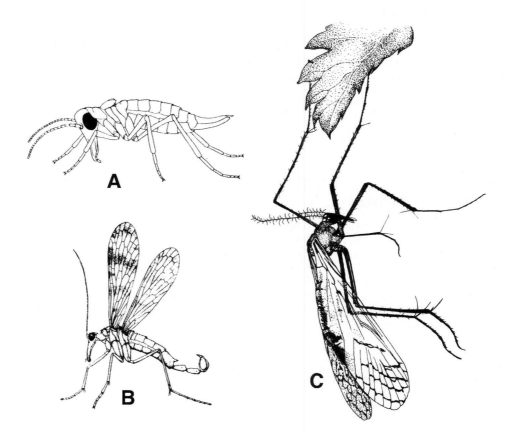

FIGURE 9.2.   Mecoptera. (A) *Boreus brumalis* (Boreidae); (B) *Panorpa helena* (Panorpidae); and (C) *Bittacus pilicornis* (Bittacidae). [A, B, from D. J. Borror, D. M. Delong, and C.A. Triplehorn, 1976, *An Introduction to the Study of Insects*, 4th ed. By permission of Brooks/Cole, a division of Thomson Learning. C, from D. W. Webb, N. D. Penny, and J. C. Marlin, 1975. The Mecoptera, or scorpionflies, of Illinois. *Bull. Ill. Nat. Hist. Surv.* **31**:251–316. By permission of the Illinois Natural History Survey.]

Boreidss are small, with hooklike wings in males used to grasp the female during mating, and scalelike reduced wings in females. They are sometimes called "snow fleas" as they may be found walking or jumping on snow patches. Only two, widely disjunct species, *Austromerope poultoni* (Western Australia) and *Merope tuber* (eastern United States), are placed in the MEROPEIDAE. These are cockroachlike in appearance as their head is largely hidden beneath the enlarged pronotum. Males have elongate genitalia possibly used in fighting for females. *Notiothauma reedi*, found in central Chile, is the only member of the NOTIOTHAUMIDAE. Similarly, *Apteropanorpa tasmanica* from Tasmania, which resembles boreids in general form and habits, is the sole representative of the APTEROPANORPIDAE. CHORISTIDAE, with only eight species, are restricted to Australia. The two remaining families, PANORPODIDAE and PANORPIDAE (Figure 9.2B), are the most specialized mecopteran groups. The former includes only two genera, *Brachypanorpa* (four species in the United States) and *Panorpodes* (five species in Japan and Korea), while the latter is the largest mecopteran family (300 species) with an essentially holarctic distribution, though it has representatives in Asia.

Literature

Because of the central position that this order occupies in any discussion of the evolution of most endopterygote orders, interest in the Mecoptera has been out of proportion to the number of extant species. Information on the biology of Mecoptera is provided by Webb *et al.* (1975), Kaltenbach (1978), and Byers and Thornhill (1983). The evolution of the order and its relationship to the other panorpoid groups are discussed by Hennig (1981), Kristensen (1981), Willmann (1987), and Whiting (2002). Keys to families of Mecoptera in North America and Australia are provided by Arnett (2000) and Byers (1991), respectively. The British species are identifiable from Plant (1997).

Arnett, R. H. Jr., 2000, *American Insects: A handbook of the Insects of America North of Mexico*, 2nd ed., CRC Press, Boca Raton, FL.

Byers, G. W., 1991, Mecoptera, in: *The Insects of Australia*, 2nd ed., Vol. 2 (CSIRO, ed.), Melbourne University Press, Carlton, Victoria.

Byers, G. W., and Thornhill, R., 1983, Biology of the Mecoptera, *Annu. Rev. Entomol.* **28**:203–228.

Hennig, W., 1981, *Insect Phylogeny*, Wiley, New York.

Hinton, H. E., 1958, The phylogeny of the panorpoid orders, *Annu. Rev. Entomol.* **3**:181–206.

Kaltenbach, A., 1978, Mecoptera (Schnabelhafte, Schnabelfliegen), in: *Handbuch der Zoologie*, Vol. IV, Insecta Lfg. **28**:1–111, de Gruyter, Berlin.

Kristensen, N. P., 1981, Phylogeny of insect orders, *Annu. Rev. Entomol.* **26**:135–157.

Plant, C. W., 1997, A key to adults of British lacewings and their allies (Neuroptera, Megaloptera, Raphidioptera and Mecoptera), *Field Stud.* **9**:179–269.

Webb, D. W., Penny, N. D., and Marlin, J. C., 1975. The Mecoptera, or scorpionflies, of Illinois, *Bull. Ill. Nat. Hist. Surv.* **31**(7):251–316.

Whiting, M. F., 2002, Mecoptera is paraphyletic: Multiple genes and phylogeny of Mecoptera and Siphonaptera, *Zool. Scrip.* **31**:93–104.

Willmann, R., 1987, The phylogenetic system of the Mecoptera, *Syst. Entomol.* **12**:519–524.

3. Diptera

SYNONYMS: none   COMMON NAMES: true flies; includes mosquitoes, midges, black flies, deer flies, horse flies, house flies

Generally minute to small soft-bodied insects; head highly mobile with large compound eyes, antennae of varied size and structure, and suctorial mouthparts; prothorax and metathorax small and fused with large mesothorax, wings present only on mesothorax, halteres present on metathorax; legs with five-segmented tarsi; abdomen with varied number of visible segments, female genitalia simple in most species, male genitalia complex, cerci present.

Larvae eruciform and in most species apodous; head in many species reduced and retracted. Pupae adecticous and obtect or exarate, the latter enclosed in a puparium.

The more than 120,000 species of Diptera described to date represent perhaps two-thirds of the world total. The order has a truly worldwide distribution, representatives occurring even in the Antarctic. More than 19,500 species have been described from North America, almost 8000 from Australia, and close to 7000 from Britain. Some of the world's commonest insects and a large number of species of veterinary and medical importance belong to this order.

Structure

The great structural diversity found in the Diptera reflects the variety of niches that the true flies have exploited.

Adult.   Adults range in size from about 0.5 mm to several centimeters and they are generally soft-bodied. The head is relatively large and highly mobile. It carries well-developed compound eyes, which in males are frequently holoptic. The antennae are of varied size and structure and are important taxonomically. In Muscomorpha-Schizophora there is a ∩-shaped ptilinal (frontal) suture that runs transversely above the antennae and extends downward on each side of them. This suture indicates the position of the ptilinum, a membranous sac that is exserted and distended at eclosion in order to rupture the puparium and assist a fly in tunneling through soil, etc. The mouthparts are adapted for sucking and are described in Chapter 3 (see Section 3.2.2 and Figures 3.14–3.16). The prothoracic and metathoracic segments are narrow and fused intimately with the very large mesothorax, which bears the single pair of membranous wings. The hind wings are extremely modified, forming halteres, small, clublike structures important as organs of balance (see Chapter 14, Section 3.3.4). In a few species halteres and wings are reduced or absent. The legs almost always have five-segmented tarsi, and in some species one or more pairs are modified for grasping prey. Primitively, there are 11 abdominal segments, but in most Diptera this number is reduced and rarely more than 4 or 5 are readily visible. Frequently, the more posterior segments (postabdomen) are telescoped into the anterior part of the abdomen (preabdomen) (see Figure 3.29). The postabdomen thus formed is used as an extensible ovipositor. The male genitalia are complex and their homologies uncertain because of the rotation of the abdomen and asymmetric growth of the individual components during the pupal stage.

In most Diptera the cibarium is strongly muscular and serves as a pump for sucking up liquids into the gut. In the bloodsucking Tabanidae and Culicidae a large pharyngeal pump is also present. The alimentary canal is, in most primitive forms, relatively unconvoluted. In Muscomorpha, however, it is much more coiled because of the increase in length of the midgut. The esophagus divides posteriorly into the gizzard and, usually, one diverticulum, the food reservoir (misleadingly called the "crop"). In Culicidae three diverticula are found. In Nematocera the midgut is a short, saclike structure; in Muscomorpha it is long and convoluted. Generally there are four Malpighian tubules that arise in pairs from a common duct on either side of the gut. In the nervous system a complete range of specialization

is seen. In primitive Nematocera three thoracic and seven abdominal ganglia occur, but all intermediate conditions are found between this arrangement and the situation in the more advanced Muscomorpha where a composite thoracoabdominal ganglion exists. In females the paired ovaries comprise a varied number of polytrophic ovarioles. In viviparous species there may be only one or two, but in the majority of oviparous flies there may be more than 100. In viviparous forms the common oviduct is dilated to form a uterus, and the accessory ("milk") glands produce a nutritive secretion. One to four spermathecae are always present. In males the testes are generally small, ovoid, and pigmented. The short, paired vasa deferentia lead into a muscular ejaculatory sac. Paired accessory glands may occur.

Larva and Pupa.   Larvae are usually elongate and cylindrical. Body segmentation is usually distinct, but in a few groups the true number of segments is masked as a result of secondary division or fusion of the original segments. True thoracic legs are always absent, though prolegs may occasionally be present on the thorax and/or abdomen. In primitive Diptera the head capsule is distinct and sclerotized (the eucephalous condition). In most species, however, it is much reduced (hemicephalous) or entirely vestigial (acephalous) (Figure 3.13). The antennae and chewing mouthparts, including horizontally moving mandibles, are well developed in Nematocera. In orthorrhaphous species antennae and maxillae may be well developed, but the mandibles are sickle-shaped and move in a vertical plane. In Muscomorpha the antennae are in the form of minute papillae and the mouthparts are reduced to a pair of curved hooks (the original mandibles). The internal structure of larvae generally resembles that of adults. Dipteran pupae are always adecticous. Pupae of nematocerous and orthorrhaphous species are obtect, whereas those of Muscomorpha are secondarily exarate and coarctate, being enclosed in a puparium, the hardened cuticle of the third larval instar.

Life History and Habits

Adult Diptera are active, mostly free-living insects that are found in all major habitats. They are predominantly diurnal and usually associated with flowers or with decaying organic matter. With the exception of a few species that do not feed as adults, flies feed on liquids (rarely, also pollen), a habit that may reflect the use of honeydew from homopterans as their ancestral energy source (Downes and Dahlem, 1987). Most Diptera feed on nectar or the juices from decaying organic matter, but a few groups are adapted for feeding on the tissue fluids of other animals especially arthropods and vertebrates. This is achieved in some species by simply cutting the skin or squeezing prey with the labella and sucking up the exuded fluid. In the majority of body-fluid feeders, however, a fine proboscis is used to pierce the skin and penetrate directly to the fluid, usually blood. The habit is usually confined to females. It is through the bloodsucking habit and the subsequent importance of these insects as vectors of disease-causing microorganisms that the order is generally considered the most important of the entire class from the medical and veterinary point of view. Although parthenogenesis is known to occur in a few species, most Diptera reproduce bisexually. Copulation is preceded in some species by an elaborate courtship. Usually females actively search for, and lay eggs directly on, the larval food source. Members of a few groups are ovoviviparous or viviparous. Egg development is normally rapid and hatching occurs in a few days.

Larvae are usually found in moist locations such as soil, mud, decaying organic matter, and plant or animal tissues, though a few are truly aquatic. The majority are liquid feeders or microphagous. Some aquatic forms trap their food in specially developed mouth brushes. Larvae of many species are of agricultural or medical importance. Usually four larval instars occur, but up to eight are found in some species, and in Muscomorpha the fourth is suppressed. Prior to pupation, larvae generally crawl to a drier location. Pupae may be naked, but those of many nematocerous and orthorrhaphous species are enclosed in a cocoon, and those of Muscomorpha are ensheathed by the puparium. The pharate adult swallows air to facilitate emergence from the pupal skin or puparium, the latter splitting lengthwise in obtect pupae; in coarctate forms, however, the adult protrudes the ptilinum to push off the anterior end of the puparium.

Phylogeny and Classification

From a structural comparison of primitive living Mecoptera and Diptera and from the relatively scarce fossil evidence, it seems probable that Diptera evolved from mecopteralike insects in the Permian period. *Permotipula patricia*, from the Upper Permian of Australia, was originally thought to be the oldest dipteran. However, it is now considered an early Mecopteran. Similarly, other four-winged forms (the Protodiptera *Permotanytarsus* and *Choristotanyderus*) from the same strata, which were originally considered to be Diptera (Riek, 1977), have been reexamined and, as a result, moved from the dipteran stem group (Willmann, 1989; Wootton and Ennos, 1989). Thus, the earliest reliable records of fossil Diptera come from the Middle Triassic, though these are quite meager. By the late Triassic, the more primitive suborder, Nematocera, had undergone a considerable radiation, possibly making use of the honeydew produced by the already abundant homopterans (Downes and Dahlem, 1987). By the Lower Jurassic, well-developed Nematocera (some assignable to recent families) and primitive orthorrhaphous forms were present. The Muscomorpha (Cyclorrhapha) evolved from orthorrhaphan stock, probably in the Late Jurassic. The great radiation of the order, and the establishment of many of the structures and habits of modern flies, took place in the Cretaceous period. This was, of course, correlated with the evolution of the flowering plants and mammals. By the Eocene period, the dipteran fauna was similar in many respects to that which survives today. Indeed, many Eocene fossils are assigned to modern genera. A possible phylogeny of the order is shown in Figure 9.3.

Classification of the modern Diptera continues to present problems, particularly the rank assignable to different groups. The difficulty arises in part because the order is extremely old; it contains many extinct groups and others that are in decline; and yet there are also groups that are still evolving at a rapid rate. Thus, while ancient families have well-established differences and are easily separated, more recent groups, notably the families and superfamilies of Muscomorpha, are sometimes little more than convenient divisions because of the vast number of species that must be considered. The differences between these groups, therefore, are relatively minor. The classification used here is adapted from the views presented in the *Manual of Nearctic Diptera* (McAlpine *et al.*, 1981–1989). In this arrangement, the Diptera are arranged in two suborders, Nematocera and Brachycera. Within the Nematocera, seven infraorders are recognized. Figure 9.3 indicates that Nematocera are monophyletic. However, it should be noted that some authors believe that this is a paraphyletic group, with the Bibionomorpha or Psychodomorpha being the sister group to the Brachycera (Yeates and Wiegmann, 1999). In the undoubtedly monophyletic Brachycera there are three infraorders; the first two, Asilomorpha and Tabanomorpha, correspond to the "Orthorrhapha" while the third, Muscomorpha, is equivalent to the "Cyclorrhapha" of

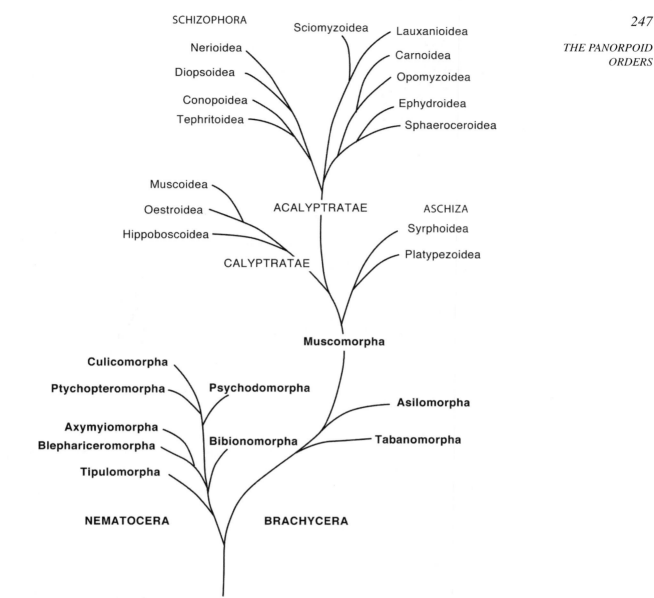

FIGURE 9.3. A suggested phylogeny of the Diptera. [After J. F. McAlpine and D. M. Wood, coordinators, 1989, *Manual of Nearctic Diptera*, Vol. 3, Agriculture Canada Monograph No. 32. By permission of the Minister of Supply and Services Canada.]

earlier systems. Within the Muscomorpha, there are two easily defined but very unequally sized sections, the probably paraphyletic Aschiza and the monophyletic Schizophora. The latter has two subdivisions, Acalyptratae and Calyptratae.

Suborder Nematocera

Most Nematocera are small, delicate flies, with 6- to 14-segmented antennae of simple structure, and 3- to 5-segmented maxillary palps. Larvae have a well-developed head and chewing mandibles that move in the horizontal plane.

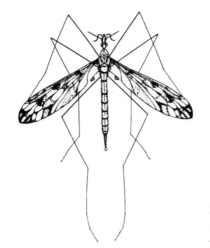

FIGURE 9.4. Tipulomorpha. A crane fly, *Tipula trivittata* (Tipulidae). [From F. R. Cole and E. I. Schlinger, 1969. *The Flies of Western North America.* By permission of the University of California Press.]

The suborder contains the oldest families of Diptera, most of which are now on the decline. Some, however, like the Culicidae, have undergone relatively recent radiations and are among the most successful modern groups.

Infraorder Tipulomorpha

Members of this group are placed in the single family TIPULIDAE (Figure 9.4) (crane flies, daddy longlegs), which, with about 14,000 species (including >1500 in North America), is the largest family of Diptera. Representatives of this worldwide family are mostly associated with moist, temperate habitats, though some occur in open meadows, rangelands, and deserts. Adults range in size from small (wingspan of 2 mm) to very large (wingspan up to 8 cm). Larvae are found in a wide variety of habitats from strictly aquatic to dry soils where they typically feed on plant material or decaying organic matter; occasionally, they become pests by feeding on seedling field crops.

Infraorder Blephariceromorpha

Almost all members of this small group are included in the widely distributed family BLEPHARICERIDAE (net-winged midges) (200 species). Adults are slender flies with long legs, and in both sexes the eyes are holoptic. In some species both sexes feed on nectar; in others females catch smaller flies and feed on their hemolymph. Adults are generally found near fast-flowing streams; the aquatic larvae attach themselves to rocks where they feed on diatoms and algae.

Infraorder Axymyiomorpha

Containing only five species in the one family AXYMYIIDAE, the taxonomic position of this group is controversial, and other authorities have included it in a variety of other nematoceran families. Species occur in North America, eastern Europe, and Siberia. Adults are stout-bodied flies that resemble Bibionidae. Larvae live in cavities excavated in rotting wood, perhaps feeding on fungi or other microorganisms.

FIGURE 9.5. Psychodomorpha. A moth fly, *Psychoda* sp. (Psychodidae). [From F. R. Cole and E. I. Schlinger, 1969, *The Flies of Western North America.* By permission of the University of California Press.]

Infraorder Psychodomorpha

About one half of the approximately 1000 species in this group are included in the PSYCHODIDAE (moth flies) (Figure 9.5), a widely distributed group of small flies recognizable by the hairy wings held rooflike over the body when at rest. Although most species do not feed as adults, some females feed on blood, including those of the genus *Phlebotomus* (sand flies), species of which are vectors of various virus- and leishmania-induced diseases. TRICHOCERIDAE (winter crane flies) (110 species, worldwide) are easily confused with the true crane flies. They carry the preference for cool, moist habitats to an extreme, and many species are common in caves and mines. Adults are often encountered in large swarms during winter. The SCATOPSIDAE form a worldwide family of about 200 species of mostly very small flies. Larvae are found in decaying organic material, both plant and animal. The ANISOPODIDAE (100 species) is a worldwide group of primitive Diptera whose larvae are found in decaying and fermenting organic matter.

Infraorder Ptychopteromorpha

This very small infraorder comprises two families, PTYCHOPTERIDAE (phantom crane flies) (60 species), found in all except the Australian and neotropical regions, and TANYDERIDAE (40 species), a mostly Australian group of crane flylike insects. In both families larvae are semiaquatic, living in the substrate at the edges of streams.

Infraorder Culicomorpha

Included in this large group of generally small and delicate flies are some extremely well-known Diptera. The group is divided into four large and three small families by McAlpine *et al.* (1981–1989). The general habits of the widespread families CULICIDAE (mosquitoes) (3000 species) (Figure 9.6) and CHAOBORIDAE (phantom midges) (75 species, often included as a subfamily of Culicidae) (Figure 9.7) present an interesting contrast. Larval mosquitoes are filter feeders that strain microorganisms from the water in which they live. Adult males do not feed, but females are voracious bloodsuckers and as such are responsible for the spread of some human and livestock diseases, for example, malaria, yellow fever, filariasis, West Nile, and equine encephalitis. (Some, incidentally, also spread the myxomatosis virus of rabbits and are, therefore, of some positive economic value.) Larval Chaoboridae, on the other hand, are predators (particularly of mosquito larvae!). The adults, however, are nectar feeders. The DIXIDAE form a small (150 species) but widely distributed group that is frequently considered a subfamily of the Culicidae, mainly on the basis of the adult wing venation and the similarity between the larval and

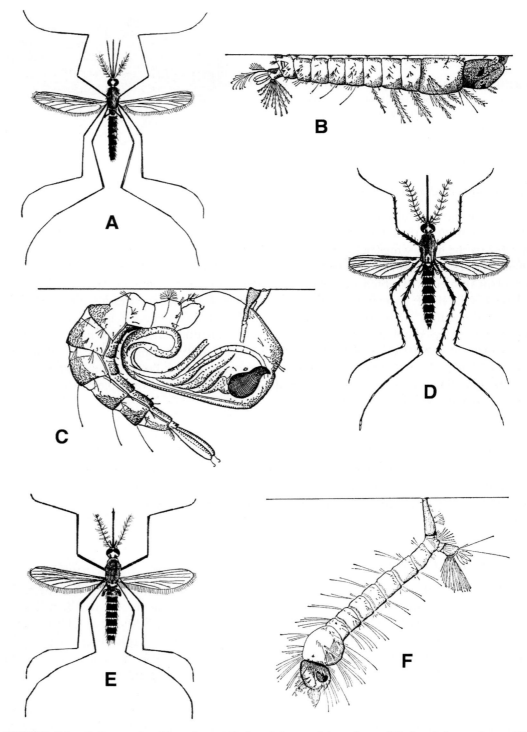

FIGURE 9.6. Culicomorpha. Mosquitoes. (A) *Anopheles quadrimaculatus;* (B) *Anopheles* sp. larva; (C) *Anopheles* sp. pupa; (D) *Aedes canadiensis;* (E) *Culex pipiens;* and (F) *Culex* sp. larva. [A, D, E, from S. J. Carpenter and W. J. LaCasse, 1955, *Mosquitoes of North America.* By permission of the University of California Press. A, E, drawn by Saburo Shibata. D, drawn by Kei Daishoji. B, C, F, from J. D. Gillett, 1971, *Mosquitos,* Weidenfeld and Nicolson. By permission of the author.]

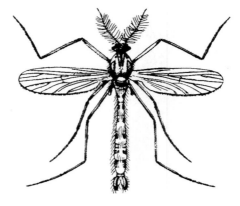

FIGURE 9.7. Culicomorpha. The clear lake gnat, *Chaoborus astictopus* (Chaoboridae). [From F. R. Cole and E. I. Schlinger, 1969, *The Flies of Western North America.* By permission of the University of California Press.]

pupal stages of the two groups. The CHIRONOMIDAE (TENDIPEDIDAE) (Figure 9.8) constitute a large, widely distributed family of more than 5000 species. Adults are small, mosquitolike flies, though they do not feed. They often form massive swarms in the vicinity of water. Larvae are aquatic and either are free-living or lie buried in the substrate, members of many species constructing a special tube. The CERATOPOGONIDAE (biting midges, punkies, no-see-ums) (Figure 9.9) form a widespread family of minute or small flies, many females of which suck the blood of vertebrates and arthropods, or prey on other insects. Members of most species, however, feed on nectar and/or pollen and render considerable benefit through cross-fertilization of the plants. Larvae occupy a variety of moist habitats, including soil, moss, under bark, and in rock pools; they may be algivorous, saprophagous, mycophagous, or predaceous. The approximately 1100 species of SIMULIIDAE (black flies, buffalo gnats) (Figure 9.10) form a widespread family, females of which attack birds, mammals, and other insects. Several species of *Simulium* are of extreme importance as vectors for the filarial nematode, *Onchocerca volvulus,* which causes onchocerciasis (river blindness) in tropical Africa, Central America, northern South America, and Yemen. Other species transmit nematode, protozoan, and viral pathogens of birds and mammals, including livestock. The larvae are found in swiftly flowing water, attached

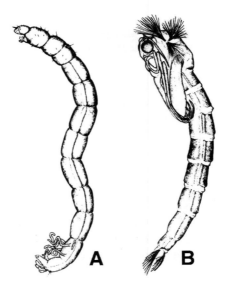

FIGURE 9.8. Culicomorpha. *Chironomus tentans* (Chironomidae). (A) Larva; and (B) pupa. [From O. A. Johannsen, 1937, Aquatic Diptera. Part IV. Chironomidae: Subfamily Chironominae, *Mem. Cornell Univ. Agri. Exp. Stn.* **210**:52 pp. By permission of Cornell University Agriculture Experimental Station.]

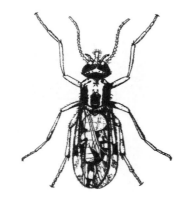

FIGURE 9.9. Culicomorpha. A punkie, *Culicoides dovei* (Ceratopogo-nidae). [From F. R. Cole and E. I. Schlinger, 1969, *The Flies of Western North America.* By permission of the University of California Press.]

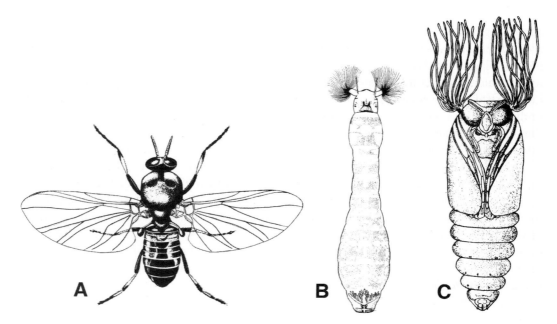

FIGURE 9.10. Culicomorpha. A black fly, *Simulium nigricoxum* (Simuliidae). (A) Female; (B) mature larva; and (C) pupa. [From A. E. Cameron, 1922, The morphology and biology of a Canadian cattle-infesting black fly, *Simulium simile* Mall. (Diptera, Simuliidae), *Bulletin #5—New Series (Technical).* By permission of Agriculture and Agri-Food Canada.]

to the substrate by an anal sucker, and are filter feeders or grazers. The THAUMALEIDAE (80 species) constitute a small, primarily holarctic family of minute midges whose affinities are uncertain. In some features its members resemble the other Culicomorpha, in others the Bibionomorpha.

Infraorder Bibionomorpha

This large and diverse group includes four major families. The BIBIONIDAE (March flies) (700 species) (Figure 9.11A) are robust, hairy flies of medium to small size. Large swarms, consisting largely of males, are seen in the Northern Hemisphere spring (hence, the common name). Larvae of this cosmopolitan group feed gregariously on roots or decaying vegetation. The MYCETOPHILIDAE (2000 species) and SCIARIDAE (500 species) are

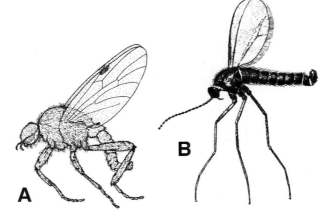

FIGURE 9.11. Bibionomorpha. (A) A March fly, *Bibio albipennis* (Bibionidae); and (B) the Hessian fly, *Phytophaga destructor* (Cecidiomyiidae). [A, from D. J. Borror and D. M. Delong, 1971, *An Introduction to the Study of Insects*, 3rd ed. By permission of Brooks/Cole, a division of Thomson Learning. B, from L. A. Swan and C. S. Papp, 1972, *The Common Insects of North America.* Copyright 1972 by L. A. Swan and C. S. Papp. Reprinted by permission of Harper & Row, Publishers, Inc.]

sometimes included in a single family of the former name. They are commonly known as fungus gnats from the observation that the larvae feed mainly on fungi and decaying plant material. Adults are commonly encountered in cool, damp situations. The CECIDOMYIIDAE (gall midges) form a very large family (4000 species) of minute flies, most of which feed, in the larval stage, on plant tissues, frequently causing the formation of galls. There are, however, saprophagous or predaceous species. Within the family are several economically important species, for example, the Hessian fly, *Phytophaga (= Mayetiola) destructor* (Figure 9.11B), whose larvae feed on wheat shoots. Many species are paedogenetic, the full-grown larvae becoming sexually mature and reproducing parthenogenetically. As the young larvae grow, they devour their parent from within. Several generations of paedogenetic larvae may develop in a season, and the larval population can thus increase enormously. Eventually, the larvae pupate normally and sexual reproduction follows.

Suborder Brachycera

Most Brachycera are rather stout flies with antennae having fewer than seven segments and, often, an arista; maxillary palps are unsegmented or two-segmented. Larvae are hemicephalous or acephalous (maggotlike), with sickle-shaped mandibles that move in the vertical plane.

Members of this suborder were traditionally placed in two subgroups, Orthorrhapha (Brachycera *sensu stricto*) and Cyclorrhapha, which were sometimes each given subordinal rank. It is now clear that the Orthorrhapha is a paraphyletic group, comprising the infraorders Tabanormorpha and Asilomorpha, and that the cyclorrhaphous forms, although monophyletic, merit only infraordinal status (Muscomorpha).

Infraorder Tabanomorpha

As constituted by McAlpine *et al.* (1981–1989), the Tabanomorpha includes seven families, three of which contain between them about 95% of the species. The largest family is the TABANIDAE (Figure 9.12A), with more than 3000 species, which includes those bloodsucking insects commonly known as horse and deer flies, clegs, March flies (in the Southern Hemisphere), and probably many other, less polite names! The bloodsuckers belong to only three genera, *Tabanus, Chrysops,* and *Haematopota,* whose evolution has

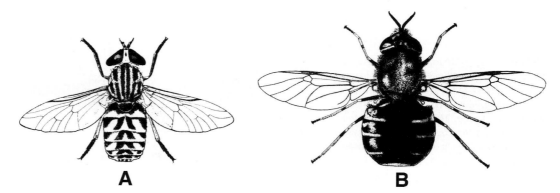

FIGURE 9.12. Tabanomorpha. (A) A horse fly, *Tabanus opacus* (Tabanidae); and (B) a soldier fly, *Odontomyia hoodiana* (Stratiomyidae). [A, from J. F. McAlpine, 1961, Variation, distribution and evolution of the *Tabanus (Hybomitra) frontalis* complex of horse flies (Diptera: Tabanidae), *Can. Entomol.* **93**:894–924. By permission of the Entomological Society of Canada. B, from F. R. Cole and E. I. Schlinger, 1969, *The Flies of Western North America.* By permission of the University of California Press.]

closely followed that of the hoofed mammals. Although they are known to be capable of transmitting various diseases both human and of livestock, tabanids cause far greater economic losses by their disturbance and irritation of livestock, resulting in lower yields of milk and meat. Only female horse flies suck blood, in the absence of which they feed, like males and like members of most tabanid species, on nectar and pollen. Larvae mostly occur in mud, decaying vegetation, and shallow water (moving or still) where they prey on other invertebrates. Another large and well-distributed family is the STRATIOMYIDAE (Figure 9.12B), containing some 1500 species, commonly known as soldier flies. The weakly flying adults are encountered among low-growing herbage and are most probably nectar feeders. Members of many species are conspicuously striped, and some species are wasp mimics. Larvae, often gregarious, are aquatic or terrestrial, occurring in decaying organic matter, in dung, or under bark. Some are scavengers, others are predaceous or phytophagous, the latter group including pests of lawns and sugarcane. The RHAGIONIDAE (snipe flies) are perhaps the most primitive Brachycera. The family, which contains more than 300 extant species, as well as fossils from the Upper Jurassic, is widely distributed, though seldom encountered, because of the secretive, solitary habits of adult flies. Members of many species are nectar feeders, others are predaceous, and females of some species suck blood. Larvae occur in damp soil, rotting wood, etc. and are believed to prey on other insects.

Infraorder Asilomorpha

The 12 families in this very large taxon fall into three well-defined superfamilies, Asiloidea, Bombylioidea, and Empidoidea. About 80% of the asiloids belong to the cosmopolitan family ASILIDAE (robber flies) (Figure 9.13A), which with about 5000 species is among the largest of the orthorrhaphous Brachycera. Adults suck the body fluids of a variety of other insects. They are powerful fliers and catch their prey on the wing, have well-developed eyes and some degree of stereoscopic vision, possess strong legs for grasping the prey and are usually hairy, especially around the face, a feature that perhaps protects them during the struggle. THEREVIDAE form a widely distributed group of about 500 species that generally resemble robber flies, though they are not predaceous, feeding instead on

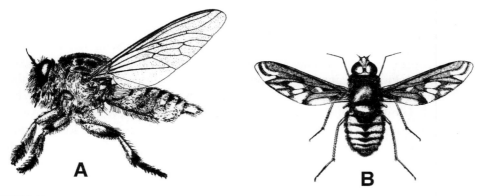

FIGURE 9.13. Asilomorpha. (A) A robber fly, *Mallophorina pulchra* (Asilidae); and (B) a bee fly, *Poecilanthrax autumnalis* (Bombyliidae). (From F. R. Cole and E. I. Schlinger, 1969, *The Flies of Western North America.* By permission of the University of California Press.]

nectar, plant exudates, etc. Therevid larvae burrow in soil, rotting bark, fungi, and rotting fruit, and are voracious predators, especially of beetle larvae and earthworms. Of the Bombylioidea, some 4000 species belong to the BOMBYLIIDAE (bee flies) (Figure 9.13B). The common name of these nectar feeders has double significance. First, the flies resemble bumblebees, and second, in many species, female flies deposit eggs at the nest entrance of a solitary bee or wasp so that the larvae may feed on the pollen, honey, and even young Hymenoptera. Larvae of other species search for and feed on grasshopper eggs, thereby playing an important role in the natural regulation of grasshopper populations. In a few species eggs are laid directly onto larvae of Lepidoptera or Hymenoptera. The NEMESTRINIDAE (250 species) and ACROCERIDAE (450) species are both widespread and ancient groups, sometimes considered to be Muscomorpha. In both families adults are found around flowers and some feed on nectar (others apparently do not feed as they have vestigial mouthparts). The larvae, which undergo heteromorphosis (see Chapter 21, Section 3.3.2), are parasitoids of grasshoppers and locusts (and may be an important factor in controlling their populations) and of spiders, respectively. The Empidoidea contain two large families of advanced Brachycera, the EMPIDIDAE (dance flies, balloon flies) and DOLICHOPODIDAE (long-legged flies). There are about 3000 species of Empididae, which are predaceous in both adult and juvenile stages. The family is largely restricted to the temperate regions of both hemispheres, and its members gain their common name from the courtship behavior of most, though by no means all, species. In many species males have an elaborate courtship display in which a female is offered a gift of food (real or imitation). Presumably this served originally as an insurance policy against the male's life. Dolichopodidae, which constitute a family of about 6000 species, appear to be a specialized offshoot of the empid line. They are generally found in cool, moist habitats, including the seashore and salt marshes. Adults and most larvae are predaceous, especially on other Diptera; larvae of a few species are phytophagous.

Infraorder Muscomorpha

The Muscomorpha are arranged in two series. The first, Aschiza, is perhaps a polyphyletic group, containing the more primitive members, which lack a ptilinal suture. The second, Schizophora, includes those flies in which a ptilinal suture is present.

In older classifications a third series, Pupiparia, was often found. McAlpine (in McAlpine *et al.*, 1981–1989), however, has presented strong reasons for considering this group as a monophyletic superfamily, Hippoboscoidea, within the calyptrate Schizophora. The majority of Schizophora (i.e., those that are winged as adults) can be arranged in two subdivisions, the Calyptratae, which contains flies that possess a calypter, or lobe, at the base of the fore wing that covers the halteres, and the Acalyptratae, whose members have no such lobe. Though this division is a natural one, that is, it represents a true evolutionary divergence, it should be recognized that (1) some groups have secondarily gained or lost the calypter, and (2) the adults of some parasitic families are wingless, though their affinities are clearly either calyptrate or acalyptrate.

Series Aschiza

Superfamily Platypezoidea

This superfamily has one large, cosmopolitan family, PHORIDAE (2700 species), and several small to very small ones, some of which are often included with the phorids. Although many phorids are free-living, fully winged flies found among low vegetation, on or near decaying organic matter, on fungi, in bird nests, etc., they seem to prefer to run rather than fly, a feature that foreshadows the brachypterous or apterous condition of the many species that live underground, as inquilines or parasites in ant and termite nests. Larvae are maggotlike, with diverse habits. Some are scavengers on fungi (and may become pests on mushroom farms), carrion, and human corpses; others are parasites of earthworms, other insects, spiders, and myriapods. PLATYPEZIDAE (250 species) form a worldwide group whose members prefer shaded woodland where they feed on nectar. Larvae are fungivorous.

Superfamily Syrphoidea

One small and one very large family make up the Syrphoidea. The PIPUNCULIDAE (big-headed flies) (400 species) are a cosmopolitan group of small humpbacked flies with large heads covered almost entirely by the compound eyes. Adults tend to be found hovering over flowers; the larvae are endoparasites of homopterans and, as such, are important natural control agents. The SYRPHIDAE (5000 species worldwide) (Figure 9.14) are the well-known hover flies. They form one of the largest and most easily recognized groups of

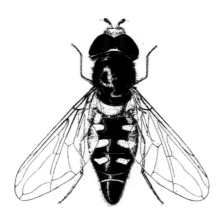

FIGURE 9.14. Syrphoidea. A hover fly, *Eupeodes volucris* (Syrphidae). [From a drawing by Charles S. Papp. By permission of the artist.]

Diptera. They are generally brightly colored, often striped, and many mimic bees or wasps. In some species there are obvious reasons for the preciseness of this mimicry, for the hover fly lays its eggs in the nests of Hymenoptera and, because of its similarity, presumably avoids detection. For other species the reason is less obvious, and no relationship is apparent between the mimic and its model. In contrast to the rather uniform, nectar-feeding habits of adult hover flies, those of larvae are extremely varied, phytophagous, zoophagous, and saprophagous species being known.

Series Schizophora

Subdivision Acalyptratae

Superfamily Conopoidea

The superfamily Conopoidea, the most primitive of the Schizophora, contains the single, widespread family CONOPIDAE (800 species) whose members typically mimic wasps and bees. Adults are nectar feeders and are especially associated with flowers of Compositae, Labiatae, and Umbelliferae. Conopids are parasites of bees and wasps, cockroaches, and calyptrate Diptera, the female catching the host in flight and depositing an egg directly on its body.

Superfamily Tephritoidea

This group includes eight families in the scheme of McAlpine *et al.* (1981–1989) of which two are large and three are of medium size. The OTITIDAE (picture-winged flies) are predominantly a north-temperate group of some 400 species. Adults are common in dense vegetation; larvae are typically saprophagous, though a few are phytophagous, including pests of onions and sugar beet. PLATYSTOMATIDAE (1000 species) are worldwide but most common in Africa, Australia, and Asia. Both adults and larvae resemble members of the previous family in morphology and habits. The cosmopolitan PYRGOTIDAE (330 species) are typically nocturnal flies that parasitize scarabaeid beetles. Females land on the beetles in flight and oviposit on the thin abdominal tergites beneath the elytra. LONCHAEIDAE (500 species worldwide) generally occur in forests; larvae feed on rotting plant material, rarely flower heads and root crowns. The largest and best known family, with some 4000 species, is the TRYPETIDAE (TEPHRITIDAE), the fruit flies, a group that includes some major agricultural pests. Their larvae feed on a variety of plant materials. They may be leaf or stem miners, gall formers, flower-inhabiting species, or fruit and seed eaters. In the latter category are the Mediterranean fruit fly, *Ceratitis capitata* (Figure 9.15), which attacks citrus and other fruits, and *Rhagoletis pomonella*, the apple maggot fly, whose larvae tunnel into apples, pears, etc.

Superfamily Nerioidea

Nerioidea (Micropezoidea) form a small group of three families, the largest of which is the MICROPEZIDAE (500 species), commonly called stilt flies because of their long legs. Members of this basically tropical family are found in wooded areas; their larvae are primarily saprophagous, though a few phytophagous species may become pests (e.g., of ginger and legumes).

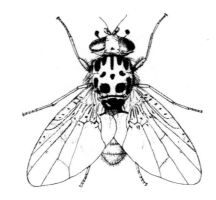

FIGURE 9.15. Tephritoidea. The Mediterranean fruit fly, *Ceratitis capitata* (Tephritidae). [From F. R. Cole and E. I. Schlinger, 1969, *The Flies of Western North America.* By permission of the University of California Press.]

Superfamily Diopsoidea

Diopsoidea (Tanypezoidea) make up another small superfamily whose members mostly fall into the PSILIDAE (200 mostly holarctic species) and DIOPSIDAE (150 mostly tropical species). Psilid larvae feed on roots and stems and a few are pests, for example, *Psila rosae* (carrot rust fly), on carrots, celery, and other root crops. Adult diopsids are called stalk-eyed flies because of the lateral extensions of the head which bear the compound eyes; their larvae are saprophagous or phytophagous (especially on Graminae), the latter occasionally becoming minor pests.

Superfamily Sciomyzoidea

Most members of this small superfamily belong to the cosmopolitan families SCIOMYZIDAE (550 species) or SEPSIDAE (240 species). Sciomyzid larvae feed on terrestrial or aquatic Mollusca, alive or dead, or on their eggs and embryos. Larvae of Sepsidae scavenge in dung, including sewage sludge, or in decaying plant or animal material.

Superfamily Lauxanioidea

Lauxanioids were formerly included in the previous superfamily. The great majority of species belong to the LAUXANIIDAE (1200 mainly tropical species), adults of which are sedentary and collect on low-growing vegetation, especially adjacent to water. Larvae are saprophagous, occurring in decaying vegetation, leaf litter, and bird nests.

Superfamily Opomyzoidea

This is possibly a polyphyletic group that, as constituted by McAlpine *et al.* (1981–1989), includes about a dozen families. Other authorities separate the families into three superfamilies, Opomyzoidea *sensu stricto*, Agromyzoidea, and Asteioidea. Most of the families are very small and have a restricted distribution. Adult CLUSIIDAE (220 species worldwide) occur around rotting logs and feed on nectar, exudates of rotting material, etc. Their larvae are found in rotting wood, and in the tunnels of termites and bark beetles. Some 1800 species of AGROMYZIDAE are known, including pests of shade trees, vegetables, and flowers. Larvae of this cosmopolitan family are mostly leaf or stem miners; some feed in seeds, bore in wood, or are gall formers. ASTEIIDAE (100 species) are widely distributed. Larvae of these tiny flies appear to be scavengers in rotting plants or fungi.

The relationships of the nine families included in the Carnoidea (Chloropoidea) remain debatable, and frequently some of the families are placed in the Opomyzoidea or Ephydroidea. The great majority of species belong to the cosmopolitan families MILICHIIDAE (300 species) and CHLOROPIDAE (2000 species). Adult milichiids are found at flowers; some species ride on spiders and predatory insects such as asilid flies and reduviid bugs, helping themselves to the juices that exude from their host's prey. Larvae are saprophagous or dung feeders, including some that live in the fungus gardens of leaf-cutting ants. Adult chloropids typically occur in vast numbers on foliage, and mostly feed on sap exudates, honeydew, etc. However, the so-called eye gnats (species of *Hippeletes* and *Siphunculina*) are attracted to wounds and secretions of the eyes, nose, lips, and skin and are vectors of conjunctivitis and skin diseases. Larvae may be saprophagous or phytophagous and some of the latter are important pests, for example, *Oscinella frit* (European frit fly) (Figure 9.16B) on cereals.

Superfamily Sphaeroceroidea

Included in the Sphaeroceroidea (Heleomyzoidea) are the cosmopolitan families SPHAEROCERIDAE (700 species, including about 120 that are brachypterous or apterous) and HELEOMYZIDAE (400 species), together with several small groups. Sphaerocerids are associated with dung, decaying plant material, fungi (occasionally becoming pests on mushroom farms), seaweed, and carrion. Larval heleomyzids, too, are found in decaying organic matter and fungi, as well as in the nests of birds and mammals, and bat caves.

Superfamily Ephydroidea

Also called Drosophiloidea, this group includes about 2600 species, almost all of which fall into two large families. The EPHYDRIDAE (shore flies, brine flies) (Figure 9.16A), with more than 1000 described species, is well known because of the remarkable variety of habitats that its members occupy. They are typically found near water, both fresh and

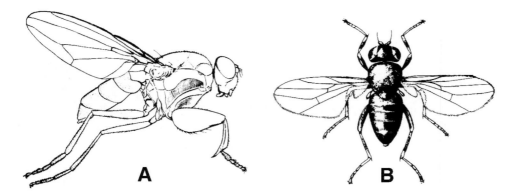

FIGURE 9.16. Ephydroidea and Carnoidea. (A) An ephydrid with raptorial forelegs, *Ochthera mantis* (Ephydridae); and (B) the European frit fly, *Oscinella frit* (Chloropidae). [A, from F. R. Cole and E. I. Schlinger, 1969, *The Flies of Western North America.* By permission of the University of California Press. B, from L. A. Swan and C. S. Papp, 1972, *The Common Insects of North America.* Copyright 1972 by L. A. Swan and C. S. Papp. Reprinted by permission of Harper & Row Publishers, Inc.]

salt, and in many species the larvae are truly aquatic. Larvae of some species feed on algae, but generally they and the adults are carnivorous or carrion feeders, sometimes almost to the point of being parasitic. Two examples, illustrating the extreme habitats in which Ephydridae are found, are *Ephydra riparia*, which is found in the Great Salt Lake of Utah (see also Chapter 18, Section 4.3), and *Psilopa (Helaeomyia) petrolei*, whose larvae live in pools of crude petroleum in California. The closely related DROSOPHILIDAE (pomace or fruit flies) are small flies generally seen in the vicinity of decaying vegetation or fruit, or near breweries and vinegar factories. The larvae are mostly fungivorous, though a few are leaf miners or prey on other insects. Various species of *Drosophila* have, of course, been extensively used for a wide range of biological research.

Subdivision Calyptratae

How the Calyptratae should be subdivided remains debatable; some authors lump all families in the Muscoidea while others, including McAlpine *et al.* (1981–1989), believe that there are three monophyletic subgroups in the subdivision, Muscoidea *sensu stricto*, Oestroidea, and Hippoboscoidea (= Pupiparia of earlier authors).

Superfamily Muscoidea

In the cosmopolitan family MUSCIDAE (3000 species) are many common pests, for example, the Australian bush fly (*Musca vetustissima*), house flies [*Musca domestica* (Figure 9.17A) and *Fannia canicularis*], and bloodsucking species such as the stable fly (*Stomoxys calcitrans*) and face fly (*Musca autumnalis*). However, the great majority of species are non-pestiferous. Adults are predaceous, saprophagous, pollenophagous, hematophagous, or feed on exudates of mammals. Females are typically oviparous, though a few are ovoviviparous or larviparous. Larvae are mostly saprophagous or dung feeders, though some are predaceous and a few are ectoparasites on birds. Most of the 1000 species of the primarily holarctic group ANTHOMYIIDAE (root maggot flies) are phytophagous in the larval stage and, as a result, many are economically important, for example, the cabbage

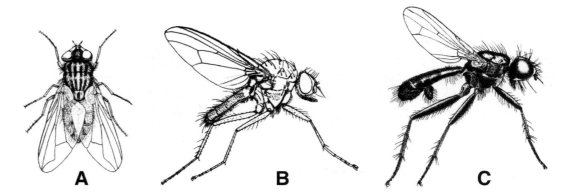

FIGURE 9.17. Muscoidea. (A) The house fly, *Musca domestica* (Muscidae); (B) the wheat bulb fly, *Hylemya coarctata* (Anthomyiidae); and (C) a dung fly, *Cordilura criddlei* (Scathophagidae). (A, from V. B. Wigglesworth, 1959, Metamorphosis, polymorphism, differentiation, *Scientific American*, February 1959. By permission of Mr. Eric Mose, Jr. B, C, from F. R. Cole and E. I. Schlinger, 1969, *The Flies of Western North America*. By permission of the University of California Press.]

root fly, *Hylemya brassicae*, and the wheat bulb fly, *H. coarctata* (Figure 9.17B). Others are saprophagous, dung feeders, or are inquilines in the burrows of solitary Hymenoptera or rodents. The small, primarily holarctic family SCATHOPHAGIDAE (dung flies) (500 species) is considered the most primitive of the Calyptratae. Adults (Figure 9.17C) are mostly predaceous on other insects, though some feed on the juice of dung. Larvae have varied habits; many are leaf and stem miners, others predaceous, and some feed on dung.

Superfamily Oestroidea

Five families are included in this very large group of Diptera. CALLIPHORIDAE (1000 species, 80% of which are restricted to the Old World) is a cosmopolitan group that includes blow flies, green- and bluebottles, and screwworm flies. Among members of the family, which may be paraphyletic (Rognes, 1997), a complete spectrum of larval feeding habits can be seen, ranging from true carrion feeders, through species that feed on exudates or open wounds of living animals, to truly parasitic forms. Calliphoridae of medical or veterinary importance include the sheep blow flies (*Lucilia* spp.), the screwworms [*Cochliomyia* (*Callitroga*) spp.] (Figure 9.18A), and bluebottles (*Calliphora* spp.), which are vectors of human diseases. Closely related to the calliphorids are the SARCOPHAGIDAE (flesh flies) (Figure 9.18B), whose larvae feed on decaying animal tissue or are true parasites of arthropods, mollusks, or annelids. Most of the 2000 species in this cosmopolitan group are viviparous, depositing first-instar larvae directly into the food source. The TACHINIDAE (Figure 9.18C,D), with some 8000 species worldwide, form the second largest family of Diptera. Without exception, the larvae are parasitic on other arthropods, mainly insects. An egg, or in the many viviparous species, a larva, is frequently deposited directly on the body of the host. Alternatively, the egg is laid on the host's food plant. The host usually dies as a result of the parasitism, and there is little doubt that tachinids play a role equal to that of many parasitic Hymenoptera in controlling the population level of certain species. Not surprisingly, some have been employed as biological control agents against pests (for examples, see Table 24.6). Included in the OESTRIDAE (bot flies and warble flies) (Figure 9.18E–G), a group of about 150 species, are four well-defined subfamilies that are often given family rank. OESTRINAE are holarctic and African flies that larviposit in the nasal and pharyngeal cavities of large herbivores; HYPODERMATINAE have a similar distribution, their hosts include rodents as well as herbivores, females oviposit on the host's skin, and the larvae develop subcutaneously; CUTEREBRINAE are restricted to the New World where they oviposit in places frequented by the host, the larvae hatching in response to radiant heat from an adjacent host and developing subcutaneously on primates, rodents, and lagomorphs; and the cosmopolitan GASTEROPHILINAE mostly oviposit on the legs or near the mouth of horses, zebras, and elephants, the larvae eventually making their way to the host's stomach or intestine where they complete development.

Superfamily Hippoboscoidea

Four families are included in this group, GLOSSINIDAE, HIPPOBOSCIDAE, STREBLIDAE, and NYCTERIBIIDAE, the last three formerly being considered to constitute the Pupipara. In all of the families the adults are bloodsucking parasites of birds or mammals, and the larvae mature, one at a time, entirely within the genital tract of the female,

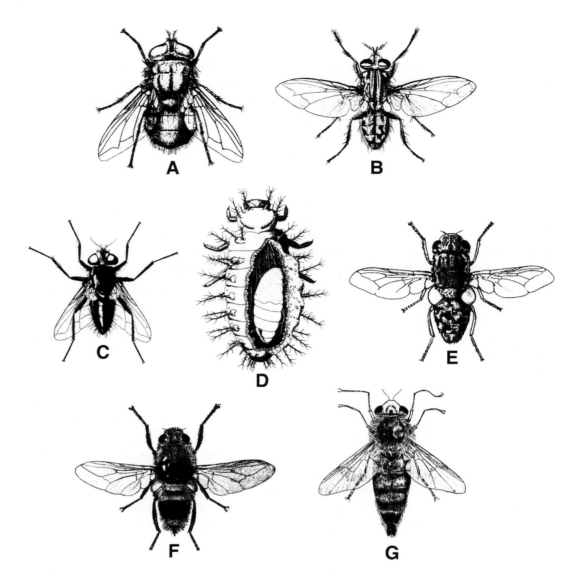

FIGURE 9.18. Oestroidea. (A) The screwworm fly, *Cochliomyia hominivorax* (Calliphoridae); (B) *Sarcophaga kellyi* (Sarcophagidae), a parasite of grasshoppers; (C) the bean beetle tachinid, *Aplomyiopsis epilachnae* (Tachinidae); (D) *A. epilachnae* larva inside bean beetle grub; (E) the sheep bot fly, *Oestrus ovis* (Oestridae); (F) the cattle warble fly, *Hypoderma bovis* (Oestridae); and (G) the horse bot fly, *Gasterophilus intestinalis* (Oestridae). [A, G, from M. T. James, 1948, The flies that cause myiasis in Man, *U.S. Dep. Agric., Misc. Publ. #631.* By permission of the U.S. Department of Agriculture. B, from L. A. Swan and C. S. Papp, 1972, *The Common Insects of North America.* Copyright 1972 by L. A. Swan and C. S. Papp. Reprinted by permission of Harper & Row Publishers, Inc. C, D, by permission of the U.S. Department of Agriculture. E, F, from A. Castellani and A. J. Chambers, 1910, *Manual of Tropical Medicine.* By permission of Bailliere and Tindall.]

being nourished by secretions of the accessory ("milk") glands. Pupation immediately follows birth and occurs off the host. Glossinidae (tsetse flies) is a very small but well-known family of about 20 species of *Glossina* from tropical Africa (excluding Madagascar). They are vectors of various species of *Trypanosoma* that cause sleeping sickness and nagana in humans and cattle, respectively. The Hippoboscidae (louse flies and keds) (Figure 9.19A) resemble tsetse flies in several ways and the two probably are sister groups. However, in

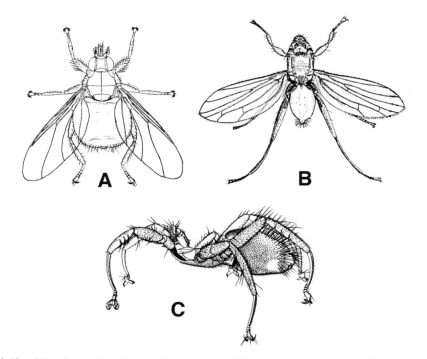

FIGURE 9.19. Hippoboscoidea. (A) *Lynchia americana* (Hippoboscidae), a parasite of owls and hawks; (B) a bat fly, *Strebla vespertilionis* (Streblidae); and (C) *Cyclopodia greefi* (Nycteribiidae). [A, from F. R. Cole and E. I. Schlinger, 1969, *The Flies of Western North America.* By permission of the University of California Press. B, from Q. C. Kessel, 1925, A synopsis of the Streblidae of the world, *J. N. Y. Entomol. Soc.* **33**:11–33. By permission of the New York Entomological Society. C, from H. Oldroyd, 1964, *The Natural History of Flies*, Weidenfeld and Nicolson. By permission of Mrs. J. M. Oldroyd.]

contrast to tsetse flies, which are strong fliers, hippoboscids rarely fly (indeed, some shed their wings after settling on a host). This group of about 330 species has a cosmopolitan distribution. They mainly parasitize birds but include some ungulates (*Melophagus ovinus*, the sheep ked is a major pest) and other mammals among their hosts. Most of the 160 species of Streblidae (bat flies) (Figure 9.19B) have wings, though these are pleated to facilitate movement through the host's fur. Females of *Ascodipteron* species shed their wings and legs and burrow into the host's skin. Most streblids are associated with colonial species of bats that roost in caves or forests and are found in tropical and subtropical regions. The Nycteribiidae (250 species) (Figure 9.19C) are wingless parasites of bats found mostly in warmer regions of the world.

Literature

There exists a massive volume of literature on Diptera, including many books, too numerous to mention specifically, on particular groups of flies or aspects of their biology. Oldroyd (1964) and Volumes 1 and 2 of the series edited by McAlpine *et al.* (1981–1989) are excellent sources of information on the biology of the group. Rohdendorf (1974), authors in Volume 3 of McAlpine *et al.* (1981–1989), Michelsen (1996), Friedrich and Tautz (1997), and Yeates and Wiegmann (1999) discuss the phylogeny and classification of the order. Griffiths (1994) [Brachycera], Oosterbroek and Courtney (1995) [Nematocera], Nagatomi

(1996) [orthorrhaphans], and Nirmala *et al.* (2001) [Calyptratae] deal with the major subordinal groups. Keys for identification of Diptera are given by Oldroyd (1949) (and other authors in Volumes 9 and 10 of the same series) and Colyer and Hammond (1951) [British species]; Cole and Schlinger (1969) and McAlpine *et al.* (1981–1989) [North American genera]; and Colless and McAlpine (1991) [Australian families and some subfamilies].

Cole, F. R., and Schlinger, E. I., 1969, *The Flies of Western North America*, University of California Press, Berkeley.

Colless, D. H., and McAlpine, D. K., 1991, Diptera, in: *The Insects of Australia*, 2nd ed., Vol. 2 (CSIRO, ed.), Melbourne University Press, Carlton, Victoria.

Colyer, C. N., and Hammond, C. O., 1951, *Flies of the British Isles*, Warne, London.

Downes, W. L., Jr., and Dahlem, G. A., 1987, Keys to the evolution of Diptera: Role of Homoptera, *Environ. Entomol.* **16**:847–854.

Friedrich, M., and Tautz, D., 1997, Evolution and phylogeny of the Diptera: A molecular phylogenetic analysis using 28S rDNA sequences, *Syst. Biol.* **46**:674–698.

Griffiths, G. C. D., 1994, Relationships among the major subgroups of Brachycera (Diptera): A critical review, *Can. Entomol.* **126**:861–880.

McAlpine, J. F., Peterson, B. V., Shewell, G. E., Teskey, H. J., Vockeroth, J. R., and Wood, D. M. (eds.), 1981–1989, *Manual of Nearctic Diptera*, Vols. 1–3, Research Branch, Agriculture Canada, Ottawa.

Michelsen, V., 1996, Neodiptera: New insights into the adult morphology and higher level phylogeny of Diptera (Insecta), *Zool. J. Linn. Soc.* **117**:71–102.

Nagatomi, A., 1996, An essay on the phylogeny of the orthorrhaphous Brachycera (Diptera), *Entomol. Mon. Mag.* **132**:95–148.

Nirmala, X., Hypsa, V., and Zurovec, M., 2001, Molecular phylogeny of calyptratae (Diptera: Brachycera): The evolution of 18S and 16S ribosomal rDNAs in higher dipterans and their use in phylogentic inference, *Insect Mol. Biol.* **10**:475–485.

Oldroyd, H., 1949, Diptera: Introduction and key to families, *R. Entomol. Soc. Handb. Ident. Br. Insects* **9**(1):49 pp.

Oldroyd, H., 1964, *The Natural History of Flies*, Weidenfeld and Nicolson, London.

Oosterbroek, P., and Courtney, G., 1995, Phylogeny of the nematocerous families of Diptera (Insecta), *Zool. J. Linn. Soc.* **115**:267–311.

Riek, E. F., 1977, Four-winged Diptera from the Upper Permian of Australia, *Proc. Linn. Soc. N.S.W.* **101**:250–255.

Rognes, K., 1997, The Calliphoridae (blowflies) (Diptera: Oestroidea) are not a monophyletic group, *Cladistics* **13**:27–66.

Rohdendorf, B., 1974, *The Historical Development of Diptera*, University of Alberta Press, Edmonton.

Willmann, R., 1989, Rediscovered: *Permotipula patricia,* the oldest known fly, *Naturwissenschaften* **76**:375–377.

Wootton, R. J., and Ennos, A. R., 1989, The implications of function on the origin and homologies of the dipterous wing, *Syst. Entomol.* **14**:507–520.

Yeates, D. K., and Wiegmann, B. M., 1999, Congruence and controversy: Towards a higher-level phylogeny of Diptera, *Annu. Rev. Entomol.* **44**:397–428.

4. Siphonaptera

SYNONYMS: Aphaniptera, Suctoria   COMMON NAME: fleas

Small, wingless, laterally compressed, jumping ectoparasites of birds and mammals; head sessile without typical compound eyes, antennae short and lying in grooves, mouthparts of piercing and sucking type; coxae large, tarsi 5-segmented; abdomen 10-segmented and bearing unsegmented cerci.

Larvae eruciform and apodous. Pupae adecticous and exarate, enclosed in a cocoon.

More than 2000 species of fleas have been described, about 90% of which are parasitic on placental mammals, especially rodents. More than 250 species occur in North America, about 90 in Australia, and 60 in Britain.

Adult. Fleas are highly modified for their ectoparasitic life, a feature that has made determination of their relationships with other Insecta difficult. The adults, which are between 1 and 10 mm in length, are highly compressed laterally and heavily sclerotized. The many hairs and spines on the body are directed posteriorly to facilitate forward movement. The head is broadly attached to the body and carries the short three-segmented antennae in grooves. Typical compound eyes are absent; however, the two "lateral ocelli" that occur are considered by some authors to be highly modified compound eyes and not homologous with the ocelli of other endopterygotes. The mouthparts are modified for piercing the host's skin and sucking blood. Mandibles are absent, the laciniae are elongate and together with the epipharynx form a piercing organ that rests in the grooved prementum. The thoracic segments are freely mobile and increase in size posteriorly. The legs are adapted for jumping and clinging to the host. The coxae are very large, and the tarsi terminate in a pair of strong claws. Ten abdominal segments occur, the last three of which are modified for reproductive purposes, especially in males, where the sternum and tergum of the ninth segment form clasping organs.

Both the cibarium and pharynx are strongly muscular for sucking up blood. The small proventriculus is fitted with cuticular rods (acanthae) that may serve to break up blood cells. The midgut (stomach) is large and fills most of the abdomen. Four Malpighian tubules arise at the anterior end of the short hindgut. The nervous system is primitive and includes three thoracic and seven or eight abdominal ganglia. As in Mecoptera, there is typically one more abdominal ganglion in males than in females. The testes are fusiform and are connected to a small seminal vesicle by means of fine vasa deferentia. The ovaries contain from four to eight panoistic ovarioles (polytrophic in Hystrichopsyllidae).

Larva and Pupa. Larvae (Figure 9.20D) are white and vermiform. They have a well-developed head that in some respects resembles that of nematocerous Diptera. The mouthparts, though modified, are of the biting type. There are 13 body segments, but the distinction between thoracic and abdominal regions is poor. Pupae are adecticous and exarate and enclosed in a cocoon. Traces of wings can be seen on the pupae of some species.

Life History and Habits

Adult fleas of both sexes are exclusively bloodsucking ectoparasites, though their association with a host is a rather loose one, that is, they spend a considerable time off the host's body. Host-parasite specificity is varied. A few species of fleas are monoxenous (restricted to a single host species) but most are polyxenous, with more than 20 potential hosts recorded for some species. Conversely, an animal may host one or many flea species. It is this lack of host specificity that makes fleas important vectors of certain diseases. The cosmopolitan rat flea, *Xenopsylla cheopsis*, for example, is responsible for transmitting bubonic plague and typhus, normally diseases of rodents, to humans. Fleas also are the intermediate hosts of the dog and rodent tapeworms that can infect humans. Interestingly, the European rabbit flea, *Spilopsyllus cuniculi*, introduced into Australia in 1966, appears to be a useful biocontrol agent by spreading the myxomatosis virus in rabbit populations. Adult fleas may survive for several months in the absence of a host and may live for more than a year when food is available.

Mating may occur on the host or in its nest, being triggered by warmth or feeding, the latter also being a prerequisite for egg maturation. In *S. cuniculi*, which is very host-specific, reproductive events are controlled by the level of hormones, especially corticosteroids, in the host's blood. Thus, the flea's breeding activity is closely linked with that of its host.

However, a similar phenomenon is not seen in polyxenous fleas. Eggs are almost never attached to the host but fall to the floor, hatching after 2–12 days.

Except for those of the Australian flea *Uropsylla tasmanica*, which live within the skin of their marsupial host, larvae are free-living and feed on organic debris. The larval stage may last several months depending on environmental conditions. After two molts the larvae pupate in a cocoon. Adult emergence is often dependent on a physical stimulus such as contact, increase in carbon dioxide level, or rise in temperature, as would be occasioned by the return of the host to its nest. Newly emerged adults may survive for several months without food.

Phylogeny and Classification

The origin and classification of fleas remain controversial, in part due to the very poor fossil record. *Saurophthirus*, from eastern Siberia, and *Saurophthirodes*, from Mongolia, are Lower Cretaceous fossils that some authors have claimed to be Siphonaptera. It is speculated that these fed on the wing membranes of flying reptiles. However, these lack some of the characters of fleas, for example, jumping hindlegs and lateral body compression. The only certain flea fossils (species of *Palaeopsylla*, *Pulex*, and *Rhopalopsyllus*), from Miocene amber, are assignable to extant families. Both the Diptera and the Mecoptera have been suggested as the sister group to the Siphonaptera, the differences arising from the interpretation of whether the common features of the groups have arisen by convergence or are genuine synapomorphies (see Chapter 2, Section 3.2, and discussions in Hennig, 1981; Kristensen, 1981). Indeed, some authors have gone so far as to suggest that the Siphonaptera should be placed *within* one of these orders (Byers 1996; Whiting, 2002). Classification of fleas is made difficult by their rather uniform habits and structure, though careful study of their host relationships has been valuable (see Traub, 1985). Depending on the authority, modern fleas are arranged in 2–5 superfamilies and 15 or 16 families. The system of Holland (1964) is followed here.

Superfamily Pulicoidea

Included in this superfamily are the families PULICIDAE and TUNGIDAE. The Pulicidae, a worldwide family of about 160 species, contains some important cosmopolitan fleas that attack humans and domestic animals, including *X. cheopsis*, *Pulex irritans* (the so-called "human flea") (Figure 9.20A) whose normal host is the pig, *Ctenocephalides canis* and *C. felis* (dog and cat fleas), and *Echidnophaga gallinacea* (the sticktight flea). Most Pulicidae have a rather loose association with their hosts, but female sticktight fleas become permanently attached in the manner of ticks. The Tungidae (Figure 9.20B) form a small (about 20 species), mainly tropical group of fleas commonly called jiggers, chigoes, or sand fleas, females of which burrow under the skin of the host, especially under the toenails or between the toes. Hosts include birds, various rodents, and occasionally humans.

Superfamily Malacopsylloidea

The two families that comprise the Malacopsylloidea were originally included in the Ceratophylloidea. The family RHOPALOPSYLLIDAE includes about 120 species of fleas parasitic on sea birds and mammals (mainly rodents). The group is mainly neotropical, but representatives are found also in Australia and southern North America. The MALACOPSYLLIDAE (two species) are found solely on armadillos in South America.

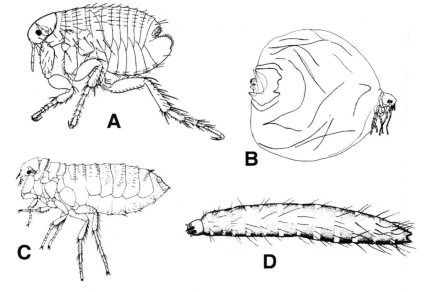

FIGURE 9.20. Siphonaptera. (A) The human flea, *Pulex irritans* (Pulicidae); (B) the female chigoe flea, *Tunga penetrans* (Tungidae); (C) the sand-martin flea, *Ceratophyllus styx* (Ceratophyllidae); and (D) larva of *Spilopsyllus cuniculi* (Pulicidae). [A, from L. A. Swan and C. S. Papp, 1972, *The Common Insects of North America.* Copyright 1972 by L. A. Swan and C. S. Papp. Reprinted by permission of Harper & Row Publishers, Inc. B–D, from R. R. Askew, 1971, *Parasitic Insects.* By permission of Heinemann Educational Books Ltd.]

Superfamily Ceratophylloidea

In Holland's (1964) scheme this very large group includes 12 families, the largest of which are mentioned below. More recent classifications break the ceratophylloids into two or three distinct superfamilies. The HYSTRICHOPSYLLIDAE (CTENOPHTHALMIDAE) constitute the largest family of fleas with about 620 species. Representatives occur throughout the world, though the group is mainly a holarctic one. Most species are parasites of rodents and shrews, though a few are found on carnivores and marsupials. Another large group of about 460 species is the CERATOPHYLLIDAE (Figure 9.20C), which includes several cosmopolitan species. Ceratophyllids are found mainly on rodents, though some occur on birds. Several species are believed to be capable of transmitting plague from rodents to humans, and others can serve as the intermediate host for the tapeworm, *Hymenolepis diminuta*. Related to the previous family are the LEPTOPSYLLIDAE, a family of about 240 species found usually on small rodents, but also known from rabbits and lynx in North America. The ISCHNOPSYLLIDAE (about 120 species) are restricted to bats, especially insectivorous forms. The PYGIOPSYLLIDAE, a group of about 160 species, has representatives in Australasia, Southeast Asia, and South America parasitic on a wide range of monotremes, marsupials, rodents, and passerine and sea birds.

Literature

Holland (1964, 1985), Askew (1971), Rothschild (1975), Traub and Starcke (1980), Marshall (1981), and Traub (1985) include a good deal of general information on fleas, especially concerning host-parasite relationships. The phylogeny of fleas is discussed by Hennig

(1981), Kristensen (1981), Byers (1996), and Whiting (2002). Lewis (1998) summarizes the world flea fauna. Keys for identification are given by Arnett (2000) [North American families], Dunnet and Mardon (1991) [Australian genera], and Smit (1957) [British species].

Arnett, R. H., Jr., 2000, *American Insects: A Handbook of the Insects of America North of Mexico*, 2nd ed., CRC Press, Boca Raton, FL.

Askew, R. R., 1971, *Parasitic Insects*, American Elsevier, New York.

Byers, G. W., 1996, More on the origin of Siphonaptera, *J. Kansas Entomol. Soc.* **69**:274–277.

Dunnet, G. M., and Mardon, D. K., 1991, Siphonaptera, in: *The Insects of Australia*, 2nd ed., Vol. 2 (CSIRO, ed.), Melbourne University Press, Carlton, Victoria.

Hennig, W., 1981, *Insect Phylogeny*, Wiley, New York.

Holland, G. P., 1964, Evolution, classification, and host relationships of Siphonaptera, *Annu. Rev. Entomol.* **9**:123–146.

Holland, G. P., 1985, The fleas of Canada, Alaska and Greenland (Siphonaptera), *Mem. Entomol. Soc. Can.* **130**:1–631.

Kristensen, N. P., 1981, Phylogeny of insect orders, *Annu. Rev. Entomol.* **26**:135–157.

Lewis, R. E., 1998, Résumé of the Siphonaptera (Insecta) of the world, *J. Med. Entomol.* **35**:377–389.

Marshall, A. G., 1981, *The Ecology of Ectoparasitic Insects*, Academic Press, New York.

Rothschild, M., 1975, Recent advances in our knowledge of the order Siphonaptera, *Annu. Rev. Entomol.* **20**:241–259.

Smit, F. G. A. M., 1957, Siphonaptera, *R. Entomol. Soc. Handb. Ident. Br. Insects* **1**(16):94 pp.

Traub, R., 1985, Coevolution of fleas and mammals, in: *Coevolution of Parasitic Arthropods and Mammals* (K. C. Kim, ed.), Wiley, New York.

Traub, R. E., and Starcke, H. (eds.), 1980, *Fleas. Proceedings of the International Conference on Fleas, Ashton Wold, Peterborough, U.K., 21–25 June 1977*, Balkema, Rotterdam.

Whiting, M., 2002, Mecoptera is paraphyletic: Multiple genes and phylogeny of Mecoptera and Siphonaptera, *Zool. Scripta* **31**:93–104.

5. Trichoptera

SYNONYMS: Phryganoidea, Phryganeida   COMMON NAME: caddisflies

Small- to medium-sized mothlike insects; head with setaceous antennae, reduced mandibulate mouthparts, and usually small compound eyes (large in some males); prothorax small, two pairs of wings almost always present and covered with hairs, held rooflike over body at rest, legs identical with five-segmented tarsi.

Larvae aquatic, usually eruciform (portable case-dwelling species) or campodeiform (fixed retreat and free-living species). Pupae mostly decticous (sometimes adecticous) and exarate.

Close to 11,000 species of Trichoptera have been described, including some 1400 species in North America, 500 in Australia, and 200 in Britain. The order has a worldwide distribution.

Structure

Adult. Caddisflies are between 1.5 and 40 mm in length. They are mothlike and usually drab in color. The setaceous antennae are always long, sometimes several times the length of the wings. The compound eyes are usually small, but in some males they are very large and almost meet at the vertex. Ocelli (three) are present in some species, absent in others. The mouthparts are reduced; the mandibles are vestigial, the maxillae are small and closely associated with the labium. The hypopharynx is well developed and in several groups modified for sucking up liquid food. The head, pro- and mesonotum have raised, wartlike areas that bear setae. The prothorax is small and ringlike; the mesothorax and metathorax are well developed. The legs are long and slender and have five-segmented tarsi. Two pairs of membranous wings are almost always present, though one or both pairs are greatly reduced in a few species. The wings are typically covered with fine hairs (denser on fore wings); however, these sometimes have a restricted distribution (e.g., along veins or in clumps) or are modified to scales. The fore and hind wings are coupled during flight. The wing venation is generalized and resembles that of some primitive Lepidoptera. There is in most species a whitish spot, the thyridium, devoid of hairs near the center of each wing. The wings are held rooflike over the body when not in use. Ten abdominal segments can be distinguished. In males the genitalia comprise a pair of claspers, a phallus usually with an extensile aedeagus and a pair of parameres, and various accessory structures derived from the 10th segment. In females of some species the terminal segments are retractile and function as an ovipositor.

Though not well known, the internal structure appears quite generalized. The gut is short and straight, and there are six Malpighian tubules. In the ventral nerve cord three thoracic and seven abdominal ganglia are found, the metathoracic and first abdominal ganglia having fused together. The testes are saclike; the ovaries contain numerous polytrophic ovarioles.

Larva and Pupa.   Larvae are generally campodeiform (free-living and fixed-retreat species) (Figures 9.22 and 9.23) or eruciform (portable case-building species) (Figure 9.24). The head is well sclerotized and carries a pair of very short antennae, mandibulate mouthparts, and two lateral clusters of ocelli. The thorax is variably sclerotized and bears well-developed legs that have an unsegmented tarsus. The forelegs are fairly short and used more for holding food and constructing the case than for walking. The abdomen is 10-segmented. The first abdominal segment of most species of Limnephiloidea has three prominent, retractile papillae that bear sensory hairs; these may enable the insect to maintain its position in the case. Prolegs are absent on all but the last abdominal segment. These have a pair of strong hooks for anchoring the insect to the case or substrate. In a few species respiration is entirely cutaneous but most species have gills. These are usually simple filamentous structures developed on the abdomen, occasionally also on the thorax. Sometimes they are arranged in groups, or they may be branched basally. Blood gills (nontracheated) occur in some species; they are usually eversible and have an osmoregulatory function.

Pupae are mostly decticous (adecticous in some Phryganeidae) and always exarate. They are aquatic, respiring by means of the larval gills or cutaneously, and have well-developed mandibles which the pharate adult uses to cut its way out of the case or cocoon.

Life History and Habits

Caddisflies, which are usually found close to fresh water, are mainly crepuscular or nocturnal, hiding by day among vegetation. They do not take solid food but are capable of sucking up nectar or water, which enables them to survive for several weeks. Coupling takes place during flight, but the insects come to rest for insemination. Eggs are laid in ribbons or masses either directly into the water or on some object immediately above it, the young larvae dropping into the water on hatching. Development may be rapid and in most species includes five larval instars. Most temperate region caddisflies have only one generation per year (univoltine), though some species are bivoltine or semivoltine. Diapause in the final larval or adult instar, rarely the egg stage, is characteristic of many species. Some larvae are free-living and construct a shelter only at pupation; others build a non-portable silken web (fixed-retreat species), which, in addition to providing them with protection, also serves to trap food. Most larvae, however, construct a portable case whose shape and the nature of the materials incorporated into it are varied but characteristic for a particular family or even genus. The case has a silken lining to which other materials both organic and inorganic are stuck. Broadly speaking, the free-living forms are predators or feed on dead animal materials, the fixed-retreat builders are filter feeders or gather organic matter trapped in their web, and the portable-case bearers may be shredders, grazers and scrapers, or piercers (obtaining their food by sucking fluids from living plants).

Prior to pupation, larvae of most case-building species attach their case to the substrate and close off each end with a perforated silken wall which allows the continued passage of water through the case for gas exchange. Some species close the anterior end with plant materials held together by silk, and a few species build an entirely new case. In most fixed-retreat species mature larvae abandon their web and construct a separate pupal enclosure similar to that of case builders; others pupate within the web. Mature larvae of many free-living species first construct a rough, dome-shaped structure that they attach to the substrate. Within this they build a discrete smooth and imperforate cocoon. Water cannot flow through the cocoon and the gas exchange occurs solely by diffusion across the fluid-filled cavity (Wiggins and Wichard, 1989). The pharate adult (often incorrectly referred to as the "pupa") (see Chapter 21, Section 3.3.3) bites its way out of the case or cocoon using the pupal mandibles and crawls or swims to a suitable emergence site. Molting takes place at the water surface and adults can fly immediately after their wings are expanded.

Phylogeny and Classification

The Trichoptera and its sister group Lepidoptera are thought to have evolved from a stem group of mecopteralike insects in the Permian. A few Trichoptera have been collected from Lower Permian deposits in the Czech Republic and the United States, and representatives of some extant families probably existed at least as early as the Jurassic. Fossilized caddis larval cases are known from the Lower Cretaceous, by which time the order had undergone a wide radiation. Within the order the evolutionary development of the larval stage has far outstripped that of the adult. Ross (1956, 1967) suggested that the order evolved in a forested habitat, through which ran small, cool-water streams. Such a habitat would not be conducive to a wide adaptive radiation of adults, but larvae, evolving in a new, relatively unexploited habitat, were able to diversify in a remarkable manner. Subsequently, representatives of many families have invaded warm, moving-water, many still-water, and even some temporary aquatic habitats (Wiggins, 1996). Curiously, perhaps, despite their wide-ranging differences in behavior and ecology, caddis larvae are quite uniform morphologically, making their identification sometimes quite difficult. By contrast, the evolution of adults has been concerned for the most part with the development of specific behavior patterns and changes in the genitalia to prevent interspecific mating, the latter making species-level identification relatively easy.

Trichopterists are agreed that the traditional arrangement whereby the extant Trichoptera were placed in two suborders, Annulipalpia and Integripalpia, is no longer tenable. Unfortunately, they have not yet reached a consensus on a more appropriate classification (see Kjer *et al.*, 2002). The central problem is the position of the Rhyacophiloidea, with some

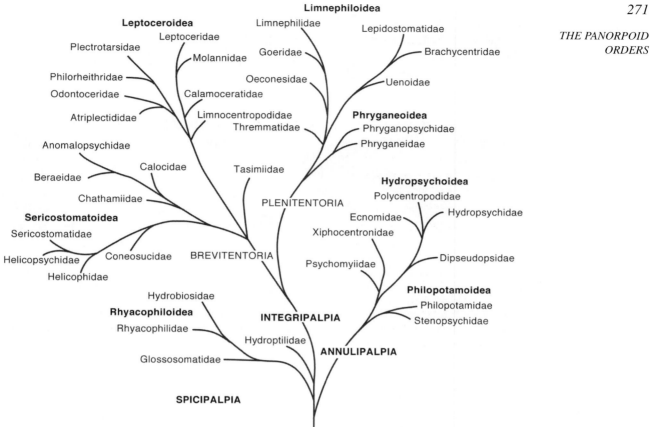

FIGURE 9.21. A suggested phylogeny of Trichoptera.

authorities suggesting that this group's affinities lie with the Annulipalpia and others taking the view that it belongs with the Integripalpia. A third point of view, supported by some morphological and molecular studies of extant species, is that the rhyacophiloids are sufficiently distinct as to constitute a separate suborder Spicipalpia (closed-cocoon makers), with the remaining species falling into the more narrowly defined suborders Annulipalpia (fixed-retreat makers) and Integripalpia (portable-case makers). In this scheme the Annulipalpia is the sister group to the Integripalpia + Spicipalpia. Whereas the Annulipalpia and Integripalpia are generally accepted as monophyletic, there are differing opinions on whether the Spicipalpia are monophyletic or paraphyletic. A possible phylogeny of the group is illustrated in Figure 9.21.

Suborder Annulipalpia

Members of this suborder are united by the following features: adults with five-segmented maxillary palps, the terminal segment being annulate, flexible, and at least twice as long as the preceding segment; females with a single anal-vaginal opening; larvae campodeiform, prognathous, and fixed-retreat makers; at maturity, larvae construct a rough shelter of stones in which they produce their perforate cocoon. The suborder includes two superfamilies, the Philopotamoidea and Hydropsychoidea.

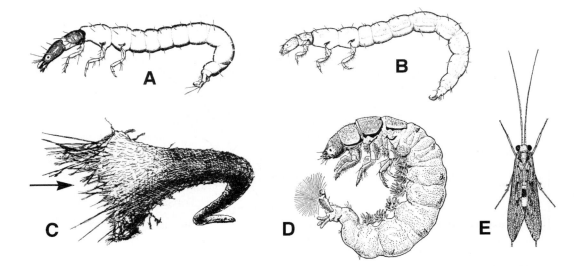

FIGURE 9.22. Annulipalpia. (A) *Chimarra* sp. (Philopotamidae) larva; (B) *Neureclipsis bimaculata* (Polycentropodidae) larva; (C) capturing net of *N. bimaculata*, arrow indicating direction of current; (D) *Hydropsyche simulans* (Hydropsychidae) larva; and (E) *H. simulans* adult. [A-C, from G. B. Wiggins, 1977, *Larvae of the North American Caddisfly Genera (Trichoptera)*. By permission of the Royal Ontario Museum. D, E, from H. H. Ross, 1944, The caddisflies, or Trichoptera, of Illinois, *Bull. Ill. Nat. Hist. Surv.* **23**:1–326. By permission of the Illinois Natural History Survey.]

Superfamily Philopotamoidea

About 90% of the 470 described species in this group are included in the widely distributed family PHILOPOTAMIDAE (Figure 9.22A), adults of which are small (6–9 mm long) and generally have brownish bodies and gray or blackish wings. Sometimes females are apterous. Larvae live in fast-flowing streams and build tubular webs whose entrance is larger than the exit. A larva stays in its web and feeds on algae and fine organic particles caught in it.

Superfamily Hydropsychoidea

Of the six families in this group, three (PSYCHOMYIIDAE, POLYCENTROPODI-DAE, and HYDROPSYCHIDAE) are quite large and common, including between them about 75% of the species. Psychomyiidae (150 species) are found in all but the neotropical region. Larvae, which are mostly found in cool, running waters, build silken tubes covered with particles of sand or detritus on rocks or logs where they feed on algae, etc. Larvae of Polycentropodidae (300 species) (Figure 9.22B) occur worldwide in both running and standing waters. Some build silken tubes or sheets covering depressions in rocks and are predaceous; others have funnel-shaped nets that filter food material (Figure 9.22C). The widespread family Hydropsychidae (900 species) (Figure 9.22D,E) includes mainly species associated with streams or rivers, though some are found along wave-washed lake shores. Larvae live in fixed retreats made from plant material or rock particles on logs or rocks, in front of which they build a cup-shaped net to trap algae, organic material, or microinvertebrates. The mesh size is characteristic of both the species and its position in the stream.

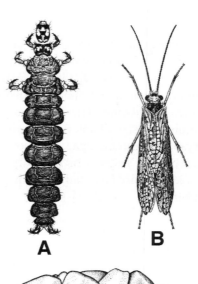

FIGURE 9.23. Spicipalpia. (A) *Rhyacophila fuscula* (Rhyacophilidae) larva; (B) *Rhyacophila fenestra* adult; and (C) *Glossosoma* sp. (Glossosomatidae) case with larva. [A, C, from G. B. Wiggins, 1977, *Larvae of the North American Caddisfly Genera (Trichoptera).* By permission of the Royal Ontario Museum. B, from H. H. Ross, 1944, The caddisflies, or Trichoptera, of Illinois, *Bull. Ill. Nat. Hist. Surv.* **23**:1–326. By permission of the Illinois Natural History Survey.]

Suborder Spicipalpia

Members of this suborder have the following characters: five-segmented maxillary palps, with terminal segment identical to others and neither annulate nor flexible; females with a single anal-vaginal opening; larvae that build an imperforate silk cocoon in which to pupate.

Superfamily Rhyacophiloidea

The family RHYACOPHILIDAE (Figure 9.23A, B) contains the most archaic members of the order. Of the 450 species in the group, more than 90% belong to the genus *Rhyacophila.* The family is widely distributed though absent from the Australian, Ethiopian, and neotropical regions. Adults are generally brownish or mottled and vary in length from 3 to 13 mm. Larvae are mainly predaceous (rarely algivorous) and found only in clear, cold-water streams. They do not construct a portable case but move freely over the stream bed, producing a continuous strand of silk that serves as an anchor line. At maturation, larvae build a rough shelter of stones in which they spin a separate cocoon. HYDROBIOSIDAE (150 species), which have frequently been considered a subfamily of Rhyacophilidae, are restricted largely to the Australian and neotropical regions. The larvae are quite similar to those of Rhyacophilidae in habits, but some species are found in warmer waters and stagnant pools. GLOSSOSOMATIDAE (400 species distributed worldwide) are little different as adults from Rhyacophilidae. The different habits of the larvae are, however, considered sufficient justification for their separation. Larvae, found in swiftly flowing water, feed on algae scraped from rocks and live in saddlelike or turtle-shaped cases (Figure 9.23C);

that is, the ventral side of the case is flat and composed of fine sand grains, and the dorsal surface is strongly convex and built of coarser material. When larvae are about to pupate, they cut away the ventral part of the case and fix the dorsal component to the substrate. Adult HYDROPTILIDAE (600 species) are hairy, minute Trichoptera (1.5–6 mm in length), often called micro-caddisflies, whose larvae are found in a variety of permanent waters worldwide. The larvae are of particular interest in that they exhibit hypermetamorphosis (see Chapter 21, Section 3.3.2). The first four instars are active, free-living larvae that feed on algae and show very little growth. Nearly all growth occurs in the last, case-making instar, which is very different in appearance from its predecessors. In many species the case has a purselike shape, and members of the family are thus called purse-case makers.

Suborder Integripalpia

Between 20 and 30 families (depending on the authority) comprise this large suborder whose species have the following common features: adult maxillary palps with five segments (fewer in many males), the apical segment not annulate; females with separate anal and vaginal openings; larvae eruciform and mostly hypognathous, living in portable cases in which pupation usually occurs. Within the suborder, two major evolutionary lines can be seen, the limnephilid and leptocerid branches of Ross (1967), equivalent to the infraorders Plenitentoria and Brevitentoria of Weaver (1984). Within the Plenitentoria are the superfamilies Phryganeoidea and Limnephiloidea; in the Brevitentoria are the Sericostomatoidea and Leptoceroidea.

Superfamily Phryganeoidea

This group of fewer than 100 species mostly belong to the holarctic and oriental family PHRYGANEIDAE. The largest caddisflies are found in this family, with lengths up to 40 mm. Larvae usually construct cases of leaf or bark pieces arranged in a spiral (Figure 9.24C) or in cylindrical sections glued end to end. The larvae are predaceous or detritus feeders found mainly in marshes and lakes, though a few species live in slow-moving streams or temporary pools.

Superfamily Limnephiloidea

By far the largest family in this group is the LIMNEPHILIDAE (1000 species) (Figure 9.24A,B), a primarily holarctic family of small to large caddisflies (7–25 mm in length). The larval case is built of plant or mineral material, its characteristics often being genus-specific. Larvae are found in moving or standing waters (including temporary and brackish pools) where they are typically detritus feeders, though a few scrape algae off rocks. LEPIDOSTOMATIDAE (250 species) are widely distributed though absent from Australia. Larvae live mainly in cool, slow-moving waters, occasionally in the littoral zone of lakes, where they feed on detritus. The BRACHYCENTRIDAE (100 species) are confined to the holarctic and oriental regions where they occur in a variety of cool, moving-water habitats. Larvae of most species are algivores; however, a few species are predaceous, using their long, spiny legs as filters with which to trap passing insects.

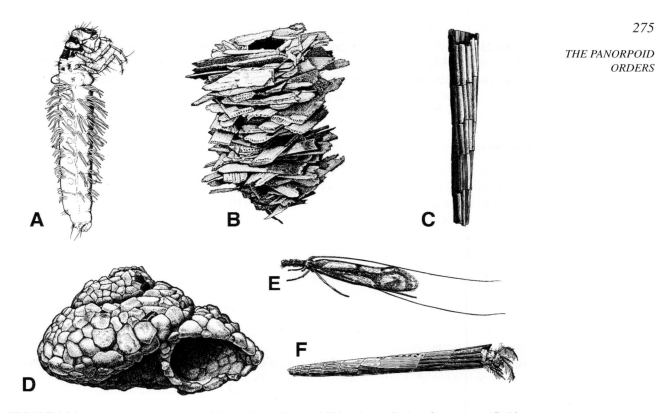

FIGURE 9.24.   Integripalpia. (A) *Limnephilus indivisus* (Limnephilidae) larva; (B) *L. indivisus* case; (C) *Phryganea cinerea* (Phryganeidae) case; (D) *Helicopsyche borealis* (Helicopsychidae) case; (E) *Triaenodes tarda* (Leptoceridae) adult; and (F) *T. tarda* larva in case. [A-D, from G. B. Wiggins, 1977, *Larvae of the North American Caddisfly Genera (Trichoptera).* By permission of the Royal Ontario Museum. E, F, from H. H. Ross, 1944, The caddisflies, or Trichoptera, of Illinois, *Bull. Ill. Nat. Hist. Surv.* **23**:1–326. By permission of the Illinois Natural History Survey.]

Superfamily Sericostomatoidea

This small superfamily, comprising only about 300 species, is split into eight families by Weaver (1984), including the HELICOPSYCHIDAE and SERICOSTOMATIDAE, each with about 100 species. Though a small group, helicopsychids are widely distributed and include some of the best-known caddis larvae because they build cases of sand grains shaped like a snail shell (Figure 9.24D). Larvae live in running waters, both cool and warm, feeding mainly on algae and organic particles scraped from rocks. Sericostomatidae are also widely distributed, though absent from Australia. Their larvae are found in still or slow-moving waters, sometimes burrowing in loose sediments. They are detritivores and carry a curved case built of rock particles.

Superfamily Leptoceroidea

Only two of the seven families recognized by Weaver (1984) in this group of about 1000 species are of any size. The LEPTOCERIDAE (800 species) occur worldwide, and the adults are easily recognized by their antennae whose length may be twice that of the body (Figure 9.24E). The larvae are mainly omnivorous and live in standing water or slowly moving rivers, building very elongate cases of fine stones or sand. In some species there is a changeover to the use of vegetable material in case construction by older instars (Figure 9.24F) when spiral cases similar to those of Phryganeidae may be built. CALAMOCERATIDAE (100 species) are mainly subtropical caddisflies whose larvae are typically detritus feeders in marshes, coastal lakes, or slow-moving streams. Larvae build cryptic cases with pieces of leaf or bark, occasionally hollowing out twigs in which to live.

Literature

Hickin (1967), Mackay and Wiggins (1979), and Wiggins (1996) provide accounts of the biology of Trichoptera, emphasizing the larval stage. Ideas on the evolution and classification of the order are discussed by Morse (1997) and Kjer *et al.* (2002) (and authors cited therein). Keys for the identification of Trichoptera are given by Wiggins (1984, 1987, 1996) and Morse and Holzenthal (1984) [larvae of North American genera], Hickin (1967) [larvae of British species], and Neboiss (1991) [adults and larvae of Australian families].

Hickin, N. E., 1967, *Caddis Larvae*, Hutchinson, London.
Kjer, K. M., Blahnik, R. J., and Holzenthal, R. W., 2002, Phylogeny of caddisflies (Insecta, Trichoptera), *Zool. Scripta* **31**:83–91.
Mackay, R. J., and Wiggins, G. B., 1979, Ecological diversity in Trichoptera, *Annu. Rev. Entomol.* **24**:185–208.
Morse, J. C., 1997, Phylogeny of Trichoptera, *Annu. Rev. Entomol.* **42**:427–450.
Morse, J. C., and Holzenthal, R. W., 1984, Trichoptera genera, in: *An Introduction to the Aquatic Insects of North America*, 2nd ed. (R. W. Merritt and K. W. Cummings, eds.), Kendall/Hunt, Dubuque, IA.
Neboiss, A., 1991, Trichoptera, in: *The Insects of Australia*, 2nd ed., Vol. 2 (CSIRO, ed.), Melbourne University Press, Carlton, Victoria.
Ross, H. H., 1956, *Evolution and Classification of the Mountain Caddis flies*, University of Illinois Press, Urbana.
Ross, H. H., 1967, The evolution and past dispersal of the Trichoptera, *Annu. Rev. Entomol.* **12**:169–206.
Weaver, J. S., 1984, Evolution and classification of the Trichoptera, part I: Groundplan of Trichoptera, in: *Proceedings of the Fourth International Symposium on Trichoptera* (J. C. Morse, ed.), Junk, The Hague.
Wiggins, G. B., 1984, Trichoptera, in: *An Introduction to the Aquatic Insects of North America*, 2nd ed. (R. W. Merritt and K. W. Cummins, eds.), Kendall/Hunt, Dubuque, Iowa.
Wiggins, G. B., 1987, Order Trichoptera, in: *Immature Insects* (P. W. Stehr, ed.), Kendall/Hunt, Dubuque, Iowa.
Wiggins, G. B., 1996, *Larvae of the North American Caddisfly Genera (Trichoptera)*, 2nd ed., University of Toronto Press, Toronto.
Wiggins, G. B., and Wichard, W., 1989, Phylogeny of pupation in Trichoptera, with proposals on the origin and higher classification of the order, *J. North Am. Benthol. Soc.* **8**:260–276.

6. Lepidoptera

SYNONYMS: none   COMMON NAMES: butterflies and moths

Insects whose body and appendages are covered with scales; head with large compound eyes and mouthparts almost always in form of suctorial proboscis; prothorax in most species reduced, two pairs of membranous wings present in almost all species with few crossveins; posterior abdominal segments much modified in connection with reproduction, cerci absent.

Larvae eruciform with well-developed head, mandibulate mouthparts, and 0–11 (usually 8) pairs of legs. Pupae in most species adecticous and obtect, in others decticous.

Lepidoptera are probably the most familiar and easily recognized of all insects. Some 200,000 species have been described, including about 11,300 from North America, 10,000 from Australia, and 2500 from Britain.

Adult.   The great majority of adult Lepidoptera (i.e., the Ditrysia, which includes about 98% of the described species) have a very constant fundamental structure, a feature that has led to some difficulty in establishing the phylogeny and classification of the order. Lepidoptera range in size from very small (wingspans of about 3 mm) to very large (wingspans of 25 cm). The entire body and appendages are covered with scales (modified hairs). The compound eyes are large and cover a major portion of the head capsule. Two ocelli are present in most species but are concealed by scales. The antennae are of varied size and structure. In most Lepidoptera mandibles are absent and the maxillae are modified as a suctorial proboscis (see Chapter 3, Section 3.2.2 and Figure 3.12). When not in use the proboscis is coiled beneath the thorax. In most species the prothorax is reduced and collarlike. The mesothorax is the larger of the pterothoracic segments, which bear large tegulae, a characteristic feature of the order. Auditory organs are present on the metathorax of Noctuoidea. Both the fore and hind wings are generally large, membranous, and covered with scales. The latter are flattened macrotrichia supported on a short, thin stalk; they may be solid (the "primitive-type" scales found in non-Glossata) or with an internal cavity that may contain pigments ("normal-type" scales of Glossata). The iridescent colors of many Lepidoptera result from the layered structure of the scale (see Chapter 11, Section 4.3). In males of some species certain scales (androconia) are modified so as to facilitate the volatilization of material produced in the underlying scent glands. In primitive Lepidoptera the wing venation is identical in the fore and hind wings (homoneurous) and resembles that of primitive Trichoptera. In advanced forms there is considerable divergence between the venations of the fore and hind wings (the heteroneurous condition). The wing-coupling apparatus of primitive Lepidoptera comprises simply the small jugum of the fore wing, which lies on top of the hind wing. Occasionally a few frenular bristles are present on the anterior part of the hind wing, which assist in coupling. In higher Lepidoptera the coupling apparatus is usually made up of the retinaculum of the fore wing and frenulum of the hind wing (Chapter 3, Section 4.3.2 and Figure 3.28). However, in certain families the frenulum has been lost, and wing coupling is achieved simply by overlapping. The humeral area of the hind wing is greatly enlarged and strengthened, and lies beneath the fore wing. This is the amplexiform system. In females of a few species wings are reduced or lost. There are 10 easily identifiable abdominal segments, though the sternum of segment 1 is missing and the 9th and 10th (sometimes also the 7th and 8th) segments are modified in relation to the genitalia. The male genitalia are complex and their homologies unclear. In females there are three basic types of genitalia: monotrysian, exoporian, and ditrysian. In the monotrysian type there is a single genital opening on fused sterna 9 and 10 that serves both for insemination and for oviposition. In the exoporian type, found in Hepialoidea and Mnesarchaeoidea, there are separate openings for insemination and oviposition though both occur on fused segment 9/10, and in Ditrysia the opening for insemination is on sternum 7 or 8, with that for egg laying on fused segment 9/10 (Figure 19.3). In most species the genital aperture is flanked by a pair of soft lobes, but these may be strongly sclerotized and function as an ovipositor in species that lay their eggs in crevices or plant tissue. A pair of auditory organs is found on segment 1 in some moths. Cerci are absent.

The anterior region of the foregut is modified as a pharyngeal sucking pump from which a narrow esophagus leads posteriorly and, in some forms, expands to form a crop. In higher Lepidoptera, however, the crop is a large lateral diverticulum. The midgut is short and straight, the hindgut longer and coiled. In most species there are six Malpighian tubules that enter the gut via two lateral ducts. Only two tubules occur in some species. The nervous system is somewhat concentrated. In the most primitive condition three thoracic and five abdominal ganglia are found. In the majority of Lepidoptera, however, there are two thoracic and four abdominal centers. In most male Lepidoptera the four testis follicles on each side are intimately fused to form a single median gonad, though the vas deferens on each side remains separate and posteriorly becomes dilated to form a seminal vesicle before opening into the ductus ejaculatorius duplex. The latter also receives one, rarely two, accessory glands. Sperm is stored in both the seminal vesicles and the duplexes, the latter opening into the common ejaculatory duct (simplex). In most species four polytrophic ovarioles occur in each ovary, which joins the common oviduct via a short lateral tube. Various accessory glands open into the oviduct. During mating, sperm (transferred in a spermatophore) is deposited in the bursa copulatrix; in monotrysian and exoporian forms, it remains here, but in Ditrysia the sperm then moves via the sperm duct to the spermatheca where it is stored till required.

Larva and Pupa. Larvae (caterpillars) are eruciform, and the three primary body divisions are easily recognizable. The head is heavily sclerotized and bears strong, biting mouthparts, six ocelli in most species, and short, three-segmented antennae. In Glossata the prementum carries a median spinneret that receives the ducts of the silk (modified salivary) glands. Three pairs of thoracic legs and, in most species, five pairs of abdominal prolegs occur; the latter are located on segments 3–6 and on segment 10. At the opposite extreme, many leaf-mining larvae are apodous. Larvae are usually equipped with some method for protecting themselves from would-be predators. They may be colored either procryptically or aposematically; they may build shelters in which to hide, or they may possess a variety of repugnatory devices, such as glands that produce an obnoxious fluid, or long irritating hairs that invest the body. The pupae of most Lepidoptera are adecticous and obtect. In lower members of the order, however, they are decticous and exarate.

Life History and Habits

Lepidoptera are found in a wide variety of habitats but are almost always associated with higher plants, especially angiosperms. Nielsen and Common (1991) suggested that the evolution of the proboscis was a major factor in the success of the order, enabling adults both to easily ingest water, hence avoid desiccation, and to obtain nectar often stored cryptically. Adults of the majority of extant species feed on nectar, the juice of overripe fruit, or other liquids. Some primitive forms feed on pollen and spores, and representatives of numerous ditrysian groups are non-feeders as adults. Some Lepidoptera are strong fliers, and some may migrate considerable distances (see Chapter 22, Section 5.2 for examples). The majority of species are cryptically colored, though many examples, especially among butterflies, of species with brilliantly colored wings are known. Balanced polymorphism (coexistence of two or more color forms), seasonal polymorphism, geographic polymorphism, or sexual dimorphism occurs in some species. In diurnal butterflies and moths visual stimuli attract males to females. However, females of nocturnal species secrete pheromones that attract males, sometimes from a considerable distance (see Chapter 13, Section 4.1). Likewise, pheromones produced by males may serve to initiate the mating response of a female. Lepidoptera almost always reproduce sexually and are oviparous, though facultative parthenogenesis and ovoviviparity occur in a few species. Egg-laying habits and the number of eggs laid are extremely varied. Some species simply drop their eggs in flight. Others are attracted to the larva's host plant by its odor and lay their eggs in a characteristic pattern either on or in it. Some species lay only a few eggs whereas others lay many thousands.

Larvae of almost all species are phytophagous, and no parts of plants remain unexploited. Because of their phytophagous habits and high reproductive rate, many species are important pests. Members of a few species are carnivorous or feed on various animal products. Larval development is usually rapid, and there may be several generations each year. However, in some species larval development requires 2 or 3 years for completion. In most Lepidoptera pupation takes place within a cocoon constructed in a variety of ways. It may be made entirely of silk or, more frequently, a mixture of silk and foreign materials. In primitive forms the pharate adult bites open the cocoon and wriggles out of it shortly before eclosion. However, eclosion frequently takes place within the cocoon, in which case the adult has various devices that facilitate escape. These include temporary spines for cutting open the cocoon and the secretion from the mouth of a special softening fluid. In many instances the cocoon itself is specially constructed to allow easy egress. Pupae of butterflies are usually naked and suspended from a small pad of silk attached to a support. This pad may represent the remains of the ancestral cocoon.

Phylogeny and Classification

Fossil Amphiesmenoptera (the stem group from which the Lepidoptera and its sister group Trichoptera evolved) are known from the Permian period. A variety of wing fragments from the Middle and Upper Jurassic, initially claimed to be those of early Lepidoptera, remain controversial, and the earliest undisputed members of the order (*Archaeolepis mane* and *Eolepidopterix jurassica*) are from the Lower and Upper Jurassic, respectively (Whalley, 1986). These early forms presumably fed on pollen and spores as adults and perhaps mosses in the caterpillar stage, much as do the living Micropterigidae. Though the ingestion of fluids, including plant exudates and nectar, cannot be ruled out, there is no indication that these early forms had a proboscis. Indeed, the earliest haustellate (proboscis-bearing) fossils do not appear until the Late Cretaceous. Central to early discussions on the evolution of the order were the Micropterigidae, which possess a combination of lepidopteran, trichopteran, mecopteran, and specialized features. According to the relative significance with which earlier authors viewed these characters, the family was placed in the Lepidoptera, Trichoptera, or in its own order Zeugloptera. More recent analyses (most notably that of Kristensen, 1984) have confirmed that the Micropterigidae are archaic Lepidoptera, though sufficiently distinct from other members of the order as to warrant separate subordinal status. The Lepidoptera appear to have remained a small group until the Cretaceous, when they underwent a remarkable radiation in conjunction with the evolution of flowering plants. Many fossils, frequently assignable to extant families and including scales, eggs, and larval fragments, are known from this period.

Historically, a variety of systems have been suggested for classifying Lepidoptera, for example, division of the group on the basis of butterflies (Rhopalocera) and moths (Heterocera), size (Macrolepidoptera and Microlepidoptera), and wing-coupling mechanism (Jugatae and Frenatae), all of which were unnatural. The system adopted here is that of Nielsen and Common (1991), which is slightly modified from the versions presented by Kristensen (1984) and Nielsen (1989). Four suborders are recognized: Zeugloptera, Aglossata, Heterobathmiina, and Glossata, the latter (proboscis-bearing forms) including more than 99% of all Lepidoptera and being divided into several infraorders. A proposed phylogeny of the order is shown in Figure 9.25.

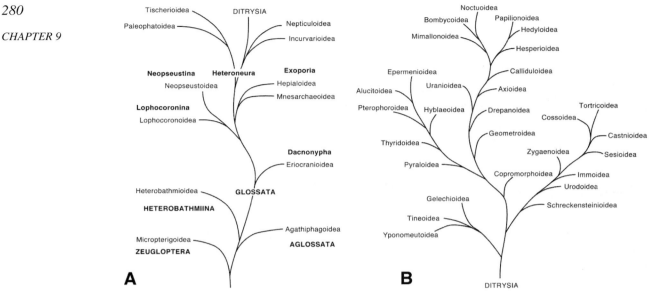

FIGURE 9.25. (A) A suggested phylogeny of Lepidoptera. (B) Superfamily relationships within the Ditrysia.

Suborder Zeugloptera

The very small, diurnal moths of the suborder Zeugloptera have the following features: functional dentate mandibles, maxilla with lacinia and free galeae, hypopharynx modified for grinding pollen grains; homoneurous wing venation, jugate wing coupling; larvae without a spinneret; and pupae decticous and exarate.

The suborder contains a single superfamily, Micropterigoidea, and one widely distributed family (MICROPTERIGIDAE) of about 100 species. Adults are metallic-colored moths with a body length of a few millimeters. They feed on pollen and fern spores. Larvae of some species feed on liverworts; others, found in rotting wood and soil, appear to be detritus feeders.

Suborder Aglossata

This suborder was erected to include two species of *Agathiphaga* (AGATHIPHAGIDAE, Agathiphagoidea) from northeastern Australia and the southwestern Pacific. These nocturnal moths, which resemble caddisflies, have large mandibles, though it is not known whether they feed. Their larvae are miners in the seeds of kauri pines (*Agathis*).

Suborder Heterobathmiina

Discovered only in the late 1970s, the handful of *Heterobathmia* species that comprise this suborder were originally included as a subfamily of Micropterigidae. Subsequent discovery of the larval stages led to the erection of a new family (HETEROBATHMIIDAE), superfamily (Heterobathmioidea), and suborder for these small, diurnal moths from temperate South America. The adults eat pollen; their larvae are leaf miners in *Nothofagus* (southern beech).

Among the features that characterize members of this suborder are: adults (after eclosion) with non-functional (often reduced) mandibles, maxilla without a sclerotized lacinia, galeae forming a proboscis; larvae with a spinneret. The suborder is split into four very small infraorders, Dacnonypha, Neopseustina, Lophocoronina, and Exoporia, and one extremely large infraorder, Heteroneura, included in which is the series Ditrysia to which about 98% of Lepidoptera belong. In Dacnonypha the proboscis is non-muscular; in the remaining groups (except Lophocoronina where proboscis muscles are absent), sometimes known as Myoglossata, the proboscis has an intrinsic musculature.

Infraorder Dacnonypha

A single superfamily, Eriocranioidea, containing three very small families, is included in this group. The ERIOCRANIIDAE (20 holarctic species) are diurnal, iridescent moths with a short proboscis. They appear to feed on sap exudates. Larvae are leaf miners in chestnut, oak, birch, and hazel. Pupation takes place in a cocoon buried in soil.

Infraorder Neopseustina

The approximately one dozen species in this infraorder are included in the single superfamily Neopseustoidea and family NEOPSEUSTIDAE. These mostly nocturnal moths, found in Southeast Asia and South America, have a short proboscis, though their feeding habits are unknown. The larvae have yet to be discovered.

Infraorder Lophocoronina

This group contains six species of *Lophocorona,* endemic to southern Australia, in the family LOPHOCORONIDAE (superfamily Lophocoronoidea). These very small, nocturnal caddisflylike moths occur in dry eucalypt forest or scrub. Their larvae are unknown but the piercing ovipositor of the female suggests that they may be borers or leaf miners.

Infraorder Exoporia

Included in this group are two superfamilies, Mnesarchaeoidea (with one family MNE-SARCHAEIDAE, endemic to New Zealand) and Hepialoidea (almost all of the 500 species of which belong to the family HEPIALIDAE). Mnesarchaeids are small, diurnal moths with a well-developed proboscis, though they are not apparently associated with flowers. The ground-dwelling larvae were discovered only in the late 1970s. Hepialidae (swift or ghost moths) are small to large, diurnal, crepuscular, or nocturnal moths whose proboscis is short or absent. The family is cosmopolitan, though about one-quarter of the species occur in Australia. Larvae live in vertical tunnels excavated either in wood, feeding on regrowth bark at the entrance, or in soil where they feed on roots or emerge to eat low-growing foliage or leaf litter. Some are important grassland pests.

Infraorder Heteroneura

This group includes the so-called "Monotrysia" (comprising the first four superfamilies outlined below) and the Ditrysia, subdivided into about 30 superfamilies. The term

Monotrysia was earlier used to define a suborder of Lepidoptera whose females are characterized by their single genital opening. It is now realized that this group is paraphyletic.

Superfamily Nepticuloidea

Also called Stigmelloidea, this superfamily includes the NEPTICULIDAE (STIGMELLIDAE) (400 species worldwide) and the OPOSTEGIDAE (50 species, mostly in tropical Asia and Australia). In the Nepticulidae are the smallest Lepidoptera. Larvae of both families are apodous miners in leaves, stems, bark, and seeds, or rarely gall formers.

Superfamily Incurvarioidea

Members of this group are split into as many as six, mostly quite small families. The IN-CURVARIIDAE (300 species worldwide) are small to minute, mostly diurnal moths whose larvae are leaf miners, stem borers, or case makers. HELIOZELIDAE (100 species) are small to very small, sometimes metallic-colored moths. Larvae mine in leaves and petioles of shrubs and trees, occasionally reaching pest proportions. The family has a cosmopolitan distribution.

Superfamily Tischerioidea

All members of this small monogeneric (*Tischeria*) group are included in the TIS-CHERIIDAE, a mostly North American family, but extending into Europe, Asia, and Africa. Larvae of these very small or minute moths are leaf miners.

Superfamily Palaephatoidea

The approximately 60 species in this group, all in the family PALAEPHATIDAE, are split more or less evenly between South America and Australia. Some species in this group of small to very small moths are sexually dimorphic. Larvae are initially leaf miners but later live in shelters formed by tying together leaves.

Series Ditrysia

All of the remaining moths and the butterflies are included in this huge but clearly monophyletic group, whose members are characterized by the separate copulatory and egg-laying openings of the female. Other possible autapomorphies are the complex musculature of the proboscis, the presence of abdominal sex-pheromone glands, and unique spermatozoa. The proposed relationships of the almost 30 superfamilies that make up the Ditrysia are shown in Figure 9.25B.

Superfamily Tineoidea

The Tineoidea are the most primitive of the tineoid group of superfamilies. Though authors split the group into as many as 10 families, the great majority of the more than 10,000 species belong to the TINEIDAE, PSYCHIDAE, and GRACILLARIIDAE. Tineoid moths are small- to medium-sized and most have narrow wings bordered with long hairs. Larvae are concealed feeders that live in portable cases, silken tubes, or mines within the food. The Psychidae (Figure 9.26A, B), a worldwide group of 6000 species, are commonly known as bagworm moths because of the portable case composed of silk and vegetable

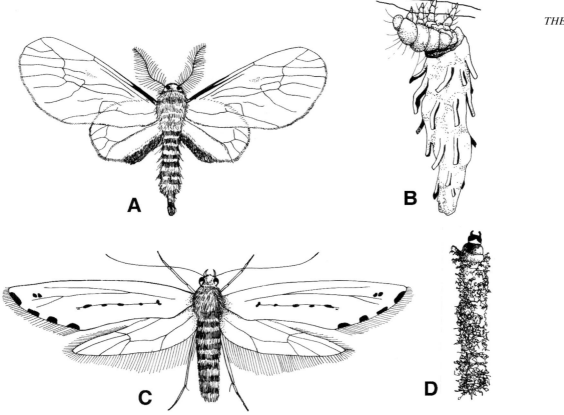

FIGURE 9.26. Tineoidea. (A) Male bagworm moth, *Thyridopteryx ephemeraeformis* (Psychidae); (B) *T. ephemeraeformis* larva with bag; (C) the case-making clothes moth, *Tinea pellionella* (Tineidae); and (D) *T. pellionella* larva in case. [A, C, D, from L. A. Swan and C. S. Papp, 1972, *The Common Insects of North America.* Copyright 1972 by L. A. Swan and C. S. Papp. Reprinted by permission of Harper & Row Publishers, Inc. B, after W. J. Holland, 1920, *The Moth Book*, Doubleday and Co., Inc.]

matter that a larva inhabits. Pupation occurs in the case. Females of many species are wingless and apodous; they do not leave the case but lay eggs directly in it. The cosmopolitan Tineidae (3000 species) includes the clothes moths *Tineola biselliella* and *Tinea pellionella* (Figure 9.26C,D), a case-bearing species, and the carpet moth *Trichophaga tapetzella.* Larvae of this family may or may not live in a portable case and feed usually on dried vegetable or animal matter. Gracillariidae (1000 species) form a cosmopolitan family of small moths recognized by their narrow, fringed wings. The larvae, which are leaf miners, undergo hypermetamorphosis. When young they are flattened and have bladelike mandibles, used to lacerate the cells on whose sap they feed. Later they metamorphose, develop normal mouthparts, and feed in the typical manner on parenchyma.

Superfamily Yponomeutoidea

The constituent families of this group were previously considered as part of the Tineoidea. There are two common families. LYONETIIDAE (300 species) are small to minute moths with narrow wings; their larvae are miners in leaves, stems, or bark, or web builders. YPONOMEUTIDAE (including the PLUTELLINAE which are often given family

FIGURE 9.27. Yponomeutoidea. The diamondback moth, *Plutella xylostella* (Yponomeutidae).

status) form a cosmopolitan group of about 1000 species of small to very small moths whose larvae are miners, web builders, or exposed feeders. Some species are widespread pests, for example, the diamondback moths, of which *Plutella xylostella* (Figure 9.27) is probably the best known for its damage to cruciferous crops.

Superfamily Gelechioidea

This large group, containing more than 13,000 described species, is subdivided into 17 families by Nielsen and Common (1991). The great majority of species are contained in four families. The OECOPHORIDAE (mallee moths) (including the Xylorictidae and Stenomatidae of other authors) contains some 7000 species, about 80% of which are Australian. Larvae of most species are external feeders on decaying organic matter, fungi, or the leaves, flowers, or seeds of angiosperms; for protection, they build portable cases of leaf fragments, spin shelters among leaves, or live in silken tubes. By feeding on leaf litter, many Australian species play a key role in energy transfer through the eucalypt ecosystem. Larvae of a few species are stem miners, or prey on soft-bodied arthropods. Though most species are small, the family includes the largest gelechioids with wingspans up to 7.5 cm. The GELECHIIDAE form a widely distributed family containing about 4000 species of small to very small moths. Their larvae show diverse habits; many are leaf folders, but others are leaf miners or borers into stems, seeds, fruit, flowers, or tubers. The family includes some cosmopolitan pests; for example, *Pectinophora gossypiella* (pink bollworm) damages cotton, *Sitotroga cerealella* (Angoumois grain moth) causes much damage to stored grains, and *Phthorimaea operculella* (potato moth) attacks potato and tobacco plants. COSMOPTE-RIGIDAE (1200 species worldwide) are miners, web builders, leaf tiers, case makers, or gall formers on stems and roots. Most of the 500 species of COLEOPHORIDAE have a holarctic distribution. Larvae of these small to very small gelechioids are almost always case bearers that feed on a variety of plants, occasionally becoming pests.

Superfamily Cossoidea

Cossoidea, which form one of the most primitive ditrysian superfamilies, show affinities with the Tortricoidea and have often been included in this group. About 90% of the 1100 species belong to the widely distributed family COSSIDAE, whose members are commonly known as goat, carpenter, or leopard moths (Figure 9.28). They are small to large, swiftly

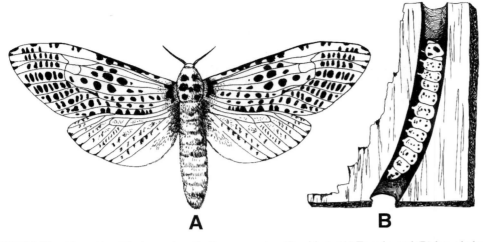

FIGURE 9.28. Cossoidea. The leopard moth, *Zeuzera pyrina* (Cossidae). (A) Female; and (B) larva in burrow. [After L. O. Howard and F. H. Chittenden, 1916, The leopard moth: A dangerous imported insect enemy of shade trees, U.S. Dept. Agr. *Farm. Bull.* **708**:12 pp.]

flying, nocturnal moths, generally grayish in color, often with dark wing spots. Larvae ("carpenter worms") are mainly wood borers, though a few live in soil and feed externally on plant roots.

Superfamily Tortricoidea

The Tortricoidea are small moths that are more common in temperate than in tropical regions. Adults are grayish or brownish, frequently mottled, moths. Larvae are concealed feeders, living in shelters formed by rolling leaves or tying several together, or as miners of fruits, seeds, bark, etc. All 4500 species are included in the family TORTRICIDAE, a group that has several major widespread pests, for example, *Cydia pomonella* (codling moth) (Figure 9.29), whose larvae mine in apples, *Grapholita molesta* (oriental fruit moth), a pest of stone fruit, and *Choristoneura* spp. (budworm moths), which are important defoliators of a variety of evergreens (see Chapter 22, Section 5.1).

Superfamily Castnioidea

The superfamily Castnioidea contains only about 160 species in one family, CASTNIIDAE, distributed in tropical America, India, Malaysia, and Australia. The adults are brightly colored and in some respects resemble nymphalid or hesperid butterflies, a feature that led some authors to mistakenly suggest that they are the ancestral group from which butterflies are derived.

Superfamily Sesioidea

About 95% of sesioids are included in two families, CHOREUTIDAE (400 species mostly in the Old World tropics) and SESIIDAE (1000 species worldwide). Choreutids are mostly diurnal moths with metallic-colored wings. Larvae are typically solitary web builders among leaves but occasionally live gregariously or are stem borers. Sesiids, commonly called clearwings, mimic wasps and bees, not only in structure but also in behavior. Their larvae

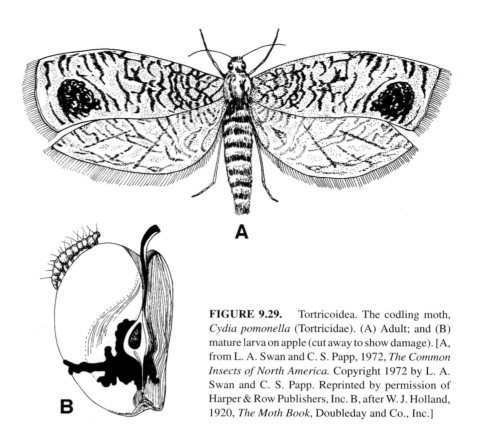

FIGURE 9.29.   Tortricoidea. The codling moth, *Cydia pomonella* (Tortricidae). (A) Adult; and (B) mature larva on apple (cut away to show damage). [A, from L. A. Swan and C. S. Papp, 1972, *The Common Insects of North America.* Copyright 1972 by L. A. Swan and C. S. Papp. Reprinted by permission of Harper & Row Publishers, Inc. B, after W. J. Holland, 1920, *The Moth Book*, Doubleday and Co., Inc.]

are typically stem borers in woody or herbaceous plants and can become pests of currants, gooseberries, raspberries, grape vines, and cucurbits.

Superfamily Zygaenoidea

Most of the nine zygaenoid families included by Nielsen and Common (1991) are very small. Zygaenoid larvae are typically stout, sluglike animals that are exposed feeders on plants; a few are parasitic on homopterans or live in ant nests. ZYGAENIDAE (burnets, foresters), a widespread group of about 800 species, are diurnal, sometimes brightly colored (and probably distasteful) moths whose larvae are leaf feeders. LIMACODIDAE (1000 species, especially tropical) have larvae that are remarkably sluglike, with the head hidden by the prothorax, long antennae, very small thoracic legs, and prolegs replaced by suckers. The larvae are often aposematically colored and have stinging hairs.

Superfamily Immoidea

This small (240 species) tropical group is placed in the single family IMMIDAE. Adults are small, stocky, nocturnal moths. Larvae are exposed feeders on foliage.

Superfamily Urodoidea

This superfamily, and its single family URODIDAE, was recently erected for about 60 species previously included in the Yponomeutoidea. Most species occur in South and Central America. Larvae are exposed feeders on broad-leaved trees.

This extremely small monofamilial (SCHRECKENSTEINIIDAE) group includes five species of *Schreckensteinia*, found in North America (one also in Europe). Larvae feed on leaves and *S. festaliella* is occasionally a minor blackberry pest.

Superfamily Copromorphoidea

Two small families make up this group, which was previously included in the Alucitoidea. The COPROMORPHIDAE (60 species) is a primarily Asian-Australian family, with a few species in Europe and North America. CARPOSINIDAE (200 species) is a more widely distributed group, though with a strong Asian and Australian component. Larvae of both families mine bark, fruit, stems, and twigs, occasionally becoming minor pests.

Superfamily Alucitoidea

This small group (130 species) is mostly placed in the family ALUCITIDAE, the many-plume moths, so-called because their wings are cleft into six or more plumelike divisions. Larvae of this widespread family are miners of buds, flowers, and shoots.

Superfamily Epermenioidea

The approximately 70 species of epermenioids are included in the widespread family EPERMENIIDAE, Like those of the previous two superfamilies, with which they had previously been included, the larvae of these very small moths, especially the early instars, are miners in leaves, seeds, fruits, and flowers, sometimes becoming external web builders when older.

Superfamily Pterophoroidea

The single family PTEROPHORIDAE that constitutes the Pterophoroidea includes about 500 species commonly known as plume moths because their wings are usually split into two or three, rarely four, plumes. The larvae are usually leaf miners when young and later become surface feeders or stem borers on low-growing, broad-leaved shrubs. Some are pests of artichokes and grape vines in North America.

Superfamily Hyblaeoidea

The 20 or so species in this group are, with one tropical American exception, from the Old World tropics. They fall into the single family HYBLAEIDAE, a group previously associated with Sesioidea, Pyraloidea, and Noctuoidea. Larvae of these stout-bodied moths are leaf tiers.

Superfamily Thyridoidea

This entirely tropical to subtropical group of about 600 species is placed in the family THYRIDIDAE. Originally included in the Pyraloidea, the thyridids were removed from this

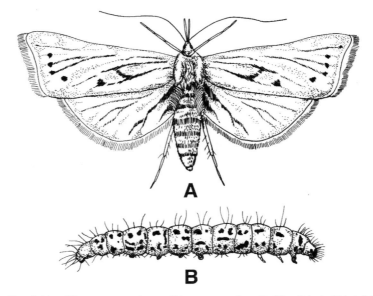

FIGURE 9.30. Pyraloidea. The sugarcane borer, *Diatraea saccharalis* (Pyralidae). (A) Adult; and (B) larva. [From L. A. Swan and C. S. Papp, 1972, *The Common Insects of North America.* Copyright 1972 by L. A. Swan and C. S. Papp. Reprinted by permission of Harper & Row Publishers, Inc.]

superfamily after the realization that they lacked abdominal tympanal organs. The larvae are miners in twigs and stems, or build leaf shelters.

Superfamily Pyraloidea

All members of this group are included in the largest lepidopteran family, the cosmopolitan PYRALIDAE (25,000 species), characterized by the presence of a tympanal organ on the first abdominal segment. Species occupy a wide range of habitats, both terrestrial and aquatic. Many are miners, using all plant parts, some build shelters among leaves or mosses, a few are case builders, and others are associated with other insects, for example, as inquilines in nests of Hymenoptera and predators of scale insects. Not surprisingly, a large number are pests. The grass moths (CRAMBINAE), whose larvae bore into stems, include *Chilo suppressalis*, the rice-stem borer, and *Diatraea saccharalis* (Figure 9.30), the sugarcane borer. The GALLERIINAE include the wax moth, *Galleria mellonella*, which lives in beehives. In the large subfamily PYRAUSTINAE, the larvae build a web among leaves and fruits. Important pests belonging to this group are *Pyrausta* (= *Ostrinia*) *nubilalis*, the European corn borer, which attacks maize, and various webworms belonging to the genera *Loxostege* and *Diaphania*. Among the PHYCITINAE are both pest and beneficial species: *Ephestia,Cadra*, and *Plodia* include species that are stored-products pests, while *Cactoblastis cactorum* is among the best-known biological control agents (see Chapter 24, Section 2.3 and Figure 24.1).

Superfamily Geometroidea

Geometroidea form an extremely large and cosmopolitan group with more than 20,000 species, about 10% of which are found in North America. Adults have abdominal tympanal organs. In most species adults and larvae are extremely well camouflaged in their

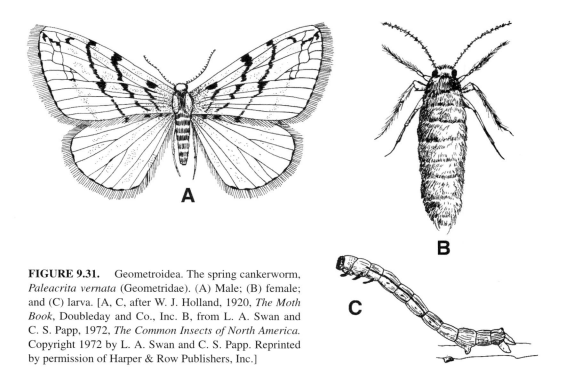

FIGURE 9.31. Geometroidea. The spring cankerworm, *Paleacrita vernata* (Geometridae). (A) Male; (B) female; and (C) larva. [A, C, after W. J. Holland, 1920, *The Moth Book*, Doubleday and Co., Inc. B, from L. A. Swan and C. S. Papp, 1972, *The Common Insects of North America.* Copyright 1972 by L. A. Swan and C. S. Papp. Reprinted by permission of Harper & Row Publishers, Inc.]

natural habitat. All geometroids belong to the family GEOMETRIDAE. Adults are generally small, slender-bodied, and with large wings that are held horizontally when the moths are resting. In some species females are apterous. Larvae, which frequently resemble twigs, have the anterior two or three pairs of prolegs reduced or absent. They are often known as geometers or inchworms because of the way they move in looping fashion. The family includes a number of important defoliators of fruit and shade trees, for example, *Paleacrita vernata*, the spring cankerworm (Figure 9.31), and *Alsophila pometaria*, the fall cankerworm.

Superfamily Drepanoidea

A group of about 1000 species with a wide distribution, the Drepanoidea are included in a single family, DREPANIDAE, by Nielsen and Common (1991). Many drepanids are known as hook-tip moths because of the sickle-shaped apical angle of the fore wings. Larvae feed on trees and shrubs, either exposed or in loosely rolled leaves.

Superfamily Uranioidea

Previously included in the Geometroidea by virtue of their tympanal organs, the 700 species of URANIIDAE have now been placed in their own superfamily following the appreciation that the organs have quite different structures in the two groups. Adults are medium to very large moths, often iridescent and with prominent "tails" on the hind wings. Larvae of this predominantly tropical family are sometimes gregarious web builders when young and feed on a variety of broad-leaved trees.

Superfamily Axioidea

The affinities of this extremely small group (10 species, in the single family AXIIDAE) of moths restricted to the Mediterranean area remain obscure. Earlier workers included them in the Geometroidea on the basis of their supposed tympanal organs. However, these organs occur on abdominal segment 7 (not segment 1 as in geometroids) and, in any event, may not have a sensory function.

Superfamily Calliduloidea

Calliduloidea form a small group of two families, CALLIDULIDAE (100 species) and PTEROTHYSANIDAE (10 species), whose members are diurnal moths with butterflylike habits (e.g., they rest with the wings folded vertically over the body). Both families are primarily oriental in distribution.

Superfamily Hedyloidea

The 35 species of the monogeneric (*Macrosoma*) family HEDYLIDAE were included originally in the geometroid group. However, Scoble (1986) has presented a strong argument for their inclusion in the "Rhopalocera" as the sister group of the Papilionoidea, though they do not possess the clubbed antennae that traditionally characterize the butterflies. The group is found in tropical America, including Cuba and Trinidad. Adults are medium-sized and mostly nocturnal, though some do fly in daylight. Larvae are leaf feeders, characteristically lying on or alongside the midrib or secondary veins.

Superfamily Hesperioidea

Hesperioidea and members of the next superfamily constitute the butterflies. This homogeneous superfamily of just over 3000 species includes the single family HESPERIIDAE (Figure 9.32), commonly known as skippers for their jerky erratic flight. Most species are diurnal, a few crepuscular. The larvae construct shelters by joining or rolling leaves, emerging

FIGURE 9.32. Hesperioidea. The common sooty wing, *Pholisora catullus* (Hesperiidae). (From L. A. Swan and C. S. Papp, 1972, *The Common Insects of North America.* Copyright 1972 by L. A. Swan and C. S. Papp. Reprinted by permission of Harper & Row Publishers, Inc.]

at night to feed. In most subfamilies larvae feed on grasses and other monocotyledonous plants; in others, dicotyledonous plants, notably legumes, are preferred.

Superfamily Papilionoidea

The approximately 14,000 species that comprise this superfamily are arranged in four families by Nielsen and Common (1991), though other authorities (e.g., Munroe, 1982) split the group up into more than a dozen families. NYMPHALIDAE (Figure 9.33A,B) form the dominant butterfly family with some 6000 described species. Members of the family are recognized by their short, functionless, hairy forelegs. Many are distasteful or mimic distasteful species. The major subfamilies are the DANAINAE (milkweed butterflies and monarchs), SATYRINAE (satyrs, wood nymphs, meadow browns, heaths, etc.), NYMPHALINAE (fritillaries, peacocks, admirals, tortoiseshells), HELICONIINAE (heliconians), and MORPHINAE (morphos). The almost 600 species of PAPILIONIDAE (swallowtails) (Figure 9.33C) are large, mainly tropical or subtropical butterflies. Adults are strikingly colored and in many species mimic Danainae. The sexes are frequently dimorphic, and among females polymorphism is common. Larvae are usually procryptically colored. They may be smooth, with a row of raised tubercles on the dorsal surface. Situated dorsally on the prothorax is an eversible osmeterium that emits a pungent odor. The family PIERIDAE (whites, sulfurs, orange tips) (1200 species) includes some very common butterflies. The family is primarily tropical, though well represented in temperate regions. Included in the genus *Pieris* are several species that are pests, especially of Cruciferae. The commonest of these is *Pieris rapae*, probably the most economically important of all butterflies. The LYCAENIDAE (blues, coppers, hairstreaks) form a widespread family of about 6000 species of small- to medium-sized butterflies. The upper surfaces of the wings are metallic blue or coppery in color, the undersides are somber and often with eyespots or streaks. Larvae are onisciform (shaped like a woodlouse). Many species are carnivorous on homopterans, and a few live in ant nests, feeding on eggs and larvae. Many of the phytophagous species are nocturnal feeders, hiding in holes by day.

Superfamily Mimallonoidea

This group of about 200 species is primarily neotropical, though a handful of species occur in North America. All species are included in the family MIMALLONIDAE, which had been considered as belonging to the Bombycoidea or Pyraloidea by earlier authors. The lack of tympanal organs rules out the latter possibility. Larvae of these nocturnal moths live in a web between two leaves when young, but older instars construct a portable case of leaves, silk, and sometimes frass.

Superfamily Bombycoidea

The outstanding feature of Bombycoidea is the reduction or absence of adult characters. Ocelli and tympanal organs are never present. Reduction or loss of the frenulum occurs, and the amplexiform method of wing coupling is developed. The proboscis is rudimentary or absent in the more specialized families. The group is split up into as many as 15 families, including the Sphingidae, which is often given its own superfamily. Relationships among the families remain unclear, that is, the Bombycoidea as presently constituted may be polyphyletic. Most families are small to extremely small, with more than 95% of the species

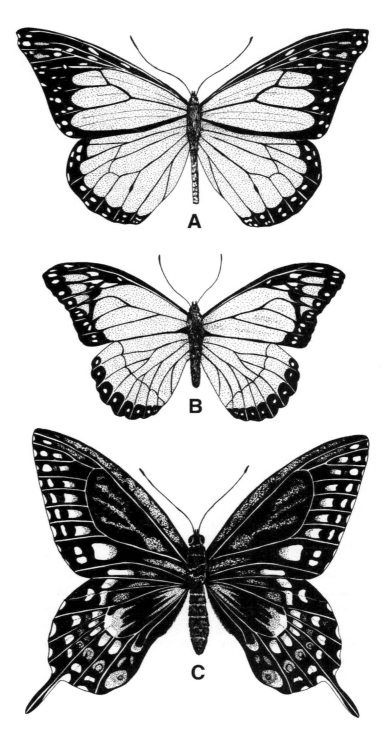

FIGURE 9.33. Papilionoidea. (A) The monarch butterfly, *Danaus plexippus* (Nymphalidae); (B) the viceroy, *Limenitis archippus* (Nymphalidae), a mimic of *D. plexippus*; and (C) the black swallowtail, *Papilio polyxenes asterius* (Papilionidae). [From L. A. Swan and C. S. Papp, 1972, *The Common Insects of North America.* Copyright 1972 by L. A. Swan and C. S. Papp. Reprinted by permission of Harper & Row Publishers, Inc.]

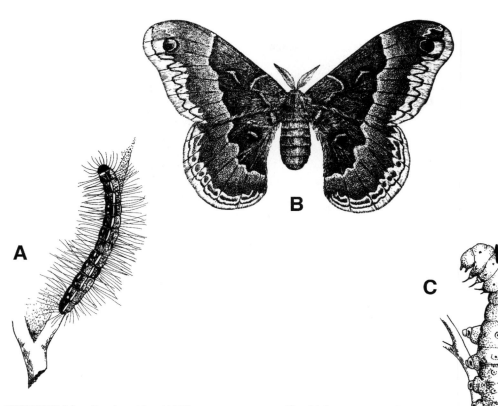

FIGURE 9.34.   Bombycoidea. (A) The eastern tent caterpillar, *Malacosoma americanum* (Lasiocampidae); (B) the promethea moth, *Callosamia promethea* (Saturniidae); and (C) *C. promethea* larva. [A, C, after W. J. Holland, 1920, *The Moth Book*, Doubleday and Co. Inc. B, from J. H. Comstock: *An Introduction to Entomology*, 9th ed. Comstock Publishing Co., Inc.]

falling into six families. The LASIOCAMPIDAE (eggars, lappet moths) is a widely distributed group of about 2200 species of medium-sized or large, cryptically colored, sexually dimorphic moths with stout bodies. Larvae are hairy and in many species gregarious, living in a communal silk nest. These "tent" caterpillars (Figure 9.34A) leave the nest to feed during the day. Some species are important defoliators. The family SATURNIIDAE (giant silkworm moths) (Figure 9.34B, C) includes some of the largest Lepidoptera with wingspans up to 25 cm *(Attacus* spp,). This group, comprising 1300 species, has a worldwide, though primarily tropical distribution. Adults are non-feeding and often have large "eyespots" on the wings used to deter would-be predators. Larvae are characterized by the scoli (spiny protuberances) on their dorsal surface and some species are major defoliators of shade trees and species of pine. Others (e.g., *Samia* and *Antheraea* spp.) produce silk of commercial value. The BOMBYCIDAE (100 species) is an Asiatic group whose larvae are covered with tufts of hair, though the family's best known representative, *Bombyx mori*, the silkworm, is naked and has a short anal horn. EUPTEROTIDAE form a family of about 400 species distributed principally in Africa and Asia. Both adults and larvae are nocturnal, the latter often living gregariously in webs and occasionally becoming pests through defoliation of shade and timber trees. The 250 species of APATELODIDAE are mainly neotropical, with a few representatives in North America, The adults are nocturnal; the larvae are exposed feeders on trees and shrubs. There are more than 1000 species of SPHINGIDAE (hawk

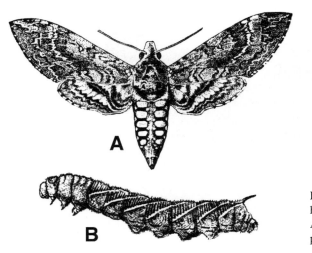

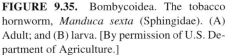

FIGURE 9.35. Bombycoidea. The tobacco hornworm, *Manduca sexta* (Sphingidae). (A) Adult; and (B) larva. [By permission of U.S. Department of Agriculture.]

moths, sphinx moths) (Figure 9.35). The family is widespread in tropical and temperate regions and contains medium-sized to large moths that generally possess a very long proboscis. Many species are capable of hovering like hummingbirds; others, with transparent wings, mimic bumblebees. Larvae are smooth and characterized by a large dorsal horn on the eighth abdominal segment, which gives them their common name of hornworms. Some species are important pests, for example, *Manduca* spp. on tomato, tobacco, and potato and *Agrius* spp. on sweet potato.

Superfamily Noctuoidea

This is easily the largest lepidopteran superfamily with close to 40,000 described species. It is a remarkably homogeneous group, and its members are characterized by their metathoracic tympanal organs (secondarily reduced in some groups); this uniformity makes the constituent families difficult to define. Using Nielsen and Common's (1991) system, four of the nine families contain the great majority of species. The family NOTODON-TIDAE (prominents) (3000 species) has often been given its own superfamily. The adults are nocturnal moths with stout bodies. Larvae are exposed feeders on trees and shrubs. When disturbed, some larvae raise the anterior and posterior ends of the body in the air and become motionless. In this attitude they vaguely resemble a twig or dead leaf. Others spray formic acid or ketones. Some species are pests through their defoliation of fruit, shade, and forest trees. About 21,000 species (including 3000 in North America) of the worldwide family NOCTUIDAE (owlet moths) have been described. These mainly nocturnal moths are procryptically colored with dark fore wings; usually they rest on tree trunks during the day. Larvae are typically phytophagous, though some prey on homopterans. They usually have four pairs of prolegs, but in some species one or more anterior pairs are reduced, and the caterpillar moves in a looping manner. Pupation in most species takes place in the ground. The family contains a large number of major pests. These include *Pseudaletia* (= *Leucania*) *unipuncta*, the armyworm, so-called because of its habit of appearing in massive numbers and marching gregariously to feed on cereal crops; *Helicoverpa zea*, the corn earworm, which feeds on maize cobs and other plants; *Trichoplusia ni*, the cabbage looper; and several other species belonging to different genera (e.g., *Agrotis*, *Prodenia*, *Feltia*) that are commonly known as cutworms (Figure 9.36) from their habit of cutting off

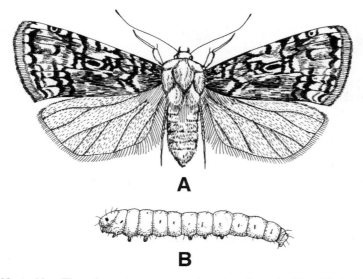

FIGURE 9.36. Noctuoidea. The pale western cutworm, *Agrotis orthogonia* (Noctuidae). (A) Adult; and (B) larva. [After C. J. Sorenson and H. F. Thornby, 1941, The pale western cutworm, *Bulletin* #297; By permission of the College of Agriculture and Agricultural Experiment Station, Utah State University.]

the plant stem level with the ground. LYMANTRIIDAE (LIPARIDAE) (tussock moths) are medium-sized moths, very similar to the noctuids, whose larvae are often densely hairy and may have osmeteria on the sixth and seventh abdominal segments. Most of the 2500 species of lymantriids occur in the Old World tropics. A few species, including *Lymantria dispar* (gypsy moth) and *Euproctis chrysorrhoea* (brown-tail moth), both introduced from Europe into North America, are important defoliators of shade and fruit trees (see Chapter 24, Section 4.3). ARCTIIDAE (tiger moths and footman moths) (11,000 species) are nocturnal or diurnal moths whose wings are often conspicuously spotted or striped. Many diurnal species are aposematic or mimic other insects including both Lepidoptera and Hymenoptera (in the latter case the arctiids may have their abdomen constricted at its anterior end). Some nocturnal species emit ultrasonic sounds that may either interfere with the echolocating ability of bats or warn the bats that the moths are distasteful. Larvae are often very hairy and known as woolly bears, or are aposematically colored and able to store secondary plant substances (notably alkaloids and cardenolides). Occasionally, species becomes pests, for example, *Hyphantria cunea* (fall webworm) which lives gregariously in large webs in apple, oak, and ash.

Literature

There is an enormous volume of literature on Lepidoptera. This deals especially with the identification and distribution of butterflies and moths, a feature that may not be surprising when the popularity of this group with insect collectors is considered. The biology of Lepidoptera is described by Common and Waterhouse (1981), Scott (1986), Douglas (1986), authors in Vane-Wright and Ackery (1989), Common (1990), Scoble (1992), Young (1997), and Leverton (2001). Discussions on the phylogeny of the order include those of Common (1975), Kristensen (1984, 1997), Nielsen (1989), Minet (1991), Heppner (1998), and Wiegmann *et al.* (2002). North American forms may be identified through the works of Dominick *et al.* (1971), Howe (1975), Scott (1986), and Arnett (2000). Mansell and Newman

(1968), Higgins and Riley (1970), Skinner (1984), and Thomas and Lewington (1991) deal with the British species. Common and Waterhouse (1981) [butterflies] and Common (1990) [moths] provide a guide to the Australian Lepidoptera.

Arnett, R. H., Jr., 2000, *American Insects: A Handbook of the Insects of America North of Mexico*, 2nd ed., CRC Press, Boca Raton, FL.

Common, I. F. B., 1975, Evolution and classification of the Lepidoptera, *Annu. Rev. Entomol.* **20**:183–203.

Common, I. F. B., 1990, *Moths of Australia*, Melbourne University Press, Carlton, Victoria.

Common, I. F. B., and Waterhouse, D. F., 1981, *Butterflies of Australia* (revised edition), Angus and Robertson, Sydney.

Douglas, M. M., 1986, *The Lives of Butterflies*, University of Michigan Press, Ann Arbor.

Dominick, R. B., Edwards, C. R., Ferguson, D. C., Franclemont, J. G., Hodges, R. W., and Munroe, E. G. (eds.), 1971, *The Moths of America North of Mexico*, Classey, Hampton, U.K.

Heppner, J. B., 1998, Classification of Lepidoptera: Part 1. Introduction, *Hol. Lepid.* **5**(Suppl. 1):1–148.

Higgins, L. G., and Riley, N. D., 1970, *A Field Guide to the Butterflies of Britain and Europe*, Collins, London.

Howe, W. H. (ed.), 1975, *The Butterflies of North America*, Doubleday, New York.

Kristensen, N. P., 1984, Studies on the morphology and systematics of primitive Lepidoptera (Insecta), *Steenstrupia* **10**:141–191.

Kristensen, N. P., 1997, Early evolution of the Lepidoptera plus Trichoptera lineage: Phylogeny and the ecological scenario, *Mem. Mus. Nat. Hist. Natur.* **173**:253–271.

Leverton, R., 2001, *Enjoying Moths*, Poyser, London.

Mansell, E., and Newman, H. L., 1968, *The Complete British Butterflies in Colour*, Rainbird, London.

Minet, J., 1991, Tentative reconstruction of the ditrysian phylogeny (Lepidoptera: Glossata), *Entomol. Scand.* **22**:69–95.

Munroe, E. G., 1982, Lepidoptera, in: *Synopsis and Classification of Living Organisms* (S.P. Parker, ed.), 2 vols., McGraw-Hill, New York.

Nielsen, E. S., 1989, Phylogeny of major lepidopteran groups, in: *The Hierarchy of Life* (B. Fernholm, K. Bremer, and H. Jörnvall, eds.), Elsevier, Amsterdam.

Nielsen, E. S., and Common, I. F. B., 1991, Lepidoptera, in: *The Insects of Australia*, 2nd ed., Vol. 2 (CSIRO, ed.), Melbourne University Press, Carlton, Victoria.

Scoble, M. J., 1986, The structure and affinities of the Hedyloidea: A new concept of the butterflies, *Bull. Br. Mus. Nat. Hist. Entomol.* **53**:251–286.

Scoble, M. J., 1992, *The Lepidoptera: Form, Function and Diversity*, Oxford University Press, London.

Scott, J. A., 1986, *The Butterflies of North America: A Natural History and Field Guide*, Stanford University Press, Stanford.

Skinner, B., 1984, *Colour Identification Guide to the Moths of the British Isles*, Viking, Harmondsworth, U.K.

Thomas, J., and Lewington, R., 1991, *The Butterflies of Britain and Ireland*, Dorling Kindersley, London.

Vane-Wright, R. I., and Ackery, P. R. (eds.), 1989, *The Biology of Butterflies*, 2nd ed., Academic Press, New York.

Whalley, P., 1986, A review of the current fossil evidence of Lepidoptera in the Mesozoic, *Biol. J. Linn. Soc.* **28**:253–271.

Wiegmann, B. M., Regier, J. C., and Mitter, C., 2002, Combined molecular and morphological evidence on the phylogeny of the earliest lepidopteran lineages, *Zool. Scripta* **31**:67–81.

Young, M., 1997, *The Natural History of Moths*, Poyser, London.

The Remaining Endopterygote Orders

1. Introduction

The six remaining endopterygote orders dealt with in this chapter are quite distinct from those that form the panorpoid complex. Of the six, the order Hymenoptera appears most isolated phylogenetically and is sometimes considered in a distinct superorder, the Hymenopteroidea, perhaps the sister group to the panorpoid complex. Except for the Strepsiptera, whose affinities remain unclear, the remaining orders are then tentatively united in a neuroptero-coleopteroid group (see Chapter 2, Section 3.2). Some authors include the Mecoptera and Raphidioptera as suborders within the order Neuroptera.

2. Megaloptera

SYNONYMS: Corydalida, Sialoidea (in order Neuroptera *sensu lato*)

COMMON NAMES: alderflies and dobsonflies

Large, soft-bodied insects; head with chewing mouthparts, elongate antennae, and large compound eyes, three ocelli present (Sialidae) or absent (Corydalidae); two pairs of identical wings with primitive venation and large number of crossveins, abdomen 10-segmented without cerci.

Larvae aquatic with chewing mouthparts and paired abdominal gills. Pupae decticous and exarate.

Representatives of this small (300 species) order are found especially in temperate regions, though their distribution is discontinuous. Some 43 species have been described from North America, about 25 from Australia, and 3 from Britain.

Structure

Adult. Adult Megaloptera are generally large insects, with members of some species having a wingspan of about 17 cm. Their prognathous head carries well-developed compound eyes, long multisegmented antennae, and chewing mouthparts [including enormously elongate mandibles in some male Corydalidae (Figure 10.1C)]. Three ocelli are present in Sialidae but absent in Corydalidae. The thoracic segments are well developed and freely movable; the pronotum is broad. All legs are similar. Four membranous wings occur, with all of the major veins and a large number of crossveins present. The wings lack

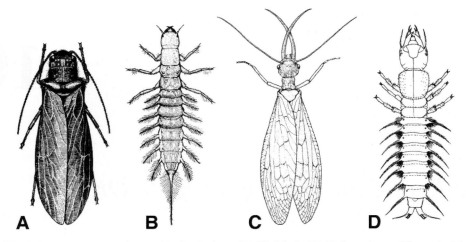

FIGURE 10.1. Megaloptera. (A) An alderfly, *Sialis mohri* (Sialidae); (B) *Sialis* sp. larva; (C) a male dobsonfly, *Corydalus cornutus* (Corydalidae); and (D) *Corydalus* sp. larva. [A, B, from H. H. Ross, 1937, Studies of Nearctic aquatic insects. I. Nearctic alderflies of the genus *Sialis* (Megaloptera, Sialidae), *Bull. Ill. Nat. Hist. Surv.* **21**(3). By permission of the Illinois Natural History Survey. D, from A. Peterson, 1951, *Larvae of Insects*, By permission of Mrs. Helen Peterson.]

a pterostigma. The wing-coupling apparatus is of the jugofrenate type. The abdomen is 10-segmented and lacks cerci.

The structure of the internal organs is poorly known. The alimentary canal has a mediodorsal food reservoir; six or eight Malpighian tubules are present; the nervous system is primitive with three thoracic and generally seven abdominal ganglia; females have a varied number of panoistic (Corydalidae) or telotrophic (Sialidae) ovarioles.

Larva and Pupa. Larvae are elongate and in some species may reach a length of 8 cm. The prognathous head is well developed and carries chewing mouthparts. The thorax bears three pairs of strong legs, the abdomen seven (Sialidae), or eight (Corydalidae) pairs of gills. Pupae are decticous, exarate, and not enclosed in a cocoon.

Life History and Habits

Adult Megaloptera are generally found in the vicinity of streams or in other cool, moist habitats where, during the day, they rest on vegetation. They probably feed very little and are generally short-lived. Reproduction appears to be entirely sexual and eggs are attached, in batches of several hundred to several thousand, to stones, vegetation, etc., usually near water.

Larvae are aquatic and predaceous. Development in most species is completed in a season, but in some large forms it may take up to 5 years. Larvae pass through 10–12 instars and when mature leave the water and burrow into soil or moss or under stones where pupation occurs. Before emergence pharate adults wriggle to the surface of the pupation medium.

Phylogeny and Classification

Fossil Megaloptera are known from as early as the Upper Permian and probably had a common ancestry with the Raphidioptera and/or Neuroptera in the Upper Carboniferous. The cladistic analysis of Aspöck *et al.* (2001) indicates that the Megaloptera and Neuroptera are sister groups. Corydalidae, including larval forms, are known from the Lower Cretaceous, and representatives of modern genera occur in Eocene Baltic amber. Extant species are placed in a single superfamily Sialoidea, which includes two families, SIALIDAE (alderflies) (Figure 10.1A,B) and CORYDALIDAE (dobsonflies) (Figure 10.1C,D). Probably less than 100 species of Sialidae occur, of which perhaps about one half have been described. The family is cosmopolitan. Corydalidae, with about 150 described species, are not found in Europe or North Africa.

Literature

The biology of Megaloptera is described by Berland and Grassé (1951). Kristensen (1991), Aspöck *et al.* (2001), and Aspöck (2002) discuss the phylogenetic position of the order. Keys to North American genera are given by Evans and Neunzig (1984), to Australian families by Theischinger (1991), and to British species by Elliott (1996) and Plant (1997).

Aspöck, U., 2002, Phylogeny of the Neuropterida (Insecta: Holometabola), *Zool. Scripta* **31**:51–55.
Aspöck, U., Plant, J. D., and Nemeschkal, H. L., 2001, Cladistic analysis of Neuroptera and their systematic position within Neuropterida (Insecta: Holometabola: Neuropterida: Neuroptera), *Syst. Entomol.* **26**:73–86.
Berland, L., and Grassé, P.-P., 1951, Super-ordre des Néuroptéroides, in: *Traité de Zoologie* (P.-P. Grassé, ed.), Vol. X, Fasc.1, Masson, Paris.
Elliott, J. M., 1996, British freshwater Megaloptera and Neuroptera: A key with ecological notes, *F. W. Biol. Assoc. Sci. Publ.* **54**:1–68.
Evans, E. D., and Neunzig, H. H., 1984, Megaloptera and aquatic Neuroptera, in: An *Introduction to the Aquatic Insects of North America*, 2nd ed. (R. W. Merritt and K. W. Cummins, eds.), Kendall/Hunt, Dubuque, IA.
Kristensen, N. P., 1991, Phylogeny of extant hexapods, in: *The Insects of Australia*, 2nd ed., Vol. I (CSIRO, ed.), Melbourne University Press, Carlton, Victoria.
Plant, C. W., 1997, A key to the adults of the British lacewings and their allies (Neuroptera, Megaloptera, Raphidioptera, and Mecoptera), *Field Stud.* **9**(1):179–269.
Theischinger, G., 1991, Megaloptera, in: *The Insects of Australia*, 2nd ed., Vol. I (CSIRO, ed.), Melbourne University Press, Carlton, Victoria.

3. Raphidioptera

SYNONYMS: Raphidiodea, Raphidioidea (in order Neuroptera *sensu lato*)   COMMON NAME: snakeflies

Large insects similar to Megaloptera but distinguished by elongate "neck"; head with chewing mouthparts, bulging compound eyes, and elongate antennae; thorax with two pairs of identical wings; abdomen 10-segmented, cerci absent, females with elongate ovipositor.

Larvae terrestrial with chewing mouthparts. Pupae decticous and exarate.

Recent members of this order, which comprises about 210 described species, are found only in the Northern Hemisphere, mostly between 35° and 50° N. About 100 species occur in Europe, mostly in the Mediterranean region but including 4 in Britain. Some 30 species are in central Asia, and 21 in southwestern North America. In the southern part of their range they are found at higher altitudes (up to 3000 m).

Structure

Adult.   Snakeflies resemble Megaloptera but may be distinguished by the elongate "neck" formed from the prothorax and narrow, posterior part of the head. The prognathous head carries a pair of bulging compound eyes, chewing mouthparts, and long, multiseg-mented antennae. Three ocelli are present (Raphidiidae) or absent (Inocelliidae). The wings are identical, and have a primitive venation of many crossveins, and a pterostigma. Females have a long hairlike ovipositor.

The internal structure is not well known. The crop has a dorsal food reservoir, and there are six Malpighian tubules. The central nervous system includes three thoracic and eight abdominal ganglia. The testes each comprise 12 follicles, the ovaries each about 40 telotrophic ovarioles.

Larva and Pupa. Larvae are elongate and have a prognathous head with chewing mouthparts. The thoracic legs are all identical. The abdomen lacks appendages. Pupae are decticous, exarate, and closely resemble adults.

Life History and Habits

Adult Raphidioptera are diurnal and may be found on flowers, foliage, tree trunks, etc. where they prey on soft-bodied arthropods, especially aphids and caterpillars; they also eat some pollen. A long courtship precedes mating, after which the female lays several hundred eggs, in batches of up to 100, in cracks in the bark of trees, etc.

Larvae occur under loose bark, especially of conifers, where they prey on other insects. Larval development usually takes 2 years or more, and there are 10 to 15 instars. Mature larvae form a cell in which to pupate. Pupae may actively move about until eclosion.

Phylogeny and Classification

The Raphidioptera probably separated from their sister group Megaloptera + Neu-roptera (Aspöck *et al.*, 2001) in the Upper Carboniferous, though fossils are not known before the Early Jurassic (supposed pre-Jurassic specimens have now been reassigned). Though Recent species occur only in the Northern Hemisphere, fossil snakeflies have been found in Early Cretaceous deposits in Brazil. Eocene Baltic amber includes species assignable to the two extant families. Recent members of the order are included in a single superfamily Raphidioidea, containing two families, RAPHIDIIDAE (155 species) (Figure 10.2) and INOCELLIIDAE (20 species). The families have similar world distribu-tions, but within each family species generally have limited distributions. North American species are restricted to three genera, *Agulla* and *Alena* (Raphidiidae) and *Negha* (Inocelli-idae) (Aspöck, 1986).

Literature

Aspöck (1975) and Aspöck (1986, 1998, 2002) describe the biology of Raphidioptera. The phylogenetic relationships of the order are discussed by Achtelig and Kristensen (1973),

FIGURE 10.2. Raphidioptera. A snakefly, *Agulla adnixa* (Raphidiidae). [From D. J. Borror, D. M. Delong, and C. A. Triplehorn, 1976, *An Introduction to the Study of Insects*, 4th ed. By permission of Brooks/Cole, a division of Thomson Learning.]

Kristensen (1991), Aspöck *et al.* (2001), and Aspöck (2002). Aspöck (1975) deals with the North American species, and Plant (1997) with the British species.

Achtelig, M., and Kristensen, N. P., 1973, A re-examination of the relationships of the Raphidioptera (Insecta), *Z. Zool. Syst. Evolutionsforsch.* **11**:268–274.

Aspöck, H., 1986, The Raphidioptera of the world: A review of present knowledge, in: *Recent Research in Neuropterology* (J. Gepp, H. Aspöck, and H. Hölzel, eds.), Graz (published privately).

Aspöck, H., 1998, Distribution and biogeography of the order Raphidioptera: Updated facts and a new hypothesis, *Acta Zool. Fenn.* **209**:33–44.

Aspöck, H., 2002, The biology of Raphidioptera: A review of present knowledge, *Acta Zool. Acad. Sci. Hung.* **48**(Suppl. 2):35–50.

Aspöck, U., 1975, The present state of knowledge on the Raphidioptera of America (Insecta, Neuropteroidea), *Pol. Pismo Entomol.* **45**:537–546.

Aspöck, U., 2002, Phylogeny of the Neuropterida (Insecta: Holometabola), *Zool. Scripta* **31**:51–55.

Aspöck, U., Plant, J. D., and Nemeschkal, H. L., 2001, Cladistic analysis of Neuroptera and their systematic position within Neuropterida (Insecta: Holometabola: Neuropterida: Neuroptera), *Syst. Entomol.* **26**:73–86.

Kristensen, N. P., 1991, Phylogeny of extant hexapods, in: *The Insects of Australia*, 2nd ed., Vol. 1 (CSIRO, ed.), Melbourne University Press, Carlton, Victoria.

Plant, C. W., 1997, A key to the adults of the British lacewings and their allies (Neuroptera, Megaloptera, Raphidioptera, and Mecoptera), *Field Stud.* **9**(1):179–269.

4. Neuroptera

SYNONYM: Planipennia   COMMON NAMES: lacewings, mantispids, antlions

Minute to large soft-bodied insects; head with chewing mouthparts, long multisegmented antennae, and well-developed compound eyes; two pairs of identical wings present, most species with primitive venation and veins bifurcated at wing margins; abdomen 10-segmented, cerci absent.

Larvae of most species terrestrial with suctorial mouthparts. Pupae decticous and exarate, enclosed in a cocoon.

The order includes about 6000 species and is represented in all world regions though is more diverse in warmer climates. About 335 species have been described from North America, 620 from Australia, and 66 from Britain.

Structure

Adult.   Adults range in size from a few millimeters to several centimeters. Members of most species are soft-bodied, weakly flying insects whose head carries a pair of well-developed compound eyes, long, multisegmented antennae, and chewing mouthparts. Ocelli are absent except in Osmylidae. In most species the legs are identical, though in Mantispidae and some Berothidae the forelegs are large and raptorial. The four wings are membranous, about equal in size, and generally have a primitive venation. In Coniopterygidae, however, the number of longitudinal and crossveins is much reduced. The abdomen is 10-segmented and lacks cerci.

Internal structure is poorly known. A dorsal crop diverticulum occurs in some groups, and there are six or eight Malpighian tubules. Three thoracic and seven abdominal ganglia are present. Females have polytrophic ovarioles.

Larva and Pupa.   Larvae have a prognathous head with suctorial mouthparts. Each mandible is sickle-shaped and grooved on the inner side. The lacinia is closely apposed to the groove, forming a tube up which soluble food can be drawn. The alimentary canal is occluded posterior to the midgut. The Malpighian tubules are secondarily attached at their tips to the rectum and secrete silk with which mature larvae spin a pupal cocoon. Pupae are decticous and exarate, with functional mandibles.

Life History and Habits

Neuroptera may be found in a wide range of ecosystems, but in most cases adults are associated with vegetation, sometimes being strongly host-plant specific (a feature related to the prey preferences of the larvae). Adults of most species prey on other arthropods, though others are omnivorous and some Chrysopidae feed on honeydew. Sex attractants (in several families) or sounds (in some Chrysopidae) produced by tapping the abdomen on the substrate are used in courtship, Neuroptera reproducing only sexually. Eggs are laid singly or in small batches, either scattered or cemented to the substrate.

Larvae show a range of feeding habits. Most are free-living predators; larvae of Ithonidae are subterranean and feed on decaying plant material; many Mantispidae are heteromorphic (Chapter 21, Section 3.3.2), the triungulins seeking the egg sacs of spiders or larvae of social Hymenoptera on which to feed; and the aquatic larvae of Sisyridae feed on freshwater sponges. There are usually three larval instars, development in most species being completed in a season. Mature larvae spin a cocoon in which to pupate. Prior to eclosion, the pharate adult chews its way out of the cocoon using the pupal mandibles.

Phylogeny and Classification

Neuroptera, like members of the previous two orders, probably had their origin in the Upper Carboniferous period, though the earliest fossils are from the Lower Permian. Representatives of the modern family Osmylidae first appeared in the Triassic, with those of other groups occurring in the Jurassic and Cretaceous. According to Aspöck *et al.* (2001), the Neuroptera are the sister group to the Megaloptera. There is still debate on some of the major subdivisions of extant Neuroptera, with respect to the families they contain and their relationships to each other. Compare, for example New's (1991) arrangement, followed here, with that proposed by Aspöck *et al.* (2001) and Aspöck (2002).

Superfamily Ithonoidea

The ITHONIDAE, which constitutes the only family in this group, contains some 15 species of mothlike insects, all but one of which are endemic to Australia. They are among the most primitive Neuroptera, and their similarities with Sialidae (Megaloptera) suggest that they left the main neuropteran stem at an early time. Some authors align the group with the Osmyloidea.

Superfamily Coniopterygoidea

The 300 or so species in this rather homogeneous group of very small Neuroptera are placed in a single family CONIOPTERYGIDAE. Their wings are covered with a white powdery exudate so that they resemble whitefly and are known as "dusty wings." Like the ithonids, they appear to be an isolated and ancient family. Coniopterygid larvae are often associated with particular types of vegetation, suggesting that they may be quite prey-specific, and they search actively for food, two features that have stimulated interest in using them as biological control agents.

Superfamily Osmyloidea

This is likely a polyphyletic group as presently constituted. About one half of its 240 species are in the family OSMYLIDAE, a largely Southern Hemisphere group. Osmylids are often found near water, their semiaquatic or aquatic larvae preying on small arthropods. SISYRIDAE (50 species) are commonly known as "spongeflies" or "spongillaflies" because their larvae are obligate predators on freshwater sponges. This worldwide family has been placed in the Mantispoidea or Hemerobioidea by some authors. DILARIDAE (40 species) form a small but widely distributed family whose biology remains largely unknown. Larvae, which are reputed to undergo up to 12 molts, live beneath the bark of recently dead trees. Members of the very small family NEVRORTHIDAE (NEURORTHIDAE) have a very disjunct distribution, being found in the western Palearctic, Asia, and Australia. Nevrorthids were previously included in the Sisyridae, mainly on the basis of the similar biology of the aquatic larvae. Aspöck *et al.*'s (2001) study indicates, however, that the similarities between the two groups are due to convergence. These authors argue that larvae of Nevrorthidae are primitively aquatic (and the family is the sister group to all other Neuroptera), whereas larvae of the Sisyridae are secondarily aquatic, the family having evolved from terrestrial ancestors.

Superfamily Mantispoidea

The two worldwide families, MANTISPIDAE (400 species) and BEROTHIDAE (70 species), that constitute this group are sometimes placed in the Osmyloidea. Adult mantispids and some berothids are characterized by their raptorial forelegs (Figure 10.3). Many larval mantispids, as noted above, are heteromorphic and feed on spiders' eggs or larvae of social Hymenoptera.

Superfamily Hemerobioidea

The great majority of species in this group are included in two families, the HEMEROBIIDAE (brown lacewings) (500 species) and CHRYSOPIDAE (green lacewings) (1950 species). Hemerobiidae have a worldwide distribution but are especially common in temperate regions. They are generally nocturnal and arboreal, often being associated with specific types of trees, where they are important predators on other small arthropods, especially homopterans. Several species have been studied for their potential as biological control agents. Among their attributes are high fecundity, short development time (up to five generations in a season), and low developmental temperature thresholds, the latter allowing them to serve as early season predators when, for example, Coccinellidae (ladybird beetles) and

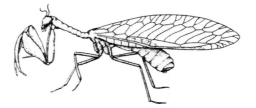

FIGURE 10.3. Mantispoidea. *Mantispa cincticornis* (Mantispidae). [From D. J. Borror, D. M. Delong. and C. A. Triplehorn, 1976, *An Introduction to the Study of Insects*, 4th ed. By permission of Brooks/Cole, a division of Thomson Learning.]

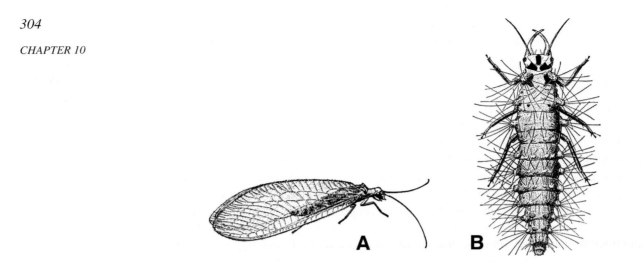

FIGURE 10.4. Hemerobioidea. A lacewing, *Chrysopa* sp. (Chrysopidae). (A) Adult and (B) larva. [By permission of the Illinois Natural History Survey.]

Chrysopidae are inactive. The Chrysopidae (Figure 10.4) are familiar insects, often found in long grass, though many species occur in shrubs and trees. Another common name for adult Chrysopidae is "stinkflies" because of their ability to produce a disagreeable odor when caught. Like hemerobiids, chrysopids have good potential as biological control agents; *Chrysopa carnea* is mass-reared in North America for augmentation of natural predator populations and in Europe for control of greenhouse pests.

Superfamily Myrmeleontoidea

Included in this group is the largest neuropteran family, the MYRMELEONTIDAE (2000 species) (Figure 10.5), a cosmopolitan group but especially common in drier regions where the soil is sandy or friable, notably the eastern Mediterranean, Asia, southern Africa, Australia, and the southern United States. Adults are typically crepuscular or nocturnal,

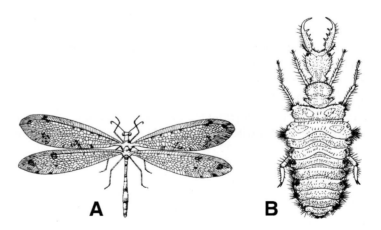

FIGURE 10.5. Myrmeleontoidea. (A) Adult antlion, *Dendroleon obsoletum* (Myrmeleontidae); and (B) *Myrmeleon* sp. (Myrmeleontidae) larva [A, from D. J. Borror, D. M. Delong, and C. A. Triplehorn, 1976, *An Introduction to the Study of Insects*, 4th ed. By permission of Brooks/Cole, a division of Thomson Learning. B, from A. Peterson, 1951, *Larvae of Insects*. By permission of Mrs. Helen Peterson.]

though there are some brightly colored diurnal species that broadly resemble damselflies. Larvae, commonly called "antlions," generally remain concealed under stones or debris, or cover themselves with particles of sand, lichen, etc., and await their prey (passing insects). Others lack such camouflage and simply hide in crevices or beneath the soil surface, actively pursuing prey that comes into the vicinity. Larvae of a few species construct pits, at the bottom of which they conceal themselves and wait for prey to fall in. The closely related ASCALAPHIDAE (350 species) is another widely distributed group whose members prefer savannah-type ecosystems. Adults of many species are broadly similar to dragonflies both in appearance and in habits; they are brightly colored, with large eyes and strong mouthparts, and catch their prey on the wing. Others are more like butterflies with broad wings and clubbed antennae. Larvae are much like antlions in their habits and form. NEMOPTERIDAE (140 species) are also associated with arid or semiarid areas of the world, except North America. Adults, which are characterized by their elongate, often filiform, hind wings, are typically nocturnal and feed on other insects and pollen. Larvae of many species are long-necked, occurring in caves, among rocks, etc., where they prey on other arthropods. Others are more like antlions, living beneath the ground surface in wait of prey.

Literature

The biology of Neuroptera is dealt with by Throne (1971), Aspöck *et al.* (1980), and New (1986, 1989). Schlüter (1986), Kristensen (1991), Aspöck *et al.* (2001), and Aspöck (2002) discuss the phylogeny of the order. Arnett (2000) gives a key to the common North American genera, New (1991) to the Australian families, and Plant (1997) to adults of the British species.

Arnett, R. H., Jr., 2000, *American Insects: A Handbook of the Insects of America North of Mexico*, 2nd ed., CRC Press, Boca Raton, FL.

Aspöck, H., Aspöck, U., and Hölzel, H., 1980, *Die Neuropteren Europas,* Vol. 1, Goecke and Evers, Krefeld.

Aspöck, U., 2002, Phylogeny of the Neuropterida (Insecta: Holometabola), *Zool. Scripta* **31**:51–55.

Aspöck, U., Plant, J. D., and Nemeschkal, H. L., 2001, Cladistic analysis of Neuroptera and their systematic position within Neuropterida (Insecta: Holometabola: Neuropterida: Neuroptera), *Syst. Entomol.* **26**:73–86.

Kristensen, N. P., 1991, Phylogeny of extant hexapods, in: *The Insects of Australia*, 2nd ed., Vol. 1 (CSIRO, ed.), Melbourne University Press, Carlton, Victoria.

New, T. R., 1986, A review of the biology of the Neuroptera Planipennia, *Neuroptera Int. Suppl.* **1**:1–57.

New, T. R., 1989, Planipennia (lacewings), in: *Handbuch der Zoologie,* Vol. IV, Insecta Lfg. **30**:1–132, de Gruyter, Berlin.

New, T. R., 1991, Neuroptera, in: *The Insects of Australia*, 2nd ed., Vol. 1 (CSIRO, ed.), Melbourne University Press, Carlton, Victoria.

Plant, C. W., 1997, A key to the adults of the British lacewings and their allies (Neuroptera, Megaloptera, Raphidioptera, and Mecoptera), *Field Stud.* **9**(1):179–269.

Schlüter, T., 1986, The fossil Planipennia—A review, in: *Recent Research in Neuropterology* (J. Gepp, H. Aspöck, and H. Hölzel, eds.), Graz (published privately).

Throne, A. L., 1971, The Neuroptera-suborder Planipennia of Wisconsin. Parts I and II, *Mich. Entomol.* **4**:65–78, 79–87.

5. Coleoptera

SYNONYMS: Eleutherata, Elytroptera   COMMON NAME: beetles

Minute to very large insects; head with chewing mouthparts and extremely varied antennae, compound eyes present or absent; prothorax large and freely movable, fore wings modified into rigid elytra which usually meet middorsally and cover most of the abdomen, hind wings membranous and usually folded beneath elytra, occasionally reduced or absent; abdomen varied.

Larvae with a distinct head and biting mouthparts, with or without thoracic legs, only rarely with prolegs. Pupae adecticous and commonly exarate, rarely obtect.

The Coleoptera are the most diverse order of insects, with more than 300,000 described species, including about 24,000 in North America, 20,000 in Australia, and 4000 in Britain. Cursory examination of their basic structure may do little to suggest why the group should be so successful, yet they have come to occupy an amazing variety of habitats, with the single exception of the sea, though many littoral species occur. The single most important structural feature contributing to the order's success has been the development of elytra (sclerotized fore wings), which protect the hind wings when these are not in use and enable the insects to occupy enclosed spaces and cryptic habitats. Other important developments associated with this mode of life include the housing of the coxal segments of the legs in cavities, flattening of the body, and an increase in the proportion of the body surface that is sclerotized, the latter being especially important in protecting the insect from predators and disease. Finally, the spiracles have become hidden in the subelytral cavity, so that water loss from the tracheal system is significantly reduced, allowing the beetles to invade arid environments. Most beetles are phyophagous, including more than 135,000 species in just two superfamilies, Chrysomeloidea and Curculionoidea. The evolution of these two groups, and their species richness, appears to have been closely tied to the evolution of the angiosperms (Farrell, 1998).

Structure

Adult. Among the living Coleoptera are some of both the largest and smallest of recent Insecta. The scarabaeid *Dynastes hercules* reaches 16 cm in length, in contrast to many Ptiliidae which are 0.5 mm or less. The head, which is primitively prognathous but sometimes hypognathous, is usually heavily sclerotized and of varied shape. Compound eyes are present or absent; occasionally they are so large as to meet both dorsally and ventrally, and in some Scarabaeoidea and Gyrinidae they are divided into upper and lower regions. Ocelli (never more than two) are absent in most species. The antennae are typically 11-segmented, but their length and form are extremely varied. The mouthparts typically are of the chewing type, but their precise structure is varied. In many species the mandibles are sexually dimorphic, being enormously enlarged and frequently branched in the male. The prothorax is the largest of the thoracic segments and is usually quite mobile. The mesothorax is small and the metathorax relatively large, except in species in which the hind wings are reduced or absent. The fore wings are modified as hard elytra that meet in the midline but are not fused except in species in which the hind wings are lacking. The metathoracic wings are membranous and typically are longer than the elytra beneath which they are folded longitudinally and transversely when not in use. The legs are usually all similar, though one or more pairs may be modified for the performance of particular functions. The number of visible abdominal segments is varied. Basically there are 10 segments, though the first is much reduced, and the last two or three are reduced and/or telescoped within the more anterior ones.

In accord with their varied diet, the alimentary canal shows a range of structure. Basically, however, it comprises a short, narrow pharynx, a widened expansion, the crop, followed by a poorly developed gizzard, a midgut that is highly varied, though usually possessing a large number of ceca, and a hindgut of varied length, Typically four or six Malpighian tubules occur, and in Cucujiformia and most Bostrychoidea a cryptonephridial arrangement exists (see Chapter 18, Sections 2.1 and 4.1). The entire range of concentration of the central nervous system is found in Coleoptera, from the primitive condition in which three thoracic and seven or eight abdominal ganglia can be distinguished to that in which all of the thoracic and abdominal ganglia are fused to form a composite structure. In males the paired testes may be simple, coiled tubular structures (Adephaga) or subdivided into a number of discrete follicles (Polyphaga). Paired vasa deferentia lead to the median ejaculatory duct, into which also open accessory glands of varied number and structure. The ovarioles of females are polytrophic (Adephaga) or acrotrophic (Polyphaga), and varied in number. A single spermatheca and its associated accessory gland enter the vagina by means of a long duct.

Larva and Pupa. The general form of beetle larvae is widely varied, though in all species the head is well developed and sclerotized, and the thoracic and abdominal segments (usually 10, rarely 9 or 8) are readily distinguishable. Thoracic legs are present or absent; abdominal prolegs are absent. Four basic larval types are found, campodeiform, eruciform, scarabaeiform, and apodous (see Chapter 21, Section 3.3.1, for further details of these types). In a few families, for example, the Meloidae, hypermetamorphosis occurs when a larva passes through all four forms during its development. Pupae are always adecticous and in most species exarate, though in Coccinellidae, most Staphylinidae, and a few other groups they are obtect.

Life History and Habits

Although they are found in large numbers in most major habitats, beetles are among the least frequently observed insects by virtue of their secretive habits. The vast majority (ca. 98%) of the world's species are terrestrial, approximately 5000 species live in fresh water, and a few species have managed to invade the littoral zone of the shore where they are submerged twice daily in seawater. Terrestrial species are most common in soil and rotting plant and animal remains, though many live on or in all parts of living plants and fungi, in dung, in nests of vertebrates and other insects, and in caves. Many species are associated with humans and live in clothes, carpets, furniture, and food. Some groups are very well adapted for survival in extremely dry situations. A wide range of adaptation for an aquatic existence is found. In some species only the larvae are aquatic, though the adults live in the vicinity of water and may be able to survive short periods of submersion. Other species are aquatic both as larvae and as adults, the latter having various devices for retaining air in a layer around their body, as their spiracles are still functional (see Chapter 15, Section 4.2). Perhaps surprisingly, there are very few truly aquatic pupae; larvae almost always leave the water to pupate or construct a submerged but air-filled cocoon. The only exception to this is in a few species of Psephenidae (Dryopoidea), where pupae respire by means of spiracular gills.

About 75% of beetle species are phytophagous in both the larval and adult stages, living in or on plants, wood, fungi, and a variety of stored products, including cereals, tobacco, and dried fruits. However, almost all Adephaga and some Polyphaga are carnivores, capturing almost anything of suitable size. Others feed on dead animal materials such as wool and leather. Species representing many families have evolved symbiotic relationships with social insects, living as inquilines in the nests of their hosts that either serve as prey or are persuaded to feed their guests. Only a few Coleoptera are parasitic. It is obvious from the above examples that many beetles will be of economic importance to humans. The majority of these are pests, though many are beneficial through their feeding on weeds or other insects such as aphids.

Most beetles reproduce sexually and are oviparous. Sex attractant and aggregation pheromones have been reported for many species, as well as complex courtship behavior. Parthenogenesis, paedogenesis, viviparity, etc., are extremely rare phenomena in this order. The number of eggs laid varies among species, as does the subsequent pattern of development. There may be one or several generations annually, or larval development may take several years. Parental care is seen in some species. Pupation usually takes place in the soil or in the food plant. In some groups a cocoon is spun.

Phylogeny and Classification

The modern consensus of opinion is that the Coleoptera were derived from some Megaloptera-like ancestor, probably in the early part of the Permian period. Evidence in support of this opinion was produced with the discovery of the Lower Permian fossil *Tshekardocoleus* (Protocoleoptera) whose elytral venation is intermediate between those of primitive existing Coleoptera and Megaloptera. The earliest undoubted beetle fossils are from the Upper Permian and belong to the suborder Archostemata, a few species of which survive to the present day. Fossil Adephaga, commonly aquatic forms, are known from the Lower Triassic, but Crowson (1960, 1981) suggested that the suborder is as old as the Archostemata. Though it appears possible that the earliest members of the suborder Polyphaga also evolved in the late Permian, fossils of this group are not known before the middle of the Triassic. Crowson (1960, 1981) speculated that a rapid radiation of the Coleoptera in the Jurassic led to the formation of three stocks: carnivorous Adephaga, wood-boring Archostemata, and Polyphaga (including the now separate Myxophaga). The Polyphaga was initially a small group of wood or fungus eaters that only later (in the Cretaceous) underwent a massive radiation correlated with the advent of the flowering plants. More than 80% of the world's species are included in this suborder. A suggested phylogeny of the Coleoptera is shown in Figure 10.6.

The classification of the order used here is based largely on that given by Crowson (1981). The suborder Archostemata is basal to the remaining suborders. Of these, the Adephaga is the sister group to the Myxophaga + Polyphaga. This arrangement is supported by the extensive analysis of Beutel and Haas (2000). However, the molecular study of Caterino *et al.* (2002), while supporting the basal nature of the Archostemata, had the Myxophaga as the sister group to the remaining two. In a recent extensive phylogenetic analysis, based on 63 hind wing characters, Kukalová-Peck and Lawrence (2004) have concluded that the Polyphaga is the sister group to the remaining Coleoptera, with the Archostemata sister to the Myxophaga and Adephaga.

Suborder Archostemata

Adult Archostemata have the following characters: wings with distal part coiled when at rest; hind wings with oblongum cell; notopleural sulcus present on prothorax; and hind coxae not immovably fixed to metasternum. Larvae have five-segmented legs with one or two claws and mandibles with a molar area. Urogomphi are absent.

Fewer than 30 species of beetles are contained in this suborder, all of them in the superfamily Cupedoidea. Three families are recognized. OMMATIDAE (six species) occur in Australia, South America, and Italy. CUPEDIDAE (21 species) are found on all continents and most large islands. Adults are short-lived and some feed on pollen; larvae are long-lived and bore in fungus-infested wood. The MICROMALTHIDAE includes only one species,

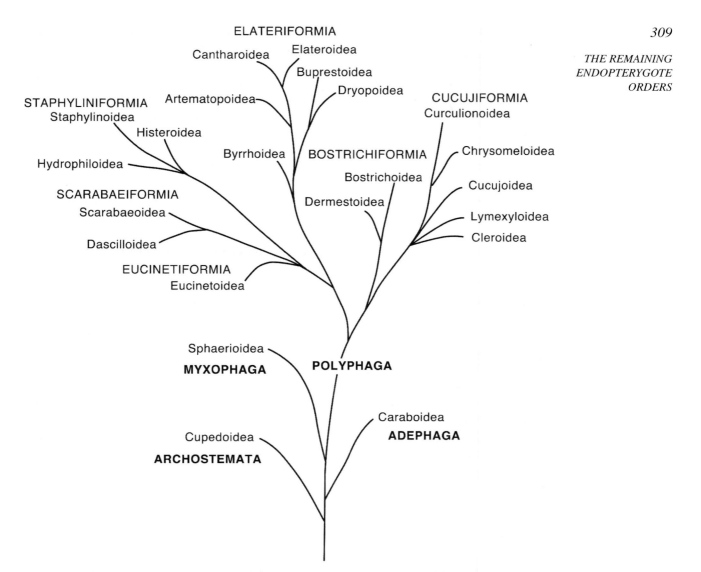

FIGURE 10.6. Proposed phylogeny of the Coleoptera. [After R. A. Crowson, 1981, *The Biology of the Coleoptera*. By permission of Academic Press, Inc., New York, and the author.]

Micromalthus debilis, which lives in rotting wood. It is native to North America but has been spread by commerce to South America, South Africa, Cuba, Hawaii, and Hong Kong.

Suborder Adephaga

The suborder Adephaga includes beetles with the following features: hind wings with an oblongum; notopleural sulcus on prothorax; hind coxae immovably attached to metasternum; testes tubular and coiled; and ovarioles polytrophic. Larvae have five-segmented legs with one or two claws. Their mandibles lack a molar area. Segmented urogomphi are present in most species.

This undoubtedly monophyletic suborder includes a single superfamily Caraboidea, most of whose members are predaceous beetles, with only a few species secondarily phytophagous, mycophagous, or algophagous. Members of the superfamily are traditionally divided into two sections, Geadephaga (terrestrial forms) and Hydradephaga (aquatic forms), an arrangement that has received support from both cladistic analysis (Beutel and Haas, 1996) and molecular studies (Shull *et al.*, 2001).

The Geadephaga contains the families RHYSODIDAE and CARABIDAE (the latter, perhaps paraphyletic, is split into many families by some coleopterists). Rhysodidae are small, black beetles that live in fungus-infected rotting wood in both the adult and larval stages. About 150 species are known, mostly from warmer areas. In contrast, about 30,000 species of Carabidae have been described from all parts of the world. About 80% of these belong to the subfamily CARABINAE (ground beetles), which, as their common name suggests, live in the soil, under stones or bark, or in logs. The elytra are frequently fused together and the wings atrophied. Both adults and larvae are carnivorous, and some species are of considerable benefit through their destruction of pest Lepidoptera. *Calosoma sycophanta* (Figure 10.7A,B) was introduced into the United States from Europe in the early 1900s for the control of the gypsy and browntail moths. Members of the subfamily CICINDELINAE (tiger beetles) (Figure 10.7C) are brightly colored, voracious predators. Larvae are typically ambush predators living in vertical tunnels in soil, rarely wood. Their flattened head forms a plug at the opening of the tunnel where they await passing prey. The subfamily comprises about 2000 species and is mainly tropical or subtropical. Among the several families of Hydradephaga are the HALIPLIDAE, DYTISCIDAE, and GYRINIDAE. The Haliplidae (Figure 10.7D) constitute a small (200 species) but widely distributed and common family of water beetles. Adults generally crawl among the green algae on which they (and the larvae) feed, though they can swim. The Dytiscidae (Figure 10.7E,F) contains about 3000 species and is especially common in the palearctic region. Both adults and larvae are predaceous. The Gyrinidae (Figure 10.7G,H), with about 700 species, includes the familiar whirligig beetles, recognized by the compound eyes split into upper and lower parts and their habit of swimming *en masse* in tight circles on the water surface. Adults feed on insects that fall onto the water surface while the larvae are bottom-dwelling predators.

Suborder Myxophaga

Myxophaga are minute beetles with clubbed antennae, a prothorax with a notopleural sulcus, and hind wings that have an oblongum and fringe of long hairs and are coiled apically. Larvae are aquatic and have mandibles with a molar area.

Crowson (1955) proposed this suborder for four families, totaling about 60 species, that were previously included in the Polyphaga. The four families are LEPICERIDAE (two species in Mexico and northern South America), TORRIDINCOLIDAE (25 species in South America, central and southern Africa, and Madagascar), MICROSPORIDAE (18 species in North and Central America, Europe, Asia, Australia, and Madagascar), and HYDROSCAPHIDAE (13 species in North America, Asia, North Africa, and Madagascar). The biology of all families is poorly known. Adults are found at stream margins, sometimes in the splash zone. They and their aquatic larvae appear to feed on algae growing on rock surfaces. A few species of Hydroscaphidae have been collected in hot springs.

Suborder Polyphaga

The beetles of the suborder Polyphaga have hind wings that lack an oblongum and are never coiled distally. A notopleural sulcus is absent from the prothorax, the hind coxae are

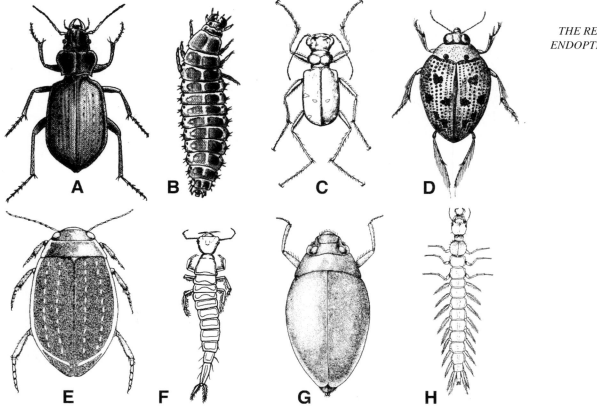

FIGURE 10.7. Caraboidea. (A) A ground beetle, *Calosoma sycophanta* (Carabidae); (B) *C. sycophanta* larva; (C) a tiger beetle, *Cicindela sexguttata* (Carabidae); (D) *Peltodytes edentulus* (Haliplidae); (E) a diving beetle, *Dytiscus verticalis* (Dytiscidae); (F) *Dytiscus* sp. larva; (G) a whirligig beetle, *Dineutes americanus* (Gyrinidae); and (H) *D. americanus* larva [A, B, from L. A. Swan and C. S. Papp, 1972, *The Common Insects of North America.* Copyright 1972 by L. A. Swan and C. S. Papp. Reprinted by permission of Harper & Row, Publishers, Inc. C, D, from R. H. Arnett, Jr., 1968, *The Beetles of the United States (A Manual for Identification).* By permission of the author. F, from E. S. Dillon and L. S. Dillon, 1972, *A Manual of Common Beetles of Eastern N. America.* By permission of Dover Publications, New York. H, from A. G. Böving and F. C. Craighead, 1930, An illustrated synopsis of the principal larval forms of the order Coleoptera, *Entomol. Am.* **XI**:1–351. Published by the Brooklyn Entomological Society. By permission of the New York Entomological Society.]

movable, the testes are not tubular and coiled, and the ovarioles are acrotrophic. The legs of larvae are either four-segmented (without tarsus) plus single claw, vestigial, or absent. The larval mandibles may or may not have a molar area.

Within the Polyphaga six major series (evolutionary lines) can be recognized (Figure 10.6). Crowson (1960) suggested that the first adaptive radiation of the Polyphaga occurred in the Triassic when three ancestral stocks had their origins: the staphyliniform, eucinetoid, and dermestoid. The former gave rise to the modern Staphyliniformia. The eucinetoid group evolved into the Eucinetiformia, Scarabaeiformia, and Elateriformia, and the dermestoid group was ancestral to the Bostrichiformia and Cucujiformia. The latter series includes more than one half of the total beetle species and can therefore be considered as the most highly evolved group within the order.

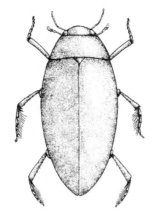

FIGURE 10.8. Hydrophiloidea. *Hydrophilus triangularis* (Hydrophilidae).

Series Staphyliniformia

Three superfamilies are included in the series Staphyliniformia: the Histeroidea, Hydrophiloidea, and Staphylinoidea. The Hydrophiloidea appear to be the most primitive group.

Superfamily Hydrophiloidea

More than 80% of the 2400 species included in the Hydrophiloidea belong to the family HYDROPHILIDAE (Figure 10.8), whose members are mainly aquatic beetles somewhat similar to Dytiscidae. However, adults are scavengers rather than predators, do not usually rest head downward at the surface of the pond, and, when swimming, move their legs alternately rather than synchronously. The larvae are predaceous. Some hydrophilids are terrestrial, but restricted to damp places, for example, in decaying plant material or dung.

Superfamily Histeroidea

Almost all of the approximately 3000 species of Histeroidea are placed in the family HISTERIDAE (Figure 10.9), both the adults and larvae of which feed on other insects. These small, usually shiny black beetles with short elytra (which leave the last one or two abdominal segments exposed) are found under bark, in rotting animal or vegetable matter including dung, or in ant nests, where they produce appeasement substances over their bodies to gain acceptance by their hosts.

FIGURE 10.9. Histeroidea. *Margarinotus immunis* (Histeridae). [From R. H. Arnett, Jr., 1968, *The Beetles of the United States (A Manual for Identification).* By permission of the author.]

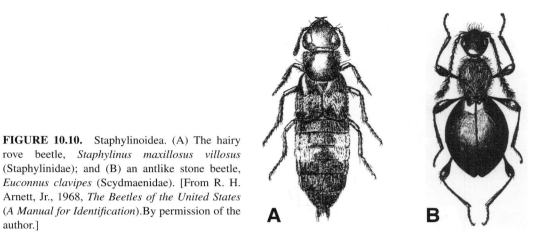

FIGURE 10.10. Staphylinoidea. (A) The hairy rove beetle, *Staphylinus maxillosus villosus* (Staphylinidae); and (B) an antlike stone beetle, *Euconnus clavipes* (Scydmaenidae). [From R. H. Arnett, Jr., 1968, *The Beetles of the United States* (*A Manual for Identification*).By permission of the author.]

Superfamily Staphylinoidea

The very large superfamily Staphylinoidea contains nearly 40,000 species, adults of which are characterized by unusually short elytra that leave about half of the abdomen visible. Some 30,000 of these species belong to the family STAPHYLINIDAE (rove beetles) (Figure 10.10A), a group of very diverse habits. Most appear to be carnivorous or saprophagous, though precise details of their feeding habits are not known. They occur in many places, for example, in decaying animal or vegetable matter, under stones or bark, on flowers, under seaweed, in moss or fungi, and in the nests of birds, mammals, and social insects. The PSELAPHIDAE (about 5000 species) closely resemble rove beetles both morphologically and in habits. They are found, moreover, in similar habitats. Adult SCYDMAENIDAE (2000 species), in contrast to those of the two families above, have fully developed elytra. These small to minute, hairy predators are somewhat antlike in appearance (Figure 10.10B), being found under stones, in humus, or sometimes in ant nests. The ANISOTOMIDAE (LEIODIDAE) form a family of about 2000 species of beetles that are found in decaying organic matter, under bark, etc. A number of species have become adapted for a cave-dwelling existence.

Series Eucinetiformia

Superfamily Eucinetoidea

About 90% of the nearly 700 species in this group belong to the SCIRTIDAE (= HELODIDAE), a worldwide family with greatest diversity in the temperate regions. Adults are usually found on vegetation near water, while larvae occupy a variety of still-water habitats and occasionally wet, rotting wood where they are filter feeders on algae, diatoms, fungi, and other organic material.

Series Scarabaeiformia

Superfamily Scarabaeoidea

Of the nearly 28,000 species in the superfamily Scarabaeoidea, about 25,000 are included in the family SCARABAEIDAE. Adults may be recognized by the lamellate terminal

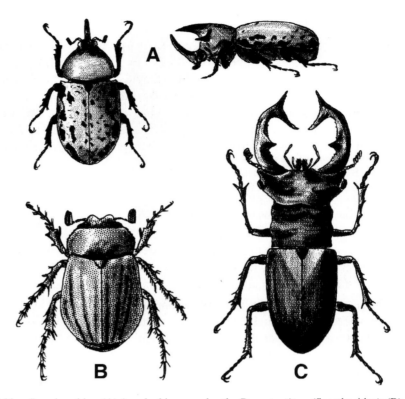

FIGURE 10.11. Scarabaeoidea. (A) A male rhinoceros beetle, *Dynastes tityus* (Scarabaeidae); (B) a May beetle, *Phyllophaga rugosa* (Scarabaeidae); and (C) the giant stag beetle, *Lucanus elaphus* (Lucanidae). [From L. A. Swan and C. S. Papp, 1972, *The Common Insects of North America.* Copyright 1972 by L. A. Swan and C. S. Papp. Reprinted by permission of Harper & Row Publishers, Inc.]

segments of their antennae. Many common beetles belong to the family, including dung beetles (SCARABAEINAE = COPRINAE) (Figure 24.3), cockchafers (May or June bugs) (MELOLONTHINAE) (Figure 10.11B), shining leaf chafers (known as Christmas beetles in Australia) (RUTELINAE), and the large and striking elephant (rhinoceros) beetles (DYNASTINAE) (Figure 10.11A). Most scarabaeids feed on decaying organic matter, especially dung, in both adult and juvenile stages, though there are many variations of this theme. Larvae of some species feed underground on plant roots and a few live in termite nests. Adults frequently feed on nectar, foliage, or fruit, or they do not feed at all. The family LUCANIDAE (stag beetles) (Figure 10.11C) includes about 1200 species in which the adults are sexually dimorphic. The mandibles of males are enormously enlarged, though the significance of this is not understood. Larvae generally feed on rotting wood. Adults are mainly nectar, occasionally foliage, feeders. The PASSALIDAE form a mainly tropical, forest-dwelling family of about 500 species. Large numbers of these beetles are often found in the same log, and it appears that the adults assist in feeding the larvae by partially chewing the rotting wood beforehand. The GEOTRUPIDAE (600 species) is a worldwide family found especially in drier regions. The beetles, which often have conspicuous horns on the head and prothorax, dig burrows that they provision with food (usually fungi, but some species use dung or decaying organic material) prior to egg laying.

Crowson (1981) included three small families in this group. DASCILLIDAE (70 species) are found in humid regions of western North America, Europe, Southeast Asia, and Australia; adults live in flowers, larvae in soil where they feed on organic material. KARUMIIDAE (10 species) occur in drier regions of North and South America, North Africa, and Asia Minor, and are associated with termites. Females are wingless. RHIPICERIDAE (50 species) live in North and South America, eastern and southern Africa, southern Europe, Asia, and Australia. Little is known of their habits though one species is a parasite of cicadas.

Series Elateriformia

Crowson (1960) suggested that the Elateriformia and Scarabaeiformia may have had a common ancestor which perhaps resembled the recent Dascillidae in general form but which was semiaquatic in habit and fed on algae. Crowson pointed out that in contrast to the Adephaga, Staphyliniformia, and Cucujiformia, in which the larval life is short and adult life long, in the Elateriformia larvae are long-lived and adults short-lived, frequently not taking any protein food. Thus, the adaptations of larvae are usually more important than those of adults. Within the series six superfamilies are recognized.

Superfamily Byrrhoidea

The 300 species of Byrrhoidea are placed in the family BYRRHIDAE (pill beetles). The family is largely restricted to temperate regions, where both the adults and larvae typically feed on mosses and liverworts. Some species camouflage themselves by mimicking mammal droppings.

Superfamily Dryopoidea

Most of the slightly more than 2100 species of Dryopoidea are subaquatic or aquatic. Adults generally live in mud or on vegetation at the margin of ponds or streams though ELMIDAE are truly aquatic, living in running water and obtaining oxygen by means of a plastron or air bubble (see Chapter 15, Section 4.2). Larvae are also aquatic and show a number of adaptations for this mode of life. Of the nine families that Crowson (1981) included in the group, the Elmidae is the largest with 700 species and occurs worldwide. PTILODACTYLIDAE (300 species) are primarily tropical; adults are found on foliage, larvae in leaf litter, rotten wood, and debris at the edges of streams. HETEROCERIDAE (300 species) form a worldwide group that live in tunnels constructed in mud or sand at the edge of streams or ponds. They feed on algae, diatoms, etc.

Superfamily Buprestoidea

The 15,000 species of Buprestoidea are placed in a single family BUPRESTIDAE (Figure 10.12). They are commonly known as "jewel beetles" because of their usually brilliant coloration. The family is particularly common in forests where the adults are found on flowers, while the larvae, which are commonly called flat-headed borers because of their

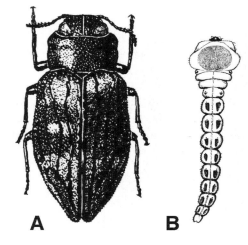

FIGURE 10.12. Buprestoidea. (A) *Chrysobothris femorata* (Buprestidae); and (B) *Chrysobothris* sp. larva. [A, from E. S. Dillon and L. S. Dillon, 1972, *A Manual of Common Beetles of Eastern N. America.* By permission of Dover Publications, New York. B, from A. D. Imms, 1957, *A General Textbook of Entomology*, 9th ed., (revised by O. W. Richards and R. G. Davies). Methuen and Co.]

expanded and flattened prothorax (Figure 10.12B), may sometimes become serious pests as they bore in wood (living or dead) or herbaceous plants where they cause galls.

Superfamily Artematopoidea

As constructed by Crowson (1981), this group includes three families, CALLIRHIP-IDAE (150 species, worldwide), ARTEMATOPIDAE (60 species, holarctic, Central and South America), and BRACHYPSECTRIDAE (3 species, western North America, southern India, and Malaysia). Other authors place these families in the Byrrhoidea or Dryopoidea, the Elateroidea, and the Cantharoidea, respectively.

Superfamily Elateroidea

About 9000 of the approximately 10,500 species of Elateroidea belong to the cosmopolitan family ELATERIDAE (Figure 10.13), the adults and larvae of which are commonly known as click beetles and wireworms, respectively. Adults are found on flowers or under bark. Larvae live in soil, litter, rotting wood, etc., and are phytophagous (often causing extensive damage to cereal crops, beans, cotton, and potatoes), saprophagous,

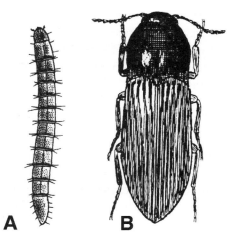

FIGURE 10.13. Elateroidea. The wheat wireworm, *Agriotes mancus* (Elateridae). (A) Larva and (B) adult. [From L. A. Swan and C. S. Papp, 1972, *The Common Insects of North America.* Copyright 1972 by L. A. Swan and C. S. Papp. Reprinted by permission of Harper & Row Publishers, Inc.]

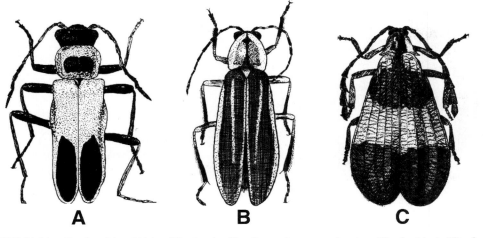

FIGURE 10.14. Cantharoidea. (A) A soldier beetle, *Chauliognathus pennsylvanicus* (Cantharidae); (B) a firefly, *Photuris pennsylvanica* (Lampyridae); and (C) a net-winged beetle, *Calopteron reticulatum* (Lycidae). [From E. S. Dillon and L. S. Dillon, 1972, *A Manual of Common Beetles of Eastern N. America.* By permission of Dover Publications, New York.]

or predaceous, obtaining their food in liquid form following extraoral digestion. The EUCNEMIDAE (1200 species) is another widespread group, especially common in the Tropics, whose members, like elaterids, make a clicking sound as they flick themselves into the air so that they are sometimes called false click beetles. Larvae occur in rotting wood and probably feed mainly on fungi or slime molds.

Superfamily Cantharoidea

Nearly 11,000 species of Cantharoidea are known, most of these being assigned to three families. The worldwide family CANTHARIDAE (Figure 10.14A) contains about 5000 species of soft-bodied, generally hairy beetles, commonly known as soldier beetles and found on flowers where they eat pollen and nectar. Larvae inhabit soil and litter where they are mostly predaceous. The approximately 2000 species of LAMPYRIDAE (fireflies) (Figure 10.14B) are renowned for their ability, in both adult and juvenile stages, to produce light. This feature has obvious sexual significance in mature insects. However, its purpose is by no means clear in the larval and pupal stages, Sivinski (1981) suggesting that it may be a form of warning coloration against predators. In adults light is produced by special cells on the abdomen; in larvae and pupae the entire body is faintly luminescent. The LYCIDAE (net-winged beetles) (3500 species) (Figure 10.14C) are brightly colored, distasteful beetles found on tree trunks or foliage. They are commonly mimicked by other insects, notably moths, flies, wasps, and other beetles. Larvae live in soil or litter, or beneath bark, where they feed on rotting vegetation and, possibly, slime molds and yeasts.

Series Bostrichiformia

As presently constituted, this is likely a paraphyletic group, and relationships among and within the higher taxa are not well understood, especially the makeup of the Dermestoidea and its relationship to the Cucujiformia.

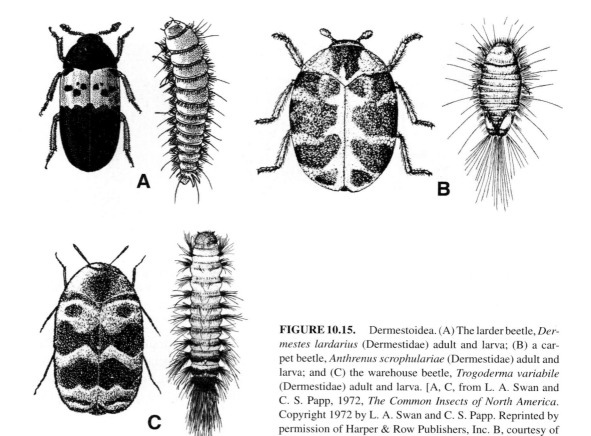

FIGURE 10.15. Dermestoidea. (A) The larder beetle, *Dermestes lardarius* (Dermestidae) adult and larva; (B) a carpet beetle, *Anthrenus scrophulariae* (Dermestidae) adult and larva; and (C) the warehouse beetle, *Trogoderma variabile* (Dermestidae) adult and larva. [A, C, from L. A. Swan and C. S. Papp, 1972, *The Common Insects of North America.* Copyright 1972 by L. A. Swan and C. S. Papp. Reprinted by permission of Harper & Row Publishers, Inc. B, courtesy of Cornell University Agricultural Experimental Station.]

Superfamily Dermestoidea

Most of the 900 or so species of Dermestoidea belong to the family DERMESTIDAE (skin and carpet beetles) (Figure 10.15), a group that contains a number of economically important, cosmopolitan species. Adults may be found on flowers, feeding on nectar and pollen, or on the larval food. Larvae are scavengers on a variety of plant and animal materials including furs, hides, wool, museum specimens, clothing, carpets, and various foods such as bacon and cheese. *Dermestes*, *Anthrenus*, *Attagenus*, and *Trogoderma* are genera that contain economically important species.

Superfamily Bostrichoidea

About one-half of the 2800 species of Bostrichoidea belong to the family ANOBIIDAE, which includes the "deathwatch" beetles, so-called because the tapping noise they make as they bore was supposed to be a sign of a future death in the house. Most species are wood borers in the larval stage, but adults leave the wood for mating purposes. A few species do not live in wood but attack stored products such as cereal products and tobacco. Economically important species include the cosmopolitan furniture beetle, *Anobium punctatum*, and the drugstore beetle or biscuit weevil, *Stegobium paniceum* (Figure 10.16A). The related family PTINIDAE (spider beetles) (Figure 10.16B), with about 450 species, includes no wood-boring forms, but only beetles associated with stored food, and dried animal

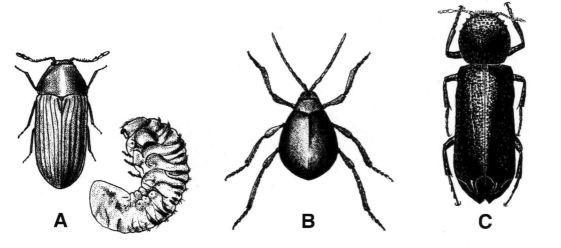

FIGURE 10.16. Bostrichoidea. (A) The drugstore beetle, *Stegobium paniceum* (Anobiidae) adult and larva; (B) the humpbacked spider beetle, *Gibbium psylloides* (Ptinidae); and (C) the apple twig borer, *Amphicerus hamatus* (Bostrichidae). [A, B, from L.A. Swan and C. S. Papp, 1972, *The Common Insects of North America.* Copyright 1972 by L. A. Swan and C. S. Papp. Reprinted by permission of Harper & Row Publishers, Inc. C, from E. S. Dillon and L. S. Dillon, 1972, *A Manual of Common Beetles of Eastern N. America.* By permission of Dover Publications, New York.]

or vegetable matter: The BOSTRICHIDAE (about 450 species) (Figure 10.16C) mainly bore in felled timber or dried wood. Adults resemble scolytids but may be distinguished by their strongly deflexed head and rasplike pronotum. Species of *Lyctus* are commonly known as powder-post beetles, as they usually reduce the dry wood in which they burrow to a powder. *Rhizopertha dominica* (the lesser grain borer) is a major pest of stored grain.

Series Cucujiformia

Within the immense but monophyletic group Cucujiformia are five superfamilies. Three of these, Cleroidea, Lymexyloidea, and Cucujoidea, are considered to be "primitive" or "lower" groups when compared with the other two, the Chrysomeloidea and Curculionoidea.

Superfamily Cleroidea

This superfamily contains about 10,000 species, more than 4000 of which belong to the mainly tropical family CLERIDAE (Figure 10.17B). Checkered beetles, as clerids are commonly known because of their striking color patterns, are hairy beetles mainly found on or in tree trunks where they prey on other insects, especially scolytid beetle larvae. A few species are found in stored products with a high fat content, for example, ham, cheese, and copra. MELYRIDAE (Figure 10.17A) (5000 species) are found worldwide but are especially diverse in the Mediterranean region. Adults occur on flowers and are frequently brightly colored. They feed on pollen and insects. Their predaceous larvae are found mostly in soil or litter and under bark. The family TROGOSSITIDAE, containing nearly 600 species, is a mainly tropical group whose members mainly prey on insects. Included in this family is the cadelle, *Tenebroides mauritanicus*, a cosmopolitan species

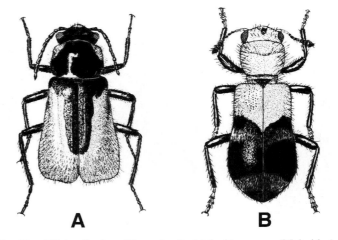

FIGURE 10.17. Cleroidea. (A) A soft-winged flower beetle, *Malachius aeneus* (Melyridae); and (B) a checkered beetle, *Enoclerus nigripes* (Cleridae). [From E. S. Dillon and L. S. Dillon, 1972, *A Manual of Common Beetles of Eastern N. America.* By permission of Dover Publications, New York.]

found in stored cereals. Although it may cause some damage to the cereals, this tends to be offset by the beneficial effect this species has by consuming other insects in the grain.

Superfamily Lymexyloidea

Lymexyloidea are a group of less than 50 species, all of which are placed in one family, LYMEXYLIDAE. Both adults and larvae are elongate insects that bore in tree stumps and logs that are beginning to decay. They feed on a fungus that grows in the tunnels, transfer of the fungus to new locations being effected by the females which carry spores in pouches near the tip of their ovipositor. Crowson (1981) included Strepsiptera (Section 6) in this superfamily, as the family Stylopidae. The weak evidence in support of this arrangement has been reviewed by Lawrence and Newton (1982).

Superfamily Cucujoidea

Though outnumbered by the Chrysomeloidea and Curculionoidea in terms of species, the Cucujoidea are the most diverse superfamily of beetles as is indicated by the large number of families (more than 50) that it contains. Most of these are relatively small, that is, contain less than 500 species, and the majority of the approximately 45,000 species in the superfamily are contained in the few very large families outlined below. Crowson (1981) subdivided the superfamily into two sections, Clavicornia and Heteromera, which other authors (e.g., Lawrence and Britton, 1991) raised to the level of superfamily, namely Cucujoidea *(sensu stricto)* and Tenebrionoidea, respectively.

Within the Clavicornia, the NITIDULIDAE (sap beetles) (Figure 10.18A) (3000 species) constitute one of the most diverse families in terms of their habits. Some species live in flowers, where they feed on pollen and nectar; others feed on fungi or ripe fruit, or are leaf miners, or prey on insects. A few species live in the nests of social Hymenoptera. Most CRYPTOPHAGIDAE (600 species) are very small, hairy beetles that feed on fungi. They are most common in moldy materials, including stored products. A few species live in the

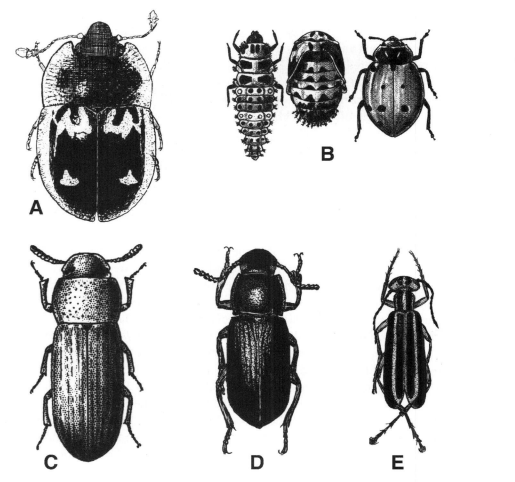

FIGURE 10.18.   Cucujoidea. (A) A sap beetle, *Prometopia sexmaculata* (Nitidulidae); (B) larva, pupa, and adult of the convergent ladybug, *Hippodamia convergens* (Coccinellidae); (C) the confused flour beetle, *Tribolium confusum* (Tenebrionidae); (D) adult of the yellow mealworm, *Tenebrio molitor* (Tenebrionidae); and (E) the striped blister beetle, *Epicauta vittata* (Meloidae). [A, D, E, from E. S. Dillon and L. S. Dillon, 1972, *A Manual of Common Beetles of Eastern N. America.* By permission of Dover Publications, New York. B, C, from L. A. Swan and C. S. Papp, *The Common Insects of North America.* Copyright 1972 by L. A. Swan and C. S. Papp. Reprinted by permission of Harper & Row Publishers, Inc.]

nests of birds, mammals, bees, and wasps. The EROTYLIDAE (2500 species) constitute a mainly South American group of beetles whose larvae feed on the fruiting bodies of fungi. The frequently brightly colored adults are usually found beneath the bark of rotting logs. The family COCCINELLIDAE (ladybugs, ladybird beetles) (Figure 10.18B), comprising about 4500 species, is a very important group of mainly brightly colored, spotted, or striped beetles. The majority of species are carnivorous as both adults and larvae on homopterans, other soft-bodied insects, and mites. Some species have been used successfully in biological control programs, the best example being the introduction of the Australian *Rodolia cardinalis* to control the cottony-cushion scale, *Icerya purchasi*, in the citrus fruit orchards of California (see Chapter 24, Section 2.3). Other coccinellids (e.g., *Epilachna* spp.) are phytophagous, however, and may cause much damage to solanaceous or bean crops. The ENDOMYCHIDAE (1100 species) are brightly colored, mainly tropical beetles found under bark or in rotting wood where they feed on fungi. CUCUJIDAE (1200 species) occur world-wide and most are found under bark or leaf litter feeding on decaying material and fungi. Others are predaceous and some (e.g., *Oryzaephilus surinamensis*, the saw-toothed grain beetle) are pests of stored products. The 900 species of LANGURIIDAE constitute a mainly tropical group of brightly colored beetles whose larvae are mostly stem borers. Others are saprophagous and a few are pests of stored products. PHALACRIDAE (600 species) are common in most parts of the world, mostly associated with flowers or vegetation. Most larvae feed on fungi. The CERYLONIDAE (650 species) are predominantly tropical in distribution where they are found in leaf litter, rotting wood, etc., and probably feed on fungi. A few species occur in bird droppings and nests of ants and mammals. Most of the 500 species of LATHRIDIIDAE, a widely distributed family, are spore feeders, though a few species have become cosmopolitan pests of stored products.

About two-thirds of the 27,000 species of Heteromera fall into the family TENEBRI-ONIDAE (darkling beetles) (Figure 10.18C,D), making it one of the largest beetle families. Adults, though generally dark in color, show a remarkable divergence of form. In contrast, larvae are extremely uniform and generally resemble wireworms. Most species are ground-dwelling insects found beneath stones or logs, but they are also found in rotting wood, fungi, nests of birds and social insects, and stored products where they may occur in enormous numbers and do considerable damage. It is to the latter group that the familiar flour beetles (*Tribolium* spp.) and mealworms (larvae of *Tenebrio* spp.) belong. Members of some species are capable of withstanding considerable periods of desiccation, and are able to absorb moisture from the air should the need arise. The COLYDIIDAE, with more than 1400 species, form a mainly tropical family of beetles that are generally asociated with rotting wood, where it is thought that they prey on the larvae of wood-boring beetles. A few species occur in ant or bee nests. The OEDEMERIDAE (1000 species) form a widely distributed family of usually brightly colored beetles that eat pollen, nectar, and sometimes insects. Many species are aposematic mimics of Lycidae and contain cantharidin (see below). Larvae feed on rotten wood and the cosmopolitan species, *Nacerdes melanura*, damages wharves, ship timbers, etc. MORDELLIDAE (1200 species) are worldwide, the adults being found in flowers, their larvae in rotten wood, stems of herbaceous plants, rarely in fungi. Adult ADERIDAE (1100 species) normally occur on foliage, while the larvae of this widespread but uncommon family live in rotten wood and leaf litter, and under bark. The MELOIDAE (blister beetles) (Figure 10.18E) form a well-known group of about 3000 species of frequently strikingly colored beetles. The family is of particular interest for the general occurrence in its larvae of heteromorphosis (Chapter 21, Section 3.3.2) and the production in the adults of cantharidin, a substance extracted from the elytra and used in certain drugs (and which causes blisters when applied to the skin). The life history is somewhat varied, but generally the early instars feed on grasshopper eggs or bee larvae and their food stores; a non-feeding, overwintering stage, the pseudopupa or coarctate larva, then follows, after which develops either another active feeding stage or a true pupal stage. Adults are phytophagous and may do much damage to solanaceous and leguminous crops. A smaller but closely related family is the RHIPIPHORIDAE (250 species), a unique family of beetles in which the larvae are at least temporarily endoparasitic, in addition to being heteromorphic. Bees, wasps, cockroaches, and, possibly, some beetles are parasitized. Adults, which are short-lived and found in flowers, usually have both elytra and wings. In the subfamily RHIPIDIINAE, however, only males have wings; females are apterous and somewhat larviform. The family thus has many parallels with the Strepsiptera and must enter into any discussion of the evolution of this group (see Section 6).

Though almost as large a group as the Cucujoidea, the superfamily Chrysomeloidea is much more uniform, being divided by Crowson (1981) into five families, two of which are very large, and two small. Both adults and larvae are phytophagous or xylophagous. The CERAMBYCIDAE (35,000 species) are commonly known as long-horned beetles (Figure 10.19A). The family, which is mainly tropical but has representatives throughout the world, contains some of the largest insects, with adults of some species more than 7 cm in length (excluding the antennae). Adults, which are brightly colored, cryptically colored, or mimic other insects, feed on flowers, foliage, or bark. Larvae generally bore into wood, though a few are restricted to the softer roots and stems of herbaceous plants. In most species the larval stage lasts for several years. The BRUCHIDAE (1500 species) are commonly known as pea and bean weevils, though they are not weevils in the true sense. Larvae feed almost exclusively in the seeds of Leguminosae, though some species attack Palmaceae (coconuts,

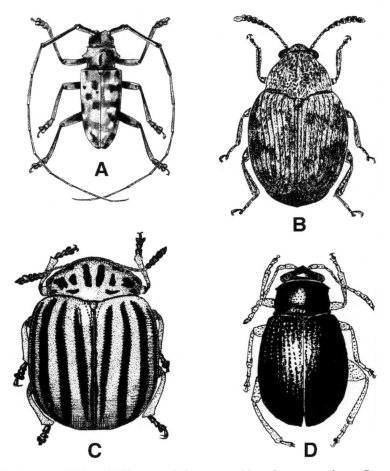

FIGURE 10.19. Chrysomeloidea. (A) The spotted pine sawyer, *Monochamus maculosus* (Cerambycidae); (B) the bean weevil, *Acanthoscelides obtectus* (Bruchidae); (C) the Colorado potato beetle, *Leptinotarsa decemlineata* (Chrysomelidae); and (D) the potato flea beetle, *Epitrix cucumeris* (Chrysomelidae). [A, from L. A. Swan and C. S. Papp, 1972, *The Common Insects of North America.* Copyright 1972 by L. A. Swan and C. S. Papp. Reprinted by permission of Harper & Row Publishers, Inc. B, from R. H. Arnett, Jr., 1968, *The Beetles of the United States* (A *Manual for Identification*). By permission of the author. C, D, from E. S. Dillon and L. S. Dillon, 1972, *A Manual of Common Beetles of Eastern N. America.* By permission of Dover Publications, New York.]

etc.). The group includes two cosmopolitan pests, the pea weevil, *Bruchus pisorum*, which feeds on growing, though not stored (i.e., dried) peas, and the bean weevil, *Acanthoscelides obtectus* (Figure 10.19B), which eats both beans and peas, either in the field or in storage. On the other hand, two Mexican species of *Acanthoscelides* have been imported into Australia for the biological control of the giant sensitive plant, *Mimosa pigra*. Some 35,000 species of CHRYSOMELIDAE (leaf beetles) have been described. Adults are frequently brightly colored beetles that feed on foliage and flowers. The larvae are varied in their habits. They may feed exposed on leaves or stems, or mine into them. Some live below ground and feed on roots. Others live in ant nests and a few are aquatic. The family includes a large number of pest species, perhaps the most infamous of which is the Colorado potato beetle, *Leptinotarsa decemlineata* (Figure 10.19C), belonging to the subfamily CHRYSOMELINAE. Most pest species, however, are in the subfamilies GALERUCINAE and ALTICINAE. For example, *Acalymma vittata,* the striped cucumber beetle, feeds in the adult stage on a variety of cucurbits and, as a larva, on the roots of various plants. This species is known to act as a vector of certain diseases that cause wilting. Flea beetles (Alticinae), so-called because of their jumping ability, attack a variety of crops, for example, canola, mustard and turnip (*Phyllotreta* spp.), and tobacco, potato, tomato, and eggplant (*Epitrix* spp.) (Figure 10.19D). Conversely, some species are important biological control agents for weeds; for example, *Longitarsus jacobaeae* (from southern Europe) for ragwort (*Senecio jacobaea*) in Australia, and *Agasicles hygrophila* (from Argentina) for alligator weed (*Alternanthera philoxeroides*) in Australia and the United States.

Superfamily Curculionoidea

Most adult Curculionoidea are easily recognized by the prolongation of the head to form a rostrum at the tip of which are located the mouthparts. It is estimated that the superfamily already includes more than 57,000 described species, about 50,000 of which belong to the family CURCULIONIDAE (weevils, snout beetles) (Figure 10.20A). Arrangement of this vast number of species into subfamilies and taxa of lower rank is perhaps the major problem for coleopteran systematists at the present time. Some authorities recognize more than 100 subfamilies, though some of these are given family status by others. Almost all weevils are phytophagous, and all parts of plants are exploited. Adults bore into seeds, fruit, and other parts of plants. Larvae usually feed within plants or externally below ground. Not surprisingly, the group includes a very large number of pests, for example, the granary and rice weevils, *Sitophilus* (= *Calandra*)*granarius* and *S. oryzae*, respectively, which attack cereal seeds, peas, and beans, and various *Anthonomus* species, of which the best known is *A. grandis*, the boll weevil, which attacks cotton. Two subfamilies that require specific mention are the PLATYPODINAE (pinhole borers) and SCOLYTINAE (bark beetles, ambrosia beetles) (Figure 10.20B), both of which are frequently given the rank of family. Members of these two groups are woodborers and, unlike the majority of weevils, adults do not have a well-developed rostrum. In both subfamilies the beetles generally live beneath the bark of the tree where they construct a characteristic pattern of tunnels (Figure 10.20C). Many species attack healthy trees, which they girdle and kill. The beetles usually feed on fungi, which they cultivate in the tunnels. In North America Dutch elm disease is transmitted by the smaller European elm bark beetle, *Scolytus multistriatus*, introduced into the United States around the 1900s, and by the native elm bark beetle, *Hylurgopinus rufipes*.

The ANTHRIBIDAE (2600 species) form a mainly tropical family, most species of which are found in rotting wood or fungi, though a few feed on seeds, etc., including the

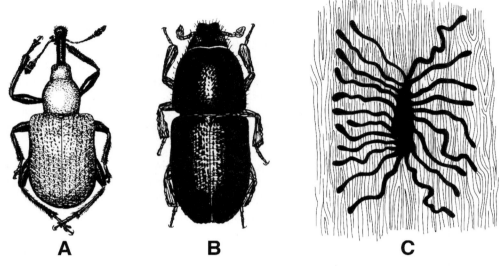

FIGURE 10.20. Curculionoidea. (A) The rose curculio, *Rhynchites bicolor* (Curculionidae); (B) the shothole borer, *Scolytus rugulosus* (Curculionidae); and (C) boring pattern of *S. rugulosus* [A, B, from E. S. Dillon and L. S. Dillon, 1972, *A Manual of Common Beetles of Eastern N. America.* By permission of Dover Publications, New York. C, from L. A. Swan and C. S. Papp, 1972, *The Common Insects of North America.* Copyright 1972 by L. A. Swan and C. S. Papp. Reprinted by permission of Harper & Row Publishers, Inc.]

coffee bean weevil, *Araecerus fasciculatus*, an important pest of coffee and cocoa beans, dried fruit, and nutmeg. The BRENTIDAE (2300 species) are mainly confined to tropical and subtropical forests where the larvae mostly bore into freshly killed trees, possibly feeding on ambrosia fungi. Others occupy the burrows of platypodines and scolytines or live in ant nests. The APIONIDAE (2200 species) constitute a widely distributed family, whose larvae mainly burrow in seeds, stems, or roots of plants, especially legumes in which they sometimes become pests.

Literature

Much general information on Coleoptera is given by Dillon and Dillon (1972), Crowson (1981), Arnett and Thomas (2001), and Arnett *et al.* (2002). The phylogeny and classification of the order are discussed by Crowson (1955, 1960, 1981), Lawrence and Newton (1982), Evans (1985), Beutel and Haas (1996, 2000), Shull *et al.* (2001), Caterino *et al.* (2002), and Kukalová-Peck and Lawrence (2004). The British species may be identified using Joy (1932) and Hodge and Jones (1995), Linnsen (1959), and the Royal Entomological Society handbooks (Crowson, 1956, and others in the series). Dillon and Dillon (1972), Headstrom (1977), Downie and Arnett (1996), Arnett (2000), Arnett and Thomas (2001), and Arnett *et al.* (2002) deal with the North American forms, while Lawrence and Britton (1991) consider the order from an Australian perspective. Lawrence (1991) provides a key to the larvae of world families and many subfamilies.

Arnett, R. H., Jr., 2000, *American Insects: A Handbook of the Insects of America North of Mexico*, 2nd ed., CRC Press, Boca Raton, FL.

Arnett, R. H., Jr., and Thomas, M. C. (eds.), 2001, *American Beetles*, Volume 1 (Archostemata, Myxophaga, Adephaga, Polyphaga: Staphyliniformia), CRC Press, Boca Raton, FL.

Arnett, R. H., Jr., Thomas, M. C., Skelley, P. E., and Frank, J. H. (eds.), 2002, *American Beetles*, Volume 2 (Polyphaga: Scarabaeoidea through Curculionoidea), CRC Press, Boca Raton, FL.

Beutel, R. G., and Haas, A., 1996, Phylogenetic analysis of larval and adult characters of Adephaga (Coleoptera) using cladistic computer programs, *Entomol. Scand.* **27**:197–205.

Beutel, R. G., and Haas, A., 2000, Phylogenetic relationships of the suborders of Coleoptera, *Cladistics* **16**:103–141.

Caterino, M. S., Shull, V. L., Hammond, P. M., and Vogler, A. P., 2002, Basal relationships of Coleoptera inferred from 18S rDNA sequences, *Zool. Scripta* **31**:41–49.

Crowson, R. A., 1955, *The Natural Classification of the Families of Coleoptera*, Nathanial Lloyd, London.

Crowson, R. A., 1956, Coleoptera. Introduction and keys to families, *R. Entomol. Soc. Handb. Ident. Br. Insects* **4**(1):59 pp.

Crowson, R. A., 1960, The phylogeny of Coleoptera, *Annu. Rev. Entomol.* **5**:111–134.

Crowson, R. A. 1981, *The Biology of the Coleoptera*, Academic Press, New York.

Dillon, E. S., and Dillon, L. S., 1972, *A Manual of Common Beetles of Eastern N. America*, Dover, New York.

Downie, N. M., and Arnett, R. H., Jr., 1996, *The Beetles of Northeastern North America*, Sandhill Crane Press, Gainsville, FL.

Evans, M. E. G., 1985, Hydradephagan comparative morphology and evolution: Some locomotor features and their possible phylogenetic implications, *Proc. Acad. Nat. Sci. Phila.* **137**(1):1–201.

Farrell, B. D., 1998, "Inordinate fondness" explained: Why are there so many beetles? *Science* **281**:555–558.

Headstrom, R., 1977, *The Beetles of America*, Barnes, Cranbury, NJ.

Hodge, P. J., and Jones, R. A., 1995, *New British Beetles—Species not in Joy's Practical Handbook*, Br. Entomol. Nat. Hist. Soc., Reading, U.K.

Joy, N. H., 1932, *A Practical Handbook of British Beetles*, 2 vols., Classey, Faringdon, U.K.

Kukalová-Peck, J., and Lawrence, J. F., 2004, Relationships among coleopteran suborders and major endoneopteran lineages: Evidence from hind wing characters, *Eur. J. Entomol.* **101**:95–144.

Lawrence, J. F., 1991, Order Coleoptera, in: *Immature Insects*, Vol. 2 (F. W. Stehr, ed.), Kendall/Hunt, Dubuque, IA.

Lawrence, J. F., and Britton, E. B., 1991, Coleoptera, in: *The Insects of Australia*, 2nd ed., Vol. 1 (CSIRO, ed.), Melbourne University Press, Carlton, Victoria.

Lawrence, J. F., and Newton, A. F., Jr., 1982, Evolution and classification of beetles, *Annu. Rev. Ecol. Syst.* **13**:261–290.

Linnsen, E. F., 1959, *Beetles of the British Isles*, 2 vols., Warne, London.

Shull, V. L., Vogler, A. P., Baker, M. D., Maddison, D. R., and Hammond, P. M., 2001, Sequence alignment of 18S ribosomal RNA and the basal relationships of adephagan beetles: Evidence for monophyly of aquatic families and the placement of Trachypachidae, *Syst. Biol.* **50**:945–969.

Sivinski, J., 1981, The nature and possible functions of luminescence in Coleoptera larvae, *Coleopt. Bull.* **35**:167–179.

6. Strepsiptera

SYNONYM: Stylopida   COMMON NAME: stylopoids

Males free living; mouthparts degenerate, antennae conspicuous and flabellate; fore wings reduced, metathorax and hind wings well developed, fore- and midlegs lacking a trochanter. Females larviform and viviparous, usually parasitoid and enclosed in puparium; with three to five secondarily segmental genital openings.

Larvae heteromorphic; pupae adecticous and exarate in males, suppressed in females except Mengenillidae.

The Strepsiptera are an order containing about 560 described species of most highly specialized insects, the males of which are free-living, the females usually parasitoid, and larvae always so in other insects. The order has been recorded from all of the major zoogeographical areas and includes more than 100 species in North America, about 30 in Australia, and 15 in Britain.

Adult Male.   Males (Figure 10.21A) are small (1.5–4.0 mm in length) and usually black or brownish. The head is very distinct with its flabellate antennae and protruding, berrylike compound eyes. The mouthparts are mandibulate but invariably reduced. The prothorax and mesothorax are small, the metathorax is very large and usually accounts for about half the body length. The fore wings are reduced and without veins. They may function like the halteres of Diptera. The hind wings are large but have few veins. The legs are weak and the first two pairs lack a trochanter. They are used for holding onto the female during copulation. The abdomen is 10-segmented, has a usually hooked aedeagus, but lacks cerci.

Adult Female.   In only one family, the Mengenillidae, are females free-living and larviform; all other female Strepsiptera are parasitoid. Only the greatly reduced head and thorax protrude from the host's body, the abdomen remaining enclosed within the last larval cuticle, the puparium, which is itself in the abdomen of the host (Figure 10.21B). Between the puparium and the true ventral surface of the female lies a flattened cavity, the brood passage. It is along this passage that insemination occurs and the first-instar larvae emerge. Secondary genital openings (median invaginations of the ventral integument) occur on abdominal segments 2–5.

In both sexes the gut is reduced and ends blindly behind the midgut; two or three Malpighian tubules occur. The tracheal system of Mengenillidae includes two thoracic

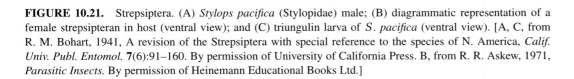

FIGURE 10.21.   Strepsiptera. (A) *Stylops pacifica* (Stylopidae) male; (B) diagrammatic representation of a female strepsipteran in host (ventral view); and (C) triungulin larva of *S. pacifica* (ventral view). [A, C, from R. M. Bohart, 1941, A revision of the Strepsiptera with special reference to the species of N. America, *Calif. Univ. Publ. Entomol.* **7**(6):91–160. By permission of University of California Press. B, from R. R. Askew, 1971, *Parasitic Insects.* By permission of Heinemann Educational Books Ltd.]

(males only) and seven abdominal spiracles; in both sexes of most other species only the first abdominal spiracle is functional. The nervous system is highly concentrated, with an anterior ganglionic mass that includes the original thoracic centers and the first two abdominal ganglia, and a posterior mass comprising the posterior abdominal ganglia. The reproductive system of males includes paired testes and tubes that lead to a common duct. In females the reproductive organs histolyse during the third larval instar, releasing the eggs into the hemocoel.

Larva and Pupa.   The first-instar larva is a very small (0.08–0.30 mm long), free-living creature known as a triungulin (Figure 10.21C). It has small antennae and reduced mouthparts, but well-formed legs and long caudal setae. Should this reach a suitable host it metamorphoses into an apodous, grublike form. Pupae are adecticous and exarate in males but suppressed in females except in Mengenillidae.

Life History and Habits

The principal hosts of Strepsiptera are Auchenorrhyncha, and Sphecoidea, Vespoidea, and Apoidea within the Hymenoptera. Other hosts include Thysanura, Orthoptera, Dictyoptera, Heteroptera, Diptera, and other aculeate Hymenoptera. Members of a given strepsipteran family usually are parasitoids of only one or a very few major insect groups; for example, Mengenillidae are known only from Thysanura. However, at the generic and specific levels the degree of host specificity is varied. It is generally high when a hymenopteran is the host, lower if the host is a homopteran. Both sexes of the host are used, and both sexes of the parasitoid may occur on the same host, in varied numbers. Curiously, in Myrmecolacidae male and female larvae use hosts from different taxa (see below). The location at which the parasitoid protrudes from the host varies but is generally specific for a particular genus.

After emergence, males, which survive for only a few hours, search actively for a virgin female. She assists by releasing an attractant pheromone. Except in Mengenillidae, copulation occurs on the host. In some species it appears that parthenogenesis is probable. The embryos develop within the female's body, which presumably receives nourishment by direct absorption through the very thin puparial cuticle. On hatching, the triungulins (usually several thousand are produced by a female) escape via the brood passage. They usually remain on the host in large numbers until an opportunity presents itself for entering a new (immature) host. The precise details of this process have rarely been observed, but it is assumed to occur in the nest in species that are parasitoids of Hymenoptera or on plants in species using other groups. The triungulins are active and, in many species, capable of jumping for distances of 2 or 3 cm using the caudal setae. They normally enter the host via the abdominal cuticle, though entry via the host's tarsi has recently been recorded (Kathirithamby, 2001). Entry is gained through a combination of enzymatic and physical activity. Fluid is secreted from the mouth and appears to partially dissolve the host's cuticle. Within the host a larva soon molts to the second stage, a more grublike form. In *Elenchus tenuicornis* third and fourth instars follow, though these do not shed the previous instar cuticles, which remain as a sheath around the larva. The final-instar larva moves to the host's integument; in Mengenillidae it leaves the host to pupate under stones, etc., but in other families (suborder Stylopidia) the larva only extrudes its anterior end outside the host before pupating (Figure 10.21B). Again, the last larval cuticle is not shed but becomes tanned and remains as a puparium within which metamorphosis occurs. Extrusion usually coincides with the final larval or pupal stage of the host. The presence of the parasitoid is not without effect on the host, which is said to be stylopized. Most noticeable are changes in the structure of the external genitalia and other external sexual characters, and atrophy of the gonads and other internal structures. Whether these changes are directly caused by the parasitoid or are merely the result of inadequate nutrition, is not known. Female hosts with parasitoids are not fertile, but males may be.

Phylogeny and Classification

When the first strepsipteran was described, in 1793, it was believed to be a hymenopteran. In 1808 it was transferred to the Diptera, and in 1813 placed in its own order Strepsiptera. Since then, strepsipterans have been considered as a separate order having affinities with the Coleoptera, Diptera, or Hymenoptera, as a superfamily within the Coleoptera-Cucujiformia, and even as a family, Stylopidae, of lymexyloid beetles (some of which are heteromorphic). Kristensen (1981) even questioned the endopterygote nature of strepsipteran development. Though it is now generally accepted that the Strepsiptera deserve ordinal status, the highly polarized debate over whether the Coleoptera or the Diptera are the sister group continues vigorously (Kathirithamby, 1989; Kinzelbach, 1990; Kukalová-Peck and Lawrence, 1993; Whiting, 1998; Huelsenbeck, 2001). Those in favor of a Strepsiptera-Coleoptera sister-group relationship cite a number of common morphological features, as well as the similar life histories of Strepsiptera and the heteromorphic beetles. Supporters of the Strepsiptera-Diptera idea argue that the supposed similarities of beetles and strepsipterans are not synapomorphies but are simply the result of convergence. They offer other morphological, as well as molecular, evidence to support their case, sometimes even suggesting that the two orders should be combined in a superorder "Halteria," reflecting the purported similarity in structure and function between dipteran halteres and the reduced fore wings of strepsipterans (reviewed by Whiting, 1998).

Fossil Strepsiptera are known mainly from Eocene and Miocene ambers and, with one exception, are assignable to extant families. Only adult males, two male puparia in a poorly preserved host, and a single supposed triungulin in the gut of an elaterid beetle have been described (Kinzelbach and Pohl, 1994).

Kinzelbach (1990, and earlier) divided the order into two suborders and nine families, eight of which contain extant species.

Suborder Mengenillidia

In this suborder males are recognized by their five-segmented tarsi that end in a pair of strong claws but no sensory spots. Females (of extant species) are free-living, larviform insects, with eyes, legs, antennae, and a single genital opening.

The suborder includes two families, MENGEIDAE, known only from the male of one species (*Mengea tertiaria*) in Baltic amber, and MENGENILLIDAE (12 species), which are parasitoids of Zygentoma and occur in the Mediterranean area, Asia, and Australia.

Suborder Stylopidia

Males have two- to four- (rarely five-)segmented tarsi, with or without claws and sensory spots. Females are parasitoid, remaining within the host with only the anterior end of the body extruded; eyes, legs, and antennae are absent.

Almost one-half of the described species of Strepsiptera belong to the family STYLOP-IDAE (275 species), a cosmopolitan group whose hosts include vespid and sphecid wasps and several families of bees. Another large and worldwide group are the HALICTOPHAGI-DAE (112 species), which are parasitoids of Auchenorrhyncha, Heteroptera, Orthoptera, Blattodea, and Diptera. CORIOXENIDAE (36 species) and ELENCHIDAE (24 species) are also widespread, being parasitoids of Heteroptera and homopterans, respectively. The 99 species of MYRMECOLACIDAE are predominantly tropical in distribution. Larval females are parasitoids of ensiferan Orthoptera and Mantidae; larval males occur in ants. The two species that comprise the family CALLIPHARIXENIDAE, from Southeast Asia, are known only as females and first-instar larvae. Their hosts are Scutelleridae (Heteroptera). BOHARTILLIDAE (one living species from a scolioid wasp in Honduras and one fossil species in Dominican amber) are known only as males.

Literature

Descriptions of the biology of Strepsiptera are given by Askew (1971) and Kathirithamby (1989). Kathirithamby (1989), Kinzelbach (1990), Kukalova-Peck and Lawrence (1993), Whiting (1998), and Huelsenbeck (2001) have discussed the evolutionary position of the group. Kathirithamby (1989) provides a key to the world's families (and some subfamilies).

Askew R. R., 1971, *Parasitic Insects*, American Elsevier, New York.

Huelsenbeck, J. P., 2001, A Bayesian perspective on the Strepsiptera problem, *Tidschr. Entomol.* **144**:165–178.

Kathirithamby, J., 1989, Review of the order Strepsiptera, *Syst. Entomol.* **14**:41–92.

Kathirithamby, J., 2001, Stand tall and they still get you in your Achilles foot pad, *Proc. R. Soc. Lond. Ser. B* **268**:2287–2289.

Kinzelbach, R., 1990, The systematic position of the Strepsiptera (Insecta), *Am. Entomol.* **36**:292–303.

Kristensen, N. P., 1981, Phylogeny of insect orders, *Annu. Rev. Entomol.* **26**:135–157.

Kinzelbach, R., and Pohl, H., 1994, The fossil Strepsiptera (Insecta: Strepsiptera), *Ann. Entomol. Soc. Am.* **87**:59–70.

Kukalova-Peck, J., and Lawrence, J. P., 1993, Evolution of the hind wing in Coleoptera, *Can. Entomol.* **125**:181–258.

Whiting, M. F., 1998, Phylogenetic position of the Strepsiptera: Review of molecular and morphological evidence, *Int. J. Insect Morphol. Embryol.* **27**:53–60.

7. Hymenoptera

SYNONYM: Vespida   COMMON NAMES: bees, wasps, ants, sawflies

Minute to medium-sized insects; head usually with well-developed compound eyes, mandibulate mouthparts (though usually adapted for sucking also); two pairs of transparent wings present in most species, fore and hind wings coupled, venation reduced; abdomen in most species markedly constricted between segments 1 and 2, with former intimately fused with metathorax; females with an ovipositor which is modified in some species for purposes in addition to egg laying.

Larvae caterpillarlike (Symphyta) or maggotlike (Apocrita). Pupae adecticous and in most species exarate, often in a cocoon.

The order Hymenoptera includes some 130,000 described species, of which about 18,000 occur in North America, 15,000 in Australia, and 6500 in the United Kingdom. In the evolution of the order, which contains some of the most advanced and highly specialized insects, emphasis has been laid not so much on structural and physiological modifications as has occurred in other orders, but on the development of complex behavior patterns. These are particularly related to provision of food for the progeny and have led ultimately to the evolution of sociality in several groups.

Structure

Adult. The hypognathous head usually is very mobile and bears very large, in some species holoptic, compound eyes. Three ocelli are usually present. The antennae contain between 9 and 70 segments and are sometimes sexually dimorphic. The mouthparts show a wide range in form from the generalized mandibulate type found in the suborder Symphyta to a highly specialized, sucking type found in the most advanced Apocrita, such as the bees (see Chapter 3, Section 3.2.2). The prothorax is small; its tergum is collarlike and fused with the large mesonotum. The pro-sternum is very small and usually can be seen only with difficulty. Two pairs of wings are generally present, with the fore wings larger than the hind pair. The venation is much reduced, and, rarely, veins are completely absent. The fore and hind wings are coupled by means of hamuli. Brachyptery or aptery occurs, for example, in ant workers and some Chalcidoidea. Hymenoptera are unique among Insecta in that a trochantellus is present on at least some legs. This is actually part of the femur, though it appears as a second segment of the trochanter. The legs are frequently specialized for particular functions, for example, digging, grasping, and carrying prey, and collection of pollen (see Chapter 3, Section 4.3.1 and Figure 3.25). In Symphyta the first abdominal segment is clearly recognizable as a part of the abdomen. In Apocrita, however, the tergum has become intimately fused with the metathorax and is distinguishable only by the presence of spiracles. In this condition it is known as the propodeum. The first abdominal sternite has disappeared entirely. In Apocrita, a marked constriction, the petiole, separates the first from the remaining abdominal segments; the latter constitute the gaster (metasoma) that normally has a posterior pair of unsegmented cerci (pygostyles), though these are reduced to pads of sensory hairs in some species and lost entirely in female Aculeata. The male terminalia are usually large and comprise the lateral parameres, the aedeagus, and a pair of ventral lobes, the volsellae. The ovipositor is well developed in females and is frequently modified for sawing, boring, piercing, or stinging, although only in the latter does it no longer participate in egg laying (Chapter 3, Section 5.2.1 and Figure 3.32).

The gut has a uniform structure throughout the order. The esophagus is narrow and long, especially when the petiole is elongate, and leads to a thin-walled crop (honey stomach) in the anterior part of the abdomen. Behind the crop is the proventriculus, which serves apparently to regulate the entry of food into the usually large stomach (ventriculus). This is followed by the ileum and rectum. The number of Malpighian tubules varies from 2, in most parasitic forms, to more than 250 in some nectar feeders. The nervous system shows various degrees of specialization; in primitive forms, three thoracic and nine abdominal ganglia occur in the ventral nerve cord, whereas in most Apocrita three thoracic and between two and six abdominal ganglia can be found. The paired testes are separate in Symphyta and a few Apocrita, but fused in other Hymenoptera. The vasa deferentia are swollen basally into vesicula seminales that lead into paired ejaculatory canals. The latter also receive the ducts of the two large accessory glands prior to forming a common tube. In females the number of polytrophic ovarioles varies from 1 to more than 100 per ovary. The two oviducts fuse to form the vagina, which also receives the duct of the median spermatheca. The vagina is swollen posteriorly to form the bursa copulatrix. Two major accessory glands occur, the venom (poison) glands and their associated sac, and the Dufour's gland. Originally, the venom glands may have produced mucus that coated and protected the eggs. In extant species they produce a variety of materials; for example, in Siricoidea (wood wasps) their secretion renders the food plant more susceptible to the fungal symbiont, and in Apocrita the secretion paralyzes or kills the prey. The secretion of Dufour's gland also has a variety of functions, including lubrication of the ovipositor, "labeling" the host (to inform conspecifics that it has already been used), and release of alarm and trail-marking pheromones (Chapter 13, Sections 4.4 and 4.5).

Larva and Pupa. Larvae of Symphyta have a distinct, well-sclerotized head, and 3 thoracic and 9 or 10 abdominal segments. The mandibulate mouthparts are well developed. If a larva is a surface feeder, as in sawflies, it usually has well-developed thoracic legs and six to eight pairs of prolegs. In boring or mining species the thoracic legs are reduced and prolegs absent. Larvae of Apocrita are apodous and resemble maggots. The head is only weakly sclerotized or much reduced and, in parasitic forms, sunk into the prothorax. Heteromorphosis is common in parasitic species. The first-instar larva is extremely varied in form, but the final instar is always maggotlike. Pupae are adecticous and in most species exarate. With the exception of certain groups (Cynipidae, Chalcidoidea, most Apoidea, and many Formicidae), a cocoon is spun in which pupation occurs.

Life History and Habits

Most adult Hymenoptera feed on nectar or honeydew and thus are found on or near flowers. A few are predaceous, while others feed on plant tissue or fungi. Through their search for nectar and pollen for feeding the larvae, the social Hymenoptera play an extremely important role in the cross-pollination of flowering plants. In addition, some parasitoid and predatory forms are important agents in the control of insects that could otherwise become pests. For these reasons the Hymenoptera are considered to be the insect order of greatest benefit to humans.

In the complexity of its members' behavior the Hymenoptera surpass all other insect orders. The development of this behavior has accompanied the evolution of parental care and, ultimately, social life in the order. Within the order the importance of the male sex shows a gradual diminution, and facultative or cyclic diploid parthenogenesis (Chapter 20, Section 8.1) becomes the rule rather than the exception in some groups of social Hymenoptera. The majority of individuals produced are females, which, however, do not mature sexually but merely serve the colony as a whole. Males are produced only for the purpose of founding new colonies; they develop from unfertilized eggs and thus are haploid.

Four broad categories of life history can be recognized in Hymenoptera, and these are paralleled by an increasing complexity of behavior patterns exhibited by females. These life histories are based on the feeding habits of the larvae. The simplest life history is found in Symphyta, whose larvae are generally foliage eaters or wood borers, and females simply deposit their eggs on a host plant. The second stage is seen in the primitive Apocrita, whose larvae are parasitoids, that is, they feed on the tissues of a host that they eventually kill. This necessitated the evolution, in females, of the ability to search out a suitable (sometimes highly specific) host on which to oviposit. In more advanced Apocrita such as the solitary wasps and bees, the beginning of parental care is seen. A female builds a special cell in which to deposit an egg. She then uses either mass provisioning, that is, the supplying of sufficient food at one time for the whole of larval development, or progressive provisioning in which food is supplied at intervals throughout larval life. Associated with this development is the evolution of the ovipositor as a sting, for paralyzing, though not killing, prey. Thus, the stage has been reached where a female comes into contact with her progeny. The fourth type of life history is found in the social Hymenoptera where a larva is fed throughout its development on food provided by the parent or another adult (which is usually sterile). The food is animal material (in wasps), or plant material such as pollen and nectar (bees), or seed, tissues, or fungi (ants). In primitive social species mass provisioning is used. At a more advanced stage progressive provisioning occurs, and there may be several egg-producing females in the colony. Gradually, division of labor evolves, and the colony then contains a single egg-laying female, the queen, and large numbers of structurally distinguishable females, the workers, which perform various duties in the colony. In some species there is a temporal separation of these duties, that is, an individual performs different duties at different ages.

Phylogeny and Classification

Hymenoptera appear to be relatively isolated from other endopterygotes, and speculations on the origin of the group have been widely divergent. In certain of their features, for example, the large number of Malpighian tubules, they resemble orthopteroid insects, and more than one author has suggested that the order evolved from Protorthoptera, implying a diphyletic origin of the endopterygotes. More frequently, on the basis of many common morphological features and habits possessed by Mecoptera and primitive Hymenoptera, it has been concluded that Hymenoptera form the sister group to the panorpoid complex. Separation of the two probably took place in the Upper Carboniferous period because the earliest fossil Hymenoptera, from the Middle Triassic, are already well developed and clearly assignable to the recent family Xyelidae. Fossils belonging to other symphytan families were present in the Upper Triassic and Jurassic periods, by which time the phytophagous Symphyta had undergone a wide radiation associated with the evolution of pteridophytes, gymnosperms, and other non-flowering vascular plants (Gauld and Bolton, 1988). A few Apocrita (Proctotrupoidea) have also been identified in Middle Jurassic deposits, but the great expansion of this suborder did not occur until the Cretaceous, in which period representatives of all extant superfamilies of Aculeata and Parasitica were present.

It is generally accepted that Apocrita evolved from Symphyta. However, most Symphyta feed on plant materials of various kinds (Jervis and Vilhelmsen, 2000), and the evolution of the primitive Parasitica from such a group has been the subject of much discussion (reviewed by Malyshev, 1968). On different occasions all the major symphytan superfamilies have been suggested as being ancestral to Apocrita. Both morphological cladistic analyses (Ronquist *et al.*, 1999; Vilhelmsen, 2001) and molecular studies (Dowton and Austin, 1994, 1999) indicate that the sister group is to be found in the Siricoidea, either the Orussidae or the Siricidae. Perhaps the former is more likely, as the larvae of modern orussids prey on the burrowing larvae of horntails and beetles (see under Siricoidea). What remains speculative, however, is how the endophagous habit of the ancestral orussid larva evolved toward facultative, then obligate zoophagy, that is, the development of the parasitoid life style. Equally fascinating is how females evolved the ability to seek out the eggs and/or larvae of their hosts. It is evident that the parasitoid habit has evolved on several separate occasions because the Parasitica is clearly a paraphyletic group (Ronquist, 1999; Ronquist *et al.*, 1999). Though members of the majority of Parasitica are parasitoids, in the Cynipoidea and Chalcidoidea many species have reverted to a phyophagous habit, becoming gall formers.

The majority of Hymenoptera have proceeded beyond this stage. In these forms (the Aculeata) the ovipositor took on a new function as a sting for paralyzing (though not killing) the prey. The advantage of this was twofold. First, the prey remained "fresh" while the larva consumed it, and second, the immobile prey could not carry the parasitoid into a different, perhaps inhospitable habitat. This phase in hymenopteran evolution is seen in the extant Chrysidoidea.

From such an ancestral group of paralyzers, the recent groups of higher Aculeata evolved. Two major lines of evolution can be distinguished. In the line that led to the solitary and social wasps (Vespoidea), the insects began to construct special cells in which to place the prey and lay their eggs, presumably to increase the chances of the offspring's survival. However, in one family of vespoids, the ants (Formicidae), an important change occurred. Primitive ants are carnivorous or feed on animal products, especially honeydew, but higher forms are secondarily phytophagous and live on plant tissue, seeds, or fungi. This dietary change was perhaps a response to the increased difficulty of obtaining sufficient animal food as the size of a colony grew. In the second major line, leading to the bees (Apoidea), there was again a trend toward a change in diet from a carnivorous one to one in which plant products, namely, pollen and nectar, were stored, also in special cells, to provide food for the developing larvae. A possible phylogeny of the Hymenoptera is shown in Figure 10.22.

The Hymenoptera have traditionally been arranged in two suborders, Symphyta (Chalastogastra) and Apocrita (Clistogastra), and while this arrangement is continued here, it

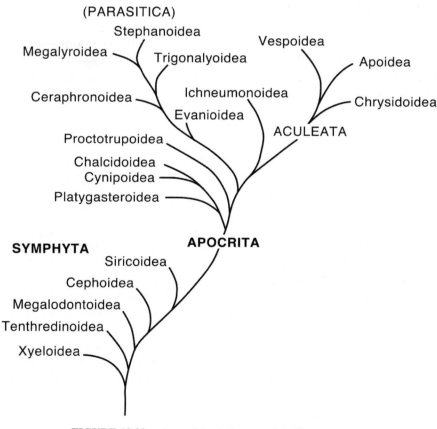

FIGURE 10.22. A possible phylogeny of the Hymenoptera.

should be noted that the Symphyta is a paraphyletic group (Ronquist, 1999; Vilhelmsen, 2001). In contrast, the monophyletic nature of the Apocrita is widely accepted. Further, the long-standing division of the Apocrita into the infraorders Parasitica and Aculeata can no longer be retained because, as noted above, the former group is paraphyletic. Within the monophyletic Aculeata, there is still disagreement as to the major taxa, some authors splitting the group into as many as seven superfamilies, while others recognize only three (see Gauld and Bolton, 1988).

Suborder Symphyta

In adults of the suborder Symphyta the abdomen is broadly attached to the thorax, and there is no marked constriction between the first and second abdominal segments. Larvae have a well-developed head, as well as thoracic, and, in most species, abdominal legs.

Superfamily Xyeloidea

Xyeloidea form a small (50 species) and extremely primitive superfamily containing the single family XYELIDAE. Adults feed on flowers and have a generalized wing venation. Larvae are found in flowers and have prolegs on all abdominal segments. The family has a holarctic distribution, with most species in North America.

Superfamily Megalodontoidea

The primitive superfamily Megalodontoidea includes two small families, the PAM-PHILIIDAE (170 species, holarctic) and MEGALODONTIDAE (45 species, palearctic). Adults feed on flowers, while larvae, which lack prolegs, live gregariously in webs or rolled leaves.

Superfamily Tenthredinoidea

Tenthredinoidea form a very large and likely paraphyletic group with diverse habits. Females have a sawlike rather than a boring ovipositor and are commonly called sawflies. More than 3000 species belong to the cosmopolitan (except Australia) family TENTHREDINIDAE. Adults are often carnivorous. Parthenogenesis is common, and larvae are usually caterpillarlike in form and habits. A few are leaf miners and apodous. Some species are economically important, for example, *Nematus ribesii*, the imported currant worm (Figure 10.23A), and *Pristiphora erichsonii*, the larch sawfly. Some 800 species of the primarily tropical and warm temperate family ARGIDAE have been described. Larvae of some species are gregarious, living under a silken cover; others are leaf miners. Females of some species protect their offspring. The PERGIDAE (400 species) are primarily found in Australia and Central and South America. Larvae are usually gregarious feeders on foliage, rarely leaf miners. Some eucalypt-feeding Australian species accumulate eucalyptus oils in a special gut diverticulum and regurgitate these when disturbed. Other families, which are not large but contain economically important species, are the mainly holarctic and oriental CIMBICIDAE (130 species, including *Cimbex americana*) (Figure 10.23B), which defoliate various broad-leaved trees, and the holarctic DIPRIONIDAE (conifer sawflies) (90 species) (Figure 10.23C), which include several of North America's most important forest pests (*Neodiprion* spp. and *Diprion* spp.).

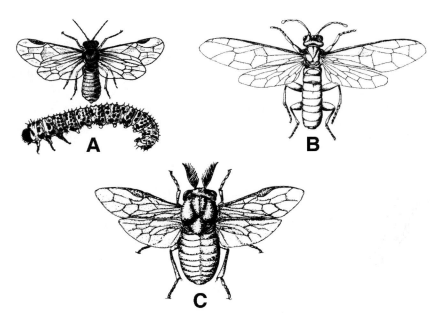

FIGURE 10.23. Tenthredinoidea. (A) The imported currant worm, *Nematus ribesii* (Tenthredinidae) adult and larva; (B) the elm sawfly, *Cimbex americana* (Cimbicidae); and (C) the redheaded pine sawfly, *Neodiprion lecontei* (Diprionidae). [A, from L. A. Swan and C. S. Papp, 1972, *The Common Insects of North America.* Copyright by L. A. Swan and C. S. Papp. Reprinted by permission of Harper & Row Publishers, Inc. B, C, from the U.S. Department of Agriculture.]

Superfamily Siricoidea

Female Siricoidea have a boring ovipositor, and larvae have reduced thoracic legs, no prolegs, and live in wood. The 250 species in this possibly paraphyletic group (Vilhelmsen, 2001) are divided approximately evenly among three families, XIPHYDRIIDAE, SIRICIDAE, and ORUSSIDAE. Adult siricids (horntails) (Figure 10.24) are large, often brightly colored insects found in temperate Northern Hemisphere forests. Larvae burrow extensively in the heartwood of both deciduous and evergreen trees and may cause much damage, especially those species that live symbiotically with rot-producing fungi. Though a cosmopolitan group, orussids (parasitic wood wasps) are principally tropical and Australian in distribution. Females have an extremely long ovipositor that is coiled within the body when not in use. The structure is used to lay the very elongate and thin eggs onto the body of horntail larvae and buprestid beetle larvae in their tunnels. Xiphydriidae (wood wasps)

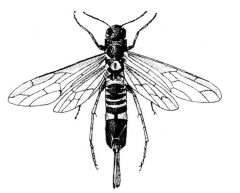

FIGURE 10.24. Siricoidea. The pigeon tremex, *Tremex columba* (Siricidae). [From H. E. Evans and M. J. W. Eberhard, 1970, *The Wasps.* By permission of the University of Michigan Press.]

FIGURE 10.25. Cephoidea. The wheat stem sawfly, *Cephus cinctus* (Cephidae). [From L. A. Swan and C. S. Papp, 1972, *The Common Insects of North America.* Copyright 1972 by L. A. Swan and C. S. Papp. Reprinted by permission of Harper & Row Publishers, Inc.]

are similar to horntails in appearance, though usually smaller. The family is cosmopolitan (except Africa). Larvae mine in branches and small trunks of angiosperm trees and shrubs.

Superfamily Cephoidea

All Cephoidea are included in the family CEPHIDAE (stem sawflies), a largely Eurasian group of about 100 species. Larvae bore into the stems of grasses and berries. Some species are of considerable economic importance, for example, *Cephus cinctus*, the wheat stem sawfly (Figure 10.25), and *Janus integer*, which bores in the stems of currants.

Suborder Apocrita

Almost all adult Apocrita have the first abdominal segment (propodeum) intimately fused with the thorax and a constriction between the first and second abdominal segments. Larvae are apodous and have a reduced head capsule.

The first 10 superfamilies were formerly included in the Parasitica, now recognized as a paraphyletic group.

Superfamily Megalyroidea

This very small group, containing about 50 species in a single family, MEGALYRIDAE, occurs mainly in the Old World tropics. Larvae of these stout-bodied wasps are parasitoids of beetle larvae under tree bark.

Superfamily Stephanoidea

Sometimes included in the Megalyroidea, the 100 species in this group are placed in the family STEPHANIDAE, which has a tropical distribution. Adults are slender wasps with highly modified hindlegs thought to be able to detect sounds created by the wood-boring beetle larvae that serve as hosts for their larvae.

Superfamily Trigonalyoidea

Trigonalyoidea form a small (100 species worldwide) group of archaic but highly specialized Apocrita whose members are all contained in the family TRIGONALYIDAE. Structurally they possess a combination of features of Parasitica and Aculeata. The trigonalyids are hyperparasites of sawfly larvae or lepidopteran caterpillars. The eggs are laid on plant tissue, which is then eaten by the sawfly larvae or caterpillars. These are then parasitized by other Hymenoptera (ichneumons) or Diptera (tachinids), which in turn become the primary host for the trigonalyids.

Superfamily Ceraphronoidea

These small (1 to 3 mm) wasps are arranged in two cosmopolitan families, MEGASPILIDAE (450 species) and CERAPHRONIDAE (360 species). The former are primary parasitoids of Coccoidea, Neuroptera, and puparia of Diptera or hyperparasites of aphid-infesting Braconidae. Ceraphronids, likewise, may be endoparasitoids (in Thysanoptera, Lepidoptera, Neuroptera, Cecidomyiidae, and puparia of Diptera) or hyperparasites (taken from cocoons of Braconidae). A few species have been collected in ant nests, where they are perhaps parasitoids of myrmecophilous Diptera.

Superfamily Ichneumonoidea

This group contains the two largest families of Hymenoptera, ICHNEUMONIDAE (ichneumon flies) (Figure 10.26A) with about 20,000 species (about 3500 in North America), and BRACONIDAE (Figure 10.26B), with some 35,000–40,000 species (2000 in North America). Ichneumonoids are mainly parasitoids of larvae and pupae of endopterygotes (all orders except Megaloptera and Siphonaptera), laying their eggs either in concealed locations on the host (the primitive condition) or in the host, when various strategies are used to avoid attack by the host's immune system. Venom may be injected into the host prior to oviposition, the poison paralyzing or killing the recipient. Ichneumonids are largely restricted to juvenile stages of endopterygotes, though a few use the eggs of pseudoscorpions and spiders, or adult spiders, as hosts. Many Braconidae are parasitoids of exopterygotes,

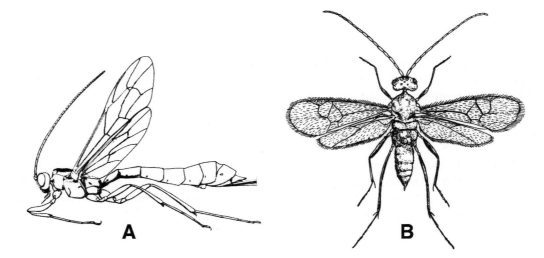

A   **B**

FIGURE 10.26. Ichneumonoidea. (A) An ichneumon fly, *Rhyssa persuasoria* (Ichneumonidae). Note that the entire ovipositor is not drawn; and (B) a braconid, *Apanteles cajae* (Braconidae). [A, reproduced by permission of the Smithsonian Institution Press from *United States National Museum Bulletin 216, Part 2 (Ichneumon-Flies of America North of Mexico: 2. Subfamilies Ephialtinae Xoridinae Acaenitinae)* by Henry and Marjorie Townes: FIGURE 302b, pages 598. Washington D.C., 1960, Smithsonian Institution. B, from R. R. Askew, 1971, *Parasitic Insects.* By permission of Heinemann Educational Books Ltd.]

notably homopterans (especially aphids), Heteroptera, Isoptera, and Psocoptera, sometimes exhibiting extremely high host specificity, but do not attack spiders. Many species are beneficial to humans because they are major natural control agents on populations of pest insects. Examples are *Bracon cephi*, a parasitoid of wheat-stem sawfly larvae, and several species of *Apanteles* and *Cotesia* that attack cabbageworms, the tobacco hornworm, clothes moth larvae, and gypsy moth caterpillars, among others. Several ichneumons and braconids have been used in the biological control of pests (see Table 24.6).

Superfamily Evanioidea

The constituent families are AULACIDAE (150 species worldwide), GASTERUPTI-IDAE (500 species, mainly tropical), and EVANIIDAE (400 species, mainly tropical but some species are cosmopolitan, living in buildings where their hosts occur). Aulacids are parasitoids of wood-boring larvae of Coleoptera and Xiphydriidae. Gasteruptiid females oviposit in nests of solitary bees and wasps; the larva eats the host's egg and then pollen or prey stored in the cell. Evaniids are exclusively parasites of cockroach oothecae, the larvae eating the eggs then pupating inside the host's egg case.

Superfamily Proctotrupoidea

Almost all of the approximately 2700 species in this group fall into the two families DIAPRIIDAE and PROCTOTRUPIDAE. Diapriidae (2300 species worldwide) mainly parasitize larvae and pupae of Diptera, especially of species living in moist habitats, though a few use larvae of beetles, ants, or termites as hosts. Larval proctotrupids (300 species worldwide) are mostly parasitoids of litter- or rotten wood-inhabiting beetle larvae, though a few species feed on larvae of mycetophilid Diptera.

Superfamily Platygasteroidea

Often included in the previous superfamily, the two large and cosmopolitan families that comprise the Platygasteroidea differ from the above group in the structure and operation of their ovipositor and in the structure of the female antennae. SCELIONIDAE (3000 species) are egg parasites of a wide range of insects, especially Lepidoptera, Hemiptera and orthopteroids, and spiders. Females of some species are phoretic; that is, they locate and are carried about on a suitable host until the latter begins to oviposit. The scelionid then leaves the host and lays her eggs among those of the host. PLATYGASTERIDAE (1100 species) are either egg or larval parasites, especially of Cecidomyiidae, including a number of pest species, for example, the Hessian fly, over which they exert major population control. Other hosts include cerambycids, homopterans, and sphecid wasps. Some species are polyembryonic (Chapter 20, Section 8.2), and females may be highly specific with respect to the host tissue in which an egg is laid.

Superfamily Cynipoidea

Most of the 2300 species so far ascribed to the Cynipoidea belong to the family CYNIP-IDAE, subfamily CYNIPINAE (gall wasps) (Figure 10.27), and are either gall makers or inquilines in already-formed galls. The host plant and the form of the gall are usually specific for a particular species of cynipid, though in some primitive species a range of host plants

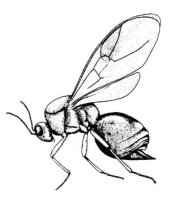

FIGURE 10.27. Cynipoidea. A gall wasp, *Diplolepis rosae* (Cynipidae). [From D. J. Borror, D. M. Delong, and C. A. Triplehorn, 1976, *An Introduction to the Study of Insects*, 4th ed. By permission of Brooks/Cole, a division of Thomson Learning.]

is chosen. In some species the life history is complex with two generations per year. The first generation comprises entirely females that reproduce parthenogenetically. The second generation, which develops in a different gall on a different plant, includes both males and females. The 800 or so species of the predominantly tropical family EUCOILIDAE (sometimes considered a subfamily of Cynipidae) are internal parasitoids of fly larvae, especially those associated with dung and rotting fruit.

Superfamily Chalcidoidea

This is a highly diverse group of mostly small to minute, parasitoid or phytophagous (usually gall-forming) insects, most of which are arranged in a number of large and important families. Several of these include species that have been used successfully as biological control agents against insect pests (for examples, see Table 24.6). The family AGAONIDAE (fig insects) contains perhaps 650 species of chalcidoids, which show some remarkable biological features. The species are restricted to living in the receptacles of species or varieties of fig, whose flowers a female pollinates in her search for an oviposition site. The Smyrna fig, a cultivated variety, requires cross-pollination with the wild fig before fruit can be formed. To facilitate this, fig growers place branches of the wild fig, from whose receptacle female fig wasps emerge, among those of the Smyrna fig. As a female searches for a suitable egg-laying site, she accidentally visits the Smyrna fig flowers, though, because they are the wrong shape, she does not oviposit in them. The TORYMIDAE (1150 species widely distributed) exhibit a wide range of life histories. Most species are parasitoids of gall-forming insects, though a few are themselves gall formers or inquilines of galls. Some are parasitoids of mantid oothecae, others develop in bee or wasp nests, and a few feed on seeds. The CHALCIDIDAE, a widespread group of about 1900 species, are parasitoids of lepidopteran, dipteran, and coleopteran larvae or pupae, or hyperparasites of tachinids or ichneumon flies. The EURYTOMIDAE (Figure 10.28A) constitute a cosmopolitan family (1400 species) of very diverse habits. Commonly its members produce galls on the stems of grasses, including cereals. Other species feed on seeds, are inquilines in the nests of bees and wasps, are parasitoids on gall-forming insects, or egg parasites of Orthoptera. Another very large family is the EULOPHIDAE (3900 species) (Figure 10.28B), a group whose members are very small parasitoids or hyperparasites. The parasitoids are important control agents of many insect pests, especially leaf-mining species of Lepidoptera, Diptera, Hymenoptera, and Coleoptera, but including scale insects, aphids, and some surface feeders. The TRICHOGRAMMATIDAE (675 species worldwide) (Figure 10.28C) are minute egg

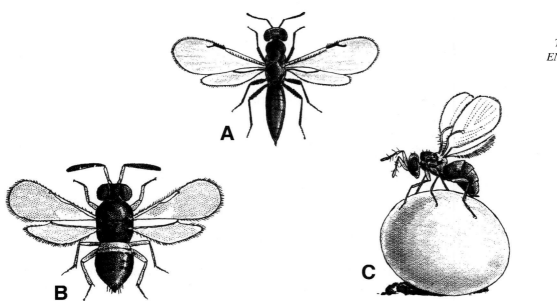

FIGURE 10.28. Chalcidoidea. (A) The wheat jointworm, *Harmolita* (= *Tetramesa*) *tritici* (Eurytomidae); (B) *Aphelinus mali* (Eulophidae), a parasitoid of the woolly apple aphid (*Eriosoma lanigerum*); and (C) ovipositing *Trichogramma minutum* (Trichogrammatidae), a parasitoid of the eggs of more than 200 species of Lepidoptera. [From L. A. Swan and C. S. Papp, 1972, *The Common Insects of North America.* Copyright 1972 by L. A. Swan and C. S. Papp. Reprinted by permission of Harper & Row Publishers, Inc.]

parasites, especially of Lepidoptera. Some species are reared in large numbers to control certain pests. The MYMARIDAE are some of the smallest insects, with adults of some species having a length of 0.21 mm! Members of this cosmopolitan family of about 1400 species are egg parasites, especially of Coccoidea and other Hemiptera. The ENCYRTIDAE (3800 species worldwide) are one of the most important groups from the perspective of biological control. They are endoparasitoids, especially of Coccoidea, though eggs, larvae, and pupae of species from many other orders, as well as spiders, are attacked. Like members of the previous family, PTEROMALIDAE (4100 species) attack a wide range of insect hosts, at all stages of their life history. The cosmopolitan APHELINIDAE (1120 species) are mainly parasitoids of aphids, scale insects, whitefly, and psyllids, for which they are important biological control agents. However, some species attack the eggs of Lepidoptera and Orthoptera, the eggs and larvae of Diptera, and even other chalcidoids. The mainly tropical to subtropical EUPELMIDAE (900 species) are predators, parasitoids, or hyperparasites of the eggs, larvae, or pupae of a wide range of insects and spiders.

Infraorder Aculeata

As noted earlier, the Aculeata is widely accepted as a monophyletic group. Earlier classifications split the infraorder into as many as eight superfamilies but, more recently, it has been appreciated that several of these were superfluous and that a system of only three superfamilies reflected the natural relationships and evolution of the infraorder (Brothers, 1975; Gauld and Bolton, 1988). The Chrysidoidea (Bethyloidea of earlier classifications) are the most primitive aculeates and form the sister group to the other members of the infraorder. The Vespoidea (including Tiphioidea, Scolioidea, Formicoidea, Pompiloidea, and Vespoidea of older systems) require further study and may be paraphyletic. The Apoidea (= Apoidea + Sphecoidea of earlier classifications) is now clearly established as a monophyletic taxon.

Superfamily Chrysidoidea

The great majority of the more than 7500 species in this group are arranged in three families, CHRYSIDIDAE, BETHYLIDAE, and DRYINIDAE, the last two being placed in a separate superfamily Bethyloidea by some earlier workers. Chrysidids (3000 species) are especially diverse in temperate deserts of both hemispheres. Their common name, cuckoo wasps, is derived from their habit of laying an egg in the cells of solitary wasps or bees. Usually, the egg does not hatch until the host larva has consumed its own food supply. Bethylidae (2200 species) are primarily tropical and subtropical wasps, females of which paralyze the host (typically beetle larvae or caterpillars, sometimes much larger than the bethylid). Some species drag the prey into a sheltered location, a habit that foreshadows the situation in the Scoliidae (digging wasps), prior to laying several eggs on it. Dryinids (1100 species worldwide) are parasitoids of immature and adult Auchenorrhyncha, especially Fulgoridae and Cicadellidae, and several species have been used as biological control agents, for example, against pests of sugarcane in Hawaii.

Superfamily Vespoidea

As presently constituted, the Vespoidea contains 10 families, though the vast majority of the approximately 24,000 described species (including 2000 in North America) fall into five large groups. The SCOLIIDAE (digging wasps) (Figure 10.29A) form a small but distinctive family (300 species) of large, hairy wasps with a cosmopolitan distribution. The female burrows into soil, rotting wood, etc., and locates a beetle larva, usually a scarabaeid, which she paralyzes, then lays an egg on it. In some species the female builds a special cell around the host larva. Female TIPHIIDAE (1500 species worldwide) also attack mainly scarabaeid larvae, though some species search out larvae of tiger beetles, cerambycids, solitary and social bees and wasps, and mole crickets. Females of about half of the species are wingless and are either transported to flowers by males, or fed by males on nectar or honeydew. MUTILLIDAE (velvet ants) (Figure 10.29B) form a large, cosmopolitan family (5000 species) of strongly sexually dimorphic wasps, the males being winged, the females wingless and somewhat resembling ants in their form and behavior. Both sexes are densely hairy and often aposematically colored. Mutillids are mainly parasitoids of both social and solitary bees and wasps, though some species oviposit on the larvae of Diptera, chrysomelid and other beetles, and Lepidoptera, and on cockroach oothecae. Members of the primarily tropical family POMPILIDAE (Figure 10.30), with some 4200 species, are commonly called spider wasps. These large, solitary, often strikingly marked wasps are parasitoids of spiders, rarely other Arachnida. Prey is paralyzed or killed and usually deposited in a nest, which may contain one to several cells, in the ground. The female lays a single egg in each cell. Some species are cleptoparasites of other pompilids; they detect the provisioned nest, dig down to it, consume the original wasp egg and lay their own in its place.

The ants (FORMICIDAE) form a worldwide, though mainly tropical, family containing between 9000 (Goulet and Huber, 1993) and 15,000 (Gauld and Bolton, 1988) described species, including about 600 in North America. They are social, with the exception of a few secondarily solitary parasitic forms. In the latter there is no worker caste, and a female deposits her eggs in the nest of a closely related species whose workers then rear the resulting

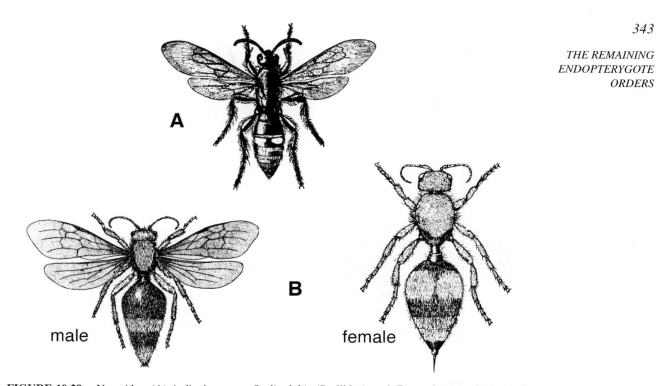

FIGURE 10.29. Vespoidea. (A) A digging wasp, *Scolia dubia* (Scoliidae); and (B) a velvet ant, *Dasymutilla occidentalis* (Mutillidae), male and female. [A, from L. A. Swan and C. S. Papp, 1972, *The Common Insects of North America.* Copyright 1972 by L. A. Swan and C. S. Papp. Reprinted by permission of Harper & Row, Publishers, Inc. B, from D. J. Borror, D. M. Delong, and C. A. Triplehorn, 1976, *An Introduction to the Study of Insects*, 4th ed. By permission of Brooks/Cole, a division of Thomson Learning.]

larvae. Polymorphism reaches the extreme in ants, where, in some species, several different forms of worker occur, each performing a particular function. Workers form the vast bulk of individuals in the nest. There may be one to several queens; males are few in number and probably produced seasonally. Except for a short period prior to swarming, all individuals are apterous. However, winged queens and males are produced in most species (not in Dorylinae) in order to found new nests. After a short mating flight, a queen alone finds a suitable nesting site, sheds her wings, and eventually begins egg laying. Except in a few primitive species, the queen does not forage for food during this initial phase of nest building but lives entirely on fat body reserves and the products of wing muscle degeneration.

FIGURE 10.30. Vespoidea. A spider wasp, *Episyron quinquenotatus* (Pompilidae). [From D. J. Borror, D. M. Delong, and C. A. Triplehorn, 1976, *An Introduction to the Study of Insects*, 4th ed. By permission of Brooks/Cole, a division of Thomson Learning.]

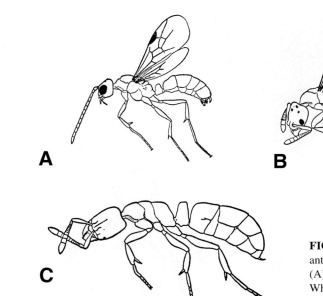

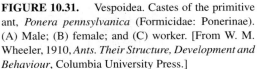

FIGURE 10.31. Vespoidea. Castes of the primitive ant, *Ponera pennsylvanica* (Formicidae: Ponerinae). (A) Male; (B) female; and (C) worker. [From W. M. Wheeler, 1910, *Ants. Their Structure, Development and Behaviour*, Columbia University Press.]

Several distinct subfamilies are recognized. The most primitive ants, which are carnivorous and form only small colonies, are the PONERINAE (Figure 10.31), which nest in the ground or rotting logs. In the ponerine ants there is little structural difference between the queen and workers, which are monomorphic. DORYLINAE are nomadic ants (army ants) of tropical regions. Like members of the previous subfamily, they are carnivorous, and in their search for food may form massive columns whose length may cover 100 m or more. The remaining ant subfamilies have solved the problem of obtaining sufficient food for the colony (which may contain several million individuals) by changing from a carnivorous to a generally phytophagous diet. Not surprisingly, in view of their enormous biomass, ants are extremely important in both energy flow and soil mixing in the ecosystem. The MYRMICINAE (Figure 10.32) form the largest and most common subfamily. Many species are harvester ants, so-called because they collect seeds that they store in the nest. Others grow fungi on decaying leaf fragments and ant excreta in special subterranean chambers. It is to this subfamily that many of the inquiline and parasitic species belong. DOLICHODERINAE and FORMICINAE, which form the second largest subfamily, are generally nectar or honeydew feeders. Workers usually have a flexible integument that stretches remarkably as food is imbibed. In the honey ants there is a distinct form of worker, the replete (Figure 10.33), which spends its life in the nest and serves as a living bottle in which food can be stored. Many species have established a symbiotic relationship with honeydew-secreting insects, mainly homopterans. In return for a copious supply of honeydew the ants move the insects to new "pasture," protect them if there is a disturbance, build shelters for them, and store their eggs during the winter.

In addition, ants have a close relationship with many other insects that actually live in their nest. The relationship ranges from one in which the inquilines are scavengers or predators and are treated with hostility by the ants, through one where the ants behave indifferently toward the visitors, to a situation in which the ants "welcome" their guests, to the extent of feeding and rearing them. In return for this hospitality, the guests appear to exude substances that are highly attractive to the ants. Frequently, two species of ants may

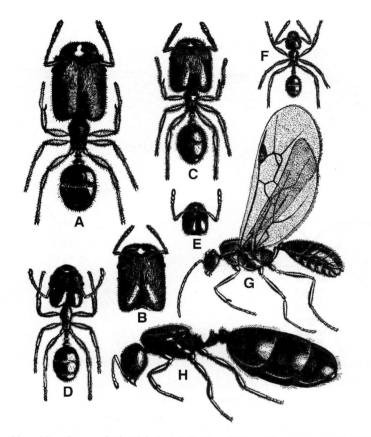

FIGURE 10.32. Vespoidea. Castes of *Pheidole instabilis* (Formicidae: Myrmicinae). (A) Soldier; (B–E) intermediate workers; (F) typical worker; (G) male; and (H) dealated female. [From W. M. Wheeler, 1910, *Ants. Their Structure, Development and Behaviour*, Columbia University Press.]

live quite amicably in close proximity. From this it is only a short step to social parasitism in which the workers of one species adopt a queen of another species. They tend the eggs laid by her and feed the larvae that develop. The host queen is killed, and eventually the nest is taken over entirely by the intruders. A relationship of a different form is shown by the slave-making ants, which capture workers of another species; these then perform domestic duties within the colony.

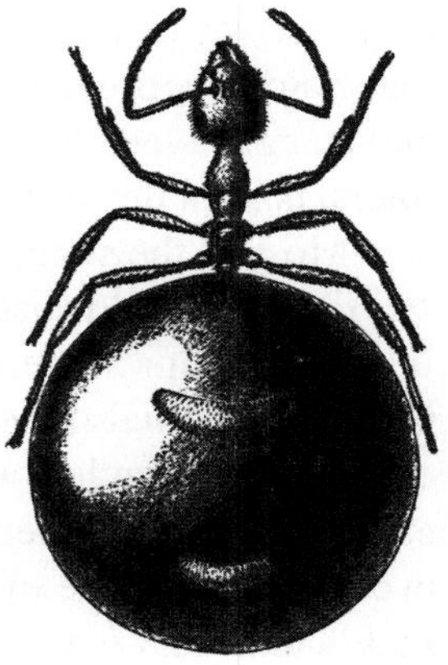

FIGURE 10.33. Vespoidea. Replete of the honey ant, *Myrmecocystus hortideorum* (Formicidae: Formicinae). [From W. M. Wheeler, 1910, *Ants. Their Structure, Development and Behaviour*, Columbia University Press.]

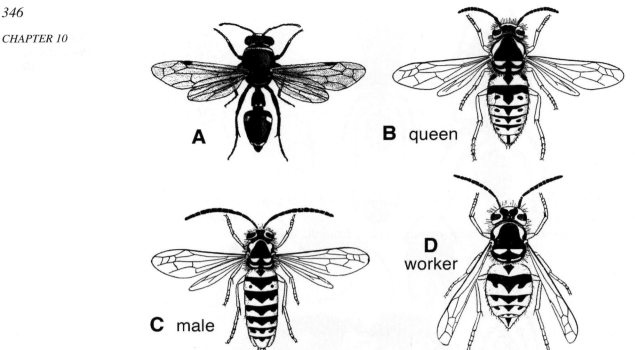

FIGURE 10.34. Vespoidea. (A) A potter wasp, *Eumenes fraterna* (Vespidae); and (B–D) a hornet, *Vespula pennsylvanica* (Vespidae) queen, male, and worker. [A, from L. A. Swan and C. S. Papp, 1972, *The Common Insects of North America.* Copyright 1972 by L. A. Swan and C. S. Papp. Reprinted by permission of Harper & Row Publishers, Inc. B–D, from E. O. Essig, 1954, *Insects of Western North America.* Reprinted with permission of the Macmillan Publishing Co., Inc. Copyright 1926 by Macmillan Publishing Co., Inc., renewed 1954 by E. O. Essig.]

With about 4000 species, the cosmopolitan (but predominantly tropical) family VESPI-DAE includes the true wasps. Both solitary and social forms occur, and in both the larvae are typically reared in specially constructed cells and fed animal material. Adults feed on nectar, honeydew, or ripe fruit. In some species, larvae are also fed on pollen and nectar. Most of the solitary forms can be arranged in two subfamilies, MASARINAE and EUMENINAE. The former is a small group of highly specialized wasps that burrow in soil or build exposed mud cells and feed their larvae on pollen and nectar. The eumenines vary in their nest-building habits. Many construct cells in wood or stems; others, the common mason or potter wasps (Figure 10.34A), build cells of mud fastened to twigs and other objects. Once a cell is provisioned (usually with larvae of Lepidoptera, Coleoptera, or Symphyta) and an egg layed, the entrance is sealed. A few eumenines are subsocial, that is, the female interacts with the larva by continuing to feed it through development.

Most of the social wasps belong to the subfamilies POLISTINAE (paper wasps) and VESPINAE (yellow jackets and hornets) (Figure 10.34B–D), the former especially diverse in the Neotropics, the latter in the holarctic and oriental regions. The polistines build annual or perennial nests of paper (masticated wood fragments mixed with saliva), usually comprising a single comb. Larvae are progressively fed with masticated insects, especially caterpillars, and sometimes honey. A colony may be founded by several females, though eventually only one of these becomes the queen, that is, the egg layer, the remaining females, although fertile, becoming workers. In Vespinae a single queen founds a colony, and

FIGURE 10.35. Apoidea. The black and yellow mud dauber, *Sceliphron caementarium* (Sphecidae). [From L. A. Swan and C. S. Papp, 1972, *The Common Insects of North America.* Copyright 1972 by L. A. Swan and C. S. Papp. Reprinted by permission of Harper & Row Publishers, Inc.]

once her first offspring have emerged and can function as workers, she does not leave the nest. Usually, vespine nests are multicombed and may include several thousand individuals. Workers are morphologically quite distinguishable from the queen.

Superfamily Apoidea

In Gauld and Bolton's (1988) system this superfamily includes only two families, SPHECIDAE and APIDAE, though most earlier classifications split each of these groups into as many as 10 distinct families. Some 8000 Sphecidae have been described worldwide, including about 1200 in North America. Commonly called mud daubers (Figure 10.35), sand wasps, thread-waisted wasps, or digger wasps, the majority of Sphecidae are solitary and mass provision their cells, though some species are subsocial or social, progressively provisioning the cells throughout the larva's life. The prey, which may be paralyzed or killed outright, comprises a wide variety of insects (both immature and adult) and arachnids. In some species a high degree of specificity exists between the wasp and its prey. A few species are cleptoparasites of other sphecids.

Among the approximately 15,000 species of Apidae (including about 3500 in North America) are both solitary and social bees. The great majority of species are solitary, the social forms being restricted to three subfamilies, Halictinae, Anthophorinae, and Apinae. Most species differ from sphecids in that they use pollen and nectar rather than animal material for feeding the larvae. However, a few forms use carrion as food and a few are cleptoparasites or social parasites on other, nest-making bees. The cleptoparasites enter the host's nest and oviposit, the larvae killing the host's egg or young and being reared by the host. Social parasite females take over the host's nest, in effect becoming queen over the host workers, which then rear her offspring.

The subfamily COLLETINAE (2000 species) is particularly common in the Southern Hemisphere, especially Australia. It contains the most primitive bees whose females construct simple nests in soil, hollow stems, or holes in wood. They are sometimes called plasterer bees from their habit of lining the nest with a salivary secretion that dries to form a thin transparent sheet.

The HALICTINAE form a large cosmopolitan subfamily (3000 species), of mostly solitary bees. In some species, however, large numbers of egg-laying females occupy the same nest site, usually a hole in the ground, though there is no division of labor within the colony. A few species are truly social; each season a single female constructs a nest and rears a brood of young, all of which develop into workers, though these may not be structurally much different from the queen. The workers care for the eggs laid subsequently, which develop into both males and females. Only fertilized females overwinter.

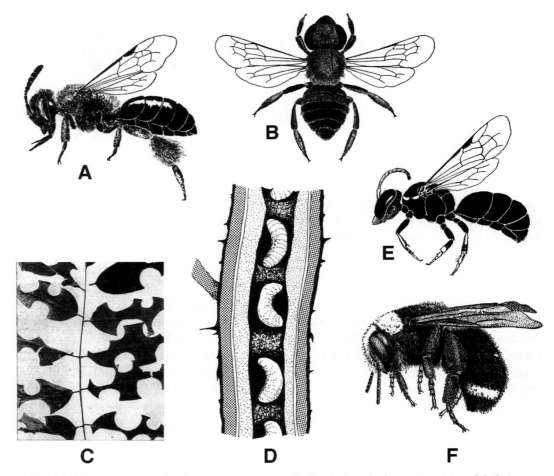

FIGURE 10.36. Apoidea: Apidae. (A) *Andrena* sp. (Andreninae); (B) a leaf-cutter bee, *Megachile latimanus* (Megachilinae); (C) the work of leaf-cutter bees; the removed portions of the leaves were used in nest building; (D) nest of the small carpenter bee, *Ceratina dupla* (Anthophorinae); (E) *Ceratina acantha* (Anthophorinae); and (F) the yellow-faced bumble bee, *Bombus vosnesenskii* (Apinae). [A–C, E, F, from E. O. Essig, 1954, *Insects of Western North America.* Reprinted with permission of the Macmillan Publishing Co., Inc. Copyright 1926 by Macmillan Publishing Co., Inc.; renewed 1954 by E. O. Essig.]

The ANDRENINAE (Figure l0.36A) are the common solitary bees of the holarctic region, with about 2000 species, including 1200 from North America. Andreninae typically nest in burrows in the ground. Often, large numbers nest together in the same "apartment," each bee with its own "suite."

The MEGACHILINAE form another very large subfamily (3000 species), which includes the familiar leaf-cutter bees (Figure 10.36B,C), females of which build nests from leaf fragments. Others, however, build nests from mud or live in burrows, under stones, and in other suitable holes. Some species are parasitic on other bees. *Megachile rotundata*, the alfalfa leaf-cutter bee, is cultured on a large scale in North America for use as an alfalfa pollinator. The weaker honey bee can pollinate this plant but experiences difficulty inforcing its way into the flower and soon learns that there are easier sources of food. The use of leaf-cutter bees can increase the yield of seed severalfold.

FIGURE 10.37. Apoidea. The honey bee, *Apis mellifera* (Apidae). (A) Queen, (B) worker, and (C) drone. [From G. Nixon. 1959, *The World of Bees*, Hutchinson. Reprinted by permission of The Random House Group Ltd.]

The ANTHOPHORINAE form a large group (4000 species) of mainly solitary or parasitic bees. Females of solitary forms frequently nest in large numbers close together, usually in holes in the ground (digger bees) or burrow into wood or plant stems (carpenter bees) (Figure 10.36D,E). In some carpenter bees there is a primitive social organization in which the queen, morphologically identical with her offspring (the workers), lays eggs and is long-lived, whereas the workers do not usually lay eggs and live for only a relatively short time.

The subfamily APINAE, a cosmopolitan group of about 1000 species, includes all of the highly social bees and a few neotropical solitary species. The subfamily includes orchid bees (tribe EUGLOSSINI), bumble bees (BOMBINI), honey bees (APINI), and stingless bees (MELIPONINI). Orchid bees are neotropical, mostly solitary species, males of which are pollinators of many orchid species. Bumble bees (Figure 10.36F) are common, large, hairy bees, found mainly in the holarctic region. The social organization of bumblebees is primitive, and workers frequently differ from the queen only in size. They do not construct a true comb but rear larvae in "pots." Often, these are sealed off after egg laying, and only older larvae are fed regularly. Only the queen overwinters, and new nests are produced annually. Some female Bombini (*Psithyrus* spp., cuckoo bumble bees) lay their eggs in the nests of other bumble bees, occasionally killing the host queen but more often living side by side. Bumble bee workers then attend to and rear the *Psithyrus* larvae in preference to those of their own species. Included in the Apini are five species of *Apis*, of which the most familiar is *A. mellifera*, the honey bee (Figure 10.37), a native of Europe and Africa, but now cosmopolitan as a result of commerce (see Chapter 24, Section 2.1). Most species of stingless bees are highly social, with well differentiated queens and workers, and a communication system comparable to that of honey bees. However, in *Melipona* and *Trigona* mass provisioning of brood cells occurs, and there is no contact between the adults and young. Young queens of *Trigona* are reared in special cells, as are those of *Apis*. However, in contrast to *Apis*, a young queen and attendants form a new nest; the old queen, being too large and heavy to fly, cannot move to a new site as occurs in *Apis*.

Literature

Primarily because of their fascinating behavior, Hymenoptera, especially the social forms, have been the subject of much literary effort, and only a sample can be given here. Good summaries of hymenopteran biology are given by Malyshev (1968) [from a phylogenetic perspective], Brown (1982), Gauld and Bolton (1988), and Goulet and Huber (1993). Doutt (1959), Askew (1971), and Gordh *et al.* (1999) have reviewed the biology of the parasitic Hymenoptera. Authors dealing with the biology of specific major groups include

Sudd (1967), Wilson (1971), and Hölldobler and Wilson (1990) [ants]; Evans and Eberhard (1970), Spradbury (1973), Edwards (1980), Ross and Matthews (1991), and O'Neill (2001) [wasps]; Stephen *et al.* (1969), Butler (1974), Michener (1974, 2000), Alford (1975), Free (1982), and Seeley (1985) [bees]. Hymenopteran phylogeny and classification are discussed by Malyshev (1968), Brothers (1975), Gauld and Bolton (1988), Goulet and Huber (1993), Ronquist (1999), and Vilhelmsen (2001). Goulet and Huber (1993) also provide a key to the world's Hymenoptera families. The British Hymenoptera may be identified from the Royal Entomological Society's handbooks (Richards, 1977, and other authors). North American and Australian families are dealt with by Arnett (2000) and Naumann (1991), respectively.

Alford, D. V., 1975, *Bumblebees*, Davis-Poynter, London.

Arnett, R. H., Jr., 2000, *American Insects: A Handbook of the Insects of America North of Mexico*, 2nd ed. CRC Press, Boca Raton, FL.

Askew, R. R., 1971, *Parasitic Insects*, American Elsevier, New York.

Brothers, D. J., 1975, Phylogeny and classification of the aculeate Hymenoptera, with special reference to Mutillidae, *Univ. Kansas Sci. Bull.* **50**:483–648.

Brown, W. L., Jr., 1982, Hymenoptera, in: *Synopsis and Classification of Living Organisms*, Vol. 2 (S. P. Parker, ed.), McGraw-Hill, New York.

Butler, C. G., 1974, *The World of the Honeybee*, 2nd ed., Collins, London.

Doutt, R. L., 1959, The biology of parasitic Hymenoptera, *Annu. Rev. Entomol.* **4**:161–182.

Dowton, M., and Austin, A. D., 1994, Molecular phylogeny of the insect order Hymenoptera: Apocritan relationships, *Proc. Natl. Acad. Sci. U S A* **91**:9911–9915.

Dowton, M., and Austin, A. D., 1999, Models of analysis for molecular datasets for the reconstruction of basal hymenopteran relationships, *Zool. Scripta* **28**:69–74.

Edwards, R., 1980, *Social Wasps: Their Biology and Control*, Rentokil, East Grinstead, U.K.

Evans, H. E., and Eberhard, M. J. W., 1970, *The Wasps*, University of Michigan Press, Ann Arbor.

Free, J. B., 1982, *Bees and Mankind*, Allen and Unwin, London.

Gauld, I., and Bolton, B. (eds.), 1988, *The Hymenoptera*, Oxford University Press, London.

Gordh, G., Legner, E. F., and Caltagirone, L. E., 1999, Biology of parasitic Hymenoptera, in: *Handbook of Biological Control: Principles and Applications of Biological Control* (T. S. Bellows and T. W. Fisher, eds.), Academic Press, San Diego.

Goulet, H., and Huber, J. T. (eds.), 1993, *Hymenoptera of the World: An Identification Guide to Families*, Agriculture Canada, Research Branch, Ottawa.

Hölldobler, B., and Wilson, E. O., 1990, *The Ants*, Springer-Verlag, Berlin.

Jervis, M., and Vilhelmsen, L., 2000, Mouthpart evolution in adults of the basal, 'symphytan', hymenopteran lineages, *Biol. J. Linn. Soc.* **70**:121–146.

Malyshev, S. M., 1968, *Genesis of the Hymenoptera and the Phases of Their Evolution* (O. W. Richards and B. Uvarov, eds.), Methuen, London.

Michener, C. D., 1974, *The Social Behavior of the Bees: A Comparative Study*, Harvard University Press, Cambridge, MA.

Michener, C. D., 2000, *The Bees of the World*, Johns Hopkins University Press, Baltimore.

Naumann, I. D., 1991, Hymenoptera, in: *The Insects of Australia*, 2nd ed., Vol. 2 (CSIRO, ed.), Melbourne University Press, Carlton, Victoria.

O'Neill, K. M., 2001, *Solitary Wasps: Behavior and Natural History*, Cornell University Press, Ithaca.

Richards, O. W., 1977, Hymenoptera. Introduction and keys to families (2nd ed.), *R. Entomol. Soc. Handb. Ident. Br. Insects* **6**(1):1–100.

Ronquist, F., 1999, Phylogeny of the Hymenoptera (Insecta): The state of the art, *Zool. Scripta* **28**:3–11.

Ronquist, F., Rasnitsyn, A. P., Roy, A., Eriksson, K., and Lindgren, M., 1999, Phylogeny of the Hymenoptera: A cladistic reanalysis of Rasnitsyn's (1988) data, *Zool. Scripta* **28**:13–50.

Ross, K. G., and Matthews, R. W. (eds.), 1991, *The Social Biology of Wasps*, Cornell University Press, Ithaca.

Seeley, T. D., 1985, *Honeybee Ecology: A Study of Adaptation in Social Life*, Princeton University Press, Princeton, NJ.

Spradbury, J. P., 1973, *Wasps: An Account of the Biology and Natural History of Social and Solitary Wasps*, Sidgwick and Jackson, London.

Stephen, W. P., Bohart, G. E., and Torchio, P. F., 1969, *The Biology and External Morphology of Bees with a Synopsis of the Genera of Northwestern America*, Agricultural Experiment Station, Oregon State University, Corvallis.

Sudd, J. H., 1967, *An Introduction to the Behaviour of Ants*, Arnold, London.

Wilson, E. O., 1971, *The Insect Societies*, Harvard University Press, Cambridge, MA.

Vilhelmsen, L., 2001, Phylogeny and classification of the extant basal lineages of the Hymenoptera (Insecta), *Zool. J. Linn. Soc.* **131**:393–442.

II

Anatomy and Physiology

II

11

The Integument

1. Introduction

The integument of insects (and other arthropods) comprises the basal lamina, epidermis, and cuticle. It is often thought of as the "skin" of an insect but it has many other functions (Locke, 1974). Not only does it provide physical protection for internal organs but, because of its rigidity, it serves as a skeleton to which muscles can be attached. It also reduces water loss to a very low level in most Insecta, a feature that has been of great significance in the evolution of this predominantly terrestrial class. In addition to these primary functions, the cuticular component of the integument performs a number of secondary duties. It acts as a metabolic reserve, to be used cyclically to construct the next stage, or during periods of great metabolic activity or starvation. It prevents entry of foreign material, both living and nonliving, into an insect. In many insects the waxy outer layer serves as a repository for contact sex pheromones (Chapter 13, Section 4.1.1). The color of insects is also a function of the integument, especially the cuticular component.

The integument is not a uniform structure. On the contrary, both its cellular and acellular components may be differentiated in a variety of ways to suit an insect's needs. Epidermal cells may form specialized glands that produce components of the cuticle or may develop into particular parts of sense organs. The cuticle itself is variously differentiated according to the function it is required to perform. Where muscles are attached or where abrasion may occur it is thick and rigid; at points of articulation it is flexible and elastic; over some sensory structures it may be extremely thin.

2. Structure

The innermost component of the integument (Figure 11.1) is the basal lamina, an amorphous but selectively porous acellular layer that is attached by hemidesmosomes to the epidermal cells. It is up to 0.5 μm thick and is produced mainly by the epidermis, though there are reports that hemocytes also participate. The chemical nature of the basal lamina is poorly understood though neutral mucopolysaccharide, glycoproteins, and collagen, similar to that of vertebrates, have been identified.

The epidermis (hypodermis) is a more or less continuous sheet of tissue, one cell thick, responsible for secreting the bulk of the cuticle. During periods of inactivity, its

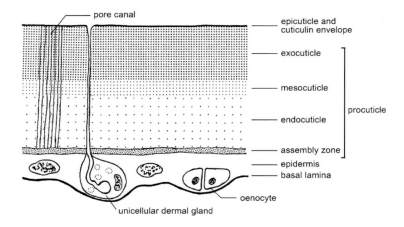

FIGURE 11.1. Diagrammatic cross-section of mature integument.

cells are flattened and intercellular boundaries are indistinct. When active, the cells are more or less cuboidal, and their plasma membranes are readily apparent; one to several nucleoli, extensive rough endoplasmic reticulum, and many Golgi complexes are evident (Locke, 1991, 1998). A characteristic feature of the apical (cuticle-facing) surface of epidermal cells are the plasma membrane plaques, specialized regions of the plasma membrane at the tips of fingerlike microvilli, from which the cuticulin envelope and new chitin fibers arise (Section 3.1). Electron microscopy has shown that, at metamorphosis, the epidermal cells develop basal processes ("feet") which can extend to become connected with the basal lamina and with other epidermal cells. When the feet shorten, the basal lamina is buckled and rearrangement of cells occurs, resulting in a change in the insect's shape, for example, from a long, thin caterpillar to a short, fat pupa (Locke, 1991, 1998). Epidermal cells also possess the ability to develop various forms of cytoskeletal extensions which can be used, for example, to draw tracheoles closer to the cell for increased oxygen supply, or to maintain intercellular contact as the cells migrate during wound healing and changes in body shape. The density of cells in a particular area varies, following a sequence that can be correlated with the molting cycle. The cells often contain granules of a reddish-brown pigment, insectorubin, which in some insects contributes significantly to their color. However, in most insects color is produced by the cuticle (Section 4.3).

Epidermal cells may be differentiated into sense organs or specialized glandular cells. Oenocytes are large, ductless, often polyploid cells, up to 100 μm in diameter. They occur in pairs or small groups and the cells of each group may be derived from one original epidermal cell. Usually they move to the hemocoelic face of the basal lamina, though in some insects they form clusters in the hemocoel or migrate and reassemble within the fat body. Oenocytes show signs of secretory activity that can be correlated with the molting cycle, and, on the basis of certain staining reactions, it has been suggested that they produce the lipoprotein component of epicuticle. In addition, ultrastructural and biochemical studies have led to the proposal that these cells produce ecdysone (Locke, 1969; Romer, 1991). They also synthesize components of the cuticular wax, including some contact sex pheromones (Blomquist and Dillwith, 1985; Schal *et al.*, 1998). Dermal glands of various types are also differentiated. In their simplest form the glands are unicellular and have a long duct that penetrates the cuticle to the exterior. More commonly, they are composed of several cells.

The gland cells again exhibit cyclical activity associated with new cuticle production, and it has been proposed that they secrete the cement layer of epicuticle.

The cuticle, which is mainly produced by the epidermal cells, usually includes three primary layers, the inner procuticle, middle epicuticle, and outer cuticulin envelope (Locke, 2001) (Figure 11.2). In older accounts of the integument the cuticulin envelope is treated as part of the epicuticle. However, Locke (1998, 2001) has argued that, because of its distinct origin, structure and functions, the cuticulin envelope should be considered separate from the epicuticle. All three primary layers are present over most of the body surface and in the cuticle that lines major invaginations such as the foregut, hindgut, and tracheae. However, the procuticle is very thin or absent, and certain components of the epicuticle may be missing, where flexibility or sensitivity is needed, for example, over sensory structures and the lining of tracheoles. Only the cuticulin envelope is universally present, except for the pores over chemosensilla (Chapter 12, Section 4.1).

The procuticle (= fibrous cuticle) forms the bulk of the cuticle and in most species is differentiated into two zones, endocuticle and exocuticle, which differ markedly in their

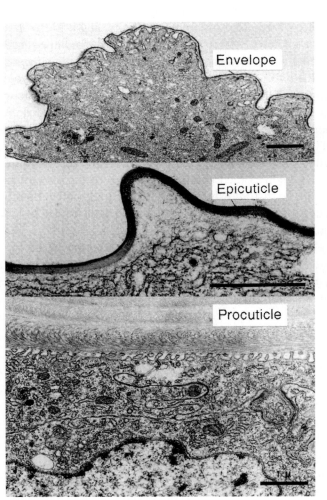

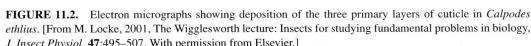

FIGURE 11.2. Electron micrographs showing deposition of the three primary layers of cuticle in *Calpodes ethlius*. [From M. Locke, 2001, The Wigglesworth lecture: Insects for studying fundamental problems in biology, *J. Insect Physiol.* **47**:495–507. With permission from Elsevier.]

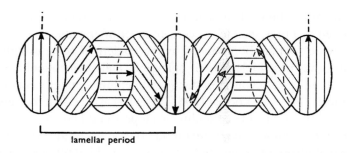

lamellar period

FIGURE 11.3. Diagram showing orientation of microfibers in lamellae of endocuticle. [From A.C. Neville and S. Caveney, 1969, Scarabaeid beetle exocuticle as an optical analogue of cholesteric liquid crystals, *Biol. Rev.* **44**: 531–562. By permission of Cambridge University Press, London.]

physical properties but only slightly in their chemical composition. In some cuticles the border between the two is not clear and an intermediate area, the mesocuticle, is visible. Adjacent to the epidermal cells a narrow amorphous layer, the assembly zone, may be seen where chitin microfibers are deposited and oriented.

The endocuticle is composed of lamellae (Figure 11.3). Electron microscopy reveals that each lamella is made up of a mass of microfibers arranged in a succession of planes, all fibers in a plane being parallel to each other. The orientation changes slightly from plane to plane making cuticle like plywood with hundreds of layers. The exocuticle is the region of procuticle adjacent to the epicuticle that is so stabilized that it is not attacked by the molting fluid and is left behind with the exuvium at molting (Locke, 1974). Not only is the exocuticle chemically inert, it is hard and extremely strong. It is, in fact, procuticle that has been "tanned" (Section 3.3). Exocuticle is absent from areas of the integument where flexibility is required, for example, at joints and intersegmental membranes, and along the ecdysial line. In many soft-bodied endopterygote larvae the exocuticle is extremely thin and frequently cannot be distinguished from the epicuticle and cuticulin envelope.

Procuticle is composed almost entirely of protein and chitin. The latter is a nitrogenous polysaccharide consisting primarily of *N*-acetyl-D-glucosamine residues together with a small amount of glucosamine linked in a β1,4 configuration (Figure 11.4). In other words, chitin is very similar to cellulose, another polysaccharide of great structural significance, except that the hydroxyl group of carbon atom 2 of each residue is replaced by an acetamide group. Because of this configuration, extensive hydrogen bonding is possible between adjacent chitin molecules which link together (like cellulose) to form microfibers. Chitin makes

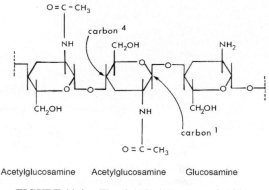

Acetylglucosamine   Acetylglucosamine   Glucosamine

FIGURE 11.4. The chemical structure of chitin.

up between 25% and 60% of the dry weight of procuticle but is not found in the epicuticle and cuticulin envelope. It is associated with the protein component, being linked to protein molecules by covalent bonds, forming a glycoprotein complex. Studies have shown that the epidermis secretes more than a dozen major proteins into the cuticle in a carefully timed sequence, probably under hormonal control (Suderman *et al.*, 2003). Interestingly, cuticular proteins of similar molecular weights have been found in a range of insect species suggesting that the chemical nature of the cuticle has been strongly conserved through evolution. The amino acid composition of the proteins determines their properties. For example, endocuticular proteins are generally rich in hydrophobic amino acids with bulky side chains and are loosely packed (not compact) molecules. This provides the endocuticle with flexibility and will also facilitate "creep" (the ability of layers to slide over each other), hence intrastadial growth in soft-bodied insects such as caterpillars. Conversely, in hard, stiff exocuticle, it is small, compact amino acids that predominate (Hepburn, 1985).

In the exocuticle, adjacent protein molecules are linked together by a quinone molecule, and the cuticle is said to be tanned (Section 3.3). The tanned (sclerotized) protein, which is known as "sclerotin," comprises several different molecules. Resilin is a rubberlike material found in cuticular structures that undergo springlike movements, for example, wing hinges, the proboscis of Lepidoptera, the hind legs of fleas (Chapter 14, Section 3.1.2.), and the wing-hinge ligament that stretches between the pleural process and second axillary sclerite (Chapter 14, Sections 3.3.1 and 3.3.3) (Neville, 1975). Like rubber, resilin, when stretched, is able to store the energy involved. When the tension is released, the stored energy is used to return the protein to its original length.

In addition to these structural proteins, enzymes also exist in the cuticle, including phenoloxidases, which catalyze the oxidation of dihydric phenols used in the tanning process (Section 3.3). These enzymes appear to be located in or just beneath the epicuticle.

A variety of pigments have been found in the cuticle (or in the epidermis) which may give an insect its characteristic color (Section 4.3). Also, in a few beetles and larvae and pupae of some Diptera, mineralized calcium (as the carbonate) is deposited, presumably to increase rigidity (Leschen and Cutler, 1994).

Certain processes occur at the surface of the cuticle after it has been formed, for example, secretion and repair of the wax layer and tanning of the outer procuticle. Thus, a route of communication must remain open between the epidermis and cuticular surface. This route takes the form of pore canals which are formed as the new procuticle is deposited (Section 3.1), and which may or may not contain a cytoplasmic process. Most often, the canals do not contain an extension of the epidermal cell but have at least one "filament" produced by the cell. Locke (1974) suggested that the filament(s) might keep a channel open in the newly formed cuticle until the latter hardens, and anchor the cells to the cuticle. In some insects the pore canals become filled with cuticular material once epicuticle formation (including tanning) is complete. The pore canals terminate immediately below the epicuticle. Running from the tips of the pore canals to the outer surface of the epicuticle are lipid-filled channels known as wax canals.

The epicuticle is a composite structure produced partly by epidermal cells and partly by specialized glands. It ranges in thickness from a fraction of a micrometer to several micrometers and generally comprises three layers. The layers are, from outside to inside, cement, wax (these are secreted outside the cuticulin envelope), and the so-called protein epicuticle. The nature of cement varies, though it is likely to be approximately similar to shellac. The latter is a mixture of laccose and lipids. The cement is undoubtedly a hard, protective layer in some insects. In others it appears to be more important as a sponge that soaks up excess wax. The latter could quickly replace that lost, for example, by surface abrasion. The wax is a complex mixture whose composition varies both among and within species, sometimes over different body regions of the same insect, and in some species seasonally. Generally, long-chain hydrocarbons and fatty acid esters predominate, though varied proportions of alcohols, fatty acids, and sterols may also occur. In some species the mixture has relatively few different components, whereas in others, for example, *Musca*, more than 100 compounds have been identified (Blomquist and Dillwith, 1985; Jacob *et al.*, 1997). According to Locke (1974), within the wax layer three regions can be distinguished. Adjacent to the cuticulin envelope is a monolayer of tightly packed molecules in liquid form that gives the cuticular surface its high contact angle with water and its resistance to water loss (but see Section 4.2.). Most wax is in the middle layer, which is less ordered and permeates the cement. The outer wax layer, which comprises crystalline wax blooms, is not present in all insects. The innermost layer of the epicuticle, the protein epicuticle, lies beneath the cuticulin envelope. It may be several micrometers thick and like the cuticulin envelope it covers almost all of the surface of the insect. It is absent from tracheoles and parts of some sense organs. It comprises dense, amorphous protein tanned in a manner similar to the protein of the exocuticle (Section 3.3) but contains no chitin.

The cuticulin envelope (about 20 nm thick) extends over the entire body surface and ectodermal invaginations, including the most minute tracheoles, but is absent from specific areas of sense organs and from the tips of certain gland cells. It may be considered the most important layer of the cuticle for the following reasons (Locke, 1974, 2001). (1) It is a selectively permeable barrier. During breakdown of the old cuticle, it allows the "activating factor" for the molting gel to move out and the products of cuticular hydrolysis to enter, yet it is impermeable to the enzymes in the molting fluid. It is permeable to waxes (as these are deposited only after the cuticulin layer has formed) and, in some insects, it permits the entry of water. (2) It is inelastic and, therefore, serves as a limiter of growth. (3) It provides the base on which the wax monolayer sits. The nature of the cuticulin envelope will therefore determine whether the wax molecules are oriented with their polar or nonpolar groups facing outward and, therefore, the surface properties of the cuticle. (4) It plays a role in determining the surface pattern of the cuticle. (5) It is resistant to abrasion and helps prevent infection. (6) It is involved in production of physical colors. Despite the importance of the cuticulin envelope, its composition is largely unknown.

3. Cuticle Formation

Formation of new cuticle (Figure 11.5) may be viewed largely as a succession of syntheses by epidermal cells, with dermal glands and oenocytes adding their products at the appropriate moment (Locke, 1974). It must be realized, however, that other, related processes such as dissolution of old cuticle are going on concurrently and that cuticle formation is partly a preecdysial and partly a postecdysial event; that is, much endocuticle formation, tanning of the outer procuticle, wax secretion, and other processes occur after the remains of the old cuticle are shed.

3.1. Preecdysis

In most species the onset of a molting cycle is marked by an increase in the volume of the epidermal cells and/or by epidermal mitoses. These events are soon followed by

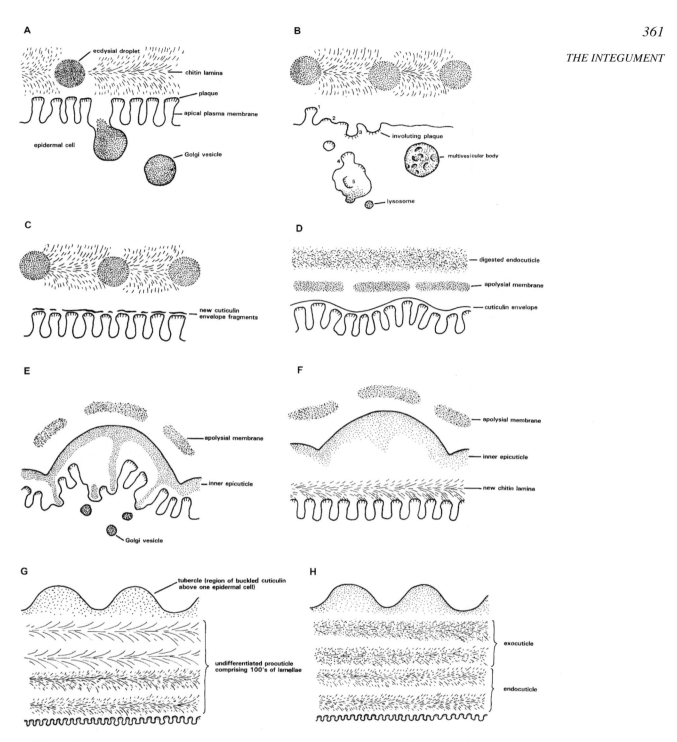

FIGURE 11.5.   Summary of cuticle formation during the molt/intermolt cycle. Individual components are not drawn to scale. The numbers in Figure 11.5B indicate the sequence of actions resulting in plaque digestion. (A) Secretion of ecdysial droplets. (B) Pinocytosis and apolysis of plasma membrane. (C) Redifferentiation of plaques and cuticulin envelope deposition. (D) Cuticulin envelope complete and digestion of old cuticle. (E) Secretion of inner epicuticle and bucking of cuticulin envelope. (F) Beginning of procuticle secretion. (G) Cuticle immediately after ecdysis. (H) Cuticle after tanning.

apolysis, the detachment of the epidermis from the old cuticle. The epidermal cells, at this time, show signs of preparation for future synthetic activity. One or more nucleoli become prominent, the number of ribosomes increases, and the ribonucleic acid content of the cells is elevated. Two components of the epidermal cells are especially important, namely, the Golgi complexes and the plasma membrane plaques, whose activities alternate to create the new cuticle. Just prior to apolysis, Golgi complex activity increases, and the vesicles produced migrate to the apical plasma membrane where they release their contents—ecdysial droplets—between the epidermal microvilli (Figure 11.5A). The ecdysial droplets contain proteinases and chitinases for cuticle digestion, though the enzymes remain in an inactive form until after formation of the new cuticulin envelope when the epidermal cells secrete an "activation factor." In *Calpodes ethlius* large quantities of an amidase are generated by the epidermis and fat body during the intermolt. The amidase (in its inactive form) accumulates in the hemolymph until the molt cycle begins, when it moves into the molting fluid and is activated, enabling precise initiation of cuticle breakdown (Marcu and Locke, 1999). Between 80% and 90% of the old cuticle is digested and may be reused in the production of new cuticle. In earlier accounts it was assumed that the molting fluid, including the breakdown products, were resorbed across the body wall. However, recent studies have demonstrated that most of the molting fluid is recovered by both oral and anal drinking, reentering the body cavity by absorption across the midgut wall (Yarema *et al.*, 2000). The exocuticle, muscle insertions, and sensory structures in the integument are not degraded by molting fluid. Thus, an insect is able to move and receive information from the environment more or less to the point of ecdysis.

After release of the ecdysial droplets, the microvilli are withdrawn and their plaques are pinocytosed and digested in multivesicular bodies (Figure 11.5B). New microvilli, with plaques at their tips, then differentiate. The first layer of new cuticle deposited is the cuticulin envelope. Minute convex patches of cuticulin appear above the plaques (Figure 11.5C), the patches eventually fusing together to form a continuous but buckled layer (Figure 11.5D). The buckling permits expansion of the cuticle after molting and is also important in the formation of annuli and taenidia in tracheae and tracheoles (Chapter 15, Section 2.1). Other buckling patterns determine the specific surface structure of scales, bristles, and microtrichia. Oenocytes are maximally active at this time, and it is possible that they are involved in cuticulin formation, perhaps by synthesizing a precursor for the epidermal cells. When the envelope is complete, it becomes tanned. The Golgi complexes then show renewed activity, their vesicles discharging their contents to form the inner (protein) epicuticle (Figure 11.5E).

Before the inner epicuticle is fully formed production and deposition of the new procuticle begin. In contrast to the epicuticle, whose layers are produced sequentially from inside to outside, the new procuticle is produced with the newest layers on the inside. Again, it is the plasma membrane plaques that are involved, new chitin fibers arising on their outer surface (Figure 11.5F,G). However, details of the mechanism by which new procuticle is produced remain sketchy. The epidermal cells contain the enzymes necessary for synthesis of acetylglucosamine from trehalose. Acetylglucosamine units perhaps are then secreted into the apolysial space, polymerization into chitin being promoted by the enzyme chitin synthetase attached to the plasma membrane plaques. Some procuticular proteins are synthesized by the epidermal cells while others are acquired from the hemolymph (Sass *et al.*, 1993; Suderman *et al.*, 2003). How the proteins become incorporated into the procuticle remains unclear.

Deposition of the wax layer of the epicuticle begins some time prior to ecdysis. For example, in *Blattella germanica* oenocytes associated especially with the integument of abdominal sternites 3–6 become major producers of hydrocarbons early in the molt cycle. The hydrocarbons are stored in fat body, then transported to the epidermis bound to lipophorin a few days before molting occurs (Schal *et al.*, 1998; Young *et al.*, 1999). The wax is secreted by the epidermal cells, probably as lipid-water liquid crystals, and passes along the pore canals to the outside. Wax production continues after ecdysis and, in some insects, throughout the entire intermolt period and in the adult stage.

3.2. Ecdysis

At the time of ecdysis, the old cuticle comprises only the original exocuticle and epicuticle. In many insects it is separated from the new cuticle by an air space and a thin ecdysial (apolysial) membrane that is formed from undigested inner layers of the endocuticle. These layers are not digested because they became tanned along with the new cuticulin envelope. Shortly before molting an insect begins to swallow air (or water, if aquatic), thereby increasing the hemolymph pressure by as much as 12 kPa. Hemolymph is then localized in the head and thorax following contraction of intersegmental abdominal muscles. In many insects these muscles become functional only at the time of ecdysis and histolyze after each molt. The local increase in pressure in the anterior part of the body causes the old cuticle to split along a weak ecdysial line where the exocuticle is thin or absent. An insect continues to swallow air or water after the molt in order to stretch the new cuticle prior to tanning.

3.3. Postecdysis

Several processes are continued or initiated after ecdysis. As noted wax secretion continues, and the major portion of the endocuticle is deposited at this time. Indeed, endocuticle production in some insects appears to be a more or less continuous process throughout the intermolt period. It is also at this time that the dermal glands release the cement.

The most striking postecdysial event, however, is the differentiation of the exocuticle, that is, the hardening of the outer procuticle (Figure 11.5H). This results from a biochemical process known as tanning (sclerotization), in which proteins become covalently bound to each other (and hence stabilized) by means of quinones. Hardening is usually accompanied by darkening (melanization), though the two may be distinct processes; that is, some species have pale but very hard cuticles. Though tanning is discussed here in the context of the cuticle, it should be noted that it also an important process in the final structure of insect egg shells, egg cases (oothecae) and protective froths, cocoons, puparia and various silk structures. Indeed, much of the basic understanding of tanning came from studies using the cockroach ootheca and the fly puparium (Andersen, 1985; Hopkins and Kramer, 1992). More recently, the cuticles of the *Manduca sexta* pupa and of locusts and grasshoppers have proved to be excellent models for study of this process. Though the details may differ, it is now possible to provide a basic scheme for the events that culminate in a tanned cuticle (Figure 11.6).

Before tanning begins, the level of the amino acid tyrosine in the hemolymph increases. The tyrosine is mostly bound to glucose, phosphate, or sulphate, forming water-soluble conjugates. This is thought to increase the amount of tyrosine that can be carried in the

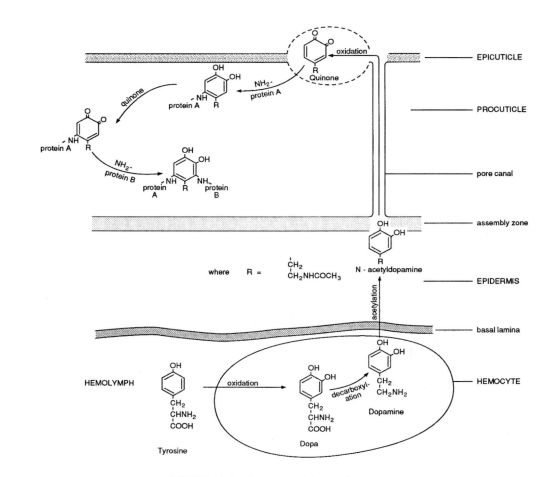

FIGURE 11.6. Summary of the tanning process.

hemolymph, to protect it from oxidation, and to prevent its use in competing metabolic pathways. In a few species the tyrosine accumulates within fat-body vacuoles, particularly as the dipeptide N-β-alanyltyrosine. The tyrosine is converted to "dopa" (dihydroxypheny-lalanine) which is then decarboxylated forming dopamine, N-β-alanyldopamine, and other catecholamines (Hopkins and Kramer, 1992; Hopkins *et al.*, 1999). Like tyrosine, the catecholamines occur as conjugates and accumulate in the hemolymph. Only when the catecholamines are taken up by the epidermal cells are the conjugates released. Within the epidermis, catecholamines are converted to the corresponding N-acetyl derivatives which then move via the pore canals to the epicuticle where they are oxidized to quinones under the influence of phenoloxidases. Two distinct phenoloxidases have been isolated from cuticle, tyrosinase, and laccase. Tyrosinase may be involved in tanning in some situations but is likely more important in wound healing (Chapter 17, Section 5.1). By contrast, the activity of laccase is closely correlated with the period of tanning, and this enzyme is therefore assumed to play the major role in the synthesis of quinone tanning agents. The quinones diffuse back into the procuticle and link protein molecules together. According to Hackman (1974), quinones may polymerize, a process that makes them larger, more effective tanning agents. The polymer will have more reactive sites and be capable of bridging larger distances, the result being that more protein molecules can be linked. The quinones combine covalently with the terminal amino group or a sulfhydryl group of the protein to form an N-catechol protein, which in the presence of excess quinone is oxidized to an *N*-quinonoid protein. The latter is able to react with the terminal amino group or sulfhydryl group of another protein to link the two molecules. In a number of species β-alanine is incorporated into the cuticle during tanning, and it is speculated that this amino acid becomes attached to proteins, leaving its amino terminus free to participate in linking the quinone and protein. In addition, the quinone can react with the ε-amino groups of lysine residues in the protein molecule to effect further binding. Autotanning is also a possibility. This is the oxidation of aromatic amino acids (e.g., tyrosine) within a polypeptide chain to form quinone compounds that can subsequently link with amino groups in adjacent chains. Though autotanning has never been demonstrated, "dopa" has been found in small amounts in certain cuticular proteins of *Manduca sexta* (Okot-Kotber *et al.*, 1994) and presumably could engage in cross-linking reactions.

The color of tanned cuticle depends on the amount of *o*-quinone that is present. When this molecule is present in small quantities the cuticle is pale; if it is in excess, and especially if it polymerizes, the cuticle when tanned is dark. In some cuticles production of the characteristic dark brown or black color (melanization) is entirely separate from tanning, although it involves tyrosine and its oxidation product dopa. Oxidation of dopa yields dopaquinone which can form an indole ring derivative, dopachrome. Decarboxylation, oxidation, and polymerization then occur to produce the pigment melanin.

As tanning occurs, the cuticle undergoes a number of physical rearrangements. Its water content decreases as a result of (1) a decline in the number of hydrophilic groups such as -NH$_2$ because of their involvement in tanning and (2) a tighter packing of the chitin and protein molecules. This leads to an overall decrease in the thickness of the cuticle.

3.4. Coordination of Events

It is essential that the complex series of events comprising a molt cycle be coordinated. Central to this coordination are hormones (Chapter 13, Section 3), though the expression of their effects can be modified by other factors, for example, nutrition and injury. Many events within a molt cycle are influenced by hormones, but still unclear is whether this influence is direct or indirect; that is, do hormones directly regulate all of these events or merely initiate a cascade of reactions? The picture is further complicated because more than one hormone may influence the same event.

β-Ecdysone affects many events throughout a molt cycle. It initiates apolysis, stimulates epidermal cell mitosis, increases chitinase activity, promotes the synthesis of select epidermal cell proteins, triggers release of tyrosine from its conjugates, and induces the synthesis of several enzymes involved in tanning. Further, the development and activity of the epidermal feet are correlated with changing hemolymph ecdysone levels in the final larval instar. Among the earliest studies of the site of action of ecdysone were those of Karlson *et al.* (see references in Neville, 1975), specifically in relation to tanning. Ecdysone stimulates synthesis of dopa decarboxylase mRNA. Karlson's group suggested that ecdysone acted on the epidermal cells and this is supported by *in vitro* studies on isolated epidermis. However, for some species, it also seems likely that the hemocytes are a target organ. More recently, the ability of β-ecdysone to induce the expression of genes for other key molecules in the molt cycle has been shown, for example, for chitinase (Zheng *et al.*, 2003). It appears that genes associated with the molt cycle differ in their sensitivity to the hormone, some being switched on at low (or decreasing) concentrations, others requiring higher (or increasing) titers for their expression. In this way, varying amounts of β-ecdysone throughout the molt cycle serve to synchronize the many steps and processes that occur. It must be noted that β-ecdysone does not act directly on genes; rather, it binds with receptors in the nuclear membrane causing release of second messengers (e.g., cyclic AMP) whose action then stimulates gene expression.

Though the modifying influence of juvenile hormone on ecdysone effects has been known for a considerable time (Chapter 21, Section 6.1), it has been shown only recently that it, too, acts on the genome. Its effect is to program the epidermis to secrete a cuticle characteristic of the juvenile stage. Like β-ecdysone, juvenile hormone binds with nuclear receptors to precipitate, via a second messenger, reactions leading to gene expression.

Bursicon, like ecdysone, affects many processes, though all are related to tanning and melanization of the cuticle. Bursicon's primary effect appears to be to increase the permeability of the hemocyte wall to tyrosine and the epidermal cell wall to dopamine. The hormone may exert this effect via a cyclic AMP-mediated system. In addition, bursicon may activate a tyrosinase in hemocytes, which catalyzes the oxidation of tyrosine to dopa.

Finally, ecdysis is regulated by an eclosion hormone secreted by the brain in silk moths (Chapter 21, Section 6.2) and wax synthesis and endocuticle deposition require the presence of the corpus allatum/corpus cardiacum complex in *Calpodes ethlius*.

4. Functions of the Integument

Most functions of the integument relate to the physical structure of the cuticle though the latter may serve as a source of metabolites during periods of starvation. The primary functions may be discussed under three headings: strength and hardness, permeability, and production of color.

4.1. Strength and Hardness

The few studies that have been carried out on the mechanical properties of insect cuticle indicate that is of medium rigidity and low tensile strength (Locke, 1974). There is, however, wide variation from this general statement; for example, the cuticles of most endopterygote larvae are extremely plastic, whereas the mandibular cuticle of many biting insects may be extremely hard, enabling them to bite through metal. Further, there is an obvious difference in properties between sclerites and intersegmental membranes, and between typical non-elastic cuticle and that which contains a high proportion of resilin.

Though the above properties indicate that the cuticle is satisfactory as a "skin" preventing physical damage to internal organs, discussion of the suitability of the cuticle as a skeletal component must include an appreciation of overall body structure (Locke, 1974). Most components of insect (and other arthropod) bodies may be considered as cuticular cylinders or spheres. Such a tubular shell (used here in the engineering sense to mean a surface-supporting structure that is thin in relation to total size) is about three times as strong as a solid rod of the same material having the same cross-sectional area as the shell (i.e., they both contain the same amount of skeletal material). The force required to distort the shell is proportional to the thickness of the shell and inversely proportional to the cross-sectional area of the whole body. Thus, in small organisms where the thickness of the shell is great relative to the cross-sectional area of the body, the use of a shell as an exoskeletal structure is quite feasible. In larger organisms the advantage of the extra strength relating to a shell type of skeleton is greatly outweighed (literally as well as metaphorically) by the massive increase in thickness of the shell that would be required and, perhaps, by the physiological problems of producing the large amounts of material required for its construction.

4.2. Permeability

For different insects there exists a wide range of materials that are potential permeants of the integument, and of factors that affect their rate of permeation. Sometimes specific regions of the integument are constructed to facilitate entry or exit of certain materials; more often the integument is structured to prevent entry or loss. At this time we shall consider only the permeability of the cuticle to water and insecticides, of which the latter may now be considered a normal hazard for most insects. The passage of gases through the integument is considered in Chapter 15.

Water. Water may be either lost or gained through the integument. In terrestrial insects, which exist in humidities that are almost always less than saturation, the problem is to prevent loss through evaporation. In freshwater forms the problem is to prevent entry related to osmosis.

In many terrestrial insects the rate of evaporative water loss is probably less than 1% per hour of the total water content of the body (i.e., of the order of 1–3 mg/cm^2 per hr for most species). Most of this loss occurs via the respiratory system, despite the evolution of mechanical and physiological features to reduce such loss (Chapter 15). Water loss through the integument (*sensu stricto*) is extremely slight, mainly because of the highly impermeable epicuticle and in particular the wax components. Early experiments demonstrated that permeability of the integument is relatively independent of temperature up to a certain point (the transition temperature), above which it increases markedly. As a result of his studies on both artificial and natural systems, Beament (1961) concluded that the initial impermeability is related to the highly ordered wax monolayer whose molecules sit on the tanned cuticulin envelope at an angle of about 25° to the perpendicular axis, with their polar ends facing inward and nonpolar ends outward. In this arrangement the molecules are closely packed and held tightly together by van der Waals forces. As temperature increases, the molecules gain kinetic energy, and eventually the bonds between them rupture. Spaces appear and water loss increases significantly. The nature of the wax and its transition temperature can be correlated with the normal niche of the insect. Insects from humid environments or that have access to moisture in their diet, for example, aphids, caterpillars, and bloodsucking insects, have "soft" waxes, with low transition temperatures. Forms from dry environments or stages with water-conservation problems, for example, eggs and pupae, are covered with "hard" wax, whose transition temperature is high (in most species above the thermal death point of the insect).

More recent studies have questioned the validity of Beament's ordered monolayer model. Evidence against it includes the observation that hydrocarbons (non-polar molecules) are the dominant component of wax, physicochemical analyses that indicate that the lipids have no preferred orientation, and mathematical calculations that show the abrupt permeability changes at the so-called transition point to be artifactual (Blomquist and Dillwith, 1985).

Some insects that are normally found in extremely dry habitats and may go for long periods without access to free water, for example, *Tenebrio molitor* and prepupae of fleas, are able to take up water from an atmosphere in which the humidity is relatively high. Originally it was believed that uptake occurred across the body surface perhaps via the pore canals. However, it has now been demonstrated that uptake occurs across the wall of the rectum (Chapter 18, Section 4.1).

In many freshwater insects, for example, adult Heteroptera and Coleoptera, the cuticle is highly impermeable because of its wax monolayer and water gain is probably 4% or less of the body weight per day. In most aquatic insects, however, the wax layer is absent. Thus, gains of up to 30% of the body weight per day are experienced, the excess water being removed via the excretory system (Chapter 18, Section 4.2).

Insecticides.   Economic motives have stimulated an enormous interest in the permeability of the integument to chemicals, especially insecticides and their solvents (Ebeling, 1974). Though, for the most part, the cuticle acts as a physical barrier to decrease the rate of entry of such materials, there is evidence that in some insects it may also bring about metabolic degradation of certain compounds, and consequently reduction of their potency. It follows that increased resistance to a particular compound may result from changes in either the structure or the metabolic properties of the integument (see also Chapter 16, Section 5.5). For most insects, the primary barrier to the entrance of insecticides is the epicuticular wax, which dissolves and retains these largely lipid-soluble materials. For the same reason, the cement layer also probably provides some protection against penetration. The procuticle offers both lipid and aqueous pathways along which an insecticide may travel, but the precise rate at which a compound moves depends on many variables, especially thickness of the cuticle, presence or absence of pore canals, and whether the latter are filled with cytoplasmic extensions or other material. It follows that the rate of penetration will vary according to the location of an insecticide on the integument. However, it has also been noted that dissolution in the wax will facilitate lateral movement of the insecticides, perhaps allowing them to reach the tracheal system and thus gain access. Thin, membranous cuticle such as occurs in intersegmental regions or covers tactile or chemosensory hairs generally provides little resistance to penetration. The tracheal system is another site of entry. The extent to which tanning of the procuticle occurs is also related to penetration rate. As the chitin-protein micelles become more tightly packed and the cuticle partially dehydrated, permeability decreases.

In addition, but obviously related to the physical features of the cuticle, the physicochemical nature of an insecticide is an important factor in determining the rate of entry. Especially significant is the partition coefficient (the relative solubility in oil and in water) of an insecticide or its solvent. In order to penetrate the epicuticular wax the material must be relatively lipid-soluble. However, in order to pass through the relatively polar procuticle and, eventually, to leave the integument to move toward its site of action, the material must be partially water-soluble. Thus, correct formulation of an insecticidal solution is an important consideration.

It should be apparent from the above discussion that few generalizations can be made. At the present time, therefore, the suitability of an insecticide must be considered separately for each species. Because of the factors that affect the entry of insecticides, a great difference usually exists between "real toxicity," that is, toxicity at the site of action, and "apparent toxicity," the amount of material that must be applied topically to bring about death of the insect. The chief feature that relates the two is obviously the "penetration velocity," that is, the rate at which material passes through the cuticle. When the rate is high, the real and apparent toxicity values will be nearly identical.

As in other animals, the color of insects serves to conceal them from predators (sometimes through mimicry), frighten or "warn" predators that potential prey is distasteful, or facilitate intraspecific and/or sexual recognition. It may be used also in thermoregulation. The color of an insect generally depends on the integument. Rarely, an insect's color may be the result of pigments in tissues or hemolymph below the integument. For example, the red color of *Chironomus* larvae is caused by hemoglobin in solution in the hemolymph. Integumental colors may be produced in two ways. Pigmentary colors are produced when pigments in the integument (usually the cuticle) absorb certain wavelengths of light and reflect others (Fuzeau-Bresch, 1972). Physical (structural) colors result when light waves of a certain length are reflected as a result of the physical features of the surface of the integument.

Pigmentary colors result from the presence in molecules of particular bonds between atoms. Especially important are double bonds such as C=C, C=O, C=N, and N=N which absorb particular wavelengths of light (Hackman, 1974; Kayser, 1985). The integument may contain a variety of pigment molecules that produce characteristic colors. Usually the molecule, known as a chromophore, is conjugated with a protein to form a chromoprotein. The brown or black color of many insects results usually from melanin pigment. Melanin is a molecule composed of polymerized indole or quinone rings. Typically, it is located in the cuticle, but in *Carausius* it occurs in the epidermis, where it is capable of movement and may be concerned with thermoregulation as well as concealment. Carotenoids are common pigments of phytophagous insects. They are acquired through feeding as insects are unable to synthesize them. Carotenoids generally produce yellow, orange, and red colors, and, in combination with a blue pigment, mesobiliverdin, produce green. Examples of the use of carotenoids include the yellow color of mature *Schistocerca* and the red color of *Pyrrhocoris* and *Coccinella*. Pteridines, which are purine derivatives, are common pigments of Lepidoptera, Hymenoptera, and the hemipteran *Dysdercus*, and produce yellow, white, and red colors. Ommochromes, which are derivatives of tryptophan, an amino acid, are an important group of pigments that produce yellow, red, and brown colors. Examples of colors resulting from ommochromes are the pink of immature adult *Schistocerca*, the red of Odonata, and the reds and browns of nymphalid butterflies. In some insects the characteristic red or yellow body color is the result of flavones originally present in the foodplant. Uric acid, the major nitrogenous excretory product of insects (Chapter 18, Section 3.1), is deposited in specific regions of the epidermis in some insects. For example, in *Dysdercus* it is responsible for the white areas of the integument.

Physical colors are produced by scattering, interference, or diffraction of light though the latter is extremely rare. Most white, blue, and iridescent colors are produced using the first two methods. White results from the scattering of light by an uneven surface or by granules that occur below the surface. When the irregularities are large relative to the wavelength of light, all colors are reflected equally, and white light results. An interference color is produced by laminated structures when the distance between successive laminae is similar to the wavelength of light that produces that particular color. As light strikes the laminae light waves of the "correct" length will be reflected by successive surfaces, and the color they produce will therefore be reinforced. Light waves of different lengths will be out of phase. Changing the angle at which light strikes the surface (or equally the angle at which the surface is viewed) is equivalent to altering the distance between laminae. In turn, this will alter the wavelength that is reinforced and color that is produced. This change of color in relation to the angle of viewing is termed iridescence. Iridescent colors are common in many Coleoptera and Lepidoptera.

4.4. Other Functions

The cuticular waxes may have important roles in preventing the entry of microorganisms and in chemical communication (i.e., they serve as semiochemicals). It has been suggested that the waxes may prevent adhesion of microorganisms or may be toxic to them. Cuticular hydrocarbons are also known to serve as contact sex pheromones, for example, in female Diptera and *Blattella*, attracting or inducing copulatory behavior in males, or serving as an aphrodisiac to keep the male in position until insemination has occurred (Schal *et al.*, 1998). In termites, the cuticular hydrocarbon blend is highly specific and serves as a species- and/or caste-recognition pheromone. (See also Chapter 13, Section 4.1.2.) Interestingly, some beetles that live in termite colonies produce the same hydrocarbon profile as the host, enabling them to remain unmolested in the nest. The species-specific nature of the lipids has been turned to advantage by some parasitic Hymenoptera who use these chemical cues (known as kairomones [Chapter 13, Section 4.2.) to locate their host (Blomquist and Dillwith, 1985).

5. Summary

The integument is a layered structure that comprises a basal lamina, epidermis, procuticle, epicuticle, and cuticulin envelope. The basal lamina contains carbohydrate and collagenlike material and is mainly a product of the epidermal cells. The epidermis is mostly a one-cell-thick layer of uniform cells, though the cells can differentiate to form dermal glands, oenocytes, or sensory structures. The procuticle includes an inner endocuticle and an outer exocuticle, both of which contain a mixture of chitin (a polymerized nitrogenous polysaccharide of fibrous form) and protein. The endocuticle is laminar, flexible, and capable of being digested by molting fluid; the exocuticle is hard, inflexible, and chemically inert as a result of tanning, the covalent linking of proteins via quinones. Exocuticle is absent from areas of the body where flexibility is required, for example, at joints and intersegmental membranes, and is very thin in soft-bodied larvae. The epicuticle includes a tanned protein layer, the proteinaceous epicuticle, which lies inside the cuticulin envelope, as well as in most terrestrial species wax and cement which sit outside the envelope. Wax is produced by oenocytes, cement by dermal glands. The cuticulin envelope, a very thin layer of unknown composition, is the single most important component of the cuticle.

Cuticle formation is a succession of syntheses by the epidermal cells, with the oenocytes and dermal glands adding their secretions at the appropriate time. After apolysis, ecdysial droplets are released and, after formation of the new cuticulin envelope, the enzymes in the droplets are activated to digest almost all of the old endocuticle. In new procuticle formation, much of the raw material from the digested endocuticle is reused. Wax deposition occurs just prior to ecdysis and, like endocuticle production, continues during intermolt. The cement layer is laid down after ecdysis. Tanning of the outer procuticle, to form the exocuticle, also takes place at this time.

The strength and hardness of the cuticle enable this layer to serve both as an exoskeleton and in the protection of the insect against physical damage and entry of pathogens. The wax layer is important in reducing water loss (entry) in terrestrial (freshwater) insects, is a barrier to insecticides, and, for some insects, contains pheromones. Color is also a function of the integument, being produced either by pigments in the epidermis or, more frequently, as a result of the structure of the cuticle.

6. Literature

General reviews of integument structure and function are given by Locke (1974), Neville (1975), Hepburn (1976, 1985), and Bereiter-Hahn *et al.* (1984). Specialized chapters include those by Ebeling (1974) and Blomquist and Dillwith (1985) [permeability of cuticle, especially the wax layer], Hackman (1974) [chemistry of cuticle], Andersen (1985) and Hopkins and Kramer (1992) [tanning], and Kayser (1985) [pigments], and by authors in the treatise edited by Binnington and Retnakaran (1991).

Andersen, S. O., 1985, Sclerotization and tanning of the cuticle, in: *Comprehensive Insect Physiology, Biochemistry and Pharmacology*, Vol. 3 (G.A. Kerkut and L.I. Gilbert, eds.), Pergamon Press, Elmsford, NY.

Beament, J. W. L., 1961, The water relations of insect cuticle, *Biol. Rev.* **36**:281–320.

Bereiter-Hahn, J., Matoltsy, A. G., and Richards, K. S. (eds.), 1984, *Biology of the Integument*, Vol. I (Invertebrates), Springer-Verlag, Berlin.

Binnington, K., and Retnakaran, A. (eds.), 1991, *Physiology of the. Insect Epidermis*, CSIRO, Melbourne.

Blomquist, G. J., and Dillwith, J. W., 1985, Cuticular lipids, in: *Comprehensive Insect Physiology, Biochemistry and Pharmacology*, Vol. 3 (G. A. Kerkut and L. I. Gilbert, eds.), Pergamon Press, Elmsford, NY.

Ebeling, W., 1974, Permeability of insect cuticle, in: *The Physiology of Insecta*, 2nd ed., Vol. VI (M. Rockstein, ed.), Academic Press, New York.

Fuzeau-Bresch, S., 1972, Pigments and color changes, *Annu. Rev. Entomol.* **17**:403–424.

Hackman, R. H., 1974, Chemistry of the insect cuticle, in: *The Physiology of Insecta*, 2nd ed., Vol. VI (M. Rockstein, ed.), Academic Press, New York.

Hepburn, H. R. (ed.), 1976, *The Insect Integument*, American Elsevier, New York.

Hepburn, H. R., 1985, Structure of the integument, in: *Comprehensive Insect Physiology, Biochemistry and Pharmacology*, Vol. 3 (G. A. Kerkut and L. I. Gilbert, eds.), Pergamon Press, Elmsford, NY.

Hopkins, T. L., and Kramer, K. J., 1992, Insect cuticle sclerotization, *Annu. Rev. Entomol.* **37**:273–302.

Hopkins, T. L., Starkey, S. R., Xu, R., Merritt, M. E., Schaefer, J., and Kramer, K. J., 1999, Catechols involved in sclerotization of cuticle and egg pods of the grasshopper, *Melanoplus sanguinipes*, and their interactions with cuticular proteins, *Archs Insect Biochem. Physiol.* **40**:119–128.

Jacob, J., Raab, G., and Hoppe, U., 1997, Surface lipids of the silverfish (*Lepisma saccharina* L.), *Z. Naturforsch.* *C***52**:109–113.

Kayser, H., 1985, Pigments, in: *Comprehensive Insect Physiology, Biochemistry and Pharmacology*, Vol. 10 (G. A. Kerkut and L. I. Gilbert, eds.), Pergamon Press, Elmsford, NY.

Leschen, R. A. B., and Cutler, B., 1994, Cuticular calcium in beetles (Coleoptera: Tenebrionidae: Phrenapetinae), *Ann. Entomol. Soc. Amer.* **87**:918–921.

Locke, M., 1969, The ultrastructure of the oenocytes in the molt/intermolt cycle of an insect, *Tissue Cell* **1**:103–154.

Locke, M., 1974, The structure and formation of the integument in insects, in: *The Physiology of Insecta*, 2nd ed., Vol. VI (M. Rockstein, ed.), Academic Press, New York.

Locke, M., 1991, Insect epidermal cells, in: *Physiology of the Insect Epidermis* (K. Binnington and A. Retnakaran, eds.), CSIRO, Melbourne.

Locke, M., 1998, Epidermis, in: *Microscopic Anatomy of Invertebrates*, Vol. 11A (Insecta) (F. W. Harris and M. Locke, eds.), Wiley-Liss, New York.

Locke, M., 2001, The Wigglesworth lecture: Insects for studying fundamental problems in biology, *J. Insect Physiol.* **47**:495–507.

Marcu, O., and Locke, M., 1999, The origin, transport and cleavage of the molt-associated cuticular protein CECP22 from *Calpodes ethlius* (Lepidoptera, Hesperiidae), *J. Insect Physiol.* **45**:861–870.

Neville, A. C., 1975, *Biology of the Arthropod Cuticle*, Springer-Verlag, Berlin.

Okot-Kotber, B. M., Morgan, T. D., Hopkins, T. L., and Kramer, K. J., 1994, Characterization of two high molecular weight catechol-containing proteins from pharate pupal cuticle of the tobacco hornworm, *Manduca sexta*, *Insect Biochem. Molec. Biol.* **24**:787–802.

Romer, F., 1991, The oenocytes of insects: Differentiation, changes during molting, and their possible involvement in the secretion of molting hormone, in: *Morphogenetic Hormones of Arthropods*, Vol. I, Part 3 (A. P. Gupta, ed.), Rutgers University Press, New Brunswick, NJ.

Sass, M., Kiss, A., and Locke M., 1993, Classes of integument peptides, *Insect Biochem. Mol. Biol.* **23**:846–857.

Schal, C., Sevala, V., Young, H. P., and Bachmann, J. A. S., 1998, Sites of synthesis and transport pathways of insect hydrocarbons: Cuticle and ovary as target tissues, *Amer. Zool.* **38**:382–393.

Suderman, R. J., Andersen, S. O., Hopkins, T. L., Kanost, M. R., and Kramer, K. J., 2003, Characterization and cDNA cloning of three major proteins from pharate pupal cuticle of *Manduca sexta*, *Insect Biochem. Mol. Biol.* **33**:331–343.

Yarema, C., McLean, H., and Caveney, S., 2000, L-Glutamate retrieved with the moulting fluid is processed by a glutamine synthetase in the pupal midgut of *Calpodes ethlius*, *J. Insect Physiol.* **46**:1497–1507.

Young, H. P., Bachmann, J. A. S., Sevala, V., and Schal, C., 1999, Site of synthesis, tissue distribution, and lipophorin transport of hydrocarbons in *Blattella germanica* (L.) nymphs, *J. Insect Physiol.* **45**:305–315.

Zheng, Y. P., Retnakaran, A., Krell, P. J., Arif, B. M., Primavera, M., and Feng, Q. L., 2003, Temporal, spatial and induced expression of chitinase in the spruce budworm, *Choristoneura fumiferana*, *J. Insect Physiol.* **49**:241–247.

12

Sensory Systems

1. Introduction

Organisms constantly monitor and respond to changes in their environment (both external and internal) so as to maintain themselves under the most favorable conditions for growth and reproduction. The structures that receive these environmental cues are sense cells, and the cues are always forms of energy, for example, light, heat, kinetic (as in mechanoreception and sound reception), and potential (as in chemoreception, the sense of smell and taste) (Dethier, 1963). The sensory structures use the energy to do work, namely, to generate a message that can be conducted to a decoding area, the central nervous system, so that an appropriate response can be initiated. The message is, of course, in the form of a nerve impulse. Sensory structures are generally specialized so as to respond to only one energy form and are usually surrounded by accessory structures that modify the incident energy.

As Dethier (1963) noted, the small size and exoskeleton of insects have had marked influence on their sensory and nervous systems. Smallness and, therefore, short neural pathways provide for a very rapid response to stimuli. However, it also means that there are relatively few axons and, therefore, a limited number of responses to a given stimulus. This has led to a situation in insects where stimulation of a single sense cell may trigger a series of responses. Further, almost all insect sense cells are *primary* sense cells, that is, they not only receive the stimulus but initiate and transmit information to the central nervous system; in other words, they are true neurons. In contrast, in vertebrates, almost all sensory systems include both a specialized (*secondary*) sense cell and a sensory neuron that transmits information to the central nervous system. The cuticle provides protection and support by virtue of its rigid, inert nature, yet sense cells must be able to respond to very subtle (minute) energy changes in the environment. Thus, only where the cuticle is sufficiently "weakened" (thinner and more flexible) will the energy change be sufficient to stimulate the cell. An insect, therefore, must strike a balance between safety and sensitivity. In contrast to mammalian skin, which has millions of generally distributed sensory structures, the surface of an insect has only a few thousand such structures, and most of these are restricted to particular regions of the body.

Two broad morphological types of sense cells are recognizable (Dethier, 1963; French, 1988), those associated with cuticle (and therefore including invaginations of the body wall) (Type I neurons) and those that are never associated with cuticle and lie on the inner

373

side of the integument, on the wall of the gut, or alongside muscles or connective tissue where they function as proprioceptors (Type II neurons) (Section 2.2). A Type I neuron and its associated cells are derived embryonically from the same epidermal cell. They and the associated cuticle form the sensillum (sense organ). All types of sensilla, with the possible exception of the ommatidia of the compound eye, are homologous and derived from cuticular hairs.

2. Mechanoreception

Insects receive and respond to a wide variety of mechanical stimuli. They are sensitive to physical contact with solid surfaces (touching and being touched); they detect air movements, including sound waves; and they have gravitational sense, that is, through particular mechanosensilla they gain information about their body position in relation to gravity. This is especially important in flying or swimming insects which are in a homogeneous medium; they receive information about their body posture and the relationship of different body components to each other, and they obtain information on physical events occurring within the body, for example, the extension of muscles in movement, the filling of the gut by food, and the stretching of the oviduct when mature eggs are present.

Information on the above is gathered by a spectrum of mechanosensilla associated with which, in most cases, are accessory structures that transform the energy of the stimulus into usable form, namely, a mechanical deformation of the sense cell's plasma membrane (French, 1988).

2.1. Sensory Hairs

The simplest form of mechanosensillum is seen in sensory hairs (sensilla trichodea) (Figure 12.1), which occur on all parts of the body but are in greatest concentration on those that frequently come into contact with the substrate, the tarsal segments of the legs, antennae, and mouthparts. Typically, they are single structures but on occasion they are found in large groups known as hair plates (Figure 12.2). In its simplest form a sensillum comprises a rigid, poreless hair set in a membranous socket and four associated cells; these are the inner sheath cell (also known as the trichogen or generative hair cell), outer sheath cell (tormogen or membrane-producing cell), neurilemma cell, which ensheathes the cell body and axon of the sensory neuron, and the sensory neuron whose dendrite often is cuticularized and includes a terminal cuticular filament (scolopale) (McIver, 1985; Keil, 1997, 1998). In addition to their generative function, the outer sheath cells have an important physiological role in maintaining the appropriate ionic and molecular environment for stimulus transduction and conduction by the dendrites. Specifically, they pump K^+ ions into the space that surrounds the tip of the sensory dendrite to facilitate generation of the receptor current (Section 2.3). A characteristic feature of the tip of sensory neurons are large numbers of microtubules. Because of their position and experiments with antimicrotubule drugs, it has been suggested that the microtubules may play a role in transduction. However, French's (1988) assessment of the evidence led him to conclude that their more likely roles are in the development and structural maintenance of the sensilla.

Within the above-generalized structure, hairs may differ widely in their detailed morphology, physiology, and function (Section 2.3). Nevertheless, they are all designed such

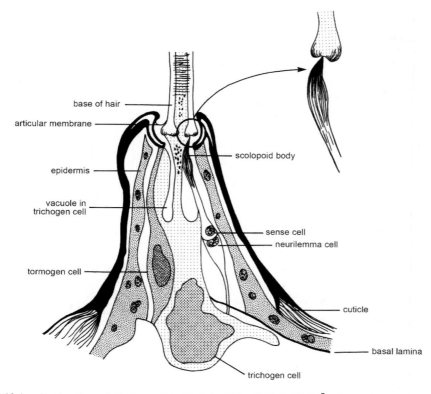

FIGURE 12.1. Section through the base of tactile hair. [After F. Hsü, 1938, Étude cytologique et comparée sur les sensilla des insectes, *La Cellule* **47**: 5–60. By permission of *La Cellule*.]

that the slightest deformation (a few nanometers) of the membrane of the sensory neuron will generate an action potential. Note also that some hairs include several neurons, but only one of these is mechanosensory, the remainder are chemosensory or thermosensory.

2.2. Proprioceptors

Proprioceptors are sense organs able to respond continuously to deformations (changes in length) and stresses (tensions and compressions) in the body. They provide an organism with information on posture and position. Five types of proprioceptors occur in insects: hair plates, campaniform sensilla, chordotonal organs, stretch receptors, and nerve nets. In common, they respond tonically and adapt very slowly to a stimulus.

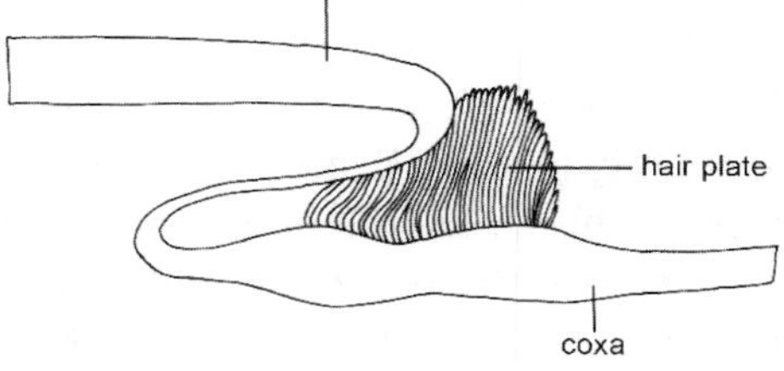

FIGURE 12.2. Hair plate at joint of coxa with pleuron. [After J. W. S. Pringle, 1938, Proprioceptors in insects. III. The function of the hair sensilla at the joints, *J. Exp. Biol.* **15**: 467–473. By permission of Cambridge University Press, London.]

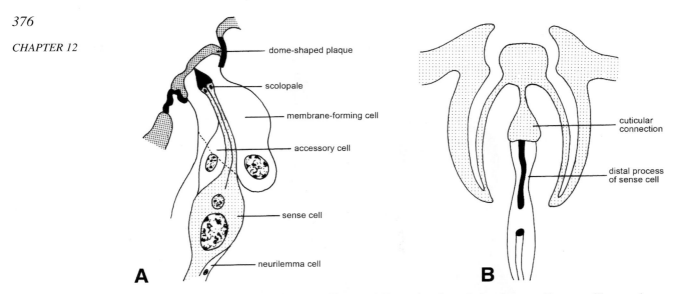

FIGURE 12.3. (A) Campaniform sensillum; and (B) section through tip of campaniform sensillum to show stiffening rod of cuticle running along cuticular plate present in some species. [After V. G. Dethier, 1963, *The Physiology of Insect Senses*, John Wiley and Sons, Inc. By permission of the author.]

A campaniform sensillum (Figure 12.3A) includes all of the components of a tactile hair with which it is homologous except for the hair shaft, which is replaced by a dome-shaped plate of thin cuticle. The plate may be slightly raised above the surrounding cuticle, flush with it, or recessed, but in all cases it is contacted at its center by the distal tip of the neuron and serves as a stretch or compression sensor. In many species the plate is elliptical and has a stiffening rod of cuticle running longitudinally on the ventral side, to which the neuron tip is attached (Figure 12.3B). The sensillum shows directional sensitivity, being stimulated by stress perpendicular to the longitudinal axis of the rod. Typically, the sensilla are arranged in groups. In *Periplaneta*, for example, the sensilla of the tibia occur in two groups, with their rod axes at right angles. During walking (Chapter 13, Section 2.3), contraction of the flexor and extensor muscles stimulates the proximal and distal groups of sensilla, respectively, which are thus important in the overall coordination of the process. In addition, when the insect is standing, the proximal sensilla whose axes are perpendicular to the axis of the tibia are continuously stimulated because of the stress in the cuticle. Information from sensilla passes to the central nervous system where it inhibits the so-called "righting reflex." When the insect is turned on its back there are no longer stresses in the cuticle, the sensilla are not stimulated, the righting reflex is not inhibited, and the insect undertakes a series of kicking movements in order to regain the standing position.

Chordotonal (scolophorous) sensilla (= scolopidia) (Figure 12.4A) are another widely distributed form of proprioceptor in insects. Unlike the sensilla discussed earlier, chordotonal sensilla lack a specialized exocuticular component, though, it should be emphasized, they are believed to be homologous with the sensillum trichodeum and other types of sensilla. A distinctive feature of scolopidia is the scolopale, an intracellular secretion of the sheath cells surrounding the dendrite of the sensory cell, that is assumed to be important in the transduction of the stimulus. They are associated with the body wall, internal skeletal structures, tracheae, and structures in which pressure changes occur. Though they are found singly, more commonly they occur in groups. Chordotonal organs exist as strands of

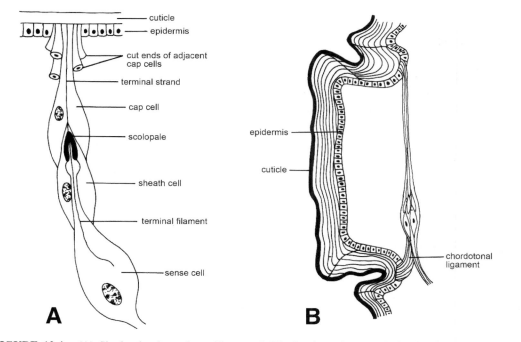

FIGURE 12.4. (A) Single chordotonal sensillum; and (B) chordotonal organ. [After V. G. Dethier, 1953, Mechanoreception, in: *Insect Physiology* (K. D. Roeder, ed.). Copyright © 1953 John Wiley and Sons, Inc. Reprinted by permission of John Wiley and Sons, Inc.]

tissue that stretch between two points. The proximal end of the sensory neuron is attached to one point by means of a ligament and the distal end is covered by a cap cell, which is attached to the second point (Figure 12.4B). Chordotonal sensilla are highly sensitive. Thus, a change in the relative position of the points that causes the strand's length to be altered by as little as 1 nm will produce bending or stretching of the dendritic membrane, hence stimulation of the sense cell. Frequently, alteration of the positions of the points is brought about as a result of pressure changes, for example, in the air within the tracheal system, in the hemolymph within the body cavity, or in aquatic insects in the water in which they are swimming. In relatively few insects chordotonal sensilla are aggregated in large numbers and capable of being stimulated by changes in external air pressure, that is, sound waves (Section 3).

Stretch receptors (Figure 12.5) comprise a multipolar neuron (Type II) whose dendrites terminate in a strand of connective tissue or a modified muscle cell, the ends of which are attached to the body wall, intersegmental membranes, and/or muscles. As the points to which the ends are attached move with respect to each other, the receptor is stimulated. Stretch receptors are probably most important in providing information to the central nervous system on rhythmically occurring events within the insect, for example, breathing movements, waves of peristalsis along the gut, and locomotion.

A peripheral system of multipolar sensory neurons is located beneath the body wall in many larvae whose cuticle is thin and flexible, or beneath the intersegmental and arthrodial membranes of insects with a rigid integument. The nerve endings are presumably stimulated by tension in the body wall or movements of joints. A similar arrangement is present in the wall of the bursa copulatrix of *Pieris rapae*, measuring the degree of stretching of this structure when a spermatophore is present (Chapter 19, Section 3.1.3), and in the

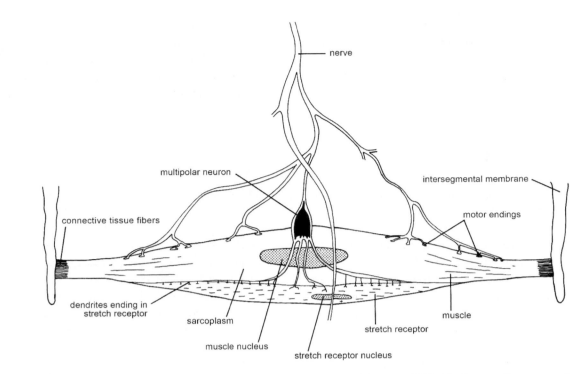

FIGURE 12.5. Stretch receptor of male mosquito. [After R. F. Chapman, 1971, *The Insects: Structure and Function*. By permission of Elsevier/North-Holland, Inc., and the author.]

alimentary canal, though the sensory neurons in this case pass their information on to the visceral nervous system (Chapter 13, Section 2.2).

2.3. Signal Detection

Detection of stimuli by mechanoreceptors is a three-step process: coupling, transduction, and encoding (French, 1988). Coupling refers to the deformation of the sensory neuron's dendritic membrane caused by movement of the hair in its socket, the inward movement of the cuticular dome in campaniform sensilla, or the stretching of the chordotonal sensillum. Coupling in Type II proprioceptors (stretch receptors and peripheral nerve nets) presumably also results in distortion of the neuronal membrane, though these systems are much less studied. Transduction is the generation, followed by its flow through the dendritic membrane, of the receptor (generator) current. It results from the stretching of the membrane and the opening of transduction (stretch-activated) channels contained therein. Studies with *Drosophila* mutants have elegantly demonstrated that the channels include specific proteins that serve as "gates" (Walker *et al.*, 2000). When the gates are opened as a result of membrane distortion, K^+ ions rush in from the extra-dendritic space, creating the receptor current. Encoding, the final step, is the transfer of information from the sensillum to the central nervous system. In common with typical neurons, this is seen as a train of action potentials induced by the receptor current.

Hairs differ in their sensitivity; long, delicate hairs respond to the slightest force, even air-pressure changes, whereas shorter, thicker spines (sometimes called sensilla chaetica) require considerable force for stimulation. Associated with this varied sensitivity are differences in the electrophysiology of the hairs. Delicate hairs typically adapt quickly, that is, rapidly lose their sensitivity to a continuously applied stimulus. More strongly built hairs, however, adapt only very slowly. Most hairs respond to a stimulus only while moving and are said to be "velocity-sensitive" and the response is "phasic." Such hairs are found on structures that "explore" the environment. The remainder responds continuously to a static deformation ("pressure-sensitive" forms with a "tonic" response). These are found usually as hair plates, at joints or on genitalia, which gather information on position with respect to gravity or posture. In these situations they are serving as proprioceptors.

3. Sound Reception

Sounds are waves of pressure detected by organs of hearing. A sound wave is produced when particles are made to vibrate, the vibration causing displacement of adjacent particles. Usually sound is thought of as an airborne phenomenon; however, it should be appreciated that sounds can pass also through liquids and solids. It will be apparent, therefore, that the distinction between sound reception and mechanoreception is not clear-cut. Indeed, many insects that lack specialized auditory organs can clearly "hear," in that they respond in a characteristic manner to particular sounds. For example, caterpillars stop all movements and contract their bodies in response to sound. If, however, their bodies are coated with water or powder, or the hairs removed, the response is abolished. Further, insects with specialized sound sensors may continue to respond to sounds of low frequency even after the specialized organs have been damaged or removed. The structures that respond to these low-frequency sound waves (see Figure 12.8) are the most delicate mechanosensilla, namely, the sensilla trichodea and, probably, chordotonal sensilla distributed over the body surface. In some species, hairs sensitive to sound may be restricted to particular areas, for example, antennae or cerci.

Among Insecta, hearing has evolved independently in at least 12 groups (Michelsen and Larsen, 1985). Though insect hearing organs include a number of common elements, their structural complexity reflects the interaction of three factors: the evolutionary history of the group, the size of the insect, and the acoustic features of the insect's environment. For example, the tympanal organs of most moths, which are sensitive only to the sounds emitted by bats preying on them, are relatively simple whereas the tympanal organs of crickets and grasshoppers tend to be complex because they need to distinguish the (equally complex) songs of conspecifics.

Broadly speaking, insect-hearing structures can be divided into two categories: near-field detectors and far-field detectors. As their names indicate, the detectors are able to perceive sounds that originate a short distance (from a few millimeters up to about 1 m) or a long distance (tens of meters), respectively. However, there are several other features unique to each type of detector. Near-field detectors are displacement receivers (activated by vibrations of adjacent air particles), are sensitive to low-frequency sound (75–500 Hz), and usually have a relatively simple structure that does not include a tympanum (Römer and Tautz, 1992). Examples of near-field detectors are the hairs on the cerci of cockroaches, on the aristae of *Drosophila*, and on the thorax of some noctuid caterpillars, as well as the specialized Johnston's organ (Section 3.1). *Drosophila* males vibrate their wings at about 330 Hz during courtship. The sound produced is picked up by the aristae of a female, provided she is within about 2 mm. If she is unreceptive, she produces her own song (at about 300 Hz) that causes the male to turn away and stop courting (Bennet-Clark and Ewing, 1970). Caterpillars of the noctuid moth *Mamestra brassicae* have eight fine thoracic hairs that show maximal sensitivity to air-borne vibrations in the 100–600 Hz range. At these frequencies, crawling caterpillars stop moving, and may squirm and lose contact with the substrate. These have been interpreted as avoidance reactions, as the frequencies correspond with those made by wing beats of caterpillar parasitoids (tachinids and ichneumonids) and predators (wasps). Apparently, the caterpillars can hear the sounds at distances up to about 70 cm (Markl and Tautz, 1975).

In contrast, far-field detectors are pressure difference receivers (are stimulated by changes in air pressure created by sound waves), are sensitive to a wide range of high frequencies (2 to over 100 kHz), almost always have a tympanum, and hence are commonly called "tympanal organs" (Section 3.2).

Specialized auditory organs comprise groups of chordotonal sensilla and associated accessory structures that enhance the sensitivity of the organ. They include Johnston's organ, tympanal organs, and subgenual organs. The first two are sensitive to only airborne vibrations, the latter mainly to vibrations in solids, though in a few species airborne sounds are detected by these structures.

3.1. Johnston's Organ

Johnston's organ, that is, one or more groups of chordotonal sensilla located in the pedicel of the antenna, is present in all adult and many larval insects and generally serves a proprioceptive function, providing information on the position of the antenna with respect to the head, the direction and strength of air or water currents, or, in back swimmers (Notonectidae), the orientation of the insect in the water. However, in male mosquitoes and chironomids, as well as in worker honey bees, the structure has become specialized to perceive sounds, notably those produced by the wings of conspecifics.

Male *Aedes aegypti* are attracted especially to sounds in the frequency range 500–550 Hz, which compare with the flight tone of females, 449–603 Hz. At these frequencies, males show a characteristic mating response. Sounds of other frequencies (including a male's flight tone, which is somewhat higher than that of a female) are also perceived by males, though these do not attract them. A male's antennae are extremely bushy, covered with long, fine hairs that vibrate in unison at certain frequencies, causing the flagellum to move in its socket, the pedicel (Figure 12.6). The bulbous pedicel accommodates a large number of sensilla that are arranged in two primary groups, the inner and outer rings. It is suggested that Johnston's organ resolves the sound into two components, a component running parallel to the flagellum, to which the inner sensilla are most sensitive, and a component perpendicular to the flagellum, which stimulates the outer sensilla. By "estimating" the relative strength of the stimulus from each component, a male is able to determine a female's position.

The waggle dance has long been recognized as the means by which foraging worker honey bees communicate information on the distance to and direction of a food source on their return to the hive (see also Section 7.1.4). However, only recently was it realized that sounds produced during the dance provide the information on distance. Low frequency (up to 500 kHz) sounds produced by wing vibration are heard by the Johnston's organ, the duration of the auditory signal being a measure of the distance to the food source (Dreller and Kirchner, 1995).

3.2. Tympanal Organs

Tympanal organs are present in some species from at least seven orders of insects (Hoy and Robert, 1996; Göpfert *et al.*, 2002): Orthoptera (fore tibiae of Tettigoniidae and Gryllidae, first abdominal segment of Acrididae), Lepidoptera (abdomen in Geometridae

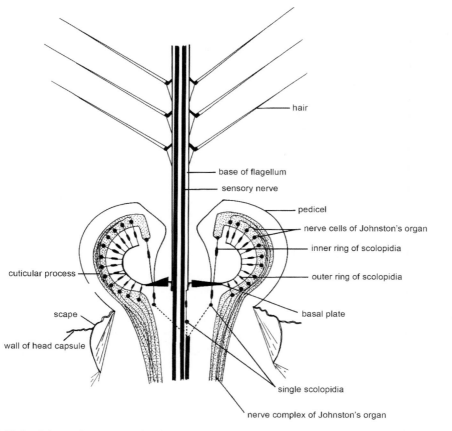

FIGURE 12.6.   Johnston's organ. [After H. Autrum, 1963, Anatomy and physiology of sound receptors in invertebrates, in: *Acoustic Behaviour of Animals* (R. G. Busnel, ed.). By permission of Elsevier/North-Holland Biomedical Press, Amsterdam.]

and Pyralidae, metathorax in Noctuidae and Notodontidae, fore or hind wing base in Nymphalidae and Hedylidae, mouthparts in Sphingidae), Hemiptera (abdomen in Cicadidae, thorax in Corixidae), Coleoptera (abdomen in Cicindelinae, cervical membrane in Scarabaeidae), Dictyoptera (metathorax in Mantodea, metathoracic leg in Blattodea), Neuroptera (wing base), and Diptera (ventral prosternum). Though their detailed structure varies, almost all tympanal organs have three common features: a cuticular membrane (the tympanum); a large tracheal air sac appressed to the membrane, the two structures forming a "drum"; and a group of chordotonal sensilla (Figure 12.7) (Yager, 1999). The tympanum is much thinner (1 μm in cicadas, 40–100 μm in some ensiferans) than the surrounding cuticle, providing the sensitivity required for sound reception.

Sound waves that strike the drum cause it to vibrate and, therefore, the sensilla to be stimulated. The range of frequency of the waves that stimulate tympanal organs is high. For example, in Acrididae, it extends from less than 1 kHz to about 50 kHz. Over this range the sensitivity of the organ varies greatly, with a maximum in the 2- to 15-kHz range (Figure 12.8). In contrast, the human ear is most sensitive to a frequency of 1–3 kHz. As pressure difference receivers, insect tympanal organs have directional sensitivity. Thus, insects with these organs can locate the source of a sound.

The functional significance of tympanal organs varies (Spangler, 1988; Hoy and Robert, 1996). In Orthoptera and Hemiptera, the ability to hear is complemented by the ability

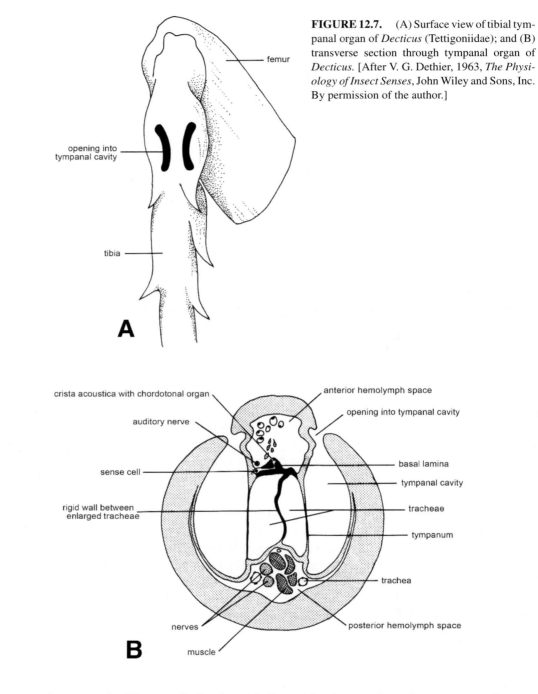

FIGURE 12.7. (A) Surface view of tibial tympanal organ of *Decticus* (Tettigoniidae); and (B) transverse section through tympanal organ of *Decticus*. [After V. G. Dethier, 1963, *The Physiology of Insect Senses*, John Wiley and Sons, Inc. By permission of the author.]

to produce sounds (Chapter 3, Section 4.3.1), and in these orders the organs are important in species aggregation and/or mate location. Experimentally, it has been shown that the tympanal organs of nocturnal Lepidoptera, Dictyoptera, Neuroptera, Orthoptera and Coleoptera are sensitive to high-frequency sounds, and it is generally assumed that this enables these insects to detect the approach of predators, principally insectivorous bats. It must be noted, however, that for most groups this has not been observed under field conditions. In Lepidoptera, for which good evidence is available, the tympanal organs are most sensitive

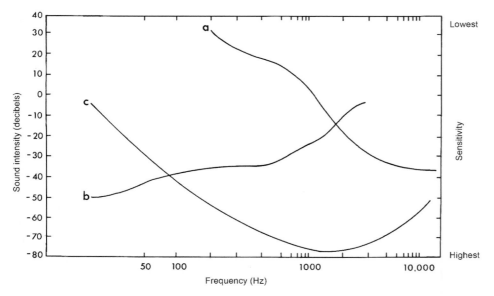

FIGURE 12.7. Sensitivity curves for (A) tympanal organ of *Locusta*, (B) cercal hairs of *Gryllus*, and (C) human ear. [After V. G. Dethier, 1953, Mechanoreception, in: *Insect Physiology* (K. D. Roeder, ed.). Copyright © 1953 John Wiley and Sons, Inc. Reprinted by permission of John Wiley and Sons, Inc.]

to frequencies in the range of 15–60 kHz, which comes within the frequency range of the sounds uttered by the bats as they echolocate. A moth's response varies according to the intensity of the sound. At low intensity (i.e., when the bat is 30 m or more distant) a moth moves away from the sound. At high intensity, a moth takes more striking action, flying an erratic course, or dropping to the ground. In a few species of Lepidoptera ultrasound is used in sexual communication (Spangler, 1988).

The phenomenon of acoustic parasitism was first reported by Cade (1975) who observed that gravid females of the tachinid fly *Ormia ochracea* locate their host, the field cricket *Gryllus integer*, by homing in on the male cricket's mating call. Since then, other tachinids and some sarcophagid flies that parasitize cicadas have been shown to use the same strategy for host location. On locating a cricket, the fly larviposits on or near it. The tachinid's ears are typical tympanal organs, located on the prosternum, and are most sensitive in the range of 4–6 kHz (the dominant frequency of the cricket's song is 4.8 kHz). Interestingly, in these nocturnally active flies males also have tympanal organs though these are insensitive to frequencies around 5 kHz. However, the ears of both male and female flies are sensitive to ultrasonic sounds (20–60 kHz). It is thought that this range of sensitivity is related to the flies' need to detect and avoid becoming prey for insectivorous bats.

3.3. Subgenual Organs

These are chordotonal organs present in the tibiae of most insects, excluding Thysanura, Hemiptera, Coleoptera, and Diptera. The organ, which comprises between 10 and 40 sensilla, generally detects vibrations in the substrate though the mechanism by which the organ is stimulated is not known. In a few species, for example, the Madagascar hissing cockroach (*Gromphadorhina portentosa*), the subgenual organs respond to airborne vibrations. Their peak sensitivity is at 1.8 kHz, which coincides with the species' courtship song.

4. Chemoreception

Chemoreception, essentially taste (contact chemoreception) and smell (distance chemoreception), is an extremely significant process in the Insecta, as it initiates some of their most important behavior patterns, for example, feeding behavior, selection of an oviposition site, host or mate location, behavior integrating caste functions in social insects, and responses to commercial attractants and repellents.

Though taste and smell are distinguished traditionally, such a distinction has no firm morphological or physiological basis. The sensilla for the two senses are structurally very similar; indeed, in some species the same structure is used for both olfaction (smell) and gustation (taste). Further, stimulation of a sensillum by either tastes or odors probably entails comparable subcellular or molecular interactions. Any difference between smell and taste is, then, a matter of degree. Smell may be defined as chemostimulation by compounds in very low concentration but volatile at physiological temperatures, and taste as chemostimulation by higher concentrations of liquids that are not volatile at physiological temperatures.

In addition to taste and smell, insects have a third method of detecting chemical stimuli, the common chemical sense. This is the response of an insect (always an avoiding reaction) to high concentrations of noxious chemicals. It is not a response caused by stimulation of normal chemosensilla, because the response is not abolished after surgical removal of the structures bearing these sensilla. It would seem to be a non-specific response of other types of sensory neurons.

4.1. Location and Structure of Sensilla

Behavioral and electrophysiological experiments have been used to establish the location and nature of chemosensilla. Organs of taste are common on the mouthparts, especially the palps, though they have been identified also on the antennae (Hymenoptera), tarsi (many Lepidoptera, Diptera, and the honey bee), ovipositor (parasitic Hymenoptera and some Diptera), and on the general body surface. The antennae are the primary site of olfactory organs and often bear many thousands of these structures. The mouthparts also carry olfactory structures in many species.

Earlier authors classified chemosensilla according to their morphology, but this system proved inadequate because structures that looked similar at the light microscope level were shown, using electron microscopy combined with electrophysiological studies, to have different structures and functions. Slifer (1970) grouped chemosensilla in two categories, "thick-walled" and "thin-walled," which, though a seemingly simple structural criterion for separation, is valid in all except a few cases, and broadly correlates with their functions as organs of taste and smell, respectively.

Thick-walled (uniporous) chemosensilla (Figure 12.9A) take the form of hairs, pegs, or papillae (Mitchell *et al.*, 1999; Ryan, 2002). Generally, they serve as taste sensilla, though some are also sensitive to strong odors. The chemosensitive hairs and pegs broadly resemble tactile hairs, though they can be distinguished with the electron microscope. Whereas tactile hairs have a sharply pointed top and are innervated by a single neuron whose dendrite terminates at the base of the hair, thick-walled chemosensory hairs have a rounded tip with a terminal pore, multineuronal innervation, and dendrites that extend along the length of the hair to terminate just beneath the pore. The dendrites are usually enclosed in a cuticular sheath. Occasionally, the thick-walled pegs may be set in pits, when they are known as sensilla coeloconica and perhaps have an olfactory function. Papillae having

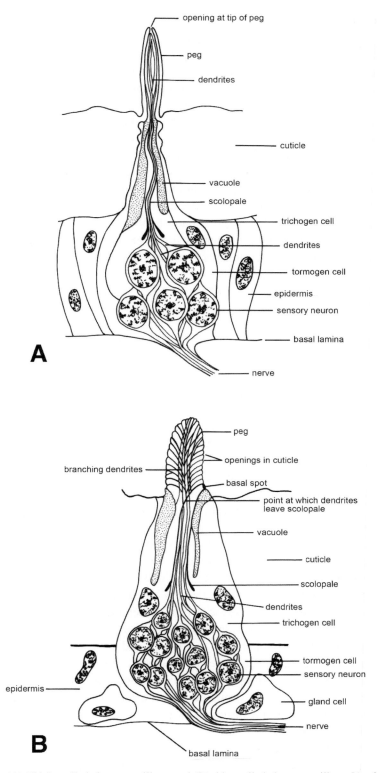

FIGURE 12.8. (A) Thick-walled chemosensillum; and (B) thin-walled chemosensillum. [A, after E. H. Slifer, J. J. Prestage, and H. W. Beams, 1957, The fine structure of the long basiconic sensory pegs of the grasshopper (Orthoptera; Acrididae) with special reference to those on the antenna, *J. Morphol.* **101**:359–397. By permission of Wistar Press. B, after E. H. Slifer, J. J. Prestage, and H. W. Beams, 1959, The chemoreceptors and other sense organs on the antennal flagellum of the grasshopper (Orthoptera; Acrididae), *J. Morphol.* **105**:145–191. By permission of Wistar Press.]

the same general features as chemosensory hairs have been observed in the food canal of aphids, on the labellum of flies, and on the cockroach hypopharynx. Some thick-walled hairs are both mechano- and chemosensory.

Thin-walled (multiporous) chemosensilla of various types (Figure 12.9B) have been described from the antennae of all species sufficiently well studied (Keil, 1999; Ryan, 2002). They have in common a thin cuticular covering (0.1–1 μm in width) perforated by many (up to several thousand) pores. Usually, beneath each pore is a small chamber from whose base numerous pore tubules connect with the sensillar sinus. The pores are filled with lipid that is transported to the cuticular surface to reduce water loss (Chapter 11, Section 2). Most sensilla receive several neurons whose dendrites are much branched and terminate beneath the pore tubules. Thin-walled chemosensilla exist as surface hairs and pegs, pegs in pits, and plate organs (sensilla placodea). The latter are roundish, flat, or domed areas of cuticle, which occur in great density on the antennae of many Hymenoptera, some Coleoptera, and some homopterans. For some time their function was controversial; however, recent electrophysiological and electron microscope studies have confirmed that they are olfactory.

4.2. Physiology of Chemoreception

Early studies, mainly behavioral, established broad parameters for the senses of smell and taste in insects and revealed some interesting comparisons between the senses in insects and those in humans. Like humans, insects appear to "recognize" the four basic taste qualities: sweet (acceptable), salty, acidic, and bitter (all non-acceptable). Further, their response to chemical stimuli varies according to their physiological state: age, sex, normal diet, and immediate history. The sensitivity of insects to taste stimuli is, broadly speaking, equal to or greater than that of humans. Sucrose, for example, can be detected by humans at a concentration of 2×10^{-2} M, by the honey bee at 6×10^{-2} M; hydrochloric acid stimulates at 1.25×10^{-3} M in humans, at 10^{-3} M (in 1 M sucrose) in the honey bee. However, as in humans, sensitivity to a substance increases if that substance has not been experienced for some time. The red admiral butterfly, *Pyrameis atalanta*, for example, fed regularly on sucrose has a tarsal threshold sensitivity of 10^{-1}–10^{-2}M. If, however, the sugar is withheld for some time sensitivity increases so that a concentration of 8×10^{-5} M will elicit a response. Chemosensilla on different parts of the body have differing sensitivity to particular chemicals; for example, in the fly, *Calliphora vomitaria*, those on the tarsi are 16 times more sensitive to sucrose than those on the labellum.

Disaccharides are more stimulating than monosaccharides. Trisaccharides are generally non-stimulating and polysaccharides never so. Of the inorganic ions, cations show increasing stimulation in parallel with their partition coefficient and ionic mobility; that is, H^+ is more stimulating than $NH_4^+ > K^+ > Ca^{2+} > Mg^{2+} > Na^+$. For anions, the situation appears more complex, and the relationship between ability to stimulate and physical properties of the ions is unclear. Organic acids stimulate in proportion to their degree of dissociation, indicating that the H^+ ion is the principal factor in stimulation. The stimulating power of non-electrolytes is usually proportional to the oil:water partition coefficient, though contradictions to this generalization occur.

Some taste sensilla are capable of being stimulated by various substances. Electrophysiological work has shown that this is possible because sensory neurons in the sensillum respond differentially to the substances. For example, in the labellar hairs of *Protophormia* there are four chemoreceptor cells and one mechanoreceptor cell (which terminates at the base of the hair). Of the chemoreceptors, one is sugar-sensitive, one is salt-sensitive, one is water-sensitive, and the fourth responds to deterrent stimuli.

Species vary in their sensitivity and response to odors, as do individuals in different physiological states. As with taste, insects are especially sensitive to odors of significance to them, and insects that are exposed to a wide range of odors have a better sense of smell (are more sensitive) than insects whose "olfactory environment" is relatively uniform. Sensitivity depends on several factors, including the number of sensilla, the number of sensory neurons, the number of branches from each dendrite, and the number of receptor sites. The last two mentioned enable summation of subthreshold responses to occur, so that a sensory cell is stimulated.

Chemoreceptor stimulation appears to involve processes analogous to those described for mechanoreceptors, namely, perireceptor activity, transduction, and encoding. Perireceptor activity refers to the collection and transport of the stimulants/deterrents to the dendritic membrane of the sensory neuron. Transduction describes the manner in which the odorant or taste molecules induce changes in the structure of the membrane, leading to the opening of ion channels, hence generation of the receptor current. Encoding is then the transfer of the stimulus from the receptor to the central nervous system by means of action potentials.

On reaching a sensillum odor molecules, which are generally lipophilic, dissolve in the wax covering the pores. To reach the dendritic membrane they must be solubilized in order to cross the fluid-filled sinus. This is achieved by attachment to odorant-binding proteins (OBPs). These may be highly specific, for example, in the sensilla of male moths where they bind pheromone molecules, or non-specific and able to bind with a range of molecules. On reaching the dendritic membrane, the odorant molecules are passed from the OPBs to receptor proteins, an action that triggers, probably via a second messenger system, the opening of ion channels and induction of the receptor current.

How taste sensilla work is not well understood. It may be supposed that organic molecules in an insect's taste repertoire reach the dendritic membrane attached to binding proteins comparable to those in olfactory sensilla. On reaching the dendrites these molecules will bind to receptor sites to induce the receptor current. It is evident that the sugar-sensitive neuron in *Protophormia* responds to more than one sugar, hence may have several specific receptor sites. The means by which ions and other inorganic components reach the dendrites is even more problematic. However, once there, these seem likely to act directly on the ion channels to induce their opening and hence generation of the receptor current.

5. Humidity Perception

Many observations on their behavior indicate that insects are able to monitor the amount of water vapor in the surrounding air. Insects actively seek out a "preferred" humidity in which to rest, or orient themselves toward a source of liquid water. The value of the preferred humidity varies with the physiological state of the insect, especially its state of desiccation. Normally, for example, the flour beetle, *Tribolium castaneum*, prefers dry conditions; however, after a few days without food and water, it develops a preference for more humid conditions. Ablation experiments have established that humidity detectors are typically located on the antennae, though they occur on the anterior sternites in *Drosophila* larvae, and surround the spiracles in *Glossina*, where they monitor the air leaving the tracheal

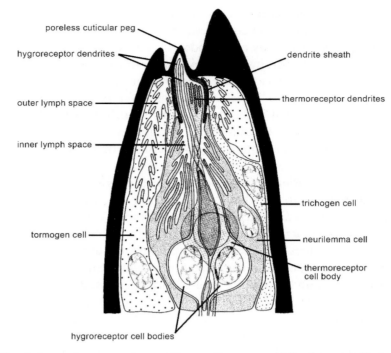

FIGURE 12.9. A thermo-/hygroreceptive sensillum on the antenna of *Bombyx mori*. [Redrawn from R. A. Steinbrecht, 1989, The fine structure of thermo-/hygrosensitive sensilla in the silkmoth *Bombyx mori*: Receptor membrane substructure and sensory cell contacts, *Cell Tiss. Res.* **255**:49–57. By permission of Springer, Berlin.]

system. Hygroreceptors exist as thin-walled, aporous hairs or pegs that are typically also thermosensitive (Altner and Loftus, 1985). Each hygroreceptor contains the dendrites of two neurons: the dendrites that penetrate the full length of the hair are hygrosensitive, whereas those that terminate near the base of the hair are temperature sensitive (Figure 12.10). Because they are poreless, hygroreceptors clearly cannot operate in the same manner as chemoreceptors, that is, by water molecules binding to receptors in the dendritic membrane. In fact, it seems that they function as mechanoreceptors, behaving like a hair hygrometer. The sensillar cuticle or a substance associated with it is hygroscopic and undergoes deformations as humidity changes, leading to physical stimulation of the dendrites. It appears that there are two types of hygroreceptors: "moist" receptors respond to increasing humidity, while "dry" receptors are stimulated by a decrease in humidity. In both cases, the stimulus is encoded as an increase in the rate of generation of action potentials.

6. Temperature Perception

This is the least understood of insect senses. Insects clearly respond to temperature in a behavioral sense, by seeking out a "preferred" temperature. For example, outside its preferred range, the desert locust, *Schistocerca gregaria*, becomes active. This locomotor activity is random but may take the insect away from the unfavorable conditions. Within the preferred range, the insect remains relatively inactive. Under field conditions, the locust will alter its orientation to the sun, raise or lower its body relative to the ground, or climb vegetation in order to keep its body temperature in the preferred range.

Parasitic insects such as *Rhodnius*, *Cimex*, and mosquitoes, which feed on mammalian blood, are able to orient to a heat source. Though the ability to sense heat is present over the entire body surface, it appears that in bloodsucking species, the antennae and/or legs are especially sensitive and probably carry specialized sensilla in the form of thick-walled hairs (Davis and Sokolove, 1975; Reinouts van Haga and Mitchell, 1975). In other species aporous, thin-walled, non-articulated hairs or pegs are both thermo- and hygroreceptive (Figure 12.10), containing at least two neurons. Occasionally, the hairs are socketed and serve also as mechanosensors. The transduction mechanism for thermoreceptors remains unclear. Encoding, however, seems to be relatively simple, with greater numbers of action potentials generated at higher temperatures, and vice versa.

Some insects, including *Rhodnius* and some fire beetles (Buprestidae), have infrared-sensitive sensilla that enable them to locate a heat source (Schmitz *et al.*, 2000, 2001). In the buprestid *Melanophila acuminata* each of the paired infrared organs, containing 50–100 sensilla, is located in a pit adjacent to the metathoracic coxae. The sensilla broadly resemble mechanosensilla; that is, they are innervated by a single neuron and are poreless. However, a unique feature is a thin cuticular lenslike structure thought to be where infrared radiation is focused. The dendrites of the neuron terminate immediately below this structure (Vondran *et al.*, 1995). In another buprestid, *Merimna atrata*, there are two pairs of ventral abdominal infrared organs. The dominant feature of the organ is a large multipolar sensory neuron whose dendrites reach two chordotonal organs in addition to the thermosensitive structure (Schmitz *et al.*, 2001). How the encoded messages from the two sensory inputs are distinguished is unclear.

7. Photoreception

Almost all insects are able to detect light energy by means of specialized photosensory structures: compound eyes, ocelli, or stemmata. In the few species that lack these structures, for example, some cave-dwelling forms, there is commonly sensitivity to light over the general body surface.

The use that insects make of light varies from a situation in which it serves as a general stimulant of activity, through one of simple orientation (positive or negative phototropism), to a state where it enables an insect to carry out complex navigation, and/or to perceive form, patterns, and colors. At all levels of complexity, however, the basic mechanism of stimulation is very likely the same; that is, the solar energy striking the photosensory cell is absorbed by pigment in the cell. The pigment undergoes a slight conformational change that causes a momentary increase in permeability of the receptor cell membrane and, thereby, the initiation of nerve impulses that travel to the optic lobes and central nervous system.

7.1. Compound Eyes

Paired compound eyes, the main photosensory system, are well developed in most adult insects and juvenile exopterygotes, but may be reduced or absent in parasitic or sedentary forms, such as lice, fleas, and female scale insects. Typically, the eyes occupy a relatively large proportion of head surface, from which they bulge out to provide a wide visual field. In dragonflies, male tabanids, and horseflies, the eyes meet in the middorsal line, the holoptic condition. In some other species, the eye is divided into readily distinguishable dorsal and ventral regions. Occasionally, these regions are physically separate, as in the male mayfly, *Cloeon*, and water beetle, *Gyrinus*.

A compound eye comprises a varied number of photosensilla, the ommatidia. The number ranges from 1 in the ant *Ponera punctatissima* to as many as 28,000 in some dragonflies. Each ommatidium (Figure 12.11) includes a light-gathering component (corneal lens plus crystalline cone); the primary sense cells (retinular cells), which collect and transduce light energy; and various enveloping (pigment) cells. The lens, a region of transparent cuticle, is produced by the primary pigment cells. Its surface is usually smooth, but in some nocturnal moths it is covered with numerous minute pimples, about 0.2 µm high, which by reducing glare improve the amount of light transmitted (equivalent to "blooming" a camera lens with an antireflection coat of magnesium fluoride). Another likely function of this antireflection layer is to improve camouflage during daylight hours, the dull surface of the moth's eye being less visible to would-be predators. In some species, for example, tabanid flies and lacewings the eyes are colored metallic gold-green due to the multilayered nature of the lens (Stavenga, 2002) (see Chapter 11, Section 4.3). The crystalline cone is a clear, hard material produced by four cells (Semper's cells). The material is typically intracellular, and the nuclei are situated around it (eucone type). In some species the material is extracellular (pseudocone type); in others there is no crystalline cone and the cells, which are transparent, occupy the area (acone type). Primitively, eight retinular cells occur beneath the crystalline cone, though one of these is usually degenerate or eccentrically located. The seven remaining cells are arranged around a central axis in most species; occasionally, they exist in two tiers of three and four cells, respectively. The mature sensory cells are unipolar, that is, lack dendrites. Instead, their inner surface is modified to form a receptive area, the rhabdomere. Collectively, the rhabdomeres form a rhabdom. By means of electron microscopy, the rhabdomeres can be seen to comprise closely packed microvilli, of diameter about 500 Å, which extend from the cell surface. In cross section the microvilli are hexagonal. The details of rhabdom construction are varied though, in general, "open" and "closed" types can be distinguished (Figure 12.11C,D). In the open type, found in Diptera and Hemiptera, individual rhabdomeres are physically separated and each serves as an optical waveguide; in the closed type, common to most insects, the rhabdomeres are wedge-shaped and closely packed around the central axis. In a closed rhabdom the rhabdomeres function collectively as a single waveguide. However, even in the most compact rhabdoms, there is extracellular space between microvilli to permit the ionic movements that are the basis of impulse transmission (Goldsmith and Bernard, 1974). The rhabdom contains visual pigments that resemble those of the vertebrate eye; that is, they are conjugated proteins called rhodopsins. The photosensitive component of the molecule, the chromophore, is retinaldehyde (retinal), the aldehyde of vitamin A, or closely related derivatives (Section 7.1.3). Surrounding the photosensitive cells are the secondary pigment cells which, like the primary pigment cells, contain granules of red, yellow, and brown/black pigments (mainly ommochromes). These pigments, especially the browns and blacks are vitally important as they strongly absorb light that enters the eye at oblique angles. Thus, the rhodopsins are activated only by light that enters an ommatidium almost parallel to its longitudinal axis (but see below). Proximally, the retinular cells narrow to form discrete axons that enter the optic lobe. Between the axons, outside the basal lamina, tracheae may occur which, in addition to their respiratory function, may serve to reflect light back along the rhabdom.

Two ommatidial types can be distinguished according to the arrangement of retinular and pigment cells (Figure 12.11A, B). In photopic (apposition) ommatidia characteristic of diurnal insects, the retinular cells span the distance between the crystalline cone and

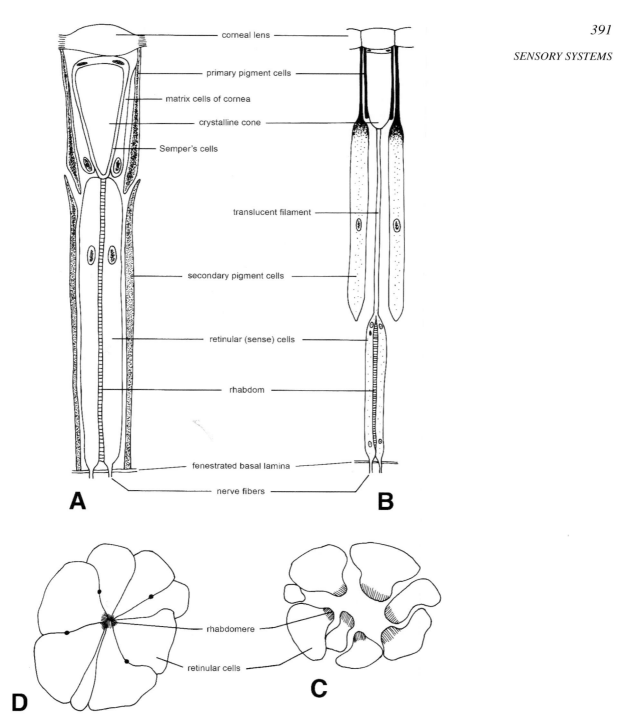

FIGURE 12.10. (A) Photopic ommatidium; (B) scotopic ommatidium; (C) open rhabdom; and (D) closed rhabdom. [A, B, after V. B. Wigglesworth, 1965, *The Principles of Insect Physiology*, 6th ed., Methuen and Co. By permission of the author. C, D, after T. H. Goldsmith and G. D. Bernard, 1974, The visual systems of insects, in: *The Physiology of Insecta*, Vol. II, 2nd ed. (M. Rockstein, ed.). By permission of Academic Press, Inc., and the authors.]

basal lamina. The secondary pigment cells lie alongside the retinular cells, and the pigment within the former does not migrate longitudinally. Scotopic (superposition) ommatidia, found in nocturnal or crepuscular species, have short retinular cells whose rhabdom is often connected to the crystalline cone by a translucent filament that serves to conduct light to the rhabdom. The secondary pigment cells do not envelop the retinular cells and their pigment granules are capable of marked longitudinal migration, allowing light from adjacent ommatidia to reach each rhabdom, enhancing rhodopsin activation.

7.1.1. Form and Movement Perception

It seems likely that early students of insect vision would assume that insects "see" in the same manner as humans; that is, a reasonably sharply defined image would form in the compound eye. Thus, early ideas on compound eye function were attempts to reconcile the observable structure of the eye with this assumed function. The classic "mosaic theory" of insect vision, introduced by Müller (1829) and expanded by Exner (1891) (cited from Goldsmith and Bernard, 1974), proposed that each ommatidium is sensitive only to light that enters at a small angle to its longitudinal axis. More oblique light rays are absorbed by pigment in the cells surrounding an ommatidium. It was assumed that little overlap existed between the visual fields of adjacent ommatidia, and thus, it was suggested, there formed in an eye an erect, mosaic image, being a composite of a large number of point source images, each formed in a separate ommatidium. The image would focus at the level of the rhabdom and its "sharpness," that is, visual acuity, would depend on the number of ommatidia per unit surface area of the eye.

Exner noted that the corneal lens and crystalline cone (which together function as a lens cylinder) appeared laminated (Figure 12.12) and suggested that this was the result of

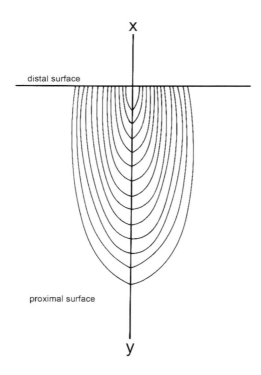

FIGURE 12.11. Lens cylinder, comprising a series of concentric lamellae of different refractive index. Refractive index is greatest along the axis *xy*. [After V. G. Dethier, 1963, *The Physiology of Insect Senses*, John Wiley and Sons, Inc. By permission of the author.]

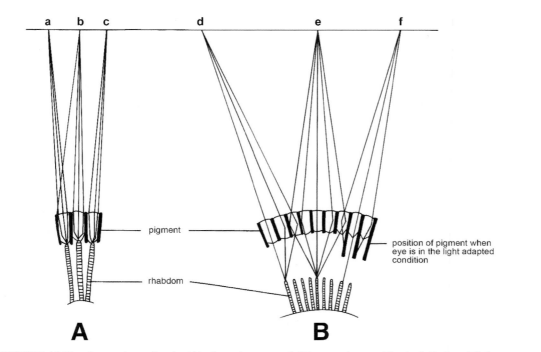

FIGURE 12.12. Image formation in (A) photopic eye; and (B) scotopic eye (B). (a–f) Paths of light rays. [After V. G. Dethier, 1963, *The Physiology of Insect Senses*, John Wiley and Sons, Inc. By permission of the author.]

a gradient of refractive index (decreasing from the axis outward). The lens cylinder would serve to bend light rays diverging from a point source back toward the axis, where they would form a point image. Exner proposed that photopic and scotopic ommatidia formed images in different ways. He suggested that in photopic eyes the lens cylinder had a focal length equal to its absolute length which would cause light rays to focus at the base of the cylinder, that is, on the upper end of the rhabdom (Figure 12.13A). The mosaic image formed in the eye would be of the "apposition" type. In scotopic eyes the focal length of the lens cylinder was supposedly one-half the absolute length of the lens cylinder with the result that light rays are brought to a focus, that is, the image is formed, some distance behind the cylinder, specifically halfway between the corneal surface and the center of curvature of the eye. According to Exner, this coincided with the tip of the rhabdom (Figure 12.13B). Further, in the dark-adapted position pigment granules aggregate at the distal end of the enveloping pigment cells. This permits light rays at a somewhat greater angle to the axis to be bent back toward the axis as they pass through adjacent lens cylinders. In other words, the point image formed on each rhabdom would be derived not from a single pencil of rays, as in the apposition type, but from a group of such pencils; that is, it would be formed by the superposition of light from a number of adjacent facets and is described therefore as a superposition image. In support of his ideas Exner managed to photograph objects, using the eye of the firefly *Lampyris* as a lens system.

During the early part of the 20th century, Exner's proposals gained acceptance and were supported by most observations that were made on insect compound eyes. However, in the 1960s, coincident with the development of more refined techniques, his ideas were subject to strong criticism, especially the concept of the lens cylinder to achieve light-ray bending, which is especially critical to the formation of a superposition image. In diurnal species there is little evidence for a gradient of refractive index in the lens and crystalline cone. By contrast, since the early 1970s, evidence has accumulated that superposition eyes invariably utilize a lens cylinder arrangement, as a result of which a single erect image is formed deep within the eye.

Although an image of sorts forms in the eye, it is now accepted that for the majority of insects an image *per se* has no physiological significance. The real function of the compound eye appears to be that of movement perception. The eye's structure is ideally suited for this function, with each ommatidium sensitive only to light shining parallel with or at very small angles to the longitudinal axis of the ommatidium. Clearly, the eye's resolution (i.e., ability to detect motion) will be related to the number of ommatidia and hence the interommatidial angle (the angle between the longitudinal axes of adjacent ommatidia). Values for ommatidial angles vary from tens of degrees in some apterygotes to 0.24° in the dragonfly *Anax junius* (Land, 1997a,b), though in many common flying insects they are in the order of 1° to 3°. Typically, eyes with the best resolution are found in predators such as Odonata, mantids, and hunting wasps, though some male Diptera which use visual cues to detect and capture mates also have high-resolution eyes. It should also be noted that resolution is often non-uniform over the whole eye. Rather, there are regions with high resolution (acute zones or "foveas") as well as regions with lower resolution, the size ratio of the two regions and the shape and position(s) of the foveas being determined by the "visual ecology" of the insect (Land 1997a,b). For example, in many predaceous species and in males that visually detect females the fovea is forward- or upward-pointing so that prey or a potential mate can be detected against the sky. Likewise, in insects that live on flat surfaces, notably water striders (Gerridae), the eyes have a narrow horizontal band of high resolution that enables the water strider to hunt for prey caught on the water surface.

Thus, a key element in insect vision is movement, more specifically, the change in relative position of the object and the insect's eyes. Many behavioral observations indicate a preference by insects for moving objects or objects with complex shapes. Bees prefer moving to stationary flowers, and they are attracted more by multistriped than by solid patterns. Dragonfly larvae will attempt to capture prey only when it is moving. Before locusts jump, they undertake "peering movements," side-to-side movements of the head that enable them to judge distance (Section 7.1.2). Stimulation of the eye by a series of changes of light intensity is known as the flicker effect. The number of stimuli per unit time to which the eye is sensitive (the flicker fusion frequency) depends on the rate at which the eye recovers from a previous stimulus and provides the basis for grouping eyes into "slow" and "fast" categories. Slow eyes have low flicker fusion frequencies and are found in more slowly moving, nocturnal insects, whereas fast eyes, with very high flicker fusion frequencies, are characteristic of fast-flying, diurnal species. As the position of the insect changes in relation to an object, the eye will receive a succession of light stimuli. Provided that the rate of change of position does not lead to a rate of stimulation that exceeds the flicker fusion frequency, the insect will, in effect, scan the object and obtain a sense of its shape. In other words, the eye translates form in space into a sequence of events in time. It follows that insects whose eyes have a high flicker fusion frequency have the best form perception. Bees, for example, are readily trained to distinguish solid shapes from striped patterns, though they cannot distinguish between two solid shapes or between two patterns of stripes. Many insects that hunt on the wing, such as dragonflies and some wasps, have excellent form perception especially species that are prey-specific. Some solitary wasps find their nest by recognizing landmarks adjacent to the entrance.

An ability to judge distance is especially important for quick-moving and/or predaceous insects. This ability may be dependent on binocular vision and/or motion parallax. In insects binocular vision is achieved when ommatidia in each eye are stimulated by the same light source, that is, when the ommatidial axes cross. For this to occur, the surface of the eyes is highly curved, which ensures a considerable overlap of their visual fields. Many predaceous species will not attack prey unless it is within catchable distance. The ability to differentiate distance appears to derive from the fact that only when certain ommatidia are stimulated is the catching reflex induced (Figure 12.14). The axes of these ommatidia cross at a point within the range of the capturing device, for example, the labium in odonate larvae, the forelegs of mantids, and the mandibles of tiger beetles. The system is not foolproof as errors in distance estimation are proportional to (a) the interommatidial angle and (b) the distance of the object in relation to the width between the eyes. To reduce possible error, the best predators have a fovea where the interommatidial angle is very small and widely set eyes, respectively. In the examples just given, the eyes are of the apposition type, with little overlap in the visual field between adjacent ommatidia and, furthermore, a fovea. By contrast, mantispids (Neuroptera) have superposition eyes where significant overlap among the visual fields of nearby ommatidia occurs. Nevertheless, when these insects feed (on bright sunny days), they appear to use a similar triangulation mechanism to that of mantids. That is, their forelegs strike when a specific set of ommatidia in each eye is stimulated by the sight of the prey. An important difference, however, between the two systems is that, in the absence of a fovea, mantispids are unable to perceive and catch prey that moves as quickly as that caught by mantids (Kral *et al.*, 2000).

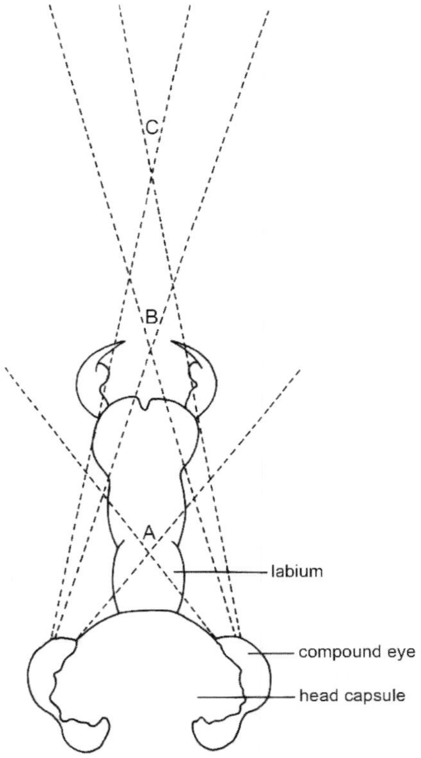

FIGURE 12.13. Distance perception in *Aeshna* larva. The insect can perceive the distance of any point that simultaneously stimulates ommatidia in both compound eyes (e.g., points A, B, and C). However, the insect extends its labium only when ommatidia whose visual axes fall between A and B are stimulated. [After V. B. Wigglesworth, 1965, *The Principles of Insect Physiology*, 6th ed., Methuen and Co. By permission of the author.]

Calculations show that even for large insects such as dragonflies the maximum distance for depth perception using binocular vision is about 20 cm. Thus, fast-moving insects employ other methods for estimating longer distances, notably motion parallax and peering (Järvilehto, 1985; Kral and Poteser, 1997). When an insect peers (Section 7.1.1), both the distance and the angle between an object and each compound eye will change. The object will move faster and by a greater angle over the nearer eye than over the farther eye. These differences are analyzed within the central nervous system to obtain a measure of the object's distance. The same technique also allows an insect to differentiate between the object and the background.

7.1.3. Spectral Sensitivity and Color Vision

Light-sensitive cells do not respond equally to all wavelengths of light; rather, they are particularly sensitive to certain parts of the spectrum. This differential sensitivity may be related to the presence in the cells of either a single visual pigment that has peak absorption at two or more wavelengths, or two or more visual pigments, each with a characteristic peak of absorption. Though the range of the spectrum to which insects are sensitive is about the same as in humans, it is shifted toward the shorter wavelengths. Many species, representative of most orders, have been shown to be very sensitive to ultraviolet light (300–400 nm) (UV) and some very significant phototaxes (behavioral responses to light) are initiated by it. For example, ants are negatively phototactic to UV and when given a choice will always congregate in a region not illuminated by UV. Many other insects are attracted by UV. Worker bees are attracted to yellow and red flowers by the pattern of UV that the latter reflect. In many species of butterflies the members of one sex have a characteristic pattern of UV-reflecting scales on the wings, invisible to would-be vertebrate predators, which facilitates intraspecific recognition and mating.

In addition to UV, the compound eye is sensitive to other wavelengths of light, though the peaks of sensitivity differ among species and even among regions of the same eye. The fly *Calliphora*, for example, shows maximum sensitivity at 470 nm (blue), 490 nm (blue-green), and 520 nm (yellow). The honey bee drone has peaks at 447 and 530 nm. All butterfly eyes have at least three distinct rhodopsins, sensitive to UV (360 nm), blue (470 nm), and green (530 nm) (White, 1985; Briscoe *et al.*, 2003). Though most insects see poorly at long wavelengths, behavioral and electrophysiological studies indicate that the eyes of some papilionid butterflies are red-sensitive (585–610 nm), in addition to having receptors sensitive to UV, blue, and green.

Although, as noted above, an insect's eye may show peaks of sensitivity to different wavelengths of light, this in itself does not constitute color vision. The latter requires that the insect has the ability to discriminate between wavelengths to which it is sensitive. Field observations and behavioral experiments have demonstrated that representatives of many orders are able to distinguish between colors. Clearly, the ability to discriminate requires at least two types of receptors (retinular cells), each of which responds to light of a particular wavelength. Further, it seems that in the insect eye, as in the human eye, the varied response of the color receptors is achieved through the possession of different visual pigments. In the rhabdom of the worker bee, each rhabdomere appears to be especially sensitive to either UV, blue, or green (Figure 12.15A). Furthermore, the bee appears to have a trichromatic color vision system similar to that of humans; that is, it can be described by a color triangle (Figure 12.15B) whose corners represent the three primary colors with points along each side being the relative stimulation of each primary color. Like vertebrates, most insects lose their ability

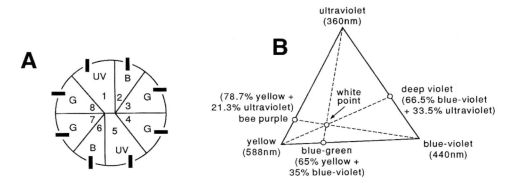

FIGURE 12.14. Color vision. (A) Fused rhabdom of worker honey bee showing that individual rhabdomeres (1–8) are sensitive to specific wavelengths [ultraviolet (UV), blue (B), or green (G)]. The orientation of the microvilli in each rhabdomere is indicated by the dark bars; and (B) color triangle for worker honey bee. Each corner corresponds approximately to the three wavelengths of maximum sensitivity. Along the base line and right side of the triangle, the color perceived is a function of the relative distance between the corners. Along the left side, the mixture is non-spectral bee purple. The hatched lines connect complementary colors that, mixed in the correct proportions, appear white because the three receptor systems are stimulated equally. [A, after R. Wehner and G. D. Bernard, 1980, Intracellular optical physiology of the bee's eye. II. Polarizational sensitivity, *J. Comp. Physiol.* **137**:205–214. B, after K. Daumer, 1956, Reizmetrische Untersuchung des Farbensehens der Bienen, *Z. Vgl. Physiol.* **38**:413–478. By permission of Springer-Verlag.]

to discriminate between colors at low light intensity though, unlike eyes of vertebrates with their rods and cones, the compound eyes of insects do not have different receptors for color and dimlight vision. An exception to this generalization are nocturnal hawkmoths which are important pollinators in tropical forests. Typically, the flowers pollinated by these moths are white (to the human eye), yet in contrast to diurnal white flowers they lack UV-reflecting patterns. Laboratory experiments with *Manduca sexta* showed that this species is attracted only to artificial flowers reflecting wavelengths greater than 400 nm, specifically in the violet and green parts of the spectrum; indeed, UV-reflecting surfaces were avoided (White *et al.*, 1994). Clearly, the visual system of nocturnal hawkmoths is able to discriminate colors at very low light intensities.

7.1.4. Sensitivity to Polarized Light

As light waves travel from their source, they oscillate sinusoidally about their longitudinal axis. The planes in which they oscillate are usually scattered randomly, through 360°, around the longitudinal axis, but under certain circumstances more waves may travel in a specific plane and the light is said to be polarized. Sunlight entering the earth's atmosphere is filtered, that is, light rays in certain planes are reflected by dust particles, etc., and the light striking the earth's surface is therefore partially polarized. Many insects, including bees, ants, crickets, locusts and some beetles, can detect and make use of the plane of polarization for navigation and orientation. The classic studies of von Frisch and his students (see von Frisch, 1967; Dyer, 2002) showed that foraging honey bees "measure" the angle between a food source, the hive, and the sun and, on returning to the hive, communicate this information to their fellows through the performance of a "waggle dance" on the honeycomb. Actual sight of the sun is not essential for the "light compass reaction," provided that a bee can see a patch of blue sky and thus determine the plane of polarized light.

In the nocturnal tenebrionid beetle *Parastizopus armaticeps*, the entire eye is able to detect polarized light. However, in many diurnal insects, especially Odonata, Hymenoptera, Coleoptera, Diptera, and Lepidoptera, the region of sensitivity is generally restricted to the dorsal part of the eye, the dorsal rim area (Labhart and Meyer, 1999). Within this region, the ommatidia are distinctly adapted to analyze the plane of polarization. The adaptations may include: (1) a fan-shaped arrangement of ommatidia within the eye, providing a wide angle of sensitivity; (2) pore canals in the corneas and irregular (as distinct from hexagonal) facets to increase light scattering; and (3) wider and shorter but very straight rhabdoms, lack of screening pigment between adjacent ommatidia, and highly specific orientation of the microvilli on the inner margins of the retinular cells, all of which increase sensitivity (Homberg and Paech, 2002). Within an ommatidium, there are two sets of rhabdomeres; the microvilli of one set are at 90° to those of the other set. The photosensitive molecules are arranged very precisely in the microvilli so as to absorb light maximally when the light waves oscillate parallel to the long axis of the microvilli. However, among different taxa the nature of the photoreceptive pigment may differ: in Orthoptera it is blue-sensitive, in Coleoptera green-sensitive, and in Hymenoptera and Diptera UV-sensitive. For additional information on the detection and use of polarized light by insects, see Wehner (1984, 1992, 1997), Rossel (1993), and Labhart and Meyer (1999).

7.2. Simple Eyes

Many adult insects and juvenile exopterygotes possess in addition to compound eyes, three simple eyes, dorsofrontal in position, known as ocelli. The larvae of endopterygotes have, as their sole photosensory structure, stemmata (lateral ocelli).

Ocelli. The structure of an ocellus is shown in Figure 12.16A. It comprises usually about 500–1000 photosensitive cells beneath a common cuticular lens (Goodman, 1970). The cells are arranged in groups of two to five cells, and, distally, each differentiates into a rhabdomere to form a central rhabdom. In contrast to the retinular cells of ommatidia, the photosensitive cells of dorsal ocelli are second-order sense cells; that is, their axons do not themselves conduct information to the central nervous system, but synapse within the ocellar nerve with axons of cells originating in the brain.

Ocellar function is poorly understood. Though an ocellus is able to form an image, it does so below the level of the rhabdom and, therefore, the image has no physiological significance. The modern view of their main function is that they serve to detect the horizon (hence, being out of focus ensures that extraneous details do not impede this function) and are thus important in maintaining stability during level flight. However, in some insects, other functions occur. For example, ocelli respond to the same wavelengths as compound eyes but are much more sensitive than compound eyes; that is, they are stimulated by very low light intensities. Thus, they may measure light intensity, and the information derived from them may be used to modify an insect's response to stimuli received by the compound eye. Painting ocelli may cause temporary reversal or inhibition of light-directed behavior, or reduce the rapidity with which an insect responds to light stimuli. Such observations suggest that ocelli act as "stimulators" of the nervous system, so that an insect detects and responds more rapidly to light entering the compound eyes. In addition, in some species ocelli appear essential for the maintenance of diurnal locomotor rhythms. Rarely, an apparent duplication of functions associated with compound eyes has been described. For example, in workers of the desert ant *Cataglyphis bicolor* the ocelli (as well as the compound eyes) determine the plane of polarized light, enabling ants with their compound eyes experimentally occluded to

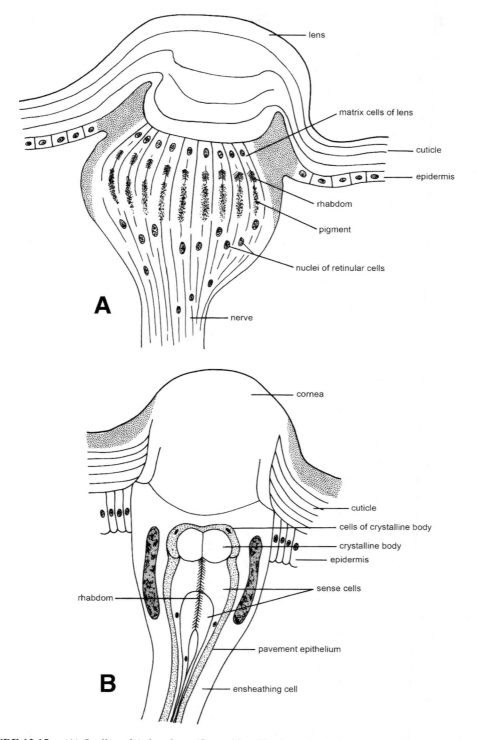

FIGURE 12.15. (A) Ocellus of *Aphrophora* (Cercopidae: Hemiptera); and (B) stemma of *Gastropacha* (Lepidoptera). [After V. B. Wigglesworth, 1965, *The Principles of Insect Physiology*, 6th ed., Methuen and Co. By permission of the author.]

successfully return to the nest after foraging (Fent and Wehner, 1985). Similarly, the escape reaction of the bloodsucking bug *Triatoma infestans* is induced equally well by stimulation of either the ocelli or the compound eyes (Lazzari *et al.*, 1998). In both these examples, it could be the two photosensors are responding to different qualities of the light stimulus (e.g., wavelength and intensity).

Stemmata. These are of three types: (1) In sawfly and beetle larvae each stemma, one on either side of the head, resembles a dorsal ocellus, that is, comprises a single cuticular lens lying beneath which are groups of photosensitive cells with a central rhabdom. (2) In larvae of Neuroptera, Trichoptera, and Lepidoptera, there are usually several laterally placed stemmata on each side of the head. Each resembles an ommatidium in that it includes both a corneal and crystalline lens and a group of retinal cells with a central rhabdom (Figure 12.16B). (3) Finally, in larvae of cyclorrhaph Diptera there are no external signs of stemmata, but a pocket of photosensitive cells occurs on each side of the pharyngeal skeleton.

Stemmata of beetle larvae and caterpillars have been best studied, and their functions appear very similar to those of compound eyes. They form a relatively focused image and are important in color vision, predator avoidance, prey capture, etc. They can also detect the plane of polarized light, though no functional significance has yet been attached to this ability. Indeed, Land (1985) raised the fascinating question of why stemmata ever came to be replaced by the structurally more complex compound eye in the adult stage.

8. Summary

Insect sensory cells are almost always primary sense cells; that is, they both receive a stimulus and conduct it to the central nervous system. Type I sense cells are associated with cuticle and accessory cells to form the sense organ (sensillum); Type II cells are not associated with cuticle and always function as proprioceptors.

Mechanosensilla include sensory hairs, hair plates, campaniform sensilla, chordotonal organs, stretch receptors, and nerve nets. Some insects can detect sound by means of fine sensory hairs (sound of low frequency) or specialized chordotonal structures, such as Johnston's organ and tympanal organs. In both mechanoreception and sound reception the stimulus leads to mechanical deformation of a sense cell's cytoplasm.

Chemosensilla occur as thick-walled (uniporous) hairs, pegs, or papillae that generally have a gustatory function, or as thin-walled (multiporous) hairs, pegs, or plate organs whose usual function is olfactory. Each chemosensillum includes several sensory cells, each of which may respond to a different stimulus. Odor and organic taste molecules reach the sensory cell membrane bound to protein molecules. At the membrane they are transferred to a receptor, an action that induces depolarization. How polar materials reach the sensory cell membrane remains unclear; however, once there, they likely act directly on ion channels to trigger impulse generation.

Generally, humidity and temperature changes are detected by dual-purpose sensilla, in the form of poreless, thin-walled hairs or pegs. Each sensillum has two multidendritic neurons, one for humidity, the other for temperature. Probably, hygroreceptors operate like mechanoreceptors, undergoing a change in form in response to changing humidity. How temperature-sensitive sensilla work remains unclear. A few insects have been shown to be sensitive to infrared heat, using thin-walled, poreless sensilla, each with a single neuron.

Insects detect light energy via compound eyes, ocelli, or stemmata, rarely a dermal light sense. Compound eyes, the chief photosensory structures, are composed of ommatidia, each of which includes a light-focusing system, photosensitive (retinular) cells, and enveloping pigment cells. Two types of ommatidia occur, the photopic characteristic of diurnal insects and the scotopic found in crepuscular and nocturnal species. Retinular cells are differentiated along their inner longitudinal axis into a rhabdomere, a series of closely packed microvilli in which are contained visual pigments. Even when an image is formed at the level of the retinular cells it has no functional significance. The primary function of compound eyes appears to be movement perception, though some appreciation of form is gained as a result of the flicker effect. The ability to perceive distance is present in some species and is based on the considerable overlap of the visual fields of the two compound eyes (up to about 20 cm in large insects) and through the use of motion parallax for greater distances. Color vision occurs in some insects and results from the presence of retinular cells with different visual pigments. Compound eyes of some species can detect the plane of polarized light, enabling these insects to carry out complex navigation behavior.

The principal function of ocelli appears to be horizon detection as a component of horizontal flight, though for some insects the ocelli may provide information on sudden changes in light intensity, serve as a general stimulus to the nervous system, initiate specific phototaxes, and detect the plane of polarized light. Stemmata have functions similar to those compound eyes.

9. Literature

Reviews dealing with particular aspects of sensory perception include those of McIver (1985), French (1988), and Keil (1997, 1998) [mechanoreception]; Michelson and Larsen (1985), Kalmring and Elsner (1985), Hoy and Robert (1996), Hoy *et al.* (1998), and Yager (1999) [sound reception]; Morita and Shiraishi (1985), Mitchell *et al.* (1999), Keil (1999) and Ryan (2002) [chemoreception]; Altner and Loftus (1985) [hygro- and thermoreception]; Horridge (1975), Land (1985), Järvilehto (1985), and White (1985) [compound eyes]; Goodman (1970, 1981) [dorsal ocelli]; Gilbert (1994) [stemmata]; and Zacharuk and Shields (1991) [sensilla of immature insects].

Altner, H., and Loftus, R., 1985, Ultrastructure and function of insect thermo- and hygroreceptors, *Annu. Rev. Entomol.* **30**:273–295.

Bennet-Clark, H. C., and Ewing, A. W., 1970, The love song of the fruit fly, *Sci. Amer.* **223** (July):85–92.

Briscoe, A. D., Bernard, G. D., Szeto, A. S., Nagy, L. M., and White, R. H., 2003, Not all butterfly eyes are created equal: Rhodopsin absorption spectra, molecular identification, and localization of ultraviolet-, blue-, and green-sensitive rhodopsin-encoding mRNAs in the retina of *Vanessa cardui, J. Comp. Neurol.* **458**:334–349.

Cade, W. H., 1975, Acoustically orienting parasitoids: Fly phonotaxis to cricket song, *Science* **190**:1312–1313.

Davis, E. E., and Sokolove, P. G., 1975, Temperature responses of antennal receptors of the mosquito, *Aedes Aegypti, J. Comp. Physiol.* **96**:223–236.

Dethier, V. G., 1963, *The Physiology of Insect Senses*, Wiley, New York.

Dreller, C., and Kirchner, W. H., 1995, The sense of hearing in honey bees, *Bee World* **76**:6–17.

Dyer, F. C., 2002, The biology of the dance language, *Annu. Rev. Entomol.* **47**:917–949.

Fent, K., and Wehner, R., 1985, Ocelli: A celestial compass in the desert ant *Cataglyphis, Science* **228**:192–194.

French, A. S., 1988, Transduction mechanisms of mechanosensilla, *Annu. Rev. Entomol.* **33**:39–58.

Gilbert, C., 1994, Form and function of stemmata in larvae of holometabolous insects, *Annu. Rev. Entomol.* **39**:323–349.

Goldsmith, T. H., and Bernard, G. D., 1974, The visual system of insects, in: *The Physiology of Insecta*, 2nd ed., Vol. II (M. Rockstein, ed.), Academic Press, New York.

Goodman, L. J., 1970, The structure and function of the insect dorsal ocellus, *Adv. Insect Physiol.* **7**:97–195.

Goodman, L. J., 1981, Organisation and physiology of the insect dorsal ocellar system, in: *Handbook of Sensory Physiology*, Vol. VII/6C (H. Autrum, ed.), Springer-Verlag, Berlin.

Göpfert, M. C., Surlykke, A., and Wasserthal, L. T., 2002, Tympanal and atympanal 'mouth-ears' in hawkmoths (Sphingidae), *Proc. R. Soc. Lond. B* **269**:89–95.

Homberg, U., and Paech, A., 2002, Ultrastructure and orientation of ommatidia in the dorsal rim area of the locust compound eye, *Arth. Struct. Dev.* 30:271–280.

Horridge, G. A. (ed.), 1975, *The Compound Eye and Vision of Insects*, Clarendon, Oxford.

Hoy, R. R., and Robert, D., 1996, Tympanal hearing in insects, *Annu. Rev. Entomol.* **41**:433–450.

Hoy, R. R., Popper, A. N., and Fay, R. R. (eds.), 1998, *Comparative Hearing: Insects*, Springer-Verlag, New York.

Järvilehto M., 1985, The eye: Vision and perception, in: *Comprehensive Insect Physiology, Biochemistry, and Pharmacology*, Vol. 6 (G. A. Kerkut and L. I. Gilbert, eds.), Pergamon Press, Elmsford, NY.

Kalmring, K., and Elsner, N. (eds.), 1985, *Acoustic and Vibrational Communication in Insects*, Paul Parey, Berlin.

Keil, T. A., 1997, Functional morphology of insect mechanoreceptors, *Microsc. Res. Tech.* **39**:506–531

Keil, T. A., 1998, The structure of integumental mechanoreceptors, in: *Microscopic Anatomy of Invertebrates*, Vol. 11 (F. W. Harrison and M. Locke, eds.), Wiley-Liss, New York.

Keil, T. A., 1999, Morphology and development of the peripheral olfactory organs, in: *Insect Olfaction* (B. Hansson, ed.), Springer-Verlag, Berlin.

Kral, K., and Poteser, M., 1997, Motion parallax as a source of distance information in locusts and mantids, *J. Insect Behav.* **10**:145–163.

Kral, K., Vernik, M., and Devetak, D., 2000, The visually controlled prey-capture behaviour of the European mantispid *Mantispa styriaca*, *J. Exp. Biol.* **203**:2117–2123.

Labhart, T., and Meyer, E. P., 1999, Detectors for polarized skylight in insects: A survey of ommatidial specializations in the dorsal rim area of the compound eye, *Microsc. Res. Tech.* **47**:368–379.

Land, M. F., 1985, The eye: Optics, in: *Comprehensive Insect Physiology, Biochemistry, and Pharmacology*, Vol. 6 (G. A. Kerkut and L. I. Gilbert, eds.), Pergamon Press, Elmsford, NY.

Land, M. F., 1997a, Visual acuity in insects, *Annu. Rev. Entomol.* **42**:147–177.

Land, M. F., 1997b, The resolution of insect compound eyes, *Isreal J. Pl. Sci.* **45**:79–91.

Lazzari, C. R., Reiseman, C. E., and Insausti, T. C., 1998, The role of the ocelli in the phototactic behaviour of the haematophagous bug *Triatoma infestans*, *J. Insect Physiol.* **44**:1159–1162.

Morita, H., and Shiraishi, A., 1985, Chemoreception physiology, in: *Comprehensive Insect Physiology, Biochemistry, and Pharmacology*, Vol. 6 (G. A. Kerkut and L. I. Gilbert, eds.), Pergamon Press, Elmsford, NY.

Markl, H., and Tautz, J., 1975, The sensitivity of hair receptors in caterpillars of *Barathra brassicae* L. (Lepidoptera, Noctuidae) to particle movement in a sound field, *J. Comp. Physiol.* **99**:79–87.

McIver, S. B., 1985, Mechanoreception, in: *Comprehensive Insect Physiology, Biochemistry, and Pharmacology*, Vol. 6 (G. A. Kerkut and L. I. Gilbert, eds.), Pergamon Press, Elmsford, NY.

Michelson, A., and Larsen, O. N., 1985, Hearing and sound, in: *Comprehensive Insect Physiology, Biochemistry, and Pharmacology*, Vol. 6 (G. A. Kerkut and L. I. Gilbert, eds.), Pergamon Press, Elmsford, NY.

Mitchell, B. K., Itagaki, H., and Rivet, M., 1999, Peripheral and central structures involved in insect gustation, *Microsc. Res. Tech.* **47**:401–415.

Reinouts van Haga, H. A., and Mitchell, B. K., 1975, Temperature receptors on tarsi of the tsetse fly, *Glossina morsitans* West., *Nature* **255**:225–226.

Römer, H., and Tautz, J., 1992, Invertebrate auditory receptors, *Adv. Comp. Environ. Physiol.* **10**:185–212.

Rossel, S., 1993, Navigation by bees using polarized skylight, *Comp. Biochem. Physiol.* **104A**:695–708.

Ryan, M. F., 2002, *Insect Chemoreception: Fundamental and Applied*, Kluwer Academic Publishers, Dordrecht.

Schmitz, H., Schmitz, A., and Bleckmann, H., 2001, Morphology of a thermosensitive multipolar neuron in the infrared organ of *Merimna atrata* (Coleoptera, Buprestidae), *Arth. Str. Dev.* **30**:99–111.

Schmitz, H., Trenner, S., Hofmann, M. H., and Bleckmann, H., 2000, The ability of *Rhodnius prolixus* (Hemiptera; Reduviidae) to approach a thermal source solely by its infrared radiation, *J. Insect Physiol.* **46**:745–751.

Slifer, E. H., 1970, The structure of arthropod chemoreceptors, *Annu. Rev. Entomol.* **15**:121–142.

Spangler, H. G., 1988, Moth hearing, defense, and communication, *Annu. Rev. Entomol.* **33**:59–81.

Stavenga, D. G., 2002, Colour in the eyes of insects, *J. Comp. Physiol. A* **188**:337–348.

Vondran, T., Apel, K. H., and Schmitz, H., 1995, The infrared receptor of *Melanophila acuminata* De Geer (Coleoptera: Buprestidae): Ultrastructural study of a unique insect thermoreceptor and its possible descent from a hair mechanoreceptor, *Tissue Cell* **27**:645–658.

von Frisch, K., 1967, *The Dance Language and Orientation of Bees*, Springer-Verlag, Berlin.

Walker, R. G., Willingham, A. T., and Zuker, C. S., 2000, A *Drosophila* mechanosensory transduction channel, *Science* **287**:2229–2234.

Wehner, R., 1984, Astronavigation in insects, *Annu. Rev. Entomol.* **29**:277–298.

Wehner, R., 1992, Arthropods, in: *Animal Homing* (F. Papi, ed.), Chapman and Hall, London.

Wehner, R., 1997, The ant's celestial compass system: Spectral and polarization channels, in: *Orientation and Communication in Arthropods* (M. Lehrer, ed.), Birkhauser, Basel.

White, R. H., 1985, Insect visual pigments and colour vision, in: *Comprehensive Insect Physiology, Biochemistry, and Pharmacology*, Vol. 6 (G. A. Kerkut and L. I. Gilbert, eds.), Pergamon Press, Elmsford, NY.

White, R. H., Stevenson, R. D., Bennett, R. R., Cutler, D. E., and Haber, W. A., 1994, Wavelength discrimination and the role of ultraviolet vision in the feeding behavior of hawkmoths, *Biotropica* **26**:427–435.

Yager, D. D., 1999, Structure, development, and evolution of insect auditory systems, *Microsc. Res. Tech.* **47**:380–400.

Zacharuk, R. Y., and Shields, V. D., 1991, Sensilla of immature insects, *Annu. Rev. Entomol.* **36**:331–354.

Nervous and Chemical Integration

1. Introduction

Animals constantly monitor both their internal and their external environment and make the necessary adjustments in order to maintain themselves optimally and thus to develop and reproduce at the maximum rate. The adjustments they make may be immediate and obvious, for example, flight from predators, or longer-term, for example, entry into diapause to avoid impending adverse conditions. The nature of the response depends, obviously, on the nature of the stimulus. Only very rarely does a stimulus act directly on the effector system; almost always a stimulus is received by an appropriate sensory structure and taken to the central nervous system, which "determines" an appropriate response under the circumstances. When a response is immediate, that is, achieved in a matter of seconds or less, it is the nervous system that transfers the message to the effector system. Such responses are usually temporary in nature. Delayed responses are achieved through the use of chemical messages (viz., hormones) and are generally longer-lasting. The nervous and endocrine systems of an individual are, then, the systems that coordinate the response with the stimulus. Semiochemicals, which constitute another chemical regulating system, coordinate behavior and development among individuals. They comprise pheromones (intraspecific coordinators) and allelochemicals (interspecific coordinators), which include kairomones and allomones.

2. Nervous System

Like that of other animals, the nervous system of insects consists of nerve cells (neurons) and glial cells. Each neuron comprises a cell body (perikaryon) where a nucleus, many mitochondria, and other organelles are located, and a cytoplasmic extension, the axon, which is usually much branched, the branches being known as neurites. Axons may be long, as in sensory neurons, motor neurons, and principal interneurons, or very short, as in local interneurons. Often, insect neurons are monopolar, lacking the dendritic tree characteristic of vertebrate nerve cells, though bipolar and multipolar neurons do occur (Figure 13.1). Motor (efferent) neurons, which carry impulses from the central nervous system, are monopolar, and their perikarya are located within a ganglion. Sensory (afferent) neurons are usually bipolar but may be multipolar, and their cell bodies are adjacent to

A

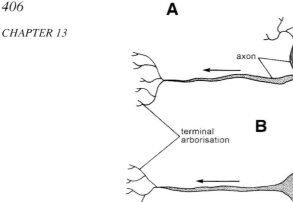

dendrites

perikaryon

axon

terminal
arborisation

B

perikaryon

dendrite

C

perikaryon

axon

axon

dendrites

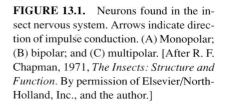

FIGURE 13.1. Neurons found in the insect nervous system. Arrows indicate direction of impulse conduction. (A) Monopolar; (B) bipolar; and (C) multipolar. [After R. F. Chapman, 1971, *The Insects: Structure and Function*. By permission of Elsevier/North-Holland, Inc., and the author.]

the sense organ. Interneurons (also called internuncial or association neurons) transmit information from sensory to motor neurons or other interneurons; they may be mono- or bipolar and their cell bodies occur in a ganglion. Interneurons may be intersegmental and branched, so that the variety of pathways along which information can travel and, therefore, the variety of responses are increased.

Neurons are not directly connected to each other or to the effector organ but are separated by a minute space, the synapse or neuromuscular junction, respectively. Impulses may be transferred across the synapse either electrically or chemically (Section 2.3). The normal diameter of axons is 5 μm or less; however, some interneurons within the ventral nerve cord, the so-called "giant fibers," have diameters up to 60 μm. These giant fibers may run the length of the nerve cord without synapsing and are unbranched except at their termini. They are well suited, therefore, for very rapid transmission of information from sense organ to effector organ; that is, they facilitate a very rapid but stereotyped response to a stimulus and for some insects are important in escape reactions (Hoyle, 1974; Ritzmann, 1984).

Neurons are aggregated into nerves and ganglia. Nerves include only the axonal component of neurons, whereas ganglia include axons, perikarya, and dendrites. The typical structures of a ganglion and interganglionic connective are shown in Figure 13.2. In a ganglion there is a central neuropile that comprises a mass of efferent, afferent, and association axons. Frequently visible within the neuropile are groups of axons running parallel, known as fiber tracts. The perikarya of motor and association neurons are normally found in clusters adjacent to the neuropile.

Surrounding the neurons are glial cells, which are differentiated according to their position and function. The peripheral glial (perineural) cells, which form the perineurium,

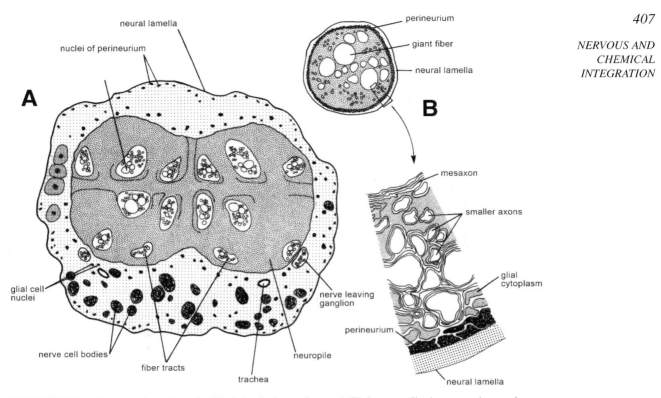

FIGURE 13.2. Cross-sections through (A) abdominal ganglion and (B) interganglionic connective to show general structure. [A, after K. D. Roeder, 1963, *Nerve Cells and Insect Behaviour.* By permission of Harvard University Press. B, after J.E. Treherne and Y. Pichon, 1972, The insect blood-brain barrier, *Adv. Insect Physiol.* **9**:257–313. By permission of Academic Press Ltd., London, and the authors.]

are very closely associated by tight junctions, forming the blood-brain barrier (Carlson *et al.,* 2000; Kretzschmar and Pflugfelder, 2002). This barrier is critical in isolating the nervous system from the hemolymph whose composition is both highly variable and inappropriate for neuronal function (see Chapter 17, Section 4). However, the barrier itself creates two potential problems, namely, obtaining adequate supplies of oxygen and nutrients for the neural elements. The former is solved by having tracheae running deeply into the ganglia, the latter by the ability of the perineural cells to transfer materials between the hemolymph and neurons. In addition, they secrete the neural lamella, a protective sheath that contains collagen fibrils and mucopolysaccharide. The lamella is freely permeable, enabling the perineural cells to accumulate nutrients from the hemolymph. The inner glial cells occur among the perikarya into which they extend fingerlike extensions of their cytoplasm, the trophospongium (Figure 13.3A). The function of these cells is to transport nutrients from perineural cells to the perikarya. Once in the perikarya, nutrients are transported to their site of use by cytoplasmic streaming.

Wrapped around each axon or groups of smaller axons are other glial (Schwann) cells (Figure 13.3B), These cells effectively isolate axons from the hemolymph in which they are bathed, However, in contrast to the situation in vertebrates, the glial cells are not compacted to form a myelin sheath but rather are loosely wound around the axons, Further, in insect nerves there are no distinct nodes of Ranvier (the regions between adjacent glial cells); hence, saltatory conduction of impulses does not occur (Section 2.3).

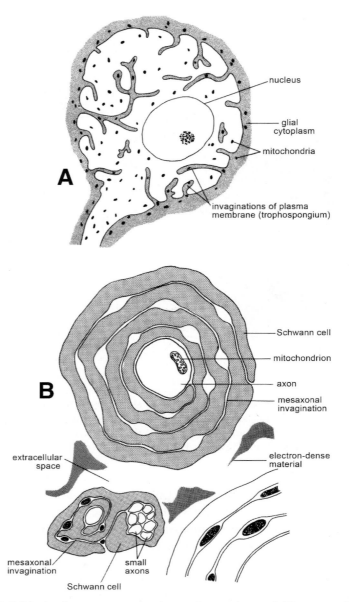

FIGURE 13.3. (A) Cell body of motor neuron showing trophospongium; and (B) cross-section through axons and surrounding Schwann cells. [A, after V. B. Wigglesworth, 1965, *The Principles of Insect Physiology*, 6th ed., Methuen and Co. By permission of the author. B, after J. E. Treherne, and Y. Pichon, 1972, The insect blood-brain barrier, *Adv. Insect Physiol.* **9**:257–313. By permission of Academic Press Ltd., London, and the authors.]

Structurally, the nervous system may be divided into (1) the central nervous system and its peripheral nerves and (2) the visceral nervous system.

2.1. Central Nervous System

The central nervous system arises during embryonic development as an ectodermal delamination on the ventral side (Chapter 20, Section 7.3). Each embryonic segment includes initially a pair of ganglia, though these soon fuse. In addition, varying degrees of

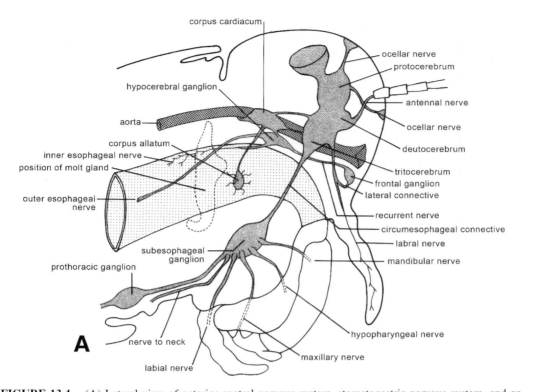

FIGURE 13.4. (A) Lateral view of anterior central nervous system, stomatogastric nervous system, and endocrine glands of a typical acridid; (B) diagrammatic dorsal view of brain and associated structures to show paths of neurosecretory axons and relationship of corpora cardiaca and corpora allata; (C) dorsal view of corpora cardiaca to show distinct storage and glandular zones; and (D,E) transverse sections through corpora cardiaca at levels a–a' and b–b', respectively. [A, after F. O. Albrecht, 1953, *The Anatomy of the Migratory Locust.* By permission of The Athlone Press. B–E, after K. C. Highnam, and L. Hill, 1977, *The Comparative Endocrinology of the Invertebrates*, 2nd ed. By permission of Edward Arnold Publishers Ltd.]

anteroposterior fusion occur so that composite ganglia result. Thus, in an adult insect the central nervous system comprises the brain, subesophageal ganglion, and a varied number of ventral ganglia.

The brain (Figure 13.4A) is probably derived from the ganglia of three segments and forms the major association center of the nervous system. It includes the protocerebrum, deutocerebrum, and tritocerebrum. The protocerebrum, the largest and most complex region of the brain, contains both neural and endocrine (neurosecretory) elements. Anteriorly it forms the proximal part of the ocellar nerves (the only occasion on which the cell bodies of sensory neurons are located other than adjacent to the sense organ), and laterally is fused with the optic lobes. Within the protocerebrum is a pair of corpora pedunculata, the mushroom bodies, so-called because of their outline in cross-section. The mushroom bodies are important association centers, receiving sensory inputs, especially olfactory and visual, and relaying the information to other protocerebral centers (Strausfeld *et al.*, 1998; Gronenberg, 2001). Further, they play a central role in learning and memory (Section 2.4), and their size can be broadly correlated with the development of complex behavior patterns. They are most highly developed in the social Hymenoptera. In worker ants, for example, they make up about one-fifth the volume of the brain. The median central body is another important association center, one function of which appears to be the coordination

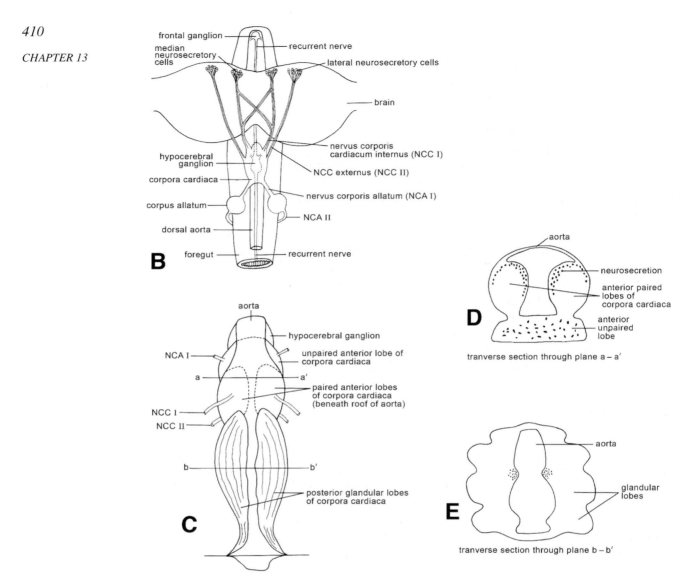

FIGURE 13.4. *(Continued)*

of segmental motor activities, for example, respiratory movements, walking, and flight. Recently, the central body and the closely associated protocerebral bridge have been shown to possess polarized light-sensitive interneurons, suggesting a role for these centers in navigation (Vitzthum *et al.*, 2002). Each optic lobe contains three neuropilar masses in which light stimuli, including those generated by polarized light, are assessed and forwarded to other brain centers.

The deutocerebrum is largely composed of the paired antennal lobes (Homberg *et al.*, 1989; Hannson and Anton, 2000). These two neuropiles include both sensory and motor neurons and are responsible for initiating both responses to antennal stimuli, especially olfactory and mechanosensory, and movements of the antennae. In species where females produce sex-pheromones the antennal lobes often show sexual dimorphism, being larger with additional interneurons in males. From the antennal lobes, interneurons convey information to association centers in both the protocerebrum and thoracic ganglia. Together with the mushroom bodies, the antennal lobes are essential in learned olfactory behavior. The transfer of mechanosensory inputs to the ventral ganglia is likely related to perception and avoidance of objects encountered during walking.

The tritocerebrum is a small region of the brain located beneath the deutocerebrum and comprises a pair of neuropiles that contain axons, both sensory and motor, leading to/from the frontal ganglion and labrum.

The subesophageal ganglion is also composite and includes the elements of the embryonic ganglia of the mandibular, maxillary, and labial segments. From this ganglion, nerves containing both sensory and motor axons run to the mouthparts, salivary glands, and neck. The ganglion also appears to be the center for maintaining (though not initiating) locomotor activity.

In most insects the three segmental thoracic ganglia remain separate. Though details vary from species to species, each ganglion innervates the leg and flight muscles (direct and indirect), spiracles, and sense organs of the segment in which it is located.

The maximum number of abdominal ganglia is eight, seen in the adult bristletail *Machilis* and larvae of many species, though even in these insects the terminal ganglion is composite, including the last four segmental ganglia of the embryonic stage. Varying degrees of fusion of the abdominal ganglia occur in different orders and sometimes there is fusion of the composite abdominal ganglion with the ganglia of the thorax to form a single thoracoabdominal ganglion. (Chapters 5–10 contain the details for individual orders.)

2.2. Visceral Nervous System

The visceral (sympathetic) nervous system includes three parts: the stomatogastric system, the unpaired ventral nerves, and the caudal sympathetic system. The stomatogastric system, shown partially in Figure 13.4, arises during embryogenesis as an invagination of the dorsal wall of the stomodeum. Generally, it includes the frontal ganglion, recurrent nerve which lies mediodorsally above the gut, hypocerebral ganglion, a pair of inner esophageal nerves, a pair of outer esophageal (gastric) nerves, each of which normally terminates in an ingluvial (ventricular) ganglion situated alongside the posterior foregut, and various fine nerves from these ganglia that innervate the foregut and midgut, and, in some species, the heart. A single median ventral nerve arises from each thoracic and abdominal ganglion in some insects. The nerve branches and innervates the spiracle on each side. In species where this nerve is absent, paired lateral nerves from the segmental ganglia innervate the spiracles. The caudal sympathetic system, comprising nerves arising from the composite terminal abdominal ganglion, innervates the hindgut and sexual organs. Nerves within the stomatogastric system both collect mechanosensory and chemical information from, and regulate the muscular activity of, the organs they supply. In the frontal ganglion, at least, the neuropile has a central pattern generator (Section 2.3) that controls rhythmic motor activity of the foregut (Ayali *et al.*, 2002).

2.3. Physiology of Neural Integration

As noted in the Introduction to this chapter, an insect's nervous system is constantly receiving stimuli of different kinds both from the external environment and from within its own body. The subsequent response of the insect depends on the net assessment of these stimuli within the central nervous system. The processes of receiving, assessing, and

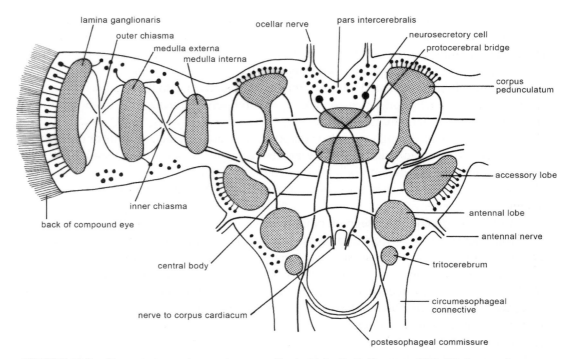

FIGURE 13.5. Cross-section to show major areas of brain. [After R. F. Chapman, 1971, *The Insects: Structure and Function.* By permission of Elsevier/North-Holland, Inc., and the author.]

responding to stimuli collectively constitute neural integration. Neural integration includes, therefore, the biophysics of impulse transmission along axons and across synapses, the reflex pathways (in insects, intrasegmental) from sense organ to effector organ, and coordination of these segmental events within the central nervous system.

Impulse transmission along axonal membranes and across synapses appears to be essentially the same as in other animals and will not be discussed here in detail. However, the absence of a myelin sheath and nodes of Ranvier precludes the phenomenon of saltatory conduction seen in vertebrates. Following the arrival of a stimulus of sufficient magnitude, an action potential is generated and the impulse travels along the axon as a wave of depolarization. The speed of impulse transmission is a function of axonal diameter so that in giant axons values of 3–7 m per sec have been recorded while in average-sized axons the speed is 1.5–2.3 m per sec. In addition to "spiking" neurons (i.e., those in which an action potential can be generated), there are in the insect central nervous system intraganglionic "non-spiking" interneurons unable to produce action potentials. Rather, the amount of neurotransmitter released at their synapses (see below) is proportional to the size of their endogenous membrane permeability changes; in other words, they release neurotransmitter (and affect the postsynaptic neuron) in a graded manner. These non-spiking interneurons may have wide importance in the initiation of rhythmic behaviors such as walking, swimming, and chewing (see below).

Transmission across a synapse, depending as it does on diffusion of molecules through fluid, is relatively slow and may take up about 25% of the total time for conduction of an impulse through a reflex arc. Rarely, when a synaptic gap is narrow (i.e., pre- and postsynaptic membranes are closely apposed), the ionic movements across the presynaptic membrane are sufficient to directly induce depolarization of the postsynaptic membrane

(Huber, 1974). Mostly, however, when an impulse reaches a synapse, it causes release of a chemical (a neurotransmitter) from membrane-bound vesicles. The chemical diffuses across the synapse and, in excitatory neurons, brings about depolarization of the postsynaptic membrane. Acetylcholine is the predominant neurotransmitter liberated at excitatory synapses, including those of interneurons and afferent neurons from mechanosensilla and taste sensilla (Homberg, 1994). 5-Hydroxytryptamine (serotonin), histamine, octopamine, and dopamine function as central nervous system excitatory neurotransmitters in specific situations on occasion. These, and other amines, have an excitatory effect when applied in low concentrations to the heart, gut, reproductive tract, etc., and it may be that they also serve as neurotransmitters in the visceral nervous system.

Sometimes a single nerve impulse arriving at the presynaptic membrane does not stimulate the release of a sufficient amount of neurotransmitter. Thus, the magnitude of depolarization of the postsynaptic membrane is not large enough to initiate an impulse in the postsynaptic axon. If additional impulses reach the presynaptic membrane before the first depolarization has decayed, sufficient additional neurotransmitter may be released so that the minimum level for continued passage of the impulse (the "threshold" level) is exceeded. This additive effect of the presynaptic impulses is known as temporal summation. A second form of summation is spatial, which occurs at convergent synapses. Here, several sensory axons synapse with one internuncial neuron. A postsynaptic impulse is initiated only when impulses from a sufficient number of sensory axons arrive at the synapse simultaneously. Divergent synapses are also found where the presynaptic axon synapses with several postsynaptic neurons. In this arrangement the arrival of a single impulse at a synapse may be sufficient to initiate impulse transmission in, say, one of the postsynaptic neurons. The arrival of additional impulses in quick succession will lead to the initiation of impulses in other postsynaptic neurons whose threshold levels are higher. Thus, synapses play an important role in selection of an appropriate response for a given stimulus.

Eventually, an impulse reaches the effector organ, most commonly muscle. Between the tip of the motor axon and the muscle cell membrane is a fluid-filled space, comparable to a synapse, called a neuromuscular junction. Again, to achieve depolarization of the muscle cell membrane and, ultimately, muscle contraction, a chemical released from the tip of the axon diffuses across the neuromuscular junction. In insect skeletal muscle, this chemical is L-glutamate; in visceral muscles, glutamate, serotonin, and the pentapeptide proctolin have all been suggested as candidate neurotransmitters.

In addition to stimulatory (excitatory) neurons, inhibitory neurons whose neurotransmitter causes hyperpolarization of the postsynaptic or effector cell membrane are also important in neural integration. When inhibition occurs at a synapse within the central nervous system, it is known as central inhibition. Central inhibition is the prevention of the normal stimulatory output from the central nervous system and may arise spontaneously within the system or result from sensory input. For example, copulatory movements of the abdomen in the male mantis, which are regulated by a segmental reflex pathway located within the terminal abdominal ganglion, are normally inhibited by spontaneous impulses arising within the brain and passing down the ventral nerve cord. In the fly *Protophormia* the stimulation of stretch receptors during feeding results in decreased sensitivity to taste caused by central inhibition of the positive stimuli received by the brain from the tarsal chemoreceptors. When inhibition of an effector organ occurs it is known as peripheral inhibition. At both synapses and neuromuscular junctions, the hyperpolarizing chemical is γ-aminobutyric acid (Homberg, 1994).

Mention must also be made of neuromodulators, a group of chemicals that can modify the effects of neurotransmitters (Orchard, 1984; Homberg, 1994). Typically, neuromodulators are released from the tip of an adjacent neuron (less commonly as a neurohormone released into the hemolymph) and act on the presynaptic or postsynaptic membrane adjacent to, but not within, the synaptic gap or neuromuscular junction. Their effects include reduction in the amount of neurotransmitter released and inhibition of the action of the neurotransmitter. Amines, especially octopamine, and some neuropeptides (e.g., proctolin) are likely to be important neuromodulators, though in many instances definitive evidence is still lacking. A probable neuromodulator of a special kind may be nitric oxide. This very short-lived, rapidly diffusing gas was discovered in nervous tissues of locusts, honey bees, and *Drosophila* in the early 1990s. Production of nitric oxide is especially rich in interneurons in the antennal and optic lobes, as well as in antennal chemosensory cells of some species, following appropriate olfactory and visual stimulation, suggesting that this unconventional neuromodulator may have roles in olfactory information processing, olfactory memory, and vision (Müller, 1997; Bicker, 1998).

In insects reflex responses are segmental, that is, a stimulus received by a sense organ in a particular segment initiates a response that travels via an interneuron located in that segment's ganglion to an effector organ in the same segment. This is easily demonstrated by isolating individual segments. For example, in an isolated thoracic segment preparation of a grasshopper, touching the tarsus causes the leg to make a stepping movement. Of course, in an intact insect such a stimulus also leads to compensatory movements of other legs to maintain balance or to initiate walking, activities that are coordinated via association centers in the subesophageal ganglion. Touching the tip of the isolated ovipositor in *Bombyx*, for example, initiates typical egg-laying movements, provided that the terminal ganglion and its nerves are intact. In other words, each segmental ganglion possesses a good deal of reflex autonomy.

Nervous activity of the type described above, which occurs only after an appropriate stimulus is given, is said to be exogenous. However, an important component of nervous activity in insects is endogenous, that is, does not require sensory input but is based on neurons with intrinsic pacemakers. Such neurons (non-spiking neurons) possess specialized membrane regions that undergo periodic, spontaneous changes in excitability (permeability) and where impulses are thereby initiated. A wide variety of motor responses are organized, in part, by endogenous activity. For example, ventilation movements of the abdomen are initiated by endogenous activity in individual ganglia. Even walking and stridulation are motor responses under partially endogenous control (Huber, 1974). An obvious question to ask, therefore, is "Why don't insects walk or stridulate continuously?" The answer is that these and all other motor responses are "controlled" by higher centers, specifically the brain and/or subesophageal ganglion. These association centers assess all information coming in via sensory neurons and, on this total assessment, determine the nature of the response. In addition, the centers coordinate and modify identical segmental activities, such as ventilation movements, so that they operate most efficiently under a given set of conditions.

Early evidence for the role of the brain and subesophageal ganglion as coordinating centers came from fairly crude experiments in which one or both centers were removed and the resultant behavior of an insect observed. More recent experiments involving localized destruction or stimulation of parts of these centers have confirmed and added significantly to the general picture obtained by earlier authors. To illustrate the complexity of coordination and control of motor activity, walking will be used as an example. This rhythmic stepping movement of each leg is controlled by a network of non-spiking neurons (called the central pattern generator and located in each half ganglion) whose endogenous activity sends signals (via motor neurons) alternately to the extensor and flexor leg muscles (Chapter 14, Section 3.2.1). The signals may be excitatory or inhibitory and, in effect, serve to switch on or off the muscles. Intraganglionic and intersegmental coordination among the central pattern generators, and ultimately leg movements, is achieved via normal interneurons. This is readily shown by cutting even one connective of the pair between adjacent ganglia when coordinated stepping is disrupted. Though the overall control of walking, that is, starting, stopping, turning, and change of speed, resides in the brain, the subesophageal ganglion is also involved. Removal of the latter, for example, sensitizes some insects so that they walk incessantly in response to even slight stimuli. In the brain the mushroom bodies and central body play major roles in the regulation of walking. Impulses originating in the mushroom bodies inhibit locomotor activity, presumably by decreasing the excitability of the subesophageal ganglion. Moreover, reciprocal inhibition may occur between the mushroom body on each side of the brain, and this is the basis of the turning response. In contrast, the central body appears to be an important excitatory system in locomotion, because its stimulation evokes fast running, jumping, and flying in some species. As yet, however, the interaction between these two cerebral association centers is not understood.

Superimposed on the central control of walking is the influence of sensory stimuli received by the insect; that is, the insect adjusts its walking pattern to suit environmental conditions such as movement uphill or downhill, along a slope, or over rough terrain. To this end, the legs are equipped with a variety of mechanoreceptors that provide information on their position, loading, and movement (Chapter 12, Section 2). In a walking Colorado potato beetle, contact between the antennae and an obstacle causes the insect to modify its body angle. The extent to which the body angle is changed is proportional to the height of the obstacle, allowing the beetle to extend the reach of the prothoracic leg so as to step up on to the obstacle. Insects that use running to escape predators receive information via other sensory pathways. For example, in the cockroach escape reaction, even the slightest air movements stimulate hairs on the cerci that are both velocity- and direction-sensitive. The information received travels via giant axons in the ventral nerve cord to the thoracic ganglia to initiate both the running and the turning away responses within 0.5 msec of the stimulus being received.

At the outset, insect behavior is dependent on the environmental stimuli received, though, as noted earlier, not all behavior patterns originate exogenously; many common patterns have a spontaneous, endogenous origin. Axons may be branched; synapses may be convergent or divergent; temporal or spatial summation of impulses may occur at synapses; neurons may be excitatory or inhibitory in their effects. Thus, an enormous number of potential routes are open to impulses generated by a given set of stimuli. The eventual routes taken and, therefore, the motor responses that follow, depend on the size, nature, and frequency of these stimuli.

2.4. Learning and Memory

The translation of sensory input into a motor response takes place within a matter of milliseconds and thereby fits well into the broad definition of "nervous control." However, another important aspect of neural physiology is learning, which, along with the related event, memory, may occupy time intervals measured in hours, days, or even years. Learning is the ability to associate one environmental condition with another; memory is the ability to store information gathered by sense organs. Within this broad definition of learning, several phenomena can be included. Habituation, perhaps the simplest form of learning, is adaptation (eventual failure to respond) of an organism to stimuli that are not significant to its well-being. For example, as noted above, a cockroach normally shows a striking escape reaction when air is blown over the cerci. If, however, this treatment is continued for a period of time, the insect eventually no longer responds to it. Conditioning is learning to respond to a stimulus that initially has no effect. Related to this is trial-and-error learning where an animal learns to respond in a particular way to a stimulus, having initially attempted to respond in other ways for which acts it received a negative reaction.

The most complex form of learning in insects, latent learning, is the ability to relate two or more environmental stimuli, though this does not confer an immediate benefit. For insects, visual and chemosensory (especially olfactory) cues are especially important in latent learning. For example, *Microplitis* spp. learn to associate color, shape, and pattern with successful oviposition; *Locusta* associates odor or visual cues with food quality; and mosquitoes learn to recognize (and return to) sites where they have successfully fed and/or oviposited (McCall and Kelly, 2002). Perhaps the best-studied example of latent learning is the recognition and use of landmarks by social Hymenoptera, enabling these insects to return to their nest or a food source. Foraging honey bees, wasps, and ants, on leaving a newly discovered food source, undertake a series of "turn-back-and-look" (TBL) manoeuvres (Lehrer, 1991; Judd and Collett, 1998; Lehrer and Bianco, 2000). In this activity the insect, when just a few centimeters from the food source, repeatedly turns and looks back at it. Likewise, inexperienced workers carry out similar learning flights on first leaving the nest to forage. It has been proposed that the TBL activity enables the insect to take a series of "snapshots" of the landmarks adjacent to the food source or nest. These pictures, memorized within the optic lobes, are then matched with the current image seen on the next trip. When the match is "exact," the insect has reached its goal.

As noted, the TBL method facilitates "close-up" landmark recognition. However, recognition of landmarks is often also used as insects move between the nest and a food source. For example, insects may learn to steer to one side of a landmark to stay on the correct path; they may go directly over landmarks that are on the flight path; and by recognizing (matching) a scene, they are able to compensate for unexpected displacement of their position provided that they have experienced the "new" position at a previous point in time (Collett, 1996).

It is not always possible to use landmark recognition to navigate between nest and food source as the terrain may be featureless. Under such conditions, path integration is employed (Collett and Collett, 2000). Essentially, path integration requires that an insect has the ability to monitor and record changes in its position over time, that is, to measure distance traveled, speed of travel, and direction traveled. The insect must then compute this information to set (or reset) a course toward the nest or food source. For many insects, path integration incorporates the insect's ability to navigate using polarized light (Chapter 12, Section 7.1.4), including a mechanism for measuring and compensating for elapsed time, necessitated by the sun's ever-changing position.

Circadian rhythms are also a form of learning. Many organisms perform particular activities at set times of the day, and these activities are initially triggered by a certain environmental stimulus, for example, the onset of darkness. Even if this stimulus is removed, by keeping an animal in constant light or dark, the activity continues to be initiated at the normal time.

Though there is no doubt that insects are able both to learn and to memorize, the physiological/molecular basis for these events is not known. However, some generalized statements can be made. The mushroom bodies are the center where complex behavior is learned, and, these structures occupy a relatively greater proportion of the brain volume in insects such as the social Hymenoptera (particularly the worker caste), which exhibit the greatest learning capacity. A group of pacemaker cells responsible for the generation of circadian rhythms in *Drosophila* have been identified in the central brain (Saunders, 1997). There is evidence that simpler forms of learning can occur in other ganglia, for example, those of the thorax. Headless insects and even individual, isolated thoracic ganglia preparations can learn to keep a leg in a certain position so as to avoid repeated electric shocks. Intact animals retain this ability for several days after the training period whereas in headless insects the retention time is only 1–2 days. However, subsequent removal of the head of insects that have learned while intact does not reduce retention time, suggesting that intact animals learn more readily than headless ones or isolated ganglia. A variety of pharmacological experiments have been undertaken in attempts to establish the molecular basis of learning and memory, including application of protein or nucleic acid synthesis-inhibiting drugs, assays of protein and nucleic acid synthesis in ganglia before and after training, measurement of cholinesterase levels, and application of cyclic AMP inhibitors. However, the results obtained are sometimes conflicting and difficult to interpret (see Eisenstein and Reep, in Kerkut and Gilbert, 1985).

3. Endocrine System

Insects, like vertebrates, possess both epithelial endocrine glands (the corpora allata and molt glands, derived during embryogenesis from groups of ectodermal cells in the region of the maxillary pouches) and glandular nerve cells (neurosecretory cells), which are found in all ganglia of the central nervous system and in parts of the visceral nervous system. Their axons terminate in storage and release sites (neurohemal organs) or run directly to their target organ. In addition, the gonads and some other structures of certain species produce hormones.

The functions of hormones are many, and discussion of these is best treated in conjunction with specific physiological systems. In this chapter, therefore, only the structure of the glands, the nature of their products, and the principles of neuroendocrine integration will be examined.

3.1. Neurosecretory Cells and Corpora Cardiaca

The best-studied neurosecretory cells are the median neurosecretory cells (mNSC) of the protocerebrum. They occur in two groups, one on each side of the midline, and their axons (which form the NCC I) pass down through the brain, crossing over en route, and normally terminate in a pair of neurohemal organs, the corpora cardiaca, where neurosecretion is stored (Figure 13.4). In some species, for example, *Musca domestica*, some neurosecretory axons do not terminate in the corpora cardiaca but pass through them to the corpora allata. In many Hemiptera-Heteroptera, the axons bypass the corpora cardiaca and, instead, terminate in the adjacent aorta wall. In aphids, some neurosecretory axons transport their product directly to the target organ. And the axons of the mNSC which produce bursicon terminate in the fused thoracoabdominal ganglion of higher Diptera and in the last abdominal ganglion of cockroaches and locusts (Highnam and Hill, 1977). The corpora cardiaca are closely apposed to the dorsal aorta into which neurosecretion and intrinsic products of the corpora cardiaca are released when the neurosecretory cell membranes are depolarized. The NCC I also contain ordinary neurons that innervate the intrinsic cells of the corpora cardiaca (see below), causing them to release their product. More than 40 years ago, it was noted that different mNSC take up characteristic stains. Further, destruction of the cells affects a wide range of physiological processes (see later chapters), leading to the proposal that they produce a variety of hormones. This was confirmed through the use of immunohistochemistry, following purification of specific neurosecretory hormones. Also in the protocerebrum are two groups of lateral neurosecretory cells (lNSC) whose axons do not cross but travel to the corpus cardiacum of the same side. However, there is almost no information on their function.

The corpora cardiaca arise as invaginations of the foregut during embryogenesis at the same time as the stomatogastric nervous system and are, in fact, modified nerve ganglia. Though their main function is to store neurosecretion, many of their intrinsic cells also produce hormones. In some species, for example, the desert locust, the neurosecretory storage zone and glandular zone (zone of intrinsic cells) are distinct (Figure 13.4C–E); in others, the neurosecretory axons terminate among the intrinsic cells.

Neurosecretory cells are also found in all of the ventral ganglia, and their axons, which contain stainable droplets, can be traced to a series of segmental neurohemal organs, the perisympathetic organs adjacent to the unpaired ventral nerve. In addition, there are many reports of multipolar neurosecretory cell bodies lying on peripheral nerves innervating the heart, gut, etc. However, it should be noted that, for both the neurosecretory cells of the ventral ganglia and those associated with peripheral nerves, only rarely has experimental evidence for their function been obtained.

Many functions have been ascribed to neurosecretory hormones and the intrinsic hormones produced by the corpora cardiaca, but for relatively few of these is there good experimental evidence. Products of the mNSC include prothoracicotropic hormone (PTTH), which activates the molt glands (Chapter 21, Section 6.1); allatotropic and allatostatic hormones, whose primary function is to regulate the activity of the corpora allata (Chapter 19, Section 3.1.3 and Chapter 21, Section 6.1); diuretic hormone, which affects osmoregulation (Chapter 18, Section 5); ovarian ecdysiotropic hormone (OEH) (formerly egg development neurosecretory hormone) (Chapter 19, Section 3.1.3); ovulation- or oviposition-inducing hormone (Chapter 19, Sections 5 and 7.2); and testis ecdysiotropin (TE) (Chapter 19, Section 3.2). Bursicon, which is important in cuticular tanning (Chapter 11, Section 3.4), has been localized in the mNSC in some species, though is principally found in the abdominal ganglia from which it is released via abdominal perivisceral organs. Neurosecretion from the mNSC also affects behavior, though in many cases this is certainly an indirect action, and is important in protein synthesis. Eclosion hormone (EH), important in ecdysis (Chapter 21, Section 6.2), is produced by neurosecretory cells in the tritocerebrum. The intrinsic cells of the corpora cardiaca produce hyperglycemic and adipokinetic hormones (AKH) important in carbohydrate and lipid metabolism (Chapter 16, Sections 5.2 and 5.3) and hormones that stimulate heartbeat rate (Chapter 17, Section 3.2), gut peristalsis, and writhing movements of Malpighian tubules. It appears, however, that the mNSC may be involved in the elaboration of these materials because extracts of these cells exert similar, though less strong, effects on these processes. Neurosecretion from the subesophageal ganglion is, in cockroaches, synthesized and released regularly and controls the circadian rhythm of locomotor activity. In many female moths, pheromone biosynthesis activating neuropeptide (PBAN) (Section 4.1) is produced in three groups of neurosecretory cells in the subesophageal ganglion (in some species also other ventral ganglia). The PBAN synthesized in the subesophageal ganglion appears to be released via the corpora cardiaca. In

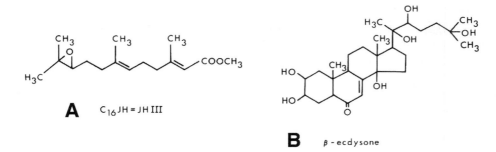

FIGURE 13.6. (A) Locust juvenile hormone ($C_{16}JH$ = JHIII); and (B) β-ecdysone.

female pupae[*] of *Bombyx,* two large neurosecretory cells in the subesophageal ganglion produce a diapause hormone which promotes the development of eggs that enter diapause (see Chapter 22, Section 3.2.3). In *Rhodnius,* diuretic hormone is produced not by the cerebral neurosecretory cells but by the hindmost group of neurosecretory cells in the fused ganglion of the thoracic and first abdominal segments.

All neurosecretory factors characterized to date are peptides (sometimes glycosylated), an observation that is entirely in keeping with those from other animals. They range in molecular weight from the tens of thousands down to a thousand or less. Examples are bursicon (M.W. about 40,000), diuretic hormone (M.W. 1500–2000), OEH (6500), TE (2500), diapause hormone (2500), and AKH (a decapeptide). The PTTH of *Drosophila* is a glycosylated polypeptide (M.W. 66,000). Curiously, the moth *Manduca sexta* produces two forms of PTTH: the smaller form (M.W. 7000) comes from the *m*NSC, whereas the larger form (M.W. 28,000) is a product of the *l*NSC. The two forms have quite different structures, yet in larvae are about equally active.

3.2. Corpora Allata

Typically the corpora allata are seen as a pair of spherical bodies lying one on each side of the gut, behind the brain (Figure 13.4A,B). However, in some species, the glands may be fused in a middorsal position above the aorta, or each gland may fuse with the corpus cardiacum on the same side. In larvae of cyclorrhaph Diptera the corpora allata, corpora cardiaca, and molt glands fuse to form a composite structure, Weismann's ring, which surrounds the aorta. Each gland receives a nerve (NCA I) from the corpus cardiacum on its own side, though the axons that form this nerve are probably those of *m*NSC, and also a nerve from the subesophageal ganglion (NCA II).

The corpora allata produce a hormone known variously as juvenile hormone, metamorphosis-inhibiting hormone, or neotenin, with reference to its function in juvenile insects (Chapter 21, Section 6.1), and gonadotropic hormone to indicate its function in adults (Chapter 19, Sections 3.1.3 and 3.2). Juvenile hormone is a terpenoid compound (Figure 13.6A) and, to date, six naturally occurring forms (JH-O, JH-I, 4-methyl-JH-I, JH-II, JH-III, and JHB_3) have been identified. In all insects investigated, except Hemiptera, Lepidoptera, and higher Diptera, only JH-III has been obtained. Though JH-III is reputedly synthesized in some Hemiptera, Numata *et al.* (1992) report that JH-I is the only form produced in the bean bug, *Riptortus clavatus*. In Lepidoptera the first five forms of JH listed

---

[*] In many Lepidoptera, including *Bombyx,* egg development begins in the pupa.

above occur, with one or two forms predominating at specific stages in the life history. In cyclorrhaph Diptera JHB$_3$ is the principal or sole juvenile hormone, with JH-III occurring as a minor component in some species (Lefevere *et al.*, 1993). In addition, a large number of related compounds have been shown to exert juvenilizing and/or gonadotropic effects. Currently, there is much interest in such compounds in view of their potential use in pest control (Chapter 24, Section 4.2).

3.3. Molt Glands

The paired molt glands generally comprise two strips of tissue, frequently branched, which are interwoven among the tracheae, fat body, muscles, and connective tissue of the head and anterior thorax. In accord with their varied position, they have been called prothoracic glands, ventral head glands, and tentorial glands, though these structures are homologous. Except in primitive apterygotes, solitary locusts (Chapter 21, Section 7), and, apparently, worker and soldier termites, the glands are found only in juvenile insects and degenerate shortly after the molt to the adult. The molt glands show distinct cycles of activity correlated with new cuticle formation and ecdysis. Their product, α-ecdysone, is a prohormone that when activated initiates several important events in this regard (see Chapter 11, Section 3.4 and Chapter 21, Section 6.1). Typically the active form is 20-hydroxyecdysone (= β-ecdysone = ecdysterone) though a large number of similar molecules with biological activity have been isolated. Ecdysones are steroids (Figure 13.6B) having the same carbon skeleton as cholesterol, which is almost certainly the natural precursor in insects.

3.4. Other Endocrine Structures

A variety of structures in insects have been proposed as endocrine glands at one time or another.

The oenocytes, which become active early in the molt cycle and again at the onset of sexual maturity in adult females, show ultrastructural similarities with steroid-producing cells in vertebrates (Locke, 1969; Romer, 1991). This has led to the suggestion that these cells may be a site for ecdysone synthesis, and a few biochemical studies support this idea (Romer, 1991; Romer and Bressel, 1994). However, their primary roles appear to be the synthesis of certain cuticular lipids and the lipoprotein layer of the epicuticle (Chapter 11, Section 2).

In contrast to those of vertebrates, the gonads of insects do not produce sex hormones that influence the development of secondary sexual characters. A large number of experiments in which insects were castrated could be cited to support this statement. The only exception to this generalization is found in the firefly *Lampyris noctiluca* (Coleoptera) where an androgenic hormone produced by the testes induces the development of male sexual characters. Thus, implantation of these organs into a female larva causes sex reversal. Ovarian tissue, however, does not produce a hormone for induction of femaleness (Naisse, 1969).

The ovaries of many insects do, however, produce hormones that affect reproductive development. Adams and others (see Adams, 1980, 1981) have demonstrated that the maturing ovaries of the house fly, *Musca domestica*, and other Diptera produce an oostatic hormone that regulates the pattern of egg maturation by inhibiting the release of OEH from the *m*NSC. In the absence of OEH, ovarian ecdysone release (see below and Chapter 19, Section 3.1.1) does not occur. After oviposition, the ovaries no longer produce oostatic hormone, and a new cycle of egg maturation begins. In contrast, the antigonadotropin produced by the abdominal perisympathetic neurosecretory organs in the bug *Rhodnius prolixus* does not act on other endocrine centers. Rather, it appears to act at the level of the follicle cells, blocking the action of juvenile hormone (Chapter 19, Section 3.1.1) (Davey and Kuster, 1981).

Reports of the existence of ecdysones in adult female insects, published in the 1950s, were largely ignored (after all, the molt glands that produce them were known to degenerate at the end of larval life), and it was not until the 1970s that the ovary was shown to be a major producer of these hormones in insects from a variety of orders (e.g., locusts, crickets, termites, mosquitoes and other Diptera, and several Lepidoptera). In some Diptera a clear role for ecdysone in vitellogenesis has been established (Chapter 19, Section 3.1.3) while in locusts the ecdysone largely accumulates in maturing eggs to be used later in the regulation of embryonic molting (Chapter 20, Section 7.2). For other species, its function remains unclear.

Evidence from a number of species indicates that the testes may also produce ecdysteroids. In the moths *Lymantria dispar* and *Heliothis virescens* testis ecdysiotropin stimulates the testis sheaths to produce ecdysteroids. In turn these trigger release of testicular factors that promote growth and development of the reproductive tract (Loeb *et al.*, 1996). Ecdysteroid production by testes, as well as by the male accessory glands and abdominal integument has also been reported in other moths, crickets, and grasshoppers (Gillott and Ismail, 1995, and references therein). However, except for a possible role in spermatogenesis (Chapter 19, Section 3.2), the function of ecdysteroids in these insects remains uncertain.

An endocrine role for the Inka cells within the epitracheal glands was originally described in *Manduca sexta* (Žitňan *et al.*, 1996). However, it has recently been confirmed that these cells occur in representatives of all major insect orders, producing pre-ecdysis- and ecdysis-triggering hormones (PETH and ETH) at the end of each developmental stage (Žitňan *et al.*, 2003) (Chapter 21, Section 6.2). In *M. sexta* there are nine pairs of segmentally arranged epitracheal glands, each attached to a large trachea immediately adjacent to a spiracle and containing a single Inka cell. However, in most insects, the Inka cells are numerous and scattered throughout the tracheal system. Both PETH and ETH are peptides, the latter composed of 26 amino acids in *M. sexta*.

4. Insect Semiochemicals

Insects interact with other members of their species, as well as with other organisms, by means of a bewildering array of so-called semiochemicals. When the interaction is intraspecific, the chemical messages are called pheromones. Chemicals involved in interspecific interaction, allelochemicals, fall into two major categories: allomones provoke a response that favors the emitter, whereas kairomones induce a response favorable to the receiver. Perhaps not surprisingly, a given semiochemical may fit into more than one of these groups; for example, pheromones emitted by prey insects may also function as kairomones by attracting predators and parasitoids. It should be understood that in interspecific interactions insects may be either the emitter or the receiver of the allelochemical signal. The latter is especially important for phytophagous insects, for which plant-released chemicals may be a major deterrent against insects attack or a specific cue by which the insect recognizes its host (Chapter 23, Section 2.3.1). In this chapter only pheromones, and kairomones and allomones released by insects will be considered.

4.1. Pheromones

Pheromones are chemicals messages produced by one individual that induce a particular behavioral, physiological, or developmental response in other individuals of the same species. Like hormones, they are produced in small quantities; indeed, they are referred to in older literature as "ectohormones," a term that may take on renewed significance following the discovery that in termites juvenile hormone apparently serves also as a pheromone that regulates caste differentiation (Section 4.1.2). Pheromones may be volatile and therefore capable of being detected as an odor over considerable distances, or they may be non-volatile, requiring actual physical contact among individuals for their dissemination. They may be highly specific, even to the extent that only a particular isomer of a substance induces the typical effect in a given species. (As a corollary, closely related species often utilize different isomeric forms of a given chemical). Pheromones are released only under appropriate conditions, that is, in response to appropriate environmental stimuli, and examples of this are given below. Thus, whereas the neural and endocrine systems coordinate the behavior, physiology, and/or development of an individual, pheromones regulate these processes within populations.

Pheromones may be arranged in rather broad, sometimes overlapping, categories based on their functions. There are sex pheromones, caste-regulating pheromones, aggregation pheromones, alarm pheromones, trail-marking pheromones, and spacing pheromones.

4.1.1. Sex Pheromones

In the term "sex pheromones" are included chemicals that (1) excite and/or attract members of the opposite sex (sex attractants), (2) act as aphrodisiacs, (3) accelerate or retard sexual maturation (in either the opposite and/or same sex), or (4) enhance fecundity and/or reduce receptivity in the female following their transfer during copulation.

Sex attractants are typically volatile chemicals produced by either male or female members of a species, whose release and detection by the partner are essential prerequisites to successful courtship and mating. At the outset, it should be realized that the term "sex attractant" is a misnomer; the chemical does not directly attract but initiates upwind orientation and movement by the recipient. Male-attracting substances are produced by virgin females of species representing many insect orders, but especially Lepidoptera and Coleoptera (for lists, see Tamaki, in Kerkut and Gilbert, 1985; Arn *et al.*, 1992). Females release their pheromones only in response to a specific stimulus. Release may be influenced by age, reproductive status, time of day, presence of host plant, temperature, and wind speed. For example, many species of moths begin "calling" (everting their pheromone-secreting glands and exuding the chemical) 1.5–2.0 hours before dawn. Other Lepidoptera are stimulated to release pheromone by the scent of the larval food plant. *Rhodnius prolixus* females release pheromone only after a blood meal, and pheromone production in *Periplaneta americana* is arrested when the ootheca is formed. It seems that in species such as *R. prolixus*, cockroaches, locusts, and beetles, which have repeated cycles of oocyte development over a period of time, pheromone production is under the control of the endocrine system, especially the corpora allata, and that the effects of stimuli such as food intake or presence of an ootheca are mediated via the neuroendocrine system. In *Musca domestica* and perhaps other Diptera, pheromone synthesis is regulated by ovarian ecdysteroids, while in many moths PBAN (Section 3.1) mediates production of sex attractant (Tillman *et al.*, 1999; Ryan 2002).

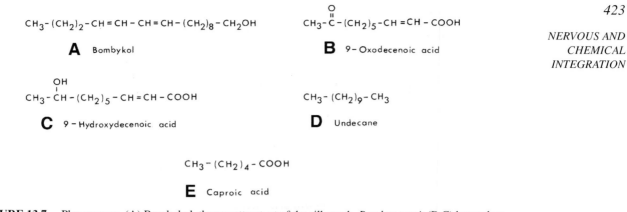

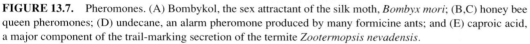

FIGURE 13.7. Pheromones. (A) Bombykol, the sex attractant of the silk moth, *Bombyx mori*; (B,C) honey bee queen pheromones; (D) undecane, an alarm pheromone produced by many formicine ants; and (E) caproic acid, a major component of the trail-marking secretion of the termite *Zootermopsis nevadensis*.

Typically, the pheromone-producing glands of female Lepidoptera are eversible sacs located in the intersegmental membrane behind the eighth abdominal stemite. In the homopteran *Schizaphis borealis* the glands are probably on the hind tibiae; in *Periplaneta* the pheromone seems to be produced in the gut and is released from fecal pellets; in the queen honey bee the mandibular glands are the source; and in the house fly sex pheromone is secreted evenly over the second through seventh abdominal segments. In Coleoptera the glands are abdominal.

The components of many male attractants have been identified, especially those of pestiferous Lepidoptera in which they appear generally to be aliphatic straight-chain hydrocarbons, alcohols, acetates, aldehydes, and ketones containing 10–21 carbon atoms (Figure 13.7A). Among Lepidoptera, species in the same family or subfamily tend to produce a "key component." For example, (Z)-11-tetradecenol or its derivative is produced by almost all Tortricinae. Usually the sex pheromone is a blend of two or more components, occurring in species-specific proportions. (Interestingly, the proportions may vary among populations of the same species in different geographic locations.) Further, only one isomeric form of a component is typically attractive in a given species. As a result, under natural conditions males respond only to the pheromone produced by females of their species. Other factors that serve to prevent interspecific attraction between males and females of species producing similar pheromones include differences in the time of day at which males are sensitive to pheromone, differences in geographic location of the species, differences in the time of year when the species are sexually mature, and the need for additional stimuli, perhaps auditory, visual, or chemical, before a male is attracted to a female. For example, the "initial" separation of two species may occur on the basis of their attraction to different host plants. Thus, in the vicinity of a host plant, the chances of being attracted to a conspecific female will be greatly increased. Nevertheless, other stimuli will be necessary to "confirm" the conspecific nature of the partner.

In many species (for lists, see Weatherston and Percy, 1977; Tamaki, in Kerkut and Gilbert, 1985; Fitzpatrick and McNeil, 1988; Birch *et al.*, 1990) it is males that produce a sex pheromone. This may function, like those of females, as a long-distance "attractant" or may trigger close-in behavior such as short-range attraction, female orientation for mating, adoption of the mating posture, and quiescence. For example, the male cockroach, *Nauphoeta cinerea*, produces seducin, which both attracts and pacifies the unmated female so that a connection can be established. The pheromone produced by the male boll weevil, *Anthonomus grandis*, attracts only females in summer; however, in the fall it serves as an aggregation pheromone, attracting both males and females. Male *Tenebrio molitor* produce both a female attractant and an antiaphrodisiac that inhibits the response of other males to the female's pheromone. The pheromone-producing glands of males are much more varied in their location than those of females. For example, pheromone is secreted from mandibular glands in ants, from glands in the thorax and abdomen of some beetles, from glands at the base of the fore wings in some Lepidoptera, from abdominal glands in other Lepidoptera and cockroaches, and from rectal glands in certain Diptera. As in females, the attractants produced by males are usually long-chain alcohols or their aldehydic derivatives.

Because of their specificity and effectiveness in very low concentrations, the use of sex attractants in pest control has great potential, an aspect that will be more fully discussed in Chapter 24, Section 4.2.

Sexual maturation-accelerating or -inhibiting pheromones are produced by many insects that tend to live in groups, including social species. These substances serve either to synchronize reproductive development so that both sexes mature together or, in social species, to inhibit the reproductive capability of almost all individuals so that their energy can be redirected to other functions. Mature male desert locusts, for example, produce a volatile pheromone that speeds up maturation of younger males and females. The pheromone is a blend of five components, of which phenylacetonitrile appears to be the most critical (Mahamat *et al.*, 2000). Conversely, gregarious juvenile desert locusts secrete a pheromone that retards the maturation of gregarious newly molted adults. This pheromone, which also serves as an aggregation pheromone for the nymphs, is a complex blend of chemicals, some produced by the epidermal cells and others associated with the feces (Assad *et al.*, 1997). The pheromones appear to operate by regulating the activity of the corpora allata in other individuals.

Each colony of social insects has only one or very few reproducing individuals of each sex. Most members of the colony, the workers, devote their effort to maintaining the colony and never mature sexually. Their failure to mature results from the production by reproductives of inhibitory pheromones. In a honey bee colony the queen produces, in the mandibular glands, a material called queen substance, 9-oxo-2-decenoic acid (Figure 13.7B), which is spread over the body during grooming to be later licked off by attendant workers. Mutual feeding among workers results in the dispersal of queen substance through the colony where the pheromone serves to stimulate foraging activity (in older workers) and "household duties" (by young workers). The glands also produce a volatile pheromone, 9-hydroxy-2-decenoic acid (Figure 13.7C), which, together with queen substance, inhibits ovarian development in workers. As the queen ages and/or the number of individuals in a colony increases, the amount of pheromone available to each worker declines, and the latter's behavior changes. The workers construct queen cells in which the larvae are fed a special diet so that they develop into new queens. The first new queen to emerge kills the others and proceeds on a nuptial flight accompanied by drones. Queen substance produced by the new queen now serves as a sex attractant for the drones and to stimulate them to mate with her. When the new queen returns to the hive from the flight, part of the colony swarms; that is, the old queen, accompanied by workers, leaves the hive to found a new colony. Again, it is queen substance, along with 9-hydroxy-2-decenoic acid, that enables workers to locate and congregate around the queen. This is another example of how the function of a pheromone may vary according to the particular environmental circumstances in which the pheromone is released.

In the lower termites, in contrast to the honey bee, there is a pair of equally important primary reproductives, the king and queen. When a colony is small, the development of additional reproductives is inhibited by means of sex-specific pheromones secreted by the royal pair. The pheromones apparently are released in the feces and transferred from termite to termite by trophallaxis. Lüscher (1972) suggested that juvenile hormone may be the pheromone that inhibits reproductive development, though how such an arrangement could act in a sex-specific manner has not been satisfactorily explained. In addition, there are indications that the king secretes a pheromone that enhances the development of female supplementary reproductives in the absence of the queen.

In some species of mosquitoes and other Diptera, Orthoptera, Hemiptera, and Lepidoptera, the male, during mating, transfers to the female via the seminal fluid, chemicals that inhibit receptivity (willingness to mate subsequently) or enhance fecundity (by increasing the rate at which eggs mature and are laid) (Gillott and Friedel, 1977; Gillott, 2003). In contrast to other pheromones, the substances that are produced in the accessory reproductive glands or their analogue are proteinaceous in nature. Both receptivity-inhibiting substances and fecundity-enhancing substances signal that insemination has occurred. The former ensure that a larger number of virgin females will be inseminated; the latter, acting as they do by triggering oviposition, increase the probability of fertilized (viable) eggs being laid (in most species unfertilized eggs are inviable). Both types of pheromones serve, therefore, to increase the reproductive economy of the species.

4.1.2. Caste-Regulating Pheromones

As noted in the previous section, in social insects very few members of a colony ever mature sexually, their reproductive development being inhibited by pheromones so that these individuals (forming the worker caste) can perform other activities for the benefit of the colony as a whole. In addition to the worker caste, there exists in termites and ants a soldier caste. The number of soldiers present is proportional to the size of the colony, a feature suggesting that soldiers regulate the numbers in their ranks by production of a soldier-inhibiting pheromone. However, the situation is made more complicated by a positive influence on soldier production (presumably pheromonal) on the part of the reproductives.

The regulation of caste differentiation in social insects is a morphogenetic phenomenon, just as are the changes from larva to larva, larva to pupa, and pupa to adult. Such changes depend on the activity of the corpora allata (level of juvenile hormone in the hemolymph) for their manifestation. It is not surprising, therefore, that the pheromones regulating caste differentiation (including the development of reproductives) exert their effect, ultimately, via the corpora allata (Chapter 21, Section 7). In lower termites, for example, soldier formation can be induced in experimental colonies by administration of juvenile hormone (through feeding, topical application, or as vapor). Thus, the soldier-inhibiting pheromone may act by inhibiting the corpora allata or by competing with juvenile hormone at its site of action (Lüscher, 1972).

4.1.3. Aggregation Pheromones

Aggregation pheromones are produced by either one or both sexes and serve to attract other individuals for feeding, mating, and/or protection. They occur in Collembola, Diptera, Hemiptera, Orthoptera, and Dictyoptera, but are especially well known in Coleoptera, particularly the bark and ambrosia beetles (Scolytinae). Larvae of the crane fly *Tipula simplex* release a pheromone in their feces that causes aggregation under cowpats and rotting wood where food and shelter occur. Likewise, Collembola aggregate in moist places in response to a pheromone to avoid desiccation and to reproduce; the latter is especially significant as these arthropods do not copulate (Chapter 19, Section 4.1). In Coleoptera the pheromones serve primarily to aggregate the beetles to a food source that may be isolated (e.g., stored grain) or require the collaborative efforts of a large number of insects to overcome host resistance (e.g., bark beetles attacking healthy trees). As noted in Section 4.1.1, juvenile locusts produce an aggregation pheromone whose function is to keep the marching swarm intact. However, gregarious adult females also produce a volatile aggregation pheromone in their egg pod froth (Saini *et al.*, 1995). The pheromone, whose most important components are acetophenone and veratrole (Rai *et al.*, 1997), attracts egg-laying females to a common site, resulting in very high egg-pod densities. In addition, it predisposes the hatchlings to take on gregarious characteristics of color, physiology, and behavior.

Aggregation pheromones are also common in a wide range of blood-feeding insects, serving to bring conspecifics together for mating, oviposition, and larviposition (McCall, 2002). For these specialized insects, the pheromones facilitate the concentration of populations in sites where mating and egg-laying can occur in relative safety (e.g., cracks and crevices in walls for bedbugs), or in locations where conditions are suitable for egg-laying. For example, the eggs of *Culex* mosquitoes, *Simulium* (black flies), and sand flies release a pheromone that triggers mass oviposition. Likewise, tsetse flies larviposit in large aggregations, especially in the dry season when suitable sites may be rare, as a result of a pheromone released as an anal exudate by the burrowing larva.

Compared to sex pheromones, relatively little work has been done to establish the chemical nature of aggregation pheromones. Those studied mostly appear to be mixtures of compounds, often including terpenoid compounds and cyclic alcohols or aldehydes, that act synergistically. Given their function, it is not surprising to discover that the aggregation pheromones of many beetles are metabolites of inhaled or ingested host-plant substances and that host odors synergize their effect (Borden, in Kerkut and Gilbert, 1985). Like sex pheromones, aggregation pheromones are typically highly specific, particular stereo and optical isomers being attractive to a given species. Beetle aggregation pheromones are rarely produced by exocrine glands. Commonly they are released in the feces though the precise sites of synthesis remain unclear. Pheromones produced from ingested material could be produced by cells in the gut wall or, as has been suggested for a few species, by gut microorganisms; those derived from host-plant vapors taken up via the tracheal system are perhaps synthesized in the hemolymph (fat body?) or Malpighian tubules. It has been suggested that some pheromones arose initially as detoxification products, their precursors being toxic to the beetles so that their modern function of promoting aggregation developed secondarily (Borden, 1982).

4.1.4. Alarm Pheromones

As their name indicates, alarm pheromones warn members of a species of impending danger. They are produced by mites and insects that live in groups, including social forms, for example, cockroaches, treehoppers, aphids, bedbugs, termites, and social Hymenoptera (Blum, in Kerkut and Gilbert, 1985; Blum, 1996). In termites it is only soldiers that produce (and respond to) the pheromone. Among Hymenoptera, hornets and honey bees, but not bumble bees, produce alarm pheromones as do almost all ants investigated. The pheromones may originate internally, being released as in treehoppers when the body wall is broken open, or in exocrine glands. Corpses (but not, living specimens) of *P. americana* release material that repels conspecifics and members of some other species. Thus, the repellent may be a cue that cockroaches use to avoid areas where others have died (Rollo *et al.*, 1995). Some Hymenoptera have more than one pheromone-producing gland; for example, in *Formica* species of ants there are mandibular glands, Dufour's glands, and poison glands. The chemical nature of alarm pheromones is highly varied but tends to be specific for each group. Mites, aphids, and termite soldiers produce terpenoid compounds, while honey bees produce a mixture of acetates, an alcohol, and a ketone. Formicine ants also produce terpenoids such as citral and citronellal and, in addition, formic acid, undecane (Figure 13.7D), and various ketones. These compounds often act synergistically. Typically, the alarm pheromones of non-social insects and mites stimulate dispersal (escape behavior). Such behavior is also seen in social species away from the nest. However, when the pheromone is released near the nest, the insects are attracted toward the source and may subsequently attack the intruder. In honey bees, for example, the pheromone is released from the sting shaft (embedded in the intruder!) causing more bees to attack.

Alarm pheromones frequently have other additional functions especially as allomones, leading Hölldobler (1984) to speculate that they evolved originally as defensive chemicals. Those of ants, for example, may also serve as defensive compounds, repelling intruders, or release digging behavior (perhaps an adaptation for excavating ants buried in a cave-in), or act as trail pheromones. In the honey bee the alarm pheromone may be released to indicate a depleted food source.

4.1.5. Trail-Marking Pheromones

Trail-marking pheromones are used by some insects to find mates, to communicate information on the location and quantity of food (and thus to recruit nest mates for food collection), and to ensure that a migrating group retains its integrity. The trails may be terrestrial, laid out on a solid substrate, or aerial, being released by a stationary insect and dependent on movement of the surrounding medium to generate a trail (in this sense, then, the sex and aggregation pheromones discussed earlier could equally be considered as trail-marking pheromones). Terrestrial trail-marking pheromones, which are laid as a solid line or as a series of spots, are produced from a variety of glands.

Trail-marking pheromones are especially well studied in social insects. In termites the sternal gland secretes the pheromone as the abdomen is dragged along or periodically pressed to the substrate. In ants trail-marking pheromones may be produced in the hindgut, Dufour's gland, poison gland, ventral glands, or on the metathoracic legs, and are released as the abdomen or limbs make contact with the substrate. Ant trail-marking pheromones are secreted as foraging workers return to the nest and serve to recruit other workers to a food source. However, they may be also used to direct other individuals to sites where nest repairs are necessary or during swarming. Most trails are relatively short-lived and fade within a matter of minutes unless continuously reinforced. Some ants, however, make trails that last for several days. Furthermore, when a source of food is good, more returning workers secrete pheromone, thereby establishing a strong trail to which more workers will be attracted. The pheromone-producing glands of the honey bee worker are distributed over the body and the pheromone is transferred during grooming to the feet. It is thereby deposited at the opening to the hive where it both attracts returning workers and stimulates them to enter. The cavity-inhabiting wasps *Vespa crabro* and *Vespula vulgaris* lay a pheromone trail between the cavity entrance and the entrance to their nest. The pheromone, which is identical to the cuticular hydrocarbons, enables returning foragers to locate the nest under the low light conditions of the cavity (Steinmetz *et al.*, 2002, 2003).

Among non-social insects, trail-marking pheromones are well known in cockroaches and gregarious caterpillars. For example, the trail-marking pheromone of *P. americana* serves to aggregate both adults and juveniles. Among Lepidoptera, trail-marking pheromones are released by gregarious caterpillars in diverse families, for example, the forest tent caterpillar (*Malacosoma disstria*) (Lasiocampidae), the pine processionary caterpillar (*Thaumetopoea pityocampa*) (Notodontidae), and *Hylesia lineata* (Saturniidae) (Fitzgerald, 2003). Trail pheromones are also produced by the gregarious larvae of the red-headed pine sawfly (*Neodiprion lecontei*) (Flowers and Costa, 2003). In these examples, the trail pheromones are used to recruit conspecifics to a feeding site, to enable foragers to return to the colony, and to maintain colony integrity, especially when the colony as a whole moves to a new location.

Knowledge of the chemistry of trail-marking pheromones is almost entirely confined to those of social insects. In termites and many ants they appear to be mixtures of long-chain fatty acids, alcohols, aldehydes, esters, or hydrocarbons (Figure 13.7E). However, within these mixtures one or two compounds are typically the major component. For example, of the 10 components identified in the trail pheromone of the ant *Mayriella overbecki* only one, methyl 6-methylsalicylate elicited trail following (Kohl *et al.*, 2000). Varying degrees of species specificity are observed. For example, species of *Solenopsis* (fire ants) have a common pheromone, and among leafcutting ants (*Atta* spp.), species will follow the trails of congenerics. In contrast, the Argentine ant, *Iridomyrmex humilis*, follows a trail of (Z)-9-hexadecenal but not of its geometric isomer, (E)-9-hexadecenal.

4.1.6. Spacing (Epideictic) Pheromones

A relatively new discovery in pheromone research are those pheromones that stimulate insects to spread out so as to maintain an optimal population density. They may be produced by immature and adult insects and may be olfactory or tactile. Spacing pheromones have been best studied in relation to oviposition. For example, after laying, the female apple maggot fly, *Rhagoletis pomonella*, releases a pheromone that deters oviposition on the fruit by other females (Prokopy, 1981). The pheromone appears to be released from the hindgut as the ovipositor is dragged over the fruit. It appears to be a water-soluble peptide that remains biologically active for several days after deposition.

Several Lepidoptera also produce oviposition-deterring pheromones that limit the number of eggs laid on a given plant (Schoonhoven, 1990); further, the presence of feeding larvae of *Pieris brassicae* inhibits egg laying, suggesting that the larvae also produce a pheromone. In the case of the flour moth, *Anagasta kühniella*, meetings between larvae result in the release of pheromone from their mandibular glands. Above a certain level of pheromone, reflecting the number of encounters and hence the density of larvae in the medium, adult females will not oviposit. However, oviposition is stimulated at low pheromone concentrations which are indicative of a medium suitable for larval development but not overcrowded.

As noted in Section 4.1.3, bark beetles produce an aggregation pheromone to ensure a "collective effort" when initially attacking new host trees. However, resident females subsequently release a spacing pheromone (verbenone) whose function is to reduce conspecific competition for egg-laying sites so that the output of each female is not compromised (Borden, 1982).

4.2. Kairomones

Olfaction is the major sense by which insects search for and locate their host, whether it is a plant or another animal. The host's odor, or less commonly its taste, is therefore serving as a kairomone for the searching insect (Chapter 23, Sections 3.3.1 and 4.2.2). Insects that prey on or parasitize other insects are especially adept at locating their host using chemical cues. These cues may include the odor of the prey insect's host plant, the odor or taste of the prey itself, and pheromones or, rarely, allomones released by the prey during the course of other activities.

Pheromones, in particular, have been exploited by predators and parasites in order to locate their host, as the following examples illustrate. The oviposition-deterring pheromone placed on fruit by the apple maggot fly (Section 4.1.6) allows the braconid parasitoid *Opius lectus* to locate its host's eggs. Likewise, the spacing pheromone released by larvae of *Anagasta kühniella* enables their predator, the ichneumonid *Nemeritis canescens*, to determine their position. Aggregation pheromones are widely used by predators for prey-finding. Thus, clerid and trogositiid beetles are strongly attracted to trees under early attack by bark beetles as a result of the aggregation pheromone that the latter emit (Haynes and Birch, in Kerkut and Gilbert, 1985). Predators or egg parasitoids often "eavesdrop" on the sex attractants released by female Lepidoptera: the predators locate and feed on the adult moths, while the parasitoids "hang around" until the moth lays its eggs soon after mating (Stowe *et al.*, 1995; Boo and Yang, 2000).

The pheromones of social insects are exploited in various ways by predaceous and parasitic insects, as well as by other social species, for example, slave-making ants. Some beetle predators, having located a pheromone trail, simply ambush passing foragers or rob them of their food. Many "guests" of ant nests, including species of beetles, flies, and Thysanura, find their way to the host colony using the ants' trail pheromones. (Further, some "guests" produce cuticular hydrocarbons identical to those of their hosts, enabling themselves to remain unrecognized while they feed on, or are fed by, the ants!)

Remarkably, in rare instances, allomones released by the host as a means of defending itself are used by some predators and parasites to locate the emitter. Thus, a number of aphid species emit (E)-β-farnesene as a defensive compound from their cornicles. However, the compound is used by searching seven-spot ladybird beetles, *Coccinella septempunctata*, as an aphid-locating cue (Al Abassi *et al.*, 2000).

4.3. Allomones

Insects release a wide array of volatile chemicals that affect the behavior of other animals, both vertebrate and invertebrate (Blum 1981; Whitman *et al.*, 1990). The great majority of these secretions are used as defensive allomones; however, a few examples of allomones used aggressively have been discovered (Blum, 1996). The chemical nature of allomones is extremely varied (see Blum, 1981). However, it has long been recognized that some allomones are chemically very similar, even identical, to alarm pheromones and sex attractants, leading to the proposal that the original role for these compounds was defensive, with the pheromonal function arising secondarily (Blum, 1996; Ryan, 2002). The allomones are typically produced in specific exocrine glands, though in a few species the allomone is sequestered within the hemolymph, to be released as a result of "reflex bleeding," that is, when hemolymph is exuded at joints and intersegmental membranes. The biosynthetic pathways are, for most allomones, not well understood, but it is evident that most insects produce these compounds endogenously. However, given the wide array of plant natural products, including many feeding deterrents, it is not surprising to discover that some specialist herbivores have evolved the ability to sequester these normally highly toxic compounds for their own use against would-be predators. A well-known example of this phenomenon is the accumulation of cardenolides by monarch butterfly caterpillars and lygaeid bugs as they feed on foliage and seeds, respectively, of milkweeds (*Asclepias* spp.). Likewise, larvae of the sawfly *Neodiprion sertifer* sequester terpenoid compounds from their host, Scots pine (*Pinus silvestris*), and use them as defensive allomones (Blum, 1981).

Allomones are occasionally used aggressively to attract prey or, in social Hymenoptera, to rob nests of other species (Blum, 1996). For example, the staphylinid beetle *Leistrophus versicolor* normally feeds on flies of dung and carrion. When these flies are unavailable, it releases an allomone that attracts Phoridae and Drosophilidae (though curiously not its normal prey). Some ants and stingless bees release allomones when they raid nests of closely related species. The allomone overrides the effect of the host species' alarm pheromones, enabling the attackers to rob or take over the nest.

5. Environmental, Neural, and Endocrine Interaction

Only very rarely is a physiological event directly influenced by environmental stimuli. Temperature changes, through their effect on reaction rate, can, in poikilotherms at least, alter the rate at which an event is occurring. However, in almost all situations stimuli are first received by sensory structures that send information to the central nervous system to be dealt with. Some of these stimuli require an immediate response, which, as noted earlier in this chapter, is achieved via the motor neuron-effector organ system. The information received by the central nervous system as a result of other stimuli, however, initiates longer-term responses mediated via the endocrine system. This information, then, must be first "translated" within the brain into hormonal language. The center for translation is the neurosecretory system. Depending on the stimulus and, presumably, the site of termination of the internuncial neurons, different neurosecreteory cells will be stimulated to synthesize and/or release their product. This material may then act directly on target organs ("one-step" or "first-order" neurosecretory control) or exert a tropic effect on other endocrine glands ("two-step" or "second-order" control).[*] A number of examples may be cited to illustrate these two levels of control. In *Rhodnius* feeding leads to stimulation of stretch receptors in the abdominal wall. Information from these receptors passes along the ventral nerve cord to the composite thoracic ganglion where the posterior neurosecretory cells are stimulated to release diuretic hormone. The latter facilitates rapid excretion of the excess water present in the blood meal (Chapter 18, Section 5). In juvenile *Rhodnius* similar information is also received by the brain whose neurosecretory cells liberate thoracotropic hormone to trigger a new molting cycle (Chapter 21, Section 6.2). A variety of stimuli may enhance the rate of egg production in the female insect. Copulation, oviposition, pheromones, photoperiod, and feeding are all stimuli that cause neurosecretory cells to release an allatotropic hormone (Chapter 19, Section 3.1.3). The latter activates the corpora allata whose secretion promotes egg development.

---

[*] In vertebrates "three-step" ("third-order") neurosecretory control is also found where hypothalamic neurosecretion exerts a tropic effect on specific cells of the anterior pituitary gland. The products of these cells are themselves "tropic" hormones which act on other epithelial endocrine glands.

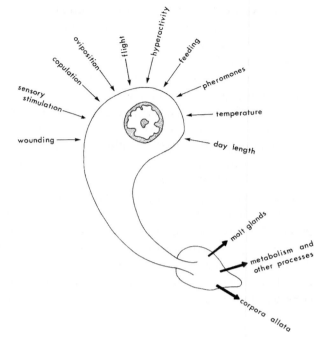

FIGURE 13.8. Diagrammatic summary of the variety of environmental and other factors that influence neurosecretory activity in insects. [After K. C. Highnam, 1965, Some aspects of neurosecretion in arthropods, *Zool. Jahrb. Abt. Allg. Zool. Physiol. Tiere* **71**:558–582. By permission of VEB Gustav Fischer Verlag.]

Figure 13.8 summarizes the variety of factors, environmental and experimental, that operate via the neurosecretory system to affect physiological events in insects.

6. Summary

The nervous system of insects, like that of other animals, comprises neurons (sensory, internuncial, and motor), synapses, and protective glial cells that wrap around neurons to effectively isolate them from the hemolymph. Neurons are aggregated to form nerves and a series of segmental ganglia. In some cases the ganglia fuse to form a composite structure. The central nervous system includes the brain, subesophageal ganglion, and a varied number of thoracic and abdominal ganglia. The visceral nervous system includes the stomatogastric system, unpaired ventral nerves, and caudal sympathetic system. Impulse transmission along an axon and across a synapse is essentially like that of other animals. Individual reflex responses are segmental, though the overall coordination of motor responses is achieved largely within the brain and/or subesophageal ganglion. Insects are able both to learn and to memorize, and, though the physiological/molecular bases for these abilities are not known, it is likely that they occur within the mushroom bodies of the brain.

The endocrine system comprises neurosecretory cells, found in most ganglia, corpora cardiaca, which both store neurosecretion and synthesize intrinsic hormones, corpora allata, molt glands, and, in some insects, the gonads. Neurosecretory hormones, which are polypeptidic, either act directly on target tissues or exert a tropic effect on other endocrine glands. In other words, the neurosecretory system "translates" neural information derived from environmental stimuli into hormonal messages that regulate a variety of physiological processes. Juvenile (gonadotropic) hormone, for which six naturally occurring forms are known, is a terpenoid compound produced by the corpora allata. Molting hormone (ecdysone) is a steroid resembling cholesterol from which it is biosynthesized.

Insects produce a wide range of semiochemicals that affect the behavior and/or physiology of other animals. The semiochemicals include pheromones, kairomones, and allomones. Pheromones induce behavioral or developmental responses in other individuals of the same species. They may be volatile, exerting their influence over considerable distance, or non-volatile when they are spread by physical contact (especially trophallaxis) among individuals. They include (1) sex pheromones, which modify reproductive behavior or development; (2) caste-regulating pheromones, which determine the proportion of different castes in colonies of social insects; (3) aggregation pheromones, which attract insects to a common feeding and/or mating site; (4) alarm pheromones, which warn of impending danger; (5) trail-marking pheromones, which provide information on the location and quantity of food available to a colony of social insects; and (6) spacing pheromones that maintain optimum population density. Kairomones and allomones exert their effect on other species, inducing effects beneficial to the receiver and emitter, respectively. Kairomones are of special significance to many predators and parasitoids, which use pheromones released by their prey as cues enabling them to locate the host. Allomones are used principally as defensive chemicals, released when the emitter is under attack; however, in a few species the allomones may be used aggressively to lure prey to the emitter or, in some ants and bees, to confuse the host, enabling the emitter to rob or take over the nest.

7. Literature

Reviews on various aspects of nervous and chemical integration are numerous, and the following list is only a small selection of publications. The structure and function of insect nervous systems are discussed by Treherne (1974), and authors in Volume 5 of Kerkut and Gilbert (1985). The neural and endocrine bases of behavior are dealt with by Markl (1974), Lindauer and Stockhammer (1974), Barton Browne (1974), and Howse (1975). Learning and memory are examined by Alloway (1972). Insect hormones are described by Novak (1975), Highnam and Hill (1977), Downer and Laufer (1983), Raabe (1982, 1989), and Nijhout (1994). Ryan (2002) and authors in Volume 9 of Kerkut and Gilbert (1985), Cardé and Minks (1997), Vander Meer *et al.* (1998), Hardie and Minks (1999), and Blomquist and Vogt (2003) review pheromones and provide lengthy bibliographies to this vast topic. Blum (1981, 1996), Haynes and Birch (in Volume 9 of Kerkut and Gilbert, 1985), and Whitman *et al.* (1990) deal with allomones and kairomones.

Adams, T. S., 1980, The role of ovarian hormones in maintaining cyclical egg production in insects, *Adv. Invert. Reprod.* **2**:109–125.

Adams, T. S., 1981, Activation of successive ovarian gonotrophic cycles by the corpus allatum in the house fly, *Musca domestica* (Diptera: Muscidae), *Int. J. Invert. Reprod.* **3**:41–48.

Al Abassi, S., Birkett, M. A., Pettersson, J., Pickett, J. A., Wadhams, L. J., and Woodcock, C. M., 2000, Response of the seven-spot ladybird to an aphid alarm pheromone and an alarm pheromone inhibitor is mediated by paired olfactory cells, *J. Chem. Ecol.* **26**:1765–1771.

Alloway, T. M., 1972, Learning and memory in insects, *Annu. Rev. Entomol.* **17**:43–56.

Arn, T., Tóth, M., and Priesner, E., 1992, *List of Sex Pheromones of Lepidoptera and Related Attractants*, 2nd ed., IOBC-WPRS, Montfavet.

Assad, Y. O. H., Hassanali, A., Torto, B., Mahamat, H., Bashir, N. H. H., and El-Bashir, S., 1997, Effects of fifth-instar volatiles on sexual maturation of adult desert locust *Schistocerca gregaria*, *J. Chem. Ecol.* **23**:1373–1388.

Ayali, A., Zilberstein, Y., and Cohen, N., 2002, The locust frontal ganglion: A central pattern generator network controlling foregut rhythmic motor patterns, *J. Exp. Biol.* **205**:2825–2832.

Barton Browne, L., 1974, *Experimental Analysis of Insect Behavior*, Springer-Verlag, Berlin.

Bicker, G., 1998, NO news from insect brains, *Trends Neurosci.* **21**:349–355.

Birch, M. C., Poppy, G. M., and Baker, T. C., 1990; Scents and eversible scent structures of male moths, *Annu. Rev. Entomol.* **35**:25–58.

Blomquist, G. J., and Vogt, R. G. (eds.), 2003, *Insect Pheromone Biochemistry and Molecular Biology*, Academic Press, New York.

Blum, M. S., 1981, *Chemical Defences of Arthropods*, Academic Press, New York.

Blum, M. S., 1996, Semiochemical parsimony in the Arthropoda, *Annu. Rev. Entomol.* **41**:353–374.

Boo, K. S., and Yang, J. P., 2000, Kairomones used by *Trichogramma chilonis* to find *Helicoverpa assulta* eggs, *J. Chem. Ecol.* **26**:359–375.

Borden, J. H., 1982, Aggregation pheromones, in: *Bark Beetles in North American Conifers: Ecology and Evolution* (J. B. Mitton and K. B. Sturgeon, eds.), University of Texas Press, Austin.

Cardé, R. T., and Minks, A. K. (eds.), 1997, *Insect Pheromone Research: New Directions*, Chapman and Hall, New York.

Carlson, S. D., Juang, J.-H., Hilgers, S. L., and Garment, M. B., 2000, Blood barriers of the insect, *Annu. Rev. Entomol.* **45**:151–174.

Collett, T. S., 1996, Insect navigation en route to the goal: Multiple strategies for the use of landmarks, *J. Exp. Biol.* **199**:227–235.

Collett, M., and Collett, T. S., 2000, How do insects use path integration for their navigation?, *Biol. Cybern.* **83**:245–259.

Davey, K. G., and Kuster, J. E., 1981, The source of an antigonadotropin in the female of *Rhodnius prolixus* Stal, *Can. J. Zool.* **59**:761–764.

Downer, R. G. H., and Laufer, H. (eds.), 1983, *Endocrinology of Insects*, Liss, New York.

Fitzgerald, T. D., 2003, Role of trail pheromone in foraging and processionary behavior of pine processionary caterpillars *Thaumatopoea pityocampa*, *J. Chem. Ecol.* **29**:513–532.

Fitzpatrick, S. M., and McNeil, J. M., 1988. Male scent in lepidopteran communication: The role of male pheromone in mating behaviour of *Pseudaletia unipuncta* (Haw.) (Lepidoptera: Noctuidae), *Mem. Entomol. Soc. Can.* **146**:131–151.

Flowers, R. W., and Costa, J. T., 2003, Larval communication and group foraging dynamics in the red-headed pine sawfly, *Neodiprion lecontei* (Fitch) (Hymenoptera: Symphyta: Diprionidae), *Ann. Entomol. Soc. Amer.* **96**:336–343.

Gillott, C., 2003, Male accessory gland secretions: Modulators of female reproductive physiology and behavior, *Annu. Rev. Entomol.* **48**:163–184.

Gillott, C., and Friedel, T., 1977, Fecundity-enhancing and receptivity-inhibiting substances produced by male insects: A review, *Adv. Invert. Reprod.* **1**:199–218.

Gillott, C., and Ismail, P. M., 1995, *In vitro* synthesis of ecdysteroid by the male accessory reproductive gland, testis and abdominal integument of the adult migratory grasshopper, *Melanoplus sanguinipes*, *Invert. Reprod. Dev.* **27**:65–71.

Gronenberg, W., 2001, Subdivisions of hymenopteran mushroom body calyces by their afferent supply, *J. Comp. Neurol.* **435**:474–489.

Hannson, B. S., and Anton, S., 2000, Function and morphology of the antennal lobe: New developments, *Annu. Rev. Entomol.* **45**:203–231.

Hardie, J., and Minks, A. K. (eds.), 1999, *Pheromones of Non-Lepidopteran Insects Associated with Agricultural Plants*, CAB International, Wallingford, U.K.

Highnam, K. C., and Hill, L., 1977, *The Comparative Endocrinology of the Invertebrates*, 2nd ed., Arnold, London.

Hölldobler, B., 1984, Evolution of insect communication, in *Insect Communication* (T. Lewis, ed.), Academic Press, New York.

Homberg, U., 1994, *Distribution of Neurotransmitters in the Insect Brain*, Gustav Fischer Verlag, Stuttgart.

Homberg, U., Christensen, T. A., and Hildebrand, J. G., 1989, Structure and function of the deutocerebrum in insects, *Annu. Rev. Entomol.* **34**:477–501.

Howse, P. E., 1975, Brain structure and behavior in insects, *Annu. Rev. Entomol.* **20**:359–379.

Hoyle, G. 1974, Neural control of skeletal muscle, in: *The Physiology of Insecta*, 2nd ed., Vol. IV (M. Rockstein, ed.), Academic Press, New York.

Huber, F., 1974, Neural integration (central nervous system), in: *The Physiology of Insecta*, 2nd ed., Vol. IV (M. Rockstein, ed.), Academic Press, New York.

Judd, S. P. D., and Collett, T. S., 1998, Multiple stored views and landmark guidance in ants, *Nature* **392**:710–714.

Kerkut, G. A., and Gilbert, L. I. (eds.), 1985, *Comprehensive Insect Physiology, Biochemistry and Pharmacology*, Pergamon Press, Elmsford, NY.

Kohl, E., Hölldobler, B., and Bestmann, H.-J., 2000, A trail pheromone component of the ant *Mayriella overbecki* Viehmeyer (Formicidae: Myrmicinae), *Naturwissenschaften* **87**:320–322.

Kretzschmar, D., and Pflugfelder, G. O., 2002, Glia in development, function, and neurodegeneration of the adult insect brain, *Brain Res. Bull.* **57**:121–131.

Lefevere, K. S., Lacey, M. J., Smith, P. H., and Roberts, B., 1993, Identification and quantification of juvenile hormone biosynthesized by larval and adult Australian sheep blow fly *Lucilia cuprina* (Diptera: Calliphoridae), *Insect Biochem. Mol. Biol.* **23**:713–720.

Lehrer, M., 1991, Bees which turn back and look, *Naturwissenschaften* **78**:274–276.

Lehrer, M., and Bianco, G., 2000, The turn-back-and-look behaviour: Bees versus robot, *Biol. Cybern.* **83**:211–229.

Lindauer, M., and Stockhammer, K. A., 1974, Social behavior and mutual communication, in: *The Physiology of Insecta*, 2nd ed., Vol. III (M. Rockstein, ed.), Academic Press, New York.

Loeb, M. J., Bell, R. A., Gelman, D. B., Kochansky, J., Lusby, W., and Wagner, R. M., 1996, Action cascade of an insect gonadotropin, testis ecdysiotropin, in male Lepidoptera, *Invert. Reprod. Devel.* **30**:181–190.

Locke, M., 1969, The ultrastructure of the oenocytes in the molt/intermolt cycle of an insect, *Tissue Cell* **1**:103–154.

Lüscher, M., 1972, Environmental control of juvenile hormone (JH) secretion and caste differentiation in termites, *Gen. Comp. Endocrinol. Suppl.* **3**:509–514.

Mahamat, H., Hassanali, A., and Odongo, H., 2000, The role of different components of the pheromone emission of mature males of the desert locust, *Schistocerca gregaria* (Forskal) (Orthoptera: Acrididae) in accelerating maturation of immature adults, *Insect Sci. Appl.* **20**:1–5.

Markl, H., 1974, Insect behavior: Functions and mechanisms, in: *The Physiology of Insecta*, 2nd ed., Vol. III (M. Rockstein, ed.), Academic Press, New York.

McCall, P. J., 2002, Chemoecology of oviposition in insects of medical and veterinary importance, in: *Chemoecology of Insect Eggs and Egg Deposition* (M. Hilker and T. Meiners, eds.), Blackwell Verlag, Berlin.

McCall, P. J., and Kelly, D. W., Learning and memory in disease vectors, *Trends Parasitol.* **18**:429–433.

Müller, U., 1997, The nitric-oxide system in insects, *Prog. Neurobiol.* **51**:363–381.

Naisse, J., 1969, Rôle des neurohormones dans la différentiation sexuelle de *Lampyris noctiluca*, *J. Insect Physiol.* **15**:877–892.

Nijhout, H. F., 1994, *Insect Hormones*, Princeton University Press, Princeton, NJ.

Novak, V. J. A., 1975, *Insect Hormones*, 2nd English ed., Chapman and Hall, London.

Numata, H., Numata, A., Takahashi, C., Nakagawa, Y., Iwatani, K., Takahashi, S., Miura, K., and Chinzei, Y., 1992, Juvenile hormone I is the principal juvenile hormone in a hemipteran insect, *Riptortus clavatus*, *Experientia* **48**:606–610.

Orchard, I., 1984, The role of biogenic amines in the regulation of peptidergic neurosecretory cells, in: *Insect Neurochemistry and Neurophysiology* (A.B. Borkovec and T. J. Kelly, eds.), Plenum Press, New York.

Prokopy, R. J., 1981, Oviposition deterring pheromone system of apple maggot flies, in: *Management of Insect Pests with Semiochemicals* (E. R. Mitchell, ed.), Plenum Press, New York.

Raabe, M., 1982, *Insect Neurohormones*, Plenum Press, New York.

Raabe, M., 1989, *Recent Developments in Insect Neurohormones*, Plenum Press, New York.

Rai, M. M., Hassanali, A., Saini, R. K., Odongo, H., and Kahoro, H., 1997, Identification of components of the oviposition aggregation pheromone of the gregarious desert locust, *Schistocerca gregaria* (Forskal), *J. Insect Physiol.* **43**:83–87.

Ritzmann, R. E., 1984, The cockroach escape response, in: *Neural Mechanisms of Startle Behavior* (R. C. Eaton, ed.), Plenum Press, New York.

Rollo, C. D., Borden, J. H., and Casey, I. B., 1995, Endogenously produced repellent from the American cockroach (Blattaria: Blattidae): Function in death recognition, *Environ. Entomol.* **24**:116–124.

Romer, P., 1991, The oenocytes of insects: Differentiation, changes during molting, and their possible involvement in the secretion of molting hormone, in: *Morphogenetic Hormones of Arthropods* (A. P. Gupta, ed.), Vol. 1, Part 3, Rutgers University Press, New Brunswick, NJ.

Romer, F., and Bressel, H. U., 1994, Secretion and metabolism of ecdysteroids by oenocyte-fat body complexes (OEFC) in adult males of *Gryllus bimaculatus* DeG (Insecta), *Z. Naturforsch. C Biosci.* **49**:871–880.

Ryan, M. F., 2002, *Insect Chemoreception: Fundamental and Applied*, Kluwer Academic Publishers, Dordrecht.

Saini, R. K., Rai, M. M., Hassanali, A., Wawije, J., and Odongo, H., 1995, Semiochemicals from froth of egg pods attract ovipositing female *Schistocerca gregaria*, *J. Insect Physiol.* **41**:711–716.

Saunders, D. S., 1997, Insect circadian rhythms and photoperiodism, *Invert. Neurosci.* **3**:155–164.

Schoonhoven, L. M., 1990, Host-marking pheromones in Lepidoptera, with special reference to two *Pieris* spp., *J. Chem. Ecol.* **16**:3040–3052.

Steinmetz, I., Sieben, S., and Schmolz, E., 2002, Chemical trails used for orientation in nest cavities by two vespine wasps, *Vespa crabro* and *Vespula vulgaris*, *Insectes Soc.* **49**:354–356.

Steinmetz, I., Schmolz, E., and Ruther, J., 2003, Cuticular lipids as trail pheromone in a social wasp, *Proc. R. Soc. Lond. B* **270**:385–391.

Stowe, M. K., Turlings, T. C. J., Lougrin, J. H., Lewis, W. J., and Tumlinson, J. H., 1995, The chemistry of eavesdropping, alarm, and deceit, *Proc. Natl Acad. Sci. U S A* **92**:23–28.

Strausfeld, N. J., Hansen, L., Li, Y., Gomez, R. S., and Ito, K., 1998, Evolution, discovery, and interpretations of arthropod mushroom bodies, *Learn. Mem., Cold Spring Harbor* **5**:11–37.

Tillman, J. A., Seybold, S. J., Jurenka, R. A., and Blomquist, G. J., 1999, Insect pheromones—An overview of biosynthesis and endocrine regulation, *Insect Biochem. Mol. Biol.* **29**:481–514.

Treherne, J. E. (ed.), 1974, *Insect Neurobiology*, Elsevier/North-Holland, Amsterdam.

Vander Meer, R. K., Breed, M. D., Espelie, K. E., and Winston, M. L. (eds.), 1998, *Pheromone Communication in Social Insects: Ants, Wasps, Bees, and Termites*, Westview Press, Boulder, CO.

Vitzthum, H., Müller, M., and Homberg, U., 2002, Neurons of the central complex of the locust *Schistocerca gregaria* are sensitive to polarized light, *J. Neurosci.* **22**:1114–1125.

Weatherston, J., and Percy, J. E., 1977, Pheromones of male Lepidoptera, *Adv. Invert. Reprod.* **1**:295–307.

Whitman, D. W., Blum, M. S., and Alsop, D. W., 1990, Allomones: Chemicals for defense, in: *Insect Defenses: Adaptive mechanisms and Strategies of Prey and Predators* (D. L. Evans and J. O. Schmidt, eds.), State University of New York Press, Albany, NY.

Žitňan, D., Kingan, T. J., Hermesman, J. L., and Adams, M. E., 1996, Identification of ecdysis-triggering hormone from an epitracheal endocrine system, *Science* **271**:88–91.

Žitňan, D., Žitňanová, I., Spalovská, I., Takáě, P., Park, Y., and Adams, M. E., 2003, Conservation of ecdysis-triggering hormone signalling in insects, *J. Exp. Biol.* **206**:1275–1289.

Muscles and Locomotion

1. Introduction

The ability to move is a characteristic of living animals and facilitates distribution, food procurement, location of a mate or egg-laying site, and avoidance of unsuitable conditions. Insects, largely through their ability to fly when adult, are among the most mobile and widely distributed of animals. Development of this ability early in the evolution of the class has made the Insecta the most diverse and successful animal group (Chapter 2, Section 3.1). However, flight is only one method of locomotion employed by insects. Terrestrial species may walk, jump, or crawl over the substrate, or burrow within it. Aquatic forms can swim in a variety of ways or run on the water surface.

In their locomotory movements, insects conform to normal dynamic and mechanical principles. However, their generally small size and light weight have led to the development of some unique structural, physiological, and biochemical features in their locomotory systems.

2. Muscles

Essentially, the structure and contractile mechanism of insect muscles are comparable to those of vertebrate skeletal (cross-striated) muscle; that is, there are no muscles in insects of the smooth (non-striated) type. Within muscle cells, the contractile elements actin and myosin have been identified, and Huxley's sliding filament theory of muscle contraction applies. Though insect muscles are always cross-striated, there is considerable variation in their structure, biochemistry, and neural control, in accord with specific functions.

Because of their small size and the variable composition of the hemolymph of insects, the neuromuscular system has some unique features (Hoyle, 1974). Being small, an insect has a limited space for muscles which are, accordingly, reduced in size. Though this is achieved to some extent by a decrease in the size of individual cells (fibers), the principal change has been a decline in the number of fibers per muscle such that some insect muscles comprise only one or two cells. Thus, to achieve a graded muscle contraction, each fiber must be capable of a variable response, in contrast to the vertebrate situation where graded muscle responses result in part from stimulation of a varied number of fibers. Similarly, the volume of nervous tissue is limited, so that there are few motor neurons for the control of muscle contraction. The hemolymph surrounding muscles may contain high concentrations of ions (especially divalent ions such as Mg^{2+}) (Chapter 17, Section 4.1.1) that could interfere with impulse transmission at synapses and neuromuscular junctions. That this does not occur is the result of the evolution of a myelin sheath that covers ganglia, nerves, and neuromuscular junctions.

2.1. Structure

Insect muscles can be arranged in two categories: (1) skeletal muscles whose function is to move one part of the skeleton in relation to another, the two parts being separated by a joint of some kind, and (2) visceral muscles, which form layers of tissue enveloping internal organs such as the heart, gut, and reproductive tract.

Attachment of a muscle to the integument must take into account the fact that periodically the remains of the old cuticle are shed; therefore, an insertion must be able to break and re-form easily. As Figure 14.1 indicates, a muscle terminates at the basal lamina lying beneath the epidermis. The muscle cells and epidermal cells interdigitate, increasing the surface area for attachment by about 10 times, and desmosomes occur at intervals, replacing the basal lamina. Attachment of a muscle cell to the rigid cuticle is achieved through large numbers of parallel microtubules (called "tonofibrillae" by earlier authors). Distally, the epidermal cell membrane is invaginated, forming numbers of conical hemidesmosomes on which the microtubules terminate. Running distad from each hemidesmosome is one, rarely two, muscle attachment fibers (= tonofibrils). Each fiber passes along a pore canal

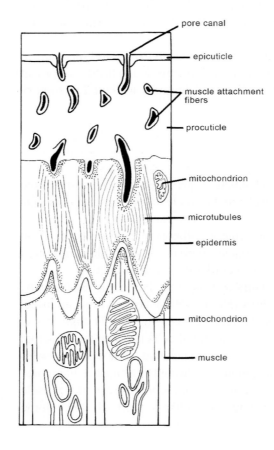

pore canal

epicuticle

muscle attachment fibers

procuticle

mitochondrion

microtubules

epidermis

mitochondrion

muscle

FIGURE 14.1. Muscle insertion. [After A. C. Neville, 1975, *Biology of the Arthropod Cuticle*. By permission of Springer-Verlag, New York.]

to the cuticulin envelope of the epicuticle to which they are attached by a special cement. As the cuticulin layer is the first one formed during production of a new cuticle (Chapter 11, Section 3.1), attachment of newly formed fibers can readily occur. Until the actual molt, however, these are continuous with the old fibers and, therefore, normal muscle contraction is possible (Neville, 1975).

Muscles comprise a varied number of elongate, multinucleated cells (fibers) (not to be confused with the muscle attachment fibers mentioned above) that may extend along the length of a muscle. A muscle is arranged usually in units of 10–20 fibers, each unit being separated from the others by a tracheolated membrane. Each unit has a separate nerve supply. The cytoplasm (sarcoplasm) of each fiber contains a varied number of mitochondria (sarcosomes). Even at the light microscope level, the transversely striated nature of muscles is visible. Higher magnification reveals that each fiber contains a large number of myofibrils (= fibrillae = sarcostyles) lying parallel in the sarcoplasm and extending the length of the cell. Each myofibril comprises the contractile filaments, made up primarily of two proteins, actin and myosin. The thicker myosin filaments are surrounded by the thinner but more numerous actin filaments. Filaments of each myofibril within a cell tend to be aligned, and it is this that creates the striated appearance (alternating light and dark bands) of the cell. The dark bands (A bands) correspond to regions where the actin and myosin overlap, whereas the lighter bands indicate regions where there is only actin (I bands) or myosin (H bands) (Figure 14.2). Electron microscopy has revealed in addition to these bands a number of thin transverse structures in the muscle fiber. Each of these Z lines (discs) runs across the fiber in the center of the I bands, separating individual contractile segments called sarcomeres. Attached to each side of the Z line are the actin filaments, which in contracted muscle are

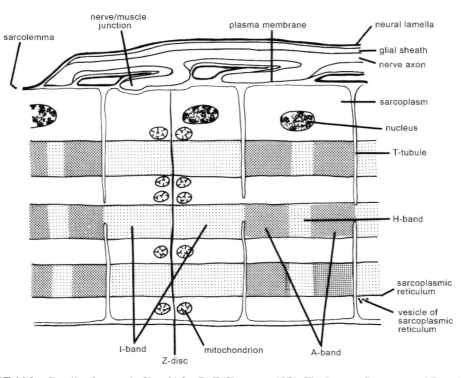

FIGURE 14.2. Details of a muscle fiber. [After R. F. Chapman, 1971, *The Insects: Structure and Function.* By permission of Elsevier/North-Holland, Inc., and the author.]

connected to the myosin filaments by a means of cross bridges present at each end of the myosin. Periodically, the plasma membrane (sarcolemma) of the muscle fiber is deeply invaginated and forms the so-called T system (transverse system). In most insect muscles the T system occurs midway between the Z line and H band; in fibrillar muscles, however, there is no regular pattern for the position of the invaginations.

Though the above description is applicable to all insect muscles, different types of muscles can be distinguished, primarily on the basis of the arrangement of myofibrils, mitochondria, and nuclei; the degree of separation of the myofibrils; the degree of development of the sarcoplasmic reticulum; and the number of actins surrounding each myosin (Figure 14.3). These include tubular (lamellar), close-packed, and fibrillar muscles, all of which are skeletal, and visceral muscles.

Leg and segmental muscles of many adult insects and the flight muscles of primitive fliers, such as Odonata and Dictyoptera, are of the tubular type, in which the flattened (lamellate) myofibrils are arranged radially around the central sarcoplasm. The nuclei are distributed within the core of sarcoplasm and the slablike mitochondria are interspersed

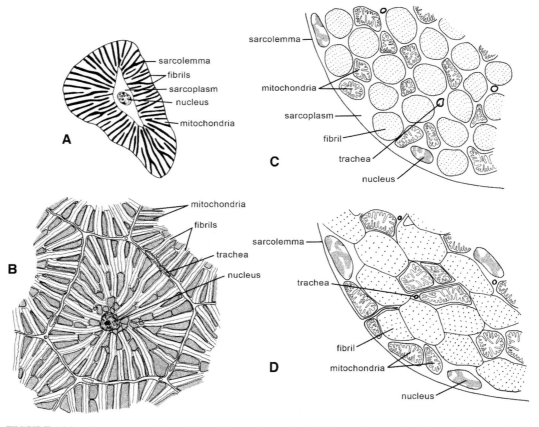

FIGURE 14.3. Transverse sections of insect skeletal muscles. (A) Tubular leg muscle of *Vespa* (Hymenoptera); (B) tubular flight muscle of *Enallagma* (Odonata); (C) close-packed flight muscle of a butterfly; and (D) fibrillar flight muscle of *Tenebrio* (Coleoptera). (Not to same scale.) [A, after H. E. Jordan, 1920, Studies on striped muscle structure. VI, *Am. J. Anat.* **27**:1–66. By permission of Wistar Press. B, C, redrawn from electron micrographs in D. S. Smith, 1965, The flight muscles of insects, *Scientific American*, June 1965, W. H. Freeman and Co. By permission of the author. D, redrawn from an electron micrograph in D. S. Smith, 1961, The structure of insect fibrillar muscles. A study made with special reference to the membrane systems of the fiber, *J. Biophys. Biochem. Cytol.* **10**:123–158. By permission of the Rockefeller Institute Press and the author.]

between the myofibrils. The body musculature of apterygotes and some larval pterygotes, the leg muscles of some adult pterygotes, and the flight muscles of Orthoptera and Lepidoptera are of the close-packed type. Here the myofibrils and mitochondria are concentrated in the center of the fiber and the nuclei are arranged peripherally. In close-packed flight muscles, the fibers are considerably larger than those of tubular flight muscles. In addition, tracheoles deeply indent the fiber, whereas in tubular muscles tracheoles simply lie alongside each fiber. It should be appreciated that the tracheoles do not actually penetrate the muscle cell membrane, that is, they are extracellular. In most insects the indirect muscles, which provide the power for flight, are nearly always fibrillar, so-called because individual fibrils are characteristically very large and, together with the massive mitochondria, occupy almost all of the volume of the fiber. Very little sarcoplasm is present, and the nuclei are squeezed randomly between the fibrils. Because of their size, there are often only a few fibrils per cell, and these are frequently quite isolated from each other by the massively indented and intertwining system of tracheoles. The presence of large quantities of cytochromes in the mitochondria gives these muscles a characteristic pink or yellow color. It should be apparent even from this brief description that fibrillar muscles are designed to facilitate a high rate of aerobic respiration in connection with the energetics of flight.

Visceral muscles differ from skeletal muscles in several ways. The cells comprising them are uninucleate, may branch, and are joined to adjacent cells by septate desmosomes. Their contractile elements are not arranged in fibrils and contain a larger proportion of actin to myosin. Nevertheless, the visceral muscles are striated (sometimes only weakly), and their method of contraction is apparently identical to that of skeletal muscles.

All skeletal muscles and many visceral muscles are innervated. The skeletal muscles always receive nerves from the central nervous system, whereas the visceral muscles are innervated from either the stomatogastric or the central nervous system. Within a particular muscle unit, each fiber may be innervated by one, two, or three functionally distinct axons. One of these is always excitatory; where two occur (the commonest arrangement), they are usually both excitatory ("fast" and "slow" axons), but may be a "slow" excitatory axon plus an inhibitory axon; in some cases all three types of axon occur. This arrangement, known as polyneuronal innervation, facilitates a variable response on the part of a muscle (Section 2.2). Each axon, regardless of its function, is much branched and, in contrast to the situation in vertebrate muscle, there are several motor neuron endings from each axon on each muscle fiber (multiterminal innervation) (Figure 14.4).

2.2. Physiology

Like those of vertebrates, insect muscles contract according to the sliding filament theory. The arrival of an excitatory nerve impulse at a neuromuscular junction causes depolarization of the adjacent sarcolemma. A wave of depolarization spreads over the fiber and into the interior of the cell via the T system. Depolarization of the T system membranes induces a momentary increase in the permeability of the adjacent sarcoplasmic reticulum, so that calcium ions, stored in vesicles of the reticulum, are released into the sarcoplasm surrounding the myofibrils. The calcium ions activate cross-bridge formation between the actin and myosin, enabling the filaments to slide over each other so that the distance between adjacent Z lines is decreased. The net effect is for the muscle to contract. Energy derived from the hydrolysis of adenosine triphosphate (ATP) is required for contraction, though its precise function is unknown. It may be used in breaking the cross-bridges, or for the active transport of the calcium ions back into the vesicles, or for both of these processes. In

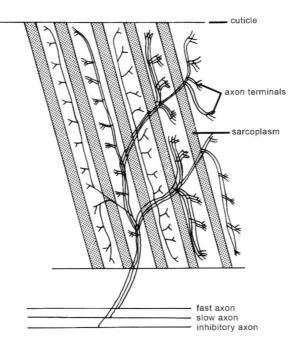

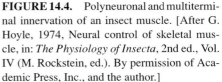

FIGURE 14.4. Polyneuronal and multiterminal innervation of an insect muscle. [After G. Hoyle, 1974, Neural control of skeletal muscle, in: *The Physiology of Insecta*, 2nd ed., Vol. IV (M. Rockstein, ed.). By permission of Academic Press, Inc., and the author.]

addition to sliding over each other, both the actin and the myosin filaments may shorten (by coiling), and in some myofibrils the Z lines disintegrate to allow the A bands of adjacent sarcomeres to overlap each other, thus enabling an even greater degree of contraction to occur.

Extension (relaxation) of a muscle may result simply from the opposing elasticity of the cuticle to which the muscle is attached. More commonly, muscles occur in pairs, each member of the pair working antagonistically to the other; that is, as one muscle is stimulated to contract, its partner (unstimulated) is stretched. Normally, the previously unstimulated muscle is stimulated to begin contraction while active contraction of the partner is still occurring (cocontraction). This is thought to bring about dampening of contraction, perhaps thereby preventing damage to a vigorously contracting muscle. Also, in slow movements, it provides an insect with a means of precisely controlling such movements (Hoyle, 1974). Muscle antagonism is achieved by central inhibition, that is, at the level of interneurons within the central nervous system (Chapter 13, Section 2.3). Thus, for a given stimulus, the passage of impulses along an axon to one muscle of the pair will be permitted, and hence that muscle will contract. However, passage of impulses to the partner is inhibited and the muscle will be passively stretched. It should be emphasized that in this arrangement the axon to each muscle is excitatory. In slow walking movements, for example, alternating stimulation of each muscle is quite distinct. At higher speeds this reciprocal inhibition breaks down, and one of the muscles remains permanently in a mildly contracted state, serving as an "elastic restoring element" (Hoyle, 1974). The other muscle continues to be alternately stimulated and thus provides the driving power for the activity.

As noted earlier, commonly muscles receive two excitatory axons, one "slow," the other "fast." These terms are somewhat misleading for they do not indicate the speed at which impulses travel along the axons, but rather the speed at which a significant contraction can be observed in the muscle. Thus, an impulse traveling along a fast axon induces a strong contraction of the "all or nothing" type; that is, a further contraction cannot be initiated until the original ionic conditions have been restored. In contrast, a single impulse from a slow axon causes only a weak contraction in the muscle. However, additional impulses arriving in quick succession are additive in their effect (summation) so that, with the slow axon arrangement a graded response is possible for a particular muscle, despite the relatively few fibers it may contain. Muscles with dual innervation use only the slow axon for most requirements; the fast axon functions only when immediate and/or massive contraction is necessary. For example, the extensor tibia muscle of the hindleg of a grasshopper is ordinarily controlled solely via the slow axon. For jumping, however, the fast axon is brought into play.

The function of inhibitory axons remains questionable. Electrophysiological work has shown that in normal activity the inhibitory axon is electrically silent, that is, shows no electrical activity, and is clearly being inhibited from within the central nervous system. During periods of great activity, impulses can sometimes be observed passing along the axon, perhaps to accelerate muscle relaxation, though normally the use of antagonistic muscles and central inhibition is adequate. Hoyle (1974) suggested that peripheral inhibition may be necessary at certain stages in the life cycle, such as molting, when central inhibition may not be possible.

3. Locomotion

3.1. Movement on or Through a Substrate

3.1.1. Walking

Insects can walk at almost imperceptibly slow speed (watch a mantis stalking its prey) or run at seemingly very high rates (try to catch a cockroach). The latter is, however, a wrong impression created by the smallness of the organism, the rate at which its legs move, and the rate at which it can change direction. Ants scurrying about on a hot summer day are traveling only about 1.5 km/hr, and the elusive cockroach has a top speed of just under 5 km/hr (Hughes and Mill, 1974).

Nevertheless, an insect leg is structurally well adapted for locomotion. Like the limbs of other actively moving animals, it tapers toward the distal end, which is light and easily lifted. Its tarsal segments are equipped with claws or pulvilli that provide the necessary friction between the limb and the substrate. A leg comprises four main segments (Chapter 3, Section 4.3.1), which articulate with each other and with the body. The coxa articulates proximally with the thorax, usually by means of a dicondylic joint and distally, with the fused trochanter and femur, also via a dicondylic joint. Dicondylic joints permit movement in a single plane. However, the two joints are set at right angles to each other and, therefore, the tip of a leg can move in three dimensions.

The muscles that move a leg are both extrinsic (having one end inserted on the wall of the thorax) and intrinsic (having both ends inserted within the leg) (Figure 14.5). The majority of extrinsic muscles move the coxa, rarely the fused trochantofemoral segment, whereas the paired intrinsic muscles move leg segments in relation to each other. Some of the extrinsic muscles have a dual function, serving to bring about both leg and wing movements. Typically, the leg muscles include (1) the coxal promotor and its antagonist, the coxal remotor, which run from the tergum to the anterior and posterior edges, respectively,

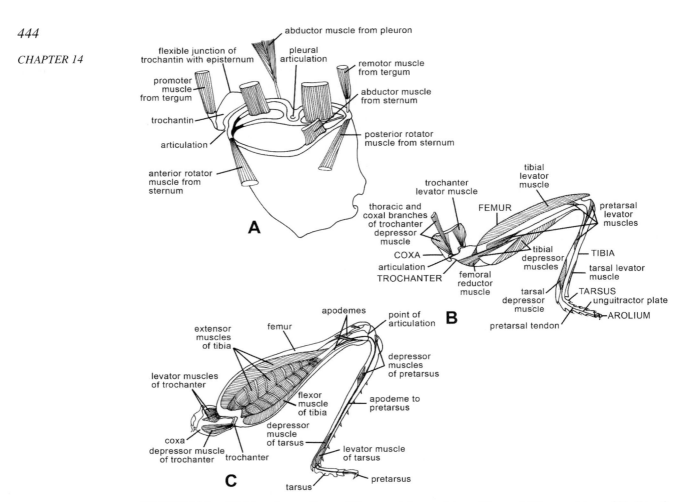

FIGURE 14.5. (A) Musculature of coxa; (B) segmental musculature of leg; and (C) musculature of hindleg of grasshopper. [A, C, from R. E. Snodgrass, *Principles of Insect Morphology.* Copyright 1935 by McGraw-Hill, Inc. Used with permission of McGraw-Hill Book Company. B, reproduced by permission of the Smithsonian Institution Press from *Smithsonian Miscellaneous Collections*, Volume 80, Morphology and mechanism of the insect thorax, Number 1, June 25, 1927, 108 pages, by R. E. Snodgrass: Figure 39, page 89. Washington, D.C., 1928, Smithsonian Institution.]

of the coxa; contraction of the coxal promotor causes the coxa to twist forward, thereby effecting protraction (a forward swing) of the entire leg; (2) the coxal adductor and abductor (attached to the sternum and pleuron, respectively), which move the coxa toward or away from the body; (3) anterior and posterior coxal rotators, which arise on the sternum and assist in raising and moving the leg forward or backward; and (4) an extensor (levator) and flexor (depressor) muscle in each leg segment, which serve to increase and decrease, respectively, the angle between adjacent segments. It should be noted that the muscles that move a particular leg segment are actually located in the next more proximal segment. For example, the tibial extensor and flexor muscles, which alter the angle between the femur and tibia, are located within the femur and are attached by short tendons inserted at the head of the tibia.

It is the coordinated actions of the extrinsic and intrinsic muscles that move a leg and propel an insect forward. In considering how propulsion is achieved, it must also be remembered that another important function of a leg is to support the body, that is, to keep

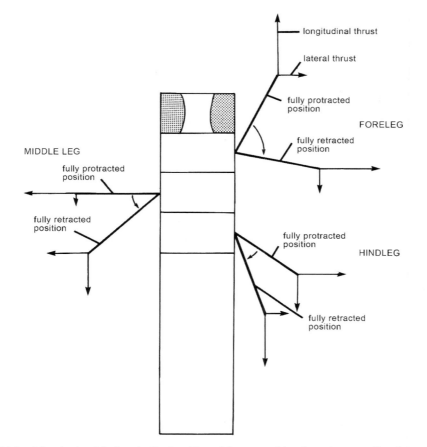

FIGURE 14.6. Magnitude of the longitudinal and lateral forces resulting from the strut effect for each leg in its extreme position. [After G. M. Hughes, 1952, The coordination of insect movements. I. The walking movements of insects, *J. Exp. Biol.* **29**:267–284. By permission of Cambridge University Press, London.]

it off the ground. In the latter situation, a leg may be considered as a single-segmented structure—a rigid strut. If the strut is vertical, the force along its length (axial force) will be solely supporting and will have no propulsive component. If the strut is inclined, the axial force can be resolved into two components, a vertical supportive force and a horizontal propulsive force. Because the leg protrudes laterally from the body, the horizontal force can be further resolved into a transverse force pushing the insect sideways and a longitudinal force that causes backward or forward motion. The relative sizes of these horizontal forces depend on (1) which leg is being considered and (2) the position of that leg. Figure 14.6 indicates the size of these forces for each leg at its two extreme positions. It will be apparent that in almost all of its positions the foreleg will inhibit forward movement, whereas the mid- and hindlegs always promote forward movement. In equilibrium, that is, when an insect is standing still, the forces will be equal and opposite. Movement of an insect's body will occur only if the center of gravity of the body falls. This occurs when the forces become imbalanced, for example, by changing the position of a foreleg so that its retarding effect is no longer equal to the promoting effect of the other legs, whereupon the insect topples forward (Hughes and Mill, 1974).

Also important from the point of view of locomotion is the leg's ability to function as a lever, that is, a solid bar that rotates about a fulcrum and on which work can be done. The fulcrum is the coxothoracic joint and the work is done by the large, extrinsic muscles.

Because of the large angle through which it can rotate and because of its angle to the body, the foreleg is most important as a lever. In contrast, the mid- and hindlegs, which each rotate through only a small angle, exert only a slight lever effect and serve primarily as struts (in the fully extended, rigid position). For the foreleg in its fully protracted position, contraction of the retractor muscle (i.e., the lever effect) will be sufficient to overcome the opposing retarding (strut) effect and, provided that the frictional forces between the ground and tarsi are sufficient, the body will be moved forward.

However, the largest component of the propulsive force is derived as a result of the leg's ability to flex and extend by virtue of their jointed nature. Flexure (a decrease in the angle between adjacent leg segments) will raise the leg off the ground so that it can be moved forward without the need to overcome frictional forces between it and the ground. In the case of the foreleg, flexure first will remove, by lifting the leg from the ground, the retarding effect as a result of its action as a strut and, second, when the leg is replaced on the substrate, will cause the body to be pulled forward. Flexure of the foreleg continues until the leg is perpendicular to the body, at which point extension begins so that now the body is pushed forward. For the mid- and hindlegs, flexure serves to bring the legs into a new forward position. Extension, as in the case of the foreleg, will push the body forward. Because the hindleg is usually the largest of the three, it exerts the greatest propulsive force.

As noted above, the horizontal axial force along each leg has a transverse as well as a longitudinal component. Thus, as an insect moves, its body zigzags slightly from side to side, the transverse forces exerted by the fore- and hindlegs of one side being balanced by an opposite force exerted by the middle leg of the opposite side in the normal rhythm of leg movements.

Rhythms of Leg Movements. Most insects use all six legs during normal walking. Other species habitually employ only the two anterior or the two posterior pairs of legs but may use all legs at higher speeds. In all instances, however, the legs are lifted in an orderly sequence (though this may vary with the speed of the insect), and there are always at least three points of contact with the substrate forming a "triangle of support" for the body. (In some species that employ two pairs of legs, the tip of the abdomen may serve as a point of support.) Two other generalizations that may be made are (1) no leg is lifted until the leg behind has taken up a supporting position and (2) the legs of a segment alternate in their movements.

In the typical hexapodal gait at low speed, only one leg at a time is raised off the ground, so that the stepping sequence is R3, R2, R1, L3, L2, L1 (where R and L are right and left legs, respectively, and 1, 2, and 3 indicate the fore-, mid-, and hindlegs, respectively). With increase in speed, overlap occurs between both sides so that the sequence first becomes R3 L1, R2, R1 L3, L2, etc., and, then, R3 R1 L2, R2 L3 L1, etc., that is, a true alternating tripodal gait.

The orthopteran *Rhipipteryx* has a quadrupedal gait, using only the anterior two pairs of legs and using the tip of the abdomen as a support. Its stepping sequence is R1 L2, R2 L1., etc. Mantids are likewise quadrupedal at low speed, using the posterior two pairs of legs (sequence R3 L2, R2 L3, etc.). At high speed, the forelegs are brought into action though the insects remain effectively quadrupedal (sequence L1 R3, L3 R2, L2 R1, etc.).

A variety of methods for turning have been observed, often in the same species. They include increasing the length of the stride on the outside of the turn, increasing the frequency of stepping on the outside of the turn, fixing one or more of the "inside" legs as pivots, and moving the legs on the inside of the turn backward.

Coordination of the movements both among segments of the same leg and among different legs requires a high level of neural activity, both sensory and motor. Like other rhythmic activities, the walking rhythm is centrally generated; that is, the endogenous activity of a network of neurons within each thoracic ganglion regulates the alternating contraction of the leg flexor and extensor muscles, hence the flexion and extension of the limb. Intersegmental coordination of leg movements is achieved principally by "central coupling," that is, by signals traveling within the central nervous system from one network to another. Starting and stopping, turning, and change of speed are controlled by the brain and subesophageal ganglion, via so-called "command neurons," though how these centers exert their control is unclear. Superimposed on this central control is the input from the insect's sense organs, especially proprioceptors on the legs themselves, which permits the insect to adjust its walking rhythm to compensate for changing environmental conditions. (See Chapter 13, Section 2.3).

3.1.2. Jumping

Jumping is especially well developed in grasshoppers, fleas, flea beetles, click beetles, and Collembola. In the first three mentioned groups, jumping involves the hindlegs, which, like those of other jumping animals, are elongate and capable of great extension. Their length ensures that the limbs are in contact with the substrate for a long time during takeoff. Extension is achieved as the initially acute angle between the femur and tibia is increased to more than 90° by the time the tarsi leave the substrate. The length and extension together enable sufficient thrust to be developed that the insect can jump heights and distances many times its body length. For example, a fifth-instar locust (length about 4 cm) may "high jump" 30 cm and, concurrently, "long jump" 70 cm.

In Orthoptera the power for jumping is developed by the large extensor tibiae muscle in the femur (Figure 14.5C). The muscle is arranged in two masses of tissue that arise on the femur wall and are inserted obliquely on a long flat apodeme attached to the upper end of the tibia. The resultant herringbone arrangement increases the effective cross-sectional area of muscle attached to the apodeme, thereby increasing the power that the muscle develops.

As the extensor tibiae contracts, all activity ceases in the motoneurons running to its antagonist, the flexor tibiae muscle, thus permitting all of the power developed to be used in extending the tibia. This inhibition results from the activity of a single, branched inhibitory interneuron. Because the apodeme is attached to the upper end of the tibia, slight contraction of the extensor muscle will cause a relatively enormous movement at the tarsus (ratio of movements 60:1 when the tibiofemoral joint is tightly flexed).

It has been calculated that for the locust to achieve the maximum thrust for takeoff, the body must be accelerated at about 1.5×10^4 cm/sec^2 over a time span of 20 msec. The force exerted by each extensor muscle is about 5×10^5 dynes (= 500 g wt) for an insect weighing 3 g (Alexander, 1968, cited from Hughes and Mill, 1974). To withstand this force, the apodeme must have a strength close to that of moderate steel. The extremely short time period over which this acceleration is developed makes it unlikely that jumping occurs as a direct result of muscle contraction. Indeed, Heitler (1974) showed that the initial energy of muscle contraction is stored as elastic energy using a cuticular locking device that holds the flexor tendon in opposition to the force developed within the extensor muscle. At a critical level, the tendon is released, allowing the tibia to rotate rapidly backward.

Likewise, in fleas, the energy of muscle contraction is first stored as elastic energy. Prior to jumping, the flea contracts various extrinsic muscles of the metathorax, which are inserted via a tendon on the fused trochantofemoral segment. This serves to draw the leg closer to the body, compressing a pad of resilin and causing the pleural and coxal walls to bend. At a certain point of contraction, the thoracic catches (pegs of cuticle) slip into notches on the sternum, thereby "cocking the system." The jump is initiated when other laterally inserted muscles contract to pull the catches out of the notches, thus allowing the stored energy to be rapidly released (Rothschild *et al.*, 1972).

Once airborne, an insect may or may not stabilize itself in preparation for landing or flight. For example, the flea beetle *Chalcoides aurata* sometimes jumps "out of control," its body rotating continuously till it hits the ground. However, it can, when necessary, control its jump by extending the wings so that a "feet-first" landing occurs (Brackenbury and Wang, 1995). Larger species such as Orthoptera use their hindlegs as rudders during the jump, facilitating an upright landing or takeoff for flight (Burrows and Morris, 2003).

3.1.3. Crawling and Burrowing

Many endopterygote larvae employ the thoracic legs for locomotion in the manner typical of adult insects; that is, they step with the legs in a specified sequence, the legs of a given segment alternating with each other. Usually, however, changes in body shape, achieved by synchronized contraction/relaxation of specific body muscles, are used for locomotion in soft-bodied larvae. In this method the legs, together with various accessory locomotory appendages, for example, the abdominal prolegs of caterpillars, are used solely as friction points between the body and substrate. Apodous larvae depend solely on peristalsis of the body wall for locomotion.

Where changes in body shape are used for locomotion the body fluids act as a hydrostatic skeleton. In other words, the insect employs the principle of incompressibility of liquids, so that contraction of muscles in one part of the body, leading to a decrease in volume, will require a relaxation of muscles and a concomitant increase in volume in another region of the body. Special muscles keep the body turgid, enabling the locomotor muscles to effect these volume changes.

Crawling in lepidopteran caterpillars is probably the best studied method of locomotion in endopterygote larvae and comprises anteriorly directed waves of contraction of the longitudinal muscles, each wave causing the body to be pushed upward and forward (Hughes and Mill, 1974; Brackenbury, 1999). Three main phases can be recognized in each wave of contraction (Figure 14.7). First, contraction of the dorsal longitudinal muscles and transverse muscles causes a segment to shorten dorsally and its posterior end to be raised so that the segment behind is lifted from the substrate. The dorsoventral muscles and leg retractor muscles then contract, lifting both feet of the segment from the substrate. Finally, contraction of the ventral longitudinal muscles, combined with the relaxation of the dorsoventral and leg retractor muscles, moves the segment forward and down to the substrate. Compared to walking, crawling and burrowing are relatively slow means of forward progression, with speeds of about 1 cm/sec for typical caterpillars. To take evasive action, for example, from a predator, caterpillars may walk backward by simply reversing the direction of peristalsis. Under extreme provocation, the caterpillar may coil up into a wheel and simply roll backward, achieving speeds up to 40 times greater than normal walking (Brackenbury, 1999). Little work has been done on the neural coordination of crawling, though it seems probable that endogenous activity within the central nervous system is responsible. However, proprioceptive stimuli undoubtedly influence the process.

Crawling or burrowing in apodous larvae is comparable to peristaltic locomotory movements found in other invertebrates, for example, mollusks and annelids. Larvae that crawl over the surface of the substrate grip the substrate with, for example, protrusible prolegs or creeping welts (transversely arranged thickenings equipped with stiff hairs) situated at

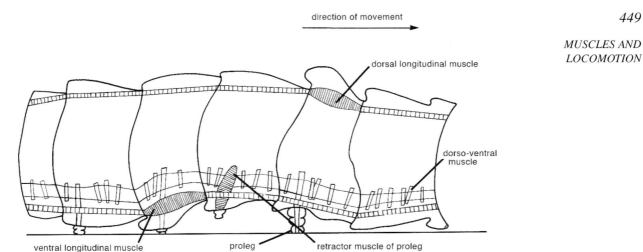

FIGURE 14.7. Phases in the passage of a peristaltic wave along the body of a caterpillar. [After G. M. Hughes, 1965, Locomotion: Terrestrial, in: *The Physiology of Insecta*, 1st ed., Vol. II (M. Rockstein, ed.). By permission of Academic Press, Inc., and the author.]

the posterior end of the body. A peristaltic wave of contraction then moves anteriorly, lengthening and narrowing the body. The anterior end is attached to the substrate while the posterior is released and pulled forward as the anterior longitudinal muscles contract. In many burrowing forms peristalsis proceeds in the opposite direction to movement, so that the narrowing and elongation begins at the anterior end and runs posteriorly. As the anterior end relaxes behind the peristlatic wave, it expands. This expansion serves both to anchor the anterior end and to enlarge the diameter of the burrow (Hughes and Mill, 1974; Berrigan and Pepin, 1995).

3.2. Movement on or Through Water

Progression on or through water presents very different problems to movement on a solid substrate. For small organisms, such as insects that live on the water surface, surface tension is a hindrance in production of propulsive leg movements. For submerged insects, the liquid medium offers considerable resistance to movement, especially for actively swimming forms.

Insects that move slowly over the surface of the water, for example, *Hydrometra* (Hemiptera), or crawl along the bottom, for example, larval Odonata and Trichoptera, normally employ the hexapodal gait described above for terrestrial species. More rapidly moving species typically operate the legs in a rowing motion; that is, both legs of the segment move synchronously. Some species do not use legs but have evolved special mechanisms to facilitate rapid locomotion.

3.2.1. Surface Running

The ability to move rapidly over the surface of water has been developed by most Gerroidea (Hemiptera), whose common names include pondskaters and waterstriders. To stay on the surface, that is, to avoid becoming waterlogged, these insects have developed various waterproofing features, especially hydrophobic (waxy) secretions, on the distal parts of the legs. However, these features considerably reduce the frictional force between the legs and water surface which is necessary for locomotion. This problem is overcome in many species by having certain parts of the tarsus, particularly the claws, penetrate the surface film and/or by having special structures, for example, an expandable fan that opens when the leg is pushed backward. In the Gerridae, however, the backward push of the legs is sufficiently strong that a wave of water is produced that acts as a "starting block" against which the tarsi can push (Nachtigall, 1974).

The functional morphology and mechanics of movement have been examined in detail in *Gerris* (Brinkhurst, 1959; Darnhofer-Demar, 1969). This insect has greatly elongated middle legs through which most of the power for movement is supplied. Some power is derived from the hindlegs, though these function primarily as direction stabilizers. The articulation of the coxa with the pleuron is such that the power derived from contraction of the large trochanteral retractor muscles is used exclusively to move the legs in the horizontal plane, that is, to effect the rowing motion. Equally, the coxal muscles serve only to lift the legs from the water surface during protraction. At the beginning of a stroke, the forelegs are lifted off the surface. The middle legs are rapidly retracted so that a wave of water forms behind the tarsi. As the legs are accelerated backward, the tarsi then push against this wave, causing the insect to move forward. After each acceleration stroke, the insect glides over the surface for distances up to 15 cm. The power developed in each leg of a segment is identical and the insect glides, therefore, in a straight line. Turning can occur only between strokes and is achieved by the independent backward or forward movement of the middle legs over the water surface.

Waterstriders of the genus *Velia* and the staphylinid beetle *Stenus* use an ingenious means of skimming across the water surface. They release a surface tension-reducing secretion behind themselves and are thus pulled rapidly forward, reaching speeds of 45–70 cm/sec *(Stenus)*.

3.2.2. Swimming by Means of Legs

Both larval and adult aquatic Coleoptera (Dytiscidae, Hydrophilidae, Gyrinidae, Haliplidae) and Hemiptera (Corixidae, Belostomatidae, Nepidae, Notonectidae) swim by means of their legs. Normally, only the hindlegs or the mid- and hindlegs are used and these are variously modified so that their surface area can be increased during the propulsive stroke and reduced when the limbs are moving anteriorly. Modifications include (1) an increase in the relative length and a flattening of the tarsus; (2) arrangement of the leg articulation, so that during the active stroke the flattened surface is presented perpendicularly to the direction of the movement, whereas during recovery the limb is pulled with the flattened surface parallel to the direction of movement; the leg is also flexed and drawn back close to the body during recovery; (3) development of articulated hairs on the tarsus and tibia that spread perpendicularly to the direction of movement during the power stroke, yet lie flat against the leg during recovery; such hairs may increase the effective area by up to five times; and (4) in *Gyrinus*, development of swimming blades on the tibia and tarsus (Figure 3.24A). These are articulated plates that normally lie flattened against each other. During the power stroke, the water resistance causes them to rotate so that their edges overlap and their flattened surface is perpendicular to the direction of movement of the leg.

In addition to the surface area presented, the speed at which the leg moves is proportional to the force developed. Thus, it is important for the propulsive stroke to be rapid, whereas the recovery stroke is relatively slow. Accordingly, the retractor muscles are well developed compared with the protractor muscles.

In the best swimmers other important structural changes can be seen, such as streamlining of the body and restriction of movement and/or change in position of the coxa. In adult *Dytiscus*, for example, which may reach speeds of 100 cm/sec when pursuing prey, the coxa is inserted more posteriorly than in terrestrial beetles and is fused to the thorax. Thus, the fulcrum for the rowing action is the dicondylic coxotrochanteral joint which operates like a hinge so that the leg moves only in one plane. Because of this arrangement, all of the muscle power can be used to effect motion in this plane (Ribera *et al.*, 1997).

Several variations are found in the rhythms of leg movements. Where a single pair of legs is used in swimming, both legs retract together. When both the midlegs and hindlegs are used, both members of the same body segment usually move simultaneously, but are in opposite phase with the legs of the other segment; that is, when one pair is being retracted, the other pair is being protracted. In adult Haliplidae and Hydrophilidae and many larval beetles all three pairs of legs are used, in a manner comparable with the tripodal gait of terrestrial insects.

Steering in the horizontal plane (control of yawing) is achieved by varying the power exerted by the legs on each side. For vertical steering (movement up or down) the non-propulsive legs become involved. These may be held out from the body in the manner of a rudder or may act as weakly beating oars. By varying the angle to the body at which the legs are placed the insect will either dive, surface, or move horizontally through the water. Most aquatic insects are quite stable in the rolling and pitching planes because of their dorsoventrally flattened body. (See Figure 14.14 for explanation of the terms yawing, pitching, and rolling.)

3.2.3. Swimming by Other Means

A variety of other methods for moving through water can be found in insects, including body curling and somersaulting found in many larval and pupal Diptera, body undulation (larval Ephemeroptera and Zygoptera), jet propulsion (larval Anisoptera), and flying (a few adult Lepidoptera and Hymenoptera) (Nachtigall, 1974).

Many midge and mosquito larvae rapidly coil the body sideways, first in one direction, then the other, to achieve a relatively inefficient form of locomotion. Chironomids, for example, lose 92% of the energy expended in the power stroke during recovery. Consequently, a 5.5-mm larva oscillating its body 10 times per second moves at only 1.7 mm/sec through the water. Mosquito larvae possess flattened groups of hairs (swimming fans) or solid "paddles" at the tip of the abdomen and are consequently more efficient and more active swimmers than chironomids.

Pupae of midges and mosquitoes somersault through the water, especially when attempting to escape predators. Interestingly, the surface-dwelling pupa of the mosquito *Culex pipiens* somersaults at a greater frequency (hence swims faster) than the bottom-dwelling pupa of the midge *Chironomus plumosus*, perhaps because there is greater predator pressure for a surface-dweller (Brackenbury, 2000).

Larvae of Zygoptera and some Ephemeroptera undulate the abdomen, which is equipped at its tip with three flattened lamellae (Zygoptera, Figure 6.11) or swimming fans (Ephemeroptera, Figure 6.4). Some ephemeropteran larvae supplement the action of the fans by rapidly folding their abdominal gills against the body.

Dragonfly larvae (Anisoptera) normally take in and expel water from the rectum during gas exchange (Chapter 15, Section 4.1). In emergencies this arrangement can be converted into a jet propulsion system for moving an insect forward at high speed (up to 50 cm/sec). Rapid contraction of longitudinal muscles causes the abdomen to shorten by up to 10%. Simultaneous contraction of the dorsoventral muscles leads to an increase in hemolymph pressure which forces water out of the rectum via the narrow anus at speeds approaching 250 cm/sec.

Female *Hydrocampa nympheata* (Lepidoptera) and adult *Dacunsa* (Hymenoptera) use their wings in addition to legs for swimming underwater. Other Hymenoptera (*Polynema* and *Limnodites*) swim solely by the use of their wings.

3.3. Flight

As noted in Chapter 2, Section 3.1, wing precursors originally had functions quite unrelated to flight. Subsequent evolution led to enlargement and perhaps articulation of these structures as they took on a new function, propulsion of an insect through the air, partly as a result of which insects were able to move into new environments to become the diverse group we know today. Despite this diversity, there is sufficient similarity of skeletal and neuromuscular structure and function to suggest that wings had a monophyletic origin (Pringle, 1974).

Examination of the form and mode of operation of the pterothorax reveals certain trends, all of which lead to an improvement in flying ability. Primitively, the power for wing movement was derived from various "direct" muscles, that is, those directly connected with the wing articulations. These muscles serve also to determine the nature of the wing beat. Even today, the direct muscles remain important power suppliers in the Odonata, Orthoptera, Dictyoptera (Blattodea), and Coleoptera. In other insect groups, efficiency is increased by separating the control of wing beat (by the direct muscles) from power production, which becomes the job of large "indirect" muscles located in the thorax.

There are important differences in the fine structure and neuromuscular physiology of the direct and indirect flight muscles. Generally, in insects that flap their wings relatively slowly (up to 100 beats/sec), each beat of the wings is initiated by a burst of impulses to the power-producing muscles, which are of the tubular or close-packed type (Figure 14.3B, C). This applies to all users of direct muscles for powering flight, plus Lepidoptera in which the indirect muscles are used. In contrast, in fliers that use indirect muscles and whose wing-beat frequency is high (up to 1000 beats/sec), muscle contraction is not in synchrony with the arrival of nerve impulses at the neuromuscular junction. Rather, the rhythm of contraction is generated within the muscles themselves, which are fibrillar (Figure 14.3D). Accordingly, the two forms of rhythm are described as synchronous (neurogenic) and asynchronous (myogenic), respectively. The use of asynchronous muscles to power flight has evolved several times within the Insecta. Its significance appears to be the facilitation of high wing-beat frequencies, thereby moving more air, so that even insects with small wings relative to their body size are efficient fliers.

3.3.1. Structural Basis

Each wing-bearing segment is essentially an elastic box whose shape can be changed by contractions of the muscles within, the changes in shape causing the wings to move up and down. The skeletal components of a generalized wing-bearing segment are shown in Figure 3.18. The essential features are as follows. Each segment contains two large intersegmental invaginations, the prephragma and postphragma, between which the dorsal longitudinal muscles stretch. The alinotum bears on each side an anterior and a posterior notal process, to which the wing is attached via the first and third axillary sclerites. The pleuron is largely sclerotized and articulates with the wing by means of the pleural wing process, above which sits the second axillary sclerite. The hinge so formed is important in wing movement because of the resilin that it contains (but see Section 3.3.3). Two other important articulating sclerites, which are usually quite separate from the sclerotized portion of the pleuron, are the anterior basalar and posterior subalar. Internally, the pleuron and sternum are thickened, forming the pleural and sternal apophyses, respectively, which brace the pterothorax. In some insects these apophyses are fused, but generally they are joined by a short but powerful pleurosternal muscle (Figure 14.8B).

The muscles used in flight may be separated into three categories according to their anatomical arrangement (Figure 14.8A, B). The indirect flight muscles include the dorsal longitudinal muscles, dorsoventral muscles, oblique dorsal muscles, and oblique intersegmental muscles. The direct muscles are the basalar and subalar muscles, and the axillary muscles attached to the axillary sclerites (including the wing flexor muscle, which runs from the pleuron to the third axillary sclerite). In the third category are the accessory indirect muscles that comprise the pleurosternal, tergopleural, and intersegmental muscles. Their function is to brace the pterothorax or to change the position of its components relative to each other. In addition, certain extrinsic leg muscles may also be important in wing movements. For example, in Coleoptera, whose coxae are fused to the thorax, the coxotergal muscles can assist the dorsoventral muscles in supplying power to raise the wings. In other species with articulated coxae the upper point of insertion of the extrinsic coxal muscles may change to the basalar or subalar. Thus, the muscles can alter both leg and wing positions, that is, they may have a dual locomotory function. For example, in *Schistocerca gregaria* the anterior and posterior tergocoxal (indirect) muscles act synergistically during flight (1) to provide power for the upstroke of the wings and (2) to draw the legs up close to the body. This is achieved through polyneuronal innervation, whereby the parts of the tergocoxal muscles that receive slow and inhibitory motor axons are responsible for the drawing up of the legs, and the muscle fibers innervated by the fast axon move the wings. In contrast, when the insect is running, the anterior and posterior tergocoxal muscles function antagonistically, effecting promotion and remotion, respectively, of the legs. In the same species, the (direct) second basalar and subalar muscles, while acting synergistically to aid the indirect muscles in the production of power for the downstroke of the wings, act antagonistically to bring about wing twisting (pronation and supination, respectively) or, when running, promotion and remotion, respectively, of the legs (Figure 14.9) (Wilson, 1962).

3.3.2. Aerodynamic Considerations

Flight occurs when the air pressure is greater on the lower than on the upper surface of a body. The wings, with their large surface area, act as aerofoils, that is, the portion of the body that is largely responsible for lift (the component of net aerodynamic force that acts perpendicular to the direction of movement). In a fixed-wing aircraft flying horizontally, lift on the wings is vertical (equal and opposite to the aircraft's weight). In insects, however, where the wings flap, twist, and deform cyclically, lift on the wings varies through the cycle.

Lift develops when air is accelerated unequally over the upper and lower surfaces of a wing. Aerofoils on fixed-wing aircraft are designed with their upper surface curved,

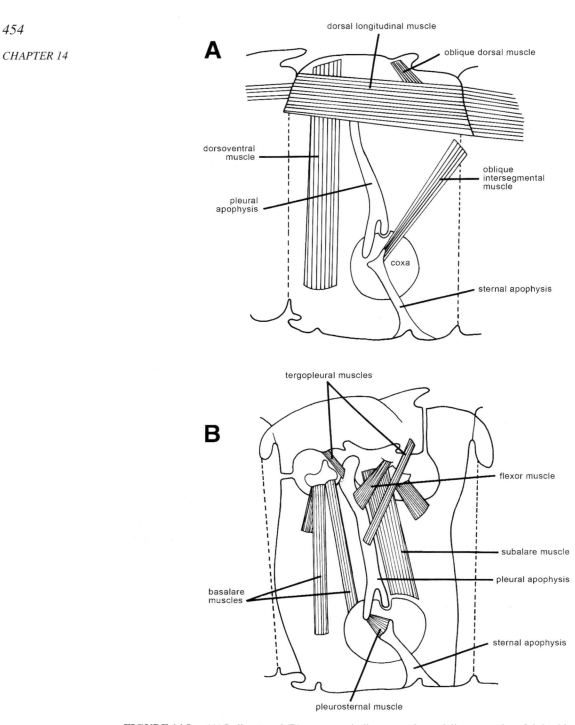

FIGURE 14.8. (A) Indirect; and (B) accessory indirect muscles and direct muscles of right side of wing-bearing segment, seen from within. [After J. W. S. Pringle, 1957, *Insect Flight*. By permission of Cambridge University Press, London.]

their lower surface flat (Figure 14.10A), so as to make use of Bernoulli's principle that the pressure exerted by flowing air is inversely related to the square of the velocity. Air flowing over the upper surface of an aerofoil travels farther and therefore has a greater velocity than the air flowing beneath. Hence, the pressure beneath the aerofoil is greater than that above

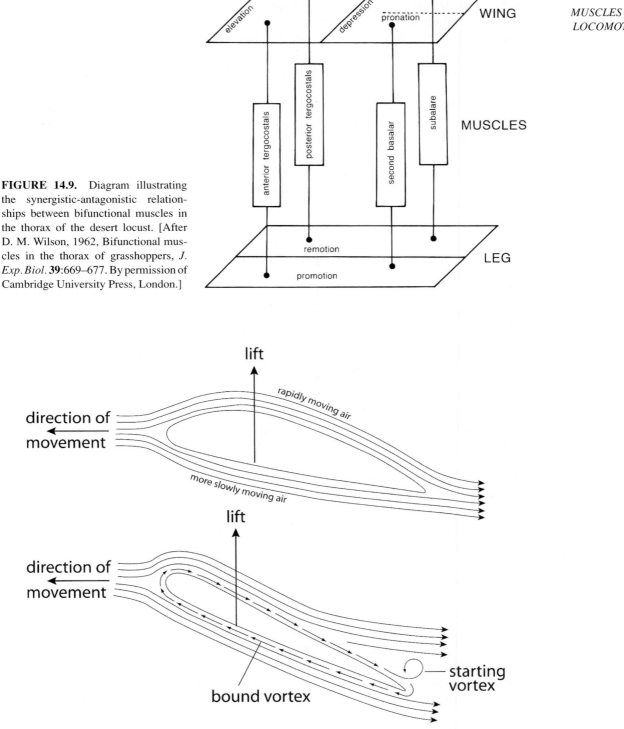

FIGURE 14.9. Diagram illustrating the synergistic-antagonistic relationships between bifunctional muscles in the thorax of the desert locust. [After D. M. Wilson, 1962, Bifunctional muscles in the thorax of grasshoppers, *J. Exp. Biol.* **39**:669–677. By permission of Cambridge University Press, London.]

FIGURE 14.10. Generation of lift as a result of different air speeds above and below an aerofoil. (A) Fixed-wing aircraft whose wing has curved upper and flat lower surface; and (B) insect wing which is of more uniform thickness. At stroke initiation, a starting vortex is developed at the wing's trailing edge, which then produces a bound vortex circulating clockwise round the wing.

and, provided that the force generated is greater than the gravitational force as a result of the body's mass, the body will be raised into the air.

In aircraft the fixed wings supply only lift, and horizontal propulsive force (thrust) is supplied by engines. In insects, wings are movable and supply thrust as well as lift. Further, insect wings are of uniform thickness; that is, they do not have curved upper and flat lower surfaces. Thus, to develop both lift and thrust during their stroke, wings both move up and down and change their angle of attack.

When an insect wing begins to accelerate through air, a starting vortex (circulation of air, as in a whirlwind) is developed at the wing's trailing edge (Figure 14.10B). This vortex, in turn, creates a second vortex (bound vortex) that moves clockwise round the wing, backward over the upper surface, and forward over the lower surface. This causes the air flow to speed up on the wing's upper surface, whereas on the lower surface air flow is slowed. These differences in air speed generate lift.

The relative values of lift and thrust will change throughout the wing beat (Figure 14.11). In the middle of a downstroke, the rapid downward movement of the wing operating in conjunction with the already moving horizontal stream of air over the wings (assuming the insect is in forward flight) will result in a positive angle of attack (i.e., the air will strike the underside of the wing) and give rise to a strongly positive lift. Concurrently, because the wing is pronated (its leading edge is pulled down), there will be a slightly positive thrust (Figure 14.11B). During an upstroke, the angle of attack becomes slightly negative (pressure of air above is greater than pressure of air below the wing) causing slightly negative lift, yet increasing the positive thrust (Figure 14.11C). The angle at which an insect holds its body in flight also results in positive lift, though this amounts to less than 5% of the total lift in the desert locust and about 20% in some Diptera.

For many years it was assumed that the lift generated by the wing movements described above was sufficient for insect flight. However, calculations showed that conventional (steady-state) aerodynamic theory, which is based on rigid wings moving at constant velocity, cannot account for the production of lift in most insects. Ellington's and Dickinson's groups (see Ellington, 1995, 1999; Ellington et al., 1996; Dickinson et al., 1999; Dickinson and Dudley, 2003; Sane, 2003) studied the aerodynamic performance of both the wings of tethered insects (bumble bees and hawk moths) and mechanically powered model wings. As a result, it is now considered that lift in most flying insects is produced by three interactive mechanisms: delayed (dynamic) stall, rapid wing rotation, and wake capture (Figure 14.12). Delayed stall operates during the main "translational" part of a stroke (both up and down), whereas the other two mechanisms occur during stroke reversal, that is, when the wings twist and reverse direction (Dickinson et al., 1999).

Delayed stall refers to the condition in which for a brief period wings can be held with a high angle of attack, thus producing a leading-edge vortex. The latter creates a region of low pressure above the wings, thus augmenting lift. The vortex appears at the beginning of the downstroke and forms a conical spiral, getting larger as it moves toward the wing tip, generating lift equivalent to about 1.5 times the weight of the hawk moth (Ellington et al., 1996). The tipward flow of the vortex also increases stability. Toward the end of a stroke, the vortex is shed as the wings twist and reverse direction.

Dickinson et al. (1999), while confirming the importance of delayed stall, showed that rapid wing rotation and wake capture also contribute to the generation of lift using their large-scale model of the Drosophila wing. Rapid circulation of air in the boundary layer is induced around a spinning object, for example, a ball. If the ball is thrown with spin,

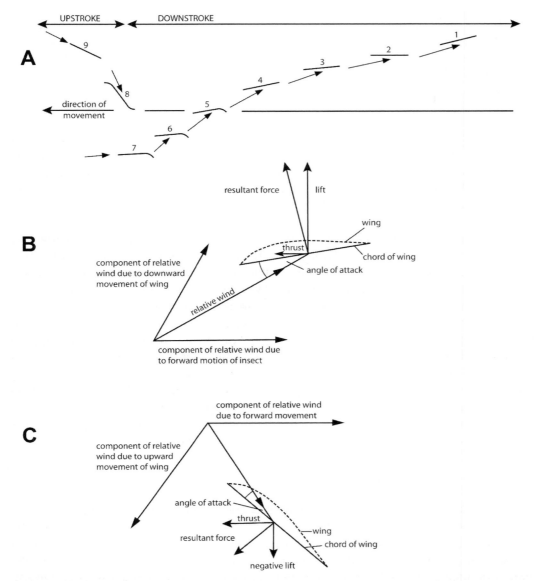

FIGURE 14.11. (A) Changes in the angle at which a wing is held in flight relative to direction of movement. Arrows indicate the angle at which air strikes the wing. Numbers indicate chronological sequence of wing positions during a stroke; and (B,C) Magnitude of lift and thrust approximately midway through downstroke and upstroke, respectively. [A, after M. Jensen, 1956, Biology and physics of locust flight. III. The aerodynamics of locust flight, *Philos. Trans. R. Soc. Lond. Ser. B* **239**:511–552. By permission of The Royal Society, London, and the author. B, C, after R. F. Chapman, 1971, *The Insects: Structure and Function*. By permission of Elsevier/North-Holland, Inc., and the author.]

air flows unequally over the top and bottom of the ball causing it to swerve. Dickinson *et al.* (1999) suggested that the transient increase in lift which occurs as a wing rotates just before the end of a stroke has a similar explanation. The key to rotation-generated lift is the timing of wing rotation. If rotation precedes stroke reversal, which is analogous to imparting backspin to the ball, positive lift is generated; if it follows stroke reversal (akin to topspin), the lift is negative. Rotation-generated lift is probably of particular importance in the fine control of flight maneuvres. The remaining contribution is derived from wake

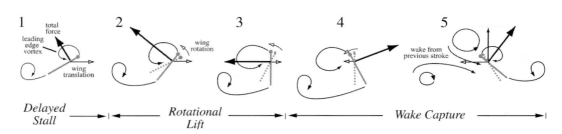

1 total force
leading edge vortex
wing translation
2 wing rotation
3
4
5 wake from previous stroke

Delayed Stall ⟶| |⟵ Rotational Lift ⟶| |⟵ Wake Capture ⟶|

FIGURE 14.12. Air flow around the wing and the resulting forces at points during a wing stroke. Delayed stall (1) results from formation of a leading edge vortex on the wing. Rotation-generated lift (2,3) occurs when the wing rapidly rotates at the end of the stroke. Wake capture (4,5) results from collision of the wing with the wake shed during the previous stroke. [Reprinted from *Encyclopedia of Insects*, M. Dickinson and R. Dudley, Flight, pages 416–426, ©2003, with permission from Elsevier.]

capture; that is, after reversing direction the wing does not move through undisturbed air but re-encounters the vortices shed from the wing tips during the previous stroke to produce a pulse of lift. Collectively, delayed stall, rotational circulation and wake capture enable insects to generate lift equivalent to several times their body weight.

A few, especially very small, insects such as thrips, whitefly, and parasitoid wasps (whose wingspans are about 1 mm or less) use a particular form of rotation-generated lift known as the "clap and fling" mechanism (Weis-Fogh, 1973) (Figure 14.13). In this system lift is generated by rotation of the wings as they separate after they have clapped together at the end of the upstroke and are flung apart as the downstroke begins. As the wings rotate, a starting vortex is formed above each wing, which serves to generate a bound vortex around each wing. As outlined above for conventionally flapping wings, the vortex causes the velocity of air passing over the top of each wing to increase, while that passing beneath the wing decreases. Thus, by Bernoulli's principle, lift is generated.

3.3.3. Mechanics of Wing Movements

All insects use the indirect tergosternal or tergocoxal muscles to raise the wing. Contraction of these muscles pulls the tergum down so that its points of articulation with the wing fall below the articulation of the wing with the pleural wing process which serves as a fulcrum (Figure 14.14A). In most insects indirect muscles are also used to lower the wing. Shortening of the dorsal longitudinal muscles causes the tergum to bow upward, raising the anterior and posterior notal processes above the tip of the pleural wing process

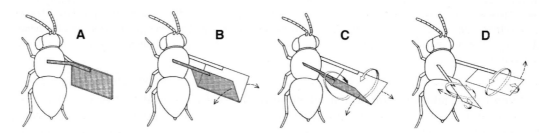

FIGURE 14.13. Clap and fling mechanism for generating lift. The wings clap together at the end of the upstroke (A), then are flung apart as the downstroke begins (B,C), creating a bound vortex (D) (greater air speed over the upper wing surface compared to the lower wing surface, thereby creating lift). [From T. Weis-Fogh, 1975, Unusual mechanisms for the generation of lift, *Sci. Amer.* **233** (November):80–87. Original drawn by Tom Prentiss. By permission of Nelson H. Prentiss.]

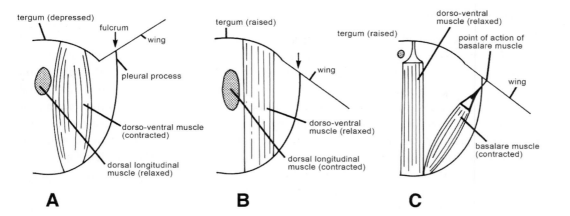

FIGURE 14.14. Diagrammatic transverse sections of thorax to show muscles used in upstroke and downstroke. (A) Use of indirect muscles to raise wing; (B) use of indirect muscles to lower wing; and (C) use of direct muscles to lower wing. [After R. F. Chapman, 1971, *The Insects: Structure and Function.* By permission of Elsevier/North-Holland, Inc., and the author.]

and, therefore, the wing to be lowered (Figure 14.14B). In Coleoptera and Orthoptera some power for a downstroke is also obtained by contraction of the basalar and subalar muscles and in Odonata and Blattodea, this power is derived entirely from contraction of these direct muscles. The points of articulation of these muscles with the wing sclerites lie outside the pleural wing process so that when the muscles contract the wing is lowered (Figure 14.14C). In all insects, however, the basalar and subalar muscles are important in wing twisting, that is, altering the angle at which the wing meets the air, thereby affecting the values of the lift and thrust generated. Contraction of the basalar muscles causes the anterior edge of the wing to be pulled down (pronation), whereas contraction of the subalar induces supination (the pulling down of the posterior edge of the wing). Contraction of the other direct muscle, the wing flexor, causes the third axillary sclerite to twist and to be pulled inward and dorsally. This pulls the vannal area of the wing (Figure 3.27) up over the body and enables the wing to fold along predetermined lines (usually the anal veins, and vannal and jugal folds). Unfolding (extension) of the wing occurs when the basalar muscle contracts.

Early descriptions of the role of muscles in wing flapping envisioned the muscles as supplying power directly to the wings to effect the wing beat. Thus, the speed with which muscles could contract would determine the speed at which a wing was lowered or raised. In turn, this determined the value of the lift generated. Not until the 1950s was it realized that a critical feature is the ability to store a large proportion of the energy released at the end of each stroke, when a wing's momentum is rapidly reduced. The momentum is reduced by (and the energy stored in) elastic structures. Then, in the following stroke the elastic energy is used to power wing movement, making the process remarkably efficient. For example, because the upstroke is aided by the pressure of the onrushing air on the underside of the wing, only a small amount of the stored elastic energy is used to raise the wing. The remainder (about 86% in the desert locust) is then used to power the following downstroke (Pringle, 1974). In insects that use synchronous muscles to generate power, the principal sites for storage of elastic energy are the lateral walls of the pterothorax and the resilin-containing hinge between the pleural wing process and second axillary sclerite. However, in insects that have fibrillar muscles with their greater elasticity (i.e., use asynchronous control of muscle contraction) the energy is mainly stored in the muscles themselves.

3.3.4. Control of Wing Movements

Like leg movements, wing flapping is a centrally generated, rhythmically repeated process; that is, each wing has its own muscles and motor neurons whose activity is initiated by specific interneurons in each segmental ganglion. In *Locusta migratoria* Robertson and Pearson (1982) identified 25 such interneurons whose spike activity or membrane potential changed rhythmically during fictive flight.* Flight is not, of course, simply the flapping of the wings. It is a complex process and at any point in time an insect is concerned with monitoring, and varying if necessary, many parameters. During flight, an insect must be able to control the frequency of wing beat, the amount of lift and thrust developed, and the direction (stability) of flight. It must also have control mechanisms for the initiation and termination of flight. The sense organs that provide the central nervous system with information on what changes are occurring in relation to flight are the compound eyes and proprioceptors strategically distributed over the body, especially on the head, wings, and legs. Responses to stimuli received by these organs are usually mediated via changes in the nature of the wing movements (particularly the degree of twisting and wing-beat frequency).

Wing-beat frequency varies widely among different insects as has been noted already in the introduction to this discussion of flight. In fliers whose wing-beat frequency is low (e.g., the desert locust, about 15–20 beats/sec) neurogenic (synchronous) control occurs. That is, there is a 1:1 ratio between wing-beat frequency and nervous input. In such fibers, therefore, wing-beat frequency can be varied by altering the rate at which nerve impulses arrive at the muscles. However, there are limits to this arrangement (maximum frequency about 100 beats/sec) because of the refractory period required to return the muscle to its resting state after each contraction. Insects whose wing-beat frequency is high, for example, the bee (190 beats/sec) and midge *Forcipomyia* (up to 1000 beats/sec), use a myogenic (asynchronous) system where the frequency of muscle contraction is much greater than that of nervous input (ratios of between 5:1 and 40:1). Such a system is possible only where there are antagonistic muscles that contract regularly and alternately. These muscles have the property of contracting autonomously when the tension developed in them as a result of stretching reaches a critical value. In the case of the flight musculature, the alternating contractions are, in a sense, self-perpetuating, though their initiation and cessation are under nervous control. There is also evidence to suggest that changes in the frequency at which nerve impulses arrive at these muscles can modify their frequency of contraction. However, as contractions are tension-dependent, their frequency can also be altered by modifying the elastic resistance in the exoskeleton. In Diptera, for example, this is achieved by contraction or relaxation of the pleurosternal muscles, which serves to move the pleural wing process closer to or farther from the tergal hinge (Figure 14.8).

As noted earlier, during a beat a wing does not make simple up and down movements, but rather twists about the vertical axis so that its tip describes an ellipse (*Schistocerca*) or figure eight (*Apis, Musca*). The size and direction of the twisting force (torque), which effectively measure the lift and thrust developed, are monitored by campaniform sensilla at the base of each wing (halteres in Diptera—see below). Input from the sensilla initiates reflex excitation of the basalar and subalar muscles, which regulate the extent to which the wing twists. This mechanism is known as the lift control reflex. *Schistocerca* employs such a reflex to control the value of lift and, in flight, holds its body at a fairly steady angle to the horizontal. Other insects vary the angle at which the body is held in flight in order to alter the lift and thrust components.

---

* Fictive flight: The locust preparation is actually a wingless, legless insect pinned in a dish! It is stimulated to "fly" by gently blowing on the frons.

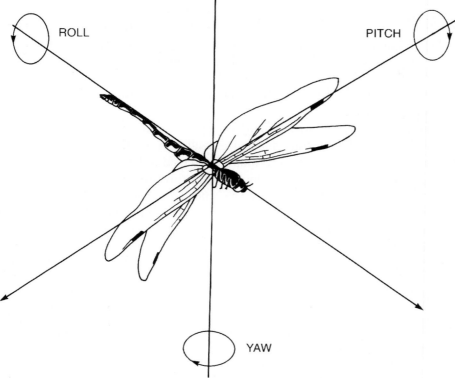

FIGURE 14.15. Diagram showing axes about which a flying insect may rotate.

The direction (and correlated with it, the stability) of flight necessitates the monitoring and regulation of movement in three dimensions (Figure 14.15): movement of the body in the horizontal, transverse axis is pitching; rotation about the vertical axis is yawing; and rotation around the longitudinal axis is rolling.

In *Schistocerca gregaria* pitching is controlled via the lift control reflex outlined above. In this species the fore and hind wings beat in antiphase. The hind wings are not capable, however, of twisting and, therefore, the value of lift generated by them is a function of the body angle (the angle at which the body is held relative to the direction of airflow). Any change in body angle, altering the lift generated by the hind wings, is compensated for by appropriate twisting of the fore wings so that the total lift produced remains constant. Other insects control pitching by varying the amplitude of the wing beat, the stroke plane (the angle at which the wings move relative to the body), or the stroke rhythm (by delaying or enhancing the moment at which pronation of the wings occurs during the beat).

The control of yawing involves sensory input from both the compound eyes and mechanoreceptors. As an insect turns, the visual field will rotate. Also, in some insects such as *Schistocerca*, this movement results in unequal stimulation of mechanosensory hairs at each side of the head. Yawing may be corrected by varying the angle of attack, the amplitude, or the frequency of beat, between the wings on each side of the body. All of these variables will alter the thrust generated.

Rolling is largely monitored as a result of certain visual stimuli. Commonly, the head is rotated so that the dorsal ommatidia receive maximal illumination, the source of which is normally directly overhead. Early and late in the day, however, when the latter is not the case, use is also made of the marked contrast in light intensity between the earth and sky at the horizon. Goodman's (1965) study of the importance of the horizon in orientation of the desert locust showed that the insect rotates its head so as to maintain the horizon horizontally across the visual field of the compound eyes. Her study also demonstrated that the dorsal light reaction overrides the "horizon response" at higher light intensity, and vice versa. Alignment of the thorax and abdomen with the head is achieved by means of proprioceptive hairs on each side of the neck. When the thorax is out of line with the head, the hairs on each side are differentially stimulated. This information induces appropriate wing twisting to compensate.

Wing movements are initiated in many insects by means of the "tarsal reflex," that is, loss of contact between the substrate and tarsi, as well as by a variety of non-specific stimuli. It seems probable that sensory input to inhibit the tarsal reflex is fed in via campaniform sensilla on the femur and trochanter which are sensitive to stresses in the cuticle in the standing insect (Chapter 12, Section 2.2). The reflex stimulates unfolding of the wings (in some species this involves the same muscles as are used in flapping the wings), stiffening of the pleural wall (contraction of the pleurosternal muscles), and raising the legs to the flying position. The tarsal reflex is not found in insects that capture prey during flight, in insects that vibrate their wings during "warm-up" prior to takeoff, in bees that employ wing movements in hive ventilation, or in beetles that must raise the elytra and unfold the wings prior to flight.

Some insects, for example, *Drosophila*, will fly until exhausted, provided their tarsi no longer make contact with the substrate. For most species, however, additional stimuli are necessary to maintain wing movements, especially airflow, which stimulates hairs on the upper part of the head, Johnston's organs in the antennae, and proprioceptors at the base of the wings. In the honey bee, flight is maintained as a result of visual stimuli received as the insect moves forward over the ground.

Visual signals experienced by the compound eyes just prior to landing stimulate leg extension. Goodman (1960), working on the fly *Lucilia sericata*, concluded that the response was based on three variables acting singly or in combination: (1) decrease in light intensity experienced by successive ommatidia as the fly approaches the substrate; (2) number of ommatidia stimulated; and (3) rate of successive stimulation of ommatidia (based on the increase in angular velocity which occurs as the fly comes closer to the substrate).

Halteres. Deserving special mention are the much modified hind wings (halteres) of Diptera (Figure 14.16), which function as gyroscopic organs of balance. The halteres vibrate in the vertical plane only, at the same frequency as the wings with which they are in antiphase. These vertical vibrations, combined with the movement of the insect through the air, cause the development of torque at the base of the halteres, which is monitored by various groups of campaniform sensilla oriented in different planes. Any change in the orientation of the insect will result in modification of the torques generated at the base of each haltere. Yawing can be detected by differential stimulation of the groups of sensilla in each haltere, whereas the detection of pitching and rolling requires the combined input of both halteres, which is assessed in the thoracic ganglion. To correct for instability in any of the planes of motion, the insect alters the extent to which each wing is twisted during beating. For a detailed discussion of the gyroscopic nature and function of halteres, see Pringle (1957) and Nalbach (1993).

3.3.5. Flight Metabolism

Insect flight is an energetically costly process, in which large quantities of substrate are oxidized in the generation of ATP. The insect body is eminently adapted to facilitate

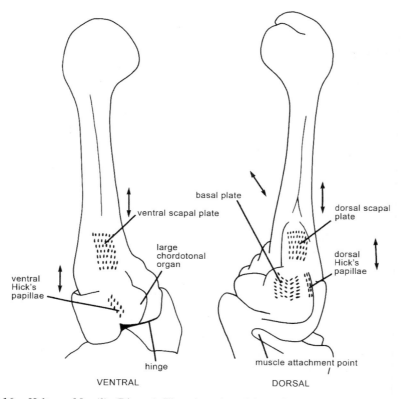

FIGURE 14.16. Haltere of *Lucilia* (Diptera). The orientation of the various groups of campaniform sensilla is indicated by double-ended arrows. [After J. W. S. Pringle, 1948, The gyroscopic mechanism of the halteres of Diptera, *Philos. Trans. R. Soc. Lond. Ser. B* **233**:347–384. By permission of The Royal Society, London, and the author.]

the production of this energy both physiologically and biochemically. The tracheal system (Chapter 16) is able to provide the large volumes of oxygen required—often more than 20 times the amount used by a resting insect. The circulatory system (Chapter 17) may store large amounts of substrate precursor molecules and have transport proteins, especially for lipids, that take energy-rich compounds to the flight muscles. And, as outlined in Section 2.1, the flight muscles are superbly equipped to deal with these materials, with many large mitochondria (occupying up to 45% of flight-muscle volume) closely apposed to the contractile elements.

Some broad ecological and behavioral correlations with the type of fuel used in flight have been suggested. Thus, insects with short-duration flights, high wing-beat frequencies, and asynchronous flight muscle, for example Diptera and Hymenoptera, use carbohydrate. Locusts, many Lepidoptera, Hemiptera, and Odonata, on the other hand, have lower-frequency wing beats, neurogenic flight muscles, and may fly for extended periods, including long-distance migrations. These insects tend to use only lipids or initially use carbohydrate, then switch to using mainly lipids within a few minutes. However, there are compounding factors (e.g., frequency of feeding and flying) that affect flight metabolism. Further, some insects use neither carbohydrate nor lipid as their flight substrate. For example, *Glossina morsitans*, *Leptinotarsa decemlineata*, and *Melolontha melolontha* use proline stored in high concentration in the hemolymph (Wheeler, 1989; Gäde and Auerswald, 1998). As hemolymph proline becomes depleted, more is formed from lipids stored in the fat body or from hemolymph alanine. Wheeler (1989) suggested that its advantages over the use of lipid substrates are its water solubility and the non-necessity of maintaining the metabolically expensive lipoprotein-carrier system.

Hormones regulate the supply of fuels to, and their use by, the flight muscles. Especially well studied is the control of flight metabolism in the migratory locust, *Locusta migratoria*, which uses carbohydrate for the first few minutes of flight but then converts to using mainly lipid. This early use of carbohydrate allows time for the hormonally controlled mobilization of lipid to gather pace (Goldsworthy, 1983; Wheeler, 1989). When flight begins in the locust, hemolymph trehalose is the substrate oxidized, and its concentration steadily decreases for about 10 minutes. During this period, the release of adipokinetic hormones (AKH) from the corpora cardiaca begins. Three forms of AKH are known, of which two seem important in flight metabolism. AKH II, which is less abundant and has a short half-life, appears to be more important at the start of flight, activating fat body glycogen phosphorylase that converts glycogen to trehalose. Soon, however, AKH III, which is 14 times more common than AKH II and has a half-life >10 times that of AKH II, becomes dominant, promoting the mobilization of the stored lipid in the fat body. Specifically, the hormone activates triacylglycerol lipase that catalyzes the conversion of stored triglycerides into diglycerides, which are released into the hemolymph for use by the flight muscles (Gäde and Auerswald, 1998; Van der Horst *et al.*, 1999).

Because of their water-insoluble nature, the lipids are transported in the hemolymph bound to proteins (lipophorins) so that high concentrations can be achieved. The proteins are also important in receptor binding (recognizing the site for uptake of the lipids) and regulating lipid usage by the muscles (Haunerland, 1997). In addition, AKH induces a switch in metabolism of the flight muscles, stimulating them to preferentially use lipid rather than carbohydrate substrate. To this end, the muscles contain high concentrations of lipases and other enzymes required for hydrolysis of lipid and degradation of free fatty acids to two-carbon fragments that can be oxidized by the Krebs cycle to carbon dioxide and water.

AKH is also responsible for the mobilization of fat body carbohydrate, as well as the conversion of hemolymph alanine to proline, which also occurs in the fat body, in species that use these substrates in flight (Gäde and Auerswald, 1998; Van der Horst *et al.*, 1999; Goldsworthy and Joyce, 2001). Other hormones, including cardioaccelerating peptides, as well as the multifunctional molecule octopamine, are also released during flight, though their roles are not always clear (Elia *et al.*, 1995; Candy *et al.*, 1997). Orchard *et al.* (1993) have proposed two roles for octopamine in flight metabolism. First, the release of octopamine within the first few minutes of flight, coincident with a small, transient increase in hemolymph lipid, suggests that octopamine acts hormonally to trigger the initial release of lipid from the fat body. *In vitro* studies have confirmed that locust fat body releases lipid when octopamine is applied. Second, anatomical and physiological evidence indicates that the release of AKH is regulated by octopaminergic neurons that terminate in the glandular lobes of the corpora cardiaca, leading to the surge in hemolymph lipid seen after about 25 minutes.

3.4. Orientation

The locomotor behavior that follows receipt of a stimulus by an insect may be described as a *kinesis* or a *taxis*. A kinesis is simply induced or enhanced activity without spatial reference to the source of the stimulus. Increased movement related to light (photokinesis) is common in many insects. For example, many butterflies fly only when the light intensity is above a certain threshold; the speed at which locusts walk is increased as the light intensity rises. A hygrokinetic response is shown by locusts whose activity is increased in moist air compared with dry. In contrast, wireworms (Elateridae) are inactive in air saturated with moisture yet become active at lower humidity. Insects that normally secrete themselves in cracks or crevices, for example, earwigs and bedbugs, exhibit stereokinesis, an inhibition of movement that occurs when a sufficient number of mechanoreceptors on the body surface are stimulated.

A taxis is orientation in response to a stimulus, the location of which can be perceived by an insect. To locate the stimulus, an insect uses either paired sense organs (compound eyes, antennae, and tympanal organs) or moves one sense organ from side to side, and compares the strength of the stimulus received by each organ or by the single organ in each position. In its simplest form, an insect's response is to move toward or away from the stimulus. For example, the bloodsucking insects *Rhodnius* and *Cimex* possess the ability to orient to a heat source (positive thermotaxis) that enables them to locate a host. Many insects exhibit chemotaxis in locating food, a mate, or a host. Locusts are attracted to a light source (positive phototaxis). A more complex form of phototaxis is where an insect orients itself and moves with its body at a constant angle to the light source. As the insect moves, therefore, it receives a constant visual stimulus. Such orientation is called menotaxis. The stimulus may be a simple point source of light, for example, the sun which forms the basis of the light-compass reaction of bees and other insects that navigate. Alternately, the stimulus may be a more complex visual pattern. Marching behavior in groups of juvenile locusts (hoppers) is based on menotaxis, involving both a simple and a complex visual stimulus. The simple stimulus, the sun, gives rise to a light-compass reaction so that the individual marches in a fixed direction. The more complex stimulus is the visual pattern created by the presence of adjacent hoppers. Thus, as one hopper moves, neighboring hoppers will also be stimulated to move in order to keep constant the visual pattern seen by each eye. This ensures that the group marches as a whole; that is, menotaxis is the basis of their gregarious behavior.

4. Summary

All insect muscles are striated, though their structural details vary according to their function. Skeletal muscles can be categorized as tubular (leg and body segment muscles, and wing muscles of Odonata and Dictyoptera), close-packed (wing muscles of Orthoptera and Lepidoptera), and fibrillar (flight muscles of most insects). All skeletal and many visceral muscles are innervated. Each muscle fiber may receive one to three functionally distinct axons, one of which is always excitatory. Excitatory axons may be subdivided into "fast" axons, whose impulses each initiate strong, rapid contractions, and "slow" axons, whose impulses individually cause a weak contraction but are additive in effect, facilitating a graded response in the muscle. Some fibers also receive an inhibitory axon. Most muscles operate in pairs, with one antagonistic to the other so that as one is stimulated to contract the other is passively stretched. Where there is no antagonistic partner, a muscle may be stretched because of the elasticity of the cuticle to which it is attached.

Most insects use six legs during walking (hexapodal gait). Others use only two pairs of legs (quadrupedal gait) but may use the tip of the abdomen as a point of support. On all occasions, however, there are at least three points of contact with the substrate, forming a triangle of support. Stepping movements are coordinated endogenously, though external stimuli clearly can affect the basic rhythm. Jumping, in most insects that use it, is a function of the hindlegs, which are elongate, muscular, and capable of great extension because of the wide angle that can be developed between femur and tibia. In order to obtain sufficient power for takeoff, the energy of muscle contraction must be stored temporarily as elastic energy. This energy, at a critical point, is suddenly and rapidly released to produce the acceleration necessary to overcome gravity. Most soft-bodied larvae that crawl over, or burrow through, the substrate depend on synchronized contraction and relaxation of muscles to effect changes in body shape, the legs and accessory locomotory appendages serving simply as points of friction between the body and substrate. In these larvae the body fluids serve as a hydrostatic skeleton.

Insects that move slowly over the surface of, or through, water typically use a hexapodal gait. More rapidly moving species, which may be streamlined, usually employ a rowing motion of the midlegs, occasionally the hindlegs. The legs are often modified to increase the surface area presented during the active stroke and their point of insertion on the body is designed so as to obtain maximum power from the stroke. Some aquatic insects swim by other means, for example, body curling, jet propulsion, or flapping the wings.

Primitively, the direct muscles (those that connect directly with the wing articulations) are used both for supplying power for flight and for controlling the nature of the wing beat. However, in most flying insects efficiency is increased by separating the supply of power (the role of large indirect muscles in the thorax) from the control of wing beat (which remains the function of the direct muscles). In fliers that have a low wing-beat frequency, control of muscle contraction is synchronous (neurogenic); that is, there is a 1:1 ratio between wing-beat frequency (= frequency of wing-muscle contraction) and the number of nerve impulses arriving at the muscle. High wing-beat frequencies are achieved by the use of fibrillar muscles and asynchronous (myogenic) control. Fibrillar muscles always operate in antagonistic pairs. Their rhythm of contraction originates endogenously and is initiated when a muscle is stretched to a critical tension, Thus, the contractions of an antagonistic pair of fibrillar muscles are self-perpetuating, though their initiation and termination are under nervous control. The indirect muscles serve to change the shape of the pterothorax, which thus acts as an elastic box, these changes in shape causing the wings to be moved up and down. Energy released at the end of each wing stroke (when a wing's momentum rapidly decreases) is stored as elastic energy, in the wall of the pterothorax, in the resilin of the wing hinge, and (in myogenic fliers) in the fibrillar muscles. This energy is then released to power the following stroke. In most insects lift is generated through a combination of delayed stall, rapid wing rotation, and wake capture. However, in some very small species a special form of rotation-generated lift, the "clap and fling" mechanism is used. In both systems the generation of vortices, creating regions of low pressure above the wings, is a critical component. Insects use a variety of substrates as energy sources for flight. Diptera and Hymenoptera metabolize trehalose; locusts, many Lepidoptera, Hemiptera, and Odonata oxidize lipid, sometimes after an initial phase of trehalose breakdown. A few insects use proline to fuel flight. Regulation of flight metabolism is under hormonal control, with adipokinetic hormone playing a major role.

Enhanced locomotor activity, which follows receipt of a stimulus, but whose direction is without spatial reference to that stimulus, is known as a kinesis. When the direction of movement is with reference to the source of the stimulus, for example, attraction to an odor and the light-compass reaction, the movement is described as a taxis.

5. Literature

The nature and properties of insect muscle are discussed by Smith (1972), by various authors in the volume edited by Usherwood (1975), and by Aidley (in Volume 5 of Kerkut and Gilbert, 1985, and in Goldsworthy and Wheeler, 1989). Walking is dealt with by Wilson (1966), Delcomyn (2004), and Hughes and Mill (1974), who also review jumping, crawling, and burrowing. Nachtigall (1974) has reviewed locomotion of aquatic insects. Pringle's (1957) monograph provides an excellent introduction to flight. More recent reviews of flight are given by Pringle (1968, 1974), Brodsky (1994), Dudley (2000), Dickinson and Dudley (2003), and Wootton (2004). Several specialist volumes cover a variety of insect flight-related topics (see Rainey, 1976; Danthanarayana, 1986; Goldsworthy and Wheeler, 1989). Wilson (1968) has examined the nervous control of flight. Several authors have contributed chapters on insect locomotion in Volume 5 of Kerkut and Gilbert (1985).

Berrigan, D., and Pepin, D. J., 1995, How maggots move: Allometry and kinematics of crawling in larval Diptera, *J. Insect Physiol.* **41**:329–337.

Brackenbury, J., 1999, Fast locomotion in caterpillars, *J. Insect Physiol.* **45**:525–533.

Brackenbury, J., 2000, Locomotory modes in the larva and pupa of *Chironomus plumosus*, *J. Insect Physiol.* **46**:1517–1527.

Brackenbury, J., and Wang, R., 1995, Ballistics and visual targeting in flea-beetles (Alticinae), *J. Exp. Biol.* **198**:1931–1942.

Brinkhurst, R. O., 1959, Studies on the functional morphology of *Gerris najas* De Geer (Hem. Het. Gerridae), *Proc. Zool. Soc. Lond.* **133**:531–559.

Brodsky, 1994, *The Evolution of Insect Flight*, Oxford University Press, Oxford.

Burrows, M., and Morris, O., 2003, Jumping and kicking in bush crickets, *J. Exp. Biol.* **206**:1035–1049.

Candy, D. J., Becker, A., and Wegener, G., 1997, Coordination and integration of metabolism in insect flight, *Comp. Biochem. Physiol. B* **117**:497–512.

Danthanarayana, W. (ed.), 1986, *Insect Flight: Dispersal and Migration*, Springer-Verlag, Berlin.

Darnhofer-Demar, B., 1969, Zur Fortbewegung des Wasserläufers *Gerris lacustris* L, auf der Wasseroberfläche, *Zool. Anz. Suppl.* **32**:430–439.

Delcomyn, F., 2004, Insect walking and robotics, *Annu. Rev. Entomol.* **49**:51–70.

Dickinson, M., and Dudley, R., 2003, Flight, in: *Encyclopedia of Insects* (R. T. Cardé and V. H. Resh, eds.), Academic Press, Amsterdam.

Dickinson, M. H., Lehmann, F.-O., and Sane S. P., 1999, Wing rotation and the aerodynamic basis of insect flight, *Science* **284**:1954–1960.

Dudley, R., 2000, *The Biomechanics of Insect Flight: Form, Function, Evolution*, Princeton University Press, Princeton, NJ.

Elia, A. J., Money, T. G. A., Orchard, I., 1995, Flight and running induce elevated levels of FMRFamide-related peptides in the haemolymph of the cockroach, *Periplaneta americana* (L.), *J. Insect Physiol.* **41**:565–570.

Ellington, C. P., 1995, Unsteady aerodynamics of insect flight, *Symp. Soc. Exp. Biol.* **49**:109–129.

Ellington, C. P., 1999, The novel aerodynamics of insect flight: Applications to micro-air vehicles, *J. Exp. Biol.* **202**:3439–3448.

Ellington, C.P., van den Berg, C., Willmott, A. P., and Thomas, A. L. R., 1996, Leading-edge vortices in insect flight, *Nature* **384**:626–630.

Gäde, G., and Auerswald, L., 1998, Insect neuropeptides regulating substrate mobilisation, *S. Afr. J. Zool.* **33**:65–70.

Goldsworthy, G., and Joyce, M., 2001, Physiology and endocrine control of flight, in: *Insect Movement: Mechanisms and Consequences* (I. P. Woiwod, D. R. Reynolds, and C. D. Thomas, eds.), CAB International, Wallingford, U.K.

Goldsworthy, G. J., 1983, The endocrine control of flight metabolism in locusts, *Adv. Insect Physiol.* **17**:149–204.

Goldsworthy, G. J., and Wheeler, C. H. (eds.), 1989, *Insect Flight*, CRC Press, Boca Raton.

Goodman, L. J., 1960, The landing responses of insects. I. The landing responses of the fly, *Lucilia sericata*, and other Calliphorinae, *J. Exp. Biol.* **37**:854–878.

Goodman, L. J., 1965, The role of certain optomotor reactions in regulating stability in the rolling plane during flight in the desert locust, *Schistocerca gregaria*, *J. Exp. Biol.* **42**:385–408.

Haunerland, N. H., 1997, Transport and utilization of lipids in insect flight muscles, *Comp. Biochem. Physiol B* **117**:475–482.

Heitler, W. J., 1974, The locust jump. Specializations of the metathoracic femoral-tibial joint, *J. Comp. Physiol.* **89**:93–104.

Hoyle, G., 1974, Neural control of skeletal muscle, in: *The Physiology of Insecta*, 2nd ed., Vol. IV (M. Rockstein, ed.), Academic Press, New York.

Hughes, G. M., and Mill, P. J., 1974, Locomotion: Terrestrial, in: *The Physiology of Insecta*, 2nd ed., Vol. III (M. Rockstein, ed.), Academic Press, New York.

Kerkut, G. A., and Gilbert, L. I. (eds.), 1985, *Comprehensive Insect Physiology, Biochemistry and Pharmacology*, Pergamon Press, Elmsford, NY.

Nachtigall, W., 1974, Locomotion: Mechanics and hydrodynamics of swimming in aquatic insects, in: *The Physiology of Insecta*, 2nd ed., Vol. III (M. Rockstein, ed.), Academic Press, New York.

Nalbach, G., 1993, The halteres of the blow fly *Calliphora*. I. Kinematics and dynamics, *J. Comp. Physiol. A* **173**:293–300.

Neville, A. C., 1975, *Biology of the Arthropod Cuticle*, Springer-Verlag, Berlin.

Orchard, I., Ramirez, J.-M., and Lange, A. B., 1993, A multifunctional role for octopamine in locust flight, *Annu. Rev. Entomol.* **38**:227–249.

Pringle, J. W.S., 1957, *Insect Flight*, Cambridge University Press, London.

Pringle, J. W. S., 1968, Comparative physiology of the flight motor, *Adv.Insect Physiol.* **5**:163–227.

Pringle, J. W. S., 1974, Locomotion: Flight, in: *The Physiology of Insecta*, 2nd ed., Vol. III (M. Rockstein, ed.), Academic Press, New York.

Rainey, R. C. (ed.), 1976, Insect flight, *Symp. R. Entomol. Soc.* **7**:287 pp.

Ribera, I., Foster, G. N., and Holt, W. V., 1997, Functional types of diving beetle (Coleoptera: Hygrobiidae and Dytiscidae), as identified by comparative swimming behaviour, *Biol. J. Linn. Soc.* **61**:537–558.

Robertson, R. M., and Pearson, K. G., 1982, A preparation for intracellular analysis of neuronal activity during flight in the locust, *J. Comp. Physiol.* **146**:311–320.

Rothschild, M., Schlein, Y., Parker, K., and Sternberg, S., 1972, Jump of the oriental rat flea *Xenopsylla cheopis* (Roths.), *Nature* **239**:45–48.

Sane, S. P., 2003, The aerodynamics of insect flight, *J. Exp. Biol.* **206**:4191–4208.

Smith, D. S., 1972, *Muscle*, Academic Press, New York.

Usherwood, P. N. R. (ed.), 1975, *Insect Muscle*, Academic Press, New York.

Van der Horst, D. J., Van Marrewijk, W. J. A., Vullings, H. G. B., and Diederen, J. H. B., 1999, Metabolic neurohormones: Release, signal transduction and physiological responses of adipokinetic hormones in insects, *Eur. J. Entomol.* **96**:299–308.

Weis-Fogh, T., 1973, Quick estimates of flight fitness in hovering animals, including novel mechanisms for lift production, *J. Exp. Biol.* **59**:169–230.

Wheeler, C. H., 1989, Mobilization and transport of fuels to the flight muscles, in: *Insect Flight* (G. J. Goldsworthy and C. H. Wheeler, eds.), CRC Press, Boca Raton.

Wilson, D. M., 1962, Bifunctional muscles in the thorax of grasshoppers, *J. Exp. Biol.* **39**:669–677.

Wilson, D. M., 1966, Insect walking. *Annu. Rev. Entomol.* **11**:103–122.

Wilson, D. M., 1968, The nervous control of insect flight and related behavior, *Adv. Insect Physiol.* **5**:289–338.

Wootton, R. J., 2004, Flight, in: *Design in Nature*, Vol. 2 (M. Collins, M. Atherton, and J. A. Bryant, eds.), WIT Press, Southampton, U.K. (In Press)

15

Gas Exchange

1. Introduction

In all organisms gas exchange, the supply of oxygen to and removal of carbon dioxide from cells, depends ultimately on the rate at which these gases diffuse in the dissolved state. The diffusion rate is proportional to (1) the surface area over which diffusion is occurring and (2) the diffusion gradient (concentration difference of the diffusing material between the two points under consideration divided by the distance between the two points). Diffusion alone, therefore, as a means of obtaining oxygen or excreting carbon dioxide can be employed only by small organisms whose surface area/volume ratio is high (i.e., where all cells are relatively close to the surface of the body) and organisms whose metabolic rate is low. Organisms that are larger and/or have a high metabolic rate must increase the rate at which gases move between the environment and the body tissues by improving (1) and/or (2) above. In other words, specialized respiratory structures with large surface areas and/or transport systems that bring large quantities of the gas closer to the site of use or disposal (thereby improving the diffusion gradient) have been developed. For most terrestrial animals prevention of desiccation is another important problem, and this has had a major influence on the development of their respiratory surfaces through which considerable loss of water might occur. Typically, respiratory surfaces of terrestrial animals are formed as invaginated structures within the body so that evaporative water loss is greatly reduced.

In insects the tracheal system, a series of gas-filled tubes derived from the integument, has evolved to cope with gas exchange. Terminally the tubes are much branched, forming tracheoles that provide an enormous surface area over which diffusion can occur. Furthermore, tracheoles are so numerous that gaseous oxygen readily reaches most parts of the body, and, equally, carbon dioxide easily diffuses out of the tissues. Thus, in most insects, in contrast to many other animals, the circulatory system is unimportant in gas transport. Because they are in the gaseous state within the tracheal system, oxygen and carbon dioxide diffuse rapidly between the tissues and site of uptake or release, respectively, on the body surface. Oxygen, for example, diffuses 3 million times faster in air than in water (Mill, 1972). Again, because the system is gas-filled, much larger quantities of oxygen can reach the tissues in a given time. (Air has about 25 times more oxygen per unit volume than water.)

The eminent suitability of the tracheal system for gas exchange is illustrated by the fact that, for most small insects and many large insects at rest, simple diffusion of gases in/out of the tracheal system entirely satisfies their requirements (but see Section 3.3). In large, active insects the gradient over which diffusion occurs is increased by means of ventilation; that is, air is actively pumped through the tracheal system.

2. Organization and Structure of the Tracheal System

A tracheal system is present in all Insecta and in other hexapods with the exception of the Protura and many Collembola. It arises during embryogenesis as a series of segmental invaginations of the integument. Up to 12 (3 thoracic and 9 abdominal) pairs of spiracles may be seen in embryos, though this number is always reduced prior to hatching, and further reduction may occur in endopterygotes during metamorphosis. Various terms are used to describe the number of pairs of functional spiracles, for example, holopneustic (10 pairs, located on the mesothorax and metathorax and 8 abdominal segments), amphipneustic (2 pairs, on the mesothorax and at the tip of the abdomen), and apneustic (no functional spiracles). The last condition is common in aquatic larvae, which are said, therefore, to have a closed tracheal system (Section 4.1).

The proportion of the body filled by the tracheal system varies widely, both among species and within the same individual throughout a stadium. In active insects whose tracheal system includes air sacs (see below) the tracheal system occupies a greater fraction of the body than in less active species. Further, in the former, the volume of the tracheal system may decrease dramatically during a stadium (e.g., in *Locusta* from 48% to 3%) as the air sacs become occluded by the increased hemolymph pressure that results from tissue growth. After ecdysis, when body volume has increased (Chapter 11, Section 3.2), the tracheal system expands because of the lowered hemolymph pressure.

2.1. Tracheae and Tracheoles

In apterygotes other than lepismatid Zygentoma, the tracheae that run from each spiracle do not anastomose either with those from adjacent segments or with those derived from the spiracle on the opposite side. In the Lepismatidae and Pterygota both longitudinal and transverse anastomoses occur, and, though minor variations can be seen, the resultant pattern of the tracheal system is often characteristic for a particular order or family. Generally, a pair of large-diameter, longitudinal tracheae (the lateral trunks) run along the length of an insect just internal to the spiracles. Other longitudinal trunks are associated with the heart, gut, and ventral nerve cord. Interconnecting the longitudinal tracheae are transverse commissures, usually one dorsal and another ventral, in each segment (Figure 15.1). Parts of the tracheal system, for example, that of the pterothorax, may be effectively isolated from the rest of the system by reduction of the diameter or occlusion of certain longitudinal trunks. This arrangement is associated with the use of autoventilation as a means of improving the supply of oxygen to wing muscles during flight (Section 3.3). Also, tracheae are often dilated to form large thin-walled air sacs that have an important role in ventilation (Section 3.3) and other functions.

Numerous smaller tracheae branch off the main tracts and undergo progressive subdivision until at a diameter of about 2–5 μm they form a number of fine branches each 1 μm or less across known as tracheoles. Tracheoles are intracellular, being enclosed within a very thin layer of cytoplasm from the tracheoblast (tracheal end cell) (Figure 15.2), and ramify throughout most tissues of the body. They are especially abundant in metabolically active tissues. Thus, in flight muscles, fat body, and testes, for example, tracheoles indent

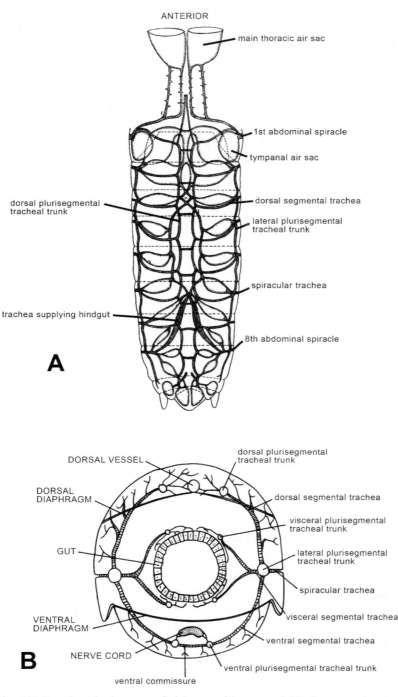

FIGURE 15.1. (A) Dorsal tracheal system of abdomen of locust; and (B) diagrammatic transverse section through abdomen of a hypothetical insect to illustrate main tracheal branches. [A, from F. O. Albrecht, 1953, *The Anatomy of the Migratory Locust.* By permission of Athlone Press. B, from R. E. Snodgrass, *Principles of Insect Morphology.* Copyright 1935 by McGraw-Hill, Inc. Used with permission of McGraw-Hill Book Company.]

individual cells, so that gaseous oxygen is brought into extremely close proximity with the energy-producing mitochondria.

In *Calpodes ethlius* caterpillars not all tracheae end in tracheoles within or on tissues. In particular, some tracheae originating from the spiracles of the eighth abdominal segment

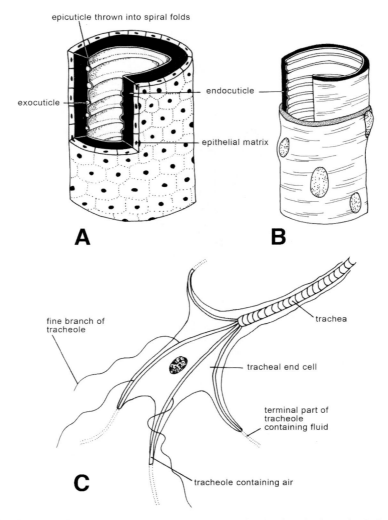

FIGURE 15.2. Structure of (A) large; and (B) small tracheae. (C) Origin of tracheole. [After V. B. Wigglesworth, 1965, *The Principles of Insect Physiology*, 6th ed., Methuen and Co. By permission of the author.]

form large, greatly branched tufts that are suspended within the hemolymph (Figure 15.3). The tracheolar tips of these tufts are connected to heart and other muscles so that the tufts are constantly moved within the hemolymph (Locke, 1998). The observation that hemocytes were abundant within the tufts led Locke to speculate that the tufts are sites of gas exchange for the blood cells, analogous to the lungs of vertebrates. Though similar tracheal tufts are found in caterpillars from other lepidopteran families, their existence and function(s) among other insect groups has not been examined.

As derivatives of the integument, tracheae comprise cuticular components, epidermis, and basal lamina (Figure 15.2). Adjacent to the spiracle, the tracheal cuticle includes, the cuticulin envelope, epicuticle and procuticle; in smaller tracheae and most tracheoles only the cuticulin envelope and epicuticle are present. Providing the system with strength yet flexibility, tracheal cuticle has internal ridges that may be either separate (annuli) or form a continuous helical fold (taenidium). In large tracheae the ridges include some procuticle, but this is absent from those of tracheoles. Taenidia are absent from, or poorly developed in, air sacs. The epicuticle of tracheae comprises the same layers as that of the integument.

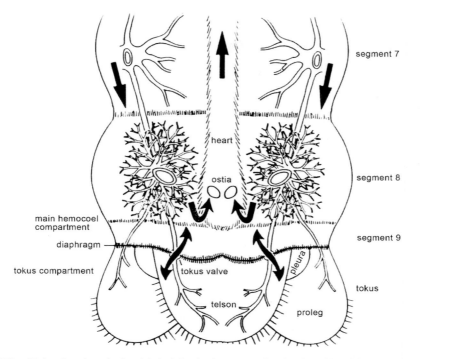

FIGURE 15.3. Tufts of tracheae in the eighth abdominal segment of *Calpodes ethlius* that perhaps serve to aerate hemocytes. Arrows indicate direction of hemolymph flow. [From M. Locke, 1998, Caterpillars have evolved lungs for hemocyte gas exchange, *J. Insect Physiol.* **44**:1–20. With permission from Elsevier.]

In the smallest tracheoles, however, only the cuticulin envelope is present and, furthermore, this contains fine pores. These two features may be associated with movement of liquid into and out of tracheoles in connection with gas exchange (Section 3.1) (Locke, 1966).

2.2. Spiracles

Only in some apterygotes do tracheae originate at the body surface. Normally, they arise slightly below the body surface from which they are separated by a small cavity, the atrium (Figure 15.4A). In this arrangement, the term "spiracle" generally includes both the atrium and the spiracle *sensu stricto*, that is, the tracheal pore. Except for those of a few insects that live in humid microclimates, spiracles may be covered, for example, by the elytra or wings in Hemiptera and Coleoptera, or are equipped with various valves for prevention of water loss. The valves may take the form of one or more cuticular plates that can be pulled over a spiracle by means of a closer muscle (Figure l5.4B–D). Opening of the valve(s) is effected either by the natural elasticity of the surrounding cuticle or by an opener muscle. Alternatively, the valve may be a cuticular lever which by muscle action constricts the trachea adjacent to the atrium (Figure l5.4E,F). In lieu of, or in addition to, the valves, there may be hairs lining the atrium or a sieve plate (a cuticular pad penetrated by many fine pores) covering the atrial pore. It is commonly assumed that an important function of these hairs and sieve plates is to prevent dust entry. However, as Miller (1974) noted, sieve plates are not better developed on inspiratory than on expiratory spiracles and several other functions can be suggested: (1) they may prevent waterlogging of the tracheal system in terrestrial species during rain, in aquatic insects, and in species that live in moist soil, rotting

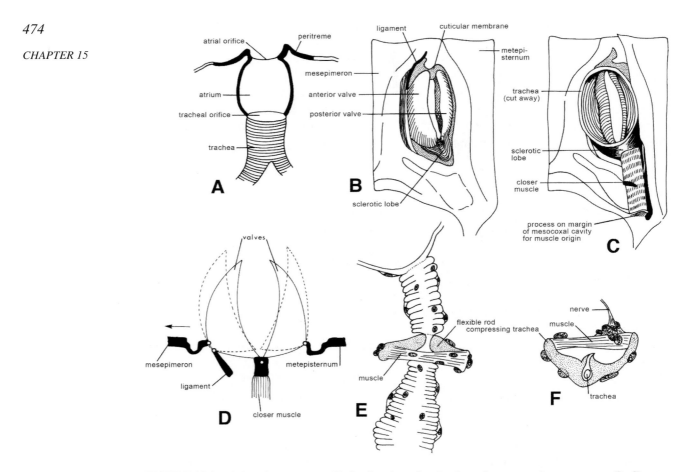

FIGURE 15.4. Spiracular structure. (A) Section through spiracle to show general arrangement; (B, C) outer and inner views of second thoracic spiracle of grasshopper; (D) diagrammatic section through spiracle to show mechanism of closure. The valve is opened by movement of the mesepimeron, closed by contraction of the muscle; (E) closing mechanism on flea trachea; and (F) section through flea trachea at level of closing mechanism. [A–C, from R. E. Snodgrass, *Principles of Insect Morphology*. Copyright 1935 by McGraw-Hill, Inc. Used with permission of McGraw-Hill Book Company. D, after P. L. Miller, 1960, Respiration in the desert locust. II. The control of the spiracles, *J. Exp. Biol.* **37**:237–263. By permission of Cambridge University Press. E. F, after V. B. Wigglesworth, 1965, *The Principles of Insect Physiology*, 6th ed., Methuen and Co. By permission of the author.]

vegetation, etc.; (2) they may prevent entry of parasites, especially mites, into the tracheal system; and (3) they may reduce bulk flow of gases through the system caused by body movements, thereby reducing evaporative water loss. This would be disadvantageous in insects that ventilate the tracheal system, and it is of interest, therefore, that those spiracles important in ventilation commonly lack a sieve plate or have a plate that is divided down the middle so that it may be opened during ventilation.

3. Movement of Gases within the Tracheal System

Gas exchange between tissues and the tracheal system occurs almost exclusively across the walls of tracheoles, for it is only their walls that are sufficiently thin as to permit a satisfactory rate of diffusion. It is necessary, therefore, to ensure that a sufficient concentration of oxygen is maintained in tracheoles to supply tissue requirements and, at the same time, that carbon dioxide produced in metabolism is removed quickly, preventing its buildup to toxic levels. In small insects and inactive stages of larger insects, diffusion of gases between the spiracle and tracheoles is sufficiently rapid that these requirements are met. However, if the spiracles are kept permanently open, the amount of water lost via the tracheal system may become important. Thus, many insects utilize discontinuous gas exchange as a means of reducing this loss. The needs of large, active insects can be satisfied only by shortening the distance over which diffusion must occur. This is achieved by active ventilation movements.

3.1. Diffusion

The absence of obvious breathing movements led many 19th century scientists to assume that insects obtained oxygen by simple diffusion. It was not, however, until 1920 that Krogh (cited in Miller, 1974) calculated, on the basis of (1) measurements of the average tracheal length and diameter, (2) measurements of oxygen consumption, and (3) the permeability constant for oxygen, that for *Tenebrio* and *Cossus* (goat moth) larvae at rest with the spiracles open, the oxygen concentration difference between the spiracles and tissues is only about 2% and easily maintainable by diffusion.

Even in large active insects that ventilate (Section 3.3), diffusion is a significant process, because the ventilation movements serve only to move the air in the larger tracheae. For example, in the dragonfly *Aeshna*, oxygen reaches the flight muscles by diffusion between the primary (ventilated) air tubes and tracheoles, a distance of up to 1 mm. Even in flight, when the oxygen consumption of the muscle reaches 1.8 ml/g/min and the difference in oxygen concentration between the primary tube and tracheoles is 5–13%, diffusion is quite adequate (Weis-Fogh, 1964).

Diffusion is also important in moving gases between the tracheoles and mitochondria of the tissue cells. Because diffusion of dissolved gases is relatively slow, the distance over which it can function satisfactorily (in structural terms, half the distance between adjacent tracheoles) is directly related to the metabolic activity of the tissue. In highly active flight muscles of Diptera and Hymenoptera, for example, it has been calculated that the maximum theoretical distance between tracheoles is 6–8 μm. In practice, tracheoles, which indent the muscle cells, are within 2–3 μm of each other, allowing a significant "safety margin" (Weis-Fogh, 1964).

In many insects, distal parts of tracheoles are not filled with air but liquid under normal resting conditions. During activity, however, the tracheoles become completely air-filled; that is, fluid is withdrawn from them only to return when activity ceases. Wigglesworth (1953) suggested that the level of fluid in tracheoles depends on the relative strengths of the capillary force drawing fluid along the tube and the osmotic pressure of the hemolymph. During metabolic activity, the osmotic pressure increases as organic respiratory substrates are degraded to smaller metabolites, causing fluid to be withdrawn from the tracheoles (perhaps via the pores mentioned earlier) and, therefore, bringing gaseous oxygen closer to the tissue cells (Figure 15.5). As the metabolites are fully oxidized and removed, the osmotic pressure will fall, and once again the capillary force will draw fluid along the tracheoles.

Though carbon dioxide is more soluble, and has a greater permeability constant, than oxygen in water and could conceivably move by diffusion through the hemolymph to leave the body via the integument, this route does not normally eliminate a significant quantity of the gas (e.g., 2–10% of the total in some dipteran larvae with thin cuticles). The great majority of carbon dioxide leaves by gaseous diffusion via the tracheal system.

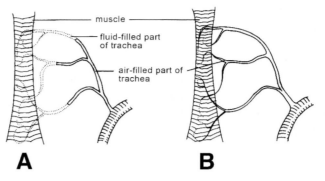

FIGURE 15.5. Changes in level of tracheolar fluid as a result of muscular activity. (A) Resting muscle; and (B) active muscle. [After V. B. Wigglesworth, 1965, *The Principles of Insect Physiology*, 6th ed., Methuen and Co. By permission of the author.]

3.2. Discontinuous Gas Exchange

Originally discovered in diapausing pupae of *Hyalophora cecropia* and other lepidopterans, the discontinuous gas exchange cycle (DGC) [formerly known as passive (suction) ventilation] is now known to occur in all ants, many bees, some wasps, several different families of beetles, cockroaches, grasshoppers and locusts (Lighton 1996, 1998). The DGC, which makes use of the high solubility of carbon dioxide in water, comprises three phases (Figure 15.6): constricted- or closed-spiracle phase, fluttering-spiracle phase, and open-spiracle phase. In the constricted-spiracle phase the spiracular valves are kept almost closed. As oxygen is used in metabolism, the carbon dioxide so produced is stored, largely as bicarbonate in the hemolymph and tissues but partially also in the gaseous state in the tracheal system. Thus, a slight vacuum is created within the tracheal system that sucks in more air. Eventually, the tracheal oxygen concentration falls to about 3.5% and the carbon dioxide concentration rises to about 4%. At this point, the low oxygen concentration induces "fluttering" (rapid opening and closing of the valves), the effect of which is to allow some outward diffusion of nitrogen and more air to flow into the tracheal system. However, carbon dioxide cannot escape and its concentration increases to about 6.5%, at which point the valves are opened and remain open between 15 and 30 minutes. During this period there is rapid diffusion of carbon dioxide out of the tracheal system and massive release of carbon dioxide from the hemolymph. When the concentration of carbon dioxide in the tracheal system falls to 3.0%, the valves reclose. Experiments in which gases of different composition are perfused through the tracheal system or over the segmental ganglia have shown that there is dual control over the opening of the valves. Hypercapnia (above normal carbon dioxide concentration) directly stimulates relaxation of the valve closer muscle (the valve opens as a result of cuticular elasticity), whereas hypoxia (insufficient oxygen) acts at the level of the central nervous system (probably the metathoracic ganglion). In diapausing *Hyalophora* pupae the periods between bursts of carbon dioxide release may be as long as 7 hours.

For diapausing pupae and certain other postembryonic stages of insects, the slight net inflow of air during the constricted-spiracle phase of the DGC may serve as a means of conserving moisture that would otherwise be lost as a result of gas exchange. However, there are many insects that show DGC but do not normally experience water-loss problems. Conversely, there are many desert insects in which DGC does not occur. This has led to the proposal that DGC evolved primarily to facilitate gas exchange in hypoxic and hypercapnic

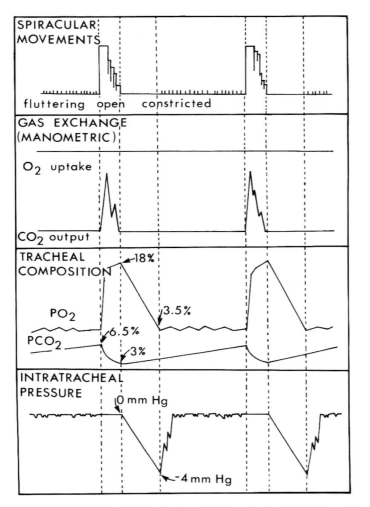

FIGURE 15.6. Discontinuous release of carbon dioxide in pupa of *Hyalophora cecropia* in relation to spiracular valve opening and closing. [After R. I. Levy and H. A. Schneiderman, 1966, Discontinuous respiration in insects. IV, *J. Insect Physiol.* **12**:465–492. By permission of Pergamon Press Ltd.]

environments (Lighton 1996, 1998). Certainly, DGC is common in some subterranean groups, notably ants and beetles that burrow in soil, dung or wood. However, there are exceptions to this generalization, perhaps the major one being termites (Shelton and Appel, 2000, 2001). Further, some species that normally use DGC may abandon it under hypoxic conditions (Chown and Holter, 2000; Chown, 2002).

3.3. Active Ventilation

By alternately decreasing and increasing the volume of the tracheal system through compression and expansion of larger air tubes, a fraction of the air in the system is periodically renewed and the diffusion gradient between tracheae and tissues kept near the maximum. These breathing (ventilation) movements may be continuous as in locusts and dragonflies, intermittent as in cockroaches, or occur only after activity as is seen in wasps. The volume changes are normally brought about by contraction of abdominal dorsoventral and/or longitudinal muscles, which increases the hemolymph pressure, thus causing the tracheae to flatten or collapse. In some species both inspiration and expiration are brought about by muscles; in others, only expiration is under muscular control, and inspiration occurs as a result of the natural elasticity of the body wall. During high metabolic activity, supplementary ventilation movements may occur. For example, the desert locust normally ventilates by means of dorsoventral movements of the abdomen but can supplement these by "telescoping" the abdomen and protraction/retraction of the head and prothorax (Miller, 1960). Remarkably, it has recently been reported that many insects showing no obvious signs of breathing have rapid cycles of tracheal compression in the head and thorax (Westneat et al., 2003). In analogy with the situation in the abdomen, tracheal compression in these regions is induced indirectly, by contraction of mandible and leg muscles. When these muscles relax, the elasticity in the taenidial rings returns the tracheae to their original shape and volume. This discovery may require reconsideration of the proposal that diffusion alone satisfies the gas-exchange requirements of small insects.

The diffusion gradient can be further improved by increasing the volume of air in the tracheal system that is renewed during each stroke (the tidal volume). This has been achieved through the development of large, compressible air sacs. However, simple tidal flow (pumping of air in to and out of all spiracles) is still somewhat inefficient because a considerable volume of air (the dead air space) remains within the system at each stroke. The size of the dead air space is greatly reduced by using unidirectional ventilation in which air is made to flow in one direction (usually anteroposteriorly) through the tracheal system. Unidirectional airflow is achieved by synchronizing the opening and closing of spiracular valves with ventilation movements. In the resting desert locust, for example, the first, second, and fourth spiracles are inspiratory, while the tenth (most posterior) is expiratory. When the insect becomes more active, the first four spiracles become inspiratory, the remainder expiratory. The spiracular valves do not form an airtight seal, however, so that a proportion of the inspired air continues to move tidally rather than unidirectionally (20% in the resting desert locust).

During flight, the oxygen consumption of an insect increases enormously (up to 24 times in the desert locust), almost entirely because of the metabolic activity of the flight muscles. To facilitate this activity, a massive exchange of air occurs in the pterothorax, made possible by certain structural features of the pterothoracic tracheal system and by changes in the body's normal (resting) ventilation pattern. As noted earlier, the tracheal system of the pterothorax is effectively isolated from that of the rest of the body by reduction in the diameter or occlusion of the main longitudinal tracheae. Autoventilation of flight muscle tracheae also occurs. This is ventilation that results from movements of the nota and pleura during wing beating, and it brings about a considerable flow of air into and out of the thoracic tracheae. During autoventilation, normal unidirectional flow, where such occurs, becomes masked by the massive increase in tidal flow in the pterothorax. To achieve this tidal flow, in the desert locust, spiracles 2 and 3 remain permanently open. Spiracles 1 and 4–10, however, continue to open and close in synchrony with abdominal ventilation so that some unidirectional flow occurs. The rate of abdominal ventilation movements also increases during flight to about four times the resting value, but these movements probably serve primarily to increase the rate of flow of hemolymph around the body, bringing fresh supplies of metabolites to the flight musculature. However, keeping the central nervous system well supplied with oxygen also appears to be important.

Autoventilation is used by many Odonata, Orthoptera, Dictyoptera, Isoptera, Hemiptera, Lepidoptera, and Coleoptera, but is of little importance in Diptera and Hymenoptera. Its significance can be broadly correlated with body size, the type of muscles used in flight (Chapter 14, Section 2.1), and the extent of movements of the thorax during flight. Odonata, for example, are generally large; movements of their thorax during wing beating are pronounced and their flight muscles are of the tubular type which lack indented tracheoles. Therefore, autoventilation is extremely important in this order. In contrast, in Hymenoptera and Diptera, movements of the thorax in flight are relatively slight so that the volume change that can be achieved in the tracheal system is not significant. However, the fibrillar flight muscles of these insects are much indented with tracheoles so that gaseous oxygen is brought close to the mitochondria. Thus, in Hymenoptera simple telescopic abdominal ventilation normally creates sufficient air exchange in the thorax. In Diptera, abdominal ventilation movements are weak or non-existent, and diffusion alone satisfies the insects' oxygen requirements in flight (Miller, 1974).

Ventilation movements are initiated within and controlled via the central nervous system. Some isolated abdominal segments, provided that they contain a ganglion, can carry out normal respiratory movements, though usually an appropriate stimulus such as hypercapnia or hypoxia is necessary to initiate the movements. The coordination of these autonomous ventilation movements and, where unidirectional air flow occurs, of spiracular valve opening and closing, is achieved by a central pattern generator (CPG) situated usually in the metathoracic or first abdominal ganglion. The nature of the CPG is not understood. However, in the desert locust, it sends bursts of impulses to each ganglion, which both excite the motor neurons to the muscles used in expiration and inhibit those going to inspiratory muscles. The activity of the CPG is modified by sensory input. For example, in the grasshopper *Taeniopoda eques* receptors in the central nervous system are sensitive to changing carbon dioxide and oxygen concentrations, which stimulate CPG activity so that ventilation is increased and decreased, respectively (Bustami *et al.*, 2002).

4. Gas Exchange in Aquatic Insects

Perhaps not surprisingly, in view of the rapid rate at which gases can diffuse within it, a gas-filled tracheal system has been retained by almost all aquatic forms in their evolution from terrestrial ancestors. Only rarely, for example, in the early larval stages of *Chironomus* and *Simulium* (Diptera), and *Acentropus* (Lepidoptera), is the system filled with liquid.

Oxygen may enter the tracheal system in gaseous form, that is, via functional spiracles (the "open" tracheal system) or may pass, in solution, directly across the body wall to the tracheal system, in which arrangement the spiracles are sealed (non-functional), and the tracheal system is said to be "closed." Aquatic insects with open tracheal systems exchange the gas within the system by periodically visiting the water surface, by obtaining gas from gas-filled spaces in aquatic plants, or through the use of a "gas gill" (a bubble or film of air that covers the spiracles, in to or out of which oxygen and carbon dioxide, respectively, can diffuse from/to the surrounding water). A significant amount of gas exchange may occur by direct diffusion across the body surface (cutaneous respiration) in larvae with an open system whose integument is thin, for example, mosquito larvae. Cutaneous respiration may entirely satisfy the requirements of insects with closed tracheal systems. However, in many species supplementary respiratory surfaces, "tracheal gills," have evolved, though these often become important only under oxygen-deficient conditions.

4.1. Closed Tracheal Systems

For small aquatic insects or those with a low metabolic rate, diffusion of gases across the body wall provides an adequate means of obtaining oxygen and excreting carbon dioxide.

In larger and/or more active forms, all or part of the body wall has a very thin cuticle and becomes richly tracheated to facilitate rapid entry and exit of gases from the tracheal system. Some tube-dwelling species show rhythmic movements that create water currents over the body so that water in the tube is periodically renewed and the thickness of the "boundary layer" (the layer of still water adjacent to the body wall) is reduced.

In many aquatic larvae there are richly tracheated outgrowths of the body wall (hindgut in dragonfly larvae) collectively known as tracheal gills. Whether these function as accessory respiratory structures was originally controversial because even without them an insect may survive perfectly well and its oxygen consumption may not change. Tracheal gills include caudal lamellae (in Zygoptera), lateral abdominal gills (in Ephemeroptera, and some Zygoptera, Plecoptera, Neuroptera, and Coleoptera), rectal gills (in Anisoptera), and fingerlike structures, often found in tufts on various parts of the body (e.g., in some Plecoptera and case-bearing Trichoptera).

Juvenile Zygoptera have three caudal lamellae (Figure 6.11) whose functions have long been controversial (Burnside and Robinson, 1995), in large part because even without them larvae survive perfectly well. Suggested uses of the lamellae include swimming, especially to avoid predators, sacrificial structures for diverting a predator's attention, and gas exchange. There is evidence that all these possibilities may be important. For example, larvae with lamellae missing are poorer swimmers than those with a full complement; further, there are significant differences in the design of lamellae between good and poor swimmers. The lamellae have a breaking joint near their base, enabling them to be shed easily when grasped by a predator. The lamellae are also highly tracheated, and experiments have demonstrated that the lamellae normally are major sites of gas exchange and in oxygen-deficient water become especially important. For example, in *Coenagrion* up to 60% of oxygen uptake may occur via the lamellae (Harnisch, 1958; cited from Mill, 1974). In certain *Enallagma* species normal larvae can survive in water with an oxygen content of only 2.4% saturation, whereas the minimum for lamellaeless larvae is 14.5% saturation. Above this value, lamellaeless larvae live apparently normally and obtain sufficient oxygen by cutaneous diffusion (Pennak and McColl, 1944).

In larval Ephemeroptera the lateral abdominal gills are paired, segmental, normally platelike or branched structures (Figures 6.4–6.7) whose size is inversely related to the oxygen content of the surrounding water. In other orders, the gills are filamentous (Mill, 1974). In burrowing mayfly larvae and larvae of other species that live in an oxygen-poor environment, the gills are an important site of gas exchange, and about 50% of an insect's oxygen requirement may be obtained via this route. In addition, their rhythmic beating facilitates cutaneous respiration by moving a current of water over the body and reducing the thickness of the boundary layer. The frequency at which the gills beat depends both on their size and on the oxygen content of the water. In species that inhabit fast-flowing water, the gills do not beat, and removal of them does not affect oxygen consumption, indicating that most gas exchange occurs cutaneously.

The anterior part of the rectum of larval Anisoptera is enlarged to form a branchial chamber whose walls bear six rows of richly tracheated gills. Water is periodically drawn into and forced out of the rectum via the anus as a result of muscular movements of the abdomen, comparable to those that effect ventilation in terrestrial species. Contraction of dorsoventral muscles in the posterior abdominal segments decreases the abdominal volume and results in water being forced out of the rectum. A muscular diaphragm in the fifth segment may prevent hemolymph from being forced anteriorly. Water enters the rectum as the volume of the abdomen is increased, as a result of the contraction of two transverse muscles, the diaphragm and subintestinal muscles, which pull the tergal walls inward and force the sterna downward. This is aided by the natural elasticity of the body wall and by relaxation of the dorsoventral muscles (Hughes and Mill, 1966; Mill, 1977). The rate of ventilation of the branchial chamber varies with the oxygen content of the water and metabolic rate of the insect. Cutaneous respiration is probably not significant in Anisoptera, whose body wall cuticle is generally thick. Rectal pumping also occurs in larval Zygoptera, though its function here is primarily in relation to ion uptake (Chapter 18, Section 4.2) as these insects lack rectal gills (Miller, 1993).

4.2. Open Tracheal Systems

Among insects that have an open tracheal system can be traced a series of stages leading from complete dependence on atmospheric air (i.e., where an insect must frequently visit the water surface to exchange the gas in its tracheal system) to the stage where an insect maintains around its body a supply of atmospheric gas into which oxygen can diffuse from the surrounding water at a sufficient rate to totally satisfy the insect's needs. Thus, the insect is completely independent of atmospheric air.

Surface-breathing insects must solve two problems. First, they must prevent waterlogging of the tracheal system when they are submerged, and second, they must be able to overcome the surface tension force at the air-water interface. Both problems are solved by having hydrofuge (water-repellent) structures around the spiracles. In many dipteran larvae special epidermal glands (perispiracular/peristigmatic glands) secrete an oily material at the entrance of the spiracle. The spiracles of some other aquatic insects are surrounded by hydrofuge hairs, which, when submerged, close over the opening but when in contact with the water surface spread out to permit exchange of air (Figure 15.7). In addition, insects may possess other modifications to the tracheal system to help cope with these problems, for example, reduction of the number of functional spiracles and restriction of the spiracles to special sites, typically at the tip of a posterior extension of the body (postabdominal respiratory siphon), as occurs in mosquito larvae (Figure 9.6F) and water scorpions (Figure 8.16D). In larvae of some Syrphidae (e.g., *Eristalis*) the siphon is very extensible, its shape and flexibility giving rise to the common name of these insects—rattailed maggots. In a few species of Coleoptera and Diptera, whose larvae live in mud, the siphon, which is rigid and pointed, is forced into air spaces in the roots of aquatic plants.

Insects have gained variable degrees of independence from atmospheric air by holding a gas store about their body. The gas store may be subelytral or may occur as a thin film

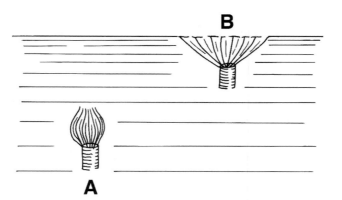

FIGURE 15.7. Hydrofuge hairs surrounding a spiracle. (A) Position when submerged; and (B) position when at water surface. [After V. B. Wigglesworth, 1965, *The Principles of Insect Physiology*, 6th ed., Methuen and Co. By permission of the author.]

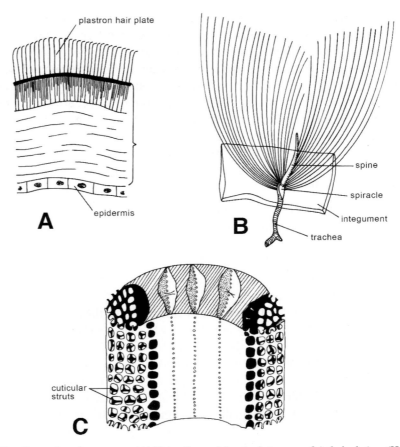

FIGURE 15.8. Examples of gas stores. (A) Hair pile on abdominal sternum of *Aphelocheirus* (Hemiptera); (B) spiracular gill of pupa of *Psephenoides gahani* (Coleoptera) comprising about 40 hollow branches; and (C) part of wall of spiracular gill branch of *P. gahani* showing cuticular struts that support the plastron. [A, after W. H. Thorpe and D. J. Crisp, 1947, Studies on plastron respiration. I. *J. Exp. Biol.* **24**:227–269. By permission of Cambridge University Press. B, C, after H. E. Hinton, 1968, Spiracular gills, *Adv. Insect Physiol.* **5**:65–162. By permission of Academic Press Ltd.]

of gas over certain parts of the body, held in place by a mat of hydrofuge hairs (Figure 15.8A). A third arrangement is for the gas to be held as a layer adjacent to the body by cuticular extensions of the body wall adjacent to the spiracles, known as spiracular gills (Figure 15.8B,C).

The degree of independence from atmospheric air (measured as the length of time that an insect is able to remain submerged between visits to the surface) depends on a number of factors. These include (1) the metabolic rate of the insect, which itself is temperature-dependent; (2) the volume, shape, and location of the gas store, which determine the surface area of the store in contact with the water; and (3) the oxygen content of the water. Factors (2) and (3) relate particularly to use of the gas store as a physical (gas) gill, that is, a structure that can take up oxygen from the surrounding water. When the oxygen used by an insect can be only partially replaced by diffusion into the gas store of oxygen from the water, the volume of the gas store will decrease and, eventually, the insect must return to the surface to renew the gas store. This is known as a temporary (compressible) gas gill. When an insect's oxygen requirements can be fully satisfied by diffusion of oxygen into the gill, whose volume will therefore remain constant, the gill is described as a permanent (incompressible) gas gill or plastron.

In a compressible gill the pressure of the gas will be equal to that of the surrounding water. The gas in the gill may be considered to include only oxygen and nitrogen, as the carbon dioxide produced in metabolism will readily dissolve in the water. Immediately after a visit to the surface, the composition of the gas in the gill will approximate that of air, that is, about 20% oxygen and 80% nitrogen and will be in equilibrium with the dissolved gases in the surrounding water. As an insect uses oxygen, a diffusion gradient will be set up so that this gas will tend to move into the gill. At the same time, the proportion of nitrogen in the gas will have increased and, therefore, nitrogen will tend to diffuse out of the gill in order to restore equilibrium. However, oxygen diffuses into the gill about three times as rapidly as nitrogen diffuses out. Thus, inward movement of oxygen will be more important than outward movement of nitrogen in restoring equilibrium. The effect of this is to prolong considerably the life of the gas store and, therefore, the duration over which an insect can remain submerged. The rate of diffusion, which is a function of the surface area of the gill exposed to the water, will obviously be greater in the case of films of gas spread over the body surface than in subelytral gas stores whose contact with the surrounding water is relatively slight. Ultimately, however, in a compressible gill the rate of oxygen use by the insect will reduce the proportion of oxygen in the gill below a critical level, and the insect will be stimulated to return to the surface.

For example, adult water beetles of the family Dytiscidae and back swimmers (*Notonecta* spp.: Hemiptera) are generally medium to large insects, which in summer, when water temperatures may approximate 20–25°C, show great activity. Under these conditions, the gas stored beneath the elytra (and also on the ventral body surface in *Notonecta*) must be regularly exchanged by visits to the water surface. However, in winter, when the water is cold (or may even be continuously frozen for several months as occurs, for example, on the Canadian prairies), the insects are more or less inactive. During this period, the gas store satisfies most or all of the insects' oxygen requirements.

In contrast, the volume of a plastron is constant but small. Hence, a plastron does not serve as a store of oxygen but solely as a gas gill. In certain adult Hemiptera (e.g., *Aphelocheirus*) and Coleoptera (e.g., *Elmis*) the plastron is held in place by dense mats of hydrofuge hairs. The hairs number about $200 \times 10^6/cm^2$ in *Aphelocheirus*, are bent at the tip, and are slightly thickened at the base. As a result, they can resist becoming flattened (which would destroy the plastron) by the considerable pressure differences that may arise between the gas in the plastron and the surrounding water as the insect moves into deeper water or uses up its oxygen supply. Spiracular gills are mostly found on the pupae of certain Diptera (e.g., Tipulidae and Simuliidae) and Coleoptera (e.g., Psephenidae) but occasionally occur on beetle larvae (Hinton, 1968). In almost all instances they include a plastron. In many dipteran pupae the gill is a long, hollowed-out structure (Figure 15.8C) that carries a plastron over its surface. The plastron is held in place by means of rigid cuticular struts and connects via fine tubes with the gas-filled center of the gill (and hence the spiracle).

Because the volume of the plastron remains constant, the nitrogen does not diffuse out. However, the rate and direction of diffusion of oxygen will depend on differences in the oxygen content of the plastron (always somewhat less than maximum because of use of oxygen by the insect) and the surrounding water. Therefore, to ensure that oxygen always diffuses into the plastron, the water around it should be saturated with the gas. If the water is not saturated with oxygen, then the latter will either diffuse into the plastron relatively slowly or may even diffuse out, leading eventually to asphyxiation of the insect. It is not surprising, then, to discover that plastron respiration is used by aquatic insects living in fast-moving streams, at the edges of shallow lakes, and in the intertidal zone. Plastron-bearing species known to inhabit water whose oxygen content may fluctuate daily, for example, that of marshes which drops greatly at night, may employ behavioral means to overcome the danger of asphyxiation, such as moving closer to the water surface or even climbing out of the water. Hinton (1968) also pointed out that the waters occupied by plastron-bearers are often subject to changes in level, leaving the insects periodically exposed to atmospheric air. However, desiccation does not present a problem to these insects because the connection between the plastron and internal tissues may be quite restricted. In addition, bulk flow of gas within the plastron is negligible, so that evaporative water loss is small.

5. Gas Exchange in Endoparasitic Insects

It is probably not surprising that endoparasitic insects, as they too are surrounded by fluid, show many parallels with aquatic insects in the way that they obtain oxygen. Most endoparasites satisfy a proportion of their requirements by cutaneous diffusion. In some first-instar larvae of Hymenoptera and Diptera the tracheal system may be liquid-filled, but generally it is gas-filled with closed spiracles and includes a rich network of branches immediately beneath the integument. Many endoparasitic forms, especially larval Braconidae, Chalcididae, and Ichneumonidae (Hymenoptera), and some Diptera, possess 'tails' filled with hemolymph or can evaginate the wall of the hindgut through the anus. It has been suggested that these structures may facilitate gas exchange, though the evidence on which this suggestion is based is generally not strong.

Endoparasites with greater oxygen requirements usually are in direct contact with atmospheric air either via the integument of the host or via the host's tracheal system. In larvae of many Chalcidoidea, for example, only the posterior spiracles are functional, and these open into an air cavity formed at the base of the egg pedicel that penetrates the host's integument (Figure 15.9A). Many larval Tachinidae (Diptera) become enclosed in a

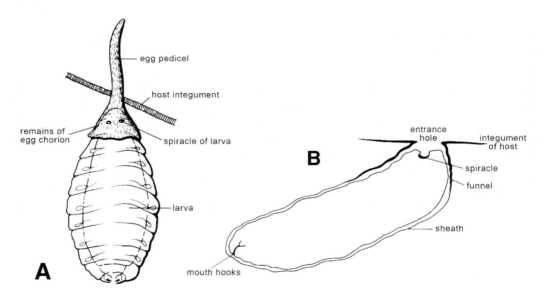

FIGURE 15.9. Respiratory systems of endoparasites. (A) Larva of *Blastothrix* (Hymenoptera) attached posteriorly to remains of egg, thereby maintaining contact with the atmosphere via the egg pedicel; and (B) larva of *Thrixion* (Diptera) surrounded by the respiratory funnel formed by ingrowth of the host's integument. [From A. D. Imms, 1937, *Recent Advances in Entomology*, 2nd ed. By permission of Churchill-Livingstone, Publishers.)

respiratory funnel produced by the host in an attempt to encapsulate the parasite (Figure 15.9B). The funnel is produced by inward growth of the host's integument or tracheal wall. Within it, the parasite attaches itself by means of mouth hooks while retaining contact with atmospheric air via the entrance of the funnel.

6. Summary

The tracheal system is a system of gas-filled tubes that develops embryonically as a series of segmental invaginations of the integument. The invaginations anastomose and branch and eventually form tracheoles across which the vast majority of gas exchange occurs. The external openings of the tracheal system (spiracles) are generally equipped with valves, hairs, or sieve plates, whose primary function is probably prevention of water loss.

Because oxygen can diffuse more rapidly in the gaseous state and is in higher concentration in air than in water, the requirements of most small insects and many large resting insects can be satisfied entirely by diffusion. Possibly to reduce water loss or as an adaptation to living in hypoxic or hypercapnic environments, some larger insects use a discontinuous gas exchange cycle (DGC). In DGC the spiracles are kept almost closed and carbon dioxide is temporarily stored, largely as bicarbonate in the hemolymph. As oxygen is used, a slight vacuum is created in the tracheal system that sucks in more air and reduces outward diffusion of water vapor. Periodically, the spiracular valves are opened for a short time when massive release of carbon dioxide occurs.

To increase the diffusion gradient between the tracheal system and tissues, many large insects ventilate the system by alternately increasing and decreasing its volume. Frequently the volume of tidal air moved during ventilation is increased through the development of large compressible air sacs, and by unidirectional air flow. Autoventilation, which relies on movements of the pterothorax during wing beating, occurs in many groups that have synchronous flight muscles.

Aquatic insects have either a closed tracheal system in which the spiracles are sealed or an open tracheal system with functional spiracles. In many insects with a closed tracheal system, cutaneous diffusion may entirely satisfy oxygen requirements. Accessory respiratory structures (tracheal gills) may be present though these are often important only under oxygen-deficient conditions. Some insects with open tracheal systems obtain oxygen by periodic visits to the water surface, or from air spaces in plants. Others hold a store of gas (gas gill) about their body. This may be either temporary, when an insect must visit the water surface to renew the oxygen content of the gill, or permanent (a plastron), when oxygen is renewed by diffusion into the gill from the surrounding medium.

For many endoparasitic insects, cutaneous diffusion is sufficient to satisfy requirements. In others there are special structural adaptations that ensure that the parasite remains in contact with atmospheric air.

7. Literature

Gas exchange is reviewed by Mill (1972, 1974, 1985), and Miller (1974). Hinton (1968) discusses spiracular gills. The regulation of gas exchange is examined by Miller (1966). Whitten (1972) and Mill (1997) provide structural details of the tracheal system.

Burnside, C. A., and Robinson, J. V., 1995, The functional morphology of caudal lamellae in coenagrionid (Odonata: Zygoptera) damselfly larvae, *Zool. J. Linn. Soc.* **114**:155–171.

Bustami, H. P., Harrison, J. F., and Hustert, R., 2002, Evidence for oxygen and carbon dioxide receptors in insect CNS influencing ventilation, *Comp. Biochem. Physiol. A* **133**:595–604.

Chown, S. L., 2002, Respiratory water loss in insects, *Comp. Biochem. Physiol. A* **133**:791–804.

Chown, S. L., and Holter, P., 2000, Discontinuous gas exchange cycles in *Aphodius fossor* (Scarabaeidae): A test of hypotheses concerning origins and mechanisms, *J. Exp. Biol.* **203**:397–403.

Hinton, H. E., 1968, Spiracular gills, *Adv. Insect Physiol.* **5**:65–162.

Hughes, G. M., and Mill, P. J., 1966, Patterns of ventilation in dragonfly larvae, *J. Exp. Biol.* **44**:317–334.

Lighton, J. R. B., 1996, Discontinuous gas exchange in insects, *Annu. Rev. Entomol.* **41**:309–324.

Lighton, J. R. B., 1998, Notes from underground: Towards ultimate hypotheses of cyclic, discontinuous gas exchange in tracheate arthropods, *Am. Zool.* **38**:483–491.

Locke, M., 1966, The structure and formation of the cuticulin layer in the epicuticle of an insect, *Calpodes ethlius*, *J. Morphol.* **118**:461–494.

Locke, M., 1998, Caterpillars have evolved lungs for hemocyte gas exchange, *J. Insect Physiol.* **44**:1–20.

Mill, P. J., 1972, *Respiration in the Invertebrates*, Macmillan, London.

Mill, P. J., 1974, Respiration: Aquatic insects, in: *The Physiology of Insecta*, 2nd ed., Vol. VI (M. Rockstein, ed.), Academic Press, New York.

Mill, P. J., 1977, Ventilation motor mechanisms in the dragonfly and other insects, in: *Identified Neurons and Behavior of Arthropods* (G. Hoyle, ed.), Plenum Press, New York.

Mill, P. J., 1985, Structure and physiology of the respiratory system, In: *Comprehensive Insect Physiology, Biochemistry and Pharmacology*, Vol. 3 (G. A. Kerkut and L. I. Gilbert, eds.), Pergamon Press, Elmsford, NY.

Mill, P. J., 1997, Tracheae and tracheoles, in: *Microscopic Anatomy of Invertebrates*, Vol. 11a (F. W. Harrison and M. L. Locke, eds.), Wiley-Liss, New York.

Miller, P. L., 1960, Respiration in the desert locust. I-III., *J. Exp. Biol.* **37**:224–236, 237–263, 264–278.

Miller, P. L., 1966, The regulation of breathing in insects, *Adv. Insect Physiol.* **3**:279–344.

Miller, P. L., 1974, Respiration-aerial gas transport, in: *The Physiology of Insecta*, 2nd ed., Vol. VI (M. Rockstein, ed.), Academic Press, New York.

Miller, P. L., 1993, Responses of rectal pumping to oxygen lack by larval *Calopteryx* splendens (Zygoptera: Odonata), *Physiol. Entomol.* 18:379–388.

Pennak, R. W., and McColl, C. M., 1944, An experimental study of oxygen absorption in some damselfly naiads, *J. Cell. Comp. Physiol.* **23**:1–10.

Shelton, T. G., and Appel, A. G., 2000, Cyclic CO_2 release and water loss in the western drywood termite (Isoptera: Kalotermitidae), *Ann. Entomol. Soc. Am.* **93**:1300–1307.

Shelton, T. G., and Appel, A. G., 2001, Carbon dioxide release in *Coptotermes formosanus* Shiraki and *Reticulitermes flavipes* (Kollar): Effects of caste, mass, and movement, *J. Insect Physiol.* **47**:213–224.

Weis-Fogh, T., 1964, Diffusion in insect wing muscle, the most active tissue known, *J. Exp. Biol.* **41**:229–246.

Westneat, M. W., Betz, O., Blob, R. W., Fezzaa, K., Cooper, W. J., and Lee, W.-K., 2003, Tracheal respiration in insects visualized with synchrotron X-ray imaging, *Science* **299**:558–560.

Whitten, M. J., 1972, Comparative anatomy of the tracheal system, *Annu. Rev. Entomol.* **17**:373–402.

Wigglesworth, V. B., 1953, Surface forces in the tracheal system of insects, *Q. J. Microsc. Sci.* **94**:507–522.

16

Food Uptake and Utilization

1. Introduction

Insects feed on a wide range of organic materials. About 75% of all species are phytophagous, and these form an important link in the transfer of energy from primary producers to second-order consumers. Others are carnivorous, omnivorous, or parasitic on other animals. In accord with the diversity of feeding habits, the means by which insects locate their food, the structure and physiology of their digestive system, and their metabolism are highly varied.

The feeding habits of insects take on special significance for humans, on the one hand, because of the enormous damage that feeding insects do to our food, clothing, and health, and, on the other, because of the massive benefits that insects provide as plant pollinators during their search for food (see also Chapter 24). In addition, because many species are easily and cheaply mass-cultured in the laboratory, they have been used widely in research on digestion and absorption, as well as in the elucidation of basic biochemical pathways, the role of specific nutrients, and other aspects of animal metabolism.

2. Food Selection and Feeding

Distinct visual, chemical, and mechanical cues act at each step of the food location and ingestion process. These steps include attraction to food, arrest of movement, tasting, biting, further tasting as ingestion begins, continued ingestion, and termination of feeding. The sensitivity of the insect to these cues varies with its physiological state. For example, a starved insect may become highly sensitive to odors or tastes associated with its normal food, and in extreme cases may become quite indiscriminate in terms of what it ingests. On the other hand, a female whose abdomen is full of eggs is normally "uninterested" in feeding.

In some plant-feeding (phytophagous) species, visual stimuli such as particular patterns (especially stripes) or colors may serve to initially attract an insect to a potential food source. Usually, however, the initial orientation, where this occurs, is dependent on olfactory stimuli. In many larval forms there appear to be no specific orienting stimuli because, under normal circumstances, larvae remain on the food plant selected by the mother prior to oviposition. In the migratory locust, on which much work has been done, olfaction is of primary importance in food location. Once the insect makes contact with the vegetation, tarsal chemosensilla initiate a reflex that results in the stoppage of movement. Sensilla on the labial and maxillary palps then taste the surface waxes of the plant, after which the locust takes a small bite. Whether feeding continues is sometimes determined by mechanosensillar responses to physical stimuli such as the hardness, toughness, shape, and hairiness of the food. More commonly, it is substances in the released sap that, by stimulating chemosensilla in the cibarial cavity, regulate the continuation or arrest of feeding (Chapman, 2003). These substances are called "phagostimulants" or "deterrents," respectively. The substances may have nutritional value to the insect or may be nutritionally unimportant ("token stimuli"). Nutritional factors are almost always stimulating in effect. Sugars, especially sucrose, are important phagostimulants for most phytophagous insects. Amino acids, in contrast, are generally by themselves weakly stimulating or non-stimulating, though may act synergistically with certain sugars or token stimuli. For example, Heron (1965) showed in the spruce budworm (*Choristoneura fumiferana*) that, whereas sucrose and L-proline in low concentration were individually only weak phagostimulants, a mixture of the two substances was highly stimulating. In addition to sugars and amino acids, other specific nutrients may stimulate feeding in a given species. Such nutrients include vitamins, phospholipids, and steroids. Token stimuli may either stimulate or inhibit feeding. Thus, derivatives of mustard oil, produced by cruciferous plants, including cabbage and its relatives, are important phagostimulants for a variety of insects that normally feed on these plants, for example, larvae of the diamondback moth (*Plutella xylostella*), the cabbage aphid (*Brevicoryne brassicae*), and the mustard beetle (*Phaedon cochleariae*). Indeed, *Plutella* will feed naturally only on plants that contain mustard oil compounds. Many secondary plant metabolites, including alkaloids, terpenoids, phenolics, and glycosides, are feeding deterrents for phytophagous insects. In a given food source there will probably be a mixture of phagostimulants and deterrents, and the balance of this sensory input, integrated through the central nervous system, determines the overall palatability of the food.

Species whose choice of food is limited are said to be oligophagous. In extreme cases, an insect may be restricted to feeding on a single plant species and is described as monophagous. Species that may feed on a wide variety of plants are polyphagous, though it must be noted that even these exhibit selectivity when given a choice. Not surprisingly, monophagous and oligophagous species are especially sensitive to the presence of deterrents in non-host plants.

In many predaceous insects, especially those that actively pursue prey, vision is of primary importance in locating and capturing food. As noted in Chapter 12 (Section 7.1.2), some predaceous insects have binocular vision that enables them to determine when prey is within catching distance. Carnivorous species, especially larval forms, whose visual sense is less well developed, depend on chemical or tactile stimuli to find prey. For example, many beetle larvae that live on or in the ground locate prey by their scent. Species parasitic on other animals usually locate a host by its scent, though tsetse flies may initially orient by visual means to a potential host. For many species that feed on the blood of birds and mammals, temperature and/or humidity gradients are important in determining the precise location at which an insect alights on a host and begins to feed.

The extent of food specificity for carnivorous insects is varied. Many insects are quite non-specific and will attempt to capture and eat any organism that falls within a given size range (even to the extent of being cannibalistic). Others are more selective; for example, spider wasps (Pompilidae), as their name indicates, capture only spiders for provisioning their nest. Parasitic insects, too, exhibit various degrees of host specificity. Thus, certain sarcophagid flies parasitize a range of grasshopper species; the common cattle grub (*Hypoderma lineatum*) is typically found on cattle or bison, rarely on horses and humans; lice are extremely host-specific, as would be expected of sedentary species.

The termination of feeding (assuming that food supply is not limiting) is largely related to the amount of food ingested and the stimulation of strategically located stretch receptors. In locusts, for example, the filling of the crop is measured by receptors at the anterior end that send signals to the brain via the stomatogastric system. In flies both crop- and esophagus-filling are important in bringing feeding to a close, while in female mosquitoes and, probably, *Rhodnius prolixus* the signals arise from stretch receptors located in the abdominal wall, which is greatly distended after a blood meal. Other factors that may play a minor role in terminating feeding are adaptation of the chemosensilla on the mouthparts and a change in the osmotic pressure or composition of the hemolymph as absorption of digested materials occurs (see chapters in Chapman and de Boer, 1995).

Apart from the specific cues outlined above that facilitate location and selection of food, there are other factors that influence feeding activity. Typically, insects do not feed shortly before and after a molt, or when there are mature eggs in the abdomen. In addition, a diurnal rhythm of feeding activity may occur, in response to a specific light, temperature, or humidity stimulus. For example, the red locust (*Nomadacris septemfasciata*) feeds in the morning and evening, and many mosquitoes feed during the early evening (though this may change in different habitats). Pupae, most diapausing insects, and some adult Ephemeroptera, Lepidoptera, and Diptera do not feed.

3. The Alimentary System

The gut and its associated glands (Figure 16.1) triturate, lubricate, store, digest, and absorb food material and expel the undigested remains. Structural differences throughout the system reflect regional specialization for performance of these functions and are correlated also with feeding habits and the nature of normal food material. The structure of the system may vary at different stages of the life history because of the different feeding habits of the larva and adult of a species. The gut normally occurs as a continuous tube between the mouth and anus, and its length is broadly correlated with feeding habits, being short in carnivorous forms where digestion and absorption occur relatively rapidly, and longer (often convoluted) in phytophagous forms. In a few species that feed on fluids, such as larvae of Neuroptera and Hymenoptera-Apocrita, and some adult Heteroptera there is little or no solid waste in the food, and the junction between the midgut and hindgut is occluded.

As Figure 16.1 indicates, food first enters the buccal cavity, which is enclosed by the mouthparts and is not strictly part of the gut. It is into the buccal cavity that the salivary glands release their products. The gut proper comprises three main regions: the foregut, in which the food may be stored, filtered, and partially digested; the midgut, which is the primary site for digestion and absorption of food; and the hindgut, where some absorption and feces formation occur.

3.1. Salivary Glands

Salivary glands are present in most insects, though their form and function are extremely varied, and they may or may not be innervated (Ribeiro, 1995). Frequently they are known

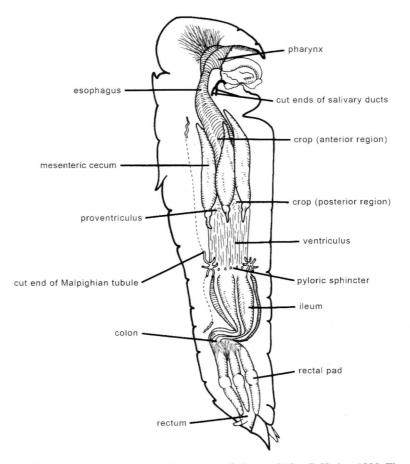

FIGURE 16.1. Alimentary canal and associated structures of a locust. [After C. Hodge, 1939, The anatomy and histology of the alimentary tract of *Locusta migratoria* L. (Orthoptera: Acrididae), *J. Morphol.* **64**:375–399. By permission of the Wistar Press.]

by other names according to either the site at which their duct enters the buccal cavity, for example, labial glands and mandibular glands, or their function, for example, silk glands and venom glands.

Typically, saliva is a watery, enzyme-containing fluid that serves to lubricate the food and initiate its digestion. Like that of humans, the saliva generally contains only carbohydrate-digesting enzymes (amylase and invertase), though there are exceptions to this statement. For example, the saliva of some carnivorous species contains protein- and/or fat-digesting enzymes only; that of bloodsucking species has no enzymes. In termite saliva there are cellulose-digesting enzymes: a β-1-4-glucanase that brings about the initial splitting of the polymer, and β-glucosidase that degrades the resulting cellobiose to glucose (Nakashima *et al.*, 2002; Tokuda *et al*, 2002). (See also Section 4.2.4.)

In the innervated glands of cockroaches and locusts, release of saliva is induced when food stimulates mechano- and chemosensilla on the mouthparts and antennae. The information travels to the subesophageal ganglion and then along aminergic or peptidergic neurons to the glands where it induces relaxation of the muscles that normally close off the opening of the salivary gland duct (Ali, 1997). In contrast, the non-innervated glands of *Calliphora erythrocephala* are stimulated to release saliva by a hemolymph factor, possibly serotonin (Trimmer, 1985).

Other substances that may occur in saliva, though having no direct role in digestion, are important in food acquisition. For example, the saliva of aphids has a viscous component, released during penetration of the stylets, which hardens to form a leakproof seal around the mouthparts. Aphid saliva also contains pectinase and peroxidase. The former facilitates penetration of the stylets through the intercellular spaces of plant tissues while the latter may inactivate toxic phytochemicals (Miles, 1999). Hyaluronidase, which breaks down connective tissue, is secreted by some insects that suck animal tissue fluids. A spectrum of compounds that assist feeding is present in the saliva of bloodsucking species. These include anticoagulants, inhibitors of platelet disintegration, pyrase (an enzyme that breaks down ADP, to prevent platelet aggregation), and vasodilators such as nitric oxide (Ribeiro, 1995; Ribeiro and Francischetti, 2003). The nitric oxide is carried to the host's skin on heme-containing proteins (nitrophorins) (Valenzuela and Ribeiro, 1998). The nitrophorins also strongly bind histamine, released by the host to induce wound healing (Weichsel *et al.*, 1998). Toxins (venoms), which paralyze or kill the prey, occur in the saliva of some assassin bugs (Reduviidae) and robber flies (Asilidae). It is also reported that substances that induce gall formation by stimulating cell division and elongation are present in the saliva of some gall-inhabiting species. Larvae of black flies and chironomid midges secrete large amounts of viscous saliva, forming nets that capture food particles.

In some species the glands have taken on functions quite unrelated to feeding, for example, production of cocoon silk by the labial glands of caterpillars and caddisfly larvae, and pheromone production by the mandibular glands of the queen honeybee.

3.2. Foregut

The foregut, formed during embryogenesis by invagination of the integument, is lined with cuticle (the intima) that is shed at each molt. Surrounding the intima, which may be folded to enable the gut to stretch when filled, is a thin epidermis, small bundles of longitudinal muscle, a thick layer of circular muscle, and a layer of connective tissue through which run nerves and tracheae (Figure 16.2). The foregut is generally differentiated into pharynx, esophagus, crop, and proventriculus. Attached to the pharyngeal intima are dilator muscles. These are especially well developed in sucking insects and form the pharyngeal pump (Chapter 3, Section 3.2.2). The esophagus is usually narrow but posteriorly may be dilated to form the crop where food is stored. In Diptera and Lepidoptera, however, the crop is actually a diverticulum off the esophagus. During storage the food may undergo some digestion in insects whose saliva contains enzymes or that regurgitate digestive fluid from the midgut. In some species the intima of the crop forms spines or ridges that probably aid in breaking up solid food into smaller particles and mixing in the digestive fluid (Figure 16.2A). The hindmost region of the foregut is the proventriculus, which may serve as a valve regulating the rate at which food enters the midgut, as a filter separating liquid and solid components, or as a grinder to further break up solid material. Its structure is, accordingly, quite varied. In species where it acts as a valve the intima of the proventriculus may form longitudinal folds and the circular muscle layer is thickened to form a sphincter. When a filter, the proventriculus contains spines that hold back the solid material, permitting only liquids to move posteriorly. Where the proventriculus acts as a gizzard, grinding up food, the intima is formed into strong, radially arranged teeth, and a thick layer of circular muscle covers the entire structure (Figure 16.2B).

Posteriorly the foregut is invaginated slightly into the midgut to form the esophageal (= stomodeal) invagination (Figure 16.3). Its function is to ensure that food enters the

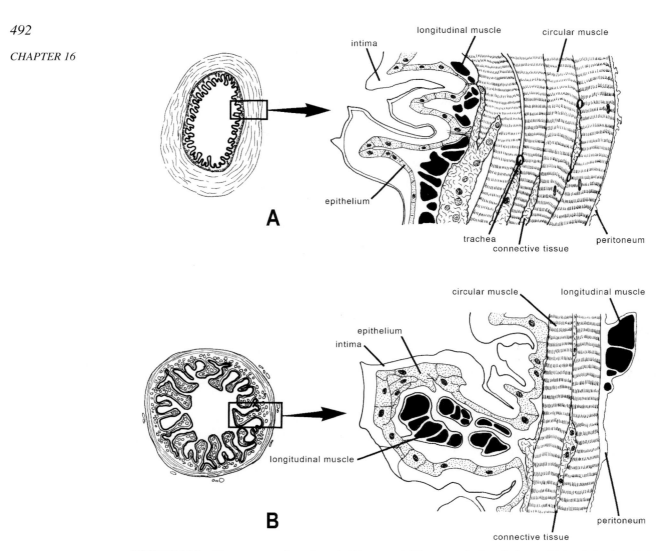

FIGURE 16.2. Transverse sections through (A) crop and (B) proventriculus of a locust. [After C. Hodge, 1939, The anatomy and histology of the alimentary tract of *Locusta migratoria* L. (Orthoptera: Acrididae), *J. Morphol.* **64**:375–399. By permission of Wistar Press.]

midgut within the peritrophic matrix. It also appears to assist in molding the peritrophic matrix into the correct shape in some insects.

3.3. Midgut

The midgut (= ventriculus = mesenteron) is of endodermal origin and, therefore, has no cuticular lining. In most insects, however, it is lined by a thin peritrophic matrix (PM) composed of proteins bound to a meshwork of chitin fibrils (Figure 16.4). Some PM proteins, the peritrophins, are heavily glycosylated like mucus in the intestine of vertebrates. The functions of the PM are to prevent mechanical damage to the midgut epithelium, to prevent entry of microorganisms into the body cavity, to bind potential toxins and other damaging chemicals, and to compartmentalize the midgut lumen, that is, to divide it into an endoperitrophic space (within the matrix) and an ectoperitrophic space (adjacent to the

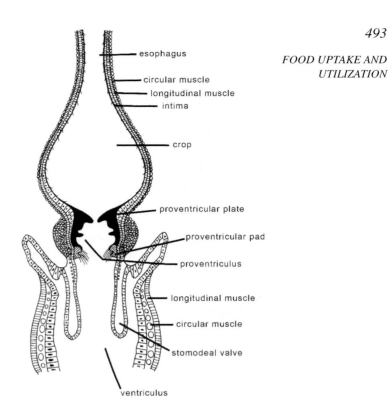

FIGURE 16.3. Longitudinal section through crop, proventriculus, and anterior midgut of a cockroach. [From R. E. Snodgrass, *Principles of Insect Morphology*. Copyright 1935 by McGraw-Hill, Inc. Used with permission of McGraw-Hill Book Company.]

midgut epithelium) (Terra, 1996; Lehane, 1997). This separation of the epithelium from the food improves digestive efficiency by segregating enzymes between the spaces and enabling some enzymes to be recycled (Section 4.2.1).

The PM is generally absent in fluid-feeding insects, for example, Hemiptera, adult Lepidoptera, and bloodsucking Diptera. However, some insects produce the PM only at certain times (e.g., female mosquitoes after a blood meal). Further, as described below, the

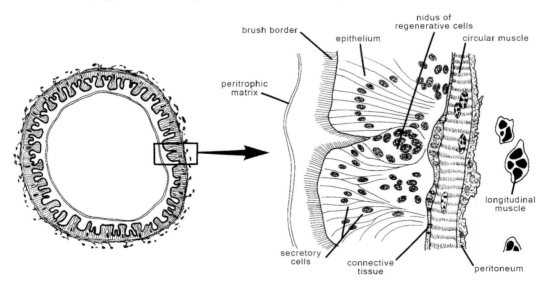

FIGURE 16.4. Transverse section through midgut of a locust. [After C. Hodge, 1939, The anatomy and histology of the alimentary tract of *Locusta migratoria* L. (Orthoptera: Acrididae), *J. Morphol.* **64**:375–399. By permission of Wistar Press.]

type of PM, whether or not a PM is produced, and the manner in which it is produced, may vary between life stages (Lehane, 1997).

The PM is formed in two principal ways. In Type I PM delamination of successive concentric lamellae occurs along the midgut (in Odonata, Ephemeroptera, Phasmida, some Orthoptera, some Coleoptera, and larval Lepidoptera). The Type II PM forms by secretion from a special zone of cells (cardia) at the anterior end of the midgut (in Diptera, Dermaptera, Isoptera, Embioptera, and some Lepidoptera). In this method the esophageal invagination presses firmly against the anterior wall of the midgut so that the originally viscous secretion of the PM-producing cells, as it hardens, is squeezed to form the tubular membrane. In Dictyoptera, other Orthoptera and Lepidoptera, Hymenoptera, and Neuroptera, a combination of both methods seems to be used. In mosquitoes, larvae produce a Type II PM, whereas the adults have a Type I PM.

The PM is made up of a meshwork of microfibrils between which is a thin proteinaceous film. The microfibrils have a constant 60° orientation to each other in Type I PM, thought to result from their secretion by the hexagonally close-packed microvilli of the epithelial cells. In Type II PM the orientation of the microfibrils is random. The PM is permeable to the products of digestion and to certain digestive enzymes released from the epithelial cells (Section 4.2.1). However, it is not permeable to other large molecules, such as undigested proteins and polysaccharides, indicating that the PM has a distinct polarity and is not merely an ultrafilter (Richards and Richards, 1977; Lehane, 1997).

The midgut is usually not differentiated into structurally distinct regions apart from the development, at the anterior end, of a varied number of blindly ending ceca, which serve to increase the surface area available for enzyme secretion and absorption of digested material. In many Heteroptera, however, the midgut is divided into three or four easily visible regions. In the chinch bug (*Blissus leucopterus*) four such regions occur (Figure 16.5). The anterior region is large and saclike, and serves as a storage region (no crop is present). The second region serves as a valve to regulate the flow of material into the third region where digestion

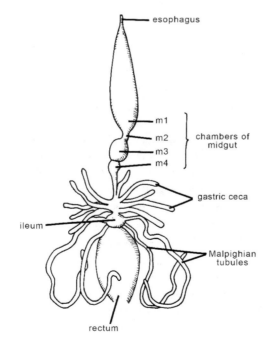

FIGURE 16.5. Alimentary canal of chinch bug (*Blissus leucopterus*) showing regional differentiation of midgut. [After H. Glasgow, 1914, The gastric caeca and the caecal bacteria of the Heteroptera, *Biol. Bull.* **26**:101–170.]

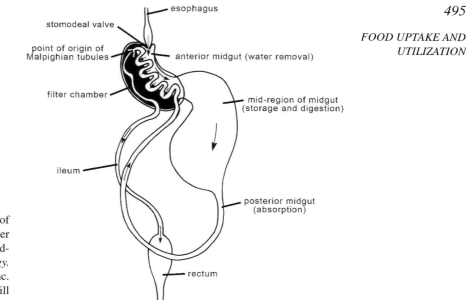

FIGURE 16.6. Alimentary canal of cercopid (Cercopoidea) showing filter chamber arrangement. [From R. E. Snodgrass, *Principles of Insect Morphology.* Copyright 1935 by McGraw-Hill, Inc. Used with permission of McGraw-Hill Book Company.]

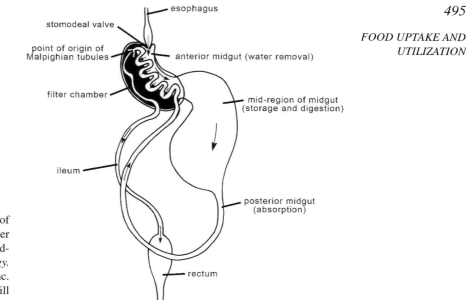

probably occurs. Ten fingerlike ceca filled with bacteria are attached to the fourth region, which may be absorptive in function. The role of the bacteria is not known.

In many homopterans, which feed on plant sap, the midgut is modified both morphologically and anatomically so that excess water present in the food can be removed, thus preventing dilution of the hemolymph. Though details vary among different groups of homopterans, the anterior end of the midgut (or, in some species, the posterior part of the esophagus) is brought into close contact with the posterior region of the midgut (or anterior hindgut), and the region of contact becomes enclosed within a sac called the "filter chamber" (Figure 16.6). Such an arrangement facilitates rapid movement of water by osmosis from the lumen of the anterior midgut across the wall of the posterior midgut and possibly also the Malpighian tubules. Thus, relatively little of the original water in the food actually passes along the full length of the midgut.

The lack of morphological differentiation within the midgut of most species is reflected in its uniform histology. Throughout its length, the mature cells lining the lumen are identical and serve to produce digestive enzymes, to absorb the products of digestion, and in some insects secrete the Type I PM. Replacement of degenerate cells occurs with the maturation and differentiation of regenerative cells found singly or in groups (nidi) near the base of the epithelium (Figure 16.4). Numerous peptide hormone-containing cells also occur in the midgut, which may play a role in modulating midgut contraction (Lange and Orchard, 1998).

In some species histological differentiation is found. For example, specialization of certain anterior cells for Type I PM production was noted earlier. In addition, differentiation into digestive and absorptive regions occurs in some species. In tsetse flies the cells of the anterior midgut are small and are concerned with absorption of water from the ingested blood. They produce no enzymes and digestion does not begin until food reaches the middle region where the cells are large, rich in ribonucleic acid, and produce enzymes. In the posterior midgut the cells are smaller, closely packed, and probably concerned with absorption of digested food. In some species different regions of the midgut are apparently adapted to the absorption of particular food materials. In *Aedes* larvae the anterior midgut is concerned with fat absorption and storage, whereas the posterior portion absorbs carbohydrates and stores them as glycogen. In larval Lepidoptera goblet cells, with a large flask-shaped central cavity, are scattered among the regular epithelial cells. They are thought to play a role in the regulation of the potassium level within the hemolymph (Chapter 18, Section 2.2).

3.4. Hindgut

The hindgut is an ectodermal derivative and, as such, is lined with cuticle, though this is thinner than that of the foregut, a feature related to the absorptive function of this region. The epithelial cells that surround the cuticle are flattened except in the rectal pads (see below) where they become highly columnar and filled with mitochondria. Muscles are only weakly developed and, usually, the longitudinal strands lie outside the sheet of circular muscle.

The hindgut usually has the following regions: pylorus, ileum, and rectum. The pylorus may have a well-developed circular muscle layer (pyloric sphincter) and regulate the movement of material from midgut to hindgut. Also, the Malpighian tubules characteristically enter the gut in this region. The ileum (Figure 16.7A) is generally a narrow tube that serves to conduct undigested food to the rectum for final processing. In some insects, however, some absorption of ions and/or water may occur in this region. In a few species production and excretion of nitrogenous wastes occur in the ileum (Chapter 18, Section 2.2). In many wood-eating insects, for example, species of termites and beetles, the ileum is dilated to form a fermentation pouch housing bacteria or protozoa that digest wood particles. The products of digestion, when liberated by the microorganisms, are absorbed across the wall of the ileum. The most posterior part of the gut, the rectum, is frequently dilated. Though for the most part thin-walled, the rectum includes six to eight thick-walled rectal pads (Figure 16.7B) whose function is to absorb ions, water, and small organic molecules (Chapter 18, Section 4). As a result, the feces of terrestrial insects are expelled as a more or less dry pellet. Frequently, the pellets are ensheathed within the PM, which continues into the hindgut.

4. Gut Physiology

The primary functions of the alimentary canal are digestion and absorption. For these processes to occur efficiently, food is moved along the canal. In some species, enzyme secretions are moved anteriorly so that digestion can begin some time before food reaches the region of absorption.

4.1. Gut Movements

Though the alimentary canal is innervated, neural control is principally associated with the opening/closing of valves that occur within the canal (see below). The rhythmic peristaltic muscle contractions that move food posteriorly through the gut are myogenic; that is, they originate within the muscles themselves rather than occurring as a result of nervous stimuli. Myogenic centers have been located in the esophagus, crop, and proventriculus, in *Galleria*, for example. In insects that form a Type II PM, backward movement of food is aided by growth of the membrane. Antiperistaltic movements also occur in some species and serve to move digestive fluid forward from the midgut into the crop.

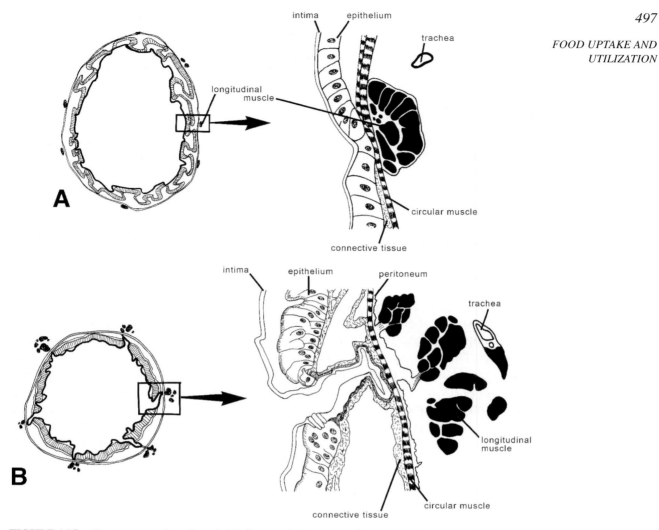

FIGURE 16.7. Transverse sections through (A) ileum and (B) rectum of a locust. [After C. Hodge, 1939, The anatomy and histology of the alimentary tract of *Locusta migratoria* L. (Orthoptera: Acrididae), *J. Morphol.* **64**:375–399. By permission of Wistar Press.]

The rate at which food moves through the gut is not uniform. It varies according to the physiological state of an insect; for example, it is greater when an insect has been starved previously or is active. The rate may also differ between sexes and with age. Another important variable is the nature of the food. Some insects are able to move some components of the diet rapidly through the gut while retaining others for considerable periods. Within the gut, food moves at variable rates in different regions.

The proventricular and pyloric valves are important regulators of food movement, though little is known about how their opening and closing are controlled. In *Periplaneta* opening of the proventriculus was shown to depend on the osmotic pressure of ingested fluid (Davey and Treherne, 1963). As the concentration is increased, the proventriculus opens less often and less widely, and vice versa. Davey and Treherne suggested that osmoreceptors in the pharynx provide information on the osmotic pressure of the food and this information travels via the frontal ganglion to the ingluvial ganglion that controls the proventriculus. However, no osmoreceptor has been located and it may be that the osmotic feedback comes from the hemolymph rather than directly from the gut as seems to be the case in *Locusta*. Distension of the foregut or, in blood-feeding species, the abdomen, is known to cause release of neurosecretion from the corpora cardiaca which enhances gut peristalsis and, hence, the rate of food passage. Localized enhancement of peristalsis may be induced by release of peptide hormones from cells in the wall of the midgut (Lange and Orchard, 1998).

4.2. Digestion

As noted above, digestion may be initiated by enzymes present in the saliva either mixed with the food as it enters the buccal cavity or secreted onto the food prior to ingestion. Most digestion is dependent, however, on enzymes secreted by the midgut epithelium. Digestion mostly occurs in the lumen of the midgut, though regurgitation of digestive fluid into the crop is important in some species. In wood-eating forms, much of the digestion is carried out by microorganisms in the hindgut (Section 4.2.4).

4.2.1. Digestive Enzymes

A wide variety and large number of digestive enzymes have been reported for insects. In many instances, however, enzymes have been characterized (and named) on the basis of their activity on unnatural substrates, that is, materials that do not occur in the normal diet of the insect. This is because many digestive enzymes, especially carbohydrases, are "group-specific"; that is, they hydrolyze any substrate that includes a particular bond between two parts of the molecule. For example, α-glucosidase splits all α-glucosides, including sucrose, maltose, furanose, trehalose, and melezitose. Further, in preparing enzyme extracts for analysis, either gut contents or midgut tissue homogenates are typically used. As House (1974) noted, the former may include enzymes derived from the food *per se*, while the latter contains endoenzymes (intracellular enzymes) that have no digestive function. Thus, reports on digestive enzyme activity must be examined cautiously.

As would be expected, the enzymes produced reflect both qualitatively and quantitatively the normal constituents of the diet. Omnivorous species produce enzymes for digesting proteins, fats, and carbohydrates. Carnivorous species produce mainly lipases and proteases; in some species these may be highly specific in action. Blow fly larvae (*Lucilia cuprina*), for example, produce large amounts of collagenase. The nature of the enzymes produced may change at different stages of the life history as the diet of an insect changes. For example, caterpillars feeding on plant tissue secrete a spectrum of enzymes, whereas nectar-feeding adult Lepidoptera produce only invertase. Interestingly, however, even in those endopterygotes in which the larvae and adults utilize the same food the properties of the enzymes change at metamorphosis. In *Tenebrio*, for example, the larval and adult trypsins and chymotrypsins differ in molecular size, substrate specificity, and kinetics, though why this should be is not clear.

Insects can digest a wide range of carbohydrates, even though only a few distinct enzymes may be produced. As noted earlier, α-glucosidase will hydrolyze all α-glucosides. Likewise, β-glucosidase facilitates splitting of cellobiose, gentiobiose, and phenylglucosides; β-galactosidase hydrolyzes β-galactosides such as lactose. In some species, however, there appear to be carbohydrate-digesting enzymes that exhibit absolute specificity. Thus, adult *Lucilia cuprina* produce an α-glucosidase, trehalase, that splits only trehalose. The normal polysaccharide-digesting enzyme produced is amylase for hydrolysis of starch, though particular species may produce enzymes for digestion of other polysaccharides. For example, firebrats and silverfish (Zygentoma), larvae of wood-boring Cerambycidae and Anobiidae (Coleoptera), as well as both lower and higher termites, have endogenous cellulases, though in most insects production of this enzyme is restricted to microorganisms present in the hindgut. Curiously, lower termites produce cellulase in the saliva, whereas in higher termites the midgut is the source of this enzyme. Scolytinae (Coleoptera) produce a hemicellulase, chitinase is reported to occur in the intestinal juice of *Periplaneta*, and some herbivorous Orthoptera produce lichenase.

As in other organisms, the protein-digesting enzymes produced by the midgut are divisible into two types: endopeptidases, which effect the initial splitting of proteins into polypeptides, and exopeptidases, which bring about degradation of polypeptides by the sequential splitting off of individual amino acids from each end of a molecule. Exopeptidases can be further categorized into carboxypeptidases, which remove amino acids from the carboxylic end of a polypeptide, and aminopeptidases, which cause hydrolysis at the amino end of a molecule. A dipeptidase also is frequently present. In some species only endopeptidases occur in the midgut lumen (specifically within the endoperitrophic space), the exopeptidases being found outside the PM or even attached to the apical plasma membrane of the epithelial cells. Some insects produce specific enzymes for the digestion of particularly resistant structural proteins. Collagenase has been mentioned already. Keratin, the primary constituent of wool, hair, and feathers, is a fibrous protein whose polypeptide components lie side by side linked by highly stable disulfide bonds between adjacent sulfur-containing amino acids, such as cystine and methionine. A keratinase has been identified in clothes moth larvae (*Tineola*) and may also occur in other keratin-digesting species, such as dermestid beetles and Mallophaga. The keratinase is active only under anaerobic (reducing) conditions and, in this context, it is interesting to note that the midgut of *Tineola* is poorly tracheated.

Dietary fats of either animal or plant origin are almost always triglycerides, that is, glycerol in combination with three fatty acid molecules. The latter may range from unsaturated to fully saturated. Lipases, which hydrolyze fats to the constituent fatty acids and glycerol, have low specificity. Therefore, the presence of one such enzyme will normally satisfy an insect's needs. In a few species, however, at least two lipases have been identified, having different pH optima and acting on triglycerides of different sizes. Fat digestion is generally somewhat slow as insects lack anything comparable to the bile salts of vertebrates that would emulsify and stabilize lipid droplets.

4.2.2. Factors Affecting Enzyme Activity

According to House (1974), three factors markedly affect digestion in insects: pH, buffering capacity, and redox potential of the gut.

The pH determines not only the activity of digestive enzymes, but also the nature and extent of microorganisms in the gut and the solubility of certain materials in the gut lumen. The latter affects the osmotic pressure of the gut contents and, in turn, the rate of absorption of molecules across the gut wall. Analyses of the pH in various regions of the gut have been made for a wide range of species, and various authors have attempted to correlate these with the feeding habits or phylogenetic position of an insect. At best, these correlations are only broadly correct, and many exceptions are known. In most insects the gut is slightly acid or slightly alkaline throughout its length. Further, the pH generally increases from foregut to midgut, then decreases from midgut to hindgut. Though the latter is true for most phytophagous species, in many omnivorous and carnivorous species the pH of the hindgut is greater than that of the midgut.

Many variables affect the pH of different regions of the gut. Generally the pH of the crop is the same as that of the food, though in some species it is consistently less than 7 because of the digestive activity of microorganisms or regurgitation of digestive juice from the midgut. The pH of the midgut differs among species but tends to be constant for a given species because of the presence in this region of buffering agents. In a few species there are local variations in pH within the midgut that can be related to changes in digestive function from one part to another. For example, in the cockroach *Nauphoeta cinerea* the pH of the anterior midgut is 6.0–7.2, which coincides with the pH optimum of the amylase found mainly in this region. In the posterior midgut, on the other hand, the pH is about 9, near the optimum for the proteinases that are active there (Elpidina *et al.*, 2001). The hindgut typically has a pH slightly less than 7, presumably resulting from the presence of the nitrogenous waste product, uric acid (Chapter 18, Section 3.2). The hindgut contents of some phytophagous species may be quite acidic as a result of the formation of organic acids from cellulose by symbiotic microorganisms.

The relatively constant pH found in different regions of the gut results from the presence in the lumen of both inorganic and organic buffering agents. In some species, inorganic ions, especially phosphates, but including aluminum, ammonium, calcium, iron, magnesium, potassium, sodium, carbonate, chloride, and nitrate, seem to offer sufficient buffering capacity. In other species organic acids, including amino acids and proteins, tend to supplement or replace the buffering effect of the inorganic ions. For some of the inorganic ions and water, active secretory or resorption mechanisms are known to regulate their concentration in the midgut. These mechanisms are also capable of inducing fluid flow, especially in the ectoperitrophic space. Specifically, secretion of ions and water across the posterior midgut epithelium simultaneously with their resorption at the anterior end establishes a forward flow of digestive fluid. This is thought to conserve nutrients and recycle enzymes.

Redox potential, which measures ability to gain or lose electrons, that is, to be reduced or oxidized, respectively, is an important factor in digestion in some insects as it affects the structure of both dietary proteins and proteolytic enzymes. The gut redox potential, which is closely linked to pH, is normally positive, indicating oxidizing (aerobic) conditions. However, in species able to digest keratin the redox potential of the midgut fluid is strongly negative. It has been suggested that such an anaerobic (reducing) environment is necessary to enable the keratinase to split the disulfide bonds (House, 1974). Subsequently, normal proteases hydrolyze the polypeptides.

4.2.3. Control of Enzyme Synthesis and Secretion

Numerous studies have shown that enzyme activity in the midgut varies in relation to food intake, though it is not always clear whether it is synthesis and/or release of the enzymes that is being controlled. In many species, including *Locusta migratoria* and *Tenebrio molitor*, enzymes are not stored in the midgut cells but are liberated immediately into the gut lumen. In other insects, for example, *Stomoxys calcitrans* and some mosquitoes, enzymes are stored (possibly in an inactive form) and released when feeding occurs. The synthesis/release of enzyme in proportion to the amount of food ingested may be regulated by secretagogue, hormonal or neural mechanisms, though there is very little evidence for the latter. Unfortunately, for most species, the evidence presented in support of one mechanism or another is equivocal. In a secretagogue system enzymes are produced in response to food present in the midgut. Presumably the amount produced is directly influenced by the concentration of food in the lumen. The best evidence for secretagogue control of enzyme activity is for mosquitoes and other blood feeders. In *Aedes* a blood meal cannulated directly into the midgut stimulates production of a proportionate amount of trypsin (Briegel and Lea, 1975). These authors showed that a variety of components of blood could serve as secretagogues. However, it is noteworthy that the amount of enzyme produced by this means was reduced in insects whose median neurosecretory cells had been removed. Where hormonal control of enzyme production has been proposed, the amount of food passing along the foregut, measured as the degree of stretching of the gut wall, is believed to result in the release from the corpora cardiaca of a proportionate amount of neurosecretion which travels via the hemolymph to the midgut cells.

At present, there is no consensus as to whether a midgut epithelial cell produces a complete package of enzymes or whether the proportions of different enzymes can vary with changes in the diet. Certainly, a secretagogue method for regulating enzyme activity could more easily account for changes in the level of specific enzymes reported to occur with alterations to the diet of some species.

4.2.4. Digestion by Microorganisms

Microorganisms (bacteria, fungi, and protozoa) may be present in the gut, but for only a few species has there been a convincing demonstration of their importance in digestion. In many insects microorganisms appear to have no role, as the insects can be reared equally well in their absence. In other species microorganisms may be more important with respect to an insect's nutrition than digestion *per se*. Where a role for microorganisms in digestion has been demonstrated, the relationship between the microorganisms and insect host is not always obligate, but may be facultative or even accidental.

Bacteria are important cellulose-digesting agents in many phytophagous insects, especially wood-eating species whose hindgut may include a fermentation pouch in which the microorganisms are housed. In other species, for example, the wood-eating cockroach *Panesthia*, bacteria in the crop are essential for cellulose digestion. In larvae of the wax moth *Galleria*, bacteria normally present in the gut undoubtedly aid in the digestion of beeswax, yet bacteriologically sterile larvae produce an intrinsic lipase capable of degrading certain wax components. Finally, many insects feed on decaying vegetation and must, therefore, ingest a large number of saprophytic bacteria which, temporarily at least, would continue their degradative activity in the gut. In this sense, therefore, though the relationship is accidental, the microorganisms are assisting in digestion.

In lower termites and some primitive wood-eating cockroaches (*Cryptocercus*), flagellate and ciliate protozoa occur in enormous numbers in the hindgut. The relationship between the insects and protozoa is mutualistic; that is, in return for a suitable, anaerobic environment in which to live, the protozoa phagocytose particles of wood eaten by the insects, fermenting the cellulose and releasing large amounts of glucose (in *Cryptocercus*) or organic acids (in termites) for use by the insects. In higher termites (Termitidae) the hindgut contains bacteria, not protozoa, but there is no evidence that the bacteria produce cellulolytic enzymes (and see Section 4.2.1).

Fungi rarely play a direct role in the digestive process of insects, though it is reported that yeasts capable of hydrolyzing carbohydrates occur in the gut of some leafhoppers (Cicadellidae). However, a mutualistic relationship has evolved between many fungi and insects whereby the fungi convert wood into a more usable form, while the insects serve to transport the fungi to new locations. Some ants and higher termites, for example, culture ascomycete or basidiomycete fungi in special regions of the nest called fungus gardens. Chewed wood or other vegetation is brought to the fungus garden and becomes the substrate on which the fungi grow, forming hyphae to be eaten by the insects. Certain wood-boring insects, for example, bark beetles (Scolytinae), inoculate their tunnels with fungal cells when they invade a new tree. The fungal mycelium that develops, along with partially decomposed wood, can then be used as food by the insects.

4.3. Absorption

The majority of absorption occurs in the midgut, especially the anterior portion, including the mesenteric ceca. A few reports have indicated that absorption of lipid materials (including the insecticides parathion and dieldrin) may occur across the crop wall, while the uptake of a range of small organic molecules occurs in the hindgut. The latter region is, however, primarily of importance as the site of water or ion resorption in connection with osmoregulation (Chapter 18, Section 4), though in insects that have symbiotic microorganisms in the hindgut it may also be an important site for absorption of small organic molecules, especially carboxylic and amino acids.

Most absorption of organic molecules across the midgut wall is passive, that is, from a higher to a lower concentration, though the rapid rate at which some molecules are absorbed suggests that special carriers facilitate their movement. The absorption rate is enhanced by a steep concentration gradient maintained between the midgut lumen and hemolymph. This may be achieved by absorption of water from the gut lumen so that the hemolymph becomes more dilute or by rapid conversion of the absorbed molecules to a more complex form. The absorption of organic molecules across the gut wall in *Schistocerca* and *Periplaneta* formed the subject of a series of papers by Treherne in the late 1950s (see Treherne, 1967, for details).

Using isotopically labeled monosaccharides, Treherne demonstrated that nearly all sugars are absorbed in the anterior region of the midgut, especially the ceca. Further, the monosaccharides are converted rapidly to the disaccharide trehalose in the fat body. Interestingly, in *Schistocerca* much of the fat body is in close proximity to the midgut wall. Of the monosaccharides studied, glucose was found to be absorbed most rapidly. Fructose and mannose are absorbed relatively slowly because of their accumulation in the hemolymph. The latter is related to the lower rate at which they are converted to trehalose. Apparently no mechanism for active uptake of monosaccharides occurs (or is necessary) in *Schistocerca* because its principal "blood sugar" is trehalose. In certain insects, for example, the honey bee, a considerable amount of glucose is normally present in the hemolymph and, in such species, active transport systems may be necessary for sugar absorption.

In *Schistocerca*, amino acids, like sugars, are apparently absorbed passively through the wall of the mesenteric ceca and anterior midgut. The observation that their absorption is passive is of interest, as it is known that the concentration of amino acids in the hemolymph is normally very high. Treherne discovered that, prior to amino acid absorption, there is rapid movement of water from the gut lumen to the hemolymph, which establishes a favorable concentration gradient for passive absorption of amino acids. In contrast, in lepidopteran larvae a suite of absorption mechanisms has been found (Turunen, 1985; Wolfersberger, 1996, 2000; Sacchi and Wolfersberger, 1996). In high midgut concentrations the amino acids either diffuse passively across the gut wall or are moved on specific carriers (facilitated diffusion). At low concentrations, active uptake of amino acids occurs across the midgut. For most neutral amino acids, a symport system operates; that is, active transport of the amino acid occurs concurrently with movement of a cation, especially K^+. Some amino acids, however, are transported in the absence of cations (the uniport system). Within the midgut, there are regional differences in uptake ability; for example, in larvae of *Manduca sexta* the symport system for leucine and proline is concentrated in the posterior midgut (Wolfersberger, 1996).

Another mechanism that may facilitate their absorption, is to convert amino acids to a more complex storage molecule. For example, when certain amino acids were fed to starved *Aedes*, glycogen rapidly appeared in some of the cells of the midgut and ceca. However, it is not known whether the glycogen was formed directly from these amino acids.

Early histochemical studies demonstrated that droplets of lipid are present in the epithelial cells of the crop and gave support to the idea that the crop was the site of lipid absorption. Though, as noted above, a few reports indicate that certain lipoidal molecules can penetrate the crop wall, Treherne's work, again involving labeled compounds, showed that, in *Schistocerca*, absorption of lipid occurs not across the crop wall but via the anterior midgut and ceca. In other insects the occurrence of lipophilic cells in the middle and posterior regions of the midgut suggests that these may be sites of lipid absorption. Absorption of lipids is again a passive process and is relatively slow compared to sugars and amino acids; its rate is, however, increased by the esterification of the absorbed materials in di- and triglycerides and other complex lipids in the midgut epithelium, a process analogous to the situation in vertebrates.

5. Metabolism

Substances absorbed through the gut wall (occasionally the integument; e.g., certain insecticides) seldom remain unchanged in the hemolymph for any length of time but are quickly converted into other compounds. Metabolism comprises all of the chemical reactions that occur in a living organism. It includes anabolism (reactions that result in the formation of more complex molecules and are, therefore, energy-requiring) and catabolism (reactions from which simpler molecules result and energy is released). Anabolic reactions include, for example, the formation of structural proteins or enzymes from amino acids, and the formation from simple sugars of polysaccharides that serve as an energy store. Many catabolic reactions have evolved for the specific purpose of producing the large quantities of energy required by the organism for performance of work.

The metabolism of insects generally resembles that of mammals, details of which can be found in standard biochemical texts. The present account, therefore, will be largely comparative in nature.

5.1. Sites of Metabolism

Chemical reactions are carried out by all living cells, though they are usually limited in number and, of course, are related to the specific function of the cell in which they occur. For example, in midgut epithelial cells, metabolism is directed largely toward synthesis of specific proteins, the enzymes used in digestion. Metabolism in muscle cells is specifically concerned with production of large amounts of energy, in the form of ATP, for the contraction process. In epidermal cells reactions leading to the production of chitin and certain proteins, the components of cuticle, are predominant. Certain tissues, however, are not so specialized and in them a multitude of biochemical reactions, involving the three major raw materials (sugars, amino acids, and lipids), are carried out. In vertebrates the liver performs these multiple functions. The analogous tissue in insects is the fat body (Kilby, 1965; Keeley, 1985).

5.1.1. Fat Body

The fat body is derived during embryogenesis from the mesodermal walls of the coelomic cavities. In other words, it is initially a segmentally arranged tissue though this becomes obscured as the hemocoel develops. Nevertheless, and contrary to what a casual examination may suggest, the fat body does have a definite arrangement in the hemocoel characteristic of the species. Typically, there are subepidermal and perivisceral layers of fat body, plus sheets or cords of cells occur in other specific locations. Thus, the fat body presents a large surface area to the hemolymph, allowing the rapid exchange of metabolites (Dean *et al.*, 1985). Contrary to what was thought originally, the fat body is not a single, uniform tissue (Haunerland and Shirk, 1995; Jensen and Børgesen, 2000). Not only are there regional and species-specific differences, but also differences between larval and adult fat body, as well as in fat body cell types, have been reported. For example, Jensen and Børgesen (2000) described 11 cell types in queens of the pharaoh ant, *Monomorium pharaonis*, based on their position, histochemistry, and ultrastructure. Groups of cells of each type are located in specific positions throughout the body. It should be stressed, however, that there is little evidence for the functions of these many histotypes.

The fat body is composed mainly of cells called trophocytes, though in some species urate cells (urocytes) and/or mycetocytes (Section 4.1.2) also can be seen scattered throughout the tissue. In embryos, early postembryonic stages, and starved insects the individual trophocytes are easily distinguishable, their nucleus is rounded, and their cytoplasm contains few inclusions. As such, they closely resemble hemocytes, with which they probably have a close phylogenetic relationship. In later larval stages and adults the trophocytes enlarge and become vacuolated. The vacuoles contain reserves of fat, protein, and glycogen. The trophocyte nuclei are proportionately large and frequently become elongate and much branched. During metamorphosis in endopterygotes, the reserves are liberated into the hemolymph. In some Diptera and Hymenoptera the majority of trophocytes also disintegrate at this time, and the fat body appears to be completely re-formed in the adult from the few cells that remain.

In a number of species uric acid accumulates in large quantities in specific cells, the urocytes, within the fat body. Cochran (1985) disputed the traditional view that this is a form of storage excretion (Chapter 18, Section 3.3) and proposed that such accumulation represents a mobile reserve of nitrogen, especially in species such as cockroaches and termites whose natural diet is deficient in this element. In cockroaches the urocytes surround mycetocytes (see below) and there is circumstantial evidence to suggest that the bacteria are intimately involved in the synthesis and utilization of the uric acid (Cochran, 1985).

5.1.2. Mycetocytes

Mycetocytes (bacteriocytes) are found in widely different groups of insects, though in common these have nutritionally poor or unbalanced diets such as wood (ants, cerambycid, and anobiid beetle larvae), phloem sap (aphids, planthoppers, mealybugs), and blood (bedbugs, tsetse flies, sucking lice). Mycetocytes are generally stated to be specialized fat body cells (e.g., Haunerland and Shirk, 1995). However, Braendle *et al.* (2003) showed that in the pea aphid, *Acyrthosiphon pisum*, some mycetocytes originate from nuclei near the posterior end of the embryo while others may have their origin as nuclei in the middle of the embryonic central syncytium. Usually, the mycetocytes form specific structures known as mycetomes (bacteriomes) distinct from the fat body. The mycetocytes contain symbiotic bacteria (rarely, yeasts) for which various roles have been proposed (Douglas, 1989, 1998; Dixon, 1998). The mycetocyte system of aphids is particularly well studied. The primary bacterial symbiont is *Buchnera aphidicola*, which is inherited transovarially (from mother to daughters via the ovary). The number of mycetocytes in the host is fixed at birth (i.e., they do not divide), though the cells grow to about four times their original size during larval development as the bacteria multiply. For aphids there is strong evidence that the bacteria supply the host with essential amino acids. The bacteria may also participate in nitrogen recycling and upgrading (converting excretory nitrogen to useful materials) as suggested above for cockroaches. In aphids the evidence is against a role for the bacteria in the synthesis of vitamins and lipids (Douglas, 1998); however, this possibility cannot be ruled out for other insect groups. Curiously, and contrary to what is generally assumed, there is no known benefit to the bacteria from this association (Douglas, 1998).

5.2. Carbohydrate Metabolism

As in other animals, simple sugars provide a readily available substrate that can be oxidized for production of energy. However, in contrast to vertebrates where glucose in the blood is the sugar of importance as an energy source, in insect hemolymph glucose and other monosaccharides usually are present only in minimal amounts. An exception to this statement is the worker honey bee whose hemolymph glucose may reach a concentration of almost 3 g/100 ml and is used as the energy source during flight. In most insects, a disaccharide, trehalose, is the immediate energy source. Trehalose consists of two glucose molecules joined through an $\alpha 1$, 1-linkage. Its level in the hemolymph is constant and in a state of dynamic equilibrium with glycogen stored in the fat body (Friedman, 1978). In this respect, therefore, the situation is comparable to that in vertebrates whose blood glucose level is in equilibrium with liver glycogen. The similarity goes further. Just as the conversion of liver glycogen to blood glucose is promoted by the hormone glucagon, which stimulates glycogen phosphorylase activity, in insect fat body the formation of trehalose from glycogen is promoted by a hyperglycemic hormone released from the corpora cardiaca. This hormone activates the phosphorylase, which removes a glucose unit from the glycogen. Because of its highly polar, polyhydroxyl nature, trehalose does not easily penetrate the muscle cell membrane. Therefore, before it can be oxidized by muscle it must first be converted into glucose by a hydrolyzing enzyme, trehalase, present in the muscle cell membrane. Hemolymph trehalose is also the source of the glucose that is converted by epidermal cells into acetylglucosamine during production of the nitrogenous polysaccharide chitin (Chapter 11, Section 3.1).

Glycogen is an important reserve substance in almost all insects and is found in high concentration in the fat body, with smaller amounts in muscle, especially flight muscle, and sometimes the midgut epithelium. In the mature bee larva, for example, glycogen makes up about one-third of the dry weight. It is produced principally from glucose and other monosaccharides absorbed from the gut following digestion. In some insects glycogen may also be synthesized from amino acids. As noted above, fat body glycogen (and perhaps also that stored in the midgut epithelium) is used to maintain a constant level of trehalose in the hemolymph. The glycerol produced as an antifreeze in the hemolymph of some insects that must withstand extremely low winter temperatures is also derived from fat body glycogen. Glycogen in muscle is used directly as an energy source, being degraded as in mammalian tissue via the glycolytic pathway, Krebs cycle, and respiratory chain, with resultant production of ATP.

Glycogen also is a significant component of the yolk in the eggs of some insects. Its use as an energy source in this situation is, however, secondary to its importance as a provider of glucose units for chitin synthesis in the developing embryo.

As in vertebrates, the major functions of the pentose cycle in insects are (1) production of reducing equivalents (as NADP) that are used, for example, in lipid synthesis, and (2) production of five-carbon sugars for nucleic acid synthesis. In addition, through its ability to interconvert sugars containing from three to seven carbon atoms, the pentose cycle can change "unusual" sugars produced during digestion into six-carbon derivatives and hence into glycogen.

5.3. Lipid Metabolism

For most insects, fats stored in the fat body are the primary energy reserve and, like those of other animals, are mostly triglycerides. They may be formed directly by combination of the fatty acids and glycerol produced during digestion or from amino acids and simple sugars. Typically, fat is stored throughout the juvenile period, especially in endopterygotes where at metamorphosis it may make up between one-third and one-half of the dry weight of an insect. Large amounts of fat also accumulate in the egg during vitellogenesis. Fats are used as an energy source during "long-term" energy-requiring events, for example, embryogenesis, metamorphosis, starvation, and sustained flight (Chapter 14, Section 3.3.5). On a weight-for-weight basis, fats contain twice as much energy as carbohydrates; they are therefore more economical to store.

In addition to the fats just described which serve solely as energy reserves or sources of carbon, many other lipids having structural or metabolic functions occur in insects. Waxes are mixtures of long-chain alcohols or acids, their esters, and paraffins. The number of carbon atoms that form the chain ranges between 12 and 36, and both unsaturated and saturated compounds have been identified. It seems that the various components of wax are synthesized from fatty acid precursors. The paraffins and, possibly, some acids are produced by oenocytes, whereas alcohol and ester synthesis occurs in the fat body. Compound lipids are fatty acids combined with a variety of organic or inorganic residues, for example, carbohydrates, nitrogenous bases, amino acids, phosphate, and sulfate. The metabolism of these lipids in insects is for the most part poorly known. Certain of them, for example, choline, a phospholipid, cannot be synthesized by insects and must be included in the diet. Likewise, sterols are essential components of the diet in almost all insects.

5.4. Amino Acid and Protein Metabolism

In growing insects a large proportion of the amino acids that result from digestion is used directly in the formation of new tissue proteins, both structural and metabolic. Within the fat body especially, but also in other tissues, a variety of transaminations also occur; that is, the amino group from an amino acid can be transferred to a keto acid to form a new amino acid. Such transaminations are especially important when an insect's diet contains insufficient amounts of particular amino acids. As in vertebrates, not all amino acids can be synthesized in insect tissues. Those that cannot must be included in the diet or provided by symbiotic microorganisms (Section 5.1.2). During starvation or when present in excess, amino acids may undergo oxidative deamination (i.e., be oxidized and simultaneously lose their α-amino group) within the fat body, resulting in the formation of the corresponding keto acids. The latter can then be further oxidized via the Krebs cycle and respiratory chain to provide energy or may be converted into carbohydrate or fat reserves. The ammonia produced during deamination is normally converted into uric acid.

The fat body is an important site of hemolymph protein synthesis, especially in the late juvenile stages of endopterygotes and in adult female insects. Storage hexamers (so-called because the proteins comprise six homologous subunits) have been characterized from more than 20 species in six orders, mainly Diptera and Lepidoptera (Telfer and Kunkel, 1991). They reach very high concentrations in the hemolymph just before metamorphosis, and are effectively serving as a store of amino acids for use in adult tissue and protein formation. If stored "individually," the amino acids would create a severe osmotic problem for the insect. In silk moth (*Bombyx*) caterpillars, for example, the hemolymph protein concentration increases sixfold from the fourth to the final instar, in preparation for spinning the cocoon, metamorphosis, and egg production. When the adult emerges, the concentration has fallen to about one-third the value at pupation. During sexual maturation in females of many species, the fat body produces vitellogenins ("female-specific" proteins). These proteins, whose synthesis is regulated by juvenile hormone or ecdysone, are accumulated in large amounts by the developing oocytes (Chapter 19, Section 3.1.1). In female insects that do not feed as adults (e.g., *Bombyx*), the storage hexamers are the source of the amino acids for vitellogenin production which occurs during the pharate adult stage (Chapter 21, Section 3.3.3). In the males of some species the fat body produces proteins that are accumulated by the accessory reproductive glands, probably for use in spermatophore production. Lipophorins are another important group of hemolymph proteins synthesized in the fat body. These, often very large, molecules are important in the absorption and transport of lipids and, additionally, serve as coagulogens in hemolymph clotting (Chapter 17, Section 4.2.2).

5.5. Metabolism of Insecticides

Nowadays insecticides may be regarded as a normal environmental hazard for insects, survival over which has been achieved through natural selection of resistant strains. Though the purpose of this section is to outline the biochemical pathways by which insecticides are rendered harmless, it should be realized that resistance can also be developed, solely or partially, as a result of physical rather than metabolic changes in a species. This is because, ultimately, the degree of resistance is dependent on the rate at which an insect can degrade the toxic material so that lethal quantities do not accumulate at the site of action. Among the physical alterations that may lead to increased resistance are (1) a decline in the permeability of the integument (achieved by increasing cuticle thickness or the extent of tanning, or by modifying the composition of the cuticle); (2) a change in the pH of the gut, resulting in a decrease in solubility and, therefore, rate of absorption of an insecticide; (3) an increase in the amount of fat stored (most insecticides are fat-soluble and, therefore, accumulate in fatty tissues in which they are ineffective); (4) a decrease in permeability of the membranes surrounding the target tissue (usually the nervous system); and (5) a change in the physical structure of the target site.

Insects have various methods for detoxifying potentially harmful substances, many of which parallel those found in vertebrates. Hydrolysis, hydroxylation, sulfation, methylation, acetylation, and conjugation with cysteine, glycine, glucose, glucuronic acid, or phosphate are examples of the methods employed (Perry and Agosin, 1974; Wilkinson,

1976). Hydroxylation and conjugation are also important in making the normally fat-soluble insecticides water-soluble so that they can be excreted. Each of these processes is enzymatically controlled, and it is not surprising to find, therefore, that metabolic resistance to insecticides most often results from qualitative or quantitative changes in the enzymes concerned so that the rate of detoxication is increased. For example, a variety of esterases bring about hydrolysis, mixed-function oxidases commonly induce hydroxylation, and glutathione S-transferase is responsible for promoting conjugation with this tripeptide. In fact, the involvement of these enzymes in insecticide detoxication appears to be an extension of a more general function. Thus, insects that encounter a broad range of naturally occurring plant-derived toxicants, for example, polyphagous Lepidoptera, have significantly higher mixed-function oxidase levels than do oligophagous species (Ronis and Hodgson, 1989). In some insecticide-resistant strains, an increase in the amount of detoxicant enzyme has been observed, which at the gene level results from either an increase in the rate of transcription (mRNA production) or gene amplification (the presence of many identical copies of the DNA coding for the enzyme) (Devonshire and Field, 1991). In others, it appears that the enzyme has changed so that it is now more specific toward its "new" substrate, the insecticide. In a few species, resistance seems to have developed as a result of an increase in the quantity, or decrease in the sensitivity, of the enzyme normally affected by the insecticide, specifically cholinesterase in the nervous system.

Mechanisms for detoxication vary among the different categories of insecticides. It is appropriate, therefore, to examine separately the metabolism of compounds in these categories. Three categories of insecticides will be considered: chlorinated hydrocarbons, organophosphates, and carbamates.

Among the chlorinated hydrocarbon insecticides are DDT, lindane (γ-BHC), chlordane, heptachlor, aldrin, and isodrin.* The chlorinated hydrocarbons act on the nervous system, preventing normal impulse transmission by binding to sodium or chloride channel proteins in the axonal membrane (Bloomquist, 1996; Zlotkin, 1999). By binding to sodium-channel proteins, DDT and its analogues enable sodium ions to diffuse readily across the membrane of excitatory neurons; thus, the membrane becomes permanently depolarized. DDT was the first synthetic insecticide to be developed and "appropriately" was the first to which insects developed resistance. Resistance to DDT and its analogues is most often the result of the presence of an enzyme that dechlorinates the compound, forming the less toxic dichloroethylene derivative, DDE. Some species, however, convert DDT into other less harmful materials such as DDA (the acetic acid derivative), dicofol (kelthane, the trichloroethanol derivative), and DDD (the dichloroethane derivative). Many comparisons of the activity of the DDT-degrading enzyme in resistant and susceptible strains of a species have shown, however, that metabolic resistance alone is often insufficient to account for the full extent of resistance. Specifically, the known maximum rate of degradation of DDT measured *in vitro* is not high enough to account for the high tolerance shown by the resistant strain. In such cases, further work has usually shown the importance also of physical resistance mechanisms of the type outlined earlier.

The cyclodiene compounds, heptachlor, aldrin, and isodrin, are of interest from several viewpoints. In themselves they are not toxic but are oxidized within an insect's tissues to the highly toxic epoxy derivatives, heptachlorepoxide, dieldrin, and endrin, respectively, a process known as "autointoxication." Because of this conversion, insects treated with

---

* These are approved common names. For the chemical names, see Perry and Agosin (1974).

these compounds show no symptoms for 1 or 2 hours after treatment, in contrast to insects treated with other insecticides that react within a matter of minutes. In contrast to DDT, these insecticides block chloride channels in inhibitory neurons, by binding to the GABA-receptor protein, causing hyperexcitation of the nervous system. The resistance shown by certain strains is not because they no longer convert an insecticide to its toxic form as might be anticipated. Further, the toxic derivatives appear to have great stability, remaining unchanged even in resistant insects for several days. Resistance is due to a simple change in the structure of the GABA-receptor protein, specifically, substitution of alanine to serine or glycine (ffrench-Constant *et al.*, 1993, 2000).

Organophosphates (e.g., parathion, malathion, diazinon, and dimethoate) bind covalently with and inhibit the action of cholinesterase, the enzyme that normally degrades acetylcholine at excitatory synapses, though there are reports that their toxicity is partially related also to inhibition of other tissue esterases. As with chlorinated hydrocarbons, resistance to organophosphates may be developed as a result of physical change, but generally is metabolic. Like cyclodienes, many organophosphates are "activated" (rendered more toxic) as a result of oxidation; for example, parathion is converted to paraoxon. Thus, resistance may be caused by a decrease in the rate of activation and/or an increase in the rate of conversion of the compound to a non-toxic form. (Apparently, resistance does not develop as a result of decreased sensitivity of the cholinesterase to an insecticide.) Many reports have shown that resistant strains are more able to carry out conjugation, especially with glutathione, or hydrolysis of the insecticide than are susceptible insects. This ability results from the presence of either greater quantities of esterifying enzymes or enzymes that, through mutation and natural selection, have become more specific for an insecticide.

Carbamates, for example, furadan, sevin, pyrolan, and isolan, are substituted esters of carbamic acid, which, like organophosphates, attack cholinesterase. Resistance to these insecticides also is very similar to that for organophosphates. Some resistance can be achieved by physical changes, but most is the result of increased rates of degradation, especially through oxidation and hydrolysis.

Shortly after the discovery that most resistance is metabolic, that is, results from increased quantities or specificity of particular enzymes that cause more rapid breakdown of an insecticide, it was realized that the phenomenon of synergism might be explored to advantage in the use of insecticides. Synergism describes the situation in which the combined effect of two substances is much greater than the sum of their separate effects. In practical terms, in the present context, it means that appropriate substances (synergists), when mixed with an insecticide, would increase the latter's effectiveness by combining with (and inhibiting) the enzymes that normally degrade the insecticide. The synergists used may be quite unrelated chemically to the insecticide but, most often, are analogues. The principle of synergism has been applied with limited success in the case of pyrethrin insecticides and DDT. For example, in the early 1950s DMC, the ethanol derivative of DDT, was found to be an effective synergist for DDT in DDT-resistant houseflies. However, perhaps not surprisingly, by 1955 the flies had developed resistance to the combination!

6. Summary

Visual, tactile, or chemical cues stimulate food location and/or selection in most insects. The stimuli may be general, for example, color, pattern, and size, or highly specific such as the particular odor or taste of a chemical. The chemicals that promote feeding (phagostimulants) may have no nutritional value for an insect.

Typically saliva lubricates and initiates digestion of the food. However, it may include compounds that act indirectly to facilitate food uptake and digestion or that have functions unrelated to feeding. The gut includes three primary subdivisions, foregut, midgut, and hindgut, and these are typically differentiated into regions of differing function. The foregut is concerned with storage and trituration of food, the midgut with digestion and absorption of small organic molecules, and the hindgut with absorption of water and ions, though some absorption of small organic molecules may occur across the hindgut wall, especially in insects with symbiotic microorganisms in their hindgut.

The digestive enzymes produced match qualitatively and quantitatively the normal composition of the diet. The enzymes may have low specificity, enabling an insect to digest a variety of molecules of a given type, or may be highly specific, for example, when a species feeds solely on a particular food. Gut fluid is buffered within a narrow pH range to facilitate digestion and absorption. Enzymes are released as soon as they are synthesized. Synthesis is regulated so that an appropriate amount of enzyme is produced for the food consumed. Microorganisms in the gut may be important in digestion, especially in wood-eating species, where they degrade cellulose.

Absorption of digestion products occurs mostly in the anterior midgut and mesenteric ceca. It is generally a passive process, though carrier molecules may be used to facilitate the process. The rate at which sugars are absorbed is linked to the rate at which they are converted to trehalose and, hence, glycogen. Lipid absorption is generally slow, with the lipids being converted into di- and triglycerides as they move through the midgut epithelium. Amino acid absorption may be preceded by absorption of water across the midgut wall to produce a favorable gradient for diffusion. However, facilitated diffusion and active transport systems are used for some amino acids in some species.

The fat body is the primary site of intermediary metabolism as well as a site for storage of metabolic reserves. In most insects, trehalose in the hemolymph is the sugar of importance as an energy reserve. Its concentration in the hemolymph is constant and is in dynamic equilibrium with glycogen stored in the fat body. Lipids in the fat body form the major energy reserve molecules and are used in long-term energy-requiring processes such as flight, metamorphosis, starvation, and embryogenesis. The fat body is important in protein metabolism, including amino acid transamination and synthesis of some specific proteins.

The development of resistance to insecticides is normally the result of increased ability of an insect to degrade the insecticides to less harmful and excretable products, but may be related also to increased physical resistance, that is, to structural changes that prevent insecticides from reaching or recognizing the site of action. Metabolic resistance normally develops through the production of more specific or greater quantities of insecticide-degrading enzymes.

7. Literature

A vast literature exists on gut physiology and metabolism, and the following is a representative selection. Food selection and the regulation of feeding are examined by Barton Browne (1975), Bernays (1985), and authors in Chapman and de Boer (1995). Authors in

Lehane and Billingsley (1996) deal with diverse aspects of midgut structure and function. Richards and Richards (1977), Terra (1996), and Lehane (1997) review the peritrophic matrix. Digestion is covered by House (1974), Terra (1990), and Terra and Ferreira (1994). Breznak (1982) and Breznak and Brune (1994) examine the role of microorganisms in the digestion of cellulosic materials. Absorption is considered by Treherne (1967), Turunen (1985), Sacchi and Wolfersberger (1996), and Turunen and Crailsheim (1996). Dean *et al.* (1985), Keeley (1985), Haunerland and Shirk (1995), and Locke (1998) review the fat body. Gilmour (1965) summarizes general insect metabolism, while Steele (1976, 1983), Keeley (1978), and Gäde (2004) review the hormonal control of metabolism. Downer's (1981) text deals with the energy metabolism of insects, emphasizing how it differs from that of other animals. The metabolism of insecticides is reviewed by Perry and Agosin (1974), and authors in Wilkinson (1976) and Kerkut and Gilbert (1985), Volume 10.

Ali, D. W., 1997, The aminergic and peptidergic innervation of insect salivary glands, *J. Exp. Biol.* **201**:1941–1949.

Barton Browne, L., 1975, Regulatory mechanisms in insect feeding, *Adv. Insect Physiol.* **11**: 1–116.

Bernays, E. A., 1985, Regulation of feeding behaviour, in: *Comprehensive Insect Physiology, Biochemistry and Pharmacology*, Vol. 4 (G. A. Kerkut and L. I. Gilbert, eds.), Pergamon Press, Elmsford, NY.

Bloomquist, J. R., 1996, Ion channels as targets for insecticides, *Annu. Rev. Entomol.* **41**:163–190.

Braendle, C., Miura, T., Bickel, R., Shingleton, A. W., Kambhampati, S., and Stern, D. L., 2003, Developmental origin and evolution of bacteriocytes in the aphid-*Buchnera* symbiosis, *Pub. Libr. Sci. Biol.* **1**:70–75.

Breznak, J. A., 1982, Intestinal microbiota of termites and other xylophagous insects, *Annu. Rev. Microbiol.* **36**:323–343.

Breznak, J. A., and Brune, A., 1994, Role of microorganisms in the digestion of lignocellulose by termites, *Annu. Rev. Entomol.* **39**:453–487.

Briegel, H., and Lea, A. O., 1975, Relationship between protein and proteolytic activity in the midgut of mosquitoes, *J. Insect Physiol.* **21**:1597–1604.

Chapman, R. F., 2003, Contact chemoreception in feeding by phytophagous insects, *Annu. Rev. Entomol.* **48**:455–484.

Chapman, R. F., and de Boer, G. (eds.), 1995, *Regulatory Mechanisms in Insect Feeding*, Chapman and Hall, New York.

Cochran, D. G., 1985, Nitrogenous excretion, in: *Comprehensive Insect Physiology, Biochemistry and Pharmacology*, Vol. 4 (G. A. Kerkut and L. I. Gilbert, eds.), Pergamon Press, Elmsford, NY.

Davey, K. G., and Treherne, J. E., 1963, Studies on crop function in the cockroach. I and II, *J. Exp. Biol.* **40**:763–773; 775–780.

Dean, R. L., Locke, M., and Collins, J. V., 1985, Structure of the fat body, in: *Comprehensive Insect Physiology, Biochemistry and Pharmacology*, Vol. 3 (G. A. Kerkut and L. I. Gilbert, eds.), Pergamon Press, Elmsford, NY.

Devonshire, A. L., and Field, L. M., 1991, Gene amplification and insecticide resistance, *Annu. Rev. Entomol.* **36**:1–23.

Dixon, A. F. G., 1998, *Aphid Ecology: An Optimization Approach*, 2nd ed., Chapman and Hall, London.

Douglas, A. E., 1989, Mycetocyte symbiosis in insects, *Biol. Rev.* **69**:409–434.

Douglas, A. E., 1998, Nutritional interactions in insect-microbial symbioses: Aphids and their symbiotic bacteria *Buchnera*, *Annu. Rev. Entomol.* **43**:17–37.

Downer, R. G. H. (ed.), 1981, *Energy Metabolism in Insects*, Plenum Press, New York.

Elpidina, E. N., Vinokurov, K. S., Gromenko, V. A., Rudenskaya, Y. A., Dunaevsky Y. E., and Zhuzhikov, D. P., 2001, Compartmentalization of proteinases and amylases in *Nauphoeta cinerea* midgut, *Arch. Insect Biochem. Physiol.* **48**:206–216.

ffrench-Constant, R. H., Rocheleau, T. A., Steichen, J. C., and Chalmers, A. E., 1993, A point mutation in a *Drosophila* GABA receptor confers insecticide resistance, *Nature* **363**:449–451.

ffrench-Constant, R. H., Anthony, N., Aronstein, K., Rocheleau, T., and Stilwell, G., 2000, Cyclodiene insecticide resistance: From molecular to population genetics, *Annu. Rev. Entomol.* **48**:449–466.

Friedman, S., 1978, Trehalose regulation, one aspect of metabolic homeostasis, *Annu. Rev. Entomol.* **23**:389–407.

Gäde, G., 2004, Regulation of intermediary metabolism and water balance of insects by neuropeptides, *Annu. Rev. Entomol.* **49**:93–113.

Gilmour, D., 1965, *The Metabolism of Insects*, Oliver and Boyd, Edinburgh.

Haunerland, N. H., and Shirk, P. D., 1995, Regional and functional differentiation in the insect fat body, *Annu. Rev. Entomol.* **40**:121–145.

Heron, R. J., 1965, The role of chemotactic stimuli in the feeding behavior of spruce budworm larvae on white spruce, *Can. J. Zool.* **43**:247–269.

House, H. L., 1974, Digestion, in: *The Physiology of Insecta*, 2nd ed., Vol. V (M. Rockstein, ed.), Academic Press, New York.

Jensen, P. V., and Børgesen, L. W., 2000, Regional and functional differentiation in the fat body of pharoah's ant queens, *Monomorium pharaonis* (L.), *Arth. Struct. Funct.* **29**:171–184.

Keeley, L. L., 1978, Endocrine regulation of fat body development and function, *Annu. Rev. Entomol.* **23**:329–352.

Keeley, L. L., 1985, Physiology and biochemistry of the fat body, in: *Comprehensive Insect Physiology, Biochemistry and Pharmacology*, Vol. 3 (G. A. Kerkut and L. I. Gilbert, eds.), Pergamon Press, Elmsford, NY.

Kerkut, G. A., and Gilbert, L. I., (eds.), 1985, *Comprehensive Insect Physiology, Biochemistry and Pharmacology*, Pergamon Press, Elmsford, NY.

Kilby, B. A., 1965, Intermediary metabolism and the insect fat body, *Symp. Biochem. Soc.* **25**:39–48.

Lange, A. B., and Orchard, I., 1998, The effects of schistoFLRFamide on contractions of the locust midgut, *Peptides* **19**:459–467.

Lehane, M. J., 1997, Peritrophic matrix structure and function, *Annu. Rev. Entomol.* **42**:525–550.

Lehane, M. J., and Billingsley, P. F. (eds.), 1996, *Biology of the Insect Midgut*, Chapman and Hall, London.

Locke, M., 1998, The fat body, in: *Microscopic Anatomy of Invertebrates*, Vol. 11A (Insecta), Wiley-Liss, New York.

Miles, P. W., 1999, Aphid saliva, *Biol. Rev.* **74**:41–85.

Nakashima, K, Watanabe, H., Saitoh, H., Tokuda, G., and Azuma, J.-I., 2002, Dual cellulose-digesting system of the wood-feeding termite, *Coptotermes formosanus* Shiraki, *Insect Biochem. Molec. Biol.* **32**:777–784.

Perry, A. S., and Agosin, M, 1974, The physiology of insecticide resistance by insects, in: *The Physiology of Insecta*, 2nd ed., Vol. VI (M. Rockstein, ed.), Academic Press, New York.

Ribeiro, J. M. C., 1995, Insect saliva: Function, biochemistry, and physiology, in: *Regulatory Mechanisms in Insect Feeding* (R. F. Chapman and G. de Boer, eds.), Chapman and Hall, New York.

Ribeiro, J. M. C., and Francischetti, I. M. B., 2003, Role of arthropod saliva in blood feeding: Sialome and post-sialome perspectives, *Annu. Rev. Entomol.* **48**:73–88.

Richards, A. G., and Richards, P. A., 1977, The peritrophic membranes of insects, *Annu. Rev. Entomol.* **22**:219–240.

Ronis, M. J. J., and Hodgson, E., 1989, Cytochrome P-450 monooxygenases in insects, *Xenobiotica* **19**:1077–1092.

Sacchi, V. F., and Wolfersberger, M. G., 1996, Amino acid absorption, in: *Biology of the Insect Midgut* (M. J. Lehane and P. F. Billingsley, eds.), Chapman and Hall, London.

Steele, J. E., 1976, Hormonal control of metabolism in insects, *Adv. Insect Physiol.* **12**:239–323.

Steele, J. E., 1983, Endocrine control of carbohydrate metabolism in insects, in: *Endocrinology of Insects* (R. G. H. Downer and H. Laufer, eds.), Liss, New York.

Telfer, W. H., and Kunkel, J. G., 1991, The function and evolution of insect storage hexamers, *Annu. Rev. Entomol.* **36**:205–228.

Terra, W. R., 1990, Evolution of digestive systems in insects, *Annu. Rev. Entomol.* **35**:181–200.

Terra, W. R., 1996, Evolution and function of insect peritrophic membrane, *Cien. Cult. Sao Paulo* **48**:317–324.

Terra, W. R., and Ferreira, C., 1994, Insect digestive systems: Properties, compartmentalization and function, *Comp. Biochem. Physiol. B* **109**:1–62.

Tokuda, G., Saito, H., and Watanabe, H., 2002, A digestive β-glucosidase from the salivary glands of the termite, *Neotermes koshuensis* (Shiraki): Distribution, characterization and isolation of its precursor cDNA by 5′- and 3′-RACE amplifications with degenerate primers, *Insect Biochem. Molec. Biol.* **32**:1681–1689.

Treherne, J. E., 1967, Gut absorption, *Annu. Rev. Entomol.* **12**:43–58.

Trimmer, B. A., 1985, Serotonin and the control of salivation in the blow fly *Calliphora*, *J. Exp. Biol.* **114**:307–328.

Turunen, S., 1985, Absorption, in: *Comprehensive Insect Physiology, Biochemistry and Pharmacology*, Vol. 4 (G. A. Kerkut and L. I. Gilbert, eds.), Pergamon Press, Elmsford, NY.

Turunen, S., and Crailsheim, K., 1996, Lipid and sugar absorption, in: *Biology of the Insect Midgut* (M. J. Lehane and P. F. Billingsley, eds.), Chapman and Hall, London.

Valenzuela, J. G., and Ribeiro, J. M. C., 1998, Purification and cloning of the salivary nitrophorin from the hemipteran *Cimex lectularius*, *J. Exp. Biol.* **201**:2659–2664.

Weichsel, A., Andersen, J. F., Champagne, D. E., Walker, F. A., and Montfort, W. R., 1998, Crystal structures of a nitric oxide transport protein from a blood-sucking insect, *Nature Struct. Biol.* **5**:304–309.

Wilkinson, C. F. (ed.), 1976, *Insecticide Biochemistry and Physiology,* Plenum Press, New York.

Wolfersberger, M. G., 1996, Localization of amino acid absorption systems in the larval midgut of the tobacco hornworm *Manduca sexta*, *J. Insect. Physiol.* **42**:975–982.

Wolfersberger, M. G., 2000, Amino acid transport in insects, *Annu. Rev. Entomol.* **45**:111–120.

Zlotkin, E., 1999, The insect voltage-gated sodium channel as target of insecticides, *Annu. Rev. Entomol.* **44**:429–455.

17

The Circulatory System

1. Introduction

The circulatory system of insects, like that of all arthropods, is of the "open" type; that is, the fluid that circulates is not restricted to a network of conducting vessels as, for example, in vertebrates, but flows freely among the body organs. An open system results from the development, in evolution, of a hemocoel rather than a true coelom. A consequence of the open system is that insects have only one extracellular fluid, hemolymph, in contrast to vertebrates, which have two such fluids, blood and lymph. The occurrence of an open system does not mean that hemolymph simply bathes the organs it surrounds because usually thin granular membranes separate the tissues from the hemolymph itself. Insects generally possess pumping structures and various diaphragms to ensure that hemolymph flows throughout the body, reaching the extremities of even the most delicate appendages. As the only extracellular fluid, it is perhaps not surprising that the hemolymph, in general, serves the functions of both blood and lymph of vertebrates. Thus, the fluid fraction (plasma) is important in providing the correct milieu for body cells, is the transport system for nutrients, hormones, and metabolic wastes, and contains elements of the immune system, while the cellular components (hemocytes) provide the defense mechanism against foreign organisms that enter the body and are important in wound repair and the metabolism of specific compounds.

2. Structure

The primary pump for moving hemolymph around the body is a middorsal vessel that runs more or less the entire length of the body (Figure 17.1). The posterior portion of the vessel has ostia (valves) and is sometimes known as the heart, whereas the cephalothoracic portion, which is often a simple tube, may be termed the aorta (Figure 17.1A). In some insects the heart is the only part that contracts, but in many others the entire vessel is contractile. The vessel is held in position by connective tissue strands attached to the dorsal integument, tracheae, gut, and other organs and by a series of paired, usually fan-shaped, alary muscles. Normally, the vessel is a straight tube, though in many species the aorta may loop vertically. Anteriorly the aorta runs ventrally to pass between the corpora cardiaca and under the brain. Generally the dorsal vessel is closed posteriorly; however, in Diplura,

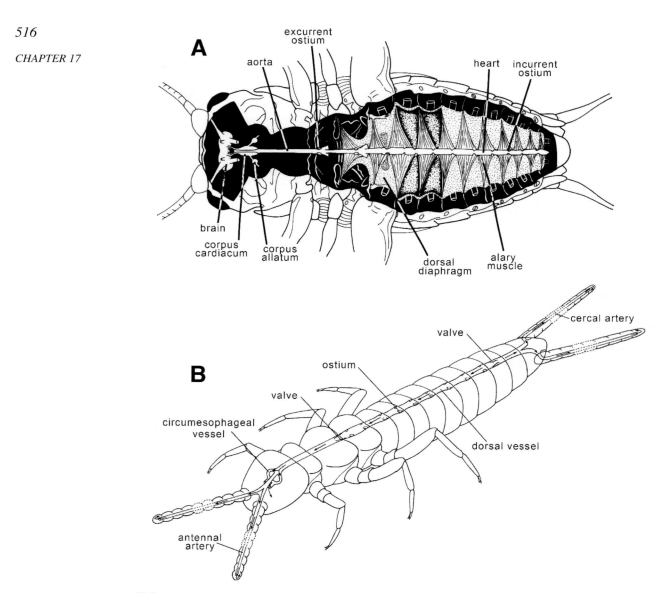

FIGURE 17.1. (A) Ventral dissection of the field cricket, *Acheta assimilis*, to show dorsal vessel and associated structures; and (B) circulatory system of *Campodea augens* (Diplura) showing anterior and posterior arteries running off the dorsal vessel. [A, after W. L. Nutting, 1951, A comparative and anatomical study of the heart and accessory structures of the orthopteroid insects, *J. Morphol.* **89**:501–597. By permission of Wistar Press. B, from a figure kindly supplied by Dr. Günther Pass.]

Archaeognatha, Zygentoma, and some Ephemeroptera the dorsal vessel connects at its rear with arteries that run along the cerci and median caudal filament (Gereben-Krenn and Pass, 2000). In Diplura an artery also supplies each antenna (Figure 17.1B), and in Dictyoptera and some Orthoptera there are pairs of segmental arteries in the abdomen (Hertel and Pass, 2002). However, except as noted, in pterygotes circulation to appendages is achieved by means of accessory pulsatile organs and septa (see below).

In most insects the dorsal vessel is well tracheated. The heart may not be innervated or may receive paired lateral nerves from the brain and/or segmental ventral ganglia. Ostia may be simple, slitlike valves or deep, funnel-shaped structures in the wall of the heart, or internal

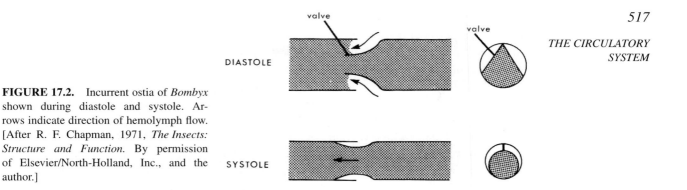

FIGURE 17.2. Incurrent ostia of *Bombyx* shown during diastole and systole. Arrows indicate direction of hemolymph flow. [After R. F. Chapman, 1971, *The Insects: Structure and Function.* By permission of Elsevier/North-Holland, Inc., and the author.]

flaps (Figure 17.2). Their position and number are equally varied. They may be lateral, dorsal, or ventral and may be as numerous as 12 pairs (in cockroaches) or as few as 1 pair (in some dragonflies). Ostia are usually incurrent, that is, they open to allow hemolymph to enter the heart but close to prevent backflow. In some orthopteroid insects, however, some ostia are excurrent. Histologically, the dorsal vessel in its simplest form comprises a single layer of circular muscle fibers, though more often longitudinal and oblique muscle layers also occur. Ultrastructural examination of the heart muscle cells reveals, however, that they contain, in addition to contractile elements, prominent Golgi complexes and vesicles, suggesting that the insect heart is secretory and, like that of vertebrates, may have a more significant role in homeostasis than just pumping hemolymph (Locke, 1989).

Assisting in directing the flow of hemolymph, especially in postlarval stages, are various diaphragms (septa) (Figure 17.3) that include both connective tissue and muscular elements. The spaces delimited by the diaphragms are known as sinuses. The pericardial septum (dorsal diaphragm) lies immediately beneath the dorsal vessel and spreads between the alary muscles. Laterally, it is attached at intervals to the terga and in most species has

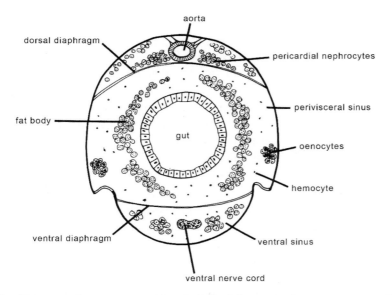

FIGURE 17.3. Diagrammatic transverse section through abdomen to show arrangement of septa. [From R. E. Snodgrass, *Principles of Insect Morphology.* Copyright 1935 by McGraw-Hill, Inc. Used with permission of McGraw-Hill Book Company.]

openings so that the pericardial sinus is in effect continuous with the perivisceral sinus. Ventrally, a perineural septum (ventral diaphragm) may occur, which cuts off the perineural sinus from the perivisceral sinus. Generally, the ventral diaphragm is restricted to the abdomen and occurs only in species whose ventral nerve cord extends into this region of the body (Miller, 1985). It is capable of performing posteriorly directed undulations and may have openings. It may receive motor nerves from segmental ganglia, which regulate the rate at which it undulates, though the undulations originate myogenically. In some insects, for example, caddisflies and cockroaches, the ventral diaphragm is reduced to a few transverse or longitudinal muscles, respectively. Frequently, there is a close physical association between the diaphragm (or its vestiges) and the nerve cord. Thus, in cockroaches and *Pseudaletia unipuncta* the actions of the longitudinal muscle remnants cause the nerve cord to oscillate laterally, bringing it into greater mix with the hemolymph and possibly improving hemolymph flow (Koladich *et al.*, 2002). Hemolymph circulation through the legs and palps of some insects is assisted by the presence of a longitudinal septum that partitions the appendage into afferent and efferent sinuses.

To further facilitate hemolymph flow, especially through appendages, accessory pulsatile organs (auxiliary hearts) commonly occur (Pass, 1998, 2000). These have been identified in the head, antennae, thorax, legs, wings, and ovipositor. In many species they are saclike structures that have a posterior incurrent ostium and an anteriorly extended vessel. In antennal pulsatile organs the vessel may run the length of the appendage but is perforated at intervals to permit exit of hemolymph. The wall of the sac may be muscular, so that constriction of the sac is the active phase, and dilation results from elasticity of the wall, or the sac may have attached to it a discrete dilator muscle, and constriction is due to the sac's elasticity. In some situations, for example, the legs of Orthoptera and Hemiptera, the accessory pulsatile organ is simply one or two small muscles that attach to the longitudinal septum. Indeed, in Hemiptera, the organ is clearly derived from a skeletal muscle, the pretarsal depressor (Figure 14.5C) (Hantschk, 1991). Contraction narrows the efferent sinus, while enlarging the afferent sinus. Valves ensure that hemolymph is pushed toward the limb tip, then back toward the body cavity. Normally, accessory hearts are quite separate from the dorsal vessel, though in some Odonata they are connected via short vessels with the aorta into which they pump hemolymph. Most accessory pulsatile organs are not innervated.

Hemopoietic organs have been described for a number of insects. For example, in *Gryllus* there are pairs of such organs, in the second and third abdominal segments, directly connected with the dorsal vessel. Like those of vertebrates, the hemopoietic organs serve both as the site of production of at least some types of hemocytes and as centers for phagocytosis. The same cells within the hemopoietic organ can carry out both of these functions, though not simultaneously; thus, during periods of infection, division of the cells to form new prohemocytes is greatly retarded.

At specific locations in the circulatory system are sessile cells, usually conspicuously pigmented, called athrocytes (Locke and Russell, 1998). They occur singly, in small groups, or form distinct lobes, and are always surrounded by a basal lamina, a feature that distinguishes them from hemocytes. In most species athrocytes are situated on the surface of the heart (occasionally also along the aorta), and these are referred to as pericardial cells. They may also be found as scattered cells in the fat body (in *Lepisma*), in clusters at the bases of legs (in *Gryllus* and *Periplaneta*), or as a garland of cells around the esophagus (in some larval Diptera). When mature they may contain several nuclei, as well as mitochondria, Golgi apparatus, and pigment granules or crystals of various colors. The cells are able to accumulate colloidal particles, for example, certain dyes, hemoglobin, and chlorophyll, which led to an early suggestion that they segregated and stored waste products (hence their alternate name of nephrocytes). The usual view is that the cells accumulate and degrade large molecules such as proteins, peptides, and pigments, and the products are then used or excreted. However, their structure includes much rough endoplasmic reticulum and well-developed Golgi complexes, characteristics of cells producing protein for export. Indeed, Fife *et al.* (1987) demonstrated that the pericardial cells synthesized and secreted several proteins into the hemolymph.

3. Physiology

3.1. Circulation

Contractions of the dorsal vessel and accessory pulsatile organs, along with movements of other internal organs and abdominal ventilatory movements (coelopulses), serve to move hemolymph around the body (Miller, 1997). In *Periplaneta* larvae, for example, circulation time is 3–6 minutes; in *Tenebrio* the time for complete mixing of injected radioisotope is 8–10 minutes. Generally hemolymph is pumped rapidly through the dorsal vessel but moves slowly and discontinuously through sinuses and appendages.

The direction of hemolymph flow in most insects is indicated in Figure 17.4A–C. Hemolymph is pumped anteriorly through the dorsal vessel from which it exits via either excurrent ostia of the heart or mainly the anterior opening of the aorta in the head. The resultant pressure in the head region forces hemolymph posteriorly through the perivisceral

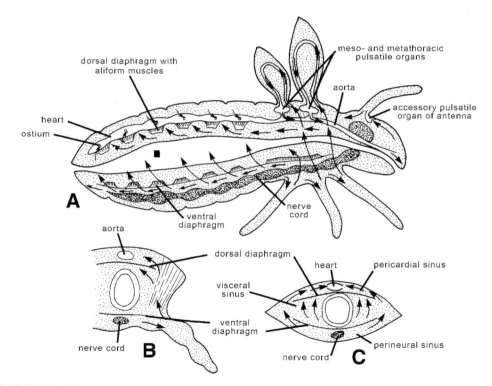

FIGURE 17.4. Diagrams showing direction of hemolymph flow. (A) Longitudinal section; (B) transverse section through thorax; and (C) transverse section through abdomen. Arrows indicate direction of flow. [After V. B. Wigglesworth, 1965, *The Principles of Insect Physiology*, 6th ed., Methuen and Co. By permission of the author.]

and perineural sinuses. Undulations of the ventral diaphragm aid the backward flow of hemolymph. Relaxation of the heart muscle results in an increase in heart volume, and, by negative pressure, hemolymph is sucked in via incurrent ostia. As noted earlier, circulation through appendages is aided by accessory pulsatile organs. In most insects hemolymph enters the wings via the anterior veins and returns to the thorax via the anal veins. Though the structure of wing pulsatile organs is varied, they always operate by sucking hemolymph out of the posterior wing veins (Pass, 1998, 2000). In some Coleoptera and Lepidoptera, tidal flow of hemolymph occurs in the wings; that is, hemolymph flows into or out of all veins simultaneously.

In apterygotes and mayflies hemolymph flow is bidirectional (Figure 17.1B). Anterior to a valve located in the heart at about the level of the eighth abdominal segment, hemolymph flows forward toward the head, while behind the valve the hemolymph is pushed backward along arteries that terminate at the tips of the cerci and median filament (Gereben-Krenn and Pass, 2000). Reversal of heartbeat may also occur and is characteristically seen in pupae and adults of Lepidoptera and Diptera.

In some actively flying insects, for example, locusts, butterflies, saturniid moths, and possibly some Hymenoptera, as well as in diapausing lepidopteran pupae, hemolymph movements are closely coordinated with the ventilation movements for gas exchange (Chapter 15, Sections 3.2 and 3.3). Abdominal pumping not only improves gas exchange within the tracheal system but also brings about tidal flow (oscillating circulation) of hemolymph. In other words, hemolymph flows back and forth between the abdomen and the anterior part of the body. Tidal flow of hemolymph into the abdomen is aided by reverse peristalsis of the dorsal vessel (Miller, 1997).

Control of circulation is especially important in large flying insects such as bumble bees, dragonflies, and night-flying moths that thermoregulate. Thermoregulation allows these insects to warm up their wing musculature at low ambient temperatures and to dissipate heat produced during flight at high temperatures. At low ambient temperatures, the heartbeat is weak and contraction of the ventral diaphragm infrequent, so that heat produced by pre-flight contraction of wing muscles is retained within the thorax. As the thoracic temperature becomes suitable for flight, heartbeat rate and amplitude increase, as do the frequency and strength of contractions of the ventral diaphragm, taking heat away from the thorax to prevent overheating (Miller, 1997).

3.2. Heartbeat

Contraction of the heart (systole) is followed, as in other animals, by a phase of relaxation (diastole) during which muscle cell membranes become repolarized. A third phase, diastasis, may follow diastole, when the diameter of the dorsal vessel suddenly enlarges because of the influx of hemolymph. Diastole in many insects seems to be passive, that is, the result of natural elasticity of the heart muscle. Though alary muscles may be quite well developed in such species, they apparently have no role in the relaxation process. They have been shown to be electrically inexcitable in locusts and cockroaches, and cutting them has no effect on the rate and strength of the heartbeat. In a few species structural integrity of the heart and alary muscles is vital, and cutting the alary muscles terminates the heartbeat.

In most pterygotes, where hemolymph flow is unidirectional, contraction of the dorsal vessel begins at the posterior end and passes forward as a peristaltic wave. Experimentally contraction can be induced at any point along the length of the vessel and individual semiisolated segments (portions of the heart with tergum still attached) continue to beat rhythmically. These observations suggest that the heartbeat is normally coordinated by a pacemaker located posteriorly. In adult *Manduca sexta* in which heartbeat reversal occurs, distinct pacemakers exist for the anteriorly and posteriorly directed contractions (Dulcis *et al.*, 2001).

Whether or not an insect heart is innervated, its beat is myogenic, that is, the beat originates in the heart muscle itself (Jones, 1977; Miller, 1985, 1997). This contrasts with the situation in Crustacea and Arachnida, which have neurogenic hearts. For innervated insect hearts, it is generally assumed that, as in vertebrates, control of the rate and amplitude of the heartbeat resides in the cardiac neurons. However, as Miller (1985) pointed out, such regulation has been demonstrated in only a few cases.

The rate at which the heart beats varies widely both among species and even within an individual under different conditions. In the pupa of *Anagasta kühniella*, for example, the heart beats 6–11 times per minute. In larval *Blattella germanica* rates of 180–310 beats/min have been recorded (Jones, 1974). Many factors affect the rate of heartbeat. Generally, there is a decline in heartbeat rate in successive juvenile stages, and in the pupal stage the heart beats slowly or even ceases to beat for long periods. In adults the heart beats at about the rate observed in the final larval stage. Heartbeat rate increases with activity, during feeding, with increase in temperature or in the presence of carbon dioxide in low concentration, but is depressed in starved or asphyxiated insects. Hormones, too, may affect heartbeat rate. Authors have reported a wide range of cardioaccelerating and cardioinhibiting factors, including juvenile hormone, neurosecretory peptides, octopamine, and 5-hydroxytryptamine. However, in many instances, an effect of these substances on the metabolism of the insect may cause the change in heartbeat rate observed.

As noted, the ventral diaphragm and accessory pulsatile organs may or may not be innervated. Thus, it may be anticipated that, as with the dorsal vessel, contraction of these structures may be controlled neurally and hormonally or by hormones alone. Pharmacological studies have shown that a range of amines and small peptides can modulate contraction of these structures. However, immunohistochemistry has identified amine- and peptide-releasing neurons terminating at these structures, tending to cloud the picture with respect to which system is regulating their activity (Hertel and Pass, 2002; Koladich *et al.*, 2002).

4. Hemolymph

Hemolymph, like the blood of vertebrates, includes a cellular fraction, the hemocytes, and a liquid component, the plasma, whose functions are broadly comparable with those found in vertebrates. Several of the features of hemolymph, however, contrast markedly with what is seen in vertebrates. First, associated with the evolution of a tracheal system, hemolymph has no gas-transporting function, except perhaps in some chironomid larvae. In addition, the composition of hemolymph (especially in the more advanced endopterygotes) is both very different from that of blood and is much more variable on a day-to-day basis. Among the trends seen in the evolution of the higher endopterygotes are substitution of organic molecules for the predominant inorganic ions (sodium and chloride), an increase in the proportion of divalent to monovalent cations, and an increase in the importance (quantitatively) of organic phosphate. Further, as noted in the previous chapter, monosaccharide sugars are generally of little importance in the hemolymph and are replaced by the disaccharide trehalose. The use of this molecule rather than the reducing sugar glucose appears to be an adaptation to overcome problems of osmotic pressure and chemical reactivity that would result if the monosaccharide were the major form of fuel in the hemolymph (Wheeler, 1989). In many insects the hemolymph osmotic pressure is held reasonably constant over a range of environmental conditions. In other species, the osmotic pressure changes in parallel with the environmental conditions, yet the body cells are able to tolerate these changes (Chapter 18, Section 4).

4.1. Plasma

4.1.1. Composition

Plasma contains a large variety of components both organic and inorganic whose relative proportions may differ greatly both among species and within an individual under different physiological conditions. Despite this variability, some general statements may be made.

In primitive orders, the predominant cation is sodium, with potassium, calcium, and magnesium present in low proportions. The major anion is chloride, though plasma also contains small amounts of phosphate and bicarbonate. These inorganic constituents are the major contributors to the hemolymph osmotic pressure (Figure 17.5A,B).

In higher orders certain trends can be observed. The relative importance of sodium decreases at the expense of potassium and, especially, magnesium. Chloride also decreases in importance and is replaced by organic anions, especially amino and carboxylic acids. Finally, the relative contribution that inorganic ions make to the hemolymph osmotic pressure declines, and organic constituents become the major osmotic effectors (Figure 17.5D,E).

Superimposed on these phylogenetic relationships may be dietary and ontogenetic considerations, especially with respect to the cationic components of hemolymph. Thus, zoophagous species generally have a larger proportion of sodium in the hemolymph, in contrast to phytophagous species where magnesium (derived from chlorophyll) and potassium are the major cations (Figure 17.5C). The ionic composition may also change with stage of development, in endopterygotes at least, though whether this is related to a change of diet which, of course, may also occur from the juvenile to the adult stage, does not seem to have been considered. For example, in the exopterygotes *Aeshna cyanea* (Odonata),

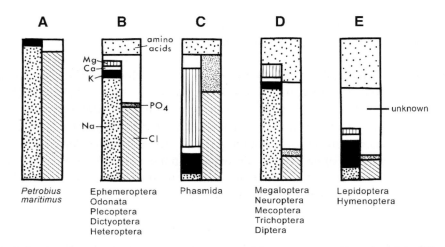

FIGURE 17.5. Relative contributions to osmotic pressure of the components of hemolymph in different insect groups. Each column represents 50% of the total osmolar concentration. [After D. W. Sutcliffe, 1963, The chemical composition of hemolymph in insects and some other arthropods in relation to their phylogeny, *Comp. Physiol. Biochem.* **9**:121–135. By permission of Pergamon Press, Elmsford, NY.]

Periplaneta americana (Dictyoptera), and *Locusta migratoria* (Orthoptera) and the endopterygote *Dytiscus* (Coleoptera), the composition of the hemolymph is similar in larvae and adults, but so, too, is the diet. In contrast, in a few endopterygote species, specifically Lepidoptera and Hymenoptera, for which data are available, larval hemolymph is of the high magnesium type, whereas adult hemolymph has a much greater sodium content. Again, however, the fact that the diet of the adult (if it feeds) is typically different from that of the larva apparently has not received consideration.

As Florkin and Jeuniaux (1974) pointed out, insects whose hemolymph contains such large quantities of magnesium and potassium must have become adapted so that physiological processes, especially neuromuscular function, can be carried out normally because these ions are detrimental. Hoyle (1954, cited in Florkin and Jeuniaux, 1974) suggested that the high magnesium-potassium type of hemolymph characteristic of phytophagous endopterygote larvae might reduce, through an effect on the nervous system, locomotor activity of larvae, so that they would tend to remain close to their food. The adult, in contrast, is usually much more active and possesses the more primitive high sodium type of hemolymph.

Organic acids are important hemolymph constituents, especially in juvenile endopterygotes. Carboxylic acids (citric, α-ketoglutaric, malic, fumaric, succinic, and oxaloacetic), which in the hemolymph are anionic, are present in large amounts and may neutralize almost 50% of the inorganic cations. They are apparently synthesized by insects (or by symbiotic bacteria), as their levels in the hemolymph are independent of diet. Whether these acids, which are components of the Krebs cycle, also have a metabolic function in the hemolymph is not known. Insects, especially endopterygotes, also characteristically have high concentrations of amino acids in their hemolymph. The proportions of amino acids vary among species and within an individual according to diet and developmental and physiological state, though glutamine, glycine, histidine, lysine, proline, and valine generally each constitute at least 10% of the total amino acid pool. The amino acids make a significant contribution to the hemolymph osmotic pressure, though whether they function as cations or as anions depends both on the pH of the fluid (usually between 6.0 and 7.0) and on the individual amino acid. In addition, some have important metabolic roles.

The hemolymph protein concentration is generally about 1% to 5% but varies with species and individual physiological states. For example, it is low in starved insects and high in females with developing oocytes. In endopterygotes the protein concentration often increases through larval life, especially in the final instar, but then declines during pupation. As noted in Chapter 16 (Section 5.4), the fat body is the major source of the many proteins found in the hemolymph though other tissues, notably midgut epithelium, epidermis, pericardial cells, and hemocytes, also contribute. For relatively few of these proteins has the function been determined. Some, for example, lysozyme, phenoloxidase, cecropins, and attacins, are important in prevention of infection. The female-specific proteins (vitellogenins) are selectively accumulated by oocytes during yolk formation. Some proteins act as transport agents for lipids and juvenile hormone. Yet others may serve simply as concentrated stores of nitrogen that can be degraded for use in growth and metabolism. Among the enzymes identified in hemolymph are hydrolases (e.g., amylase, esterases, proteases, and trehalase), dehydrogenases and oxidases important in carbohydrate metabolism, and tyrosinase.

The principal carbohydrate in the hemolymph of most insects is the disaccharide trehalose, which serves as a source of readily available energy (Chapter 16, Section 5.2). Monosaccharides and polysaccharides normally occur in only small amounts. Glycerol and sorbitol are sometimes present in high concentration in the hemolymph of overwintering stages where they serve as antifreezes.

Free lipids seldom occur in high concentrations in hemolymph, except in some species after feeding, during flight, or at metamorphosis when they are being transported to sites of use or storage.

Other constituents of hemolymph include phosphate esters, urates, and traces of other nitrogenous waste products, amino sugars such as acetylglucosamine produced during digestion of the cuticle, pigments (often conjugated with protein), and hormones, which may be transported to their sites of action in combination with protein.

4.1.2. Functions

Apart from the specific functions of particular components, which were outlined in the above consideration of its composition, the plasma has some important general functions. It serves as the medium in which nutrients, hormones, and waste materials can be transported to sites of use, action, and disposal, respectively. It is an important site for the storage, usually temporary, of metabolites. Plasma is the source of cell water, and during periods of desiccation its volume may decline at the expense of water entering the tissues. By virtue of some components (proteins, amino acids, carboxylic acids, bicarbonate, and phosphate) it is a strong buffer and resists changes in pH that might occur as a result of metabolism. As a liquid, it is also used to transmit pressure changes from one part of the body to another. It is used hydrostatically, for example, to maintain the turgor necessary for movement in soft-bodied insects, split the old exocuticle during ecdysis, expand appendages after ecdysis, evert structures such as the penis, and extend the labium of larval Odonata. The plasma also has an important thermoregulatory function in many actively flying insects. Various structural and physiological features of the circulatory system have evolved that allow heat to be retained in the thoracic region (at lower ambient temperatures) or to be easily transported to the abdomen and hence dissipated when ambient temperatures are high (Miller, 1985).

4.2. Hemocytes

4.2.1. Origin, Number, and Form

The embryonic origin of hemocytes is considered in Chapter 20, Section 7.5. In postembryonic stages of many species, hemocytes are produced in discrete hemopoietic organs; in other species multiplication of hemocytes takes place in the hemocoel *per se*, either as the cells circulate or as they rest on the surface of tissues.

In many species all or nearly all of the hemocytes are in circulation; in some species very few hemocytes circulate, the great majority remaining loosely attached to tissue surfaces. In adult mosquitoes there are no circulating hemocytes (Jones, 1977). Hemocyte counts are normally highly variable within the same species, as well as differing among species. Thus, in *Periplaneta americana*, counts ranging from 45,000–120,000 hemocytes/μl may be measured. As the insect has about 170 μl of hemolymph, it will have between 7 and 20 million circulating cells.

According to Crossley (1975), there is evidence that the number of circulating hemocytes may depend on the hormone titer of the hemolymph. Both ecdysone and juvenile hormone are said to stimulate an increase in the number of circulating hemocytes, especially plasmatocytes. In some instances, the increase seems to be related to increased mobility of preexisting cells rather than an increased rate of cell multiplication, though how hormones bring about this effect is not known.

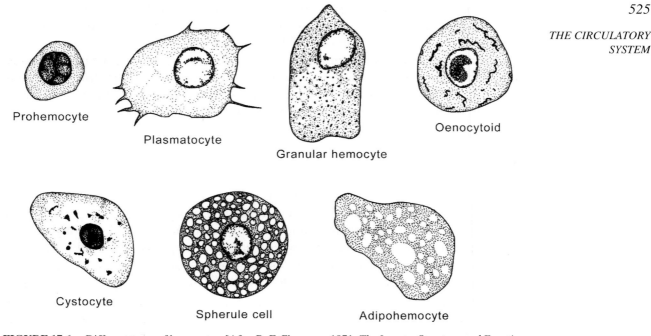

Prohemocyte

Plasmatocyte

Granular hemocyte

Oenocytoid

Cystocyte

Spherule cell

Adipohemocyte

FIGURE 17.6. Different types of hemocytes. [After R. F. Chapman, 1971, *The Insects: Structure and Function.* By permission of Elsevier/North-Holland, Inc., and the author.]

Though several types of hemocytes have been recognized, which differ in size, stainability, function, and cytology (including fine structure) (Figure 17.6), their classification and relationships have proven difficult. This difficulty stems partly from the natural structural variability and multifunctional nature of some hemocytes both within and among species, and partly from the differences in methodology and criteria used to distinguish different hemocytes (Arnold, 1974; Crossley, 1975; Jones, 1977). Notwithstanding these difficulties, it is apparent that three types of hemocytes are common to almost all insects, though one or more additional types may also occur in a given species. In this account, the scheme of Arnold (1974) is followed.

The three types common to most insects are prohemocytes, plasmatocytes, and granular hemocytes (granulocytes). Prohemocytes (stem cells) are small (10 μm or less in diameter), spherical, or ellipsoidal cells whose nucleus fills almost the entire cytoplasm. They are frequently seen undergoing mitosis and are assumed to be the primary source of new hemocytes and the type from which other forms differentiate. Plasmatocytes (phagocytes) are cells of variable shape and size, with a centrally placed, spherical nucleus surrounded by well vacuolated cytoplasm. In the cytoplasm are a well-developed Golgi complex and endoplasmic reticulum, as well as many lysosomes. The cells are capable of amoeboid movement and are phagocytic. Granulocytes are usually round or disc-shaped, with a relatively small nucleus surrounded by cytoplasm filled with prominent granules. In some species they are amoeboid and phagocytic which, together with the occurrence of intermediate forms, suggests that they may be derived from plasmatocytes. More often, they are non-motile and appear to be involved in intermediary metabolism.

Other types of hemocytes include adipohemocytes, oenocytoids, spherule cells, and cystocytes. As their name indicates, the adipohemocytes are cells whose cytoplasm normally contains droplets of lipid. In addition to lipid droplets, the cytoplasm may have non-lipid vacuoles and granules that contain carbohydrate material. The cells, which occasionally are phagocytic, are considered by some authors to be a form of granulocyte. Oenocytoids are spherical or ovoid cells with one, occasionally two, relatively small, eccentric nuclei. They are almost never phagocytic, as the absence of lysosomes, small Golgi complexes, and poorly developed endoplasmic reticulum attest. The oenocytoids are fragile cells that easily lyse, leading some authors to suggest that they may be a type of cystocyte. Spherule cells are readily identifiable cells whose central nucleus is often obscured by the mass of dense spherical inclusions occupying most of the cytoplasm. Though reported from a range of orders, they are especially common in higher Diptera and Lepidoptera, and a variety of functions have been proposed for them, including phagocytosis, uptake and transport of materials, synthesis of some blood proteins, and a role in bacterial immunity. Antibody-binding studies support the view that spherule cells may be a form of granulocyte (Gardiner and Strand, 1999). Cystocytes (coagulocytes) are spherical cells in whose small central nucleus the chromatin is so arranged as to give the nucleus a "cartwheel-like" appearance. The cytoplasm contains granules that, when liberated from these fragile cells, cause the surrounding plasma to precipitate. Thus, the cells, which are again a specialized kind of granulocyte, play a major role in hemolymph coagulation.

4.2.2. Functions

The major functions of hemocytes are endocytosis, nodule formation, encapsulation, and coagulation. For the first three of these a key element is the ability to distinguish between foreign (including altered self) and self. Hemocytes probably also have a variety of metabolic and homeostatic functions.

Endocytosis. This is a process whereby a cell plasma membrane folds around a substance that is thus ingested by the cell without rupture of the membrane. It includes pinocytosis ("cell drinking") in which the material ingested is in solution and phagocytosis where the material engulfed is particulate. Once it reaches the interior of the cell, the membrane-bound vesicle containing the substance fuses with enzyme-containing lysosomes and degradation of the substance ensues. In insects the primary phagocytic cells are plasmatocytes, though, as noted above, other types of hemocytes may also engulf material. Like other endocytotic cells, hemocytes are apparently selective with regard to what they ingest though the basis of this selectivity is not well known.

Endocytosis, especially phagocytosis, is important in both metamorphosis and defense against disease, as well as in routine cleaning up of dead or damaged cells. In some insects (including Lepidoptera and higher Diptera), during metamorphosis, phagocytic hemocytes invade many larval tissues and bring about their rapid histolysis. However, invasion does not occur (at least in muscle which has been well studied) until autolysis has begun in tissue cells. Autolysis is believed to be under hormonal control and presumably makes tissue appear foreign to the hemocytes. In other insects (including Coleoptera and mosquitoes) the hemocytes never invade larval tissues during metamorphosis, and the tissues disappear strictly by autolysis. A variety of microorganisms (viruses, bacteria, fungi, rickettsias, and protozoa) are known to be phagocytosed by hemocytes, and in most instances insects have excellent resistance to infection as a result. However, for some viruses, phagocytosis is not followed by digestion and therefore does not prevent infection. Rather, the virus uses this process as a means of entering a cell prior to replication. Furthermore, habitual protozoan and fungal parasites often escape phagocytosis, presumably because they are not recognized as foreign by the hemocytes.

Nodule formation.   This occurs in response to invasion of the body by large quantities of particulate matter (especially bacterial aggregates) that cannot be removed effectively by phagocytosis. Typically, nodule formation takes place in two phases. Within minutes of the foreign matter arriving in the body cavity, it is surrounded either by granulocytes or by specific plasma proteins and by material discharged from coagulocytes. This initial layer, which is often melanized, includes "plasmatocyte-spreading peptide" that induces plasmatocytes to aggregate and adhere to a foreign surface. Thus, in the second phase, which takes several hours to complete, plasmatocytes surround the melanized core, become flattened, and form a multicellular sheath comparable to that seen after encapsulation (Lavine and Strand, 2002). Plasmatocyte-spreading peptide may not be the only hemocyte-aggregating factor: evidence indicates that eicosanoids (polyunsaturated fatty acid derivatives), synthesized and secreted by hemocytes, induce cell aggregation and nodule formation (Miller and Stanley, 2001). Though the innermost plasmatocytes may carry out a limited amount of phagocytosis, nodules generally persist for the duration of the insect's life.

Encapsulation.   Encapsulation is essentially nodule formation on a larger scale. After the invader is initially coated with a thin layer of granulocytes or plasma proteins, layers of plasmatocytes surround it (Nappi, 1975; Lavine and Strand, 2001, 2002). This reaction on the part of hemocytes forms an insect's most important defense mechanism against metazoan endoparasites such as nematodes and insects. Typically, a capsule is 50 or more cells thick and includes, especially in the layers adjacent to the encapsulated object, intercellular secretions, in particular, mucopolysaccharide. Melanin may be present also in the capsule, though whether hemocytes are involved in its production is not known. Humoral (cell-free) encapsulation is particularly common in some species of Diptera, all characterized by having low numbers of circulating hemocytes. Plasma phenoloxidase is involved in the formation of the capsule sheath, perhaps by promoting tanning, analogous to that seen in the formation of the exocuticle (Chapter 11, Section 3.3), and melanin is usually deposited. An encapsulated organism almost always dies, as a result of starvation, asphyxiation, and perhaps poisoning by quinones, antibacterial peptides, hydrogen peroxide and nitric oxide. It should be stressed, however, that there is little evidence to support any of these mechanisms (Vass and Nappi, 2001; Lavine and Strand, 2002).

Recognition of Foreignness and Altered Self.   Recognition of an object as foreign or altered self is a key event in phagocytosis, nodule formation and encapsulation, yet how this takes place remains largely a mystery. The limited evidence suggests that there are some similarities to the innate immune system of vertebrates. Thus, for biological foreign objects there appear to be pattern-recognition receptors (PRR) both on the surface of hemocytes and in the plasma. As their name indicates, the PRR recognize, then bind with particular molecular patterns on the surface of the foreign organism. They also recognize altered self, specifically changes to the basal lamina that sits beneath the epidermis (Figure 11.1) and covers all internal organs (Salt, 1970; Lavine and Strand, 2002). This is important at metamorphosis, when juvenile tissues are histolyzed, when an insect is wounded, and when microorganisms penetrate the midgut epithelium. The nature of PRR is poorly known. Several proteins have been suggested as candidate humoral PRR, including hemolin, gram-negative bacteria recognition protein, and lectins (= hemagglutinins, agglutinins). In common, these are large, highly specific proteins with multivalent capacity to bind with membrane-surface carbohydrates, notably lipopolysaccharides, peptidoglycans and β-1, 3-glucans. Even less is known about hemocyte-surface PRR, though there are indications that these may, in fact, be membrane-bound forms of the humoral PRR (Lavine and Strand, 2002).

Experimentally, it has been demonstrated that hemocyte and humoral receptors also recognize and bind to abiotic objects, specifically, nylon, latex, and chromatography beads. Abiotic targets are recognized by their physicochemical characteristics, for example, surface charge, bead structure, and functional groups (Lackie, 1981; Lavine and Strand, 2001).

For biotic systems recognition may involve two distinct processes. First, there may be specific attraction of hemocytes to the foreign object (chemotaxis), though some authors suggest that hemocytes and foreign objects will meet randomly as a result of normal hemolymph circulation. And second, there is physical contact between hemocytes and the foreign material, central to which are the PRR. As noted, the hemocytes may have PRR built in to their plasma membrane and thus are directly able to recognize the foreign matter, or they may recognize the foreign material only after PRR in the plasma have become attached to it (Dunn, 1986). Either way, the interaction between PRR and foreign material promotes the next step in the sequence, namely, phagocytosis, nodule formation, or encapsulation.

Coagulation. Studies on hemolymph coagulation in insects lag behind those on vertebrate blood clotting for which a mass of physiological and biochemical information is available. According to Gregoire (1974), this is related largely to the extremely rapid, even instantaneous nature of hemolymph coagulation, which has hampered studies on the relative roles of hemocytes and plasma and the biochemical events that occur.

It is now accepted that all forms of coagulation involve the participation of a specific form of hemocyte, the coagulocyte, though there is still controversy as to whether other hemocyte types may also play a role in clot formation. Three basic patterns have been described on the basis of microscopical observations. Among orthopteroid insects, most Hemiptera, some Coleoptera, and some Hymenoptera, coagulation is initiated by a sometimes "explosive" discharge of small pieces of cytoplasm from the coagulocytes, though the cells remain intact. As a result, plasma surrounding the cells forms granular precipitates which gradually increase in size and density. In some Scarabaeidae (Coleoptera) and larval Lepidoptera, the coagulocytes extrude long, threadlike pseudopodia-like processes that interweave with and stick to each other to form a cytoplasmic network in which other hemocytes become trapped. Concurrently, the plasma itself gels, forming transparent elastic sheets between the processes. In some homopterans, other Coleoptera, for example, Tenebrionidae, and other Hymenoptera, clotting appears to combine the features of the first two types.

In some insects [some Heteroptera, Curculionidae (Coleoptera), and Neuroptera] hemocytes morphologically identical to coagulocytes are present, but these do not release material to induce clotting. They do, however, accumulate at wounds, and Crossley (1975) suggested that they may release bacteriostatic substances.

Limited progress has been made in understanding the biochemistry of coagulation (Bohn, 1986; Theopold *et al.*, 2002). Early studies showing that insect plasma lacked, for example, prothrombin, thrombin, and thromboplastin, suggested that in insects this process must be very different from its counterpart in mammals. However, recent work has shown that a number of similarities do exist. For example, in insects as in mammals there are both plasma- and cellular clotting proteins (coagulogens), the latter being released as the cells rupture. Interestingly, in many insects the plasma coagulogens are lipophorins, high-molecular-weight lipoproteins synthesized in the fat body that transport lipids and hormones. The hemocyte coagulogen is a glycoprotein. Other similarities include the involvement of calcium ions, the formation of membranous vesicles (microparticles), and enzyme-controlled cross-linking of the plasma and hemocyte coagulogens.

A possible scheme for hemolymph coagulation is given in Figure 17.7. Coagulation is initiated when coagulocytes rupture, a step that requires calcium ions. This releases

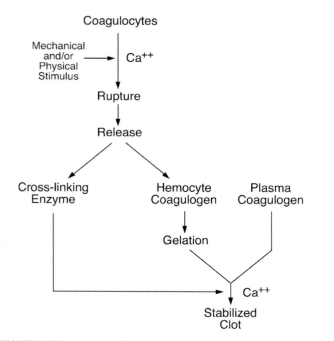

FIGURE 17.7. Proposed scheme for hemolymph clotting in insects.

hemocyte coagulogen, the cross-linking enzyme, and microparticles. These are bits of coagulocyte plasma membrane that have turned inside out, exposing negatively charged phospholipids. It is thought that the phospholipids may be important in the elimination of the dead or damaged cells, perhaps by attracting phagocytic hemocytes to the site of the wound. The hemocyte coagulogen swells on entering the plasma because of hydration and forms a gel, the basic framework of the clot. The clot is then stabilized and strengthened as the plasmacoagulogen becomes cross-linked to the gel. This step is enzyme-controlled and also requires calcium. It is possible that the cross-linking enzyme is part of the phenoloxidase enzyme complex, promoting tanning of the proteins, as occurs in the formation of exocuticle (Chapter 11, Section 3.3). This speculation is based on the observation by Li *et al.* (2002) that the insect coagulation pathway has components that also occur in the prophenoloxidase cascade (see Section 5.2).

   Metabolic and Homeostatic Functions.   Hemocytes have been implicated in a variety of metabolic and homeostatic functions, though for most of these convincing evidence is not available. Largely on the basis of electron microscopy and histochemical studies, in which the cells have been shown to contain, for example, glycogen, mucopolysaccharide, lipid, and protein, it has been suggested that hemocytes are important in storage of nutrients and their distribution to growing tissues, formation of connective tissue, synthesis of chitin, and maintenance of hemolymph sugar level. Crossley (1975) noted that certain amino acids, including glutamate, can be actively accumulated by hemocytes, suggesting that the cells might be important in hemolymph amino acid homeostasis. The ability to accumulate and store glutamate (thereby effectively removing it from the hemolymph) may be of special significance in view of this substance's role as a transmitter at the neuromuscular junction (Chapter 13, Section 2.3). Certain hemocytes (spherule cells in Diptera, oenocytoids in other insects) contain enzymes for metabolism of tyrosine, derivatives of which are important in tanning and/or darkening of cuticle (Chapter 11, Section 3.3) and have a bacteriostatic effect.

Various authors also have suggested that the hemocytes may have a significant role in detoxicating poisons, though again convincing evidence is lacking.

5. Resistance to Disease

Resistance to disease may be considered to include two components, prevention of entry of the disease-causing organisms and rendering harmless organisms that do manage to reach the body cavity. The insect cuticle, covering the entire body surface, tracheal system, foregut and hindgut, and the peritrophic matrix lining the midgut are major obstacles to the entry of such organisms.[*] A considerable number of potentially dangerous microorganisms must be ingested during feeding, yet normally these have no detrimental effect on the insect. What little work has been done suggests that specific substances in the gut deal with these organisms. For example, bacteria (especially gram-positive forms) appear to be digested by lysozyme. Though this enzyme is normally produced in the midgut, in mosquitoes the salivary glands are its source, perhaps to provide early protection against bacteria that enter the alimentary canal in ingested nectar (Moreira-Ferro *et al.*, 1998). Substances with antiviral activity are also known to occur in the gut, though their nature and mode of action have not been studied.

Hemolymph plays a major role in an insect's resistance to disease. It is important, through its role in wound healing, in preventing entry of pathogenic organisms, and in the destruction of organisms that manage to enter the body cavity.

5.1. Wound Healing

Wounding leads to a rapid increase in the number of circulating hemocytes and their aggregation in the vicinity of the wound. The nature of the factors that cause mobilization and aggregation are unknown, though it has been proposed that the damaged tissue cells release chemicals (wound hormones) that attract hemocytes. At the site of damage, hemocytes may become involved in various ways. The coagulocytes may initiate clotting; the plasmatocytes and perhaps other types of hemocytes may phagocytose or encapsulate dead cells and foreign organisms that have entered the wound; and hemocytes may arrange themselves in sheets to form a scaffold on which damaged tissue can regenerate.

5.2. Immunity

Immunity in the present discussion may be defined as the ability of an insect to resist the pathogenic effects of microorganisms that have gained entry into the body cavity. As in vertebrates, two forms of immunity may be distinguished, innate (natural) and acquired (induced) (Gillespie, *et al.*, 1997; Vilmos and Kurucz, 1998; Vass and Nappi, 2001; Lavine and Strand, 2002).

Innate immunity refers to resistance produced by factors already present in an organism, that is, prior to any stimulation resulting from appearance of a pathogen. Innate immunity in insects appears to be principally a cellular phenomenon, comprising phagocytosis, nodule formation, and encapsulation on the part of the hemocytes. However, the plasma may

[*] Some fungi enter the host via the integument (Chapter 23, Section 4.2.4).

contain powerful agglutinating and lytic substances, including lysozyme, phenoloxidase, and lectins. Lysozyme, which is produced by phagocytic hemocytes, destroys the murein skeleton of the bacterial cell wall. Phenoloxidase, a highly reactive enzyme, is stored as prophenoloxidase, which is activated by a variety of materials, including the carbohydrates in the cell wall of invading microorganisms. The enzyme appears to have two functions: first, its oxidation products may be toxic to the invaders, and second, it has an as yet unclear role in the recognition of foreignness. Lectins, as noted earlier, also facilitate recognition of foreign material.

Acquired immunity is resistance that is developed only when a pathogen enters a host; that is, the pathogen (or toxins produced by it) stimulates the development of this resistance. The existence of the phenomenon was first demonstrated by Paillot and Metalnikov, working independently, in 1920 (cited from Whitcomb *et al.*, 1974). These authors found that insects previously vaccinated with killed bacteria were immune to live pathogens injected subsequently. The immunity that develops is different, however, from the antigen-antibody system of vertebrates in two major respects. First, the immunity is short-lived, lasting only a matter of days, indicating that the cells that produce the immunogen (the substance that reacts against the pathogen) have no "memory," as do the antibody-producing lymphoid cells in vertebrates. Second, the response is non-specific, which is in marked contrast to the highly specific interaction between antibody and antigen in vertebrates.

Relatively little progress was made in understanding the insect immune response for some 40 years. In the 1960s circumstantial evidence accumulated for the idea that the immunogens might not be proteins, since they were apparently heat stable, dialyzable, and unaffected by trypsin, but rather phenolic compounds. It is now appreciated, however, that this corresponds with the transient activation of the prophenoloxidase system following infection and is not the induced immune response *per se*. At the beginning of the 1980s, insect immunology entered a new era with the discovery and isolation of the antibacterial peptides known as cecropins from the plasma of *Hyalophora cecropia*. Since then, some 60 antibiotic peptides, including attacins, lysozymes, defensins, and diptericins, have been confirmed as inducible immunogens for a variety of Lepidoptera, Diptera, Coleoptera, Hymenoptera, and Hemiptera (Boman and Hultmark, 1987; Vilmos and Kurucz, 1998). In common, these peptides are positively charged, enabling them to bind to the negatively charged bacterial surface. In a given species the antibiotic peptides exist in several slightly different forms that collectively provide the host with broad-spectrum coverage against bacteria. Further, the peptides act synergistically by targeting different parts of the bacterial cell envelope. Thus, attacins and diptericins attack gram-negative bacteria, inhibiting the synthesis of the cell membrane during division. Defensins kill gram-positive bacteria by inducing ion-channel formation in the membrane, hence lysis of the cell. Cecropins also cause the formation of ion channels in the membrane, but act against both gram-positive and gram-negative bacteria. Lysozyme, which also attacks both types of bacteria, hydrolyzes the bacterial cell wall (Vilmos and Kurucz, 1998). Antifungal peptides have also been described for a few species.

Many details of the immune response (the chain of events leading to immunity) remain unclear. However, it seems that some initial phagocytic action by hemocytes is necessary, producing peptidoglycan fragments (by lysis of the bacterial cell walls). These then trigger antibacterial peptide synthesis, principally in the fat body. After infection, the response is very rapid, immune-specific RNA being detectable within a few hours. The titer of immunogens rises to a peak after a few days (e.g., 7–8 days for cecropins in *Hyalophora*), then declines to undetectable levels after a further week.

Boman and Hultmark (1987) compared the immune mechanisms of insects and vertebrates. They noted that insects have relatively good physical protection from potential invaders by way of the cuticle and peritrophic matrix, whereas vertebrates offer greater access via the mucous linings of the digestive, respiratory, excretory, and reproductive systems. The link between the duration of an immune response and the life span and reproductive rate was also considered. Thus, in insects, with generally short lives but high fecundity, the immune response is very rapid, being based on RNA and protein synthesis, and is broad spectrum in nature. In contrast, in vertebrates, which have a long life span and produce relatively few offspring, the immune response is based on cell proliferation and is much longer lasting and highly specific. Of course, there are exceptions to every rule, and it is interesting that work on *Periplaneta americana*, which for an insect is relatively long-lived, indicates the existence in this species of a specific, long-lived "memory" type of immune response, functionally similar to that of vertebrates (Dunn, 1986, 1990).

Finally, it should be noted that the development of an immune response is not without costs to the host. The costs include not only those associated with developing and maintaining the potential for an immune response, but also longer-term responses. For example, successful defense against parasitoid attack by larvae of *Drosophila* leads to a reduction in adult size and number of eggs produced, as well as greater susceptibility to pupal parasitoids. Further, larvae whose immune system has been challenged have reduced competitive ability when food is limited, compared to unchallenged larvae (Kraaijeveld *et al.*, 2002).

5.2.1. Resistance to Host Immunity

Though humans have not solved the mystery of how hemocytes recognize foreign matter, many parasites and pathogens obviously have because on entering a host they do not elicit a reaction, with disastrous consequences for the host. Boman and Hultmark (1987) distinguished between passive and active resistance to host immunity.

In passive resistance the pathogen's or parasite's mechanism for avoiding attack is already in place on entry into the host. For example, many parasitic Hymenoptera oviposit with great precision in specific tissues of the host so that the egg does not trigger the immune response. Similarly, some nematodes, on entering the insect's body, immediately migrate to a ganglion where they grow rapidly and undisturbed (Vinson, 1990). Several mechanisms are used to prevent an encapsulation response. For example, the eggs of some parasitoids are coated with material that prevents hemocytes from sticking to them (Eslin and Prévost, 2000). Another tactic is for the eggs to be covered by a fibrous coat that is not recognized as foreign (Hu *et al.*, 2003). In a slight variation on this theme, the eggs and larvae of the ichneumonid wasp *Venturia* (= *Nemeritis*) *canescens* are covered by a glycoprotein layer that binds the host's plasma lipophorin, which thereby camouflages the foreign egg surface (Kinuthia *et al.*, 1999).

Active resistance refers to active steps taken by the invader when the host's system attempts to attack it. One strategy is for the parasitoid to deposit its eggs before the host's immune system is very responsive. For example, the braconid wasp *Asobara tabida* preferentially oviposits in first- or second-instar larvae of *Drosophila* in which there are few circulating hemocytes. Growth of the parasitoid larva is very rapid, which further reduces the likelihood of efficient encapsulation. Other active resistance mechanisms include the introduction into the host, at the time of egg deposition, of (1) compounds that suppress production of humoral immunogens such as the antibacterial peptides, (2) substances that degrade or inhibit activity of the immunogens, or (3) chemicals that inhibit mobilization or multiplication of hemocytes. For example, the parasitic nematode *Heterorhabditis bacterio-phora* releases a proteinase that specifically degrades cecropins in the wax moth *Galleria mellonella* (Jarosz, 1998). However, the majority of evidence for these mechanisms has come from studies on hymenopteran parasitoids.

Except in some braconids and ichneumonids (see below), hymenopteran venom, produced by the female accessory glands (Figure 19.1) and injected at oviposition, may attack host granulocytes, causing them to lyse or lose their ability to spread over a foreign surface. The venom also interferes with the host's endocrine and metabolic systems. As a result of the disruption to its hormone levels, the host can no longer molt, though it continues to feed. The venom also interferes with the normal process of storage of nutrient reserves by the fat body. Collectively, these two effects enable the parasitoid larva to grow and mature rapidly in a food-rich environment, without the risk that the host will metamorphose prematurely (Vinson, 1990; Coudron, 1991; Strand and Pech, 1995).

Some braconid and ichneumonid wasps, on the other hand, have entered into a remarkable symbiosis with a virus as their strategy for resisting host attack. As these wasps oviposit, they inject a protein together with numerous particles of polydnavirus, so-called because these complex viruses have up to 28 complete double-stranded circles of DNA (Beckage, 1997; Shelby and Webb, 1999). The ovarian protein temporarily blocks encapsulation by causing the granulocytes and plasmatocytes to lyse, giving the polydnavirus time to invade the host's tissues where viral protein synthesis begins. By about 24 hours after oviposition, when the ovarian protein is no longer effective, sufficient viral protein has been produced that lysis of hemocytes continues and encapsulation does not occur. The polydnavirus also has two other major effects on the host's immune response. It inhibits the prophenoloxidase cascade, preventing production of melanin and quinones, and it significantly reduces the production and release of antibacterial peptides in the fat body. Like venom, polydnavirus also inhibit the host's normal development and metabolism. Specifically, synthesis of molting hormone and juvenile hormone esterase are inhibited, while production of juvenile hormone is enhanced. Collectively these actions prevent the host from growing and molting normally (see Chapter 21, Section 6.1). Simultaneously, the virus disrupts the synthesis of storage hexamers and carbohydrates, presumably allowing the parasitoid easier access to nutrients. It should be stressed that production of viral particles occurs only in the female parasitoid, specifically in the calyx region of the reproductive tract (Figure 19.1). Curiously, however, the polydnavirus genome is part of the genome of both male and female wasps. Further, some polydnavirus genes are identical to those that code for venom proteins, raising the fascinating question of whether the polydnavirus had an independent evolutionary origin or has always been an integral part of the wasp's genome (Beckage, 1997).

6. Summary

The insect circulatory system includes a contractile middorsal vessel, divisible into a posterior heart, equipped with valves (ostia) through which hemolymph enters the vessel, and an anterior aorta that delivers hemolymph to the head region. Only rarely are there arteries for conduction of hemolymph to specific body regions. Muscular diaphragms and, in most species, accessory pulsatile organs assist in directing the hemolymph around the body. The heart may or may not be innervated, but in any event the heartbeat is always myogenic. Control of heartbeat frequency and amplitude is complex and influenced by neuronal and neurohormonal factors. Except in primitive insects where hemolymph flow is bidirectional, the heartbeat is typically seen as a forwardly moving wave of peristalsis that begins near the posterior end of the dorsal vessel. However, reverse peristalsis is common, especially in pupae and adults, and may play a role in the tidal flow of hemolymph in association with the ventilation movements for gas exchange.

Hemolymph is composed of cellular components, hemocytes, and a non-cellular fraction, plasma. In primitive insects, inorganic ions, especially sodium and chloride, are the major osmotic effectors in hemolymph, but, through evolution, the tendency is for sodium and chloride to be replaced by magnesium, potassium, and organic anions, and for organic components to become the dominant contributors to osmotic pressure. However, diet may considerably influence the composition of plasma. Plasma, especially of larval endopterygotes, contains high and relatively constant concentrations of amino and carboxylic acids. Other organic constituents of plasma include enzymes and other proteins, trehalose and other carbohydrates, hormones, and lipids, but, with the exception of trehalose, their concentration is variable. Apart from the specific functions of its components, plasma is important as a transport system, a source of cell water, a hydrostatic system for transmitting pressure changes from one part of the body to another, and for some species in thermoregulation.

In many species virtually all of the hemocytes are in circulation; in some other species most hemocytes do not normally circulate through the body but remain loosely attached to tissue surfaces to be mobilized under specific conditions such as wounding, entry of foreign organisms, and ecdysis. Several types of hemocytes occur of which three, prohemocytes, plasmatocytes, and granulocytes, are common to most insects. Prohemocytes are the stem cells from which other types differentiate. Plasmatocytes are the principle phagocytic hemocytes and, together with granulocytes, are nodule-forming and encapsulating agents. Granulocytes are probably also important in intermediary metabolism. In addition, in many species, cystocytes are present which participate in hemolymph coagulation.

Hemolymph plays an important role in insect immunity to disease. In insects innate immunity appears to be principally a hemocyte-based phenomenon and includes wound healing, coagulation, and destruction of invading organisms by phagocytosis, nodule formation and encapsulation for which the ability to distinguish between self and foreign (including altered self) is critical. Some innate immunity also results from agglutinating and lytic substances in the plasma. In keeping with their generally short life history and high fecundity, for most species acquired immunity is short-lived and non-specific. The immunogens are peptides and include cecropins, defensins, attacins, diptericins, and lysozymes. The first two substances alter permeability of the bacterial membrane, attacins and defensins interfere with cell membrane synthesis, and lysozymes degrade the murein in the cell wall.

Many parasitoids resist the immune response of their host, either by passive or active means. Passive resistance can result from the invader being in a location where it does not trigger an immune response or having a surface coat that is not recognized as foreign. Active resistance mechanisms include immobilization and destruction of the host's hemocytes and prevention of immunogen formation or action. In addition, the parasitoid disrupts the host's endocrine balance and modifies its metabolism, preventing normal host growth and channeling resources for the parasitoid's benefit.

7. Literature

Jones (1977) and Miller (1985) give general descriptions of the circulatory system. Pass (1998, 2000) reviews accessory pulsatile structures. The forms and functions of hemocytes are discussed by Arnold (1974), Crossley (1975), Jones (1975), Gupta (1985), and authors in the volume edited by Gupta (1979). Florkin and Jeuniaux (1974) and Mullins (1985) deal with hemolymph composition, and Grégoire (1974), Bohn (1986), and Theopold *et al.* (2002) with coagulation. Insect immunity is reviewed by Salt (1970), Gotz and Boman (1985), Dunn (1986, 1990), Boman and Hultmark (1987), Gillespie *et al.* (1997), Trenczek (1998), Vilmos and Kurucz (1998), Vass and Nappi (2001), Lavine and Strand (2002), and authors in Brey and Hultmark (1998). Vinson (1990), Beckage (1997), and Shelby and Webb (1999) examine how parasites deal with their host's immune system.

Arnold, J. W., 1974, The hemocytes of insects, in: *The Physiology of Insecta*, 2nd ed., Vol. V (M. Rockstein, ed.), Academic Press, New York.

Beckage, N., 1997, The parasitic wasp's secret weapon, *Sci. Amer.* **277**(November):82–87.

Bohn, H., 1986, Hemolymph clotting in insects, in: *Immunity in Invertebrates* (M. Brehélin, ed.), Springer-Verlag, Berlin.

Boman, H. G., and Hultmark, D., 1987, Cell-free immunity in insects, *Annu. Rev. Microbiol.* **41**:103–126.

Brey, P. T., and Hultmark, D. (eds.), 1998, *Molecular Mechanisms of Immune Responses in Insects*, Chapman and Hall, London.

Coudron, T. A., 1991, Host-regulating factors associated with parasitic Hymenoptera, in: *Naturally Occurring Pest Bioregulators* (P. A. Hedin, ed.), American Chemical Society, Washington, DC.

Crossley, A. C., 1975, The cytophysiology of insect blood, *Adv. Insect Physiol.* **11**:117–221.

Dulcis, D., Davis, N. T., and Hildebrand, J. G., 2001, Neuronal control of heart reversal in the hawkmoth *Manduca sexta, J. Comp. Physiol. A* **187**:837–849.

Dunn, P. E., 1986, Biochemical aspects of insect immunology, *Annu. Rev. Entomol.* **31**:321–339.

Dunn, P. E., 1990, Humoral immunity in insects, *BioScience* **40**:738–744.

Eslin, P., and Prévost, G., 2000, Racing against host's immunity defenses: A likely strategy for passive evasion of encapsulation in *Asobara tabida* parasitoids, *J. Insect Physiol.* **46**:1161–1167.

Fife, H. G.,Palli, S. R., and Locke. M., 1987, A function for pericardial cells in an insect, *Insect Biochem.* **17**:829–840.

Florkin, M., and Jeuniaux, C., 1974, Hemolymph: Composition, in: *The Physiology of Insecta*, 2nd ed., Vol. V (M. Rockstein, ed.), Academic Press, New York.

Gardiner, E. M. M., and Strand, M. R., 1999, Monoclonal antibodies bind distinct classes of hemocytes in the moth *Pseudoplusia includens, J. Insect Physiol.* **45**:113–126.

Gereben-Krenn, B.-A., and Pass, G., 2000, Circulatory organs of abdominal appendages in primitive insects (Hexapoda: Archaeognatha, Zygentoma and Ephemeroptera), *Acta Zool.* **81**:285–292.

Gillespie, J. P., Kanost, M. R., and Trenczek, T., 1997, Biological mediators of insect immunity, *Annu. Rev. Entomol.* **42**:611–643.

Gotz, P., and Boman, H. G., 1985, Insect immunity, in: *Comprehensive Insect Physiology, Biochemistry and Pharmacology*, Vol. 3 (G.A. Kerkut and L. I. Gilbert, eds.), Pergamon Press, Elmsford, NY.

Grégoire, C., 1974, Hemolymph coagulation, in: *The Physiology of Insecta*, 2nd ed., Vol. V (M. Rockstein, ed.). Academic Press, New York.

Gupta, A. P. (ed.), 1979, *Insect Hemocytes*, Cambridge University Press, London.

Gupta, A. P., 1985, Cellular elements in the hemolymph, in: *Comprehensive Insect Physiology, Biochemistry and Pharmacology*, Vol. 3 (G. A. Kerkut and L. I. Gilbert, eds.), Pergamon Press, Elmsford, NY.

Hantschk, A., 1991, Functional morphology of accessory circulatory organs in the legs of Hemiptera, *Int. J. Insect Morphol. Embryol.* **20**:259–273.

Hertel, W., and Pass, G., 2002, An evolutionary treatment of the morphology and physiology of circulatory organs in insects, *Comp. Biochem. Physiol. A* **133**:555–575.

Hu, J., Zhu, X.-X., and Fu, W.-J., 2003, Passive evasion of encapsulation in *Macrocentrus cingulum* Brischke (Hymenoptera: Braconidae), a polyembryonic parasitoid of *Ostrinia furnacalis* Guenée (Lepidoptera: Pyralidae), *J. Insect Physiol.* **49**:367–375.

Jarosz, J., 1998, Active resistance of entomophagous rhabditid *Heterorhabditis bacteriophora* to insect immunity, *Parasitology* **117**:201–208.

Jones, J. C., 1974, Factors affecting heart rates in insects, in: *The Physiology of Insecta*, 2nd ed., Vol. V (M. Rockstein, ed.), Academic Press, New York.

Jones, J. C., 1975, Forms and functions of insect hemocytes, in: *Invertebrate Immunity:Mechanisms of Invertebrate Vector-Parasite Relations* (K. Maramorosch and R. E. Shope, eds.), Academic Press, New York.

Jones, J. C., 1977, *The Circulatory System of Insects*, Thomas, Springfield, IL.

Kinuthia, W., Li, D., Schmidt, O., and Theopold, U., 1999, Is the surface of endoparasitic wasp eggs and larvae covered by a limited coagulation reaction?, *J. Insect Physiol.* **45**:501–506.

Koladich, P. M., Tobe, S. S., and McNeil, J. N., 2002, Enhanced haemolymph circulation by insect ventral nerve cord: Hormonal control by *Pseudaletia unipuncta* allatotropin and serotonin, *J. Exp. Biol.* **205**:3123–3131.

Kraaijeveld, A. R., Ferrari, J., and Godfrey, H. C. J., 2002, Costs of resistance in insect-parasite and insect-parasitoid interactions, *Parasitology* **125**(Suppl.):S71–S82.

Lackie, A. M., 1981, Immune recognition in insects, *Dev. Comp. Immunol.* **5**:191–204.

Lavine, M. D., and Strand, M. R., 2001, Surface characteristics of foreign targets that elicit an encapsulation response by the moth *Pseudoplusia includens*, *J. Insect Physiol.* **47**:965–974.

Lavine, M. D., and Strand, M. R., 2002, Insect hemocytes and their role in immunity, *Insect Biochem. Molec. Biol.* **32**:1295–1309.

Li, D., Scherfer, C., Korayem, A. M., Zhao, Z., Schmidt, O., and Theopold, U., 2002, Insect hemolymph clotting: Evidence for interaction between the coagulation system and the prophenoloxidase cascade, *Insect Biochem. Mol. Biol.* **32**:919–928.

Locke, M., 1989, Secretion by insect heart muscle cells, *J. Insect Physiol.* **35**:53–56.

Locke, M., and Russell, V. W., 1998, Pericardial cells or athrocytes, in: *Microscopic Anatomy of Invertebrates*, Vol. 11B (F. Harrison and M. Locke, eds.), Wiley, New York.

Miller, J. S., and Stanley, D. W., 2001, Eicosanoids mediate microaggregation reactions to bacterial challenge in isolated insect hemocyte preparations, *J. Insect Physiol.* **47**:1409–1417.

Miller, T. A., 1985, Structure and physiology of the circulatory system, in: *Comprehensive Insect Physiology, Biochemistry and Pharmacology*, Vol. 3 (G. A. Kerkut and L. I. Gilbert, eds.). Pergamon Press, Elmsford, NY.

Miller, T. A., 1997, Control of circulation in insects, *Gen. Pharmacol.* **29**:23–38.

Moreira-Ferro, C. K., Daffre, S., James, A. A., and Marinotti, O., 1998, A lysozyme in the salivary glands of the malaria vector *Anopheles darlingi*, *Insect Mol. Biol.* **7**:257–264.

Mullins, D. E., 1985, Chemistry and physiology of the hemolymph, in: *Comprehensive Insect Physiology, Biochemistry and Pharmacology*, Vol. 3 (G. A. Kerkut and L. I. Gilbert, eds.), Pergamon Press, Elmsford, NY.

Nappi, A. J., 1975, Parasite encapsulation in insects, in: *Invertebrate Immunity: Mechanisms of Invertebrate Vector-Parasite Relations* (K. Maramorosch and R. E. Shope, eds.), Academic Press, New York.

Pass, G., 1998, Accessory pulsatile organs, in: *Microscopic Anatomy of Invertebrates*, Vol. 11B (F. Harrison and M. Locke, eds.), Wiley, New York.

Pass, G., 2000, Accessory pulsatile organs: Evolutionary innovations in insects, *Annu. Rev. Entomol.* **45**:495–518.

Salt, G., 1970, *The Cellular Defence Reactions of Insects*, Cambridge University Press, London.

Shelby, K. S., and Webb, B. A., 1999, Polydnavirus-mediated suppression of insect immunity, *J. Insect Physiol.* **45**:507–514.

Strand, M. R., and Pech, L. L., Immunological basis for compatibility in parasitoid-host relationships, *Annu. Rev. Entomol.* **40**:31–56.

Theopold, U., Li, D., Fabbri, M., Scherfer, C., and Schmidt, O., 2002, The coagulation of insect hemolymph, *Cell. Molec. Life Sci.* **59**:363–372.

Trenczek, T., 1998, Endogenous defense mechanisms of insects, *Zoology, Jena* **101**:298–315.

Vass, E., and Nappi, A. J., 2001, Fruit fly immunity, *BioScience* **51**:529–535.

Vilmos, P., and Kurucz, E., 1998, Insect immunity: Evolutionary roots of the mammalian innate immune system, *Immunol. Lett.* **62**:59–66.

Vinson, S. B., 1990, How parasites deal with the immune system of their host: An overview, *Arch. Insect Biochem. Physiol.* **13**:3–27.

Wheeler, C. H., 1989, Mobilization and transport of fuels to the flight muscles, in: *Insect Flight* (G. J. Goldsworthy and C. H. Wheeler, eds.), CRC Press, Boca Raton.

Whitcomb, R. P., Shapiro, M., and Granados, R. R., 1974, Insect defense mechanisms against microorganisms and parasitoids, in: *The Physiology of Insecta*, 2nd ed., Vol. V (M. Rockstein, ed.), Academic Press, New York.

18

Nitrogenous Excretion and Salt and Water Balance

1. Introduction

Enzymatically controlled reactions occur at the optimum rate within a narrow range of physical conditions. Especially important are the pH and ionic content of the cell fluid, as these factors readily affect the active site on an enzyme. As the conditions existing within cells and tissues are necessarily dependent on the nature of the fluid that bathes them—in insects, the hemolymph—it is the regulation of this fluid that is important. By regulation is meant the removal of unwanted materials and the retention of those that are useful, to maintain as nearly as possible the best cellular environment. Regulation is a function of the excretory system and is of great importance in insects because they occupy such varied habitats and, therefore, have different regulatory requirements. Terrestrial insects lose water by evaporation through the integument and respiratory surfaces and in the process of nitrogenous waste removal. Brackish-water and saltwater forms also lose water as a result of osmosis across the integument; in addition, they gain salts from the external medium. Insects inhabiting fresh water gain water from and lose salts to the environment. The problem of osmoregulation is complicated by an insect's need to remove nitrogenous waste products of metabolism, which in some instances are very toxic. This removal uses both salts and water, one or both of which must be recovered later from the urine.

2. Excretory Systems

2.1. Malpighian Tubules—Rectum

The Malpighian tubules and rectum, functioning as a unit, form the major excretory system in most insects. Details of the rectum are given in Chapter 16, Section 3.4, and only the structure of the tubules is described here.

The blindly ending tubules, which usually lie freely in the hemocoel, open into the alimentary canal at the junction of the midgut and hindgut (Figure 18.1A). Typically they enter the gut individually but may fuse first to form a common sac or ureter that leads into the gut. Their number varies from two to several hundred and does not appear to be

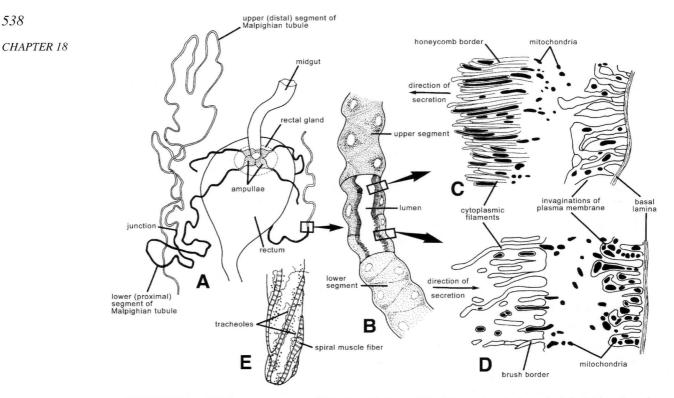

FIGURE 18.1. (A) Excretory system of *Rhodnius*. Only one Malpighian tubule is drawn in full; (B) junction of proximal and distal segments of a Malpighian tubule of *Rhodnius*. Part of the tubule has been cut away to show the cellular differentiation; (C, D) sections of the wall of the distal and proximal segments, respectively, of a tubule; and (E) tip of Malpighian tubule of *Apis* to show tracheoles and spiral muscles. [A, B, E, after V. B. Wigglesworth, 1965, *The Principles of Insect Physiology*, 6th ed., Methuen and Co. By permission of the author. C, D, from V. B. Wigglesworth and M. M. Saltpeter, 1962, Histology of the Malpighian tubules in *Rhodnius prolixus* Stal. (Hemiptera), *J. Insect Physiol.* **8**:299–307. By permission of Pergamon Press Ltd.]

closely related to either the phylogenetic position or the excretory problems of an insect. Malpighian tubules are absent in Collembola, some Diplura, and aphids; in other Diplura, Protura, and Strepsiptera there are papillae at the junction of the midgut and hindgut. With the tubules are associated tracheoles and, usually, muscles (Figure 18.1E). The latter take the form of a continuous sheath, helical strips, or circular bands and are situated outside the basal lamina. They enable the tubules to writhe, which ensures that different parts of the hemolymph are exposed to the tubules and assists in the flow of fluid along the tubules.

A tubule is made up of a single layer of epithelial cells, situated on the inner side of a basal lamina (Figure 18.1B–D). In many species where the tubules have only a secretory function (Section 3.2) the histology of the tubules is constant throughout their length and basically resembles that of the distal part of the tubule of *Rhodnius* (Figure 18.1C). The inner (apical) surface of the cells takes the form of a brush border (microvilli). The outer (basal) surface is also extensively folded. Both of these features are typical of cells involved in the transport of materials and serve to increase enormously the surface area across which transport can occur. Numerous mitochondria occur, especially adjacent to or within the folded areas, to supply the energy requirements for active transport of certain ions across the tubule wall. In many species various types of intracellular crystals occur which are presumed to represent a form of storage excretion (Section 3.3). Adjacent cells are closely apposed near their apical and basal margins, though not necessarily elsewhere.

In some insects (e.g., *Rhodnius*), two distinct zones can be seen in the Malpighian tubule (Figure 18.1C, D). In the distal (secretory) zone the cells possess large numbers of closely packed microvilli, but very few infoldings of the basal surface. Mitochondria are located near or within the microvilli. In the proximal (absorptive) part of the tubule the cells possess fewer microvilli, yet show more extensive invagination of the basal surface. The mitochondria are correspondingly more evenly distributed. In the flies *Dacus* and *Drosophila*, where pairs of Malpighian tubules unite to form a ureter prior to joining the gut, the ultrastructure of the ureter resembles that of the proximal part of the *Rhodnius* tubule, suggesting that the ureter may be a site of resorption of materials from the urine.

Yet other species have even more complex Malpighian tubules in which up to four distinct regions may be distinguished on histological or ultrastructural grounds. On the basis of the structural features of their cells, these regions have been designated as secretory or absorptive, though it must be emphasized that physiological evidence for these proposed functions is largely lacking. For a survey of insects whose tubules show regional differentiation and a discussion of tubule function in such species, see Jarial and Scudder (1970).

A cryptonephridial arrangement of Malpighian tubules is found in larvae and adults of many Coleoptera, some larval Hymenoptera and Neuroptera, and nearly all larval Lepidoptera (Figure 18.2). Here the distal portion of the Malpighian tubules is closely apposed to the surface of the rectum and enclosed within a perinephric membrane. The system is particularly well developed in insects living in very dry habitats, and in such species its function is to improve water resorption from the material in the rectum (Section 4.1).

2.2. Other Excretory Structures

Even in insects that use the rectum as the primary site of osmoregulation, the ileum may nonetheless be a site for water or ion resorption. In other species where the rectum is unimportant in osmoregulation, serving only to store urine and feces prior to expulsion, the ileum often takes on this role.

In a few insects the labial glands may function as excretory organs. In apterygotes that lack Malpighian tubules the glands can accumulate and eliminate dyes such as ammonia carmine and indigo carmine from the hemolymph, but there is no evidence that they can deal similarly with nitrogenous or other wastes. The labial glands of saturniid moths excrete copious amounts of fluid just prior to emergence from the cocoon, and it may well be that the primary function of the glands is to reduce hemolymph volume and hence body weight, which, in such large flying insects, needs to be kept as low as possible. The midgut of silkmoth larvae actively removes potassium from the hemolymph, thus protecting the tissues from the very high concentration of potassium ions present in the leaves eaten by these insects.

In a few insects it appears that the Malpighian tubules, though present, play no part in nitrogenous excretion. In *Periplaneta americana*, for example, uric acid is not found in the tubules but does occur in small amounts in the hindgut, which may excrete it directly from the hemolymph. In *P. americana* much uric acid is stored in urate cells in the fat body, and the major form of excreted nitrogen in this species is ammonia. How this reaches the hindgut lumen in *P. americana* is unclear. However, in the flesh fly *Sarcophaga bullata*, ammonia, the primary excretory product, is actively secreted as ammonium ions into the lumen across the anterior hindgut wall.

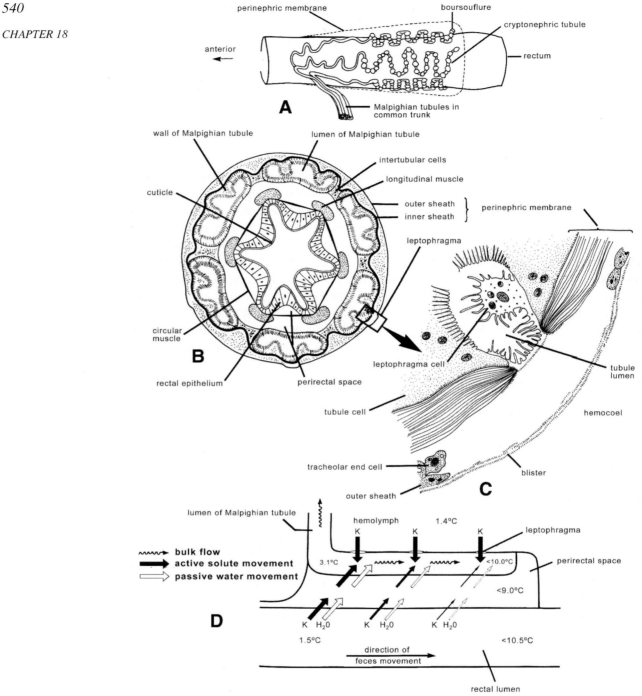

FIGURE 18.2. Cryptonephridial arrangement of Malpighian tubules in *Tenebrio* larva. (A) General appearance. Note that only three of the six tubules are drawn fully and that in reality the tubules are much more convoluted and have more boursouflures than are shown; (B) cross section through posterior region of cryptonephridial system; (C) details of a leptophragma; and (D) diagram illustrating proposed mode of operation of system. Solid arrows indicate movements of potassium, hollow arrows indicate movements of water. Numbers indicate osmotic concentration (measured as freezing-point depression) of fluids in different compartments. [After A. V. Grimstone, A. M. Mullinger, and J. A. Ramsay, 1968, Further studies on the rectal complex of the mealworm *Tenebrio molitor* L. (Coleoptera, Tenebrionidae), *Philos. Trans. R. Soc. Lond. Ser. B* **253**:343–382. By permission of the Royal Society, London, and Professor J. A. Ramsay.]

In males of some species of cockroaches, for example, *Blattella germanica*, a considerable amount of uric acid (as much as 5% of the live weight of the insect) is found in the utriculi majores (part of the accessory reproductive gland complex). The uric acid becomes part of the wall of the spermatophore and is, in a sense, "excreted" during copulation.

3. Nitrogenous Excretion

3.1. The Nature of Nitrogenous Wastes

In nitrogenous wastes structural complexity, toxicity, and solubility go hand in hand. The simplest form of waste (ammonia) is highly toxic and very water-soluble. It contains a high proportion of hydrogen that can be used in production of water. It is generally found as the major excretory product, therefore, only in those insects that have available large amounts of water, for example, larvae and adults of freshwater species. Nonetheless, exceptions are known, the best examples being the larvae of meat-eating flies and *P. americana* under certain dietary regimes. Generally, however, in insects, as in other terrestrial organisms, water must be conserved, and more complex nitrogenous wastes are produced, which are both less toxic and less soluble. In the egg and pupal stage the problem is accentuated because water lost cannot be replaced, and nitrogenous wastes must remain in the body in the absence of a functional excretory system. Most insects, then, excrete their waste nitrogen as uric acid. This is only slightly water-soluble, relatively non-toxic, and contains a smaller proportion of hydrogen compared with ammonia.

However, uric acid is not the only form of nitrogenous waste. Usually traces of other materials (especially the related compounds allantoin and allantoic acid) can be detected, and in many species one of these has become the predominant excretory product (Bursell, 1967). Urea is rarely a major constituent of insect urine, usually representing less than 10% of the nitrogen excreted. Traces of amino acids can be found in the excreta of many insects, but their presence should be regarded as accidental loss rather than deliberate excretion by an insect (Bursell, 1967). Only occasionally has the excretion of particular amino acids been authenticated; for example, the clothes moth *Tineola* and the carpet beetle *Attagenus* excrete large amounts of the sulfur-containing amino acid cystine. Although in tsetse flies uric acid is the primary excretory product, two amino acids, arginine and histidine, are important components of the urine. These make up about 10% of the protein amino acids in human-blood; because their nitrogen content is high, it is probably uneconomical to degrade them, and they are therefore excreted unchanged (Bursell, 1967). The amino acids voided in honeydew by plant-sucking Hemiptera must be considered as largely fecal and not metabolic waste products. Because of the large amount of water taken in by aphids, it has been suggested that they might produce ammonia as their nitrogenous waste. Indeed, uric acid, allantoin, and allantoic acid cannot be detected in their excreta. However, ammonia makes up only 0.5% of the total nitrogen excreted, which has led to the suggestion that it is used (and detoxified) by symbiotic bacteria in mycetomes (Chapter 16, Section 5.1.2). Table 18.1 contains selected examples to show the variety of nitrogenous wastes produced by insects.

As can be seen in Figure 18.3, uric acid and the other nitrogenous waste products are derived from two sources, nucleic acids and proteins. Degradation of nucleic acids is of minor importance; most nitrogenous waste comes from protein breakdown followed by synthesis of hypoxanthine from amino acids. The biochemical reactions that lead to synthesis of this purine appear to be similar to those found in other uric acid-excreting organisms (Bursell, 1967; Barrett and Friend, 1970).

TABLE 18.1. Nitrogenous Excretory Products of Various Insects[a,b]

	Uric acid	Allantoin	Allantoic acid	Urea	Ammonia	Amino acids
Odonata						
Aeshna cyanea (larva)	0.08	—	0.00	—	1.00	—
Dictyoptera/Phasmida						
Periplaneta americana	1.00	0.00	0.00	—	—	—
Blatta orientalis	0.64	0.64	1.00	—	—	—
Dixippus morosus	0.69	1.00	0.44	—	—	—
Hemiptera						
Dysdercus fasciatus	0.00	1.00	0.00	0.26	—	0.24
Rhodnius prolixus	1.00	—	—	0.33	—	Trace
Coleoptera						
Melolontha vulgaris	1.00	0.00	0.00	—	—	—
Attagenus piceus	0.72	—	—	1.00	0.57	0.50
Diptera						
Lucilia sericata	1.00	0.30	—	—	0.30	—
Lucilia sericata (pupa)	1.00	0.00	—	—	0.15	—
Lucilia sericata (larva)	0.05	0.02	—	—	1.00	—
Lepidoptera						
Pieris brassicae	1.00	0.04	0.01	—	—	—
Pieris brassicae (pupa)	1.00	0.03	0.05	—	—	—
Pieris brassicae (larva)	0.28	0.16	1.00	—	—	—

[a]From Bursell (1967), after various authors.
[b]The quantity of nitrogen excreted in the different products is expressed as a proportion of the nitrogen in the predominant end product.

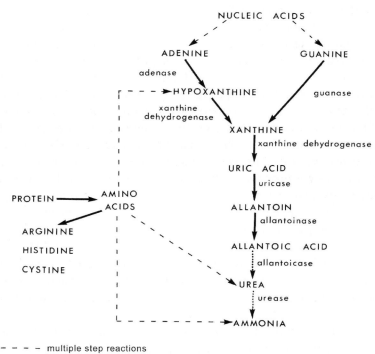

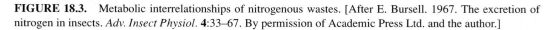

FIGURE 18.3. Metabolic interrelationships of nitrogenous wastes. [After E. Bursell. 1967. The excretion of nitrogen in insects. *Adv. Insect Physiol.* **4**:33–67. By permission of Academic Press Ltd. and the author.]

In addition to the enzymes for uric acid synthesis there are also uricolytic enzymes that catalyze degradation of this molecule in many insects (Figure 18.3). Uricase has a wide distribution within the Insecta. Active preparations of allantoinase have been obtained from many species, but the distribution of this enzyme appears to be rather restricted compared with uricase. Although there are reports that indicate the occurrence of allantoicase and urease in tissue extracts from a few insects, their presence should not be regarded as having been established unequivocally. In other words, when urea and ammonia are produced in significant amounts, they are probably derived in a manner other than by the degradation of uric acid. The existence of an ornithine cycle for urea production, such as is found in vertebrates, has not been proved conclusively, even though the constituent molecules of the cycle (arginine, ornithine, and citrulline) and the enzyme arginase have been identified in several species (Cochran, 1975). Cochran (1985) suggested that urea is merely a by-product of the biochemical conversion of arginine to proline, used in flight metabolism (Chapter 14, Section 3.3.5). Similarly, the way in which ammonia is produced (especially in those insects in which it is a major excretory molecule) is poorly understood. It is generally assumed to result from deamination of amino acids, but the precise way in which this occurs remains unclear.

It has been suggested that the most primitive state was that in which the complete series of uricolytic enzymes was present, and ammonia was the excretory material. As insects became more independent of water, selection pressures led to loss of the terminal enzymes and production of more appropriate excretory molecules. This simple view should be regarded with caution. Thus, in some caterpillars, diet can affect the nature of the nitrogenous waste. In certain insects substantial quantities of a particular nitrogenous waste molecule are produced, yet the appropriate enzyme in the uricolytic pathway has not been demonstrated, and vice versa; that is, the effects of other metabolic pathways may override the uricolytic system. In many insects (especially endopterygotes) the predominant nitrogenous excretory product changes during development. For example, in the mosquito *Aedes aegypti* urea is the principal nitrogenous waste in the (aquatic) larvae, while uric acid becomes dominant in pupae and adult females (von Dungern and Briegel, 2001). In *Pieris brassicae* (Lepidoptera) the major excretory product in the pupa and adult is uric acid; in the larva this compound constitutes only about 20% of the nitrogenous waste, allantoic acid being the predominant end product (Table 18.1). Indeed, in some Lepidoptera, the ratio of uric acid to allantoin may fluctuate widely from day to day (Razet, 1961, cited from Bursell, 1967). Of great interest will be determination of factors that stimulate inhibition or activation (degradation or synthesis?) of uricolytic enzymes so that the most suitable form of nitrogenous waste is produced under a given set of conditions.

3.2. Physiology of Nitrogenous Excretion

Uric acid is produced in the fat body and/or Malpighian tubules (occasionally the midgut) and released into the hemolymph. How the highly insoluble uric acid is transported in the hemolymph remains unclear though the most likely means seems to be as the sodium or potassium salt, or in combination with specific carrier proteins (Cochran, 1985). The uric acid is secreted into the lumen of the tubules as the sodium or potassium salt, along with other ions, water, and various low-molecular-weight organic molecules. In a typical insect, for example *Dixippus*, secretion occurs along the entire length of the tubule. No resorption of materials takes place across the tubule wall, and urate leaves the tubule in solution. In the rectum resorption of water and sodium and potassium ions occurs, and the pH of the fluid

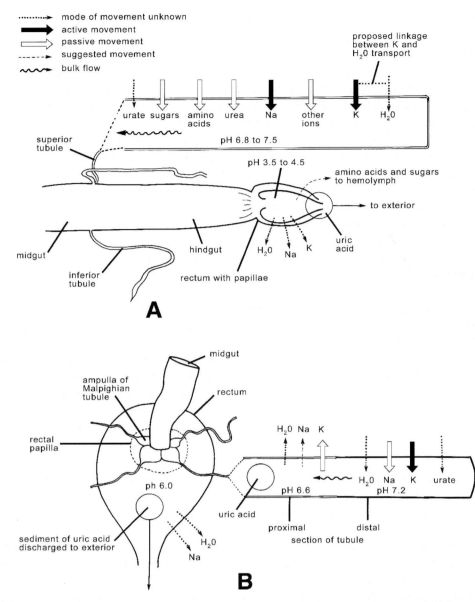

FIGURE 18.4. Movements of water, ions, and organic molecules in the excretory systems of (A) *Dixippus* and (B) *Rhodnius*. [After R. H. Stobbart and J. Shaw, 1974, Salt and water balance: Excretion, in: *The Physiology of Insecta*, 2nd ed., Vol. V (M. Rockstein, ed.). By permission of Academic Press, Inc. and the authors.]

decreases from 6.8–7.5 to 3.5–4.5. The combined effect of water resorption and pH change is to cause massive precipitation of uric acid. Useful organic molecules such as amino acids and sugars are also resorbed through the rectal wall. The Malpighian tubule-rectal wall excretory system thus shows certain functional analogies with the vertebrate nephron. The excretion of uric acid in *Dixippus* is summarized in Figure 18.4A.

In *Rhodnius*, whose tubules show structural differentiation along their length, the process of excretion is basically the same as in *Dixippus*. However, in *Rhodnius* only the distal portion of the tubule is secretory and resorption of water and cations begins in the proximal part. Slight change in pH occurs (from 7.2 to 6.6) as the fluid passes along the tubule and this is sufficient to initiate uric acid precipitation. Further water and salt resorption occurs in the rectum (pH 6.0), causing precipitation of the remaining waste (Figure 18.4B).

Although allantoin is the major nitrogenous waste in many insects, its mode of excretion appears to have been studied in only one species, *Dysdercus fasciatus* (Hemiptera) (Berridge, 1965). This insect is required, because of its diet, to excrete large quantities of unwanted ions (magnesium, potassium, and phosphate). This, combined with the insect's inability to actively resorb water from the rectum, results in the production of a large volume of urine. Because no resorption or acidification occurs which could cause precipitation of uric acid, this molecule is no longer used as an excretory product. Thus, allantoin, which is 10 times more soluble than uric acid (yet of equally low toxicity), is preferred. However, the insect does not possess a mechanism for actively transporting this molecule from the hemolymph to tubule lumen; that is, allantoin only moves passively across the wall of the tubule. It is therefore maintained in high concentration in the hemolymph to achieve a sufficient rate of diffusion into the tubule. Whether a similar mechanism occurs in other allantoin-excreting insects remains to be seen. It may be significant that many other allantoin producers are herbivorous and have the problem of removing large quantities of unwanted ions.

The physiological mechanisms for excretion of other nitrogenous wastes are poorly understood. Aquatic insects are presumed to excrete ammonia in very dilute urine, whereas larvae of meat-eating flies such as *Lucilia cuprina* and *S. bullata* produce highly concentrated, ammonia-rich excreta, apparently by actively transporting ammonium ions across the anterior hindgut wall. Urea probably moves passively into the Malpighian tubules and becomes concentrated in the hindgut because of its inability to permeate the cuticular lining as water resorption occurs.

3.3. Storage Excretion

An alternative strategy to the removal of wastes through the Malpighian tubule-rectum system used by some insects is storage excretion, the retention of the wastes in "out of the way places" within the body. In *Dysdercus*, for example, uric acid is deposited permanently in the epidermal cells of the abdomen, forming distinct, white transverse bands (Berridge, 1965). Adult Lepidoptera convert much of their waste nitrogen into pteridines that are stored in the integument, eyes, or wing scales, giving the insects their characteristic color patterns (Chapter 11, Section 4.3).

At other times storage of urate occurs even when the tubules are working normally and may be regarded as a supplementary excretory mechanism for occasions when the tubules cannot cope with all the waste that is being produced. In the larval stages of many species uric acid crystallizes out in ordinary fat body cells and epidermis, even though the Malpighian tubules are functional. It appears that this is caused by the metabolic activity of the cells themselves (i.e., they are not accumulating uric acid from the hemolymph), and crystallization occurs by virtue of the particular conditions (pH, ionic content, etc.) existing in the cells. During the later stages of pupation the crystals disappear, the uric acid apparently having been transferred to the meconium (the collective wastes of pupal metabolism, released at eclosion) via the excretory system. It is worth noting that in many species the Malpighian tubules are entirely reconstituted during the pupal stage. Thus, storage of uric acid in fat body and epidermal cells is of great importance at this time. Yet other insects, notably termites and cockroaches, retain large quantities of uric acid in special cells (urocytes) within the fat body. However, as Cochran (1985) pointed out, this is not a form of storage excretion but an important means of conserving nitrogen in these insects whose normal diet is severely nitrogen deficient (Chapter 16, Section 5.1.1).

Temporary storage of other materials may also take place. Calcium salts (especially carbonate and oxalate) are found in the fat body of many plant-eating insect larvae. During metamorphosis they are released and dissolved, to be excreted via the Malpighian tubules in the adult. Dyes present in food are often accumulated in fat body cells where they appear to become associated with particular proteins. These proteins are then transferred to the egg during vitellogenesis and the dyes subsequently "excreted" during oviposition.

Nephrocytes (Chapter 17, Section 2) accumulate a variety of substances, especially pigments, and their name is derived from the mistaken idea that storage excretion is one of their major functions. As Locke and Russell (1998) pointed out, nephrocytes are involved in the metabolism of hemolymph macromolecules.

4. Salt and Water Balance

Salt and water balance involves more than simply the control of hemolymph osmotic pressure; the relative proportions of the ions that contribute to this pressure must be maintained within narrow limits. The osmotic pressure of the hemolymph is generally within the same limits as that of the blood of other organisms, but it can be increased considerably under specific conditions (by the addition, for example, of glycerol, which serves as an antifreeze during hibernation). Regulation of the salt and water content is obviously related to the nature of the external environment. Insects in different habitats face different osmotic problems. Nevertheless, these problems have been solved using the same basic mechanism, namely, the production of a "primary excretory fluid" in the Malpighian tubules followed by differential resorption from or secretion into this fluid when it reaches the rectum. For clarity the problems of insects living on land, in fresh water, or in brackish or salt water are considered separately. However, considerable similarity in the solution of these problems will be seen.

4.1. Terrestrial Insects

Terrestrial insects appear able to regulate their hemolymph osmotic pressure over a wide range of conditions. For example, in *Tenebrio* the hemolymph osmotic pressure varies only from 223 to 365 mM/l (measured as the equivalent of a sodium chloride solution) over a range of relative humidity from 0% to 100% (Marcuzzi, 1956, cited in Stobbart and Shaw, 1974). In starving *Schistocerca* there is only a 30% difference in hemolymph osmotic pressure between animals kept in air at 100% relative humidity and given only tap water and those kept in air at 70% relative humidity and given saline (osmotic pressure equivalent to 500 mM/l sodium chloride) to drink (Phillips, 1964a).

In terrestrial insects water is lost (1) by evaporation across the integument, although this is considerably reduced by the presence of the wax layer in the epicuticle (Chapter 11, Section 2); (2) during respiration through the spiracles [many insects possess devices both physiological and structural for reducing the loss (Chapter 15, Section 2.2)]; and (3) during excretion. Despite these adaptations, insects that inhabit extremely dry environments may become greatly dehydrated. For example, some desert beetles can survive the loss of 75% of their body water. The critical factor for these beetles is to maintain the intracellular water concentration by using the water in the hemolymph; in other words, the hemolymph volume is reduced. To avoid the potential osmotic problems that this withdrawal of water creates, osmotically active particles can be excreted or rendered inactive; for example, ions are chelated and amino acids are polymerized into peptides. The strategy used appears to be correlated with the insects' diet: carnivorous species, whose food contains abundant sodium, tend to excrete the excess ions. The diet of herbivores, by contrast, is deficient in sodium, so these species use chelation as a means of retaining the sodium within the body (Pedersen and Zachariassen, 2002).

The major source of water for most terrestrial insects is obviously food and drink. Some insects may eat excessively solely for the water content of the food. Where sufficient water cannot be obtained by drinking or in food, the insect must obtain it by other means. One source is the water produced during metabolism. Absorption of water vapor from the atmosphere is a method employed by a few insects (e.g., *Thermobia* and *Tenebrio*) that are normally found in very dry conditions. Interestingly, the site of absorption is the rectum, which, as is noted below, is the site of uptake of liquid water in other terrestrial and saltwater insects.

Small amounts of ions are lost from the body via the excretory system, and these are readily made up by absorption across the midgut wall. Indeed, in terrestrial insects the usual problem is removal of unwanted ions present in the diet. The food often contains ions in concentrations that are widely different from those of the hemolymph. It is probable that these ions enter the hemolymph passively in the same proportions as they occur in the diet, and excesses are subsequently expelled via the excretory system. In other words, the midgut does not act as a selectively permeable barrier to the entry of ions (Stobbart and Shaw, 1974).

The role of the Malpighian tubules and rectum was investigated by examination of the ionic composition of the fluids within them and, more recently, by the use of radioisotopes to measure the direction and rate of movement of individual ions. The studies of Ramsay in the 1950s (see reviews for references) revealed that the fluid in the tubules is isosmotic with the hemolymph (Table 18.2) but has a very different ionic composition. Particularly obvious is the difference in potassium ion concentration, which is several times higher in the tubule fluid than in the hemolymph. The sodium ion concentration is usually lower in the fluid than in the hemolymph, as is the case with most other ions (except phosphate). The tubule fluid, which is produced continuously, also contains a number of low-molecular-weight organic molecules, for example, amino acids and sugars; thus, it is broadly comparable with the glomerular filtrate of the vertebrate kidney, though it is not produced by hydrostatic pressure. The high potassium concentration in the tubule fluid and the demonstration that the rate at which tubule fluid is formed depends on the hemolymph potassium concentration led Ramsay to suggest that the active transport of potassium ions is fundamental to the production and flow of the fluid. Bloodsucking insects that have just fed are exceptional in that both potassium and sodium ions are actively transported into the tubule lumen. This modification to the basic plan is necessitated by the heavy sodium chloride load in the plasma fraction of the vertebrate host's blood and by the need to remove as rapidly as possible excess water taken into the body as a result of feeding. Active cation transport is accompanied by the movement of anions (principally chloride) to maintain electrical neutrality and by the flow of water into the tubule lumen by osmosis (Pannabecker, 1995).

Most other ions and organic molecules appear to enter the tubule fluid passively. However, active transport of sulfate and of some dyes and toxic compounds (e.g., alkaloids) has been demonstrated in some insects, notably those species that encounter these molecules in their natural diet.

It is clear that, as the tubule fluid and hemolymph are isosmotic, the tubules are not directly concerned with regulation of hemolymph osmotic pressure. Ramsay, using isolated

TABLE 18.2. The Osmotic Pressure and Concentration (mM/l) of Some Ions in the Hemolymph (H), Malpighian Tubule Fluid (MT), and Rectal Fluid (R) in Insects from Different Habitats[a]

Habitat	Species (stage and conditions)	Fluid	Osmotic pressure (≡NaCl solution)	Ions		
				Na$^+$	K$^+$	Cl$^-$
Terrestrial	*Schistocerca gregaria* (adult, water-fed)	H	214	108	11	115
		MT	226	20	139	93
		R	433	1	22	5
	Dixippus morosus (adult, feeding)	H	171	11	18	87
		MT	171	5	145	65
		R	390	18	327	—
	Rhodnius prolixus (adult, 19–29 hr after meal)	H	206	174	7	155
		MT	228	114	104	180
		R	358	161	191	—
Salt water	*Aedes detritus* (larvae, in seawater)	H	157	—	—	—
		MT	—	—	—	—
		R	537	—	—	—
	Aedes detritus (larvae, in distilled water)	H	97	—	—	—
		MT	—	—	—	—
		R	56	—	—	—
Fresh water	*Aedes aegypti* (larvae, in distilled water)	H	138	87	3	—
		MT	130	24	88	—
		R	12	4	25	—

[a] Data mainly from Stobbart and Shaw (1974).

tubules, showed that the isosmotic condition is retained over a wide range of external concentrations. Indirectly, however, the tubules are important in regulation, as the rate at which ions and water are excreted from the body is the difference between their rate of secretion into the tubule lumen and their rate of resorption by the rectum.

As noted in Section 3.2, in some insects the tubules show regional differentiation, secretion taking place in the distal part and resorption beginning in the proximal part of the tubule. In most species, however, resorption occurs mainly in the rectum, though the ileum may also modify the fluid. In the rectum major changes occur in the osmotic pressure and composition of the urine (Table 18.2). Generally, the urine becomes greatly hypertonic to the hemolymph, but when much water is available a hypotonic fluid may be excreted.

In the rectum, water is resorbed against a concentration gradient; that is, it is an active process and energy is expended. Phillips (1964a) showed that in *Schistocerca* the rate of water movement across the rectal wall is independent of the rate of salt accumulation. The rate at which water is resorbed depends on the osmotic gradient across the wall, and, as the gradient increases during resorption, the point is reached at which the rate of active accumulation is balanced by the rate of passive diffusion back into the rectal lumen; that is, the concentration of the rectal fluid reaches a maximum value. However, this value varies according to the water requirements of the insect. For example, locusts that have been kept in a dry environment and given strong saline to drink have a rectal fluid whose osmotic pressure is about twice that of insects with access to tap water. The physiological basis of this increased ability to concentrate the urine is not known.

The precise mechanism of water uptake is still unclear. Though models have been proposed in which water *per se* is actively transported across the rectal wall, there is now direct evidence that water movements occur as a result of active movements of inorganic ions, especially sodium, potassium, and chloride (secondary transport of water) (Phillips, 1977). Fine-structural studies of the rectal wall of *Calliphora* (Berridge and Gupta, 1967), *Periplaneta* (Oschman and Wall, 1969) and other insects, and the elegant work of Wall and Oschman (1970) and Wall *et al.* (1970), who used micropuncture to obtain fluid samples from the subepithelial sinus and intercellular spaces in the rectal epithelium, led to the following scheme for water absorption from the rectum (Figure 18.5). Ions are actively secreted into the intercellular space between the highly convoluted plasma membranes of adjacent epithelial cells so that local pockets of high salt content are formed. Thus, an osmotic gradient is developed down which water flows from the rectal lumen to the intercellular spaces via the cytoplasm of the epithelial cells. Water may also enter the intercellular spaces directly from the rectal lumen if the apical septate junctions are leaky as has been proposed. The entry of water into the spaces produces a hydrostatic pressure that forces the ions and water toward the hemolymph. As the fluid moves through the larger (inner) intercellular spaces and subepithelial sinus, active resorption of ions occurs across the epithelial cell membrane. However, relatively little water moves into the cells because the spaces have a low surface area/volume ratio (i.e., the plasma membrane of the cells is not convoluted in these regions as it is in the distal intercellular spaces).

Ramsay's early work on *Rhodnius* and *Dixippus* provided a strong indication that the rectum is also capable of resorbing salts, and this has been confirmed by Phillips (1964b) in *Schistocerca*. This author showed that sodium, potassium, and chloride ions are accumulated

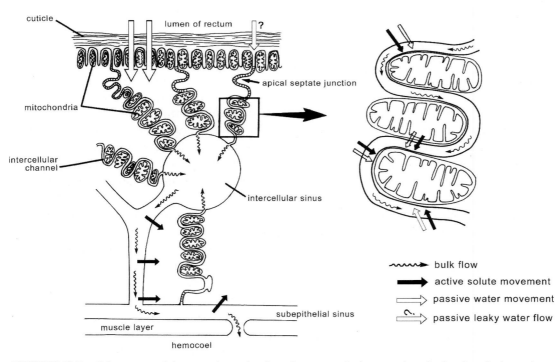

FIGURE 18.5. Scheme to explain water absorption from the rectum. Active secretion of solute into the intercellular channels induces passive movement of water into the channels from the epithelial cells and, perhaps, directly from the rectal lumen. The intercellular fluid thus formed flows toward the hemocoel, and, as it moves through the sinuses, solute is actively resorbed by the cells for recycling. For further details, see text. [After S. H. P. Maddrell, 1971, The mechanisms of insect excretory systems, *Adv. Insect Physiol.* **8**:199–331. By permission of Academic Press, Ltd. and the author.]

against a concentration gradient and independently of the movement of water. Furthermore, the rate of accumulation of these ions depends on their concentrations in the rectal fluid and the hemolymph. In this way the requirements of the insect can be satisfied. In water-fed locusts ions are resorbed from the rectum as quickly as they arrive in the fluid from the tubules, and low rectal concentrations are found (Table 18.2). At the other extreme, in saline-fed animals, the rates of resorption are low and a greatly hyperosmotic fluid is produced.

The cryptonephridial arrangement of Malpighian tubules increases the power of the rectal wall to resorb water against high concentration gradients. The system is particularly well developed in insects that inhabit dry environments, enabling such species not only to extract the maximum amount of water from their feces but also, under fasting conditions, to take up water from moist air (relative humidity at least 88%) by holding open their anus (Machin, 1980). According to Ramsay (1964), the perinephric membrane is impermeable to water, and, under dry conditions, the osmotic pressure of the perinephric cavity is raised mainly because of the presence of potassium chloride. Thus, the concentration gradient across the rectal wall is reduced, facilitating water uptake. Ramsay suggested that potassium and chloride ions are actively transported into the lumen of the perirectal tubules (which contain many mitochondria), with an accompanying movement of water. Resorption of ions and water occurs across the wall of the parts of the tubule bathed in hemolymph. An apparent lack of mitochondria in the perinephric membrane and leptophragma cells led Ramsay to conclude that ions move passively across the membrane in order to balance those removed by the tubule. However, the fine-structural and experimental study of Grimstone *et al.* (1968) has shown this conclusion to be wrong. These authors found that the leptophragma cells have a normal complement of mitochondria. Active transport of potassium ions occurs across the cells, which are, however, impermeable to water. Chloride ions follow passively. The scheme is summarized in Figure 18.2D.

4.2. Freshwater Insects

In freshwater insects water enters the body osmotically despite the relatively impermeable cuticle (Chapter 11, Section 4.2) and must be removed, and salts will be lost from the body and must be replaced if the hyperosmotic condition of the hemolymph is to be maintained. Freshwater insects can regulate their hemolymph osmotic pressure successfully to the point at which the external environment becomes isosmotic with the hemolymph (Figure 18.6). This is achieved by the production of urine that is hypoosmotic to the hemolymph. Beyond this point, regulation breaks down because freshwater insects are not able to produce hyperosmotic urine; that is, they cannot resorb water against a concentration gradient or excrete excess ions (Stobbart and Shaw, 1974). In other words, they become osmoconformers, their hemolymph osmotic pressure closely paralleling that of the external medium.

As in terrestrial insects, the Malpighian tubules produce a fluid that is isosmotic with the hemolymph but of different ionic composition. Particularly obvious is the great difference in the potassium ion concentration between the two solutions, as a result of active transport (Table 18.2). When the primary urine enters the rectum, resorption of ions occurs. The osmotic pressure of the fluid that finally leaves the body is much lower than that of the hemolymph but not as low as would be expected from knowledge of the extent of ion resorption in the rectum. This is because large quantities of ammonium ions appear in the rectal fluid. These ions cannot be detected in the Malpighian tubules, and it is presumed that they are secreted directly across the rectal wall, as occurs in larvae of *S. bullata* (Section 2.2).

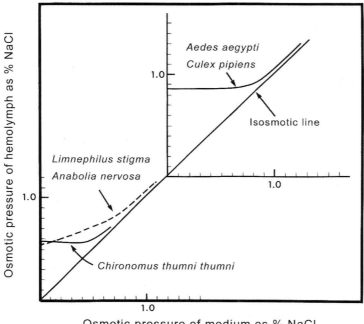

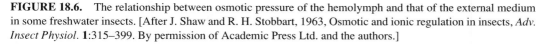

FIGURE 18.6. The relationship between osmotic pressure of the hemolymph and that of the external medium in some freshwater insects. [After J. Shaw and R. H. Stobbart, 1963, Osmotic and ionic regulation in insects, *Adv. Insect Physiol.* **1**:315–399. By permission of Academic Press Ltd. and the authors.]

In freshwater insects, food is the usual source of ions that are absorbed through the midgut wall. However, in some species ions are accumulated through other parts of the body, for example, the gills of caddisfly larvae, the rectal respiratory chamber of dragonfly, damselfly and mayfly larvae, the anal gills of syrphid larvae, and the anal papillae of mosquito and midge larvae. The role of the anal papillae in ionic regulation has been particularly well studied. In mosquito larvae a pair of papillae is located on each side of the anus (Figure 18.7A). They communicate with the hemocoel and are well supplied with tracheae. Their walls are a one-cell-thick syncytium (perhaps an adaptation to eliminate intercellular leakage) and covered with a thin cuticle (Figure 18.7B). Mosquito larvae can accumulate chloride, sodium, potassium, and phosphate ions against large concentration gradients using the papillae. The ability to accumulate ions varies with the habitat in which an insect is normally found. Thus, *Culex pipiens*, which is found in contaminated water, is less efficient at collecting ions than *Aedes aegypti*, which typically lives in fresh rainwater pools. Indeed, normal larvae of the latter species can maintain a constant hemolymph sodium concentration when the sodium concentration of the external medium is only 6 μM/l. The hemolymph sodium concentration under these conditions is only about 5% below the normal level (Shaw and Stobbart, 1963).

4.3. Brackish-Water and Saltwater Insects

Brackish water may be defined as water whose osmotic concentration is in the range 300 mOsm (about 1.1% sodium chloride) (the osmotic concentration of the hemolymph) to 1000 mOsm (about 3.5% sodium chloride) (the concentration of normal seawater), with salt water having osmotic concentrations greater than those of natural seawater. The definitions

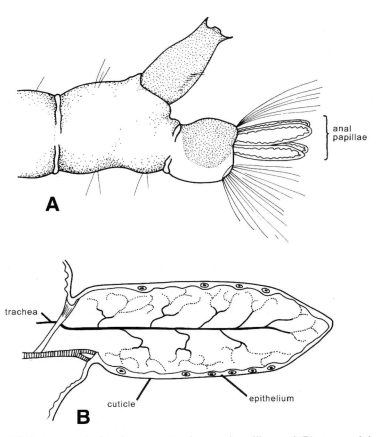

FIGURE 18.7. (A) Posterior end of *Aedes aegypti* to show anal papillae; and (B) structural details of a single anal papilla. [A, after V. B. Wigglesworth, 1965, *The Principles of Insect Physiology*, 6th ed., Methuen and Co. By permission of the author. B, after V. B. Wigglesworth, 1933, The effect of salts on the anal glands of the mosquito larva, *J. Exp. Biol.* **10**:1–15. By permission of Cambridge University Press.]

are not entirely arbitrary, as they tend to describe habitats occupied by particular species. For example, larvae of the mosquito *Culiseta inornata* are found in a variety of brackish waters, including tidal estuaries and some inland ponds, but cannot survive in natural seawater. In contrast, larvae of *Aedes taeniorhynchus*, which occur in coastal salt marshes, have a maximum saline tolerance equivalent to 10% sodium chloride. Even more remarkable are larvae of *Ephydra cinerea* that are found in the Great Salt Lake of Utah where the salinity may exceed the equivalent of 20% sodium chloride (i.e., about six times that of normal seawater).

The habitat occupied by brackish-water and saltwater insects can vary widely in ionic content and osmotic pressure. During periods of warm, dry weather the salinity may increase several fold. Conversely, after heavy rains or the melting of snow in spring, the salinity may approach that of fresh water. It is not surprising, therefore, to find experimentally that such insects can regulate their hemolymph osmotic pressure over a wide range of external salt concentrations (Figure 18.8). Larvae of *Aedes detritus* and *Ephydra riparia*, inhabitants of salt marshes, can survive in media containing the equivalent of 0 to about 7–8% sodium chloride. Over this range of concentrations the hemolymph osmotic pressure changes by only 40–60%.

When their external medium is dilute (i.e., its osmotic pressure is less than that of the hemolymph), both brackish-water and saltwater insects osmoregulate to keep their

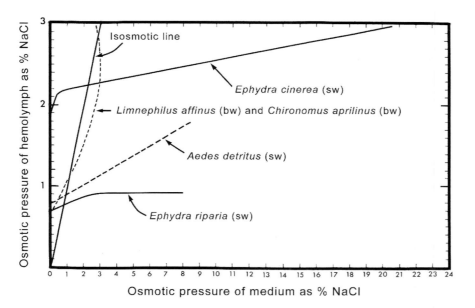

FIGURE 18.8. The relationship between osmotic pressure of the hemolymph and that of the external medium in some saltwater (sw) and brackish-water (bw) larvae. [After J. Shaw and R. H. Stobbart, 1963, Osmotic and ionic regulation in insects, *Adv. Insect Physiol.* **1**:315–399. By permission of Academic Press Ltd. and the authors.]

hemolymph osmotic pressure more or less constant. Like freshwater species, they produce dilute urine and actively resorb salts through the rectal wall. Mosquitoes from these habitats may also take up ions via the anal papillae. However, when the external osmotic pressure rises above that of the hemolymph, leading to loss of water from the body by osmosis, saltwater and brackish-water species employ different strategies.

Saltwater mosquitoes, on which most experimental work has been done, counteract the loss of water by drinking the external medium; for example, *A. taeniorhynchus* larvae ingest 240% of their body weight per day (Bradley, 1987). The water taken in, and the ions it contains, are then absorbed across the midgut wall, a process that may serve to concentrate food in the midgut prior to digestion (Phillips *et al.*, 1978). Thus, the problem for these insects is to rid themselves of the excess ions that enter the hemolymph. As in insects from other habitats, the Malpighian tubules produce a potassium-rich fluid isosmotic with the hemolymph, and their main function appears to be the secretion of sulfate. Formation of hyperosmotic urine occurs as a result of active secretion of ions across the wall of the rectum, which in these species is divided into anterior and posterior segments (Figure 18.9). The posterior segment appears to be the more important, secreting sodium, chloride, magnesium, and potassium ions into the rectal lumen. The role of the anterior segment is less clear, though exchange of bicarbonate for chloride ions occurs here, as may the resorption of selected inorganic and organic solutes (especially in larvae in dilute media). As a result of these activities, saltwater mosquitoes are able to strongly regulate their hemolymph osmotic pressure and ionic content over a wide range of external concentrations (Figure 18.8).

In contrast to saltwater forms, brackish-water insects become osmoconformers in external media more concentrated than their hemolymph; that is, their hemolymph osmotic pressure increases approximately parallel with that of the surrounding medium (Figure 18.8). Unlike the situation in freshwater species, however, where osmoconformation is largely because of increases in hemolymph inorganic ion levels (mainly those of sodium

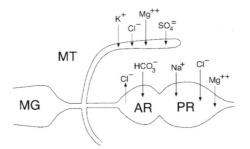

FIGURE 18.9. Excretory system of saltwater mosquito larvae. Known pathways of active ion transport are shown. AR, anterior rectal segment; MG, midgut; MT, Malpighian tubule; PR, posterior rectal segment. [After T. J. Bradley, 1987, Physiology of osmoregulation in mosquitoes, *Annu. Rev. Entomol.* **32**:439–462. Reproduced, with permission, from the Annual Review of Entomology, Volume 32 ©1987 by Annual Reviews, Inc.]

and chloride), in brackish-water species the hemolymph inorganic ion levels remain more or less constant and the increased osmotic pressure is the result of the addition of organic components. In *Culex tarsalis* larvae acclimated to 600 mOsm seawater, for example, the concentrations of amino acids (especially proline and serine) and trehalose were several fold greater than those of insects acclimated to 50 mOsm seawater (Bradley, 1987). The use of these molecules to prevent osmotic dehydration is interesting in view of their somewhat parallel role as cryoprotectants in cold-hardy insects (Chapter 22, Section 2.4.1).

5. Hormonal Control

As in other systems with a homeostatic role, the activities of the excretory system, including both production of nitrogenous waste and osmoregulation, need to be regulated to suit the specific but changing requirements of the insect. This coordination is effected by hormones.

Though there is strong circumstantial evidence for involvement of hormones in the synthesis of uric acid, neither the nature of the factor(s) involved nor the site of action is known. Changes in the rate of uric acid production, for example, occur at specific stages in an insect's life and these can be correlated with fluctuations in the levels of juvenile hormone and ecdysteroids. Similarly, removal of the corpora allata or corpora cardiaca may markedly affect uric acid production though, again, it is uncertain whether the response is direct or indirect, for example, via an influence on amino acid or protein metabolism.

In contrast, the occurrence of both diuretic and antidiuretic hormones is firmly established (Phillips, 1983; Spring, 1990; Coast, 1998a, 2001; Gäde, 2004). In most species, neurosecretory cells in the brain produce these hormones that are then stored in the corpora cardiaca, though there are many reports of diuresis-modifying factors in extracts from other ganglia in the ventral nerve cord. Almost all identified osmoregulatory hormones are peptides, though in some insects (e.g., *Rhodnius*, locusts and crickets) serotonin also appears to be an important diuretic factor.

One or more diuretic hormones are released after feeding in many terrestrial insects. In *Rhodnius*, which takes intermittent large blood meals, stretching of the abdominal wall brings about hormone release (Maddrell, 1964). In *Schistocerca, Dysdercus*, and other insects that feed more or less continuously, it is probably the stretching of the foregut that causes release of hormone (Mordue, 1969; Berridge, 1966). Diuretic hormones appear to act primarily on the Malpighian tubules, stimulating them to secrete potassium ions at a greater rate, thereby creating an enhanced flow of water across the tubule wall (Pilcher, 1970). This primary action of diuretic hormone will affect both osmoregulation and the excretion of uric acid. In some insects, for example *Calliphora* and *Schistocerca*, diuretic hormone has a dual action, causing accelerated secretion through the tubules and a slowing down of water resorption through the rectal wall. In *Rhodnius* and other insects that have two diuretic factors, it is believed that these work synergistically so that a small change in the amount of either will induce a large change in the rate of production of the primary fluid. In other words, this arrangement endows the excretory system with great sensitivity, enabling the Malpighian tubules to react quickly to changes in water load within the body.

An ancillary effect of diuretic peptides and serotonin is to stimulate contraction of the muscles on the outside of the tubules, enhancing their writhing movements. This will reduce the thickness of the unstirred layer of hemolymph adjacent to the tubule, as well as improving fluid flow within the tubule (Coast, 1998b).

The nature and mode of action of antidiuretic factors are less well understood, and until recently it was thought that they act only at the level of the rectum, enhancing fluid uptake. One possible means of achieving this would be through stimulation of ion secretion as proposed in the scheme for water resorption outlined earlier (Figure 18.5). A chloride transport-stimulating peptide (CTSP) (molecular weight 8000) isolated from the corpora cardiaca of *Schistocerca* stimulates resorption of these ions from the rectal lumen (Spring, 1990; Phillips and Audsley, 1995), but whether it promotes water resorption has not been determined. Curiously, a second peptide (ion transport peptide [ITP]) from the same source and having a similar molecular weight (8500) has been identified (Phillips *et al.*, 1998). ITP also stimulates chloride ion resorption, but in the ileum, and the consensus is that they are different hormones.

For a few insects, including *Rhodnius* and *Tenebrio*, the antidiuretic hormone acts on the Malpighian tubules, inhibiting the active transport of potassium ions across the tubule wall (Eigenheer *et al.*, 2002).

A further aspect of hormonal control is that some insects produce both diuretic and antidiuretic hormones simultaneously after feeding, an action that may appear to be counterproductive. Proux *et al.* (1984) suggested, however, that this action may enhance fluid cycling to clear metabolic wastes from the hemolymph.

Because, presumably, there is a direct relationship between the amount of food consumed, the amount of hormone released, and the quantity of water removed across the tubules or resorbed from the rectal lumen, it is difficult to see how such a simple arrangement will work other than in insects such as *Rhodnius* whose food has a constant water content. The physical stimulus of "stretching" alone would not provide a precise enough mechanism for water regulation in insects whose food differs in water content. Some other control system must therefore operate. Perhaps an insect can monitor the water content of its food or, alternatively, the resorptive power of the rectal wall may be controlled directly by the hemolymph itself. Phillips (1964b) showed that the rate at which sodium and potassium ions are resorbed is dependent on the ionic concentration of the hemolymph.

Very little work has been done on the hormonal control of salt and water balance in aquatic insects. However, as a result of his experiments on *Aedes aegypti*, Stobbart (1971) suggested that accumulation of sodium ions across the wall of the anal papillae is under endocrine control. In the water boatman, *Cenocorixa blaisdelli*, and larvae of the dragonfly, *Libellula quadrimaculata*, extracts of the head and thoracic ganglia, respectively, stimulate Malpighian tubule secretion (Spring, 1990).

6. Summary

The removal of nitrogenous wastes and maintenance of a suitable hemolymph salt and water content are two closely linked processes. In most insects, the predominant nitrogenous waste is uric acid, which is removed from the hemolymph via the Malpighian tubules as the soluble sodium or potassium salt. Precipitation of uric acid occurs usually in the rectum as a result of resorption of ions and water from, and acidification of, the urine. Allantoin and allantoic acid are excreted in quantity by some insects and may be the major nitrogenous waste. Urea is of little significance as a waste product, and ammonia is generally produced only in aquatic species.

Insects are usually able to regulate the salt and water content of the hemolymph within narrow limits. In all insects, a primary excretory fluid, isosmotic with hemolymph but differing in ionic composition, is produced in the Malpighian tubules. Production of tubule fluid is driven by active transport of potassium ions. When this fluid reaches the posterior rectum, it is modified according to an insect's needs. In terrestrial insects selective resorption of ions and/or water occurs. Freshwater species osmoregulate by producing hypoosmotic urine from which useful materials have been resorbed. In a hyperosmotic medium they become osmoconformers, their hemolymph osmotic pressure paralleling that of the medium. Brackish-water and saltwater insects have excellent ability to osmoregulate over a wide range of environmental conditions. In dilute media they behave much like freshwater species, forming hypoosmotic urine and resorbing useful components. In media with osmotic pressures greater than that of hemolymph, saltwater species drink excessively and produce hyperosmotic urine by secreting ions across the rectal wall; in contrast, brackish-water insects osmoconform by increasing the concentration of amino acids and trehalose in the hemolymph.

Both diuretic and antidiuretic hormones are known. The former stimulate Malpighian tubule fluid production and may inhibit water resorption from the rectum; antidiuretic hormones mostly appear to act only by stimulating water resorption from the rectal lumen; however, in a few species the antidiuretic factor inhibits potassium ion transport (hence formation of the primary excretory fluid) in the Malpighian tubule.

7. Literature

For additional information, see Maddrell (1971, 1980), Stobbart and Shaw (1974), Phillips (1981), and Bradley (1985) [general]; Bursell (1967) and Cochran (1975, 1985) [nitrogenous waste excretion]; and Phillips (1977) [salt and water balance]. Hormonal regulation of excretion is considered by Phillips (1983), Spring (1990), Phillips and Audsley (1995), Coast (1998a), and Gäde (2004). Machin (1980) reviews the uptake of water from the atmosphere.

Barrett, F. M., and Friend, W. G., 1970, Uric acid synthesis in *Rhodnius prolixus*, *J. Insect Physiol.* **16**:121–129.

Berridge, M. J., 1965, The physiology of excretion in the cotton stainer, *Dysdercus fasciatus* Signoret. III. Nitrogen excretion and excretory metabolism, *J. Exp. Biol.* **43**:511–521.

Berridge, M. J., 1966, The physiology of excretion in the cotton stainer, *Dysdercus fasciatus* Signoret. IV. Hormonal control of excretion, *J. Exp. Biol.* **44**:553–566.

Berridge, M. J., and Gupta, B. L., 1967, Fine-structural changes in relation to ion and water transport in the rectal papillae of the blow fly, *Calliphora*, *J. Cell Sci.* **2**:89–112.

Bradley, T. J., 1985, The excretory system: Structure and physiology, in: *Comprehensive Insect Physiology, Biochemistry and Pharmacology*, Vol. 4 (G. A. Kerkut and L. I. Gilbert, eds.), Pergamon Press, Elmsford, NY.

Bradley, T. J., 1987, Physiology of osmoregulation in mosquitoes, *Annu. Rev. Entomol.* **32**:439–462.

Bursell, E., 1967, The excretion of nitrogen in insects, *Adv. Insect Physiol.* **4**:33–67.

Coast, G. M., 1998a, The regulation of primary urine production in insects, in: *Recent Advances in Arthropod Endocrinology* (G. M. Coast and S. G. Webster, eds.), Cambridge University Press, Cambridge.

Coast, G. M., 1998b, The influence of neuropeptides on Malpighian tubule writhing and its significance for excretion, *Peptides* **19**:469–480.

Coast, G. M., 2001, The neuroendocrine regulation of salt and water balance in insects, *Zoology, Jena* **103**:179–188.

Cochran, D. G., 1975, Excretion in insects, in: *Insect Biochemistry and Function* (D. J. Candy and B. A. Kilby, eds.), Chapman and Hall, London.

Cochran, D. G., 1985, Nitrogenous excretion, in: *Comprehensive Insect Physiology, Biochemistry and Pharmacology*, Vol. 4 (G. A. Kerkut and L. I. Gilbert, eds.), Pergamon Press, Elmsford, NY.

Eigenheer, R. A., Nicolson, S. W., Schegg, K. M., Hull, J. J., and Schooley, D. A., 2002, Identification of a potent antidiuretic factor acting on beetle Malpighian tubules, *Proc. Natl Acad. Sci. USA* **99**:84–89.

Gäde, G., 2004, Regulation of intermediary metabolism and water balance of insects by neuropeptides, *Annu. Rev. Entomol.* **49**:93–113.

Grimstone, A. V., Mullinger, A. M., and Ramsay, J. A., 1968, Further studies on the rectal complex of the meal worm *Tenebrio molitor* L. (Coleoptera, Tenebrionidae), *Philos. Trans. R. Soc. Lond. Ser. B* **253**:343–382.

Jarial, M. S., and Scudder, G. G. E., 1970, The morphology and ultrastructure of the Malpighian tubules and hindgut of *Cenocorixa bifida* (Hung.) (Hemiptera, Corixidae), *Z. Morphol. Tiere* **68**:269–299.

Locke, M., and Russell, V. W., 1998, Pericardial cells or athrocytes, in: *Microscopic Anatomy of Invertebrates* (F. Harrison and M. Locke, eds.), Wiley-Liss, New York.

Machin, J., 1980, Atmospheric water absorption in arthropods, *Adv. Insect Physiol.* **14**:1–48.

Maddrell, S. H. P., 1964, Excretion in the blood-sucking bug, *Rhodnius prolixus* Still. III. The control of the release of the diuretic hormone, *J. Exp. Biol.* **41**:459–472.

Maddrell, S. H. P., 1971, The mechanisms of insect excretory systems, *Adv. Insect Physiol.* **8**:199–331.

Maddrell, S. H. P., 1980, The control of water relations in insects, in: *Insect Biology in the Future* (M. Locke and D. S. Smith, eds.), Academic Press, New York.

Mordue, W., 1969, Hormonal control of Malpighian tube and rectal function in the desert locust, *Schistocerca gregaria, J. Insect Physiol.* **15**:273–285.

Oschman, I. L., and Wall, B. J., 1969, The structure of the rectal pads of *Periplaneta americana* L. with regard to fluid transport, *J. Morphol.* **127**:475–510.

Pannabecker, T., 1995, Physiology of the Malpighian tubule, *Annu. Rev. Entomol.* **40**:493–510.

Pedersen, S. A., and Zachariassen, K. E., 2002, Sodium regulation during dehydration of a herbivorous and a carnivorous beetle from African dry savannah, *J. Insect Physiol.* **48**:925–932.

Phillips, J. E., 1964a, Rectal absorption in the desert locust, *Schistocerca gregaria* Forskill. I. Water, *J. Exp. Biol.* **41**:14–38.

Phillips, J. E., 1964b, Rectal absorption in the desert locust, *Schistocerca gregaria* Forskill. II. Sodium, potassium and chloride, *J. Exp. Biol.* **41**:39–67.

Phillips, J. E., 1977, Excretion in insects: Function of gut and rectum in concentrating and diluting the urine, *Fed. Proc.* **36**:2480–2486.

Phillips, J. E., 1981, Comparative physiology of insect renal function, *Am. J. Physiol.* **241**:R241–R257.

Phillips, J. E., 1983, Endocrine control of salt and water balance: Excretion, in: *Endocrinology of Insects* (R. G. H. Downer and H. Laufer, eds.), Liss, New York.

Phillips, J. E., and Audsley, N., 1995, Neuropeptide control of ion and fluid transport across locust hindgut, *Amer. Zool.* **35**:503–514.

Phillips, J. E., Bradley, T. I., and Maddrell, S. H. P., 1978, Mechanisms of ionic and osmotic regulation in saline-water mosquito larvae, in: *Comparative Physiology—Water, Ions and Fluid Mechanics* (K. Schmidt-Nielson, L. Bolis, and S. H. P. Maddrell, eds.), Cambridge University Press, London.

Phillips, J. E., Meredith, J., Audsley, N., Ring, M., Macins, A., Brock, H., Theilmann, D., and Littleford, D., 1998, Locust ion transport peptide (ITP): Function, structure, cDNA and expression, in: *Recent Advances in Arthropod Endocrinology* (G. M. Coast and S. G. Webster, eds.), Cambridge University Press, Cambridge.

Pilcher, D. E. M., 1970, The influence of the diuretic hormone on the process of urine secretion by the Malpighian tubules of *Carausius morosus, J. Exp. Biol.* **53**:465–484.

Proux, B., Proux, J., and Phillips, J. E., 1984, Antidiuretic action of a corpus cardiacum factor (CTSH) on long term absorption across locust recta *in vitro, J. Exp. Biol.* **113**:409–421.

Ramsay, J.A., 1964, The rectal complex of the mealworm *Tenebrio molitor* L. (Coleoptera, Tenebrionidae), *Philos. Trans. R. Soc. Lond. Ser. B* **248**:279–314.

Shaw, J., and Stobbart, R. H., 1963, Osmotic and ionic regulation in insects, *Adv. Insect Physiol.* **1**:315–399.

Spring, J. H., 1990, Endocrine regulation of diuresis in insects, *J. Insect Physiol.* **36**:13–22.

Stobbart, R. H., 1971, The control of sodium uptake by the larva of the mosquito *Aedes aegypti* (L.), *J. Exp. Biol.* **54**:29–66.

Stobbart, R. H., and Shaw, J., 1974, Salt and water balance: Excretion, in: *The Physiology of Insecta*, 2nd ed., Vol. V (M. Rockstein, ed.), Academic Press, New York.

von Dungern, P., and Briegel, H., 2001, Protein catabolism in mosquitoes: Ureotely and uricotely in larval and imaginal *Aedes aegypti*, *J. Insect Physiol.* **47**:131–141.

Wall, B. J., and Oschman, J. L., 1970, Water and solute uptake by the rectal pads of *Periplaneta americana*, *Am. J. Physiol.* **218**:1208–1215.

Wall, B. J., Oschman, J. L., and Schmidt-Nielson, B., 1970, Fluid transport: Concentration of the intercellular compartment, *Science* **167**:1497–1498.

III

Reproduction and Development

19

Reproduction

1. Introduction

As was discussed in Chapter 2 (Section 4.1), an important factor in the success of the Insecta is their high reproductive capacity, the ability of a single female to give rise to many offspring, a relatively large proportion of which may reach sexual maturity under favorable conditions. As reproduction is almost always sexual in insects, there arise within insect populations large numbers of genetic combinations, as well as mutations, which can be tested out in the prevailing environmental conditions. As these conditions change with time, insects are able to adapt readily, through natural selection, to a new situation. Over the short term their high reproductive capacity enables insects to exploit temporarily favorable conditions, for example, availability of suitable food plants. The latter requires that both the timing of mating, egg production, and hatching, and the location of a suitable egg-laying site must be carefully "assessed" by an insect.

Like other terrestrial animals insects have had to solve two major problems in connection with their reproductive biology, namely, the bringing together of sperm and egg in the absence of surrounding water and the provision of a suitable watery environment in which an embryo can develop. The solution to these problems has been the evolution of internal fertilization and an egg surrounded by a waterproof cover (chorion), respectively. The latter has itself created two secondary problems. First, because of the generally impermeable nature of the chorion, structural modifications have had to evolve to ensure that adequate gaseous exchange can occur during embryonic development. Second, the chorion is formed while an egg is still within the ovarian follicle, that is, prior to fertilization, which has necessitated the development of special pores (micropyles) to permit entry of sperm.

2. Structure and Function of the Reproductive System

The external structure of male and female reproductive systems has been dealt with in Chapter 3 (Section 5.2.1), so that only the structure of internal reproductive organs will be described here.

2.1. Female

Functions of the female reproductive system include production of eggs, including yolk and chorion formation, reception and storage of sperm, sometimes for a considerable period, and coordination of events that lead to fertilization and oviposition.

Though details vary, the female system (Figure 19.1) essentially includes a pair of ovaries from each of which runs a lateral oviduct. The lateral oviducts fuse in the midline, and the common oviduct typically enters a saclike structure, the vagina. In some species the vaginal wall evaginates to form a pouchlike structure, the bursa copulatrix, in which

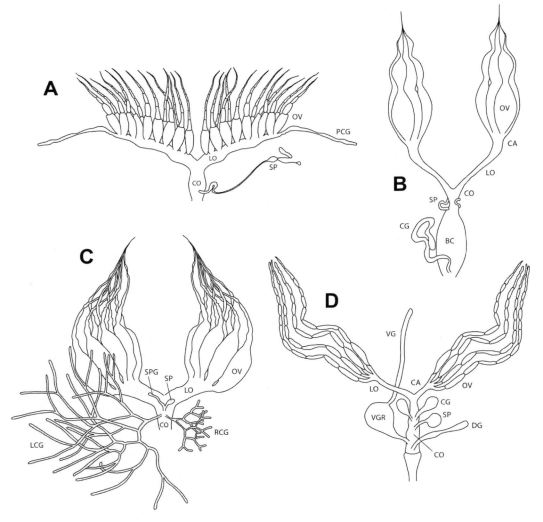

FIGURE 19.1. Representative female reproductive systems (not to scale). (A) *Melanoplus sanguinipes* (Orthoptera); (B) *Rhodnius prolixus* (Hemiptera); (C) *Periplaneta americana* (Dictyoptera); and (D) *Nasonia vitripennis* (Hymenoptera). Abbreviations: BC, bursa copulatrix; CA, calyx; CG, collateral (accessory) glands; CO, common oviduct; DG, Dufour's gland; LCG, left colleterial gland; LO, lateral oviduct; OV, ovariole; PCG, pseudocolleterial gland; RCG, right colleterial gland; SP, spermatheca; SPG, spermathecal gland; VG, venom gland; VGR, venom gland reservoir. [A, C, D, from C. Gillott, 2002, Insect accessory reproductive glands: Key players in production and protection of eggs, in: *Chemoecology of Insect Eggs and Egg Deposition* (M. Hilker and T. Meiners, eds.), By permission of Blackwell Verlag, Berlin; B, from R. P. Ruegg, 1981, Factors influencing reproduction in *Rhodnius prolixus* (Insecta: Hemiptera), Ph.D. Thesis, York University, Canada.]

spermatophores and/or seminal fluid is deposited during copulation. Also connected with the vagina are the spermatheca in which sperm are stored and various accessory glands. In some species part of the spermatheca takes the form of a diverticulum, the spermathecal gland. It is noteworthy that the ovaries themselves lack innervation though the ductal components of the system receive nerves from the terminal abdominal ganglion (Sections 5 and 7.2).

The ovaries are usually dorsolateral to the gut, and each comprises a number of tubular ovarioles ensheathed by a network of connective tissue in which numerous tracheoles and muscles are embedded. The number of ovarioles per ovary, though approximately constant within a species, varies widely among species. For example, in some viviparous aphids and in dung beetles there is one ovariole per ovary in contrast to the more than 2000 ovarioles per ovary in some higher termite queens. The wall of each ovariole includes an outer epithelial sheath and an inner acellular, elastic layer, the tunica propria. Each ovariole (Figure 19.2)

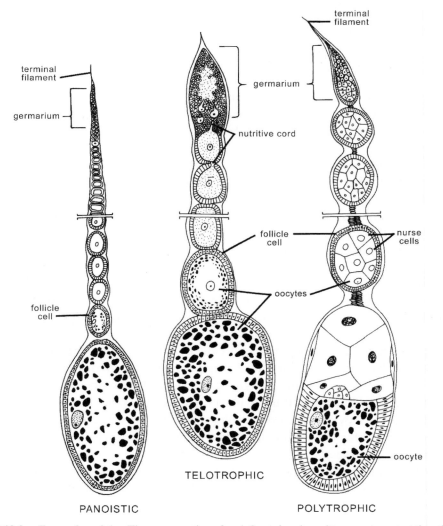

FIGURE 19.2. Types of ovarioles. The upper portion of each figure is enlarged to a greater extent than the lower in order to make details of germarial structure clear. [After A. P. Mahowald, 1972, Oogenesis, in: *Developmental Systems: Insects*, Vol. I (S. J. Counce and C. H. Waddington, eds). By permission of Academic Press Ltd., and the author.]

consists of a terminal filament, germarium, vitellarium, and pedicel (ovariole stalk). The terminal filaments may fuse to form a sheet of tissue attached to the dorsal body wall or dorsal diaphragm by which an ovary is suspended within the abdominal cavity. Within the germarium, oogonia, derived from primary germ cells, give rise to oocytes and, in some types of ovarioles, also to nutritive cells (see below). As oocytes mature and enter the vitellarium they tend in most insects to become arranged in a linear sequence along the ovariole. Each oocyte also becomes enclosed in a one-cell-thick layer of follicular epithelium derived from mesodermal prefollicular tissue located at the junction of the germarium and vitellarium. As its name indicates, the vitellarium is the region in which an oocyte accumulates yolk, a process known as vitellogenesis (Section 3.1.1). Normally vitellogenesis occurs only in the terminal oocyte, that is, the oocyte closest to the lateral oviduct, and during the process the oocyte's volume may increase enormously, for example, by as much as 10^5 times in *Drosophila*. Each ovariole is connected to a lateral oviduct by a thin-walled tube, the pedicel, whose lumen is initially occluded by epithelial tissue. This plug of tissue is lost during ovulation (movement of a mature oocyte into a lateral oviduct) and replaced by the remains of the follicular epithelium that originally covered the oocyte. Ovarioles may join a lateral oviduct linearly, as in some apterygotes, Ephemeroptera and Orthoptera, or, more often, open confluently into the distal expanded portion of the oviduct, the calyx.

Three types of ovarioles can be distinguished (Figure 19.2). The most primitive type, found in Thysanura, Paleoptera, most orthopteroid insects, Siphonaptera, and some Mecoptera, is the panoistic ovariole in which specialized nutritive cells (trophocytes) are absent. Trophocytes occur in the two remaining types, the polytrophic and telotrophic ovarioles, which are sometimes grouped together as meroistic ovarioles. In polytrophic ovarioles, several trophocytes (nurse cells) are enclosed in each follicle along with an oocyte. The trophocytes and oocyte originate from the same oogonium. Polytrophic ovarioles are found in most endopterygotes, and in Dermaptera, Psocoptera, and Phthiraptera. In Hemiptera and Coleoptera telotrophic (acrotrophic) ovarioles occur in which the trophocytes form a syncytium in the proximal part of the germarium and connect with each oocyte by means of a trophic cord.

The lateral oviducts are thin-walled tubes that consist of an inner epithelial layer set on a basal lamina and an outer sheath of muscle. In many species they include both mesodermal and ectodermal components. In almost all insects they join the common oviduct medially beneath the gut, but in Ephemeroptera the lateral oviducts remain separate and open to the exterior independently. The common oviduct, which is lined with cuticle, is usually more muscular than the lateral oviducts. Posteriorly, the common oviduct is confluent with the vagina that, as noted above, may evaginate to form the bursa copulatrix. In some species the bursa forms a diverticulum off the oviduct. In nearly all Lepidoptera the bursa is physically distinct from the oviduct and opens to the outside via the vulva (Figure 19.3). A narrow sperm duct connects the bursa with the oviduct and forms the route along which the sperm migrate to the spermatheca.

Usually a single spermatheca is present in which sperm are stored, though in some higher Diptera up to three such structures occur. The spermatheca and the duct with which it joins the bursa are lined with cuticle. The cuticle overlays a one-cell-thick layer of epithelium whose cells are glandular and assumed to secrete nutrients for use by the stored sperm. Typically, also, the cells have a much folded apical plasmalemma, adjacent to which are many mitochondria, features characteristic of cells involved in ion exchange (compare the structure of Malpighian tubule and rectal epithelial cells, Chapter 18, Sections 2.1

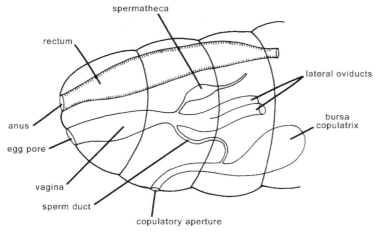

FIGURE 19.3. Reproductive system of female Lepidoptera-Ditrysia. [After A. D. Imms, 1957, *A General Textbook of Entomology*, 9th ed. (revised by O. W. Richards and R. G. Davies), Methuen and Co.]

and 4.1). These features may indicate that the sperm stored in the spermatheca require an ionic milieu different from that of the surrounding hemolymph.

Various accessory glands (also called collateral or colleterial glands) may be present and usually open into the bursa. However, in Acrididae (Orthoptera) the glands, known as pseudocolleterial glands, are anterior extensions of the lateral oviducts (Figure 19.1A). Normally, there is one pair of glands, which secrete materials that form a protective coating around the eggs or stick the eggs to the substrate during oviposition. Less commonly the glands produce antibacterial substances that coat the eggs, toxic egg protectants, and oviposition-stimulating or oviposition-deterring pheromones (Gillott, 2002). In some species the glands may be structurally distinct bi- or multipaired structures, each pair presumed to have discrete functions. In Hymenoptera, the glands are single, not paired, and produce the venom used in the sting, secrete trail- or oviposition site-marking pheromones, or lubricate the ovipositor valves (Figure 19.1D).

2.2. Male

Functions of the male reproductive system include production, storage, and, finally, delivery to the female of sperm. In some species, the system produces substances transferred during copulation that regulate female receptivity and fecundity (Gillott, 1995, 2003; Wolfner, 1997, 2002). An additional function may be to supply the female with nutrients that can be incorporated into developing oocytes, thereby increasing the rate and number of eggs produced.

The male system includes paired testes (in Lepidoptera these fuse to form a single median organ), paired vasa deferentia and seminal vesicles, a median ejaculatory duct, and various accessory glands (Figure 19.4). The testes, which lie either above or below the gut, comprise a varied number of tubular follicles bound together by a connective tissue sheath. The follicles may open into the vas deferens either confluently or in a linear sequence. The wall of each follicle is a layer of epithelium set on a basal lamina. Within the follicles several zones of development can be readily distinguished (Figure 19.5). The distal zone is the germarium in which spermatogonia are produced from germ cells. In Orthoptera,

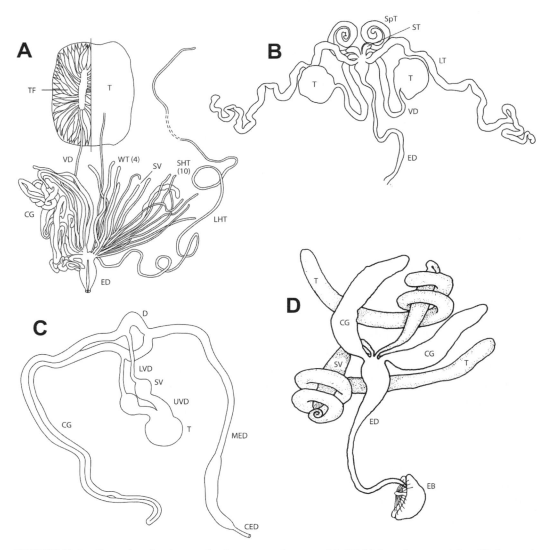

FIGURE 19.4. Examples of male reproductive systems (not to scale). (A) *Melanoplus sanguinipes* (Orthoptera); (B) *Lytta nuttalli* (Coleoptera); (C) *Anagasta kühniella* (Lepidoptera); and (D) *Drosophila melanogaster* (Diptera). In *M. sanguinipes* 16 pairs of tubules make up each collateral gland (CG). There are 4 white tubules (WT), 10 short hyaline tubules (SHT), and a long hyaline tubule (LHT). The 16th tubule serves as a seminal vesicle (SV). In *L. nuttalli* there are three tubules in each collateral gland; a spiral tubule (SpT), short tubule (ST), and a long tubule (LT). *Other abbreviations*: CED, cuticular ejaculatory duct; D, duplex; EB, ejaculatory bulb; ED, ejaculatory duct; LVD, lower vas deferens; MED, mesodermal ejaculatory duct; T, testis; TF, testis follicle; UVD, upper vas deferens; VD, vas deferens. (A, from C. Gillott, 2002, Insect accessory reproductive glands: Key players in production and protection of eggs, in: *Chemoecology of Insect Eggs and Egg Deposition* (M. Hilker and T. Meiners, eds.). By permission of Blackwell Verlag, Berlin; B, from G. H. Gerber, N. S. Church, and J. G. Rempel, 1971, The anatomy, histology, and physiology of the reproductive systems of *Lytta nuttalli* Say (Coleoptera: Meloidae). I. The internal genitalia, *Can. J. Zool.* **49**:523–533. By permission of the National Research Council of Canada; C, from a diagram supplied by Dr. J. G. Riemann; D, from E. Kubli, 1996, The *Drosophila* sex-peptide: A peptide pheromone involved in reproduction, *Adv. Dev. Biochem.* **4**:99–128. By Permission of JAI Press, Inc.)

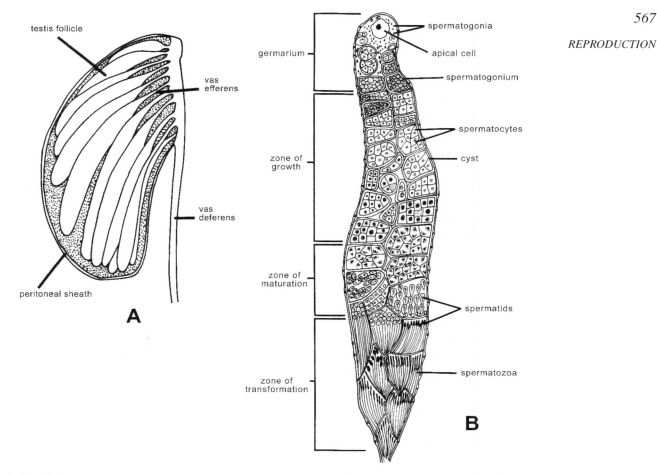

FIGURE 19.5. (A) Section through testis to show arrangement of follicles; and (B) zones of maturation in testis follicle. [A, from R. E. Snodgrass, *Principles of Insect* Morphology. Copyright 1935 by McGraw-Hill, Inc. Used with permission of McGraw-Hill Book Company. B, after V. B. Wigglesworth, 1965, *The Principles of Insect Physiology*, 6th ed., Methuen and Co. By permission of the author.]

Dictyoptera, Hemiptera, and Lepidoptera a prominent apical cell is also present whose presumed function is to supply nutrients to the spermatogonia. As each spermatogonium moves proximally into the zone of growth, it becomes enclosed within a layer of somatic cells, forming a "cyst." Within the cyst, the cell divides mitotically to form a varied number (usually 64–256) of spermatocytes. In the zone of maturation, the spermatocytes undergo two maturation divisions, so that from each spermatocyte four haploid spermatids are formed. In the proximal part of the follicle, the zone of transformation, spermatids differentiate into flagellated spermatozoa. At this time the cyst wall normally has ruptured, though often the sperm within a bundle (spermatodesm) remain held together by a gelatinous cap that covers their anterior end. This cap may be lost as the sperm enter the vas deferens or persist until the sperm have been transferred to the female.

In Lepidoptera two types of sperm occur. Pyrene (nucleate) sperm are those that fertilize eggs, while apyrene (anucleate) sperm are speculated to have several functions, including assisting in the movement of pyrene sperm from the testes to the seminal vesicles, providing nourishment to the pyrene sperm, and destroying sperm from previous matings (Silberglied *et al.*, 1984). In each species within the *Drosophila obscura* complex two size classes of nucleated sperm are produced, which differ in head and tail lengths. Snook and Karr (1998) confirmed that only the long-sperm type fertilize eggs, though the function(s) of the short sperm remain unidentified.

Sperm are moved from the testes to their site of storage (normally, the seminal vesicles) by peristaltic contractions of the vas deferens. The seminal vesicles are dilations of the vasa deferentia. Their walls are well tracheated and frequently glandular, which may indicate a possible nutritive function. In Acrididae (Orthoptera) sperm are stored in a pair of highly modified accessory gland tubules (Figure 19.4A). In many Lepidoptera the migration of sperm follows a circadian rhythm (Giebultowicz *et al.*, 1989). Typically sperm are released from the testes into the upper vasa deferentia shortly before or just after dark, and then are moved into the seminal vesicles during the next light phase. However, they quickly leave this site, being moved into the duplex region of the reproductive tract (Figure 19.4C), where they remain until the next copulation occurs. The timing of sperm movement is such that the sperm produced each day move into the duplex a few hours after the male's daily period of receptivity to female pheromone. This ensures that when the male next has an opportunity to mate, a substantial amount of new sperm will be available for insemination.

The vasa deferentia enter the anterior tip of the ejaculatory duct, an ectodermally derived tube lined with cuticle whose walls normally are heavily muscularized. Posteriorly, the ejaculatory duct may run through an evagination of the body wall, which thus forms an intromittent organ. In insects that form a complex spermatophore, subdivision of the ejaculatory duct into specialized regions may occur. In Ephemeroptera no ejaculatory duct is present, and each vas deferens opens directly to the exterior.

The accessory glands may be either mesodermal (mesadenia) or ectodermal (ectadenia) in origin and are connected with either the lower part of the vasa deferentia or the upper end of the ejaculatory duct. In some species considerable morphological and functional differentiation of the glands occurs. Essentially, however, their secretions may contribute to the seminal fluid and/or form the spermatophore. In some species the glands produce substances that, when transferred to the female during insemination, cause increased egg production (Section 3.1.3) and/or decreased receptivity (willingness to mate subsequently) (Gillott, 1988, 2003).

3. Sexual Maturation

Most male insects eclose (emerge as adults) with mature sperm in their seminal vesicles. Indeed, in a few insects, for example, some Lepidoptera, Plecoptera, and Ephemeroptera, both egg and sperm production occur in the final larval or pupal instar to enable mating and egg laying to take place within a few hours of eclosion. Generally, however, after eclosion, a period of sexual maturity is required in each sex during which important structural, physiological, and behavioral changes occur. This period may extend from only a few days up to several months in species that have a reproductive diapause (Section 3.1.3).

3.1. Female

Among the processes that occur as a female insect becomes sexually mature are vitellogenesis, development of characteristic body coloration, maturation of pheromone-producing glands, growth of the reproductive tract, including accessory glands, and an increase in receptivity. These processes are controlled by the endocrine system whose activity, in turn, is influenced by various environmental stimuli.

569

REPRODUCTION

3.1.1. Vitellogenesis

As noted above, vitellogenesis occurs, by and large, only in the terminal oocyte within an ovariole, yet in many species the process is highly synchronized among ovarioles and between ovaries; that is, the eggs are produced in batches. Why vitellogenesis does not occur to any great extent in the more distal oocytes is unclear, though various suggestions have been made. One suggestion is that the terminal oocyte, as the first to mature, that is, to become capable of vitellogenesis, simply outruns the competition. In other words, once the oocyte begins vitellogenesis and growth, its increasing surface area enables it to capture virtually all of the available nutrients. This, however, cannot be the complete answer because in many female insects vitellogenesis in the penultimate oocyte appears to be inhibited even after the terminal oocyte has completed its yolk deposition and become chorionated, provided that the mature egg is not laid. Two explanations have been proposed. Adams and co-workers (see Adams, 1970, 1981) showed, in *Musca domestica* at least, an ovary containing mature eggs produces an oostatic hormone that prevents release of the ovarian ecdysiotropic hormone necessary for vitellogenesis (Section 3.1.3). In contrast, in *Rhodnius prolixus* and *Locusta migratoria* an antigonadotropic hormone is produced by the abdominal perisympathetic organs and thoracic ganglia, respectively, when the ovariole contains a mature egg. The function of this hormone, it is proposed, is to block the action of juvenile hormone on the follicle cells (Section 3.1.3), again preventing vitellogenesis (Huebner and Davey, 1973; Davey *et al.*, 1993). Remarkably, the metacestode stage of the rat tapeworm, *Hymenolepis diminuta*, which infects female mealworm beetles (*Tenebrio molitor*), also produces an antigonadotropin. This acts similarly to that of *Rhodnius*, thus allowing the parasite to make use of resources originally intended for egg production (Hurd, 1998).

As yolk appears, it is seen to be made up almost entirely of roundish granules or vacuoles known as yolk spheres. Within the yolk spheres, protein, lipid, or carbohydrate can be detected. The membrane-bound protein yolk spheres are most abundant, followed by lipid droplets that are not membrane-bound. Relatively few glycogen-containing yolk spheres are usually present. Small amounts of nucleic acids are normally detectable, but these are not within the yolk spheres. The source of some of these materials is different in the various types of ovarioles.

In all ovarioles, however, almost all yolk protein is extraovarian in origin. In most insects the proteins are accumulated from the hemolymph. The source of these proteins, as was noted in Chapter 16, Section 5.4, is the fat body, which, during vitellogenesis, synthesizes* and releases large quantities of a few specific proteins (vitellogenins or female-specific proteins) that are selectively accumulated by the terminal oocytes. The higher Diptera are a notable exception in that the follicular epithelium is also a major source of yolk proteins; indeed, in the stable fly, *Stomoxys calcitrans*, all yolk protein production occurs here (Kelly, 1994).

Shortly before vitellogenesis, a space appears between the follicle cells and the terminal oocytes, and intercellular spaces develop in the follicular epithelium (patency), so that the oocytes become bathed in hemolymph. The tunica propria appears to be freely permeable to all solutes within the hemolymph. Electron microscopic and other studies have shown

---

* In insects that have fully developed eggs at eclosion, the proteins are synthesized (and stored) by the fat body during larval development, to be released during the pupal stage when vitellogenesis occurs.

that extensive pinocytosis occurs in the oocyte plasma membrane, resulting in the accumulation of yolk protein in membrane-bound vacuoles. Vitellogenins are large (molecular weight 200,000–650,000), conjugated proteins containing both lipid and carbohydrate components. The latter are oligosaccharides and probably important as the agents by which the oocyte plasmalemma recognizes the vitellogenins prior to pinocytosis. Though most yolk protein is produced in the fat body, small contributions may be made by follicular epithelial cells, nurse cells where present, or the terminal oocyte. Studies have shown, for example, that in some species active RNA synthesis occurs in the follicular epithelium during vitellogenesis, especially the early stages, and that isotopically labeled amino acids are first taken up by follicle cells to appear later in protein spheres within the oocytes. In telotrophic and polytrophic ovarioles some protein may be transferred during early vitellogenesis to the oocyte from the nurse cells. However, the latter appear to be more important as suppliers of nucleic acids to the developing oocyte, and several autoradiographic and electron microscopic studies have shown the movement of labeled RNA or ribosomes down the trophic cord in telotrophic ovarioles or across adjacent nurse cells into the oocyte in polytrophic ovarioles. It is presumed that this RNA is then associated with protein synthesis within the oocyte. In addition to the RNA derived from nurse cells, RNA may also be produced by the oocyte nucleus for use in protein synthesis. In several species, active incorporation of labeled RNA precursors into the oocyte nucleus has been observed to occur early in vitellogenesis, concomitant with the accumulation of protein (non-membrane-bound) adjacent to the nuclear envelope.

Studies on the accumulation of lipid, carbohydrate, and other components of yolk are few. Though lipid may make up a considerable proportion of the yolk, its source remains doubtful in most species. An apparent association between the Golgi apparatus and the accumulation of lipid led early authors to suggest that the lipid was synthesized by the oocyte *per se*, though this has not been confirmed. Another suggestion that requires further study is that the follicle cells contribute lipid to the oocyte. In the polytrophic ovariole of *Drosophila*, the nurse cells supply lipids to the oocyte, though this apparently is not the case in *Culex*. In telotrophic ovarioles, lipid may be derived both from the nurse cells (early in vitellogenesis) and from the follicle cells.

Glycogen usually can be detected only in small amounts and, in meroistic ovarioles, after degeneration of the nurse cells. In *Apis* and *Musca*, labeled glucose injected into the hemolymph is rapidly accumulated by oocytes in late vitellogenesis and apparently converted to glycogen (Engelmann, 1970).

In the German cockroach, *Blattella germanica*, the terminal oocytes accumulate large amounts of hydrocarbons during vitellogenesis. The hydrocarbons apparently are synthesized by the epidermis of the abdominal sternites, then transported to the ovaries bound to a hemolymph lipophorin. Later, they become incorporated into the ootheca (Fan *et al.*, 2002).

3.1.2. Vitelline Membrane and Chorion Formation

When vitellogenesis is completed, the vitelline membrane and, later, the chorion (eggshell) are formed. Though some early observations suggested that the vitelline membrane was produced by the oocyte itself, perhaps as a modification of the existing plasmalemma, recent studies covering a range of insect orders have confirmed that the vitelline membrane is secreted by the follicle cells. The nature of the membrane varies both among species (correlated with the egg's environment) and regionally over the egg surface. For example, the vitelline membrane of *Drosophila melanogaster* is perforated in the "collar"

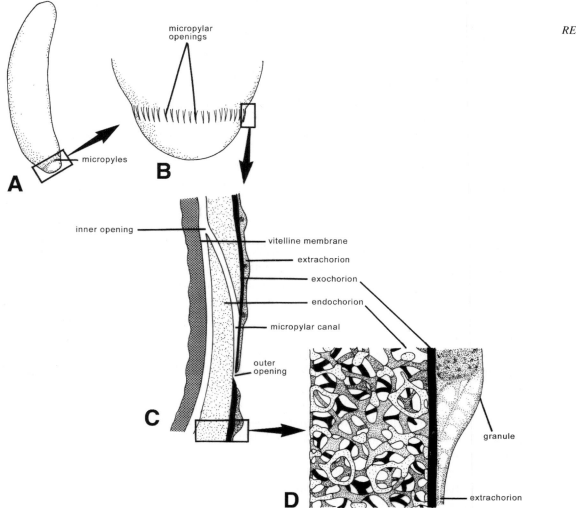

FIGURE 19.6. Egg of *Locusta*. (A) General view; (B) enlargement of posterior end; (C) section through chorion along micropylar axis; and (D) details of chorion structure. [A, after R. F. Chapman, 1971, *The Insects: Structure and Function*. By permission of Elsevier/North-Holland, Inc., and the author. B, C, after M. L. Roonwal, 1954, The egg-wall of the African migratory locust, *Locusta m. migratorioides* R&F. (Orthoptera: Acrididae), *Proc. Natl. Inst. Sci. India* **20**:361–370. By permission of the Indian National Science Academy. D, after J. C. Hartley, 1961, The shell of acridid eggs, Q. *J. Microsc. Sci.* **102**:249–255. By permission of Cambridge University Press.]

region which ruptures to facilitate hatching, while in the eggs of parasitic Hymenoptera the membrane is extremely thin to permit uptake of nutrients from the host's hemolymph. In *D. melanogaster* the follicle cells secrete wax immediately after production of the vitelline membrane. The wax layer is found in eggs of many species and, like that of the cuticle, prevents desiccation. A wax layer is not found, however, in eggs that are normally laid in humid or wet microclimates or that take up water from the environment prior to embryonic development.

The chorion is usually secreted entirely by the follicle cells and comprises two main layers, an endochorion adjacent to the vitelline membrane and an exochorion (Figure 19.6). In some insects, for example, Acrididae, the shell takes on a third layer, the extrachorion,

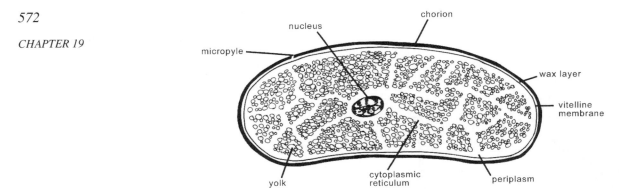

FIGURE 19.7. Diagrammatic sagittal section through an egg at oviposition. [After R. F. Chapman, 1971, *The Insects: Structure and Function.* By permission of Elsevier/North-Holland, Inc., and the author.]

as an oocyte moves through the common oviduct. Interestingly, though the follicle cells are mesodermal derivatives, the chorion is cuticlelike in nature and contains layers of protein and lipoprotein (but not chitin), some of which are tanned by polyphenolic substances released by the cells.

The chorion is not produced as a uniform layer over the oocyte. For example, in some species a ring of follicle cells near the anterior end of the oocyte secrete no exochorion, so that a line of weakness is created at this point for ease of hatching. Also, certain follicle cells appear to have larger than normal microvilli which, when withdrawn after chorion formation, leave channels (micropyles) to permit entry of sperm (Figure 19.6C,D). The aeropyles (air canals) appear to be formed in a similar way. In eggs of many species, both terrestrial and aquatic, the aeropyles connect with a network of minute air spaces within the endochorion (the plastron), which improves gas exchange between the oocyte and the atmosphere. The plastron in eggs of terrestrial species usually serves to prevent desiccation caused by evaporation; for aquatic insects (or terrestrial species whose eggs become immersed temporarily in water, for example, after rain), the plastron prevents waterlogging of the egg while facilitating gas exchange (see Chapter 15, Section 4.2). Like the vitelline membrane, the chorion varies in thickness among species. In the cecropia moth, *Hyalophora cecropia*, it is 55–60 μm, providing a rigid protective coat, while in parasitic Hymenoptera it is very thin (e.g., *Habrobracon juglandis*, 0.17 μm) and flexible to permit stretching of the egg during its passage along the very narrow ovipositor.

The internal structure of a mature (i.e., chorionated) egg is shown diagrammatically in Figure 19.7.

3.1.3. Factors Affecting Sexual Maturity in the Female

Environmental. A number of environmental factors have been observed to influence the rate at which a female becomes sexually mature, for example, quantity and quality of food consumed, population density and structure, mating, photoperiod, temperature, and humidity. For most factors there is evidence that they exert their influence by modifying the activity of the endocrine system, though availability of food and temperature also have obvious direct effects on the rate of egg development.

Many reports indicate that the qualitative nature of food may have a marked effect on the number of eggs matured. However, it is often not clear whether the observed differences in egg production are related to differences in palatability (more palatable foods might be eaten in larger quantities) or to differences in the nutritive value of the food. For most insects, dietary proteins are essential for maturation of eggs. Many Diptera, for example, may survive for several weeks on a diet that contains carbohydrate but no protein, yet will mature no eggs. Equally, different proteins may have different nutritional values related to their amino acid composition and, possibly, to their digestibility. Carbohydrates, too, are important, especially in the provision of energy for synthesis of yolk components. As noted above, lipids may be a significant component of the yolk, and therefore, an essential part of the diet, though some may be formed from carbohydrates. Water, vitamins, minerals, and, for some species, specific growth substances are also necessary constituents of the diet if maximum egg production is to be achieved.

Quantitative influences of feeding are much more easily documented. Anautogenous mosquitoes and many other bloodsucking insects, for example, do not mature eggs until they have fed. Furthermore, the number of eggs produced is, within limits, proportional to the quantity of blood ingested. However, it must be noted that in endopterygotes many of the materials to be used in egg production are laid down during larval development, and therefore the nutrition of the larva must also be considered when attempting comparisons.

In addition to its direct function of providing raw materials for egg maturation, feeding also has an important indirect effect, namely, to stimulate endocrine activity. In continuous feeders (those that take a series of small meals) stretching of the foregut as food is ingested results in information being sent via the stomatogastric nervous system to the brain/corpora cardiaca complex where it stimulates synthesis and release of neurosecretion. In occasional feeders such as *Rhodnius* and anautogenous mosquitoes, which require a single, large meal in order to mature a batch of eggs, stretching of the abdominal wall is the stimulus, sent to the brain via the ventral nerve cord, which triggers endocrine activity. In both arrangements the degree of endocrine activity and, in turn, the number of eggs developed, is proportional (other things being equal) to the amount of food ingested.

In some insects, especially gregarious species, population density and structure may influence sexual maturation. In some species of *Drosophila*, *Locusta migratoria*, and *Nomadacris septemfasciata* (the red locust), for example, the greater the population density, the slower is the rate of egg maturation. Though there are some obvious potential reasons for this, such as interference with feeding, other effects of this stress, perhaps manifest through a decrease in endocrine activity, may also be important. By contrast, in the desert locust, *Schistocerca gregaria,* crowded females mature eggs faster than isolated individuals, probably as a result of increased endocrine activity in the former group. In addition, in *S. gregaria* and *S. paranensis* (the Central American locust) there is evidence that egg development in females is promoted in the presence of older males, which are believed to secrete a maturation-accelerating pheromone (Chapter 13, Section 4.1.1). Conversely, in colonies of social insects, secretion of a maturation-inhibiting pheromone by the queen prevents development of the reproductive system of other females, the workers. Again, the effects of these pheromones are probably mediated through the endocrine system, though the evidence for this proposal is mostly circumstantial.

Mating is, for many species, a most important factor in sexual maturation. For example, in some bloodsucking Hemiptera and some cockroaches almost no eggs develop in virgin females. In other insects eggs of virgin females mature more slowly than those of mated females, and many are eventually resorbed if mating does not take place. The stimulus given to a female may be physical or chemical in nature, but in either case its effect may be to enhance endocrine activity. In some cockroaches, for example, mechanical stimulation of the genitalia or the presence of a spermatophore in the bursa results in information being sent to the brain via the ventral nerve cord, followed by activation of the corpora allata. In representatives of many insect orders male accessory gland secretions are transferred to the female during mating. These chemicals, which move to a range of sites within the female's body (Lay *et al.*, 2004), may serve either as signals, triggering subsequent steps in the reproductive process, or as nutrient contributions enabling females to produce more eggs. For example, in *Drosophila*, many species of mosquitoes, and some Lepidoptera peptidic materials from the male accessory glands stimulate egg development. Thus, in *Drosophila* one of the actions of the so-called "sex peptide" is to promote juvenile hormone synthesis and hence increased vitellogenesis (Wolfner, 1997, 2002; Kubli, 2003). In tettigoniids, some butterflies, and *Photinus* fireflies the spermatophore, either eaten or digested within the reproductive tract, provides nutrients that increase the number of eggs produced (see also Section 4.3.1) (Boggs, 1995; Gwynne, 2001, Rooney and Lewis, 2002).

Like all metabolic processes, egg development is affected by temperature and occurs at the maximum rate at a specific, optimum temperature, whose value presumably reflects the normal temperature conditions experienced by a species during reproduction. On each side of this optimum egg maturation is decreased, in the normal temperature-dependent manner of enzymatically controlled reactions. Sometimes superimposed on this basic effect, however, are more subtle effects of temperature, of both a direct and an indirect nature. For example, in *Locusta migratoria* regular temperature fluctuations (provided these are not too extreme) appear to stress the insect, causing release of neurosecretion and enhanced rates of development. In some other species mating occurs only within a certain temperature range, yet, as noted earlier, may have an important influence on egg development. It follows that, in this situation, temperature can have an important indirect effect on maturation.

Few direct observations have been made on the effects of humidity on egg maturation, though it is known that humidity may determine whether or not oviposition occurs. In many species eggs are laid only when the relative humidity is high (80% to 90%), and oviposition is increasingly retarded as the environment becomes drier. Engelmann (1970) suggested a possible explanation for this is that, as increasing amounts of water are lost from the body by evaporation, insufficient remains for use in egg development whose rate is therefore decreased.

Photoperiod, the earth's naturally recurring alternation of light and darkness, is probably the best-studied environmental factor that influences egg maturation. The effect of photoperiod is long-term (seasonal) and serves to correlate egg development with the availability of food, suitable egg-laying conditions, and/or suitable conditions for the eventual development of the larvae. Implicit in this statement is the idea that an insect, by having its reproductive activity seasonal in nature, is able to overcome adverse environmental conditions. Commonly, an insect survives these adverse conditions by entering a specific physiological condition known as diapause, whose onset and termination are induced by changes in daylength (sometimes acting in conjunction with temperature). (For a general discussion of diapause, see Chapter 22, Section 3.2.3.) Essentially diapause is a phase of arrested development, and, in the context of adult (reproductive) diapause, this means that the eggs do not mature. In different species diapause may be induced by increasing day length (number of hours of light in a 24-hour period), which enables an insect to overcome hot and/or dry summer conditions (estivation), or by decreasing day length prior to the onset of cold winter conditions (hibernation).

For example, in the Egyptian tree locust, *Anacridium aegyptium*, reproductive diapause is induced by the decreasing day lengths experienced in fall and is maintained for about

4 months during which no eggs are produced. Termination of diapause, brought about by increasing day lengths in spring, is correlated with renewed availability of oviposition sites and food for the juvenile stages. Likewise, hibernation in newly eclosed adult Colorado potato beetles (*Leptinotarsa decemlineata*) is induced by the short day lengths of fall (and also by a lack of food at this time). At the onset of diapause, the beetles become negatively phototactic and bury themselves under several inches of soil. Termination of diapause in *L. decemlineata* is unlikely to be induced by increasing day length (given the beetles' position); rather, diapause may have terminated simply as a result of the passage of time or by a temperature cue as soil warms in the spring (Hodek, 2002, and personal communication).

Estivation is seen in *Schistocerca gregaria* and *Hypera postica,* the alfalfa weevil. Sexual maturation in *S. gregaria* is retarded by long and promoted by short day lengths. This observation can be correlated with the availability of food and oviposition sites in its natural habitat, arid areas of Africa and Asia, where in the summer the weather is hot and dry, but rain falls intermittently during the winter. *H. postica* adults emerge in late spring and undergo reproductive diapause before laying eggs in late summer and fall. Low winter temperatures prevent the eggs from hatching until the following spring when new alfalfa foliage on which the larvae feed has begun to appear.

The effects of photoperiod on egg maturation are mediated via the endocrine system, though for many species the evidence for this statement is largely circumstantial, for example, differences in the histological appearance of the endocrine glands in diapausing and non-diapausing insects. In diapausing adult insects the corpora allata are small, and the neurosecretory system is typically full of stainable material, which are taken to indicate inactivity of these glands. When diapause is terminated and egg development begins, the corpora allata increase in volume and the amount of stainable material in the neurosecretory system decreases. In the beetle *Galeruca tanaceti* autoradiographic studies have shown that in post-diapause beetles the rate of incorporation of labeled cystine into neurosecretory cells (taken to be a measure of their synthetic activity) is high compared to that of estivating insects. A limited amount of experimental work supports these histological correlations. For example, in *L. decemlineata* maintained under long-day (non-diapause) conditions, allatectomy, treatment with precocene (which destroys the corpora allata), or cautery of the brain neurosecretory cells mimics the diapause-inducing effects of short days, that is, stimulates digging behavior, arrests yolk deposition, and causes oxygen consumption to decrease. Conversely, treatment of diapausing beetles with juvenile hormone or its mimics causes the beetles to leave the soil, and stimulates feeding activity and egg maturation (Denlinger, 1985).

Endocrine.   The endocrine control of sexual maturation in female insects, especially the control of oocyte development, continues to be among the most intensely studied areas of insect physiology. Yet, despite the wealth of literature that has resulted, some major aspects of hormonal control remain unclear. It is apparent, however, that among the Insecta the relative importance of the various endocrine centers in reproduction may differ, as might be anticipated in a group of such diverse habits (Belles, 1998). The following account is therefore generalized, though the major points of contention and differences among insects will also be outlined (see also Figure 19.8).

The two principal hormonal components involved are juvenile hormone (JH) and cerebral neurosecretory factors, though in some insects ecdysone, oostatic hormone, or antigonadotropic hormone also are important.

The importance of the corpora allata in egg development first became apparent in 1936 when Wigglesworth and Weed-Pfeiffer (cited in de Wilde and de Loof, 1974b) demonstrated independently that in *Rhodnius* and *Melanoplus*, respectively, allatectomy (removal of the

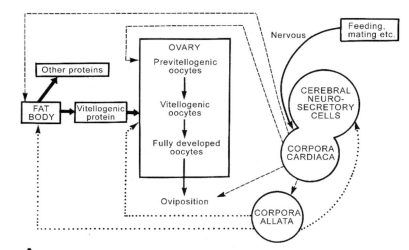

A *Schistocerca, Locusta, Anacridium*

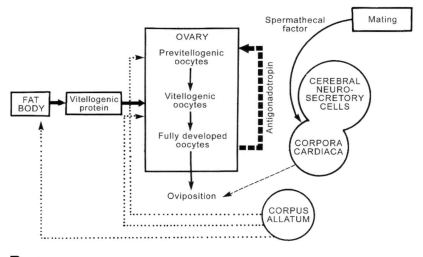

B *Rhodnius*

FIGURE 19.8. Endocrine control of egg development. (A) *Schistocerca gregaria* and other locusts; (B) *Rhodnius prolixus*; (C) *Aedes aegypti* and other mosquitoes; and (D) *Sarcophaga bullata*. [After K. C. Highnam and L. Hill, 1977, *The Comparative Endocrinology of the Invertebrates*, 2nd ed.. By permission of Edward Arnold Publishers Ltd.]

corpora allata) prevented vitellogenesis. Since this date, many authors, using allatectomy, followed by replacement therapy (implantation of "active" glands from other insects, or treatment with JH or its mimics), have confirmed the importance of JH as a gonadotropic factor in most insects. However, in the flesh fly, *Sarcophaga bullata*, vitellogenesis occurs only when the median neurosecretory cells are present and is, apparently, independent of JH (Figure 19.8D). In mosquitoes, JH controls only the previtellogenic growth of the primary follicles and is not required for vitellogenesis (Figure 19.8C) (Dhadialla and Raikhel, 1994; Klowden, 1997).

Originally, it was believed that JH probably triggered the synthesis of yolk precursors in the follicle cells, which then passed these materials on to developing oocytes. However,

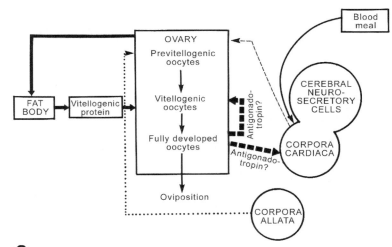

C *Aedes* and other mosquitoes

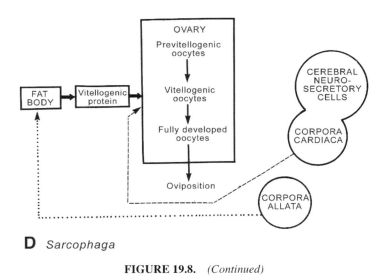

D *Sarcophaga*

FIGURE 19.8. *(Continued)*

subsequent studies have shown that in most species yolk is mainly of extraovarian origin, especially, the fat body (see below). In the ovary, JH has two effects: early in vitellogenesis it "primes" the follicular epithelial cells, while later it regulates the permeability of the follicular epithelial layer. In *Rhodnius* and *Locusta*, for example, JH causes the follicular epithelial cells to shrink slightly, resulting in the development of prominent intercellular channels (patency) (Davey *et al.*, 1993; Davey, 1996). Vitellogenins have been shown to move along these channels, to be accumulated pinocytotically by the oocyte during vitellogenesis. The precise way in which JH stimulates the follicle cells to shrink remains unclear, though an effect on the arrangement of microtubules within the cells has been demonstrated. (Evidence from other non-insectan systems strongly implicates the involvement of these organelles in the regulation of cell shape.)

As was noted in Section 3.1.1, for most insects the source of most yolk components is extraovarian, specifically the hemolymph, which serves as a reservoir for materials synthesized in the fat body. In the early to mid-1960s, evidence was collected that suggested a hormone from the median neurosecretory cells of the brain regulated protein synthesis in the fat body (references in Highnam and Hill, 1977). Only later came the realization that the neurosecretion does not act directly on the fat body, but is allatotropic in effect; that is, it stimulates the corpora allata to produce JH, which then promotes vitellogenesis in the fat body in addition to its effects on the ovary (Figure 19.8A). In some insects, for example *Rhodnius*, neurosecretion also has other important functions in the reproductive process (see Section 7.2) (Figure 19.8B) (Davey, 1997).

A different arrangement occurs in mosquitoes in which neurosecretion [ovarian ecdysteroidogenic hormone (OEH)] is released for only the first few hours after the blood meal (anautogenous species such as *A. aegypti*) or after eclosion (autogenous species, for example, *A. atropalpus,* that do not need a blood meal to mature eggs), and, as noted earlier, the corpora allata are not important in vitellogenesis (Klowden, 1997). In these insects OEH acts directly on the ovary, stimulating it to release ecdysone, which regulates vitellogenin synthesis in the fat body (Hagedorn *et al.*, 1975) (Figure 19.8C). Of particular interest is the involvement of ecdysone, a hormone traditionally associated with molting and produced by glands that disappear in all but a very few adult insects. Ecdysone has now been detected in adults of many insect species and, in higher Diptera also is involved in vitellogenesis (Hagedorn *et al.*, 1975; Hagedorn, 1985; Yin and Stoffolano, 1994). It is now clear that the follicle cells are the major source of the ecdysone. However, ovariectomy does not lead to the complete depletion of the hormone in the hemolymph, and other sources, notably oenocytes, have been suggested. Relatively huge (microgram) amounts of ecdysone occur in the ovaries of *Locusta migratoria*, 99% of which is in the terminal oocytes in the conjugated (inactive) form. During embryogenesis, free ecdysone is periodically released, the peaks of release corresponding with bouts of embryonic cuticle production and hatching (Chapter 20, Section 7.2).

Among the Lepidoptera, which show a wide range of reproductive strategies, the endocrine control of vitellogenesis and metamorphosis are often closely linked (Ramaswamy *et al.*, 1997). Four patterns have been identified within the order. In those species in which egg development is completed in the larval or early pupal stage, vitellogenesis is triggered by the release of prothoracicotropic hormone. This stimulates the molt glands to produce ecdysone, which initiates vitellogenesis in the fat body and may also induce patency in the follicular epithelium. In some other Lepidoptera (Pyralidae) where vitellogenesis occurs later in the pupal stage, it is the declining level of ecdysone that apparently promotes the process. In a third group, of which *Manduca sexta* is an example, the dependence on ecdysone is lost: vitellogenesis occurs in the pharate adult in the absence of ecdysteroids, though what other hormonal factor(s) regulate the process are unclear. Finally, in the Lepidoptera that develop eggs only after eclosion and a period of sexual maturation (for example, butterflies and noctuid moths), JH serves both to promote vitellin production in the fat body and the development of patency in the follicular epithelium, as occurs in insects in general.

An aspect of egg maturation that requires much more study is the interrelationship of the endocrine glands, ovary, and fat body. That is, how is endocrine activity increased or decreased during each cycle of egg development so that the system operates most efficiently? For most insects, the neurosecretory system acts as the center for translating external stimuli into endocrine language. As a result of these stimuli, production and release of allatotropic, allatostatic, and other neurosecretory peptides will be altered, causing, in turn, changes in the level of production of JH, ecdysone, etc. However, endocrine gland activity can be modified in other ways, as is obviously the situation in insects whose neurosecretory system is not involved in egg development. For example, in the allatic nerves (Figure 13.4) there are non-neurosecretory axons that conceivably can carry neural information from the brain to activate or inhibit the corpora allata. Another possibility is that the corpora allata respond directly to the nutritional milieu in which they are bathed. In an actively feeding insect whose hemolymph contains large quantities of nutrients, the glands could be expected to be active to facilitate production and accumulation of yolk. As oocytes mature, feeding activity declines, so that there are fewer nutrients in the hemolymph, which might lead to a reduction in corpus allatum activity. An important question is, therefore, "What causes feeding activity to decline as vitellogenesis nears completion?" To date, there is no clear answer to this question, though some authors have suggested that distension of the oviducal or abdominal wall by mature eggs might inhibit feeding behavior via the central nervous system.

A further possibility that may account for the cyclic nature of egg production involves a common endocrine principle, namely, feedback inhibition. There is evidence that a high level of circulating JH may inhibit synthesis and release of neurosecretion. In turn, this will result in a decline in corpus allatum activity, a drop in the level of JH and, eventually, a renewal of neurosecretory activity.

A relatively unexplored aspect of these interrelationships is whether the ovary may function as an endocrine structure whose secretion may competitively bind with a hormone receptor, for example, that for JH, or may inhibit production of hormone by a gland. As noted earlier in this chapter and in Chapter 13 (Section 3.4), oostatic and antigonadotropic hormones are produced in *Musca*, and *Rhodnius* and *Locusta*, respectively, and it will be interesting to see whether comparable situations are widespread among Insecta.

3.2. Male

Compared to that of females, sexual maturation in male insects has received much less attention and, consequently, is relatively poorly understood.

In species in which males after eclosion live for only a brief time, spermatogenesis is completed during the late larval and/or pupal stages. Presumably, also, in such short-lived species, when sperm is transferred to the female in a spermatophore, the accessory glands (which produce the spermatophore components) must become active in the juvenile stages. In many male insects, especially those that mate frequently, spermatogenesis continues at a low rate after eclosion. During sexual maturation, other events may also occur, such as synthesis and accumulation of accessory gland secretions for use in spermatophore or seminal fluid formation. Pheromone-producing glands may become functional, and the male may develop characteristic behavior patterns and coloration. All of these processes appear to be influenced by the endocrine system.

Involvement of the endocrine system in sperm production is now reasonably clear (Dumser, 1980). Several early workers observed that precocious adult males, obtained by removing the corpora allata from larvae, had testes that contained sperm capable of fertilizing eggs and concluded that JH inhibited differentiation of sperm. Conversely, other authors had noted that in diapausing larvae or pupae differentiation within the testis ceased and suggested that ecdysone promoted this process, an effect that has been confirmed in several species following the commercial availability of this hormone. If, however, the hormones regulate differentiation, then an adult insect, which was thought to lack ecdysone but has active corpora allata, would presumably present an unfavorable environment for spermatogenesis. Yet, as noted above, this process occurs after eclosion in some species. Dumser and Davey (1975 and earlier), working on *Rhodnius*, resolved this paradox by proposing that the hormones affect only the rate of spermatogonial mitosis and not differentiation *per se.* They suggested that in the testes there is a basal (endogenous) rate of mitosis that occurs even in the absence of hormones. Ecdysone increases the rate of division, whereas JH depresses it, though never below the basal level. Differentiation of the germ cells follows division, but its rate is constant (not directly affected by hormones) and species-specific. Thus, in early juvenile stages, the rate of division will be low, because of the presence in the hemolymph of JH as well as ecdysone. However, in the final larval and/or pupal stages when the corpora allata are inactive (Chapter 21, Section 6.1), ecdysone will accelerate spermatogonial division (reflected in the enormous growth of the testes seen at this time), which results in production of large numbers of mature sperm. In adults, spermatogenesis continues at the basal rate, despite the presence of JH in the hemolymph. The demonstration that ecdysteroids are produced by the testicular sheath and other tissues (Hagedorn, 1985; Loeb *et al.*, 1996) provides, of course, a ready explanation for the continuation of spermatogenesis in mature males of some species.

The male accessory reproductive glands are fully differentiated at eclosion. During sexual maturation they become active and increase greatly in size as a result of the synthesis and accumulation of materials used in spermatophore and/or seminal fluid formation (Section 4.3.1). In some species, for example, the cockroach *Leucophaea maderae*, secretory activity of the accessory glands apparently is not under JH control; thus, allatectomized adult males are able to produce spermatophores throughout their life. In contrast, in many other species (including other cockroaches [Schal *et al.*, 1997]) there is a clear correlation between the onset of corpus allatum activity and development of secretory activity in the accessory glands. Further, accessory glands of allatectomized males of these species remain small and show only weak secretory activity, effects that can be reversed by treatment with JH (Gillott, 1988; Gillott and Gaines, 1992). Though Engelmann (1970) suggested that the effect of JH on the secretory activity of the accessory glands may be indirect, stemming from a general endocrine control of protein metabolism, other authors, for example, Odhiambo (1966), Gillott and Friedel (1976), and Gillott (1996) argued that JH has a specific role, namely, to control the synthesis of specific accessory gland proteins. Indeed, studies of the production of accessory gland secretion in the male migratory grasshopper, *Melanoplus sanguinipes*, revealed some interesting parallels with the process of vitellogenesis (Friedel and Gillott, 1976; Gillott, 1996). Male *M. sanguinipes* are promiscuous insects. They may copulate several times on a single day and, on each occasion, transfer several spermatophores. Such promiscuity requires either extremely active accessory glands or the participation of other tissue(s) in the production of the necessary materials. It was shown that the fat body produces specific proteins that are accumulated by the accessory glands. Removal of the accessory glands led to accumulation of protein in the fat body and hemolymph. Furthermore, both synthesis of these proteins in the fat body and their accumulation by the accessory glands were prevented by allatectomy, an effect that could be reversed by treating operated insects with a JH mimic. Several other demonstrations of extraglandular synthesis of accessory gland proteins have been made, suggesting that the phenomenon may be a common occurrence among Insecta (Figure 19.9).

Ecdysteroids have been reported in adult males of a range of species and may also be involved in regulating accessory gland protein synthesis. For example, *in vitro* experiments with the accessory glands of *M. sanguinipes* have demonstrated that β-ecdysone can promote or inhibit the production of specific proteins. Further, the level of ecdysteroids does vary during sexual maturation and is increased briefly after mating (Gillott, 1996).

The development and endocrine control of sexual behavior in male insects have been studied in detail in relatively few species, mostly Acrididae. As Engelmann (1970) noted,

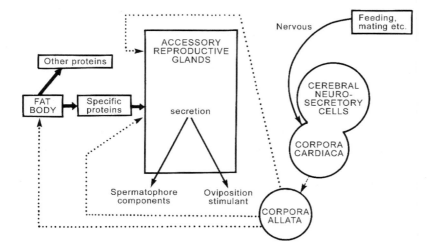

FIGURE 19.9. Endocrine relationships in male insects. Note that some of these relationships may not exist in all species.

the best-known species in this regard is the desert locust *S. gregaria,* which begins to show elements of sexual behavior between 6 and 12 days after eclosion. Paralleling the onset of this behavior is a change in body color, from the grayish-brown-pink of newly emerged insects to the uniform yellow of mature males, and the beginning of secretion from epidermal glands of a maturation-accelerating pheromone. The work of Loher, Odhiambo, and Pener (references in Pener, 1974, 1983) established that the changes are controlled by JH in this species. For example, allatectomy shortly after eclosion leads to retention of immature coloration and a lack of desire to copulate, effects that may be reversed by implantation of corpora allata from mature individuals. Similar procedures have demonstrated JH-controlled sexual behavior in the males of some other species. However, in male migratory locusts, *L. migratoria* ssp. *migratorioides,* the corpora allata control only the development of the mature body color; mating behavior is controlled primarily by specific median neurosecretory cells in the brain. The control exerted by the cells is direct, that is, not mediated through the corpora allata.

4. Mating Behavior

Mating behavior, which can be considered as the events that encompass and relate to the act of mating, may be subdivided into four components, some or all of which may be observed in a given species. The components are (1) location and recognition of a mate, (2) courtship, (3) copulation, and (4) postcopulatory behavior. The primary function of mating behavior is to ensure the transfer of sperm from male to female, though as is discussed below, there are additional functions of this behavior that optimize the reproductive economy of the species (Cade, 1985; Gillott *et al.*, 1991).

4.1. Mate Location and Recognition

A prerequisite to internal fertilization for almost all terrestrial animals, insects included, is the coming together of male and female so that sperm can be transferred directly into a female's reproductive tract. Interestingly, however, in some apterygotes indirect sperm transfer occurs. In some Collembola, for example, males deposit packages of sperm onto the substrate, though at the time no females may be in the vicinity. Females apparently find the sperm packages by chance and take them up into the reproductive tract. In Thysanura, also, droplets of sperm are placed on the substrate, though only when a female is present.

Among pterygote insects, a variety of stimuli alone or in combination may signal the location of a potential mate. These may be visual, olfactory, auditory, or tactile. Visual cues are used mostly, though not solely, by many diurnal species. Movement, color, form, and size may attract one individual to another, though final determination of the suitability of the proposed partner is usually made by tactile or chemical stimuli. Many male Diptera, for example, are attracted randomly to dark objects of a given size range but do not make copulatory movements unless they receive the correct tactile or chemical stimulus on contact. Male butterflies may fly toward objects whose movements follow a particular sequence and which are of a particular color pattern. Again, however, final contact and attempts to mate depend on the odor of the object. The best-known nocturnal insects that use a visual cue to locate (or attract) a mate are fireflies (Lampyridae: Coleoptera), both sexes or only females of which may produce light by means of bioluminescent organs located on the posterior abdominal segments. In some species a female "glows" continuously to attract males that land in her general vicinity, then locate her exact position by olfaction. In others, a male in flight flashes regularly (!), and a female responds (begins to flash) when he comes within a certain range. The time interval between the partners' flashes is critical and determines whether a male is attracted to a responding female. Again, final contact is made by means of touch and/or smell.

Pheromones (Chapter 13, Section 4.1.1) are probably the most common signal used by insects in mate location. In contrast to visual stimuli, volatile chemical attractants are both highly specific and capable of exerting their effect over a considerable distance (in some Lepidoptera, several kilometers). Pheromones are usually produced by females and are employed by both diurnal and nocturnal species.

Auditory stimuli, like pheromones, can exert their effect over some distance and may be species-specific. Because transmission of sound is not seriously impeded by vegetation, auditory stimuli are particularly useful cues for insects that live among grass, etc., for example, many Orthoptera. Among the Gryllidae, only males produce sounds that serve to orient and attract potential mates from distances up to 30 m in some species. In some Acrididae, in contrast, both males and females produce sounds that attract the opposite sex. Males of certain species of cicadas, katydids, and crickets sing "aggression songs" that serve to keep the males a minimum distance apart (i.e., spread the males over the widest area), thereby reducing interference during courtship and mating. In some species conspecific males sing only during the interval of silence in a neighboring male's song. This may facilitate mate location by the female because of the reduction in background noise.

"Accidental" sex attraction occurs when all members of a species are attracted by the topography or scent of a particular site or when individuals of both sexes produce aggregation sounds (as in cicadas) or pheromones (as in bark beetles).

4.2. Courtship

Recognition and attraction is immediately followed, in some insects, by copulation. In other species, however, copulation is preceded by more or less elaborate forms of courtship for which several functions have been suggested (Manning, 1966).

Males of some species, including the migratory locust (*Locusta migratoria*), some damselflies (Figure 6.9), and the monarch butterfly, will mount or grasp an available female when the opportunity arises, yet may not mate for some time, specifically, until the female is about to oviposit (Simmons, 2001). This phenomenon, known as precopulatory mate guarding, ensures that the mounted male's sperm will be used to fertilize the eggs (but see Section 4.3.1). Male desert locusts (*Schistocerca gregaria*) take the process of precopulatory mate guarding a step further: under crowded conditions, they also secrete a courtship-inhibiting pheromone (phenylacetonitrile) to protect their mate from rivals (Seidelmann and Ferenz, 2002).

Specialized courtship behavior may prevent interspecific mating, especially between closely related species, though Manning (1966) argued that there is little evidence to support this proposal. In species where a female is normally aggressive (even predaceous to the point of consuming a potential mate), courtship may serve to appease her so that she becomes willing to copulate. A well-known example of such appeasement is found in the dance flies (Empididae: Diptera), so-called because in many species a male has an elaborate courtship dance in which he presents a female with a silken ball. In some species the ball contains prey that is actually eaten by the female during copulation. In others, it contains nutritionally useless material, though by the time a female discovers this it is too late! Finally, in some species the procedure becomes entirely ritualized, and a male simply presents his partner with an empty ball.

Males of many species use liquid or pheromonal secretions as aphrodisiacs to pacify their partner. In many male Orthoptera, Dictyoptera, and Coleoptera glands on the thorax, head, antennae, palps, or elytra release a secretion that pacifies the female when ingested. Many male Lepidoptera produce pheromones that serve as arrestants, inhibiting the female's movement and stimulating her to adopt the mating position (Birch *et al.,* 1990).

As an antithesis to female appeasement, courtship may also be a necessary prerequisite for bringing a male into a suitable state for insemination. In other words, the process serves to synchronize the behavior of the pair, thereby increasing the chance of successful sperm transfer. In some species, courtship has a very obvious function, namely, bringing the partners together in the correct physical relationship for insemination. For example, in his courtship display, the male cockroach *Byrsotria fumigata* raises his wings to expose on the metanotum a gland whose secretion is attractive to the female. At the same time, the male turns away from the female, who, in order to reach the secretion and feed on it, must mount the male from the rear. In this way the female comes to take up the appropriate position for copulation.

Courtship may also be very important in sexual selection, that is, the choice of a mate by the female. For example, in the firefly *Photinus ignitus* male flash duration and size of the lantern (flashing apparatus) are correlated with the size of the male's spermatophore. Females, which do not feed as adults, prefer males with longer flashes and larger lanterns, perhaps to gain the benefits of a larger spermatophore, including more sperm and greater nutrient content (Cratsley and Lewis, 2003). Similarly, female bush crickets (*Ephippiger ephipigger*) prefer the songs of younger males, perhaps related to their superior insemination ability compared to that of older individuals (Ritchie *et al.*, 1995).

4.3. Copulation

Receptivity (willingness of a female to copulate) depends not only on a male's efforts to seduce her but also on the female's physiological state. Age and the presence of semen in the spermatheca are the two most important factors governing receptivity, though external influences may also be important. For many species an obvious correlation exists between receptivity and state of egg development in virgin females, and it appears that the level of circulating JH governs receptivity. Many females mate only once or a few times, and after mating become unreceptive to males for a varied length of time. The switching off of receptivity has been related to the presence of semen in the spermatheca. In some species it appears that stretching of the spermathecal wall may lead to nervous inhibition of receptivity. In others, inhibition is pheromonal, the seminal fluid containing a "receptivity-inhibiting" substance that either directly, or by causing the spermatheca to liberate a hormone into the hemolymph, acts on the brain to render the female unreceptive (Gillott, 1988, 2003; Wolfner, 1997).

In many species, mating occurs only at a certain time of the day. For example, certain fruit flies (*Dacus* spp.) mate only when the light intensity is decreasing. Other species have built-in circadian rhythms of mating.

Some Lepidoptera mate only in the vicinity of the larval food plant, the odor of the plant stimulating release of sex attractant by the female.

During copulation, some of the behavioral elements introduced during courtship may be continued, presumably to keep the female pacified until insemination is completed. Sometimes pacification continues after insemination and is thought to prevent the female from ejecting or eating the spermatophore until the sperm have migrated from it.

4.3.1. Insemination

Insemination is the transfer from male to female of sperm and seminal fluid. Although the latter has some obvious functions such as the protection of sperm and the provision of a nourishing fluid in which the sperm can be moved to their site of storage, it also contains chemicals that may modulate the female's postcopulatory behavior and physiology to the male's advantage.

As noted earlier, indirect sperm transfer occurs in some apterygotes, but in almost all Pterygota, sperm is transferred during copulation directly to the female reproductive tract. Primitively, sperm are enclosed in a special structure, a spermatophore, which may be formed some time before copulation (in Gryllidae and Tettigoniidae) or, more often, as copulation proceeds. Spermatophores are not produced by males of many species of endopterygotes, especially higher Diptera, or by some male Hemiptera; rather, the semen is transferred to the spermatheca via an intromittent organ.

Spermatophore Production. Production of a spermatophore has been studied in relatively few species, and it is difficult, therefore, to generalize. Essentially, however, the structure is formed by secretions of the accessory glands and sometimes also the ejaculatory duct. The secretions from different gland components may mix or remain separate, so that the wall of the spermatophore is formed of a series of layers surrounding a central mass of sperm. In most katydids and some other Ensifera the spermatophore comprises a sperm-containing sac and a large sperm-free mass (spermatophylax) that is eaten by the female after copulation (see below).

Gerber (1970) proposed four general methods of spermatophore formation, which form a distinct evolutionary series. In the most primitive method (first male-determined method), found in many orthopteroid species, the spermatophore is complex and formed either at the anterior end of the ejaculatory duct or within the male copulatory organ. After transfer to a female, the spermatophore is usually held between her external genital plates and only its anterior, tubelike portion enters the vagina or bursa. In the second male-determined method, the spermatophore has a less complex structure and is formed within a special sper-

matophore sac of the copulatory organ. The thin-walled sac is everted into the bursa, which therefore essentially determines the final shape and size of the spermatophore. After spermatophore formation, the sac is withdrawn. *Rhodnius,* some Coleoptera, and some Diptera form spermatophores in this way. The two remaining methods are female-determined. In the first, seen in Trichoptera, Lepidoptera, some Diptera, some Coleoptera, and a few Hymenoptera, male accessory gland secretions empty directly into the vagina or bursa in a definite sequence, either before or after transfer of sperm which they encapsulate. The spermatophore takes up the shape of the genital duct. The spermatophore produced by the second female-determined method is the least complex. Male accessory gland secretions are produced concurrently with or immediately after sperm transfer and often do not encapsulate the sperm; rather, they harden to form a mating plug that prevents backflow and loss of semen (and also, further mating). This method is seen in mosquitoes, the honey bee, and some Lepidoptera. Gerber (1970) speculated that the next step in the evolutionary sequence would be complete loss of the spermatophore and, concurrently, the development of a more elongate penis for depositing sperm close to the spermatheca.

The number of spermatophores formed, the length of time the spermatophore remains with a female, and its fate are varied. In some species a male produces a single spermatophore during each copulation, and this may remain in the female's genital tract or between the genital plates for several hours to ensure complete evacuation of semen. Empty spermatophores are generally discarded. However, as noted earlier, they may be eaten by the female, or partially to completely digested within her genital tract, thus providing a nutrient contribution toward egg production.

Precisely how sperm move from the spermatophore to the spermatheca is for most species unclear, In some species the anterior end of the spermatophore is open to facilitate the escape of sperm. Where the spermatophore completely encloses the sperm, its wall may be either ruptured by spines protruding from the wall of the bursa or digested by secretions of the bursa wall or accessory glands. Transfer of sperm into the spermatheca is achieved normally as a result of rhythmic contractions of the reproductive tract. In *Rhodnius* the contractions are promoted by a substance contained within the seminal fluid and produced in the male's accessory glands. In some species sperm may migrate actively into the spermatheca, possibly in response to a chemotactic stimulus released by the storage organ.

Insemination Without a Spermatophore. In species where spermatophores are not used in insemination, the penis may be rigid and erection achieved by means of muscles. In such instances the organ penetrates only a short distance into the female's genital tract. Alternatively and more commonly, the penis is thin-walled and erected by hydrostatic pressure, either of the hemolymph or of fluid contained within a special reservoir off the ejaculatory duct. It extends far into the genital tract of the female and terminates adjacent to the spermatheca.

In Odonata, where a unique mode of sperm transfer is found, the male copulatory organ is a tubular structure formed on the third abdominal sternum. Prior to mating, a male transfers sperm to the copulatory organ by coiling the abdomen ventrally. During mating (which occurs in flight), the female, held around the head or prothorax by the male's terminal claspers, brings the tip of her abdomen into contact with the copulatory organ (Figure 6.9). Remarkably, in damselflies and some dragonflies the male's copulatory organ, which is equipped with horns and/or spines, not only transfers sperm to the female but first removes sperm deposited in a previous mating (Waage, 1979, 1986). In other dragonflies the copulatory organ has an inflatable head that is used to push previously deposited sperm to the rear of the female's storage organ before insemination (Simmons, 2001). Removal of old sperm prior to insemination also occurs in some beetles.

Hemocoelic Insemination. This most unusual form of insemination, used by Cimicoidea (Hemiptera) and Strepsiptera, refers to the injection of sperm into the body cavity of a female from which they migrate to specialized storage sites, conceptacula seminales (not homologous with the spermathecae of other Insecta), adjacent to the oviducts. Comparative studies of the phenomenon in Cimicoidea have led authors to propose a possible evolutionary sequence (Hinton, 1964; Carayon, 1966). Primitively, the penis is placed in the vagina but penetrates its wall, thereby injecting semen into the body cavity. At a more advanced stage, the penis penetrates the integument, though not at any predefined site, and sperm are still injected into the hemolymph. Next, the site of penetration becomes fixed, and beneath it a special structure, the spermalege, develops to receive sperm. However, the sperm must still migrate via the hemolymph to the conceptacula seminales. At the most advanced level, insemination into a spermalege occurs, and sperm move to the conceptacula along a solid core of cells. What is the functional significance of this arrangement? In all forms of hemocoelic insemination a proportion of the sperm are phagocytozed either by hemocytes or by cells of the spermalege. As a result, it has been suggested that in Cimicoidea hemocoelic insemination is a method of providing nutrients to the female, enabling her to survive for longer periods in the absence of suitable food. It should be remembered that many Cimicoidea are semiparasitic or parasitic, and the chances of locating a host are slight. Interestingly, some species are apparently homosexual; that is, males inseminate other males, enabling the recipients to resist starvation for longer periods, while reducing the donors' own viability, surely a truly noble and altruistic act!

Modulators of Female Behavior and Physiology. Included in the seminal fluid are chemicals secreted by the accessory glands (rarely other parts of the male reproductive system) that bring about profound changes in the postcopulatory behavior and physiology of the female (Gillott, 1988, 2003; Wolfner, 1997, 2002; Kubli, 2003). These changes work to ensure that the male sires at least a portion of the eggs laid They include induction of refractoriness (unwillingness of the female to remate), reduction in female attractiveness, plugging of the female reproductive tract, acceleration of egg development, stimulation of egg laying, and displacement or incapacitation of sperm from previous matings. Other suggested effects of the seminal fluid chemicals are modification of host-seeking behavior in bloodsucking species (which typically mate near or on the host), change in the female's circadian rhythmicity, and a decrease in the female's life span.

4.4. Postcopulatory Behavior

In many species characteristic behavior follows copulation. This may be a continuation of the events that occur during copulation, for example, antennation or palpation of the female by the male, or feeding by the female on special secretions (nuptial gifts) offered by the male. As a result of these actions, the female remains passive, enabling sperm to be evacuated from the spermatophore.

Postcopulatory mate guarding has also been identified in species belonging to many orders (Simmons, 2001). Its purpose is to ensure complete transfer of sperm or to enable the female to oviposit undisturbed. It includes both contact and non-contact forms of behavior. Two forms of contact guarding occur: continued genital contact, seen, for example, in some terrestrial bugs, phasmids, and tettigoniids; and mounting, when the male simply remains on the female, as in beetles, many pond skaters, some grasshoppers, and some flies. In non-contact guarding, seen in crickets, most odonates, and other flies, the male disengages from the female but remains close by ready to repel other males. The duration and intensity of postcopulatory mate guarding may vary even within a species, according to the intensity of competition from other males (Simmons, 2001).

5. Ovulation

Ovulation, the movement of an egg from the ovary into the lateral oviduct, may or may not be well separated in time from the actual process of egg laying (oviposition). In most insects, especially those that lay their eggs singly (e.g., *Rhodnius*), oviposition immediately follows ovulation. However, in insects that deposit eggs in batches or are viviparous (Chapter 20, Section 8.3) the two events may occur several days apart. For example, in *Schistocerca gregaria* the eggs accumulate in the lateral oviducts for a week prior to being laid.

Ovulation is induced by a neurosecretory factor from the brain. In *Rhodnius*, for example, median neurosecretory cells produce a myotropic peptide that is stored in the corpora cardiaca. Mating is an important (though not the only) stimulus for release of the peptide, which induces contractions of the ovarian muscles. However, as mating may occur well in advance of ovulation, it cannot be the only factor involved. It appears that mating triggers periodic releases of ecdysone from the ovary (coincident with the maturation of an egg), and it is this hormone that effects the release of the peptide from the corpora cardiaca (Davey, 1985b).

6. Sperm Use, Entry into the Egg, and Fertilization

6.1. Sperm Use

Compared to mammals where about 10 million sperm are ejaculated to fertilize one or a few eggs, insects show remarkable economy in this regard. For example, the queen bee spermatheca contains about 5 million sperm, yet she produces on average at least 200,000 fertilized eggs during her 4-year life span. In other words, an average of about 25 sperm per egg are released from the spermatheca. This suggests that a very precise control mechanism exists for sperm release though almost nothing is known about the process.

Females of some species are essentially monogamous; that is, they mate once, then become unreceptive or unattractive to males as a result of chemicals from the successful male's seminal fluid (Section 4.3.1). In such females, therefore, the question of sperm competition (which male's sperm will be used to fertilize the eggs) does not arise. Most female insects, however, mate several to many times (though they may become unreceptive for some time after each mating), receiving new sperm (and sometimes nutrients) on each occasion. In these species sperm competition will occur, especially if sperm from the various donors mix in the storage organs. It is clear, however, that sperm use in multiply mated insects is non-random; that is, strategies have evolved to ensure that some sperm take precedence over others. These strategies may be either male- or female-driven.

From the male's perspective the simplest strategy is to be the most recent mate, as his ejaculate will thus tend to be closest to the opening of the spermatheca; indeed, in most insects the "last in, first out" method of sperm precedence occurs. However, strengthening of this positional advantage may be achieved by various means, including sperm stratification [pushing the existing sperm to the rear of the storage organ (Section 4.3.1)], sperm loading (injecting a high number of sperm relative to the number already there), and sperm incapacitation (killing or inhibiting the activity of sperm already there by components of the seminal fluid) (Simmons, 2001). It should be stressed that there is currently little evidence for the occurrence of the second and third mechanisms. Other strategies, for which the evidence is more compelling, include sperm removal prior to ejaculation (Section 4.3.1) and sperm flushing whereby previously deposited sperm are washed out by the most recent ejaculate.

Females may also exert selection over which male's sperm they use to fertilize their eggs. For example, they may discard sperm soon after ejaculation has occurred (i.e., before it reaches the storage organ). Failure to transport the sperm to storage or fertilization sites has also been reported for some species. Rapid remating with a "better" male and reduction in the number of eggs deposited are other mechanisms that females may use to control which male's sperm they use (Eberhard, 1996).

6.2. Sperm Entry into the Eggs

In almost all insects sperm enter the eggs as the latter pass through the common oviduct during oviposition. However, in Cimicoidea and some scale insects sperm enter the eggs in the ovary, and in Strepsiptera sperm entry occurs as the eggs float within the hemocoel. Two problems associated with the entry of sperm are (1) release of sperm from the spermatheca in synchrony with movement of an egg through the oviduct and (2) location by sperm of the micropyles. The solution to the first problem remains unclear, though in some species the presence of muscle in the wall of the spermatheca and its duct indicates that sperm are squeezed toward the oviduct rather than moving of their own accord. In *Periplaneta,* it has been suggested that a nervous pathway might exist between sensory hairs found within the oviduct and the spermathecal muscle that could synchronize these events. Other insects have spermathecae with rigid walls, and in these a chemotactic stimulus for induction of sperm movement seems more likely.

Various mechanisms ensure that sperm can locate and enter the micropyles. In *D. melanogaster*, whose egg has but a single micropyle, the egg is precisely oriented as it moves along the oviduct so that the micropylar region directly faces the opening of the ventral receptacle, which contains the sperm. Furthermore, movement of an egg along the oviduct may stop briefly at this point. Where an egg's orientation is less precise, there may be a large number of micropyles. For example, in *Periplaneta* up to 100 of these funnel-shaped structures occur in a cluster at the cephalic end of the egg. In *Rhodnius* the micropyles lie within a groove that encircles the anterior end of the egg. As an egg moves along the oviduct, the groove comes to lie opposite the openings of the paired spermathecae (Davey, 1965). Actual entry of sperm into the micropyles, whose diameter at the inner end may be only a fraction of a micrometer, may result from release of a chemical attractant by the oocyte, though there is little evidence to support this suggestion.

Polyspermy (the entry of two or more sperm into an egg) is common in insects, though only one sperm normally undergoes subsequent transformation into a pronucleus, and the remainder degenerate.

6.3. Fertilization

Fertilization, that is, the fusion of male and female pronuclei, does not occur until after oviposition. Indeed, completion of the meiotic divisions of the oocyte nucleus, which give rise to three polar body nuclei and the female pronucleus, is inhibited until after an egg is laid. Whether or not entry of sperm is responsible for removal of this inhibition is uncertain. Pronuclear fusion may occur at a relatively fixed site to which both pronuclei migrate or may occur randomly within an oocyte, depending on the rate and direction of movement of the pronuclei. The polar body nuclei normally migrate to the periphery of the oocyte and eventually degenerate.

7. Oviposition

Oviposition (egg laying) is an extremely important phase in an insect's life history for, unless it is carried out at the correct time and in a suitable location, the chances of the eggs developing and of the larvae reaching adulthood are slim. Except in normally partheno-genetic species, unfertilized eggs are generally inviable, and it is important, therefore, that oviposition occurs only after mating. The eggs when laid must be protected from desiccation and predation. Further, because larvae are relatively immobile, it is necessary to lay eggs close to or even on/in the larva's food. This is especially true when the food is highly specialized and/or in limited supply as, for example, in many parasitic species.

Conversely, it is desirable that the female not expend energy searching for and testing potential oviposition sites until she is ready to lay. Accordingly, oviposition behavior characteristically does not begin until the eggs are more or less mature and may be induced by the hormone balance in the female at this time.

7.1. Site Selection

Only a few examples are known of insects that apparently show no site selection behavior. Phasmids simply release their eggs, which fall among the dead vegetation beneath the host plant, though even this may be adaptive, as the eggs presumably will be hidden and protected from heat, desiccation, and predation. Females of many species attach their eggs, either singly or in batches, to an appropriate surface (often the food source) using secretions of the accessory glands. Such species typically lack an ovipositor. Other insects lay their eggs in crevices, or in plant or animal tissues, typically using an ovipositor formed by modification of either the terminal abdominal segments *per se* or the appendages of these segments (Chapter 3, Section 5).

The initial location of an egg-laying site may be more or less specific, often depending on rather general visual or chemical stimuli that tend to attract the female. For example, cabbage butterflies (*Pieris* spp.), which are attracted to blue or yellow objects when sexually immature, show a preference for green when ready to lay. Mosquitoes are attracted to dark or colored waters and to sites where fermenting materials provide olfactory signals. For grasshoppers and locusts, the texture, moisture content, and salt content of the soil are important site-selection cues, detected by sensory receptors at the tip of the abdomen. Many parasitoids are attracted initially to a likely host site by the odor and color of the host's food plant.

The final choice of an oviposition site depends on more specific stimuli. Thus, continuing the examples cited above, *Pieris* is attracted by the odor of oil of mustard, which is released by cabbage and its relatives. Even then, the female may not lay if she detects the oviposition-deterring pheromone that coats the eggs deposited by a prior female (Schoonhoven, 1990). *Culex* mosquitoes assess a potential egg-laying site using tarsal sensilla that determine water quality (salinity, pH, and amount and type of organic matter present). Oviposition-aggregation pheromone released from previously deposited egg rafts may also be an important oviposition cue (McCall, 2002). Likewise, the egg-pod froth (Section 7.3) of desert locusts releases an aggregation pheromone that attracts conspecific females to a common oviposition site (Saini *et al.*, 1995). Having located suitable host habitat, parasitoids seek out a potential host using an array of sensory cues: sight, touch, smell, and taste. Once a potential host is found, the female may probe it with her ovipositor. In this way she recognizes the host and, moreover, determines whether or not it is already parasitized (Steidle and van Loon, 2002).

7.2. Mechanics and Control of Oviposition

Compared to ovulation, oviposition is much more complex in terms of both its mechanics and its regulation, the latter being influenced by both internal and external variables as outlined in Section 7.1. Expulsion of eggs via the genital pore results from rhythmic peristaltic contractions of the walls of the oviducts. Concurrently, muscles may move the ovipositor, where present, so that a suitable egg-laying cavity is formed, and sensory input from mechanoreceptors on the ovipositor may be important where precise positioning of eggs in the substrate is required. Removal of the terminal abdominal ganglion or section of the ventral nerve cord anterior to the ganglion has demonstrated the importance of neural control in many species. In other species myotropic (oviposition-stimulating) hormones and/or pheromones may also be involved. Where ovulation is followed immediately by oviposition, as in *Rhodnius*, it seems likely that the ovulation-inducing hormone will also trigger peristalsis of the oviduct muscles. And there is evidence from species with temporally separated ovulation and oviposition (e.g., grasshoppers and locusts) for the production, by median neurosecretory cells, of a myotropic hormone. The action of this hormone is to enhance the frequency and amplitude of contractions of the oviducal muscles.

Release of a myotropic hormone or initiation of motor impulses in species where egg laying is controlled neurally depends on sensory input received by the central nervous system, and examples were given above of chemical and physical information on the oviposition site acquired in this way. However, another important consideration for a gravid female is whether or not she has been inseminated, since, as noted above, unfertilized eggs are normally inviable. In some insects, such as the cockroaches *Pycnoscelus* and *Leucophaea*, filling the spermatheca with semen stimulates neural pathways that terminate in the muscles used in oviposition. In *Rhodnius* filling the spermatheca causes this structure to release a hormone that, directly or indirectly, triggers oviposition. In a number of Diptera and Orthoptera, and other insects, there is evidence that semen contains a pheromone (fecundity-enhancing substance) that stimulates release of a myotropic hormone by the brain (Gillott and Friedel, 1977; Gillott, 2003).

7.3. Oothecae

Many orthopteroid insects, for example, locusts and grasshoppers, mantids, and cockroaches, surround their eggs, which are laid in batches, with a protective coat, the ootheca. The ootheca is produced from protein secretions of the accessory glands that are tanned in a manner similar to those of the exocuticle (Chapter 11, Section 3.3). The resulting structure is thought to prevent desiccation and, perhaps, parasitism. The egg pod of Acrididae consists of eggs surrounded by a hard, frothy mass (Figure 19.10A). The egg "cocoon" of

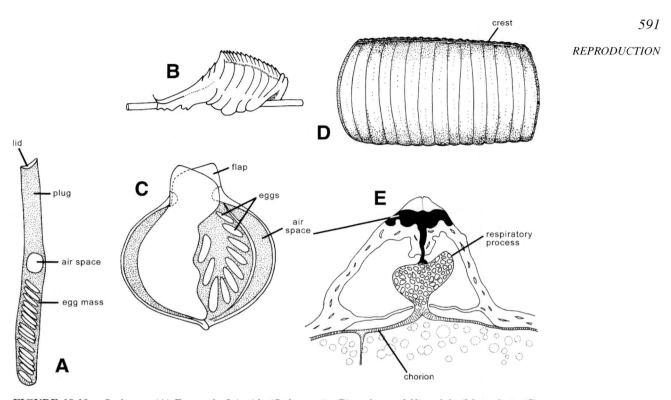

FIGURE 19.10. Oothecae. (A) Egg pod of *Acrida* (Orthoptera); (B) ootheca of *Hierodula* (Mantodea); (C) transverse section through ootheca of *Hierodula*; (D) ootheca of *Blattella* (Blattodea); and (E) transverse section through crista of *Blattella* ootheca. [A, after R. F. Chapman and I. A. D. Robertson, 1958, The egg pods of some tropical African grasshoppers, *J. Entomol. Soc. South. Afr.* **21**:85–112. By permission of the Entomological Society of Southern Africa. B, C, after Kershaw, J. C., 1910, The formation of the ootheca of a Chinese mantis, *Hierodula saussurii*, Psyche **17**:136–141. E, after V. B. Wigglesworth, 1965, *The Principles of Insect Physiology*, 6th ed., Methuen and Co. By permission of the author.]

mantids is also a hard but vacuolated sheath (Figure 19.10B,C), and, in these examples, there is clearly no hindrance to gas exchange. In contrast, the ootheca of cockroaches is a hardened shell (Figure 19.10D) formed by tanning of the proteins secreted from the accessory glands and for the most part is virtually impermeable to gases. Accordingly, the ootheca contains an air-filled cavity along the crista that connects with the exterior via small pores (Figure 19.10E). Oothecae are restricted to the lower insect orders, and Davey (1965) speculated that their absence in higher insects might represent an economy measure that permits members of these groups to produce more eggs.

8. Summary

Almost all species of insects use internal fertilization and lay eggs that contain much yolk. The female reproductive system basically includes paired ovaries and lateral oviducts, a common oviduct, bursa copulatrix, spermatheca, and accessory glands. Each ovary includes ovarioles, which may be panoistic (lacking nurse cells) or meroistic, where nurse cells either are enclosed in each follicle with an oocyte (polytrophic type) or form a syncytium in the germarium and connect with an oocyte via a trophic cord (telotrophic type).

The male system includes paired testes, vasa deferentia and seminal vesicles, a median ejaculatory duct, and accessory glands. Each testis is composed of follicles in which zones of development of germ cells occur.

In most insects a period of sexual maturation is required after eclosion. In females maturation may include vitellogenesis (formation of yolk), development of characteristic body coloration, maturation of pheromone-producing glands, growth of the reproductive tract, and an increase in receptivity. Sexual maturation is affected by quality and quantity of food eaten, population density and structure, mating, temperature, humidity, and photoperiod. These factors exert their influence by modifying endocrine activity. The two primary endocrine components in most insects are the cerebral neurosecretory system and corpora allata. The neurosecretory system liberates various peptides, including allatotropic and allatostatic hormones that regulate corpus allatum activity. The corpora allata release juvenile hormone (JH), which has both gonadotropic effects (priming the follicular epithelium, then stimulating its differentiation so that it becomes permeable to proteins) and a metabolic effect (promoting synthesis of vitellogenins in the fat body.) However, in mosquitoes and probably other Diptera, JH acts only during previtellogenesis and it is ecdysone, synthesized by the follicle cells in response to ovarian ecdysteroidogenic hormone, that promotes vitellogenesis in the fat body. In Lepidoptera vitellogenesis is regulated by ecdysone in species that develop eggs in the larval and/or pupal stages, while JH takes on this role in species that develop eggs as adults.

Spermatogenesis occurs principally in the last juvenile instar, under endocrine control, but may continue at a reduced rate in adult males. Sexual maturation in adult males is largely devoted to synthesis and accumulation of accessory gland secretions, development of pheromone-producing glands and mature body coloration, and courtship behavior. As in females, maturation is controlled hormonally, especially by juvenile hormone, though ecdysteroids may be important in some species.

Mating behavior serves to ensure that sperm is transferred from male to female under the most suitable conditions, and, perhaps, to prevent interspecific mating. It includes mate location and/or recognition, courtship, copulation, and postcopulatory behavior. Mate location and recognition are achieved through visual, chemical, auditory, and tactile stimuli. Courtship may synchronize the behavior of male and female and appease a normally aggressive female. Copulation may depend on the receptivity of a female and may occur only under specific conditions, for example, at a set time of the day, near the food plant, or immediately after feeding.

Primitively, sperm are transferred in a spermatophore produced from secretions of the accessory glands and formed within the male genital tract. At a more advanced stage, the spermatophore is formed in the female genital tract and may simply serve as a plug to prevent loss of semen. Some species do not form spermatophores but use a penis for depositing sperm in the female genital tract. Sperm normally reach the spermatheca as a result of peristaltic movements of the genital tract, though in some species active migration may occur in response to a chemical attractant. Seminal fluid includes a variety of chemicals that may induce major changes in the behavior and physiology of the mated female. These changes, which are designed to ensure paternity of the eggs laid, include: induction of refractoriness, reduction in attractiveness, plugging of the reproductive tract, acceleration of egg development, stimulation of egg laying, and displacement or incapacitation of sperm from previous matings.

Both male- and female-driven postinsemination strategies have evolved to ensure that some sperm are used in preference to others for egg fertilization. Males primarily take the "last in, first out" approach (i.e., sperm from the most recent mating will be used), though examples of stratification (packing), loading, incapacitation, and removal of sperm are known. Females may discard sperm, not transport them to storage sites, or simply remate quickly.

In almost all insects sperm enter eggs via micropyles as the eggs move down the common oviduct during oviposition. Release of sperm from the spermatheca is closely synchronized with movements of the eggs. Eggs may be precisely oriented in the oviduct or the micropyles may be arranged in a cluster or a groove to ensure location of the micropyles by sperm. Polyspermy is common, though usually only one sperm undergoes transformation into a pronucleus. Fertilization (fusion of male and female pronuclei) does not occur until after eggs are laid.

Insects may show great selectivity in their choice of oviposition sites. Eggs may be attached to surfaces by secretions of the female accessory glands or buried using an ovipositor. Eggs may be covered with an ootheca, again formed from accessory gland secretions, which may prevent desiccation and/or parasitism.

9. Literature

General introductions to insect reproduction are given by Davey (1965, 1985a,b), Engelmann (1970), and de Wilde and de Loof (1974a). Highnam and Hill (1977), Koeppe *et al.* (1985), Hagedorn (1985), Raabe (1986), Nijhout (1994), and Belles (1998) discuss the endocrine control of reproduction. Reviews of other aspects of insect reproduction are given by Phillips (1970), Baccetti (1972), and Jamieson (1989) [sperm]; Dumser (1980) [spermatogenesis]; Cade (1985), Bailey and Ridsdill-Smith (1991), and Gillott *et al.* (1991) [sexual behavior]; King (1970), Mahowald (1972), Telfer (1975), and King and Büning (1985) [ovarian development]; Margaritis (1985) [vitelline membrane and chorion;]; Chen (1984), Gillott (1988, 2003), and Kaulenas (1992) [male accessory reproductive glands]; and Hinton (1981) [eggs].

Adams, T. S., 1970, Ovarian regulation of the corpus allatum in the house fly, *Musca domestica, J. Insect Physiol.* **16**:349–360.

Adams, T. S., 1981, Activation of successive ovarian gonotrophic cycles by the corpus allatum in the house fly, *Musca domestica* (Diptera: Muscidae), *Int. J. Invert. Reprod.* **3**:41–48.

Baccetti, B., 1972, Insect sperm cells, *Adv. Insect Physiol.* **9**:315–397.

Bailey, W. J., and Ridsdill-Smith, J. (eds.), 1991, *Reproductive Behaviour of Insects*, Chapman and Hall, London.

Belles, X., 1998, Endocrine effectors in insect vitellogenesis, in: *Recent Advances in Arthropod Endocrinology* (G. M. Coast and S. G. Webster, eds.), Cambridge University Press, Cambridge.

Birch, M. C., Lucas, D., and White, P. R., 1990, Scents and eversible scent structures of male moths, *Annu. Rev. Entomol.* **35**:25–58.

Boggs, C. L., 1995, Male nuptial gifts: Phenotypic consequences and evolutionary implications, in: *Insect Reproduction* (S. R. Leather and J. Hardie, eds.), CRC Press, Boca Raton.

Cade, W, H., 1985, Insect mating and courtship behaviour, in: *Comprehensive Insect Physiology, Biochemistry and Pharmacology*, Vol. 9 (G. A. Kerkut and L. I. Gilbert, eds.), Pergamon Press, Elmsford, NY.

Carayon, J., 1966, Traumatic insemination and the paragenital system, *The Thomas Say Foundation, Publ.* **7**:81–166.

Chen, P. S., 1984, The functional morphology and biochemistry of insect male accessory glands and their secretions, *Annu. Rev. Entomol.* **29**:233–255.

Cratsley, C. K., and Lewis, S. M., 2003, Female preference for male courtship flashes in *Photinus ignitus* fireflies, *Behav. Ecol.* **14**:135–140.

Davey, K. G., 1965, *Reproduction in the Insects*, Oliver and Boyd, Edinburgh.

Davey, K. G., 1985a, The male reproductive tract, in: *Comprehensive Insect Physiology, Biochemistry and Pharmacology*, Vol. 1 (G. A. Kerkut and L. I. Gilbert, eds.), Pergamon Press, Elmsford, NY.

Davey, K. G., 1985b, The female reproductive tract, in: *Comprehensive Insect Physiology, Biochemistry and Pharmacology*, Vol. 1 (G. A. Kerkut and L. I. Gilbert, eds.), Pergamon Press, Elmsford, NY.

Davey, K. G., 1996, Hormonal control of the follicular epithelium during vitellogenin uptake, *Invert. Reprod. Develop.* **30**:249–254.

Davey, K. G., 1997, Hormonal controls on reproduction in female Heteroptera, *Arch. Insect Biochem. Physiol.* **35**:443–453.

Davey, K. G., Sevala, V. M., and Gordon, D. R. B., 1993, The action of juvenile hormone and antigonadotropin on the follicle cells of *Locusta migratoria*, *Invert. Reprod. Develop.* **24**:39–45.

Denlinger, D. L., 1985, Hormonal control of diapause, in: *Comprehensive Insect Physiology, Biochemistry and Pharmacology*, Vol. 8 (G. A. Kerkut and L. I. Gilbert, eds.), Pergamon Press, Elmsford, NY.

Dhadialla, T. S., and Raikhel, A. S., 1994, Endocrinology of mosquito vitellogenesis, in: *Perspectives in Comparative Endocrinology* (K. G. Davey, R. E. Peter, and S. S. Tobe, eds.), National Research Council of Canada, Ottawa.

de Wilde, J., and de Loof, A., 1974a, Reproduction, in: *The Physiology of Insecta*, 2nd ed., Vol. I (M. Rockstein, ed.), Academic Press, New York.

de Wilde, J., and de Loof, A., 1974b, Reproduction-Endocrine control, in: *The Physiology of Insecta*, 2nd ed., Vol. I (M. Rockstein, ed.), Academic Press, New York.

Dumser, J. B., 1980, The regulation of spermatogenesis in insects, *Annu. Rev. Entomol.* **25**:341–369.

Dumser, J. B., and Davey, K. G., 1975, The *Rhodnius* testis: Hormonal effects on germ cell division, *Can. J. Zool.* **53**:1682–1689.

Eberhard, W. G., 1996, *Female Control: Sexual Selection by Cryptic Female Choice*, Princeton University Press, Princeton.

Engelmann, F., 1970, *The Physiology of Insect Reproduction*, Pergamon Press, Elmsford, NY.

Fan, Y., Chase, J., Sevala, V., and Schal, C., 2002, Lipophorin-facilitated hydrocarbon uptake by oocytes in the German cockroach *Blattella germanica*, *J. Exp. Biol.* **205**:781–790.

Friedel, T., and Gillott, C., 1976, Extraglandular synthesis of accessory reproductive gland components in male *Melanoplus sanguinipes*, *J. Insect Physiol.* **22**:1309–1314.

Gerber, G. H., 1970, Evolution of the methods of spermatophore formation in pterygotan insects, *Can. Entomol.* **102**:358–362.

Giebultowicz, J. M., Riemann, J. G., Raina, A. K., and Ridgway, R. L., 1989, Circadian system controlling release of sperm in the insect testes, *Science* **245**:1098–1100.

Gillott, C., 1988, Arthropoda-Insecta, in *Reproductive Biology of Invertebrates*, Vol. III (Accessory Sex Glands) (K. G. Adiyodi and R. G. Adiyodi, eds.), Wiley, New York.

Gillott, C., 1995, Insect male mating systems, in: *Insect Reproduction* (S. R. Leather and J. Hardie, eds.), CRC Press, Boca Raton.

Gillott, C., 1996, Male insect accessory glands: Functions and control of secretory activity, *Invert. Reprod. Dev.* **30**:199–205.

Gillott, C., 2002, Insect accessory reproductive glands: Key players in production and protection of eggs, in: *Chemoecology of Insect Eggs and Egg Deposition* (M. Hilker and T. Meiners, eds.), Blackwell Verlag, Berlin.

Gillott, C. 2003, Male accessory gland secretions: Modulators of female reproductive physiology and behavior, *Annu. Rev. Entomol.* **48**:163–184.

Gillott, C., and Friedel, T., 1976, Development of accessory reproductive glands and its control by the corpus allatum in adult male *Melanoplus sanguinipes*, *J. Insect Physiol.* **22**:365–372.

Gillott, C., and Friedel, T., 1977, Fecundity-enhancing and receptivity-inhibiting substances produced by male insects: A review, *Adv. Invert. Reprod.* **1**:199–218.

Gillott, C., and Gaines, S. B., 1992, Endocrine regulation of male accessory gland development and activity, *Can. Entomol.* **124**:871–886.

Gillott, C., Mathad, S. B., and Nair, V. S. K., 1991, Arthropoda-Insecta, in: *Reproductive Biology of Inverte brates*, Vol. V (Sexual Differentiation and Behaviour) (K. G. Adiyodi and R. G. Adiyodi, eds.), Oxford and IBH, New Delhi.

Gwynne, D. T., 2001, *Katydids and Bush-Crickets:Reproductive Behavior and Evolution of the Tettigoniidae*, Cornell University Press, Ithaca, NY.

Hagedorn, H. H., 1985, The role of ecdysteroids in reproduction, in: *Comprehensive Insect Physiology, Biochemistry and Pharmacology*, Vol. 8 (G. A. Kerkut and L. I. Gilbert, eds.), Pergamon Press, Elmsford, NY.

Hagedorn, H. H., O'Connor, J. D., Fuchs, M. S., Sage, B., Schlaeger, D. A., and Bohm, M. K., 1975, The ovary as a source of α-ecdysone in an adult mosquito, *Proc. Natl. Acad. Sci. U S A* **72**:3255–3259.

Highnam, K. C., and Hill, L., 1977, *The Comparative Endocrinology of the Invertebrates*, 2nd ed., Arnold, London.

Hinton, H. E., 1964, Sperm transfer in insects and the evolution of haemocoelic insemination, *Symp. R. Entomol. Soc.* **2**:95–107.

Hinton, H. E., 1981, *Biology of Insect Eggs*, 3 vols., Pergamon Press, Elmsford, NY.

Hodek, I., 2002, Controversial aspects of diapause development, *Eur. J. Entomol.* **99**:163–173.

Huebner, E., and Davey, K. G., 1973, An antigonadotropin from the ovaries of the insect *Rhodnius prolixus* Stal, *Can. J. Zool.* **51**:113–120.

Hurd, H., 1998, Parasite manipulation of insect reproduction: Who benefits? *Parasitology* **116**(Suppl.):S13–S21.

Jamieson, B. G. M., 1989, *The Ultrastructure and Phylogeny of Insect Spermatozoa*, Cambridge University Press, London.

Kaulenas, M. S., 1992, *Insect Accessory Reproductive Structures. Function, Structure and Development*, Springer, Berlin.

Kelly, T. J., 1994, Endocrinology of vitellogenesis in *Drosophila melanogaster*, in: *Perspectives in Comparative Endocrinology* (K. G. Davey, R. E. Peter, and S. S. Tobe, eds.), National Research Council of Canada, Ottawa.

King, R. C., 1970, *Ovarian Development in Drosophila melanogaster*, Academic Press, New York.

King R. C., and Büning, J., 1985, The origin and functioning of insect oocytes and nurse cells, in: *Comprehensive Insect Physiology, Biochemistry and Pharmacology*, Vol. 1 (G. A. Kerkut and L. I. Gilbert, eds.), Pergamon Press, Elmsford, NY.

Klowden, M. J., 1997, Endocrine aspects of mosquito reproduction, *Arch. Insect Biochem. Physiol.* **35**:491–512.

Koeppe, J. K., Fuchs, M., Chen, T. T., Hunt, L.-M., Kovalick, G. E., and Briers, T., 1985, The role of juvenile hormone in reproduction, in: *Comprehensive Insect Physiology, Biochemistry and Pharmacology*, Vol. 8 (G. A. Kerkut and L. I. Gilbert, eds.), Pergamon Press, Elmsford, NY.

Kubli, E., 2003, Sex-peptides: Seminal peptides of the *Drosophila* male, *Cell. Mol. Life Sci.* **60**:1689–1704.

Lay, M., Loher, W., and Hartmann, R., 2004, Pathways and destination of some male gland secretions in female *Locusta migratoria migratorioides* (R&F) after insemination, *Archs. Insect Biochem. Physiol.* **55**:1–25.

Loeb, M. J., Bell, R. A., Gelman, D. B., Kochansky, J., Lusby, W., and Wagner, R. M., 1996, Action cascade of an insect gonadotropin, testis ecdysiotropin, in male Lepidoptera, *Invert. Reprod. Dev.* **30**:181–190.

Mahowald, A. P., 1972, Oogenesis, in: *Developmental Systems: Insects*, Vol. I (S. J. Counce and C. H. Waddington, eds.), Academic Press, New York.

Manning, A., 1966, Sexual behaviour, *Symp. R. Entomol. Soc.* **3**:59–68.

Margaritis, L. H., 1985, Structure and physiology of the eggshell, in: *Comprehensive Insect Physiology, Biochemistry and Pharmacology*, Vol. 1 (G. A. Kerkut and L. I. Gilbert, eds.), Pergamon Press, Elmsford, NY.

McCall, P. J., 2002, Chemoecology of oviposition in insects of medical and veterinary importance, in: *Chemoecology of Insect Eggs and Egg Deposition* (M. Hilker and T. Meiners, eds.), Blackwell Verlag, Berlin.

Nijhout, H. F., 1994, *Insect Hormones*, Princeton University Press, Princeton.

Odiambo, T. R., 1966, Growth and the hormonal control of sexual maturation in the male desert locust, *Schistocerca gregaria* (Forskål), *Trans. R. Entomol. Soc.* **118**:393–412.

Pener, M. P., 1974, Neurosecretory and corpus allatum controlled effects on male sexual behaviour in acridids, in: *Experimental Analysis of Insect Behaviour* (L. Barton Browne, ed.), Springer, Berlin.

Pener, M. P., 1983, Endocrine research in orthopteran insects, *Occ. Pap. Pan Am. Acridol. Soc.* **1**:42 pp.

Phillips, D. M., 1970, Insect sperm: Their structure and morphogenesis, *J. Cell Biol.* **44**:243–277.

Raabe, M., 1986, Insect reproduction: Regulation of successive steps, *Adv. Insect Physiol.* **19**:29–154.

Ramaswamy, S. B., Shu, S., Park, Y. I., and Zeng, F., 1997, Dynamics of juvenile hormone-mediated gonadotropism in the Lepidoptera, *Arch. Insect Biochem. Physiol.* **35**:539–558.

Ritchie, M. G., Couzin, I. D., and Snedden, W. A., 1995, What's in a song? Female bush crickets discriminate against the song of older males, *Proc. Roy. Soc. Lond. Ser. B* **262**:21–27.

Rooney, J., and Lewis, S. M., 2002, Fitness advantage from nuptial gifts in female fireflies, *Ecol. Entomol.* **27**:373–377.

Saini, R. K., Rai, M. M., Hassanali, A., Wawiye, J., and Odongo, H., 1995, Semiochemicals from froth of egg pods attract ovipositing female *Schistocerca gregaria*, *J. Insect Physiol.* **41**:711–716.

Schal, C., Holbrook, G. L., Bachmann, J. A. S., and Sevala, V. L., 1997, Reproductive biology of the German cockroach, *Blattella germanica*: Juvenile hormone as a pleiotropic master regulator, *Arch. Insect Biochem. Physiol.* **35**:405–426.

Schoonhoven, L. M., 1990, Host-marking pheromones in Lepidoptera, with special reference to two *Pieris* spp., *J. Chem. Ecol.* **16**:3040–3052.

Seidelmann, K., and Ferenz, H. J., 2002, Courtship inhibition pheromone in desert locusts, *Schistocerca gregaria*, *J. Insect Physiol.* **48**:991–996.

Silberglied, R. E., Shepherd, J. G., and Dickinson, J. L., 1984, Eunuchs: The role of apyrene sperm in Lepidoptera, *Am. Nat.* **123**:255–265.

Simmons, L. W., 2001, *Sperm Competition and Its Evolutionary Consequences in Insects*, Princeton University Press, Princeton.

Snook, R. R., and Karr, T. L., 1998, Only long sperm are fertilization-competent in six sperm-heteromorphic *Drosophila* species, *Curr. Biol.* **8**:291–294.

Steidle, J. L. M., and van Loon, J. J. A., 2002, Chemoecology of parasitoid and predator oviposition behaviour, in: *Chemoecology of Insect Eggs and Egg Deposition* (M. Hilker and T. Meiners, eds.), Blackwell Verlag, Berlin.

Telfer, W. H., 1975, Development and physiology of the oocyte-nurse cell syncytium, *Adv. Insect Physiol.* **11**:223–319.

Waage, J. K., 1979, Dual functionality of the damselfly penis: Sperm removal and transfer, *Science* **203**:916–918.

Waage, J. K., 1986, Evidence for widespread sperm displacement ability among Zygoptera (Odonata) and the means for predicting its presence, *Biol. J. Linn. Soc.* **28**:285–300.

Wolfner, M. F., 1997, Tokens of love: Functions and regulation of *Drosophila* male accessory gland products, *Insect Biochem. Mol. Biol.* **27**:179–192.

Wolfner, M. F., 2002, The gifts that keep on giving: Physiological functions and evolutionary dynamics of male seminal proteins in *Drosophila*, *Heredity* **88**:85–93.

Yin, C.-M., and Stoffolano, J. G., Jr., 1994, Endocrinology of vitellogenesis in blow flies, in: *Perspectives in Comparative Endocrinology* (K. G. Davey, R. E. Peter, and S. S. Tobe, eds.), National Research Council of Canada, Ottawa.

20

Embryonic Development

1. Introduction

Embryonic development begins with the first mitotic division of the zygote nucleus and terminates at hatching. Not surprisingly, in view of their diversity of form, function, and life history, insects exhibit a variety of embryonic developmental patterns, though certain evolutionary trends are apparent. Eggs of most species contain a considerable amount of yolk. In exopterygote eggs there is such a preponderance of yolk that the egg cytoplasm is readily obvious only when it forms a small island surrounding the nucleus. In eggs of endopterygotes, the yolk:cytoplasm ratio is much lower than that of exopterygotes and the cytoplasm can be seen as a conspicuous network connecting the central island with a layer of periplasm lying beneath the vitelline membrane. This trend toward reduction in the relative amount of yolk in the egg, carried to an extreme in certain parasitic Hymenoptera and viviparous Diptera (Cecidomyiidae), whose eggs are yolkless and receive nutrients from their surroundings, has some important consequences. Broadly speaking, the eggs of endopterygotes are smaller (size measured in relation to the body size of the laying insect) and develop more rapidly than those of exopterygotes. The increased quantity of cytoplasm leads to the more rapid formation of more and larger cells at the yolk surface that facilitates the formation of a larger embryonic area from which development can take place. Compared with that of exopterygotes, development of endopterygotes is streamlined and simplified. There has been, as Anderson (1972b, p. 229) put it, "reduction or elimination of ancestral irrelevancies," which when taken to an extreme, seen in the apocritan Hymenoptera and cyclorrhaph Diptera, results in the formation of a structurally simple larva that hatches within a short time of egg laying. However, superimposed on this process of short-circuiting may be developmental specializations associated with an increasing dissimilarity of juvenile and adult habits.

2. Cleavage and Blastoderm Formation

As it moves toward the center of an egg after fusion, the zygote nucleus begins to divide mitotically. The first division occurs at a predetermined site, the cleavage center (Figure 20.1), located in the future head region, which cannot be recognized morphologically but which appears to become activated either when sperm enter an egg or when an egg is laid. Early divisions are synchronous, and as nuclei are formed and migrate

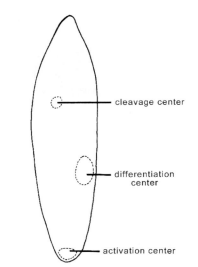

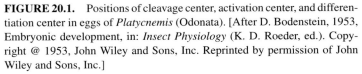

FIGURE 20.1. Positions of cleavage center, activation center, and differentiation center in eggs of *Platycnemis* (Odonata). [After D. Bodenstein, 1953, Embryonic development, in: *Insect Physiology* (K. D. Roeder, ed.). Copyright @ 1953, John Wiley and Sons, Inc. Reprinted by permission of John Wiley and Sons, Inc.]

through the yolk toward the periplasm, each becomes surrounded by an island of cytoplasm (Figure 20.2A). Each nucleus and its surrounding cytoplasm are known as a cleavage energid. In eggs of endopterygotes and possibly exopterygotes, but not those of apterygotes, the energids remain interconnected by means of fine cytoplasmic bridges.

The rate at which nuclei migrate to the yolk surface and the method of colonization are varied. In eggs of some species nuclei appear in the periplasm as early as the 64-energid state (after six divisions); in others, nuclei are not seen in the periplasm until the 1024-energid stage. In eggs of most endopterygotes and in those of paleopteran and hemipteroid exopterygotes, the periplasm is invaded uniformly by the energids. However, in eggs of orthopteroid insects the periplasm at the posterior pole of the egg receives energids first, after which there is progressive colonization of the more anterior regions.

In eggs of most insects not all cleavage energids migrate to the periphery but continue to divide within the yolk to form primary vitellophages, so-called because in most species they become phagocytic cells whose function is to digest the yolk (Figure 20.2B). In eggs of Lepidoptera, Diptera, and some orthopteroid insects, however, all of the energids migrate to the periplasm and only later do some of their progeny move back into the yolk as secondary vitellophages (Figure 20.2F). Secondary vitellophages are also produced in eggs of other insects to supplement the number of primary vitellophages. So-called tertiary vitellophages are produced in eggs of some cyclorrhaph Diptera and apocritan Hymenoptera from the anterior and posterior midgut rudiments.

After their arrival at the periplasm, the energids continue to divide, often synchronously, until the nuclei become closely packed (the syncytial blastoderm stage), after which cell membranes form by radial infolding, then tangential expansion of the original egg plasmalemma (the uniform blastoderm stage) (Figure 20.2C–F). From the resulting monolayer of cells develop all of the cells of the larval body, except in a few species where vitellophages or yolk cells contribute to the formation of the midgut (Section 7.4).

3. Formation and Growth of Germ Band

The next stage is blastoderm differentiation, giving rise to the embryonic primordium (an area of closely packed columnar cells from which the future embryo forms) and the

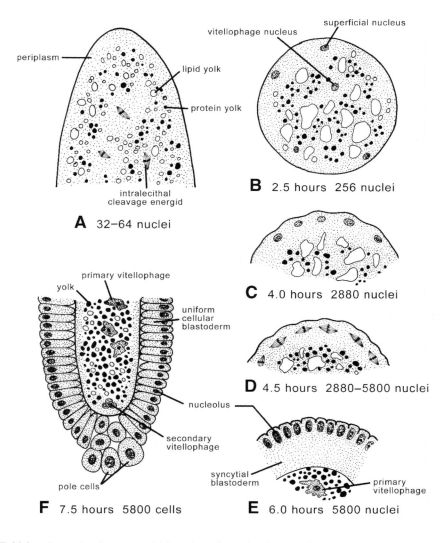

FIGURE 20.2. Stages in cleavage and blastoderm formation in egg of *Dacus tryoni* (Diptera). (A) Frontal section through anterior end during 6th division; (B) transverse section after 8th division; (C) transverse section after 12th division; (D) transverse section during 13th division; (E) transverse section at syncytial blastoderm stage; and (F) frontal section through posterior end after formation of uniform cellular blastoderm. [After D. T. Anderson, 1972b, The development of holometabolous insects, in: *Developmental Systems: Insects*, Vol. I (S. J. Counce and C. H. Waddington, eds.). By permission of Academic Press Ltd., and the author.]

extra-embryonic ectoderm from which the extra-embryonic membranes later differentiate (Figure 20.3). For more than a century, attempts have been made to explain how the body pattern of an insect is determined. Following the classic experiments of the German embryologist Seidel in the late 1920s, it was widely believed that differentiation was controlled by two centers (Counce, 1973; Heming, 2003). As energids move toward the posterior end of the egg, they interact with a so-called "activation center" (Figure 20.1), and differentiation subsequently occurs. Seidel's experiments showed that neither an energid nor the activation center alone could stimulate differentiation. It was presumed that the center is caused to release an unidentified chemical that diffuses anteriorly. This diffusion is seen morphologically as a clearing and slight contraction of the yolk. As the chemical reaches

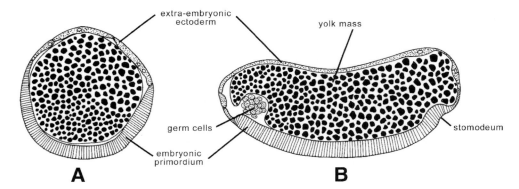

extra-embryonic
ectoderm

yolk mass

germ cells

embryonic
primordium

stomodeum

A   **B**

FIGURE 20.3. Diagrammatic transverse (A) and sagittal (B) sections of egg of *Pontania* (Hymenoptera) to show differentiation of blastoderm into embryonic primordium and extra-embryonic ectoderm. Note also the germ (pole) cells at the posterior end. [After D. T. Anderson, 1972b, The development of holometabolous insects, in: *Developmental Systems: Insects*, Vol. I (S. J. Counce and C. H. Waddington, eds.). By permission of Academic Press Ltd., and the author.]

the future prothoracic region of the embryo (the "differentiation center") (Figure 20.1), the blastoderm in this region gives a sharp twitch and becomes slightly invaginated. Blastoderm cells aggregate within this invagination and differentiate into the embryonic primordium. (Later in embryogenesis, other processes, for example, mesoderm formation and segmentation, begin at the differentiation center and spread anteriorly and posteriorly from it.)

An alternate view for the cause of embryonic differentiation is the "gradient hypothesis," which had its origins at the end of the 19th century but then fell out of favor after Seidel's pioneering work (Sander, 1984, 1997; Lawrence, 1992). Essentially, the hypothesis proposes that a chemical produced at each end of an egg diffuses throughout the egg, producing two gradients of concentration (Figure 20.4). Cells within the egg then "recognize" their position within the egg by the relative concentrations of the chemical and differentiate accordingly. Initial support for the existence of chemical gradients in eggs came from experiments in which eggs either were ligatured at various distances along their length and at varied times after embryonic development began or were centrifuged, thereby disrupting the proposed gradient. Recently, the application of genetic and molecular techniques to the study of pattern development in *Drosophila* has given further support to the idea of gradients. Thus, a modern interpretation of Seidel's differentiation center is that it is a "commitment center"; that is, it is the point at which blastoderm cells are committed to following a particular path of differentiation by virtue of their position within the gradients (Heming, 2003).

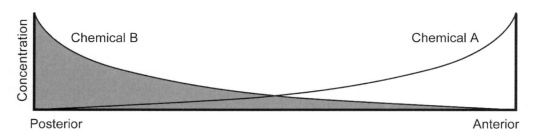

Chemical B   Chemical A

Posterior   Anterior

FIGURE 20.4. Diagrammatic representation of the gradient hypothesis. A chemical produced at each end of an egg diffuses lengthwise, forming two gradients of concentration. At any point along the length of the egg, the relative concentration of the two chemicals provides positional information to cells.

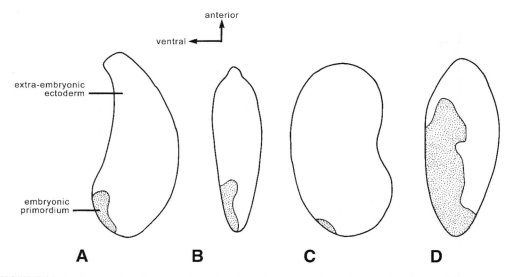

FIGURE 20.5. Form and position of embryonic primordium in exopterygotes. (A) *Periplaneta*; (B) *Platycnemis*; (C) *Zootermopsis*; and (D) *Notonecta*. [After D. T. Anderson, 1972a, The development of hemimetabolous insects, in: *Developmental Systems: Insects*, Vol. I (S. J. Counce and C. H. Waddington, eds.). By permission of Academic Press Ltd., and the author.]

As a result of the differing amounts of yolk that exopterygote and endopterygote eggs contain, important differences occur in the formation of the embryonic primordium. In exopterygote eggs where there is initially little cytoplasm, the embryonic primordium is normally relatively small, and its formation depends on the aggregation and, to some extent, proliferation of cells. In these eggs it usually occupies a posterior midventral position (Figure 20.5A–D). In contrast, in endopterygote eggs with their greater quantity of cytoplasm, the primordium forms as a broad monolayer of columnar cells that occupies much of the ventral surface of the yolk (Figure 20.6A,B). In other words, the primordium in endopterygote eggs does not require to undergo much increase in size, as is necessary in eggs of exopterygotes, so that tissue differentiation can occur directly and embryonic growth more rapidly. At its extreme, seen in eggs of some Diptera and Hymenoptera, the primordium occupies both ventral and lateral areas of the egg, with the extra-embryonic ectoderm covering only the dorsal surface (Figure 20.6C).

The shape of the primordium is varied, though in most insects the anterior region is expanded laterally as a pair of head lobes (= protocephalon), behind which is a region of varied length, the protocorm (postantennal region) (Figure 20.5). In eggs of Paleoptera, hemipteroid insects, and some orthopteroid species, the protocorm is semilong and at its formation includes the mouthpart-bearing segments, the thoracic segments, and a posterior growth region from which the abdominal segments arise. In eggs of other orthopteroid insects the postantennal region consists initially of only the growth zone. Though the protocorm in most endopterygote embryos is long, it also includes a posterior growth zone from which rudimentary abdominal segments proliferate. As the embryonic primordium elongates and begins to differentiate, it becomes known as the germ band. During elongation and differentiation, the abdomen grows around the posterior end and forward over the dorsal surface of the egg (Figure 20.7). In eggs of some higher endopterygotes (Hymenoptera-Apocrita and Diptera-Muscomorpha), there is no posterior growth zone and the abdominal segments arise directly from the primordium.

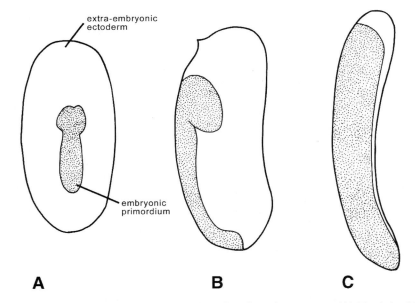

extra-embryonic
ectoderm

embryonic
primordium

A   B   C

FIGURE 20.6.   Form and position of embryonic primordium in endopterygotes. (A) *Tenebrio*; (B) *Sialis*; and (C) *Pimpla*. [After D. T. Anderson, 1972a,b, The development of hemimetabolous insects, and The development of holometabolous insects, in: *Developmental Systems: Insects*, Vol. I (S. J. Counce and C. H. Waddington, eds.). By permission of Academic Press Ltd., and the author.]

It is during the differentiation and elongation of the germ band that the primordial germ cells first become noticeable in most endopterygote eggs, though in those of some Coleoptera they are distinguishable even as the syncytial blastoderm is forming. They are largish, rounded cells in a distinct group at the posterior pole of the yolk, and accordingly are referred to as pole cells (Figure 20.3). In eggs of Dermaptera, Psocoptera, Thysanoptera, and homopterans also, the germ cells differentiate early at the posterior end of the primordium. In those of most exopterygotes, however, they are not apparent until gastrulation or somite formation has occurred.

As the germ band elongates and becomes broader, segmentation and limb-bud formation appear externally and are accompanied internally by mesoderm and somite formation. Growth of the germ band may occur either on the surface of the yolk (superficial growth) as seen in eggs of Dictyoptera, Dermaptera, Isoptera, some other orthopteroid insects, and all endopterygotes (Figure 20.7), or by immersion into the yolk (immersed growth) as occurs in eggs of Paleoptera, most Orthoptera, and hemipteroid insects (Figure 20.8). Immersion of the germ band (anatrepsis) forms the first of a series of embryonic movements, collectively known as blastokinesis. The reverse movement (katatrepsis), which brings the embryo back to the surface of the yolk, occurs later (see Section 6). Anatrepsis has developed secondarily (i.e., superficial growth is the more primitive method) and convergently among those exopterygotes in which it occurs. Its functional significance is, however, not clear (Anderson, 1972a; Heming, 2003).

4. Gastrulation, Somite Formation, and Segmentation

As the embryonic primordium begins to increase in length, its midventral cells sink inward to form a transient, longitudinal gastral groove (Figure 20.9A). The invaginated

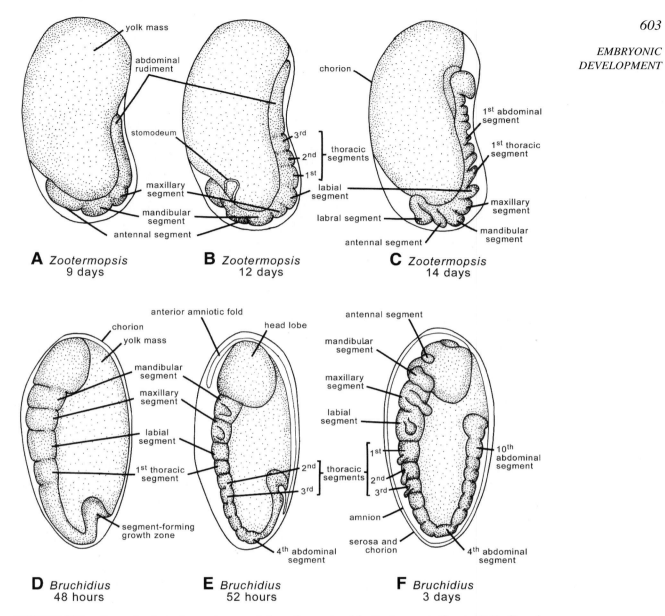

FIGURE 20.7. Stages in elongation and segmentation of germ band in *Zootermopsis* (Isoptera) (A–C) and *Bruchidius* (Coleoptera) (D–F). [After D. T. Anderson, 1972a,b, The development of hemimetabolous insects, and the development of holometabolous insects, in: *Developmental Systems: Insects*, Vol. I (S. J. Counce and C. H. Waddington, eds.). By permission of Academic Press Ltd., and the author.]

cells soon separate from the outer layer, which closes to obliterate the groove. It is from the anterior and posterior points of closure of the gastral groove that the stomodeum and proctodeum, respectively, develop. The outer layer can now be distinguished as the embryonic ectoderm. The invaginated cells, which proliferate and spread laterally, form the mesoderm (Figure 20.9B,C) except adjacent to the developing stomodeum and proctodeum where they become the anterior and posterior midgut rudiments, respectively. The mesodermal cells become concentrated into paired longitudinal tracts which soon separate into

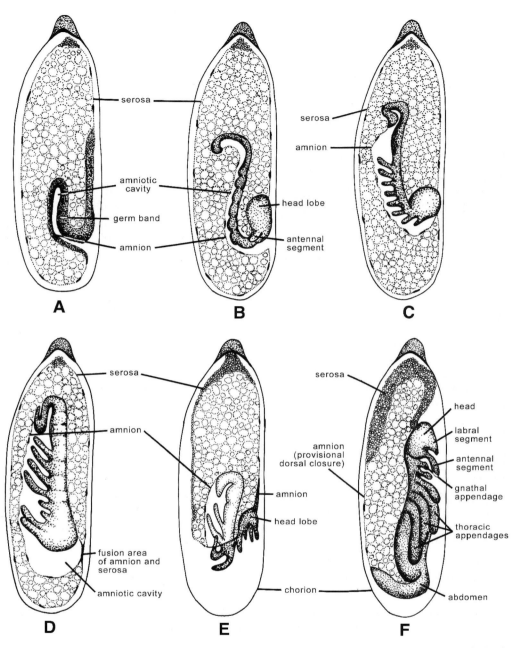

FIGURE 20.8. Early embryonic development in *Calopteryx* to show anatrepsis and katatrepsis. [A–E, after O. A. Johannsen and F. H. Butt, 1941, *Embryology of Insects and Myriapods.* By permission of McGraw-Hill Book Co., Inc. F, After R. F. Chapman, 1971, *The Insects: Structure and Function.* By permission of Elsevier/North-Holland, Inc., and the author.]

segmental blocks, leaving only a thin longitudinal strip, the median mesoderm, from which hemocytes later differentiate. From these segmental blocks, paired hollow somites usually arise (Figure 20.9E). Somite formation is initiated and occurs more or less simultaneously in the gnathal and thoracic segments, spreading anteriorly and posteriorly after gastrulation takes place. Formation of the coelom (the cavity within a somite) may occur in one of two ways, by internal splitting of a somite or by median folding of the lateral part of each somite.

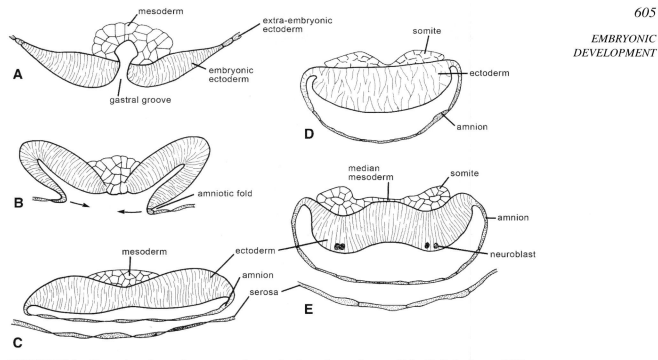

FIGURE 20.9.   Formation of gastral groove, somites, and embryonic membranes. [After D. T. Anderson, 1972a, The development of hemimetabolous insects, in *Developmental Systems: Insects*, Vol. I (S. J. Counce and C. H. Waddington, eds.). By permission of Academic Press Ltd., and the author.]

In embryos of a given species, one or both methods may be seen in different segments. For example, internal splitting of the somites occurs in all segments of embryos of Phasmida, most hemipteroid insects, and most endopterygotes, and in the abdominal segments of *Locusta* embryos. Median folding is the method used in all segments in embryos of Odonata, Dictyoptera, and Mallophaga, and in the gnathal and thoracic segments of those of *Locusta*, and some Coleoptera, Lepidoptera, and Megaloptera. In exopterygote embryos, all somites usually develop a central cavity, though this may be only temporary. Among endopterygotes, members of more primitive orders retain a full complement of somites in their embryos and the latter usually develop a coelom. In embryos of some species, however, cavities may not form, and somite formation may be suppressed in the head segments. In embryos of Diptera and Hymenoptera, no distinct head somites appear, and in those of some Muscomorpha and Apocrita, somite formation is entirely suppressed, so that mesodermal derivatives are produced directly from a single midventral mass.

5. Formation of Extra-Embryonic Membranes

Simultaneously with gastrulation and somite formation, two extra-embryonic membranes, the amnion and serosa, develop from the extra-embryonic ectoderm (Figure 20.9). Cells at the edge of the germ band proliferate and the tissue formed on each side folds ventrally to give rise to the amniotic folds. These meet and fuse in the ventral midline to form inner and outer membranes, the amnion and serosa, respectively, the former enclosing a central fluid-filled amniotic cavity. Many authors have suggested that such a cavity would provide space in which an embryo could grow and also prevent physical damage. Anderson (1972a) considered, however, that these functions are redundant and that the cavity must have an as yet unidentified function. Another possibility is that the amnion and its cavity are used to store wastes, which are thus kept separate from the yolk. The general method of amnion and serosa formation outlined above is found in all insect embryos (with some modification where immersion of the germ band into the yolk occurs) except those of Muscomorpha and Apocrita, in which, it will be recalled, the embryonic primordium covers most of the yolk surface. In these, embryonic membranes are greatly reduced or lost. In embryos of Apocrita the extra-embryonic ectoderm separates from the edge of the primordium and grows ventrally to form the serosa; that is, amniotic folds are not formed. In embryos of Muscomorpha neither an amnion nor a serosa forms, and the extra-embryonic ectoderm covers the yolk until definitive dorsal closure occurs (see below).

After the embryonic membranes form, the serosa in most insect eggs secretes a cuticle that is often as thick as the chorion. For several species, production of the serosal cuticle is closely synchronized with a peak of molting hormone in the egg (see Section 9).

6. Dorsal Closure and Katatrepsis

When germ band elongation and segmentation are complete, limb buds develop, the embryonic ectoderm grows dorsolaterally over the yolk mass, and internally organogenesis begins. This phase of growth is ended abruptly as the extra-embryonic membranes fuse and rupture and the germ band reverts to its original (pre-anatreptic) position (in most exopterygotes) or shortens (endopterygotes).

In embryos of most insects, the amnion and serosa fuse in the vicinity of the head, and the combined tissue then splits to expose the head and rolls back dorsally over the yolk (Figure 20.10A). As a result, the serosa is reduced to a small mass of cells, the secondary dorsal organ, and the amnion becomes stretched over the yolk, forming the provisional dorsal closure (Figure 20.10B). In some endopterygote embryos, variations of this process can be seen. In those of Nematocera (Diptera) and Symphyta (Hymenoptera), for example, it is the amnion that ruptures and is reduced, leaving the serosa intact. As noted above, in eggs of Apocrita only a serosa is formed, and this persists until definitive dorsal closure occurs, and in those of Muscomorpha no extra-embryonic membranes develop, and the yolk remains covered by the extra-embryonic ectoderm until definitive dorsal closure.

Except in dictyopteran embryos where the germ band remains superficial and ventral during elongation, extensive movement of the germ band now occurs in exopterygote eggs which serves (1) to bring an immersed germ band back to the surface of the yolk and (2) to restore the germ band to its pre-anatreptic orientation, that is, on the ventral surface of the yolk with the head end facing the anterior pole of the egg. This movement, the reverse of anatrepsis, is known as katatrepsis (Figure 20.8).

At the beginning of provisional dorsal closure, the germ band of most endopterygotes is quite long so that, although its anterior end is ventral, its posterior component passes round the posterior tip of the yolk and forward along the dorsal side (Figure 20.7F). During closure, the germ band shortens and broadens rapidly so that its posterior end now comes to lie near the posterior end of the egg (Figure 20.11A).

Definitive dorsal closure, that is, the enclosing of the yolk within the embryo, then occurs. It is achieved in all insect embryos by a lateral growth of the embryonic ectoderm, which gradually replaces the amnion or, rarely, the serosa (Figures 20.10C and 20.11B).

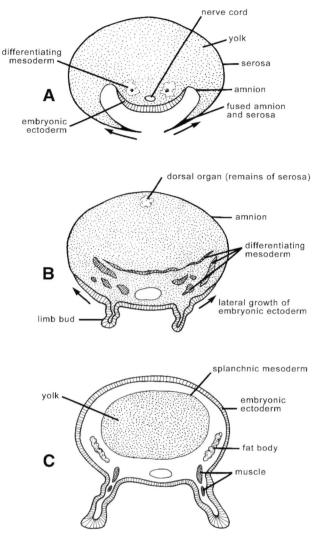

FIGURE 20.10. Diagrammatic representation of dorsal closure. (A) Initial fusion of amnion and serosa and beginning of rolling back; (B) provisional dorsal closure with amnion covering yolk; and (C) definitive dorsal closure with yolk enclosed within embryonic ectoderm.

7. Tissue and Organ Development

7.1. Appendages

Paired segmental evaginations of the embryonic ectoderm appear on the thoracic, antennal, and gnathal segments while the abdominal part of the germ band is still forming (see Figure 20.7). Their subsequent growth results from proliferation of the ectoderm as a single layer of cells and of mesodermal cells within. The cephalic and thoracic limbs ultimately differentiate into their specific form, except in eggs of secondarily apodous species where they soon shorten or become reduced to epidermal thickenings. In embryos of Muscomorpha, the thoracic appendages never develop beyond the epidermal thickening stage.

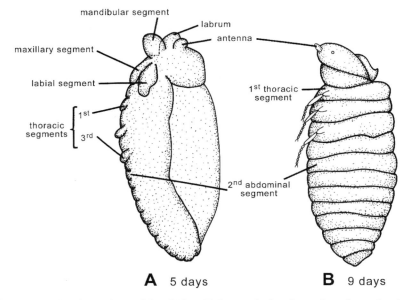

FIGURE 20.11. (A) Five-day embryo of *Bruchidius* (Coleoptera) after shortening of germ band. Compare this figure with Figure 20.7F; and (B) embryo of *Bruchidius* at hatching stage (9 days). [After D. T. Anderson, 1972b, The development of holometabolous insects in: *Developmental Systems: Insects*, Vol. I (S. J. Counce and C. H. Waddington, eds.). By permission of Academic Press Ltd., and the author.]

In Paleoptera and orthopteroid insects, 11 pairs of abdominal appendages evaginate before provisional dorsal closure. In most hemipteroid embryos, no sign of abdominal limbs is evident, though in those of Hemiptera and Thysanoptera appendages develop on the first and last abdominal segments. Ten pairs of abdominal evaginations develop in most endopterygote embryos. The fate of the abdominal appendages varies, and some or all of them may disappear before embryonic development is completed. The first (most anterior) pair disappears after blastokinesis in embryos of Paleoptera and some orthopteroid insects, but remains as glandular pleuropodia in those of Dictyoptera, Phasmida, Orthoptera, Hemiptera, and some Coleoptera and Lepidoptera. The function of the pleuropodia is uncertain, though some authors have suggested that in orthopteran embryos they secrete chitinase that brings about dissolution of the serosal cuticle. The pleuropodia are resorbed or discarded before or shortly after hatching. The appendages of the second through seventh abdominal segments are resorbed, except in some endopterygotes where they persist as larval prolegs. Pairs 8–10 may differentiate into the external genitalia or disappear, while the last pair either persists as cerci or disappears.

7.2. Integument and Ectodermal Derivatives

Soon after definitive dorsal closure, the outer embryonic ectoderm differentiates into epidermis, which in embryos of most insects then secretes the first instar larval cuticle. In some insects, however, one or more embryonic cuticles are produced which may be shed before or at hatching. As in larvae, production of the cuticles in embryos appears to be regulated by molting hormone (Section 9).

External sensilla generally develop from a dividing precursor epidermal cell, whose daughter cells then differentiate to form the sensory neuron and accessory cells (Chapter 12). The axon of the sensory neuron finds the appropriate interneurons within the central nervous system by growing along the surface of pioneer neurons (Section 7.3) or neurons of previously formed sensilla (Heming, 2003). Similarly, both compound and simple eyes develop from groups of epidermal cells, each of which divides and differentiates to form the photoreceptive and accessory components of the light-sensitive structures.

Imaginal discs and histoblasts, from which many adult tissues are derived at metamorphosis in higher Diptera and Hymenoptera (Chapter 21, Section 4.2), can be recognized soon after germ-band formation. They are groups of cells that separate from the ectoderm in characteristic numbers, sizes and shapes, at specific sites in the body (Heming, 2003).

Concurrently with the formation of abdominal appendages a number of ectodermal invaginations develop, from which differentiate endoskeletal components, various glands, the tracheal system, and certain parts of the reproductive tract (for the latter, see Section 7.6). From ventrolateral invaginations at the junctions of the antennal/mandibular segments and the mandibular/maxillary segments are derived the anterior and posterior arms of the tentorium. Paired mandibular apodemes differentiate from invaginations near the bases of the mandibles. The apodemes of the trunk region arise from intersegmental invaginations in the thorax and abdomen.

Salivary glands develop from a pair of invaginations near the bases of the labial appendages. When the appendages fuse the invaginations merge to form a common salivary duct that opens midventrally on the hypopharynx.

The corpora allata develop from a pair of ventrolateral invaginations at the junction of the mandibular/maxillary segments. Initially, they exist as hollow vesicles, though these fill in as they move dorsally to their final position adjacent to the stomodeum. The molt glands also originate as paired ventral ectodermal invaginations, usually on the prothoracic segment. Although the endocrine glands arise before katatrepsis, at this time they are nonsecretory, and maternally derived hormones (especially ecdysteroids) stored in the egg are used in embryonic endocrine regulation (Section 9). Other invaginations on the head may give rise to specialized exocrine glands on the mandibles or maxillae.

Elements of the tracheal system can be seen first as paired lateral invaginations on each segment from the second thoracic to the eighth (ninth in a few Thysanura) abdominal. However, not all of these invaginations develop completely into tracheae. Those that do, bifurcate and anastomose with branches from adjacent segments and from their opposite partner of the same segment. The cells differentiate as tracheal epithelium and then secrete a cuticular lining. After cuticle secretion but before hatching, gas is secreted into the tracheal system. Some of the invaginated ectodermal cells differentiate into oenocytes. These may remain closely associated with the tracheal system, form definite clusters in specific body regions, or become embedded as single cells in the fat body.

7.3. Central Nervous System

Soon after somite formation has commenced, specialized ectodermal cells on each side of the midventral line, the neuroblasts (Figure 20.9E), begin to proliferate, resulting in the formation of paired longitudinal neural ridges separated by a neural groove. As proliferation occurs, the cells move slightly inward so that they become separated from the ectoderm. They then begin to divide vertically, unequally and repeatedly, the small daughter cells eventually developing into ganglion cells from which both neurons and glial cells differentiate (Figure 20.12). Remarkably, the number and arrangement of neuroblasts is highly conserved across the Insecta: each half-ganglion has 30 or 31 neuroblasts from which all neurons are produced (Thomas et al., 1984). However, the number of neurons in

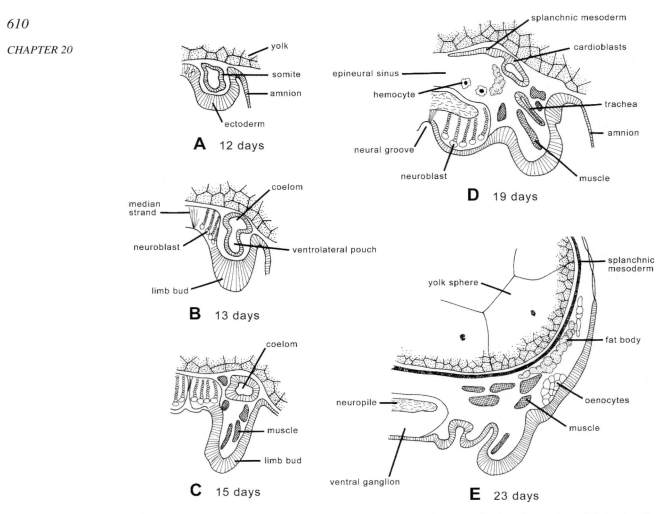

FIGURE 20.12. Transverse sections to show development of nervous tissue and mesodermal derivatives in prothoracic segment of *Tachycines* (Orthoptera). In A–D, which are prekatatreptic stages, the serosa is omitted. [After D. T. Anderson, 1972a, The development of hemimetabolous insects, in: *Developmental Systems: Insects*, Vol. I (S. J. Counce and C. H. Waddington, eds.). By permission of Academic Press Ltd., and the author.]

a ganglion can be widely varied; for example, in a thoracic ganglion of an aptergygote there are about 1500 neurons, whereas in the thoracic ganglia of grasshoppers and flies there are about 3000 and 4000 neurons, respectively, reflecting the increased demands on the nervous system associated with the evolution of flight (Truman and Ball, 1998).

With the onset of segmentation, neuroblasts in the intersegmental regions become less active, so that paired segmental swellings, the future ganglia, now become apparent. As the neurons form, their cell bodies become arranged peripherally around the central axis (neuropile). Subsequent growth of the axons leads to formation of longitudinal connectives, transverse commissures, and to motor nerves innervating a variety of effector organs. As noted above, sensory nerves form as a result of the inward growth of axons from peripheral sense cells.

An important aspect of the development of the nervous system is how developing neurons find other neurons so that the correct connections are made within the insect's body. The first neurons to arise (from the central nervous system) are the central pioneer neurons. Their axons grow along predetermined paths when their tips recognize particular cues on cell surfaces within the embryo. Thus, a scaffold of axonal pathways is initially erected, which later-developing axons are then able to follow, being guided by specific recognition signals on the surfaces of the pioneer neurons (Goodman, 1984; Goodman and Bastiani, 1984). Peripheral pioneer neurons, originating from ectoderm at the tips of appendages, have also been identified. These grow toward the central nervous system, serving later as guides for motor axons innervating effectors, especially muscles, and the sensory axons from integumental sensilla. The peripheral pioneer axons die when the necessary connections have been established (Heming, 2003).

As embryogenesis continues, fusion of ganglia occurs in the head region to form the brain and subesophageal ganglion and at the posterior end of the abdomen where ganglia from segments 8–11 form a composite structure. In embryos of species belonging to different orders of Insecta, varying degrees of fusion of other ganglia may subsequently occur.

7.4. Gut and Derivatives

As noted earlier, the stomodeum and proctodeum arise at the anterior and posterior ends of the gastral groove, respectively. Both develop as hollow invaginated tubes, the stomodeum slightly earlier than the proctodeum, and as they differentiate into the subdivisions of the foregut and hindgut, respectively, various associated structures arise.

On the roof of the stomodeum, evaginations or thickenings give rise to the frontal ganglion, hypocerebral ganglion, intrinsic cells of the corpora cardiaca, and ingluvial ganglia. At its anterior end, the proctodeum develops pouches, which are the rudiments of the Malpighian tubules. (In many insects additional tubules develop during larval life.)

The anterior and posterior midgut rudiments that appeared at gastrulation begin to proliferate at about the time of provisional dorsal closure to form the midgut. Each rudiment proliferates a pair of strands that grow caudad or cephalad, respectively, between the nerve cord and yolk. After strands from each rudiment meet in the middle of the embryo, they grow laterally and dorsally so as to eventually enclose the yolk mass. The cells then differentiate as midgut epithelium. Prior to this, in embryos of some species, the vitellophages form a temporary "yolk sac" around the yolk mass.

7.5. Circulatory System, Muscle, and Fat Body

The heart, aorta, musculature, fat body, lining of the hemocoel, and some components of the reproductive system are derived from the somites and median mesoderm formed after gastrulation. As noted earlier, the median mesoderm gives rise to hemocytes. In most embryos, each somite becomes hollow and forms three interconnected chambers, the anterior, posterior, and ventrolateral pouches. The latter grows into the adjacent ectodermal limb bud and breaks up to form intrinsic limb muscles (Figure 20.12). The splanchnic walls (i.e., those facing the yolk) of the two remaining pouches spread round the gut, forming the gut musculature, some fat body, and part of the reproductive system (Section 7.6). The somatic walls (those facing the ectoderm) of the anterior and posterior pouches give rise to extrinsic limb muscles, dorsal and ventral longitudinal muscles, and more fat body. For each muscle in an insect there is a single founder (pioneer) cell that serves as a base to which other progenitor muscle cells (myoblasts) become attached. As each muscle develops, it grows toward a specific epidermal insertion site (Chapter 14, Section 2.1).

The breaking up of the somite walls into discrete tissues means that in insects as in other arthropods there is no true coelom. Rather, the latter merges with the epineural sinus (the space between the dorsal surface of the embryo and the yolk) and is correctly called a mixocoel (hemocoel). From mesodermal cells at the dorsal junction of the somatic and splanchnic walls of the labial to the tenth abdominal somites, a sheet of cardioblasts develops. As the mesoderm grows around the gut, the sheets on each side become apposed to form the heart. Other somatic mesoderm cells adjacent to the cardioblasts differentiate as alary muscles, pericardial septum, and pericardial cells. The aorta develops from the median walls of the antennal somites, which become apposed and grow posteriorly to meet the heart.

In embryos of insects where the somites remain solid or are not formed as discrete segmental structures, the mesoderm still gives rise to the same components.

7.6. Reproductive System

The reproductive system includes both mesodermal and ectodermal components. In female exopterygotes, the vagina and spermatheca develop after hatching as midventral ectodermal invaginations of the seventh or eight abdominal segment. In males, the ejaculatory duct and ectadenes (ectodermal accessory glands) are formed from a similar midventral invagination of the ectoderm of the ninth or tenth abdominal segment.

The paired genital ducts and mesadenes (mesodermal accessory glands) arise in exopterygotes from mesoderm of the splanchnic walls of certain abdominal somites which first thickens than hollows out to form coelomoducts. Some of these soon disappear, but those of the seventh and eighth somites (in females) or ninth and tenth somites (in males) enlarge to form the ducts and/or accessory gland components. In endopterygotes, the genital ducts are formed during postembryonic development. In *Drosophila* and other muscomorph Diptera the reproductive system (excluding the gonads) develops from a single or pair of imaginal discs during metamorphosis (Chapter 21, Section 4.2).

Development of the gonads varies, though two related trends can be seen, namely, earlier segregation of the primordial germ cells and restriction of these cells to fewer abdominal segments. In the most primitive arrangement, seen in some thysanuran and orthopteran embryos, the germ cells do not become distinguishable until they appear in the splanchnic walls of several abdominal somites. In *Locusta* embryos, for example, they are found initially in the somites of abdominal segments 2–10, though they remain only in segments 3–6. Eventually they fuse longitudinally to form a compact gonad on each side. Such a segmental arrangement is presumably primitive as it is seen also in adult Annelida, Onychophora, Myriapoda, and non-insectan apterygotes (Anderson, 1972a).

In embryos of Dictyoptera, Phasmida, Embioptera, and Heteroptera the germ cells become apparent early in gastrulation. Nevertheless, they still become associated with the splanchnic mesoderm of several anterior abdominal segments.

In embryos of Dermaptera, Psocoptera, Thysanoptera, homopterans, and endopterygotes, the germ cells differentiate as the blastoderm forms (see Figure 20.3B). After somite formation, they migrate to abdominal segments 3 and 4 in exopterygote embryos or abdominal segments 5 and 6 in endopterygote embryos where they divide into left and right halves and become surrounded by splanchnic mesoderm.

8. Special Forms of Embryonic Development

The great majority of insect species are bisexual and females lay eggs that contain a considerable amount of yolk. However, in some species, males may be rare and females may produce viable offspring from unfertilized eggs (parthenogenesis). In another form of asexual reproduction, polyembryony, which is characteristic of some parasitic Hymenoptera and Strepsiptera, several embryos develop from one fertilized egg. In other insects, fertilized eggs may be retained within the female reproductive tract for varied periods of time so that a young insect may hatch from the egg almost as soon as or even before the latter is laid (viviparity). In a few species paedogenesis may occur where mature larvae are able to produce, parthenogenetically and usually viviparously, a further generation of young.

8.1. Parthenogenesis

The ability of unfertilized eggs to develop is common to many insect species and in some is the normal mode of reproduction under certain conditions. In all insects except Lepidoptera and Trichoptera, the female is the homogametic sex (i.e., having two X sex chromosomes) and the male, heterogametic (XY or XO). Unfertilized eggs, therefore, will contain only X chromosomes. However, whether they contain one or two such chromosomes and, therefore, the sex of parthenogenetic offspring, depends on the behavior of the chromosomes during meiosis in the oocyte nucleus (Suomalainen, 1962; White, 1973).

Two forms of female-producing parthenogenesis (thelytoky) are known. In some species no meiotic division occurs during oogenesis. Therefore, offspring are diploid and female (ameiotic or apomictic parthenogenesis) and will have the same genetic makeup as the mother, unless mutation or insertion of transposable elements occurs (Heming, 2003). In meiotic (automictic) parthenogenesis, the typical reduction division is followed by nuclear fusion so that a diploid chromosome complement is retained. Again, therefore, the offspring are female but they have a different genetic make up from their mother.

Haploid parthenogenesis (arrhenotoky), where the oocyte nucleus undergoes meiosis that is not followed by nuclear fusion, is of relatively rare occurrence, though typical of Hymenoptera, Thysanoptera, and some homopterans and Coleoptera. It results in the production of males. In Hymenoptera, haploid parthenogenesis is facultative; that is, a female determines whether or not an egg will be fertilized. In the honey bee, for example, a queen normally lays fertilized eggs that develop into workers (diploid females). However, under certain conditions, for example, when the hive is crowded, and the workers construct larger than normal drone cells on the honeycomb, she will lay unfertilized eggs from which haploid males develop, as a preliminary to swarming.

Parthenogenesis, producing in most species female offspring, may confer two advantages. In a species whose population density may be (temporarily) low, the ability of an isolated female to reproduce parthenogenetically may ensure survival of her genotype until the population density increases and males are again likely to be encountered. More often, however, parthenogenesis is employed as a mechanism that provides a rapid mode of reproduction, to enable a species to take full advantage of temporarily ideal conditions. Thus, a parthenogenetic female, who does not require to locate, or be located by, a male can devote her time and energy to egg production. Further, all her offspring are female, so that her maximum reproductive potential can be realized. The disadvantage of parthenogenesis is that the genotype of successive generations remains more or less constant so that adaptation of a species to changing environmental conditions is very slow. To counteract this, many species alternate one or more parthenogenetic generations with a normal sexual generation. Aphids, for example, reproduce for most of the year by ameiotic parthenogenesis (Figure 8.8). However, toward fall (and affected by changing environmental conditions) there occurs, during maturation of some oocytes, a separation of the two X chromosomes, one of which migrates to the polar body and is destroyed. From such eggs (with an XO constitution) males will develop. As spermatogenesis occurs in these individuals, spermatocytes containing either one or no X chromosome are produced. However, the latter do not mature, so that only sperm with an X chromosome result. Therefore, the overwintering eggs produced as a result of mating will have an XX sex chromosome complement and give rise the following spring only to females.

8.2. Polyembryony

Polyembryony, the development of more than one embryo from one egg, is known to be a normal occurrence in about 30 species of parasitic Hymenoptera (mostly Encyrtidae, Platygasteridae, and Braconidae) and one species of Strepsiptera (Ivanova-Kasas, 1972). In these insects it is always associated with either parasitism or viviparity and is presumed to have evolved in conjunction with the abundance of food offered by these two modes of life. Characteristically, the eggs of polyembryonic species are minute and devoid of yolk. Because they depend on an external (host or maternal) source of nutrients the chorion, which is initially thin and permeable, soon disappears. Further, in Hymenoptera, the serosa becomes modified for the uptake of nutrients and is known as a "trophamnion."

Both the number of embryos formed and the point in development at which they become discernible vary. In *Platygaster hiemalis*, a parasite of Hessian fly larvae, for example, at the four-cell stage, the cells may separate into two groups so that twin embryos are formed. In contrast, in the chalcidid *Litomastix truncatellus*, which parasitizes larvae of the moth genus *Plusia*, formation of embryos does not begin until the 220- to 225-blastomere stage. At this stage, certain of the blastomeres become spindle-shaped and fuse to form a syncytial sheath that divides the remaining blastomeres into groups, the primary embryonic masses. In due course, secondary, tertiary, etc., embryonic masses form so that the final number of potential embryos may exceed 1000. The early development of *Litomastix* is summarized in Figure 20.13. Eventually the polygerm (the total embryonic mass within the trophamnion) disintegrates, and each embryo develops into a larva. The larvae feed within a host until all the usable parts are consumed and then pupate. At this point the host is nothing more than a cuticular bag full of parasites (Figure 20.14).

8.3. Viviparity

Viviparity, the retention of developing offspring within the maternal genital tract, is found in a range of complexity within the Insecta. It is seen in different forms in species from several orders, but among Diptera the entire range of variation may occur.

In its simplest form (ovoviviparity) the eggs retain their full complement of yolk for nourishment of the embryo. The eggs may be retained within the mother for a varied period of time but usually are laid just before they hatch. Such an arrangement is seen in many Tachinidae (Figure 20.15) where the first-instar larvae actually escape from the chorion during oviposition. Associated with retention of the eggs, which presumably affords them greater protection, is a trend toward production of fewer of them. As fewer eggs are produced each can acquire more yolk so that larvae can hatch at more advanced stages of development. For example, many Sarcophagidae (flesh flies) produce only 40–80 eggs but are larviparous; that is, larvae hatch from the eggs while the latter are still within the reproductive tract. At the extreme, the number of eggs that mature simultaneously is reduced to one, as, for example, in *Hylemya strigosa* (Anthomyiidae) and *Termitoxenia* sp. (Phoridae). In *Hylemya*

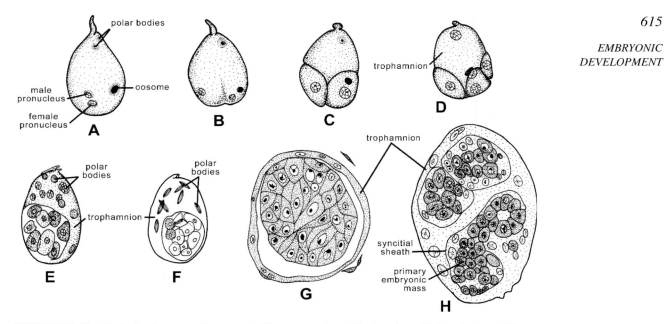

FIGURE 20.13. Early development of *Litomastix* (Hymenoptera). (A) Fertilization; (B) first cleavage; (C) two-cell stage; (D–F) next stages; (G) formation of spindle cells; and (H) formation of secondary embryonic masses. [After O. M. Ivanova-Kasas, 1972, Polyembryony in insects, in: *Developmental Systems: Insects*, Vol. I (S. J. Counce and C. H. Waddington, eds.). By permission of Academic Press Ltd., and the author.]

the larva that emerges from a newly laid egg molts immediately to the second instar; in *Termitoxenia*, whose egg is relatively larger, it is a third-instar larva that emerges from an egg and it pupates within a few minutes.

In truly viviparous species, developing offspring obtain their food from the mother. Accordingly, the structures of the maternal reproductive system and egg are modified to facilitate this exchange. As in ovoviviparity, the trend is toward reduction of the number of embryos being developed simultaneously.

Some aphids, Psocoptera, and Dermaptera (*Hemimerus*) show pseudoplacental viviparity (Hagan, 1951). Eggs of these insects contain little or no yolk and lack a chorion. They develop within the ovariole, where the follicle cells supposedly supply at least some nourishment to the embryo. (In species with meroistic ovarioles, the nurse cells are also important). In *Hemimerus*, for example, follicle cells adjacent to the anterior and posterior ends of an oocyte proliferate and become connected with the embryonic membranes forming pseudoplacentae. Later, the follicle cells degenerate because, it is assumed, they are supplying nutrients to the developing embryo (Figure 20.16).

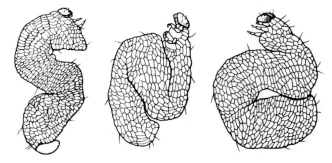

FIGURE 20.14. Caterpillars parasitized by *Litomastix*. [From R. R. Askew, 1971, *Parasitic Insects*. By permission of Heinemann Educational Books Ltd.]

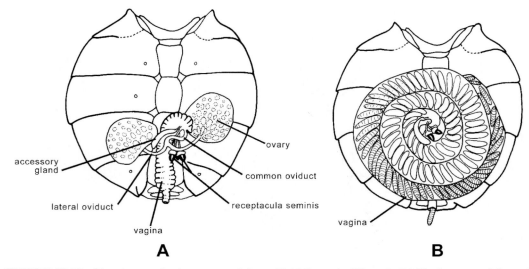

FIGURE 20.15. Female reproductive system of the tachinid *Panzeria* (Diptera). (A) Newly emerged fly; and (B) mature female, with greatly enlarged vagina forming a brood chamber. An egg containing a fully formed embryo is being laid, [After V. B. Wigglesworth, 1965, *The Principles of Insect Physiology*, 6th ed., Methuen and Co. By permission of the author.]

In *Glossina* spp. and pupiparous Diptera adenotrophic viviparity occurs. In this arrangement, an egg is normal, that is, contains yolk and possesses a chorion, yet is retained within the expanded bursa, the so-called uterus. Embryonic development is, therefore, correctly described as ovoviviparous. However, after hatching, the larva remains within the uterus and feeds on secretions (uterine milk) of the enormous accessory glands that ramify through the abdomen (Figure 20.17). One larva at a time develops and pupation occurs shortly after birth.

In hemocoelic viviparity, used by Strepsiptera and some paedogenetic Cecidomyiidae (Diptera), oocytes are released from the ovarioles into the maternal hemocoel. In Strepsiptera fertilization occurs within the maternal body cavity, and, during embryonic development,

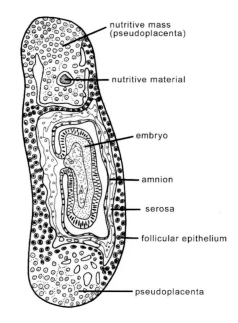

FIGURE 20.16. Longitudinal section through ovarian follicle of *Hemimerus* (Dermaptera) to show pseudoplacentae. (After V. B. Wigglesworth, 1965, *The Principles of Insect Physiology*, 6th ed., Methuen and Co. By permission of the author.]

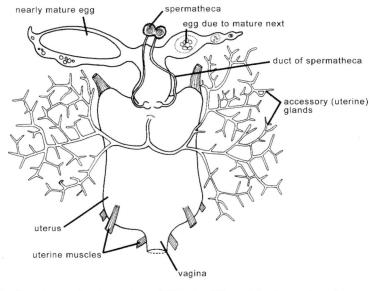

FIGURE 20.17. Female reproductive system of *Glossina* (Diptera) to show enlarged accessory glands. Note also that only one egg at a time is maturing. [After V. B. Wigglesworth, 1965. *The Principles of Insect Physiology*, 6th ed., Methuen and Co. By permission of the author.]

nutrients are absorbed directly from the hemolymph. After hatching, the larvae escape from the female's body via the genital pores. In some cecidomyiids, oocytes develop parthenogenetically. Initially, development occurs within the ovarioles, but the larvae on hatching escape into the hemocoel. The larvae remain within the mother and feed on her tissues until just prior to pupation when they exit via the body wall.

8.4. Paedogenesis

Though not actually a form of embryonic development, paedogenesis, that is, precocious sexual maturation of juvenile stages, is conveniently mentioned here. Paedogenesis is usually associated with both parthenogenesis and viviparity, and, probably, is best shown in certain Cecidomyiidae, though it is known to occur also in some Chironomidae (Diptera), Coleoptera, and Hemiptera. In some cecidomyiids, the oocytes develop viviparously in the hemocoel of the last larval or pupal instar. In some chironomids, embryonic development begins in female pupae. In the viviparous hemipteran *Hesperoctenes* female larvae that have been inseminated hemocoelically develop embryos within their ovarioles. In an extreme situation seen in some aphids, development of young may begin in the mother while she herself is still in here own mother's reproductive system!

9. Factors Affecting Embryonic Development

Temperature is probably the single most important environmental variable affecting embryonic development. For eggs of most species, there are upper and lower temperature limits, outside which development is greatly retarded or completely inhibited. Within these limits, however, an inverse but linear relationship exists between temperature and time required to complete development; that is, the total heat requirement (temperature above minimum required X duration of exposure to this temperature) is constant for a given species. This heat requirement is typically measured in degree-days. Outside these developmental limits, yet within the limits of viability, an egg may survive but does not develop. Under these conditions, it is said to be quiescent and in this state may survive for a considerable length of time. In a quiescent state, an egg is always ready to take advantage of favorable conditions, even if only temporary, to continue its development. However, quiescence is a relatively sensitive developmental state; that is, outside certain temperature limits, an egg will be killed. For many species, therefore, which exist in habitats exposed to climatic extremes, especially of temperature but also of precipitation, a more resistant state of developmental arrest, diapause, has evolved to permit their survival. Diapause is discussed at greater length in Chapter 22 (Section 3.2.3), though in the present context it is worth noting that diapause may occur at different stages of embryonic development and with varied strength in different species. In all instances, however, it is characterized by a cessation of morphogenesis and a considerable lowering of the metabolic rate. Also, the water content of an egg is often low at this time. In *Bombyx*, diapause, which begins in overwintering eggs almost as soon as they are laid, is extremely strong; that is, even when eggs are experimentally maintained at 15°C to 20°C from the time of laying they will not develop. Development begins only after they have been exposed to a temperature of about 0°C for several months. In eggs of the damselfly *Lestes congener*, diapause is also strong but does not commence until after anatrepsis. Diapause in eggs of the grasshopper *Melanoplus differentialis* also occurs after anatrepsis but is weak. Should the temperature to which eggs are exposed be maintained at summer levels (around 25°C), some of the eggs will develop directly, though more slowly than those that have undergone chilling. In eggs of some insects, for example, certain mosquitoes *(Aedes* spp.) and the damselfly *Lestes disjunctus*, embryonic development is almost completed before diapause is initiated.

Through an effect on parthenogenesis, temperature may also affect the sex ratio of the offspring. For example, in Hymenoptera higher temperatures often favor production of haploid males. In some bisexual species, extreme high or low temperatures may disrupt the normal sex chromosome distribution that occurs during meiosis, so that a preponderance of males or females results.

Water is another important requirement and in eggs of many species must be acquired from the external environment before embryonic development can begin. When it is available to an egg in insufficient quantity, the embryo becomes quiescent or remains in diapause (though this was not induced by the lack of moisture). Some species can obtain sufficient water from moisture in the air. For example, eggs of the beetle *Sitona*, when kept at 20°C and 100% relative humidity, hatch in 10.5 days; at the same temperature but only 62% relative humidity, development takes twice as long. In other species contact of the egg with liquid water is necessary for continued development. Such is the case in the damselfly eggs mentioned above which pass the winter in snow-covered, dried-out *Scirpus* stems and do not continue their development until the stems become waterlogged following the spring thaw.

Though the presence of ecdysonelike molecules and juvenile hormone was first reported some 40 years ago, their sources and roles in embryonic development are only slowly being clarified. As noted above, the source of these hormonal factors, at least until after katatrepsis, is maternal; that is the compounds are deposited as conjugates within the egg prior to oviposition. For species such as locusts and grasshoppers that produce several embryonic cuticles, there is a clear correlation between peaks of free ecdysone and bouts of cuticle synthesis. In other species the onset and termination of embryonic diapause is associated with changes in free ecdysone levels. Juvenile hormone seems to be involved in embryonic development in a manner similar to that in larvae, namely, the qualitative expression of cuticle structure; however, little experimental evidence supports this conjecture. Some studies have identified juvenile hormone in quite early stages of embryogenesis (i.e., well before cuticulogenesis) suggesting that it has other roles. One such role may be the regulation of ectodermal growth leading to dorsal closure. Though neurosecretory cells and corpora cardiaca have been identified in descriptions of embryonic development, and specific products have been assayed in a few species, their roles in embryogenesis remain unknown.

10. Hatching

To escape from the egg, a larva must break through the various membranes that surround it. These include the chorion, vitelline membrane, and, in eggs of some species, serosal cuticle. Further, in many exopterygotes and some endopterygotes a newly hatched larva is surrounded by embryonic cuticle that also must be shed before the insect is truly free.

The general mechanism of hatching is as follows. An insect first swallows amniotic fluid,* followed usually by air, or water, which diffuses into the egg. The abdomen is then contracted to force hemolymph into the head and thorax, which enlarge and cause the egg membranes to rupture. To facilitate rupture the chorion may have predetermined lines of weakness that run longitudinally or transversely, the latter separating an anterior egg cap from the more posterior portion of the egg. In many species, egg bursters, in the form of hard cuticular spines or plates, or thin eversible bladders, may develop on the head, thorax, or abdomen. In Acrididae and those Hemiptera in which pleuropodia develop, it is believed that these glands secrete chitinase that dissolves the serosal cuticle as an aid to hatching. Larvae of Lepidoptera simply eat their way out of the egg.

Where an embryonic cuticle is present, this may be shed concurrently with the other enclosing membranes, or may ensheath a larva until it has completely escaped from the egg, as in Odonata, Orthoptera, and some Hemiptera. In these insects the embryonic cuticle underwent apolysis some time prior to hatching, and the first-instar larval cuticle is already formed beneath. Thus, the insect hatches as a pharate first-instar larva. In Orthoptera and endophytic Odonata, the embryonic cuticle presumably protects a larva until it reaches the surface of the substrate in which the egg was laid. In other species, however, its function is unclear. It is shed a few minutes after a larva has reached the surface, a process called the intermediate molt.

11. Summary

Most insects are oviparous and therefore lay eggs that contain much yolk. However, there is an evolutionary trend toward reduction of the yolk:cytoplasm ratio to permit more rapid embryonic development.

Cleavage begins at a predetermined site, the cleavage center, and early divisions are synchronous. Most energids migrate through the yolk to the periplasm and form the blastoderm;

---

* This fluid is no longer in the amniotic cavity whose membranes were destroyed during dorsal closure.

some remain in the yolk as vitellophages that supply nutrients to the embryo. Posteriorly moving energids receive a signal at the activation center, which stimulates differentiation of part of the blastoderm into the embryonic primordium. Differentiation begins at a predetermined site, the differentiation (commitment) center, located in the region of the future prothorax. The embryonic primordium of exopterygotes is usually small and grows by aggregation and proliferation of cells, whereas that of most endopterygotes is large to permit rapid tissue differentiation and embryonic growth. Elongation and differentiation of the primordium (now known as the germ band) occur, and externally segmentation and appendage formation are obvious; internally somites form and mesoderm differentiates. Simultaneously, in embryos of most species the amnion and serosa develop from proliferating extra-embryonic cells at the margins of the germ band. Anatrepsis, movement of the germ band into the yolk core, occurs in eggs of most exopterygotes at this time.

At the end of germ band formation the amnion and serosa fuse, then break in the head region, and the combination rolls back dorsally over the yolk which is left covered by only the amnion (provisional dorsal closure). Katatrepsis now takes place in eggs with immersed germ bands, so that the embryo is returned to the yolk surface with its head facing the anterior pole of the egg. Embryonic ectoderm now extends around the yolk to replace the amnion (definitive dorsal closure).

Paired segmental appendages develop from evaginations of the embryonic ectoderm but may become reduced or disappear. Shortly after definitive dorsal closure, the embryonic ectoderm differentiates into epidermis and secretes a cuticle. Specific epidermal cells differentiate into external sensilla and eyes, and in some species form imaginal discs and histoblasts. Invaginations of the ectoderm give rise to the endoskeleton, tracheal system, salivary glands, corpora allata, molt glands, exocrine glands, and, in females, the vagina and spermatheca, in males, the ejaculatory duct and ectadenes. The foregut and hindgut develop from ectodermal invaginations at the anterior and posterior ends, respectively, of the gastral groove. The midgut is formed from anterior and posterior midgut rudiments that grow toward each other and on meeting extend dorsolaterally to enclose the yolk. The central nervous system arises from neuroblasts in the midventral line. The stomatogastric nervous system develops from evaginations in the roof of the stomodeum.

The heart, aorta, septa, muscle, fat body, paired genital ducts, and mesadenes are mesodermal derivatives. Gonads arise from primordial germ cells that become enclosed in mesoderm.

Parthenogenesis, the development of unfertilized eggs, may be ameiotic (no meiosis in oocyte nucleus) or meiotic (meiosis is followed by nuclear fusion), both of which result in diploid (female) offspring, or haploid (meiosis is not followed by nuclear fusion) from which males arise.

Polyembryony, the formation of more than one embryo in a single, small, yolkless egg, is restricted to a few parasitic or viviparous Hymenoptera and Strepsiptera.

Viviparity occurs in several forms. Ovoviviparity is retention of yolky eggs in the genital tract. In true viviparity developing offspring receive their nourishment directly from the mother. In pseudoplacental viviparity the follicle cells and embryonic membranes become closely apposed, and nourishment appears to be derived largely from the degeneration of follicle cells and from trophocytes. In adenotrophic viviparity, eggs are yolky, but larvae are retained in the uterus and feed on secretions of the accessory glands. Hemocoelic viviparity is where embryos receive nutrients directly from the hemolymph.

Paedogenesis is precocious sexual maturation of juvenile stages and is normally associated with parthenogenesis and viviparity.

Within species-specific limits the rate of embryonic development is inversely related to temperature. Outside these limits, an embryo may survive but not develop; that is, it is quiescent. Survival of an embryo at extreme temperatures may be achieved through diapause. Eggs of many species must take up water from the environment before embryonic development can begin. Ecdysone and juvenile hormone are involved in the regulation of embryogenesis though their precise roles remain unknown. In respect of embryonic cuticle formation they appear to work as they do in postembryonic development.

At hatching, hemolymph is forced into the head and thorax as a result of abdominal muscle contraction. As the anterior end of the embryo increases in volume, the chorion is split. Hatching may be facilitated by lines of weakness in the chorion, by egg bursters or eversible bladders on the head or thorax, or by secretion of pleuropodial chitinase that dissolves the serosal cuticle.

12. Literature

Johannsen and Butt (1941), Anderson (1972a,b, 1973), Jura (1972), Counce (1973), Sander *et al.* (1985), and Heming (2003) give general descriptions of insect embryogenesis. Works dealing with specific aspects of embryonic development include those by Hagan (1951) and Retnakaran and Percy (1985) [viviparous insects], Ivanova-Kasas (1972) and Retnakaran and Percy (1985) [polyembryony), White (1973) [parthenogenesis and sex determination], Matsuda (1976) [embryogenesis of abdomen, gonads, and germ cells], and Hoffmann and Lagueux (1985) [endocrine aspects]. Lawrence (1976, 1992) and Sander (1984, 1997) review experimental embryogenesis, especially in relation to pattern formation.

Anderson, D. T., 1972a, The development of hemimetabolous insects, in: *Developmental Systems: Insects*, Vol. 1 (S. J. Counce and C. H. Waddington, eds.), Academic Press, New York.

Anderson, D. T., 1972b, The development of holometabolous insects, in: *Developmental Systems: Insects*, Vol. 1 (S. J. Counce and C. H. Waddington, eds.), Academic Press, New York.

Anderson, D. T., 1973, *Embryology and Phylogeny in Annelids and Arthropods*, Pergamon Press, Elmsford, NY.

Counce, S. J., 1973, The causal analysis of insect embryogenesis, in: *Developmental Systems: Insects*, Vol. 2 (S. J. Counce and C. H. Waddington, eds.), Academic Press, New York.

Goodman, C. S., 1984, Landmarks and labels that help developing neurons find their way, *BioScience* **34**:300–307.

Goodman, C. S., and Bastiani, M. J., 1984, How embryonic nerve cells recognize one another, *Sci. Am.* **251**(June):58–66.

Hagan, H. R., 1951, *Embryology of the Viviparous Insects*, Ronald Press, New York.

Heming, B. S., 2003, *Insect Development and Evolution*, Cornell University Press, Ithaca, NY.

Hoffmann, J. A., and Lagueux, M., 1985, Endocrine aspects of embryonic development in insects, in *Comprehensive Insect Physiology, Biochemistry and Pharmacology*, Vol. 1 (G. A. Kerkut and L. I. Gilbert, eds.), Pergamon Press, Elmsford, NY.

Ivanova-Kasas, O. M., 1972, Polyembryony in insects, in: *Developmental Systems: Insects*, Vol. 1 (S. J. Counce and C. H. Waddington, eds.), Academic Press, New York.

Johannsen, O. A., and Butt, F. H., 1941, *Embryology of Insects and Myriapods*, McGraw-Hill, New York.

Jura, C., 1972, Development of apterygote insects, in: *Developmental Systems: Insects*, Vol. 1 (S. J. Counce and C. H. Waddington, eds.), Academic Press, New York.

Lawrence, P. A. (ed.), 1976, Insect development, *Symp. R. Entomol. Soc.* **8**:240 pp.

Lawrence, P. A., 1992, *The Making of a Fly: The Genetics of Animal Design*, Blackwell, Oxford.

Matsuda, R., 1976, *Morphology and Evolution of the Insect Abdomen*, Pergamon Press, Elmsford, NY.

Retnakaran, A., and Percy, J., 1985, Fertilization and special modes of reproduction, in: *Comprehensive Insect Physiology, Biochemistry and Pharmacology*, Vol. 1 (G. A. Kerkut and L. I. Gilbert, eds.), Pergamon Press, Elmsford, NY.

Sander, K., 1984, Embryonic pattern formation in insects: Basic concepts and their experimental foundations, in: *Pattern Formation: A Primer in Developmental Biology* (G. M. Malacinski, ed.), Macmillan, New York.

Sander, K., 1997, Pattern formation in insect embryogenesis: The evolution of concepts and mechanisms, *Int. J. Insect Morphol. Embryol.* **25**:349–367.

Sander, K., Gutzeit, H. O., and Jackle, H., 1985, Insect embryogenesis: Morphology, physiology, genetical and molecular aspects, in: *Comprehensive Insect Physiology, Biochemistry and Pharmacology*, Vol. 1 (G. A. Kerkut and L. I. Gilbert,eds.), Pergamon Press, Elmsford, NY.

Suomalainen, E., 1962, Significance of parthenogenesis in the evolution of insects, *Annu. Rev. Entomol.* **7**:349–366.

Thomas, J. B., Bastiani, M. J., Bate, M., and Goodman, C. S., 1984, From grasshopper to *Drosophila*: A common plan for neuronal development, *Nature* **310**:203–207.

Truman, J. W., and Ball, E. E., 1998, Patterns of embryonic neurogenesis in a primitive wingless insect, the silverfish, *Ctenolepisma longicaudata*: Comparison with those seen in flying insects, *Develop. Genes Evol.* **208**:357–368.

White, M. J. D., 1973, *Animal Cytology and Evolution*, 3rd ed., Cambridge University Press, London.

21

Postembryonic Development

1. Introduction

During their postembryonic growth period insects pass through a series of stages (instars) until they become adult, the time interval (stadium) occupied by each instar being terminated by a molt. Apterygotes continue to grow and molt as adults, periods of growth alternating with periods of reproductive activity. In these insects structural differences between juvenile and adult instars are slight, and their method of development is thus described as ametabolous. Among the Pterygota, which with rare exceptions do not molt in the adult stage, two forms of development can be distinguished. In almost all exopterygotes the later juvenile instars broadly resemble the adult, except for their lack of wings and incompletely formed genitalia. Such insects, in which there is some degree of change in the molt from juvenile to adult, are said to undergo partial (incomplete) metamorphosis, and their development is described as hemimetabolous. Endopterygotes and a few exopterygotes have larvae whose form and habits, by and large, are very different from those of the adults. As a result, they undergo striking changes (complete metamorphosis), spread over two molts, in the formation of the adult (holometabolous development). The final juvenile instar has become specialized to facilitate these changes and is known as the pupa (see also Chapter 2, Section 3.3).

In insect evolution increasing functional separation has occurred between the larval phase, which is concerned with growth and accumulation of reserves, and the adult stage, whose functions are reproduction and dispersal. Associated with this trend is a tendency for an insect to spend a greater part of its life as a juvenile, which contrasts with the situation in many other animals. Thus, in apterygotes, the adult stage may be considerably longer than the juvenile stage. Furthermore, feeding (in the adult) serves to provide raw materials both for reproduction and for growth. In exopterygotes and primitive endopterygotes adults may live for a reasonable period, but this is not usually as long as the larval phase. Feeding in the adult stage is primarily associated with reproductive requirements, though in some insects it provides nutrients for an initial, short "somatic growth phase" in which the flight muscles, gut, and cuticle become fully developed. Many endopterygotes live for a relatively short time as adults and may feed little or not at all because sufficient reserves have been acquired during larval life to satisfy the needs of reproduction.

2. Growth

2.1. Physical Aspects

Growth in insects and other arthropods differs from that of mammals in various respects. In insects growth is almost entirely restricted to the larval instars, though in some species there is a short period of somatic growth in newly enclosed adults when additional cuticle may be deposited, and growth of flight muscles and the alimentary canal may occur. As a consequence, the length of the juvenile stage is considerably longer than that of the adult. An extreme example of this is seen in some mayfly species whose aquatic juvenile stage may require 2 or 3 years for completion, yet give rise to an adult that lives for only a few hours or days. Growth in many animals is discontinuous or cyclic; that is, periods of active growth are separated by periods when little or no growth occurs. Nowhere is discontinuous growth better seen than in arthropods, which must periodically molt their generally inextensible cuticle in order to significantly increase their size (volume). It should be appreciated, however, that, though increases in volume may be discontinuous, increases in weight are not (Figure 21.1). As an insect feeds during each stadium, reserves are deposited in the fat body, whose weight and volume increase. In a hard-bodied insect this increase in volume may be compensated for by a decrease in the volume occupied by the tracheal system or by extension of the abdomen as a result of the unfolding of intersegmental membranes. In many endopterygote larvae, of course, the entire body is largely covered with extensible cuticle, and body size increases almost continuously (but see below).

For many insects grown under standard conditions the amount of growth that occurs is predictable from one instar to the next; that is, it obeys certain "growth laws." For example, Dyar's law, based on measurements of the change in width of the head capsule which occurs at each molt, states that growth follows a geometric progression; that is, the proportionate increase in size for a given structure is constant from one instar to the next. Mathematically expressed, the law states $x/y = $ constant (value usually 1.2–1.4), where $x = $ size in a given instar and $y = $ size in previous instar (Figure 21.1). Thus, when the size of a structure is plotted logarithmically against instar number, a straight line is obtained, whose gradient is constant for a given species (Figure 21.2). In those insects where it applies Dyar's law can be used to determine how many instars there are in the life history. However, so many factors

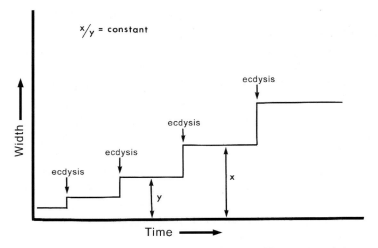

FIGURE 21.1. Change in head width with time to illustrate Dyar's law.

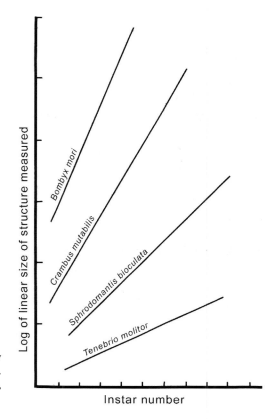

FIGURE 21.2. Head width plotted logarithmically against instar number in various species. [After V. B. Wigglesworth, 1965, *The Principles of Insect Physiology*, 6th ed., Methuen and Co. By permission of the author.]

affect growth rates and the frequency of ecdysis that the law is frequently inapplicable. In any event, the law requires that the interval between molts remains constant, but this is rarely the case.

As winged insects grow, they change shape; that is, the relative proportions of different parts of the body change. This disproportionate growth, which is not unique to insects, is described as "allometric" (heterogonic, disharmonic). In other words, each part has its own growth rate, expressed by the equation $y = bx^k$ (y = linear size of the part, x = linear size of the standard (e.g., body length), b = initial growth index (y intercept), and k = allometric coefficient). Normally, allometric growth is expressed as a log-log plot, when k is the gradient of the slope (Figure 21.3).

Growth laws do not apply in situations where the number of instars is variable. This variability may be a natural occurrence, especially in primitive insects such as mayflies that have many instars. In addition, females that are typically larger than males may have a greater number of instars than males. Variability may also be induced by environmental conditions. For example, rearing insects at abnormally high temperature often increases the number of instars, as does semistarvation. In contrast, in some caterpillars crowding leads to a decrease in the number of molts.

Caterpillars and probably other larvae whose cylindrical body is covered with a thin integument rely on hydrostatic (hemolymph) pressure to maintain the rigidity necessary for locomotion. However, this presents a problem with respect to their body form during growth. Mechanically, in a cylinder under internal pressure the hoop stress (around the body) is twice the axial (lengthwise) stress. Thus, a caterpillar theoretically should become proportionately fatter as it grows, much like a balloon when inflated. That it does not do

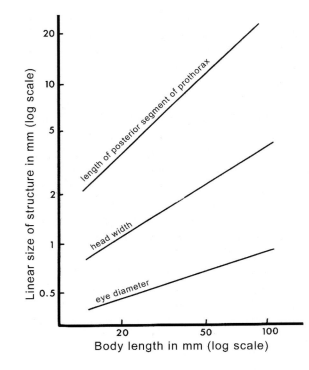

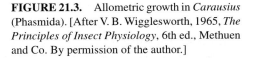

FIGURE 21.3. Allometric growth in *Carausius* (Phasmida). [After V. B. Wigglesworth, 1965, *The Principles of Insect Physiology*, 6th ed., Methuen and Co. By permission of the author.]

so is due to the occurrence of axial pleats (transverse cuticular folds) that reduce the axial stress by unfolding as the insect enlarges (Carter and Locke, 1993).

2.2. Biochemical Changes during Growth

Like the physical changes noted above, biochemical changes that occur during postembryonic development may also be described as allometric. That is, the relative proportions of the various biochemical components change as growth takes place. These changes are especially noticeable in endopterygotes during the final larval and pupal stages. At hatching, the fat content of a larva is typically low (less than 1% in the caterpillar *Malacosoma*, for example) and remains at about this level until the final larval stadium when fat is synthesized and stored in large quantity, reaching about 30% of the dry body weight. Though fat is the typical reserve substance in most insects, members of some species store glycogen. Again, this usually occurs in small amounts in newly hatched insects, but its proportion increases steadily through larval development, and at pupation glycogen may be a significant component of the dry weight (one-third in the honey bee). Like fat, glycogen is stored in the fat body.

In contrast, the proportions of water, protein, and nucleic acids generally decline during larval development. However, this is often not the situation in larvae that require large amounts of protein for specific purposes, for example, spinning a cocoon. In *Bomybyx mori*, for example, the hemolymph protein concentration increases sixfold in late larval development, and about 50% of the total protein content of a mature larva is used in cocoon formation. The great increase in concentration of hemolymph protein often can be accounted for almost entirely by synthesis, in the fat body, of a few specific proteins. In the fly *Calliphora stygia*, for example, the protein "calliphorin" makes up 75% (about 7 mg) of the hemolymph protein by the time a mature larva stops feeding. The calliphorin is used in the pupa as a major source of nitrogen (in the form of amino acids) for formation of adult tissues and as a source of the energy required in biosynthesis. Thus, at eclosion (emergence of the adult), the hemolymph calliphorin content has fallen to 0.03 mg, and, 1 week after emergence, the protein has entirely disappeared.

During metamorphosis some of the above trends may be reversed. The proportions of fat and/or glycogen decline as these molecules are utilized in energy production. In *Calliphora* the fat content decreases from 7% to 3% of the dry weight through the pupal period. In the honey bee, which mainly uses glycogen as an energy source, the glycogen content drops to less than 10% of its initial value as metamorphosis proceeds. For most insects there is little change in the net protein content during pupation, though major qualitative changes occur as adult tissues develop. In members of a few species a significant decline in total protein content occurs during metamorphosis as protein is used as an energy source. The moth *Celerio*, for example, obtains only 20% of its energy requirements in metamorphosis from fat, the remaining 80% coming largely from protein.

Superimposed on the overall biochemical changes from hatching to adulthood are changes that occur in each stadium, related to the cyclic nature of growth and molting. Factors to be considered include the phasic pattern of feeding activity throughout the stadium, synthesis of new and degradation of old cuticle, and net production of new tissues (though some histolysis also occurs in each instar).

Measurement of oxygen consumption shows that it follows a U-shaped curve through each stadium with maximum values being obtained at the time of molting. The maxima are correlated with the great increase in metabolic activity at this time, associated especially with the synthesis of new cuticle and formation of new tissues. In *Locusta* larvae there are significant decreases in the carbohydrate and lipid contents of the fat body and hemolymph at ecdysis, probably correlated with the use of these substrates to supply energy (Hill and Goldsworthy, 1968). Conversely, as feeding restarts after a molt, these materials are again accumulated.

Changes in the amount of protein in the fat body and hemolymph of *Locusta* are also cyclical, with maximum values occurring in the second half of each stadium (Hill and Goldsworthy, 1968). The early increase in protein content is related to renewed feeding activity after the molt. Feeding activity reaches a peak in the middle of the stadium, providing materials for growth of muscles (and presumably other tissues, though these were not studied by Hill and Goldsworthy) and for the synthesis of cuticle. Excess material is stored in the fat body and hemolymph. In the second half of the stadium feeding activity declines, and this is followed by a decrease in the level of protein in the hemolymph and fat body. Hill and Goldsworthy (1968) suggested that the latter probably reflects the use of protein in the synthesis of new cuticle. However, recycled protein from the old cuticle may account for most (about 80% in *Locusta*) of the protein content of the new cuticle.

3. Forms of Development

Through insect evolution there has been a trend toward increasing functional and structural divergence between juvenile and adult stages. Juvenile insects have become more concerned with feeding and growth, whereas adults form the reproductive and dispersal phase. This specialization of different stages in the life history became possible with the introduction into the life history of a pupal instar, though the latter's original function was probably related specifically to evagination of the wings and development of the wing musculature (Chapter 2, Section 3.3).

In modern insects three basic forms of postembryonic development can be recognized, described as ametabolous, hemimetabolous, and holometabolous, according to the extent of metamorphosis from juvenile to adult (Figure 21.4).

3.1. Ametabolous Development

In Thysanura (and other primitive hexapods), which as adults remain wingless, the degree of change from juvenile to adult form is slight and is manifest primarily in increased body size and development of functional genitalia. Juvenile and adult apterygotes inhabit the same ecological niche, and the insects continue to grow and molt after reaching sexual maturity. The number of molts through which an insect passes is very high and variable. For example, in the firebrat, *Thermobia domestica,* between 45 and 60 molts have been recorded.

3.2. Hemimetabolous Development

Exopterygotes usually molt a fixed number of times, but, with the exception of Ephemeroptera, which pass through a winged subimago stage, never as adults. In species where the female is much larger than the male, she may undergo an additional larval molt. The number of molts is typically 4 or 5, though in some Odonata and Ephemeroptera whose larval life may last 2 or 3 years a much greater and more variable number of molts occurs (e.g., 10–15 in species of Odonata, 15–30 in most Ephemeroptera).

In almost all exopterygotes the later juvenile instars broadly resemble the adult, except that their wings and external genitalia are not fully developed. Early instars show no trace of wings, but, later, external wing buds arise as sclerotized, non-articulated evaginations of the tergopleural area of the wing-bearing segments. Wings develop within the buds during the final larval stadium and are expanded after the last molt. Other, less obvious, changes that occur during the growth of exopterygotes include the addition of neurons, Malpighian tubules, ommatidia, and tarsal segments, plus the differentiation of additional sensilla in the integument. This mode of development is described as hemimetabolous and includes a partial (incomplete) metamorphosis from larva to adult.

3.3. Holometabolous Development

Holometabolous development, in which there is a marked change of form from larva to adult (complete metamorphosis), occurs in endopterygotes and a few exopterygotes, for example, whiteflies (Aleurodidae: Hemiptera), thrips (Thysanoptera), and male scale insects (Coccidae: Hemiptera). Perhaps the most obvious structural difference between the larval and adult stages of endopterygotes is the absence of any external sign of wing development in the larval stages. The wing rudiments develop internally from imaginal discs that in most larvae lie at the base of the peripodial cavity, an invagination of the epidermis beneath the larval cuticle, and are evaginated at the larval-pupal molt (see Section 4.2 and Figure 21.11).

As noted above, the evolution of a pupal stage in the life history has made holometabolous development possible. The pupa is probably a highly modified final juvenile instar (Chapter 2, Section 3.3) which, through evolution, became less concerned with feeding and building up reserves (this function being left to earlier instars) and more

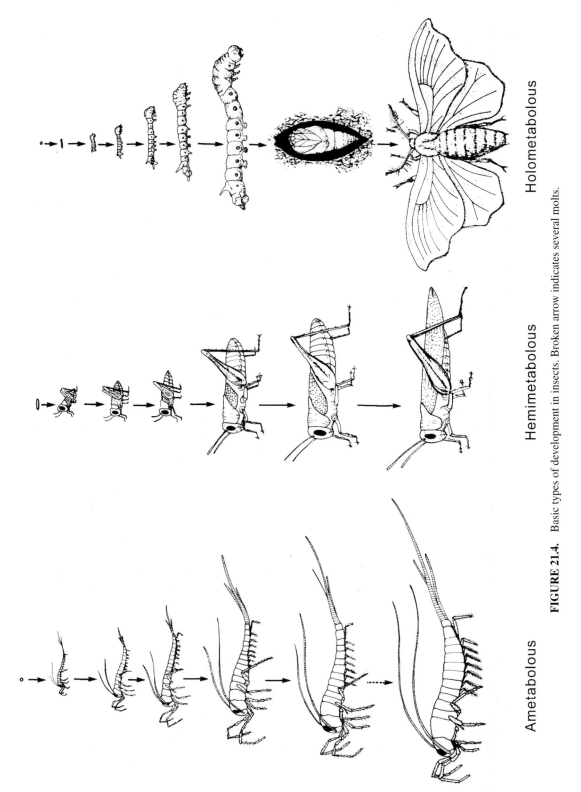

FIGURE 21.4. Basic types of development in insects. Broken arrow indicates several molts.

Holometabolous

Hemimetabolous

Ametabolous

specialized for the breakdown of larval structures and construction of adult features. In other words, the pupa has become a non-feeding stage; it is generally immobile as a result of histolysis of larval muscles; it broadly resembles the adult and thereby serves as a mold for the formation of adult tissues, especially muscles.

3.3.1. The Larval Stage

Among endopterygotes the extent to which the larval and adult habits and structure differ [and therefore the extent of metamorphosis (Section 4.2)] is varied. Broadly speaking, in members of more primitive orders the extent of these differences is small, whereas the opposite is true, for example, in the Hymenoptera and Diptera. Endopterygote larvae can be arranged in a number of basic types (Figure 21.5). The most primitive larval form is the oligopod. Larvae of this type have three pairs of thoracic legs and a well-developed head with chewing mouthparts and simple eyes. Oligopod larvae can be further subdivided into (1) scarabaeiform larvae (Figure 21.5A), which are round-bodied and have short legs and a weakly sclerotized thorax and abdomen, features associated with the habit of burrowing into the substrate, and (2) campodeiform larvae (Figure 21.5B), which are active, predaceous surface-dwellers with a dorsoventrally flattened body, long legs, strongly sclerotized thorax and abdomen, and prognathous mouthparts. Scarabaeiform larvae are typical

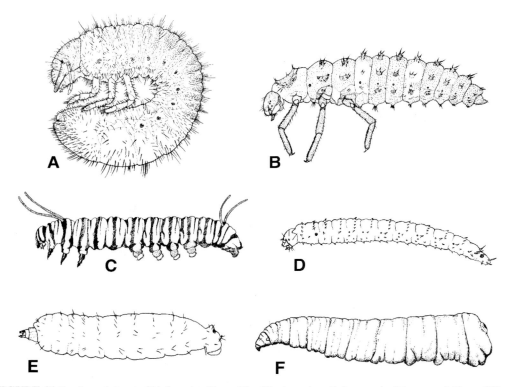

FIGURE 21.5. Larval types. (A) Scarabaeiform (*Popillia japonica*, Coleoptera); (B) campodeiform (*Hippodamia convergens*, Coleoptera); (C) eruciform (*Danaus plexippus*, Lepidoptera); (D) eucephalous (*Bibio* sp., Diptera); (E) hemicephalous (*Tanyptera frontalis*, Diptera); and (F) acephalous (*Musca domestica*, Diptera). [A–E, from A. Peterson, 1951, *Larvae of Insects.* By permission of Mrs. Helen Peterson. F, from V. B. Wigglesworth, 1959, Metamorphosis, polymorphism, differentiation, *Scientific American*, February 1959. By permission of Mr. Eric Mose, Jr.]

of the Scarabaeidae and other beetle families; campodeiform larvae occur in Neuroptera, Coleoptera-Adephaga, and Trichoptera.

Polypod (eruciform) larvae (Figure 21.5C) have, in addition to thoracic legs, a varied number of abdominal prolegs. The larvae are generally phytophagous and relatively inactive, remaining close to or on their food source. The thorax and abdomen are weakly sclerotized in comparison with the head, which has well-developed chewing mouthparts. Eruciform larvae are typical of Lepidoptera, Mecoptera, and some Hymenoptera [sawflies (Tenthredinidae)].

Apodous larvae, which lack all trunk appendages, occur in various forms in many endopterygote orders but in common are adapted for mining in soil, mud, or animal or plant tissues. The variability of form concerns the extent to which a distinct head capsule is developed. In eucephalous larvae (Figure 21.5D), characteristic of some Coleoptera (Buprestidae and Cerambycidae), Strepsiptera, Siphonaptera, aculeate Hymenoptera, and more primitive Diptera (suborder Nematocera), the head is well sclerotized and bears normal appendages. The head and its appendages of hemicephalous larvae (Figure 21.5E) are reduced and partially retracted into the thorax. This condition is seen in crane fly larvae (Tipulidae: Nematocera) and in the larvae of orthorraphous Diptera. Larvae of Diptera-Muscomorpha are acephalous (Figure 21.5F); no sign of the head and its appendages can be seen apart from a pair of minute papillae (remnants of the antennae) and a pair of sclerotized hooks believed to be much modified maxillae.

Frequently a larva in the final instar ceases to feed and becomes inactive a few days before the larval-pupal molt. Such a stage is known as a prepupa. In some species, the entire instar is a non-feeding stage in which important changes related to pupation occur. For example, in the prepupal instar of sawflies, the salivary glands become modified for secreting the silk used in cocoon formation.

3.3.2. Heteromorphosis

In most endopterygotes the larval instars are more or less alike. However, in some species of Neuroptera, Coleoptera, Diptera, Hymenoptera, and in all Strepsiptera, a larva undergoes characteristic changes in habit and morphology as it grows, a phenomenon known as heteromorphosis (hypermetamorphosis). In such species several of the larval types described above may develop successively (Figure 21.6). For example, blister beetles (Meloidae) hatch as free-living campodeiform larvae (planidia, triungulins) that actively search for food (grasshopper eggs and immature stages, or food reserves of bees or ants). At this stage the larvae can survive for periods of several weeks without food. Larvae that locate food soon molt to the second stage, a caterpillarlike (eruciform) larva. The insect then passes through two or more additional larval instars, which may remain eruciform or become scarabaeiform. Some species overwinter in a modified larval form known as the pseudopupa or coarctate larva, so-called because the larva remains within the cuticle of the previous instar. The pseudopupal stage is followed the next spring by a further larval feeding stage, which then molts into a pupa.

3.3.3. The Pupal Stage

The pupa is a non-feeding, generally quiescent instar that serves as a mold in which adult features can be formed. For many species it is also the stage in which an insect survives adverse conditions by means of diapause (Chapter 22, Section 3.2.3). The terms "pupa" and "pupal stage" are commonly used to describe the entire preimaginal instar. This is,

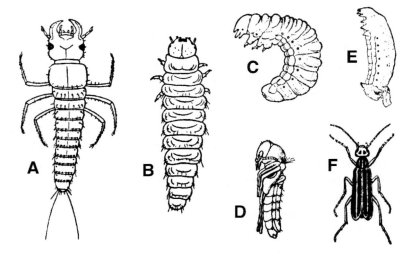

FIGURE 21.6. Heteromorphosis in *Epicauta* (Coleoptera). (A) Triungulin; (B) caraboid second instar; (C) final form of second instar; (D) coarctate larva; (E) pupa; and (F) adult. [From J. W. Folsom, 1906, *Entomology: With Special Reference to Its Biological and Economic Aspects*, Blakiston.]

strictly speaking, incorrect because for a varied period prior to eclosion, the insect is a "pharate adult," that is, an adult enclosed within the pupal cuticle. The insect thus becomes an adult immediately after apolysis of the pupal cuticle and formation of the adult epicuticle (Chapter 11, Section 3.1). The distinction between the true pupal stage and the pharate adult condition becomes important in consideration of so-called "pupal movements," including locomotion and mandibular chewing movements (used in escaping from the protective cocoon or cell in which metamorphosis took place). In most instances these movements result from the activity of muscles attached to the adult apodemes that fit snugly around the remains of the pupal apodemes (Figure 21.7).

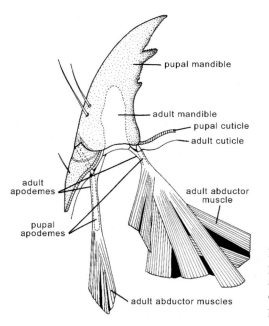

FIGURE 21.7. Section through mandible of a decticous pupa to show adult apodemes around remains of pupal apodemes. [After H. E. Hinton, 1946, A new classification of insect pupae, *Proc. Zool. Soc. Lond.* **116**:282–328. By permission of the Zoological Society of London.]

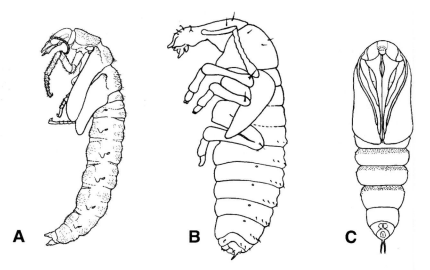

FIGURE 21.8. Pupal types. (A) Decticous (*Chrysopa* sp., Neuroptera); (B) exarate adecticous (*Brachyrhinus sulcatus*, Coleoptera); and (C) obtect adecticous (*Heliothis armigera*, Lepidoptera). [From A. Peterson, 1951, *Larvae of Insects.* By permission of Mrs. Helen Peterson.]

Pupae are categorized according to whether or not the mandibles are functional and whether or not the remaining appendages are sealed closely against the body (Figure 21.8). Decticous pupae, found in more primitive endopterygotes [Neuroptera, Mecoptera, Trichoptera, and Lepidoptera (Zeugloptera and Dacnonypha)], have well-developed, articulated mandibles (moved by the pharate adult's muscles) with which an insect can cut its way out of the cocoon or cell. Decticous pupae are always exarate; that is, the appendages are not sealed against the body so that they may be used in locomotion. Some neuropteran pupae, for example, can crawl and some pupae of Trichoptera swim to the water surface prior to eclosion. Adecticous pupae, whose mandibles are non-functional and often reduced, may be either exarate or obtect. In the latter condition the appendages are firmly sealed against the body and are usually well sclerotized. Adecticous exarate pupae are characteristic of Siphonaptera, brachycerous Diptera, most Coleoptera and Hymenoptera, and Strepsiptera. In nematocerous Diptera, Lepidoptera (Heteroneura), and in a few Coleoptera and Hymenoptera, pupae are of the adecticous obtect type.

In muscomorph Diptera at the end of the final larval stadium the cuticle becomes thickened and tanned. The tanned cuticle is not shed but remains as a rigid coat (puparium) around the insect. A few hours after pupariation the larval epidermis apolyses so that a pharate pupal instar is formed within the puparium, serving as in other endopterygotes as the mold for adult tissues.

An immobile pupa is vulnerable to attack by predators or parasites and to severe changes in climatic conditions, particularly as the pupal stadium may last for a considerable time. To obtain protection against such adversities the pupa typically has a thick, tanned cuticle. Also, in many species it is enclosed within a cocoon or subterranean cell constructed by the previous larval instar. The cocoon may comprise various kinds of extraneous material, for example, soil particles, small stones, leaves or other vegetation, or may be made solely of silk. In some endopterygotes the pupa is exposed (not surrounded by a protective cocoon) but obtains additional protection by taking on the color of its surroundings. Many parasitic species remain within, and are thus protected by, the host's body in the pupal stage.

4. Histological Changes During Metamorphosis

Though we have distinguished, in the preceding account, between hemimetabolous development (where partial metamorphosis occurs in the molt from larva to adult) and holometabolous development (in which metamorphosis is striking and requires two molts, larval-pupal and pupal-imaginal, for completion), the distinction is primarily useful in discussions of insect evolution. In a physiological sense the difference between partial and complete metamorphosis is a matter of degree rather than kind. Indeed, as is described in Section 6.1, the endocrine basis of growth, including molting and change of form, is common to all insects.

4.1. Exopterygote Metamorphosis

In most exopterygotes the larval and adult forms of a species occupy the same habitat, eat the same kinds of food (though specific preferences may change with age), and are subject to the same environmental conditions. Accordingly, most organ systems of a juvenile exopterygote are smaller and/or less well-developed versions of those found in an adult and simply grow progressively during larval life to accommodate changing needs. Even larval Odonata and Ephemeroptera that are aquatic and possess transient adaptive features such as gills or caudal lamellae broadly resemble the adult stage. The system that undergoes the most obvious change at the final molt is the flight mechanism. In the last larval instar wings develop within the wing buds as much folded sheets of integument, and, concurrently, the articulating sclerites differentiate. Direct flight muscle rudiments are present in larval instars and are attached to the integument at points corresponding to the future locations of the sclerites. Some of these (bifunctional muscles) may be important in leg movements during larval life (Chapter 14, Section 3.3.1). Like the direct flight muscles, the indirect flight muscles grow progressively through larval life but remain unstriated and non-functional until the adult stage.

4.2. Endopterygote Metamorphosis

In more primitive endopterygotes such as Neuroptera and Coleoptera, as in exopterygotes, a good deal of progressive development of organ systems occurs during larval life so that metamorphosis, relatively speaking, is slight and concerns, again, mainly the flight mechanism. At the opposite extreme, seen in many Diptera and Hymenoptera, most larval tissues are histolyzed, with adult tissues being formed anew, often from specific groups of undifferentiated cells, the imaginal discs and abdominal histoblasts. The imaginal discs occur as thickened regions of epidermis whose cells remain embryonic; that is, in the larval instars their differentiation is suppressed by the hormonal milieu existing in the insect at this time. At metamorphosis striking changes occur in the concentration of certain hormones, as a result of which the cells can multiply and differentiate into specific adult organs and tissues (Figure 21.9). Experiments in which cells have been selectively destroyed by X-irradiation have shown that formation of imaginal discs occurs very early in embryogenesis and at specific sites. Furthermore, each imaginal disc differentiates in a predetermined manner. During larval development, the discs grow exponentially in relation to general body growth and, typically, come to lie within an invagination, the peripodial cavity, beneath the cuticle (Figure 21.11). In contrast to the imaginal discs, the abdominal histoblasts, which

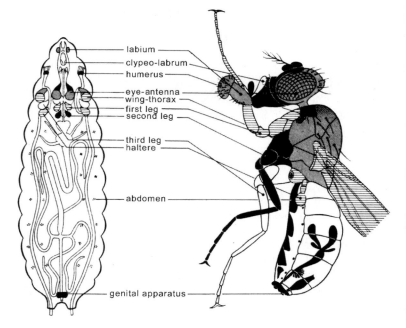

labium
clypeo-labrum
humerus
eye-antenna
wing-thorax
first leg
second leg
third leg
haltere
abdomen
genital apparatus

FIGURE 21.9.   Imaginal discs of *Drosophila* and their derivatives in the adult body. [From H. Wildermuth, 1970, Determination and transdetermination in cells of the fruitfly, *Sci. Prog.* (*Oxford*) **58**:329–358. By permission of Blackwell Scientific Publications.]

are groups of loosely associated cells in the larval integument, do secrete larval cuticle. At metamorphosis, under hormonal influence, they divide and differentiate into the adult abdominal epidermis, fat body, oenocytes, and some muscles.

With the evolution of imaginal discs the way was open for the development of a larva whose form is highly different from that of the adult, and capable of existing in a different habitat from that of the adult, thus avoiding competition for food and space.

To clarify the histological changes that occur in endopterygote metamorphosis, the various organ systems will be considered separately

Epidermal cells carried over from the larval stage produce the cuticle of most adult endopterygotes. However, in Hymenoptera-Apocrita and Diptera-Muscomorpha the larval epidermis is more or less completely histolyzed and replaced by cells derived from imaginal discs and histoblasts. In Muscomorpha, histolysis of the larval epidermis does not occur until after pupariation.

Appendage formation is also varied. In lower endopterygotes formation of adult mouthparts, antennae, and legs begins early in the final larval stadium from larval epidermis. Certain predetermined areas of the epidermis thicken, then proliferate and differentiate so that, at pupation, the basic form of the adult appendages is evident. During the pupal stadium the final form of the adult appendages is expressed (Figure 21.10). In contrast, where the larval appendages are very different from those of the adult, or are absent, the adult structures develop from imaginal discs that undergo marked proliferation and differentiation in the last larval instar and are evaginated from the peripodial cavity at the larval-pupal molt. Wings are formed in all endopterygotes from imaginal discs. In most species their early development is similar to the development of paired segmental appendages outlined above; that is, the wing rudiments form in a peripodial cavity and become everted at the larval-pupal molt (Figure 21.11). The forming wing bud in the peripodial cavity is initially a hollow,

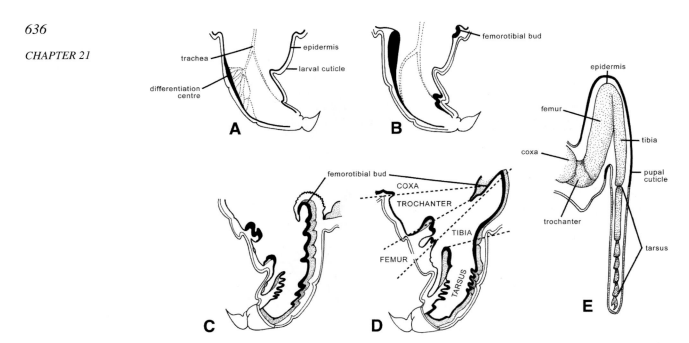

FIGURE 21.10. Sections through leg of *Pieris* (Lepidoptera) to show development of adult appendage. (A) Leg of last-instar larva 3 hours after ecdysis; (B) same as (A) but 1 day after ecdysis; (C) same as (A) but 3 days after ecdysis; (D) leg at beginning of prepupal stage showing presumptive areas of adult leg; and (E) leg of pupa. [After C.-W. Kim, 1959, The differentiation center inducing the development from larval to adult leg in *Pieris brassicae* (Lepidoptera), *J. Embryol. Exp. Morphol.* **7**:572–582. By permission of Cambridge University Press.]

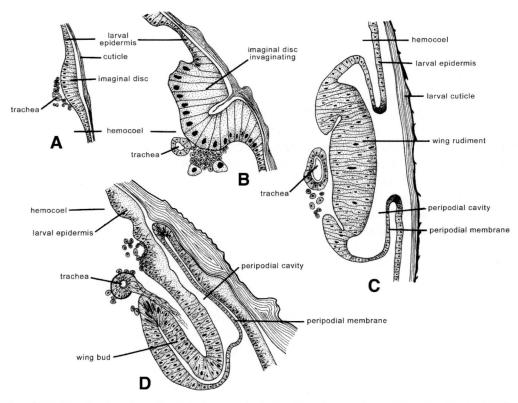

FIGURE 21.11. Sections through developing wing bud of first four larval instars of *Pieris* (Lepidoptera). [After J. H. Comstock, 1918, *The Wings of Insects*, Comstock.]

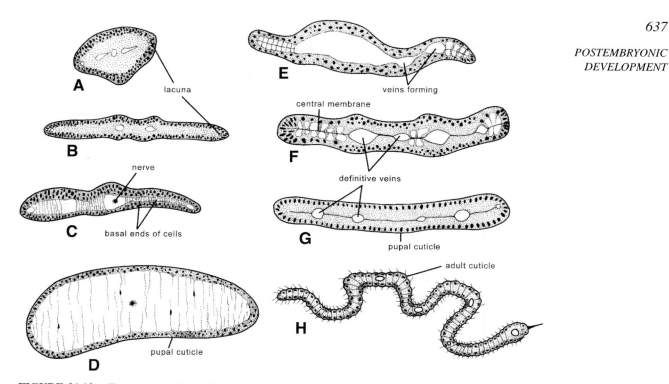

FIGURE 21.12. Transverse sections of developing wing of *Drosophila*. (A, B) Successive stages in pharate pupa; (C–G) stages in pupa; and (H) pharate adult. [After C. H. Waddington, 1941, The genetic control of wing development in *Drosophila*, *J. Genet.* **41**:75–139. By permission of Cambridge University Press.]

fingerlike structure, but this becomes flattened so that the central cavity is more or less obliterated, leaving only small lacunae (Figure 21.12A, B). A nerve and trachea that have been associated with the imaginal disc now grow along each lacuna occasionally branching in a predetermined pattern. At the larval-pupal molt, hemolymph pressure forces the sides of a wing bud apart so that there is sufficient space within it for the development of an adult wing (Figure 21.12C, D). During the pupal stadium extensive proliferation of the epidermal cells occurs within the wing bud, as a consequence of which the epidermis becomes folded and closely apposed over most of the wing surface. The epidermal layers remain separate adjacent to the nerve and trachea, forming the definitive wing veins (Figure 21.12E–H).

The gut of endopterygotes typically changes its form markedly during metamorphosis. In Coleoptera the foregut and hindgut undergo relatively slight modification, this being achieved by the activity of larval cells. In higher endopterygotes these regions are partially or entirely renewed from groups of primordial cells located at the junctions of the foregut and midgut and midgut and hindgut, and adjacent to the mouth and anus. The larval midgut of all endopterygotes is fully replaced as a result of the activity of either regenerative cells from the larval midgut, or undifferentiated cells at the junction of the midgut and hindgut, or both. In either arrangement, the histolyzed larval cells eventually are surrounded by adult tissue. To protect the insect from potential pathogens in the absence of a peritrophic matrix, the differentiating pupal midgut epithelium releases a mixture of antibacterial peptides into the gut lumen (Russell and Dunn, 1996).

Larval Malpighian tubules may be retained in some adult Diptera, but in other endopterygotes they are partially or completely replaced at metamorphosis from special cells located either along the length of each tubule or at the anterior end of the hindgut.

Generally, the larval tracheal system is carried over to the adult with little modification, except as required by the development of new tissues such as flight muscles and reproductive organs. However, in Diptera, considerable replacement of the larval system occurs from scattered cells in the larval tracheae.

The central nervous system of most endopterygotes becomes more concentrated at metamorphosis because of shortening of the interganglionic connectives and the forward migration of ganglia. It is likely that the surrounding glial cells are responsible for this process as their ensheathing cytoplasm is filled with microtubules oriented lengthwise along the connectives. Within a ganglion, some neurons may enlarge while others differentiate from neuroblasts. Many others are phagocytozed by glial cells so that only a few (5% to 10% in *Drosophila*, for example) are carried over to the adult. These are nevertheless important as they serve as a scaffold on which adult sensory neurons can reach and connect with the central nervous system (Williams and Shepherd, 2002). In some moths clusters of immature neurons, each surrounded by a giant glial cell, migrate along the interganglionic connectives to the next posterior ganglion (Cantera *et al.*, 1995). The giant glial cells, which span the interganglionic connectives, probably drag the neuron clusters between ganglia. Breakdown and rebuilding of the perineurium and neural lamella are necessitated by these changes. In insects that as adults make extensive use of associative learning (Chapter 13, Section 2.4) the mushroom bodies undergo large-scale reorganization during metamorphosis, with extensive degradation of the larval structures and establishment of the adult form from neuroblasts carried over from the embryo (Farris *et al.*, 1999).

During metamorphosis the muscular system undergoes considerable modification especially in connection with flight. Typically, many larval muscles histolyze, though the point at which histolysis begins is varied and somewhat dependent on their function. For many muscles histolysis begins in the final larval instar and continues in the pupa. Other muscles, however, have particular functions at the larval-pupal molt and beyond and do not histolyze until later. As an extreme example, special eclosion muscles that facilitate the adult's escape from the puparium differentiate during the pupal stage but disappear within a few hours of emergence. The muscles of an adult insect arise in several ways: (1) from larval muscles that remain unchanged, (2) from partially histolyzed and reconstructed larval muscles, (3) from previously inactive imaginal nuclei within the larval muscles, (4) from myoblasts that previously adhered loosely to the surface of the larval muscles, or (5) from rudimentary, non-functional fibers present in the larva (Whitten, 1968). Methods 3–5 are increasingly important in higher endopterygotes. Further, the presence of motor neurons is essential for the proliferation and correct organization of the new muscles (Kent *et al.*, 1995).

The extent of histolysis of the fat body is quite varied and depends on the extent of metamorphosis. In more primitive endopterygotes where metamorphosis is relatively slight, much of the larval fat body is carried over unchanged into the adult stage. However, in muscomorph Diptera, for example, the larval tissue is completely broken down, and the adult fat body is formed from mesenchyme cells associated with imaginal discs.

The heart and aorta are normally not histolyzed and continue to contract in the pupa. However, in muscomorph Diptera, contraction stops midway through the pupal stadium, the larval heart muscle cells break down, and new contractile elements differentiate from myoblasts. The dorsal and ventral diaphragms together with the accessory hearts are formed at metamorphosis, apparently from myoblasts associated with existing neurons.

Rarely, both ectodermal and mesodermal rudiments of the reproductive system become distinguishable in late larval instars. However, in the great majority of endopterygotes the entire process of differentiation of the reproductive system occurs at metamorphosis, either from groups of undifferentiated cells or, in the case of the muscomorph Diptera and higher Hymenoptera, from genital imaginal discs.

5. Eclosion

For exopterygotes adult emergence (eclosion) consists solely of escape from the cuticle of the previous instar. Many endopterygotes must, in addition, force their way out of the cocoon or cell in which pupation occurred and, in some species, to the surface of the substrate in which they have been buried. Some aquatic species that pupate under water have special devices to enable the adult to reach the water surface.

For adults of many species, emergence is triggered by environmental factors, especially temperature and photoperiod, or is entrained as a circadian rhythm (Myers, 2003) (see also Chapter 22, Sections 2.3 and 3.1.1). As a result, emergence of populations of adults is highly synchronized, this being of particular importance in species whose adult life is short.

Eclosion is accomplished in a manner similar to larval-larval molts. A pharate adult swallows air to increase its body volume and, by contraction of abdominal muscles, forces hemolymph anteriorly. As the hemolymph pressure increases, the pupal cuticle splits along an ecdysial line on the thorax and/or head. In obtect pupae the pupal mouth is sealed over, but the adult swallows air that enters the pupal case via the tracheal system. Some adult spiracles remain in contact with those of the pupa, whereas others become separated so that a channel is open along which air can move into the pupal case.

Among more primitive endopterygotes an insect escapes from its cocoon or cell as a pharate adult using the mandibles of the decticous pupa to force an opening in the wall. Pharate adults of some species also have backwardly facing spines on the pupal cuticle, which enable them to wriggle out of the cell and through the substrate. Many primitive Lepidoptera and Diptera, whose pupae are adecticous, also escape as pharate adults, frequently making use of special spines (cocoon cutters) on the pupal cuticle. In higher Lepidoptera, adults may shed the pupal cuticle while still in the cocoon. In such species the cocoon may possess an "escape hatch" or part of it may be softened by special salivary secretions. Further, for those species that pupate in soil, the adult cuticle may become temporarily plasticized to facilitate tunneling to the surface. In muscomorph Diptera an eversible membranous sac on the head, the ptilinum, can be expanded by hemolymph pressure. This enables an adult to push off the tip of the puparium and tunnel to the surface of the substrate in which it has been buried. Adult Coleoptera, Hymenoptera, and Siphonaptera leave the pupal cuticle while in the cocoon or cell, then use their mandibles or cocoon cutters to cut their way out. In some species this is the sole function of the mandibles, which, like cocoon cutters, are shed after emergence.

6. Control of Development

Despite the apparently wide differences in the pattern of development seen in Insecta, the physiological system that regulates growth, molting, and metamorphosis is common to all members of the class, namely, the endocrine system. Variations in the relative levels of different hormones in an insect's body determine the nature and extent of tissue differentiation that is expressed at the next molt. In other words, it is the hormone balance that determines, in a holometabolous insect, for example, whether the next molt is larval-larval,

larval-pupal, or pupal-adult. Hormones also coordinate the sequence of events in the growth and molting cycle and in some species ensure that an adult emerges when environmental conditions are suitable. The hormones act by regulating genetic activity. In a particular hormonal milieu, the genes that are active are responsible for expression of larval characters; under other hormonal conditions genes for pupal or imaginal features are activated.

Many environmental factors can modify developmental patterns. Some of these factors, for example, temperature, may act directly to affect development; most factors, however, exert their effect indirectly via the endocrine system.

6.1. Endocrine Regulation of Development

Postembryonic development is controlled by three endocrine centers: the brain-corpora cardiaca complex, corpora allata, and molt glands (see Chapter 13, Section 3, for a description of their structure and products). A molt cycle is initiated when, as a result of appropriate signals (see Section 6.2), the median neurosecretory cells of the brain release ecdysiotropin [prothoracicotropic hormone (PTTH)], which stimulates molting hormone (ecdysone) (MH) production by the molt glands. In all insects studied there is a major peak of MH in the hemolymph during the second half of each molt cycle, and it is this MH surge [in reality, the MH is first converted to the biologically active form, 20-hydroxyecdysone (20-HE), probably at its target site] that initiates the various physiological events constituting a molt cycle (Figure 21.13). In addition, in the final larval instar of holometabolous species one or more smaller hemolymph MH peaks precede the major peak and are thought to be responsible for reprogramming tissues for pupal rather than larval syntheses.

The corpora allata produce juvenile hormone (JH). The regulation of corpus allatum activity is complex (Tobe and Stay, 1985). Both allatotropic (corpus allatum-stimulating) and allatostatic (corpus allatum-inhibiting) neurosecretory factors have been reported, while in some insects the brain exerts direct neural control over the gland. It should also be noted that the hemolymph contains highly active esterases with the potential for degrading free JH. Thus, the concentration of circulating JH is determined not only by the secretory activity of the corpora allata but also by these hemolymph esterases.

JH can exert an influence on development only in the presence of MH, that is, after a molting cycle has begun. It is the concentration of circulating JH during one or more critical periods of the stadium that determines the nature of the succeeding molt. If the concentration of JH is above a threshold value* during the critical period, the next molt will be larval-larval (for this reason, JH has been described as the "*status quo*" hormone). When there is little or no circulating JH, an adult will appear at the next molt (Figure 21.13).

This scenario applies to hemimetabolous insects, which have a single critical period during each stadium and, of course, lack a pupal instar. In holometabolous insects there are two critical periods in the last larval stadium. In the first the absence of JH programs the development of pupal characters. In the second, which is just before the larval-pupal molt, a sharp increase in JH concentration occurs that prevents premature differentiation of imaginal discs. In the pupal stadium the absence of JH in the critical period permits the expression of adult characters. In species that show phenotypic polymorphism (Section 7) there may be several extra JH-sensitive critical periods for expression of the various forms (Nijhout, 1994).

---

* The absolute concentration appears unimportant, provided that it is above or below a threshold range. However, the threshold range will vary among species.

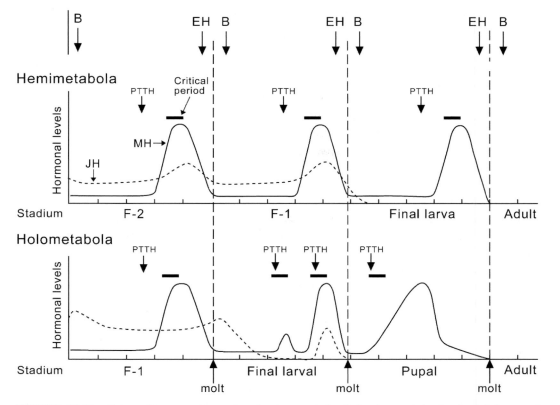

FIGURE 21.13. Schematic comparison of endocrine control of development in hemimetabolous and holometabolous insects. Pulses of prothoracicotropic hormone (PTTH) trigger synthesis and release of MH. Levels of JH determine the nature of the molt: when JH is present during a critical period, a larval-larval molt occurs; if no JH is present, the next molt will be larval-adult (Hemimetabola), or larval-pupal or pupal-adult (Holometabola). In the final larval stage of Holometabola there are two critical periods: in the first (no JH present) the switch to pupal development occurs; in the second, there is a pulse of JH that prevents premature differentiation of imaginal discs. Just prior to, or immediately after, each molt, release of eclosion hormone (EH) or bursicon (B), respectively, occurs (see also Figure 21.14). JH and MH levels are not drawn to the same scale. Numbers on the horizontal axes indicate the percent duration of each stadium. Other abbreviations: F-1, penultimate larval stadium; F-2, antepenultimate larval stadium. [Original, based on data from several sources.]

Apart from permitting the expression of adult characters, the low concentration of JH during the final stadium has another major effect: it leads to degeneration of the molt glands, which disappear within a few days of eclosion in most insects, exceptions being apterygotes, which, as noted earlier, continue to grow and molt as adults and solitary locusts whose corpora allata apparently do not become completely inactive at the final molt. Though degeneration of the molt glands is of critical importance in the life history of an insect, when viewed in the perspective of metamorphosis the phenomenon becomes simply an example of apoptosis (programmed cell death). In other words, the molt glands, like many other structures in juvenile insects, especially endopterygotes, are larval tissues whose structural well-being is dependent on JH. In the absence of JH, at metamorphosis, they histolyze. It remains unclear how JH affects the molt glands. However, a very early sign of apoptosis in the glands is nuclear DNA cleavage; this is prevented by the experimental application of JH (Dai and Gilbert, 1998).

The precise modes of action of the developmental hormones remain unclear. One of the earliest observable effects of PTTH is renewed RNA synthesis in the molt glands. This effect appears to be achieved, as for other peptide hormones, through a second messenger system, namely, cyclic AMP and calcium ions.

By contrast, 20-HE and JH are lipophilic and thus able to move through the cell membrane to the nuclear membrane where they bind to specific receptors. In *Drosophila* and *Manduca* the receptor for 20-HE has been identified and characterized as a protein heterodimer (Hall, 1999; Riddiford *et al.*, 1999). Its two components are EcR (ecdysone receptor) and USP (ultraspiracle*), both of which may exist in slightly different forms and concentrations. The 20-HE binds to the EcR part of the dimer, and while the function of USP remains unclear, both components of the dimer are necessary to transport the hormone to the chromosomes for gene activation (Lezzi *et al.*, 1999). Precisely how 20-HE works is not known; the most widely accepted view is that the hormone, like the steroid hormones of vertebrates, induces a gene-activation cascade. The 20-HE activates so-called "early" genes that encode regulatory proteins. The latter, in turn, modulate the activity of "middle," and eventually "late," genes whose transcriptional products carry out the appropriate tissue-specific process (Riddiford, 1985; Doctor and Fristrom, 1985). The pathways of gene activation that are followed depend on the 20-HE concentration and receptor isoform; as well, they will be species- and tissue-specific, resulting in the numerous effects induced by this hormone.

The site and mode of action of JH are less clear, a statement that is perhaps not surprising given its multifunctional role in insects. However, with respect to postembryonic development where it qualitatively modifies the effects of 20-HE, JH is generally assumed to act at the same level; that is, it influences gene activation and messenger RNA synthesis. Elucidation of its mode of action has been hampered by the inability to characterize its receptors. The proposal of Jones and Sharp (1997) that USP is a receptor for JH in *Drosophila* is intriguing but requires extension to other insects. One suggestion for the manner in which JH works is that it may alter the conformation of the chromatin so that 20-HE can activate only larva-specific genes.

6.2. Factors Initiating and Terminating Molt Cycles

Compared with the enormous volume of literature on the endocrine interactions that regulate growth and molting, relatively little is known about the external factors that initiate or terminate molting cycles.

A number of environmental variables have been shown to affect growth and molting, and some of these clearly exert their effect via the neurosecretory system, that is, they stimulate or depress the synthesis/release of hormones from the brain. Probably the best studied of these variables are feeding and photoperiod.

In the blood feeders *Rhodnius* and *Cimex* and in some sap-feeding Hemiptera, for example, *Oncopeltus*, whose food is highly consistent in nature, engorgement causes distension of the abdominal wall and initiates a molt cycle. Information from stretch receptors in the wall passes along the ventral nerve cord to the brain where neurosecretory cells are activated to release PTTH. In insects that feed more or less continuously through the stadium, also, the intake of food is probably an important stimulus for the release of neurosecretion. *Locusta migratoria*, for example, has stretch receptors in the wall of the pharynx, which are stimulated as food passes through the foregut. Information from the receptors reaches the

---

* The name derives from the fact that the gene (*usp*) producing this protein also has other developmental roles. When it malfunctions, in *Drosophila*, the larva develops an extra pair of posterior spiracles.

brain-corpora cardiaca complex via the stomatogastric nervous system (Clarke and Langley, 1963).

In continuous feeders it is often necessary for an insect to achieve a minimal nutritional status or body weight before a new molt cycle is initiated. This may be especially important for species whose diet is variable. In other words, during each stadium there is an initial period of obligate feeding, which results in acquisition of the minimal nutritive requirements (and release of sufficient PTTH to trigger a molt cycle), followed by a phase of facultative feeding, the nutritive contribution from which gives rise to larger larvae.

Photoperiod is another important environmental factor in the regulation of growth and molting, particularly in relation to diapause, a more or less prolonged condition of arrested development, which enables insects to survive periods of adverse conditions (Chapter 22, Section 3.2). Members of most species studied enter diapause when the daily amount of light to which they are exposed falls below a certain value (usually 14–16 hours). In diapause, an insect is physiologically "turned-off"; generally, it does not feed or move actively, and its metabolic rate is abnormally low. These effects result from inactivity of the endocrine system. In a manner that is not clear, short day lengths lead to reduced neurosecretory activity that, in turn, results in inactivity of molt glands and corpora allata. Conversely, diapause is terminated as the day length increases beyond a certain point in spring, because of renewed endocrine activity. In members of some species, however, the neurosecretory system must be exposed to low temperatures for a critical length of time during diapause before it can respond to increasing day length (Chippendale, 1977).

In some insects, the "feel" of the surroundings is important for continued normal development. For example, larvae of the wheatstem sawfly, *Cephus cinctus*, will not pupate if removed from the cavity at the base of the stem. Larvae of the squash fly, *Zeugoducus depressus*, live in the cavity of squash where the carbon dioxide concentration is initially about 4% to 6%. Pupation is delayed by this concentration of gas and will not occur until the level falls to about 1%, some 6 months later. This delay serves to synchronize the emergence of adult flies with the opening of the squash flowers (in which eggs are laid) the following season.

Still relatively unexplored are the changes of endocrine activity and other events that bring a molting cycle to a close with the shedding of the old cuticle. Certainly negative feedback pathways exist so that when the concentration of circulating hormone reaches a critical level, the activity of the gland producing it is depressed. The pathway may be direct, that is, the hormone itself depresses glandular activity. Alternatively, circulating ecdysone and JH may inhibit the activity of the PTTH- and allatotropic hormone-producing cells, respectively. A third possibility is that hormone levels are monitored by chemoreceptors, which send the information via sensory neurons to the brain. Reduction in activity of the molt glands and/or corpora allata might then be brought about via the nerves to the glands.

The complex behaviors that enable an insect to escape from the old exuvium are coordinated by the interplay of several hormones, including 20-HE, eclosion hormone (EH), pre-ecdysis-triggering hormone (PETH), ecdysis-triggering hormone (ETH), crustacean cardioactive peptide (CCAP), and bursicon (Truman, 1985, 1990, 1992; Reynolds, 1986; Horodyski, 1996; Myers, 2003). Only when the 20-HE level falls below a threshold value can molting occur, for two reasons. First, the target tissues for EH only acquire their competence to respond and, second, EH release only occurs at very low 20-HE concentrations. EH is produced in the Chinese oak silk moth, *Antheraea pernyi*, and other saturniid moths by neurosecretory cells in the ventral part of the brain. In larval instars these cells release EH at neurohemal organs on the hindgut; however, in the pharate adult the neurosecretory cells

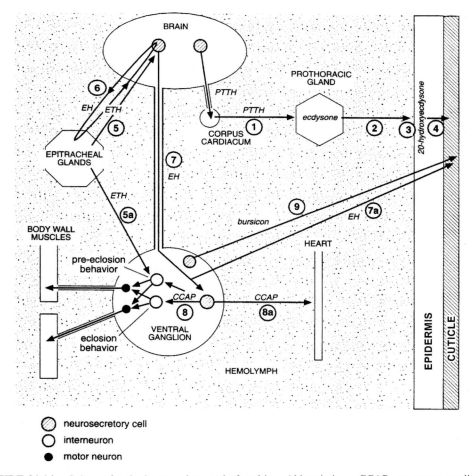

neurosecretory cell

interneuron

motor neuron

FIGURE 21.14. Scheme for the hormonal control of molting. Abbreviations: CCAP, crustacean cardioactive peptide; EH, eclosion hormone; ETH, ecdysis-triggering hormone; PETH, pre-ecdysis-triggering hormone. [After R. F. Chapman, 1998, *The Insects: Structure and Function* (4th ed.). Reprinted with the permission of Cambridge University Press.]

are restructured and terminate in the corpora cardiaca. For about 25 years after its discovery in the early 1970s, EH was thought to be the only hormone involved in the initiation of the various behaviors and physiological events that encompass the shedding of the exuvium. However, it is now apparent that release of EH is but one step in a complex hormonal cascade (Žitňan *et al.*, 1996; Žitňan and Adams, 2000).

Žitňan and Adams (2000) have proposed the following model for the control of molting, which includes three phases: pre-ecdysis I, pre-ecdysis II, and ecdysis (Figure 21.14). The behavioral sequence is initiated when low levels of EH are released. The EH induces PETH and ETH secretion from the Inka cells (Chapter 13, Section 3.4). PETH acts on the abdominal ganglia to trigger pre-ecdysis I; that is, it causes motor neurons to fire, bringing about strong dorsoventral muscle contractions in the body wall. ETH has two effects. First, by positive feedback it causes further, massive release of EH (and, in turn, more ETH), levels of which peak about 1 hour before molting occurs. Second, it induces (also by acting on the abdominal ganglia) pre-ecdysis II (strong posterioventral and proleg muscle contractions). The high level of EH has both direct and indirect effects related to molting. First, EH released from the tips of neurosecretory axons within the abdominal ganglia stimulates other neurons to release CCAP, which controls the initiation of the third phase, ecdysis. CCAP switches off pre-ecdysis I and II, then activates motor neurons that control the swallowing of air, heartbeat, and skeletal muscle functions. Together, these actions lead to the splitting of the old cuticle, wriggling free, and expansion of the new cuticle and wings. In addition, EH released into the hemolymph causes parts of the cuticle to plasticize, stimulates the cement-producing (Verson's) glands to discharge, and is a signal for release of bursicon. As described in Chapter 11 (Section 3.4), bursicon has important roles in tanning of the cuticle. It also appears to be involved in cuticle plasticization and the degeneration of specific muscles and neurons. Other processes in which bursicon may play a role include postecdysial cuticle deposition, postecdysial diuresis, and tracheal air-filling (Reynolds, 1986). In some insects that must wriggle to the surface of the substrate in which they pupated, proprioceptive stimuli are also important. These inhibit early release of bursicon, so that premature tanning of the adult cuticle is avoided.

The above model is based largely on data from experiments with Lepidoptera, especially *Manduca sexta*. However, both EH and ETH have been shown to occur in larvae from a wide range of insect orders (Truman *et al.*, 1981; Žitňan *et al.*, 2003), suggesting that this mechanism for hormonal control of ecdysis has been conserved across the Class.

7. Polymorphism

Polymorphism, the existence of several distinct forms of the same life stage of an organism, though not a common phenomenon in insects, occurs in representatives of several widely different orders. The phases of locusts (Orthoptera) and some caterpillars (Lepidoptera), castes of social insects (Isoptera and Hymenoptera), wing polymorphism in crickets (Orthoptera), aphids and other Hemiptera, and color polymorphism of mimetic butterflies (Lepidoptera) are examples of insect polymorphism. Though these examples refer only to difference of form, it should be appreciated that the physiology, behavior, and ecology of these forms are also different (Applebaum and Heifetz, 1999).

Polymorphism, like differentiation, has a genetic basis. In some examples of polymorphism, such as the color forms of butterflies and moths, the genetic system is relatively little influenced by short-term changes in environmental conditions. This so-called "genetic (obligate) polymorphism" includes transient polymorphism, the situation in which a trait is spreading through a population (e.g., melanism in the peppered moth, *Biston betularia*), and balanced polymorphism where a trait is maintained at a constant frequency in the population by opposing selection pressures [e.g., mimicry by the (edible) viceroy butterfly of the (distasteful) monarch butterfly (Figure 9.15)]. At the opposite extreme is "phenotypic (facultative) polymorphism" (now commonly referred to as "polyphenism") in which the development of characters is greatly influenced by changing environmental conditions and is manifest within a few generations (e.g., caste polymorphism, phase polymorphism, and wing polymorphism). The changing environmental conditions exert their influence via the neuroendocrine system; in other words, it is changes in the hormonal milieu that lead to polyphenism, particularly changes in the level of JH, as will be shown in the examples described below (Nijhout and Wheeler, 1982).

Aphid polymorphism is a complex phenomenon for, in addition to extensive structural polymorphism (some species include as many as eight distinct forms) (Figure 21.15) and the physiological polymorphism that accompanies it, there is also "temporal" or "successive"

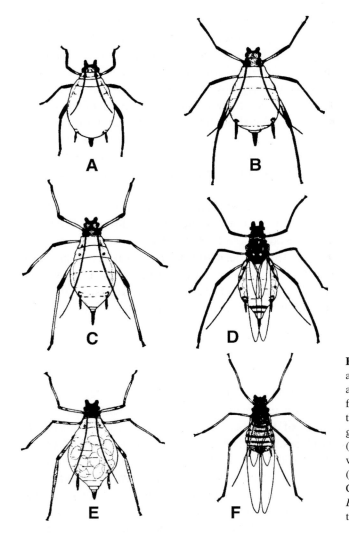

FIGURE 21.15. Forms of the vetch aphid. *Megoura viciae*. The fundatrix (A), a parthenogenetic female that emerges from the overwintering egg, produces the fundatrigeniae (B) from which many generations of wingless (C) or winged (D) virginoparae develop. In late summer, wingless egg-laying oviparae (E) and males (F) are formed. [From A. O. Lees, 1961, Clonal polymorphism in aphids, *Symp. R. Entomol. Soc.* **1**:68–79. By permission of the Royal Entomological Society.]

polyphenism in which these structural and physiological features gradually change from generation to generation (Lees, 1966; Hardie and Lees, 1985). Aphids reproduce parthenogenetically (and in some species paedogenetically) for a large part of the year, giving rise to large numbers of wingless individuals that can exploit the rich food supplies available in spring and summer. However, to avoid starvation caused by overcrowding, to move to nutritionally more valuable food sources, and to reproduce sexually, it is necessary for winged individuals (alates) to develop.

The development of alates is influenced by many environmental factors, for example, photoperiod, temperature, population density, and water content of food plants, all of which ultimately appear to bring about changes in endocrine activity. Thus, treatment of presumptive gynoparae (a winged migrant form) of the bean aphid (*Aphis fabae*) with JH mimics the effects of long-day conditions, causing them to molt to wingless adults and to show a preference for tic bean (*Vicia fabae*), their normal summer host. Conversely, topical application of precocene (which destroys the corpus allatum) to winged adults of other species stimulates production of winged progeny. In some species these environmental factors act directly on an individual to modify its form, whereas in others the effect is not seen until the following generation. For example, when first- or, to a lesser extent, second-instar larvae of the green peach aphid (*Myzus persicae*) are kept under crowded conditions, they develop into winged forms. In contrast, in the vetch aphid (*Megoura viciae*) crowding young larvae does not induce wing development either in the larvae or in their progeny. In this species, sensitivity to crowding develops in the fourth (final) instar and is retained through adulthood. Thus, when older larvae or wingless adults are crowded, their progeny are winged. Experiments have indicated that the stimulus is not visual or chemical but due to repeated physical contact between individuals. Interestingly, ants that tend aphids for their honeydew have a "tranquilizing" effect on the aphids. This effect apparently leads to reduction in the amount of physical contact between the aphids, which thus remain apterous, to the ants' obvious advantage. It is presumed that the crowding stimulus operates via the brain, which somehow reduces corpus allatum activity. The nature of the link between the brain and corpus allatum is not known but is likely to be hormonal because in species such as *M. viciae* the crowding stimulus is received by the maternal brain, but its effect is made apparent in the progeny. In other words, it is the corpora allata of the developing embryos within the mother whose activity is modified, yet there is no nervous connection between these glands and the mother's brain (Lees, 1966).

In colonies of social insects, individuals fall into a number of functionally and, usually, structurally distinct castes. In lower termites, for example, there is a pair of primary reproductives (king and queen) which found the colony, supplementary (replacement) reproductives which develop as the colony reaches a certain size and eventually take over the reproductive function, soldiers, nymphs (juveniles with wing buds from which primary reproductives develop), and larvae (juveniles that lack wing buds). Each caste contains members of both sexes. The number of individuals belonging to each caste is normally maintained as a fixed proportion of the total number of insects in the colony by means of inhibitory pheromones secreted by already differentiated individuals (Chapter 13, Sections 4.1.1 and 4.1.2). When the concentration of a pheromone falls below a certain level, resulting from, for example, growth of the colony or death of the pheromone-producing individuals, inhibition no longer occurs, and differentiation of new individuals restores the correct proportion. The relationship of the castes and the course of development in a lower termite, *Zootermopsis angusticollis*, are indicated in Figure 21.16. Young larvae pass through several progressive molts (i.e., grow and differentiate) until they become pseudergates (false workers) comparable with the true workers of higher termites. Pseudergates may undergo additional progressive molts to form specific castes, or may molt without differentiation (stationary molts) (Yin and Gillott, 1975).

The inhibitory pheromones influence caste differentiation by modifying the activity of the endocrine system, especially the corpora allata. The extensive studies of Lüscher and his associates in the 1950s led to the development of an elaborate hypothesis for hormonal control (see Hardie and Lees, 1985). This included the suggestion that the corpora allata produced three distinct hormones: JH, for control of juvenile → juvenile and juvenile → adult differentiation (as in non-polymorphic insects); soldier-inducing hormone that promoted soldier formation; and gonadotropic hormone, which was produced in adult females to control oocyte development.

As a result of subsequent work by Lüscher's group and other authors, especially experiments involving application of JH analogues, it was realized that the corpora allata do not produce more than one hormone. Rather, all developmental possibilities result from variations in the concentration of circulating JH at critical times during a stadium. Nijhout

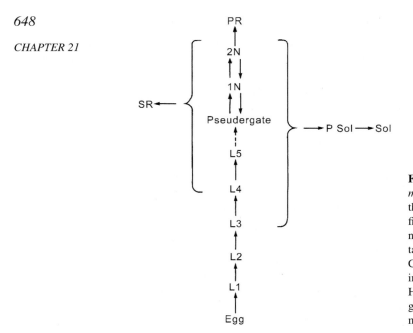

FIGURE 21.16. Course of development in *Zootermopsis angusticollis* (Isoptera). Broken arrow indicates the potential for several molts. Abbreviations: L1–L5, first to fifth instar larvae; 1N, 2N, first- and second-stage nymphs; PR, primary reproductive; SR, supplementary reproductive; PSol, presoldier; Sol, soldier. [After C.-M. Yin and C. Gillott, 1975, Endocrine activity during caste differentiation in *Zootermopsis angusticollis* Hagen (Isoptera): A morphometric and autoradiographic study, *Can. J. Zool.* **53**:1690–1700. By permission of the National Research Council of Canada.]

and Wheeler (1982) presented a model for JH-mediated control of caste differentiation in termites (Figure 21.17). Their proposal includes three critical periods in which JH is influential: the first of these is when sexual characters are determined; in the second period the JH level determines whether or not adult non-sexual characters differentiate; and in the third soldier characters are regulated. It must be stressed that the JH levels presented in the model are hypothetical; that is, they have never been measured experimentally.

In contrast to the situation in aphids and termites, whose different forms are clearly distinct (discontinuous polymorphism), in some locusts a spectrum of slightly different forms (continuous polymorphism) can be obtained by varying the population density over a number of generations. At opposite ends of the spectrum are the solitary form (phase),

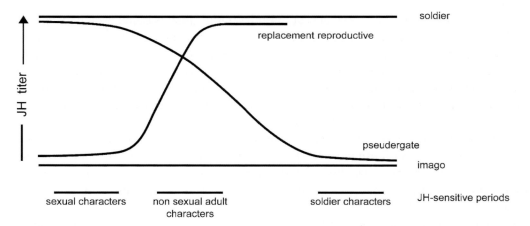

FIGURE 21.17. Proposed changes in hemolymph juvenile hormone level during caste differentiation in lower termites. In the larval stadium there are three critical periods when hormone level determines the path of differentiation to be followed. [From Nijhout, H. F., and Wheeler, D. E., 1982, Juvenile hormone and the physiological basis of insect polymorphisms, *Quart. Rev. Biol.* **57**:109–133. By permission of University of Chicago Press.]

produced when the population density remains continuously low for a long period of time, and the gregarious phase, typical of high-density populations. It is not feasible here to describe in detail the many differences that occur between the two phases. Suffice it to say there are significant differences in their structure, color, physiology (growth rates, occurrence of diapause, reproductive capacity), and behavior (feeding activity, migratory habits), which enable populations to develop and reproduce at the optimum rate according to the habitat in which they occur. For example, female solitary locusts, for which food is abundant, are larger, may lay up to five times more eggs, and have slightly shorter wings than gregarious females. Because of the greater population density, gregarious females must have the capacity to migrate to new food supplies (see also Chapter 22, Section 5.2) for which smaller body size and larger wings will be advantageous. In addition, they accumulate more fat (used as an energy source during migration) than solitary females, an aspect that may be correlated with production of fewer eggs (Pener, 1991; Pener and Yerushalmi, 1998).

Some experimental evidence suggests that the influence of population density and other environmental factors on phase determination is exerted via hormones, especially JH. For example, adults with some features of solitary locusts can be produced by implanting corpora allata or by applying JH to gregarious larvae. Solitary adults are structurally more juvenile than gregarious adults, and, as noted above, solitary females lay more eggs than their gregarious counterparts, both of which features are related to greater levels of JH in the hemolymph. Because of the above-normal levels of hemolymph JH, the molt glands of solitary individuals do not degenerate at eclosion. However, they do not produce ecdysone so never trigger a further molt. In the general model for hormonal control of polyphenism proposed by Nijhout and Wheeler (1982) and Nijhout (1994) phase change would occur as a result of the JH levels during specific critical periods of a stadium. However, other authors, notably Pener (1991), Pener and Yerushalmi (1998), and Breuer *et al.* (2003), while accepting that JH can induce certain characteristics of the solitary phase, have argued that JH is not the primary causal factor in phase determination; that is, changing JH levels are simply a result of other, as yet unknown, factors that induce phase change.

8. Summary

Among the evolutionary trends that may be seen in insect postembryonic development are (1) increasing separation of the processes of growth and accumulation of reserves (functions of the juvenile stage) from reproduction and dispersal, which are functions of the adult stage; (2) the spending of a greater proportion of the insect's life in the juvenile stage; and (3) an increasing degree of difference between larval and adult habits and form. The latter has been accompanied by modification of the final larval instar into a pupa in which the considerable changes from larval to adult form can occur.

Insects may be arranged in three basic groups in terms of the pattern of postembryonic development that they display. Apterygotes are ametabolous; that is, the changes from juvenile to adult form are very slight. Adults continue to molt, and the number of instars is both large and variable. Almost all exopterygotes are hemimetabolous. Juveniles broadly resemble adults and undergo only a partial metamorphosis. The number of instars is generally four or five and constant for a species. The major event in exopterygote metamorphosis is the full development of wings and genitalia. Internal organs grow progressively through larval life. Endopterygotes and a very few exopterygotes are holometabolous. Juveniles and adults are normally strikingly different and major changes (complete metamorphosis) occur in the pupa. In primitive endopterygotes most organs grow progressively during larval life, and metamorphosis consists mainly of the development of the flight mechanism. In most endopterygotes considerable differentiation of adult tissues occurs during metamorphosis, often from imaginal discs, groups of cells that remain embryonic through larval life, probably because of the hormonal milieu in juvenile instars.

Several types of endopterygote larvae may be distinguished: oligopod (including scarabaeiform and campodeiform larvae), which are most primitive; polypod (eruciform); and apodous (including eucephalous, hemicephalous, and acephalous larvae).

Pupae may be decticous (having functional mandibles) and exarate (appendages not sealed against the body), or adecticous and exarate or obtect (appendages sealed against the body). For protection, a pupa may be enclosed within a cocoon or cell, or may be heavily sclerotized and/or camouflaged, or in cyclorrhaph Diptera remain inside the cuticle of the last larval instar (the puparium).

Adult emergence (eclosion) is achieved by swallowing air to increase the body volume and thus split the pupal cuticle. When an insect pupates in a cocoon, etc., it may chew or tear its way out, either as a pharate adult or after eclosion, using mandibles, spines, or cocoon cutters. The cocoon of some species is equipped with an escape hatch.

Hormones regulate development. A molt cycle is initiated with the release of prothoracicotropic hormone from the brain, which stimulates release of ecdysone from the molt glands. The nature of a molt is determined by the concentration of JH at a critical period in the stadium. When the JH concentration is above a threshold value, a larval-larval molt follows; at below-threshold concentrations the molt is larval-pupal or pupal-adult.

Polymorphism is the existence of several distinct forms of the same stage of a species. It may have a genetic basis (as in transient and balanced polymorphism) or be induced by changing external conditions (polyphenism), whose effects are manifest via the endocrine system, specifically the concentration of JH at critical periods. Examples of polyphenism are caste differentiation in social insects, seasonal polymorphism in aphids, and the phases of locusts.

9. Literature

Postembryonic development, especially its endocrine control, continues to be a strong area of research, and literature on the subject is abundant. The following list is therefore highly selective. Sehnal (1985) provides basic information on insect growth and life cycles, as well as references to important early papers. Highnam and Hill (1977) and Wigglesworth (1985) review the endocrine regulation of postembryonic development from an historical perspective, thereby providing a basis for chapters of a more specialized nature in Volume 8 of the treatise edited by Kerkut and Gilbert (1985). Nijhout (1994), Gilbert *et al.* (1996), and Heming (2003) provide recent summaries of the endocrinology of postembryonic development. Chippendale (1977) and Denlinger (1985) have reviewed the hormonal control of diapause. Aspects of polymorphism are considered in the volumes edited by Kennedy (1961), Lüscher (1976), and Gupta (1991), and by Nijhout and Wheeler (1982), Retnakaran and Percy (1985), Hardie and Lees (1985), Pener (1991), Pener and Yerushalmi (1998), and Applebaum and Heifetz (1999). Imaginal discs are discussed in a series of papers edited by Ursprung and Nöthiger (1972), as well as by Oberlander (1985) and Doctor and Fristrom (1985).

Applebaum, S. W., and Heifetz, Y., 1999, Density-dependent physiological phase in insects, *Annu. Rev. Entomol.* **44**:317–341.

Breuer, M., Hoste, B., and De Loof, A., 2003, The endocrine control of phase transition: Some new aspects, *Physiol. Entomol.* **28**:3–10.

Cantera, R., Thompson, K. S. J., Hallberg, E., Nässel, D. R., and Bacon, J. P., 1995, Migration of neurons between ganglia in the metamorphosing insect nervous system, *Roux's Arch. Develop. Biol.* **205**:10–20.

Carter, D., and Locke, M., 1993, Why caterpillars do not grow short and fat, *Int. J. Insect Morphol. Embryol.* **22**:81–102.

Chippendale. G. M., 1977, Hormonal regulation of larval diapause, *Annu. Rev. Entomol.* **22**:121–138.

Clarke, K. U., and Langley, P. A., 1963, Studies on the initiation of growth and moulting in *Locusta migratoria migratorioides* R. and F. IV. The relationship between the stomatogastric nervous system and neurosecretion, *J. Insect Physiol.* **9**:423–430.

Dai, J.-D., and Gilbert, L. I., 1998, Juvenile hormone prevents the onset of programmed cell death in the prothoracic glands of *Manduca sexta, Gen. Comp. Endocrinol.* **109**:155–165.

Denlinger, D. L., 1985, Hormonal control of diapause, in: *Comprehensive Insect Physiology, Biochemistry and Pharmacology*, Vol. 8 (G. A. Kerkut and L. I. Gilbert, eds.), Pergamon Press, Elmsford, NY.

Doctor, J. S., and Fristrom, J. W., 1985, Macromolecular changes in imaginal discs during postembryonic development, in: *Comprehensive Insect Physiology, Biochemistry and Pharmacology*, Vol. 2 (G. A. Kerkut and L. I. Gilbert, eds.), Pergamon Press, Elmsford, NY.

Farris, S. M., Robinson, G. E., Davis, R. L., and Fahrbach, S. E., 1999, Larval and pupal development of the mushroom bodies in the honey bee, *Apis mellifera, J. Comp. Neurol.* **414**:97–113.

Gilbert, L. I., Rybczynski, R., and Tobe, S. S., 1996, Endocrine cascade in insect metamorphosis, in: *Metamorphosis: Postembryonic Reprogramming of Gene Expression in Amphibian and Insect Cells*, 3rd ed. (L. I. Gilbert, J. R. Tata, and B. G. Atkinson, eds.), Academic Press, San Diego.

Gupta, A. P. (ed.), 1991, *Morphogenetic Hormones of Arthropods: Roles in Histogenesis, Organogenesis, and Morphogenesis*, Rutgers University Press, New Brunswick, NJ.

Hall, B. L., 1999, Nuclear receptors and the hormonal regulation of *Drosophila* metamorphosis, *Am. Zool.* **39**:714–721.

Hardie, J., and Lees, A. D., 1985, Endocrine control of polymorphism and polyphenism, in: *Comprehensive In sect Physiology, Biochemistry and Pharmacology*, Vol. 8 (G. A. Kerkut and L. I. Gilbert, eds.), Pergamon Press, Elmsford, NY.

Heming, B. S., 2003, *Insect Development and Evolution*, Cornell University Press, Ithaca, NY.

Highnam, K. C., and Hill, L., 1977, *The Comparative Endocrinology of the Invertebrates*, 2nd ed., Arnold, London.

Hill, L., and Goldsworthy, G. J., 1968, Growth, feeding activity, and the utilisation of reserves in larvae of *Locusta, J. Insect Physiol.* **14**:1085–1098.

Horodyski, F. M., 1996, Neuroendocrine control of insect ecdysis by eclosion hormone, *J. Insect Physiol.* **42**:917–924.

Jones, G., and Sharp, P. A., 1997, Ultraspiracle: An invertebrate nuclear receptor for juvenile hormones, *Proc. Natl. Acad. Sci. U S A* **94**:13499–13503.

Kennedy, J. S. (ed.), 1961, Insect polymorphism, *Symp. R. Entomol. Soc.* **1**:115 pp.

Kent, K. S., Consoulas, C., Duncan, K., Johnston, R. M., Luedeman, R., and Levine, R. B., 1995, Remodelling of insect neuromuscular systems during insect metamorphosis, *Am. Zool.* **35**:578–584.

Kerkut, G. A., and Gilbert, L. I. (eds.), 1985, *Comprehensive Insect Physiology, Biochemistry and Pharmacology*. Pergamon Press, Elmsford, NY.

Lees, A. D., 1966, The control of polymorphism in aphids, *Adv. Insect Physiol.* **3**:207–277.

Lezzi, M., Bergman, T., Mouillet, J.-F., and Henrich, V. C., 1999, The ecdysone receptor puzzle, *Arch. Insect Biochem. Physiol.* **41**:99–106.

Lüscher, M. (ed.), 1976, *Phase and Caste Determination in Insects*, Pergamon Press, Elmsford, NY.

Myers, E. M., 2003, The circadian control of eclosion, *Chronobiol. Int.* **20**:775–794.

Nijhout, H. F., 1994, *Insect Hormones*, Princeton University Press, Princeton, NJ.

Nijhout, H. F., and Wheeler, D. E., 1982, Juvenile hormone and the physiological basis of insect polymorphisms, *Quart. Rev. Biol.* **57**:109–133.

Oberlander, H., 1985, The imaginal discs, in: *Comprehensive Insect Physiology, Biochemistry and Pharmacology*, Vol. 2 (G. A. Kerkut and L. I. Gilbert, eds.), Pergamon Press, Elmsford, NY.

Pener, M. P., 1991, Locust phase polymorphism and its endocrine relations, *Adv. Insect Physiol.* **23**:1–79.

Pener, M. P., and Yerushalmi, Y., 1998, The physiology of locust phase polymorphism: An update, *J. Insect Physiol.* **44**:365–377.

Retnakaran, A., and Percy, J., 1985, Fertilization and special modes of reproduction, in: *Comprehensive Insect Physiology, Biochemistry and Pharmacology*, Vol. I (G. A. Kerkut and L. I. Gilbert, eds.), Pergamon Press, Elmsford, NY.

Reynolds, S. E., 1986, Endocrine timing signals that direct ecdysial physiology and behavior, in: *Insect Neurochemistry and Neurophysiology, 1986* (A. B. Borkovec and D. B. Gelman, eds.), Humana Press, Clifton, NJ.

Riddiford, L. M., 1985, Hormone action at the cellular level, in: *Comprehensive Insect Physiology, Biochemistry and Pharmacology*, Vol. 8 (G. A. Kerkut and L. I. Gilbert, eds.), Pergamon Press, Elmsford, NY.

Riddiford, L. M., Hiruma, K., Lan, Q., and Zhou, B.-H., 1999, Regulation and role of nuclear receptors during larval molting and metamorphosis of Lepidoptera, *Am. Zool.* **39**:736–746.

Russell, V., and Dunn, P. E., 1996, Antibacterial proteins in the midgut of *Manduca sexta* during metamorphosis, *J. Insect Physiol.* **42**:65–71.

Sehnal, F., 1985, Growth and life cycles, in: *Comprehensive Insect Physiology, Biochemistry and Pharmacology*, Vol. 2 (G. A. Kerkut and L. I. Gilbert, eds.), Pergamon Press, Elmsford, NY.

Tobe, S. S., and Stay, B., 1985, Structure and regulation of the corpus allatum, *Adv. Insect Physiol.* **18**:305–432.

Truman, J. W., 1985, Hormonal control of ecdysis, in: *Comprehensive Insect Physiology, Biochemistry and Pharmacology*, Vol. 8 (G. A. Kerkut and L. I. Gilbert, eds.), Pergamon Press, Elmsford, NY.

Truman, J. W., 1990, Neuroendocrine control of ecdysis, in: *Molting and Metamorphosis* (E. Ohnishi and H. Ishozaki, eds.), Springer-Verlag, Berlin.

Truman, J. W., 1992, The eclosion hormone system of insects, *Prog. Brain Res.* **92**:361–374.

Truman, J. W., Taghert, P. H., Copenhaver, P. F., Tublitz, N. J., and Schwartz, L. M., 1981, Eclosion hormone may control all ecdyses in insects, *Nature* **291**:70–71.

Ursprung, H., and Nothiger, R. (eds.), 1972, *The Biology of Imaginal Disks*, Springer-Verlag, Berlin.

Whitten, J., 1968, Metamorphic changes in insects, in: *Metamorphosis* (W. Etkin and L. I. Gilbert, eds.), Appleton-Century-Crofts, New York.

Wigglesworth, V. B., 1985, Historical perspectives, in: *Comprehensive Insect Physiology, Biochemistry and Pharmacology*, Vol. 7 (G. A. Kerkut and L. I. Gilbert, eds.), Pergamon Press, Elmsford, NY.

Williams, D. W., and Shepherd, D., 2002, Persistent larval sensory neurons are required for the normal development of the adult sensory afferent projections in *Drosophila*, *Development* **129**:617–624.

Yin, C.-M., and Gillott, C., 1975, Endocrine activity during caste differentiation in *Zootermopsis angusticollis* Hagen (Isoptera): A morphometric and autoradiographic study, *Can. J. Zool.* **53**:1690–1700.

Žitňan, D., and Adams, M. E., 2000, Excitatory and inhibitory roles of central ganglia in initiation of the insect ecdysis behavioural sequence, *J. Exp. Biol.* **203**:1329–1340.

Žitňan, D., Kingan, T. J., Hermesman, J. L., and Adams, M. E., 1996, Identification of ecdysis-triggering hormone from an epitracheal endocrine system, *Science* **271**:88–91.

Žitňan, D., Žitňanová, I., Spalovská, I., Takáč, P., Park, Y., and Adams, M. E., 2003, Conservation of ecdysis-triggering hormone signalling in insects, *J. Exp. Biol.* **206**:1275–1289.

IV

Ecology

22

The Abiotic Environment

1. Introduction

The development and reproduction of insects are greatly influenced by a variety of abiotic factors. These factors may exert their effects on insects either directly or indirectly (through their effects on other organisms) and in the short- or long-term. Light, for example, may exert an immediate effect on the orientation of an insect as it searches for food, and may induce changes in an insect's physiology in anticipation of adverse conditions some months in the future. Another abiotic factor to which insects are now routinely subjected (deliberately or otherwise) are pesticides. Apart from the obvious effect of lethal doses of such chemicals, pesticides may have more subtle, indirect effects on the distribution and abundance of species, for example, alteration of predator-prey ratios and, in sublethal doses, changes in fecundity or rates of development.

Under natural conditions organisms are subject to a combination of environmental factors, both biotic and abiotic, and it is this combination that ultimately determines the distribution and abundance of a species. Frequently, the effect of one factor modifies the normal response of an organism to another factor. For example, light, by inducing diapause (Section 3.2.3), may make an insect unresponsive to (unaffected by) temperature fluctuations. As a result, an insect is not harmed by abnormally low temperatures, but nor does it become active in temporary periods of warmer weather that may occur in the middle of winter.

2. Temperature

2.1. Effect on Development Rate

The body temperature of insects, as poikilothermic animals, normally follows closely the temperature of the surroundings. Within limits, therefore, metabolic rate is proportional to ambient temperature. Consequently, the rate of development is inversely proportional to temperature (Figure 22.1). Outside these temperature limits the rate of development no longer bears an inversely linear relationship to temperature, because of the deleterious effects of extreme temperatures on the enzymes that regulate metabolism, and eventually temperatures are reached (the so-called upper and lower lethal limits) where death occurs.

655

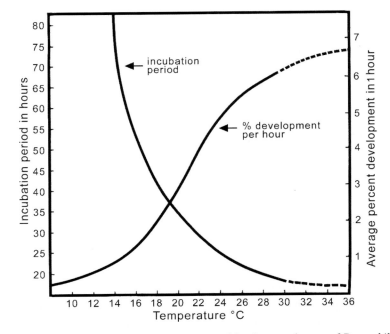

FIGURE 22.1. Relationship between temperature and rate of development in eggs of *Drosophila melanogaster* (Diptera). The two curves represent different ways of expressing this relationship, each being the reciprocal of the other. [After H. G. Andrewartha, 1961, *Introduction to the Study of Animal Populations*, University of Chicago Press. By permission of the author.]

Within the range of linearity the product of temperature multiplied by time required for development will be constant. This constant, known as the thermal constant or heat budget, is commonly measured in units of degree-days. This relationship will hold even when the temperature fluctuates, provided that the fluctuations do not exceed the range of linearity.

The temperature limits outside which development ceases and the rate of development at a given temperature vary among species, two seemingly obvious points that were apparently overlooked in some early attempts at biological control of insect pests. A predator that, on the basis of laboratory tests and short-term field trials, had good control potential was found to exert little or no control of the pest under natural conditions. Further study showed this to be related to the differing effects of temperature on development, hatching, and activity between the pest and its predator.

A broad correlation exists between the temperature limits for development and the habitat occupied by members of a species. For example, many Arctic insects that overwinter in the egg stage complete their entire development (embryonic + postembryonic) in the temperature range 0°C to 4°C, whereas in the Australian plague grasshopper, *Austroicetes cruciata*, development ceases below 16°C. This means that the distribution of a species will be limited by the range of temperature experienced in different geographic regions, as well as by other factors. However, the distribution of a species may be significantly greater than that anticipated on the basis of temperature data for the following reasons: (1) temperature adaptation may occur, that is, genetically different strains may evolve, each capable of surviving within a different temperature range; (2) the temperature limits of development may differ among developmental stages [this also serves as an important developmental synchronizer in some species (Section 2.3)]; and (3) the insect may have mechanisms for surviving extreme temperatures (Section 2.4).

Because of the ameliorating effects of the water surrounding them, aquatic insects are not normally exposed to the temperature extremes experienced by terrestrial species. Further, because ice is a good insulator, development may continue through the winter in some aquatic species in temperate climates, though air temperatures render development of terrestrial species impossible. Indeed, through evolution there has been a trend in some insects (e.g., species of Ephemeroptera and Plecoptera) to restrict their period of growth to the winter, passing the summer as eggs in diapause. Such species, whose developmental threshold is usually only slightly above 0°C, appear to gain at least two advantages from this arrangement. First, through the winter there is an abundance of food in the form of rotting vegetation, yet relatively little competition for it. Second, they are relatively safe from predators (fish) which are sluggish and feed only occasionally at these temperatures (Hynes, 1970b). Such a life cycle may also allow some species to inhabit temporary or still bodies of water that dry up or become anaerobic during summer.

2.2. Effect on Activity and Dispersal

Through its effect on metabolic rate, temperature clearly will affect the activity of insects. Many of the generalizations made above with regard to the influence of temperature on development have their parallel in relation to activity. Thus, there is a range of temperature within which activity is normal, though this range may vary among different strains of the same species. The temperature range for activity is correlated with a species' habitat; for example, in the Arctic, chironomid larvae are normally active in water at 0°C, and adults can fly at temperatures as low as 3.5°C (Downes, 1964).

By affecting an insect's ability to fly temperature may have a marked effect on a species' dispersal and, therefore, its distribution. Further, because flight is of such importance in food and/or mate location and, ultimately, reproduction, temperature is of great consequence in determining the abundance of species. Insects use various means of raising their body temperature to that at which flight is possible even when the ambient temperature is low. For example, they may be darkly colored so as to absorb solar radiation, or they may bask on dark surfaces, again using the sun's heat. Some moths and bumblebees beat their wings while at rest and simultaneously reduce hemolymph circulation in order to increase the temperature of the thorax (Chapter 17, Section 3.1). A dense coat of hairs or scales covers the body of some insects, which, by its insulating effect, will retard loss of heat generated or absorbed.

In extremely cold climates these physiological, behavioral, or structural features may no longer be sufficient to enable flight to occur, especially in a larger-bodied, egg-carrying female. Thus, different temperature-adaptation strategies are employed, some of which are exemplified especially well by Arctic black flies (Simuliidae: Diptera). Typical adult temperate-climate species are active insects that mate in flight, and females mayfly considerable distances in search of a blood meal necessary for egg maturation. In contrast, females of Arctic species seldom fly. Their mouthparts are reduced and eggs mature from nutrients acquired during larval life. Mating occurs on the ground as a result of chance encounters close to the site of adult emergence. In two species parthenogenesis has evolved, thereby overcoming the difficulty of being found by a male (Downes, 1964).

Temperature change, through its effect on the solubility of oxygen in water, may markedly modify the activity and, ultimately, the distribution and survival of aquatic insects. Members of many aquatic species are restricted to habitats whose oxygen content remains relatively high throughout the year. Such habitats include rivers and streams that are normally well oxygenated because of their turbulent flow and lower summer temperature, and high-altitude or -latitude ponds and lakes, which generally remain cool through the summer. Alternatively, as noted in Section 2.1, the life cycle of some species is such that the warmer (oxygen-deficient) conditions are spent in a resistant, diapausing, egg stage.

2.3. Temperature-Synchronized Development and Emergence

Many species of insects have highly synchronized larval development (all larvae are more or less at the same developmental stage) and/or synchronized eclosion, especially those that live in habitats where the climate is suitable for growth and reproduction for a limited period each year. Synchronized eclosion increases the chances of finding a mate. It may also increase the probability of finding suitable food or oviposition sites, or of escaping potential predators. Synchronized larval development also may be related to the availability of food, and in some situations it may be necessary in order to avoid interspecific competition for the same resource. For certain carnivorous species, such as Odonata, synchronized development may help reduce the incidence of cannibalism among larvae.

Perhaps not surprisingly in view of its effects on rate of development and activity, temperature is an important synchronizing factor in the life of insects. Its importance may be illustrated by reference to the life history of *Coenagrion angulatum*, which, along with several other species of damselflies (Odonata: Zygoptera), is found in or around shallow ponds on the Canadian prairies (Sawchyn and Gillott, 1975). For these insects the season for growth and reproduction lasts from about mid-May to mid-October. For the remaining 7 months of the year C. *angulatum* exists as more or less mature larvae, which, between about November and April, are encased in ice as the ponds freeze to the bottom. (The larvae themselves do not freeze, as the ice temperature seldom falls more than a few degrees Celsius below zero as a result of snow cover.) In C. *angulatum* both larval development and eclosion are synchronized by temperature. Synchronized development is achieved (1) by means of different temperature thresholds for development in different instars, that is, younger larvae can continue to grow in the fall after the growth of older larvae has been arrested by decreasing water temperatures, and (2) by a photoperiodically induced diapause. Thus, samples collected in mid-September include larvae of the last seven instars, whereas those from early October are composed almost entirely of larvae of the last three instars. Conversely, after the ice melts the following April, younger larvae can continue their development earlier than their more mature relatives, so that by mid-May more than 90% of the larvae are in the final instar. After their release from the ice larvae migrate into shallow water at the pond margin whose temperature parallels that of the air. Emergence occurs when the air temperature is 20°C to 21°C (and the water temperature is about 12°C). It begins normally during the last week of May and reaches a peak within 10 days. Emergence of C. *angulatum* follows that of various chironomids and chaoborids (Diptera), which form the main food of the adult damselflies during the period of sexual maturation. The development and emergence of other damselfly species that inhabit the same pond are also highly synchronized but occur at different times of the growing season. This enables the species to occupy the same pond and make use of the same resources, yet avoid interspecific competition. This is discussed further in Chapter 23 (Section 3.2.1).

Though unpredictable on a day-to-day basis, temperature does have a regular seasonal pattern that controls the onset and termination of diapause in some species. Temperature is the primary diapause-inducing stimulus for some subterranean species [e.g., some ground beetles (Carabidae)], wood- and bark-inhabiting species, and pests of stored products that live in darkness. It also is the major cue for diapause induction in some insects living near the equator where changes in photoperiod are too small to act as signals of seasonal change (Tauber *et al.*, 1986; Denlinger, 1986). Temperature can also exert a strong influence on diapause and other photoperiodically controlled phenomena, as is discussed below (Section 3.2).

2.4. Survival at Extreme Temperatures

In many tropical areas climatic conditions are suitable for year-round development and reproduction in insects. In other areas of the world, the year is divisible into distinct seasons, in some of which growth and/or reproduction is not possible. One reason for this arrest of growth and/or reproduction may be the extreme temperatures that occur at this time and are potentially lethal to an insect. In many instances shortage of food would also occur under these conditions.

To avoid the detrimental effects of periods of moderately low (down to freezing) or high temperature, and to ensure that development and reproduction occur at favorable times of the year, insects use an array of behavioral and physiological mechanisms (Danks, 2001, 2002). First, the life history of many species is arranged so that the period of adverse temperature is passed as the immobile, non-feeding egg or pupa. Second, prior to the advent of adverse conditions [and it should be realized that the token stimulus that triggers this behavior is not, in itself, adverse (see Section 3.2)], an insect may actively seek out a habitat in which the full effect of the detrimental temperature is not felt. For example, it may burrow or oviposit in soil, litter, or plant tissue, which acts as an insulator. Third, it may enter diapause where its physiological systems are largely inactive and resistant to extremes of temperature.

2.4.1. Cold-Hardiness

Cold-hardiness refers to an insect's ability to adapt to and survive low temperatures. Some insects are "chill-intolerant," that is, suffer lethal injury even at temperatures above 0°C. Others are "chill-tolerant," though a period of gradual temperature acclimation (hardening) may be required for tolerance to develop (Bale, 1993, 1996; Sømme, 1999). For insects in environments that experience temperatures below 0°C, an additional problem presents itself, namely, how to avoid being damaged by freezing of the body cells. The formation of ice crystals within cells causes irreversible damage to and frequently death of an organism (1) by physical disruption of the protoplasm and (2) by dehydration, reduction of the liquid water content that is essential for normal enzyme activity. Insects that survive freezing temperatures are described as either freezing-susceptible or freezing-tolerant. Freezing-susceptible species are those whose body fluids have a lower freezing point and may undergo supercooling. Freezing-tolerant (= freezing-resistant = frost-resistant) species are ones whose extracellular body fluids can freeze without damage to the insect.

In both groups, two or three types of cryoprotectants (substances that protect against freezing) are produced. Cryoprotectants identified to date fall into three categories: (1) ice-nucleating agents (proteins), produced only in freezing-tolerant species; (2) low-molecular-weight polyhydroxyl substances such as proline, glycerol, sorbitol, mannitol, threitol, sucrose and trehalose; and (3) thermal-hysteresis or antifreeze proteins (Duman and Horwath, 1983; Lee, 1991; Bale, 2002). Typically, insects produce two or more polyhydroxyls. This may be because they are toxic at higher concentrations, an effect that can be avoided by the use of a multicomponent system.

To appreciate the mode of action of these cryoprotectants, it is necessary to understand the process of freezing. When water is cooled the speed at which individual molecules move decreases, and the molecules aggregate. As cooling continues there is an increased probability that a number of aggregated molecules will become so oriented with respect to each other as to form a minute rigid latticework, that is, a crystal. Immediately this crystal (nucleator) is formed the rest of the water freezes rapidly as additional molecules bind to the solid frame now available to them. Freezing of a liquid does not always depend on the formation of a nucleator, but can be induced by foreign nucleating agents such as dust particles or, in the present context, particles of food in the gut or a rough surface such as that of the cuticle.

In freezing-susceptible species cold-hardiness is attained in a two-step process (Bale, 2002). In the first step behavioral and physiological activities occur that collectively reduce the insect's chance of freezing. These may include emptying the gut of food and overwintering as a non-feeding pupa, hibernating in dry locations, building structures that prevent contact with moisture, reducing body water content, and increasing fat content. Collectively, these processes may lower the supercooling point to $-20°C$. In the second step polyhydroxyls and antifreeze proteins are produced. These molecules not only increase the concentration of solutes in the body fluid so that the freezing point is depressed, but by their chemical nature they considerably improve the insect's supercooling capacity; that is, the body fluids remain liquid at temperatures much below their normal freezing point. Because of their hydroxyl groups, the cryoprotectants are capable of extensive hydrogen bonding with the water within the body. The binding of the water has two important effects with respect to supercooling. First, it greatly reduces the ability of the water molecules to aggregate and form a nucleating crystal, and second, even if an ice nucleus is formed, the rate at which freezing spreads through the body is greatly retarded because of the increased viscosity of the fluid.

A remarkable degree of supercooling can be achieved through the use of cryoprotectants. In the overwintering larva of the parasitic wasp *Bracon cephi*, for example, glycerol makes up 25% of the fresh body weight (representing a 5-Mole concentration) and lowers the supercooling point of the hemolymph to $-47°C$. Perhaps a disadvantage to the use of supercooling as a means of overwintering is that the probability of freezing occurring increases both with duration of exposure and with the degree of supercooling so that, for example, an insect might freeze in 1 minute at $-19°C$ but survive for 1 month at $-10°C$. Thus, to ensure survival an insect must have the ability to remain supercooled at extreme temperatures for significant periods of time, even though the average temperatures to which it is exposed may be $10°C$ to $15°C$ higher. In other words, it may have to produce much more antifreeze in anticipation of those extremes than would be judged necessary on the basis of the average temperature.

The alternative method, employed by freezing-tolerant species, is to permit (be able to withstand) a limited amount of freezing within the body. Freezing must be restricted to the extracellular fluid, as intracellular freezing damages cells. Ice formation in the extracellular fluid, which is accompanied by release of heat (latent heat of fusion), will therefore reduce the rate at which the body's tissues cool as the ambient temperature falls. Thus, it will be to an insect's advantage to have a large volume of hemolymph (and there is evidence that this is characteristic of pupae) and to be able to tolerate freezing of a large proportion of the water within it. Two subsidiary problems accompany the freezing-tolerant strategy: it is necessary (1) to prevent freezing from extending to the cell surfaces (and hence into the cells) and (2) to prevent damage to cells as a result of dehydration. As water in the extracellular fluid freezes, the osmotic pressure of the remaining liquid will increase, so that water will be drawn out of the cells by osmosis.

Freezing tolerance is generally found in insects living in extremely cold environments. The general strategy used by freezing-tolerant species is to synthesize ice-nucleating proteins in late fall/early winter (i.e., at temperatures above $-10°C$) that initiate freezing of extracellular fluids. This early induction of ice formation is advantageous because the rate of ice formation is less than at lower temperatures, thus allowing water to move out of cells to maintain osmotic equilibrium and reduce the likelihood of intracellular freezing (Baust and Rojas, 1985). Through the winter both intra- and extracellular polyhydroxyls are generated. With their ability to bind extensively with water the extracellular cryoprotectants will retard the rate at which freezing spreads, while the intracellular cryoprotectants will hold water within cells, to counteract the outwardly pulling osmotic force. It has also been suggested that the cryoprotectants may bind with plasma membranes to reduce their permeability to water. The role of the antifreeze proteins in freezing-tolerant insects is less clear (Bale, 2002). An early suggestion was that they may protect insects from freezing in early fall, before the ice-nucleating agents have been synthesized. A more likely function is that they prevent "secondary recrystallization" (refreezing) in the spring, when polyhydroxyls are being degraded under the influence of rising temperatures, yet the insect must be safeguarded against unexpected freezing temperatures.

Of interest is the evolutionary selection of glycerol as the dominant cryoprotectant because in high concentration this molecule is toxic at above-freezing temperatures. Storey and Storey (1991) suggested that at least three factors have been critical. First, two molecules of glycerol are produced from each molecule of its precursor hexose phosphate, important where colligative properties are concerned. Second, the synthesis of a triol (3-carbon-containing polyol) from a 6-carbon precursor conserves the carbon pool compared to synthesis of 4- or 5-carbon polyols (when the extra carbons are lost as carbon dioxide). Third, the pathways for glycerol synthesis and breakdown already exist in the fat body as part of lipid metabolism. Insects that use glycerol have biochemical pathways for synthesizing it in increasing amounts as the temperature falls progressively below $0°C$ and, equally, for degrading it when the temperature increases. Such has been shown to be the case in *Pterostichus brevicornis*, an Arctic carabid beetle that overwinters as a freezing-tolerant adult. In *P. brevicornis* glycerol synthesis begins when an insect is exposed to a fall temperature of $0°C$, and by the following December-January the concentration of this molecule may reach or exceed 30 g/100 ml, sufficient to enable an insect to withstand the $-40°C$ to $-50°C$ temperatures to which it may be exposed at this time. Conversely, as temperatures increase toward $0°C$ with the advent of spring, the glycerol concentration falls and the cryoprotectant disappears from the hemolymph by about the end of April, coincident with the return of above-freezing average temperatures (Baust and Morrissey, 1977). A comparable situation is observed in *Eurosta solidaginis*, a gall-forming fly that overwinters as a freezing-tolerant third-instar larva. The larva has a three-phase cryoprotectant system that comprises glycerol, sorbitol, and trehalose. Production of the molecules begins somewhat above $0°C$ but is probably triggered by declining temperatures. At temperatures below $0°C$, production of glycerol and sorbitol is greatly enhanced. With the return of warm weather in spring, the concentration of the three molecules rapidly declines (Baust and Morrissey, 1977).

Cold-hardiness and overwintering diapause (Section 3.2.3) frequently occur together, and the question of whether the phenomena are physiologically related has been widely debated (Denlinger, 1991). As noted above, studies have correlated the synthesis of cryoprotectants with lowered temperatures, and vice versa. However, only a handful of examples are known in which insects exposed to short days (but not low temperatures) develop increased cold tolerance, presumably by synthesizing cryoprotectants (Saunders, 2002). Denlinger (1991) concluded that, given the diversity of overwintering strategies found among insects, generalization was not possible. Thus, in some species cold-hardiness occurs in the absence of diapause; in others, diapause and cold-hardiness may occur coincidentally or may be physiologically linked (regulated by the same signals). According to Pullin (1996) there is increasing evidence that the production of polyhydroxyls is linked to the great suppression of metabolic rate which accompanies diapause.

3. Light

Light exerts a major influence on the ability of almost all insects to survive and multiply. A well-developed visual system enables insects to respond immediately and directly to light stimuli of various kinds in their search for food, a mate, a "home," or an oviposition site, and in avoidance of danger (Chapter 12, Section 7). But light influences the biology of many insects in another manner which stems from the earth's rotation about its axis, resulting in a regularly recurring 24-hour cycle of light and darkness, the photoperiod.* Because the earth's axis is not perpendicular to the plane of the earth's orbit around the sun, and because the orbit varies throughout the year, the relative amounts of light and darkness in the photoperiod change seasonally and from point to point over the earth's surface.

Photoperiod influences organisms in two ways: it may either induce short-term (diurnal) behavioral responses which occur at specified times in the 24-hour cycle, or bring about long-term (seasonal) physiological responses which keep organisms in tune with changing environmental conditions. In both situations, however, a key feature is that the organisms that respond have the ability to measure time. In short-term responses the time interval between the onset of light or darkness and commencement of the activity is important. For seasonal responses, the absolute day length (number of hours of light in a 24-hour period) is usually critical, though in some species it is the day-to-day increase or decrease in the light period that is measured. In other words, organisms that exhibit photoperiodic responses are said to possess a "biological clock," the nature of which is unknown, though its effects in animals are frequently manifest through changes in endocrine activity.

3.1. Daily Influences of Photoperiod

Various advantages may accrue to members of a species through the performance of particular activities at set times of the photoperiod. It may be advantageous for some insects to become active at dawn, dusk, or through the night when ambient temperatures are below the upper lethal limit, chances of predation are reduced, and the rate of water loss through the cuticle is lessened by the generally greater relative humidity that occurs at these times. For other insects, in which visual stimuli are important, activity during specific daylight hours may be advantageous; for example, food may be available for only a limited part of the day, or conversely, other, detrimental factors may restrict feeding to a specific period. For many species it is clearly beneficial for their members to show synchronous activity, as this will increase the chance of contact between sexes. "Activity" in this sense is not

---

* As Beck (1980) noted, some authors use this term to describe the light portion of a light-dark cycle (i.e., synonymously with day length).

restricted to locomotion, however. For example, in many species of moths, it is by and large only the males that exhibit daily rhythms of locomotor activity. The females are sedentary, but, in their virgin condition, have daily rhythms of calling (secretion of male-attracting pheromones) that enable males to locate them.

3.1.1. Circadian Rhythms

In a few species daily rhythms of activity are triggered by environmental cues and are therefore of exogenous origin. For example, the activity of the stick insect *Carausius morosus* is directly provoked by daily changes in light intensity. However, in most species these rhythms are not simply a response to the onset of daylight or darkness; that is, dawn or dusk do not act as a trigger that switches the activity on or off. Rather, the rhythms are endogenous (originate within the organism itself) but are subject to modification (regulation) by photoperiod and other environmental factors. That the rhythm originates internally may be demonstrated by placing the organism in constant light or darkness. The organism continues to begin its activity at approximately the same time of the 24-hour cycle, as it did when subject to alternating periods of light and darkness. Because the rhythm has an approximately 24-hour cycle, it is described as a circadian rhythm. When the rhythm is not influenced by the environment, that is, when environmental conditions are kept constant, the rhythm is described as "free-running." When environmental conditions vary regularly in each 24-hour cycle, and the beginning of the activity occurs at precisely the same time in the cycle, the rhythm is "entrained." For example, if a cockroach begins its locomotor activity 2 hours after darkness, this activity is said to be photoperiodically entrained. The role of photoperiod is therefore to adjust (phase set) the endogenous rhythm so that the activity occurs each day at the same time in relation to the onset of daylight or darkness. Though photoperiod is probably the most important regulator of circadian rhythms in insects, other environmental factors such as temperature, humidity, and light intensity, as well as physiological variables such as age, reproductive state, and degree of desiccation or starvation may modify behavior patterns. Photoperiodically entrained daily rhythms are known to occur in relation to locomotor activity, feeding, mating behavior (including swarming), oviposition, and eclosion, examples of which are given below.

Many examples are known of insects that actively run, swim, or fly during a characteristic period of the 24-hour cycle, this activity usually occurring in relation to some other rhythm such as feeding or mate location. In *Periplaneta* and other cockroaches activity begins shortly before the anticipated onset of darkness, reaches a peak some 2–3 hours after dark, and declines to a low level for the remaining period of darkness and during most of the light period (Figure 22.2A). *Drosophila robusta* (Figure 22.2B) flies actively during the last 3 hours of the light phase but is virtually inactive for the rest of the 24-hour period. Male ants of the species *Camponotus clarithorax* are most active during the first few hours of the light period but show little activity at other times (Figure 22.2C). The above examples show a well-defined single peak (unimodal rhythm) of activity. Other species, however, have bimodal or trimodal rhythms. For example, females of the silver-spotted tiger moth, *Halisidota argentata*, show two peaks of flight activity during darkness, the first shortly after darkness begins, the second about midway through the dark period (Figure 22.3A). Males of this species, in contrast, have a trimodal rhythm of flight activity (Figure 22.3B) (Beck, 1980).

Rhythmic feeding activity is apparent in larvae of some Lepidoptera, for example, *H. argentata*, which feed almost exclusively during darkness. Female mosquitoes, too,

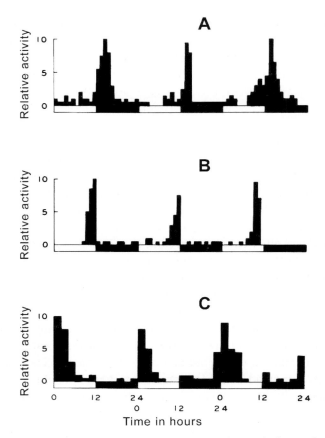

FIGURE 22.2. Locomotor activity rhythms in insects, illustrating photoperiodic entrainment. (A) *Periplaneta*; (B) *Drosophila*; and (C) *Camponotus*. [From S. D. Beck, 1968, *Insect Photoperiodism*. By permission of Academic Press, Inc., and the author.]

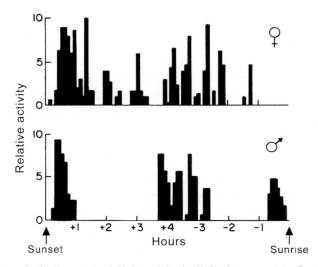

FIGURE 22.3. Photoperiodically entrained flight activity in *Halisidota argentata* (Lepidoptera). [From D. K. Edwards, 1962, Laboratory determinations of the daily flight times of separate sexes of some moths in naturally changing light, *Can. J. Zool.* **40**:511–530. By permission of the National Research Council of Canada.]

show peaks of feeding activity either at dawn or dusk, or during both of these periods, though there is some argument with regard to whether feeding activity is endogenous or simply a direct response to a particular light intensity.

Several good examples may be cited to illustrate the importance of photoperiod in entraining daily endogenous rhythms of mating behavior. Many virgin female Lepidoptera begin to secrete male-attracting pheromones shortly after the onset of darkness and are maximally receptive to males about midway through the dark period. Equally, males show maximum excitability to these pheromones in the early part of the dark period. The males of certain ant species undertake mating flights at characteristic times within the light period, typically near dawn or dusk. Mosquitoes and other Nematocera form all-male swarms that females enter for insemination. Formation of these swarms, which occurs both at dawn and at dusk, is an endogenous rhythm, entrained by photoperiod, though temperature and light intensity are also involved (Beck, 1980).

For some insect species, egg laying has been shown to be a photoperiodically entrained endogenous rhythm. In the mosquitoes *Aedes aegypti* and *Taeniorhynchus fuscopennatus*, for example, oviposition is concentrated in the period immediately after sunset and before dawn, respectively. In other mosquitoes, however, no oviposition rhythm exists and egg laying appears to be dependent on light intensity.

Many examples are known of insects that molt to adults during a characteristic period of the day. Many tropical Odonata exhibit mass eclosion during the early evening and are able to fly by the following morning. Corbet (1963) suggested that this minimizes the effects of predators such as birds and other dragonflies that hunt by sight. In temperate climates where nighttime temperatures are generally too low for emergence, there may be a switch to emergence during certain daylight hours. In more rigorous climates temperature appears to override photoperiod* as a factor regulating emergence, which occurs opportunistically at any time of the day provided the ambient temperature is suitable (Section 2.3). Species of Ephemeroptera and Diptera also have daily emergence patterns, which may be associated with immediate mating and oviposition. Though many insect species are known that have a daily emergence rhythm, for only a few of these, mainly Diptera, is experimental evidence available that proves the endogenous nature of the rhythm. In contrast to the previously described rhythmic processes, emergence occurs but once in the life of an insect and results in the appearance of a very different developmental stage, the adult. Nevertheless, this single event, like daily repeated processes, is an endogenous rhythm, entrained by environmental stimuli, especially photoperiod, that exert their effect in earlier developmental stages. For example, populations of many *Drosophila* species emerge at maximum rates 1–2 hours after dawn on the basis of photoperiodic entrainment either in the larval or pupal stage. Thus, if a culture of *Drosophila* larvae of variable ages is maintained in darkness from the egg stage except for one brief period of light (a flash lasting as little as 1/2000 of a second is sufficient) the adults will emerge at regular 24-hour intervals, based on the onset of the light period being equivalent to dawn; that is, the beginning of the light period serves as the reference point for entraining the insects' emergence rhythm (Beck, 1980).

Physiological and molecular studies of light-regulated circadian rhythms have revealed that, although there is a basic operating system, this may occur in a variety of forms (Saunders, 2002). The basic system comprises three components: an input pathway, which includes a photoreceptor; an oscillator ("clock" or pacemaker) that receives the entraining

---

* In the Arctic summer there are, of course, 24 hours of light per day, and photoperiod cannot serve as an entraining factor for diurnal rhythms.

signal from the photoreceptor; and the output pathway that connects the clock to the effector structures for the rhythms.

In cockroaches, crickets and some beetles the compound eyes are the photoreceptors, and a clock is located near the medulla regions of each optic lobe (Figure 13.5). However, in flies, moths and other beetles neither the compound eyes nor the ocelli are essential as photoreceptors for circadian rhythms. These insects may use several photoreceptors for entrainment, including groups of neurons within the central brain region. In flies the clock is located in the lateral neurons (located near the border of the optic lobe and brain) or their equivalent, while in moths the clock lies deep in the central part of the brain. Output pathways vary in their nature, depending on the rhythmic activity that is being controlled. For example, in cockroaches the clock connects with interneurons that run to the thoracic ganglia where locomotor activity is regulated. By contrast, the eclosion rhythm in moths is triggered by eclosion hormone whose release is controlled by the clock in the brain.

Though behavioral rhythms such as locomotor activity and eclosion are regulated by a central clock, it is evident that many other circadian rhythms operate independently; that is, many organs and tissues possess their own clock (Giebultowicz, 2000, 2001). This is readily shown by separating a structure from the rest of the body and observing that it retains its rhythmicity of function. To date, peripheral clocks have been reported for gonads, Malpighian tubules, endocrine glands, epidermis, and some sense organs.

Molecular studies, mainly using *Drosophila* mutants, have identified at least 10 genes in the central brain oscillator that are involved in circadian locomotor and eclosion rhythms. Two of these genes, *period* and *timeless*, and their protein products (PER and TIM, respectively) are responsible for the actual timing mechanism, showing circadian activity and production, respectively. A third gene, *cryptochrome*, codes for a photosensitive protein that modulates the action of PER and TIM, thereby resetting the clock. Other genes are on the output pathway (i.e., downstream of the clock), and their gene products induce manifestation of the circadian rhythm. Homologous genes to *period* have been detected in the central oscillators of moths and cockroaches, as well as in organs with peripheral clocks (Giebultowicz, 2000), indicating that there may be a common molecular basis for all circadian clocks. For further details of the molecular components of circadian clocks, see Giebultowicz (2000, 2001), Stanewsky (2002), and Zordan *et al.* (2003).

3.2. Seasonal Influences of Photoperiod

Photoperiod affects a variety of long-term physiological processes in insects and, in doing so, allows a species to (1) exploit suitable environmental conditions and (2) survive periods when climatic conditions are adverse. Some of the ways in which species are enabled to exploit a suitable environment include being in an appropriate developmental stage as soon as the suitable conditions appear, and growing or reproducing at the maximum rate while conditions last. Obviously, to survive adverse conditions, members of a species must already be in an appropriate physiological state when the conditions develop. In other words, the physiological or behavioral changes are induced by cues that are not in themselves adverse. Thus, among the processes known to be affected by photoperiod are the nature (qualitative expression) and rate of development, reproductive ability and capacity, synchronized adult emergence, induction of diapause, and possibly cold-hardiness. Several of these processes are closely related and are therefore affected simultaneously. Other environmental factors, especially temperature, may modify the effects of photoperiod.

3.2.1. Nature and Rate of Development

In some species larval growth rates are affected by photoperiod. For some species growth is accelerated under long-day conditions (when there are 16 or more hours of light in each 24-hour cycle) and inhibited in photoperiods that contain 12 or fewer hours of light; for other species, the converse is true. Often the effect of photoperiod on growth rate is correlated with the nature of diapause induction; that is, species that grow more slowly under short-day conditions tend also to enter diapause as a result of short days. However, it should be noted that the growth rate of many species that enter a photoperiodically controlled diapause is not affected by photoperiod.

Exposure to different photoperiods such as occur in different seasons may result in the development of distinct forms of a species, that is, polyphenism. The physiological (endocrine) basis of polyphenism has been outlined in Chapter 21 (Section 7), and the present discussion is restricted to a consideration of its induction by photoperiod. Sometimes the forms that develop are so strikingly different that they were described originally as separate species. Beck (1980) cited as an example the European butterfly *Araschnia levana*, described originally as two species, *A. levana* and *A. prorsa*, but which is now known to be a seasonally dimorphic species. Caterpillars reared under long-day conditions metamorphose into the non-diapausing, black-winged (prorsa) form; when they have developed at short day lengths the caterpillars emerge as red-winged (levana) adults that overwinter in diapause. This example shows a typical feature of most dimorphic Lepidoptera, namely, that one form is characteristically found in summer and is non-diapausing, whereas the alternate form is the diapausing, overwintering stage.

Photoperiodically influenced polyphenism is also seen in the seasonal occurrence of normal-winged, brachypterous, and/or apterous forms of species of Orthoptera and Hemiptera. But perhaps the best-known example of the effects of photoperiod on development is that of polyphenism in aphids. The life cycle of aphid species (see Figure 8.8) is complex and varied but shows beautifully how an insect takes full advantage of suitable conditions for growth and reproduction. A key feature of the life cycle is the occurrence within it of wingless, neotenic females that reproduce viviparously and parthenogenetically. In many species the offspring are entirely female. This combination of features enables aphids to reproduce rapidly and build up massive populations in the spring and summer when weather conditions are good and food is abundant. As a result of the crowding that results from this reproductive activity, winged migratory forms develop, and a part of the population moves on to alternate host plants. From these migratory forms several more generations of female aphids (alienicolae) are produced (again through viviparity and parthenogenesis), which may be winged or apterous. Eventually the alienicolae give rise to winged sexuparae (all female) that migrate back to the original host plant, and whose progeny may be either winged males or wingless females (oviparae). These reproduce sexually and lay eggs that pass the winter in diapause on the host plant. The following spring each egg gives rise to a female individual, the "stem mother" or fundatrix, normally wingless, that reproduces asexually, and from which several generations of neotenic females (fundatrigeniae) arise. There are many variants of this generalized life cycle, most often through its simplification; that is, one or more of the life stages is omitted as, for example, in species that do not alternate hosts when migrants and whose offspring do not appear as distinct forms. Indeed, in some species sexual forms have never been described and reproduction appears to be strictly parthenogenetic.

The development of these seasonally occurring aphid forms is influenced by a variety of environmental factors, including day length. Crowding is the major factor that influences production of summer migrants, whereas the shorter days of late summer and early fall induce development of sexuparae and oviparae. For some species there is a critical day length for induction of oviparous forms. In *Megoura viciae* (Figure 21.15), for example, which does not alternate host plants (i.e., it has no migrant form, and the oviparae are produced directly from fundatrigeniae), the critical day length is 14 hours 55 minutes at 15°C. At greater day lengths continuous production of viviparous, parthenogenetic females occurs; when the day length is below this critical value oviparae are produced.

In some species production of males also is induced by short days, though temperature and maternal age exert a strong influence. For example, in the pea aphid *Acyrthosiphon pisum* male offspring are not produced by young females or by females reared under long-day conditions. Old females reared at short day lengths and temperatures from 13°C to 20°C produce a large proportion of males. Outside this temperature range the proportion of males declines.

3.2.2. Reproductive Ability and Capacity

The effects of photoperiod on reproductive processes are almost all indirect, that is, result from other photoperiodically induced phenomena, especially adult diapause (see below). By its effect on the nature of development, as in aphids, photoperiod may indirectly modify the fecundity of a species. Beck (1980) noted one example of an apparently direct effect of photoperiod on fecundity. In *Plutella xylostella*, the diamondback moth, egg production in individuals reared under long-day photoperiods averaged 74 eggs/moth, whereas egg production under short-day conditions was only half this value.

3.2.3. Diapause

Beck (1980, p. 119) described diapause as "a genetically determined state of suppressed development, the expression of which may be controlled by environmental factors." It is a physiological state in which insects can survive cyclic, usually long, periods of adverse conditions, unsuited to growth and reproduction, including high summer or low winter temperatures, drought, and absence of food. In other words, it includes both hibernation (overwintering) and estivation (summer dormancy). Insects enter diapause usually some time in advance of the adverse conditions and terminate diapause after the conditions have ended. In other words, natural selection has favored the development of a safety margin against prematurely unseasonal conditions. Furthermore, the factor that leads to the induction of diapause (most often photoperiod) is not in itself an adverse condition. Thus, diapause differs markedly from quiescence, which is a temporary form of dormancy, usually induced directly by the arrival of adverse conditions.

Occurrence and Nature. Diapause may occur at any stage of the life history, egg, larva, pupa, or adult, though this stage is usually species-specific. Only rarely does diapause occur at more than one stage in the life history of a species. Such is typically the case in species that require 2 or more years in which to complete their development. For example, in the cockroach *Ectobius lapponicus*, which has a 2-year life history, the first winter is passed as a diapausing egg, the second as a quiescent second- or third-instar larva or as a diapausing fourth-instar larva.

Anticipation of the arrival of adverse conditions means that the environmental stimuli that induce diapause must exert their influence at an earlier stage in development. Thus, egg diapause is the result of stimuli that affect the parental generation. These stimuli act on the female parent either in the adult stage or, more often, during her embryonic or larval development. In *Bombyx mori*, for example, the day length experienced by developing female embryos determines whether or not these insects will lay eggs that enter diapause. Specifically, exposure of embryos to long day lengths results in females that lay diapausing eggs, and vice versa. For *B. mori* good evidence exists for the production of a diapause hormone by females exposed during embryogenesis to long-day conditions. This hormone, synthesized in the subesophageal ganglion, has as its target organ the ovary, which is caused to produce diapause eggs. Whether this scheme is applicable to the induction of egg diapause in other species is not known.

Diapause may occur at any larval stage, though the instar in which it is present is usually characteristic for a species. In many species it occurs in the final instar and, upon termination, is immediately followed by pupation. In this situation it is referred to as prepupal diapause. In the induction of larval diapause environmental stimuli normally exert their influence at an earlier larval stage, though species are known in which the environmental stimulus is given in the egg stage or during the previous generation. For example, in the pink bollworm, *Pectinophora gossypiella*, the photoperiod experienced by the eggs, as well as that during larval life, is important in determining whether or not prepupal diapause occurs. In *Nasonia vitripennis*, a parasitic hymenopteran, induction of larval diapause is dependent on the age of the female parent at oviposition, as well as on the photoperiod and temperature to which she is exposed early in adult life.

The pupa is the stage in which a large number of species enter diapause. The environmental signal that induces diapause is generally given during larval development, though for some species the influence is exerted in the parental generation. In the Chinese oak silkworm, *Antheraea pernyi*, for example, the last two larval instars are sensitive to photoperiod, whereas in the horn fly, *Haematobia irritans*, pupal diapause results when the female parent has been exposed to short day lengths.

Diapause may also occur in adult insects when it is known as reproductive diapause. Either young adults or larval instars are the stages sensitive to environmental stimuli. Newly emerged adult Colorado potato beetles (*Leptinotarsa decemlineata*), when subjected to short day lengths, will enter diapause. In the boll weevil, *Anthonomus grandis*, short-day conditions experienced by larvae will induce diapause in the adult stage.

A few examples of intraspecific variation in the diapausing instar are known, especially in species that occupy a wide geographic range. In the pitcher-plant mosquito, *Wyeomyia smithii*, northern North American populations overwinter as third-instar larvae whereas those in the southern United States diapause one instar later.

Mansingh (1971) subdivided diapause and the events surrounding it into a number of phases (Figure 22.4). This arrangement is convenient for a description of the sequence in which various processes occur, though it must be realized that these phases normally are not clearly separated in time but merge gradually with one another. In the preparatory phase environmental factors induce changes in metabolic activity in anticipation of diapause, resulting generally in accumulation of reserves, especially fats, but including carbohydrates and in some hibernating species cryoprotectants. During this phase the metabolic rate remains normal. Entry into the first phase (induction phase) of diapause is signaled by great declines in metabolic rate and, in postembryonic stages, activity of the endocrine system. For example, in diapausing *Hyalophora cecropia*, a saturniid moth, the rate of oxygen consumption (a measure of metabolic rate) is only about 2% of the prediapause value; in larval European corn borers (*Ostrinia nubilalis*), which have a "weak" diapause (see below), the rate of oxygen consumption falls to about one-quarter of the prediapause level. In the

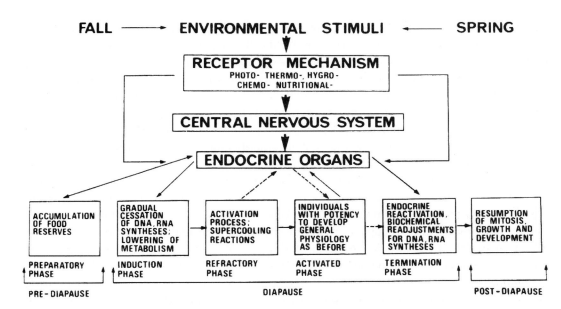

FALL ⟶ **ENVIRONMENTAL STIMULI** ⟵ **SPRING**

FIGURE 22.4. Phases before, during, and after diapause in overwintering insects. Probable (solid arrows) and possible (broken arrows) relationships between the environment, endocrine system, and the various phases are indicated. [From A. Mansingh, 1971, Physiological classification of dormancies in insects, *Can. Entomol.* **103**:983–1009. By permission of the Entomological Society of Canada.]

induction phase continued production of certain reserves, notably cryoprotectants in overwintering species, probably occurs. What causes the decline in metabolic rate associated with the beginning of diapause is uncertain. It may be related to the continued effects of environmental stimuli, or it may result from the changed metabolism of an insect. Collectively, the behavioral and physiological changes that occur during the preparatory and induction phases make up the so-called "diapause syndrome." The refractory phase (phase of diapause development), which follows induction, is perhaps the least understood aspect of the diapause condition. Originally, it was proposed that for overwintering insects diapause development would occur only during a period of exposure to low temperature. Such chilling was thought to be necessary for breakdown of the diapause-inducing or growth-inhibiting substances produced earlier, or for reactivation of specific systems (e.g., the endocrine system) important in postdiapause development. Tauber *et al.* (1986) and Hodek (2002) summarized the evidence that, in fact, no such chilling is required; rather, the refractory phase, like other phases of diapause, is maintained as a result of species-specific temperature and/or photoperiod requirements. Of course, under natural conditions, species may be subject to low winter temperatures, but these serve only to prevent premature development, to prevent breakdown of metabolites important in cold-hardiness and to synchronize postdiapause development. The refractory phase is followed by the activated phase, a period in which insects are capable of terminating diapause but do not do so because of prevailing environmental conditions (especially low temperature). Certain authors (e.g., Hodek, 2002) consider that once insects reach this stage, when their dormancy is (often) simply temperature-dependent, they must be considered as being quiescent, that is, no longer in diapause. Mansingh (1971), however, pointed out that, although insects in this phase are capable of continued development, several aspects of their physiology are similar to those of the refractory phase, for example, the greatly depressed respiratory rate and presence of cryoprotectants. He believed therefore that activated insects should be considered to be still in diapause. In Mansingh's scheme the final phase of diapause is the termination phase, which occurs when environmental conditions become favorable for development. In this phase the metabolic rate returns to normal, the endocrine system once more becomes active, body tissues again become capable of nucleic acid and protein synthesis, and any cryoprotectants present gradually disappear. As a result of these changes postdiapause development can begin.

In view of the varying degrees of severity of climatic conditions that insects in different geographic regions may encounter, it is perhaps not surprising to find that the intensity (duration and stability) of diapause varies. This variability, which is both interspecific and intraspecific, is manifest as a broad spectrum of dormancy that ranges from a state virtually indistinguishable from quiescence ("weak" or "shallow" diapause) to one of great stability ("strong" or "intense" diapause) in which an insect can resist extremely unfavorable conditions. In each situation the strength of diapause is precisely adjusted through natural selection to provide an insect with adequate protection against the adverse conditions, yet to continue growth and reproduction as soon as an amenable climate returns. Broadly speaking, insects from less extreme climates show weak diapause [called oligopause by Mansingh (1971)], in which development may not be completely suppressed; the insects may continue to grow slowly (and even molt) and feed when conditions permit during the period of generally adverse climate. In weak diapause the induction phase is relatively short, since the biochemical adjustments that an insect makes in order to cope with the adverse conditions are relatively simple. As a corollary of this, insects that overwinter in weak diapause are not, for example, very cold-tolerant. The refractory phase is short so that the activated phase is entered relatively soon after diapause has begun, and diapause is quickly terminated when environmental conditions return to normal. Conversely, in strong diapause, which is the rule in insects from severe climates, there is a lengthy induction phase, after which development is fully suppressed. The refractory phase usually lasts for several weeks or months, and the activated phase usually does not begin until diapause is more than half over. The termination phase is relatively slow, normally spanning 2 or 3 weeks after the return of suitable climatic conditions. Frequently insects that overwinter in strong diapause are very cold-hardy.

Diapause was formerly subdivided into facultative and obligate diapause. Facultative diapause described the environmentally controlled diapause of bivoltine and multivoltine species (having two or more generations per year) in which the members of certain generations had no diapause in their life history. Obligate diapause referred to the diapause found in univoltine species (those with one generation per year) in which every member of the species undergoes diapause. It was incorrectly assumed that in univoltine species diapause was not induced by environmental factors. However, experimental work on a number of univoltine species has revealed that in these species diapause is environmentally controlled. Further study may well demonstrate that this is always the case and render invalid the distinction between obligate and facultative diapause.

Induction, Maintenance, and Termination.   Various factors may influence the course of diapause. Photoperiod is especially important in the induction of diapause, though ambient temperatures, population density, and diet during the preparatory and induction phases may influence the incidence (proportion of individuals entering diapause) and intensity of dormancy. For many species photoperiod is also important, though temperature plays an increasing role, either alone or in combination with photoperiod, in diapause maintenance. Examples of both hibernating and estivating species that require a specific photoperiod to terminate diapause are also known.

Temperature and moisture have also been suggested as key factors in the termination of diapause, especially for some overwintering forms. However, it is often unclear whether these factors are serving as token stimuli for terminating diapause or are enabling postdiapause development to begin (Denlinger, 1986; Tauber *et al.*, 1986). According to Hodek (2002) and others, diapause typically terminates naturally simply by the passage of time; that is, no specific temperature or moisture stimulus is required and any effects of these factors is exerted on the quiescent phase that follows diapause.

In contrast to circadian clocks for which, as noted earlier, there is now a reasonable understanding of their mechanism, the photoperiodic clock that regulates insect seasonality remains largely a "black box." Despite extensive study no single hypothesis predominates; rather, a series of complex models have been proposed, details of which are beyond the scope of this text. Most of the models incorporate two interdependent components: a circadian-type clock system that measures the number of hours of light (or darkness) in each day, and a counter that sums the number of cycles. Adjustments are made for variations in temperature, latitude, food, etc. The information gathered is stored in a "memory link" until a "critical value" is reached, at which point an effector (e.g., the neuroendocrine system) is activated. For further information, refer to Takeda and Skopik (1997), Vaz Nunes and Saunders (1999), Tauber and Kyriacou (2001), and Saunders (2002).

For the great majority of insects that exhibit a photoperiodically induced diapause it is the absolute day length that is critical rather than daily changes in day length. Most insects studied to date show a long-day response to photoperiod (Figure 22.5A). That is, when reared under long-day conditions, they show continuous development, whereas at short day lengths diapause is induced. Between these extremes is a critical day length at which the incidence of diapause changes abruptly. Examples of insects that show a long-day response are the Colorado potato beetle, *L. decemlineata*, and the pink bollworm, *P. gossypiella*. In

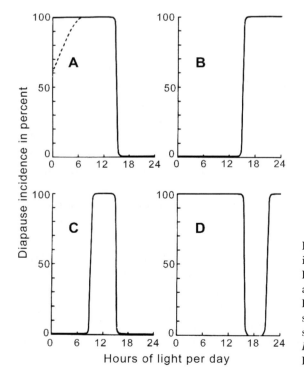

FIGURE 22.5. Different types of diapause incidence-day length relationships in insects. (A) Long-day; (B) short-day; (C) short-day-long-day; and (D) long-day-short-day. The hatched line in Figure 22.5A indicates that in some long-day species diapause incidence is less than 100% at very short day lengths. [From S. D. Beck, 1968, *Insect Photoperiodism*. By permission of Academic Press, Inc., and the author.]

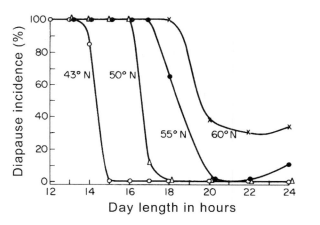

FIGURE 22.6. Effect of photoperiod on diapause incidence in *Acronycta rumicis* (Lepidoptera) populations from different northern latitudes. [From S. D. Beck, 1968, *Insect Photoperiodism*. By permission of Academic Press, Inc., and the author.]

a number of species, including the silkworm, *Bombyx mori*, diapause is induced when the day length is long, while at short day lengths development is continuous. Such insects are said to show a short-day response (Figure 22.5B). The European corn borer, *O. nubilalis*, and the imported cabbage worm, *Pieris brassicae*, have a short-day-long-day response to photoperiod; that is, the incidence of diapause is low at short and long day lengths, but high at intermediate day lengths (14–16 hours of light per day) (Figure 22.5C). The ecological significance of such a response is unclear, since under natural conditions, insects would already be hibernating when the day length was short. A few northern species of Lepidoptera behave in the opposite manner, namely, show a long-day-short-day response to photoperiod (Figure 22.5D). All photoperiods except those with 16–20 hours of light per day induce diapause. Again, however, the ecological value of such a response is uncertain.

The precise value of the critical day length for a species varies with latitude (Figure 22.6). For example, the sorrel dagger moth, *Acronycta rumicis*, studied in Russia by Danilevskii (1961), is a long-day insect which, near Leningrad (latitude about 60°N), has a critical day length of about 19 hours. In more southerly populations the critical day length is gradually reduced and is, for example, only 15 hours on the Black Sea coast (43°N).

In the dragonfly *Anax imperator* and perhaps a few other insects, diapause is induced by daily changes in day length rather than by absolute number of hours of light per day. *Anax* larvae that have entered the final instar by the beginning of June are able to metamorphose the same year. Those that reach the final instar after this date enter diapause and do not emerge until the following spring. It seems that larvae are able to determine the extent by which the day length increases. When the daily increment is 2 minutes or more per day larvae can develop directly, whereas at smaller increases or decreases in day length diapause is induced (Corbet, 1963).

Temperature may profoundly modify or overrule the normal effect of photoperiod on diapause induction. For example, the critical photoperiod depends on the particular (constant) temperature at which insects are maintained: in *A. rumicis* a 5°C difference in temperature results in a 1-hour difference in the critical day length. At extreme values the effects of temperature may overcome those of photoperiod with reference to induction of diapause. In long-day insects exposure to constant high temperature may completely prevent diapause induction regardless of photoperiod. Conversely, in short-day insects high temperature induces diapause, even under long-day conditions.

In nature temperatures normally fluctuate daily about a mean value. This daily fluctuation (thermoperiod) also may modify the influence of photoperiod according to whether or not it is in phase with the light-dark cycle. For example, in *A. rumicis* the incidence of diapause is increased by low nighttime temperatures and vice versa, though at very long day lengths (18 hours or more) temperature has little effect.

In most species studied diet influences the induction of diapause only slightly or not at all. In *P. gossypiella*, for example, the incidence of diapause induction may be increased by feeding the larvae on cotton seeds whose water content is low and/or oil content high, provided that the day length is not much greater than the critical value. Host-plant maturity can be correlated with the onset of diapause in a number of species though the chemical basis for this remains unknown. For some predaceous species (e.g., the convergent lady beetle, *Hippodamia convergens*), prey density is inversely correlated with the incidence of diapause.

For most insects termination of hibernation or estivation is not under photoperiodic control but occurs, under natural conditions, with the return of suitable temperatures for development. In a few species, however, exposure to appropriate photoperiods will terminate diapause. For example, in adult *Leptinotarsa decemlineata* and pupae of *H. cecropia* and *P. gossypiella* long-day conditions terminate diapause. Conversely, in some *Limnephilus* species (Trichoptera) that estivate as adults, diapause is ended by short day lengths.

In some species, especially those that overwinter in the egg stage or in a partially dehydrated condition, contact with liquid water is necessary for continued (i.e., postdiapause) development and activity. In diapausing larvae of *O. nubilalis*, for example, whose water content falls by midwinter to about 50% of the prediapause level, uptake of water (by drinking) is essential before the insect can continue its development (Beck, 1980). *Lestes congener*, a damselfly found on the Canadian prairies, oviposits in late summer in dry, dead stems of *Scirpus* (bulrush). The eggs begin to develop immediately but only to the end of anatrepsis and then enter diapause. Continued development in the spring will not begin until the eggs are wetted, regardless of temperature. Wetting is achieved under natural conditions as the level of the water rises during snow melt and also as a result of wind action, which causes ice movements and subsequent breaking and submersion of the plant stems (Sawchyn and Gillott, 1974a). As a result of such observations, it has frequently been claimed that water/moisture is a factor that terminates diapause. However, as noted above, in most instances it is unclear whether this is correct or whether in fact diapause has already ended; in other words, the insect is now quiescent but requires water for its continued development (Hodek, 2003).

4. Water

Water, an essential constituent of living organisms, is obviously an important determinant of their distribution and abundance. Active organisms must retain within their body a certain proportion of water in order for metabolism to occur normally. Deviation from this proportion for any length of time may result in injury or death. For some terrestrial insects, especially those from regions with striking dry and wet seasons, moisture may also serve as a token stimulus for seasonally regulated processes (but see below).

4.1. Terrestrial Insects

For terrestrial organisms, the problem generally is to reduce water loss from the body, which occurs as a result of surface evaporation and during excretion of metabolic wastes. Surface evaporation is especially important in small organisms, including insects whose surface area is relatively large in relation to body volume. That insects have been able to solve this problem is one of the main reasons for their success as a terrestrial group. Not only do insects generally possess a highly impermeable cuticle (Chapter 11, Section 4.2) and various devices for reducing water loss from the respiratory system (Chapter 15, Section 3), but they also have an efficient method of excretion, that is, one that uses a minimum of water for urine production (Chapter 18, Section 4.1). Such water loss as does occur is normally made up by drinking or from water in the food, though active members of a few species from very dry habitats are able to take up water from moist air should the opportunity arise, or use water produced in metabolism.

As dormant insects are largely unable to acquire water from their surroundings, they typically have a "prevention is better than cure" strategy; that is, they use behavioral or physiological mechanisms to reduce water loss (Danks, 2000). Behavioral means include spending the dormant period in cocoons, in soil or leaf litter, under bark, and in groups (e.g., ladybird beetles). Examples of physiological strategies are reducing the size of the spiracular opening, increasing the thickness of the cuticle, especially the wax layer, altering the composition of the wax to raise the transition temperature (Chapter 11, Section 4.2), increasing the osmotic pressure of the hemolymph by synthesizing cryoprotectants, and by significantly reducing metabolic rate. Some insects, nevertheless, lose considerable body water during diapause, and in hibernating species this is often correlated with the production of cryoprotectants (Section 2.4.1) (Block, 1996). For example, yellow woolly bear caterpillars (*Diacrisia virginica*) enter diapause as mature larvae weighing about 600 mg. During diapause their weight falls to about 200 mg, mainly as a result of the loss of water. However, the water loss is achieved by decreasing the hemolymph volume, enabling intracellular water to be kept at a physiologically suitable level.

For estivating insects, especially those that live in areas with distinct wet and dry seasons, moisture may act as an important stimulus for continued development and activity, in much the same manner as photoperiod and temperature regulate the seasonal ecology of many temperate species (Tauber *et al.*, 1998; Hodek, 2003). Though there is currently a lack of evidence for the importance of moisture as a seasonal regulator, Tauber *et al.* (1998) hypothesize that moisture may affect life cycles in three ways: (1) it may serve to induce, maintain, or terminate diapause; (2) it may be a developmental modulator, for example, by controlling rates of growth, maturation, and reproduction, as well as feeding and locomotion; and (3) It may be an important behavioral cue for such processes as molting, mating, and egg laying.

The importance of water is not restricted to postembryonic stages. During embryogenesis, also, the correct proportion of water must be present within the egg. Again, the primary problem is to prevent water loss (unlike postembryonic stages of most species, eggs cannot move in search of water or into habitats where loss is reduced!). To facilitate this a female may lay eggs in batches rather than singly, oviposit in a moist medium, and surround the eggs with protective material (Chapter 19, Section 7). In addition, the eggshell (chorion) is highly impermeable to water. As a result the egg is very resistant to desiccation and is frequently the stage in which periods of drought are overcome. Moisture may also be an important cue that triggers postdiapause embryonic development and hatching, as described in Section 3.2.3.

In view of the importance of water, it is not surprising to find that many terrestrial insects behave in a characteristic manner with respect to moisture in the surrounding air or substrate. The response may have immediate survival value for the individual concerned or may confer a long-term advantage on the individual and, ultimately, on the species. The ability to recognize and respond to slight differences in relative humidity enables an insect to move into a region of preferred humidity. Not only does this have immediate survival value, but because other individuals of the species will tend to respond similarly it may also increase the chances for perpetuation of the species. Some insects seek out sites with a preferred humidity, in which to enter diapause. Though this behavior is of no immediate value to the insect, it increases the chances of survival of the dormant stage. Similarly, female grasshoppers about to oviposit dig "test holes" with their ovipositor to determine the moisture content (and probably other physical and chemical features) of the soil. Eggs are normally laid in moist soil, and a female may retain the eggs in the oviducts for some time if she does not immediately find a suitable site. Again, this behavior has no immediate value to the female but certainly increases the eggs' chances of survival.

Thus far, the discussion has emphasized the harmful effects of too little water and the mechanisms by which terrestrial insects avoid this problem. On occasions, however, too much moisture may be equally detrimental to insects' survival. The effects of excessive moisture may be direct (namely, causing drowning) but more often are indirect. For example, insects that normally develop cold-hardiness partially as a result of dehydration may be less cold-hardy and therefore less capable of surviving low temperatures of winter if this has been preceded by a wet fall. Wet conditions may also affect a species' food supply. However, the most important way in which excessive moisture affects insect populations is by stimulating the development and spread of pathogenic microorganisms (bacteria, protozoa, fungi, and viruses). For example, in the summer of 1963 in Saskatchewan (Canada) the weather was abnormally humid, with above-average rainfall in some areas of the province. These conditions appeared ideal for the fungus *Entomophthora grylli*, which underwent a widespread epizootic, causing high mortality in populations of several species of grasshoppers, especially *Camnula pellucida*, the clear-winged grasshopper, and to a lesser extent *Melanoplus bivittatus* (two-striped grasshopper) and *M. packardii* (Packard's grasshopper). Such was the effect of the fungus on *C. pellucida* that by the fall of 1963 its proportion in the grasshopper species complex had fallen to 7% compared with 64% the previous year (Pickford and Riegert, 1964).

Finally, the beneficial effects of snow on the survival of insects must be noted. Snow is an excellent insulator and in extremely cold climates serves to reduce considerably the rate of heat loss from the substrate. Thus, the substrate remains considerably warmer than the air above the snow. For example, with an air temperature of –30°C and a snow depth of 10 cm, the temperature of soil about 3 cm below its surface is about –9°C. In the absence of snow the soil temperature at this depth is only a degree or two higher than that of the air. This means that species with only limited cold-hardiness may be able to survive the winter in cold climates provided that there is ample snow cover. In other words, because of snow a species may be able to extend its geographical range into areas with low winter temperatures. In Saskatchewan, for example, the damselflies *Lestes disjunctus* and *L. unguiculatus* overwinter as eggs (in diapause) laid in emergent stems of *Scirpus*. The eggs can tolerate exposure to temperatures as low as –20°C and remain viable. Below this temperature mortality increases significantly (Sawchyn and Gillott, 1974b). At Saskatoon, where this study was carried out, the *mean* temperature for January is, however, about –22°C, though the temperature frequently falls well below this value (the record low being about –48°C!). Field collection of eggs throughout the winter showed that, whereas the viability of eggs from beneath the snow remained near 100%, no eggs collected from exposed stems survived. Thus, the insulating effect of snow is essential to the survival of these species in this region of Canada. In addition, the snow cover may also prevent desiccation.

4.2. Aquatic Insects

The most important features of the surrounding medium that affect the distribution and abundance of aquatic insects appear to be its temperature, oxygen content, ionic content, and rate of flow. The influence of temperature on development and activity (through its effect on oxygen content) has already been outlined in Sections 2.1 and 2.2.

The ability of insects to regulate both the total ionic concentration and the level of individual ions in the hemolymph is a major determinant of their distribution. Typical freshwater insects are restricted to waters of low ionic content because, although they are capable of excreting excess water that enters their body osmotically, they have no mechanism for removing excess ions that enter the body when the insect is in a saline medium; that is, they cannot produce a hyperosmotic urine (Chapter 18, Section 4.2). Further, members of some species may be unable to colonize some freshwater habitats because these contain certain ions such as Mg^{2+} and Ca^{2+} in too high a concentration.

In contrast, members of many species that normally inhabit saline environments appear to be able to regulate their hemolymph osmotic pressure and ionic content over a wide range of external salt concentrations. In other words, they can produce hyperosmotic urine when it is necessary, in a saline medium, to excrete excess ions, or hypoosmotic urine, when in fresh water, to remove excess water from the body (Chapter 18, Section 4.3). As they are normally found only in saline habitats, it must be assumed that their distribution is governed by other environmental factors.

The insect fauna of an aquatic habitat may be correlated with the speed at which the water is moving. Insects in still or slowly moving water are not prevented from moving, for example, in search of food or to the surface for gaseous exchange. In contrast, rheophilic species (those that live in swiftly moving streams or rivers) have evolved structural, behavioral, and physiological adaptations to survive in these habitats. Among the structural adaptations that may be found in rheophilic insects are flattening or streamlining of the body, and the development of friction discs or hydraulic suckers (Hynes, 1970a,b). Flattening may take on differing significance among species, though ultimately its function is to enable insects to avoid being washed downstream by the current. In members of some species, which live on exposed surfaces, flattening enables them to remain within the boundary layer, a thin layer of almost static water covering the substrate. For members of most species flattening is associated with their cryptic habit, permitting them to live under stones, in cracks, crevices, etc. Streamlining, too, is a modification mainly used by insects to avoid currents by burrowing into the substrate, though members of a few streamlined species, for example, most species of *Baetis* and *Centroptilum* (mayflies), do live on exposed surfaces and are able to swim against quite strong currents (Hynes, 1970a,b).

The major physiological adaptation of rheophilic species is related to gas exchange. Because of the danger of being washed downstream, insects in moving water cannot come to the surface to obtain oxygen; they rely on oxygen dissolved in the medium. Through evolution, members of rheophilic species have become adapted to a medium with high oxygen content and conduct most or all gas exchange directly across the body wall. Further, they depend on the water current to renew the oxygen supply at their body surface. As a result, in many species, gills, if present, are reduced, and the ability to ventilate, by flapping the gills or undulating the abdomen, has been lost.

Their relative inability to move because of the current has been paralleled, in many rheophilic insects, by the evolution of devices that enable them to obtain food passively; that is, they depend on the current to bring food (especially microorganisms and detritus) to them. These devices include the nets built by many trichopteran larvae, fringes of hairs on the forelegs and/or mandibles of some larval Plecoptera, the fans on the premandibles of black fly larvae, and the sticky strings of saliva produced by the chironomid *Rheotanytarsus* (Hynes, 1970a,b).

An important factor in the distribution of aquatic insects, and one that is related to the extent of water movement, is the substratum. Many species of stream insects are characteristically associated with particular types of substratum. For some insects the significance of this association is easily understood. For example, water pennies [larvae of Psephenidae (Coleoptera)], found in fast-moving waters, require largish rocks to which they can become attached. Similarly, larval Blephaoceridae (Diptera) need smooth rocks, not covered with silt or algal growth, to which to attach their suckers. And some Leuctridae (Plecoptera) require gravel of the correct texture in which to burrow.

5. Weather

Because of their weight and relatively large surface area/volume ratio, insects may be profoundly affected by weather, especially by temperature, wind, and rain. Weather is a major factor limiting the abundance of many insect species, especially close to the edge of their range. Its effect may be both direct and indirect. For example, by altering the rate of evaporation of water from the body surface wind may be important in the water relations of the insect. Flight activity (whether or not flight occurs, the direction of movement, and the distance traveled) is also directly related to the strength and direction of the wind. Wind action may also exert indirect effects on insects, for example, by causing erosion of soil or snow so that the insects (or their eggs) are exposed to predators, extremes of temperature, or desiccation. Temperature has both obvious direct effects on development rate (Section 2.1) and less easily quantified indirect effects, for example, on a species' host plants, pathogens, and parasitoids, and is thus a key factor in insect population dynamics. Rain probably exerts its influence on most insect populations only indirectly, notably by affecting the availability and quality of food or the incidence of disease. However, it can occasionally have specific, direct effects. For example, through the formation of temporary pools it provides egg-laying sites for some mosquitoes and it is an important factor in termination of larval diapause for some species in semiarid, tropical climates. In other tropical species, for example, the desert locust, *Schistocerca gregaria*, which has an adult (reproductive) diapause through the dry season, the arrival of rain serves as a cue for copulation, dispersal, and oviposition (Denlinger, 1986).

Through its effect on the flight activity of winged insects and because insects by virtue of their weight are easily transported on wind currents, wind is an important factor in dispersal, the movement away from a crowded habitat so that scattering of a population results. Though a good deal of insect dispersal is of no obvious benefit, for some species the dispersal is adaptive, that is, confers a long-term advantage on the species by transferring some adult members to new breeding sites. Because of its advantageous nature, physiological, structural, and behavioral features that facilitate adaptive dispersal (= migration) will become fixed in a population through natural selection.

5.1. Weather and Insect Abundance

To illustrate the key role of weather, especially temperature, as a limiter of insect populations, it will be useful to refer to two specific examples, both important forest pests of eastern North America, namely, the fall webworm (*Hyphantria cunea*) and the spruce budworm (*Choristoneura fumiferana*). In eastern Canada, which is the northern limit of the range for *H. cunea*, the species is univoltine (has one generation per year), and summer temperatures play a major role in limiting population density through a variety of direct and indirect effects. First, late summer temperatures near or above optimum (about 32°C) permit rapid larval development, high larval survival rates, and formation of large pupae from which adults with high fecundity will emerge the following spring. Warm weather in the spring will allow pupae to complete metamorphosis and the adults to emerge early enough that their offspring can grow and pupate before the following fall. In addition, the following indirect influences operate. In cooler years larval development is prolonged, and larvae must therefore feed on older, less nutritious foliage; as a result, larval survival and the fecundity of the females that emerge are much reduced. This effect on survival is apparently cumulative; thus, if the offspring of these females have to feed on older foliage because of a second cool summer, very few survive beyond the larval stage. Finally, under cooler conditions the life cycles of *H. cunea* and one of its major parasitoids, the ichneumonid *Sinophorus validus*, are more closely synchronized so that more hosts will be infected.

Outbreaks of *C. fumiferana*, which despite its common name is primarily a pest of balsam fir (*Abies balsamea*), generally result from a sequence of warm, dry years. Such conditions, while obviously directly conducive to the insect's development, are much more influential over abundance through their stress effect on the host plant. Normally, mature balsam firs flower in alternate years; however, under these stress conditions they flower heavily each year. Ovipositing *C. fumiferana* lay more eggs on taller, more exposed trees which also produce more flowers. First-instar larvae, which hatch in late summer, seek out the cuplike remains of the flowers. After building a silken hibernaculum, they molt and enter diapause. Under these conditions the larval survival rate is high. In early spring the larvae begin to feed on the flowers that emerge before the vegetative buds and provide nourishing food for the larvae whose development is more rapid on flowers than on buds and needles. Also, the new foliage of flowering trees has a greater amino acid content than that of non-flowering trees and old foliage, a feature that leads to formation of larger pupae and more fecund females. Figure 22.7 summarizes the interactions that lead to population outbreaks in *C. fumiferana*.

5.2. Migration

Johnson (1969, p. 8) described migration as "essentially a transference of adults of a new generation from one breeding habitat to others." Implicit in this statement is the idea that migration takes place because the current habitat is already, or will soon become, unsuitable. In many species migration is preceded by highly synchronized adult emergence and begins shortly after the molt to the adult. In other words, it occurs in adults that are sexually immature or in reproductive diapause. In a sense, therefore, migration forms part of a species' development just as do mating and oviposition. The apparent incompatibility between reproductive development and migration led to the concept of the oogenesis-flight syndrome as a general phenomenon (Johnson, 1969). However, this viewpoint no longer seems tenable, given the many demonstrations of interreproductive migration (i.e., the migration of mature females that had already laid a proportion of their eggs) (Gatehouse and Zhang, 1995).

Given its central role in development and reproduction, it is not surprising that juvenile hormone (JH) also plays a major part in the initiation of migratory behavior (McNeil *et al.*, 1995; Dingle, 2001). In most insects JH stimulates egg development (Chapter 19,

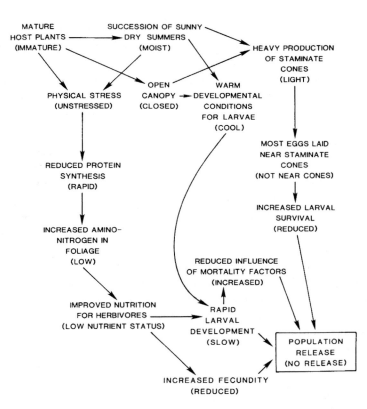

FIGURE 22.7. Interactions leading to population outbreaks in the spruce budworm (*Choristoneura fumiferana*). Terms in parentheses relate to populations under non-outbreak conditions. [From P. W. Price, *Insect Ecology*, 2nd ed. Copyright © 1984 by John Wiley and Sons, Inc. Reprinted by permission of John Wiley and Sons, Inc.]

Section 3.1.3). However, early experiments in which JH or its mimics were applied to insects indicated that the hormone also promotes migratory flight, which is seemingly at odds with the oogenesis-flight syndrome. Careful analyses have shown that the JH level is key: low levels affect neither activity; intermediate concentrations trigger migratory behavior; and high levels induce egg development (Rankin, 1991). Note, however, that in Lepidoptera, which produce a suite of JH homologues, it may be a specific blend of JH forms that induces migratory flight (McNeil *et al.*, 1995). The level of JH may also affect longevity in relation to migratory ability. For example, in the monarch butterfly (see below) the southbound migratory adults live from August/September to March of the following year. During this time, they are in reproductive diapause and their JH level is low. By contrast, adults of the summer generation live only about 2 months, reproduce, and have high JH levels (Herman and Tatar, 2001). In species with migratory and non-migratory populations the effects of JH may be expressed as morphological differences (polyphenism), especially with respect to the flight system. Thus, high JH levels can be correlated with short-winged or wingless non-migratory populations with high fecundity, whereas lower levels are found in fully winged forms capable of migration but with lower fecundity. Interestingly, the differing levels of JH seem not to be the result of altered corpus allatum activity, rather changes in the level of the degrading enzyme JH-esterase (Dingle, 2001).

The form of migration varies widely among species. Some of the variables are the proportion of the population that migrates, whether migration occurs in every generation or only in certain generations, the distance traveled, and the nature of the migratory movements (wind-dependent or wind-independent, feeding en route or proceeding directly). For example, the swarming flights of social insects, such as ants and termites, involve only a fraction of a colony's population, may be completed in a matter of minutes, and may take the migrating individuals only a few meters from the original colony. In contrast, the migrations of locusts are undertaken by all members of a population and may cover several thousand kilometers. The migrations extend over a number of weeks and are interspersed with short periods of feeding activity.

The nature of the navigational cues used by migrating insects is not well understood. Species that are wind-dependent respond to weather disturbances in which temperature, wind speed and direction, and light intensity may change rapidly. In the "correct" combination these variables trigger a mass take off, and the direction taken is wind-determined. When the combination changes (e.g., at lower temperatures and wind speeds), the swarm will land.

Species that are wind-independent, that is, migrate within the "flight boundary layer" under their own power, require an orientation system in order to reach a predetermined goal. For some diurnal species, a light-compass reaction, similar to that used by bees, exists. Other diurnal migrants may use the earth's magnetic field or local landmarks as cues (Srygley and Oliveira, 2001). On the basis of somewhat tenuous evidence, it has been suggested that the well-known migrant, the monarch butterfly (see below), might use geomagnetic cues for orientation. However, recent experiments appear to have shown this suggestion to be unwarranted; rather, the monarchs use a light-compass system (Mouritsen and Frost, 2002). Day-flying species that rely on wind currents for their movements typically also show positive phototropism. Their upward flight takes them out of the boundary layer (the shallow layer of relatively still air adjacent to the earth's surface) and into horizontally moving airstreams. Such insects also may make use of thermals, columns of warm rising air, to gain height.

Insects from a wide range of orders, but especially Ephemeroptera, Trichoptera, Diptera, and Lepidoptera, show nocturnal flight activity that can be correlated with the phase of the moon. Such species may use polarized light reflected from the moon's surface for navigational purposes. For other nocturnal migrants that show a common orientation even under dark and completely overcast conditions, other cues have been suggested, including the earth's magnetic field, infrared energy perception, and stellar navigation (Danthanarayana, 1986).

An important component in migration, whatever cues are used for direction finding, is the ability to compensate for varying conditions, notably the changing position of the sun or moon and crosswinds that may blow the insects off course. Thus, an integral part of the light-compass response and possibly astronavigation is the ability to time-compensate. To compensate for wind-drift effect, insects appear to use a ground reference mechanism (estimation of the relative motion of landmarks beneath their flight path), which causes them to turn into the wind until the cross-track component of their upwind speed is equal and opposite to the crosswind drift (Srygley and Oliveira, 2001).

5.2.1. Categories of Migration

Johnson (1969) suggested that all forms of migration may be arranged in three major categories, though there is gradation both within and between each of them. In the first category are included species that as adults, migrate from the emergence site to a new breeding site where they oviposit, then die. Johnson includes in this group species such as the house fly, all of whose members, at emergence, leave the old habitat and disperse randomly. After a period of maturation, they seek out new breeding sites. The category also contains species that seasonally produce populations of weakly flying, migratory individuals, for example, aphids, termites, and ants. Migrations are largely wind-dependent and may occur in every direction from the emergence site. The migration time and distance are usually short, and once individuals reach a suitable breeding site, they remain there for the rest of their lives. Indeed, on reaching such a site, some species characteristically shed their wings.

Also placed in the first category, but having migrations that are of a much grander scale, are the migratory locusts. Desert locusts, *Schistocerca gregaria*, for example, may travel thousands of kilometers as they move from one breeding area to another as each becomes unsuitable because of drought (Figure 22.8). Like those of aphids, etc., the migrations of desert locusts are wind-dependent. Because they inhabit fairly dry regions, breeding in the desert locust is synchronized with the arrival of a rainy season. As rain comes to different parts of the inhabited area at different times of the year, adults migrate in order to continue breeding activity. It is the winds on which locusts migrate that also bring rain to the new areas. During spring, breeding occurs in north and northwest Africa (in conjunction with rain in the Mediterranean region), in the Middle East across to Pakistan, and to the south in East Africa. In the latter regions, local seasonal rains occur at this time. As northern Africa and the Middle East become dry in early summer, locusts migrate southward on prevailing winds to an area that runs across Africa, from east to west, lying just south of the Sahara desert, then northward across southern Arabia and into Pakistan. This area is closely associated with the Inter-Tropical Convergence Zone (ITCZ), where hot, northbound air from the equatorial region meets cooler air flowing south. The mixing of these air masses within the ITCZ results in the production of rain and a reduction of wind speed so that the locusts again become earthbound. Locusts from East Africa, south of the ITCZ, move north and east on prevailing winds to be deposited in southern Arabia and India. These summer migrations may take locust swarms several hundred or even thousands of kilometers in a relatively short time. In contrast, the fall and winter movements that bring locusts to their spring breeding sites generally consist of a number of shorter migrations made over a longer period of time, mainly because the air temperatures at this time of the year are only intermittently suitable for migration. The northward migration results largely from cyclonic weather disturbances that move eastward across Africa every few days. These disturbances bring with them warm southerly winds on which locusts may be carried. As the winds push northward they mix with cooler air, which results in rainfall and temporary cessation of migration. Successive waves of warm air gradually bring the locusts to their spring habitat. The above summary of the annual movements of locusts is of necessity extremely simplified. Nevertheless, it shows how migration, based on wind currents, has evolved as an integral part of locusts' life history to facilitate year-round breeding activity through the exploitation of temporarily suitable habitats.

Finally, Category 1 includes some species (mainly Lepidoptera) whose migrations are independent of wind currents. That is, the insects do not rely on wind to make the migrations, though their movements are undoubtedly influenced by wind speed and direction. A much-studied example of a species that migrates under its own power is the great southern white butterfly, *Ascia monuste* (Pieridae), that migrates up and down the Florida coastline. In Florida the species breeds year-round but not in all localities simultaneously. Periodically populations comprising immature females and males of all ages make migratory flights over distances up to 150 km or more to new areas where *Batis maritima* (maritime salt wort),

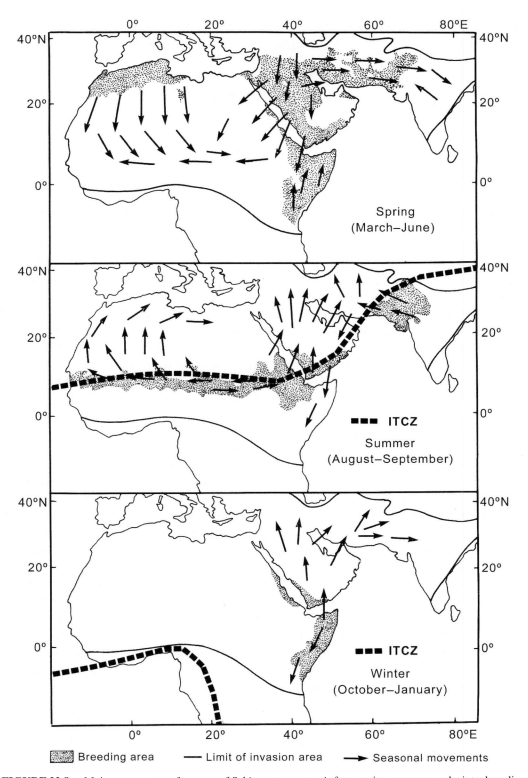

FIGURE 22.8. Major movements of swarms of *Schistocerca gregaria* from spring, summer, and winter breeding areas, in relation to the position of the Inter-Tropical Convergence Zone (ITCZ). [From Z. Waloff, 1966, *Antilocust Memoir* 8. Crown copyright 1966. Reproduced with permission of the Controller of Her Britannic Majesty's Stationery Office.]

the primary host plant, is abundant. In contrast to the situation in locust migration, it is not the arrival of adverse conditions that stimulates migration in *A. monuste*. The migrations occur from areas where food and oviposition sites are still abundant. Though the sun has been suggested as a reference point by which the butterflies orient themselves during flight, local cues are also important. For example, the insects may closely follow the shoreline, roads, railway tracks, or telephone lines.

In his second category Johnson (1969) included species whose migration is in two parts, an emigration to feeding sites where sexual maturation occurs, followed by a return flight to the original (or a similar) site of emergence where the insects oviposit. Many Odonata, for example, do not remain in the area of the pond from which they emerge but migrate to nearby woods or hedgerows to prey on other insects until mature. They then return to water, mate, and lay eggs. Some species regularly return to the feeding habitat between each oviposition period. Some species of mosquitoes also have a two-part migration, first to find a host on which to feed and later to locate an egg-laying site. In some cases the initial part of the migration is considerably longer than the second. Like that of Odonata, the migratory flight of some mosquitoes is wind-independent. For most migratory species, however, wind determines the direction and distance traveled.

The third category includes species that again have a two-part migration. The initial migratory flight takes the species to suitable hibernating or estivating sites where they enter diapause, after which they return to the region in which they emerged and reproduce. Within this type of migration three subcategories can be recognized. In the first, the sites of diapause are within the general breeding area of the species. Species that adopt this arrangement include the Colorado potato beetle, *Leptinotarsa decemlineata*, and the corn thrips, *Limothrips cerealium*.

Belonging to the second subcategory are species that migrate to a climatically different region prior to diapause. Especially common is migration between warmer lowland areas where the insects have emerged and mountainous regions, either to avoid summer heat or to overwinter. Such migrations are seen in various noctuid moths, for example, the army cutworm, *Euxoa auxiliaris*, in Montana, which moves southwest to the Rocky Mountains, and in some coccinellid beetles, such as the convergent lady beetle *Hippodamia convergens*, in northern California. Adult beetles first appear in early May and soon most migrate, using prevailing winds, to mountain canyons in the Sierra Nevada range where they aggregate under stones, litter, etc., and enter diapause. Diapause lasts for about 9 months, and the following February and March adults, again windborne, return to the valleys where a new generation is produced. Breeding activity is thus closely correlated with mass emergence of spring-breeding aphids. (Interestingly, as a result of human agricultural activities, aphids are now available on a year-round basis, and some populations of *H. convergens* no longer migrate but through the summer produce several generations of progeny, the last one of which overwinters in diapause at the breeding site.)

Migrations included in the third subcategory differ from those of the second subcategory only in terms of the distance covered, especially during prediapause movements. The classic example of a species in this group is the monarch butterfly, *Danaus plexippus*, whose remarkable migration (Figure 22.9) is intimately associated with the abundant *Asclepias* (milkweeds) found in North America. Two major populations of monarchs exist. The smaller western population (thought to have evolved from that east of the Rockies) spends the summer in the valleys and coastal areas of British Columbia, Oregon, Washington, Nevada, and California, and migrates each fall to at least 38 permanent overwintering sites along the Californian coast. The eastern population, whose summer range extends from the Great

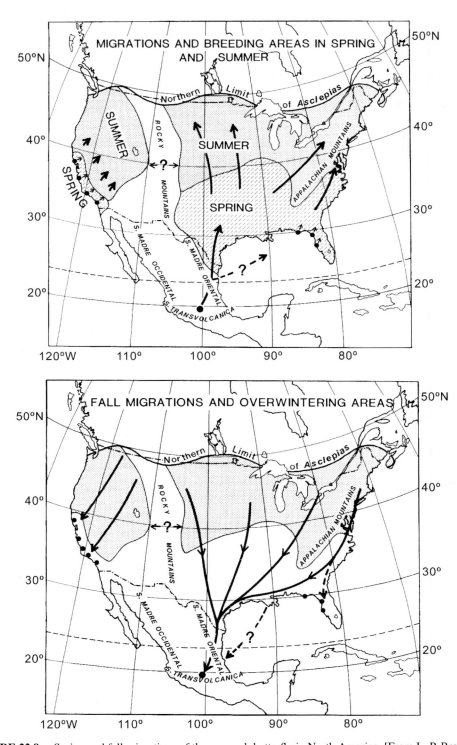

FIGURE 22.9. Spring and fall migrations of the monarch butterfly in North America. [From L. P. Brower and S. B. Malcolm, 1991, Animal migrations: Endangered phenomena, *Am. Zool.* **31**:265–276. By permission of the American Society of Zoologists.]

Plains to the eastern seaboard, migrates on prevailing winds southward and westward during September to November to some 10 permanent overwintering sites in the Transvolcanic Range of central Mexico. At these cool and moist sites, which are dominated by stands of the oyamel fir (*Abies religiosa*), extremely dense and spectacular clusters of monarchs are able to spend the winter until the milkweed flora of the north reappears the following spring. Through the winter the monarchs remain in reproductive diapause, but about the end of February egg development and mating begin and the butterflies initiate a relatively rapid remigration northward (again using prevailing winds), reaching the southern United States in late March and early April.

For many years there was debate over whether the overwintered monarchs were responsible for establishing the spring generation over the entire range of the eastern population (the "single sweep hypothesis") or whether the overwintered butterflies produced offspring only in the southern United States and then died, the offspring continuing the northward movement (the "successive brood hypothesis") (Brower and Malcolm, 1991). This controversy was resolved in favor of the successive brood hypothesis by analyzing the cardenolides found in the body of the butterflies. The cardenolides, which are sequestered by the monarch larvae as they feed on the milkweeds and serve to deter would-be vertebrate predators, are characteristic for each *Asclepias* species and can thus be used to determine where the butterflies lived as larvae. Essentially, the study showed that overwintered, early spring migrants to the southern United States had cardenolides characteristic of *A. syriaca*, a northern species, which they would have acquired as larvae the previous summer. However, the monarchs that reached North Dakota east to Massachusetts in May and June had cardenolide patterns typical of *Asclepias* species found only in the southern United States (Malcolm *et al.*, 1991). Through the summer period, there may be several non-migratory generations of monarchs, only the last generation of adults entering reproductive diapause and developing migratory behavior. Tragically, the overwintering sites of both eastern and western populations of the monarch are under severe pressure as a result of numerous and varied human activities. Especially significant is illegal deforestation in the Mexican national parks where the eastern population's overwintering sites are located (Brower *et al.*, 2002). Logging not only reduces the monarch's habitat, but by fragmenting the forest it also opens the area to localized climate changes, especially freezing temperatures, which increase the butterfly's mortality. Brower and Malcolm (1991) (p. 270) concluded that the monarch's "eastern North American migratory phenomenon is now threatened with extinction and will probably be destroyed within 10–20 years."*

As the above examples demonstrate, migration can take many forms yet its common purpose is to improve a species' ability to survive and multiply. For most species, wind supplies the power for migration, and their physiology and behavior have so evolved as to make best use of this.

6. Summary

The distribution and abundance of insects are markedly affected by temperature, photoperiod, water, and weather.

---

* Brower (personal communication, May 2004) considered the situation "more than critical. It is disastrous," mainly due to the federal government's failure to protect the forest. For more information, visit *www.MonarchWatch.org*.

As poikilotherms, insects have a metabolic rate that within species- and stage-specific limits is proportional to temperature. Their rate of development within these limits is inversely proportional to temperature. Outside these limits insects will survive, but their development is retarded or prevented. The temperature extremes for survival are known as the upper and lower lethal limits. Survival at extreme ambient temperatures may be accomplished by (1) behavioral means, such as burrowing or ovipositing in a substrate, and/or (2) entering a physiologically dormant condition (diapause). At below-freezing temperatures insects may also become freezing-tolerant, that is, capable of withstanding freezing of their extracellular fluids, or, when they are freezing-susceptible, become supercooled. In both arrangements, polyhydroxyl cryoprotectants, thermal-hysteresis proteins, and (for freezing-tolerant species) ice-nucleating proteins are important. In species from habitats whose climate is suitable for development and/or reproduction for a limited period each year, temperature may be an important synchronizer of development and/or eclosion.

Photoperiod, the naturally occurring 24-hour cycle of light and darkness, exerts both short-term and long-term effects on behavior and physiology, which keep insects in tune with changing environmental conditions. In a few species changing light intensity triggers daily activities; that is, they arise exogenously. However, in most species, diurnal (circadian) rhythms of activity, for example, locomotor activity, feeding, mating behavior, oviposition, and eclosion, originate endogenously and are phase set by photoperiod.

By responding to seasonal changes in photoperiod insects can exploit suitable environmental conditions for development and reproduction and survive periods when climatic conditions are adverse. Among the long-term processes affected by photoperiod are the nature and rate of development, reproductive ability and capacity, synchronized eclosion, diapause, and possibly cold-hardiness.

Diapause is a genetically determined state of suppressed development. It may occur at any stage of the life history, though this is usually species-specific. Photoperiod exerts its influence at a stage earlier than the one in which diapause occurs, thus ensuring safety against prematurely unseasonal weather. Induction of diapause is, in almost all species, a response to the absolute day length (number of hours of light in a 24-hour cycle) rather than daily differences in the day length. For a species, there is a critical day length at which the incidence of diapause (proportion of individuals that enter diapause) changes markedly. Long-day insects develop continuously at all day lengths above the critical day length (usually about 16 hours of light per day) but enter diapause at shorter day lengths. In short-day insects, development is continuous at day lengths below the critical value (usually about 12 hours). In short-day-long-day insects, development is continuous at short and long day lengths, but at intermediate day lengths (about 14–16 hours of light per day) the incidence of diapause is high. Long-day-short-day insects develop continuously within a narrow range of day lengths (16–20 hours of light per day) and enter diapause at all other day lengths. The value of the critical day length for a species may change with temperature, latitude, and food availability.

Water is an important determinant of the distribution and abundance of insects. A problem for most terrestrial species is to reduce water loss from the integument and tracheal system and in excretion. In postembryonic stages this is achieved by means of a relatively impermeable cuticle, valves and/or hairs that reduce water vapor movement out of the tracheae, production of highly concentrated urine, as well as by selection of more humid microclimates. Eggs are covered with a cuticle-like chorion and may be laid in an ootheca and/or substrate. For some species in cold climates snow cover may be important as an insulator and in preventing desiccation.

In addition to temperature and light, important abiotic factors affecting the distribution and abundance of aquatic insects are oxygen content, ionic content, and rate of movement of the surrounding water.

Because of their size, insects may be greatly affected by weather, especially temperature, wind, and rain. Wind affects the rate of water loss from the body and is an important agent of dispersal in many terrestrial species. Dispersal by wind may be adaptive, that is, advantageous to a species that can migrate to new breeding sites. The form of migration varies among species, in terms of the proportion of the population that migrates, whether migration occurs in all or selected generations, the distance traveled, and whether the migration is continuous or intermittent.

7. Literature

Information on the effects of abiotic factors on the distribution and abundance of insects is provided by Varley *et al.* (1973), Price (1984), and authors in the volume edited by Huffaker and Gutierrez (1999). Saunders (1974, 1981, 2002), Beck (1980), and Danks (2003) discuss insect photoperiodism while Beck (1983, 1991) reviews thermoperiodism. Circadian rhythms are examined by Giebultowicz (2000, 2001), Saunders (2002), Stanewsky (2002), and Homberg *et al.* (2003). Tauber and Tauber (1976), Jungreis (1978), Denlinger (1986, 2002), Tauber *et al.* (1986), Zaslavski (1988), Nechols *et al.* (1999), and Danks (1987, 2001, 2002) review insect seasonality, including diapause. Insect migration and dispersal are considered by Johnson (1969), Wehner (1984), Danthanarayana (1986), Dingle (1989), and authors in the volumes edited by Drake and Gatehouse (1995) and Woiwod *et al.* (2001). Duman and Horwath (1983), Baust and Rojas (1985), Bale (1996, 2002), Sømme (1999, 2000), and authors in the volume edited by Lee and Denlinger (1991) review insect cold-hardiness. Abiotic factors that affect the distribution of aquatic insects are discussed by Hynes (1970a,b), Macan (1974), and by authors in the volumes edited by Merritt and Cummins (1978) and Resh and Rosenberg (1984). Cloudsley-Thompson (1975) reviews the adaptations of insects to arid environments.

Bale, J. S., 1993, Classes of insect cold hardiness, *Func. Ecol.* **7**:751–753.

Bale, J. S., 1996, Insect cold hardiness: A matter of life and death, *Eur. J. Entomol.* **93**:369–382.

Bale, J. S., 2002, Insects and low temperatures: From molecular biology to distributions and abundance, *Phil. Trans. R. Soc. B* **357**:849–862.

Baust, J. G., and Morrissey, R. E., 1977, Strategies of low temperature adaptation, *Proc. XV Int. Congr. Entomol.*, pp. 173–184.

Baust, J. G., and Rojas, R. R., 1985, Insect cold hardiness: Facts and fancy, *J. Insect Physiol.* **31**:755–759.

Beck, S. D., 1980, *Insect Photoperiodism*, 2nd ed., Academic Press, New York.

Beck, S. D., 1983, Insect thermoperiodism, *Annu. Rev. Entomol.* **28**:91–108.

Beck, S. D., 1991, Thermoperiodism, in: *Insects at Low Temperature* (R. E. Lee, Jr. and D. L. Denlinger, eds.), Chapman and Hall, New York.

Block, W., 1996, Cold or drought—The lesser of two evils for terrestrial arthropods, *Eur. J. Entomol.* **93**:325–339.

Brower, L. P., and Malcolm, S. B., 1991, Animal migrations: Endangered phenomena, *Am. Zool.* **31**:265–276.

Brower, L. P., Castilleja, G., Peralta, A., Lopez-Garcia, J., Bojorquez-Tapia, L., Diaz, S., Melgarejo, D., and Missrie, M., 2002, Quantitative changes in forest quality in a principal overwintering area of the monarch butterfly in Mexico, 1971–1999, *Cons. Biol.* **16**:346–359.

Cloudsley-Thompson, J. L., 1975, Adaptations of Arthropoda to arid environments, *Annu. Rev. Entomol.* **20**:261–283.

Corbet, P. S., 1963, *A Biology of Dragonflies*, Quadrangle Books, Chicago.

Danilevskii, A. S., 1961, *Photoperiodism and Seasonal Development of Insects* (Engl. transl., 1965), Oliver and Boyd, Edinburgh.

Danks, H. V., 1987, *Insect Dormancy: An Ecological Perspective*, Biological Survey of Canada, Ottawa.

Danks, H. V., 2000, Dehydration in dormant insects, *J. Insect Physiol.* **46**:837–852.

Danks, H. V., 2001, The nature of dormancy responses in insects, *Acta Soc. Zool. Bohem.* **65**:169–179.

Danks, H. V., 2002, The range of dormancy responses in insects, *Eur. J. Entomol.* **99**:127–142.

Danks, H. V., 2003, Studying insect photoperiodism and rhythmicity: Components, approaches and lessons, *Eur. J. Entomol.* **100**:209–221.

Danthanarayana, W., 1986, *Insect Flight: Dispersal and Migration*, Springer-Verlag, Berlin.

Denlinger, D. L., 1986, Dormancy in tropical insects, *Annu. Rev. Entomol.* **31**:239–264.

Denlinger, D. L., 1991, Relationship between cold hardiness and diapause, in: *Insects at Low Temperature* (R. E. Lee, Jr. and D. L. Denlinger, eds.), Chapman and Hall, New York.

Denlinger, D. L., 2002, Regulation of diapause, *Annu. Rev. Entomol.* **47**:93–122.

Dingle, H., 1989, The evolution and significance of migratory flight, in: *Insect Flight* (G. J. Goldsworthy and C. H. Wheeler, eds.), CRC Press, Boca Raton.

Dingle, H., 2001, The evolution of migratory syndromes in insects, in: *Insect Movement: Mechanisms and Consequences* (I. P. Woiwod, D. R. Reynolds, and C. D. Thomas, eds.), CAB International, Wallingford, UK.

Downes, J.A., 1964, Arctic insects and their environment, *Can. Entomol.* **96**:280–307.

Drake, V. A., and Gatehouse, A. G. (eds.), 1995, *Insect Migration: Tracking Resources through Space and Time*, Cambridge University Press, Cambridge.

Duman, J., and Horwath, K., 1983, The role of hemolymph proteins in the cold tolerance of insects, *Annu. Rev. Physiol.* **45**:261–270.

Gatehouse, A. G., and Zhang, 1995, Migratory potential in insects: Variation in an uncertain environment, in: *Insect Migration: Tracking Resources through Space and Time* (V. A. Drake and A. G. Gatehouse, eds.), Cambridge University Press, Cambridge.

Giebultowicz, J. M., 2000, Molecular mechanism and cellular distribution of insect circadian clocks, *Annu. Rev. Entomol.* **45**:769–793.

Giebultowicz, J. M., 2001, Peripheral clocks and their role in circadian timing: Insights from insects, *Phil. Trans. R. Soc. Lond. B* **356**:1791–1799.

Herman, W. S., and Tatar, M., 2001, Juvenile hormone regulation of longevity in the migratory monarch butterfly, *Proc. R. Soc. Lond. B* **268**:2509–2514.

Hodek, I., 2002, Controversial aspects of diapause development, *Eur. J. Entomol.* **99**:163–173.

Hodek, I., 2003, Role of water and moisture in diapause development, *Eur. J. Entomol.* **100**:223–232.

Homberg, U., Reischig, T., and Stengl, M., 2003, Neural organization of the circadian system of the cockroach *Leucophaea maderae*, *Chronobiol. Int.* **20**:577–591.

Huffaker, C. B., and Gutierrez, A. P. (eds.), 1999, *Ecological Entomology*, 2nd ed., Wiley, New York.

Hynes, H. B. N., 1970a, *The Ecology of Running Waters*, University of Toronto Press, Toronto.

Hynes, H. B. N., 1970b, The ecology of stream insects, *Annu. Rev. Entomol.* **15**:25–42.

Johnson, C. G., 1969, *Migration and Dispersal of Insects by Flight*, Methuen, London.

Jungreis, A. M., 1978, Insect dormancy, in: *Dormancy and Developmental Arrest* (M. E. Clutter, ed.), Academic Press, New York.

Lee, R. E., Jr., 1991, Principles of insect low temperature tolerance, in: *Insects at Low Temperature* (R. E. Lee, Jr. and D. L. Denlinger, eds.), Chapman and Hall, New York.

Lee, R. E., Jr., and Denlinger, D. L. (eds.), 1991, *Insects at Low Temperature*, Chapman and Hall, New York.

Macan, T. T., 1974, *Freshwater Ecology*, 2nd ed., Wiley, New York.

Malcolm, S. B., Cockrell, B. J., and Brower, L. P., 1991, Spring recolonization of eastern North America by the monarch butterfly: Successive brood or single sweep migration? in: *Biology and Conservation of the Monarch Butterfly* (S. B. Malcolm and M. P. Zalucki, eds.), Natural History Museum of Los Angeles County, Los Angeles.

Mansingh, A., 1971, Physiological classification of dormancies in insects, *Can. Entomol.* **103**:983–1009.

McNeil, J. N., Cusson, M., Delisle, J., Orchard, I., and Tobe, S. S., 1995, Physiological integration of migration in Lepidoptera, in: *Insect Migration: Tracking Resources through Space and Time* (V. A. Drake and A. G. Gatehouse, eds.), Cambridge University Press, Cambridge.

Merritt, R. W., and Cummins, K. W. (eds.), 1978, *An Introduction to the Aquatic Insects of North America*, Kendall/Hunt, Dubuque, Iowa.

Mouritsen, H., and Frost, B. J., 2002, Virtual migration in tethered flying monarch butterflies reveals their orientation mechanisms, *Proc. Natl. Acad. Sci.* **99**:10162–10166.

Nechols, J. R., Tauber, M. J., Tauber, C. A., and Sasaki, M., 1999, Adaptations to hazardous seasonal conditions: Dormancy, migration, and polyphenism, in: *Ecological Entomology*, 2nd ed. (C. B. Huffaker and A. P. Gutierrez, eds.), Wiley, New York.

Pickford, R., and Riegert, P. W., 1964, The fungous disease caused by *Entomophthora grylli* Fres., and its effects on grasshopper populations in Saskatchewan in 1963, *Can. Entomol.* **96**:1158–1166.

Price, P. W., 1984, *Insect Ecology*, 2nd ed., Wiley, New York.

Pullin, A. S., 1996, Physiological relationships between insect diapause and cold tolerance: Coevolution or coincidence, *Eur. J. Entomol.* **93**:121–129.

Rankin, M. A., 1991, Endocrine effects on migration, *Amer. Zool.* **31**:217–230.

Resh, V. H., and Rosenberg, D. M. (eds.), 1984, *The Ecology of Aquatic Insects*, Praeger, New York.

Saunders, D. S., 1974, Circadian rhythms and photoperiodism in insects, in: *The Physiology of Insecta*, 2nd ed., Vol. II (M. Rockstein, ed.), Academic Press, New York.

Saunders, D. S., 1981, Insect photoperiodism-the clock and the counter: A review, *Physiol. Entomol.* **6**:99–116.

Saunders, D. S., 2002, *Insect Clocks*, 3rd ed., Elsevier Science B.V., Amsterdam.

Sawchyn, W. W., and Gillott, C., 1974a, The life history of *Lestes congener* (Odonata: Zygoptera) on the Canadian prairies, *Can. Entomol.* **106**:367–376.

Sawchyn, W. W., and Gillott, C., 1974b, The life histories of three species of *Lestes* (Odonata: Zygoptera) in Saskatchewan, *Can. Entomol.* **106**:1283–1293.

Sawchyn, W. W., and Gillott, C., 1975, The biology of two related species of coenagrionid dragonflies (Odonata: Zygoptera) in Western Canada, *Can. Entomol.* **107**:119–128.

Sømme, L., 1999, The physiology of cold hardiness in terrestrial arthropods, *Eur. J. Entomol.* **96**:1–10.

Sømme, L., 2000, The history of cold hardiness research in terrestrial arthropods, *CryoLetters* **21**:289–296.

Srygley, R. B., and Oliveira, E. G., 2001, Orientation mechanisms and migration strategies within the flight boundary layer, in: *Insect Movement: Mechanisms and Consequences* (I. P. Woiwod, D. R. Reynolds, and C. D. Thomas, eds.), CAB International, Wallingford, UK.

Stanewsky, R., 2002, Clock mechanisms in *Drosophila*, *Cell Tiss. Res.* **309**:11–26.

Storey, K. B., and Storey, J. M., 1991, Biochemistry of cryoprotectants, in: *Insects at Low Temperature* (R. E. Lee, Jr. and D. L. Denlinger, eds.), Chapman and Hall, New York.

Takeda, M., and Skopik, S. D., 1997, Photoperiodic time measurement and related physiological mechanisms in insects and mites, *Annu. Rev. Entomol.* **42**:323–349.

Tauber, E., and Kyriacou, B. P., 2001, Insect photoperiodism and circadian clocks: Models and mechanisms, *J. Biol. Rhythms* **16**:381–390.

Tauber, M. J., and Tauber, C. A., 1976, Insect seasonality: Diapause maintenance, termination, and postdiapause development, *Annu. Rev. Entomol.* **21**:81–107.

Tauber, M. J., Tauber, C. A., and Sasaki, M., 1986, *Seasonal Adaptations of Insects*, Oxford University Press, London.

Tauber, M. J., Tauber, C. A., Nyrop, J. P., and Villani, M. G., 1998, Moisture, a vital but neglected factor in the seasonal ecology of insects: Hypotheses and tests of mechanisms, *Environ. Entomol.* **27**:523–530.

Vaz Nunes, M., and Saunders, D. S., 1999, Photoperiodic time measurement in insects: A review of clock models, *J. Biol. Rhythms* **14**:84–104.

Varley, G. C., Gradwell, G. R., and Hassell, M. P., 1973, *Insect Population Ecology: An Analytical Approach*, Blackwell, Oxford.

Wehner, R., 1984, Astronavigation in insects, *Annu. Rev. Entomol.* **29**: 277–298.

Woiwod, I. P., Reynolds, D. R., and Thomas, C. D. (eds.), 2001, *Insect Movement: Mechanisms and Consequences*, CAB International, Wallingford, U.K.

Zaslavski, V. A., 1988, *Insect Development, Photoperiodic and Temperature Control*, Springer-Verlag, Berlin.

Zordan, M., Sandrelli, F., and Costa, R., 2003, A concise overview of circadian timing in *Drosophila*, *Front. Biosci.* **8**:d870–d877.

23

The Biotic Environment

1. Introduction

This chapter will deal with the biotic environment of insects, which is composed of all other organisms that affect insects' ability to survive and multiply. In other words, the interactions of insects with other organisms (of the same and other species) will be discussed. Food is the most obvious and important biotic factor, and insects are involved in a wide spectrum of trophic relationships with other organisms, both living and dead. As the majority of insects feed on plant material in one form or another, they are key components in the flow of energy through the ecosystem. However, other interactions are known that, though not as easily recognized as feeding, are nonetheless important regulators of insect distribution and abundance.

2. Food and Trophic Relationships

Insects have evolved diverse feeding habits that allow them to exploit virtually every naturally occurring organic substance. Among their adaptations are specialized ingestive and digestive systems, the ability to detoxify or physically avoid toxins produced by the host, mutualistic relationships between the insect and microorganisms, and life-history strategies that result in temporal avoidance of resource-poor situations (including those resulting from interspecific competition) or times when the host's toxins are abundant. Thus, insects participate in an array of trophic interactions as herbivores, predators, parasites, parasitoids, detritivores, and prey in both terrestrial and freshwater ecosystems (Figures 23.1 and 23.2). Food may be an important limiter of insect population growth; it may also affect the distribution and the dispersal of species over time (Price, 1997).

2.1. Quantitative Aspects

Though the amount of food available might be considered as an important regulator of insect abundance, it has been found in natural communities that populations do not normally use more than a small fraction of the total available food. This is primarily because other

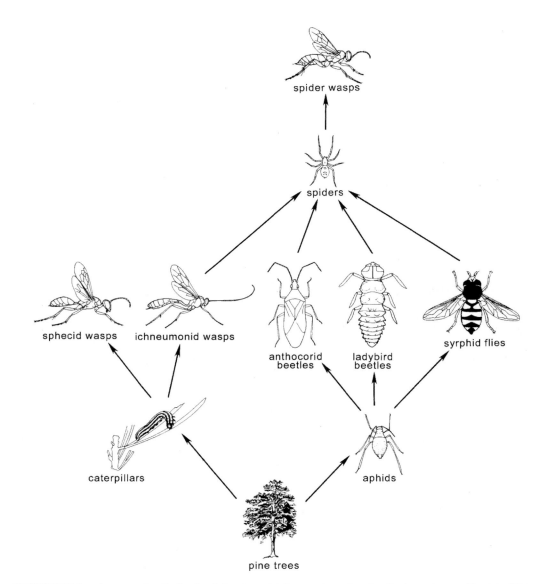

FIGURE 23.1. An example of a food web in a terrestrial ecosystem, showing the importance of insects. [From P. W. Price, The concept of the ecosystem, in: *Ecological Entomology* (C. B. Huffaker and R. L. Rabb, eds.). Copyright © 1984 by John Wiley and Sons, Inc. Reprinted by permission of John Wiley and Sons, Inc.]

components of the environment especially weather but including, for example, predators, parasites, and pathogens, usually have a significant adverse effect on growth and reproduction. Other features of insects may, however, be important in this regard. Many species, especially plant feeders, are polyphagous. Thus, when the preferred food plant is in limited quantity, alternate choices can be used. Among endopterygotes, larvae and adults of a species may eat quite different kinds of food, and in some species such as mosquitoes the food of the adult female differs from that of the adult male.

Two situations may occur in which the quantity of food limits insect distribution and abundance. In the first, there is no absolute shortage of food, but only a proportion of the total is available to a species. Thus, there is said to be a "relative shortage" of food. Various reasons may account for the food not being available. (1) The food may be concentrated

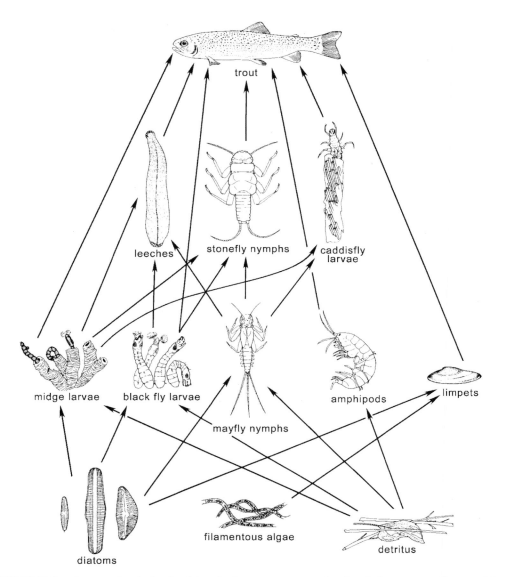

FIGURE 23.2. An example of a food web in a freshwater ecosystem, showing the importance of insects. [From P. W. Price, The concept of the ecosystem, in: *Ecological Entomology* (C. B. Huffaker and R. L. Rabb, eds.). Copyright © 1984 by John Wiley and Sons, Inc. Reprinted by permission of John Wiley and Sons, Inc.]

within a small area so that it is available to relatively few insects. As an interesting example of this Andrewartha (1961) cited the Shinyanga Game-Extermination Experiment in East Africa in which, over the course of about 5 years, the natural hosts of tsetse flies were virtually exterminated over an area of about 800 square miles. At the end of this period one small elephant herd and various small ungulates remained in the game reserve. However, almost no tsetse flies could be found, despite the fact that, collectively, the mammals that remained could supply enough blood to feed the entire original population of flies. The distribution of the food was now so sparse that the chance of flies obtaining a meal was practically nil. (2) The food may be randomly distributed but difficult to locate. Thus, only a fraction of the individuals searching ever find food. Such is probably the situation in many parasitic or hyperparasitic species whose host is buried within the tissues of plants or other animals. (3) A proportion of the food may occur in areas that for other reasons are not normally visited by the consumer so that, in effect, it is not available.

In the second situation, food may become a limiting factor in population growth when a species' numbers are not kept in check by other influences, especially natural enemies. This may happen, for example, when a species is accidentally transferred (often as a result of human activity) from its original environment to a new geographic area where its natural enemies are absent. Under these conditions, the population may grow unchecked, and its final size is limited only by the amount of food available. Occasionally, even in a species' natural habitat, food may limit population growth, for example, when weather conditions are favorable for development of a species but not for development of those organisms that prey on or parasitize it.

2.2. Qualitative Aspects

The nature of the food available may have striking effects on the survival, rate of growth, and reproductive potential of a species, and much work has been done on insects in this regard. For example, of the insect fauna associated with stored products, the saw-toothed grain beetle, *Oryzaephilus surinamensis*, can survive only on foods with a high carbohydrate content such as flour, bran, and dried fruit, whereas species of spider beetles, *Ptinus* spp., and flour beetles, *Tribolium* spp., have no such carbohydrate requirement and are consequently cosmopolitan, occurring in animal meals and dried yeast, in addition to plant products. For some phytophagous insects, a combination of plants of different kinds appears necessary for survival and/or normal rates of juvenile development. In the migratory grasshopper, *Melanoplus sanguinipes*, for example, a smaller percentage of insects survive from hatching to adulthood, and the development of those that do survive is slower when the grasshoppers are fed on wheat (*Triticum aestivum*) alone compared with wheat plus flixweed (*Descurainia Sophia*) or dandelion (*Taraxacum officinale*) (Pickford, 1962).

Both the rate of egg production and the number of eggs produced may be markedly affected by the nature of the food available. Many common flies, for example, species of *Musca*, *Calliphora*, and *Lucilia*, may survive as adults for some time on a diet of carbohydrate. However, for females to mature eggs a source of protein is essential. Pickford (1962) showed that *M. sanguinipes* females fed a diet that included wheat and wild mustard (*Brassica kaber*) or wheat and flixweed produced far more eggs (579 and 467 eggs per female, respectively) than females fed on wheat (243 eggs/♀), wild mustard (431 eggs/♀), or flixweed (249 eggs/♀), alone. These differences in egg production resulted largely from variations in the duration of adult life, though differences in rate of egg production were also evident. For example, percent survival of females fed wheat plus mustard after 1, 2, and 3 months was 93%, 60%, and 13%, respectively. These females produced, on average, 8.4 eggs/female per day. The corresponding figures for females fed on wheat alone were 87%, 27%, and 0% survival over 1, 2, and 3 months, respectively, and 4.6 eggs/female per day. The metabolic basis for these differences was not determined.

3. Insect-Plant Interactions

3.1. Herbivores

Though insects feed on plants from all of the major taxonomic groups, the greatest number of herbivorous species feed on angiosperms with which they have been coevolving since the Cretaceous period (Chapter 2, Section 4.2). All parts of a plant may be exploited as a result of the activities of grazing, sucking, and boring insects. As might be anticipated in view of the length of time over which this coevolution has occurred, some of the relationships between herbivorous insects and angiosperms are extremely intimate and refined, though essentially the relationships have a common theme. Insects gain energy (food) at the expense of plants, whereas plants attempt to defend themselves (conserve their energy) or at least to obtain something in return for the energy that insects take from them. Though the theme remains constant through time, the relationships themselves are always changing as a result of natural selection. Insects strive to improve their energy-gathering efficiency (most often by concentrating on energy in a particular form and from a restricted source and by specialization of the method used to collect the energy) while plants concurrently improve their defenses. Most authors, for example, Price (1997) view this relationship as "constant warfare" between insects and plants, which forms the basis of their coevolution. Other authors such as Owen and Wiegert (1987) believe that herbivory is a form of mutualism. They point out that, in analogy with pruning, mowing, and similar activities carried out by humans, a frequent effect of insects grazing on newly formed plant tissues is to stimulate the plant to produce more branches and, eventually, more reproductive structures and seed; in other words, the plant is making an adaptive, mutualistic response to the herbivore.

The most common method used by plants as defense against insects (and other herbivorous animals) is production of toxic metabolites. Plants produce a wide array of such chemicals in secondary metabolic pathways (i.e., those not used for generation of major components such as proteins, nucleic acids, and carbohydrates). Particular types of secondary plant compounds are commonly restricted to specific plant families, for example, glucosinolates in Brassicaceae (crucifers), cardenolides (mainly cardiac glycosides) in Asclepiadaceae (milkweeds), and cucurbitacins in Cucurbitaceae (Panda and Khush, 1995; Schoonhoven *et al.*, 1998). Moreover, the compounds often accumulate within specific tissues or areas of the plant, for example, trichomes (terpenes), the wax layer (phenolics), vacuoles (alkaloids), and seeds (non-protein amino acids) (Bernays and Chapman, 1994). The reproductive parts of plants, which represent concentrated stores of energy, are especially attractive to herbivores and often serve as a sink for secondary metabolites. For example, *Hypericum perforatum* (Klamath weed) produces the toxic quinone hypericin. The concentration of hypericin is 30 µg/g wet weight in the lower stem, 70 µg/g in the upper stem, and 500 µg/g in the flower (Price, 1997). Typically, the toxins are chemically combined with sugars, salts, or proteins to render them inactive while in storage. When the plant tissue is damaged, enzymes release the toxin from its conjugate, allowing a localized effect at the site of the wound (Bernays and Chapman, 1994).

The evolutionary origin of these secondary metabolites remains a matter of speculation. An early view was that the chemicals arose as waste products of a plant's primary metabolism, and the plant, being unable to excrete the molecules, simply retained them within its tissues. This idea is now considered unlikely given the highly complex nature of some of these compounds and, therefore, the amount of energy required for their synthesis. A more likely possibility is that originally the metabolites were simply short-lived intermediates in normal biochemical pathways within plants and/or provided a means of storing chemical energy for later use by the plant. In other words, the original function(s) of these compounds may have been unrelated to the occurrence of herbivores. An example of such a compound might be nicotine produced by the tobacco plant *(Nicotiana* spp.). Radioisotope studies have shown that, although about 12% of the energy trapped in photosynthesis is used for nicotine production, the nicotine has a relatively short half-life, 40% of it being converted to other metabolites (possibly sugars, amino acids, and organic acids) within 10 hours.

Thus, animals adapted to feeding on plants that produce toxins will be at a considerable advantage over animals that are not. Among herbivores, insects show the greatest ability to cope with the toxins. In part, this arises from the enormous period of time over which coevolution of insects and plants has occurred, but it is also related to insects' high reproductive rate and short generation time, which facilitate rapid adaptation to changes in the host plant. Through evolution, many insect species have not only developed increasing tolerance to a host plant's toxins but are now attracted by them. In other words, such insects locate food plants by the scent or taste of their toxic substance and frequently are restricted to feeding on such plants. For example, certain flea beetles, *Phyllotreta* spp., and cabbage worms, *Pieris* spp., feed exclusively on plants such as Cruciferae that produce glucosinolates (mustard oil). Colorado potato beetles, *Leptinotarsa decemlineata*, and various hornworms, *Manduca* spp., feed only on Solanaceae, the family that includes potato (*Solanum tuberosum*) (produces solanine), tobacco (*Nicotiana* spp.) (nicotine), and deadly nightshade (*Atropa belladonna*) (atropine) (Price, 1997).

The method most often used to overcome the potentially harmful effects of these chemicals is to convert them into non-toxic or less toxic products. Especially important in such conversions is a group of enzymes known as mixed-function oxidases (polysubstrate monooxygenases), which, as their name indicates, catalyze a variety of oxidation reactions (Schoonhoven *et al.*, 1998). The enzymes are located in the microsome fraction* of cells and occur in particularly high concentrations in fat body and midgut. Perhaps unsurprisingly, it is these same enzymes that are often responsible for the resistance of insects to synthetic insecticides (Chapter 16, Section 5.5.).

Some insects are able to feed on potentially dangerous plants as a result of either temporal or spatial avoidance of the toxic materials. For example, the life history of the winter moth, *Operophtera brumata*, is such that the caterpillars hatch in the early spring and feed on young leaves of oak (*Quercus* spp.), which have only low concentrations of tannins, molecules that complex with proteins to reduce their digestibility. Though weather conditions are suitable and food is still apparently plentiful later in the season, a second generation of winter moths does not develop because by this time large quantities of tannins are present in the leaves. Spatial avoidance is possible for many Hemiptera whose delicate suctorial mouthparts can bypass localized concentrations of toxin in the host plant. Some aphids feed on senescent foliage where the concentration of toxin is less than that of younger, metabolically active tissue (Price, 1997).

Price (1997) proposed that at least four advantages may accrue to an insect able to feed on potentially toxic plants. First, competition with other herbivores for food will be much reduced. Second, the food plant can be located easily. Related to this, as members of a species will tend to aggregate on or near the food plant, the chances of finding a mate will be increased. Third, if an insect is able to store the ingested toxin within its tissues, it may gain protection from would-be predators. Many examples of this ability are known, especially among Lepidoptera (Blum, 1981; Nishida, 2002). Thus, most of the insect fauna associated with milkweeds are able to store the cardenolides produced by these plants. These substances, at sublethal levels, induce vomiting in vertebrates. Other well-known examples are pyrrolizidine alkaloids sequestered by arctiid moths, and cucurbitacins

---

* The microsome fraction is obtained by differential high-speed centrifugation of homogenized cells and consists of fragmented membranes of endoplasmic reticulum, ribonucleoproteins, and vesicles.

accumulated by cucumber beetles. In Lepidoptera the chemicals are accumulated by the caterpillar stage and are transferred at metamorphosis to the adult. Further, in some species the female endows her eggs with the toxin so that they, too, are protected (Blum and Hilker, 2002). Most insect species that sequester toxins from their host plant are aposematically (brightly and distinctly) colored, a feature commonly indicative of a distasteful organism and one that makes them stand out against the background of their host plant. On sampling such insects, a would-be vertebrate predator discovers their unpalatability and quickly learns to avoid insects having a particular color pattern. Remarkably, a few insect predators have evolved tolerance to the plant-produced toxins stored by their insect prey and are, themselves, unpalatable to predators further up the food chain (Eisner et al., 1997). The fourth advantage to be gained by tolerance to these plant products is protection against pathogenic microorganisms. For example, cardiac glycosides in the hemolymph of the large milkweed bug, *Oncopeltus fasciatus*, have a strong antibacterial effect. Also, cucurbitacins sequestered by the adult female cucumber beetle, *Diabrotica undecimpunctata howardi*, provide antifungal protection for her eggs and offspring (Tallamy et al., 1998).

The channeling of energy into production of toxic or repellent substances is the most often used method by which plants may obtain protection, though others are known. A few plants expend this "energy of protection" on formation of structures that prevent or deter feeding, or even harm would-be feeders. For example, passion flowers (*Passiflora adenopoda*) have minute hooked hairs that grip the integument of caterpillars attempting to feed on them. The hairs both impede movement and tear the integument as the caterpillars struggle to free themselves so that the insects die from starvation and/or desiccation (Gilbert, 1971). Leguminous plants have evolved a variety of physical (as well as chemical) mechanisms to protect their seeds from Bruchidae (pea and bean weevils). These include production of gum as a larva penetrates the seed pod so that the insect is drowned or its movements hindered, production of a flaky pod surface that is shed, carrying the weevil's eggs with it, as the pod breaks open to expose its seeds, and production of pods that open explosively so that seeds are immediately dispersed and, therefore, not available to females that oviposit directly on seeds (Center and Johnson, 1974).

3.2. Insect-Plant Mutualism

Not all insect-plant relationships are of the "constant warfare" type just discussed. For a large number of insect and plant species, an interaction of mutual benefit has evolved. Thus, some insects live in close association with plants, protecting them in return for food. For example, the bull's-horn acacias (*Acacia* spp.) are host to colonies of ants (*Pseudomyrmex* spp.) that live within the swollen, hollow stipular thorns and feed on nectar (produced in petioles) and protein (in Beltian bodies at the tips of new leaves) (Figure 23.3). In return, the aggressive ants guard the plants against herbivores and suppress the growth of nearby, potentially competitive plants by chewing their growing tips (Hölldobler and Wilson, 1990).

A mutualistic relationship of a very different kind is that in which the host supplies food to insects, in return for which the insects provide the transport system for dispersal of pollen, seeds, and spores. Though their importance as pollinators for higher plants has been extensively studied (Kevan and Baker, 1983, 1999; Schoonhoven et al., 1998), it must be emphasized that some insects are essential for spore dispersal in some mosses and many fungi (see Chapter 16, Section 4.2.4), as well as transporting seeds of angiosperms. The success (importance) of insects as pollinators compared with pollinators from other groups such as birds and bats is presumably a result of their much longer evolutionary association

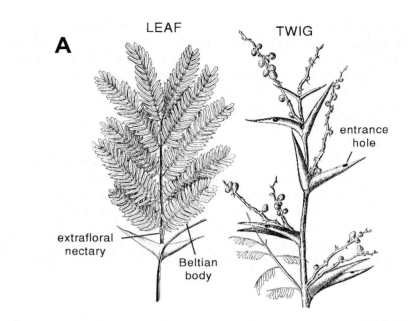

FIGURE 23.3. Mutualism between bull's-horn acacia and ants. (A) Acacia leaf and twig showing extrafloral nectary, hollow thorns, and leaflets with Beltian bodies at tips; (B) enlarged view of hollow thorn with entrance hole of ant nest; and (C) close-up view of Beltian bodies and ant visitor. [A, redrawn from W. M. Wheeler, 1910, *Ants. Their Structure, Development and Behaviour*. Columbia University Press. B, C, photographs courtesy of Dan L. Perlman.]

with plants. Most of the modern insect orders were well established by the time the earliest flowering plants appeared about 225 million years ago. Thus, insects were able to gain a considerable head start as pollinators over birds and bats, the earliest fossil records for which date back about 150 and 60 million years, respectively (Price, 1997).

To achieve effective cross-pollination, two important factors must be taken into consideration in an evolutionary sense. First, plants must produce precisely the right amount of

FIGURE 23.3. *(Continued)*

nectar to make an insect's visit energetically worthwhile, yet stimulate visits to other plants, and second, plants of the same species must be easily recognized by an insect. If too much energy is made available by each plant, then insects need visit fewer plants and the extent of cross-pollination is reduced. If a plant produces too little food (to ensure that an insect will visit many plants), there is a risk that the insect will seek more accessible sources of food, Natural selection determines the precise amount of energy that each plant must offer to an insect, and this amount depends on a number of factors. The amount of energy gained by an insect during each visit to a flower is related to both quantity and quality of available food. Thus, until recently, it was considered that many adult insects obtained their carbohydrate requirements from nectar and their protein requirements from other sources such as pollen, vegetative parts of the plant (as a result of larval feeding), or other animals. Baker and Baker (1973) showed, however, that the nectar of many plants contains significant amounts of amino acids, so that insects can concentrate their efforts on nectar collection. This not only increases the extent of cross-pollination by inducing more visits to flowers, but may also lead to economy in pollen production, as pollen becomes less important as food for the insects. The amount of nectar produced is a function of the number of flowers per plant. Hence, for plants with a number of flowers blooming synchronously, it is important that each flower produces only a small amount of nectar and pollen, so that an insect must visit other plants to satisfy its requirements.

More nectar is produced by plant species whose members typically grow some distance apart, so that it is still energetically worthwhile for an insect species to concentrate on these plants. Related to this, insects that forage over greater distances are larger species such as bees, moths, and butterflies whose energy requirements are high. When nectar is produced in large amounts, it is typically accessible only to larger insects that are strong enough to gain entry into the nectar-producing area or have sufficiently elongate mouthparts. This ensures that nectar is not wasted on smaller insects lacking the ability to carry pollen to other members of the plant species.

Temperature also affects the amount of nectar produced, as it is related to the energy expended by insects in flight and to the time of day and/or season. For example, in temperate regions and/or at high altitudes, flowers that bloom early in the day or at night, or early or late in the season, when temperatures may not be much above freezing, must provide a large enough reward to make foraging profitable at these temperatures. An alternative to production of large amounts of nectar by individual flowers is for plants that bloom at lower temperatures to grow in high density and flower synchronously (Heinrich and Raven, 1972).

Beyond a certain distance between plants, however, the amount of nectar that an insect requires to collect at each plant (in order to remain "interested" in that species) exceeds the maximum amount that the plant is able to produce. Thus, the plant must adopt a different strategy. Among orchids, for example, about one half of the species produce no nectar, but rely on other methods to attract insects, especially deception by mimicry. The flowers may resemble (1) other nectar-producing flowers, (2) female insects so that males are attracted and attempt pseudocopulation, (3) hosts of insect parasitoids, or (4) insects that are subsequently attacked by other territorial insects. These somewhat risky methods of attracting insects are offset by the evolution of highly specific pollen receptors (so that only pollen from the correct species is acquired) and a high degree of seed set for each pollination (Price, 1997).

It is important for both plants and insects that insects visit members of the same plant species. The chances of this occurring are greatly increased (1) when the plant species has a restricted period of bloom, in terms of both season and/or time of day; (2) where members of a species grow in aggregations, though this is counterbalanced by a restriction of gene flow if pollinators work within a particular plant population; and (3) when the flowers are easily recognized by an insect which learns to associate a given plant species with food. Recognition is achieved as a result of flower morphology (and related to this is accessibility of the nectar and pollen), color, and scent. The advantage to an insect species when its members can recognize particular flowers is that, through natural selection, the species will become more efficient at gathering and utilizing the food produced by those flowers.

The degree of influence that these variables exert is manifest as a spectrum of intimacy between plants and their insect pollinators. At one end of the spectrum, the plant-insect relationship is non-specific; that is, a variety of insect species serve as pollinators for a variety of plants. Neither insects nor flowers are especially modified structurally or physiologically. At the opposite extreme, the relationship is such that a plant species is pollinated by a single insect species. Structural features of the pollinator precisely complement flower morphology; the plant's blooming period is synchronized with the life history and diurnal activity of the insect; and, where present, nectar is produced in exactly the right quantity and quality to satisfy the insect's requirements.

Harvester ants (those that use seeds as food) are important seed dispersers, resulting from accidental loss of the seeds as they transport them back to the nest or by failure to use the seeds before they germinate. This activity partially compensates for the damage caused to the plant by the ants' seed predation. This mutualistic relationship has been taken to a new level of sophistication by myrmecochorous plants, which produce attractive appendages (elaiosomes) on their seeds and chemicals to induce the ants to transport the seeds without damaging them (Figure 23.4). The elaiosomes are rich in nutrients and form the food of the ants, while the seed itself is discarded. Though examples of myrmecochory are known worldwide, it seems to be a phenomenon of habitats that are nutrient-poor (especially those deficient in phosphorus and potassium), notably the dry schlerophyll regions of Australia and South Africa where more than 90% of the 3100 known species of myrmecochorous plants are found. It has been speculated that plants in these habitats use myrmecochory because energetically it is far less expensive than the production of the larger fruits preferred by vertebrates (Beattie, 1985; Hölldobler and Wilson, 1990).

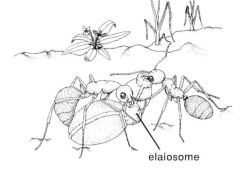

FIGURE 23.4. Workers of *Formica podzolica* from the northern United States gathering violet (*Viola nuttallii*) seeds. Note the elaiosomes that will later be eaten by colony members. [From A. J. Beattie. 1985, *The Evolutionary Theory of Ant-Plant Mutualisms.* By permission of Cambridge University Press.]

3.3. Detritivores

Feeding on detritus, essentially the remains of dead animals and plants together with the microorganisms that bring about their decay, is a very old habit among Insecta, and most orders include some detritivorous species (Southwood, 1972). Detritus-feeding insects are especially significant in aquatic ecosystems where large quantities of dead plant matter may accumulate annually. In streams and rivers the source of this material is largely overhanging vegetation. In still waters the major contributor to detritus is phytoplankton, though in shallow areas emergent vegetation may make a significant contribution. In terrestrial ecosystems insects are generally unimportant as detritivores, this role falling predominantly to other arthropods, notably Collembola and oribatid mites. Two notable exceptions are termites, the majority of which feed on wood litter, and the Australian Oecophoridae and Tortricidae (Lepidoptera) many of whose larvae feed on eucalyptus litter (Common, 1980). The foliage of eucalyptus contains substantial amounts of phenolics (including tannins) and essential oils, rendering it unpalatable to many herbivores, and is also extremely deficient in nitrogen. Thus, the ability of these moth larvae to utilize this resource gives them a key role in energy flow and matter recycling in this specialized ecosystem.

According to Anderson and Cargill (1987), some 45% of an estimated 10,000 species of aquatic insects in North America ingest some detritus. The major groups of detritivores are in the orders Trichoptera, Diptera, Ephemeroptera, Plecoptera, and Coleoptera, and their eating habits may be categorized as shredding and gouging, scraping, filter feeding, and deposit collection (Figure 23.2). The shredders, gougers, and scrapers inevitably will ingest the living saprophytic microorganisms on the surface of the dead vegetation, as well as the plant tissue itself, and a key question is the relative importance of these as food. For some detritivores, there is evidence that the microorganisms provide all of the nutrients with the relatively indigestible plant tissue simply passing unchanged through the gut. In other species, dead plant tissues are the main source of energy though microorganisms may supply some essential components. Generally, only about 10% of the ingested material is assimilated, the rest being defecated. However, the breaking up of the material into smaller particles as it passes along the alimentary tract is in itself important, the increase in surface area facilitating microbial activity.

Detritus may vary considerably in its nutrient availability and in its content of feeding deterrents, and a major role of the microorganisms is to "condition" the material, for example, by softening the tissues, by chemically converting the contents into a form that can be used by the insects, and by detoxifying the deterrent compounds. The availability of detritus as food may also vary seasonally, typically becoming maximal in the fall following leaf drop and the death of emergent vegetation and phytoplankton and again in spring as temperatures rise permitting renewed microbial activity. Thus, many detritivores (especially shredders) exhibit their greatest growth rates in the fall and early winter, before entering a phase of arrested development until spring.

4. Interactions between Insects and Other Animals

Interactions between insects and other animals (including other members of the same species) take many forms, though most are food-related. Insects may be predators (which require more than one prey individual in order to complete development), parasitoids, or parasites (which need only one host to complete development). Parasitoids differ from parasites in that they ultimately kill their host, which is typically another arthropod. Alternatively, insects may serve as prey or host for other animals. Insects may also enter into mutualistic relationships with other species. In another form of interaction, insects may compete either with other members of the species or with other animals for the same resource, for example, food, breeding or egg-laying sites, overwintering sites, or resting sites. Between opposite sexes of the same species, the interaction may be for a very obvious purpose, propagation of the species.

4.1. Intraspecific Interactions

The nature and number of interactions among members of the same species will depend on the density of the population. These interactions may be either beneficial or harmful. It follows that there will be an optimal range of density for a given population, within which the net effect of these interactions will be most beneficial. Outside this range of density, that is, when there is underpopulation or overpopulation (crowding), the net result of these interactions will be less than optimal for perpetuation of the species. Animals have evolved various regulatory mechanisms that serve either to maintain this optimal density or to alter the existing density so as to bring it within the optimal range. In the discussion that follows it will be seen how these mechanisms operate in some insects.

4.1.1. Underpopulation

Probably the most obvious detrimental effect of underpopulation is the increased difficulty of locating a mate for breeding purposes. For most species, mate location requires an active search on the part of the members of one sex, which under conditions of underpopulation might present a special problem for weakly flying insects. To alleviate this, many species have evolved highly refined mechanisms (e.g., production of pheromones, sounds, or light) that facilitate aggregation or location of individuals of the opposite sex (Chapter 19, Section 4.1). The relatively slight chance of finding a mate is offset to some extent by the fact that one mating may suffice for fertilization of all eggs a female may produce; that is, sperm may remain viable for a considerable period, and a female may produce a large number of eggs.

On some occasions the effect of non-specific predators is much greater when prey density is lower than normal because the chance of an individual prey organism being eaten is increased. For example, in the period 1935–1940 the population density of the Australian plague grasshopper, *Austroicetes cruciata*, was high. However, drought conditions in the winter of 1940 resulted in a great shortage of grasshopper food and a decline in population numbers. Only a few small areas of land remained moist enough to support growth of grass, and surviving grasshoppers congregated in these areas. But so did the birds that normally fed on the insects and they reduced the population density to an extremely low level (Andrewartha, 1961).

Lower than normal densities may also have a serious effect in species that modify their environment, for example, social insects. The temperature and humidity within the nest are normally quite different from those outside, being regulated by the activity of members of the colony. If a proportion of the population is removed or destroyed, those individuals that remain may no longer be able to keep the temperature and humidity at the desired level and the colony may die. Another example is the lesser grain borer, *Rhizopertha dominica* (Coleoptera), which is a serious pest of stored grain in the United States. In damaged (cracked) grain beetles can survive and reproduce even at low population density. However, in sound grain only cultures whose density is quite high will survive because the insects themselves, through their chewing activity, can cause sufficient damage to the grain that it becomes suitable for reproduction. The nature of this suitability is unknown.

Below a certain level of population, the so-called "threshold density," the chances of survival for a population are slim because of the unlikelihood of a meeting between insects of opposite sex and in reproductive condition. This is important in two areas of applied entomology, namely, quarantine service and biological control. Quarantine regulations are designed so that for a given pest the number of individuals entering a country over a period of time is sufficiently low that the chances of the pest establishing itself are very slim. In biological control programmes that use insects as the controlling agents, experience shows that it is better to release the insects in a restricted area, especially if they are limited in number, rather than distributing them sparsely, in order to improve the chances of establishing a breeding population.

Populations of many insect species may be considered self-regulating; that is, should the density of the population fall below normal (though not below threshold) it will, over time, return to its original level. There may be various reasons for this. As a species' density falls, its predators may experience greater difficulty in finding food so that they migrate elsewhere or produce fewer young. As a result of the decline in predator density, a larger proportion of the prey species may survive to reproduce. If this continues for several generations, the original population density may be reestablished. Another possibility is that with a decrease in density, there will be a greater choice of oviposition and, perhaps, resting sites. Selection of the best of these sites will again increase the chances of survival of an insect or its progeny and lead to a population increase. Some species have rather more specific mechanisms for overcoming the disadvantages of underpopulation. For example, females of some species use facultative parthenogenesis in the absence of males, which serves not only to maintain continuity of the population, but, as the progeny are generally all female, any offspring that do find a mate can make a substantial contribution to the next generation. In the desert locust adult females in the solitary phase (Chapter 21, Section 7) live longer, so that the chance of encountering a male is increased and, further, produce up to four times as many eggs compared to gregarious females.

4.1.2. Overpopulation

As population density rises beyond the normal level, members of a species will increasingly compete with each other for such resources as oviposition sites, overwintering sites, resting places, and, occasionally, food. Such competition may itself have a regulatory effect, as a proportion of the population will have to be satisfied with less than optimal conditions. Thus, if oviposition sites are marginally suitable, few or no progeny may result. In less than adequate overwintering sites, insects may die if the weather is severe. If insects cannot find proper resting places, their chances of discovery by predators or parasitoids are increased, as are their chances of dying because of unfavorable weather conditions. As noted earlier (Section 2.1) food is seldom limiting, though under unusual circumstances it may become so.

In addition to the general regulating mechanisms just mentioned, some insects regulate population density in more specific ways. Migration (Chapter 22, Section 5.2) is a means by which a species may reduce its population density. In such species, it is crowding that induces the necessary physiological and behavioral changes that put an insect into migratory condition. Crowding may also lead to reduction in fecundity. For example, in *Schistocerca gregaria* gregarious females lay fewer eggs than solitary females because (1) their ovaries contain fewer ovarioles, (2) a smaller proportion of the ovarioles produce oocytes in each ovarian cycle [as a result of (1) and (2), each egg pod contains fewer eggs], and (3) they have fewer ovarian cycles (Kennedy, 1961). Likewise, in the migratory grasshopper, *Melanoplus sanguinipes*, mating frequency, which is a function of population density, is inversely related to number of eggs produced and longevity (Pickford and Gillott, 1972).

A few species employ a very obvious means of reducing crowding, namely, cannibalism of either the same or a different life stage. In larval Zygoptera, for example, cannibalism of earlier instars is common under crowded conditions. For species whose larvae inhabit small and/or temporary ponds with limited food resources, cannibalism may be important in ensuring that at least a proportion of the population reaches the adult stage. Another consequence is that stragglers are eliminated, which results in greater synchrony of adult emergence (see also Chapter 22, Section 2.3). In the confused flour beetle, *Tribolium confusum*, and some other beetle species, all of whose life stages are spent in grain or its products, egg cannibalism occurs. Adults eat any eggs they find, and, therefore, the higher the adult population density, the greater the number of eggs consumed.

In some species of insects that inhabit a homogeneous environment, population density is regulated by making the environment less suitable for growth. For example, *T. confusum* larvae and adults "condition" the flour in which they live. As a result, a smaller proportion of the larvae survive to maturity, and the duration of the larval stage is increased. The nature of this conditioning is not known.

In some species, regulation of population density is achieved by having individuals that dominate others so that the reproductive capacity of the latter is either reduced or totally suppressed. This is seen most clearly in social Hymenoptera where one individual, the queen, dominates the other members of the colony, which are mostly female. In more primitive species, dominance is achieved initially by physical aggression, though in time the subordinates recognize the queen by scent and consequently avoid her. In highly social forms, such as the honey bee, dominance is asserted entirely through the release of pheromones.

In many species, dominance has taken on another form, namely, territoriality, the defense of a particular area. The size of the area defended (territory) may vary, but not below a minimum value, so that a maximum population density is attained. Territoriality is shown by insects in a number of orders, both primitive and advanced, and is typically associated with some aspect of reproduction. Most often males establish territories and defend them against other males, usually by chasing and fighting but occasionally by nonaggressive means such as chirping in some Orthoptera. Females enter males' territories for mating and oviposition. Among Odonata, where mating immediately precedes egg laying, males also protect females as they oviposit.

Territoriality with respect to food availability may be seen in some species, especially parasitoids and social insects. For example, female ichneumons, braconids, chalcidids, and scelionids (Hymenoptera) may mark the host either chemically or physically as they oviposit so that other females of the species do not lay in the same host. Such behavior ensures that the offspring will have adequate food for complete development. (The marks may also be the means by which hyperparasites locate a host!) Social insects defend both their nest and foraging sites against members of other colonies.

4.2. Interspecific Interactions

4.2.1. Competition and Coexistence

An important form of interaction is when an insect competes with other organisms for the same resources. Grasshoppers, sheep, and rabbits all eat grass, and if this is in short supply the presence of the mammalian herbivores will have a very obvious effect on the distribution and abundance of grasshoppers living in the same area. However, as noted earlier, food is seldom a limiting factor as far as the abundance of animals is concerned, and the other requirements of these three species are so different that the species can coexist perfectly well. The collection of requirements that must be satisfied in order for a species to survive and reproduce under natural conditions is described as a niche. Thus, a niche includes both physical and biotic requirements, and its complexity varies with the environment in which a species finds itself. For example, as noted in Chapter 22, Section 3.2.3, the critical day length for induction of diapause in a species may vary with latitude. Equally, with reference to biotic requirements, the complexity of a niche will differ according to the number and nature of other species utilizing the same resources. The more closely two species are related, the more nearly identical will be their requirements (i.e., their niche), and the greater will be the degree of competition between them where the two species coexist. Normally, in this situation the less well-adapted species becomes extinct or restricted to areas where it can again compete favorably with the other species as a result of different environmental conditions, a phenomenon known as competitive exclusion or displacement. In the absence of competition, a species' niche will be broader (less complex); that is, a species' requirements will be less stringent and form the so-called "fundamental" niche. Conversely, the niche occupied by a species that coexists with others is known as the "realized" niche.

A well-documented example of competitive exclusion in insects involves three species of chalcidid, belonging to the genus *Aphytis*, which are parasitoids of the California red scale, *Aonidiella aurantii*, found on citrus fruits. In the early 1900s, the golden chalcidid, *Aphytis chrysomphali*, was accidentally introduced into southern California, probably along with red scale on nursery stock imported from the Mediterranean region, though it is a native of China. During the next 50 years, *A. chrysomphali* spread along with its host throughout the citrus-growing area and exerted a reasonable degree of control over red scale, particularly in the milder coastal areas. However, in 1948, a second species, also Chinese, *A. lingnanenis*, was introduced in the hope of obtaining even better control of the pest. During the 1950s, *A. lingnanensis* gradually displaced *A. chrysomphali* so that, by 1961, the latter was virtually extinct, being restricted to a few small areas along the coast. However, *A. lingnanensis* was ineffective as a control agent of red scale in the inland citrus-growing areas around San

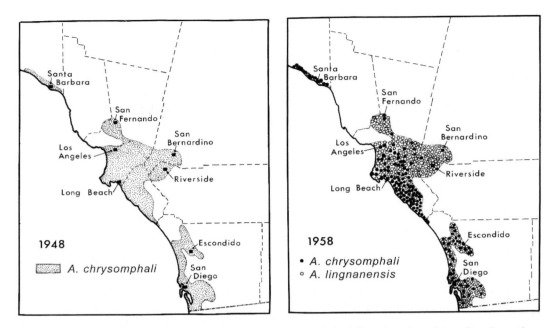

FIGURE 23.5. Changes in the distribution of *Aphytis chrysomphali*, *A. lingnanensis*, and *A. melinus* in southern California between 1948 and 1965. [After P. DeBach and R. A. Sundby, 1963, Competitive displacement between ecological homologues, *Hilgardia* **34**:105–166. By permission of Agricultural Sciences Publications, University of California; and P. DeBach, D. Rosen, and C. E. Kennet, 1971, Biological control of coccids by introduced natural enemies, in: *Biological Control* (C. B. Huffaker, ed.). By permission of Plenum Publishing Corporation and the authors.]

Fernando, San Bernadino, and Riverside, where annual climatic changes are greater. It was found, for example, that periods of cool weather (18°C or less for 1 or 2 weeks) or several nights of hard frost caused high mortality of all stages. Even light overnight frosts (−1°C for 8 hours) killed sperm in the spermathecae of females, which did not mate again, and rendered males sterile. Also, exposure of females to a temperature of 15°C for 24 hours led to an increase in proportion of male progeny. These factors caused a reduction in "effective progeny production" (number of female offspring produced per female) from 21.4 to 4.5 and a resultant inability of the species to control red scale populations. Consequently, a third species, *A. melinus*, was introduced from India and Pakistan in 1956 and 1957. This species rapidly displaced *A. lingnanensis* from these inland areas, and by 1961 virtually the entire population of chalcidids in these areas was made up of *A. melinus* (DeBach and Rosen, 1991). Changes in the distribution of the three *Aphytis* species between 1948 and 1965 are summarized in Figure 23.5.

Competitive displacement does not always occur, however, because closely related organisms have evolved mechanisms that enable them to occupy almost but not quite the same niche. These mechanisms include habitat selection (spatial selection), microhabitat selection, temporal (diurnal and seasonal) segregation, and dietary differences. Two or more of these mechanisms may operate simultaneously to prevent competition between species.

Spatial segregation is shown by the distribution of *A. lingnanensis* and *A. melinus* in southern California, which DeBach and Rosen (1991) considered to have stabilized, with *A. lingnanensis* occupying the milder (less climatically extreme) coastal districts and *A. melinus* the interior. Spatial separation is also seen in larval damselflies (Zygoptera), which

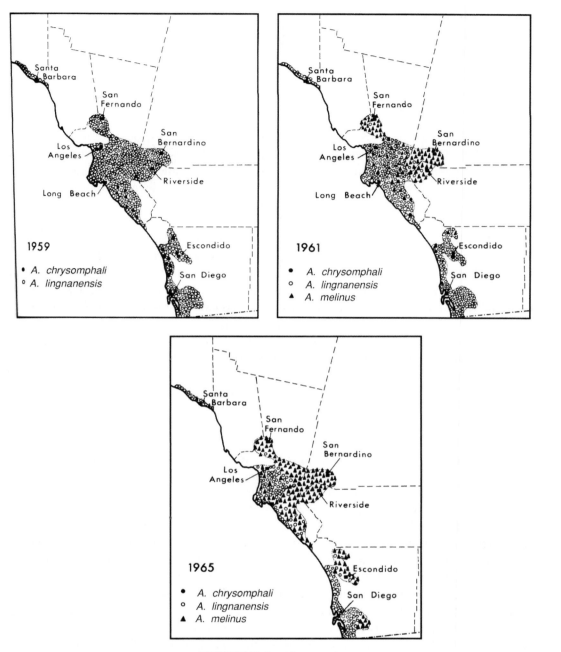

FIGURE 23.5. (*Continued*)

inhabit prairie ponds. For example, two species of Coenagrionidae, *Coenagrion resolutum* and *Enallagma boreale*, hatch, develop, and emerge as adults almost synchronously. However, the species can coexist because larval *E. boreale* are restricted to deep, open water, while *C. resolutum* occurs in shallow water with emergent vegetation (Sawchyn and Gillott, 1975).

Price (1997) provided several examples of microhabitat selection that enable closely related species to coexist. In England, two species of Psocoptera, *Mesopsocus immunis* and *M. unipunctatus*, coexist on larch twigs with no readily obvious differences in their biology.

Studies revealed, however, that the species oviposit in different microhabitats. *M. immunis* prefers to oviposit in the axils of dwarf side shoots, whereas *M. unipunctatus* selects girdle scars and leaf scars. The congeneric flea beetles, *Phyllotreta cruciferae* and *P. striolata*, are concentrated on different parts of their food plant, *Brassica oleracea* (cabbage and relatives). The former species shows a preference for sunny locations and occurs largely on the upper surface of top and middle leaves. *P. striolata* is concentrated on the underside of leaves, especially those near the base of the plant.

Diurnal segregation is shown by two species of *Andrena*, *A. rozeni* and *A. chylismiae*, solitary bees that forage on evening primrose (*Oenothera clavaeformis*) whose flowers remain in bloom for less than a day. Flowers open in late afternoon and are visited by *A. rozeni* between about 1600 and 1900 hours. *A. chylismiae* is an early morning forager and visits flowers between 0500 and 0800 hours, that is, just before they wilt.

Clear-cut seasonal segregation is shown by the damselflies studied by Sawchyn and Gillott (1974a,b, 1975), who were able to arrange the damselflies into three types according to their seasonal biology. Type A species, which includes *Coenagrion resolutum*, *Enallagma boreale*, and other Coenagrionidae, overwinter in diapause as well-developed larvae and emerge highly synchronously between the last week of May and mid-June. Sexual maturation takes about 1 week and the oviposition period extends to the end of July. Females lay eggs in the submerged parts of floating plants. Embryogenesis is direct and requires less than 3 weeks; half-grown larvae may be collected before the end of July and mature larvae by mid-September. Included in Type B are three species of *Lestes*: *L. unguiculatus*, *L. disjunctus*, and *L. dryas*, which overwinter in diapause as well developed embryos. Eggs hatch synchronously during early May, but the very young larvae are not preyed on by the larger larvae of Type A species either because they are too small, that is, outside the range of prey size, or because the Type A larvae have ceased to feed in preparation for the final molt. Type B larvae develop rapidly, and synchronized adult emergence begins in early July and is completed within 2 weeks. Adult maturation requires 16–18 days, and females oviposit in green emergent stems of *Scirpus* (bulrush), which may relate to the requirement of water for embryogenesis. Adults are not normally seen after the end of August, though in mild years they may survive into October. In Type C is included one species, *Lestes congener*, which is characterized by the lateness of its seasonal chronology. *L. congener* overwinters in diapause at an early (preblastokinetic) stage of embryogenesis. Embryonic development continues in the spring after the eggs are wetted and hatching occurs at the end of May. However, the young larvae are too small to serve as prey for the Type B species. Larval development is rapid in *L. congener* so that synchronized emergence begins in late July and continues for about 3 weeks. The much larger larvae of *L. congener* generally do not eat larvae of Type A species. Sexual maturation in *L. congener* takes about 3 weeks. Oviposition begins in mid-August and copulating adults may be seen until early October. Female *L. congener* oviposit only in dry stems of *Scirpus*, a feature associated with the lack of prediapause embryonic development observed in this species.

Thus, the occurrence of seasonal segregation between types and of microhabitat segregation (e.g., deep versus shallow water for larvae, and oviposition in floating vegetation, or emergent green or dry stems in adults) both between and within types, enables a number of species of Zygoptera to coexist and make use of the rich food supply (in the form of *Daphnia*, *Diaptomus*, and dipteran larvae) which is found in prairie ponds.

Dietary differences also enable closely related species to coexist. For example, larvae of the caddisflies *Pycnopsyche gentilis* and *P. luculenta* coexist in woodland streams in

Quebec because the former prefers fallen leaves, whereas *P. luculenta* feeds on submerged twigs or, if these are not available, on detritus or well-rotted leaves (MacKay and Kalff, 1973).

4.2.2. Predator-Prey Relationships

It will be abundantly clear that the distribution and abundance of a species will be greatly affected by those organisms that use it as food and that the reverse is also true, namely, that the distribution and abundance of prey will determine the distribution and abundance of predators.

Most Insecta feed on plant material in one form or another, that is, are primary consumers, and therefore play a major role in the flow of energy stored in plants to higher trophic levels. However, another large group, probably about 10% of known species, feed on other animals, especially insects. Some of these are typical predators or parasites, but the majority are parasitoids that belong especially to the Tachinidae (Diptera), Strepsiptera, and so-called "parasitic" Hymenoptera (Chapter 10, Section 7). A parasitoid may be defined as "an insect that requires and eats only one animal in its life span, but may be ultimately responsible for killing many" (Price, 1997, p. 141). Typically, a female parasitoid deposits a single egg or larva on each host, which is then gradually eaten as the offspring develops. Adult parasitoids are free-living and either do not feed or subsist on nectar and/or pollen. Thus, a parasitoid differs from a typical predator, which feeds on many organisms during its life, and a parasite, which may feed on one to several host individuals but does not kill them. However, as Price (1997) pointed out, the distinction between predator and parasitoid is not always clear. For example, a bird that captures insects as food for its offspring is comparable with a parasitoid that lays its egg on a freshly killed or paralyzed host. Further, a predator and parasitoid face the same problem, namely, location of prey (host), and may solve the problem in an identical manner. Of course, from the prey's point of view, it matters not whether the aggressor is predator or parasitoid; for either, it must take appropriate steps to avoid being eaten! In the final analysis, the population dynamics of predator-prey and parasitoid-host relationships will be identical, and it is therefore appropriate to discuss these relationships under the same heading. In the remainder of this section, therefore, the terms "predator" and "prey" should be taken to include "parasitoid" and "host," respectively, except where specifically stated otherwise.

First, what strategies are employed by prey species in order to reduce the chances of their members being eaten? Probably, the most obvious strategy is for insects to avoid detection. This they may do in various ways—by burrowing into a substrate, which frequently also serves as food, by hiding, for example, on the underside of leaves, by becoming active for a restricted period of the day, or through camouflage where their color pattern merges with the background on which they normally rest, or they precisely resemble a twig or leaf of their food plant. Other prey species have evolved other protective mechanisms that depend on initial recognition of the prey by the predator for their effectiveness. Such mechanisms include being distasteful, a feature usually accompanied by aposematic (warning) coloration so that a predator soon learns to recognize that species are distasteful. Related to this is Müllerian mimicry in which distantly related, distasteful species resemble each other, so that if a predator recognizes their pattern of coloration all species are protected. Another form of mimicry is Batesian, in which an edible species (the mimic) comes to resemble a distasteful species (the model) (Figure 9.33). The success of this method of avoiding predation relies on the probability of the predator selecting the distasteful model rather than the edible mimic; that is, the population density of the model must greatly outweigh that of the mimic. Another chemical method of defense is to secrete obnoxious liquid or vapor whose odor repels predators. Other species release poisons that, on contact with skin or when injected by means of spines, hairs, or sting, injure or kill the attacker (Blum, 1981).

Some insects, especially species of butterflies, practice intimidation displays aimed at frightening would-be (vertebrate) predators. The butterflies normally rest with their wings closed vertically above the body. On being disturbed, the butterflies rapidly open their wings to reveal a striking color pattern, often including large "eyespots," intended to evoke prompt retreat of the aggressor.

Predators use a variety of stimuli in order to locate prey. Some may attempt to capture and eat anything that moves within a certain size range and employ only simple visual or mechanical cues for detection of prey. Most species are, however, relatively prey-specific (feed on only a few or a single species of prey), and prey location is therefore a much more elaborate process. For many of these more specialized predators, the first step is location of the prey's habitat, and this is often achieved as a result of attraction to odors released from the food of the prey. For example, females of the ichneumon fly, *Itoplectis conquisitor*, a parasitoid, are attracted by the odor of pine oil, especially that of Scots pine (*Pinus sylvestris*), on which one of its preferred hosts, caterpillars of the European pine shoot moth, *Rhyacionia buoliana*, are found. For some predators, attraction is greater after the food of the prey has been damaged by the prey. The pteromalid *Nasonia vitripennis*, for example, is attracted to meat, especially when this has been contaminated by the parasitoid's hosts, various muscid flies. Similarly, the ichneumon *Nemeritis* (= *Venturia*) *canescens* is more attracted to oatmeal contaminated by its host, larvae of the Mediterranean flour moth, *Ephestia kühniella*, than to clean oatmeal (Vinson, 1975).

Having been attracted to the habitat of its prey, a predator must now specifically locate the prey. For many species this involves a systematic search, though this behavior is initiated only after receipt of an appropriate signal that indicates the likelihood of prey in the immediate area. Again, such signals are usually chemical in nature and include odors from the prey's feces or from the damaged tissues of the prey's food plant. Pheromones released by the host are often used by parasitoids to locate the host (Powell, 1999; Powell and Poppy, 2001; and see Chapter 13, Section 4.2). Final location of prey is commonly achieved by means of its taste, rarely by its odor (effective only over a very short distance). Some parasitoids locate hosts by the vibrations or sounds the latter make as they burrow through the substrate.

In some species, location of prey is followed immediately by feeding or, in parasitoids, oviposition or larviposition. In others, additional stimuli must be received before prey is deemed acceptable. These appear to be especially critical in parasitoids for which selection of hosts of the correct age (judged by size, color, shape, texture, or taste) may by important. For example, some parasitoids accept only hosts above a certain size; cylindrical host shape improved acceptance in the ichneumon *Pimpla instigator*; hairiness of the host (by preference, caterpillars of the gypsy moth, *Lymantria dispar*) is an important determinant of acceptability in the braconid *Apanteles melanoscelus*; and female *Itoplectis conquisitor* probe host larvae with their ovipositor but will lay eggs only if the host's hemolymph has a suitable taste. Movement may be an important stimulant or deterrent of acceptability. Some parasitoids oviposit only if a larval host moves, whereas movement of an embryo within an egg (indicative of the egg's age) may inhibit oviposition by egg parasitoids (Vinson, 1975, 1976). A special feature of many parasitoids is their ability to discriminate between non-parasitized and parasitized hosts on the basis of physical markings, odors, or tastes

FIGURE 23.6.   Ant feeding on honeydew exuded from the anus of a treehopper (Membracidae). [Photograph courtesy of Dan L. Perlman.]

left by the original parasitoid as it oviposited or larviposited. Such marks render a host unacceptable to the parasitoids that locate it subsequently and ensure that the parasitoid larva, on hatching, has sufficient food for its complete development. Other parasitoids leave trail-marking pheromones as they search for hosts, which inhibit researching of an area and thereby facilitate dispersal of the species.

When given a choice a predator may consistently select a particular species for prey. However, in its natural habitat its survival does not normally depend on availability of that species, and in its absence other species are acceptable as prey. Special mention is made of this point, as it has sometimes been overlooked in attempts at biological control of pests. Some attempts have failed because the introduced predator reduced the pest to low density and then died out because alternate prey was not available in the new habitat. As a result, the pest was able to rebound to an economically important level. In other words, secondary prey species form an important reservoir for the predator at times when the density of the primary prey is low.

4.2.3. Insect-Insect Mutualisms

Mutualistic relationships between insects are relatively uncommon, an important exception being that in which some species of ants tend, protect, and sometimes transport to new locations other insects, notably various homopterans (aphids, psyllids, and membracids) and the larvae of many lycaenid butterflies, in return for which the ants feed on fluids produced by their associate. In homopterans the fluid is the honeydew defecated in large amounts through the anus (Figure 23.6); in myrmecophilous lycaenids, however, special "Newcomer's glands" on the dorsum of the seventh abdominal segment produce an enriched sugar solution (Hölldobler and Wilson, 1990).

5. Insect Diseases

In insects, as well as in almost all other organisms, the great majority of individuals (80% to 99.99% of those born) never survive to reproduce. Fifty percent or more of this mortality may be related to predation, while the remainder results from unsuitable weather conditions, perhaps starvation, and especially disease. Diseases may be subdivided into non-infectious and infectious categories. The former includes those that result from physical or chemical injury, nutritional diseases (caused by deficiencies of specific nutrients), genetic diseases (inherited abnormalities), and physiological, metabolic, or developmental disturbances. Infectious diseases, which can be spread rapidly within a population of organisms, are caused by microorganisms, including viruses, rickettsias, spirochetes, bacteria, fungi, protozoa, and nematodes. Though infections of many of these pathogens may be directly fatal, other pathogens simply weaken an insect, rendering it more susceptible to predation, parasitism, or other pathogens, to chemical and other means of control, or altering its growth rate and reproductive capacity. Most of the time in natural populations, the effects of pathogens are not readily obvious. This is described as the enzootic stage. On occasion, however, conditions are such that the pathogens can reproduce and spread rapidly to decimate the host population. This is known as the epizootic phase, and the outbreak is described as an epizootic, comparable to an epidemic within a population of humans. Study of the factors that lead to epizootics, epizootiology, is of interest not only from a purely ecological perspective but also in light of the potential use of microorganisms in the biological control of insect pests.

This section will outline the important factors in outbreaks of infectious diseases and survey the major groups of insect pathogens.

5.1. Epizootics

Essentially, there are four primary components in the development of an epizootic: the pathogen population, the host population, an efficient means of pathogen transmission, and the environment, all of which are closely interrelated.

Key features of a pathogen are its virulence (disease-producing power), infectivity (capacity to spread among hosts), and ability to survive. Clearly, pathogens (or specific strains of a pathogen) that have both high virulence and high infectivity are the ones that most often cause epizootics, though the susceptibility of the host is also important. Some pathogens may be highly virulent but of low infectivity and, as a result, have a low potential for causing epizootics. *Bacillus thuringiensis*, for example, though pathogenic for many Lepidoptera, seldom causes an epizootic under natural conditions because of its poor powers of dispersal. Indeed, inability to disperse and limited capacity to survive outside a host are probably the main reasons why epizootics occur relatively rarely. Dispersal may be effected either by abiotic or biotic agents in the environment, including wind, rain, running water, snow, host organisms (both healthy and infected), and their predators (both vertebrate and invertebrate) or parasites. Host organisms may disperse the pathogen as a result of defecation, regurgitation, oviposition (i.e., the pathogen occurs either on or within the eggs), disintegration of the body after death, or cannibalism. Predators commonly distribute pathogens via their feces, though some insect parasitoids transfer the microorganisms via their ovipositor when they either sting the host or lay an egg on or in it. Pathogens may survive in either the host or the environment, sometimes for considerable periods. Those that survive in the environment typically have a highly resistant resting stage, such as spores (bacteria, fungi, and protozoa), inclusion bodies (viruses), or cysts (nematodes).

Members of the host population pick up pathogens as a result of physical contact with contaminated surfaces, or eating contaminated food (including cannibalism), or receive pathogens directly from the mother via transovarian transmission. Contact with a pathogen, however, does not necessarily result in ill effects for the host, as the latter has various means of defending itself (Chapter 17, Section 5). Further, even when some members of a population are susceptible to a pathogen, an epizootic does not always follow because of the difficulty of dispersal of the pathogen referred to above and because other members of the population may have varying degrees of resistance.

Density, distribution, and mobility of hosts are important factors in the development of an epizootic. Generally, epizootics are more likely to occur at high densities, even distribution, and high mobility of hosts, as the chances of dispersal of the pathogen are greater under these conditions. On occasion, an epizootic may develop at low host density, as a result of widely dispersed but long-lived pathogens that remain from a previous high-density outbreak. Even at high host-population density, an epizootic may not develop if the host population has a discontinuous distribution and/or poor mobility.

The importance of the environment, physical and biotic, in the dispersal and survival of pathogens has already been noted. Environmental factors are important in other ways in relation to epizootics. For example, factors that induce stress in an insect, especially extremes of temperature, high humidity, and inadequate food, may lower its resistance to a pathogen.

In conclusion, it is clear that whether or not an epizootic occurs depends on a variety of conditions relating to pathogen, host, and environment. Only when knowledge of all of these factors is available for a given host-pathogen interaction can an accurate forecast of a potential disease outbreak be made. Such knowledge is of critical importance in determining the success or otherwise of biological control using pathogenic microorganisms.

5.2. Types of Pathogens

It is natural that the best-known pathogens are those that cause epizootics in economically important insects and show potential for use in microbial control of pest species. In this short account, the major features of some of these pathogens will be outlined, as a basis for the discussion on microbial control presented in Chapter 24, Section 4.3.1.

5.2.1. Bacteria

Many species of bacteria have been shown experimentally to be highly pathogenic should they gain access to the body cavity. However, under natural conditions, the majority of these never cause epizootics (even though their infectivity is high) because of the barrier presented by the integument and lining of the gut. Further, the gut may be unsuitable for survival and multiplication of the bacteria because of its pH or redox potential, or because of the presence within it of antibacterial substances or antagonistic microorganisms. Infection of an individual insect occurs when the integument or gut is physically damaged. The commonest route of invasion appears to be via the midgut. Bacteria are sometimes able to slip between the peritrophic matrix and the midgut epithelium at either the anterior or the posterior end of the midgut, or through ruptures in the matrix caused by the passage of food along the gut.

The subsequent mode of action of the bacteria, that is, how they cause a pathological condition, is varied. Invasion and destruction of the midgut epithelium is typically the first step, and for some bacteria this may be their only activity, their host then dying of starvation. More often, bacteria not only destroy the midgut epithelium but then grow rapidly in the hemolymph and other tissues, causing massive septicemia. Others act by toxemia; that is, they liberate toxins that kill the host's cells. Other bacteria become pathogenic only when they are able to enter the hemocoel via lesions in the midgut epithelium caused by other microorganisms such as protozoa, viruses, and nematodes.

Bacteria known to be naturally pathogenic in insects can be arranged in two groups, the non-spore-formers and the spore-formers.* Included among the non-spore-forming bacteria is *Serratia marcescens*, varieties of which attack a range of insect species and whose presence is recognized by the reddish color of the dead host. Outbreaks of *S. marcescens* are common in high-density laboratory cultures of insects. According to Bailey (1968), at least five species of bacteria are involved in European foulbrood disease of larval honey bees, though a non-spore-former, *Streptococcus pluton*, is the causative agent. The remaining species are secondary pathogens or saprophytes.

The spore-formers are the most important group from the point of view of epizootics and for their potential importance in biological control, largely because they remain viable for a considerable time outside their host. Further, some species (the so-called crystalliferous bacteria) are highly pathogenic in some insects should their sporangia be ingested because the sporangia include, in addition to a spore, a crystalline structure, the parasporal body that contains various toxic proteins (the δ-endotoxins).

Two of the better known, spore-forming, non-crystalliferous bacteria are *Bacillus cereus*, which has been isolated from a range of host species, and *B. larvae*, which is the cause of American foulbrood in bee larvae. Only young larvae (up to 55 hours old) are susceptible, suggesting that only in early larval life are gut conditions suitable for spore germination. Despite the availability of antibiotics for treatment of American foulbrood, it continues to be a common disease. Though this is partially related to the long-lived nature of *B. larvae* spores (especially in honey), its appearance is frequently the result of poor management on the part of beekeepers who expose contaminated yet unoccupied supers.† Bees from adjacent hives visit the supers to steal honey and in doing so spread the disease.

Of the crystalliferous species, *B. thuringiensis*, *B. popilliae*, and *B. lentimorbus* have been the subject of considerable research, pure and applied, during the past 50 years. Less well known but with some economic potential is *B. sphaericus*.

B. thuringiensis includes a large number of highly pathogenic subspecies [Beegle and Yamamoto (1992), list 36] which attack Lepidoptera and representatives of other orders (notably Diptera and Coleoptera), including a wide array of economically important species (Glare and O'Callaghan, 2000). Fortunately, honey bees and many hymenopterous parasitoids are not directly susceptible though, through their effect on the host, the parasitoids may die within the dead host or have reduced reproductive potential (Vinson, 1990). The key factor that determines host susceptibility is the pH of the midgut. The parasporal body of most subspecies of *B. thuringiensis* (but not *B. t. israelensis* which attacks Diptera) is solubilized by alkaline gut juices and broken down by gut proteases into smaller, toxic fragments. The δ-endotoxins bind with specific glycoprotein receptors in the plasmalemma of the gut epithelial cells, causing disruption of ion and amino acid transport and cytolysis (Gill *et al.*, 1992). Two other classes of toxins are produced by *B. thuringiensis*, thuringiensin (β-exotoxin), which is a nucleotide and inhibits RNA synthesis, and a group of miscellaneous toxins that include phospholipase C and the "louse factor," so-called because it (rather than the δ-endotoxins or thuringiensin) is responsible for the toxicity of *B. thuringiensis* against the sheep body louse (*Bovicola ovis*) (Gough *et al.*, 2002).

Extensive field testing showed that *B. thuringiensis* has desirable attributes for successful control of selected insect pests, and commercial preparations are now available for

---

* This is not a taxonomic arrangement but one of convenience used by insect pathologists.

† Supers are the boxes, without top or bottom, that contain the frames of honeycomb. Placed one on top of the other, they constitute the hive.

a wide range of applications, making it easily the most successful microbial pest control agent. This aspect is taken up again in Chapter 24 (Section 4.3.1).

B. popilliae and *B. lentimorbus* cause "milky disease" in larvae of some scarabaeid beetles, including the Japanese beetle, *Popillia japonica*, an important pest in the eastern United States, that feeds on roots of grasses, vegetables, and nursery stock. Mass-produced spore preparations of *B. popilliae* are marketed for control of this pest.

The first pathogenically active strain of *B. sphaericus* was isolated in 1965 from mosquito larvae (Charles *et al.*, 1996). Though many strains of this bacterium are known, not all are pathogenic. Strains isolated from larvae of *Culex* and *Anopheles* spp. are generally highly active, whereas those from *Aedes* are only weakly pathogenic or non-toxic. There has been some interest in developing *B. sphaericus* as a microbial pesticide, and three commercial formulations are available, though its narrow host range is a significant drawback (Khetan, 2001).

5.2.2. Rickettsias

Only a few species of rickettsias are pathogenic in insects, though some infect pests and may have potential as biological control agents, for example, *Rickettsiella popilliae*, which attacks Japanese beetles. However, almost no work has been done to test this possibility perhaps, as St. Julian *et al.* (in Bulla, 1973) suggested, because some rickettsias of insects are also pathogenic in mammals.

5.2.3. Viruses

According to David (1975), more than 700 species of insects are known to be susceptible to viral diseases. Of these, about 80% are Lepidoptera, 10% Diptera, and 5% Hymenoptera. Many of these species are economically important, and much research has been and is being conducted to determine the value of viruses in biological control.

Viruses may be arranged in two categories, according to whether the nucleic acid they contain is in the form of DNA or RNA. Viruses in both groups use raw materials in host cells for replicating their nucleic acids, and, by producing viral mRNA, they use the host's ribosomes for synthesis of viral proteins, thereby causing disruption of the host cell's metabolism. The DNA viruses (baculoviruses) include nucleopolyhedroviruses (NPVs), granuloviruses (GVs), and iridescent viruses (IVs), while among the RNA-containing viruses the best known are the cypoviruses (CPVs). The characteristics of these various forms are given in Steinhaus (1963, Vol. 1), by David (1975), and in standard texts on virology or microbiology.

Like bacteria, viruses mainly infect insects when ingested on contaminated food or during cannibalism, though some may be transmitted on the surface of or within eggs (Rothman and Myers, 2000). Baculoviruses have rod-shaped viral particles enclosed within a protective proteinaceous sheath, forming an occlusion body. On reaching the midgut, the occlusion body dissolves in the alkaline gut juice, releasing the virus particles. Initially, these infect the midgut epithelial cells, replicate, and then enter the hemocoel where the infection spreads to the fat body, hemocytes, and, to a lesser extent, other tissues. In the final stages of infection, more occlusion bodies are produced, sometimes in massive numbers (10^9 per corpse) to be released as the corpse decays or is eaten.

Some viruses are relatively short-lived outside the insect host and their survival from year to year requires that at least some members of the host population survive. Others, such as NPVs and CPVs, are able to survive outside the host for a considerable time under suitable conditions (e.g., in soil, in leaf litter, on tree bark, and in cadavers). Cabbage looper NPV, for example, may persist for 9 years in soil. Viruses are relatively stable within the temperature and humidity range that they normally experience. However, they are sensitive to sunlight, especially its shorter (ultraviolet) wavelengths, and are inactivated by a few hours of continuous exposure. Thus, those that are disseminated on the host's food plant may have limited viability. In commercial preparations viability is increased by additives that screen out ultraviolet radiation. Viruses are also sensitive to pH and need to be maintained in conditions close to neutrality. Again, in commercial preparations such sensitivity may be partially overcome by the inclusion of buffers.

The best-studied (and most-commercialized) viruses are the baculoviruses, which are restricted to invertebrate (mostly insect) hosts and are structurally dissimilar to mammalian or plant viruses (Moscardi, 1999; Battu *et al.*, 2002). Of these, NPVs have been isolated from numerous Lepidoptera and from representatives of other orders, whereas GVs apparently are restricted to Lepidoptera. Though the majority of pest Lepidoptera are susceptible to NPVs and/or GVs under experimental conditions, for relatively few of these have the viruses been used successfully in integrated pest management (see Table 24.8). In rare cases, for example, *Gilpinia hercyniae* (European spruce sawfly) in Canada, NPVs exert a strong degree of natural control so that populations of this potential pest do not reach epizootic levels.

IVs have been isolated mainly from Diptera, including mosquitoes and black flies. Though highly infective when injected experimentally into the hemocoel, IVs do not appear to have much future as biological control agents in Diptera, at least, because they are rapidly destroyed when ingested. Further, the peritrophic matrix of dipteran larvae does not have gaps like that of larval Lepidoptera (David, 1975).

Almost all of the approximately 150 species of insect from which CPVs have been isolated are Lepidoptera, and a number of these are pests. Relatively little work appears to have been carried out on the potential of CPVs as control agents, perhaps because many of the pests in which they occur are also susceptible to the better-known NPVs and because they are relatively slow-acting.

5.2.4. Fungi

Studies on entomogenous fungi, including their potential as control agents, tend to be overshadowed by the enormous volume of work being carried out on bacteria and viruses. However, the first attempt at microbial control (by Metschnikoff in 1879 against larvae of the wheat cockchafer, *Anisoplia austriaca*) used the green-muscardine fungus, *Metarhizium anisopliae*, and the first large-scale program in the United States, which covered almost the whole of Kansas, began in about 1890 against the chinch bug, *Blissus leucopterus*, using the fungus *Beauveria bassiana* (Glare and Milner, 1991).

Unlike bacteria, viruses, and protozoa, fungi normally enter insects via the integument rather than the gut, whose conditions (especially pH) are unsuitable for fungal spore germination. They are the principal pathogens of sucking insects, which cannot ingest pathogens whose route of entry is the gut wall. They are also important disease-causing agents of Coleoptera, which seem especially resistant to viral and bacterial diseases (Hajek and St. Leger, 1994). Apart from suitable temperature and pH (in soil-dwelling fungi), high humidity or even liquid water is essential for spore germination, an observation that accounts for the frequent occurrence of fungal epizootics during periods of rainy or humid weather. After germination, access through the cuticle is probably achieved through the secretion of chitinase and proteinase enzymes by elongating hyphae. Initially, the hyphae that penetrate colonize the epidermis but then spread to other specific body tissues. Some tissues are not attacked until after the host's death when the fungus becomes saprophytic. Death may result from the mere physical presence of a numerous hyphae, which disrupt tissues or inhibit hemolymph circulation. Alternatively, fungal secretions may cause histolysis or be toxic to the host.

At the end of an epizootic a fungus may survive in various ways. It may persist at low incidence in remaining members of the host population or it may infect other, less susceptible, species. It may enter a facultatively saprophytic phase, for example, in soil. Or, most commonly, it may produce spores that can survive outside the host for some time. The longevity of spores varies among species and in relation to weather conditions. Low humidity and temperatures a few degrees above freezing point appear to result in greatest longevity. Spores are also susceptible to sunlight, though whether this is due to ultraviolet radiation is uncertain.

About 700 species of fungi are pathogenic in insects. Of these only about 10 species are very common and can cause epizootics (Hajek and St. Leger, 1994). Naturally, therefore, these are the best studied, with emphasis on their potential as microbial control agents. Selections of *Metarhizium anisopliae*, which has over 200 host species, mostly soil inhabitants, *Beauveria bassiana*, and *Verticillium lecanii* have been commercialized (Khachatourians *et al.*, 2002), though their impact in the bio-pesticide market is not yet significant. *Coelomomyces* species are aquatic fungi that are obligate pathogens of arthropods; they have complex life cycles that include both intermediate and final hosts. The former are ostracod or copepod crustacea, while the latter are mosquitoes. Under natural conditions, *Coelomomyces* species are probably important regulators of mosquito population density. However, large-scale production of these fungi has not yet been achieved, limiting their potential in microbial control (Ferron *et al.*, 1991). The genus *Entomophthora* includes more than 100 species of fungi, many of which cause epizootics among grasshoppers, aphids, caterpillars, mosquitoes, and houseflies. Though *Entomophthora* species are usually highly specific with reference to their insect host, many are facultative parasites and are amenable to culture on a mass scale.

5.2.5. Protozoa

Protozoa are one of the most diverse groups of insect pathogens, and the diseases they cause range from benign to highly virulent. Hosts typically acquire the pathogens as they ingest food, though some pathogenic protozoa may be passed on to subsequent generations in the host's eggs or via parasitoids. In some species the midgut epithelium is the first tissue attacked, and only later are other tissues, especially fat body, invaded. In other species the protozoa migrate through the midgut wall to specific tissues in the hemocoel. Disease is caused by the general debilitating effect of the protozoa as they reproduce, by toxins released from the protozoa, or as a result of secondary invasions of viruses or bacteria. Protozoa are disseminated as spores or cysts, some of which may survive for long periods outside the host.

Most disease-causing protozoa in insects are Sporozoa, including Coccidia, Gregarinia, and especially Microsporidia for which more than 200 host species are known. The protozoa probably play an important role in regulation of insect populations, though in only a few instances has this been authenticated. Species of the microsporidian *Nosema* are agents for a number of well-known insect diseases, two examples of which are pébrine disease of silkworms caused by *N. bombycis* and nosema disease of honey bees (pathogen *N. apis*). Other species of *Nosema* known to be highly pathogenic include *N. lymantriae* [primary host, the gypsy moth (*Lymantria dispar*)], *N. polyvora* [imported cabbageworm (*Pieris brassicae*)], *N. pyrausta* [European corn borer (*Ostrinia nubilalis*)], and *N. locustae* (almost 60 species of Orthoptera) (Maddox, 1987).

Though some pathogenic protozoa would seem to be good candidates for biological control agents, research into this possibility is proceeding slowly, primarily because, with few exceptions, the pathogens cannot be cultured outside their host. As a result, obtaining sufficient spore material for testing is somewhat slow.

5.2.6. Nematodes

Though obviously not microorganisms *sensu stricto*, nematodes are generally included in this term in discussions of insect pathogens. Nematodes enter into a variety of interactions with insects, including parasitism, and pathogenesis, and are important natural regulators of insect populations (Kaya *et al.*, 1993).

Most insect-associated nematodes are parasitic and in most instances do not directly cause death but, rather, protracted larval development, abnormal morphology (including wing shortening), and reduced fecundity. Many of the effects noted can be attributed to debilitation of the host as the nematode feeds on its tissues, especially fat body. Other effects result from more specific activities on the part of the parasite. For example, some morphological abnormalities probably result from endocrine imbalances, induced perhaps by toxins released by the parasite. Sterility results in some insects because the parasite selectively feeds on the gonads.

Pathogenic nematodes are found in only two small, monogeneric families, Steinernematidae (14 recognized species of *Steinernema*) and Heterorhabditidae (5 species of *Heterorhabditis*) (Kaya *et al.*, 1993). Members of these families are unique in that: (1) they are the only nematodes with the ability to carry and introduce symbiotic bacteria into the host insect's body cavity; (2) they are the only insect pathogens with a host range that includes most insect orders and families; and (3) they can be cultured on a large scale in artificial media (Poinar, 1990). The bacteria (*Xenorhabdus* spp.) are carried in the intestine of the nematode and are released into the hemocoel when the infective (juvenile) nematode enters the insect through the mouth, anus, spiracles, or body wall if the cuticle is thin. The bacteria propagate rapidly, typically killing the host insect within 48 hours as a result of septicemia. However, studies have also shown that some nematodes are pathogenic even in the absence of the bacteria, indicating that the nematodes produce toxins. The nematodes feed on the bacteria, mature, and produce up to 10^5 offspring that emerge as infective juveniles (Akhurst and Dunphy, 1993; Kaya and Gaugler, 1993).

Soil is the normal environment for entomopathogenic nematodes, which would therefore be expected to have a restricted host range and limited potential as microbial pest-control agents. However, under laboratory conditions they are capable of infecting a very wide range of insects, including mosquitoes, black flies, grasshoppers, caterpillars, weevils, and ants, opening the way to the possibility of developing nematode-based management systems for certain above-ground pests.

6. Summary

The biotic environment of insects is composed of all other organisms that affect insects' survival and multiplication.

Food is not normally an important regulator of insect abundance because other environmental factors have a significant adverse effect on insect growth and reproduction. In addition, many insects are polyphagous, and larvae and adults may eat different foods. In two situations the quantity of food may be limiting: (1) when only a proportion of the total food is available and (2) when insect population density is not kept in check by other factors. The nutritional quality of food also markedly affects survival, rate of growth, and fecundity of insects.

The evolution of diverse feeding habits has allowed insects to exploit virtually all organic carbon sources. The majority of species are herbivorous and, as the prey of other animals, are key elements in the flow of energy from primary producers to second-order consumers. Others are predators, parasites, pathogens, detritus feeders (especially in aquatic ecosystems), or may enter into a variety of mutualistic relationships with plants or other insects. Through evolution complex interactions between plants and insects have developed based on the theme that insects feed on (gain energy from) plants, while plants attempt to defend themselves (conserve this energy) or obtain service (most often cross-pollination, rarely protection) from insects in exchange. Some plants protect themselves by producing toxins. However, some insects have become able to cope with these toxins and may even accumulate them for protection against predators and, possibly, microorganisms. Other insects have become adapted to feeding on parts of plants that lack toxins (spatial avoidance) or when plants have a low toxin content (temporal avoidance). To obtain effective cross-pollination: (1) plants must produce the correct amount of nectar to maintain insects' "interest," yet stimulate visits to other plants of the same species; and (2) insects must be able to recognize members of the same plant species. The quantity of nectar produced in each flower depends on factors such as number of flowers per plant, synchronous or asynchronous blooming of flowers on a plant, population density of plants (average distance between plants), air temperature when flowers are blooming, and the size and foraging capacity of the pollinators.

Interactions between insects and other animals may be intraspecific or interspecific. Intraspecific interactions include those related to underpopulation and those that result from overpopulation. Populations are generally self-regulating, that is, maintain their density within a suitable range. As a prey species' density falls, predators may experience greater difficulty in locating food and may migrate or produce fewer progeny. Thus, more prey may survive to reproduce, leading to restoration of the original population density. In some species females produce more eggs or reproduce parthenogenetically in underpopulated conditions. Overpopulation results in competition for resources such as oviposition, overwintering, and resting sites, and, occasionally, food, and renders a greater proportion of the population susceptible to predation, the effects of weather, and disease. When overcrowded, part of a population may migrate, cannibalism may increase, females may lay fewer eggs, and the rate of larval development may be reduced. Some species regulate breeding population density by territoriality.

When two species that coexist (live in the same habitat) require a common resource, they are said to compete for that resource. The more closely related are the species, the more nearly identical will be their total requirements (niche), and the greater will be the competition between the species, leading ultimately to the competitive exclusion (displacement) of one species from that habitat. To avoid displacement closely related species evolve mechanisms that make their niches sufficiently different that both can occupy the same habitat. The mechanisms include spatial segregation, microhabitat selection, temporal (diurnal or seasonal) segregation, and dietary differences.

Insects that are preyed on by other organisms may reduce their chances of being eaten by burrowing into a substrate, hiding in vegetation, becoming active during a restricted period, or through camouflage. Some species are aposematically colored and distasteful. Distasteful species may resemble each other (Müllerian mimicry), so that when a predator learns their pattern of coloration, all species are protected. An edible species (the mimic) may resemble an inedible (distasteful) species (the model) to avoid detection (Batesian mimicry), though this method requires that the population density of the model be much greater than that of the mimic.

Most insect predators, parasitoids, and parasites are relatively or highly prey-(host-)specific. They may find suitable prey in a sequence of steps: (1) location of the prey's habitat, often by the odor of the prey's food (especially if this has been damaged by the prey); (2) search for and location of prey, stimulated by specific odors, for example, that of the prey's feces; and (3) acceptance of prey, which may be dependent on its size, color, shape, texture, or taste.

Diseases caused by pathogenic microorganisms, particularly bacteria, viruses, fungi, protozoa, and nematodes, are important regulators of insect populations. Often the incidence of disease in a population is low and the disease is said to be in the enzootic stage. When conditions are such that a disease can spread rapidly through a population, the disease is described as in the epizootic stage. The occurrence of an epizootic depends on a pathogen's virulence, infectivity, and viability, on the host's density, distribution, and mobility, and on abiotic factors such as temperature, humidity, light, and wind. The normal route of entry of pathogens is via the midgut, though fungi commonly enter via the integument. How microorganisms cause a pathological condition is varied. Bacteria may damage the midgut epithelium, causing starvation, or may invade other tissues causing septicemia and/or liberating toxins. Viruses disrupt the metabolism of the host's cells. Fungi may physically disrupt tissues, or may secrete histolyzing or toxic substances. Protozoa have a generally debilitating effect and may release toxins. Nematodes may be parasitic or pathogenic. Parasitic species typically cause protracted or morphologically abnormal larval development, or reduced fecundity. Some feed on selected organs, for example, the gonads, and thereby exert specific effects on the host. Pathogenic nematodes release their mutualistic bacteria into the host's body cavity, causing rapid septicemia and death of the host.

7. Literature

Matthews and Kitching (1984), Price (1997), Huffaker and Gutierrez (1999), Speight *et al.* (1999), and Schowalter (2000) discuss the interactions of insects with other organisms. The ecology of aquatic insects is considered in the volume edited by Resh and Rosenberg (1984). Aspects of insect-plant interactions are reviewed by Schoonhoven *et al.* (1998) [general] and in the volumes edited by Wallace and Mansell (1975) [biochemical interactions], Gilbert and Raven (1975) [insect-plant coevolution], Strong *et al.* (1984) [insect-plant communities], Beattie (1985) and Huxley and Cutler (1991) [ant-plant interactions], and Panda and Khush (1995) [host-plant resistance to insects]. Blum (1981), Bowers (1990), Blum and Hilker (2002), and Nishida (2002) review the nature and use by insects of plant allelochemicals. The trophic relationships of insects are discussed in Slansky and Rodriguez (1987). The biology of insect parasites and parasitoids is dealt with by Vinson (1975, 1984, 1985), Vinson and Iwantsch (1980), Powell (1999), Powell and Poppy (2001), and in several of the papers edited by Price (1975).

A large volume of literature on insect pathogens has developed during the past decade, especially dealing with those organisms having control potential. Chapters in Fuxa and Tanada (1987) and Beckage *et al.* (1993) [all pathogenic groups], Gaugler and Kaya (1990) and Bedding *et al.* (1993) [nematodes], and Arora *et al.* (1991) [fungi], as well as the reviews by Rothman and Myers (2000) [viruses], Ferron (1978) and Hajek and St. Leger (1994) [fungi], Henry (1981) [protozoa], and Kaya and Gaugler (1993) [nematodes], provide comprehensive treatments.

Akhurst, R. J., and Dunphy, G. B., 1993, Tripartite interactions between symbiotically associated entomopathogenic bacteria, nematodes, and their insect hosts, in: *Parasites and Pathogens of Insects*, Vol. 2 (N. Beckage, B. A. Federici, and S. N. Thompson, eds.), Academic Press, New York.

Anderson, N. H., and Cargill, A. S., 1987, Nutritional ecology of aquatic detritivorous insects, in: *Nutritional Ecology of Insects, Mites, Spiders, and Related Invertebrates* (F. Slansky, Jr. and J. G. Rodriguez, eds.), Wiley, New York.

Andrewartha, H. G., 1961, *Introduction to the Study of Animal Populations*, University of Chicago Press, Chicago.

Arora, D. K., Ajello, L., and Mukerji, K. G. (eds.), 1991, *Handbook of Applied Mycology*, Vol. 2, Dekker, New York.

Bailey, L., 1968, Honey bee pathology, *Annu. Rev. Entomol.* **13**:191–212.

Baker, H. G., and Baker, I., 1973, Amino-acids in nectar and their evolutionary significance, *Nature* **241**:543–545.

Battu, G. S., Arora, R., and Dhaliwal, G. S., 2002, Prospects of baculoviruses in integrated pest management, in: *Microbial Pesticides* (O. Koul and G. S. Dhaliwal, eds.), Taylor and Francis, London.

Beattie, A. J., 1985, *The Evolutionary Ecology of Ant-Plant Mutualisms*, Cambridge University Press, London.

Beckage, N. E., Federici, B. A., and Thompson, S. N. (eds.), 1993, *Parasites and Pathogens of Insects*, Vols. 1 and 2, Academic Press, New York.

Bedding, R., Akhurst, R., and Kaya, H. (eds.), 1993, *Nematodes and the Biological Control of Insect Pests*, CSIRO, East Melbourne, Australia.

Beegle, C. C., and Yamamoto, T., 1992, History of *Bacillus thuringiensis* Berliner research and development, *Can. Entomol.* **124**:587–616.

Bernays, E. A., and Chapman, R. F., 1994, *Host-Plant Selection by Phytophagous Insects*, Chapman and Hall, New York.

Blum, M. S., 1981, *Chemical Defences of Arthropods*, Academic Press, New York.

Blum, M. S., and Hilker, M., 2002, Chemical protection of insect eggs, in: *Chemoecology of Insect Eggs and Egg deposition* (M. Hilker and T. Meiners, eds.), Blackwell Verlag, Berlin.

Bowers, M. D., 1990, Recycling plant natural products for insect defense, in: *Insect Defenses* (D. L. Evans and J. O. Schmidt, eds.), SUNY Press, Albany, NY.

Bulla, L. A., Jr. (ed.), 1973, Regulation of insect populations by microorganisms, *Ann. N.Y. Acad. Sci.* **217**:243 pp.

Center, T. D., and Johnson, C. D., 1974, Coevolution of some seed beetles (Coleoptera: Bruchidae) and their hosts, *Ecology* **55**:1096–1103.

Charles, J.-F., Nielsen-LeRoux, C., and Delécluse, A., 1996, *Bacillus sphaericus* toxins: Molecular biology and mode of action, *Annu. Rev. Entomol.* **41**:451–472.

Common, I. F. B., 1980, Some factors responsible for imbalances in the Australian fauna of Lepidoptera, *J. Lepid. Soc.* **34**:286–294.

David, W. A. L., 1975, The status of viruses pathogenic for insects and mites, *Annu. Rev. Entomol.* **20**:97–117.

DeBach, P., and Rosen, D., 1991, *Biological Control by Natural Enemies*, 2nd ed., Cambridge University Press, London.

Eisner, T., Goetz, M. A., Hill, D. E., Smedley, S. R., and Meinwald, J., 1997, Firefly "femmes fatales" acquire defensive steroids (lucibufagins) from their firefly prey, *Proc. Natl. Acad. Sci.* **94**:9723–9728.

Ferron, P., 1978, Biological control of insect pests by entomogenous fungi, *Annu. Rev. Entomol.* **23**:409–442.

Ferron, P., Fargues, J., and Riba, G., 1991, Fungi as microbial insecticides against pests, in: *Handbook of Applied Mycology*, Vol. 2 (D. K. Arora, L. Ajello, and K. G. Mukerji, eds.), Dekker, New York.

Fuxa, J. R., and Tanada, Y. (eds.), 1987, *Epizootiology of Insect Diseases*, Wiley, New York.

Gaugler, R., and Kaya, H. K. (eds.), 1990, *Entomopathogenic Nematodes in Biological Control*, CRC Press, Boca Raton.

Gilbert, L. E., 1971, Butterfly-plant coevolution: Has *Passiflora adenopoda* won the selectional race with heliconiine butterflies? *Science* **172**:585–586.

Gilbert, L. E., and Raven, P. H., 1975, *Coevolution of Animals and Plants*, University of Texas Press, Austin.

Gill, S. S., Cowles, E. A., and Pietrantonio, P. V., 1992, The mode of action of *Bacillus thuringiensis* endotoxins, *Annu. Rev. Entomol.* **37**:615–636.

Glare, T. R., and Milner, R. J., 1991, Ecology of entomopathogenic fungi, in: *Handbook of Applied Mycology*, Vol. 2 (D. K. Arora, L. Ajello, and K. G. Mukerji, eds.), Dekker, New York.

Glare, T. R., and O'Callaghan, M., 2000, *Bacillus thuringiensis: Biology, Ecology and Safety*, Wiley, Chichester, U.K.

Gough, J. M., Akhurst, R. J., Ellar, D. J., Kemp, D. H., and Wijffels, G. L., 2002, New isolates of *Bacillus thuringiensis* for control of livestock ectoparasites, *Biol. Control* **23**:179–189.

Hajek, E. J., and St. Leger, R. J., 1994, Interactions between fungal pathogens and insect hosts, *Annu. Rev. Entomol.* **39**:293–322.

Heinrich, B., and Raven, P. H., 1972, Energetics and pollination ecology, *Science* **176**:597–602.

Henry, J. E., 1981, Natural and applied control of insects by protozoa, *Annu. Rev. Entomol.* **26**:49–73.

Hölldobler, B., and Wilson, E. O., 1990, *The Ants*, Springer-Verlag, Berlin.

Huffaker, C. B., and Gutierrez, A. P. (eds.), 1999, *Ecological Entomology*, 2nd ed., Wiley, New York.

Huxley, C. R., and Cutler, D. F. (eds.), 1991, *Ant-Plant Interactions*, Oxford University Press, London.

Kaya, H. K., and Gaugler, R., 1993, Entomopathogenic nematodes, *Annu. Rev. Entomol.* **38**:181–206.

Kaya, H. K., Bedding, R. A., and Akhurst, R. J., 1993, An overview of insect-parasitic and entomopathogenic nematodes, in: *Nematodes and the Biological Control of Insect Pests* (R. Bedding, R. Akhurst, and H. Kaya, eds.), CSIRO, East Melbourne, Australia.

Kennedy, J. S.,1961, Continuous polymorphism in locusts, *Symp. R. Entomol. Soc.* **1**:80–90.

Kevan, P. G., and Baker, H. G., 1983, Insects as flower visitors and pollinators, *Annu. Rev. Entomol.* **28**:407–453.

Kevan, P. G., and Baker, H. G., 1999, Insects on flowers, in: *Ecological Entomology*, 2nd ed. (C. B. Huffaker and A.P. Gutierrez, eds.), Wiley, New York.

Khachatourians, G. G., Valencia, E. P., and Miranpuri, G. S., 2002, *Beauveria bassiana* and other entomopathogenic fungi in the management of insect pests, in: *Microbial Pesticides* (O. Koul and G. S. Dhaliwal, eds.), Taylor and Francis, London.

Khetan, S. K., 2001, *Microbial Pest Control*, Dekker, New York.

MacKay, R. J., and Kalff, J., 1973, Ecology of two related species of caddis fly larvae in the organic substrates of a woodland stream, *Ecology* **54**:499–511.

Maddox, J. V., 1987, Protozoan diseases, in: *Epizootiology of Insect Diseases* (J. R. Fuxa and Y. Tanada, eds.), Wiley, New York.

Matthews, E. G., and Kitching, R. L., 1984, *Insect Ecology*, 2nd ed., University of Queensland Press, St. Lucia, Queensland.

Moscardi, F., 1999, Assessment of the application of baculoviruses for control of Lepidoptera, *Annu. Rev. Entomol.* **44**:257–289.

Nishida, R., 2002, Sequestration of defensive substances from plants by Lepidoptera, *Annu. Rev. Entomol.* **47**:57–92.

Owen, D. F., and Wiegert, R. G., 1987, Leaf eating as mutualism, in: *Insect Outbreaks* (P. Barbosa and J. C. Schultz, eds.), Academic Press, New York.

Panda, N., and Khush, G. S., 1995, *Host Plant Resistance to Insects*, CAB International, Wallingford, U.K.

Pickford, R., 1962, Development, survival and reproduction of *Melanoplus bilituratus* (Wlk.) (Orthoptera: Acrididae) reared on various food plants, *Can. Entomol.* **94**:859–869.

Pickford, R., and Gillott, C., 1972, Coupling behaviour of the migratory grasshopper, *Melanoplus sanguinipes* (Orthoptera: Acrididae), *Can. Entomol.* **104**:873–879.

Poinar, G. O., Jr., 1990, Biology and taxonomy of Steinernematidae and Heterorhabditidae, in: *Entomopathogenic Nematodes* in *Biological Control* (R. Gaugler and H. K. Kaya, eds.), CRC Press, Boca Raton.

Powell, W., 1999, Parasitoid hosts, in: *Pheromones of Non-Lepidopteran Pests Associated with Agricultural Plants* (J. Hardie and A. K. Minks, eds.), CAB International, Wallingford, U.K.

Powell, W., and Poppy, G., 2001, Host location by parasitoids, in: *Insect Movement: Mechanisms and Consequences* (I. P. Woiwod, D. R. Reynolds, and C. S. D. Thomas, eds.), CAB International, Wallingford, U.K.

Price, P. W. (ed.), 1975, *Evolutionary Strategies of Parasitic Insects and Mites*, Plenum Press, New York.

Price, P. W., 1997, *Insect Ecology*, 3rd ed., Wiley, New York.

Resh, V. H., and Rosenberg, D. M. (eds.), 1984, *The Ecology of Aquatic Insects*, Praeger, New York.

Rothman, L. D., and Myers, J. H., 2000, Ecology of insect viruses, in: *Viral Ecology* (C. J. Hurst, ed.), Academic Press, San Diego.

Sawchyn, W. W., and Gillott, C., 1974a, The life history of *Lestes congener* (Odonata: Zygoptera) on the Canadian prairies, *Can. Entomol.* **106**:367–376.

Sawchyn, W. W., and Gillott, C., 1974b, The life histories of three species of *Lestes* (Odonata: Zygoptera) in Saskatchewan, *Can. Entomol.* **106**:1283–1293.

Sawchyn, W. W., and Gillott, C., 1975, The biology of two related species of coenagrionid dragonflies (Odonata: Zygoptera) in Western Canada, *Can. Entomol.* **107**:119–128.

Schoonhoven, L. M., Jermy, T., and van Loon, J. J. A., 1998, *Insect-Plant Biology: From Physiology to Evolution*, Chapman and Hall, London.

Schowalter, T. D., 2000, *Insect Ecology: An Ecosystem Approach*, Academic Press, San Diego.

Slansky, F., Jr., and Rodriguez, J. G. (eds.), 1987, *Nutritional Ecology of Insects, Mites, Spiders, and Related Invertebrates*, Wiley, New York.

Southwood, T. R. E., 1972, The insect/plant relationship—An evolutionary perspective, *Symp. R. Entomol. Soc.* **6**:3–30.

Speight, M. R., Hunter, M. D., and Watt, A. D., 1999, *Ecology of Insects: Concepts and Applications*, Blackwell, Oxford.

Steinhaus, E. A. (ed.), 1963, *Insect Pathology—An Advanced Treatise,* Vols. 1 and 2, Academic Press, New York.

Strong, D. R., Lawton, J. H., and Southwood, R., 1984, *Insects on Plants: Community Patterns and Mechanisms,* Blackwell, Oxford.

Tallamy, D. W., Whittington, D. P., Defurio, F., Fontaine, D. A., Gorski, P. M., and Gothro, P. W., 1998, Sequestered cucurbitacins and pathogenicity of *Metarhizium anisopliae* (Moniliales: Moniliaceae) on spotted cucumber beetle eggs and larvae (Coleoptera: Chrysomelidae), *Environ. Entomol.* **27**:366–372.

Vinson, S. B., 1975, Biochemical coevolution between parasitoids and their hosts, in: *Evolutionary Strategies of Parasitic Insects and Mites* (P. W. Price, ed.), Plenum Press, New York.

Vinson, S. B., 1976, Host selection by insect parasitoids, *Annu. Rev. Entomol.* **21**: 109–133.

Vinson, S. B., 1984, Parasitoid-host relationships, in: *Chemical Ecology of Insects* (W. J. Bell and R. T. Cardé, eds.), Chapman and Hall, London.

Vinson, S. B., 1985, The behavior of parasitoids, in: *Comprehensive Insect Physiology, Biochemistry and Pharmacology*, Vol. 8 (G. A. Kerkut and L. I. Gilbert, eds.), Pergamon Press, Elmsford, NY.

Vinson, S. B., 1990, Potential impact of microbial insecticides on beneficial arthropods in the terrestrial environment, in: *Safety of Microbial Insecticides* (M. Laird, L. A. Lacey, and E. W. Davidson, eds.), CRC Press, Boca Raton.

Vinson, S. B., and Iwantsch, G. F., 1980, Host suitability for insect parasitoids, *Annu. Rev. Entomol.* **25**:397–419.

Wallace, J. W., and Mansell, R. L. (eds.), 1975, Biochemical interaction between plants and insects, *Recent Adv. Phytochem.* **10**:425 pp.

24

Insects and Humans

1. Introduction

This final chapter will focus on those insects that humans describe, in their economically minded way, as beneficial or harmful, though it should be appreciated from the outset that these constitute only a very small fraction of the total number of species. It must also be realized that the ecological principles governing the interactions between insects and humans are no different from those between insects and any other living species, even though humans with their modern technology can modify considerably the nature of these interactions.

Of an estimated 5–10 million species of insects, probably not more than a fraction of 1% interact, directly or indirectly, with humans. Perhaps some 10,000 constitute pests that, either alone or in conjunction with microorganisms, cause significant damage or death to humans, agricultural or forest products, and manufactured goods. Worldwide food and fiber losses caused by pests (principally insects, plant pathogens, weeds, and birds) are generally estimated at about 40%, of which 12% are attributable to insects and mites. These figures do not include postharvest losses, estimated to be about 20%, and occur despite the application of about 3 million tonnes of pesticide (worth more than US$31 billion, including about US$9 billion of insecticide) (Pimentel, 2002). In the United States alone, crop losses related to insect damage rose from 7% to 13% in the period 1945–1989, despite a tenfold increase in the amount of insecticide used (>120,000 tonnes each year) (Pimentel *et al.*, 1992).

On the other hand, the value of benefits derived from insects is severalfold that of losses as a result of their pollinating activity, their role in biological control, and their importance as honey, silk, and wax producers. That insects do more good than harm probably would come as a surprise to laypersons whose familiarity with insects is normally limited to mosquitoes, houseflies, cockroaches, various garden pests, etc., and to farmers who must protect their livestock and crops against a variety of pests. If asked to prepare a list of useful insects, many people most likely would not get further than the honey bee and, perhaps, the silkmoth, and would entirely overlook the enormous number of species that act as pollinating agents or prey on harmful insects that might otherwise reach pest proportions.

Humans have long recognized the importance of insects in their well-being. Insects and/or their products have been eaten by humans for thousands of years. Production of silk from silkmoth pupae has been carried out for almost 5000 years. Locust swarms, which originally may have been an important seasonal food for humans, took on new significance as humans turned to a farming rather than a hunting existence. However, with rare exceptions, for example, the honey bee and silkmoth whose management is relatively simple and labor-intensive, until recently humans neither desired nor were able, because of a lack of basic knowledge as well as technology, to attempt large-scale modification of the environment of insects, either to increase the number of beneficial insects or to decrease the number of those designated as pests.

Several features of recent human evolution have made such attempts imperative. These include a massive increase in population, a trend toward urbanization, increased geographic movement of people and agricultural products, and, associated with the need to feed more people, a trend toward monoculture as an agricultural practice.

The relatively crowded conditions of urban areas enable insects parasitic on humans both to locate a host (frequently a prerequisite to reproduction) and to transfer between host individuals. Thus, urbanization facilitates the spread of insect-borne human diseases such as typhus, plague, and malaria whose spectacular effects on human population are well documented. For example, in the sixth century A.D. plague was responsible for the death of about 50% of the population in the Roman Empire, and "Black Death" killed a similar proportion of England's population in the mid-1300s (Southwood, 1977).

An increasing need to produce more and cheaper food led, through agricultural mechanization, to the practice of monoculture, the growing of a crop over the same large area of land for many years consecutively. However, two faults of monoculture are (1) the ecosystem is simplified and (2) as the crop plant is frequently graminaceous (a member of the grass family, including wheat, barley, oats, rice, and corn), the ecosystem is artificially maintained at an early stage of ecological succession. By simplifying the ecosystem, humans encourage the buildup of populations of the insects that compete with them for the food being grown. Further, as the competing insects are primary consumers, that is, near the start of the food chain, they typically have a high reproductive rate and short generation time. In other words, populations of such species have the potential to increase at a rapid rate.

A massive increase in human geographic movements and a concomitant increase in trade led to the transplantation of a number of species, both plant and animal, into areas previously unoccupied by them. Some of these were able to establish themselves and, in the absence of normal regulators of population (especially predators and parasitoids), increased rapidly in number and became important pests. Sometimes, as humans colonized new areas, some of the cultivated plants that were introduced proved to be an excellent food for species of insects endemic to these areas. For example, the Colorado beetle, *Leptinotarsa decemlineata*, was originally restricted to the southern Rocky Mountains and fed on wild Solanaceae. With the introduction of the potato by settlers, the beetle had an alternate, more easily accessible source of food, as a result of which both the abundance and distribution of the beetle increased and the species became an important pest. Likewise, the apple maggot, *Rhagoletis pomonella*, apparently fed on hawthorn until apples were introduced into the eastern United States (Horn, 1976).

2. Beneficial Insects

Insects may benefit humans in various ways, both directly and indirectly. The most obvious of the beneficial species are those whose products are commercially valuable. Considerably more important, however, are the insects that pollinate crop plants. Other beneficial insects are those that are used as food, for biological control of pest insects and plants, in medicine and in research. For some of these useful species, humans modify their environment so as to increase their distribution and abundance in order to gain the benefits.

2.1. Insects Whose Products Are Commercially Valuable

The best-known insects in this category are the honey bee (*Apis mellifera*), silkworm (*Bombyx mori*), lac insect (*Laccifer lacca*), and pela wax scale (*Ericerus pela*).

The honey bee originally occupied the African continent, most of Europe (except the northern part), and western Asia, and within this area the usefulness of its products, honey and beeswax, has been known for many thousands of years. Though the discovery of sugar in cane (in India, about 500 B.C.) and in beet (in Europe, about 1800 A.D.) (Southwood, 1977) led to a decline in the importance of honey, it is nevertheless still a very valuable product.

Bee management was probably first carried out by the ancient Egyptians. Honey bees were brought to North America by colonists in the early 1600s, and today honey and beeswax production is a billion dollar industry. In 2001 world honey production was estimated at about 1.25 million tonnes and a value of about US$4 billion. China is the world's largest honey producer, accounting for almost 20% of the total; the United States lies in third place (behind the former USSR), producing about 100,000 tonnes (with a value of about US$330 million) (*www.beekeeping.com/databases/honey-market/world_honey.htm*). Beeswax is produced at the rate of about 1 kg for every 50–100 kg of honey; its value per kilogram varies between one and three times that of honey. There is a significant world trade, perhaps worth about US$10 million annually, in pollen which is used not only by beekeepers to supplement the reserves in the hive but also in the health-food industry. Other products that are collected include propolis (bee glue), venom (used to desensitize patients with severe allergies to bee stings), and royal jelly which is added to certain food supplements (Gochnauer, in Pimentel, 1991, Vol. 2).

Good bee management aims to maintain a honey bee colony under optimum conditions for maximum production. Management details vary according to the climate and customs of different geographical areas but may include (1) moving hives to locations where nectar-producing plants are plentiful, (2) artificial feeding of newly established, spring colonies with sugar syrup in order to build up colony size in time for the summer nectar flow, (3) checking that the queen is laying well and, if not, replacing her, (4) checking and treating colonies for diseases such as foulbrood and nosema, and (5) increasing the size of a hive as the colony develops, in order to prevent swarming.

Silk production has been commercially important for about 4700 years. The industry originated in East Asia and spread into Europe (France, Italy, and Spain) after eggs were smuggled from China to Italy in the sixth century A.D. The production of silk remains a labor-intensive industry, making production costs high. In 1988 world silk production totaled about 67,000 tonnes, with raw silk fetching about US$50 per kilogram. By the end of 1998, production had increased slightly, to 72,000 tonnes, though the price of raw silk had fallen to about US$26 per kilogram due to competition from cheaper synthetic fibers. China is the leading producer, with about 70% of the world total, and India has passed Japan as the world's second largest producer (Feltwell, 1990; *www.tradeforum.org/news/fullstory*).

The lac insect is a scale insect endemic to India and Southeast Asia that secretes about itself a coating of lac, which may be more than 1 cm thick. The twigs on which the insects rest are collected and either used to spread the insects to new areas or ground up and heated in order to separate the lac. The lac is a component of shellac, though its importance has declined considerably with the development of synthetic materials.

The pela wax scale has been used in China for commercial production of "China wax," principally in Sichuan province, for more than 1000 years. It is the second-instar males that produce economically valuable wax. These are carefully managed in large aggregations (200 per cm^2, and extending a distance of 1.0–1.5 m along a branch) that produce a coating of wax some 5–10 mm thick. Wax production peaked in the early 1900s and in a good year was more than 6000 tonnes. Most of this wax was used in the manufacture of candles. Starting in the 1940s, with the coming of electricity and discovery of other waxes (notably paraffin wax), interest in China wax production declined, and currently about 500 tonnes is harvested yearly. It is used for a variety of industrial, pharmaceutical, and horticultural purposes, for example, the manufacture of molds for precision instruments, insulation of electrical cables and equipment, production of high-gloss, tracing, and wax paper, as an ingredient of furniture and automobile polishes, coating candies and pills, and as a grafting agent for fruit trees (Qin, in Ben-Dov and Hodgson, 1994).

2.2. Insects as Pollinators

As was noted in Chapter 23, Section 3.2, an intimate, mutualistic relationship has evolved between many species of insects and plants, in which plants produce nectar and pollen for use by insects, while the latter provide a transport system to ensure effective cross-pollination. Though some crop plants are wind-pollinated, for example, cereals, a large number, including fruits, vegetables, and field crops such as clovers, rape, and sunflower, require the service of insects. In addition, ornamental flowers are almost all insect-pollinated.

The best known, though by no means the only, important insect pollinator is the honey bee, and it is standard practice in many parts of the world for fruit, seed, and vegetable producers either to set up their own beehives in their orchards and fields or to contract this job out to beekeepers. For example, in California about 1.4 million hives are rented annually to augment natural pollination of almonds (about 50% of the hives), alfalfa, melons, and other fruits and vegetables (Pimentel *et al.*, 1992). Under such conditions, the value of bees as pollinators may be up to 140 times their value as honey producers. Using this factor, it is estimated that the *increased* value of crops attributable to honey bee pollination in the United States is about US$15 billion each year [see Robinson *et al.* (1989) for a detailed analysis]. This estimate does not take into account the value of other, natural pollinators of crops, nor obviously has a value been placed on the importance of *all* pollinators of non-crop plants, which are vital to species diversity and as food for wildlife.

2.3. Insects as Agents of Biological Control

It is only relatively recently that humans have gained an appreciation of the importance of insects in the regulation of populations of potentially harmful species of insects and plants. In many instances, this appreciation was gained only when, as a result of human activity, the natural regulators were absent, a situation that was rapidly exploited by these species whose status was soon elevated to that of pest. In the first three examples given below (taken from DeBach and Rosen, 1991), none of the organisms is a pest in its country of origin because of the occurrence there of various insect regulators. The discovery of these regulators, followed by their successful culture and release in the area where the pest occurs, constitutes biological control (Section 4.3).

Among the best-known examples of an introduced plant pest are the prickly pear cacti (*Opuntia* spp.) taken into Australia as ornamental plants by early settlers. Once established, the plants spread rapidly so that by 1925 some 60 million acres of land were infested, mostly in Queensland and New South Wales. Surveys in both North and South America, where *Opuntia* spp. are endemic, revealed about 150 species of cactus-eating insects, of which about 50 were judged to have biological control potential and were subsequently sent to Australia for culture and trials. Larvae of one species, *Cactoblastis cactorum*, a moth, brought from Argentina in January, 1925, proved to have the required qualities and within 10 years had virtually destroyed the cacti (Figure 24.1). Perhaps the most remarkable feature of this success story is that only 2750 *Cactoblastis* larvae were brought to Australia, of which only 1070 became adults. From these, however, more than 100,000 eggs were produced, and in February–March of 1926 more than 2.2 million eggs were released in the field! Additional releases, and redistribution of almost 400 million field-produced eggs until the end of 1929, ensured the project's success.

The classical example of an insect pest brought under biological control is the cottony-cushion scale, *Icerya purchasi*, which was introduced into California, probably from Australia, in the 1860s. Within 20 years, the scale had virtually destroyed the recently established, citrus-fruit industry in southern California. As a result of correspondence between American and Australian entomologists and of a visit to Australia by an American entomologist, Albert Koebele, two insect species were introduced into the United States as biological control agents for the scale. The first, in 1887, was *Cryptochaetum iceryae*, a parasitic fly, about which little is heard, though DeBach and Rosen (1991) consider that it had excellent potential for control of the scale had it alone been imported. However, the abilities of this species appear to have been largely ignored with the discovery by Koebele of the vedalia beetle, *Rodolia cardinalis*, feeding on the scale. In total, only 514 vedalia were brought into the United States, between November 1888 and March 1889, to be cultured on caged trees infested with scale. By the end of July 1889, the vedalia had reproduced to such an extent that one orchardist, on whose trees about 150 of the imported beetles had been placed for culture, reported having distributed 63,000 of their descendants since June 1! By 1890, the scale was virtually wiped out. Similar successes in controlling scale by means of vedalia or *Cryptochaetum* have been reported from more than 60 countries (Hokkanen, in Pimentel, 1991, Vol. 2).

A third example of an introduced pest being brought under control by biological agents is the winter moth, *Operophtera brumata*, which, though endemic to Europe and parts of Asia, was accidentally introduced into Nova Scotia in the 1930s. Its initial colonization was slow, and it did not reach economically significant proportions until the early 1950s, and by 1962 it had spread to Prince Edward Island and New Brunswick. The larvae of the winter moth feed on the foliage of hardwoods such as oak and apple. Though more than 60 parasites of the winter moth are known in western Europe, only 6 of these were considered to be potential control agents and introduced into eastern Canada between 1955 and 1960. Two of these, *Cyzenis albicans*, a tachinid, and *Agrypon flaveolatum*, an ichneumonid, became established, but between them they brought the moth under control by 1963. Embree (in Huffaker, 1971) noted that the two parasites are both compatible and supplementary to each other. When the density of moth larvae is high, *C. albicans*, which is attracted to, and lays its eggs near, feeding damage caused by the larvae, is a more efficient parasite than *A. flaveolatum*. However, once in the vicinity of damage, it does not specifically seek out winter moth larvae. Thus, at lower density, it wastes eggs on non-susceptible defoliators such as caterpillars of the fall cankerworm, *Alsophila pometaria*. Hence, at low host

FIGURE 24.1. (A) *Cactoblastis cactorum* caterpillars feeding on cactus pad; and cactus-infested pasture before (B) and after (C) release of *Cactoblastis*. [From D. F. Waterhouse, 1991, Insects and humans in Australia, in: *The Insects of Australia*, 2nd ed., Vol. 1 (CSIRO, ed.), Melbourne University Press. By permission of the Division of Entomology, CSIRO.]

densities, *A. flaveolatum* is more effective because it oviposits specifically on winter moth larvae.

The three examples described above indicate one method whereby the importance of biological control can be demonstrated, namely, by introduction of potential pests into areas where natural regulators are absent. Another way of demonstrating the same phenomenon is to destroy the natural regulators in the original habitat, which enables potential pests to undergo a population explosion. This has been achieved frequently through the use of non-selective insecticides. For example, the use of DDT against the codling moth, *Cydia pomonella*, in the walnut orchards of California, led to outbreaks of native frosted scale, *Lecanium pruinosum*, which was unaffected by DDT, whereas its main predator, an encyrtid, *Metaphycus californicus*, suffered high mortality (Hagen *et al.*, in Huffaker, 1971). Another *Lecanium* scale, *L. coryli*, introduced from Europe in the 1600s, is a potentially serious pest of apple orchards in Nova Scotia but is normally regulated by various natural parasitoids (especially the chalcidoids *Blastothrix sericea* and *Coccophagus* sp.) and predators (especially mirid bugs). Experimentally it was clearly demonstrated in the 1960s that application of DDT destroyed a large proportion of the *Blastothrix* and mirid population, and this was followed in the next two years by medium to heavy scale infestations. Recovery of the parasite and predators was rapid, however, and by the third year after spraying the scale population density had been reduced to its original value (MacPhee and MacLellan, in Huffaker, 1971).

2.4. Insects as Human Food

As noted in the previous chapter, insects play a key role in energy flow through the ecosystem, principally as herbivores but also as predators or parasites, which may themselves be consumed by higher-level insectivorous vertebrates. In turn, some of these vertebrates, notably freshwater fish and game birds, are eaten by humans. Moreover, in many parts of the world, insects (including grasshoppers and locusts, beetle larvae, caterpillars, brood of ants, wasps and bees, termites, cicadas, and various aquatic species) historically played, and continue to have, an important part as a normal component of the human diet (DeFoliart, 1992, 1999).

Aboriginal people of the Great Basin region in the southwestern United States traditionally spent much time and effort harvesting a variety of insects, principally crickets, grasshoppers, shore flies (Ephydridae) (especially the pupae), caterpillars, and ants (adults and pupae) though bees, wasps, stoneflies, aphids, lice, and beetles were also consumed on an opportunistic basis. Some of the insects were eaten raw though most were baked or roasted prior to being consumed; further, large quantities, especially of grasshoppers and crickets, were dried and ground to produce a flour that was stored for winter use (Sutton, 1988).

In parts of southeastern Australia the aboriginals would seasonally gorge themselves on bogong moths (*Agrotis infusa*) which estivate from December through February in vast numbers in high-altitude caves and rocky outcrops in the Southern Tablelands (Figure 24.2). Some tribes would make an annual trek over a considerable distance (up to 200 km) to take advantage of this seasonal food source, returning each year to the same area (Flood, 1980).

In some African countries (including Botswana, South Africa, Zaire, and Zimbabwe) there is a thriving trade in mopanie caterpillars (*Gonimbrasia belina*), and when these are in season, beef sales may show a significant decline. A similar preference for insects over meat is shown by the Yupka people of Colombia and Venezuela (Ruddle, 1973). Insects are also eaten in many Asian countries; indeed, giant water bugs (*Lethocerus indicus*) and

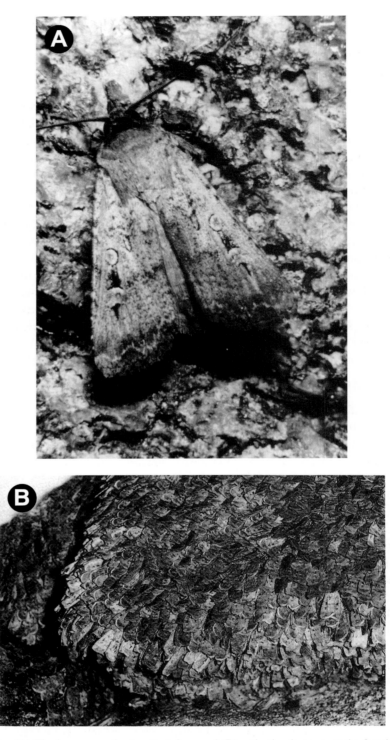

FIGURE 24.2. (A) The bogong moth, *Agrotis infusa*; and (B) estivating bogong moths forming a scalelike pattern on a cave wall. Aboriginals harvested the moths in vast numbers by dislodging them with sticks and collecting them in nets or bark dishes held beneath. [A, photograph by J. Green. B, from D. F. Waterhouse, 1991, Insects and humans in Australia, in: *The Insects of Australia*, 2nd ed., Vol. 1 (CSIRO, ed.), Melbourne University Press. By permission of the Division of Entomology, CSIRO.]

pupae of the silkmoth (*Bombyx mori*) are exported to Asian community food stores in the United States from Thailand and South Korea, respectively. Mexico also used to ship food insects to the United States, namely ahuahutle (Mexican caviar—the eggs of various aquatic Hemiptera) and maguey worms (caterpillars of *Aegiale hesperiaris*, found on agave). Shipment of ahuahutle to North America no longer occurs because of lake pollution, though it can still be found in many markets and restaurants in Mexico and is exported to Europe as bird and fish food. Maguey worms are commonly eaten in Mexico and are exported as gourmet food to North America, France and Japan (DeFoliart, 1992, 1999). However, tourists who visit Mexico are probably more familiar with another caterpillar, the red agave worm (*Comadia redtenbachi*), seen in bottles of mezcal!

There has been some increase in interest in the potential of insects as food, including discussion of the subject at international conferences. However, most North Americans and Europeans have not yet been educated to the delights of insects, despite the efforts of authors such as Taylor and Carter (1976), DeFoliart (1992, 1999), and Berenbaum (1995) to increase the popularity of insects as food. The western world's bias against eating insects has two negative impacts. First, it may be seen as a missed opportunity. Compared to livestock, insects are much more efficient at converting plant material into animal material with high nutritional value. With relatively little research, industrial-scale mass production of food insects should be possible. Second, as less-developed areas of the world become increasingly westernized, their populations may be expected to eat fewer insects. This could lead to nutritional problems in areas where the economy is already marginal (DeFoliart, 1999).

2.5. Soil-Dwelling and Scavenging Insects

By their very habit the majority of soil-dwelling insects are ignored by humans. Only those that adversely affect our well-being, for example, termites, wireworms, and cutworms, normally "merit" our attention. When placed in perspective, however, it seems probable that the damage done by such pests is greatly outweighed by the benefits that soil-dwelling insects as a group confer. The benefits include aeration, drainage, and turnover of soil as a result of burrowing activity. Many species carry animal and plant material underground for nesting, feeding, and/or reproduction, which has been compared to ploughing in a cover crop.

Many insects, including a large number of soil-dwelling species, are scavengers; that is, they feed on decaying animal or plant tissues, including dung, and thus accelerate the return of elements to food chains. In addition, through their activity they may prevent use of the decaying material by other, pest insects, for example, flies. Perhaps of special interest are the dung beetles (Scarabaeidae), most species of which bury pieces of fresh dung for use as egg-laying sites (Figure 24.3). Generally, the beetles are sufficiently abundant that a pat of fresh dung may completely disappear within a few hours, thus reducing the number of dung-breeding flies that can locate it. Furthermore, the chances of fly eggs or larvae surviving within the dung are very low because the dung is ground into a fine paste as the beetles or their larvae feed. Likewise, the survival of the eggs of tapeworms, roundworms, etc., present in the dung producer, is severely reduced by this activity.

In Australia there are an estimated 22 million cattle and 162 million sheep that collectively produce 54 million tonnes of dung (measured as dry weight) each year! The cattle dung especially provides food and shelter for many insects, including the larvae of two fly pests, the introduced buffalo fly (*Haematobia irritans exigua*) in northern Australia and the native bush fly (*Musca vetustissima*) in southeastern and southwestern areas of the country. Further, because of the generally dry climate, the dung soon dries and may remain

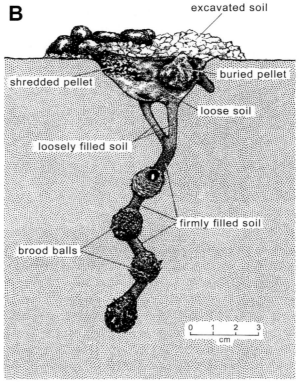

FIGURE 24.3. (A) A dung beetle, *Sisyphus rubrus*, with its ball of dung which is rolled away from the dung pad and then buried. This southern African species was introduced into Australia in 1973; and (B) diagrammatic section through nest of the Australian native dung beetle *Onthophagus compositus*, which colonizes the dung of kangaroos, wallabies, and wombats. [A, photograph by J. Green. By permission of the Division of Entomology, CSIRO. B, from G. F. Bornemissza, 1971, A new variant of the paracopric nesting type in the Australian dung beetle, *Onthophagus compositus*, *Pedobiologia* **11**:1–10. By permission of Gustav Fischer Verlag.]

unchanged for a considerable time so that the fiber and nutrients are unavailable to maintain soil fertility and texture. Rank herbage grows around each dung pat, and this is not normally eaten by cattle. Thus, at any time, dung pats render a significant proportion (estimated at about 20%) of all pasture potentially unusable (Waterhouse, 1974). Although Australia has more than 320 indigenous species of dung beetles, almost all of these use only the dung of native marsupials, especially kangaroos, wallabies, and wombats, and, furthermore, are restricted to forest and woodland habitats. Only a few species of the native *Onthophagus* have adapted to using cattle dung (Waterhouse and Sands, 2001).

In 1963, it was decided to initiate a program of biological control of dung, and in 1967, after extensive research, various species of tropical southern African dung beetles were released in northern Australia. These species had been carefully selected from regions climatically similar to northern Australia and because they were known to be effective processors of the dung of large native ruminants. The results were spectacular, the beetles rapidly multiplied and spread over wide distances, while simultaneously achieving complete or partial disposal of dung for much of the year (Waterhouse, 1974). Over the next 15 years, about 50 additional species of dung beetles, plus a few species of histerid beetles that prey on fly eggs, larvae, and puparia, were imported not only from southern Africa but also from southern Europe and Asia, each having features appropriate to a particular region of Australia. Of the 50-odd species released, 25 dung beetles and 3 histerids have established breeding populations in the field, though numbers (and effectiveness of dung processing) may be highly varied according to the species, locality, season, and weather conditions (Waterhouse and Sands, 2001). All of the species established in northern regions are common in all except the winter months, and through the summer almost complete dispersal of dung occurs. Further, there is some evidence that such intense activity results in a regional suppression of buffalo fly numbers.

In the cooler southeast and southwest of Australia the introduced species are most active in the summer and autumn months when their dung dispersal may substantially reduce bush fly abundance. However, their activity is generally low in winter and spring, the latter being the period in southwestern Australia when massive populations of bush flies develop (Doube *et al.*, 1991). Thus, in 1989, three spring-active species were imported from Spain; one of these, *Bubas bison*, established itself quickly though populations of the other two species took longer to increase because of their complex breeding behavior (Creagh, 1993).

In terms of dung disposal, the Australian dung beetle project has been a major success, saving farmers the costs of harrowing, accelerating the release of nutrients into the soil, and reducing the availability of fly breeding sites. No comprehensive study of the impact of the introduced beetles on the pest fly problem has been undertaken. However, Waterhouse and Sands (2001) provide examples to show that the beetles may significantly reduce fly populations at a local level, especially when rainfall, which prolongs the beetles' breeding activity, is favorable. A compounding factor in any attempt to estimate the beetles' impact is the ease with which the flies are carried on wind currents from regions where they have bred successfully. This tends to mask local effects of dung beetle activity on fly numbers. In all probability the dung beetle system will become but one component of an integrated program for fly population management, with other strategies being used when fly populations peak.

2.6. Other Benefits of Insects

Their relatively simple food and other requirements, short generation time, and high fecundity enable many insects to be reared cheaply and easily under laboratory conditions and, consequently, make them valuable in teaching and research. Even at the pre-college level, these attributes, plus their remarkable diversity of form and habits, make insects an important resource both in and outside the classroom (Matthews *et al.*, 1997). The fruit fly, *Drosophila melonagaster* with its array of mutants, is familiar to all who take an elementary college genetics class, though it must also be appreciated that the insect continues to have an important role in advanced genetic research. Studies on other insects have provided us with much of our basic knowledge of animal and cell physiology, particularly in the areas of nutrition, metabolism, endocrinology, and neuromuscular physiology. Investigations into the population dynamics of some pest insects, especially forest species, led to the formulation of some important concepts in population ecology (Gillott, 1985).

With the development of microbial resistance to many antibiotics, there has been a revival in the use of maggot therapy, the use of fly larvae to clean wounds and promote healing (Sherman *et al.*, 2000). Maggot therapy has been used for centuries in some societies and probably developed as a result of casual observations that the larvae of some myiasis-causing flies had beneficial effects on infected wounds. Myiasis, the infestation of animal tissues (living or dead) by maggots, appears to have evolved in some Dipteran families that were originally saprophagous, that is, bred in carrion. Currently, it is mostly seen in three families: Oestridae (all 150 species), Sarcophagidae, and Calliphoridae (about 80 species in total) (Chapter 9, Section 3). However, most of these species are unsuitable for use in maggot therapy because they feed on healthy tissue, are highly host-specific, and have other disadvantages. Of the 10 or so species of "medicinal maggots," the most common are larvae of the greenbottle fly, *Lucilia sericata*. Curiously, in the United Kingdom, continental Europe, and New Zealand, this fly is a major sheep pest, causing "strike," which may be fatal in heavy infestations (Sherman *et al.*, 2000).

Many insects give us pleasure through their aesthetic value. Because of their beauty, certain groups, especially butterflies, moths, and beetles, are sometimes collected as a hobby. Some are embedded in clear materials from which jewelry, paperweights, bookends, place mats, etc., are made. Others are simply used as models on which paintings and jewelry are based.

3. Pest Insects

Since humans evolved, insects have fed on them, competed with them for food and other resources, and acted as vectors of microorganisms that cause diseases in them or in the organisms that they value. However, as was noted in the Introduction, the impact of such insects increased considerably as the human population grew and became more urbanized. Urbanization presented easy opportunities for the dissemination of insect parasites on humans and the diseases they carry. Large-scale and long-term cultivation of the same crop over an area facilitated rapid population increases in certain plant-feeding species and the spread of plant diseases. Modern transportation, too, encourages the spread of pest insects and insect-borne diseases. Further, as described in Section 2.3, some of the attempts at pest eradication have backfired, resulting in even greater economic damage.

3.1. Insects That Affect Humans Directly

A large number of insect species may be external, or temporary internal, parasites of humans. Some of these are specific to humans, for example, the body louse (*Pediculus humanus*) and pubic louse (*Phthirus pubis*), but most have a varied number of alternate

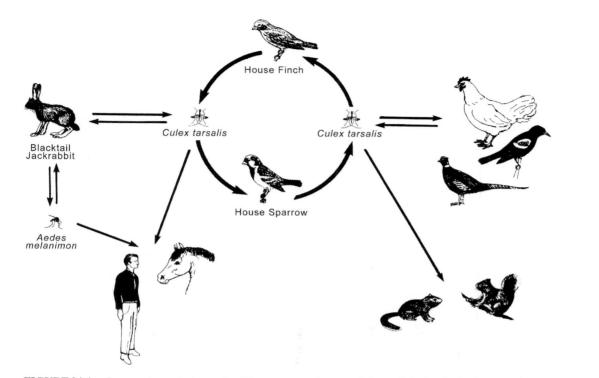

FIGURE 24.4. Summer transmission cycle of the western equine encephalomyelitis virus in the Central Valley of California. The mosquito, *Culex tarsalis*, is the primary vector of the virus, and house finches and house sparrows the primary amplifying hosts (hosts in which the virus multiplies). Secondary (less important) amplifying hosts include other passerine birds, chickens, and pheasants. Another transmission cycle involves blacktail jackrabbits, which are sometimes bitten by *C. tarsalis*, and *Aedes melaniman*. Humans and horses, as well as ground squirrels, tree squirrels, and some other wild mammals become infected but do not contribute significantly to virus amplification. [From J. L. Hardy, 1987, The ecology of western equine encephalomyelitis virus in the Central Valley of California, 1945–1985, *Am. J. Trop. Med. Hyg.* **37**(Suppl.):18S–32S. By permission of the American Society of Tropical Medicine and Hygiene.]

hosts which compounds the problem of their eradication. With rare exceptions, for example, some myiasis-causing flies, insect parasites are not fatal to humans. In large numbers, insect parasites may generally weaken their hosts, making them more susceptible to the attacks of disease-causing organisms. Or the parasites, as a result of feeding, may cause irritation or sores which may then become infected.

But by far the greatest importance of insects that parasitize humans is their role as vectors of pathogenic microorganisms (including various "worms") some well-known examples of which are given in Table 24.1. The pathogen is picked up when a parasitic insect feeds and may or may not go through specific stages of its life cycle in the insect. Bacteria and viruses are directly transmitted to new hosts, an insect serving as a mechanical vector, whereas for protozoa, tapeworms and nematodes, an insect serves as an intermediate host in which an essential part of the parasites' life cycle occurs (Figure 24.4). In the latter arrangement the insect is known as a biological vector.

A pathogen may reside (and multiply) in alternate vertebrate hosts that are immune to or only mildly infected by it. For example, the bacterium *Pasteurella pestis*, which causes bubonic plague (Black Death), is endemic in wild rodent populations. However, in domestic rats and humans, to which it is transmitted by certain fleas, it is highly pathogenic. Similarly, in South America, yellow fever virus, transmitted by mosquitoes, is found in monkeys though these are immune to it. Such alternate hosts are thus important reservoirs of disease.

TABLE 24.1. Examples of Insects That Serve as Vectors for Diseases of Humans and Domestic Animals[a]

Insect vector	Pathogen	Disease	Host	Distribution
ANOPLURA				
Pediculus humanus (body louse)	*Rickettsia prowazekii* (rickettsian)	Epidemic typhus (Brills' disease)	Humans, rodents	Worldwide
	Pasteurella tularensis (bacterium)	Tularemia	Humans, rodents	N. America, Europe, the Orient
HEMIPTERA				
Rhodnius spp. (assassin bugs)	*Trypanosoma cruzi* (protozoan)	Chagas' disease	Humans, rodents	S. America, Central America, Mexico, Texas
DIPTERA				
Phlebotomus spp. (sand flies)	*Leishmania donovani* (protozoan)	Kala-azar (Dumdum fever)	Humans	Mediterranean region, Asia, S. America
	L. tropica	Oriental sore	Humans	Africa, Asia, S. America
	L. brazilliensis	Espundia	Humans	S. America, Central America, N. Africa, southern Asia
	(virus)	Pappataci fever (Sand fly fever)	Humans	Mediterranean region, India, Sri Lanka
Anopheles spp. (mosquitoes)	*Plasmodium vivax* (protozoan)	Malaria	Humans	Worldwide in tropical, subtropical, and temperate regions
	P. malariae	Malaria	Humans	
	P. falciparum	Malaria	Humans	
Aedes spp. (mosquitoes)	(virus)	Yellow fever	Humans, monkeys, rodents	American and African tropics and subtropics
	(virus)	Dengue	Humans	Worldwide in tropics and subtropics
	(virus)	Encephalitis	Humans, horses	N. America, S. America, Europe, Asia
	Wucheria bancrofti (nematode)	Filariasis (Elephantiasis)	Humans	Worldwide in tropics and subtropics
Culex spp. (mosquitoes)	(virus)	Dengue	Humans	Worldwide in tropics and subtropics
	(virus)	Encephalitis	Humans, horses	N. America, S. America, Europe, Asia
	Wucheria bancrofti (nematode)	Filariasis (Elephantiasis)	Humans	Worldwide in tropics and subtropics
Tabanus spp. (horse flies)	*Bacillus anthracis* (bacterium)	Anthrax	Humans, other animals	Worldwide

Chrysops spp. (deer flies)	*Pasteurella tularensis* (bacterium)	Tularemia	Humans, rodents	N. America, Europe, the Orient
	Loa loa (nematode)	Loiasis (Calabar swelling)	Humans	Africa
Glossina spp. (tsetse flies)	*Trypanosoma rhodesiense* (protozoan)	Sleeping sickness	Humans, other animals	Equatorial Africa
	T. gambiense	Sleeping sickness	Humans, other animals	Equatorial Africa
	T. brucei	Nagana	Cattle, wild ungulates	Equatorial Africa
SIPHONAPTERA				
Xenopsylla cheopsis (oriental rat flea)	*Pasteurella pestis* (bacterium)	Bubonic plague (Black death)	Humans, rodents	Worldwide
Xenopsylla spp.	*Rickettsia typhi* (rickettsian)	Endemic (murine) typhus	Humans, rodents	Worldwide
Xenopsylla cheopsis	*Hymenolepis nana* (cestode)	Tapeworm	Humans	Europe and N. America
	H. diminuta (cestode)	Tapeworm	Humans	Worldwide
Nosopsyllus fasciatus (northern rat flea)	*Pasteurella pestis* (bacterium)	Bubonic plague (Black death)	Humans, rodents	Worldwide
	Rickettsia typhi (rickettsian)	Endemic (murine) typhus	Humans, rodents	Worldwide
	Hymenolepis diminuta (cestode)	Tapeworm	Humans	Worldwide
Ctenocephalides canis (dog flea)	*Dipylidium caninum* (cestode)	Tapeworm	Humans, dogs, cats	Worldwide
	Hymenolepis nana (cestode)	Tapeworm	Humans	Europe, N. America
Pulex irritans (human flea)	*Hymenolepis nana* (cestode)	Tapeworm	Humans	Europe, N. America

[a]Data from various sources.

Transmission of human disease-causing microorganisms is not, however, entirely the domain of parasitic insects. Many insects, especially flies, may act as mechanical vectors, contaminating human food as they rest or defecate on it, with pathogens picked up during contact with feces or other organic waste. Examples of such insects and the diseases transmitted by them are listed in Table 24.1.

A third category of insects that directly affect humans includes those that may bite or sting when accidental contact is made with them, for example, bees, wasps, ants, some caterpillars (with poisonous hairs on their dorsal surface), and blister beetles. Normally, the effect of the bite or sting is temporary and nothing more than skin irritation, swelling, or blister formation. Bee stings, however, may cause anaphylaxis or death in some sensitive individuals.

3.2. Pests of Domesticated Animals

A range of insect parasites may cause economically important levels of damage to domestic animals. The majority of these parasites are external and include bloodsucking flies (e.g., mosquitoes, horse flies, deer flies, black flies, and stable flies), biting and sucking lice, and fleas. Other parasites are internal for part of their life history, for example, bot, warble, and screwworm flies, which as larvae live in the gut (horse bot), under the skin (warble and screwworm of cattle), or in head sinuses (sheep bot). Other examples are given in the chapters that deal with the orders of insects. In addition to insects, other arthropods are also important livestock pests, especially various mites and ticks. Kunz *et al.* (in Pimentel, 1991, Vol. 1) estimated that direct losses caused by insect and tick pests of livestock were almost $3 billion each year in the United States. Insecticide treatment is routine for some pests, and in the absence of treatment, losses caused by these pests (e.g., cattle grubs, *Hypoderma bovis* and *H. lineatum*) might be as much as ten times greater. Indirect losses such as poor breeding performance by bulls, reduced conception rate by cows, and labor costs of treating livestock are not included in the above estimate.

Generally the effect of such parasites is to cause a reduction in the health of the infected animal. In turn, this results in a loss of quality and/or quantity of meat, wool, hide, milk, etc., produced. When severely infected by parasites, an animal may eventually die. In addition to their own direct effect on the host, some parasites are vectors of livestock diseases, examples of which are included in Table 24.1.

3.3. Pests of Cultivated Plants

Damage to crops and other cultivated plants by insects is enormous; in the United States alone losses in potential production are estimated at 13% and have a value of about US$30 billion annually, despite the application of more than 100,000 tonnes of insecticide. Remarkably, about three quarters of the insecticide used is applied to 5% of the total agricultural land, especially that growing row crops such as cotton, corn, and soybean (Pimentel *et al.*, in Pimentel, 1991, Vol. 1). Damage is caused either directly by insects as they feed (by chewing or sucking) or oviposit, or by viral, bacterial, or fungal diseases, for which insects serve as vectors. Especially important as "direct damagers" of plants are Orthoptera, Lepidoptera, Coleoptera, and Hemiptera (see the chapters that deal with these orders for specific examples of such pests). Several hundred diseases of plants are known to be transmitted by insects (for examples see Table 24.2) including about 300 that are caused by viruses (Eastop, 1977). Especially important in disease transmission are Hemiptera,

TABLE 24.2. Examples of Plant Diseases Transmitted by Insects[a]

	Disease	Important hosts	Vectors	Distribution
Viruses	Alfalfa mosaic	Alfalfa, tobacco, potato, beans, peas, celery, zinnia, petunia	Aphids (at least 16 spp.) incl. *Acyrthosiphon primulae, A. solani, Aphis craccivora, A. fabae, A. gossypii, Macrosiphum euphorbiae, M. pisi, Myzus ornatus, M. persicae, M. violae*	Worldwide
	Barley yellow dwarf	Barley, oat, wheat, rye, wild and tame grasses	Numerous aphids incl. *Macrosiphum granarium, M. miscanthi, Myzus circumflexus, Rhopalosiphum padi, R. maidis*	North America, Australia, Denmark, Holland, UK
	Bean common mosaic	Beans	Aphids (at least 11 spp.) esp. *Aphis rumicis, Macrosiphum pisi, M. gei*	Worldwide
	Beet yellows	Sugarbeet, spinach	Aphids esp. *Aphis fabae, Myzus persicae*	Wherever sugarbeet is grown
	Cauliflower mosaic	Cauliflower, cabbage, Chinese cabbage	Aphids esp. *Brevicoryne brassicae, Rhopalosiphum pseudobrassicae, Myzus persicae*	Europe, USA, New Zealand
	Dahlia mosaic	Dahlia, zinnia, calendula	Aphids esp. *Myzus persicae, Aphis fabae, A. gossypii, Macrosiphum gei, Myzus convolvuli*	Wherever dahlias are grown
	Lettuce mosaic	Lettuce, sweet pea, garden pea, endive, aster, zinnia	Aphids esp. *Myzus persicae, Aphis gossypii, Macrosiphum euphorbiae*	Europe, USA (esp. California), New Zealand
	Pea mosaic	Garden pea, sweet pea, broadbean, lupin, clovers	Aphids: *Acyrthosiphon pisum, Myzus persicae, Aphis fabae, A. rumicis*	Europe, USA, New Zealand, Australia, Japan
	Potato virus Y	Potato, tobacco, tomato, petunia, dahlia	Aphids esp. *Myzus persicae, M. certus, M. ornatus, Macrosiphum euphorbiae*	UK, France, USA
	Soybean mosaic	Soybean	Aphids incl. *Myzus persicae, Macrosiphum pisi*	Wherever soybean is grown
	Sugarcane mosaic	Sugarcane, com, sorghum, other tame and wild grasses	Numerous aphids incl. *Rhopalosiphum maidis, Aphis gossypii, Schizaphis graminum, Myzus persicae*	Wherever sugarcane is grown
	Tomato spotted wilt	Tomato, tobacco, dahlia, pineapple	Thrips: *Thrips tabaci, Frankliniella schultzeri, F. fusca, F. occidentalis*	Africa, Asia, Australia, Europe, North and South America

(Continued)

TABLE 24.2. (*Continued*)

	Disease	Important hosts	Vectors	Distribution
	Turnip yellow mosaic	Turnip, cauliflower, Chinese cabbage, kohlrabi, cabbage, Broccoli	Flea beetles (*Phyllotreta* spp.); Mustard beetle (*Phaedon cochleariae*); Grasshoppers (*Leptophyes punctatissima, Chorthippus bicolor*); Earwig (*Forficula auricularia*)	UK, Germany, Portugal, North America
Mycoplasmas	Aster yellows	Aster, celery, carrot, squash, cucumber, wheat, barley	Numerous leafhoppers incl. *Gyponana hasta, Scaphytopius acutus, S. irroratus, Macrosteles quadrilineatus, Paraphlepsius apertinus, Texananus* (several species)	Worldwide
	Clover phyllody	Most clovers	Leafhoppers incl. *Aphrodes albifrons, Macrosteles cristata. M. quadrilineatus, M. viridigriseus. Euscelis lineolatus, E. plebeja*	USA, UK
	Corn stunt	Corn	Leafhoppers esp. *Dalbulus elimatus. D. maidis. Graminella nigrifrons*	North and Central America
Bacteria	Stewart's bacterial wilt	Corn	Corn flea beetle (*Chaetocneme pulicara*); toothed flea beetle (*C. denticulata*)	USA
	Cucurbit wilt	Cucumber, muskmelon	Cucumber beetles (*Diabrotica vittata, D. duodecimpunctata*)	USA, Europe, South Africa, Japan
	Potato blackleg	Potato	Seedcorn maggot (*Hylemya cilicrura*), *H. trichodactyla*	North America
	Fire blight	90 spp. of orchard trees and ornamentals, esp. apple, pear, quince	Wide range of insect vectors, esp. bees, wasps, flies, Ants, aphids	North America, Europe
Fungi	Dutch elm disease	Elm	Elm bark beetles, esp. *Scolytus multistriatus, S. scolytus, Hylurgopinus rufipes*	Asia, Europe, North America
	Ergot	Cereals and other grasses	About 40 spp. of insects Esp. flies, beetles, aphids	Worldwide

[a] Data from various sources.

particularly leafhoppers and aphids. Three aspects of the behavior of these insects facilitate their role as disease vectors: (1) they make brief but frequent probes with their mouthparts into host plants; (2) as the population density reaches a critical level, winged migratory individuals are produced; and (3) in many species, winged females deposit a few progeny on each of many plants, from which new colonies develop. On the basis of their method of transmission and viability (persistence in the vector), viruses may be arranged in three categories. The non-persistent (stylet-borne) viruses are those believed to be transmitted as contaminants of the mouthparts. Such viruses remain infective in a vector for only a very short time, usually an hour or less. Semipersistent viruses are carried in the anterior regions of the gut of a vector, where they may multiply to a certain extent. Vectors do not normally remain infective after a molt, presumably because the viruses are lost when the foregut intima is shed. Persistent (circulative or circulative-propagative) viruses are those that, when acquired by a vector, pass through the midgut wall to the salivary glands from where they can infect new hosts. Such viruses may multiply within tissues of a vector, which retains the ability to transmit the virus for a considerable time, in some instances for the rest of its life. Persistent viruses, in contrast to those in the first two categories, may be quite specific with respect to the vectors capable of transmitting them (Hull, 2002).

3.4. Insect Pests of Stored Products

Almost any stored material, whether of plant or animal origin, may be subject to attack by insects, especially species of Coleoptera (larvae and adults) and Lepidoptera (larvae only). Among the products that are frequently damaged are grains and their derivatives, beans, peas, nuts, fruit, meat, dairy products, leather, and woolen goods. In addition, wood and its products may be spoiled by termites or ants. Again, readers should refer to the appropriate chapters describing these groups for specific examples.

Estimates of the worldwide postharvest losses of foodstuffs (especially stored grains) may be as high as 20%, of which about one-half is attributable to insects and the rest to microorganisms, rodents, and birds. Even in well-developed countries such as the United States, Canada, and Australia where storage conditions are more adequate and pesticide treatment is available, losses of 5% to 10% are estimated for stored grains. Given a worldwide estimate for the production of wheat, coarse grains, and rice as about 1.5 billion tonnes in 1981–1982, perhaps as much as 150 million tons may have been lost as a result of insect damage. The nature of the damage caused by stored products pests varies. Grain and other seed pests not only eat economically valuable quantities of food, but cause spoilage by contamination with feces, odors, webbing, corpses, and shed skins, and by creating heat and moisture damage that permits the growth of microorganisms (Wilbur and Mills, in Pfadt, 1985). Pests of household goods such as clothing and furniture principally cause damage by spoilage, for example, by tunneling, defecating, and creating odors.

4. Pest Control

As will be apparent from what was said above, pests are organisms that damage, to an economically significant extent, humans or their possessions, or that in some other way are a source of annoyance to humans. Implicit in the above description are value judgments that may vary according to who is making them, as well as where and when they are being made. Nevertheless, in a given set of circumstances, there will be an economic injury (annoyance) threshold, measured in terms of a species' population density, above which it is desirable (profitable) to take control measures that will reduce the species' density. As the margin between economic injury threshold and actual population density widens, the desirability

TABLE 24.3. Principal Control Methods in Relation to Ecological Strategies of Pests[a]

	r pests	Intermediate pests	*K* pests
Control method:	Pesticides———————→		
		Biological control	
			←———————Cultural control
			←———————Genetic control
Examples (with important features):	*Schistocerca gregaria* (desert locust) Fecundity/X = 400 eggs Generation time = 1–2 months Migratory, defoliates many crops *Aphis fabae* (black bean aphid) Fecundity/X = 100 eggs Generation time = 1–2 weeks Feeds on wide range of plants *Musca domestica* (house fly) Fecundity/X = 500 eggs Generation time = 2–3 weeks Feeds on organic waste *Agrotis ipsilon* (black cutworm) Fecundity/X = 1500 eggs Generation time = 1–1.5 months Feeds on seedlings of most crops	(Most deciduous forest pests, fruit pests, and some vegetable pests)	*Oryctes rhinoceros* (rhinoceros beetle) Fecundity/X = 50 eggs Generation time = 3–4 months Feeds on apical growing points of coconuts *Glossina* spp. (tsetse fly) Fecundity/X = 10 eggs Generation time = 2–3 months Feeds on narrow range of hosts *Cydia pomonella* (codling moth) Fecundity/X = 40 eggs Generation time = 2–6 months Larvae feed on apple and some other fruits *Melophagus ovinus* (sheep ked) Fecundity/X = 15 eggs Generation time = 1–2 months External parasite of sheep

[a]Data from Conway (1976) and Southwood (1977).

(profitability) of control increases. Pest control is, then, essentially a sociological problem—a matter of economics, politics, and psychology.

A range of methods is available for the control of insect pests. Each of these methods has its advantages and disadvantages and these must be balanced against each other in determining which (combination of) method(s) is most appropriate in a given instance. Some of these methods are spectacular but short-term and will be appropriate, for example, where massive outbreaks of pests are relatively sudden, yet unpredictable and temporary. Others are more slowly acting but relatively permanent in effect, and may be used for pests that are more or less permanent but whose populations are relatively stable.

Conway (1976) and Southwood (1977) proposed that pests can be arranged in a spectrum according to their "ecological strategies" and that the principal (best) method of control is based on their position in the spectrum (Table 24.3). At either end of the spectrum are the so-called "*r* pests" and "*K* pests," with the "intermediate pests" in between.

The *r* pests are characterized by their potentially high rates of population increase (resulting from the high fecundity and short generation time), well-developed powers of dispersal (migration) and ability to locate new food sources, and rather general food preferences. These features enable *r* pests to colonize temporarily suitable habitats, in which there is typically little interspecific competition for the resources available. Because *r* pests may occur in such large but unpredictable numbers and rapidly change their location, predators (of which there may be many) have relatively little effect on their population. Further, although like other organisms *r* pests are subject to disease, the latter is slow to take effect, by which time significant damage may have been done. Finally, because of their high reproductive potential, *r* pests are able to tolerate mass mortality and rapidly regenerate their original density. Hence, biological control, which is a relatively slow but long-term method, is of little use against *r* pests. For such pests specific insecticides, which can be stored for application at short notice, continue to be the most important tool in their control. Included in the *r*-pest group are the "classic" pests: locusts, aphids, mosquitoes, and house flies (Table 24.3).

K pests, on the other hand, have lower fecundity and longer generation time, poor ability to disperse, relatively specialized food preferences, and are found in habitats that remain stable over long periods of time. Under natural conditions, insects with the features of *K* strategists seldom become pests. If, however, probably as a result of human activity, their niche is expanded (e.g., their food plant becomes an important crop), or if they can occupy a new niche (e.g., feeding on domestic cattle rather than wild ungulates) they may become a pest. Once established, such pests are often difficult to eradicate over the short term, for example, through the use of insecticides. Insecticides are frequently not feasible tools because the *K* pests attack the fruit rather than the foliage of crop plants, or because the cost is prohibitive in view of the low density of the pest population. (In some instances, however, where even at low population density a pest may cause considerable damage, for example, codling moth on apple, insecticidal control may be profitable.) Nor is biological control an appropriate method because *K* pests have few natural enemies, a feature probably related to their low density under natural conditions. For *K* pests, the best methods of control are those that disturb their habitat, for example, the breeding of resistant strains of plant(s) or animal(s) attacked by the pests, and cultural practices. Examples of *K* pests are given in Table 24.3.

The majority of pests are classified as intermediate pests in the Conway scheme because they exhibit a mixture of the features of *r* and *K* pests. For some of these, with a relatively high reproductive potential, insecticidal control may be necessary under certain conditions, and conversely, for pests approaching the *K* end of the spectrum, cultural control sometimes may be adequate. However, the most important feature of intermediate pests is the relatively large number of natural enemies that they have. These enemies, under normal circumstances, are important regulators of the pest population. In addition, intermediate pests are frequently foliage- or root-damaging pests, for example, spruce budworm and some scale insects, and, therefore, the economic injury threshold is reasonably high; that is, a fair amount of damage can be tolerated without economic loss. Hence, for these pests, biological control would appear to be the single most appropriate method of control, which can be supplemented as necessary with insecticidal and other methods. The latter is, in other words, an integrated control program.

With these general considerations in mind, it is now appropriate to consider in more detail the methods available for pest control.

4.1. Legal Control

Also known as regulatory control, legal control is based primarily on the old adage "Prevention is better [in this instance, cheaper] than cure." Legal control is the enactment of legislation to prevent or control damage by insects (Rohwer, in Pimentel, 1991, Vol. 1). It includes, therefore, establishment of quarantine stations at major ports of entry into an area. Usually the stations are located at international borders, though in some instances domestic quarantines are necessary, for example, when certain parts of a country are widely separated from the rest (Hawaii and continental United States). At quarantine stations people and goods are inspected to prevent the accidental introduction of potential insect pests and plant and animal diseases. Prior to the introduction of quarantine legislation in the United States in the early 1900s (Plant Quarantine Act of 1912) a number of insect species had been accidentally introduced and become established as plant pests, for example, the cottony-cushion scale discussed in Section 2.3. Though quarantine has severely reduced the number of insect introductions, on average 11 exotic species are still added annually to the insect fauna of the United States (for a total of more than 800 for the period 1920–1980). About 35% of the important pests in the United States are introduced species and include pink bollworm (*Pectinophora gossypiella*), citrus blackfly (*Aleurocanthus woglumi*), Egyptian alfalfa weevil (*Hypera brunneipennis*), face fly (*Musca autumnalis*), cereal leaf beetle (*Oulema melanopus*), and Russian wheat aphid (*Diuraphis noxia*) (Sailer, 1983). As an adjunct to quarantine, many countries (or areas within countries) have legislation that requires international or interstate shipments of animals or plants, or their products, to be certified as disease- or insect-free by qualified personnel prior to shipment.

Also part of legal control is the setting up of surveillance systems for monitoring the insect population in a given area so that, should an outbreak occur, it can be dealt with before it has a chance to spread. Such surveillance is an important duty of state/provincial entomologists, in cooperation with local agriculture representatives and crop and livestock producers.

Another aspect of legal control, and one that has become increasingly important, is the licensing of insecticides and the establishment of (1) regulations regarding their use and (2) monitoring systems to assess their total impact on the environment. For example, in the United States the Environmental Protection Agency is responsible for assessing the effectiveness of pesticides, as well as their possible hazardous effects on humans, wildlife, and other organisms, including bees, other pollinating species, and beneficial parasitoids. As noted earlier (Section 2.3), indiscriminate use of insecticides can result in greatly increased rather than decreased pest damage.

4.2. Chemical Control

The use of chemicals either to kill or to repel insect pests is the oldest method of pest control. Fronk (in Pfadt, 1985) notes that the Greeks used sulfur against pests almost 3000 years ago and the Romans used asphalt fumes to rid their vineyards of insect pests. The Chinese used arsenic compounds against garden pests before 900 A.D., though arsenic was not used in the Western world until the second half of the 17th century.

Until about 1940, insecticides belonged to two major categories, the "inorganics" and the "botanicals." Among the inorganic insecticides are arsenic and its derivatives (arsenicals), including Paris green (copper acetoarsenite), which was the first insecticide to be used on a large scale in the United States—against Colorado potato beetle in 1865. Other inorganics include fluoride salts (developed at about the end of the 19th century, following the realization that toxic residues were left by arsenicals), sulfur, borax, phosphorus, mercury salts, and tartar. These inorganic insecticides were typically sprayed on the pest's food plant or mixed with suitable bait. In other words, all are "stomach poisons" that require ingestion and absorption to be effective. Thus, they were unsatisfactory pesticides for sucking insects for which "contact poisons," absorbed through the integument or tracheal system, are necessary.

The "botanicals" are organic contact poisons produced by certain plants in which they serve as protectants against insects (Chapter 23, Section 3.1). Among the earliest to be used were (1) nicotine alkaloids, derived from certain species of *Nicotiana*, including *N. tabaca* (tobacco) (family Solanaceae); (2) rotenoids extracted from the roots of derris (*Derris* spp.) and cubé (*Lonchocarpus* spp.); and (3) pyrethroids, produced by plants in the genus *Pyrethrum* (*Chrysanthemum*) (family Compositae).

Because of their high mammalian toxicity, nicotine has been completely superseded by synthetics, while rotenone use is now mainly restricted to control of some sucking pests on crops and pests of pets and livestock. The pyrethroids, when first available commercially, were an important group of insecticides for use in the home, as livestock sprays, and against stored-product, vegetable, or fruit pests, primarily because of their low toxicity to mammals. Initially, a major disadvantage of pyrethroids was their photolabile nature (instability in light) and the need, therefore, to reapply them frequently made them expensive to use. Thus, they were largely replaced by cheaper, synthetic insecticides in the 1940s, an important consequence of which was that relatively few insects became resistant to them. This feature, in conjunction with the development of several photostable synthetic pyrethroids (e.g., cypermethrin, permethrin, fenvalerate, and deltamethrin), has led to a resurgence in the importance of these compounds which now account for about one third of world insecticide use (Elliott *et al.*, 1978; Leahy, 1985; Pickett, 1988). Unfortunately, but not surprisingly, paralleling this increased usage has been a major increase in arthropod resistance to pyrethroids, from fewer than 10 species in 1970 to over 80 in 2003 (Metcalf, 1989; Georghiou and Lagunes-Tejeda, 1991; Resistant Arthropod Database, 2004).

Though two synthetic organic insecticides had been commercially available prior to 1939 (dinitrophenols in Germany, first used in 1892; organic thiocyanates in the United States from 1932 on), it is generally acknowledged that this is the year in which the synthetic organic insecticide industry took off. After several years of research for a better mothproofing compound, Müller, who worked for Geigy AG in Switzerland, discovered the value of DDT as an insecticide. In the next few years, production of DDT began at the company's plants in the United Kingdom and United States, although because of the Second World War, knowledge of DDT was kept a closely guarded secret. In early 1944, DDT was first used on a large scale, in a delousing program in Naples where typhus had recently broken out. Some 1.3 million civilians were treated with DDT, and within 3 weeks the epidemic was controlled (Fronk, in Pfadt, 1985). Later that year the identity of the "miracle cure" was revealed, and the world soon became convinced that with DDT (and other recently developed insecticides) pest insects would become a thing of the past. In 1948 Müller was awarded a Nobel Prize, though, interestingly, the first example of insect resistance to DDT had been reported 2 years earlier!

Through the 1940s and into the 1960s, much research was carried out in Western Europe and the United States for other insecticides as effective as DDT. Initially, the search focused on other chlorinated hydrocarbons, including lindane, chlordane, aldrin, dieldrin, endrin, and heptachlor, and more than 8 billion pounds were produced between 1950 and 1990 (Casida and Quistad, 1998). The use of chlorinated hydrocarbons in the western world was severely reduced, beginning in the 1970s, following the development of resistance and the recognition of the health hazards that these highly persistent insecticides pose (see below). Some, however, remain the major insecticides in some developing countries. Though discovered in 1937, organophosphates such as TEPP, diazinon, dichlorvos, parathion, and malathion did not come to prominence as insecticides till the mid-1960s. They continue to play a massive role, especially in agricultural pest control, with about one half of the top 20 sales list being organophosphates (Casida and Quistad, 1998). Carbamates, for example, sevin, isolan, and furadan, originated in the 1940s but their impact on the insecticide scene was not seen until the late 1960s. They remain important with 4 representatives in the top 20 insecticides sold. For details of structure, physical properties, formulation, lethal doses, usage, etc., consult Fronk, in Pfadt (1985), Volume 12 of Kerkut and Gilbert (1985), Hassall (1990), and Szmedra, in Pimentel (1991), Vol. 1.

The search for suitable synthetic insecticides continues today, though at a somewhat reduced rate because the profitability of such ventures for industrial concerns has greatly diminished, for a variety of interrelated reasons. The primary reasons are (1) the time and cost of discovery, development, and registration of an insecticide, estimated at an average of 7 years and US$35–45 million, with an additional US$55–65 million for the cost of a production plant [by comparison, the total cost of developing a new drug is US$360 million (Casida and Quistad, 1998)]. Only 1 in 15,000–20,000 candidate chemicals ever reaches the marketing stage; (2) the relatively short "life expectancy" of an insecticide because of the development of resistance by its target organisms and/or its becoming an environmental hazard. In contrast to the situation 20–25 years ago, companies now seek to maximize sales within 3–5 years after registration; and (3) a general unwillingness of government agencies to grant registrations for use of new insecticides (and, for that matter, other types of chemical pesticides), as a result of pressure from environmentalists, special interest groups, and the general public. As a result, some of the largest companies in the chemical industry have either considerably reduced or abandoned research into the development of new insecticides (Brown, 1977; DeBach and Rosen, 1991; Dent, 2000).

Paralleling research into new synthetic insecticides has been the discovery of several additional groups of naturally occurring compounds with insecticidal activity, for example, the avermectins, spinosyns, and azadirachtins. Avermectins, a mixture of natural products from the soil actinomycete *Streptomyces avermitilis*, were discovered in the 1970s during screening tests for natural antihelminthic compounds (Lasota and Dybas, 1991). It was also observed that they were potent insecticides and acaricides, and subsequently ivermectin and abamectin were registered for use with domestic animals and some household pests. Ivermectin is also used to control onchocerciasis in West Africa and Latin America. Though avermectins are quickly degraded in ultraviolet light and are strongly bound to soil particles, two features that improve their safety against non-target organisms, there are situations in which they may pose problems. For example, a large proportion of the avermectins administered to livestock pass unchanged out of the body in the feces. These residues remain in the dung pat for several weeks, to exert their toxic effects on a spectrum of dung-using insects, both harmful (e.g., various flies) and beneficial (e.g., dung beetles). The situation is particularly of concern in Australia where, as noted in Section 2.5, a range of exotic dung-beetle species have been introduced to deal with the massive amounts of dung produced by domestic cattle and sheep. Avermectins have been shown to exert a large number of detrimental effects on growth and reproduction of these beetles, though these effects can be significantly lessened by careful selection of the times when livestock are treated (i.e., when the dung beetles are less active). However, the adoption of a sustained-release bolus for delivery of the avermectins in livestock may have a major influence on dung-beetle populations (Strong, 1992; Herd *et al.*, 1993).

Spinosyns, discovered in the late 1980s, are produced by another actinomycete, *Saccharopolyspora spinosa*. Like avermectins, spinosyns are easily degraded by ultraviolet light and soil microorganisms. They have low toxicity to mammals, birds, and non-target (beneficial) insects. Commercial preparations are available for control of a range of pests on cotton, vegetables, turf, and ornamentals (Crouse and Sparks, 1998).

Azadirachtin is the major component of the oil extractable from the seeds and leaves of the neem (margosa) tree (*Azadirachta indica*), an evergreen endemic to southern and southeastern Asia, but now found also in Africa, the United States, and Australia. It has been shown to exert a variety of potentially useful effects including inhibition of settling, oviposition, and feeding behaviors, interference with both embryonic and postembryonic development, reduction in number of eggs matured, and mortality, and it is effective on a wide range of pest insects, especially phytophagous species. Because azadirachtin is primarily a feeding poison for juvenile phytophagous insects, it shows great selectivity in the sense that the pests' natural enemies (adult parasitoids and predators) are unaffected by it. It is also relatively non-toxic to vertebrates. Offsetting these advantages are its limited persistence in the field, its slow-acting nature, and, of course, the potential for the development of resistance. Nevertheless, it may become a useful adjunct to already available pesticides in underdeveloped countries, and a few commercial preparations are available (Schmutterer, 1990; Isman *et al.*, 1991; Gahukar, 1995; Beckage, in Rechcigl and Rechcigl, 2000).

Three major problems have arisen as a result of the massive use of insecticides over the past 70 years. First, many insects and mites have developed resistance to one or more of the chemicals (Tables 24.4 and 24.5). (See Chapter 16, Section 5.5 for a discussion

TABLE 24.4. Number of Species of Insects and Mites in Which Resistance to One or More Chemicals Has Been Documented[a]

Year	Species
1908	1
1928	5
1938	7
1948	14
1954	25
1957	76
1960	137
1963	157
1965	185
1967	224
1975	364
1980	428
1984	447
1989	504
2003	536

[a] Data from Georghiou and Taylor (1977), Metcalf (1989), Georghiou and Lagunes-Tejeda (1991), and Resistant Arthropod Database (2004).

TABLE 24.5. Number of Species of Arthropods with Reported Cases of Resistance Through 2003[a]

	Pesticide Group[b]								Importance[c]				
	DDT	Cyclod.	OCL	OP	Carb.	Pyr.	Fum.	Other	Med./Vet.	Agr.	Benef.	Other	Total
Acari	14	15	16	59	16	10		44	22	41	12		75
Araneae				1					1				1
Coleoptera	24	45	39	28	12	8	14	7		68	3	2	73
Copepoda								1	1				1
Dermaptera	1	1								1			1
Dictyoptera	2	3	1	2	1	1		2	3				3
Diptera	129	107	20	89	21	21		14	148	22	2	10	182
Ephemeroptera	2											2	2
Hemiptera	7	10	8	5	1	1			4	14			18
Homopterans	13	7	6	42	20	12	2	6	1	57			58
Hymenoptera	3	3	1	9	4	2			1	3	10		14
Lepidoptera	45	34	25	46	21	19	2	23		82			82
Neuroptera	1			1	1	1					1		1
Phthiraptera	4	5	2	3	2	2			9				9
Siphonaptera	8	6	2	4	2	1			9				9
Thysanoptera	2	3	1	2	2	3		2		7			7
Total	256	239	121	291	103	81	18	99	198	296	28	14	536
%	47.8	44.6	22.6	54.3	19.2	15.1	3.4	18.5	37.0	55.2	5.2	2.6	

[a] From Resistant Arthropod Database (2004)
[b] Cyclod., cyclodiene; OCL, other chlorohydrocarbons; OP, organophosphate; Carb., carbamate; Pyr., pyrethroid; Fum., fumigant.
[c] MedVet., medical and veterinary pests; Agr., agricultural pests; Benef., predators, parasitoids, honey producers, etc.

of the mechanism of resistance.) *Quadraspidiotus perniciosus* (San Jose scale) is credited with being the first recorded species to develop resistance to an insecticide (lime sulfur) in 1908, less than 30 years after use of the insecticide began. The house fly was the first species resistant to a synthetic insecticide (DDT in 1946), and by 1989 the number of species showing some resistance had reached more than 500, just over half of which are agricultural pests. Interestingly, between 1989 and 2003 the number of resistant species increased by only 32 (Resistant Arthropod Database, 2004). Several reasons may be suggested for this slow down in the appearance of resistant species: (1) Better use of insecticides as a result of both integrated pest management and resistance management; (2) highly diverse chemistry (and hence distinct modes of action) of currently used insecticides; (3) reduced funding for (hence less study of) resistance research; and (4) some degree of resistance has already developed in most of the world's worst arthropod pests (which number about 550 species) (Mark E. Whalon, pers. comm.).

Depending on the method of resistance developed by a species against an insecticide, species frequently have resistance to other chemicals of the same group (class resistance) or even to chemicals of other groups (cross resistance). For example, the green peach aphid (*Myzus persicae*) is resistant to 69 insecticides, while the house fly, German cockroach, and Colorado potato beetle are each resistant to more than 40 insecticides (Resistant Arthropod Database, 2004).

Numerous potential tactics for the management of resistance have been proposed, one or more of which may be appropriate in a specific pest situation. They include (1) increasing use of non-insecticidal strategies (i.e., an integrated approach to pest control; see Section

4.6); (2) reduction in the amount of insecticide applied (in conjunction with a raising of the economic injury threshold and more appropriate timing of insecticide application); (3) increasing the dose applied so that even potentially resistant genotypes are killed; (4) use of insecticide mixtures (the assumption being that the resistance mechanisms for each insecticide will be different and will not occur together in individuals); (5) use of rotations (alternation of the insecticides used) so that resistance to an insecticide decreases during the intervals between its use; (6) the use of synergists which depress the rate of detoxication; and (7) development of new forms of insecticides (but see previous paragraphs) (Georghiou, 1983; Roush, in Pimentel, 1991, Vol. 2; Denholm and Rowland, 1992).

The second problem associated with insecticide use is one already mentioned in Section 2.3, namely, the non-specificity of action of these chemicals, with the result that beneficial as well as pest species are destroyed. (Indeed, some were developed precisely because of their "general purpose" nature!) As pest species typically can recover from insecticide application more rapidly than their natural enemies (because of their greater reproductive potential), they rebound with even greater force, necessitating additional insecticidal treatment and increasing costs to the user.

The third problem is the potential health hazard, both direct and indirect, of many of the synthetic insecticides to humans, livestock, and wildlife. The World Health Organization estimated that, on a worldwide basis, about 3 million humans each year are hospitalized due to exposure to pesticides (herbicides, fungicides and insecticides), and about 220,000 persons die, almost all in developing countries (Jeyaratnam, 1990). It should be noted, however, that two-thirds of these fatalities are suicides (Palmborg, in Pimentel, 2002). Indirectly, many millions more, as well as wildlife, receive minute daily doses in their food and drink or in the air they breathe. Indeed, a feature of insecticides originally considered beneficial, namely, their highly persistent (indestructible) nature in the environment, is now realized to be a major detriment to their safety. For example, DDT is highly stable and only slowly degraded in the presence of sunlight and oxygen. Thus, a single spraying in a house or barn may remain effective up to a year, and even outdoor applications (on foliage) may be stable through an entire growing season. Unfortunately, this stability is retained following ingestion or absorption of DDT by living organisms. Thus, DDT tends to be stored in fatty tissue, because of its lipid solubility, and is concentrated as it is transferred from organism to organism in food webs. Recognition of the phenomenon of bioconcentration (biological magnification) via food webs and observation of harmful effects of insecticides in the terminal members of food chains (especially predatory birds) led to enormous public outcry against insecticides. As a result, governments have been forced to examine carefully the balance between the benefits gained and the risks entailed in the use of insecticides, and where necessary enact legislation to protect human interest. One result of this was the banning in 1972 of DDT use in the United States, in all except a very few situations where benefits clearly outweighed risks (Whittemore, 1977). Since then, the list of insecticides whose registration has been fully or partially canceled, suspended pending review, or modified (e.g., by imposing requirements for protective clothing, changes in application method, or application by a certified person) in the United States has grown considerably and now includes virtually all uses of the chlorinated hydrocarbons (Szmedra, in Pimentel, 1991, Vol. 1).

Despite this rather gloomy picture, it must be strongly emphasized that synthetic insecticides have saved and will continue to save millions of human lives and billions of dollars' worth of food and organic manufactured goods. They are still by far the principal method for control of insect pests, and a good investment for users, with a return of about US$4 for each dollar invested, and in their absence losses would increase by a further 10%. However, there is a realization among users that insecticides are probably more valuable (i.e., cheaper and having a longer period of service) when used selectively, in conjunction with other methods such as biological control, rather than in a "blanket" manner as in the past. This is associated with an appreciation that actual extermination of a pest is almost never achievable, or is even necessary in most situations; that is, a certain amount of pest damage can be tolerated without suffering economic loss. In North America the total amount of insecticide applied is now declining slightly, reflecting improved application technology that permits lower dose rates, improved formulations, more informed application, and a decline in the number of hectares treated as a result of declining farm economy and greater use of non-pesticide control strategies. However, on a worldwide basis insecticide use continues to increase, especially among the less developed nations in Africa and Central and South America, where this remains the best method of pest control.

The chemicals discussed above mainly operate on the principle of control through rapid death of (most) members of a pest population. More recently, however, great interest has been shown in other chemicals that, though widely different in nature, are collectively known as insectistatics because they suppress insects' growth and reproduction (Levinson, 1975). They include (1) substances that inhibit chitin synthesis or tanning, rendering an insect more susceptible to microbial (especially fungal) infection or preventing normal activity because the muscles do not have a firm structural base; (2) antagonists or analogues of essential metabolites (e.g., essential amino acids and vitamins) whose effect is to prolong larval life and/or retard egg production, frequently resulting in death; (3) insect growth regulators (IGRs), substances with juvenile-hormone or ecdysonelike activity; and (4) sex attractants. The ability to purify and sequence insect neuropeptides (Chapter 13, Section 3.1) and their corresponding nucleotides has led to speculation about the potential use of these molecules or their analogues in pest control. Safety may be a concern, however, as a significant degree of homology exists between insect and vertebrate peptides (Kelly *et al.*, 1994).

Some 600 compounds are known to mimic juvenile hormone to various degrees and in different species. Among the effects of these IGRs are (1) interference with embryogenesis, followed by death, at IGR doses about 1/1000th the value of conventional ovicides; (2) abnormal development of the integument in postembryonic stages, leading to inability to molt properly and impaired sensory function (hence inability to locate food, mates, oviposition sites, etc.); (3) improper metamorphosis of internal organs or external genitalia, causing sterility and/or inability to mate; (4) interference with diapause, so that an insect becomes seasonally maladjusted; and (5) abnormal polymorphism in aphids (Staal, 1975). Because these effects take some time to manifest themselves, IGRs are less valuable against rapidly growing larval pests. However, a number of commercial preparations are now available for insects that are long-lived pests and/or pests in the adult stage, for example, fenoxycarb (livestock flies), methoprene (fleas, mosquitoes, stored product pests), hydroprene (cockroaches), and kinoprene (homopteran pests of greenhouses and ornamentals) (Staal, 1987). As insects presumably do not become resistant to their own hormones, it was suggested that pests would not develop resistance to these juvenile-hormone analogues. Unfortunately, this has not proven to be the case, many examples of cross resistance to juvenile-hormone analogues in insects resistant to conventional insecticides being reported in the literature (Sparks and Hammock, 1983; Beckage, in Rechcigl and Rechcigl, 2000).

IGRs that mimic molting hormone have been investigated less intensely, probably due to their greater structural complexity and relative instability (Dhadialla *et al.*, 1998). Nevertheless, commercial products are available, for example, tebufenozide and halofenozide, which are used against lepidopteran pests and pests of turf grass and ornamentals, respectively. Significantly, tebufenozide is generally non-toxic to non-Lepidoptera, including a range of predators and parasitoids (Dhadialla *et al.*, 1998).

With the discovery and identification of pheromones (Chapter 13, Section 4.1) and their mimics (more than 1000 are now known), there were high hopes for the development of new, effective and environmentally safe methods of pest control. As initially envisaged, the pheromones, especially sex attractants and aggregation pheromones, potentially might be used in the following pest management situations: (1) for monitoring pest population density; (2) as lures to attract pests into traps; and (3) for permeating the environment, so that individuals are unable to locate mates. For the first of these possibilities, pheromones have been an outstanding success and are now an integral component of many pest management programs, with commercial preparations available for more than 250 species. For some species, especially Coleoptera, the pheromone is used in association with a synergist kairomone (Chapter 13, Section 4.2); for example, the pheromone of the western pine beetle, *Dendroctonus brevicomis*, is used together with myrcene, a volatile material produced by the host tree. The second and third possibilities, the "direct" control options, on the other hand, have met with much less success, principally because of the high costs involved for the user but also because of reluctance on the part of industry to produce materials with such specificity of action. Nonetheless, commercial mass-trapping systems are available for 19 species (including 12 Lepidoptera) and mating-disruption formulations for almost 30 species (all Lepidoptera) (Ridgway *et al.*, 1990; Shorey, in Pimentel, 1991, Vol. 2; Cardé and Minks, 1995; Suckling and Karg, in Rechcigl and Rechcigl, 2000). The majority of these are used against pests of high-value crops, including stone fruits, berries, apples, grapes, and tomatoes.

4.3. Biological Control

Biological control, in the sense used here, may be described as the regulation of pest populations by natural enemies (parasites, predators, and pathogens). Essentially, it includes four strategies, depending on the nature of the control agent and the pest: (1) natural (passive) control in which the agent is an endemic species and the pest is either endemic or introduced; (2) augmentative control in which populations of an endemic control agent are increased, either by cultural means in the field or by mass-rearing and release; (3) classical biological control (perhaps the best known form) where an exotic control agent is imported to control an introduced, *coevolved* pest; and (4) neoclassical biological control in which the agent is an exotic species brought in to control a native pest (Lockwood, 1993). This latter strategy is, in fact, a specific form of a wider approach known as the "new association" in which the control agent is an exotic species that has not coevolved with the pest (which may be native or introduced) (Hokkanen and Pimentel, 1984, 1989; Hokkanen, in Pimentel, 1991, Vol. 2). Interwoven among all of these strategies is "conservation biological control" (Barbosa, 1998). This entails the use of tactics that alter the habitat of the control agent so as to improve its survival, reproduction, etc., leading to a greater chance of success. The tactics are wide-ranging and include methods that affect the control agent either directly (e.g., providing it with alternate food sources and nest sites) or indirectly (e.g., adjusting cultivation practices to improve the microclimate, increasing ecosystem diversity, and using insecticides that are non-persistent or benign to the control agent).

As in the history of chemical control, several early successes with biological control led some scientists to believe that this method might be *the* one to solve the world's pest problems. However, with an increase in knowledge of ecology, specifically predator-prey and host-parasite relationships, and in the length of the list of failures in biological control projects, this view has been markedly tempered. It is now realized that for some pest species, biological control will not work, and that for some others, biological control has been or will be a highly effective method. Between these two extremes, and forming the majority of pests, are those for which biological control will be an effective tool when used in conjunction with other methods of pest control, that is, as a component of integrated pest management. For many years, biological control was a "poor relation" to chemical control, even though several outstanding successes of biological control were recorded before the advent of synthetic insecticides. However, with the increasing appreciation of the problems caused by synthetic insecticides, biological control began to receive a greater share of the attention of applied entomologists, industrial concerns, and government agencies.

The importance of naturally occurring biological control must be emphasized. DeBach and Rosen (1991, p. 102) suggested that "upon it rests our entire ability to successfully grow crops, because without it, the *potential* pests would overwhelm us." These authors estimated that 99% or more of potential pests are under natural control. However, by its very effect, namely, the prevention of species from becoming sufficiently abundant to be designated as pests, it is easily overlooked. Only by very careful study of ecosystems in equilibrium or by disturbance of ecosystems can its value be appreciated. Two common methods of ecosystem disruption are (1) transfer of a species from its original habitat where it is not a pest to a new habitat where, in the absence of natural enemies, it flourishes and becomes a pest, and (2) indiscriminate use of broad-spectrum insecticides that decimate both pest and natural enemy populations. As noted above, the pest normally recovers more rapidly than its predators or parasitoids and becomes even more destructive than before. Examples of both forms of disruption were given in Section 2.3. DeBach and Rosen (1991) list more than 20 cases of naturally occurring control of homopteran pests (mostly scale insects), and further examples involving pest insects of other orders are given in the papers by Hagen *et al.*, Rabb, and MacPhee and MacLellan, in Huffaker (1971).

Augmentative biological control was carried out in China at least 2300 years ago. The Chinese collected from the wild (or bought) colonies of the tree-nesting ant *Oecophylla smaragdina* which they placed in their citrus trees to control caterpillars and wood-boring beetles. In addition, they placed bamboo runways from tree to tree to facilitate the ants' movements. Apparently, this practice is still used in Burma and perhaps parts of China (DeBach and Rosen, 1991). Augmentative biological control in more recent times came to the forefront as a means of controlling pests that had become resistant to chemical insecticides (van Lenteren, 2000, 2003). Its main uses today are for control in field crops attacked by only a few pest species and in greenhouses where there is a spectrum of pests that are controlled by a variety of natural enemies (Kogan *et al.* and Parrella *et al.*, respectively, in Bellows and Fisher, 1999). Two strategies are used in this form of control. Inundative-release is where the biocontrol agents are released in large numbers to obtain rapid control (i.e., as a "living" insecticide), in systems where viable populations of the agents cannot be maintained and only one generation of the pest occurs. Seasonal inoculation, the release of several control agents simultaneously on a seasonal basis, is used where several pest generations occur. This strategy allows for both immediate control and the build up of the control agent complex for control throughout the crop growing season. Seasonal inoculation is especially useful in greenhouse operations where the crop, plus pests and natural enemies, are removed at the end of each growing season (van Lenteren, 2003). As of 2000, more than 125 control agents (including insects, mites, nematodes, and microorganisms) were being produced commercially, with a value of about US$50 million. In addition, biocontrol agents are produced and sold by many state- or farmer-supported organizations (van Lenteren, 2003).

The first known example of classical biological control occurred in the 18th century, when mynah birds imported from India were successfully used in the control of red locusts (*Nomadacris septemfasciata*) in Mauritius. However, full appreciation of the potential that insect parasitoids and predators might have as control agents developed only in the middle of the 19th century, following careful studies of the biology of such insects in the early 1800s. During the latter half of the 19th century, many well-known entomologists, both in Europe and North America, studied biological control and extolled its virtues, though these were largely unheeded until cottony-cushion scale was spectacularly controlled by the vedalia beetle in 1888–1889 (for this and other examples, see Section 2.3). In the late 1800s, based on studies of insect-diseases (especially those of the silkworm), several authorities, including Pasteur and Auduoin (France), Bassi (Italy), Metschnikoff (Russia), and Le Conte (United States), suggested that microorganisms might be used in the control of pest species (DeBach and Rosen, 1991; Federici, in Bellows and Fisher, 1999). However, for reasons outlined below, the widespread use of microorganisms for biological control did not materialize for almost a further 100 years (Section 4.3.1).

The theory behind the "new associations" approach, including neoclassical control, is that the pest has not coevolved with the biocontrol agent and therefore the former will not have an effective defense against the latter. In other words, neoclassical control should result in a higher success rate compared to classical biological control. Analysis of some 600 cases of biological control using exotic agents led Hokkanen and Pimentel (1989) to conclude that the success rate for new associations was about twice that of classical control (though in both situations there were a significant number of failures).

Biological control, when successful, generally has several advantages over control by insecticides. First, it is persistent; that is, once a control agent is established, it will exert a continuing influence on the population density of the pest. (Augmentative biocontrol agents are exceptions to this generalization and must be reapplied in the manner of chemical insecticides.) Second, in part related to its persistence, biological control is cheap because one application of the control agent is usually sufficient. Furthermore, the control agent is "ready-made"—it does not have to go through an extended and costly phase of research and development, though determination of the most suitable control agent(s) may take some time. As an example of the cheapness of biological control, DeBach and Rosen (1991) noted that the cottony-cushion scale project in 1888–1890 cost less than US$5000, yet it has saved the California citrus-fruit industry millions of dollars each year since. Because of escalating costs of insecticides, biological control may be the only (or principal) method of pest control in underdeveloped countries. Third, biological control does not endanger humans or wildlife through pollution of the environment.

A fourth advantage that has been claimed for biological control is its specificity toward the pest, though there are now known to be some major exceptions to this statement in the areas of classical and neoclassical biological control of both weeds and insect pests. These projects suffered from insufficient assessment of the risks associated with release of the biocontrol agent, in particular, failure to carry out sufficiently broad host-specificity studies (Louda *et al.*, 2003; Sheppard *et al.*, 2003). For example, *Cactoblastic cactorum*, introduced so successfully into Australia for control of exotic *Opuntia* species (Section 2.3), was released on the Caribbean island of Nevis in 1957 as a control agent for native weedy *Opuntia*. However, it was soon found on some of the rarer (non-weedy) *Opuntia* species, whose numbers have been significantly reduced. Further, the moth has now spread throughout the Caribbean, both naturally and by introductions, and has reached Florida, where it is threatening some of the native *Opuntia* species, including some that are endangered (McFadyen, 1998). The tachinid fly *Compsilura concinnata* was introduced into North America in 1906 as a control agent for the gypsy moth (*Lymantria dispar*) and the brown-tail moth (*Euproctis chrysorrhea*). Despite the knowledge that the fly is multivoltine (whereas the moths are univoltine) and would therefore use non-target species to complete its annual cycle, releases were continued until the mid-1980s. Surveys have shown that the fly attacks >200 species of Lepidoptera and Hymenoptera, including giant silkworms (Saturniidae), some of which are now endangered or extirpated in some states (Louda *et al.*, 2003).

Two other advantages that have been suggested in the past are now debatable, namely, that biological control does not stimulate "genetic counterattack" by the pests and that it does not result in the growth to pest status of species that are economically unimportant (Huffaker, 1971; Howarth, 1991).

If biological control offers so many advantages over insecticides, why has it, in most instances, come only a distant second to these compounds in terms of usage? What are its disadvantages? Perhaps the main one is psychological because users like to see immediate results (profits?) for their efforts. In biological control it may take some time (even years) for a control agent to subdue a pest, by which time a user's patience (and profit margin) have worn thin, especially under the considerable and continuous advertising pressure of insecticide producers. Further, the new equilibrium density that a pest attains when controlled biologically is almost certainly higher than the density immediately after insecticide treatment, which again may be unsatisfactory as far as a user is concerned, especially if the crop being grown is of high unit value, for example, fresh fruit. Consumers, too, are involved here, as they have become accustomed to "blemish-free" produce and may not buy even slightly damaged material. Other disadvantages have appeared after the introduction of biological control agents, including extinction of non-target species, enhancement of target pest populations as a result of secondary outbreaks, development of new pests, and effects on human health. A widely accepted tenet of ecological predator-prey theory is that the predator (i.e., biological control agent) cannot cause the extinction of its prey (i.e., pest). However, Howarth (1991) provided numerous examples of both target and non-target animals (including many insects) that have become extinct as a result of the direct or indirect effects of biological control agents. In other situations, introduction of an exotic control agent has led to displacement of already present natural agents, without significant gains in pest control: a good example of this is the displacement (increasing rarity) of several native species of ladybird beetle in North America, following the introduction of the Old World species *Coccinella septempunctata* (7-spotted ladybird beetle) (Louda *et al.*, 2003). In some cases, an introduced control agent has itself become a pest as a result of changing its diet to a useful plant either in preference to the weed or following the demise of the weed for whose control it was imported. There have also been reports that microbial insecticides can infect or cause allergic reactions in humans.

Biological control has not always been successful, though estimates of the degree of success vary. The analysis by DeBach and Rosen (1991) indicated a (partial plus complete) success rate, up to 1988, of about 40% for 416 insect pest species, including 75 species that were completely controlled. On the other hand, Hokkanen and Pimentel (1984) cited a figure of 1 in 7 attempts as the rate for partially and fully successful projects combined, while the rate for fully successful projects is only 1 in 18. In some situations biological control of a pest has failed despite numerous introductions; for example, in Canada none of the 26 parasitoids and predators introduced for European corn borer (*Pyrausta nubilalis*) control has exerted a significant effect, and in the northeastern United States only 15 of 120 introduced control agents for European gypsy moth (*Lymantria dispar*) have become established, and the moth remains a major pest (DeBach and Rosen, 1991). Though there are many specific reasons for the failure of individual projects, undoubtedly the single most important factor has been the lack of research support for biological control. Not surprisingly, in view of the "unprofitability" of successful (long-term) biological control using insect agents, industry has shown little interest in the method. And generally, neither universities nor government agencies have been willing to invest funds and labor in research in this area. In Germany the funding for research in biological control is only about 1% of that awarded for research into new pesticides. Conversely, those countries or states with the best success rates for biological projects (most notably Hawaii and California) are those in which significant research support has been provided. Furthermore, when success is achieved, the benefit:cost ratio is about 30:1 compared to only 4:1 for chemical insecticides (DeBach and Rosen, 1991).

The greatest success in biological control has been achieved through the use of insect agents, especially parasitoids but including many herbivorous species, and some examples are given in Table 24.6. Classical biological control has been primarily successful against pests of perennial crops such as orchards, vineyards, forests and rangelands, and weeds, where long-term interaction between the control agent and pest can occur without the ecological disturbances associated with management of annual crops (Dent, 2000).

Insects have also been used successfully in the control of weeds, notably in Australia, Canada, and the United States (for examples, see Julien 1992; Goeden and Andres, in Bellows and Fisher, 1999; Mason and Huber, 2001; and McFadyen, 2003). Early successes were principally against introduced weeds though, more recently, attention has also been turned to the control of native species (Harris, 1991). DeBach and Rosen (1991) provided figures to suggest that biological control of weeds has been relatively successful, with 48% of 267 projects achieving a measurable degree of success. In all, 49 of 125 weed species have been controlled, the great majority by importation of exotic insects. As with the biological control of insect pests, estimates of benefit:cost ratios vary. For example, a return of 2.3:1 was calculated for control of Noogoora burr (*Xanthium occidentale*) in Queensland, and 15:1 for tansy ragwort (*Senecio jacobaeae*) in Oregon. In Sri Lanka control of the water weed salvinia (*Salvinia molesta*) by the Brazilian weevil *Cyrtobagous singularis* was achieved with a benefit:cost ratio of 1675:1, principally because the agent had previously been tested and used in Australia (McFadyen, 2003).

4.3.1. Microbial Control

Initially, biological control using microorganisms (microbial control), specifically viruses, bacteria, protozoa, fungi, and nematodes, did not meet with the success that had been anticipated from knowledge of their pathogenic effects (Burges and Hussey, 1971; Bulla, 1973; Angus, 1977; Burges, 1981; Kurstak, 1982). This stemmed largely from the nature of the control agents, many of which are obligate pathogens, thus relatively short-lived outside the host, are relatively slow-acting, have poor powers of dispersal, and are

TABLE 24.6. Examples of Successful Biological Control Projects Using Insects as Control Agents[a]

Pest	Primary control agent[b] and source	Location and date of project
Icerya purchasi (cottony-cushion scale)	*Rodolia cardinalis* (vedalia beetle) (Australia)	California (1888–1889)
Perkinsiella saccharicida (sugarcane leafhopper)	*Paranagrus optabilis* (mymarid) (Australia) *Cytorhinus mundulus* (mirid) (Australia)	Hawaii (1904–1920)
Opuntia spp. (prickly pear cactus)	*Cactoblastis cactorum* (moth) (Argentina)	Australia (1920–1925)
Levuana iridescens (coconut moth)	*Ptychomyia remota* (tachinid) (Malaysia)	Fiji (1925)
Aleurocanthus woglumi (citrus blackfly)	*Eretmocerus serius* (aphelinid) (Malaysia)	Cuba (1930), other Caribbean islands and Central American countries (1931 on)
	Amitus hesperidum (platygasterid) Prospaltella opulenta (aphelinid) *P. clypealis* (aphelinid) (India and Pakistan)	Mexico (1949)
Nezara viridula (green vegetable bug)	*Trissolcus basalis* (scelionid) (Egypt and Pakistan)	Australia (1933–1962), New Zealand (1949–1952), Fiji and other Pacific islands (1941–1953)
	T. basalis and *Trichopoda pennipes* var. *pilipes* (tachinid) (Antigua and Monserrat)	Hawaii (1962–1965)
Planococcus kenyae (coffee mealybug)	*Anagyrus* nr.*kivuensis* (encyrtid) (Uganda)	Kenya (1938)
Pseudococcus citriculus (citriculus mealybug)	*Clausenia purpurea* (encyrtid) (Japan)	Israel (1939–1940)
Hypericum perforatum (Klamath weed)	*Chrysolina quadrigemina* (chrysomelid beetle) (Australia)	California (1944–1946) and other western states
Dacus dorsalis (oriental fruit fly)	*Opius oophilus* (braconid) (Philippines and Malaysia)	Hawaii (1947–1951)
Lepidosaphes beckii (purple scale)	*Aphytis lepidosaphes* (eulophid) (China)	California (1948 on), Texas, Mexico, Greece, Brazil, Peru (1952–1968)

TABLE 24.6. (*Continued*)

Pest	Primary control agent[b] and source	Location and date of project
Antonina graminis (rhodesgrass scale)	*Neodusmetia sangwani* (encyrtid) (India)	Texas, Florida, Brazil (1949 on)
Operophtera brumata (winter moth)	*Cyzenis albicans* (tachinid) (Europe) *Agrypon flaveolatum* (ichneumon) (Europe)	Nova Scotia (1955–1960)
Chrysomphalus aonidum (Florida redscale)	*Aphytis holoxanthus* (eulophid) (Hong Kong)	Israel, Lebanon, Florida, Mexico, South Africa, Brazil, Peru (1956 on)
Chrysomphalus dictyospermi (dictyospermum scale)	*Aphytis melinus* (eulophid) (California)	Greece (1962)
Oulema melanopus (cereal leaf beetle)	*Anaphes flavipes* (mymarid) *Tetrastichus julis* (eulophid) *Lemophagus curtus* and *Diaparsis temporalis* (ichneumonids) (Europe)	Midwestern United States (1966–1972)
Aleurothrixus floccosus (woolly whitefly)	*Amitus spiniferus* (platygasterid) (Mexico) *Cales noacki* (aphelinid) (Chile)	California (1967 on), France, Spain, and North Africa (1970 on)
Agromyza frontella (alfalfa blotch leafminer)	*Dacnusa dryas* (braconid) *Chrysocaris punctifacies* (eulophid) (Europe)	Eastern United States (1977 on)
Unaspis yanonensis (arrowhead scale)	*Physcus fulvus* (aphelinid) *Aphytis yanonensis* (eulophid) (China)	Japan (1980 on)
Phenacoccus manihoti (cassava mealybug)	*Epidinocarsis lopezi* (encyrtid) (South America)	16 African countries (1981 on)

[a] Data from DeBach and Rosen (1991).

[b] In many projects, a variety of control agents were introduced and became established, but assist in control of the pest to only a minor degree.

active only in certain environmental conditions. Added to these problems were technical difficulties that hampered research in this area, especially the inability to mass-produce microorganisms on a year-round basis for laboratory study as well as field trials. This has been mainly overcome through the use of artificial insect diets and tissue culture, though these methods are not, by and large, satisfactory for large-scale commercial production because of their labor-intensive nature and high costs. Another major technical problem that required solution was the development of suitable protectants against ultraviolet light to which most microorganisms are especially sensitive. Oddly, it was one of their disadvantages, namely, poor viability outside their host (hence the need to reapply them, much like chemical insecticides), that spurred industry into research and, ultimately, development of a number of commercial preparations. Their advantages include safety to humans and wildlife compared to conventional insecticides, specificity (and, therefore, safety to beneficial insects), biodegradability, and lower registration costs (Laird *et al.*, 1990; Roberts *et al.*, in Pimentel, 1991, Vol. 2; Federici, in Bellows and Fisher, 1999; Flexner and Belnavis, in Rechcigl and Rechcigl, 2000; Khetan, 2001; Koul and Dhaliwal, 2002).

Not surprisingly, in view of the diversity of both pests and their pathogens, there are many methods for regulating insect populations with microorganisms though Harper (1987) and Fuxa (1987) cited three general approaches: introduction, augmentation, and conservation. In the first approach, introduction of the pathogens into the pest's ecosystem may be inoculative or inundative. Following inoculative introduction (colonization), the pathogen exerts more or less permanent regulation of the pest population; in other words, this is a form of classical biological control. Good examples of introduced pathogens that work in this way are: (a) *Bacillus popilliae* and *B. lentimorbus*, which cause milky disease in the Japanese beetle, *Popillia japonica*, an important pest of lawn and other grasses in the United States; (b) a nucleopolyhedrovirus, accidentally introduced into Canada in the early 1940s, that has kept populations of the European spruce sawfly (*Gilpinia hercyniae*) below the economic threshold level since the initial epizootic; (c) a nucleopolyhedrovirus that exerts good control over the European pine sawfly (*Neodiprion sertifer*) in Canada; and (d) the nematode *Deladenus siricidicola*, released in Australia in 1970 for control of an introduced European woodwasp *Sirex noctilio*.

Inundative introductions are those in which an exotic pathogen exerts relatively rapid, but temporary, suppression of the pest population. In other words there is no recycling of the pathogen in the pest's ecosystem and reapplication of the pathogen is necessary if further control is required. Thus, this form of microbial control is analogous to control with a conventional chemical insecticide. The use of *Bacillus thuringiensis* for control of lepidopterous and coleopterous pests and mosquitoes is a prime example of inundative introduction.

Augmentation is the placing of additional amounts of a naturally occurring pathogen into the ecosystem to increase disease prevalence. As Harper (1987) noted, this is useful when natural epizootics are asynchronous with pest outbreaks or when the incidence of disease is too low to be economically valuable. In fact, many of the currently registered entomopathogens (see below) are utilized in this way, for example, the nucleopolyhedroviruses used to control *Heliothis* spp. on cotton, *Orgyia pseudotsuga* (Douglas fir tussock moth), and *Lymantria dispar* (gypsy moth), the microsporidian *Nosema locustae* against rangeland grasshoppers, and the fungus *Hirsutella thompsonii* against the citrus rust mite (*Phyllocoptruta oleivora*).

Conservation refers to the enhancement of naturally occurring microbial pest regulation as part of an overall pest management strategy. It may include, for example, choosing a fungicide against a plant disease that has least impact on the pest entomopathogen, reducing the amount of chemical pesticide applied so that host pest populations do not fall below those required for development of natural epizootics, timely irrigation so as to provide the humidity necessary to precipitate a fungal epizootic, and minimum tillage cultivation to avoid the destruction of pathogens in the upper soil layer.

Microbial products constituted about 2% of the world pesticide market by the mid-1980s. This share was expected to increase considerably as the technological problems referred to earlier were overcome and dissatisfaction with chemical insecticides continued to increase (Roberts, in Pimentel, 1991, Vol. 2). However, this has not been the case, perhaps in large part due to the development of cheaper and safer chemical insecticides, especially synthetic pyrethroids.

The potential of bacteria as microbial control agents remains largely untapped, with only five species (*Bacillus popilliae, B. lentimorbus, B. thuringiensis, B. sphaericus,* and *Serratia entomophila*) available commercially. Commercialization of the two milky disease-causing species (*B. popilliae* and *B. lentimorbus*) is of interest because they are obligate pathogens, which has necessitated their production in living hosts either collected from the field or grown in laboratory culture and harvesting the spores from the cadavers. *B. sphaericus* is registered for mosquito control and is especially effective against species of *Culex. S. entomophila* is available in New Zealand for control of the grass grub *Costelytra zealandica*.

B. thuringiensis (Bt) is by far the most successful microbial control agent, with several million kilograms being produced annually. Commercial formulations of a number of its subspecies (notably *kurstaki, israelensis,* and *tenebrionis*) are available for more than 100 species, mostly lepidopterous and coleopterous pests of forest and agricultural (including stored products) pests, and for mosquitoes and black flies (Table 24.7). The great advantage of Bt is that it can be cultured outside its hosts by liquid fermentation using cheap sources of protein such as soy meal. As noted earlier, however, Bt is short-lived, and several applications may be necessary each season. Furthermore, several examples of pests developing some degree of resistance to Bt are now known (Beegle and Yamamoto, 1992; Tabashnik, 1994; Koul and Dhaliwal, 2002). Currently, considerable effort is being devoted to genetically engineering Bt so as to improve its persistence, specificity, and toxicity. Among the possibilities being explored are adding more or improving the toxin genes, engineering the toxin gene into other organisms, including other bacteria that are more resistant to weathering and host plants whose foliage will thus produce the toxin (Sheck, in Pimentel, 1991, Vol. 2; Khetan, 2001). For additional information on Bt, see also Margalit and Dean (1985), Lüthy (1986), van Frankenhuyzen (1990), Milner (1994), and Metz (2003).

Though Table 24.8 would suggest that viruses (especially nucleopolyhedroviruses) are important microbial control agents, their commercialization and use by producers has been slow, primarily for three reasons; they are too expensive to produce, too slow to act, and too host-specific (Payne, 1986; Roberts *et al.* in Pimentel, 1991, Vol. 2; Hunter-Fujita *et al.*, 1998; Federici, in Bellows and Fisher, 1999; Moscardi, 1999; Khetan, 2001; Koul and Dhaliwal, 2002). A notable exception is the nucleopolyhedrovirus (AgNPV) that infects the velvetbean caterpillar (*Anticarsia gemmatalis*). AgNPV is applied annually to more than 1 million hectares in Brazil (and to smaller areas in some other South American countries) for control of this pest in soybean (Moscardi, 1999). Economically, viruses generally have not yet been able to compete with chemical insecticides, primarily because of their host specificity and their obligate parasitism. Because they must be ingested to be effective, it takes several days before their lethal effect is manifest, during which the pest may do considerably more damage. Other disadvantages include their sensitivity to sunlight (though in soil or litter they can survive from season to season) and, apparently, the existence of significant resistance by pests to them. Advantages of viruses include their specificity (especially in integrated pest management), environmental safety, compatibility with other

TABLE 24.7. Some Insect Pests of Agriculture and Forestry with Potential for Management by Formulations of *Bacillus thuringiensis* δ-Endotoxin[a]

Cruciferous crops	*Pieris rapae* (imported cabbageworm), *Plutella xylostella* (diamondback moth), *Trichoplusia ni* (cabbage looper)
Cotton and tobacco	*Heliothis virescens* (tobacco budworm)
Corn	*Ostrinia nubilalis* (European corn borer)
Potatoes	*Leptinotarsa decemlineata* (Colorado potato beetle)
Soybean, canola, forage crops	*Anticarsia gemmatalis* (velvetbean caterpillar), *Mamestra configurata* (bertha armyworm), *Spodoptera exigua* (beet armyworm), *S. frugiperda* (fall armyworm)
Tobacco	*Manduca quinquemaculata* (tomato hornworm), *M. sexta* (tobacco hornworm)
Tree fruits	*Archips argyrospilus* (fruit-tree leafroller), *Argyrotaenia volutinana* (red-banded leafroller), *Operophtera brumata* (winter moth), *Spilonota ocellana* (eye-spotted bud moth)
Stored grains	*Cadra cautella* (almond moth), *Plodia interpunctella* (Indian meal moth)
Forests and ornamentals	*Choristoneura fumiferana* (spruce budworm), *C.occidentalis* (western spruce budworm), *Dioryctria amatella* (southern pine coneworm), *Lambdina fiscellaria fiscellaria* (hemlock looper), *Lymantria dispar* (gypsy moth), *Malacosoma americanum* (eastern tent caterpillar), *M. disstria* (forest tent caterpillar), *Pryganidia californica* (California oakworm), *Pyrrhalta luteola* (elm leaf beetle)

[a]From Roberts *et al.*, in Pimentel (1991, Vol. 2).

biological control agents, and relative simplicity of their genome. The latter has facilitated genetic engineering to improve their usefulness: in particular, recombinant DNA technology has been used to incorporate toxins, resulting in a 20% to 40% reduction in killing time (Bonning and Hammock, 1996; Federici, in Bellows and Fisher, 1999; Bonning *et al.*, 2003).

As was noted in Chapter 23, Section 5.2.4, it was a fungus *(Metarhizium anisopliae)* that was the control agent in the first attempt at microbial control. Generally speaking, however, the potential of fungi as pest control agents has not been widely exploited, perhaps for three main reasons: inconsistency of test results (probably because of a lack of understanding of the factors that regulate spore germination and infectivity), a perceived impracticability of using fungi because, under natural conditions, their epizootics are heavily dependent on weather conditions, and early concern about their safety because some fungi that are insect pathogens are also known to affect vertebrates (Roberts, in Bulla, 1973). Other disadvantages include inactivation by sunlight and the potential for incompatibility with other components of pest control, notably fungicides for plant diseases and some

TABLE 24.8. Viruses Used Successfully for Insect Pest Management[a]

Insect host	Type of virus[b]	Host plant	Control strategy[c]	Location
Agrotis segetum	GV	Row crops	IE	Pakistan, Denmark
Anticarsia gemmatalis	NPV	Soybeans	IE	USA
A. gemmatalis	NPV	Soybeans	IA	Brazil
Cephalcia abietis	NPV	Trees	IE	Austria
Choristoneura fumiferana	NPV	Trees	IE	Canada
Dendrolimus spectabilis	CPV	Trees	IA	Japan
Erinnyis ello	GV	Cassava	A	Brazil
Gilpinia hercyniae	NPV	Trees	IE	Canada
Heliothis armigera	NPV	Cotton	IA	China
Hyphantria cunea	GV	Trees	IE	Yugoslavia
Lymantria dispar	NPV	Trees	IA	Canada. USA, Russia
L. dispar	NPV	Trees	IE	Sardinia
Lymantria monarcha	NPV	Trees	IE	Denmark. Sweden
Malacosoma disstria	NPV	Trees	IE	Canada
Neodiprion lecontii	NPV	Trees	IE	Canada
Neodiprion sertifer	NPV	Trees	IE	Canada, England, USA
N. sertifer	NPV	Trees	IA	Canada, England, Finland, Russia
Neodriprion swainei	NPV	Trees	IE	Canada
Orgyia pseudotsugata	NPV	Trees	IA	Canada, USA
Oryctes rhinoceros	BV	Oil and coconut palms	IA	South Pacific Islands
Pieris rapae	GV	Row crops	A	China
Pseudoplusia includens	NPV	Soybeans	IE	USA
Spodoptera exigua	NPV	Row crops	A	Thailand
Spodoptera littoralis	NPV	Crops	IE	Crete
Spodoptera sunia, S. exigua	NPV	Cotton, vegetables	IA	Guatemala
Trichoplusia ni	NPV	Cotton	IE	Colombia
T. ni	NPV[d]	Cotton	IA	Guatemala
Trichoplusia orichalcea	NPV	Soybeans	A	Zimbabwe
Wiseana spp.	NPV	Pastures	EM	New Zealand

[a] After Roberts *et al.*, in Pimentel (1991, Vol. 2).

[b] BV, non-occluded baculovirus; CPV, cypovirus; GV, granulovirus; NPV, nucleopolyhedrovirus.

[c] A, augmentation; EM, environmental manipulation; IA, inoculative augmentation; IE, introduction-establishment.

[d] *Autographa californica* NPV.

chemical insecticides. Advantages of fungi include their specificity and, for most species, their facultatively parasitic nature, allowing them to be mass-produced by fermentation. Because fungi attack their host by cuticular penetration (rather than having to be ingested), they may become especially useful as control agents for plant-sucking insects. Eight species of fungi were registered worldwide as of 1998 (Table 24.9), and several more are under intensive study. It would appear that, with improvements in mass-production, persistence, virulence, and formulation (leading to improved germination), fungi will have a major role to play in specific integrated pest management programs (Zimmerman, 1986; Ferron *et al.*, 1991; Roberts *et al.*, in Pimentel, 1991, Vol. 2; Butt *et al.*, 2001; Khetan, 2001; Khachatourians *et al.*, in Koul and Dhaliwal, 2002; Alves *et al.*, 2003).

Of the various groups of microorganisms with potential for pest control, the protozoa are the least studied despite the knowledge that many species (especially Microsporidia) cause regular natural epizootics among pest insects (Table 24.10) (Kellen, 1974). A major

TABLE 24.9. Some Fungi Used or with Good Potential as Microbial Control Agents for Insects[a]

Fungus	Target insects	Status[b]
Aschersonia aleyrodis	Whitefly	A—Russia; B—elsewhere
Beauveria bassiana	Pine caterpillar	A—China
	Potato beetle	A—Russia; B—elsewhere
	Corn borers	A—China, USA, Europe
	Codling moth	A—Russia
	Grasshoppers, whitefly, thrips, mealybugs, aphids	A—USA
Beauveria brongniarti	White grubs, cockchafers	A—Europe
Conidiobolus obscurus	Aphids	B
C. thromboides	Aphids	B
Culicinomyces clavisporus	Mosquitoes	B
Entomophaga grylli	Grasshoppers	B
Lagenidium giganteum	Mosquitoes	A—USA
Metarhizium anisopliae	Spittle bugs	A—Brazil
	Cockroaches, termites	A—USA
Metarhizium flavoride	Grasshoppers, locusts	A—Africa
Nomuraea rileyi	Velvetbean caterpillar	B
Paecilomyces fumosoroseus	Greenhouse pests, diamondback moth	A—USA; B—Europe
Verticillium lecanii	Aphids, whitefly	A—Europe
	Greenhouse pests	A—South America

[a] Data from various sources.
[b] A, registered and in production; B, intensively studied.

reason for this has been the difficulty of diagnosing and identifying protozoan pathogens compared to other groups. Other deterrents are the inability to culture the protozoa outside their host, their slow-acting nature, difficulties with long-term storage, and safety concerns (some Microsporidia have a rather broad host range and may even grow in vertebrate tissue culture). Currently, only one protozoan, *Nosema locustae*, is registered in the United States, for use against rangeland grasshoppers and Mormon crickets. It may not always be lethal, though infected insects feed less and produce fewer eggs. Other species of *Nosema* and another microsporidian *Vairimorpha necatrix* are being tested against a variety of forest and agricultural lepidopterous pests. Protozoan pathogens may be of greatest value when used in conjunction with conventional insecticides, the latter providing the quick mortality

TABLE 24.10. A Partial List of Microsporidia Known To Act as Biological Control Agents in Populations of Insect Pests[a]

Microsporidian	Host
Nosema scolyti	Several bark beetles
Unikaron minitum	*Dendroctonus frontalis*
Nosema sp.	*Pissodes strobi*
N. fumiferanae	*Choristoneura fumiferana*
N. lymantriae	*Lymantria dispar*
N. tortricis	*Tortrix viridana*
N. locustae	Many species of grasshopper
Pleistophora oncoperae	*Oncopera alboguttata*
N. pyrausta	*Ostrinia nubilalis*
Amblyspora sp.	*Aedes cantator*

[a] After Roberts *et al.* in Pimentel (1991, Vol. 2).

expected by users and the former providing long-term protection. Some studies have shown that pests are more susceptible to insecticides (thus, less need be applied) when treated simultaneously with a protozoan pathogen (Henry, 1981; Wilson, 1982; Roberts *et al.*, in Pimentel, 1991, Vol. 2).

As noted in Chapter 23 (Section 5.2.6), the numerous species of nematodes that attack insects fall into two major categories: parasites and pathogens. Generally, there has been little interest in determining the potential of parasitic species as biological control agents, although more than 30 years ago the importance of the facultatively parasitic *Deladenus siricidicola* (Neotylenchidae) as a control agent for the wood wasp *Sirex noctilio* was recognized (Bedding, 1984, 1993). This European wasp, introduced into New Zealand in the early 1900s, Australia in the 1950s, and South America in the 1980s, is a major pest of pines, especially Monterey pine (*Pinus radiata*), a major timber species. The free-living stage of *D. siricidicola* can be mass-produced on fungi grown *in vitro*. It is then injected into holes bored in infected pines, where it is able to locate and infect wood wasp larvae. With this method, parasitism rates up to 90% were obtained, reducing the number of trees killed to zero over a 4-year period in Tasmania.

Over the past 25 years, there has been an exponential increase in interest in the commercialization of pathogenic nematodes against insect pests, specifically the Steinernematidae and Heterorhabditidae which possess a number of attributes relevant to pest control, including a very wide host range, high virulence, ease of large-scale production in artificial media, durability of the infective stage, and safety (hence, the United States Environmental Protection Agency has exempted them from government registration requirements). As well, the infective larval stage is mobile and can actively seek out its insect host. Disadvantages are those typical of microbial agents, especially sensitivity to sunlight, desiccation, and high temperature, as well as poor storage qualities. Thus, nematode products find their greatest use in situations where the pathogens are protected from environmental extremes, for example, against soil pests and pests living in cryptic habitats. Commercial formulations of at least eight species of pathogenic nematodes are available worldwide, though Mráček (in Upadhyay, 2003) lists more than a dozen species that have been used in small and large-scale biological control programs. From an economic perspective, nematodes rank second to bacteria in importance as microbial pesticides. Among the pests for which they are used are termites, cutworms, flea larvae, mole crickets, white grubs, various soil-dwelling fly larvae, and root weevils, as well as slugs and plant nematodes (Poinar, 1986; Gaugler and Kaya, 1990; Roberts *et al.*, in Pimentel, 1991, Vol. 2; Bedding *et al.*, 1993; Flexner and Belnavis, in Rechcigl and Rechcigl, 2001; Mráček, in Upadhyay, 2003). Experimentally it has been demonstrated that many insect species which do not normally encounter pathogenic nematodes are nevertheless susceptible to them. Thus, technological developments such as the addition of antidesiccants to the formulation, genetic selection of more desiccation-resistant strains, and genetic engineering should improve the pest spectrum against which nematodes can be used (Liu *et al.*, 2000; Brey and Hashmi, in Upadhyay, 2003).

In conclusion, it is evident that pathogenic microorganisms will play an increasing role in insect pest management strategies in the foreseeable future. Both technological developments, aimed at improving the production and delivery of these bioinsecticides, and genetic manipulations to increase their viability, virulence, specificity, and other useful features will be vital to their success (Harrison and Bonning, in Rechcigl and Rechcigl, 2000). The majority of bioinsecticides will be used in a manner analogous to the chemicals that they will (partially) replace, and caution must be used, for example, to prevent the development of resistance. Scattered examples of resistance to representatives of all of the major groups of pathogenic microorganisms can be cited; indeed, this would be expected in view of the long coevolutionary relationship between the pest and its pathogen. However, as Briese (1986) noted, the consequences of altering this natural balance by the increasing use of microorganisms require careful study.

4.4. Genetic Control

Methods for genetic control fall into two broad categories: (1) those by which pests are rendered less capable of reproduction and (2) those in which resistance is increased in the organism attacked by the pest. Included in this category is the genetically engineered introduction of microbial toxin genes into plants.

A variety of genetic mechanisms are potentially applicable for regulation of a pest's reproductive capability, including dominant lethality [the basis of the sterile insect release method (SIRM)], inherited partial sterility, autosomal translocations, male-linked translocations, and hybrid sterility [see reviews by Proverbs (1969), Whitten and Foster (1975), Whitten (1985), Steffens (1986), Bartlett, in Pimentel, 1991, Vol. 2; Robinson, in Rechcigl and Rechcigl, 2001)]. To date, however, only one of these, the SIRM, has been used on a full scale; the rest are under examination in laboratories or in field trials, or still at the theoretical stage.

Knipling (1955) first proposed the idea of releasing sterile males into wild populations of the pest so as to reduce total fecundity. He suggested that successive releases of sterile males over a number of generations would lead, cumulatively, to eradication of the pest. To be successful SIRM requires that the pest can be mass-reared and sterilized, usually by irradiation though some chemicals can produce the same effect. The irradiation does not cause sperm inactivation; rather, it causes chromosomes to break, resulting in chromosome imbalance and death of the zygote.

SIRM has been used with striking success against the screwworm fly (*Cochliomyia hominivorax*), which has been eliminated from large areas of Central and North America, including Curaçao, the United States, Mexico, Puerto Rico, Guatemala, El Salvador, Belize, and Honduras (Graham, 1985; Wyss, 2000). Authorities are confident that it is only a matter of time (and political agreement) before the pest is fully eliminated from Central America. SIRM was also used to eradicate screwworm fly in Libya, following its accidental introduction in 1988. Some 1.3 billion irradiated pupae from Mexico were subsequently released, and the last case of screwworm was reported in 1991 (Lindquist *et al.*, 1992).

SIRM has also been used successfully against other pests, for example, Mediterranean fruit fly (medfly) (*Ceratitis capitata*) in California, Mexico, and Chile, melon fly (*Dacus cucurbitae*) in Japan and Taiwan, and Mexican fruit fly (*Anastrepha ludens*) in California. In the latter case a barrier zone of sterile flies in Baja, California and the Tijuana area prevents invasion by the pest. Other SIRM projects, including those against onion fly, mosquitoes, tsetse flies, pink bollworm, gypsy moth, corn earworm, tobacco hornworm, and boll weevil, are being carried out with varying degrees of success (Robinson, in Rechcigl and Rechcigl, 2001; Krafsur, 2002).

The advantages of SIRM are its specificity, the permanency of its effect (though it may take several years to achieve this), and the fact that it does not pollute the environment. An important disadvantage is its limited applicability. For example, it is not feasible to use SIRM on pests that appear sporadically or in high density; the latter would make it very difficult to achieve the necessary high ratio of sterile:wild males in the field. Initially, there were technical problems related to the mass production and/or release of only male individuals. Arguments against the concurrent release of sterile females include "dilution" of the effectiveness of the sterile males by competing with normal females for the males' attention and increasing the size of the female population in species where females are blood-feeders and, therefore, disease vectors. Genetic sexing now allows for the mass production of male-only batches for release (Krafsur, 2002). Females are either eliminated early in development (e.g., by incorporation of a sex-linked lethal mutant into their genome) or separated as distinctly colored pupae in species where they are useful, for example, as breeding stock or hosts for rearing parasitoids.

Chemosterilization is an alternative to sterilization by irradiation in situations where the latter is not practical. For example, it may prove to be useful for species that cannot easily be mass-reared; that is, the pest has to be treated directly in the field. It may be possible to combine the chemosterilants with sex pheromones on female "mimics" to which males would be attracted and attempt to mate, the physical contact thus leading to a male receiving a dose of the chemical. However, the use of chemosterilants has been limited, principally because of concerns about environmental safety.

The use of irradiation or chemosterilants to achieve sterility (known as induced sterility) may not always be feasible because of health hazards, costs, etc., and a number of other methods are being investigated that cause inherited sterility. This is sterility that results either from matings between genetically or cytoplasmically incompatible partners, with production of inviable zygotes, or because gametogenesis is abnormal in one or both sexes.

The development of plant varieties capable of resisting attack by pests is a well-known method of pest control. Probably the classic example of plant resistance is the resistance of wheat varieties to attack by the Hessian fly (*Mayetiola destructor*). Resistance of "Underhill" wheat was recorded in New York in 1782 (Gallun *et al.*, 1975), and several resistant varieties are now used regularly in the United States, with enormous annual savings.

Basically, increased resistance can be achieved by introducing either physical or biochemical changes in plants. Painter (1951) categorized plant resistance to insect attack as non-preference (now called antixenosis to reflect the fact that resistance is a property of the plant, not the resultant behavior of the insect), antibiosis, and tolerance (though these terms are not mutually exclusive). Antixenosis includes mechanisms that deter colonization of the plant by insects, for example, varieties of wheat with "hairy stems" that confer resistance to cereal leaf beetle (*Oulema melanopus*). Antibiosis refers to those features that have adverse effects on the pest's growth, reproduction, and survival. Resistance to the Hessian fly, for example, is related to the presence of antibiotics in the wheat tissues that cause death of the larvae. Plant tolerance describes the extent to which plants can sustain insect damage without loss of vigor or productivity. It is a complex response but often includes compensatory growth; for example, some genotypes of maize resistant to western corn rootworm (*Diabrotica virginifera*) produce greater root volume when fed upon by the pest, in contrast to susceptible varieties whose root volume is reduced by as much as 20%. Other examples of insect-resistant varieties of plants are: decreased amino acid and increased sugar levels in peas (giving poor development and reproduction in pea aphid, *Acyrthosiphon pisum*); asparagine deficiency in rice (causing reduced fecundity in brown planthopper, *Nilaparvata lugens*); beans with hooked hairs (that impale cowpea aphid, *Aphis craccivora*, and potato leafhopper, *Empoasca fabae*); tomatoes with sticky hairs (that trap potato aphid, *Macrosiphum euphorbiae*, and greenhouse whitefly, *Trialeurodes vaporariorum*); and low levels of cucurbitacins in various cucurbits (that deter feeding of spotted cucumber beetle, *Diabrotica unidecimpunctata howardi*). All told, plant varieties resistant to more than 100 pest species are known (DeBach and Rosen, 1991).

The advantages of using resistant varieties include greater crop yield, relatively low cost of development, and absence of side effects (environmental damage, etc.). In addition, pests that feed on these plants may be more susceptible to disease, adverse weather conditions, and insecticides, which can therefore be used in lesser amounts. Disadvantages include the length of time required to develop a resistant variety, ordinarily 10–15 years with crop plants and even longer with trees. Related to this, an extensive program of screening must be carried out to ensure that the varieties are totally satisfactory. Thus, resistance may vary according to the form of the insect, for example, between apterous and winged aphids. Or increased resistance to one pest may result in decreased resistance to another. For example, glossy non-waxy Brussels sprouts are resistant to cabbage aphid (*Brevicoryne brassicae*) but are susceptible to some other pests including green peach aphid (*Myzus persicae*), a major vector of viral diseases. Another problem may be psychological, that is, to persuade a grower that a new variety is superior to the one previously used. [See reviews by Maxwell and Jennings (1980), Smith (1989), Tingey and Steffens, in Pimentel (1991) (Vol. 1), Kogan (1994), Panda and Kush (1995), Dent (2000), and Smith, in Rechcigl and Rechcigl (2000).]

The advent of genetic engineering has engendered considerable speculation, even fear (especially in Europe), about the modification of plants to render them more pest-resistant. The two principal concerns are whether genetically engineered plants are safe for humans and whether they will harm the environment. Nevertheless, in the United States commercial production of corn, cotton and potato genetically engineered to include the gene for production of the Bt endotoxins is now a reality. In 2001 more than one quarter of all corn grown in the United States was Bt-inclusive (Shelton *et al.*, 2002). For these three crops, the need to introduce new pest control technology was imperative. Thus, European corn borer, which tunnels in the plant stalks, is inaccessible to insecticidal sprays. By contrast, pink bollworm (on cotton) and Colorado potato beetle are resistant to most insecticides registered for use against them (Gatehouse and Gatehouse, in Rechcigl and Rechcigl, 2000; Kumar, 2003). Other annual crops and trees are in the developmental stage with respect to insertion of Bt genes into their genome. Especially noteworthy is the development of Bt-expressing rice as part of the strategy for controlling the several stem-boring pests of this major crop.

Not all genetic engineering is directed towards producing "insecticidal" plants. As noted in Chapter 23 (Section 3.1), many plants possess secondary metabolites that deter insects from feeding or ovipositing on them. Thus, it should be possible to transfer the genes responsible into crop plants. Another possibility is the incorporation of a pest insect's own genes for sex pheromone production into the host plant. The idea behind this suggestion is that the host plant, by slowly releasing pheromone, would disrupt the mating behavior of the pest. This proposal is perhaps not quite so farfetched as it may sound because certain plants are known to produce compounds not much different from some insect sex pheromones (Dawson *et al.*, 1989). Other schemes for producing transgenic plants include the incorporation of genes for inhibitors of the insect's digestive enzymes (such as naturally occur in seeds of some legumes) and lectins, which can significantly reduce development and fecundity.

Though genetic engineering may appear to be a panacea for pest control, comparable to the development of chemical insecticides some decades ago, it is fraught with the same problem, namely, the potential for development of resistance. As noted in Section 4.3.1, examples of resistance to Bt applied as sprays are already known. Further, studies on Bt-engineered crop cultivars indicate a "high risk for rapid adaptation' by insects unless careful strategies are put in place to manage these valuable assets (Gould, 1998).

4.5. Cultural Control

Cultural control, the use of various agricultural practices to make a habitat less suitable for reproduction and/or survival of pests, is a long-established method of pest control. Cultural control aims, therefore, to reduce rather than eradicate pest populations and is typically used in conjunction with other control methods. However, in some instances, cultural practices alone may effect almost complete control of a pest, as occurs with the tobacco hornworm (*Manduca sexta*) in North Carolina and the pink bollworm (*Pectinophora gossypiella*) on cotton in central Texas.

The agricultural practices used either may have a direct effect on the pest or may act indirectly by stimulating population buildup of a pest's predators or parasites, or by making plants and animals more tolerant of pest attack. An essential prerequisite for effective cultural control is detailed knowledge of a pest's life history so that its most. susceptible stages can be determined. For crops, important agricultural practices include (1) crop rotation to prevent buildup of pest populations; (2) planting or harvesting out of phase with a pest's injurious stage(s), which is especially important against species that have a limited period of infestation or for plants with a short period of susceptibility; (3) use of trap crops on which a pest will concentrate, making its subsequent destruction easy; (4) soil preparation, so as to bury or expose a pest, or increase the crop's strength so that it can more easily tolerate a pest; (5) clean culture, the removal, destruction, or ploughing under of crop remains, in or under which pests may hibernate; and (6) crop diversity, that is, reversal of the current practice of monoculture (growing a single crop over a wide area). Studies have shown that damage to mixed crops or diverse natural vegetation is much less than that done to plants in monoculture. In mixed crops pests may experience greater difficulty in locating their host plant, while the occurrence of diversity may improve the resources available for parasitoids and predators of the pest. [The chapters by Stem, Cromartie, and Brust and Stinner in Pimentel (1991, Vol. 1) provide excellent discussions of the various aspects of cultural control of crop pests.]

Among the strategies available for cultural control of livestock pests are the use of barriers, habitat alteration, pasture rotation, and removal or restriction of alternate hosts (Steelman, in Pimentel, 1991, Vol. 1). Barriers are of two types: (1) areas cleared of vegetation that prevent the pest from invading the area occupied by livestock; this has been used successfully against tsetse flies in east Africa; and (2) areas containing specific vegetation to inhibit or impede entry of pests; for example, along the east coast of the United States, the planting of bushes and grasses around the breeding grounds of the horse fly *Tabanus nigrovittatus* largely prevented it reaching upland areas occupied by livestock. Examples of habitat alteration are: the selective clearing of low undergrowth adjacent to rivers to allow entry of hot dry winds from the adjacent savannah, thus destroying the humid microclimate preferred by tsetse flies; water impoundment, or drainage of marshes, to destroy mosquito breeding sites; and drainage, harrowing, and liming of pastures to eliminate sheep ticks (*Ixodes ricinus*). Pasture rotation is the practice of grazing livestock alternately in adjacent pastures, allowing enough time to lapse to ensure that pests such as ticks starve to death. A modification of this strategy is to move livestock among pastures that have seasonal variations in pest levels. Removal or restriction of alternate hosts is sometimes necessary when livestock and wildlife, especially game animals, live side by side. Thus, destruction of game and the building of game-proof fences have been widely used in east Africa to reduce tsetse fly populations, and tall fences have also been effective in reducing the density of ticks in areas of Tennessee by reducing deer populations.

4.6. Integrated Pest Management

Integrated pest management (IPM), is "a pest population management system that utilizes all suitable techniques either to reduce pest populations and maintain them at levels below those causing economic injury, or to so manipulate the populations that they are prevented from causing such injury" (van den Bosch *et al.*, in Huffaker, 1971, p. 378). A key aspect of this definition is the idea of holding populations at or below the level at which they cause significant damage; that is, it emphasizes that some level of damage can be tolerated and that the pest population does not have to be eradicated. (See Myers *et al.*, 1998, for a discussion of the costs, benefits and successes of pest-eradication programs.) In IPM the techniques employed must be compatible and the system must be flexible to accommodate changes in an ecosystem. It is an approach to pest control that arose, of necessity, with the realization that, for almost all pests, existing methods individually were not capable of exerting permanent control, gave rise to harmful side effects, or created, by their very use, new pest problems.

IPM is not a new concept. It was envisioned by C. V. Riley, former Chief Entomologist in the United States, in the 1890s and was used in the 1920s in California citrus groves (DeBach and Rosen, 1991). However, it was not in most instances given serious consideration until the use of insecticides alone had increased to the point at which an entire crop-growing operation became unprofitable (Kogan, 1998). (The importance of insecticides as environmental hazards did not enter into consideration until much later.) As has been noted previously, a major fault of pesticide use has been its indiscriminate nature. Insecticides were not used only as necessary (a decision based on pest population density) but on a regular (calendar) basis as a form of "preventive medicine" or "insurance." This approach to the use of pesticides, referred to by Doutt and Smith (in Huffaker, 1971) as "the pesticide syndrome," led rapidly to the development of resistance in the pests against which it was aimed, to the use of larger doses or of different insecticides (to which also the pests soon became resistant!), to a rise to pest status of previously innocuous species, and ultimately to greatly elevated costs and decreased profits.

Dent (2000) recognized three distinct phases in the development of an IPM program: problem definition, research, and implementation (Figure 24.5) and suggested that the success of a program depends largely on the time and effort given to the first of these. Clearly, the more information that can be collated for the definition phase, the easier it will be to identify the research needs and implement the management strategy that will provide the maximum economic and socially acceptable returns. In addition to basic "scientific" information on a pest's ecology and current methods of control, there are important socioeconomic aspects to be considered; that is, losses in yield must be placed in the context of the farming system in which they occur, and any changes in strategy must be formulated in consultation with, and meet with the approval of, farmers. In many early projects, relatively little information was available on which to base IPM strategies, and these were developed largely by intuition, despite which they were remarkably successful. Nowadays, the considerably greater body of information generally available is stored in computers for use in constructing and testing mathematical models of the agroecosystem. Once developed, such models can be used to examine the effects of varying one or more of the factors that influence the population density of a pest and, therefore, to determine the optimum method of reducing and maintaining this density below the economic injury threshold.

The following are some of the questions that might be asked in the development of an IPM strategy for the insect pests in a particular agroecosystem (remember that insects

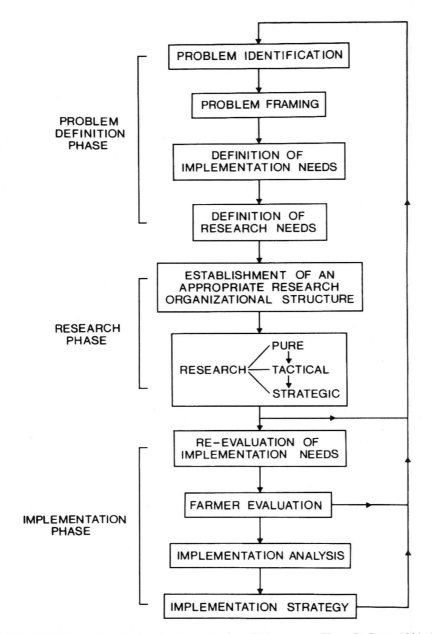

FIGURE 24.5. The three phases in the development of an IPM program. [From D. Dent, 1991, *Insect Pest Management.* By permission of CAB International, Wallingford, U.K.]

typically are only one component of the pest problem, and that any strategy for their management must be compatible with those developed for other pests and with overall socioeconomic goals). (1) What is the economic injury threshold? It has been observed frequently that insecticide users were maintaining pest population density at levels much below the economic injury threshold, increasing their operational costs in both the short and the long term. (2) What are the best parasitoids and predators to use and how might their abundance be increased? (3) Are some of the pests amenable to control by microbial insecticides? (4) What are the best insecticides to use, when should they be applied, and what are suitable doses? These aspects should be considered not only for the pest complex but also with respect to the beneficial insects in the system and thus require knowledge of the specificity of action, stability, and possible side effects of the insecticides. (5) Are suitable cultural procedures available? (6) Are resistant strains of the crop available? (7) Do the farmers already have the resources, human and technological, to adopt the new strategy? (8) Will the new strategy be acceptable to farmers, both economically and socially? Culture, tradition, and religion are important factors in determining farming practice. Thus, changes in planting and harvesting dates may be resisted if they interfere with other customs. Likewise, a suggested change in the cultivars used may not be acceptable if the new cultivars' taste, texture, or yield is different. It will be apparent from this sample of questions that IPM is interdisciplinary in nature and may require the collaboration of experts in widely different areas, for example, entomologists, plant breeders, chemists, toxicologists, meteorologists, computer modelers, economists, anthropologists, and sociologists.

Many examples of pest management might be cited that differ in complexity, geographic location, and crop pest being controlled. In its simplest (and rarest) form, pest management involves only one method of control and, strictly speaking, cannot be described as IPM. Thus, biological control is entirely satisfactory for control of sugarcane pests in Hawaii and coconut pests in Fiji (DeBach and Rosen, 1991). More often, two or more control methods are employed. For example, at the height of the pesticide syndrome, apple growers in Nova Scotia used a battery of fungicides and insecticides (especially broad-spectrum synthetics) against apple pests. Far from achieving the desired effects, this led to even greater problems. By about 1950 earlier pests such as codling moth (*Cydia pomonella*), eye-spotted bud moth (*Spilonota ocellana*), and oystershell scale (*Lepidosaphes ulmi*) had become even more serious, and other species, for example, European red mite (*Panonychus ulmi*) and fruit-tree leafroller (*Archips argyrospilus*) had become pests because of destruction of their natural enemies. The need to reduce the escalating costs of this program stimulated intensive study of apple orchard ecology, including the effects of pesticides on the natural enemy complex. The outcome of this work was the development of an integrated control program that utilized fewer but more specific insecticides (as well as different fungicides, some of which had killed some natural enemies) and allowed for recovery of many of the natural enemies. By 1955 the mite and scale insect had practically disappeared, and the density of the eye-spotted bud moth had fallen significantly.

As a final example, the IPM program for cotton pests in the San Joaquin Valley, California, will be outlined (van den Bosch *et al.*, in Huffaker, 1971; Anonymous, 1984; El-Zik and Frisbie, in Pimentel, 1991, Vol. 3; P. B. Goodell, pers. comm.). This region produces about 10% of the cotton grown in the United States. The usual picture emerges. As with many other crops, cotton growing was initially a small-scale agricultural activity, not requiring an organized program of crop protection. With the development of synthetic insecticides, a massive increase in acreage devoted to cotton growing occurred. Insect pest outbreaks were easily dealt with through the regular but infrequent application of an organochlorine insecticide. With time, however, it became necessary to increase the application frequency and dose of insecticide to combat population resurgences and increased resistance in the two major pests, the corn earworm (*Helicoverpa zea*) and the western tarnished plant bug (*Lygus hesperus*), as well as outbreaks of secondary pests such as cabbage looper (*Trichoplusia ni*), beet armyworm (*Spodoptera exigua*), salt-marsh caterpillar (*Estigmene acraea*), pink bollworm (*Pectinophora gossypiella*), tobacco budworm (*Heliothis virescens*), and spider mites (Tetranychidae) following destruction of their natural enemies.

As the use of organochlorines was discontinued, growers changed to using organophosphates or carbamates, but insect resistance to these also developed rapidly. At the peak of this "crisis phase," almost 50% of all insecticides applied to field crops across the United States were being used on cotton, and control often required 10–20 applications of insecticide per growing season. Not surprisingly, at this point cotton could no longer be grown profitably, many marginal farmers went out of business, and overall there was a massive decline in the acreage devoted to this crop. It was only then that development of a cotton IPM program began.

Early in the development of the integrated program, it was realized that no reliably established economic injury thresholds existed for either of the major pests and that, for *Lygus*, time of insecticide application was critical. For *Lygus*, it was determined that the generally accepted injury threshold of 10 bugs/50 net sweeps was invalid; rather, the major period of importance was June 1–July 20 and only if the density exceeded 10 bugs/50 net sweeps in this period should insecticides be applied. After July 20, *Lygus* densities of 20 or more bugs/50 net sweeps would not affect the quantity or quality of cotton produced, provided that the plants had flowered normally in June and early July.

It was noted that corn earworms seldom became pests except in fields treated for *Lygus* outbreaks (using the old value for economic injury threshold) where the corn earworms' natural enemies (the bugs *Geocoris pallens* and *Nabis americoferus* and the lacewing, *Chrysopa carnea*) were destroyed by insecticide. With the new threshold for economic injury due to *Lygus*, tied in with the plant's seasonal development, the use of insecticides was reduced, especially after late July, and the earworms' natural enemies survived. Further reduction in insecticide usage came about with the raising of the economic injury threshold for corn earworms from 4 earworms/plant to 15 *treatable* earworms (first- and second-instar larvae)/plant. This arose, in part, from realization that larvae often feed on surplus buds, flowers, and small bolls that are not picked. Thus, by the 1980s only one or two applications of insecticide per season were required. This has since grown to three or four applications per season, due to the emergence of new pests and the inclusion of new, narrow-spectrum insecticides in the spray regime (see below). Somewhat offsetting the higher costs of these applications has been a switch in the type of cotton grown: Pima (*Gossipium barbadense*), worth about US$2.20 per kilogram, has replaced the traditional Acala Uplands (*G. hirsutum*) (US$1.30 per kilogram) in about 25% of the San Joaquin Valley cotton-growing area.

Cultural control is also used to reduce the effects of *Lygus* on the cotton crop. *Lygus* prefers alfalfa (*Medicago sativa*) to cotton and will remain on the former plant when given a choice. Traditionally, however, alfalfa fields were solid cut (all of the field cut at once) provoking mass migration of *Lygus* into adjacent cotton fields. Alfalfa growers were persuaded to practice strip cutting to prevent such a migration, particularly following the discovery that strip cutting also reduces pest problems in the alfalfa itself by favoring *Aphidius smithi*, a parasitoid of the pea aphid (*Acyrthosiphon pisum*). Also, because alfalfa is cut at 28-day intervals, it is possible to stagger the cutting cycles in adjacent fields. An earlier proposal that cotton growers plant strips of alfalfa through their fields did not become practical because of difficulty in managing both crops. Currently, GIS technology is being used as part of a landscape-management program, designed to characterize areas with a high potential for *Lygus* outbreaks.

It was also noted that safflower could be a major source of *Lygus*, the bugs moving into cotton as the former plant began to flower. Careful monitoring of the bug populations in the safflower, and a single insecticide treatment when 70% of the insects are third- to fifth-instar nymphs, largely eliminated the threat to adjacent cotton fields. Monitoring by means of light traps and pheromone traps (using gossyplure, which has been particularly valuable for assaying pink bollworm movements into the San Joaquin Valley) is also routinely employed to determine the relative abundance of adult pest Lepidoptera.

Hagen and Hale (1974) suggested another potential cultural method, namely, the use of food sprays in attempts to prolong the seasonal effectiveness of natural enemies whose populations frequently decline after midseason. In experimental cotton plots adult *Chrysopa* were attracted to the food sprays and produced more eggs, resulting in significantly less injury to plants by corn earworms. In the mid-1990s, a commercial product, Envirofeast, was made available, and though this is not used in the San Joaquin Valley, it has been incorporated into some cotton IPM systems elsewhere.

Notwithstanding the increased importance placed on biological and cultural control methods, cotton growers in the San Joaquin Valley continue to apply insecticides as part of the IPM system. In part this is due to a change in the pest spectrum. Specifically, honeydew-producing homopterans, especially aphids and whiteflies, have become significant mid- and late-season pests. The feces of these insects can severely reduce the quality of the lint, creating "sticky" cotton. Cotton producers can choose from a battery of insecticides, including "traditional" synthetics and more recently developed insect growth regulators, avermectins, spinosyns, and microbials. For insecticide-resistance management, the insecticides can be arranged in groups according to their mode of action. Cotton growers are encouraged to follow a spray strategy based on the insecticides' mode of action; specifically, insecticides from the same group (having the same mode of action) should not be used in consecutive applications, or more than twice per season. Other guidelines stress the need for regular scouting (preferably twice-weekly), treatment only when pest threshold densities are reached, application of insecticides at the recommended rate, and whenever possible incorporation of biological and cultural controls.

Genetically engineered cotton, containing the Bt toxin, is available, though has found little use in the San Joaquin Valley where Lepidoptera are minor pests. It is, however, in widespread use where bollworms, beet army worm and other Lepidoptera are the primary pests; for example, in Mississippi more than 85% of the cotton grown is Bt-transgenic.

IPM has resulted in substantial savings, both financial and in terms of pollution, through the greatly decreased use of insecticides. For example, in the San Joaquin Valley, the pre-IPM insecticide-based control program for cotton pests cost about twice as much per hectare as the integrated program. From 1976 to 1982, the average quantity of insecticide applied per hectare to cotton decreased by 72% for the United States as a whole. In part this was due to the introduction of synthetic pyrethroids, which initially provided very effective control, though by the mid-1980s pyrethroid resistance had been documented in the tobacco budworm (Luttrell, 1994). Mainly, however, it was the result of introducing the cotton IPM program, which led to a decrease from 60% to 36% in the proportion of hectares treated. As a consequence, cotton's share of the United States field crop insecticide market dropped from 49% to 24%), while the benefit:cost ratio has been estimated at 4.6:1 (Frisbie and Adkisson, 1986).

Despite these remarkable statistics, many crop growers and government officials remain convinced that the unilateral use of insecticides is best. Persuading these individuals that to continue with this approach is not only economically unsound but will lead to long-term, perhaps irreversible deterioration of environmental quality is perhaps the greatest challenge that scientists have faced.

5. Summary

Only a very small proportion of insect species interact directly or indirectly with humans. Most of these are beneficial and relatively few insects (about 0.1% of all species) constitute pests, which alone or in combination with microorganisms cause damage, disease, or death to humans, crops, livestock, and manufactured goods.

Among the insects that directly benefit humans are those whose products, notably honey and silk, are commercially valuable, that serve as food, educational, research, and medical resources, or that simply give pleasure. Indirectly, humans gain immensely from the activities of insects, especially as pollinators, as agents for biological control of insect pests and weeds, and as soil-dwellers and scavengers.

Pest insects may be classified according to their ecological strategies as r pests, K pests, or intermediate pests. The r pests have potentially high rates of population increase, well developed ability to disperse and find new food sources, and general food preferences. Though they have a large number of natural enemies, biological control would not be effective against r pests because of their reproductive potential. The use of insecticides is the best method for the control of r pests. At the opposite end of the spectrum are K pests, which have relatively low fecundity, poor powers of dispersal, specific food preferences, and few natural enemies. They occur in relatively stable habitats and are best controlled by methods that render their habitat less stable, for example, cultural practices and by the breeding of resistant strains of the organisms they attack. Intermediate pests form a continuum between r and K pests and are normally held below the economic injury threshold by natural enemies. Biological control is the primary method for control of these pests, supplemented as appropriate by chemical, genetic, and/or cultural methods.

Legal control aims to prevent or control pest damage through legislation. It includes the setting up of quarantine stations, systems for monitoring pest populations, and mechanisms for the certification of disease-free plants and animals.

Chemical control traditionally has been the use of naturally occurring or synthetic chemicals to kill pests. It has been the major method of pest control for about 90 years but has created three serious problems: (1) a great increase in the resistance of pests to the chemicals, (2) the death of many beneficial insects as a result of the chemicals' non-specific activity, and (3) pollution of the environment. Insectistatics (insect growth regulators) have found use in specific situations where rapid "knock down" is not critical, and sex attractants and aggregation pheromones now play major roles in monitoring systems for estimating pest populations.

Biological control is the regulation of pest populations by natural enemies. For insect pests, parasitoids, predaceous insects, and microorganisms are the major control agents. Biological control (including natural control, augmentation, classical control, and neoclassical control) offers several advantages over control by insecticides: (1) the control agent is ready-made (does not have to be developed or go through an extensive registration process), cheap to produce and apply, and persistent (many microbial insecticides are exceptions to this generalization—and see point 3); (2) it does not endanger humans or wildlife through pollution of the environment; and (3) the method does not stimulate rapid genetic counterattack by pests (though examples of developing resistance to microbial insecticides are known). Its main disadvantages are its slowness of effect and the fact that the final (equilibrium) pest population density is normally higher than that achieved after insecticide application. Other problems identified from some biological control projects include extinction (of both the original pest and non-pest species), enhancement of the target pest population as a result of secondary outbreaks, and change to pest status for the original control agent.

Microbial control (the use of pathogenic bacteria, viruses, nematodes, fungi, and protozoa) is playing an increasingly important role in the control of insect pests. Strategies for the use of microorganismal agents include introduction, augmentation, and conservation. The advantages of microbial control (safety to humans and wildlife, specificity, biodegradability, and low registration costs) have been somewhat offset by the slow-acting nature, low powers of dispersal, and mass-production problems of the agents. Genetic engineering should remove some of these disadvantages.

Methods of genetic control include (1) those that render pests less capable of reproduction, for example, the sterile insect release method, use of chemosterilants, and sterility resulting from mating incompatibility or abnormal gametogenesis, and (2) those in which resistance is increased in the organisms attacked by pests. Genetic engineering is starting to play a major role in improving plant resistance to insect pests.

Cultural control includes the long-established agricultural practices that make habitats less suitable for pests. The methods used may either directly affect a pest, stimulate an increase in density of a pest's natural enemies, or make the organisms on which a pest feeds more tolerant of attack.

Integrated pest management (IPM) is a combination of methods for reducing and maintaining pest populations below the economic injury threshold. There are three phases in the development of an IPM strategy: problem definition, research, and implementation, of which the first is the most important. To be most effective, IPM requires the input of as much information as possible, not only about the agroecosystem, but also about the socioeconomic framework of the farming system in which the pest problem occurs. Thus, the collaboration of experts from a wide range of disciplines is necessary. If conducted properly, IPM leads to considerable financial saving and a great improvement in environmental quality.

6. Literature

The literature dealing with economic entomology is extensive, and the following list is merely representative. Descriptions of economically important insects (mainly North American species) may be found in standard textbooks of applied entomology, for example, Pfadt (1985), Davidson and Lyon (1987), Olkowski *et al.* (1991), and Metcalf and Metcalf (1993). Volume 12 of Kerkut and Gilbert's (1985) treatise is devoted entirely to insect control, and in the volumes edited by Pimentel (1991, 2002) all aspects of insect-pest management in agriculture are considered. Insecticide resistance is discussed by Georghiou and Taylor (1977), Georghiou and Saito (1983), Metcalf (1989), Hassall (1990), Roush and Tabashnik (1990), Georghiou (1990), and Denholm and Rowland (1992). The economic and environmental costs of chemical pesticides are examined by Pimentel *et al.* (in Pimentel, 1991, Vol. 1) and Pimentel *et al.* (1992). Coats (1994) looks at the risks from natural versus synthetic insecticides.

Biological control is the subject of the volumes edited by Huffaker (1971), Huffaker and Messenger (1976), Ridgway and Vinson (1977), Hoy and Herzog (1985), Franz (1986), Baker and Dunn (1990), Barbosa (1998), and Bellows and Fisher (1999), and the texts of van den Bosch *et al.* (1982), Waterhouse and Norris (1987), Croft (1990), and DeBach and Rosen (1991).

Deacon (1983), Khetan (2001), and authors in the volumes edited by Burges and Hussey (1971), Burges (1981), Kurstak (1982), Anderson and Canning (1982), Maramorosch and Sherman (1985), Baker and Dunn (1990), Gaugler and Kaya (1990), Bedding *et al.* (1993), Hunter-Fujita *et al.* (1998), Rechcigl and Rechcigl (2000), Koul and Dhaliwal (2002), Metz (2003), and Upadhyay (2003) discuss the potential and use of microorganisms as insect pest control agents.

The genetic bases for management of insect pest populations are reviewed by Whitten and Foster (1975), several authors in Hoy and McKelvey (1979), Whitten (in Kerkut and Gilbert, 1985), and Bartlett (in Pimentel, 1991, Vol. 2), while plant resistance to insect pests is examined by authors in Maxwell and Jennings (1980), and by Singh (1986), Smith (1989), and Panda and Khush (1995).

Kogan (1998), Dent (2000), and authors in Apple and Smith (1976), Frisbie and Adkisson (1986), Burn *et al.* (1987), and Metcalf and Luckmann (1994) discuss IPM. The use of pheromones in IPM is considered in Mitchell (1981), Kydonieus and Beroza (1982), Jutsum and Gordon (1989), and Ridgway *et al.* (1990).

Alves, S. B., Pereira, R. M., Lopes, R. B., and Tamai, M. A., 2003, Use of entomopathogenic fungi in Latin America, in: *Advances in Microbial Control of Insect Pests* (R. K. Upadhyay, ed.), Kluwer Academic/Plenum Publishers, New York.

Anderson, R. M., and Canning, E. U. (eds.), 1982, *Parasites as Biological Control Agents*, Symp. Br. Soc. Parasitol., Vol. 19, Cambridge University Press, London.

Angus, T. A., 1977, Microbial control of arthropod pests, *Proc. XV Int. Congr. Entomol.*, pp. 473–477.

Anonymous, 1984, *Integrated Pest Management for Cotton in the Western Region of the United States*, University of California, Division of Agriculture and Natural Resources, Publication 3305.

Apple, J. L., and Smith, R. F. (eds.), 1976, *Integrated Pest Management*, Plenum Press, New York.

Baker, R. R., and Dunn, P. E. (eds.), 1990, *New Directions in Biological Control: Alternatives for Suppressing Agricultural Pests and Diseases*, Liss, New York.

Barbosa, P. (ed.), 1998, *Conservation Biological Control*, Academic Press, San Diego.

Bedding, R. A., 1984, Nematode parasites of Hymenoptera, in: *Plant and Insect Nematodes* (W. R. Nickle, ed.), Dekker, New York.

Bedding, R. A., 1993, Biological control of *Sirex noctilio* using the nematode *Deladenus siricidicola*, in: *Nematodes and the Biological Control of Pests* (R. Bedding, R. Akhurst, and H. Kaya, eds.), CSIRO, East Melbourne, Victoria.

Bedding, R., Akhurst, R., and Kaya, H. (eds.), 1993, *Nematodes and the Biological Control of Pests*, CSIRO, East Melbourne, Victoria.

Beegle, C. C., and Yamamoto, T., 1992, History of *Bacillus thuringiensis* Berliner research and development, *Can. Entomol.* **124**:587–616.

Bellows, T. S., and Fisher, T. W. (eds.), 1999, *Handbook of Biological Control: Principles and Applications of Biological Control*, Academic Press, San Diego.

Ben-Dov, Y., and Hodgson, C. J. (eds.), 1994, *Soft Scale Insects: Their Biology, Natural Enemies and Control*, Elsevier/North-Holland, Amsterdam.

Berenbaum, M. R., 1995, *Bugs in the System*, Addison-Wesley, Reading, MA.

Bonning, B. C., and Hammock, B. D., 1996, Development of recombinant baculoviruses for insect control, *Annu. Rev. Entomol.* **41**:191–210.

Bonning, B. C., Boughton, A. J., Jin, H., and Harrison, R. L., 2003, Genetic enhancement of baculovirus insecticides, in: *Advances in Microbial Control of Insect Pests* (R. K. Upadhyay, ed.), Kluwer Academic/Plenum Publishers, New York.

Briese, D. T., 1986, Host resistance to microbial control agents, in: *Biological Plant and Health Protection* (J. M. Franz, ed.), Fischer Verlag, Stuttgart.

Brown, A. W. A., 1977, Epilogue: Resistance as a factor in pesticide management, *Proc. XV Int. Congr. Entomol.*, pp. 816–824.

Bulla, L. A., Jr. (ed.), 1973, Regulation of insect populations by microorganisms, *Ann. N.Y. Acad. Sci.* **217**:243 pp.

Burges, H. D. (ed.), 1981, *Microbial Control of Pests and Plant Diseases 1970–1980*, Academic Press, New York.

Burges, H. D., and Hussey, N. W. (eds.), 1971, *Microbial Control of Insects and Mites*, Academic Press, New York.

Burn, A. J., Coaker, T. H., and Jepson, P. C. (eds.), 1987, *Integrated Pest Management*, Academic Press, New York.

Butt, T. M., Jackson, C., and Magan, N. (eds.), 2001, *Fungi as Biocontrol Agents: Progress, Problems and Potential*, CAB International, Wallingford, U.K.

Cardé, R. T., and Minks, A. K., 1995, Control of moth pests by mating disruption: Successes and constraints, *Annu. Rev. Entomol.* **40**:559–585.

Casida, J. E., and Quistad, G. B., 1998, Golden age of insecticide research: Past, present, or future?, *Annu. Rev. Entomol.* **43**:1–16.

Coats, J. R., 1994, Risks from natural versus synthetic insecticides, *Annu. Rev. Entomol.* **39**:489–515.

Conway, G., 1976, Man versus pests, in: *Theoretical Ecology: Principles and Applications* (R. M. May, ed.), Blackwell, Oxford.

Creagh, C., 1993, Dung beetles make their mark, *Ecos* **75**(Autumn 1993):26–29.

Croft, B. A., 1990, *Arthropod Biological Control Agents and Pesticides*, Wiley, New York.

Crouse, G. D., and Sparks, T. C., 1998, Naturally derived materials as products and leads for insect control: The spinosyns, *Rev. Toxicol.* **2**:133–146.

Davidson, R. H., and Lyon, W. F., 1987, *Insect Pests of Farm, Garden, and Orchard*, 8th ed., Wiley, New York.

Dawson, G. W., Hallahan, D. L., Mudd, A., Patel, M. M., Pickett, J. A., Wadhams, L. J., and Wallsgrove, R. M., 1989, Secondary plant metabolites as targets for genetic modification of crop plants for pest resistance, *Pestic. Sci.* **27**:191–201.

Deacon, J. W., 1983, *Microbial Control of Plant Pests and Diseases*, Van Nostrand Reinhold, New York.

DeBach, P., and Rosen, D., 1991, *Biological Control by Natural Enemies*, 2nd ed., Cambridge University Press, London.

DeFoliart, G., 1992, Insects as human food, *Crop Prot.* **11**:395–399.

DeFoliart, G., 1999, Insects as food: Why the Western attitude is important, *Annu. Rev. Entomol.* **44**:21–50.

Denholm, I., and Rowland, M. W., 1992, Tactics for managing pesticide resistance in arthropods: Theory and practice, *Annu. Rev. Entomol.* **37**:91–112.

Dent, D., 2000, *Insect Pest Management*, 2nd ed., CAB International, Wallingford, U.K.

Dhadialla, T. S., Carlson, G. R., and Le D. P., 1998, New insecticides with ecdysteroidal and juvenile hormone activity, *Annu. Rev. Entomol.* **43**:545–569.

Doube, B. M., Macqueen, A., Ridsdill-Smith, T. J., and Weir, T. A., 1991, Native and introduced dung beetles in Australia, in: *Dung Beetle Ecology* (I. Hanski and Y. Cambefort, eds.), Princeton University Press, Princeton.

Eastop, V. F., 1977, Worldwide importance of aphids as virus vectors, in: *Aphids as Virus Vectors* (K. F. Harris and K. Maramorosch, eds.), Academic Press, New York.

Elliott, M., Janes, N. F., and Potter, C., 1978, The future of pyrethroids in insect control, *Annu. Rev. Entomol.* **23**:443–469.

Feltwell, J., 1990, *The Story of Silk*, Alan Sutton, Stroud, Gloucestershire.

Ferron, P., Fargues, J., and Riba, G., 1991, Fungi as microbial insecticides against pests, in: *Handbook of Applied Mycology*, Vol. 2 (D. K. Arora, L. Ajello, and K. G. Mukerji, eds.), Dekker, New York.

Flood, J., 1980, *The Moth Hunters*, Australian Institute of Aboriginal Studies, Canberra.

Franz, J. M. (ed.), 1986, *Biological Plant and Health Protection*, Fischer Verlag, Stuttgart.

Frisbie, R. E., and Adkisson, P. L. (eds.), 1986, *Integrated Pest Management on Major Agricultural Systems*, Texas Agricultural Experiment Station, MP-1616.

Fuxa, J. R., 1987, Ecological considerations for the use of entomopathogens in IPM, *Annu. Rev. Entomol.* **32**:225–251.

Gahukar, R. T., 1995, *Neem in Plant Protection*, Agri-Horticultural Publishing House, Nagpur, India.

Gallun, R. L., Starks, K. J., and Guthrie, W. D., 1975, Plant resistance to insects attacking cereals, *Annu. Rev. Entomol.* **20**:337–357.

Gaugler, R., and Kaya, H. K. (eds.), 1990, *Entomopathogenic Nematodes in Biological Control*, CRC Press, Boca Raton.

Georghiou, G. P., 1983, Management of resistance in arthropods, in: *Pest Resistance to Pesticides* (G. P. Georghiou and T. Saito, eds.), Plenum Press, New York.

Georghiou, G. P., 1990, Overview of insecticide resistance, in: *Managing Resistance to Agrochemicals: From Fundamental Research to Practical Strategies*, American Chemical Society, Washington, D.C.

Georghiou, G. P., and Lagunes-Tejeda, A., 1991, *The Occurrence of Resistance to Pesticides in Arthropods*, FAO, Rome.

Georghiou, G. P., and Saito, T. (eds.), 1983, *Pest Resistance to Pesticides*, Plenum Press, New York.

Georghiou, G. P., and Taylor, C. E., 1977, Pesticide resistance as an evolutionary phenomenon, *Proc. XV Int. Congr. Entomol.*, pp. 759–785.

Gillott, C., 1985, The value of fundamental research in entomology, *Can. Entomol.* **117**:893–900.

Gould, F., 1998, Sustainability of transgenic insecticidal cultivars: Integrating pest genetics and ecology, *Annu. Rev. Entomol.* **43**:701–726.

Graham, O. H. (ed.), 1985, *Symposium on Eradication of the Screwworm from the United States and Mexico*, *Misc. Publ. Entomol. Soc. Amer.* **62**

Hagen, K. S., and Hale, R., 1974, Increasing natural enemies through use of supplementary feeding and non-target prey, in: *Proceedings of the Summer Institute on Biological Control of Plant Insects and Diseases* (F. G. Maxwell and F. A. Harris, eds.), University Press of Mississippi, Jackson.

Harper, J. D., 1987, Applied epizootiology: Microbial control of insects, in: *Epizootiology of Insect Diseases* (J. R. Fuxa and Y. Tanada, eds.), Wiley, New York.

Harris, P., 1991, Classical biocontrol of weeds: Its definition, selection of effective agents, and administrative political problems, *Can. Entomol.* **123**:827–849.

Hassall, K. A., 1990, *The Biochemistry and Uses of Pesticides: Structure, Metabolism, Mode of Action and Uses in Crop Protection*, 2nd ed., VCH, New York.

Henry, J. E., 1981, Natural and applied control of insects by Protozoa, *Annu. Rev. Entomol.* **26**:49–73.

Herd, R., Strong, L., and Wardhaugh, K. (eds.), 1993, *Environmental Impact of Avermectin Usage in Livestock*, Elsevier, Amsterdam.

Hokkanen, H., and Pimentel, D., 1984, New approach for selecting biological control agents, *Can. Entomol.* **116**:1109–1121.

Hokkanen, H., and Pimentel, D., 1989, New associations in biological control: Theory and practice, *Can. Entomol.* **121**:829–840.

Horn, D. J., 1976, *Biology of Insects*, Saunders, Philadelphia.

Howarth, F. G., 1991, Environmental impacts of classical biological control, *Annu. Rev. Entomol.* **36**:485–509.

Hoy, M. A., and Herzog, D. C. (eds.), 1985, *Biological Control in Agricultural IPM Systems*, Academic Press, New York.

Hoy, M. A., and McKelvey, J. J. (eds.), 1979, *Genetics in Relation to Pest Management*, The Rockefeller Foundation, New York.

Huffaker, C. B. (ed.), 1971, *Biological Control*, Plenum Press, New York.

Huffaker, C. B., and Messenger, P. S. (eds.), 1976, *Theory and Practice of Biological Control*, Academic Press, New York.

Hull, R., 2002, *Matthews' Plant Virology*, 4th ed., Academic Press, San Diego.

Hunter-Fujita, F. R., Entwistle, P. F., Evans, H. F., and Crook, N. E., 1998, *Insect Viruses and Pest Management*, Wiley, New York.

Isman, M. B., Koul, O., Amason, J. T., Stewart, J., and Salloum, G. S., 1991, Developing a neem-based insecticide for Canada, *Mem. Entomol. Soc. Can.* **159**:39–47.

Jeyaratnam, J., 1990, Acute pesticide poisoning: A major global health problem, *World Health Stat. Quart.* **43**:139–144.

Julien, M. H. (ed.), 1992, *Biological Control of Weeds: A World Catalogue of Agents and Their Target Weeds*, CAB International, Wallingford, U.K.

Jutsum, A. R., and Gordon, R. F. S. (eds.), 1989, *Insect Pheromones in Plant Protection*, Wiley, New York.

Kellen, W. R., 1974, Protozoan pathogens, in: *Proceedings of the Summer Institute on Biological Control of Plant Insects and Diseases* (F. G. Maxwell and F. A. Harris, eds.), University Press of Mississippi, Jackson.

Kelly, T. J., Masler, E. P., and Menn, J. J.,1994, Insect neuropeptides: Current status and avenues for pest control, in: *Natural and Derived Pest Management Agents* (P. A. Hedin, J. J. Menn, and R. M. Hollingworth, eds.), American Chemical Society, Washington, D.C.

Kerkut, G. A., and Gilbert, L. I. (eds.), 1985, *Comprehensive Insect Physiology, Biochemistry and Pharmacology*, Pergamon Press, Elmsford, NY.

Khetan, S. K., 2001, *Microbial Pest Control*, Dekker, New York.

Knipling, E. F., 1955, Possibilities of insect control or eradication through the use of sexually sterile males, *J. Econ. Entomol.* **48**:459–462.

Kogan, M., 1994, Plant resistance in pest management, in: *Introduction to Insect Pest Management*, 3rd ed. (R. L. Metcalf and W. Luckmann, eds.), Wiley, New York.

Kogan, M., 1998, Integrated pest management: Historical perspectives and contemporary developments, *Annu. Rev. Entomol.* **43**:243–270.

Koul, O., and Dhaliwal, G. S. (eds.), 2002, *Microbial Biopesticides*, Taylor and Francis, London.

Krafsur, E. S., 2002, The sterile insect technique, in: *Encyclopedia of Pest Management* (D. Pimentel, ed.), Dekker, New York.

Kumar, P. A., 2003, Insect pest resistant transgenic crops, in: *Advances in Microbial Control of Insect Pests* (R. K. Upadhyay, ed.), Kluwer Academic/Plenum Publishers, New York.

Kurstak, E. (ed.), 1982, *Microbial and Viral Pesticides*, Dekker, New York.

Kydonieus, A. F., and Beroza, M. (eds.), 1982, *Insect Suppression with Controlled Release Pheromone Systems*, CRC Press, Boca Raton.

Laird, M., Lacey, L. A., and Davidson, E. W. (eds.), 1990, *Safety of Microbial Insecticides*, CRC Press, Boca Raton.

Lasota, J. A., and Dybas, R. A., 1991, Avermectins, a novel class of compounds: Implications for use in arthropod pest control, *Annu. Rev. Entomol.* **36**:91–117.

Leahy, J. P. (ed.), 1985, *The Pyrethroid Insecticides*, Taylor and Francis, London.

Levinson, H. Z., 1975, Possibilities of using insectistatics and pheromones in pest control, *Naturwissenschaften* **62**:272–282.

Lindquist, D. A., Abusowa, M., and Hall, M. J. R., 1992, The New World screwworm in Libya: A review of its introduction and eradication, *Med. Vet. Entomol.* **6**:2–8.

Liu, J., Poinar, G. O., Jr., and Berry, R. E., 2000, Control of insect pests with entomopathogenic nematodes: The impact of molecular biology and phylogenetic reconstruction, *Annu. Rev. Entomol.* **45**:287–306.

Lockwood, J. A., 1993, Environmental issues involved in biological control of rangeland grasshoppers (Orthoptera: Acrididae) with exotic agents, *Environ. Entomol.* **22**:503–518.

Louda, S. M., Pemberton, R. W., Johnson, M. T., and Follett, P. A., 2003, Nontarget effects—The Achilles' heel of biological control? Retrospective analyses to reduce risk associated with biocontrol introductions, *Annu. Rev. Entomol.* **48**:365–396.

Lüthy, P., 1986, Insect pathogenic bacteria as control agents, in: *Biological Plant and Health Protection* (J. M. Franz, ed.), Fischer Verlag, Stuttgart.

Luttrell, R. G., 1994, Cotton pest management: Part 2. A US perspective, *Annu. Rev. Entomol.* **39**:527–542.

Maramorosch, K., and Sherman, K. E. (eds.), 1985, *Viral Insecticides for Biological Control*, Academic Press, New York.

Margalit, J., and Dean, D., 1985, The story of *Bacillus thuringiensis* var. *israelensis* (B.t.i.), *J. Am. Mosq. Control Assoc.* **1**:1–7.

Mason, P. G., and Huber, J. T. (eds.), 2001, *Biological Control Programmes in Canada, 1981–2000*, CAB International, Wallingford, U.K.

Matthews, R. W., Flage, L. R., and Matthews J. R., 1997, Insects as teaching tools in primary and secondary education, *Annu. Rev. Entomol.* **42**:269–289.

Maxwell, F. G., and Jennings, P. R. (eds.), 1980, *Breeding Plants Resistant to Insects*, Wiley, New York.

McFadyen, R. E. C., 1998, Biological control of weeds, *Annu. Rev. Entomol.* **43**:369–393.

McFadyen, R. E. C., 2003, Biological control of weeds using exotic insects, in: *Predators and Parasitoids* (O. Koul and G. S. Dhaliwal, eds.), Taylor and Francis, London.

Metcalf, R. L., 1989, Insect resistance to insecticides, *Pestic. Sci.* **26**:333–358.

Metcalf, R. L., and Luckmann, W. H. (eds.), 1994, *Introduction to Insect Pest Management*, 3rd ed., Wiley, New York.

Metcalf, R. L., and Metcalf, R. A., 1993, *Destructive and Useful Insects: Their Habits and Control*, 5th ed., McGraw-Hill, New York.

Metz, M. (ed.), 2003, *Bacillus thuringiensis: A Cornerstone of Modern Agriculture*, Food Products Press, New York.

Milner, R. (ed.), 1994, Special issue: *Bacillus thuringiensis*, *Agri. Ecosys. Env.* **49**:1–112.

Mitchell, E. R. (ed.), 1981, *Management of Insect Pests with Semiochemicals: Concepts and Practice*, Plenum Press, New York.

Moscardi, F., 1999, Assessment of the application of baculoviruses for control of Lepidoptera, *Annu. Rev. Entomol.* **44**:257–289.

Myers, J. H., Savoie, A., and van Randen, E., 1998, Eradication and pest management, *Annu. Rev. Entomol.* **43**:471–491.

Olkowski, W., Daar A., and Olkowski, H., 1991, *Common-sense Pest Control*, The Taunton Press, Newtown, CT.

Painter, R. H., 1951, *Insect Resistance in Crop Plants*, Macmillan Co., New York.

Panda, N., and Khush, G. S., 1995, *Host Plant Resistance to Insects*, CAB International, Wallingford, U.K.

Payne, C. C., 1986; Insect pathogenic viruses as pest control agents, in: *Biological Plant and Health Protection* (J. M. Franz, ed.), Fischer Verlag, Stuttgart.

Pfadt, R. E. (ed.), 1985, *Fundamentals of Applied Entomology*, 4th ed., Macmillan Co., New York.

Pickett, J. A., 1988, Chemical pest control - the new philosophy, *Chemistry in Britain* February 1988:137–142.

Pimentel, D. (ed.), 1991, *CRC Handbook of Pest Management in Agriculture*, 2nd ed., Vols. I–III, CRC Press, Boca Raton.

Pimentel, D. (ed.), 2002, *Encyclopedia of Pest Management*, Dekker, New York.

Pimentel, D., Acquay, H., Biltonen, M., Rice, P., Silva, M., Nelson, J., Lipner, V., Giordano, S., Horowitz, A., and D'Amore, M., 1992, Environmental and economic costs of pesticide use, *BioScience* **42**:750–760.

Poinar, G. O., Jr., 1986, Entomophagous nematodes, in: *Biological Plant and Health Protection* (J. M. Franz, ed.), Fischer Verlag, Stuttgart.

Proverbs, M. D., 1969, Induced sterilization and control of insects, *Annu. Rev. Entomol.* **14**:81–102.

Rechcigl, J. E., and Rechcigl, N. A. (eds.), 2000, *Biological and Biotechnological Control of Insect Pests*, Lewis Publishers, Boca Raton, FL.

Resistant Arthropod Database, 2004, *www.pesticideresistance.org*, Michigan State University, East Lansing.

Ridgway, R. L., and Vinson, S. B. (eds.), 1977, *Biological Control by Augmentation of Natural Enemies*, Plenum Press, New York.

Ridgway, R. L., Silverstein, R. M., and Inscoe, M. N. (eds.), 1990, *Behavior-modifying Chemicals for Insect Management*, Dekker, New York.

Robinson, W. E., Nowogrodzki, R., and Morse, R. A., 1989, The value of honey bees as pollinators of U.S. crops, Parts I and II, *Am. Bee J.* **129**:411–423, 477–487.

Roush, R. T., and Tabashnik, B. E. (eds.), 1990, *Pesticide Resistance in Arthropods*, Chapman and Hall, London.

Ruddle, K., 1973, The human use of insects: Examples from the Yupka, *Biotropica* **5**:94–101.

Sailer, R. I., 1983, History of insect introductions, in: *Exotic Plant Pests and American Agriculture* (C. L. Wilson and C. L. Graham, eds.), Academic Press, New York.

Schmutterer, H., 1990, Properties and potential of natural pesticides from the neem tree, *Azadirachta indica,Annu. Rev. Entomol.* **35**:271–297.

Shelton, A. M., Zhao, J.-Z., and Roush, R. T., 2002, Economic, ecological, food safety, and social consequences of the deployment of Bt transgenic plants, *Annu. Rev. Entomol.* **47**:845–881.

Sheppard, A. W., Hill, R., DeClerck-Floate, R. A., McClay, A., Olckers, T., Quimby, P. C., and Zimmermann, H. G., 2003, A global review of risk-benefit-cost analysis for the introduction of classical biological control agents against weeds: A crisis in the making?, *Biocontrol News Info.* **24**:91N–108N.

Sherman, R. A., Hall, M. J. R., and Thomas, S., 2000, Medicinal maggots: An ancient remedy for some contemporary afflictions, *Annu. Rev. Entomol.* **45**:55–81.

Singh, D. P., 1986, *Breeding for Resistance to Diseases and Insect Pests*, Springer-Verlag, Berlin.

Smith, C. M., 1989, *Plant Resistance to Insects: A Fundamental Approach*, Wiley, New York.

Southwood, T. R. E., 1977, Entomology and mankind, *Proc. XV Int. Congr. Entomol.*, pp. 36–51.

Sparks, T. C., and Hammock, B. D., 1983, Insect growth regulators: Resistance and the future, in: *Pest Resistance to Pesticides* (G. P. Georghiou and T. Saito, eds.), Plenum Press, New York.

Staal, G. B., 1975, Insect growth regulators with juvenile hormone activity, *Annu. Rev. Entomol.* **20**:417–460.

Staal, G. B., 1987, Juvenoids and anti-juvenile hormone agents as IGRs, in: *International Pest Management; Quo Vadis? An International Perspective* (V. DeLucchi, ed.), Parasitis, Geneva.

Steffens, R. J., 1986, Autocidal control, in: *Biological Plant and Health Protection* (J. M. Franz, ed.), Fischer Verlag, Stuttgart.

Strong, L., 1992, Avermectins: A review of their impact on insects of cattle dung, *Bull. Entomol. Res.* **82**:265–274.

Sutton, M. Q., 1988, *Insects as Food: Aboriginal Entomophagy in the Great Basin*, Ballena Press Anthropological Papers No. 33, Ballena Press, Menlo Park, CA.

Tabashnik, B. E., 1994, Evolution of resistance to *Bacillus thuringiensis,Annu. Rev. Entomol.* **39**:47–79.

Taylor, R. L., and Carter, B. J., 1976, *Entertaining with Insects*, Woodbridge Press, Santa Barbara, CA.

Upadhyay, R. K. (ed.), 2003, *Advances in Microbial Control of Insect Pests*, Kluwer Academic/Plenum Publishers, New York.

van den Bosch, R., Messenger, P. S., and Gutierrez, A. P., 1982, *An Introduction to Biological Control*, Plenum Press, New York.

van Frankenhuyzen, K., 1990, Development and current status of *Bacillus thuringiensis* for control of defoliating forest insects, *For. Chron.* **66**:498–507.

van Lenteren, J. C., 2000, Measures of success in biological control of arthropods by augmentation of natural enemies, in: *Measures of Success in Biological Control* (G. Gurr and S. Wratten, eds.), Kluwer Academic Publishers, Dordrecht.

van Lenteren, J. C., 2003, Need for quality control of mass-produced biological control agents, in: *Quality Control and Production of Biological Control Agents: Theory and Testing Procedures* (J. C. van Lenteren, ed.), CAB International, Wallingford, U.K.

Waterhouse, D. F., 1974, The biological control of dung, *Sci. Am.* **230**(April):100–109.

Waterhouse, D. F., and Norris, K. R., 1987, *Biological Control: Pacific Prospects*, Inkata Press, Melbourne.

Waterhouse, D. F., and Sands, D. P. A., 2001, *Classical Biological Control of Arthropods in Australia*, ACIAR Monograph No. 77, CSIRO Publishing, Melbourne.

Whittemore, F. W., 1977, The evolution of pesticides and the philosophy of regulation, *Proc. XV Int. Congr. Entomol.*, pp. 714–718.

Whitten, M. J., 1985, The conceptual basis for genetic control, in: *Comprehensive Insect Physiology, Biochemistry and Pharmacology*, Vol. 12 (G. A. Kerkut and L. I. Gilbert, eds.), Pergamon Press, Elmsford, NY.

Whitten, M. J., and Foster, G. G., 1975, Genetical methods of pest control, *Annu. Rev. Entomol.* **20**:461–476.

Wilson, G. G., 1982, Protozoans for insect control, in: *Microbial and Viral Pesticides* (E. Kurstak, ed.), Dekker, New York.

Wyss, J. H., 2000, Screwworm eradication in the Americas: Overview, in: *Area-wide Control of Fruit Flies and Other Insect Pests* (K.-H. Tan, ed.), Penerbit Universiti, Sains, Malaysia.

Zimmerman, G., 1986, Insect pathogenic fungi as pest control agents, in: *Biological Plant and Health Protection* (J. M. Franz, ed.), Fischer Verlag, Stuttgart.

Index

Page numbers in **boldfaced** type indicate illustrations.